ENCYCLOPEDIA OF
MARINE SCIENCE

CONTRIBUTORS

Contributors

Craig Bonn	Beaufort Laboratory, National Oceanic and Atmospheric Administration
Kyle R. Dedrick	Marine Information Resources Corporation
Matthew Keene	Duke Marine Laboratory
David L. Porter	Johns Hopkins University Applied Physics Laboratory
Elijah W. Ramsey III	U.S. Geological Survey, National Wetlands Research Center
Royce Randlett, Jr.	Helix Mooring Systems
Robert C. Smoot, IV	McDonogh School
Bradley Weymer	Texas A&M University, Ocean Drilling Program

Photo Research

Glenn R. Williams Barbara Williams	Tryon Palace Historic Sites & Gardens

ENCYCLOPEDIA OF
MARINE SCIENCE

C. REID NICHOLS

PRESIDENT AND OCEANOGRAPHER
MARINE INFORMATION RESOURCES CORPORATION
ELLICOTT CITY, MARYLAND

AND

ROBERT G. WILLIAMS, PH.D.

SENIOR SCIENTIST
MARINE INFORMATION RESOURCES CORPORATION

ADJUNCT INSTRUCTOR
CRAVEN COMMUNITY COLLEGE
NEW BERN, NORTH CAROLINA

Facts On File
An imprint of Infobase Publishing

Dedicated to Gerhard Neumann, Ph.D. (1911–1996),
Willard J. Pierson, Ph.D. (1922–2003),
James Edward Paquin, Ph.D. (1940–2004),
and Franklin E. Kniskern (1940–2008)

ENCYCLOPEDIA OF MARINE SCIENCE

Facts On File, Inc.
An imprint of Infobase Publishing
132 West 31st Street
New York NY 10001

Library of Congress Cataloging-in-Publication Data

Nichols, C. Reid.
Encyclopedia of marine science / authors, C. Reid Nichols and Robert G. Williams
[and others].
p. cm.
Includes bibliographical references and index.
ISBN-13: 978-0-8160-5022-2 (alk. paper)
ISBN-10: 0-8160-5022-8 (alk. paper)
1. Marine science—Encyclopedias. I. Williams, Robert G. II. Title.
GC9.N53 2009
551.4603—dc22 2007045166

Facts On File books are available at special discounts when purchased in bulk quantities
for businesses, associations, institutions, or sales promotions. Please call our Special
Sales Department in New York at (212) 967-8800 or (800) 322-8755.

You can find Facts On File on the World Wide Web at http://www.factsonfile.com

Text design by James Scotto-Lavino
Illustrations by Melissa Ericksen

Printed in the United States of America

MP Hermitage 10 9 8 7 6 5 4 3 2

This book is printed on acid-free paper and contains 30 percent
postconsumer recycled content.

CONTENTS

Appendixes:

ACKNOWLEDGMENTS

This encyclopedia could not have been produced without the efforts of the contributors. With support from Frank Darmstadt and Melissa Cullen-Dupont, the authors were able to complete this work despite complications arising from the overseas deployments of Col. C. Reid Nichols, USMC Reserve. In addition, a number of people have reviewed particular entries, and the authors are thankful for their many valuable suggestions: Dr. Bill Boicourt, Helen Dawson, Patrick Dennis, John Hoyt, Robert Freeman, Jim Hall, Benjamin Nichols, Caitlin Nichols, Libby Nichols, and Cindy Nichols. Dr. Carrie Root took an interest in reading every entry, and her comments improved the overall quality of this encyclopedia. Barbara Williams and Glenn Williams worked tirelessly on the design of numerous figures and tables. The authors also thank Henry Rasof, who provided the initial impetus for this encyclopedia.

—C. Reid Nichols and
Robert G. Williams, Ph.D.

INTRODUCTION

Because of the interdisciplinary nature of marine science, the authors decided that a subject approach is more appropriate than a discipline-based one. *Encyclopedia of Marine Science* contains approximately 600 entries, arranged alphabetically to enable the reader or researcher to find major marine science topics and concepts easily. The authors describe the essence or fundamental character of each headword as a word picture. The index also provides a detailed listing, which supports research across many disparate topics relevant to oceans, marginal seas, and estuaries. For many closely related marine science topics, entries include cross-references to related topics and links to online information. There are more than 150 black-and-white photographs, charts, or diagrams that complement selected entries.

The authors have considerable experience working with other scientists on projects directly relevant to the coastal ocean. They have both taught oceanography courses and have worked together as oceanographers at the National Oceanic and Atmospheric Administration and on a variety of projects for industry relevant to waves, tides, and currents. Their backgrounds and interests in operational oceanography make the *Encyclopedia of Marine Science* particularly applicable to operators who consider environmental factors in their decision making. Several contributors to the encyclopedia come from industry, academia, and government. Various marine technologists have supported this work through technical reviews and by providing background materials. Consequently, some entries include exciting essays about marine technology and suggestions for further reading. References include the name of the author, title of the work, and year of publication. Print and Web resources, many of which were used in completing this work, are either provided within the entries or in an appendix. The full name of a contributor author and his/her professional affiliation can be found in the list of contributors in the front matter. In selecting the topics, the authors made every effort to remain unbiased and to provide the reader with a representative and contemporary view of marine science.

In this work, the authors have attempted to compile a comprehensive alphabetical treatment of marine science. This work is unique because it greatly expands upon the many important terms that are generally found in oceanography or oceanology encyclopedias. This work provides descriptive definitions for many terms that could be found only by reading numerous Earth science reference books and glossaries. The scope of this volume includes applied science, engineering, and technology disciplines. The *Encyclopedia of Marine Science* has been written for high school and college students, mariners, teachers, scientists, engineers, librarians, and anyone with an interest in the field. This material is intended for use by a range of readers, students to professionals, young and old, and everyone who sails the seas or just walks the beach.

A NOTE ON MEASUREMENTS AND CONVERSIONS

Most recently acquired data in the marine sciences are given in the International System of units, or SI system, from the French *Système international d'unités*. In the United States, measurements are often given in U.S. Customary units, formerly known as the English System, especially in engineering projects, weather reports and forecasts, and in reports by the media. However, in order to maintain competitiveness in the current global economy, the United States government continues to promote the SI system. Most scientific journals require that all units be given in the SI system. In view of the interdisciplinary and often international nature of most major marine science projects, consistency in data reporting is a matter of major concern for marine scientists.

In this encyclopedia, measurements are given in SI and U.S. Customary units, with the U.S. Customary unit given first, and the SI unit following in parentheses, such as 300 miles (500 km). Example units commonly used in marine science are listed below:

SI Base Units from Which All Others Are Derived

Quantity	Name	Abbreviation
amount of Substance	mole	Mol
electric current	ampere	A
length	meter	m
luminous intensity	candela	cd
mass	kilogram	kg
temperature	kelvin	K
time	second	s

SI Derived Units Used in Marine Science

Quantity	Name	Abbreviation
acceleration	meters per second squared	m/s^2
plane angle	radian	rad
solid angle	steradian	sr
Celsius temperature	degree Celsius	°C

Quantity	Name	Abbreviation
electric capacitance	farad	F
electric charge	coulomb	C
electric conductance	siemens	S
electric potential difference	volt	V
electric resistance	ohm	Ω
energy, work, quantity of heat	joule	J
force	newton	N
frequency	hertz	Hz
inductance	henry	H
power	watt	W
pressure, stress	pascal	Pa
speed, velocity	meter per second	m/s
magnetic flux	weber	Wb
magnetic flux density	tesla	T
radioactivity	becquerel	Bq

Conversion factors are given below to convert from SI units to U.S. Customary units.

CONVERSION FACTORS

Some important constants:

Acceleration due to gravity at sea level at an arbitrary value of 45.5 degrees of latitude 1 g = 9.80665 m s^{-2} = 32.17405 ft s^{-2}

Base of Natural Logarithms (Euler's number) e = 2.718281828 (and more . . .)

Density of air at 15°C at sea level = 1.275 kg/m^3 = 0.079596 pounds/ft^3

Density of freshwater at 4°C = 1,000 kg/m^3 = 62.43 pounds/ft^3

Density of seawater at 0°C = 1,025 kg/m^3 = 63.99 pounds/ft^3

Density of ice = 917 kg/m^3 = 57.2 pounds/ft^3

Equatorial radius of the Earth = 6,378.137 km = 3,963.19 statute miles

Normal standard seawater has a salinity of 35 Practical Salinity Units (PSU) in accordance with the Practical Salinity Scale of 1978 (see SALINITY).

Speed of sound in air at 15°C and 1 ATM = 340.4 m/s = 761.45 miles per hour (mph)

Speed of sound in freshwater at 20°C = 1,481 m/s = 3,312.90 mph

Specific Gravity (density of a fluid/density of freshwater)

freshwater = 1.0

seawater = 1.025

Specific weight (weight/volume)

9,810 N/m^3 = 62.4 lb/ft^3 (freshwater at 10°C)

10,062 N/m^3 = 64 lb/ft^3 (seawater at 0°C)

Standard atmosphere = 1,013.25 millibars (mb) = 14.7 pounds/in^2

Angle and Angular Speed

1 rad = 57.30 deg

1 deg = 1.745 × 10^{-2} rad

1 rev/min = 0.1047 rad/s

1 rad/s = 9.549 rev/min

Area

hectare = 2.471054 acres

km^2 = 0.38610 mi^2

Density

1 kg/m^3 = 10^{-3} kg/L = 10^{-3} g/cm^3

Energy

1 Btu = 252 cal = 1.054 kJ

1 J = 1 W s = .10197 kg m = 10^7 erg

1 cal = 4.1840 J

Force

1 N = 105 dyn = .10197 kg-force = .22481 lb-force

1 kg Wt = 2.20462 lb Wt

Length

2.54 cm = 1 in

1 m = 3.2808 ft

1.8288 m = 1 fathom

1 km = .62137 mi (statute)

1 mm = 0.0394 in

1 nanometer (nm) = 0.001 micrometers = 3.937 × 10^{-8} in

Mass

1 kg = 6.852 × 10^{-2} slug = 2.20462 lb at 1 g

Power

1 W = 1.341 × 10^{-3} hp = 0.7376 ft lb/s

Pressure

1 atmosphere (atm) = 1.0332×10^4 kg m^{-2} = 1,013.2 mb = 2116.2 lb ft^2 = 14.7 psi = 76 cm Hg = 1.0132×10^5 Pa = 197.1 dB SPL (Swing from total vacuum to 2 atm)

1 Pa = 1 N m^{-2} = 93.98 dB SPL = .000145 pound per square inch (psi)

Temperature

0°C = 273.15 K

100°C = 212 °Fahrenheit (°F)

72°F = 22.2°C = 295.4 K

°F = 9 / 5 × °C + 32

°C = 5/9 (°F - 32)

K = °C + 273.15

Velocity

1 m/s = 3.6 km/hour = 2.2369 mph

0.5144 m/s = 1 knot = 1.1515 mph

Volume

1 barrel (water) = 31.1 gallons

1 barrel (crude oil) = 42 gallons = 315 pounds

1 liter = 0.2642 gal

1 m^3 = 1,000 L = 35.315 ft^3

Volume Flow

0.0281317 m^3/s = 1 ft^3/sec

6.30888 = 10^{-5} m^3/s = 1 gal/min

Nautical Units

Some units have been traditional in the maritime environment. Conversion factors for these units are given below:

1 barrel × 7.2 = 1 metric ton of crude oil

1 fathom = 6 feet = 1.8288 m

1 nautical mile = 1.1508 statute mile = 1,852 m

1 knot = 1 nautical mile per hour = .5148 m/s = 51.48 cm

1 league = 3 nautical miles = 5.56 km

1 displacement ton = 35 cubic feet = 1 long ton = cubic meter of seawater

1 long ton = 2,240 lbs = 1.016 metric tons

1 metric ton = 1,000 kg = 2,205 lbs = 7.2 barrels of crude oil

1 short ton = 2,000 lbs = 0.907 metric tons

1 Sverdrup = 10^6 m^3/s

32 points of a compass = 360 degrees; 1 point of a compass = 11.25 degrees

Commonly Used Units in Marine Science Literature

Parameter	Units	Comment
air temperature	°C	Absolute zero is -459.67°F (-273.15°C)
atmospheric pressure	mb	millibars
concentration	mg/l	Parts per million is an outdated expression of mg/l.
day of year	DoY	A running count of days up to 365 or 366
density	kg/m^3	Seawater is denser than freshwater
depth	m	Average depth of ocean is 16,000 feet (5,000 m)
heading	degrees	Clockwise from true North
height	m	Especially important when talking about surface roughness and waves
Julian day	JD	Running count of days since January 1, 4713 B.C.E.

Parameter	Units	Comment
latitude	degrees	+ north, - south
liquid precipitation	mm	Linear depth in millimeters (mm)
longitude	degrees	+ west, - east
particle size	(pa) μm	Diameter (d) in micrometers (μm); ultrafine ($d_{pa} \leq 0.1$ μm), fine (0.1 μm $<d_{pa} \leq 2.5$ μm), coarse (2.5 μm $< d_{pa} \leq 10.0$ μm), and supercoarse ($d_{pa} >10.0$ μm)
radiation	W/m^2	Heat flux density
relative humidity	percent	100 percent is total saturation
seawater salinity	PSU	Dissolved salt content, ratio of conductivity of seawater and a standard potassium chloride solution
seawater temperature	°C	Measure of heat or kinetic energy
solid precipitation	cm	Linear depth in centimeters (cm)
specific humidity	g/kg	Ratio of water vapor to dry air in a particular volume
speed	m/s	Scalar rate at which an object covers distance
time	minutes	A fundamental scalar quantity
velocity	m/s	Vector rate of change of position with time

For additional information on the above units or conversation factor, please refer to the following sources:

Cleave Books. "A Dictionary of Units." Available online. URL: http://www.cleavebooks.co.uk/dictunit/index.htm. Accessed October 17, 2007.

Nelson, Robert A. *Guide for Metric Practice.* Available online. URL: http://www.physicstoday.org/guide/metric.pdf. Accessed September 14, 2007.

The NIST Reference on Constants, Units, and Uncertainty. Available online. URL: http://physics.nist.gov/cuu/Constants/index.html. Accessed September 14, 2007.

A

able-bodied seamen Entry-level mariners, men or women, capable of performing as efficient deck hands with certifiable abilities in basic seamanship are classified as able-bodied seamen. Their usual responsibilities include vessel maintenance and housekeeping, assisting in loading and unloading cargo, and standing watches on the wheel and lookout. Routine tasks include chipping rust, buffing surfaces, cleaning and painting, and assisting in the maintenance of the engine room. Important marine safety duties involve firefighting and lifeboat operations.

Obtaining an able-bodied seaman license usually involves written tests, proof of certain training courses, and at-sea experiences. Some able-bodied seamen are graduates from an accredited maritime academy or training facility. They are the skilled crew working under the supervision of the mate, engineer, and captain. Able-bodied seamen receive room, board, and medical coverage in accordance with maritime laws and pay from shipping firms, marine science organizations, and government agencies that own and operate research vessels.

Able-bodied seamen are employed aboard oil-field support vessels, seismographic boats, dive boats, research vessels, and offshore tugs. Life onboard the ship can be very rigorous. For this reason, able-bodied seamen must meet certain physical requirements such as climbing a 12-foot (3.7 m) ladder, lifting weights of approximately 100 pounds (45.4 kg), and swinging on a rope with no foot support. Living accommodations vary with ship and company. Most able-bodied seamen receive a single stateroom or share cabin facilities with others. Meals are provided on a specific schedule and are taken in a crew mess facility.

Aboard research vessels or survey ships, able-bodied seamen operate WINCHES and other forms of deck equipment in support of marine scientists. They become intimately involved in the deployment and recovery of oceanographic instruments such as CURRENT METERS or samples such as finfish from otter trawls. Aboard ships such as the 238-foot (72.5 m), 1,978-gross ton R/V *Maurice Ewing*, able-bodied seamen have participated in worldwide deep-sea oceanographic research with scientists affiliated with Columbia University's LAMONT-DOHERTY EARTH OBSERVATORY. The R/V *Maurice Ewing* was in service from 1988 to 2005 and has been replaced by the R/V *Marcus G. Langseth*.

See also BOAT; MARINE TRADES; NETS.

Further Reading

Office of Marine Operations, the Lamont-Doherty Earth Observatory. Available online. URL: http://www.ldeo. columbia.edu/res/fac/oma/langseth/index.html. Accessed January 5, 2007.

Welcome to the Marine Technology Advanced Education (MATE) Center, Monterey Peninsula College, Monterey, California. Available online. URL: http://www. marinetech.org. Accessed August 17, 2004.

abyssal hills Abyssal hills are topographic volcanic mounds found along the seaward margins of most abyssal plains and originating from the spreading of mid-ocean ridges. They rise less than 0.6 mile (.97 km) above the seafloor and have widths from 0.6 to 6.0 miles (.97–10 km). Abyssal hills generally decrease in height moving away from the ridges due to SEAFLOOR SPREADING, erosion, and deposition. The abyssal hills merge into the flat and nearly level deposit of sediments on the deep-ocean floor known as the abyssal plains (*see* ABYSSAL PLAIN).

Abyssal hills and plains make up only about one-third of the floor of the ATLANTIC and INDIAN OCEAN basins, but about three-quarters of the PACIFIC OCEAN basin. In many locations, bottom currents have transported fine-grained sediments from coastal margins to abyssal hill regions, where they have filled in the valleys between the hills. These topographic features are actually formed near the axis of the mid-ocean ridges in groups.

In this deep-sea benthic (ocean floor) environment that is well below the photic zone there is no sunlight, and hence no photosynthesis taking place. Ambient temperatures are cold, and pressures are high. Dissolved oxygen levels are surprisingly high, on the order of 4–5 milliliters per liter due to the low temperature and low oxygen demand of deep-sea creatures.

The mid-19th-century biologist Sir Edward Forbes (1815–54) hypothesized that the deep ocean was devoid of life below about 1,800 feet (548.6 m). He classified this region as the "azoic zone" following his analysis and interpretation of bottom dredge samples taken aboard the HMS *Beacon* during 1841. The azoic zone theory was abandoned following evidence collected during research studies such as the HMS *Challenger* Expedition from 1872 to 1876, led by Sir Charles Wyville Thomson (1830–82), which found living organisms in the deep sea, using bottom dredges. Subsequent deep-sea expeditions of the 20th century found living creatures on the ocean floor in increasing abundance and diversity as larger numbers of samples were taken with sophisticated equipment. Deep-sea cameras have taken many photographs of benthic (bottom-dwelling) organisms in the ocean basins.

Today it is known that creatures living in the abyssal hill province either depend on sinking organic material from the photic zone or are supported by chemosynthesis-based food webs (chemosynthesis is the chemical metabolism of bacteria that form the base of the food chain) in locations near hydrothermal vents (*see* HYDROTHERMAL VENTS).

Further Reading

Bathymetry and Topography, National Geophysical Data Center. Available online. URL: http://www.ngdc.noaa.gov/mgg/bathymetry/relief.html. Accessed January 7, 2007.

abyssal plain The abyssal plain includes the ocean floor at depths from 16,400 to 23,000 feet (4,998.7–7,010 m) with slopes that are less than 1:1,000. The abyssal plain, which is considered to be the flattest region on Earth, is the region between the continental rise and the ABYSSAL HILLS. It is one of the three major physiographic provinces within the ocean basin; the others being the oceanic ridges and the deep-sea trenches that often bound the abyssal plain. The plain is formed by the accretion of sediments that cover the rugged volcanic rocks of the ocean crust. These graded beds of silt and sand are normally derived from TURBIDITY currents that emanate from the continental rise. Sediment thickness therefore increases from the mid-ocean ridge to the continental rise. The mid-ocean ridges rise hundreds of feet (meters) above the surrounding ocean floor and stretch approximately 37,000 miles (59,545.7 km) across the Atlantic, Indian, and Pacific basins. The deepest parts of the ocean are the trenches, which are subduction (tectonic plate plunging under an adjacent plate down into the mantle) zones running along the edge of destructive margins (*see* PLATE TECTONICS). The depth at the deep-sea TRENCHES can be from 9,400 to 16,400 feet (2,865–4,998.7 m).

Early mariners used lead lines, such as a hemp line with a weight on it or a piano wire with a cannonball, to measure depths. Given the time required and the difficulty of making these measurements, few were made. It was not until the widespread application of acoustic fathometers to deep-sea sounding following World War II that the features of ocean basins were revealed. Highly accurate fathometers coupled with paper charts, called precision depth recorders, or PDRs made possible detailed bathymetric maps of the world's oceans. The abyssal plains were first identified during the 1947 and 1948 mid-Atlantic ridge expeditions and the 1948 Swedish deep-sea expedition in the INDIAN OCEAN. Researchers aboard the WOODS HOLE OCEANOGRAPHIC INSTITUTION'S R/V *Atlantis* found that the sediment layer on the floor of the Atlantic was much thinner than originally thought. Echo sounding and submarine photography were used to conduct this deep-sea research. Today, not only soundings, but mappings of bottom and sub-bottom structures are made using sonar along with observations by still and video cameras supplemented by deep sediment core samples to allow geological oceanographers to make very detailed charts of the ocean floor.

Marine geologists drilling into the deep ocean floor have found accumulations of sediments in the abyssal plain to be 1.8–3.0 miles (2.9–4.8 km) deep. They have found that deep-sea sediments (*see* SEDIMENT) are composed of various materials such as black, red, or brown terrigenous (eroded from the land surface) materials, biogenous (remains of living creatures) materials such as shells and skeletons, and hydrogenous (water-absorbing) minerals such as

MANGANESE NODULES that precipitate from seawater through chemical reactions. Similarly, ocean explorers make visual inspections of the abyssal plain with remotely operated vehicles to find shipwrecks such as RMS *Titanic,* found by Robert Ballard in 1986.

Further Reading

Ballard, Robert D. *The Eternal Darkness: A Personal History of Deep-Sea Exploration.* Princeton, N.J.: Princeton University Press, 2000.

Charton, Barbara. *A to Z of Marine Scientists.* New York: Facts On File, 2003.

MMS, Minerals Management Service, U.S. Department of the Interior. Available online. URL: http://www.mms.gov. Accessed January 6, 2007.

OceanWorld, Texas A&M University. Available online. URL: http://www.oceanworld.tamu.edu. Accessed January 7, 2007.

abyssal zone   The extensive physiographic province described as the bottom of the deep ocean is classified as the abyssal zone. In these high-pressure and low-temperature regions, depths are generally from 11,500 to 18,000 feet (3,505–5,486.4 m) and temperatures range from 32° to 39°F (0°–3.9°C). The abyssal zone includes the substrates (sub-bottom sediment layers) and supports many species of bottom-dwelling life. Even though the number of organisms per unit area is low compared to the continental shelves, the biological diversity is high. Cost-effective technologies have not been invented to sustain commercial fishing, or to extract minerals in this zone.

Features common to abyssal zones include the mid-ocean ridges that contain hydrothermal vents (*see* HYDROTHERMAL VENTS). These vents have been classified based on their temperatures and precipitates as warm-water vents, white smokers, and black smokers. Chemosynthetic (chemical metabolizing) bacteria thrive near the hydrothermal vents because of the large amounts of hydrogen sulfide and other minerals from which they derive their energy. These bacteria form the bottom of the food web and are eaten by giant tubeworms, specialized species of crabs, and shrimp.

The deep ocean is an active system, where conditions are maintained by complex interactions among biological, chemical, geological, and physical processes. The rates of these processes may be relatively slow, but they result in a state of dynamic balance. The ecosystems they support may be damaged by humans, but seem to be resilient to both natural and anthropogenic changes. It is imperative that the knowledge base for these regions be increased. Many proposals have been developed to use the abyssal depths for waste disposal, including nuclear waste; opponents argue that leakage of toxic waste into the deep sea would produce an environmental disaster of global proportions.

Further Reading

Ocean Explorer, National Oceanic and Atmospheric Administration, U.S. Department of Commerce. Available online. URL: http://oceanexplorer.noaa.gov. Accessed January 7, 2007.

OceanLink: All about the ocean. Available online. URL: http://oceanlink.island.net/index.html. Accessed January 7, 2007.

abyssopelagic zone   The vertical province over the ABYSSAL PLAIN is the abyssopelagic zone. These bottom depths are found seaward of the continental rise and generally extend from 11,500 feet (3,505 m) depth to the seafloor. The only deeper zone is the hadopelagic zone, which includes deep-sea trenches and canyons. The depth at the deep-sea trenches can be from 16,400 to 9,400 feet (4,998.7–2,865 m). Temperatures in the abyssopelagic zone range from 32° to 39°F (0°–3.9°C) and pressures are greater than 350 atmospheres. Abyssal waters are quite distinct from shallow seas found on the continental shelves. Sunlight is completely absent, and CURRENTS and turbulence are relatively weaker than those measured in shallower surface and intermediate-depth waters. Despite this incredibly harsh environment, there is life.

Abyssopelagic organisms live and feed in open waters between 11,500 and 19,685 feet (3,505–6,000 m) deep. At 19,685 feet (6,000 m) depth, the pressure is a crushing 600 atmospheres. These deep waters support many species of invertebrates such as basket stars and deep-water squid. Most species are scavengers called detritivores, living on the constant rain of dead creatures from the upper sunlit part of the ocean (the detritus). Organic aggregates in the water column are also called marine snow. There are predators, however. The deep-sea medusa, a relative of the jellyfish, captures small crustaceans, larva, and organic debris with its tentacles and digests it in its central cavity. Fish have evolved special features such as large mouths, elastic jaws, and bioluminous lures to attract prey in order to survive in the deep. The abyssal grenadier (*Coryphaenoides armatus*) or smooth-scale rattail is found at those depths, worldwide.

BIOLUMINESCENCE is extremely important to many creatures of the deep sea. It occurs in a wide range of organisms from bacteria to vertebrates and occurs on land as well as in the sea. Within deep and bottom waters, glowing and flashing creatures create their own light in order to survive. Two major chemicals are required for bioluminescence. Luciferin produces the light when the enzyme luciferase is present to drive or catalyze the reaction. Bioluminescence in the abyssopelagic zone is credited with aiding navigation, attracting prey, confusing predators, communicating, and attracting mating partners.

There are at present no cost-effective technologies to utilize abyssopelagic resources. This is the realm of bottom-water masses and currents, a possible future source for energy. In recent years, only Japan has been developing vehicles capable of transporting humans to the deepest parts of the ocean.

Further Reading

Ballard, Robert D. *The Eternal Darkness: A Personal History of Deep-Sea Exploration.* Princeton N.J.: Princeton University Press, 2000.

JAMSTEC, Japan Agency for Marine-Earth Science Technology. Available online. URL: http://www.jamstec.go.jp. Accessed January 6, 2007.

OceanWorld, Texas A&M University. Available online. URL: http://www.oceanworld.tamu.edu. Accessed January 7, 2007.

U.N. Atlas of the Oceans, United Nations Foundation. Available online. URL: http://www.oceansatlas.org. Accessed January 6, 2007.

academic programs   Academic programs in marine science are presented from grade through graduate school, and students participating in these programs can obtain professional recognition of completion by means of certificates through diplomas. In Washington, D.C., the Consortium of Ocean Leadership represents more than 95 of the nation's academic institutions, aquaria, nonprofit research institutes, and federal research laboratories with the common goal of promoting and enhancing ocean research and education. The Consortium of Ocean Leadership was formed during 2007 through the merging of the Consortium for Oceanographic Research and Education (CORE) and the Joint Oceanographic Institution (JOI). This nonprofit organization hosts the popular academic competition known as the National Ocean Sciences Bowl. Marine science programs provide students opportunity to explore this fascinating career area while testing their sea legs during at-sea experiments. Many students visit aquaria, trawl the ocean bed, take bottom samples with corers, collect water samples for analysis, interpret oceanic weather information, and learn the fundamentals of marine navigation. Viable programs in marine science expose students to all aspects of the ocean. Instruction relates to the application of the classical scientific fields to the marine environment. Typical programs offer classes in introductory ocean-

ography, coastal oceanography, marine biology, marine geology, marine chemistry, marine mammology, marine aquarium science, and provide opportunities for participation in research projects. The MARINE ADVANCED TECHNOLOGY EDUCATION CENTER at Monterey Peninsula Community College offers opportunities for hands-on, internship-based experience in a variety of topics such as designing, building and operating REMOTELY OPERATED VEHICLES. Marine science departments usually have research vessels, small craft, oceanographic instruments, a "wet" marine laboratory, wave tanks, and a marine operations facility with access to the coastal ocean. At a community college, students can earn an associate of arts degree in a topic such as biology while emphasizing marine studies during their first two years. They may then transfer to a baccalaureate program in marine biology or oceanography at one of several private or state universities. Students anticipating transfer to a four-year college should consult admission counselors for curriculum planning assistance. The bachelor of science degree programs are meant for students planning to continue with graduate studies in marine science, or for those pursuing technical careers. Graduate degree programs leading to the master of science and doctor of philosophy degrees usually show that the student has successfully completed basic research in marine and atmospheric chemistry, marine biology and fisheries, marine geology and geophysics, meteorology, physical oceanography, and applied marine physics. Many students can gain part-time employment at laboratories such as the United States Army Corps of Engineers' Field Research Facility, government agencies such as the NATIONAL OCEANIC AND ATMOSPHERIC ADMINISTRATION, and marine science or oceanographic companies as they complete their studies. These experiences help tie theoretical classroom discussions and laboratory experiments to real-world marine science applications.

Further Reading

American Geophysical Union. Available online. URL: http://www.agu.org. Accessed January 6, 2007.

ASLO: Advancing the science of limnology and oceanography, American Society of Limnology and Oceanography. Available online. URL: http://www.aslo.org. Accessed January 6, 2007.

Consortium for Ocean Leadership. Available online. URL: oceanleadership.org. Accessed on February 25, 2008.

Marine Technology Society. Available online. URL: http://www.mtsociety.org. Accessed January 6, 2007.

National Research Council. *50 Years of Ocean Discovery: National Science Foundation 1950–2000.* Washington, D.C.: National Academy Press, 2000.

Nichols, Charles R., David L. Porter, and Robert G. Williams. *Recent Advances and Issues in Oceanography.* Westport, Conn.: Greenwood, 2003.

The Oceanography Society. Available online. URL: http://www.tos.org. Accessed January 6, 2007.

Acoustical Society of America (ASA)

The Acoustical Society of America is a professional society that promotes the study of acoustics and the interests of acousticians. Acoustics is a branch of physics concerned with sound and sound waves. Acoustic scientists come from many diverse fields, including physics, engineering, computer science, meteorology, oceanography, and biology.

The society was founded on May 10–11, 1929, with a charter membership of about 450. The ASA joined, in 1931, with three other societies to found the American Institute of Physics to unite physics-related groups and to provide publishing and meeting facilities.

Since its inception, the society has promoted the development of acoustic standards for terminology, measurement, and criteria for the study of noise and vibration. The society has been a leading advocate in the development of the theory and applications of underwater sound in such diverse areas as SONAR, underwater communications, underwater location, marine bio-acoustics, geophysical studies of the ocean floor and overlying sediments, the remote sensing of temperature, water velocity, and wave height.

Papers on these and many other fields of acoustics are presented at biannual meetings of the Society. Peer-reviewed papers, new patents, and meeting information are provided in the *Journal of the Acoustical Society of America,* published monthly. The journal was first published in 1929.

The ASA recognizes achievements in acoustics by a series of awards. Awards, in the field of underwater sound, include the Pioneers of Underwater Acoustics Medal and the Silver Medal in Acoustical Oceanography and Underwater Acoustics. The A. B. Wood Medal and Prize of the Institute of Acoustics is awarded to young scientists for distinguished contributions in the applications of acoustics; preferences are given to candidates whose studies involve marine science. Albert Beaumont Wood (1890–1964) was a pioneer in underwater acoustics and is credited with designing the first directional hydrophone.

The ASA awards fellowships and grants each year, including minority fellowships, and the Medwin Prize in Acoustical Oceanography. The Medwin prize was established from a grant by Herman Medwin and his wife, and recognizes effective use of sound in the discovery and understanding of physical and biological parameters in the sea. Professor Medwin was a leading researcher for many years at the U.S. Navy Postgraduate School in Monterey, California. Today, there are more than 7,000 members of Acoustical Society of America worldwide.

Further Reading

Acoustical Society of America. "A History of the Acoustical Society of America." *Journal of the Acoustical Society of America* 26 (1951): 874–905.

Acoustical Society of America. "An Introduction to the Acoustical Society of America." Available online. URL: http//www.asa.aip.org/intro.html. Accessed May 5, 2004.

acoustic Doppler current profiler (ADCP)

An instrument for measuring water CURRENT that use the transmissions of high-frequency (kilohertz to megahertz) acoustic signals in the water is called an acoustic Doppler current profiler. The current is measured by means of a Doppler shift in the backscattered echo (*see* BACKSCATTER) resulting from plankton, suspended sediments, and bubbles, all assumed to be moving at the mean speed of the water. A Doppler shift occurs in sound waves whenever the source, the receiver, or the target is in motion. In this case, the "sources" are the moving objects, which scatter the sound back to the transmitter/receiver, called a "transducer." Usually, the transmitter and receiver are in the same location. By the Doppler principle, there is a slight increase in frequency for objects moving toward the receiver and a slight decrease in frequency for objects moving away from the receiver. Christian Johann Doppler (1803–53) was the first investigator to relate the frequency of a source to its velocity relative to an observer in a scientific paper. In an ADCP, the changes in frequency are proportional to the speeds of the target objects. By accurately detecting the time differences between the transmitted and received pulses, the time delays are related to range, making possible the measurement of current speeds at increasing ranges from the transducer. The current velocities are then sorted into "range bins," permitting the construction of a profile of current velocity versus range.

Generally, ADCPs use multiple transducer heads, the most common being two pairs in orthogonal planes, known as the "Janus configuration" after the Roman god Janus, who had two faces looking front and back. By beaming the acoustic signals in at least three directions, the total current vector is obtained. The fourth beam adds redundancy and improves accuracy. The individual current components are combined by vector addition, giving the resultant

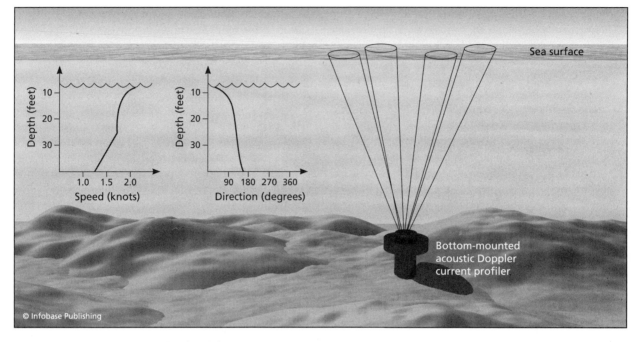

An acoustic Doppler current profiler (ADCP) deployed on the seafloor

velocity vector. The angles of the beams to the transducer axis are typically 20 to 30 degrees.

ADCPs operated from ships or buoys, or towed platforms must also measure the velocity, pitch, roll, and yaw of the supporting platform so that these motions can be removed from the beam velocities prior to computing the water velocity profiles. Scientists and engineers have developed algorithms that record vessel position and heading and then subtract out the unwanted vessel motions from the current meter measurements. Such capabilities are very important, especially in supporting drill ships. To ensure safety at sea, vessels may record positions using several systems such as GLOBAL POSITIONING SYSTEM, multiple gyrocompasses, and magnetic compasses. Shipboard computer systems may integrate data from the various measurement systems. Redundant systems are particularly important so that a faulty sensor can be bypassed. Intercomparison tests with other current meters can be completed to assess the validity of ADCP measurements.

The range of a sound beam in water is inversely proportional to frequency, such that lower frequencies give longer ranges. However, the lower the frequency, the poorer the resolution, so that for a given range one wants to use as high a frequency as possible. ADCPs are available in several operating frequencies, from 75 kilohertz with an 1,800-foot (548.6 m) range to 1.2 megahertz, with a 45-foot (13.7 m) range. The resolution of the 75-kilohertz units is on the order of several feet (m), whereas the resolution of the 1.2 megahertz units can be as fine as several inches (tenths of a meter). Because the uncertainty in a single measurement or "ping" is quite high, it is necessary to average many pings to obtain a single velocity measurement. This characteristic is typical of most sonar and radar systems.

ADCPs are deployed from ships, tow bodies, moored BUOYS, submersibles, REMOTELY OPERATOR VEHICLES (ROVs), bottom platforms, and oil drilling rigs. They are usually deployed in the vertical mode (axis of the transducer vertical), so that the beams scan upward or downward, thus obtaining a vertical profile of currents. However, they are sometimes employed in the horizontal (axis of the transducer horizontal) mode, so that the beams scan outward horizontally to obtain a cross current or cross channel profile. This mode of operation is usually possible only in deep water, because in shallow water reflections from the surface and bottom would dominate the received signal.

These instruments are an outgrowth of the Doppler speed logs, which have been used for many years to measure ship speed over the bottom by transmitting an acoustic pulse down to the ocean floor, which is reflected back to the transmitter and passed to electronic circuitry, which measures the ship's speed by means of the Doppler shift in the received signal, in a fashion similar to the police Doppler radar guns for determining vehicle speed on the highway. Research was performed to extract water velocity information by modifying the speed logs

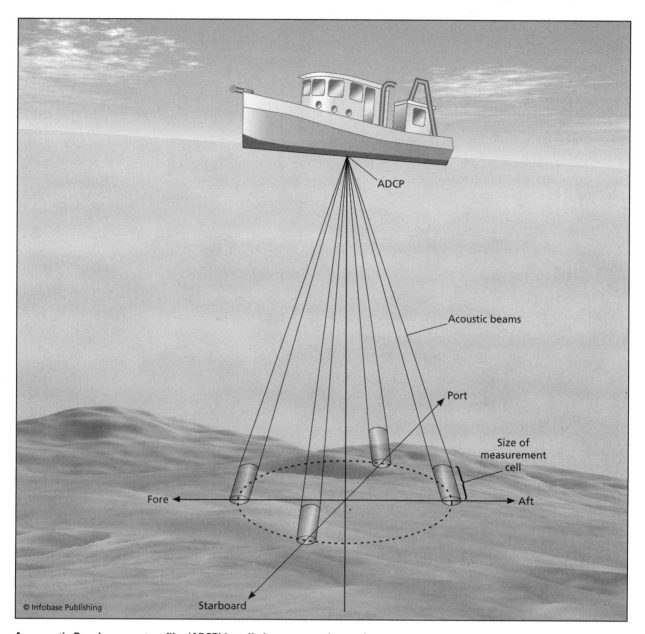

An acoustic Doppler current profiler (ADCP) installed on a research vessel

during the 1960s and 1970s. By the 1980s, water velocity measurements were being routinely measured by acoustic Doppler current meters. The profiling capability of the instruments gave rise to the name acoustic Doppler current profiler (ADCP).

Steady improvement in these instruments has resulted in multiple capabilities for measuring the ocean's dynamical properties. ADCPs are now available for making measurements in the ocean in water depths as great as 19,685 feet (6,000 m), for profiling across navigation channels, for deployment in hazardous areas such as bottom fishing grounds, and in measuring the directional wave spectrum by adding a vertically directed acoustic

beam sensing wave height above the transducer, and by integrating the directional particle velocities from the Janus beams. Self-recording ADCPs can be deployed in the deep ocean to 19,685-feet (6,000 m) depths and recovered one year later with the data recorded internally on solid-state devices. In shallow deployments from moored buoys or bottom platforms, the data are usually transmitted by satellite or direct radio link to receiving stations on shore and fed to computers for real-time presentation for navigation, as in the case of the NOAA PORTS® system (*see* essay, NOAA's PHYSICAL OCEANOGRAPHIC REAL-TIME SYSTEM). ADCPs are also available in miniature versions to be used in

An acoustic Doppler current profiler (ADCP) mounted on a catamaran for a NOAA survey of Chesapeake Bay   *(NOAA/NOS—Rich Bourgerie)*

channels, streams, and aqueducts. They can easily be deployed by divers.

The ADCP is generally recognized by oceanographers as being the premier current measuring instrument of our time. It has revolutionized the measurement of water velocities in lakes, rivers, and the sea.

Recovery of an RD Instruments acoustic Doppler current profiler (ADCP) from a deep-sea deployment   *(RD Instruments)*

Further Reading

Gordon, R. Lee. *Acoustic Doppler Current Profiler Principles of Operation: A Practical Primer.* San Diego, Calif.: RD Instruments, 1996.

acoustic release   A mechanism designed to hold a MOORING in place until released by a coded acoustic signal from a command console onboard a recovery vessel is called an acoustic release. Upon release from the ANCHOR attached to the bottom, the buoyancy of the floats supporting the mooring line will propel the mooring to the surface. In this scenario, the anchor is generally expendable. There are situations when salvage divers are used to retrieve the anchor. Usually, the mooring supports oceanographic instruments for measuring CURRENT, TEMPERATURE, SALINITY, and other variables. For subsurface deep-ocean moorings well below diver depth, acoustic releases are especially important. The instruments on these moorings usually record data internally on either magnetic tape or solid-state recorders and must be recovered to download the data. On a typical operation, the buoys are deployed for several months before being retrieved by a research vessel. Deep moorings usually employ acoustic releases in pairs, so that if one fails to release, the second (backup) responds.

Acoustic releases are generally preferable to timed releases that release the mooring at a preset time, because if the mooring is released prematurely, no one will be there to recover it. Acoustic releases provide more control in instrument recoveries than inexpensive timed releases. The least sophisticated timed releases rely on galvanic action to release transmitters, trigger the uncovering of sediment traps, underwater deployment of drogue-chutes, or to serve as a backup to expensive electronically timed or acoustic releasing devices. Although acoustic releases are most common for deep-ocean moorings, they are also used for subsurface moorings in shallow water when recovery by divers is inconvenient or impractical. A concern in shallow water is that marine growth may foul the release mechanism, since the mooring will be in the photic zone, and barnacles and plant life may cover the release mechanism. To avoid this problem, the linkages of the instrument are coated with anti-fouling paint or compound prior to deployment.

Acoustic releases are also used to retrieve arrays of acoustic sources and receivers and oil drilling and recovery gear. They typically have acoustic operating frequencies on the order of tens of kilohertz with 30 hertz (Hz) being frequently used. A hertz is a unit of frequency equivalent to one cycle per second. A research vessel recovering a mooring equipped with acoustic releases usually positions itself as close as possible over the mooring using GLOBAL POSITION-ING SYSTEM (GPS) navigation. Once positioned to retrieve the mooring, the marine technician will send the acoustic signal to release the mooring, so that the ship will be able to use its deck gear to retrieve it. Ranges of the acoustic transmitter on the recovery vessel are on the order of a mile (1.6 km) or so. Prior to deployment, surface marker buoys would have been installed on the mooring line painted in international orange to improve the visibility of the mooring when it breaks the surface. In the event of high winds and heavy seas, the mooring will be difficult to see from the recovery ship. To make recovery easier, an acoustic transponder may be placed in the mooring line for location while it is still underwater, and a radio transponder for location on the sea surface.

The signals sent by the recovery vessel to release the mooring have been coded so that spurious ambient noise or acoustic signals from other vessels will not release it prematurely. In the commercial sector, recovery vessels may use acoustic releases to retrieve arrays of acoustic sources, receivers, fishing nets, and oil drilling gear.

acoustic tomography A technique for studying the ocean by measuring the travel time of sound between multiple known transmitters and receivers is known as acoustic tomography. Changes in sound speed are related to changes in the physical properties of seawater such as TEMPERATURE, SALINITY, and PRESSURE. Thus, sound is a useful tool to remotely sense vast regions of the ocean. The name "tomography" comes from the Greek words *tomos,* meaning "to slice," and *graphein,* meaning "to write." A similar method used to image the human body is called X-ray tomography.

Sensing of ocean temperature from satellites has been a boon to oceanographers, filling in the blanks of vast areas of the ocean, such as the South Pacific, for which there is little data. However, satellite measurements provide information only about the sea surface. Acoustic tomography has the potential of providing information on the internal temperature structure of the ocean, essential to understanding ocean circulation and air-sea interaction.

By deploying multiple acoustic sources/receivers equipped with very accurate clocks, the changes in sound speed over a large number of acoustic paths provide slices of the ocean's acoustic structure of a region, typically around 100 miles (several hundred kilometers) on a side. The equations for sound speed as a function of temperature, salinity, and pressure can then be used to solve for one of these variables, usually the temperature, in terms of the variations in the acoustic signals.

It is also possible to estimate the current speed by this method. If an acoustic source and receiver are deployed in the direction of the CURRENT, the travel time of a sound pulse will be less in the direction of the current and greater in the direction opposite the current. The difference in travel times yields the speed of the current.

In addition to covering large areas hundreds of miles (several hundred kilometers) on a side, acoustic tomography offers the advantage of averaging out the omnipresent short period fluctuations in ocean properties, which cause "noise" in the measurements, and make interpretation difficult.

A University of California oceanographic acoustic tomography experiment called Acoustic Thermometry of Ocean Climate or "ATOC" began in 1995 to better provide comprehensive measurements of the variability in ocean temperatures. The study involves the use of low-frequency rumbling sources, i.e., frequencies of less than 1,000 hertz (kilohertz) off the coasts of Hawaii and California, with receivers sited around the PACIFIC OCEAN. Declassified U.S. Navy SOund SUrveillance System (SOSUS) stations in the northeast Pacific were used to record many of the signals. SOSUS consists of bottom-mounted hydrophone arrays connected to shore facilities by undersea communication cables. These arrays are found on CONTINENTAL SLOPES and SEAMOUNTS at optimal locations for undistorted long-range acoustic propagation. Other ATOC receivers were located near the Big Island of Hawaii and just north of the equator near Christmas Island. Since the speed of sound is proportional to the temperature of the water as noted above, sound speed can be used to infer the average temperature or temperature trends in the Pacific. These data taken over extended time and space scales could be used to detect GLOBAL WARMING. Fine-resolution analysis of shorter time and space scales, on the order of weeks and tens of miles (km), could identify eddies transporting water of anomalous temperatures.

Acoustic thermography tomography has raised concerns among many environmental activists, since these experiments employ powerful sonic impulses that are intended to be received across ocean basins. Environmentalists have been concerned that the high levels of acoustic power required by tomography could interfere with communications by whales, or even cause hearing loss or other adverse conditions. Researchers such as marine mammologists are evaluating the response of MARINE MAMMALS and other marine organisms to man-made sounds, and to provide information needed to direct policies for long-term protection and conservation of marine species. The NATIONAL OCEANIC AND ATMOSPHERIC ADMINISTRATION's National Marine Fisheries Service has delayed ATOC until the impact of low-frequency noise on marine mammals is better understood.

Further Reading

NPAL, North Pacific Acoustic Laboratory, Scripps Institution of Oceanography, University of California, San Diego. Available online. URL: http://npal.ucsd.edu/npal04/. Accessed July 21, 2008.

adaptive management Natural resource management where decisions are made as part of an on-going and science-based process is called adaptive management. Observed information deficiencies and data gaps need to be identified to improve monitoring efforts. Statistical validity will generally increase over the course of monitoring as information deficiencies are corrected. Descriptive STATISTICS and new understandings developed over long observational periods are used to refine subsequent monitoring efforts. The resulting scientific findings are evaluated against societal needs to provide the basis for adapting resource management rules.

Adaptive management is a central theme for coastal zone management. Managers of marine protected areas such as sanctuaries or estuarine reserves monitor the results of actions in order to develop a knowledge base that may indicate the need to change a particular course of action. Monitoring is intended to assess the effectiveness of management action in maintaining or enhancing ecosystem integrity and fitness. This management approach applies the concept of experimentation to the design and implementation of natural resource strategies by testing clearly formulated hypotheses about the response of a particular system such as the Channel Islands National Marine Sanctuary in Southern California to management decisions. Hypotheses may be tested empirically using spatial or temporal controls to evaluate resource responses or through science-based models.

Adaptive management is much more than "learning by doing," its data gathering and analyses focused on management expectations and tests of the real efficacy of management practices. This methodology provides data to compare alternative policies, prescriptions, and practices. Spatial statistics and sampling issues are important as decision makers apply numerical and statistical methods to understand environmental data that is collected from various sources. Adaptive management also ensures that there is collaboration among the public, managers, and scientists. This approach integrates the interests of managers, scientists, and public users by generating information that can feed future research and stimulate management actions benefiting the ecosystem. It is a cycle involving planning, action, monitoring, evaluation, and adjustments.

As an example, researchers during the late 1980s used population matrix models and empirical data to determine that the best plan to increase population growth rates of loggerhead sea turtles (*Caretta caretta*). Results indicated the modification of shrimp nets to prevent the capture of large juvenile turtles. Today, Turtle Excluder Devices or TEDs are added to trawl nets to allow trapped turtles to escape before they drown. The TED consists of a metal frame with vertical bars about five inches (12.7 cm) apart that is fastened to the funnel of the trawl net. A flap of net covers an opening, which turtles can swim through to escape capture. Shrimp trawlers are required by law to use these devices and face stiff penalties if caught without them.

Further Reading

National Marine Sanctuaries, National Oceanic and Atmospheric Administration. Available online. URL: http://www.sanctuaries.nos.noaa.gov. Accessed January 6, 2007.

NOAA Coastal Services Center, National Oceanic and Atmospheric Administration. Available online. URL: http://www.csc.noaa.gov. Accessed January 6, 2007.

aerial survey Aerial survey operations provide a reliable and economical means to produce topographical maps and bathymetric charts. Sensors are carefully mounted on manned aircraft and unmanned aerial vehicles. Some employ precision gyro-stabilized image-capture systems. Aerial surveying includes cataloguing, archiving and making images accessible, creating maps and records based on aerial images, and linking the images with information derived from other sources. Commercially available software provides image processing capabilities and the ability to store features in a database to rapidly build geospatial products.

Typical aerial platforms include fixed-wing aircraft such as DeHavilland Twin Otters, helicopters such as Bell 206 Jet Rangers, and unmanned aerial vehicles (UAV) such as the U.S. Department of Defense's Pioneer UAV system. An onboard GLOBAL POSITIONING SYSTEM (GPS) receiver records the camera's precise location as it gathers remotely sensed data. The camera's location is used to georeference individual data elements, such as pixels and video frames. Surveyors collect data for a host of requirements from inspecting work or collecting raw data for maps or charts. Products support resource exploration, environmental impact assessment, and mapping near-shore reefs, coastal vegetation, and pollution such as mine wastes.

Some aircraft use aerial survey cameras such as digital color and multispectral (imaging in several wavelength bands) cameras or lasers such as light detection and ranging (LIDAR) systems to collect high-resolution hydrographic data. Applications of multispectral imagery over bodies of water include detecting pollutant plumes, monitoring CORAL REEFS, finding shipwrecks, locating mines, and tracking MARINE MAMMALS. Bathymetric LIDAR can be used to accurately measure shallow depths and underwater structures. Information can be presented either graphically on topographical maps or in digital form as digital maps.

Viewing the Earth's surface from the air provides unobtrusive access, fast data acquisition, high accuracy, and lower field operation costs than ship surveys. Aerial surveys are widely accepted as an effective means of mapping large areas. Aerial survey products are used to preserve species, manage marine sanctuaries and reserves, track ships, and provide relief assistance. An example aerial survey system is the Scanning Hydrographic Operational Airborne LIDAR Survey or SHOALS that was developed by the U.S. Army. SHOALS has typically been used to survey important navigation areas, such as ports and harbors, shore protection structures, and coral reefs worldwide.

Further Reading
National Geospatial-Intelligence Agency. Available online. URL: http://www.nga.mil. Accessed January 6, 2007.
Office of Coast Survey. National Oceanic and Atmospheric Administration. Available online. URL: http://www.chartmaker.ncd.noaa.gov. Accessed January 6, 2007.
Spatial Data Branch, Operations Division, Mobile District, U.S. Army Corps of Engineers. Available online. URL: http:///gis.sam.usace.army.mil. Accessed February 19, 2008.
United States Geological Survey, Science Topics. Available online. URL: http://www.usgs.gov/science. Accessed January 6, 2007.

aerographer's mate A Navy aerographer's mate—also called an "AG"—collects, analyzes, and forecasts information on the weather and ocean conditions, using a variety of data sources, equipment, and instruments for the planning of naval expeditionary operations.

These men and women are trained in the physical sciences of meteorology and oceanography. Apprentice-level AGs (those who have been practicing naval meteorology and oceanography for one to four years) master the use of meteorological and oceanographic instruments to monitor environmental conditions such as barometric PRESSURE, air TEMPERATURE, humidity, wind speed and direction, sea temperature, SALINITY, and WAVES. They prepare and analyze atmosphere and ocean-surface "synoptic" charts in order to provide forecasters with a complete "snapshot" of environmental conditions over large geographic areas. They collect data on upper-level atmospheric conditions, using instruments attached to weather balloons to help forecasters determine the best flight levels for aircraft. Some measurements are used to determine wind and air density in order to increase the accuracy of antiaircraft artillery, naval gunfire, and the delivery of precision weapons by aircraft. AGs often make use of climatological information, automated wind and cloud height sensors, satellite imagery, and numerical prediction model runs on supercomputers located at the Fleet Numerical Meteorology and Oceanography Center in Monterey, California. At the National/Naval Ice Center in Suitland, Maryland, AGs work together with NOAA and U.S. Coast Guard staffs to provide operational global, regional, and tactical scale sea ice analyses and forecasts tailored to meet the requirements for polar ICEBREAKING ships and submarines. They also provide specialized meteorological and oceanographic support to Department of Defense agencies.

Those AGs who are the most skilled in meteorological and oceanographic data collection and analysis, and who demonstrate the most leadership potential, are chosen to acquire more advanced training and education in meteorology and oceanography in order to earn the distinguished title of "forecaster." This is the single most important accomplishment of the AG's career and is quite difficult to attain. Forecasting is the most crucial skill required of AGs when supporting naval operations. Forecasters not only provide information about current weather and ocean conditions, but they are also best suited to predict future weather and ocean effects and, more important, their influence on specific operations, troops, and equipment. Their forecasts help naval leaders decide whether an airfield remains open or shuts down, whether a multimillion-dollar exercise is cancelled or continued, and sometimes whether lives are lost or saved. Forecasters may also provide ships with "Optimum Track Ship Routing" (OTSR) services, wherein they combine their understanding of weather and ocean forecasting with navigation to provide ships' captains with the safest and most fuel-efficient paths across large ocean areas. Through the Department of Labor Military Apprentice Program, an AG can be certified as a practicing "meteorologist" by both the Department of Defense and the Department of Commerce. Unlike civilian meteorologists, however, AGs are also trained to interpret forecasts with regard to operational requirements of military missions.

A typical apprentice level AG performs weather technician duties with minimal supervision aboard COMBATANT SHIPS such as aircraft carriers and at shore stations such as regional naval meteorology and oceanography centers. A successful AG will often complete a 20-year career and serve several times at sea and at overseas duty stations. Overseas duty stations might include Rota, Spain; Keflavik, Iceland; Guantanamo Bay, Cuba; Yokusuka, Japan; and Bahrain.

Much like navy Meteorology and Oceanography (METOC) officers, the most experienced, skilled, and senior AGs become supervisors, spending less time performing practical meteorology and oceanography tasks and more time leading and managing a team of METOC technicians and forecasters. Together, AGs and METOC OFFICERS support navy and marine corps missions by synthesizing and interpreting atmospheric, electromagnetic, surf, and hydroacoustic information, and translating this information into operationally relevant decision aids for military commanders. They typically present this information as weather and oceanographic briefings to the officers and sailors assigned to U.S. Navy ships and planning staffs.

Further Reading

Naval Weather Service Association. Available online. URL: http://www.navalweather.org. Accessed January 6, 2007.

The Office of the Oceanographer & Navigator of the U.S. Navy. Available online. URL: http://www.oceanographer.navy.mil. Accessed January 6, 2007.

agnathans Agnathans are a group of jawless fishes thriving during the late Ordovician (about 450 million years ago) and Silurian (435 to 408 million years ago) periods of the Paleozoic era. The Greek word roots for the class Agnatha are *a* ("without"), *gnathos* ("jaw"), and *an* ("resembling"). Agnathans were primarily represented by ostracoderms, which were small fish covered by a bony external skeleton of solid shields or scales. Ostracoderms became extinct by the end of the Devonian (360 million years ago) period. The only modern survivors of the Agnatha are the hagfishes (Myxiniformes) and lampreys (Petromyzontiformes).

Hagfishes and lampreys (Cyclostomata) have eel-like bodies and are found in temperate waters of the Northern and Southern Hemispheres and some deep waters of the Tropics. Scientists have identified about 50 different species of hagfish and 38 species of lamprey. Unlike jawed vertebrates (Gnathostomata), which have gills that open through slits in the side of the body, hagfishes and lampreys have gill pouches that open through small pores. Both lack dermal armor, paired fins, and internal ossification (bones). The modern Agnatha have a single median nostril and a round mouth with a complex tonguelike structure bearing keratinized teeth.

While lampreys and hagfishes have many apparent similarities, there are significant differences that distinguish the two fishes. Hagfishes do not have a larval phase and spend their entire lives in the marine environment. Dead or dying fish and bottom-dwelling invertebrates in the deep sea serve as the primary food sources. The hagfishes have the unique ability to tie themselves into a knot and use their bodies to create leverage to pull flesh from fish and escape from predators. They have relatively large eggs with strong casings that are anchored to the substrate by filaments. Relatively little is known about the spawning behavior of these scavengers.

In contrast to hagfishes, lampreys are found in both fresh and salt waters and do have a larval phase. The larvae feed on algae, detritus, and microorganisms found in streams and rivers. The larval stage lasts at least three years. Metamorphosis takes three to six months and starts when the larvae are approximately 5 inches (120 mm) long. Once metamorphosis is complete, the larvae swim to sea, where feeding and growth rates increase dramatically. Lampreys feed on either blood or muscle tissue. Adults feed by attaching their mouths to a host organism and using their complex muscular tongues to rasp away flesh. When the adult phase ends, the lampreys stop feeding and return upriver to spawning grounds.

Scientists initially used the term *cyclostome* (round mouth) to classify lampreys and hagfishes because of the obvious similarities between the two groups; however, most scientists now consider hagfishes and lampreys only distantly related. Today, hagfishes are regarded as an early derivative of the first vertebrates of the Cambrian (more than 500 million years ago), whereas lampreys and gnathostomes share a number of significant characteristics that suggest they are more closely related. These characteristics include the presence of neural arches along a rodlike notochord that supports the body during development, large eyes with associated eye muscles, osmotic control of body fluids, and nervous regulation of the heart.

Paleontologists are using the ostracoderm fossil record to sequence the evolution of hagfishes, lampreys, and jawed vertebrates. By studying ostracoderm evolution, scientists believe they can discover information about the evolutionary pathway of the nasal structure, the lateral line, sense of smell, the brain, and other vertebrate characteristics.

The Agnathan fossil record is providing information about the common ancestor of all known vertebrates. Recent improvements in preparing and observing fossils suggest that continued excavations in Early/Lower Ordovician and Cambrian deposits could uncover significant information relevant to the early evolution of vertebrates.

Lampreys were introduced to the North American Great Lakes region in the 19th century. By the middle of the 20th century, large lamprey populations had decimated many native and SPORT FISH stocks and threatened the survival of many economic and recreational activities in the Great Lakes region. For example, in the middle of the 20th century, lamprey populations reduced annual trout production in Lake Superior from 2,000 to 200 tons (1.8 million kilograms to 180,000 kilograms) in less than a decade. Since then, chemical controls have reduced lamprey populations in most areas of the Great Lakes by 90 percent, and successful fishery restoration efforts are in progress.

Koreans consume more than 5 million pounds of hagfish annually, and lamprey is a popular fish served with gourmet dishes in Portugal and Spain. Intense pressures on some hagfish and lamprey fisheries have left them on the brink of collapse. Native lampreys are also an important food source for Native Americans in the northwest United States, but dammed rivers and waterfront development obstruct lamprey migrations and threaten the survival of localized lamprey populations.

See also FISH.

Further Reading

Forey, Peter L., and Phillipe Janvier. "Agnathans and the Origin of the Jawed Vertebrates." *Nature* 361 (1993): 129–134.

MarineBio.org, Houston, Texas. Available online. URL: http://ww.marinebio.org. Accessed January 7, 2007.

PALAEOS: The Trace of Life on Earth. Available online. URL: http://www.palaeos.com. Accessed March 16, 2007.

Potter, Ian C. "Jawless Fishes." In *Encyclopedia of Fishes*, 2nd ed., edited by John R. Paxon and William N. Eschmeyer. San Diego, Calif.: Academic Press, 1998.

Raloff, Janet. "Lamprey: A tasty treat from prehistory?" Available online. URL: http://www.sciencenews.org/sn_arch/8_10_96/food.htm. Accessed July 15, 2003.

Agulhas Current The Agulhas Current is a very swift, warm current flowing southwestward along the southeast coast of Africa. The current begins where the South Equatorial Current of the INDIAN

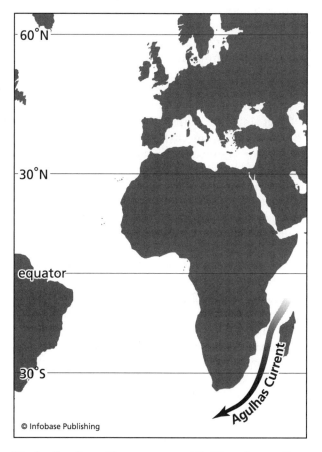

The Agulhas Current is a warm current that flows in a southwesterly direction, along the coast of East Africa.

OCEAN meets the African continent. This southward flow divides around the island of Madagascar and recombines south of Madagascar. Between the African continent and the island of Madagascar the flow into the Agulhas Current is known as the Mozambique Current. South of 30° South latitude, the width of the Algulhas Current proper becomes narrow and well defined, extending seaward about 60 miles (96.6 km) from the coast. Recent observations suggest that at times there is an undercurrent below the Agulhas Current flowing toward the northeast at about 0.6–0.8 knots (0.3–0.4) m/s. When the Agulhas Current reaches the tip of South Africa (Cape Agulhas), the current branches, with the western branch flowing around the Cape of Good Hope northward along the southwest coast of the continent into the BENGUELA CURRENT. Most of the current turns sharply eastward, and then southeast, or "retroflects" (loops back on itself) to flow into the Antarctic Circumpolar Current.

The velocity of the Agulhas Current varies with location, season, and depth. Speeds in the Agulhas Current average around 3.1 knots (1.6 m/s) and in some locations are greater than five knots (2.5 m/s) for many months, especially during the time of the southwest monsoon. The water within the current has distinct TEMPERATURE and SALINITY characteristics. The source water of the Agulhas Current is warm, flowing from the tropical Indian Ocean. The temperature of the current cools as it flows southward. Within the current, a narrow tongue of water, warm Agulhas water, extends to a depth of about 490 feet (149.4 m), with a temperature of about 68°F (20°C). There is an envelope of identifiable water around the Agulhas water with a temperature of about 62°F (17°C). The salinity at the boundary of the warm and cool Agulhas water is about 35.6 psu (practical salinity units). Salinity is determined from empirical relationships between temperature and the conductivity ratio of the collected sample to International Association for the Physical Sciences of the Ocean standard seawater. Volume transport is very large, being about 70 Sverdrups (70×10^6 m³/s) near 31° South increasing to about 100 Sverdrups (100×10^6 m³/s) at the tip of South Africa. A Sverdrup is equal to about 264 million gallons per second (1 million cubic meters per second) and is named after the Norwegian oceanographer and Arctic explorer Harald U. Sverdrup (1888–1957).

Off Cape Agulhas in the retroflection region, the current forms many ringlike eddies, many of which are carried around the Cape of Good Hope into the Benguela Current. Recent satellite, infrared, and microwave images show scalloped edges of the current and the formation of the rings and eddies, as well as current filaments. The exchange of heat and salt between the ATLANTIC OCEAN and the Indian Ocean by rings and eddies is important to the climatic balance of the region.

The Agulhas Current is perhaps most famous for its extremely high waves, which have caused the destruction of ocean liners and supertankers (see ROGUE WAVES). Strong southwest winds oppose the current, increasing wave heights; the steady progression of cold fronts and high waves from the Antarctic region further drives up wave heights. In future, with modern measurement technologies, sufficient information will be made available to forecast the severe seas that have proved so dangerous to shipping.

Further Reading

Beal, Lisa M., and Harry L. Bryden. "The velocity and vorticity structure of the Agulhas Current at 32°S." *Journal of Geophysical Research* 104, C3 (1999): 5,151–5,176.

Donohue, Kathleen A., Eric Firing, and Lisa M. Beal. "Comparison of the three velocity sections of the Agulhas Current and the Agulhas Undercurrent." *Journal of Geophysical Research* 105, C12 (2000): 28,585–28,593.

International Association for the Physical Sciences of the Oceans. Available online. URL: http://www.olympus.net/IAPSO. Accessed January 6, 2007.

Large Marine Ecosystems of the World, LME #30: Agulhas Current. Available online. URL: http://www.edc.uri.edu/lme/text/agulhas-current.htm. Accessed January 13, 2007.

Ocean Surface Currents, University of Miami, Rosenstiel School of Marine and Atmospheric Science. Available online. URL: http://oceancurrents.rsmas.miami.edu. Accessed February 7, 2004.

Shillington, Frank A., and Eckart H. Schumann. "High Waves in the Agulhas Current." *Mariners Weather Log* 37 (Fall 1993): 24–28.

U.N. Atlas of the Oceans, United Nations Foundation. Available online. URL: http://www.oceansatlas.org. Accessed January 6, 2007.

Alaska Current   The Alaska Current is a northeastward-flowing continuation of the Aleutian Current, entering the Gulf of Alaska and heading north northeastward, turning counterclockwise to follow the shoreline contours of Canada and Alaska around the Gulf of Alaska. The Alaska Current eventually flows southwest along Kodiak Island and the Alaska Peninsula, where it is known as the Alaska Stream. It eventually flows back into the Aleutian Current. The eastward flowing Aleutian Current transports subarctic water from the northern KUROSHIO and OYASHIO CURRENTS to the region between the North Pacific

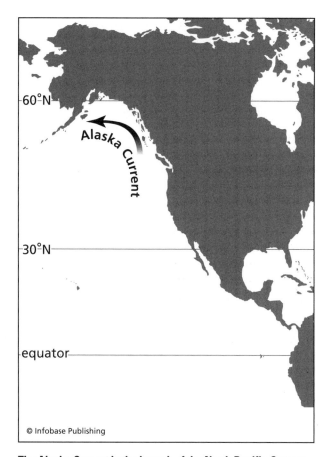

The Alaska Current is the branch of the North Pacific Current that carries warm water to the coasts of northwestern Canada and Alaska.

Current and the Aleutian Islands. Prior to reaching the Aleutian Islands, the Aleutian Current divides, with a branch going northward into the Bering Sea, and a branch going south of the Aleutian Islands. This latter branch divides again off the North Pacific coast of North America, with one branch flowing southward as the CALIFORNIA CURRENT, and one branch flowing northward as the Alaska Current.

Even though the waters of the Alaska Current are cold, it is categorized as a warm current, because its waters are warmer than those of its surroundings. Winter temperatures in the Gulf of Alaska can be as low as 32°F (0°C). The Alaska Current is colder and less saline at the surface than in the deepest part of the current. The water maintains hydrostatic stability because the increase of SALINITY with depth compensates for the increase of TEMPERATURE with depth, which ordinarily would make the water column unstable. This type of structure, with both temperature and salinity increasing with depth is often found in subarctic waters. The Alaska Current undergoes seasonal variations, with the strongest flow in winter and the weakest flow in summer.

Further Reading

U.N. Atlas of the Oceans, United Nations Foundation. Available online. URL: http://www.oceansatlas.org. Accessed January 6, 2007.

albedo Albedo is the fraction of light reflected from a surface compared to the incident light upon it. Ordinary albedo is a dimensionless number measuring the surface's brightness. If all the incident light is reflected, then the albedo is 1.0. Fresh fallen snow or the tops of clouds have an albedo of nearly 1.0. Glacier surfaces have albedoes that can range from nearly 1.0 to 0.1, depending upon whether the surface is dry snow or debris-covered ice, respectively. For this reason, it is important for researchers as well as those recreating in snow- and ice-covered regions to protect their eyes from the Sun's rays to avoid photokeratitis or snow blindness.

Bond albedo (named after American astronomer George P. Bond, 1825–65) is different from normal albedo. It is defined as the percentage of incident solar radiation reflected by a planet and is a measure of the planet's energy balance. The bond albedo of Earth is about 0.33, that of the Moon is about 0.12, and that of Venus is about 0.76. This parameter is of particular use for astronomers and remote sensing scientists. For example, scientists measure changes in sunlight reflected off Earth and onto the Moon as a tool to assess Earth's climatic patterns. Leonardo da Vinci (1452–1519) called this phenomenon Earthshine.

It is important when measuring the albedo to specify the WAVELENGTH. For example, suppose a light bulb shone for equal times on a white sheet of paper and a black sheet of paper. The white paper's TEMPERATURE would increase by a small amount. However, the black paper's temperature would increase by a large amount. The white paper has a larger albedo. It reflects more of the incoming energy. The black paper has a smaller albedo and absorbs more of the energy, causing it to heat up. If the light bulb were replaced by a heating element emitting infrared instead of visible light, both pieces of paper would warm to the same temperature. In the infrared range (750 mm to 1 mm), both pieces of paper have the same albedo. Infrared light lies between visible and microwave portions of the electromagnetic spectrum.

Albedo is an important parameter in modeling the response of the Earth's climate to incoming energy from the Sun. GLOBAL WARMING could lead to more clouds, increasing the Earth's albedo, which would reflect more energy away. The planet would then begin to cool. On the other hand, in polar regions changes in the landscape such as the spread of vegetation could increase the speed of climate change by altering snow cover and reducing albedo.

Further Reading
Canada Center for Remote Sensing. Available online. URL: http://www.ccrs.nrcan.gc.ca. Accessed January 6, 2007.
European Space Agency. Available online. URL: http://www.esa.int/esaCP/index.html. Accessed January 6, 2007.
Goddard Space Flight Center, National Aeronautics and Space Administration. Available online. URL: http://www.nasa.gov/centers/goddard/home/index.html. Accessed January 6, 2007.
Netherlands Environmental Assessment Agency. Available online. URL: http://www.mnp.nl/en/index.html. Accessed January 6, 2007.

algae Although there are many exceptions, algae are commonly regarded as aquatic organisms that are photosynthetic oxygenic autotrophs typically smaller and less structurally complex than land plants. In contrast to most terrestrial plants, algae are nonflowering and do not have roots, leafy shoots, or highly organized tissues for transporting water, sugars, and nutrients. Prokaryotic (lacking an organized nucleus) algae belong to the kingdom Monera, and eukaryotic algae (with an organized nucleus) are members of the kingdom Plantae. The fossil record indicates that blue-green algae existed 3 billion years ago and were probably the first organisms to carry chlorophyll *a* (a pigment creating the brighter hues of bluish-green). Algology (from the Latin *alga,* sea wrack) and phycology (from the Greek *phykos,* seaweed) refer to the study of algae. Algologists recognize 16 phyletic classes of algae based on differences in reproduction, cellular structure, pigmentation, and mobility.

Because algae evolved to tolerate a wide range of conditions, algae can be found almost anywhere on the planet. Most species of algae are aquatic or native to the tidal zone in either fresh or salt water, but some algae have adapted to survive in the terrestrial environment. In part, algae are successful because they tolerate a broad spectrum of physical conditions, such as dissolved oxygen and carbon dioxide concentrations, pH, solute concentrations, water turbidity, and TEMPERATURE. They can exist in the water column (planktonic), attached to the bottom (benthic), or on the surface of water (neustonic). Red algae have been observed in tropical waters 820 feet (249.9 m) beneath the ocean surface. Algae can be found on stones (epilithic), in sand and mud (epipelic), on plants (epiphytic), and on animals (epizoic). They inhabit the driest deserts, Antarctic rocks, and the waters beneath polar ice sheets. There are species of algae called cyanobacteria that can tolerate temperatures in hot springs of up to 162°F (72°C). Various terrestrial algae have evolved to coexist with fungi to form organisms known as lichens.

The body of some algae can be as small as bacteria (1 to 1.5μm), while others, like the giant kelps, can reach lengths of 200 feet (60.9 m). The primary types of algal body (thallus) include unicellular (single cell), colonial (groups of cells), filamentous (chain of cells), membranous (sheets of cells), and tubular structures. Algae may have a means of locomotion or not. Scientists often classify algae as blue-green, red, brown, green, or yellow-green depending on differences in pigmentation.

Variation in algal pigment is often associated with differences in storage products and cellular organization. The primary purpose of pigment in phototrophic or photoauxotrophic algae is to collect light energy for photosynthesis. ZOOXANTHELLAE, photosynthetic yellow-brown algae, live symbiotically inside the cells of reef-building corals and provide nutrients to corals for growth and reproduction. Other algae are heterotrophic and must either absorb organic compounds from their surroundings (osmotrophic) or ingest solid food particles (phagotrophic). Both sexual and asexual reproduction is common among algae, and like other algal characteristics, means of reproduction can be variable depending on species and habitat.

Humans have harvested nearly 500 different species of seaweed, a form of algae, over the last few thousand years. Each year the Japanese cultivate and harvest 350,000 tons of seaweed known as Nori or *Porphyra.* Nori, which is high in vitamins and minerals, is an important ingredient in food and medicine in many Asian countries and is becoming more popular in other parts of the world. The Chinese consume an estimated 100 million pounds (45 million kg) of seaweed annually. Seaweed extracts serve as thickeners and gelling agents in ice creams, toothpaste, sauces, and drinks. Seaweed or products containing seaweed are commonly used to enhance soil fertility and livestock diets.

Marine microalgae play a large role in biogeochemical cycling of carbon, nitrogen, sulfur, phosphorous and other elements, and long-term marine phytoplankton growth and sedimentation has resulted in significant oil and limestone deposits. Both freshwater and marine algal plankton are aquatic primary producers that are necessary for the sustained health of fisheries and ecologically important bird and mammal populations. Cyanobacteria are nitrogen fixing, meaning they have the ability to take nitrogen out of the atmosphere (diatomic N_2)

and release it into the ground in a form (ammonium ion) that is useful to other organisms.

Algal blooms are high concentrations of microscopic aquatic algae that usually form as a response to nutrient pollution. Human activities, such as improper sewage treatment and use of agricultural fertilizers, are a major cause of algal blooms that produce toxins responsible for poor water quality, substantial fish kills, and illness or death in vertebrates, including humans.

Because algae are sensitive to various physical and chemical pollutants, they are used as biomonitors to assess toxicity levels of detergents, pesticides, and other effluents. Algae also play an important part in some effluent treatment systems by filtering out inorganic nutrients while adding oxygen. Space researchers believe that algae may eventually assist life-support systems by removing CO_2 and adding O_2 to the atmosphere of spacecraft and planets.

See also KELP; PLANKTON; PRIMARY PRODUCTION.

Further Reading

Bold, Harold C., and Michael J. Wynne. *Introduction to the Algae,* 2nd ed. Englewood Cliffs, N.J.: Prentice Hall, 1985.

Grahm, Linda E., and Lee W. Wilcox, *Algae.* Upper Saddle River, N.J.: Prentice Hall, 2000.

The Harmful Algae Page. Available online. URL: http://www.whoi.edu/redtide. Accessed January 6, 2007.

Thomas, David N. *Seaweeds.* Washington, D.C.: Smithsonian Institution Press, 2002.

Wetlands, U.S. Environmental Protection Agency. Available online. URL: http://www.epa.gov/OWOW/wetlands/wqual/algae.html. Accessed January 6, 2007.

alkalinity Alkalinity is the measure of a solution's resistance to changes in pH or the "potential of hydrogen." It is the capacity of water to neutralize an acid or the measure of how much acid can be added to a liquid without causing a significant change in pH. The intensity of alkalinity is reflected on the pH scale by values ranging from 7 to 14. Values below 7.0 indicate the intensity of acidity, and 7.0 indicates neutral. Water that produces neither an acid nor alkaline reaction is neutral because there are an equal number of free hydrogen and hydroxide ions. Examples of common acidic-to-alkaline solutions are acids such as lemon juice or vinegar to bases such as soap or baking soda.

Many marine organisms thrive at specific pHs, while others can tolerate a range of values. As aquarists know, expensive marine fish are sensitive to pH changes, which are usually well controlled or changed very gradually. For example, the Golden Butterfly

(*Chaetodon semilarvatus*) requires high water quality and a pH from 8.1 to 8.4. These popular aquarium fish are commonly captured from the RED SEA. In the ocean, the pH is usually around 8.0, that is, slightly alkaline. For organisms such as coral, they effectively remove bases such as calcium and carbonate from the water column in order to build calcium carbonate skeletons. In the aquarium as in the natural ocean, adding a buffer such as oyster shells can raise alkalinity. The concentration of calcium carbonate can be determined to calculate the alkalinity of the water.

Oyster shells are very important to an estuary. They help to raise alkalinity by adding calcium carbonate, but also provide great habitat for benthic organisms such as worms, crustaceans, and fish. In estuaries, oyster reefs develop in favorable locations where live oysters grow on the empty shells of dead oysters. Oyster shells also have rough surfaces that greatly increase the surface area where other sessile creatures such as sea squirts, barnacles, and anemones can attach.

Further Reading

Ohrel, Ronald L., Jr., and Kathleen M. Register. *Volunteer Estuary Monitoring: A Methods Manual,* 2nd ed. EPA-842-B-06-003, March 2006. Available online. URL: http://www.epa.gov/nep/monitor. Accessed July 21, 2008.

Snyder, Carl H. *The Extraordinary Chemistry of Ordinary Things,* 4th ed. Hoboken, N.J.: John Wiley, 2003.

altimeter A device for measuring the distance from the instrument to a reference surface below is called an altimeter. Altimeters can be carried, mounted in an aircraft, or flown in an orbiting satellite. The most common type of altimeter is found in aircraft and is used to measure the height of an aircraft above the ground. In this case, the altimeter is a barometer that is set to zero at the air pressure on the ground. Other types of altimeters can use laser beams to measure the distance to the surface below. These laser altimeters have been used in aircraft to survey the terrain below to an accuracy of centimeters. They have been used to chart glaciers, monitor beach migrations, and map the debris field at "ground zero" after terrorist attacks of September 11, 2001, in the United States. Laser altimeters are also used on board orbiting satellites to map the sea surface height, the land's topography, and the polar ice caps. The biggest hindrance to using a laser altimeter is the Earth's cloud cover. The lasers are incapable of penetrating through the clouds and fog.

Another type of altimeter is the radar altimeter. The radar altimeter is capable of penetrating through the clouds and fog that shroud large portions of the Earth's oceans. Radar altimeters are used on aircraft and on spacecraft. Some space-borne altimeters operate at a frequency of 36 gigahertz. If the position of the satellite is known precisely, the sea surface height can be measured to an accuracy of about two centimeters. The circular footprint (the intersection of the beam with the ground) of a radar altimeter has a radius of about 6.2 miles (9.9 km) and is sampled about every .4 miles (.64 km) along the satellite's ground track.

A prototype radar altimeter was flown on *Skylab*, the first U.S. space station, in the early 1970s. The first highly successful and scientifically useful satellite altimeter was flown on the U.S. Navy's *Geosat* satellite. This satellite was launched in the 1980s to measure the shape of the Earth's geoid (shape of the Earth approximated as an ellipsoidal solid of revolution whose surface coincides with mean sea level). The geoid is important in orbit considerations for launching missiles from submarines; hence the U.S. Navy's interest. Furthermore, sea surface topography yields information on ocean circulation, since changes in sea surface height produce pressure gradients (*see* PRESSURE GRADIENT) that drive ocean currents. However, once this data was declassified in the 1990s, it provided a trove of scientific data and the first comprehensive and standardized charts of the Earth's bathymetry (topography of the seafloor).

Because *Geosat*'s altimeter was able to measure the height of the sea surface with an accuracy of two inches (5 cm), it was sufficiently accurate to provide data from which maps of the GULF STREAM were constructed from the analysis of the sea surface topography, and by applying the geostrophic equation (*see* GEOSTROPHY). From these data, even mesoscale (order of tens of kilometers in the ocean, such as eddies) features of the ocean were identified. (Ten kilometers equal 6.2 miles.)

The TOPEX/Poseidon altimeter, a joint NATIONAL AERONAUTICAL AND SPACE ADMINISTRATION (NASA) and Centre National d'Etudes Spatiales (CNES in France) project, was launched in the early 1990s and has operated nearly perfectly for more than 10 years. The *Jason-1* altimeter has recently joined it. This pair of satellite altimeters is providing oceanographers with a treasure of data that allows measurements on space scales from hundreds to thousands of kilometers. (One hundred kilometers equal 62 miles.)

New space-borne altimeters are in the works or proposed to measure the ice caps and the ocean bathymetry with hitherto unparalleled accuracy. Altimeters have been and will continue to be an important tool for marine scientists.

Further Reading

CNES, Centre National d'Etudes Spatiales. Available online. URL: http://www.cnes.fr. Accessed January 6, 2007.

Jet Propulsion Laboratory, California Institute of Technology. Available online. URL: http://www.jpl.nasa.gov. Accessed January 6, 2007.

Laboratory for Satellite Altimetry, Oceanic Research and Applications Division, National Environmental Satellite, Data and Information Service, National Oceanic and Atmospheric Administration. Available online. URL: http://ibis.grdl.noaa.gov/SAT. Accessed January 7, 2007.

ambient noise   Ambient noise is the level of acoustic background noise in a given location, such as a deep ocean basin, the coastal ocean, or an ESTUARY, emanating from a variety of sources. These signals are actually pressure changes or compressions and dilations moving through the water. The main contributors to ambient noise in the ocean include sources such as shipping noise (sounds from propellers and engines), drilling rigs, underwater pipelines, dredging, and biologics (sounds from whales, fish, and even smaller organisms). Ambient noise from the interaction of the ocean with the atmosphere includes sounds from the wind, WAVES, and precipitation. The seafloor itself also passes seismic noise into the water above. These combined sound signals, which make up the broadband noise, are important because they contain information about the oceanic environment. *Broadband* is the term describing acoustic energy propagating over a wide range of frequencies. In deep water, sources of natural ocean noise are often events occurring at the air-sea interface, while in shallow water, biologics such as snapping shrimp (*Synalpheus stimpsoni, Alpheus pacificus,* and *Alpheus randalli*) can easily dominate the soundscape. Snapping shrimp live in burrows, and they make a snapping sound with their claws. Shallow water ambient noise may also contain important information about the geoacoustic parameters of the seabed. Gaining information on the surface features of the seafloor such as SEAMOUNTS and ridges is important for navigation by submarines. Some researchers use arrays of hydrophones and transponders deployed from research platforms such as the Floating Instrument Platform (FLIP) to listen to background noise in the ocean. These signals are important in understanding phenomena from wind and waves to earthquakes and TSUNAMIS.

Acoustical oceanography is an important tool for measuring many processes. Investigators attempt to record, study, and identify the many sounds associated with the noise field in the ocean. As sound propagates through the water, its energy is spread over an increasing area, and the intensity of the signal decreases with distance as this spreading occurs. A propagation-loss diagram is used to represent this effect along a given ray path, which represents the sound's travel. When the intensity of the sound expressed in decibels falls below the level of the ambient noise, the signal will not be audible. There will always be work for acousticians, since ambient noise in the ocean is continuously present at varying levels and comprises acoustic energy from many sources. For example, rainfall over the ocean plays a pivotal role in atmospheric dynamics, air-sea interactions, the oceanic salt budget, and climate variability on a wide range of temporal and spatial scales. Precipitation over the ocean is one of the most difficult measurements to make reliably because of its highly episodic nature. Today, at-sea acoustical methods provide a reasonable means of making quantitative rainfall measurements. Such "acoustic rain gauges" have been included in ARGO floats. Beginning in 2000, an international program called "Argo" began with the deployment of approximately 3,000 neutrally buoyant free-drifting floats that transmit TEMPERATURE and SALINITY profiles in the upper 6,500 feet (1,981.2 m) of the ocean when they surface.

Further Reading

Argo: Part of the integrated global observation strategy. Available online. URL: http://www.argo.ucsd.edu. Accessed January 6, 2007.

National Research Council. *Environmental Information for Naval Warfare*. Washington, D.C.: National Academies Press, 2003.

Urick, Robert J. *Principles of Underwater Sound*. New York: McGraw Hill, 1983.

American Bureau of Shipping (ABS)   The American Bureau of Shipping is a preeminent ship classification society that determines structural and mechanical capabilities of ships and other marine structures. ABS is a nonprofit corporation with world headquarters in Houston, Texas, and serves the public interest as well as the needs of the marine industry by completing inspection and analysis services. ABS has offices or representatives working out of principal seaports. According to the ABS mission statement, "the primary purpose of ABS is to determine the structural and mechanical fitness of ships and other marine structures for their intended purpose. It does this through a procedure known as classification."

ABS promotes the security of life, property, and the natural environment primarily through the development and verification of standards for the design, construction, and operational maintenance of marine-related facilities. Vessels that have been "classed" are listed on the ABS Web site along with detailed information, such as type, size, construction material, machinery, fuel used, and date and type of last survey. Classification helps to ensure that ships are designed, built, and operated in accordance with internationally recognized safety standards. Classification rules cover a large variety of equipment and systems essential for the safe and reliable operation of a ship. The U.S. Coast Guard (USCG) does not require that any vessel be classed with the ABS. However, there is an April 27, 1982, Memorandum of Understanding between the USCG and ABS stating, "Unless specified elsewhere, copies of ABS approval letters, stamped plans, pass certificates, and other ABS documents provided to the Officer in Charge, marine Inspection, will be deemed sufficient for verification of compliance with USCG requirements."

ABS establishes industry standards through technical rules and guides, "for the design, construction, and operational maintenance of marine vessels and (offshore) structures." These rules and guides are based on scientific principles, including disciplines such as naval architecture, marine and ocean engineering, and oceanography. Examples of ABS rules are as follows:

- Rules for Building and Classing Steel Vessels under 61 Meters (200 Feet in Length)

- Rules for Building and Classing Offshore Installations: Part 1, Structures

- Rules for Certification of Cargo Containers, 1983

- Rules for Certification of Cargo Containers, 1987/with Marine Container Chassis Test Report

- Rules for Building and Classing Steel Vessels, 1988

ABS Guides cover numerous topics from environmental stewardship to risk management. Examples of ABS Guides are as follows:

- Prevention of Air Pollution from Ships

- Certification of Oil Spill Recovery Equipment

- Marine Safety, Quality & Environmental Management

- Risk Assessment Applications for Marine & Offshore Oil & Gas Industries

More than 60 rules and guides are listed online from the ABS Web site; other literature of direct benefit to the marine industry includes ABS Advisory Notes and Manuals.

ABS obtains input from industry trade shows, conferences, and professional society meetings, which provide forums for the presentation of technical information on marine science topics, offshore technologies, and naval architecture. ABS scientists, engineers, and architects are at the forefront in developing risk-management and risk-assessment strategies for the safe design, construction, and operation of innovative new vessels such as natural gas carriers (LNG tankers). In addition, certification applies to vessel components from bilge pumps and diesel engines to winches and cranes.

Further Reading

"ABS Company—Overview." Available online. URL: http://www.eagle.org/company/overview.html. Accessed May 5, 2003.

International Association of Classification Societies. Available online. URL: http://www.iacs.org.uk/index1.htm. Accessed January 6, 2007.

National Vessel Documentation Center, U.S. Coast Guard, U.S. Department of Homeland Security. Available online. URL: http://www.uscg.mil/hq/g-m/vdoc/nvdc.htm. Accessed January 6, 2007.

Office of Operating and Environmental Standards, U.S. Coast Guard, U.S. Department of Homeland Security. Available online. URL: http://www.uscg.mil/hq/g-m/mso/index.htm. Accessed January 6, 2007.

American Fisheries Society (AFS) The American Fisheries Society, founded in 1870 as the American Fish Culture Association, is the oldest and largest professional society representing fisheries scientists. Today, the American Fisheries Society has more than 9,000 members.

During the Civil War reconstruction period, aquaculturists in the northeastern United States began to attribute declines in local fish stocks to poaching, pollution, soil erosion, dam construction, and general habitat alteration. To address the problem, they formed the American Fish Culture Association, and the organization's goal was to restock local waters with fish from their hatcheries. Interestingly, from 1874 to 1882 the second president of the association was Robert Barnwell Roosevelt (1829–1906), the uncle of the 26th president of the United States, Theodore Roosevelt (1858–1919). In 1900, President Roosevelt signed a law to construct the second federal fisheries laboratory in the United States at Beaufort, North Carolina.

The association later changed its name to the American Fisheries Society, and their current goals are to increase the visibility and understanding of the fisheries profession, increase the influence of fisheries science, enhance visibility and public perception of AFS, and increase public appreciation of the benefits of scientifically managed fisheries. The overall mission of AFS is to improve conservation and sustainability of fishery resources and aquatic ecosystems by advancing fisheries and aquatic science and promoting the development of fisheries professionals.

AFS uses a variety of tools to accomplish their mission. AFS publishes four well-respected fisheries research journals that are available online: *Transactions of the American Fisheries Society, North American Journal of Fisheries Management, North American Journal of Aquaculture,* and *The Journal of Aquatic Animal Health.* AFS also publishes books that cover topics in biology, aquaculture, management, habitat, water quality, professional tools, policy, socioeconomic, and fisheries technology. Their books have helped to standardize the names of fish as well as fisheries procedures (e.g. fish kill counting techniques and fish disease diagnosis).

AFS aligns the expertise of its members to create discipline groups called sections, which specialize in areas such as bioengineering, fish health, marine fisheries, and native peoples' fisheries. AFS has a number of programs dedicated to professional certification, international affairs, public affairs, and public information. These resources are applied to produce opinions on various issues (such as aquatic species introduction) and publicize executive and legislative branch decisions that affect fisheries resources. Annual meetings draw thousands of participants and cover topics such as invasive species, stock assessments, and the management of recovering fish stocks.

Further Reading

American Fisheries Society. Available online. URL: http://www.fisheries.org. Accessed July 15, 2003.

Berry, Charles R., Jr. "The American Fisheries Society: An AIBS Member Society Promoting Fisheries Science and Education Since 1870." *BioScience* 52.8 (2002): 758(3).

American Geophysical Union (AGU) The American Geophysical Union is a professional society that promotes the interests of geophysics and geophysicists. Geophysicists are scientists who study the Earth and its surrounding environment in space. The disciplines of geophysics include geology, meteorology, oceanography, space, and planetary physics. The AGU was founded in 1919 by the National Research Council, and functioned for more than half a century as an unincorporated affiliate of the National Academy of Sciences. In 1972, the union was incorporated in the District of Columbia, and membership was opened to scientists and students from all nations. The AGU is one of the member societies of the American Institute of Physics.

AGU's mission is:

1. to promote the scientific study of Earth and its environment in space and to disseminate the results to the public.

2. to promote cooperation among scientific organizations involved in geophysics and related disciplines.

3. to initiate and participate in geophysical research programs.

4. to advance the geophysical disciplines through scientific discussion, publication, and dissemination of information.

AGU has defined four fundamental areas of geophysics as:

1. Atmospheric and ocean sciences, including such topics as climate change, weather, air and water quality, ocean circulation, WAVES, the ocean floor, REMOTE SENSING, biological, and chemical processes.

2. Solid Earth sciences, including such topics as structure and dynamics of the Earth's interior, tectonic plates, earthquakes, volcanoes, and mineral resources.

3. Hydrologic sciences, including such topics as water supply and quality, groundwater contamination, waste disposal, droughts, floods, erosion, and desertification.

4. Space sciences, including such topics as structure and evolution of planets, moons, asteroids, comets, gravitational and magnetic fields, solar storms, space weather, cosmic rays, and X-ray bursts.

AGU is an individual membership society open to those engaged in research in the geosciences. There are currently more than 35,000 members, of which about 7,000 are students. AGU members include outstanding scientists from academia, government, and industry.

The union administers grant programs that provide free memberships to scientists from developing countries. Numerous awards are presented to scientists for outstanding achievements in their fields of endeavor.

AGU provides a choice of premier journals dealing with the four major geophysical subject areas, which are published monthly. The AGU journal dealing with marine science is called *Oceans and Atmospheres*. All members receive a weekly newspaper of recent scientific findings and meeting announcements, student opportunities, job announcements, and so forth, as well as the American Institute of Physics publication *Physics Today*. The union also publishes numerous specialized books as well as electronic resources on the World Wide Web.

AGU is governed by a council that consists of elected officers of the union, and from each of the 11 scientific sections. Generally, two meetings of the union are held every year, as well as other specialized meetings, such as the Ocean Sciences Meeting, which is held every other year, and small, highly focused meetings and workshops. AGU strongly encourages interdisciplinary research and publishes articles on biogeochemistry and biogeophysics.

Further Reading
American Geophysical Union homepage. Available online. URL: http://www.agu.org. Accessed October 16, 2003.

American Meteorological Society (AMS) The American Meteorological Society is a professional society that promotes the development and dissemination of information and education on the atmospheric and related oceanic and hydrological sciences. Atmospheric scientists include climatologists, meteorologists, atmospheric physicists and chemists; oceanic and hydrological scientists include oceanographers, ocean engineers, hydrologists, geologists, and geophysicists. Geophysicists are scientists who study the Earth and its surrounding environment in space. Charles Franklin Brooks (1891–1958) of the Blue Hill Observatory in Milton, Massachusetts, founded the American Meteorological Society (AMS) in 1919. The first members came primarily from the U.S. Weather Bureau and the U.S. Army Signal Corps. Their first publication was the *Bulletin of the American Meteorological Society,* which is still published. In 1944–45 Professor Carl-Gustaf Arvid Rossby (1898–1957), then president of the society, organized the publication of the *Journal of Meteorology,* which is now the *Journal of Applied Meteorology,* and the *Journal of the Atmospheric Sciences.* Carl-Gustav Rossby is best remembered for his studies of large-scale motions.

Waves in the ocean and atmosphere that propogate as a result of the CORIOLIS FORCE bear his name (*see* ROSSBY WAVES).

The role of the society, to serve as a scientific and professional organization for the atmospheric and related sciences, has continued to the present day. AMS now publishes in print and online nine highly respected journals as well as an abstract journal, and the *Bulletin of the American Meteorological Society,* which is sent to all members. The present membership is more than 11,000 professionals, professors, students, and weather enthusiasts.

The AMS administers two professional certification programs, the Radio and Television Seal of Approval and the Certified Consulting Meteorological (CCM) programs. It offers a large number of undergraduate scholarships and graduate fellowships to support students pursuing careers in the atmospheric, oceanic, hydrologic, and related sciences. The society offers several prestigious awards to scientists for significant achievements in advancing knowledge of the atmosphere and the oceans.

The AMS sponsors and organizes more than a dozen scientific conferences every year. It has published more than 50 monographs, books, and educational materials of all types. For the marine sciences, the society publishes the *Journal of Physical Oceanography,* which reports the results of research related to the physics of the ocean and to processes operating at its boundaries, and the *Journal of Atmospheric and Oceanic Technology,* which contain papers on the instrumentation and methodology used in the atmospheric and oceanic sciences. The *Journal of Climate* reports the results of climate research on large-scale variability of the atmosphere, the ocean, and the land surface. *Earth Interactions* is published online at irregular intervals reporting research results in the Earth sciences, especially interdisciplinary work.

The AMS also supports local chapters whose members include anyone interested in weather and climate, as well as professional atmospheric and oceanic scientists. These chapters typically hold meetings every month, offer featured speakers, provide weather information to local officials, and hold annual elections for officers.

Further Reading
American Meteorological Society homepage. Available online. URL: http://www.ametsoc.org. Accessed October 16, 2003.

American Society of Limnology and Oceanography (ASLO)

The American Society of Limnology and Oceanography is a professional association promoting the interests of limnology, oceanography, and related sciences. Membership in ASLO provides marine scientists on-the-job learning opportunities, chances to interact and network, and exchange information on topics ranging from the latest scientific discoveries in limnology and oceanography to job opportunities. The association publishes the acclaimed journal *Limnology and Oceanography,* holds several interdisciplinary meetings per year, and conducts special symposia.

The society attracts members who are interested in studying the oceans, inland waters, fisheries, estuaries, and wetlands. ASLO members exchange information across the range of aquatic science and collaborate in order to further investigations dealing with these subjects. Research topics include methods to control invasive species, impacts from depletion of the ozone layer, and environmental restoration.

Limnologists focus their work on the physical and chemical characteristics of freshwater lakes, ponds, and marshes as well as the plants and animals that live within them. Oceanographers are interested in the physical properties of the ocean, such as CURRENTS and WAVES (physical oceanography), the chemistry of the ocean (chemical oceanography), the geology of the seafloor (marine geology), and the organisms that are found within the oceans (marine biology and biological oceanography). Related sciences include disciplines such as coastal engineering, fisheries biology, and hydrology. ASLO is one of the few marine science associations embracing all these disciplines.

ASLO provides educational resources for students and educators, both specific to limnology and oceanography, and for general science. Resources range from computer-based training to research opportunities for undergraduates. ASLO has established the Minorities in the Aquatic Sciences program to increase participation of underrepresented minorities in the sciences. Minority students from the United States have been selected to participate in minority-focused workshops and the annual ASLO meeting with funding from agencies such as the National Science Foundation and the NATIONAL OCEANIC AND ATMOSPHERIC ADMINISTRATION.

Further Reading
ASLO: American Society of Limnology and Oceanography. Available online. URL: http://www.aslo.org/. Accessed May 5, 2003.

amphibious warfare Amphibious warfare refers to the use of specially trained, equipped, and sea-based Navy and Marine Corps units to perform missions ranging from humanitarian assistance and disaster relief to small-scale contingencies and major

The amphibious warfare vessel USS *Cleveland* deploys Marines in the combat zone. *(U.S. Navy)*

theater war, normally conducted in littoral (intertidal) and coastal regions from sea, land, or forward bases. Critical equipment includes vessels such as transport ships, specialized aircraft such as vertical/ short take off and landing (V/STOL) jets, helicopters, tilt-wing propeller-driven aircraft, lighters, elevated causeways, and waterborne craft such as amphibious tractors and hovercraft. Near the coast, a wide diversity of environmental factors can affect military operations. Military commanders and their staffs rely on meteorological and oceanographic forecasts during the planning, embarkation, rehearsal, movement, and assault phases of amphibious warfare operations. They must know whether environmental conditions match critical thresholds for weapon systems and permit the movement of equipment and personnel through shallow water and across the landing BEACHES to inland objectives.

To ensure that operators have essential information, support personnel must obtain timely information that includes:

- climatological and historical data

- weather parameters such as cloud ceiling height, visibility, and winds

- sea, swell, and surf conditions

- sea surface TEMPERATURE

- astronomical data (for example: sunrise/ sunset, moonrise/ moonset, and percent of illumination)

- tidal data that illustrates water fluctuations at local anchorages and the extent of beach at various phases of tide

- longshore and rip CURRENTS

- character of surf zone (for example: beach profiles, surf zone width, nearshore currents, breaker height, and breaker type)

- degree of exposure of potential obstacles in the surf zone

- bearing strength of sediments

- wave oscillation in harbors

- hydrographic data for inshore navigation of landing craft

- treacherous regions, such as inlets to bays and harbors

- locations of SANDBARS, reefs, and shifting channels.

In order to meet many challenges facing amphibious warfare, marine scientists have contributed to the development of innovative technologies such as the amphibious vehicles the Expeditionary Fighting Vehicle and other high-speed vessels, as well as Rapidly Installed Breakwaters, modular causeways, and high-resolution imaging systems that detect important features such as waterlines and shoals. Hovercraft are especially important to moving personnel and equipment from ships to shore. The design was patented by Sir Christopher Sydney Cockerell (1910–99) in 1955. State-of-the-art measuring systems are also used to provide real-time information on conditions such as winds, WAVES, and currents. Such systems are similar to the ocean observatories that are being implemented around the United States as part of the Integrated Ocean Observing System. Tactical decision aids are employed that help planners estimate rates of mine drift, burial, and scour. These resources enable operational forces to accomplish their diverse military operations, which include tasks from ship-to-shore movement and mine countermeasures to identifying exits off the beach and routes for cross-country movement.

Further Reading

Clancy, Tom. *Marine: A Guided Tour of a Marine Expeditionary Unit.* New York: Berkley Books, 1996.

Coastal and Hydraulics Laboratory, Engineering Research and Development Center, U.S. Army Corps of Engineers. Available online. URL: http://chl.erdc.usace.army.mil. Accessed January 6, 2007.

Direct Reporting Program Manager Advanced Amphibious Assault, Expeditionary Fighting Vehicle. Available online. URL: http://www.efv.usmc.mil. Accessed January 6, 2007.

Marine Corps Warfighting Laboratory, Marine Corps Combat Development Command, U.S. Marine Corps. Available online. URL: http://www.mcwl.quantico.usmc.mil. Accessed January 6, 2007.

National Research Council. *Environmental Information for Naval Warfare.* Washington, D.C.: The National Academies Press, 2003.

Welcome to USS *Cleveland!* Available online. URL: http://www.cleveland.navy.mil/default.aspx. Accessed July 21, 2008.

amphidromic point   A point of zero amplitude of the observed or constituent tide is classified as an amphidromic point. It is a node of the tide wave around which the tide wave rotates counterclockwise in the Northern Hemisphere (*see* TIDE). The semi-diurnal, or twice daily tide completes this cycle in about 12.4 hours. The pattern of the tide is shown by plotting cotidal lines, or isolines of equal phase of the tide wave around the amphidromic point. The cotidal lines can be labeled with degrees of phase, or in time intervals in hours, of either solar time or lunar time, with respect to the average interval between the Moon's passage over the Greenwich Meridian and the time of local high water. The cotidal lines are traditionally labeled by Roman numerals.

An amphidromic point and the cotidal lines that radiate from it like the spokes of a wheel are known as an amphidromic system. Amphidromic systems are found in ocean basins, the basins of marginal seas, the Great Lakes, and large embayments and estuaries. These circulation patterns are a result of the Earth's rotation, which imposes the CORIOLIS EFFECT on the tidal motion, such that the tide wave rotates counterclockwise around the basin. The time of one cycle of the wave is determined by the period of the observed tide or tidal constituent such as the 12.4–hour semi-diurnal tide.

The Englishman William Whewell (1794–1866) developed the first plot of an amphidromic system for the North Sea in 1836, although his (correct) interpretation of the tidal circulation would not be widely accepted for many decades. The term *amphidromic,* from the Greek for "running around," was coined by Rollin A. Harris (1863–1918) of the U.S. Coast and Geodetic Survey (a predecessor organization of the present U.S. NATIONAL OCEANIC AND ATMOSPHERIC ADMINISTRATION, NOAA). Harris helped design a tide-prediction machine that computed 37 tidal constituents. He produced the first world map of cotidal lines.

See also WATER LEVEL FLUCTUATIONS.

Further Reading

Cartwright, David E. *Tides: A Scientific History.* Cambridge: Cambridge University Press, 1999.

anadromous   Anadromous fish species spend portions of their life cycle in both fresh and salt waters, entering freshwater from the ocean to breed. *Anadromous* is a term describing fish such as the American shad (*Alosa sapidissima*), which ascends

Although this old fish ladder is charming, it must be replaced by a new, more efficient ladder to help anadromous fish move past the dam at Parker River, Massachusetts. *(NOAA, Louise Kane)*

estuaries and rivers from the ocean to spawn. The American shad is indigenous to the Atlantic coast from the St. Lawrence River to Florida and enters freshwaters where it was born to spawn.

Fisheries biologists plan and execute long-term fish tagging programs to understand fish migrations and to manage fish stocks. Some researchers tag hatchery-reared fry with tetracycline, which stains the fish's otolith. An otolith is one of the small bones in the internal ear of vertebrates and in the auditory organs of many invertebrates. During capture of anadromous fish, if there is a tetracycline stain on the otolith, then the tiny ear stones are further analyzed like tree rings with microscopes. By understanding release time and locations of the fry and capture time and locations of the mature fish, investigators are able to create a picture of the species' migration patterns. Besides otoliths, scales, fins, and vertebrae are some of the other anatomical structures most frequently collected from finfish to determine age. Investigators have found that American shad enter estuaries such as the Chesapeake Bay after four to six years at sea to spawn in low-salinity tributaries such as the Susquehanna River.

Other familiar examples of anadromous fish species include Pacific trouts and salmon of the genus *Oncorhynchus*. Rainbow (*Oncorhynchus mykiss*) and cutthroat (*Oncorhynchus clarki*) trout are capable of migrating, or at least adapting to seawater. Their anadromy is either a genetic adaptation or simply an opportunistic behavior. The fish known as steelhead trout refers to the anadromous form of a rainbow trout; it bears the same scientific name as a rainbow trout. Thus, the steelhead is a rainbow trout that migrates to the ocean as a juvenile and returns to freshwater as an adult to spawn.

Chinook (*Oncorhynchus tshawytscha*), coho (*Oncorhynchus kisutch*), sockeye (*Oncorhynchus nerka*), chum (*Oncorhynchus keta*), pink (*Oncorhynchus gorbuscha*), masu (*Oncorhynchus masou*), and amago (*Oncorhynchus rhodurus*) are all Pacific salmon that die approximately one week after spawning. The dead salmon become food for birds such as bald eagles (*Haliaeetus leucocephalus*) and North American osprey (*Pandion haliaetus carolinensis*), as well as mammals such as raccoons (*Procyon lotor*) and mink (*Mustela vison*). Scientists have employed acoustic and video technologies to document and estimate the numbers of salmon and steelhead passing through fish ladders such as the one on the Lower Granite Dam on the Snake River in Washington. Engineers design fish ladders so that each step provides a gradual increase in height and a place for the fish to rest.

See also FISH; MARINE FISHES.

Further Reading

Groot, Cornelis, and Leo Margolis, eds. *Pacific Salmon Life Histories.* Vancouver: University of British Columbia Press, 1991.

Hagerman National Fish Hatchery, U.S. Fish and Wildlife Service-Pacific Region. Available online. URL: http://www.fws.gov/hagerman/lower_granite_dam.html. Accessed January 6, 2007.

McClane, Albert J. *McClane's Field Guide to Saltwater Fishes of North America.* New York: Holt, Rinehart & Winston, 1978.

McPhee, John. *The Founding Fish.* New York: Farrar, Straus and Giroux, 2002.

anchors Devices that are usually designed to dig into the seafloor and are attached to ships or buoys by chain and lines are called anchors. The traditional parts of an anchor are the shank, flukes, stock, arms, and crown. The shank is the component of the anchor that transmits the tension force to the flukes, which are projected surfaces designed to bite into the ground. The stock is a crossbar at the top of the shank, and the crown is the bottom portion of the anchor where the arms are united. Depending on size and type of vessel, the weather, and bottom conditions, different types of ground tackle will be used to hold position. On board a vessel, the anchors, the anchor line and chain, shackles, windlass, and any other gear used to anchor the vessel are called, "ground tackle." Mariners define the line and chain attached to the anchor as rode. *Scope* is the term that describes the ratio of rode paid out to the depth of the water plus the height to the bow chock. In general, greater scopes increase the holding power by contributing to a more horizontal pull on the anchor. The amount of

Marine Embedment Anchors: A Twist on Anchors

In times of fair weather, people all over the world are drawn to the ocean, and since the beginning of time, man has engineered and built objects and vessels to use in the ocean for commercial, research, and recreational purposes. But in times of stormy weather, the ocean can present a very harsh environment for property such as boats, floating docks, and fixed piers. Therefore, anchoring has become a science and an art in the marine industry. There are in fact many ways to secure structures to the seafloor, which has different soil or bottom compositions. Some traditional examples include mushroom and deadweight anchors.

Marine embedment anchors are intended to help people contend with stormy weather and to minimize damage that so often results. Therefore, the primary advantage of marine embedment anchors over traditional mooring anchors and mooring piles is the increased holding capacity. These anchors are intended as permanent mooring anchors, as indicated by their weight, shape, and installation. Viable marine embedment anchors not only deliver more holding power, but they must deliver that increased holding at competitive costs.

Helix anchors are marine embedment anchors that are screwed into harbor bottoms with hydraulic equipment, which allows their installation in resisting soils. Firm soils deliver the best holding, but one is able to compensate for softer soils by using larger anchors or by screwing the anchors deeper into the softer soils. Because the soils are so critical to the performance of anchors, one measures the resistance created by the soils during installation. Based on that resistance, one can accurately approximate the holding that will be delivered. If the resistance is low because the soils are soft, shaft extensions can be added during installation that allow one to screw the anchor deeper into the soils in search of firmer soils. Alternatively, one can remove (unscrew) the first anchor and replace it with a larger Helix anchor better suited for the softer soils. A larger Helix anchor would have additional or larger diameter disks welded to the same central shaft. This adaptation provides more anchor surface area and holding power when working with the weaker soils or soft sediments. Ability to match anchors with different soil conditions makes it possible to deliver exceptional holding in wide varieties of soil types (bottom compositions).

Helix anchors have been installed in water depths from several feet (m) to more than 100 feet (30.5 m) deep. An installation can be done by divers using submersible installation equipment, or it can be done from surface vessels working down through the water column without any diver. There are advantages to each system—depending upon the local situation and readily available equipment. Prior to any deployment, an oceanographic survey considers how factors such as bottom composition, tides, waves, and icing will support installation planning for a marine embedded anchor. The installation is planned in details and executed with precision.

Anchors that are protected by hot-dipped galvanizing and built on a solid central shaft offer a longer life in the marine environment that do traditional mooring anchors. The galvanized shaft on a Helix high-load

rode (chain) used by a ship when anchoring is normally five to seven times the depth of water.

The sea area in relation to factors such as winds, WAVES, and bottom that is suitable for anchoring vessels is called an anchorage. Good holding ground for anchoring is marked on nautical charts where bottom compositions are usually mud and sand or mud and clay. Anchoring also depends on factors such as depth, where one has to know WATER LEVEL FLUCTUATIONS to remain clear of hazards. Low-tide shoals can ground a ship, while CURRENT speeds or winds may cause anchors to drag. Depending on the currents, weather, and bottom conditions, different amounts of rode may be paid out. The main anchor should hold in up to 30 knots (15.43 m/s) of wind. Adverse anchoring conditions such as heavy weather require more scope, additional chain at the anchor, or possibly a heavier anchor or even multiple anchors. Some vessels have a "storm anchor" that would be designed to hold at wind speeds up to 42 knots (21.61 m/s). Maintaining position during severe weather is difficult since as the wind speed doubles, holding requirements can quadruple. Other tasks requiring multiple anchors range from salvage crews trying to maintain position over a sunken wreck to oceanographers measuring current speed and direction from a boat. Using multiple anchors to keep a boat from swinging back and forth requires detailed planning and ready crews to prevent tangled or broken lines and fouled propellers. It should also be noted that no anchor is 100 percent retrievable.

There are many types of anchors, which come in various shapes and sizes from lightweight grapnel anchors to 40,000-pound (18,143.7 kg) stockless anchors. The traditional styles of anchors are Admiralty, Grapnel, Navy, Danforth, Plow, and Mushroom. Admiralty anchors have good holding power and consist of a stock, a shank, which carries two square shaft anchor is 1.75 inches square (44.5 mm sq), which is a significant mass when compared to alternative anchors. Some deadweight anchors such as stacks of old railroad wheels or mushroom anchors may be used for permanent moorings, but when holding power and efficiency are of concern, planning considerations should consider holding power, depth of settlement, and potentials for dragging during heavy weather. It is also important for stewardship of the submarine environment to note that a properly installed marine embedment anchor has minimal impact on the ocean bottom, both during installation and throughout its life. This is especially important in environmentally sensitive areas such as reefs because the anchors cause almost no disturbance during installation, and they do not drag or slide around the bottom, causing damage during bad weather.

The Helix anchor marine embedment anchor is used for a variety of important marine applications. Primarily it is used to secure boat moorings (called mooring balls in some regions), floating marine docks, aquaculture fish pens as well as other grow-out and hatchery equipment, and scientific or research equipment. Helix Mooring Systems has installed anchors across the United States and in more than 20 maritime nations to meet the needs of a variety of customers, ranging from aquaculturists and marine scientists to mariners and naval forces. Example applications include mooring fish pens in Canada and Taiwan; providing storm mooring systems for charter boats in Saint Martin; mooring U.S. Coast Guard and U.S. Navy docks in Washington and Guam; mooring public and private marina docks in Panama, Michigan, Ohio, Nevada, New York, Florida, Washington, and Maine; installing community mooring fields in Massachusetts, California, Florida, and Antigua; and mooring boundary markers, navigation aids, and scientific equipment in Louisiana, Connecticut, and New Jersey. Helix anchors have also been used in lieu of traditional mooring anchors and cruising anchors to protect fragile environments, including eel grass habitats and coral reefs. Marine embedded anchors are playing an important role in restoring and safeguarding submerged aquatic vegetation beds and coral reefs.

Helix anchors are high-load square shaft anchors that have also supported components of ocean observatories such as LEO-15, which stands for Longterm Ecosystem Observatory at a depth of 49 feet (15 m). It is operated by Rutgers University and is part of the U.S. Integrated Ocean Observing System. Long-term moorings benefit from well-placed and correctly installed marine embedment anchors.

For smaller vessels or inland waterway buoys, Helix Mooring Systems offers a smaller and lower-load round shaft anchor that is suitable for installation in remote locations. These anchors, relying on the same holding concepts as high-load anchors, can be installed by a diver without any special equipment.

Further Reading

Helix Mooring Systems. Available online. URL: www.helixmooringsystems.com. Accessed July 21, 2008.

—**Royce Randlett, Jr.**, president
Helix Mooring Systems
Belfast, Maine

arms, and a crown. On the arms are two flat broad parts called the flukes, which terminate in points called the pea or bill. Grapnel or grappling anchors have four arms and limited holding power, but are ideal for grabbing onto bottom structures such as rocks when other anchors would drag. The Navy anchors use weight and two grabbing flukes to hold a vessel. Danforth anchors work best with small boats and are characterized by long, narrow, twin pivoting flukes at one end of the relatively long shank. The plow anchor has no stock and is hard to handle but, when pulled on, digs into soft bottoms. Mushroom anchors are commonly used for permanent moorings since they sink deeply into the sediments over time. Marine professionals such as naval architects often make the selection of ground tackle for larger vessels.

Marine scientists often deploy moored arrays of instruments to measure parameters and processes occurring in the water column and the marine boundary layer. Some of these moorings are long-term and are used to telemeter information to ground stations located at research laboratories or operational sites such as Coast Guard stations. The oceanographic and meteorological instruments are fixed to mooring cables and buoy platforms or masts. Some of the important hardware that is used in designing oceanographic moorings includes anchors, chain, wire rope, synthetic line, SHACKLES, SWIVELS, pear rings, hooks, subsurface floats, acoustic releases, surface buoys, radios, lights, and antennas. These moorings depend on well-designed anchoring systems such as two stockless anchors in tandem, deadweight anchors composed of cement blocks or clumps of old train wheels, or innovative square shaft anchors that auger into resisting bottom soils. Marine technicians often design, build, and deploy instrument moorings using the ship's crane and a quick release. The design of the moorings and the

Railroad wheels are often used as anchors when deploying oceanographic buoys.   *(NOAA)*

distribution of the instruments are based on the parameters to be measured as well as factors such as mooring tensions, depths, currents, waves, and winds. Technicians may mark distances along mooring lines by sliding numbered slips of plastic between strands, attach instruments directly to the mooring with combinations of shackles, pear rings, and swivels, and increase the safety factor by preventing shackle screw pins from working loose by wiring the pin to the shackle.

Structures in deep waters such as in the North Sea require innovative mooring solutions. Wave tanks and numerical models are often used to analyze hydrodynamic loads and responses of fixed and floating structures in different sea states. Offshore structures may utilize multiple anchor points and various chain lengths attached to complicated and massive drag, deadweight, pile, or embedded plate anchors. Offshore structures and big ship anchor types should be independently tested for holding power and be classified by standards organizations such as the AMERICAN BUREAU OF SHIPPING.

Mariners may also try to fix themselves in a permanent current such as the GULF STREAM. They use sea anchors that are cone-shaped frameworks covered with material such as canvas and weighted at the bottom to follow the current. These drogues or sea anchors can range from buckets to modified parachutes and also reduce a boat's drift before the wind. Sea anchor uses extend from helping fishermen stay on fishing grounds longer to helping mariners safely ride out a storm. A sea anchor is typically set off the bow of a boat so that the bow points into the wind and rough waves.

See also MOORING; SCOPE.

Further Reading

Maloney, Elbert S. *Chapman Piloting: Seamanship & Small Boat Handling,* 62nd ed. New York: Hearst Marine Books, 1996.
Marine Embedment Anchors from Helix Mooring Systems. Available online. URL: www.helixmooringsystems.com. Accessed July 21, 2008.

anemometer   An anemometer is a meteorological instrument that measures either the total wind speed or the speed of one or more components of the wind vector. Anemometers may be classified according to the type of sensor used. The most commonly used, and perhaps the oldest, is the cup anemometer, which is a mechanical instrument with a vertical axis of rotation that usually consists of three or four hemispherical or conical cups arranged symmetrically about the rotation axis. The rate of rotation of the cups is a measure of the wind speed. In gusty conditions, the cup anemometer overestimates the wind speed, because inertia carries the cups forward after the occurrence of a short pulse of high wind speed. A voltage proportional to wind speed usually transfers the rate of rotation to a visual display or recorder. An

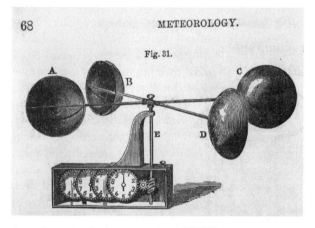

An early mechanical anemometer   *(NOAA)*

analog-to-digital converter samples the data for storage on a digital computer system. Quality instruments must come with a certificate of calibration, which was performed in a wind tunnel. Cup anemometers are usually used in conjunction with wind vanes to measure the direction of the wind, which is also transmitted to a display or recorder as the output of a compass by an electromagnetic signal. Often, a propeller rotating on a horizontal axis mounted on a wind vane, in which the whole instrument rotates into the wind direction, is used in place of the cup/wind vane combination.

For research purposes in meteorology it is often more desirable to measure orthogonal north and east components of the wind vector. Instruments employed for this purpose include Pitot tube, hot-wire or hot-film, and sonic anemometers. The Pitot-tube anemometer measures the wind speed by means of PRESSURE changes. Hot-wire or hot-film anemometers operate on the principle that a heated wire or thin film of conducting material on the surface of an insulating rod is cooled in proportion to the speed of the flow past the element. The wind speed is actually measured by the amount of electrical current needed to maintain the wire at a constant temperature. The data are recorded as a time varying current or voltage. The sonic anemometer measures the components of the wind vector by measuring the wind effect on the transit times of sound waves transmitted over known acoustic paths. Sonic anemometers can be configured in orthogonal triads to measure the three components of the wind speed, and thus determine the total wind vector. Hot-wire/hot-film anemometers and sonic anemometers have much shorter time constants than mechanical cup or propeller type instruments, and are thus more suitable for measuring wind gusts and atmospheric turbulence than cup anemometers.

The location of the anemometer vis-à-vis terrain or structural features that could block and distort the airflow around the anemometer is an important consideration in any installation. The location should be as free from structural interference as possible. The ideal exposure height of an anemometer is usually above obstructions, such as nearby trees, houses, or ship structures. If the data are acquired as part of a research project, or official weather data, national or international agreements specify the height as 32.8 feet (10 m) above ground or sea level, so that the data can be integrated with other wind measurements to produce a map of the regional wind flow.

The systematic recording of wind direction was begun in Italy in 1650 and in Britain in 1667. The speed of the wind was first observed by measuring the rate at which light objects were carried along by the air. This type of measurement, where a small, light object acts as a tracer of the wind motion, is called Lagrangian flow measurement. Measurement of the wind at a fixed point is called Eulerian flow measurement.

See also MARINE METEOROLOGIST; OCEANO-GRAPHIC INSTRUMENTS.

Further Reading

Armagh Observatory, College Hill, Northern Ireland. Available online. URL: http://www.arm.ac.uk/home.html. Accessed January 6, 2007.

Stevermer, Amy L. *Recent Advances and Issues in Meteorology.* Westport, Conn.: Oryx Press, 2002.

angular momentum Angular momentum is a measure of the spin of a body or a particle. Just as linear momentum is conserved, angular momentum is conserved. A body that is set in motion will spin. If that body interacts with another body and gives up some of its spin, that other body will begin to spin, such that the sum of the angular momentum at the beginning of the interaction and at the end are the same. Linear momentum is the product of the mass and velocity of a particle and is a conserved quantity. Similarly, angular momentum, which is the product of the mass, velocity, and distance from the point it is rotating or orbiting about, is a conserved quantity in the absence of external torques or other forces. A bicycle wheel when it is rotating has an angular momentum associated with it. The conservation of the angular momentum can be used to solve the motion of that wheel subject to the forces, such as turning the handlebars, on it. Similarly, air and water particles have an associated angular momentum. The angular momentum of those particles will be conserved. Another familiar example is an ice skater going into a spin, who spins faster and faster as he or she pulls the arms in close to the body.

Angular momentum describes how the Earth orbits steadily about a fixed axis in space such as the Sun. The Earth in its orbit changes its distance from the Sun and speed with which it rotates about the Sun. However, the product of its speed, mass, and distance from the Sun remains constant. Thus, the Earth's angular momentum is conserved.

This same concept applies to the fluids of the oceans and the atmosphere. Permanent ocean current systems such as gyres conserve their angular momentum unless a torque, such as winds at the surface, act upon them. Oceanographers and meteorologists measure the angular momentum of spinning masses of air or water in order to better understand ocean currents and Earth's winds. These same conservation laws play a large role in physics-based numerical models that are used to forecast ocean circulation and global wind patterns.

Further Reading
Newton, David E. *Recent Advances and Issues in Physics.* Phoenix, Ariz.: Oryx Press, 2000.

Antarctica   Antarctica is a pristine polar and generally uninhabited continent slightly less than 1.5 times the size of the United States, sometimes referred to in the vernacular as "the South Pole." Antarctica is the coldest continent, lying 600 miles (965.6 km) from South America and 1,600 miles (2,574.9 km) from Australia. Air temperatures in the high inland regions fall below -110°F (-79°C) in winter and rises to only -20°F (-29°C) in summer. The warmest coastal regions may reach the freezing point in summer but drop well below it in winter. During winter, more than half of the surrounding ocean freezes, which essentially doubles the size of the continent. Ice analysts at the National Ice Center have been mapping this Antarctic SEA ICE routinely since 1973 in support of U.S. Navy, U.S. Coast Guard, and U.S. Department of Commerce missions, and will soon release a complete digital record of these sea ice maps. Antarctica is predominately composed of a thick continental ice sheet and approximately 2 percent barren rock. Mountain ranges reach an elevation of 16,404 feet (4,999.9 km). The length of the coastline is approximately 11,165 miles (17,968.3 km). The SOUTHERN OCEAN surrounds the continent of Antarctica and is clearly delimited by the Antarctic Convergence, now called the Antarctic Polar Frontal Zone (APFZ). This polar ocean front is formed where cold Antarctic waters meet warmer waters to the north. Interestingly, organisms such as

Antarctica is a frozen and windswept continent. Forbidding snow and ice cover cliffs in the Lemaire Channel, 65 degrees 06 minutes S latitude, 64 degrees 00 minutes W longitude.   *(NOAA Corps—Rear Admiral Harley D. Nygren)*

krill (*Euphausia superba*) cannot pass north of this polar front.

The Antarctic ice cap contains approximately 70 percent of the world's freshwater and 90 percent of the world's ice. Huge ICEBERGS calve or break off each year from the floating ice shelves. During February 2002, an ice analyst from the National Ice Center studying space imagery of the Ross Sea located an iceberg roughly 11 nautical miles (20.4 km) long and four nautical miles (7.4 km) wide, and covering an area of approximately 58.24 square statute miles (150.84 km²). In summer, ice-free coastal areas include parts of southern Victoria Land, Wilkes Land, and Ross Island in McMurdo Sound. Natural resources include iron ore, chromium, copper, gold, nickel, platinum, coal, and hydrocarbons. Commercial fisheries include crustaceans such as krill and Antarctic king crab (*Paralomis spinosissima*), and finfish such as Patagonian toothfish (*Dissostichus eleginoides*) and mackerel icefish (*Champsocephalus gunnari*). Krill form the base of the marine food chain, and its predators include fish, squid, seabirds, and MARINE MAMMALS. The icefish have no hemoglobin or red blood pigment. Seabirds such as the emperor penguin (*Aptenodytes forsteri*) with flipper-shaped wings for swimming and black-browed albatross (*Diomedea melanophris*) with 12-feet (3.7 m) wingspans inhabit Antarctic regions. The albatross populations are declining in part due to long-line fishing, which unintentionally hooks diving seabirds as fishermen reel out or in a line many miles long with thousands of baited hooks. As the line moves through the water, the hooked birds are drowned. Marine mammal populations include elephant (*Mirounga leonina*), weddell (*Leptonychotes weddellii*), crabeater (*Lobodon carcinophagus*), leopard (*Hydrurga leptonyx*) seals and humpback (*Megaptera novaeangliae*), minke (*Balaenoptera acutorostrata*), and killer (*Orcinus orca*) whales. The Commission for Conservation of Antarctic Marine Living Resources regulates commercial fishing activities. This body tries to ensure that approved fishing activities do not harm animals such as penguins, seals, and whales.

An international body of scientists view Antarctica as a natural laboratory to study phenomena from sea level rise to polar ecosystems. Scientists study clues relevant to climate change that are recorded in the average 6,000-feet (1,828.8 m) thick ice sheet. For example, trapped bubbles in the ice hold an archive of atmospheric gases that provide indicators of global pollution. Other studies have shown evidence for ozone depletion in the upper atmosphere. Recent evidence for receding ice sheets and their contribution to world sea level rise are vital

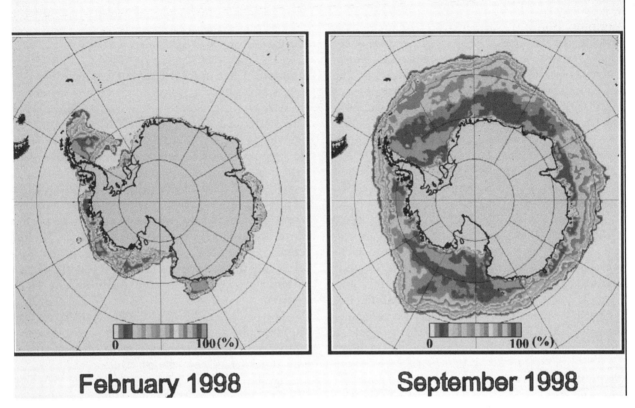

Antarctic Sea Ice

February 1998   September 1998

The variation in summer-winter sea ice coverage in the Antarctic continent   *(NOAA)*

to the understanding of global climate change. Several researchers have hypothesized that the kingdom of Atlantis lies below the Antarctic Ice Sheets, a place where people may have lived when there was no ice. The United States operates the Palmer Station, located on the southwestern rocky coast of Anvers Island off the Antarctica Peninsula, primarily for biological studies of birds, seals, and the Antarctic ecosystem. Palmer's laboratories complement research conducted onboard the R/V *Laurence M. Gould*. Besides research, the *Laurence M. Gould* resupplies and transports personnel between Palmer Station and South American ports. A burgeoning new tourist industry uses ice-strengthened oceanographic research vessels that have been converted to comfortable, safe, small group, cruise ships. These passengers visit locations such as the Palmer Station.

During the early 1980s, scientists discovered large deposits of natural resources such as coal, natural gas, and offshore oil reserves. Many factors prohibit mining natural resources in Antarctica. The 1991 Protocol on Environmental Protection to the Antarctic Treaty placed a 50-year moratorium on the exploitation of Antarctica's natural resources. It

is also technically challenging to drill for oil in Antarctica due to the extensive ice shield. Furthermore, the depth of the continental shelf ranges from about 1,200 feet (365.8 m) to 2,600 feet (792.5 m), much greater than the global mean. If oil exploration were ever permitted, new technologies would need to mitigate or prevent discharges of sludge from oil tankers, natural oil seeps, and blowouts. If oil were to suddenly escape from a well in an uncontrolled, continuous eruption, it would be extremely difficult to cap or to drill relief wells in a timely manner. Sustainable development of resources to include oil in Antarctica requires the maintenance of the 1959 Antarctic Treaty and establishing a legal framework to minimize environmental degradation in addition to holding polluters strictly liable for their damages.

See also GLOBAL WARMING; MARGINAL SEA.

Further Reading

Large Marine Ecosystem of the World, LME #61: Antarctica. Available online. URL: http://www.edc.uri.edu/lme/text/antarctic.htm. Accessed January 13, 2007.

National Geographic. *Satellite Atlas of the World*. Washington, D.C.: National Geographic Society, 1998.

U.N. Atlas of the Oceans, United Nations Foundation. Available online. URL: http://www.oceansatlas.org. Accessed January 6, 2007.

Antarctic Bottom Water (ABW)

Antarctic Bottom Water is the coldest, densest WATER MASS in the oceans that fills the world's major ocean basins, and is hence the water mass that constitutes the abyss. The density of Antarctic Bottom Water is on the order of 1.028 kg/m³; its TEMPERATURE is about 32.54°F (0.3°C), its SALINITY on the order of 34.7 practical salinity units (psu). The top of the ABW layer mixes slightly with the North Atlantic Deep Water (see ATLANTIC OCEAN) rendering both its temperature and its salinity slightly higher than in the interior of the ABW layer.

Antarctic Bottom Water is formed at the surface of the WEDDELL SEA and the Ross Sea. At the surface of these water/ice seas, the water begins to freeze leaving the salt in the still liquid water below, increasing its salinity and lowering its freezing point. Thus the freshwater freezes first; the remaining liquid water becomes more saline, hence denser, and begins to sink all the way to the bottom, where it spreads northward to fill the major ocean basins of the world. This process is known as deep winter CONVECTION and is also known as thermohaline circulation. At the time of formation, the temperature of this water mass is on the order of 35°F (-1.7°C), but it increases as it mixes with the adjacent oceanic water masses. The Antarctic Bottom Water is the deepest water mass in the Atlantic and southern regions of the Indian and Pacific Oceans; it is found at depths greater than 13,000 feet (3,962.4 m). The bottom topography has a considerable affect on the flow of bottom waters. In the eastern South Atlantic at about 30°S latitude, the Walfisch Ridge blocks the northward movement of ABW. Hence, at the same latitudes above that, the bottom water is colder in the western South Atlantic than in the eastern.

In the Atlantic Ocean, the ABW is overlain by the North Atlantic Deep Water, which is formed during the Northern Hemisphere winter in the Greenland and Norwegian Seas. In the Pacific and Indian Oceans, there is no comparable source of deep water. The water just above the ABW is called Pacific Common Water, and is a mixture of various Atlantic water masses with the water of the Antarctic Circumpolar Current (see INDIAN OCEAN; PACIFIC OCEAN).

Further Reading

Knauss, John A. *Introduction to Physical Oceanography,* 2nd ed. Upper Saddle River, N.J.: Prentice Hall/Pearson, 1996.

Neumann, Gerhard, and Williard J. Pierson, Jr. *Principles of Physical Oceanography.* Englewood Cliffs, N.J.: Prentice Hall, 1966.

Polar Challenges: Adventures and Research. International Polar Foundation. Available online. URL: http://www.antarctica.org. Accessed January 7, 2007.

Antarctic Circumpolar Current (ACC)

A major subpolar permanent CURRENT driven by persistent westerly winds is called the Antarctic Circumpolar Current. It isolates Antarctica from warm southward-flowing western boundary currents while flowing clockwise around the continent. These cold waters with an average surface temperature of 28°F (-2.2°C) directly affect the present ice cap and glacial climate. As this current approaches the South American continent, part of the WATER MASS is deviated northward as the PERU (HUMBOLDT) CURRENT, which brings cold Antarctic water to the coast of Chile and Peru. The Antarctic current also flows northward as a subsurface current between Australia and New Zealand.

The region of the ACC has the wildest seas in the world, whipped up by strong and persistent westerly winds that circle the globe, providing an exceedingly long fetch (see FETCH). During the days of sailing ships before the building of the Panama Canal, it was necessary to sail around Cape Horn to travel between Atlantic and Pacific ports, the part of the trip that terrified ship's crews. Exploration of the Antarctic continent was not really possible before the 20th century. The region off South Africa where the ACC and the Agulhas Current meet is notorious for generating rogue waves (see ROGUE WAVES) that have sent supertankers to the bottom.

The Antarctic Circumpolar Current is the only current in the world ocean that completely circles the globe, having a total length of about 14,400 miles (23,174.5 km). The region of strongest flow of the current is the Antarctic Polar Frontal Zone (APFZ), formerly known as the ANTARCTIC CONVERGENCE. Jets of current can occur in the APFZ with speeds as high as two knots (1.0 m/s). However, the average speeds of the Antarctic Circumpolar Current are low, ranging from about 0.8 knots (0.4 m/s) south of the Antarctic (Polar) Front to more than 0.4 knots (0.2 m/s) north of the Front. Because the current is very deep, extending all the way to the ocean floor, about 13,123 feet (3,999.9 m) below the surface, its volume transport is very high. Estimates are on the order of 130 Sverdrups (130 × 10⁶ m³/s), making it the greatest current in the world ocean. Oceanographers believe that friction with the bottom topography is the reason the current speed is not greater. A balance between the wind stress at the surface and the frictional stress at the bottom governs the velocity structure of this current. The current between the surface and bottom layers of frictional influence is believed to be in geostrophic balance (*see* GEOSTROPHY). Both cyclonic and anticyclonic eddies have been observed in the current, which transport cold water northward and warmer water southward.

Between the Antarctic Circumpolar Current and the Antarctic continent, the Polar Current flows westward opposite to the ACC. Since these currents flow in opposite directions, the region between them is a region of diverging Ekman transports (*see* EKMAN SPIRAL) and hence a region of diverging surface waters leading to an upwelling of subsurface waters rich in nutrients. The result is an increase in productivity that is unmatched anywhere else in the world ocean. Dense populations of zooplankton such as krill form the basis for a food chain that includes penguins, leopard seals, orcas, and giant blue whales. Unfortunately, overfishing and whaling decimated this fishery in the early and mid 20th century.

The Antarctic Circumpolar Current and its cool Peru Current branch skirting the western shores of South America northward to the equator, together with the prevailing on-shore winds, exert a strong cooling influence over the coastal regions of all the western countries of this continent except Colombia. These cool and nutrient-rich waters support an immense fish population, which in turn attracts great numbers of seabirds. Species of plankton, fish, benthic organisms, mammals, and birds have sophisticated adaptations for survival in extreme cold. To better understand this extreme cold region, Antarctic waters are regarded as a Large Marine Ecosystem.

A branch of the Antarctic Circumpolar Current extends to the Galápagos Islands, as the Peru (Humboldt). The islands were forever made famous by Charles Darwin's book *On the Origin of Species by Means of Natural Selection*, which was published in 1859. On the east coast of South America, the climate at the Falkland Islands is temperate because the Antarctic Circumpolar Current cools the surrounding sea.

A recently discovered feature of the Antarctic region is the Antarctic Circumpolar Wave. Careful analysis of records of atmospheric PRESSURE, wind stress, barometric pressure, and SEA ICE extent has revealed a wave in the atmosphere and the ocean having two complete cycles around the Antarctic continent. The wave moves at about the speed of the Antarctic Circumpolar Current. Meteorologists and oceanographers believe that it is generated by the ENSO phenomenon, that is, the EL NIÑO and its atmospheric partner the Southern Oscillation. Anomalies in ocean surface temperature due to the Antarctic Circumpolar Wave have been observed to travel northward along the Pacific coast of South America and the Atlantic coast of Africa as far as the equatorial oceans. The details of these motions are not known at the present time.

Recent advances in offshore oil and gas production have been accomplished in Arctic regions. There is a temptation to assume that such successes can be duplicated in Antarctica. While techniques for polar exploitation will be of some use, it should be stressed that offshore Arctic drilling is a recent shallow water innovation. Working in deep water Antarctic regions would be considerably different. Large Antarctic ICEBERGS colliding with oil rigs would completely destroy those structures. For example, placement of drilling gear on the sea bottom would provide no protection down to a water depth of at least 650 feet (198 m) because icebergs would scour the sea bottom. Technology innovations to mitigate this problem might include emplacement of "blowout preventers" beneath the seafloor to avoid scour, icebreaking tankers, and towing icebergs.

Due to extreme conditions, polar environments possess unique values for research. Public interest in permanent currents such as the Antarctic Circumpolar Current remains high due its influence on maritime commerce, climate, biodiversity, the stability of ice sheets, and potential natural resources. Scientists study the Antarctic region to learn more about polar flora and fauna, the thermohaline (*see* THERMOHALINE CIRCULATION) component of the global oceanic flow, and to determine whether or not GLOBAL WARMING will cause the amount of glacial ice in Antarctica to diminish significantly, exacerbating global sea-level rise.

Further Reading

Charton, Barbara. *A to Z of Marine Scientists*. New York: Facts On File, 2003.

Large Marine Ecosystems Information Portal, LMEs of the Polar Oceans. Available online. URL: http://www./me.noaa.gov/Portal/jsp/LME_PO.jsp. Accessed February 29, 2008.

Ocean Surface Currents, University of Miami, Rosenstiel School of Marine and Atmospheric Science. Available online. URL: http://oceancurrents.rsmas.miami.edu. Accessed February 7, 2004.

U.N. Atlas of the Oceans, United Nations Foundation. Available online. URL: http://www.oceansatlas.org. Accessed January 6, 2007.

Antarctic Convergence A belt of converging surface water spanning the Earth roughly between 50 and 60 degrees South latitude is called the Antarctic Convergence. The Antarctic Convergence is also known as the Antarctic Polar Front, and the region around it as the Antarctic Polar Frontal Zone. The front is actually a boundary between TEMPERATURE and SALINITY in the ocean. Since WATER MASSES have specific temperature, salinity, and chemical content, this is analogous to air masses in the atmosphere, where adjacent air masses tend to remain distinct. The temperature change at the surface going across the front is generally on the order of 35°–37°F (1.7°–2.8°C). Surface water moves northward from the edge of the Antarctic continent and meets sub-Antarctic water, where, being colder and denser, it sinks and spreads out equatorward above the heavier North Atlantic Deep Water in the South Atlantic basin, and above the ANTARCTIC BOTTOM WATER in the PACIFIC and INDIAN OCEAN basins. The Antarctic Intermediate Water, a major water mass of the world ocean, spreads out as far as 35°N in the North Atlantic Ocean.

The Antarctic Convergence occurs as a result of Ekman transport (*see* EKMAN SPIRAL AND TRANSPORT) of Antarctic surface water moving eastward in the ANTARCTIC CIRCUMPOLAR CURRENT (the Ekman transport in the Southern Hemisphere is 90° to the left of the current direction) and the THERMOHALINE CIRCULATION (the temperature-salinity distribution). The cold, low salinity water spilling off the glaciers of Antarctica sets up the latter circulation.

Strong current jets and meanders mark the Antarctic Polar Frontal Zone and cold eddies that transport cold Antarctic surface water northward. Not much is known about these features, but an increasing database is becoming available from the oceanographic measurements of ships resupplying research stations on Antarctica and from aircraft and satellites. These cold nutrient-rich waters provide an ideal environment for many types of microscopic plants called PHYTOPLANKTON. Just south of the Convergence, ZOOPLANKTON such as Antarctic krill (*Euphausia superba*) feed on the abundant phytoplankton. Krill are then eaten by many organisms, including fish, squid, sea birds, and MARINE MAMMALS. The fish and sea birds attract predators such as seals, toothed whales, dolphins, and porpoises.

aphotic zone   The aphotic zone includes the deeper portions of the ocean's water column where visible sunlight does not penetrate with enough intensity to carry on photosynthesis. Depths in the aphotic zone are usually greater than 656 feet (199.9 m). Above the aphotic zone is the upper part of the water column, which is penetrated by visible light and classified as the rich and diverse photic zone. The compensation depth, where photosynthesis and respiration are balanced, forms an approximate boundary between the photic and aphotic zones.

The deeper aphotic zone is almost completely in darkness because scattering and absorption have completely attenuated the incoming solar radiation. Scattering of light is what causes the ocean to appear blue. The water molecules are smaller than the atmosphere's nitrogen molecules, thereby scattering radiation of higher frequencies than those scattered by the air. As the sunlight penetrates the water, the longer WAVELENGTH, red to orange colors, are absorbed first, the greens and yellows next, and the shorter wavelength blue to violets are absorbed last. The absorption of long wave or infrared radiation results in rapid heating of the upper layers of the water column by incident light.

Johann Heinrich Lambert (1728–77) was one of the first scientists to investigate the passage of light through various media; he published a book on the topic in 1758. He found that light attenuation with depth increases exponentially in a mathematical relationship now known as Lambert's Law. Today, optical properties of water such as TURBIDITY are very important in understanding satellite images since the oceans contain various light-attenuating materials. Turbidity is the qualitative measure of solid particles that are suspended in water. These particles cause the water to become cloudy or even opaque in extreme cases. The colors that are seen in imagery depend on which colors are reflected and which are absorbed in the water. Thus, water quality affects the depth of the aphotic zone and the ability to image the ocean with remote sensing. Conversely, remotely sensed images are used to infer water quality.

Generally speaking, the biomass or the total mass of all living organisms in a given area decreases with depth and distance from the coast. The aphotic zone is home to many different types of bizarre species of marine animals. Many of the organisms in the aphotic zone are detritus or filter feeders. Detritus is accumulated organic debris from dead organisms. As a survival mechanism, many plankton stay in the aphotic zone during daylight hours, then rise to eat at night. The main predators are midwater and deepwater fish such as the hatchetfish (*Sternoptyx obscura*), daggertooth (*Anotopterus pharao*), and viper fish (*Chauliodus macouni*). Many of these predators tend to have large heads and jaws and flexible bodies, which allows fish like the black swallower (*Chiasmodon niger*) and the gulper eel (*Eurypharynx pelecanoides*) to eat large fish. Some like the lantern fish (*Stenobrachius leucopsarus*) have large eyes and light-producing organs called photophores. The lantern fish may follow prey from the depths to the surface during the night. SONAR operators first noticed this diurnal migration and labeled it as the DEEP SCATTERING LAYER. A carnivorous mollusk, the giant squid (*Architeuthis dux*), also inhabits the aphotic zone. Characteristics of these large invertebrates are known only from stranded specimens or from their remains recovered from the stomachs of their predators, the sperm whales (*Physeter macrocephalus*). Sperm whales are the largest of the toothed whales, made famous by Herman Melville's *Moby-Dick*.

Further Reading

Ocean Explorer, National Oceanic and Atmospheric Administration, U.S. Department of Commerce. Available online. URL: http://oceanexplorer.noaa.gov. Accessed January 7, 2007.

aquaculture   The process of farming marine organisms including fish, mollusks, crustaceans, and aquatic plants is called aquaculture. Aquatic farmers provide some sort of intervention in the mass rearing process to enhance production, such as regular stocking, feeding, protection from predators, and culminating with harvesting. Tasks such as monitoring aquatic animal health, routine testing of water quality parameters, and the production of live fish foods is essential to an aquaculture operation. Besides caring for, feeding, and monitoring various species of aquatic animals, personnel involved in aquaculture have individual or corporate ownership of the stock being cultivated.

Commercial aquaculture ventures need to implement profit-maximizing and cost-efficient management strategies. When farmers receive live aquatic animals such as channel catfish (*Ictalurus punctatus*) that are packed in closed bags at high densities, the receiving water used for acclimation of the

Experimental shrimp-raising ponds at the Kaohsiang Marine Laboratory aquaculture facility in Taiwan   *(NOAA/Sea Grant, Dr. James P. McVey)*

incoming stocks should match the TEMPERATURE, SALINITY, and pH of the shipping water. Farmers might grow the small catfish for up to two years before they weigh a marketable one or two pounds (0.45 or 0.91 kg). Specialists indicate that it takes approximately two pounds of feed to grow a one-pound fish. Catfish feed consists of soybean meal with some corn and rice ingredients.

Marine scientists involved with mollusk aquaculture operations such as oyster farming in CHESAPEAKE BAY develop mathematical models of oyster production that analyze new activities and constraints based on experimental oyster reef data. The researcher's analyses might focus on strategies to produce larger or disease-resistant Eastern oysters (*Crassostrea virginica*). In some cases, mesh oyster shell bags containing spat are moved from an unfavorable oyster bed to a more favorable oyster bed based on salinity conditions. Spat are young oyster larvae that have metamorphosed into juvenile oysters. Not-for-profit organizations such as the Oyster Recovery Partnership are working with University of Maryland's Center for Environmental Science at Horn Point to restore the ecological and economic benefits of oysters in Chesapeake Bay. Creating habitat through the rehabilitation of historically productive oyster bars and seeding those bars with hatchery disease-free oysters produced at the Horn Point Laboratory are helping to increase oyster populations.

Farm-raised crayfish and shrimp have proven to be successful enterprises around the globe. The crayfish capital of the world is considered to be Louisiana, where the harvest comes from Mississippi basin wetlands. Two of the most abundant species are Red Swamp (*Procambarus clakii*) and White River (*Procambarus acutus*) crayfish. Farming crayfish involves filling and draining ponds on a yearly cycle. Since the crayfish eat detritus, the ponds are drained to grow natural vegetation such as rice that becomes food for the crayfish when the ponds are refilled during the fall. The crayfish grow during fall and winter flooded seasons and reach a marketable size during the spring.

In Saudi Arabia, the Saudi Fisheries Company has received acclaim for its ecologically safe warm-water shrimp aquaculture operations that raise thousands of tons of tiger (*Penaeus monodon*) and white (*Penaeus indicus*) shrimp annually. Shrimp eggs or larvae are either gathered from the natural environment or grown in hatcheries after being taken from female brood stock. Both of these Red Sea species are raised to maturity in shallow ponds that are approximately 6.56 feet (2 m) deep. Outflow from the farm's rearing ponds are treated biologically without any chemical or medical substances and monitored for purity prior to being returned to the sea.

There are various commercial and academic efforts aimed at raising and harvesting marine vegetation from drifting PHYTOPLANKTON to beds of SUBMERGED AQUATIC VEGETATION or SAVs. Saltwater tanks can serve as the medium for the culture of phytoplankton, microalgae that produce oxygen as a by-product of photosynthesis. Sea life shows great promise toward development of new pharmaceuticals, and some marine algae are already used as growth stimulants, treatment of esophagitis, and as a substrate for making dental impressions. SAVs such as eelgrass (*Zostera marina*), redhead grass (*Potamogeton perfoliatus*), wild celery (*Vallisneria americana*), sago pondweed (*Potamogeton pectinatus*), and widgeon grass (*Ruppia maritima*) are rooted, vascular plants that grow completely underwater or just up to the water's surface. They provide critical habitat for shellfish such as the blue crab (*Callinectes sapidus*) and other organisms, are food for a wide variety of waterfowl, and prevent erosion. Dr. Steve Ailstock, the director of the Environmental Center at Anne Arundel Community College in Maryland, teaches various people and organizations proven techniques for growing and restoring SAV beds and wetlands. He also advocates the importance of understanding water quality and the conditions that engender success in any wetland's restoration effort that requires the introduction of starter SAVs.

There are many benefits and some risks relative to aquaculture or farming of the sea. Aquaculture is gaining in popularity due to rapidly increasing fish demand and declining marine harvests. Many governmental organizations are looking for the marine aquaculture industry to help restore depleted populations. However, the production of fish and crustaceans can generate high concentrations of pollution if the facility is not properly designed. For example, particulate organic wastes from fecal material and uneaten food or soluble, inorganic excretory waste can pollute adjacent coastal waters. Extreme cases result in sediments and bottom waters becoming completely anoxic or contributing to the eutrophication of coastal waters. These waste materials cause the water body to become rich in dissolved nutrients. The ensuing environment favors plant life and culminates in low levels of dissolved oxygen that stress marine animals. Consumers of aquaculture products need to have absolute confidence that farmed seafood are safe and high quality. They also need to determine that the farms are not creating environmental damage.

See also MARICULTURE.

Further Reading

Food and Agriculture Organization of the United Nations (FAO). "FAO Fisheries Technical Paper. No. 500. Rome: Food and Agriculture Organization of the United Nations, 2006." Available online. URL: http://www.fao.org/docrep/009/a0874e/a0874e00.htm. Accessed March, 9, 2008.

NOAA Aquaculture Program. National Ocean Service. National Oceanic and Atmospheric Administration, U.S. Department of Commerce. Silver Spring, Maryland. Available online. URL: http://aquaculture.noaa.gov/. Accessed March 9, 2008.

World Aquiculture Society. Available online. URL: http://was.org/Main/Default.asp. Accessed March 9, 2008.

aquarium An aquarium is a container for displaying, studying, or maintaining marine or freshwater aquatic organisms and systems. Aquariums range in size from just a few liters to more than a million gallons and can be made of glass, acrylic sheet, fiberglass, or reinforced concrete. Filters, lights, pumps, and temperature control devices are important components of modern aquariums.

Accumulation of wastes from aquatic organisms living in aquariums can be toxic and is disposed of by aquarium water systems that are open, closed, or semi-closed. Open systems constantly renew the water in the system from an outside source, while closed systems constantly recycle water and only occasionally renew water. A semi-closed system recycles water but is always connected to an outside source. High water quality is the primary requirement for maintaining aquatic systems in aquariums. Chemical methods, biological filtration, and replicating the natural system are the primary means of sustaining water quality. To maintain the health of aquatic organisms, aquarists may also need to consider water clarity, dissolved wastes, temperature, tank décor, disease treatment, nutrition, visitor traffic, reflections, and acoustics.

Aquariums have been valuable to humanity for thousands of years. The Sumerians supplemented food supplies by maintaining fish in artificial ponds more than 4,500 years ago. According to records, ancient Egypt and Assyria also kept fish, and the Chinese were breeding carp for food by 1000 B.C. The ancient Romans kept fish for food and entertainment by circulating fresh seawater through aquariums.

Philip Gosse, a 19th-century British naturalist, was the first to use the term *aquarium*. His work and a better scientific understanding of the relationships between plants, animals, and oxygen resulted in the establishment of the first public aquarium in London, England, in 1853. This was followed by many public and commercial aquarium openings in Europe and the United States.

Today, public aquariums are a valuable educational resource. Thousands of students visit aquariums each year, and aquariums are often a major tourist attraction in the world's largest cities. The goal of many public aquariums is to inform visitors about topics of aquatic biodiversity conservation, traditional fisheries, AQUACULTURE, and endangered species. Famous public aquariums in the United States include the Aquarium for Wildlife Conservation in Brooklyn, New York; the National Aquarium in Baltimore, Maryland; Marineland of Florida in Marineland, Florida; the Monterey Bay Aquarium in Monterey, California; and the Waikiki Aquarium in Honolulu, Hawaii.

Private aquariums are a major source of enjoyment for hobbyists around the world. Approximately 10 million households in the United States maintain an aquarium. Despite the high cost and time required to care for an aquarium, the demand for ornamental fish and invertebrates has been increasing in recent decades.

In 1998, 60 countries reported significant coral bleaching (death) events due to GLOBAL WARMING, destructive fishing practices, and pollution. Many conservationists believe that the aquarium trade is also partially responsible for the degradation of many reef communities. Each year, corals and coral reef specimens important to the aquarium trade are added to CITES, the Convention on International Trade in Endangered Species.

Research aquariums provide an opportunity to study aquatic systems in a more controlled setting. Scientists are studying the effects of rapid changes in environmental conditions, coral propagation, breeding behavior, and various types of freshwater and marine processes. As marine conservation garners more attention, aquarium research may help restore life and genetic diversity to some marine ecosystems.

See also CORAL REEF ECOLOGY; ENDANGERED SPECIES; NUTRIENT CYCLES.

Further Reading

Adey, Walter H., and Karen Loveland. *Dynamic Aquaria.* San Diego, Calif.: Academic Press, 1998.

Carlson, Bruce A. "Organism Responses to Rapid Change: What Aquaria Tell Us about Nature." *American Zoologist* 39 (1999): 44.

Okinawa Churaumi Aquarium, Japan. Available online. URL: http://www.kaiyahaku.com/eu. Accessed June 2, 2008.

archaeologist An archaeologist is a scientist with a strong background in disciplines such as anthro-pology, history, earth science, and general engineering, whose work involves unearthing and studying objects that have survived over the ages. The archaeologist attempts to reconstruct and interpret the past by analyzing, dating, and comparing artifacts through the analysis of material remains and their historical contexts. Archaeology skills include surveying, remote sensing of archeological sites, excavation and data retrieval, assessing the condition and integrity of specimens, and the conduct of basic field and laboratory procedures. Archaeologists are interested in discoveries that expand understandings of the past rather than those explorations that make a few treasure hunters rich. The science of underwater or marine archaeology is especially important in understanding historical aspects of colonial life and sea battles. Marine archaeology has been popularized by television series that display artifacts as they are found on the ocean bottom and after special cleaning, transformed into clay pots, tableware, revolvers, and ultimately as museum holdings.

MARINE ARCHAEOLOGISTS tend to be high-technology underwater discoverers who use state-of-the-art SONAR systems to locate shipwrecks and their associated artifacts. Once found, wrecks must be mapped and explored, and artifacts such as cannons, wine bottles, ceramics, and coins must be retrieved with extreme care. As is the case with surveyors, remote sensing scientists, and coastal engineers, marine archaeologists must understand geophysics in order to understand coastal changes such as shoreline retreat rates affecting historical sites such as Fort Oregon along North Carolina's barrier islands. Due to inlet migration along the Outer Banks, Fort Oregon is presently located beneath the sands in Oregon Inlet, the passage connecting coastal lagoons to the ATLANTIC OCEAN. Marine archeologists use an array of tools from old manuscripts such as dispatches and hand-drawn maps to modern tools such as Geographic Information Systems in order to understand the location of seaports such as Port Royal, forts such as Fort Fisher, or shipwrecks such as the Spanish galleon *Nuestra Señora de Atocha*. The ensuing excavations find and document underwater remains from the seafloor in order to learn about past human life and activities. The use of underwater archaeology is especially important in understanding the United States's early history, much of which has occurred near the coastal zone. For example, relics from the Civil War confirm that the fall of Fort Fisher during winter 1865 provided Union forces important access to the intercoastal waterways and the port city of Wilmington, North Carolina.

Often times, the recovery and preservation of artifacts from a shipwreck such as the SS *Central*

America, which sank during a hurricane in 1857 off the Carolina Capes, requires the use of REMOTELY OPERATED VEHICLES, SCUBA diving, and specialized excavation equipment. Remotely Operated Vehicles or ROVs are small, unmanned submersibles usually equipped with lights and a video camera and attached to a research ship or dock and operated by a marine technician from the surface. The preservation of artifacts is one of the most important considerations for the successful recovery of material from a marine archaeological site. The conservation of such artifacts is time consuming and expensive. Without conservation, the artifacts will perish and the historic data will be lost, a loss to both present and future archaeologists. The discovery of underwater artifacts and in particular treasure ships has been well documented, especially on television, and has been the basis for legislation to help preserve our cultural heritage from treasure hunters. Whether a sunken ship or an Indian gravesite, archaeological research and scientific methods are imperative to preserving historically important artifacts that are worthy of protection.

Further Reading

Kinder, Gary. *Ship of Gold in the Deep Blue Sea.* New York: Atlantic Monthly Press, 1998.

Arctic The northern polar region of the Earth that includes the Arctic Ocean and adjacent areas of the Eurasian and North American continents is called the Arctic. Some people consider it to be the area inside the Arctic Circle, which generally refers to 66° 32' N (often taken as 66.5°N) latitude. Along this line, the Sun does not set on the day of the summer solstice, and does not rise on the day of the winter solstice. Also, there are areas of land that contain discontinuous and continuous permafrost around the Arctic Circle. The Arctic Ocean is actually made up of several seas, including the Greenland Sea, the Norwegian Sea, the Barents Sea, the Kara Sea, the Laptev Sea, the East Siberian Sea, the Chukchi Sea, and the Beaufort Sea. Only the Beaufort and Norwegian Seas have depths greater than 1,640 feet (499.9 m). The bottom of the Arctic Ocean is made up of several basins separated by features such as the Nansen and Alpha Cordilleras and the Lomonosov Ridge. The Nansen Cordillera is an extension of the North Atlantic mid-ocean ridge, where seafloor spreading occurs.

In 1998, polar scientists spent 31 days onboard USS *Hawkbill,* a U.S. Navy submarine involved in a Scientific Ice Expeditions (SCICEX) cruise. They used Seafloor Characterization and Mapping Pods (SCAMP) to conduct bathymetric surveys in the Arctic Basin. They used gravity, sidescan, swath bathymetry, and chirp sub-bottom SONAR data to study the Arctic Ocean basin geology. The survey focused on the slowly spreading Gakkel Ridge (*see* PLATE TECTONICS), but scientists also collected data during the submarine's cross-Arctic transit that encountered the Alpha-Mendeleev Ridge, the Lomonosov Ridge, and the Chukchi Cap. Scientists collected the first detailed measurements of these features, significantly improving understanding of Arctic basin geology. SCAMP was one of the most complicated civilian instruments ever installed on a U.S. Navy submarine. It was funded by multiple private and government organizations. SCAMP represents a significant effort by the U.S. science community to examine the nature, origin, and evolution of the Arctic basins. The data collected by SCAMP while crossing the Arctic Ocean provides a geophysical cross section of the entire Arctic Ocean from the North American continent to the Nansen Basin. Bathymetry collected by SCAMP shows the Gakkel ridge as having sharp bathymetric relief typically in the 3,300–5,000-foot (1,005.8–1,524 m) depth range with portions as shallow as 2,000 feet (609.6 m) and regions down to about 17,000 feet (5,181 m) in depth. The ridge is defined by gravitational variations and a 13,000–20,000-foot (3,962.4–6,096 m) wide valley at about 16,000 feet (4,876.8 m), with steep slopes 3,300–5,000 feet (1,005.8–1,524 m) in height. Strong acoustic returns from sidescan sonar systems indicate an absence of sediment along the valley floor. This is typical of young volcanic regions. Trackline bathymetry shows a ridge crest composed of multiple distinct peaks rather than the more uniform ridge crest suggested by the ETOPO5 database. ETOPO5 stands for Earth Topography Five-Minute Grid, a gridded database of worldwide elevations with a resolution of five minutes of latitude and longitude. Sidescan and sub-bottom data indicate a decrease in sediment cover to the west and increasingly steeper slopes dropping into the Canadian Basin to the south.

Climate in the Arctic is harsh. The far northern coniferous forests that are found in places such as Siberia and Alaska are called "taiga." The flat or undulating region that is covered with lichen, sedges and grasses, mosses, and low shrubs is called "tundra." Along the coast there are lagoons and BEACHES where sea ice is absent at least part of the year, while the Arctic Ocean itself is continuously covered with ice whose thickness varies between 0.3 and 33 feet (0.1 and 10.1 m). About one third of the Arctic Ocean is shallow, that is, continental shelf, while the rest is more than 3,300 feet (1,005.8 m) deep. The Arctic Ocean is covered in sea ice, as opposed to glacial ice or "ice of land origin," though occasionally ICEBERGS may break away from coastal glaciers and get caught in Arctic Ocean circulations, such as the Beaufort Gyre or the Transpolar Drift Stream. The seasonal variation in ice cover is found in the Bering Sea and in the Sea of Okhotsk. Interestingly, the GULF STREAM's warm waters prevent sea ice from occurring between Norway and Svalbard, even during winter. Arctic Ocean water temperatures are around 32 to 33.8°F (0 to 1.0°C), but the salinity increases slightly with depth, which is important in the formation of sea ice. Because of its salt content, the Arctic Ocean water freezes at about 28.8°F (-1.8°C). Unlike freshwater, which is most dense at 39.2°F (4°C), seawater is most dense and sinks just prior to reaching its freezing point. Thus, in order to form sea ice, the Arctic Ocean requires a great deal more heat to be extracted by overlying cold air and radiative processes than is required for, say, a freshwater lake's surface to freeze. Because the Arctic is essentially a desert environment, precipitation and evaporation play a very small role in the surface SALINITY of the Arctic Ocean, unlike the rest of the world's oceans. Instead, the surface salinity of the Arctic Ocean is influenced mostly by river runoff from Siberia and Canada, which keeps the surface salinity relatively low. Meanwhile, higher saline, higher density seawater is supplied from the Atlantic Ocean beneath the lower salinity surface water. Because the Arctic Ocean is a relatively closed system, the general ocean circulation as well as the thickness of the pack ice in the Arctic is controlled by this density stratification. This is a defining characteristic of the Arctic Ocean and partly what makes it an important part of Earth's climate as a whole.

There are very few living species in the Arctic Ocean. There are only about 30 fish species that belong strictly to Arctic waters, as opposed to thousands of species throughout the world's oceans. The strong density stratification in Arctic waters probably has something to do with the Arctic ocean's inability to mix thoroughly, such as through upwelling events, which would bring much needed nutrients to the surface from the deep. Moreover, the production of sea ice at the surface of the Arctic Ocean actually serves to pump more salinity into the subsurface waters through the process of "brine rejection," thus enhancing the density stratification and suppressing upwelling events. Moreover, thick pack ice cover essentially prevents the entry of sunlight into surface water, which would help photosynthetic processes, and thus the Arctic Ocean's ecosystem as a whole. Surprisingly, the same brine pockets that form in solid sea ice during rapid ice growth that enhance the density stratification in the Arctic Ocean as a whole, and greatly prevent the proliferation of life forms, actually serve as miniature worlds for a few species of epontic algae. Polar scientists refer to epontic organisms as those attached to the ice or living between ice crystals.

Marine scientists working in Arctic regions focus on understanding the mechanics of sea ice conditions, the calculation of ice forces, effects of climate on off-shore structures, the performance of ships in sea ice, climate change, and the survival strategies of arctic flora and fauna. Research tools include maps derived from satellite imagery, numerical model analyses, ice core studies, and data from instrumented BUOYS. The International Arctic Buoy Program provides a sustained network of drifting buoys in the Arctic Ocean to provide real-time meteorological and oceanographic data to support a range of users from basic researchers to mariners. In 1952, the U.S. Air Force's Alaska Air Command organized Project ICICLE to establish a weather station on an ice island in order to conduct geophysical research. Since then, several drifting Arctic research stations have been established to facilitate research by scientists from many countries and research institutions. One of the United States's newest and most technologically advanced icebreakers is also a scientific research vessel. It is called the U.S. Coast Guard cutter *Healy (WAGB–20)*. *Healy* provides more than 4,200 square feet (390 m²) of scientific laboratory space, numerous electronic sensor systems, oceanographic winches, and accommodations for up to 50 scientists. *Healy* is designed to break 4.5 feet (1.4 m) of sea ice continuously at three knots (1.5 m/s) and can operate in temperatures as low as -50°F (-45.6°C). The science community provided invaluable input on the ship's laboratory design and science capabilities. As a Coast Guard cutter, *Healy* is also a capable platform for supporting logistics, search and rescue, ship escort, environmental protection, and enforcement of laws and treaties.

Basic researchers use Arctic data to study topics from Arctic climate and climate change to tracking the source and fate of samples. More operational users apply the data to tasks like weather and sea ice forecasting to support vessel navigation. Recent research and public interest in the Arctic has centered on the larger question of global climate change, sometimes called global warming. Several researchers have proposed that the polar regions are the most sensitive to small changes in global climate, which is characterized largely by air temperature wind speeds and directions and ocean circulation. Thus, if global climate change, or even long-term climate cycles are to be identified, the polar regions, including the Arctic, provide some of the first clues in the form of changing ice thickness and coverage.

See also ALBEDO; GLOBAL WARMING; SEA ICE.

Further Reading

Alaska Ocean Observing System. Available online. URL: http://ak.aoos.org/arc/. Accessed February 29, 2008.

International Arctic Buoy Program, Polar Science Center, Applied Physics Laboratory, University of Washington. Available online. URL: http://iabp.apl.washington.edu. Accessed January 7, 2007.

International Arctic Research Center, University of Alaska Fairbanks. Available online. URL: http://www.iarc.uaf. edu. Accessed January 7, 2007.

National Geographic. *Atlas of the World*, 7th ed. Washington, D.C.: National Geographic Society, 1999.

NOAA, Arctic Theme Page. Available online. URL: http://www.arctic.noaa.gov. Accessed October 16, 2003.

U.N. Atlas of the Oceans, United Nations Foundation. Available online. URL: http://www.oceansatlas.org. Accessed January 6, 2007.

USCGC HEALY (WAGB 20). Available online. URL: http://www.uscg.mil/pacarea/healy. Accessed March 17, 2007.

Arthropoda The largest phylum in the Animal Kingdom, which includes invertebrate aquatic animals. The arthropods are joint-legged animals that constitute the largest assemblage of species within the animal kingdom. There have been more than 750,000 species described by science; this is more than three times the number of all other animal species combined. This group includes many familiar forms such as horseshoe crabs (e.g., *Limulus polyphemus*), spiders, mites, ticks, scorpions, and the sea spiders known collectively as the Chelicerates. The other major subphylum includes the crustaceans, insects, centipedes, millipedes, and other lesser known forms, which constitute the Mandibulates. The insects, for example, contain the largest number of species of any animal group.

The adaptive capabilities of the arthropods have enabled them to occupy virtually every niche and habitat of marine, freshwater, and terrestrial environments—inhabiting the deepest ocean trenches, the highest mountain peaks, and virtually every habitat in between. In the evolution of arthropods, the development of the chitin-protein exoskeleton allows for articulation of the body and movement, while periodic molting of the exoskeleton allows for growth of the organism. The exoskeleton also provides the arthropods with protection against predation; it protects the more delicate body segments housed within the exoskeleton itself. The annelids or segmented worms are believed to be the ancestors of the arthropods due to a shared characteristic known as metamerism, where the body is divided into similar parts or segments. Nervous system structure and determinate cleavage also link the two groups. Metamerism has been lost or greatly reduced in modern forms of arthropods due to differentiation of the appendages to perform other important physi-ological functions. Sensory mechanisms are highly variable among groups but usually involve some specialization within the exoskeleton, allowing the organism to monitor its surrounding environment. There are usually two separate sexes with internal fertilization being the most common reproductive strategy. Various appendages are used for the transfer of sperm from male to female depending upon the species.

Many species of arthropods, most important the insects, carry diseases that infect mankind and cause millions of deaths annually. Mosquitoes, lice, fleas, bedbugs, and flies are vectors of human diseases such as malaria, elephantiasis, yellow fever, sleeping sickness, typhus, bubonic plague, typhoid fever, and dysentery among others. Many species of arthropods contribute greatly to the well-being of mankind. For example, species such as lobster, crab, and shrimp are commercially important and rank among the highest in terms of dollar value when compared to other commercially important species of fishes and invertebrates. The lobster, crab, and shrimp fisheries generate millions of dollars in sales annually. These shellfish make a significant contribution to the local and regional economy, which contributes to the national economy. Exports of these products to other countries also have a significant effect upon the nation's economy. Other benefits include the detection of bacteria that can cause diseases in humans and domesticated livestock through the utilization of blood from the horseshoe crab. An enzyme found in the blood of the horseshoe crab is used by industry to detect bacteria in pharmaceuticals and medical devices. The Limulus Amino Lysate (LAL) test has become an industry standard and is required by the Food and Drug Administration to test these products for the presence of nuisance or disease-causing bacteria.

See also MARINE CRUSTACEAN.

Further Reading
Barnes, R. D. *Invertebrate Zoology.* Philadelphia: Saunders College; Holt, Rinehart, and Winston, 1980.

MarineBio.org, Houston, Texas. Available online. URL: http://marinebio.org. Accessed January 7, 2007.

Ocean Biogeographic Information System. Available online. URL: http://www.iobis.org. Accessed January 7, 2007.

University of California Museum of Paleontology. "Introduction to the Arthropoda." Available online. URL: http://www.ucmp.berkeley.edu/arthropoda/arthropoda.html. Accessed October 7, 2007.

Atlantic Ocean The Atlantic Ocean is the world's second-largest water body. The Atlantic is subdivided into two ocean basins: the North Atlantic to the north of the equator, and the South Atlantic to the south of the equator. The Mid-Atlantic Ridge, part of the great mid-ocean ridge system that winds around the globe further divides both the North Atlantic and South Atlantic Oceans into east and west basins. The western boundary of the Atlantic Ocean is formed by North and South America; the eastern boundary by Europe and Africa; the southern boundary by the ANTARCTIC CIRCUMPOLAR CURRENT; the northern boundary by the Arctic Ocean. There is some controversy over the exact northern boundaries of the Atlantic.

Beside the Mid-Atlantic Ridge, major features of the Atlantic include large islands, such as the United Kingdom, Iceland, and Greenland, which are continental, and small islands, such as Bermuda and the Azores, which are oceanic volcanic in origin. A great island chain, including the Windward Islands, Puerto Rico, Hispaniola, and Cuba, which forms the southwestern boundary of the North Atlantic Ocean also defines the north and east boundaries of the Caribbean Sea, an important MARGINAL SEA. The floor of the Atlantic supports some SEAMOUNTS, but they are not nearly so numerous as in the PACIFIC OCEAN. The Atlantic has a few deep ocean trenches, including the Puerto Rican Trench, and the Romanche Trench. Technical data on the Atlantic Ocean basin and those of its marginal seas are provided in the appendices.

Generally speaking, the continental shelves of North America and northwestern Europe are wide and shallow; those of Africa are narrow. The CONTINENTAL SHELVES of South America are narrow in the north, and wide in the south, off the coast of Argentina. Marine geologists classify margins with distinct continental shelves, slopes, and rises as Atlantic-type margins. These margins are generally passive in contrast to the more seismically active Pacific-type margins.

The Atlantic Ocean was formed when the supercontinent Pangaea began to split apart about 170 million years ago, along what is now the Mid-Atlantic Ridge. Deep convective forces within Earth's mantle continued to force the crust apart as magma welled up from a rift in the middle of the ridge, forming new crust and thrusting the land masses to be called Eurasia and North America apart, as well as separating Africa from South America. As the separation continued over millions of years, the rift valley became wider and wider, and filled with water from the adjacent oceans. This process is still continuing; volcanic island formation can be seen near Reykjavik, Iceland, and the Azores located on the Mid-Atlantic Ridge. Exploration of the Mid-Atlantic Ridge during the mid 20th century led to the devel-

opment of the dominant theory of geophysics and geology: PLATE TECTONICS, which views the Earth's crust and upper mantle as floating on a plastic mantle that can flow under conditions of great heat, coming from convective currents deep within the Earth. French and American scientists working from deep submersible craft have observed this formation of crust under the ocean.

Several important adjacent seas border the North Atlantic Ocean, including the MEDITERRANEAN, the Caribbean, the GULF OF MEXICO, the Greenland, and the Norwegian. The South Atlantic has no adjacent seas. Important rivers discharging into the Atlantic include the Hudson, the Amazon, the Río de la Plata, the Congo, and the Rhine. The mixture of freshwater from land drainage and salt water is commonly called brackish water and is a key element for ESTUARIES such as CHESAPEAKE BAY.

Two large circulating wind systems, one in the Northern Hemisphere and one in the Southern Hemisphere, drive the circulation of the Atlantic Ocean. The winds in the Northern Hemisphere are the Northeast Trade Winds, blowing from the northeast to the southwest approximately between latitudes 10°N and 25°N latitude, and the mid-latitude westerlies generally north of latitude 30°N. The ocean responds to the wind force by forming a large GYRE, a great rotating lens of water spanning the entire North Atlantic basin, formed by the westward flow of the North Equatorial Current, the northeastward flow of the GULF STREAM, the eastward flow of the North Atlantic Current, and the southward flow of the CANARY CURRENT, which completes the gyre by joining the North Equatorial Current off the coast of Northwest Africa. The Sargasso Sea is located in the west central part of the North Atlantic basin just east of the Gulf Stream and in many ways corresponds to the center of the gyre. In the South Atlantic, the corresponding winds are the Southeast Trade Winds and the mid-latitude westerlies. The corresponding ocean currents are the South Equatorial Current, the Brazil Current, the SOUTH ATLANTIC CURRENT, and the BENGUELA CURRENT, which flows northward along the coast of Africa to complete the gyre by joining the South Equatorial Current. The Brazil Current, the Southern Hemisphere counterpart of the Gulf Stream, is a much weaker current. The Northern Hemisphere gyre rotates clockwise, and the Southern Hemisphere rotates counterclockwise as would be expected from consideration of the CORIOLIS EFFECT.

These currents are really a climatological average; ocean currents are in fact extremely variable and are characterized by numerous meanders, jets, and eddies. However, it is possible to make valid generalizations about the major currents of the Atlantic. On the western boundary, the Gulf Stream is narrow and swift, while on the eastern boundary the Canary and Benguela Currents are broad and slow.

Other major currents include a complex system of equatorial currents including the Equatorial Countercurrent, which is a return flow of water eastward between the North and South Equatorial Currents, and the Equatorial Undercurrent, which is a subsurface current with a high-speed, high-salinity core usually right at the equator centered at a depth of about 492 feet (149.9 m). An important feature of the equatorial current system is that it is not symmetric about the equator, but rather displaced northward, so that the Southeast Trade Winds reach over the equator into the Northern Hemisphere. The extent of the displacement is seasonal, with the greatest northward position of the currents being in Northern Hemisphere summer. This effect increases the eastward flow of the Equatorial Countercurrent. It is not completely known why this occurs, but it is believed to be a consequence of the greater size of landmasses in the Northern Hemisphere and the direct exposure of the South Atlantic to the effects of the Antarctic Circumpolar Current.

Polar currents sending cold water into the Atlantic include the LABRADOR CURRENT and the East Greenland Current in the north, and the Falkland (Malvinas) current in the south, which branches northward along the coast of Argentina from the Antarctic Circumpolar Current.

Surface salinities in the mid-latitude regions of the Atlantic tend to be on the order of 36.00 practical salinity units (psu), with higher salinities in the western tropical and equatorial regions. In fact, in the waters east of Puerto Rico in the North Atlantic, and off Brazil in the South Atlantic, salinities as high as 37.00 psu or higher are found. Salinities are lower in the eastern Atlantic, being on the order of 35.5 psu off northwest Africa. Surface salinities in the Arctic and Antarctic regions are much lower, of magnitude 33.00 to 34.00 psu. Surface temperatures are on the order of 78–79°F (25.6–26°C) in equatorial regions ranging down to slightly below freezing in polar and Antarctic waters. The seasonal variation of temperature is on the order of 47°F (8°C) at mid latitudes in the North Atlantic and on the order of 41°F (5°C) in the South Atlantic. The WATER MASSES of the Atlantic include ANTARCTIC BOTTOM WATER (ABW), which fills the ocean basins below about 13,000 feet (3,962.4 m) depth and North Atlantic Deep Water (NADW), which is generally found at depths below 6,600 feet (2,011.7 m). Ant-

arctic Intermediate Water (AIW) is found between depths of a few hundred feet (meters) and 13,000 feet (3,962.4 m). The surface waters found above that are known as North Atlantic Central Water, and South Atlantic Central Water. An equatorial water mass of high temperature and relatively high salinity lies between the two central water masses. The deep-water masses, originating in the Antarctic and Arctic regions, are extremely cold and constitute the oceanic "stratosphere." The central water masses are relatively warm and constitute the oceanic "troposphere," in analogy with the atmosphere. These terms were first used by German oceanographer Georg Adolf Otto Wüst (1890–1977) on the basis of his analysis of data from the *Meteor* expedition (1925–27). A complicating feature of the Atlantic Ocean is the inflow of warm, salty water from the Mediterranean Sea, which is quite dense because of its high salinity. This water spreads out at a depth of around 3,300 feet (1,005.8 m) and gradually mixes with the surrounding waters.

The weather in the North Atlantic Ocean during winter is notoriously bad. The Icelandic Low, a semipermanent feature of the pressure distribution in the Greenland-Iceland-Scotland region produces strong winter winds, causing high seas and a procession of storms traveling eastward off the North American continent toward northern Europe. It is just through this region that the major shipping lanes from Europe to America must pass. The cold and dry winter air from the North American continent passing over the warmer water of the NORTH ATLANTIC CURRENT produces dense fog, especially over the Grand Banks, adding to the hazards faced by the mariners and fishermen.

The climate of northwestern Europe is greatly influenced by the Gulf Stream system, which transports warm water from the Gulf Stream eastward to the British Isles, France, Germany, and the Scandinavian countries to make their climates much warmer than they would otherwise be. Some scientists are concerned that GLOBAL WARMING might alter the circulation of the North Atlantic Ocean, and cut off the Gulf Stream from Europe, resulting in a much colder climate for northwest Europe and northeastern North America. Such large-scale air-sea interactions are topics of major research by 21st-century meteorologists and oceanographers. If a major change in the circulation of the North Atlantic Ocean occurs as a result of global warming, governments must take action to avoid an environmental catastrophe. Large-scale air-sea interactions are currently being observed, such as the Pacific Ocean's El Niño–Southern Oscillation (ENSO), that produces global changes in weather and climate. The North Atlantic Ocean experiences an effect called the Arctic Oscillation (AO), which is related to the North Atlantic Oscillation (NAO). The AO oscillates between warm and cold states of sea surface temperature. During the warm period, there is abnormally low pressure over the polar latitudes and higher pressure over the mid latitudes. Under these conditions, winters in the eastern United States are warmer, whereas in Greenland winters are colder than normal. During the cold period, there is abnormally high pressure at the poles and lower pressure at mid latitudes. Under these conditions, the eastern United States and western Europe experience colder than normal winters, while Greenland is warmer. These changes occur on a time scale of decades and appear to be related to freezing or thawing of the ice cover of the Arctic Ocean. During the warm period, North Atlantic water penetrates farther north and melts ice, so it is thinner; during the cold period, the North Atlantic water retreats, and the ice thickens. At present, a warm period prevails, but small changes in atmospheric variables such as temperature could produce a major effect in the Arctic, which could then amplify over the North Atlantic and bordering nations.

Further Reading

Aguado, Edward, and James E. Burt. *Understanding Weather and Climate,* 3rd ed. Upper Saddle River, N.J.: Prentice Hall/Pearson, 2004.

Earle, Sylvia A. *National Geographic Atlas of the Ocean: The Deep Frontier.* Washington, D.C.: National Geographic, 2004.

Erickson, Jon. *Marine Geology: Exploring the New Frontiers of the Ocean,* rev. ed. New York: Facts On File, 2003.

National Geographic. *Atlas of the World,* 7th ed. Washington, D.C.: National Geographic Society, 1999.

Ocean Surface Currents, University of Miami, Rosenstiel School of Marine and Atmospheric Science. Available online. URL: http://oceancurrents.rsmas.miami.edu. Accessed February 7, 2004.

U.N. Atlas of the Oceans, United Nations Foundation. Available online. URL: http://www.oceansatlas.org. Accessed January 6, 2007.

atoll An island landform composed of coralline material shaped as a ring or partial ring in an ocean or marginal sea is called an atoll. These low islands contain a LAGOON and lack volcanic or continental rock in their surface geology. Research writings and presentations during 1842 by Charles Robert Darwin

An atoll protected by an encircling barrier reef *(NOAA)*

(1809–82) reported that the formation of atolls occurs in stages. His theory indicates that reefs first form near landmasses as fringing reefs on the shores of newly formed volcanic islands. Then, as the islands begin to sink, reef growth begins to form a barrier reef. The circular island becomes an atoll when the volcanic island disappears due to weathering under the ocean surface leaving a central lagoon. Observed tectonic processes and an improved understanding of SEAFLOOR SPREADING have substantiated Darwin's theory.

The reefs that make up an atoll are limestone formations produced mostly by living stony corals (*Madreporaria*) that are common in shallow, tropical marine waters. The normal range for coral growth is from 30°N to 30°S latitude. Reef-forming corals do not thrive below 100-feet (30.5 m) depths, nor where water temperatures fall below 72°F (22°C). Sediments trapped on the lagoon side of atolls eventually support vegetation such as coconut palms (*Cocos nucifera*), pandanus plants, breadfruit (*Artocarpus altilis*), and taro (*Colocasia esculenta*). Given the low-lying character of these islands, atolls are very prone to the effects of wave slamming from TSUNAMI and STORM SURGE from hurricanes or typhoons.

Geological oceanographers specializing in geomorphology or the scientific study on how landforms are formed on the Earth show particular interest in atolls. Some atolls have a completely enclosed lagoon where ocean waters reach the lagoon through a few narrow channels in the coral ring when coastal water levels are particularly high. Other atolls have a largely submerged coral ring where there is ongoing mixing between open-ocean and lagoon waters. Complex coral patterns may also form a reticulated lagoon with numerous shallow interconnected pools. The lagoons provide anchorage and economic resources such as pearls, seaweed, and other marine by-products that are used for pharmaceuticals. Mining natural resource such as coral to process for lime removes important habitat for local marine species and weakens natural coastal storm defenses.

Anthropogenic impacts on coral atolls include damage from increased coastal development, anchoring, water pollution, smothering from sedimentation, bleaching from rising sea surface temperatures, and flooding from sea level rise. One of the most famous atolls, the Bikini atoll, is found in the northwest equatorial Pacific's Marshall Islands. The United States has been monitoring, studying, and cleaning up these islands since conducting a series of nuclear tests called Operations Crossroads and Castle during the late 1940s and the mid 1950s. People are still trying to mitigate the negative social and physical consequences of these experiments with nuclear weapons in the Marshall Islands.

Further Reading

Barrett, Paul H., and Freeman, R. B., eds. *The Works of Charles Darwin, vol. 7: The Geology of the Voyage of H.M.S. Beagle, Part 1: Structure and Distribution of Coral Reefs.* New York: University Press, 1987.

Charton, Barbara. *A to Z of Marine Scientists.* New York: Facts On File, 2003.

CoRIS, NOAA's Coral Reef Information System, National Oceanic and Atmospheric Administration, U.S. Department of Commerce. Available online. URL: http://www.coris.noaa.gov. Accessed January 7, 2007.

Spalding, Mark D., Corrina Ravilious, and Edwin P. Green. *The World Atlas of Coral Reefs.* Berkeley: University of California Press, 2001.

Thurman, Harold V., and Elizabeth A. Burton. *Introductory Oceanography,* 9th ed. Upper Saddle River, N.J.: Prentice Hall, 2001.

attenuation   Attenuation is the reduction in intensity of a transmission, as it gets farther away from its source, due to absorption by the medium through which it is traveling. Absorption is the conversion of mechanical energy (sound) to heat energy. For example, the voice of a speaker becomes fainter with increasing distance. This happens because the sound is spreading and being attenuated by the air through which it travels. Both sound and light are attenuated as they travel through the ocean. Light is attenuated as it propagates through the ocean, reaching depths of 328 to 656 feet (100 to 200 m) in the ocean, but only a fraction of that in turbid waters. Absorption is greatest for the long wavelengths of light and somewhat less for shorter wavelengths, such as the blue-green wavelengths. Approximately 99 percent of natural light is gone by 328 feet (100 m); all the light is gone below 3,281 feet (1,000 m). In the ocean, high-frequency (short WAVELENGTH) sounds are greatly attenuated or weakened, while low-frequency sounds experience little attenuation.

Sound attenuation depends strongly on the sound's frequency. For a 10-kilohertz sound source the attenuation loss in the upper portion of the ocean (for a sea surface temperature of 68°F (20°C), pH of 8.0, and salinity near 35 psu) is about 0.7 decibels per kilometer. This is a very small rate of attenuation. Sound in the atmosphere (for an air temperature of 68°F (20°C), 14.7 pounds per square foot (1,013.25 mb), and 50 percent relative humidity), at the same frequency, has an attenuation loss of 160 decibels per kilometer. The corresponding speed of sound in air is 769.5 miles per hour (1,238.4 km/h), while the speed of sound in seawater is

3,403.5 miles per hour (5,477.4 km/h). The speed of sound in the water is a function of TEMPERATURE, PRESSURE, and SALINITY.

The attenuation of light in the ocean can be measured very accurately using highly sophisticated sensors, or it can be crudely measured using a SECCHI DISK (a white-and-black disk 8–12 inches (20–30 cm) in diameter). Light attenuation is important because it defines the food chain. Photosynthesis, the process that is the first link in the food chain, is totally dependent upon the depth that light can reach. Water with high attenuation allows photosynthesis to occur only very near the surface. Similarly, sound attenuation is important for animals that communicate over long distances, such as whales, and is very important for SONAR operators listening for approaching ships and submarines. Charts showing sound and light attenuation for the world's oceans are available in general reference books.

See also APHOTIC ZONE; LIGHT; LIGHT AND ATTENUATION; SONAR.

Further Reading

A Brief History of the Use of Sound in Ocean Exploration. United States Geological Survey, Woods Hole Science Center. Available online. URL: http://woodshole.er.usgs. gov/operations/sfmapping/soundhist.htm. Accessed January 7, 2007.

Jensen, Finn B., William A. Kuperman, Michael B. Porter, and Henrik Schmidt. *Computational Ocean Acoustics.* AIP Series in Modern Acoustics and Signal Processing. New York: Springer-Verlag, 2000.

Jensen, John R. *Remote Sensing of the Environment: An Earth Resource Perspective.* Upper Saddle River, N.J.: Prentice Hall, 2000.

National Physical Laboratory, Welcome to Kaye and Laby online. Available online. URL: http://www.kayelaby.npl. co.uk/general_physics/2_4/2_4_1.html. Accessed January 17, 2007.

Urick, Robert J. *Principles of Underwater Sound.* New York: McGraw Hill, 1983.

Autonomous Underwater Vehicles Autonomous Underwater Vehicles or "AUVs" are untethered submersible robots capable of navigating and carrying out prescribed missions without an onboard pilot. These unmanned, self-propelled vehicles are usually used to collect oceanographic data or to image the ocean bottom along preprogrammed courses. The AUVs are generally launched and recovered by a ship such as a research vessel that has a crane. The lack of tethers makes AUVs different from Remotely Operated Vehicles or ROVs, which are connected by a communications cable to a mother ship. *Jason Jr.* was the well-known ROV controlled from the small manned Deep Submergence Vehicle *Alvin* during Dr. Robert Ballard's discovery and exploration of the HMS *Titanic. Jason Jr.* was maneuvered inside the wreckage, and video imagery was provided to viewers around the world. The autonomy of AUVs makes them an ideal instrument platform for clandestine military hydrographic survey operations or as a mobile instrument suite to collect data in hard-to-access regions such as ice-covered seas.

Some of the earliest AUVs were developed to find dangerous mines or to collect bathymetric data in areas inaccessible by hydrographic survey ships. Similar AUVs have been used to address scientific questions. Their payloads consist of numerous sensors that can be "plugged in" or easily removed from the AUV. Users such as MARINE ARCHAEOLOGISTS, oceanographers, and mine countermeasure personnel have to carefully select their payloads because there is little room onboard the vehicle. Typical marine science payloads include cameras, conductivity-temperature-depth or CTD gauges, ACOUSTIC DOPPLER CURRENT PROFILERS (ADCPs), sensors to detect dissolved gases such as methane, transmissometers, sidescan sonars, and sediment profiling sonars. Such vehicles serve the research community by being able to make 4-d maps of features such as a methane plume above a cold seep in the deep ABYSSAL ZONE. Total darkness, extreme water PRESSURE, and cold TEMPERATURES characterize the ABYSSOPELAGIC ZONE—a perfect environment for a state-of-the-art AUVs. The ensuing surveys of a cold seep can be studied to improve human understanding of the global methane budget. They show how these greenhouse gases enter the ocean, such as from gas bubbles on the ocean floor.

Reliable navigation and positioning systems are essential for the effective use of AUVs. Navigation software uses initial fixes from the GLOBAL POSITIONING SYSTEM and readings from such sensors as fluxgate compasses for heading, inclinometers for pitch and roll, pressure cells to measure vehicle depth, and SONAR to detect and avoid obstacles. Most AUVs will dead reckon with the help of an inertial navigation unit (INU) that consists of gyroscopes and accelerometers used to provide information on attitude, heading, position, inertial velocity, and acceleration information. Measures such as inertial velocity can be calculated by integrating the signal from an accelerometer. This combination of sensors is expected to provide position with a high degree of accuracy.

The ongoing development of AUVs provides an excellent example of military research being applied to meet civilian needs. Today's AUVs come in various shapes and sizes. Most are torpedolike submersibles several meters in length and capable of reaching speeds of three to four knots (1.5 to 2.1 m/s). Ranges are dependent on batteries, but are usually on the order of 60 miles (96.6 km). Their maximum diving depth is on the order of 500 fathoms (.57 miles .91 km). Other recently developed AUVs are based on biomimetic control, actuator, and sensor architectures with highly modularized components. Biomimetic refers to human-made devices or systems that imitate nature. The actuators convert electrical signals into actions and are one of today's most significant research and development areas in microtechnology. An example of a biomimetic class of AUVs would be an eight-legged ambulatory vehicle that moves among the benthos like a lobster or crab.

Operational AUVs in the future will have long ranges (on the order of hundreds of miles) and increased maximum payloads. Such vehicles will support global climate studies by being able to measure processes that range from upwelling at the equator to downwelling at the poles. They can become an integral part of the emerging Global Ocean Observing System and perform unprecedented surveys efficiently under sea ice in the Arctic or Antarctic.

See also REMOTELY OPERATED VEHICLE; SUBMERSIBLE.

Further Reading

Ocean Explorer, National Oceanic and Atmospheric Administration, U.S. Department of Commerce. Available online. URL: http://oceanexplorer.noaa.gov. Accessed January 7, 2007.

OceanWorld, Texas A&M University. Available online. URL: http://oceanworld.tamu.edu. Accessed January 7, 2007.

B

backscatter The energy, whether electromagnetic, mechanical, or acoustical, that scatters off a surface back to its source is called backscatter. Consider the beam of a flashlight illuminating a spot on the surface of a lake at night. The beam hits a spot on the water's surface and is observable. The illuminated spot is the visible light that is backscattered from the spot to the observer's eyes. Another observer on the other side of the lake also sees the spot illuminated. That is the visible light that is reflected forward from the illuminated spot to their eyes. Backscattered energy is a very important phenomenon in electromagnetic, mechanical, and acoustical remote sensing.

An acoustic source, such as a bat or a submarine, can send out acoustic waves. The energy that is reflected back at the source from a target, such as a moth or another submarine, respectively, is called a reflection or echo. Reflections follow Snell's Laws (after Willebrord Snel van Royen (1580–1626) a Dutch astronomer and mathematician); that is, the angle of incidence equals the angle of reflection. This is true if the surface is a perfect reflector; however, surfaces usually are not perfect. Backscatter is usually referred to as the energy that returns to the source excluding the reflected energy. Thus, it is important to use this word correctly. A radar (electromagnetic energy) on a ship that is viewing another ship sees a huge amount of energy reflected back from another ship. Some of the RADAR energy hits the sea surface, and most of that is forward scattered or reflected forward. However, some is backscattered to the ship's radar. This backscattered energy is called "sea clutter."

Navigational radars rely on the reflected signal. Other radars rely on only the backscattered signal.

For example, a Synthetic Aperture Radar (SAR) mounted on spacecraft or aircraft sends out a signal where most of the energy is reflected at the sea surface away from the source. However, a small amount is backscattered to the radar. The backscattered energy is a function of the roughness of the surface below. If the surface water is flat, all the energy is scattered forward, none is backscattered, and the water appears black. If the wind is blowing and there are a lot of capillary waves on the surface, then more energy is backscattered to the radar, and that area shows up light. In summary, light and dark on radar imagery are relative measures of radar reflectivity.

Further Reading

Center for Remote Sensing, University of Delaware, Welcome to the Ocean Internal Wave Online Atlas. Available online. URL: http://newark.cms.udel.edu/crs/crs.html. Accessed August 26, 2004.

Fisheries Acoustic Research Group, Northeast Fisheries Science Center. Available online. URL: http://www.nefsc.noaa.gov/femad/ecosurvey/acoustics. Accessed January 7, 2007.

WHSC Sidescan Sonar systems, Woods Hole Science Center, United States Geological Survey. Available online. URL: http://woodshole.er.usgs.gov/operations/sfmapping/sonar_interp.htm. Accessed January 7, 2007.

bacterial plankton (bacterioplankton) Bacterioplankton comprise a group of diverse prokaryotic single-celled microorganisms found in marine realms. Their structure is simple with a loop of deoxyribonucleic acid (DNA) encased in a cell membrane and cell wall. Prokaryotic organisms lack a nuclear membrane, while eukaryotes have a distinct nucleus. The marine prokaryotes include the bacterial plankton, which mostly drift with the CURRENT unable to control their position in the water column. Others such as *Streptomyces coelicolor* dwell on or in the seabed and are a part of the BENTHOS. They are especially important for human and veterinary medicine because of their use in producing natural antibiotics. Bacterial plankton tend to be either photosynthetic cyanobacteria or nonphotosynthetic heterotrophic bacteria.

Many of the bacterial plankton are only now being fully understood. Most people have seen dense mats of bluegreen bacteria of the genera *Microcystis* that turn pond and lake waters turbid and green. These blooms usually result from large quantities of nutrients entering shallow lakes and ponds from runoff and the feces of ducks and geese. The purple sulfur bacteria (*Chromatiaceae* and *Ectothiorhodospiraceae*), which have the ability to oxidize hydrogen sulfide (H_2S) and store elemental sulfur inside the cells, are anaerobic bacteria that appear as red masses in salt marshes. Green nonsulfur bacteria are fairly well represented by *Chloroflexus, Thermomicrobium, Oscillochloris,* and *Heliothrix,* four major genera that thrive in neutral or alkaline hot springs. Some of the other more commonly encountered genera of bacterioplankton include *Bacillus, Bacterium, Corynebacterium, Micrococcus, Mycoplana, Norcardia, Pseudomonas, Sarcina, Spirillum, Streptomyces,* and *Vibrio.*

During the last two decades, investigators have learned much about chemosynthesis through studies of deep hydrothermal vent communities (*see* HYDROTHERMAL VENTS), locations where bacterioplankton at the base of the food chain obtain their energy by oxidizing sulfur-rich fluid spewing from vents. Invertebrate animals such as tubeworms (*Riftia pachyptila*), hydrothermal vent crabs, (*Bythograea thermydron*), and giant clams (*Calytogena magnifica*) contain large populations of symbiotic bacteria that provide them with organic nutrients. Higher up the food chain, bathydemersal pug-nose eels (*Thermarces cerberus*) hide in the clumps of tube worms and feed on small invertebrates such as the vent snail (*Cyathermia naticoides*).

There is an abundance of bacteria in marine waters, in freshwaters, underneath ice, and in the sediments of the deep ocean TRENCHES. Only a small number of possibly many thousands of species have been named. They are found thriving at the air-sea interface, in the water column, and in the HADAL ZONE. Their abundance is dependent on factors such as water conditions and protozoan grazing. In the photic zone, the bacterioplankton are impaired by ultraviolet radiation, short WAVELENGTH energy from the Sun that ranges from 295 to 320 nm in length. The filter-feeding Calanoid Copepod (*Calanoides carinatus*), which thrives in nutrient-rich upwelling waters, is believed to graze on bacterioplankton during its naupliar or early hatching stage (*see* COPEPODS). In salt marshes, bacteria are responsible for methane and hydrogen sulphide odors. Some bacterioplankton produce toxins that are potentially dangerous to animals and human health. Investigators studying fish kills have reported lesions and ulcers infected with bacteria in estuaries and coastal waters. Bacteria and viruses present in water due to human pollution cause illnesses. Eating oysters, mussels, and scallops that were harvested from waters with unsafe concentrations of *Escherichia coli* can lead to mild intestinal disorders all the way to acute gastroenteritis. Bacterioplankton serve the important task of degrading organic waste and recycling nutrients back into the food chain.

Further Reading
American Society for Microbiology. Available online. URL: http://www.asm.org. Accessed March 17, 2007.

MarineBio.org. Available online. URL: http://marinebio.org. Accessed January 7, 2007.

Monterey Bay Aquarium Research Institution. Available online. URL: http://www.mbari.org. Accessed January 7, 2007.

barge A barge is a long, flat-bottomed boat designed to transport cargo on inland waterways. These vessels are usually without engines or crew accommodations. Barges can be lashed together and either pushed or pulled by TUGBOATS. Barges that are used for carrying goods between ship and shore are known as lighters. Other custom-designed barges may stay moored to a pier for use as a machine shop, laboratory, classroom, and for ceremonies. In naval vernacular, a powerboat reserved for use by an admiral is called a barge.

Barges provide an ideal floating platform for different types of instruments and experiments. Barges such as the Research Barge *Robert E. Hayes* provide a crane, deck, and laboratory space for scientists to study marine birds. The 70-foot (21 m) RB *Robert E. Hayes* has 75-foot (23 m) hydraulic legs, which lift the vessel out of the water. Other unique vessels such as *Flip,* which is home-ported at the SCRIPPS INSTITUTION OF OCEANOGRAPHY, are able to operate either horizontally or vertically. The floating instrument platform on *Flip* floods its ballast tanks in order to gain a vertical orientation in the water column.

A barge being towed to a worksite in Buzzards Bay, Massachusetts. Buzzards Bay is 28 miles (45 km) long, averages about 8 miles (12 km) in width, has a mean depth of 36 feet (11 m), and is located between the westernmost part of Cape Cod and the Elizabeth Islands. *(U.S. Coast Guard, PA2 Matthew Belson)*

Marine geologists and petroleum exploration scientists might use a box core, piston core, or even a drilling derrick from a barge to investigate the seafloor. The box core takes relatively undisturbed samples of the seafloor by lowering a hollow steel-walled rectangle to the bottom via a winch cable. The piston core device consists simply of a weight stand mounted above a length of stainless steel core barrel that is lowered to the bottom, penetrates into the sediment, and is returned to the ship's deck, where the sediment core is promptly removed from the core barrel. The well-known jack-up barge hosts a drilling derrick and extendable legs that actually jack the vessel out of the water to become a drilling platform.

Multidisciplinary teams of marine scientists and engineers have worked together to develop ideas on how to build elaborate floating structures on the high seas that are composed of interconnected barges. An example of barge technology from the late 1990s that received global attention involved construction plans for sea-based heliports in Okinawa, Japan. One potential project called "megafloat" was a multistory, nearly mile-long floating heliport anchored in deep water. Another design was called the "quick installation platform," which involved anchoring a lesser heliport in shallow water. During this time, the OFFICE OF NAVAL RESEARCH spearheaded a study to evaluate the feasibility of modular interconnected barges to provide flight, maintenance, supply, and other forward logistics support operations for U.S. and Allied forces. By 2001, the investigators concluded that the "Mobile Offshore Base" concept was cost prohibitive and alternatives such as aircraft carriers and roll-on/roll-off ships were less expensive.

Further Reading

The American Waterways Operators. Available online. URL: http://www.americanwaterways.com. Accessed January 8, 2007.

barrier islands Long, very narrow islands generally parallel to the coast separated from the mainland by ESTUARIES (bays, LAGOONS, sounds) and tidal marshes are classified as barrier islands. They are divided into separate islands by tidal inlets that connect the estuaries to the ocean. They are composed mainly of sand organized into BEACHES, dunes, and tidal flats. Regions common to the barrier island include the offshore zone, which is always under water; the inshore zone, also called the shoreface, where the breakers usually form; the foreshore, where waves break at high water; the backshore, where waves break only during storms; and the upland region, where dunes begin. The dunes can consist of only one line of dunes on very narrow barrier islands, or there can be several rows of dunes separated by low areas called swale. The more lines of dunes and the higher the dunes, the more the barrier island will be protected from attack by ocean waves and storm winds. If the dune field is sufficiently developed, there may be adequate shelter for plants to grow into thickets, and even small "maritime" forests, which provide further protection to the islands. On the bay or lagoon sides, tidal flats and marshes complete the barrier island setting. The range of the tide largely determines whether the landward boundary of the barrier islands will be a lagoon or salt marsh. Lagoons tend to border microtidal areas where tidal ranges less than 6.5 feet (2 m) prevail. Areas in which mesotides prevail, tides ranging between 6.5 and 13.1 feet (2 to 4 m), tend to be bordered by salt marshes extending from the barrier island to the land. The Outer Banks of North Carolina are examples of barrier islands in a microtidal region; the barrier islands of Georgia exemplify those common to a mesotidal region.

Barrier islands separate and protect the mainland from the eroding effects of the sea. They are the dominant landform on most of the Atlantic and Gulf coasts of the United States, the largest chain of barrier islands in the world. Well-known examples of barrier islands include Cape Cod, Massachusetts, the Outer Banks of North Carolina, and Padre Island, Texas. They are found in about one-fifth of the coastlines of the world. Barrier islands are a very dynamic environment, with enormous amounts of their sand being transported by the ocean and atmosphere, especially during major storms. The islands can be battered off and on all winter long by midlatitude cyclonic storms called nor'easters. In the summertime, the relatively tranquil environment can be interrupted by the occasion of a HURRICANE striking on or near the coast. Severe hurricanes are capable of opening new inlets and closing existing ones, thus dramatically altering the configuration of the coastline.

Dunes are the principal natural protectors of barrier islands. For dune development to succeed, there must be plant growth to protect the dune from the wind and slow the rate of errosion, thus stabilizing the dunes. In the southeast, sea oats (*Uniola paniculata*) are very rugged and can withstand the salt-laden air of the foredune region to grow and protect this most exposed and vulnerable line of dunes. In the Northeast, American beach grass (*Ammophila breviligulata*) performs this job. If there is a secondary line of dunes, more sea oats, American beach grass, and salt meadow hay (*Spartina patens*) help to anchor the system. Thickets develop behind these beach grasses, including marsh elders (*Iva frutescens*), red cedars (*Juniperus virginiana*) and wax myrtles (*Myrica cerifera*) in the southeast United States. If there is a third line of dunes called the tertiary line, small maritime forests of pines (*Pinus taeda, Pinus rigida*), live oaks (*Quercus virginiana*), and other rugged trees can develop in the Southeast, and black oak (*Quercus velutina*), beach plum (*Prunus maritima*), and others in the Northeast. Although these trees and plants are usually bent and stunted, their canopy protects the sand from even very strong winds, and produce roots and organic soil to help stabilize the barrier islands.

On the lagoon or bay side of the barrier islands, tidal flats and salt marshes provide shelter and food to the young of many marine animals, including game fish such as striped bass (*Morone saxatilis*), which are commonly called rockfish south of New Jersey, and red drum (*Sciaenops ocellatus*), the North Carolina state saltwater fish. These coastal lagoons are important nursery areas, which are essential to maintain ocean fish stocks. In the estuaries, commercially important finfish prey on species such as Atlantic menhaden (*Brevooria tyrannus*), bay anchovy (*Anchoa mitchilli*), gizzard shad (*Dorosoma cepedianum*), blue crabs (*Callinectes sapidus*), and grass shrimp (*Palaemonetes vulgaris*).

There are many theories as to how barrier islands have formed. In the Northeast of the United States, most barrier islands formed from longshore currents extending existing sand spits. Subsequent storms then broke through weak points in the spit to form inlets. Monomoy Island southeast of Cape Cod was formed in this way. The Outer Banks of North Carolina may have formed when flooding of low areas behind coastal dunes occurred, resulting from the rise in sea level accompanying the melting of glaciers at the end of the last ice age. Barrier islands in the GULF OF MEXICO may have been built up by the shoaling of SANDBARS. Barrier islands are also built at the mouths of large rivers by sand deposition. Notable examples are found at the mouth of the Mississippi River in Louisiana.

Studies of the Outer Banks of North Carolina show that barrier islands are very dynamic systems. Since the time when Sir Walter Raleigh's expedition set foot on the Outer Banks in 1585, new inlets were being driven through the islands during hurricanes and severe nor'easters, while other inlets were being closed due to the infilling of sand during the relative calms following stormy periods. A new inlet was begun on Hatteras Island by Hurricane Isabel in September 2003. This hurricane also caused significant destruction to coastal structures, such as fishing piers, where uplift damage and scour occurred from waves. Waves actually lift the decking from the piling and beams.

Another dynamic feature of barrier islands is the westward (landward) retreat of barrier islands on the east coast of the United States. This westward movement has been confirmed by loss of private property and public works to the sea. There are 19th- and early to mid 20th-century cottages on shoreline roads on the eastern shore of Cape Cod that are now underwater. Cape Hatteras Light was recently moved 1,600 feet (488 m) west (inland), in a spectacular engineering feat, to save it from destruction by the encroaching ocean. Quantitative measurements of shoreline retreat were made by means of movements from reference markers in the sand called benchmarks, and from images taken from aircraft and satellites. These movements are a consequence of the global rise in SEA LEVEL. As the sea rises, its storm waves and winds can attack the higher portion of the dunes and transport their sand westward, through overwash, that is, storm waves topping the dunes. Occasionally, even a "blowout" can occur, in which the sand of a dune is blown westward into a fan-shaped mound, covering all the vital organic material that may have been accumulating over many years. The movement of sand accumulating on the landward side of the barrier island completes the process. In summary, sand is taken away from the ocean side of the barrier island and deposited on the lagoon or bay side, thus moving the whole island westward.

When trees, plants, or dunes are removed, the dunes are destabilized, facilitating the movement of sand, the erosion of beaches, and finally the migration of the entire island landward. Barrier islands are very attractive places to spend summer vacations. Millions of tourists travel to them every year, for swimming, surfing, fishing, boating, or just relaxing. The popularity of these islands has resulted in a great deal of development that frequently interferes with the natural way of maintaining the beaches. Hotels, condominiums, highways, and parking lots all block the natural movement of sand to renourish the beaches and dunes. Construction of groins to prevent beach erosion has been counterproductive. Particularly destructive to dunes are off-road vehicles that scrape off the salt crust of the dunes, kill the vegetation, and dig holes and ruts in the sand. They

are also destroying important animal habitats for marine birds, crabs, sand fleas, and the like. One compromise that allows people to enjoy the recreational benefits of barrier islands without significantly destroying them has been the creation of national parks, such as Cape Hatteras National Seashore.

Further Reading
Cape Hatteras National Seashore, National Park Service. Available online. URL: http://www.nps.gov. Accessed January 8, 2007.
Dean, Cornelia. *Against the Tide: The Battle for America's Beaches.* New York: Columbia University Press, 1999.
Frankenberg, Dirk. *The Nature of North Carolina's Southern Coast: Barrier Islands, Coastal Waters and Wetlands.* Chapel Hill: The University of North Carolina Press, 1997.
Geology Field Notes, Cape Cod National Seashore, Massachusetts, National Park Service. Available online. URL: http://www2.nature.nps.gov/geology/parks/caco. Accessed January 8, 2007.
Padre Island National Seashore, National Park Service. Available online. URL: http://www.nps.gov/archive/pais/myweb2/index.htm. Accessed January 8, 2007.
Virginia Coast Reserve Long-Term Ecological Research, University of Virginia, Department of Environmental Sciences, Charlottesville, Virginia. Available online. URL: http://atlantic.evsc.virginia.edu. Accessed February 19, 2008.

barrier reef Barrier reefs are long, narrow coral reefs, roughly parallel to the shore and separated from land by a LAGOON of considerable depth and width. It may lie a great distance from a continental coast and may be segmented by passes or channels. Reef-building corals require a hard substrate, adequate sunlight, tropical waters greater than 68°F (20°C), and low concentrations of suspended sediments. According to Charles Darwin (1809–82), barrier reefs were first fringing reefs that developed along the shores of a volcanic island. As the island moves into deeper water by processes that we understand today as SEAFLOOR SPREADING, the coral grows upward. Through time biological forces such as reef growth and physical forces such as wave action contribute to the development of a lagoon between the reef and the land. An atoll forms when land surrounded by a barrier reef subsides below sea level.

The seaward edge of the barrier reef (fore reef) is often characterized by a spur-and-groove system (buttresses and channels) formed by wave energy and irregular coral growth. The spur-and-groove system acts to dissipate wave energy, thus reducing damage to the reef. Whereas the fore reef may be steep or very irregular, the back edge of a barrier reef slopes gradually into a shallow lagoon that separates the land from the reef. Mixes of sediments originating from land and reef and intermittent patch reefs characterize the lagoon floor.

The world's largest coral reef, the Great Barrier Reef, is situated in the Coral Sea. This assembly of individual reefs and islands stretches for approximately 1,200 miles (2,000 km) along the northeastern coast of Queensland, Australia. Like most healthy coral reefs, the Great Barrier Reef supports thousands of species of plants and animals and draws hundreds of thousands of tourists annually. Besides the tropical Pacific, barrier reef systems are found along the Caribbean and Atlantic coasts of Belize, Mexico, Guatemala, and Honduras. Barrier reefs in Belize are 124 miles (200 km) long. As sea levels rise, many new barrier reefs are in the process of evolving from fringing reefs.

See also ATOLL; CORAL REEFS; FRINGING REEF; REMOTE SENSING.

Further Reading

Barnes, Richard S. K., and Roger N. Hughes. *An Introduction to Marine Ecology,* 3rd ed. Oxford: Blackwell Science, 1999.

Bonem, Rena, and Joe Strykowski. *Palaces under the Sea: A Guide to Understanding the Coral Reef Environment.* Crystal River, Fla.: Star Thrower Foundation, 1993.

CoRIS, NOAA's Coral Reef Information System, National Oceanic and Atmospheric Administration, U.S. Department of Commerce. Available online. URL: www.coris.noaa.gov. Accessed January 7, 2007.

International Coral Reef Action Network. Available online. URL: http://www.icran.org. Accessed January 8, 2007.

Spalding, Mark D., Corrina Ravilious, and Edwin P. Green. *The World Atlas of Coral Reefs.* Berkeley: University of California Press, 2001.

bathymetry   Water depths referenced to sea level or a tidal datum such as mean lower low water are bathymetry. In contrast, topography is the elevation of land relative to SEA LEVEL. The bathymetry of the ocean is as varied as that of the land. There are continental shelves covering more that 15 percent of the ocean, stretching out from 6.2 miles (10 km) to more than 62 miles (100 km) from shore with depths less than 660 feet (200 m). The deep ocean floor makes up more than 40 percent of the ocean area. The average depth of the world ocean is 12,238 feet (3,730 m). The Mariana Trench, the deepest place in the ocean, is almost 36,090 feet (11,000 m) deep. This means that if you took Mount Everest and placed it in the bottom of the Mariana Trench it would still be covered by more than 6,562 feet (2,000 m) of water!

The following are three methods to measure bathymetry:

- The first is to measure directly by dropping a weighted cable (lead line) to the bottom and measuring the length of cable.

- The second is to use acoustic signals and measure the travel time of a reflected signal from the ocean floor below.

- The third is to use a satellite altimeter to measure the slope of the oceans surface caused by the gravity of the solid earth below and infer the bathymetry.

Other less traditional methods include using lidar or wave kinematics. However, the use of lidar is dependent upon the atmospheric penetration as well as the optical quality of the water.

Bathymetric information is important for basic research, such as crustal dynamics, as well as in marine operations, such as laying deep-sea telecommunication cables, or in supplying accurate navigation charts for safe navigation. One of the most important measurements needed in determining accurate bathymetry is the mean sea level. Mean sea level is used as the zero reference for bathymetry as well as for the Earth's topography. Remote sensing scientists using aerial and satellite imagery may attempt to estimate bathymetry in shallow water. They require bathymetric surveys to test the sensor's performance. Hyperspectral imagery and bathymetric lidar are particularly useful in clear water. Remote sensing is also important in determining the waterline, the land-sea interface at time of imaging.

Tide gauges ranging from simple staffs bolted to a pier to air-acoustic tide gauges that measure the sea height and transmit it via satellite back to the data centers have been used to determine the mean sea level along the continents. Tide gauge records are some of the longest-recorded and most accurate geophysical records on the face of the Earth. The shoreline on a map is actually based on mean high water, which is an average of the high waters in an approximately 19-year tidal record. Recent advances in space-borne radar altimeters and tide modeling have allowed investigators to estimate the mean sea level over much of the world's oceans.

Accurate bathymetry information is important to the mariner. Nautical bathymetric charts give the water depth for low tide (the obvious choice for safety) and give the height of overhead transmission lines that might cross a river for high tide (again, the obvious choice for safety). Tide information is important for commerce in determining how much ships have to be loaded or unloaded before they can enter some seaports. Oil tankers, for example, must offload some of their oil to BARGES (a dangerous and costly procedure) before they can navigate up an estuary such as Delaware Bay and offload at piers. The bathymetric charts, the stage of the tide, and the draft of the vessel are some of the factors that determine how much oil is offloaded.

Bathymetric maps give insight to the geological processes that take place in the formation of the seabed. Smooth-spreading areas such as the Pacific Rise and rough-spreading areas such as the Mid-Atlantic Ridge are caused by different spreading speeds. Large areas, some the size of Oklahoma, have not been surveyed due to the high cost of ship operations. In the future, satellites or even instruments on the *International Space Station* may be able to map the bathymetry of the ocean with a horizontal resolution of approximately 3.73 miles (6 km). When this occurs, we will know the ocean's bathymetry just as well as the topography of Mars, Venus, and our Moon.

Further Reading

Bathymetry and Topography, National Geophysical Data Center. Available online. URL: http://www.ngdc.noaa.gov/mgg/bathymetry/relief.html. Accessed January 7, 2007.

Brown, Joan, Angela Colling, Dave Park, John Phillips, Dave Rothery, and John Wright. *The Ocean Basins: Their Structure and Evolution.* New York: Pergamon Press, 1989.

General Bathymetric Chart of the Oceans (GEBCO), International Hydrographic Bureau. Available online. URL: http://www.ngdc.noaa.gov/mgg/gebco/gebco.html. Accessed January 8, 2007.

Sandwell, David T., Sarah T. Gille, and Walter H. F. Smith, eds. "Bathymetry from Space: Oceanography, Geophysics, and Climate, Geoscience Professional Services." Available online. URL: http://fermi.jhuapl.edu/bathymetry/bathy_from_space_workshop.pdf. Accessed January 8, 2007.

Smith, Walter H. F., and David T. Sandwell. "Global sea floor topography from satellite altimetry and ship depth soundings." *Science* 277 (1997): 1,956–1,962.

United Nations Atlas of the Oceans. Available online. URL: http://www.oceansatlas.org/index.jsp. Accessed January 8, 2007.

bathypelagic zone The zone of the deep ocean basins not in direct contact with the bottom extending from about 3,300–12,000 feet (1,000–4,000 m) is called the bathypelagic zone. This is a zone of absolute darkness, except for the illumination of bioluminescent creatures. Water temperatures range from about 33°F (2°C) near the bottom of the zone, to close to 43°F (6°C) near the top of the zone. Pressures are crushing, ranging from 100 to 400 atmospheres of PRESSURE. Nevertheless, the creatures that inhabit the zone have adapted to the pressure, just as humans have adapted to the pressure of one atmosphere, or 14.7 pounds (6.7 kg) per square inch on their bodies. However, if the creatures of the bathypelagic zone are brought to the surface in nets, their bodies explode when subjected to low pressure. Thus, information about who they are and their habits must be gleaned from direct observations from deep submersibles, underwater cameras, REMOTELY OPERATED VEHICLES, and the like.

Creatures inhabiting the zone include cephalopods, such as the maroon-colored deep-sea squid (*Bathyteuthis abyssicola*), giant squid (*Architeuthis dux*), amphipods, siphonophores, and red shrimp. Some of the predators of the deep-sea squid include deep-diving Cuvier's beaked whales (*Ziphius cavirostris*) and Antarctic bottlenose whales (*Hyperoodon planifrons*). The predator of the giant squid is the well-known sperm whale (*Physeter macrocephalus*). These toothed whales have the remarkable ability to dive to depths of thousands of fathoms (kilometers) to hunt their prey. Other creatures include comb jellies (*ctenophores*), gulper eels (*Saccopharynx ampullaceus*), grenadiers or rat-tailed fish (e.g., *Macrourus bairdii, Macrourus berglax, Coelorhynchus carminatus*), deep-sea anglerfish (e.g., *Melanocetus johnsoni*), and pteropods. The gelatinous mollusks or pteropods that are sometimes called "sea butterflies" gracefully swim the depths grazing on planktonic prey. Many of these fish have bioluminescent lures (*see* BIOLUMINESCENCE) on their body to attract prey, or mates. Several, such as the gulper eel, are able to devour prey as big as themselves by expanding their jaws and stomachs. Although most of the fish and other creatures that live their entire lives in these depths are quite small, they surely deserve the title of "sea monsters" from their appearance and behavior. An example of both is the *vampyroteuthis*, whose name means "vampire squid from hell."

Further Reading

Byatt, Andrew, Alastair Fothergill, and Martha Holmes. *The Blue Planet: A Natural History of the Oceans.* New York: DK Publishing, 2001. (First published by BBC Worldwide Limited, London, 2001, to accompany the television series *The Blue Planet,* produced by the BBC.)

OceanExplorer, National Oceanic and Atmospheric Administration, U.S. Department of Commerce. Available online. URL: http://oceanexplorer.noaa.gov. Accessed January 8, 2007.

Bay of Bengal   The Bay of Bengal is located east of India and the island of Sri Lanka and west of the Malay Peninsula. It extends triangularly approximately 1,300 miles (2,090 km) into the northeastern part of the Indian Ocean and has a width of approximately 1,000 miles (1,610 km). The Bay of Bengal is known for the Ganges and Brahmaputra River Delta, mangrove swamps, Royal Bengal tigers (*Panthera tigris tigris*), three species of crocodiles (*Crocodilus palustris, Crocodilus porosus,* and *Gavialis gangeticus*), monsoon rains, and destructive cyclones that cause great loss of life along the bay's coast through flooding.

The bay receives sediments from many large rivers such as the Ganges, Brahmaputra, Mahanadi, Godavari, and Krishna. The meandering Ganges has its source in the Himalayas, its mouth at the Bay of Bengal, and is considered to be a "holy river." Similarly, the Brahmaputra River has its source in the Himalayas, flows into Bangladesh, and meets the Ganges to form the Ganges and Brahmaputra River Delta, one of the world's largest deltas. Sediments, composed of rock material eroded from the Himalayas and deposited from the many rivers have turned the Bay into a shallow sea. The seasonal wind field, tides, and the amount of river discharge influences the river plumes and the ensuing deposition of sediment load. The delta is therefore an extensive region of sediment dispersal and sedimentation. These riverine inputs also make the Bay of Bengal an important source of freshwater.

The bay's concave shoreline acts to funnel tidal energies, which are semidiurnal (*see* TIDE). Tidal-bores are reported in the Bay of Bengal and tidal ranges are on the order of ca. 6.5 feet (2 m). As the tide propagates into the delta, salty water from the Arabian Sea is carried inland, mixing with the freshwater to create brackish conditions in the rivers, that is, to a level that is dependent on the flow of freshwater.

Circulation in the Bay of Bengal is forced by the Asian MONSOON. GYRES develop in synchronization with monsoon winds. For example, during the period from January to October a poleward-flowing current known as the East Indian Current develops and is replaced from late October through late December by the southward-flowing East Indian Winter Jet, which returns water into the Arabian Sea. These

wind variations and circulations induce upwelling near the coasts from the East Indian Current and piling up of surface waters along the coasts from the East Indian Winter Jet. Weather in the Bay of Bengal is also influenced by the monsoon. The northeast monsoon occurs from December to April and the southwest monsoon from June to October. Tropical CYCLONES occur from May to June and from October to November. This complicated combination of deltaic flats, shallow basin, high tidal ranges, monsoons, and cyclones can result in damaging storm surges. Winds cause the seawater to pile up on the coast, which leads to coastal flooding in highly populated areas. Because the land is so low, and the population density so high, large surges from tropical cyclones result in great loss of life.

Further Reading

Bay of Bengal Program: Inter-Governmental Organization, Tamil Nadu, India. Available online. URL: http://www.bobpigo.org. Accessed January 13, 2007.

Institute of Fundamental Studies, Kandy, Sri Lanka. Available online. URL: http://www.ifs.ac.lk. Accessed January 13, 2007.

Large Marine Ecosystems of the World #34: Bay of Bengal. Available online. URL: http://www.edc.uri.edu/lme/text/bay-of-bengal.htm. Accessed January 13, 2007.

The National Institute of Oceanography, Dona Paula, Goa, India. Available online. URL: http://www.nio.org. Accessed January 13, 2007.

beach   Most marine scientists and engineers view the beach as a zone of unconsolidated material extending from the mean low water line (a vertical datum that is determined by averaging all observed low waters over a 19-year period) to the place where there is a distinct change in material or physiographic form, such as the location of permanent vegetation. Depending on the coastal zone, going from the swash toward the hinterlands, the beach may transition to dunes where sand-binding sea oats (*Uniola paniculata*) and prickly pear cactus (*Opuntia cactaceae*) are common or cliffs that have been cut by the action of storm waves. The beach is further divided into the foreshore and backshore. The foreshore is the beach region experiencing the uprush and backrush of waves in response to the periodic rise and fall of the tide. The backshore extends from the foreshore or the high tide line to the region of dunes that are impacted by waves during storms, especially when STORM SURGE occurs concurrently with high tides.

The confluence of air, land, and sea is a dynamic realm. Beaches receive their sand from the weathering of inland rock that reaches the shore through

A typical East Coast open beach—Delaware   *(NOAA)*

streams and rivers. Temperature differences over land and water influence land and sea breeze circulations. Winds can remove loose material from a beach. During HURRICANES (called typhoons in the western Pacific and cyclones in the Indian Ocean) destructive storm surges and high breakers can remove large amounts of sand from the beach. Such forces contribute to the beaches shape, which may be straight, convex such as a peninsula, or concave such as a pocket beach. As the land slopes down to the sea bottom, incoming waves feel the bottom and their heights increase while wave lengths decrease. Eventually, the unstable waves break and release their energy as spilling, plunging, collapsing, or surging surf.

The character of the surf is usually related to beach slope. Spilling breakers dissipate energy evenly across a flat gradient and are characterized by patches of foam or bubbles that spill down the front face of the wave. Plunging breakers, sought after by surfers, dissipate energy quickly over gentle to intermediate beach gradients. Their crest curls over, forming a concave front that plunges toward the wave's trough. When plunging surf propagates across steeper slopes, they may become arrested by strong offshore winds, which causes the front face to collapse instead of forming a tubelike wall of water. On very steep beach slopes, such as when mountainous terrain descends into the ocean, surging surf is easily identified because the wave's front face is deformed but manifests little or no foam as it slides up the beach face. Waves coming ashore at an angle and breaking contributes to the development of nearshore currents and offshore bars. Currents moving parallel to the shore are called LONGSHORE CURRENTS, and the currents moving perpendicular to the shore are known as RIP CURRENTS. Beachside communities are often compelled to spend large amounts of money nourishing beaches that have lost large quantities of sand. Often, human interventions to restrict the movement of sand by means of groins, jetties, breakwaters, and the like fail to perform as expected and result in considerable damage to the environment.

The slopes and grain sizes found on the beach are indicative of the ocean's strength at that location. Beaches with fine sand (grain sizes around 0.125 mm) tend to be flatter and less energetic than beaches having very coarse sand (grain sizes around 1 mm) and steeper gradients. The drift of particles that are carried by longshore currents can be picked up in the seaward flows of rip currents. The plume of the rip current can be seen just beyond the surf zone. The surf zone extends from the swash seaward to where the waves are breaking as surf. The wave climate of a region indicates how the to- and-fro movement of sand alternately builds up and tears down beach structures. Sediment transport contributes to the buildup of beaches by the smaller waves that generally occur in summer and to the retreat of the shoreline by larger, more erosive storm waves, generally during hurricanes in summer and winter storms. The large waves transport sand offshore to build sandbars, which then provide some protection to the beach from further erosion.

See also OPEN COAST BEACH; SANDBAR; SEDIMENT TRANSPORT; SURF.

Further Reading

Beaches, U.S. Environmental Protection Agency. Available online. URL: http://www.epa.gov/beaches/basicinfo. html. Accessed January 11, 2007.

Brown, Joan, Angela Colling, Dave Park, John Phillips, Dave Rothery, and John Wright. *Waves, Tides, and Shallow-Water Processes,* Oxford: Pergamon Press, 1994.

Coastal Inlet Research Program. Coastal and Hydraulics Laboratory, U.S. Army Research and Development Center, U.S. Army Corps of Engineers. Available online. URL: http://cirp.wes.army.mil/cirp/cirp.html. Accessed January 9, 2007.

beach composition   Beach composition is unconsolidated material along the shore that is dependent on the source of the sediment and on processes such as erosion and deposition. The nature of the biological and physical processes that shape the coast will also contribute to the composition of the beach. Materials may arrive from the mountains through streams and rivers, from the ocean as accumulations of the remains of marine organisms, from nearby beaches by LONGSHORE CURRENTS, and through the atmosphere by winds.

Depending on the biological and physical character of a particular shore, beach materials may be deposited in areas that are from a few feet to thousands of feet (meters to kilometers) wide. Where biological processes dominate, beaches may be composed of shell fragments and carbonate skeletal debris. Where physical processes such as high-energy waves dominate, beaches may be sandy where the waves erode sand and rocky where the waves attack a cliff. Eroded materials are then transported from high-energy locations by currents and deposited in low-energy areas, which results in well-sorted beach deposits that are devoid of clay-sized particles.

Marine scientists, especially coastal engineers, classify sediments according to particle size in order to determine rates of sediment transport. At the beach, marine scientists may also be interested in parameters such as shear strength and soil moisture. Field instruments that can help characterize the beach include cone penetrometers, lightweight deflectometers, ovens, scales, and sieves. The Udden-Wentworth scale defines size ranges for particle diameters ranging from 0.0005 mm (clay) to 256 mm (boulders). The grain sizes and the velocity profile of water over the seabed are needed in order to determine the force necessary to lift a stationary sediment grain off the seabed. Thus, erosion, transport, and deposition of sediments are related to water flow in nonlinear ways. Higher-velocity currents are needed to pick up and transport larger particles except in the case of cohesive and flat clay particles. In contrast, larger particles will settle out at higher velocities than the smaller ones. The computation of the amount and distance that beach material can be transported over a given time is important for maintaining a sand budget. Such calculations are used in developing shore protection strategies that include the installation of revetments, bulkheads, JETTIES, GROINS, and BREAKWATERS.

See also SEDIMENT; SEDIMENT TRANSPORT.

Further Reading

Beach Nourishment: A Guide for Local Government Officials, Coastal Services Center, National Oceanic and Atmospheric Administration. Available online. URL: http://www.csc.noaa.gov/beachnourishment/index.htm. Accessed January 11, 2007.

Surfrider Foundation. Available online. URL: http://www.surfrider.org. Accessed January 11, 2007.

beach protection   Beach protection—also referred to as shore protection—includes methods that are aimed at controlling shoreline erosion and recession. Beach or shoreline property is often protected when processes such as wave action, strong CURRENTS, and flooding threaten beaches, buildings, or the economic well-being of a community. Beach protection

measures may also be implemented in the aftermath of a severe storm in order to save and restore valuable beach property. Strategies to protect beaches and shoreline are simple in comparison to the complicated forces that combine to reshape the coast. Most strategies involve constructing structures to limit the damage, implementing costly beach nourishment programs, abandoning or moving endangered buildings, or a combination of the above. Erosion and sedimentation are natural processes, and when some interest is threatened, the level of protection against erosion should be balanced against the damage that might occur without protection.

Shore-forms and coasts are the result of biological and physical processes: the shore being the strip of land between the lowest tides and the highest reaches for storm waves, while coasts extend inland from the maximum level of impacting storm waves to the farthest extent of marine features such as dunes, coastal floodplains, mangroves, marshes and tideflats. Healthy mangrove stands and CORAL REEFS are examples of biological processes shaping coasts when favorable water temperatures and clarity exist. Marshes and reefs protect inland areas and are important nursery areas for many species. Anthropogenic factors such as the removal of the foredune ridge along a BARRIER ISLAND to build a beachfront hotel may lead to rapid beach loss since the local sand budget is corrupted. Physical forces such as waves and water level fluctuations batter the shore, while winds and currents carry sand up and down the shore. The term *WATER LEVEL FLUCTUATION* includes variable processes such as SEA LEVEL rise, TIDES, and STORM SURGES. The complex interaction of these biological and physical forces occurs at the land-sea interface. Shoreline changes caused

Beach protection by means of a revetment (low wall of rocks). Revetments, bulkheads, and seawalls protect the land but do not promote beach buildup. Groins and breakwaters protect the land by promoting beach buildup. *(NOAA)*

by human activity have led to asphalting and cementing long stretches of beach to protect valuable structures from washing into the ocean. The consequences observed directly from aerial imagery or by observing a vacation beach are shoreline changes, the natural result when erosion is not balanced by accretion.

If a survey and assessment has been made to quantify the rates of erosion that threaten coastal property, then a shore protection strategy can be designed to help mitigate the erosion or to save properties from destruction. Goals for a beach protection strategy might be to widen the beach berm, build dune fields, and improve surfing conditions. The strategies should consider extra erosion that might be caused by altering the natural flow of waves or currents. One strategy might involve the stabilization of river mouths or tidal inlets by constructing shoreline structures. JETTY construction involves placing with a crane, large slabs of granite, armor stone, or riprap into a narrow and elongated structure perpendicular to the shore at an inlet. The interconnected structure prevents longshore drift from filling in the inlet, provides added protection for navigation, and a platform for fishing. Another narrow and linear structure installed perpendicular to the beach is a GROIN. This structure interrupts the LONGSHORE CURRENT and traps longshore drift to build up a section of beach. According to Cornelia Dean in *Against the Tide: The Battle for America's Beaches,* the long-term effects of jetties and groins is almost always destructive to beaches. The erosion is blocked on one side of the groin, but sand is eroded from the other side, thus protecting part of the beach at the expense of other parts of the shore. A second strategy for shore protection involves the nourishment of the beaches and dunes to provide an adequate storm buffer. This method would rely on sediment being dumped, pumped or placed on the beaches by some mechanical means. If dune fields are built, it is essential that the restoration effort include the introduction of native dune vegetation such as sea oats (*Uniola paniculata*). NOAA Sea Grant recommends that approximately 64 sea oat seedlings be planted for each 10-foot (3.05 m) by 10-foot (3.05 m) area of dune. A final strategy for shore protection might involve the abandonment or removal of a building that is damaged or threatened by continuing erosion. During 1999, the Cape Hatteras Lighthouse, a national monument, was moved inland on heavy I-beam rails to protect it from beach erosion at Cape Hatteras, North Carolina. This move of the largest brick structure ever transported on rails was an engineering feat in response to the loss of beach near the lighthouse.

Further Reading

Cronin, William B. *The Disappearing Islands of the Chesapeake.* Baltimore, Md.: The Johns Hopkins University Press, 2005.

Dean, Cornelia. *Against the Tide: The Battle for America's Beaches,* New York: Columbia University Press, 1999.

Jacksonville District, U.S. Army Corps of Engineers. Available online. URL: http://www.saj.usace.army.mil. Accessed January 11, 2007.

Benguela Current   The Benguela Current is a cold eastern boundary CURRENT flowing northward along the southwest coast of Africa, between about 35°S and 15°S. The Benguela Current forms the eastern part of the South Atlantic GYRE, and is thus an eastern boundary current. These are currents that are generally broader, more diffuse, and slower flowing than western boundary currents. Mean speeds less than 0.40 knots (0.2 m/s) have been recorded. In the southern part, the current has been observed to be about 120 miles (200 km) wide, but to widen considerably

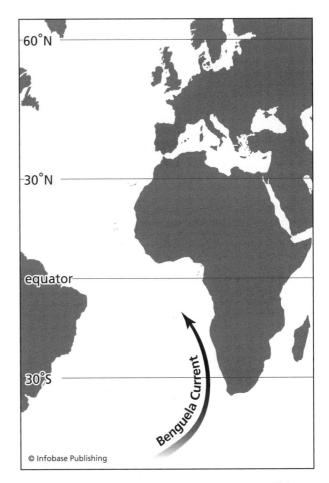

The Benguela Current is a cold current that flows parallel to the southwest African coast; the current produces upwelling and enhanced biological productivity.

as it flows northward. A major source of water for this current comes from the AGULHAS CURRENT, which flows southwestward along the southeast coast of Africa, then branches off at Cape Agulhas with one branch flowing south and southeast into the Antarctic Circumpolar Current, and the other branch flowing westward and then northward into the Benguela Current. Another source for the Benguela Current is the Antarctic Circumpolar Current flowing eastward across the South Atlantic Ocean. The Benguela Current also entrains water from the South Atlantic as it flows northward. The current is generally found within 100 miles (160 km) of the coast and is a region of upwelling due to the offshore EKMAN TRANSPORT resulting from the Southeast Trade Winds. The northern part of the Benguela Current flows into the South Equatorial Current.

Volume transport estimates are generally around 20 Sverdrups (20×10^6 m³/s) or less. Temperatures in the Benguela Current range from about 59°F (15°C) in the region of the Cape of Good Hope to about 72°F (22°C) off Cape Frio, Namibia. Current speeds are generally less than 0.50 knots (0.25 m/s).

The upwelling regions of the Benguela Current support a significant fishing industry. On occasion, climatological conditions are unfavorable to upwelling, and warm water from the tropical South Atlantic moves in, causing a condition known as a "Benguela Niño" after the El Niño in the PERU CURRENT. When these changes to the circulation occur, the dissolved oxygen content of surface waters is reduced. In the northern part of the current especially, the lack of oxygen causes a large increase in sulfate-reducing bacteria that produce hydrogen sulfide, which in turn results in large fish kills. The hydrogen sulfide layer, which initially forms above the bottom, can rise to within 215 feet (65 m) of the sea surface, where it is quite deadly to pelagic life-forms.

EDDIES and rings formed in the Agulhas Current off the southern tip of Africa often propagate northward with the Benguela Current and may persist for months. The Benguela Current has been observed to be fairly steady and well defined near the coast, with more variable flow perturbed by eddies from the Agulhas Current farther offshore. The larger eddies may vary over a large range and contain most of the flows energy. The eddy dissipates as energy and is successively transferred to smaller eddies, and is then eliminated by viscous dissipation in the smallest eddies. Viscosity is a measure of fluid's resistance to flow.

Cool surface temperature anomalies travel northward from the Antarctic region into the Benguela Current due to the Antarctic Circumpolar Wave (see ANTARCTIC CIRCUMPOLAR CURRENT).

Further Reading

Large Marine Ecosystems of the World, LME # 29: Benguela Current. Available online. http://www.edc.uri.edu/lme/text/benguela-current.htm. Accessed January 13, 2007.

Ocean Surface Currents, University of Miami, Rosenstiel School of Marine and Atmospheric Science. Available online. URL: http://oceancurrents.rsmas.miami.edu. Accessed February 7, 2004.

Benioff zone The region of earthquake foci located where a downward-moving oceanic crustal plate interacts with a continental plate is named after Hugo Benioff (1899–1968) of the United States. The Benioff zone is also called the Wadati-Benioff zone in honor of seismologist Kiyoo Wadati (1902–95) of Japan. Their investigations of deep seismicity (see SEISMIC WAVES) during the period from 1935 to 1940 independently recognized that the patterns of earthquakes in the PACIFIC OCEAN were inclined 30–80° from the horizontal and extended more than 100 miles (160.9 km) into the Earth from the ocean trenches and island arcs. These island arcs are generally located at the borders of the Pacific Plate stretching from New Zealand, along the eastern edge of Asia, north across the Aleutian Islands of Alaska, and south along the coasts of North and South America. This region comprises more than 75 percent of the world's active and dormant volcanoes and is often called the RING OF FIRE.

The Benioff zones usually lie beneath the surface where slip along the subduction thrust fault or by slip on faults within the plate that is being pulled into Earth's mantle results in earthquakes. Faults are fractures in Earth's surface where there can be movement of one or both sides relative to the other. Movement along the fault can cause earthquakes or release underlying magma to the surface. Alternating patterns of rock magnetism were discovered in surface seafloor rocks on either side of the mid-ocean ridges during the 1960s. Harry Hess's (1906–69) seafloor-spreading hypothesis was accepted thanks to discoveries such as the Benioff zones and patterns of magnetic striping on the seafloor. These combined geophysical discoveries confirmed the conveyor belt model, where oceanic crustal plates are destroyed in the subductions zones, beneath the ocean trenches, in synchronization with new seafloor being added at the mid-ocean ridges.

Further Reading

Rothery, Dave. *The Ocean Basins: Their Structure and Evolution,* 2nd ed. Prepared for the Open University Oceanography Course Team, the Open University, Milton-Keynes. Oxford: Butterworth-Heinemann, 1998.

United States Geological Survey. *Science for a Changing World.* Available online. URL: http://www.usgs.gov/. Accessed April 28, 2004.

benthos The organisms that live on, at, or in the bottom of the sea are the benthos. Included in this group are ranges of animals from the bacteria in the mud to the lobsters that scuttle across the surface of the bottom. The word *benthos* derives from the Greek "depths of the sea." The benthos actually stretches from the tide pools on the shore that children play in to the abyssal depths where neither sunlight nor humans venture. This bottom-dwelling COMMUNITY includes plants, mollusks, arthropods, and others from all levels of the aquatic food web. Most benthos feed on food as it floats by or scavenge for food on the ocean floor. There are many ways to divide the benthos up into groups. It can be done by type of creature, by where they live, or by their size.

Plant benthos, such as algae and aquatic plants are called phytobenthos. Animals or consumers such at protozoans and metazoans are referred to as zoobenthos. Bacteria and fungi are called benthic microflora. This last group is responsible for decomposing and recycling all of the essential nutrients on the ocean floor.

The benthos may also be divided up into the three communities that they inhabit. These communities are (1) infauna, (2) epifauna, and (3) demersal. The infauna is made up of the bacteria, plants, and animals that live in the sediment. This group includes diatoms, which are considered primary producers, and mollusks and worms, which are primary consumers. These communities have limited mobility and are exposed to environmental changes such as low dissolved oxygen levels and chemical contaminants. For this reason the infauna are often used to indicate the environmental health of an ecosystem. Tiny interstitial organisms such as nematodes and COPEPODS may actually inhabit the space between adjacent particles (such as sand grains) within the sediments. Nematodes are nonsegmented worms approximately 1/500 of an inch (.05 mm) in diameter and 1/20 of an inch (1.3 mm) in length. Benthic copepods are small crustaceans that are generally from 1/125 of an inch (0.2 mm) to 1/10 of an inch (2.5 mm) in length. The epifauna are the bacteria, plants, and animals that are attached to or live on the surface of the sediment. The epifauna make up the group that is most familiar to casual observers. They include all the plants and animals on the bottom, attached to pilings, in tidal pools, and moving about the bottom, such as barnacles, crabs, starfish, snails, and oysters, just to mention a few. The demersal are the bottom-feeding or bottom-dwelling fish.

These animals typically feed on the benthic infauna and epifauna. These fish include tautog (*Tautoga onitis*), left- and right-eyed flounders (family Bothidae and Pleuronectidae), sea robins (family Triglidae), and rays (family Dasyatidae), to name a few. Commercial fishing vessels catch these fish by towing trawls along the seafloor (benthic trawling).

The animals of the benthos are divided into three groups by approximate size. The three groups are (1) the microbenthos, (2) the meiobenthos, and (3) the macrobenthos. The sizes ranges are less than or equal to .1 mm, larger than 0.1mm and smaller than 1 mm, and greater than or equal to 1 mm, respectively. The macrobenthos, including polychaete worms and bivalves, are the largest (>1mm) and most extensively studied category of benthos. Nematode worms and small crustaceans known as harpacticoid copepods are common examples of the meiobenthos (0.1–1mm), while the microbenthos (<1mm) include the diatoms, bacteria, and ciliates.

The success of benthic species primarily depends upon TEMPERATURE, latitude, SALINITY, and depth. Benthic fauna are typically most diverse in tropical and temperate waters. Benthic communities in shallow waters (fringing communities) receive sufficient light to sustain diverse populations of epifauna such as flowering plants, algae, corals, and photosynthetic diatoms, which provide food and shelter to the associated benthos. In fact, bottom-dwelling algae are responsible for 2 percent of primary production in the ocean. Below the photic zone, more organisms rely on PLANKTON for food. Plankton are relatively small plants and animals that live in the water column. Plankton and marine snow, aggregates of cells, mucus, and microalgae ranging in size from 1/50 of an inch (0.5 mm) to 2/5 of an inch (1 cm), are constantly falling through the water column. The plankton and marine snow that reach the bottom provide necessary nutrients to benthic communities in even the deepest parts of the ocean (greater than 33,000 feet or 10,058 m).

Suspension feeders such as bivalves, ophiuroids, and crinoids extend a passive filter and extract food particles floating in the water column. Deposit feeders such as holothurians, echinoids, and gastropods often consume the material that suspension feeders collect but cannot use. Deposit feeders ingest living components of detrital aggregate, then excrete the inert components to be recolonized by bacteria, fungi, mollusks, and other meiofauna. Benthic predators and scavengers, such as fish, starfish, large crustaceans, and gastropods, are typically larger than other members of the bottom-dwelling community.

Benthic fauna, such as vestimentiferans and some bivalve mollusks, living near HYDROTHERMAL

VENTS have adapted to live in water with temperatures of approximately 662°F (350°C). Most of these organisms live in association with chemoautotrophic bacterial symbionts that oxidize reduced compounds released by fumaroles (holes often found in volcanic areas). The remaining species living near the vents are predatory or feed on bacteria. The constant introduction of plankton and marine snow to the bottom results in significant buildup of organic matter that bacteria aerobically decompose on the sediment surface. At greater sediment depth, oxygen becomes less plentiful, and eventually there is no oxygen available for aerobic respiration. The redox discontinuity layer is a gray transition layer that separates the aerobic sediments from the black anaerobic environment. The thiobiota are organisms that live within the gray layer and metabolize sulfur. Many of the infauna living below the redox discontinuity depth survive by maintaining oxygenated water currents that run through or into their burrows. Some bivalves hide from predators by burrowing below the redox layer and extending a siphon toward the sediment surface that collects oxygen and food. Organisms that use anaerobic metabolic processes (anaerobes) below the redox layer release hydrogen sulfide and other toxic reduced ions and molecules.

Human coastal communities have harvested benthic organisms, such as clams, scallops, crabs, and flounder, for tens of thousands of years, and today the economies and public health of many regions of the world rely on these fisheries. Fishing efficiency and effort is increasing with increasing demand; however, many benthic fisheries have failed, and many more are on the verge of collapse. Bottom trawling is a major commercial fishing activity that may significantly decrease benthic community biodiversity. Researchers are studying the effects of trawling and agricultural runoff on the benthos. Scientists are also investigating the effects of exotic invasive benthic species on benthos biodiversity and health.

See also ALGAE; BIVALVE; MOLLUSKS; PHOTIC ZONE.

Further Reading

Barnes, Richard S. K., and Roger N. Hughes. *An Introduction to Marine Ecology,* 3rd ed. Oxford: Blackwell Science, 1999.
Gray, John S. *The Ecology of Marine Sediments.* Cambridge: Cambridge University Press, 1981.
International Association of Meiobenthologists. Available online. URL: www.meiofauna.org/. Accessed July 21, 2008.

berm A relatively flat and steplike sandy area or shoulder formed by material that is deposited by wave action between the BEACH face and dunes is the berm. The berm is where beachgoers set up their camps for sunbathing and quick access to the surf. The astute architect digging a sand castle will notice layers of different size sand grains that were deposited during different wave regimes. Larger waves tend to produce steeper slopes and coarser sands, while milder waves produce flatter beaches and finer sands. As a sandcastle's moat deepens to the depth of the winter beach, the digger will notice that the winter beach slope is greater than the summer beach. In fact, summer sand that beachcombers stroll on was only brought onshore since the end of last winter.

The beach face is the sloping region directly exposed to wave uprush between mean low and high tides. Mole crabs (*Emerita talpoida*), which are sometimes called sand fleas, like to bury themselves in wet sand and are found in abundance along the foreshore or intertidal zone. They are an important food source for endangered bird species such as the piping plover (*Charadrius melodus*) and game fish such as red drum (*Sciaenops ocellatus*). The berm is above the beach face and runs alongside the backshore. It is the habitat for crustaceans such as ghost crabs (*Ocypode quadrata*) that eat dead plant and animal material found on the beach. The ghost crabs make their home in holes near or well above the high-tide line. Horseshoe crabs (*Limulus polyphemus*) move into shallow water to spawn and lay their eggs directly on the beach. They are harvested for fertilizer and biomedical research. The backshore is above mean high tide and is influenced by the ocean only during STORM SURGES or unusually high tides. Landward of the beach berm one finds either coastal dunes or steep cliffs. The dunes are formed from sand deposited by the wind or by storm overwash. Over time, the dune fields are stabilized by grass such as sea oats (*Uniola paniculata*), which may grow to heights of six feet (2 m).

Beaches are harsh dynamic environments subject to breaking waves, changing tides, storm winds, salt spray, surges, and extreme temperature changes. They are also subject to human activities such as construction of beach houses, beach nourishment to reestablish lost beaches, and shore protection to control erosion from wave actions and longshore currents (*see* BEACH PROTECTION). Beach nourishment involves depositing new sediment by artificial means to build up a beach in an effort to protect property, prevent storm damage, and reduce flooding. Shore protection structures such as JETTIES, GROINS, and BREAKWATERS are installed to control the effects of waves and currents.

A berm being constructed for beach protection *(NOAA)*

Further Reading

Coastal and Hydraulics Laboratory, U.S. Army Corps of Engineers. Available online. URL: http://www.oceanscience.net/inletsonline. Accessed August 10, 2004.

The Ocean Beach, Natural History of the Georgia Coast, Marine Education Center and Aquarium. Available online. URL: http://www.uga.edu/aquarium/nature/beach.html. Accessed January 12, 2007.

Bermuda Institute for Ocean Sciences In 1903, Harvard University, New York University, and the Bermuda Natural History Society established the Bermuda Biological Station for Research (BBSR), which later changed its name to the Bermuda Institute for Ocean Sciences on September 5, 2006. The originators of BBSR realized that the Bermuda ATOLL provided a strategic site for a research station given its location in the center of the ATLANTIC OCEAN, its branching, fire, and soft corals, and close proximity to the deep sea. Today, the Bermuda Institute of Ocean Sciences is an independent organization focused on the study of marine and atmospheric systems. The institute includes classrooms, library, and laboratory space, as well as boats such as a 16-foot (4.9 m) skiff, two 32-foot (9.8 m) workboats, and the 41-foot (12.5 m) RV *Stommel*. For research in the deep ocean, the institute uses the regional 115-foot (35.1 m) RV *Weatherbird II*, a University-National Oceanographic Laboratory System or UNOLS vessel, along with heavy-duty deck gear, meteorological and oceanographic instruments, and REMOTELY OPERATED VEHICLES. The ship's master and crew routinely support scientists on projects ranging from Conductivity, Temperature, and Depth (CTD) casts to deploying and recovering MOORING arrays. UNOLS' vessels belong to an organization of 63 academic institutions and national laboratories; they represent the United States's oceanographic research fleet.

Resources at the Bermuda Institute of Ocean Sciences support basic and applied research that are in keeping with international science objectives. Some programs and their associated measuring systems have been integrated to form what is called the Sargasso Sea Ocean/Atmosphere Observatory (S2O2). This observatory is like a wheel, where the tire can be viewed as participating researchers concerned with improved understanding of dynamically coupled atmospheric and oceanographic processes common to the Sargasso Sea. Multidisciplinary scientists are able to observe temporally and spatially variable atmospheric, physical, biogeochemical, and biological phenomena through S2O2 observations. BIOS serves as the hub where data is collected and distilled into important information that can be used by decision makers. The spokes provide structure and include coastal facilities such as the 80-foot (23 m) high walk-up sampling tower erected at Tudor Hill, state-of-the-art instrumentation and ship equipment provided by RV *Weatherbird II,* transects of data collected from the CMV *Oleander,* a remote sensing laboratory that receives and processes imagery from NOAA and NASA satellites, and networked weather stations and radiosonde data provided through the Bermuda Weather Service. The CMV *Oleander* is a container ship outfitted with instruments such as an ACOUSTIC DOPPLER CURRENT PROFILER that routinely surveys the continental shelf, GULF STREAM, and Sargasso Sea as it sails between Port Elizabeth, New Jersey, and Bermuda. The rim is composed of many programs that collect and archive information on environmental processes. Examples include the Bermuda Atlantic Time Series Study, the Oceanic Flux Program, the Bermuda Bio-Optics Project, the AErosol RObotic NETwork, the Moored *In Situ* Trace Element Serial Sampling Program, and the CArbon Retention In A Colored Ocean Program.

The institute is also focused on meeting community and industry needs through its research efforts. For example, the Risk Prediction Initiative applies climate research advances to meet the information needs of the insurance industry. One very important program, owing to increased frequency in HURRICANES, involves the improved assessment, prediction, and management of risks posed by tropical CYCLONES. Investigators are also studying Bermuda cedar (*Juniperus bermudiana*) tree rings to help quantify Bermuda's past climate and possibly correlate this data with other data holdings to characterize trends in the North Atlantic climate system, including fluctuations in hurricane activity and the North Atlantic Oscillation.

The Bermuda Institute for Ocean Sciences is a valuable repository for oceanographic data. Professor Henry Melson Stommel (1920–92) from the WOODS HOLE OCEANOGRAPHIC INSTITUTION began a time series of hydrographic measurements at Bermuda in 1954 that continues to this day. Today, this series of oceanographic observations is one of the longest for the deep ocean. Professor Stommel is most remembered for his contributions in describing the general circulation of the ocean. These long-term time series are essential for studying currents such as the Gulf Stream or the fate of deep-water masses such as North Atlantic Deep Water, which has long

period internal waves and a unique temperature and salinity signature.

Further Reading

Bermuda Institute of Ocean Sciences. Available online. URL: http://www.bios.edu. Accessed March 18, 2008.

Sargasso Sea Ocean Observatory. Available online. URL: http://www.bbsr.edu/Labs/hanselllab/s2o2/s2o2mainpage.html. Accessed June 15, 2007.

University-National Oceanographic Laboratory System, Moss Landing, Calif. Available online. URL: http://www.unols.org. Accessed January 12, 2007.

biological oceanographers   Biological oceanographers are scientists who study marine life in the ocean and nearby coastal or estuarine regions, and the relations between the two. They usually have advanced degrees, having completed specialized courses on the ocean's physical, chemical, and geological processes, which affect biological systems. They are interested in understanding the evolution of living organisms, their food, life cycle, and how they interact with the ocean environment. Biological oceanographers design and execute experiments to identify population controls and temporal variations for the ocean's organisms. A typical study might investigate the composition of a WATER MASS and the degree to which biological systems affect the seawater's composition. A water mass is a body of water that can be identified by its TEMPERATURE and SALINITY, usually identified on a Temperature-Salinity diagram.

As is the case with biologists in general, biological oceanographers tend to be either theoretical or experimental scientists. Theoretical biological oceanographers develop mathematical models to explain interactions among the biological, physical, and chemical aspects of marine systems. Experimental scientists use oceanographic instruments to collect water samples and species for analysis in the laboratory. They also observe *in situ* animal behavior by descending the ocean depths in deep submergence vehicles, listening to and tracking them via underwater SONAR, tagging them, and using remote devices to monitor their migration patterns. Long-term data archives are used to investigate the cycling of carbon and nitrogen through ocean PLANKTON ecosystems by observing the fate of the organisms and particulate organic materials as they settle from the ocean's surface to the deep sea. Their instruments can be deployed from well-equipped research vessels, moored BUOYS, aircraft, and deep submersibles. Biological oceanographers use instruments ranging from low-technology SECCHI DISKS, which measures visibility underwater, Van Dorn bottles, which collect water samples, and thermometers to measure water temperatures, to high-technology ocean color satellite sensors on Earth satellites to measure the abundance of phytoplankton and the concentration of dissolved and particulate material at the ocean surface. Dr. William G. Van Dorn patented the Van Dorn bottle during 1956. He was both a long-term yachtsman and an oceanographer with SCRIPPS INSTITUTION OF OCEANOGRAPHY from 1947 to 1987.

Biological oceanographers are somewhat different from marine biologists who focus their research on a particular organism, including their physiology and predator-prey relationships. Marine biology is a common undergraduate program for students entering biological oceanography graduate programs. Biological oceanographers normally have at least a master's degree. University research positions normally require a doctorate followed by postdoctoral research. With only a bachelor's degree in marine biology, positions are generally limited to the technician level. Fully trained biological oceanographers study biological systems in order to fully understand a class of marine systems. They go to the biological system in ships or submersibles to collect pertinent data to characterize the biosystem, as well as the marine environment.

biological reference point

A particular value of stock size, catch, fishing effort, or fishing mortality used as a goal in fisheries management is called a biological reference point. The stock defines a grouping of fish based on genetic relationship, geographic distribution, or movement patterns. Catch refers to the total number or poundage of fish captured from an area over some period of time. Benchmarks such as biological reference points allow fisheries management personnel to assess fisheries status based on abundance of the stock or the fishing mortality rate. This term is similar to carrying capacity, which represents the maximum population that a specific ecosystem can support indefinitely without deterioration of the character and quality of the resource. In order to avoid overfishing, fisheries scientists will estimate the amount of harvest or yield that can be removed from a population without causing the stock to be depleted given average environmental conditions. This concept is called SUSTAINABLE YIELD.

Further Reading
Food and Agriculture Organization of the United Nations: Helping to Build a World Without Hunger. Available online. URL: http://www.fao.org. Accessed January 13, 2007.

NOAA Fisheries: Office of Science and Technology. Available online. URL: http://www.st.nmfs.gov. Accessed January 13, 2007.

bioluminescence

The word *bioluminescence* can be broken down into its Greek and Latin roots to mean *bio* (Greek for life) and *lumin* (Latin for light). In its simplest form, bioluminescence refers to the emission of light by living plants and animals. Some scientists also refer to biologically produced light as "phosphorescence." On the East Coast of the United States, especially near water, people are familiar with lightning bugs such as *Photinus pyralis,* that glow either to attract mates or to convince potential predators that they taste bad. The light-emitting chemical reaction consists of luciferin (a substrate) combined with luciferase (an enzyme), ATP (adenosine triphosphate) and oxygen. In the ocean, this phenomenon of emitting light is extremely common since bioluminescence occurs in many species of fish, shrimp, squid, and PLANKTON. Swimmers in ESTUARIES and the ocean are familiar with comb jellies such as the sea walnut (*Mnemiopsis leidyi*), which are often called "phosphorous," which often phosphoresce emitting a blue-green glow when disturbed.

Many marine biologists study the role of bioluminescence in nocturnal predator-prey interactions. In the plankton, short luminescent flashes emitted by phytoplankton such as dinoflagellates attract small fish to potential ZOOPLANKTON food sources. These food sources could be the very COPEPODS that disturbed the dinoflagellates, causing rapid flashes that investigators time between 0.1 to 0.5 seconds. This type of interaction affects the feeding behavior of copepods and reduces the numbers of dinoflagellates that are consumed. At dark depths from approximately 984 to 13,123 feet (300 to 4,000 m), nekton such as the female deep-sea anglerfish (*Linophryne*

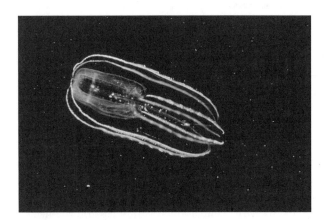

Bioluminescence exemplified by the bioluminescent lobate ctenophore *(NOAA)*

arborifer) sport glowing lures dangling on a fin ray over its huge jaw. The wiggling lure hosts millions of light-producing bacteria that causes the "lure" to light up. It looks like a worm to unwary prey such as small fish and crustaceans. On the water surface, bioluminescent dinoflagellates are generally the cause for light displays seen in breaking waves, the splashes from a swimmer, or in boat wakes. For this reason, many military organizations attempt to understand the behaviors of bioluminescent organisms in order to avoid compromising covert operations. Some U.S. Navy pilots have reported that illumination from ship wakes was so brilliant that it aided the flight navigator in locating their home aircraft carrier.

Industry and fields such as biomedical science extensively exploit bioluminescence in exciting efforts focused on enhancing the quality of human life. Glow sticks commonly seen at evening outdoor concerts, sports events, or the circus provide a good example of corporate application for bioluminescent technologies. These products are having real applications for safety, recreation, and even in commercial fishing. In *Aequorea victoria*, an abundant jellyfish found along the Pacific Coast of North America, coupled with aequorin and green fluorescent protein cause emitted light to appear green. Biomedical researchers are studying the luminescent protein aequorin and the fluorescent molecule known as green fluorescent protein or GFP. These materials have been extracted and used as an *in vivo* marker. (The term *in vivo* is Latin for happening within a living organism.) Thus, the GFP is used to obtain an image of processes going on within a living cell. Other examples include the use of lucerferin and lucerferase in medical research as possible treatments for cancer, cystic fibrosis, multiple sclerosis, and heart disease.

Further Reading

Haddock, Steven H. D., Carrie M. McDougall, and James F. Case. "The Bioluminescence Web Page." Available online. URL: http://www.lifesci.ucsb.edu/~biolum. Accessed March 11, 2004.
HBOI Bioluminescence, Harbor Branch Oceanographic Institution, Ft. Pierce, Fl. Available online. URL: http://www.biolum.org. Accessed January 13, 2007.
McClinton, Jack. "Splendor in the Dark." *Discover* 25, no. 5 (May 2004) 51–57.

bivalve   A bivalve is one of several classes of soft-bodied mollusca that are protected by two hinged shells or valves. The adductor muscles and the hinge between the two halves of the external shell help keep the shell closed. They have a foot that can be used to burrow, creep, or to attach, depending on the species. Filter feeders such as scallops, oysters, and clams are important in cycling organic matter from the water column. Many species of bivalves also make up important commercial fisheries. The term *fishery* refers to all the activities involved in catching a species of fish (shellfish) or group of species. Live markets for bivalves such as oysters include seafood distributors, retailers, restaurants, and processors. Statistics by organizations such as the National Marine Fisheries Service report landings of MOLLUSKS as pounds of meats and therefore exclude the shell weight. Scallops such as the Atlantic bay scallop (*Argopecten irradians concentricus*) stay in underwater grass beds on a soft, shallow seafloor. Scallops move in spurts by quickly opening and closing their shells to propel them in an erratic path through the water. Oysters such as the Olympia (*Ostrea lurida*) attach themselves to rocks, pilings, and old oyster shells that are found in ESTUARIES and bays. The oyster also has the ability of changing its sex. They are usually only a few inches long, but can reach sizes of three feet (1 m). Clams are invertebrate animals that burrow under the seafloor and range in size from the small coquina clams (*Donax fossor*), which live in the swash zone to the Indo-Pacific giant clam (*Tridacna gigas*), which prefers waters between 30 to 60 feet (9.1 to 18.3 m deep). The swash zone is the shallow portion of the beach favored by beachcombers where there is a landward flow of water from breaking waves. It is becoming increasingly more common for retailers to sell farm-raised oysters and clams. In the United States, key AQUACULTURE areas include Chesapeake Bay, Maine, the Gulf Coast, Long Island, and the Pacific Northwest.

Further Reading

Carnegie Museum of Natural History. "Section of Mollusks: About Mollusks." Available online. URL: http://www.carnegiemnh.org/mollusks/about.htm. Accessed January 13, 2007.
Chesapeake Bay Program: Americas Premier Watershed Partnership, Annapolis, Md. Available online. URL: http://www.chesapeakebay.net. Accessed January 13, 2007.
San Francisco Estuary Institute, Oakland, Calif. Available online. URL: http://www.sfei.org. Accessed January 13, 2007.
Washington Department of Fish and Wildlife. "Priority Habitats and Species." Available online. URL: http://www.wdfw.wa.gov/hab/phsinvrt.htm. Accessed January 13, 2007.

Black Sea   The Black Sea is a marginal or inland sea with an area of approximately 160,000 square miles (414,400 km²) and a large basin at a depth of approximately 1,200 fathoms (7,200 feet, [2,195 m]). Turkey to the south, Georgia to the east, Russia and Ukraine to the north, and Romania and Bulgaria to the west, border the Black Sea. It contains the world's largest dead zone or anoxic deep layer below depths of 42 to 50 fathoms (252 to 300 feet [77 to 91 m]).

See also MARGINAL SEAS; MEDITERRANEAN SEA.

Further Reading

Large Marine Ecosystems of the World # 62: Black Sea. Available online. URL: http://www.edc.uri.edu/lme/text/black-sea.htm. Accessed September 13, 2007.

University of Delaware College of Marine Studies. "A Black Sea Journey." Available online. URL: http://www.ocean.udel.edu/blacksea/research. Accessed January 13, 2007.

boats The U.S. Coast Guard classifies vessels less than 65 feet (19.8 m) in length as boats. However, the term *boats* is fairly indefinite and means only that the vessel is smaller than a ship. Some classify boats as any vessel that can be hoisted aboard a ship, and ships are generally large oceangoing vessels of more than 500 displacement tons. Boats are specially designed by naval architects to meet the user's needs, needs ranging from rowing out to a moored cabin cruiser in an ESTUARY to circumnavigating the globe. Boats are used for work such as long-lining to catch fish or towing side-scan SONAR to make charts. Yachtsmen may race in sailboats or cruise in powerboats. Sportsmen use high-speed boats such as hydroplanes to race across calm water bodies. Sailors may patrol regions on hydrofoils or use a launch to take personnel ashore from a warship. Since 70 percent of the world is covered by water, boats are an essential component of transport and commerce for maritime countries such as the United States.

Boats float because their volume is greater then the water they displace, and their performance in moving through the water is dependent on hull type and propulsive power. The hull is the structural body of the boat that rests in the water. The hull influences the amount of resistance that a boat generates as speed is increased. For this reason, hulls are further categorized as displacement, semi-displacement, and planing, based on their ability to transition from displacement mode to planing mode. Sailboats and trawlers are examples of boats with displacement hulls, while speedboats that get up on top of the water are examples of boats with planing hulls. Each hull type can have many sub-types, e.g., popular planing hulls include the v-hull of a Grady White or the cathedral hull of a Boston Whaler.

Other design factors that influence a boat's utility are speed, range, ride quality, and seaworthiness. Speed is related to variables such as hull type, power unit, load, and sea condition. Stability is affected by the boat's beam or width and factors such as structural modifications, cargo, catch, passengers, and fuel. The powerboat's range is predicted based on its speed and the local wind and sea conditions. Ride quality relates to the manner in which the boat rolls, pitches, and heaves in response to waves and wind. A seaworthy boat can handle waves comfortably, weather storms, and bring passengers and crew back to land safely.

Boats move through the water by various means from rowing with oars and sailing with sails to powering with inboard and outboard gas or diesel engines. Outboard engines are mounted outside the hull, while inboard engines are inside the hull. The engine's power is converted to thrust by the propellers, which usually have three, four, or five blades. On boats such as sailboats or inboards, a tiller or wheel turns the rudder in order to steer the boat. This action turns the boat by directing the flow of water passing over the rudder's surfaces. Outboard engines are steered by turning the wheel or engines and redirecting the discharge screw current that is generated by the turning propellers.

Boats are a very important platform for marine science. Boats supporting scientific operations are outfitted with deck gear such as cranes and WINCHES, computers, radios, and lab space for scientists. They are used to deploy corers that sample the sediments, bottles that sample the seawater, nets that sample organisms from minute PLANKTON to large fish, and balloons that carry radiosondes that measure the atmosphere. Hydrographers in a survey boat might determine depths in a channel, while coastal engineers in a workboat measure bed profiles under a BRIDGE to assess scour. Some boats are maintained in PORTS AND HARBORS for first responders to rapidly deploy booms that help minimize the effects of a marine spill. Scientists involved in operating boats usually hold a certification or have completed a boating safety course. At a minimum, certification means that the researcher demonstrated proficiency with the vessel and knowledge of the published U.S. Coast Guard Navigation Rules or "Rules of the Road."

Further Reading

National Association of State Boating Law Administrators. Available online. URL: http://www.nasbla.org. Accessed January 12, 2007.

United States Coast Guard Auxiliary. Available online. URL: http://nws.cgaux.org. Accessed January 12, 2007.

United States Coast Guard, Navigation Rules, International-Inland, COMDTINST M16672.2D, U.S. Department of Homeland Security. Available online. URL: http://www.navcen.uscg.gov/mwv/navrules/navrules.htm. Accessed January 13, 2007.

United States Power Squadrons. Available online. URL: http://www.usps.org. Accessed January 13, 2007.

bottom currents   Bottom currents are found in the deep ocean just above the bottom. Different types of bottom currents include CONTOUR CURRENTS, deep-sea density currents, TURBIDITY CURRENTS, and UNDERCURRENTS.

Contour currents cling to the contours of the ocean floor, especially in regions of rapid depth change, such as the CONTINENTAL SLOPE. These currents were first identified by steep temperature gradients, and later measured directly by self-recording current meters. They play a significant role in the erosion of the CONTINENTAL RISE and CONTINENTAL SHELF. They are called deep geostrophic currents by physical oceanographers.

A deep-sea circulation driven by density differences resulting from the TEMPERATURE-SALINITY structure of the oceans is found above the ocean floor at depths as great as 13,100 feet (4,000 m). Deep currents are found flowing southward below the GULF STREAM into the Southern Hemisphere, and below the BRAZIL CURRENT and into the ANTARCTIC CIRCUMPOLAR CURRENT, east of Argentina. The Antarctic Circumpolar Current flows eastward around the Antarctic continent, and is an extremely deep current, flowing from the surface all the way to the bottom. Northward flow along the bottom branches off the Antarctic Circumpolar Current into the INDIAN OCEAN east of the Republic of South Africa, and into the PACIFIC OCEAN east of New Zealand. The Pacific flow continues into the North Pacific Ocean, where it turns east toward the Americas.

The southward flow in the North Atlantic Ocean is quite concentrated, being pushed against the American continental margin by the CORIOLIS FORCE. Currents have been found as large as almost a knot (0.51 m/s). This current forms a southward pathway for the recently formed North Atlantic Deep Water (NADW), sinking down from the surface of the Greenland and Norwegian Seas.

Small-scale bottom currents have been found in the central parts of ocean basins that seem to have been induced by strong meteorological events on the surface. Such currents are sometimes referred to as underwater storms. They are often identified by patterns of suspended sediment movements appearing in photographs of the bottom, as well as side-scanning sonar records.

Turbidity currents or underwater landslides can occur, especially down submarine canyons, in which large amounts of sediment are transported from the continents to the ocean floor. These currents can be quite rapid, on the order of knots and have snapped telephone/telegraph cables. They generally result from seismic disturbances.

Other types of bottom currents include undercurrents flowing counter to eastern boundary surface currents, such as the Davidson Current flowing beneath the CALIFORNIA CURRENT, and the Gunther Current flowing beneath the PERU (HUMBOLDT) CURRENT. These deep currents form part of a benthic boundary layer, which exhibits turbid water from suspended sediment and turbulent flow induced by rough topographic features of the bottom.

See also THERMOHALINE CIRCULATION; WATER MASS.

Bouguer anomaly   Since gravity is not equal around the Earth, the fundamental gravity anomaly arising from underlying geology after latitude and elevation corrections have been applied to measured gravity data is called the Bouguer anomaly. These deviations provide important information regarding the density of rocks in the vicinity of where gravity measurements have being conducted and information about the undersea topography. For example, Harry Hess (1906–69) while collaborating with the Dutch physicist Felix Andries Vening Meinesz (1887–1966) on an investigation of the gravity anomalies of deep ocean TRENCHES hypothesized that the trenches were SUBDUCTION ZONES, where an oceanic plate is plunging down into Earth's mantle.

Gravity measurements are made with a gravimeter and the units of measurement are in gals. Named after the astronomer Galileo Galilei (1564–1642), the gal is the basic unit for gravity and equals 1 cm s^{-2}. A gravimeter is essentially a weight on a spring that measures variations in the acceleration due to gravity. Maps are useful in showing Bouguer anomaly contours or gravity values. The resulting contours highlight very small changes in gravity created by changes in subsurface rock types or elevations. This anomaly is named for Pierre Bouguer (1698–1758), a French mathematician who demonstrated that gravitational attraction decreases with altitude.

See also AERIAL SURVEY; HYDROGRAPHIC SURVEY; REMOTE SENSING.

Further Reading

Kennett, James P. *Marine Geology.* Englewood Cliffs, N.J.: Prentice Hall, 1982.

U.S. Geological Survey. *Geophysical Data Grids for the Conterminous United States, Fact Sheet 078–95.* Available online. URL: http://pubs.usgs.gov/fs/fs–0078–95/FS078–95.html. Accessed January 13, 2007.

Wybraniec, Stanislaw, Shaohua Zhou, Hans Thybo, Rene Forsberg, Edward Perehuc, Michael Lee, Gleb D. Demianov, and Vladimir N. Strakhov. *New Map Compiled of Europe's Gravity Field, Geological Institute, University of Copenhagen, Denmark.* Available online. URL: http://solid_earth.ou.edu/readings/europe_gravity.html. Accessed January 13, 2007.

box corer An instrument used to take a relatively undisturbed vertical sample of the sediment at the bottom of a body of water is called a box corer. Cores are usually taken when a simple grab sample of the bottom is inadequate. A Box corer is lowered from a research vessel to the bottom, when it strikes the bottom; a steel box is driven into the sediment, taking a sample of the material. As is the case for a conventional PISTON CORER, weights can be added to the frame of the box corer to increase the force with which it is driven into the sediments. A spade lever then moves to cut the sediment core, while forming the bottom of the box. After the sample has been

Recovery of a box core from the deep ocean *(NOAA)*

taken, the corer and core are winched back aboard the research vessel. Once the corer is back on deck, the cores are usually separated from the corer by electrolytic means, reducing the friction of the samples against the walls of the corer and allowing them to slide out. A typical length of core for a box corer is 0.98 to 1.6 feet (0.3–0.5 m). The box is typically about 0.98 feet (0.3 m) on a side.

The box corer does not compress the sediments as does a conventional corer, and it retains the benthic (ocean bottom) fauna in the core in their natural locations. Scientists can then study the spatial distribution of the organisms and the relation to their immediate environment. Visual data on the organisms can be logged as soon as the core is on deck. Box cores are often x-rayed to examine the layering of the sediments, and also to reveal burrows of benthic creatures. Living creatures can then be recovered alive for further study if desired, or preserved in an undamaged condition. Box corers are adaptations of instruments originally used by geologists on land for taking undisturbed soil samples for study in the laboratory. Small oceanographic laboratories on limited budgets have made box corers from inexpensive materials on hand for use in shallow, inshore waters.

Further Reading

Mapping Systems, Coastal and Seafloor Mapping, Western Coastal and Marine Geology, USGS. Available online. URL: http://www.walrus.wr.usgs.gov/mapping/sampling.html. Accessed January 13, 2007.

Woods Hole Oceanographic Institution, Ocean and Climate Change Institute. Available online. URL: http://www.whoi.edu/page.do?pid=7399. Accessed September 14, 2007.

Brazil Current A warm current formed where the South Equatorial Current reaches the eastern tip of South America at Ponta do Calcanhar, offshore of the region between Natal and Recife, and splits, with one part flowing northward into the GUYANA (North Brazil) CURRENT, and the other part flowing southward, is known as the Brazil Current. The flow into the Brazil Current at its northern terminus is about 17 Sverdrups (17×10^6 m³/s), with a speed of about 1 knot. At about 35°S to 40°S in the vicinity of the mouth of the Río de la Plata, the Brazil Current meets the cold northward flowing Falkland (Malvinas) Current coming up as a branch of the ANTARCTIC CIRCUMPOLAR CURRENT. The combined currents turn eastward to flow into the South Atlantic Current.

Although it is a western boundary current, forming the western branch of the South Atlantic GYRE,

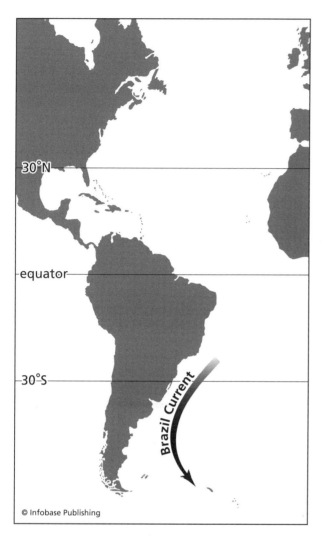

The Brazil Current is a warm current that flows southward parallel to the coast of Brazil.

Seasonal variations occur in the Brazil Current, with the current being found farther offshore in the Southern Hemisphere winter. The Brazil Current extends farther south in summer and retreats northward in winter, a movement believed to be related to the seasonal displacement of the INTERTROPICAL CONVERGENCE ZONE (ITCZ) and the South Atlantic gyre.

Further Reading

Instituto Oceanográfico da Universidade de São Paulo. Available online. URL: http://www.io.usp.br. Accessed July 21, 2008.

Ocean Surface Currents, University of Miami, Rosenstiel School of Marine and Atmospheric Science. Available online. URL: http://oceancurrents.rsmas.miami.edu. Accessed February 7, 2004.

U.N. Atlas of the Oceans, United Nations Foundation. Available online. URL: http://www.oceansatlas.org. Accessed January 6, 2007.

breakwater   A coastal structure that protects a BEACH, harbor, anchorage, or basin from the full impact of waves. Traditional breakwaters are linear structures placed offshore to dissipate the energy of incoming waves. As the wave energy is dissipated, drifting beach material settles behind the breakwater. The amount of deposition depends on unique site characteristics such as nearshore currents, wave climate, sediment composition, and the type of breakwater. Nearshore currents that run parallel to the shore are caused by waves approaching the shore at an angle. They are called "longshore" or "littoral currents." Nearshore currents flowing seaward in response to water being piled up against the shore by incoming waves and wind are called "rip currents" (*see* RIP CURRENT). The average condition of wave heights, periods, and directions at a location, over a period of years, is known as wave climate. Depending on the above factors, breakwaters help attenuate wave heights and may be fixed or floating, attached to the beach or detached.

The coast is the region where air, land, and sea meet. It is a very dynamic zone associated with large populations and commercial centers. When one looks at water movements alone, it is easy to see how vertical fluctuations such as TIDES, STORM SURGES, and breaking waves can aggravate erosion and flooding. The horizontal movements of water that include tidal currents, rip currents, littoral or longshore currents, and hydraulic currents also contribute to sediment transport and erosion or accretion. These impacts to the shore are critical since sand is as important to many benthic life-forms as air and water are to people. Coastal engineers and support

and the counterpart of the GULF STREAM, its speeds are much lower, typically being less than 2 knots (1.0 m/s), with a mean speed of about 0.6 knots (1.2 m/s). Another difference is that the density of the Brazil Current changes in the direction of flow, not perpendicular to the flow direction, as does the Gulf Stream. Its volume transport has been estimated at about 17 Sverdrups (17×10^6 m³/s). Transport varies seasonally and increases with distance downstream, becoming as high as 20 Sverdrups (20×10^6 m³/s) south of 30°S. It is a rather shallow current, with depths on the order of 1,980 feet (600 m). Below the Brazil Current, Antarctic Intermediate Water is moving northward, eventually flowing into the Northern Hemisphere. Because of the high SALINITY of its source water in the South Equatorial Current, the Brazil Current is very saline, generally ranging from 36–37 psu (practical salinity units).

Breakwater for Timbalier Island in Louisiana. The island is part of a chain of barrier islands that front Timbalier Bay and provides a line of defense against waves from the Gulf of Mexico. *(NOAA, Erik Zobrist)*

scientists such as oceanographers and geologists assist property owners as well as state and local governments in developing structural and nonstructural measures for shore protection and storm damage reduction.

Breakwater design depends on what is being protected and the nature of waves, tides, and shallow water processes that are common to the area. They are usually rubble features with a toe, core, and layers of armor stone sited parallel to the shore to provide protection against wave action. There is usually a layer of graded fine stones underlying a breakwater to prevent the natural bed material from being washed away. The toe is the lowest part of the sloping breakwater that transitions the structure to the seabed. The core provides an inner and less permeable portion of the breakwater. The outer layers comprise armor stones, large quarried stones, or specially shaped interlocking concrete blocks. During construction, armor stones are usually brought to the site in BARGES.

The United States Corps of Engineers conducts coastal research and develops new designs for structures such as breakwaters. One such recent development is the RAPIDLY INSTALLED BREAKWATER or "RIB" for ocean deployments. This V-shaped floating breakwater is unique as it spreads wave fronts apart based on geometric spreading and coupled deflections of wave motion. The moored breakwater floats at the surface and is oriented into approaching waves to provide a sheltered area inside the "V" and for some distance in the lee. Mooring loads are minimized because the structure spreads and reflects incoming wave energy. This breakwater has shown promise to military planners that need to offload ships in areas that lack a natural seaport. These

operations called "Logistics Over-The-Shore" or "LOTS" are sea state limited because they involve the at-sea use of ship-based cranes and stevedore crews. The RIB operates to reduce wave heights in its lee and to provide suitable sea states for the ships, lighters, craft, and barges that are involved in LOTS operations. The RIB system has been demonstrated at numerous sites including the Field Research Facility in Duck, North Carolina (see FIELD RESEARCH FACILITY). The Field Research Facility or "FRF" is internationally known for its 1,840-foot (560 m) pier that extends from the BARRIER ISLAND into the Atlantic Ocean. The FRF has also been designated as an ocean observatory where one can study changing shallow-water waves, coastal winds, tides, and nearshore currents in real time.

Further Reading

Coastal Structures, Coastal and Hydraulics Laboratory, Engineer Research and Development Center, U.S. Army Corps of Engineers. Available online. URL: http://chl. erdc.usace.army.mil/chl.aspx?p=s&a=ResearchAreas;2 3. Accessed January 13, 2007.

Dean, Cornelia. *Against the Tide.* New York: Columbia University Press, 1999.

bridges Civil engineering structures that provide passage across obstacles such as rivers, ESTUARIES, railroads, and valleys are bridges. Scientists and engineers involved in the construction of bridges base the initial design on the obstacle's features. The three major types of bridges are arch, beam, and suspension. There are many variations on these types such as cantilever and cable-stayed bridges. One example of a fixed bridge is the Pont du Gard aqueduct that was built by the Romans before the birth of Christ. This arch bridge stands near Nîmes, France, and the stone bridge's weight is carried outward along the curve of the arches to the abutments at each end. The world's longest bridge, the Lake Pontchartrain Causeway, is a beam bridge that is almost 24 miles (39 km) long. Lake Pontchartrain is an oval-shaped brackish water body located in southeastern Louisiana, covering an approximate area of 630 square miles (1,630 km²) with an average depth of 13 feet (4 m). A dual suspension bridge crosses CHESAPEAKE BAY with a shore-to-shore length of 4.3 miles (6.9 km). This is one of the world's most scenic overwater structures that Marylander's call the Bay Bridge. Its actual name is the William Preston Lane, Jr., Memorial Bridge, and the main span has a horizontal navigational clearance of 1,500 feet (457 m), a vertical navigational clearance of 186 feet (57 m), and is utilized by oceangoing ships to and from the Port of Baltimore. Bridges may also be movable such as swing,

The Golden Gate Bridge spans the inlet to San Francisco Bay. *(NOAA)*

floating, or draw bridges. Some major components are abutments, anchorage blocks, bents, columns, dampeners, deck, piers, pilings, pontoons, spans, stays, struts, suspension cables, trusses, and turntables.

The construction and maintenance of bridges over water bodies requires support of a multidisciplinary group of scientists, engineers, and technicians. Beginning with the design, the engineering team ensures that the flow of wind and water do not set up resonant vibrations that will shake the structure apart. Scale models of the bridge are tested in wind tunnels and flumes, while others evaluate the structure, using numerical models and simulations. Marine scientists might compute tidal prisms to help evaluate forces on the bridge bents and the potential for scour around pilings. The tidal prism is an estimate for the amount of water passing under the bridge over a tidal cycle. Waves also need to be modeled since they have many effects on bridges. The tremendous wave forces generated by storms can cause dangerous vibrations and shearing forces. Other marine science applications would utilize bridges as platforms for instruments like microwave RADAR to measure surface CURRENTS in the main navigation channels. ACOUSTIC DOPPLER CURRENT PROFILERS have also been deployed under the main spans of bridges such as the Sunshine Skyway, a cable-stayed bridge crossing Tampa Bay in Florida. These devices measure currents in the water column and provide vital information for planning docking operations for large vessels (*see* PORTS AND HARBORS). Other sensors are deployed to measure strain and even air gap. Not knowing bridge clearance can be disastrous for recreational and commercial boaters.

The environmental characteristics of a particular location have important effects on bridges. In North Carolina, there are usually between three and 11 major inlets along the dynamic barrier islands

called Outer Banks. In fact, Oregon Inlet was opened during a HURRICANE in 1846 and has moved south approximately two miles as a result of erosion and accretion. In 1963, the Herbert C. Bonner Bridge was built over Oregon Inlet. Sediment transport and inlet migration continue causing many maintenance challenges for the North Carolina's Department of Transportation. Armor stone was placed along the southern shore in order to protect the bridge and slow the rate of shoreline retreat. In October 1990, the dredge *Northerly Island* collided with the Herbert C. Bonner Bridge during a northeaster storm, causing severe damage to several of the spans. Hatteras Island to the north could be accessed only by boat or plane for many weeks while emergency construction was under way to replace the island's only highway link to the mainland. Maintenance dredging to keep Oregon Inlet open to navigation as well as frequent bridge bed profile surveys will be necessary for years to come. In 2003, Hurricane Isabel began opening a new inlet on Hatteras Island just north of Hatteras Village. The U.S. Army Corps of Engineers and other federal and state agencies are working to close this incursion, which has destroyed the major road on the island, Highway 12.

Further Reading

Confederation Bridge, Prince Edward Island, Canada. Available online. URL: http://www.confederationbridge. com. Accessed January 13, 2007.

Cooper, James D. "World's Longest Suspension Bridge Opens in Japan." *Public Roads* 62, no. 1 (July/August 1998). Available online. URL: http://www.tfhrc.gov/ pubrds/julaug98/worlds.htm. Accessed January 13, 2007.

Humber Bridge. Available online. URL: http://www. humberbridge.co.uk. Accessed January 13, 2007.

Pont du Gard Web site. Available online. URL: http://www. pontdugard.fr. Accessed January 13, 2007.

buoys   Buoys are usually anchored structures used for marking positions on the water, to alert mariners of hazards, to indicate changes in bottom contours or navigation channels, for MOORING vessels, or to provide a platform for instruments and communications equipment. There are many types of buoys, and they can all be used as guideposts for navigation. They are given specific color, shapes, numbers, and light characteristics in accordance with international agreements. Navigation buoys provide information for safe passage, while instrumented buoys have a payload of sensors that collect oceanographic, acoustic, and meteorological data. They provide information ranging from the side on which a vessel should pass to real-time winds, water levels, and CURRENTS.

Oceanographic buoy deployed for detection of environmental conditions favorable for the development of El Niño *(NOAA)*

Buoy design depends on the intended deployment location and purpose. For example, subsurface buoys deployed in the deep ocean provide taut moorings for instruments such as current meters and thermistor chains. Subsurface buoys are usually favored by oceanographers because they provide a more stable platform for instruments such as current meters than do surface buoys. This is so because ocean surface wave motions decrease exponentially with depth. Ocean engineers have spent considerable time developing computer programs to remove the unwanted affects of buoy motion from wave and current data.

Most buoys are fixed to their location by specific mooring designs based on hull shape, location, and water depth. The principal types include: a long tubular *spar* buoy; a conical *nun* or *nut* buoy; a cylindrical *can* buoy; a *dan* buoy carrying a pole with a flag or light on it; a *torroidal* buoy with a superstructure for mounting electronics; a *discus* buoy with oceanographic instruments; and a *bell* buoy, which rings mechanically or by the action of waves. Smaller buoys with all-chain moorings are typically found in shallow coastal waters, while large buoys deployed in the deep ocean utilize a combination of chain, nylon, and buoyant polypropylene materials for anchoring. Moorings consist of various types of marine hardware that fixes the buoy to the bottom as well as providing communication links for sophisticated buoys with instrument payloads.

Buoys provide information on water movement and sea conditions. Drifting buoys, both surface and subsurface, are used to track currents. Buoy data are

often sent to shore via satellite. Mariners can also assess how a buoy is riding in the water in order to estimate current speeds and directions. Instrumented buoys such as the Triaxys™ directional wave buoy

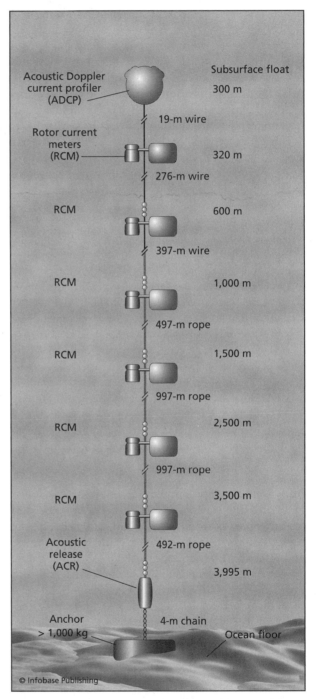

A typical deep-sea oceanographic taut-wire mooring using a subsurface float supporting acoustic Doppler and rotor current meters, which frequently contain electronic thermometers and salinometers. The buoy is retrieved by sending a coded acoustic signal to trigger the anchor release, allowing the buoy to float to the surface.

use special rubber cords within the mooring to allow unrestricted buoy motions in the waves. Radios transmit the measured wave information to ground stations. The National Weather Service's National Data Buoy Center operates and maintains a network of buoy and bottom-mounted Coastal Marine Automated Network (C-MAN) stations that are critical for collecting data that is used in weather forecasting. Information from these coastal data buoys provides a measure of safety for offshore fisherman and other mariners that are sailing in the shipping channels. Other organizations are developing networks of data buoys that telemeter meteorological and oceanographic information to users in real time as ocean observatories. Systems such as the Carolinas Coastal Ocean Observing and Prediction System are helping to reduce the impact of hurricanes that have caused extreme flooding in North Carolina.

Further Reading

Berteaux, Henri O. *Buoy Engineering*. New York: John Wiley, 1976.

Caro-COOPS: Carolinas Coastal Ocean Observing and Prediction System. Available online. URL: http://www.carocoops.org/. Accessed July 21, 2008.

Global Drifter Program, NOAA Atlantic Oceanographic and Meteorological Laboratory, Miami, Fla. Available online. URL: http://www.aoml.noaa.gov/phod/dac/gdp.html. Accessed January 13, 2007.

National Data Buoy Center, NOAA. Available online. URL: http://www.ndbc.noaa.gov. Accessed January 13, 2007.

Swedish Meteorological and Hydrological Institute. Available online. URL: http://www.smhi.se. Accessed January 13, 2007.

buoy tenders Specially configured vessels designed to service and position buoys, beacons, ranges, and leading lights that indicate the navigable channel by their position, shape, coloring, numbering, sound, and light characteristics are known as buoy tenders. These "aids to navigation" along with collision avoidance regulations or what is commonly referred to as "Rules of the Road" combine to provide directions to vessels and the prevention of collisions, groundings, and other safety mishaps. The U.S. Coast Guard is responsible for the placement and maintenance of aids to navigation in U.S. waters.

The United States Coast Guard classifies vessels less than 65 feet (19.8 m) in length as boats. Aids to Navigation Boats (ANB) specially designed for the inland waters of the United States generally range in size from 21 to 64 feet (6.4 to 19.5 m). Some of these such as the 55-foot (16.8 m) ANB are often employed to service aids found in the coastal ocean. Vessels

Small buoy tender for servicing navigation buoys in the Great Lakes, home-ported in Muskegon, Michigan *(U.S. Coast Guard, PA1 Harry C. Craft III)*

longer than 65 feet (19.8 m) are classified as Coast Guard Cutters and usually have accommodations for crew to live on board. These ships are usually equipped with launches or motorboats such as rigid hull inflatable boats, surf rescue boats, or landing craft. The Coast Guard further classifies its cutters that work on aids to navigation as river (WLR), coastal (WLM), inland (WLI), and seagoing (WLB) buoy tenders. Buoy tenders have black hulls and use state-of-the-art positioning such as the GLOBAL POSITIONING SYSTEM (GPS) to service and precisely set floating aids to navigation.

The buoy tenders sail from their homeport through stretches of shoals, ledges, currents, and ever-changing weather patterns, to ensure that their operating area's all-important buoys are on station and working properly. They install, operate, and maintain federally owned and selected privately owned aids to navigation. Some tenders even clear ice-laden domestic waters and help maintain operational lighthouses. ICEBREAKING in U.S. waters would be conducted for search and rescue or a similar emergency situation, prevention of flooding caused by ice, or to facilitate safe navigation. Other important collateral duties involve operation of the vessel of opportunity skimming system or the Spilled

Oil Recovery System for use in mitigating marine spills such as crude oil, servicing NOAA's data buoys, law enforcement, and performing search and rescue missions. While working, buoy tenders are granted special navigation privileges, and the presence of day shapes or special lights indicates their status as "vessel restricted in her ability to maneuver."

The men and women who work aboard buoy tenders use deck gear, muscle, and special handling skills to move chain links the size of footballs while setting thousand-pound buoys. Crew members know how to handle sinkers, lines of chain called shot, as well as the correct lines, hooks, swivels, shackles, and releases that safely maneuver buoys from the deck to the sea or vice versa. They are a team that is familiar with their operating area and have demonstrated skills in marlinspike seamanship, basic navigation and boat handling, WINCH and crane operations, survival and safety, and even skills such as watch standing and communications. Marlinespike seamanship, the art of handling and working with all kinds of lines and rope, is a fundamental prerequisite for the crew. Today's crews are also trained in seaport security as well as marine environmental response and protection.

Further Reading

Canadian Coast Guard Fleet, Ottawa, Canada. *Commissioned Vessels, Aircraft, and Hovercraft.* Available online. URL: http://www.ccg-gcc.gc.ca/fleet-flotte/main_e.htm. Accessed January 13, 2007.

United States Coast Guard. *Aircraft, Boats, and Cutters, Information on United States Coast Guard Resources.* Available online. URL: http://www.uscg.mil/datasheet. Accessed January 13, 2007.

C

cabin cruisers Powerboats that contain a cabin and are in excess of 26-feet (7.9 m) long are cabin cruisers. They should not be confused with cuddy cabins that have below deck accommodations in the bow for stowage and overnighting. Many cabin cruisers usually have either a deep V or a displacement hull. The body of the boat minus any features such as the deck is called the hull, which is usually made of fiberglass or wood, although some hulls are made of polyethylene, epoxy resins, carbon, or kevlar fibers. Cabin cruisers contain a head, galley, and may be equipped with several berths. On board the boat, a head is the compartment that contains a marine toilet, the galley is the kitchen, and a berth is a place to sleep. Cabin cruisers are generally powered by one to three inboard gasoline or diesel engines and usually travel at speeds up to 25 knots (12.9 m/s).

A Hatteras Yacht's 50-foot (15-m) cabin cruiser built at the Hatteras Yachts New Bern, North Carolina, yard *(Hatteras Yachts)*

Cabin cruisers have played important roles in society from entertainment to research. These are the most popular boats in yachting, and the are generally used for fishing or cruising in ESTUARIES, intercoastal waterways, and the coastal ocean. (The 1960s television comedy *Gilligan's Island* centered on the escapades of castaways from the cabin cruiser SS *Minnow,* caught in a violent storm during a sightseeing tour.) At institutions such as Oregon State University (OSU), scientists need platforms supporting coastal research aimed at understanding WAVES, flooding, fisheries, the fate of pollutants, and myriad other shallow-water processes. OSU is currently using the 29-foot- (8.8 m) long Research Vessel (R/V) *Kalipi* for nearshore oceanographic research that often involves diving operations. The R/V *Kalipi* can be trailered, launched, and cruises at about 20 knots (10.3 m/s) to rapidly investigate coastal phenomena.

Cabin cruisers share navigable waters with commercial and military vessels of all types. All water craft, yachts, and vessels must comply with the same "Rules of the Road" and understand standard aids to navigation for proper and safe conduct on the water. Yachtsmen and women should receive training to ensure safe coastal, offshore, and ocean passage. Learning to navigate properly is one major way to avoid boating accidents. Organizations such as the United States Power Squadron and the United States Coast Guard Auxiliary provide navigation and safety courses. Boat insurance is also necessary to protect the cabin cruiser and liability exposures associated with boat ownership.

See also BOATS; NAVIGATION; NAVIGATION AIDS.

Further Reading

Bowditch, Nathaniel. *The American Practical Navigator: An Epitome of Navigation.* Bethesda, Md.: National Geospatial-Intelligence Agency, 2002.

Maloney, Elbert S. *Chapman: Piloting, Seamanship and Boat Handling,* 64th ed. New York: Sterling Publishing, 2003.

calcite compensation depth (**carbonate compensation depth**) The calcite compensation depth is called the CCD by most marine scientists. It is the depth in the ocean below which material composed of calcium carbonate, $CaCO_3$, is dissolved and does not accumulate on the seafloor. In other words, at this depth, carbonate particles are dissolving faster than they can be deposited. It is the location where the amount of calcium carbonate delivered to the seafloor is equal to the amount removed by dissolution. The depth of the CCD is a function of the THERMOCLINE depth, the rate of increase of dissolution with depth, and the rate of supply of carbonate and noncarbonate materials to the sediment. In the PACIFIC OCEAN, the average depth of the CCD is 2.8 miles (4.5 km), while in the ATLANTIC OCEAN, it is approximately 3.1 miles (5 km). Only above the CCD can carbonate materials be deposited, because below the CCD they dissolve and do not reach the seafloor. Early marine scientists interested in the geographic distribution of different seafloor sediments, particularly the locations of white oozes, were the first to understand the CCD. Oozes are biogenous seafloor sediments containing more than 50 percent of the remains of microscopic sea animals and plants or nannofossils. For example, calcareous ooze might contain more than 50 percent of the shells of coccoliths, pteropods, and some forams. Coccolithophores are a type of PHYTOPLANKTON with a gelatinous covering overlapped by calcite plates called coccoliths. Organisms such as corals, ALGAE and diatoms pull carbon dioxide (CO_2) from the seawater in order to make their shells, a process that is considered by some as Earth's CO_2 filter. Pteropods are microscopic free-floating gastropods with aragonite shells. Aragonite was first found in Aragon, Spain, and has the same chemical makeup as calcite but is slightly harder. Forams are a protozoan that secrets a calcareous shell or test containing calcium carbonate. The famous pink sands of Bermuda get much of their color from the pink to red-colored shells or tests of foraminifera.

See also SEDIMENT; SEDIMENT TRANSPORT.

Further Reading

Integrated Ocean Drilling Program, Washington, D.C. Available online. URL: http://www.iodp.org. Accessed March 20, 2007.

National Climatic Data Center: NOAA Satellite and Information Service. NOAA Paleoclimatology Program. Available online. URL: http://www.ncdc.noaa.gov/paleo. Accessed March 20, 2007.

California Current The California Current is a broad, slow-moving eastern boundary current forming the eastern limb of the North Pacific GYRE (the circulation formed by the California Current, the North Equatorial Current, the Kuroshio Current, the North Pacific Current). The current can be delineated from the waters off Vancouver Island, British Columbia, Canada, to Baja California, Mexico. It is on the order of 600 miles (1,000 km) in width. The waters of the California Current are cool, about 48°F (9°C) in the northern portion, and about 79°F (26°C) in the southern portion. The California Current is a region of vigorous UPWELLING in the spring and summer due to the prevailing northwest winds which lead to an EKMAN TRANSPORT away from the coast. In the summer, the cool waters interacting with warm air produce the famous San Francisco fog.

The mean speed of the current is on the order of 0.50 knots (0.25 m/s). A countercurrent flows inshore of the California Current. This countercurrent is

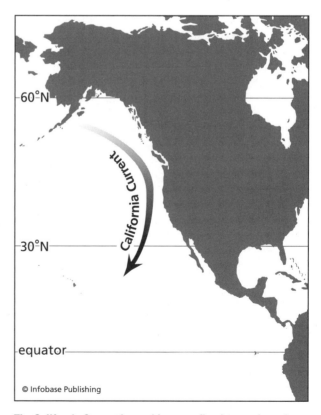

The California Current is a cold current flowing southward parallel to the coast of northwestern North America.

strongest below about 500 feet (150 m); it flows toward the north at about .50 knots (0.25 m/s).

On the whole, the California Current is a poorly defined current readily influenced by the North Pacific wind regime. Variability is strongly influenced by interactions between the climatological subtropical high-pressure region over the North Pacific Ocean and the atmospheric thermal low over California-Nevada. The region is subject to large seasonal variations.

In spring and summer, the stress on the sea surface from northerly winds produces southward oceanic flow. The associated Ekman transport induces a circulation away from the coast in near-surface layers with consequent upwelling of cold, salty, nutrient-rich water from below. During this season, an undercurrent flows northward at depths below about 650 feet (200 m). During the summer, the cool California Current also contributes to fog near the coast.

In fall and early winter, northerly winds weaken, with the predominant wind coming at times from the southwest. These southwest winds produce a northward-flowing current inshore along the coast known as the Davidson Current. Offshore, the mean flow continues toward the south. When the upwelling ceases in early fall, the core of the undercurrent moves toward the surface, merging with the Davidson Current. The Davidson Current cuts off further upwelling, deepens the surface mixed layer, and increases surface temperature The end of the Davidson Current period in late winter or spring and the beginning of the upwelling period can occur quite suddenly. The undercurrent found in spring and summer may in fact be the subsurface continuation of the Davidson Current.

Volume transport in the California Current is on the order of 10–12 Sverdrups ($10–12 \times 10^6$ m³/s) toward the south. The mean speed of the current ranges from about .23–.50 (.12–.25 m/s) but speeds over one knot (.50 m) have been observed. The highest observed speeds in the California Current occur in summer after several months of northerly winds have been acting on the surface.

Both cold-core and warm-core EDDIES are common features in the California Current region. They produce high current speeds compared to the surrounding waters. These flows may be observed on satellite images, which depict sea-surface temperatures.

The upwelling regions of the California Current produce an abundance of marine life, ranging from seals and orcas to large schools of commercially valuable fish. Intensive fishing in the past has resulted in strict controls regulating the tuna, anchovetta, and other fisheries.

Further Reading

Large Marine Ecosystems of the World #3, California Current. Available online. URL: http://www.edc.uri.edu/lme/text/california-current.htm. Accessed March 19, 2007.

National Oceanic and Atmospheric Administration. California Current Ecosystems Program, Southwest Fisheries Science Center. Available online. URL: http://nmml.afsc.noaa.gov/CaliforniaCurrent/calcurr.htm. Accessed March 19, 2007.

Canary Current The Canary Current is southeastern extension of the NORTH ATLANTIC CURRENT, which flows southward past Spain and Portugal, forming the eastern portion of the North Atlantic GYRE. The other currents of the gyre are the North Equatorial Current, the GULF STREAM and the North Atlantic Current. The Canary Current flows past the Canary Islands and then divides, with the western branch flowing southwest into the North Equatorial Current, and the eastern branch flowing southeastward into the Guinea Current, which flows eastward along the African coast into the Gulf of Guinea. The Canary Current is the eastern boundary current of the North Atlantic Ocean; it exhibits typical characteristics of eastern boundary currents being broad, diffuse, and of relatively slow speeds. It is the Atlantic counterpart of the CALIFORNIA CURRENT in the PACIFIC OCEAN. Speeds in the Canary Current are on the order of about 0.2 knot (.1 m/s) with a volume transport on the order of 16 Sverdrups (16×10^6 m³/s).

The width of the current is about 600 miles (1,000 km); the depth is about 1,500 feet (500 m). Surface waters are cool containing water upwelled off the coast of North Africa. EDDIES of a spatial scale of 60–180 miles (100–300 km) are found along the shoreward edge of the current. Eddies may form, especially as the current accelerates as it passes through the area between the Canary Islands and the coast.

The flow of the Canary Current is seasonal, with the greatest strength of the current being in the Northern Hemisphere winter, corresponding to the strongest northeast trade winds. During this time, the core of the Canary Current may have speeds as high as 2.5 feet per second (1.48 knots or 0.76 m/s) as it passes through the Canary Islands. As the trade winds weaken in spring, the Canary Current weakens and reaches a minimum speed in late summer and early autumn. These seasonal variations coincide with the seasonal variations of the North Equatorial Current into which it flows. Eddies and meanders form frequently in this region. The trade winds blowing over the Canary Current and the North Equatorial Current once carried Columbus's small fleet from Spain to the New World. During the late 18th and early 19th centuries, sailing ships trading at North Atlantic ports took advantage of the circulation of winds and currents, sailing west to the Americas in the North Equatorial Current, then north in the Gulf Stream, east across the Atlantic in the North Atlantic Current, and then south again in the Canary Current.

Further Reading

GIWA, Regions and Network, Subregion 41: Canary Current, Global International Waters Assessment, Kalmar, Sweden. Available online. URL: http://www.giwa.net/areas/area41.phtml. Accessed March 21, 2007.

Gyory, Joanna, Arthur J. Mariano, and Edward H. Ryan. "The Canary Current." Available online. http://oceancurrents.rsmas.miami.edu/atlantic/canary.html. Accessed April 1, 2004.

Large Marine Ecosystems of the World #27: Canary Current. Available online. URL: http://www.edc.uri.edu/lme/text/canary-current.htm. Accessed March 19, 2007.

capillary waves   Capillary waves are very small water waves for which the restoring force is capillary action. The most visually apparent ocean waves are called surface gravity waves because their restoring force is the pull of gravity. Capillary waves have as their restoring force the tension of the water surface. This is the same surface force, called capillary attraction, which pulls water up very thin tubes or bends the meniscus of a fluid in a test tube. Surface tension affects only very short WAVELENGTH waves, less than 0.7 inches (1.7 cm). Their amplitudes are only a fraction of a centimeter. The range of waves from 0.7 inches (1.7 cm) to about two inches (5.0 cm) in wavelength is acted upon by both gravity and surface tension. For waves more than 2 inches (5.0 cm) in wavelength, the restoring force is entirely gravity.

Capillary waves are similar to gravity waves in that they are generated by the wind blowing over the surface of the water. However, in most respects capillary waves are very different from gravity waves. Deep-water gravity waves travel faster the longer their wavelength, whereas capillary waves travel slower the longer their wavelength. Deep-water gravity waves can circumscribe the globe, whereas capillary waves die out almost immediately after the wind stops blowing.

Most of the oceans are covered by capillary waves; they are absent only when there is no wind at all and the sea is glassy. Under these conditions, a gust of wind approaching can be observed because it has a distinctive patch of capillary waves in its wake, which is visible as dark, roughened patches appar-

ently moving faster than any of the surface waves. Those who sail the ocean call these cat's paws.

See also RADAR; WAVES.

Further Reading

Capillary Action, Water Science for Schools, USGS. Available online. URL: http://ga.water.usgs.gov/edu/capillaryaction.html. Accessed March 19, 2007.

Sandia National Laboratories: Synthetic Aperture Radar Applications. Available online. URL: http://www.sandia.gov/RADAR/sarapps.html. Accessed March 19, 2007.

Wright, John, Angela Colling, and Dave Park (prepared for the Course Team). *Waves, Tides and Shallow-Water Processes*. Milton Keynes, U.K.: The Open University in association with Butterworth/Heinemann, 2001.

carbon cycle   The carbon cycle accounts for movements of carbon through reservoirs such as the atmosphere, the biosphere, the oceans, and the sediments. Carbon is one of the elements that make up the universe. Human bodies, animals, plants, and the food they eat all contain carbon. Carbon, just like water, goes through a cycle. One of the largest storage deposits of carbon on Earth is in the oceans. The oceans store vast quantities of the element and help mitigate the changes that occur to climate caused by the introduction of carbon, specifically CO_2 (carbon dioxide), into the atmosphere. Diamonds and graphite are carbon in its pure form. Carbon can combine with oxygen to form carbon monoxide, CO, or carbon dioxide, CO_2, or it can combine with hydrogen to form sugar and starches. The proteins and fats in human bodies, as well as the cellulose in plant walls are all formed with carbons. Life is a carbon-based life form. Animals and plants get carbon from the food they eat and from the air they intake, respectively. Animals use carbon to make energy. Sugar and starch are broken down, and CO_2 is given off in the process. This is called respiration. Plants take in CO_2 and get the carbon from the air that they use in the process of photosynthesis to make sugar, which contains carbon. This is the beginning of the carbon cycle. One cycle could be that a plant takes in CO_2 and produces sugars. These plants are eaten by an animal, which exhales CO_2 after the sugar has been "burned up" in the body. This CO_2 is then used by plants again in the process of photosynthesis, thereby completing this simple cycle. In actuality, the carbon cycle is more complicated. One of the big research areas for global climate change is tracking the carbon cycle and estimating its impact on adding to the greenhouse gases (H_2O, CO_2, and so forth) and increasing the average temperature of Earth. A more complicated example of the carbon cycle would be: A dinosaur expels CO_2 after eating some plants. The CO_2 is absorbed by ALGAE, which breaks down the CO_2 and uses the carbon to form a carbohydrate that forms the cell wall. The algae die and sink to the bottom where they form a deep-water OOZE. This ooze is covered over with silt, compressed, and over time turns into petroleum, carbon in another form. Years later, this petroleum is pumped to the surface and shipped to a refinery. There it is turned into gas and used to power vehicles. The gas is burned, giving off CO_2, which finds its way into the upper atmosphere, where it forms a barrier that traps the long-wave radiation from Earth, thereby heating up the atmosphere and contributing to—GLOBAL WARMING.

Since the beginning of the Industrial Revolution, the amount of CO_2 put into the atmosphere has increased dramatically. Humans, through burning of coal, gas, and wood, are putting more CO_2 into the atmosphere than the plants can use. The oceans can actually absorb CO_2 and store it in the ocean. The estimates of the amount of CO_2 in the carbon cycle include the huge amounts that reside in the oceans. Satellites are being designed to measure the amount of carbon in the atmosphere. Research vessels ply the seas, measuring the amount of carbon absorbed into the ocean and the depth to which it is stored. So far it appears that the ocean has absorbed so much carbon that it has been able to mitigate the greenhouse effects on the atmosphere. Tracking the carbon and making estimates of how it will affect our climate in the future is an important and very difficult research topic.

See also PHOTIC ZONE; PRIMARY PRODUCTION.

Further Reading

Office of Oceanic and Atmospheric Research, NOAA. "The Global Carbon Cycle." Available online. URL: http://www.oar.noaa.gov/climate/t_carboncycle.html. Accessed March 19, 2007.

cartographer   Scientists who measures, maps, and charts Earth's surface are called cartographers. Cartographers perform geographical research and compile data to actually produce maps of the land and charts of the sea. They collect, analyze, and interpret data that are derived from disparate sources, ranging from census surveys to satellite images. Spatial data such as latitude, longitude, elevation, depth and distance are merged or fused with nonspatial data or attributes such as population density, land-use patterns, annual precipitation levels, and demographics. Cartographers prepare the resultant maps or charts in either digital or graphic form, using information provided by geodetic surveys and imagery. Cartogra-

A NOAA cartographer develops a bathymetric chart.   *(NOAA)*

phers also use creativity to vitalize an otherwise lifeless map by applying an artist's knowledge of color, balance, contrast, and pattern.

As is the case with Earth scientists, cartographers study landmasses, water bodies, and atmosphere and then develop decision aids that help determine how Earth's surface can be used for specific purposes. For example, hydrographic surveys are used to determine shorelines, bathymetry, bottom type, and many other marine features. With the development of the GLOBAL POSITIONING SYSTEM and Geographic Information Systems, cartographers need specific education to deal with digital data and to develop strong computing skills. These combined systems are capable of assembling, integrating, analyzing, and displaying data identified according to location. Mapping scientists who use computers to combine and analyze various forms of geographic information are sometimes called geographic information specialists.

Cartographers usually have a degree in a field such as engineering, forestry, geography, or a physical science. Many universities and community colleges have developed certificate programs in order for geography students to demonstrate to prospective employers that they have specific practical skills in Geographic Information Systems, beyond the standard theoretical sciences commonly associated with university study. Certification in Geographic Information Systems is becoming more accepted and even required among professional engineering communities, as certification in computer-aided design (CAD) drawing and land survey are essential to many engineering careers, especially civil engineering.

Major universities, government laboratories, and private industries employ cartographers. In the United States, federal government employers include the U.S. Geological Survey (USGS), the Bureau of Land Management, the Army Corps of Engineers, the Forest Service, the NATIONAL OCEANIC AND ATMOSPHERIC ADMINISTRATION, the National Geospatial Intelligence Agency, and the Federal Emergency Management Agency. Highway departments and urban planning and redevelopment agencies employ most cartographers working in state and local government. In the private sector, construction firms, mining and oil exploration companies, and utilities employ surveyors, cartographers, photogrammetrists, and surveying technicians.

To keep abreast of recent developments in their field, cartographers must spend considerable time reading current technical journals, participating in scientific meetings and workshops, and in taking courses in new technologies. More and more, in order to customize computer systems to specific environmental applications, cartographers who work with Geographic Information Systems must be able to serve as developers, applying programming skills in various macro languages such as ARC Macro Language and Avenue as well as programming languages such as JAVA and Visual BASIC. They can expand their resources by partnering with scientists and computer professionals from other organizations and by making use of existing spatial databases available on the World Wide Web.

Further Reading

British Cartographic Society. Available online. URL: http://www.cartography.org.uk. Accessed March 19, 2007.

National Geospatial-Intelligence Agency. Available online. URL: http://www.NGA.mil. Accessed March 20, 2007.

National Oceanic and Atmospheric Administration, Office of Coast Survey, National Ocean Service. Available online. URL: http://chartmaker.ncd.noaa.gov. Accessed March 19, 2007.

North American Cartographic Information Society. Available online. URL: http://www.nacis.org. Accessed March 19, 2007.

U.S. Geological Survey. Available online. URL: http://www.usgs.gov. Accessed March 20, 2007.

catadromous stocks   Supplies of wild fish species that spend most of the life cycle in freshwater but migrate to salt water to breed are catadromous stocks. Freshwater eels (*Anguillidae*) are one of the most common families of catadromous fish. Anadromous fish, such as salmon and lamprey, live the majority of the life cycle in salt water and migrate to freshwater to spawn.

Families of fish that have many diadromous members, species that migrate between freshwater and saltwater systems, are typically ancient. Diadromy increases with increasing differences in water productivity between connected river and ocean systems. In the Tropics, freshwater systems tend to be more productive than saltwater systems, so catadromy is more common in warmer waters where freshwater food resources generally exceed those in the sea. Researchers suggest that migrating between habitats improves survivorship and reproductive success.

An endogenous biological clock initiates migration between systems. Spawning grounds may be thousands of miles away, and body protein and lipid reserves accumulate to support long-distance swimming. During migration, fish enter the water column when the tide is going in the migratory direction and hold onto the bottom when it changes (selective tidal stream transport). Fish may also undergo morphological changes, such as eye enlargement, color changes, and head elongation, during migration.

Catadromous fish must cope with osmotic water loss when migrating to salt water. Migrating fish handle the loss by decreasing urine production and increasing water that is passively reabsorbed. They also have chloride cells on the primary lamellae of the gills, and these, along with Na^+, K^+, and ATPase, are part of the bronchial mechanisms of osmoregulation that make it possible for fish to migrate between saltwater and freshwater systems.

Because catadromous fish depend on the integrity of two environments for growth and reproductive success, they may be more fragile than single-habitat species. Barriers to migration, such as dams and water flow regulation appear to have the most detrimental effect on catadromous stocks because fish must inhabit suboptimal regions of the river during growth. Pollution that inhibits osmotic and ionic regulatory performance, invasive species competing for resources, and high sediment concentrations that reduce food availability, habitat diversity, and growth rates are other examples of threats to catadromous stocks. For example, the range of the commercially important catadromous African mullet, *Myxus capensis*, has been severely reduced by development that caused severe erosion in freshwater feeding areas.

Freshwater eels, such as the American and European eels, are currently the most commercially valuable catadromous fish species, but stock status is not well understood and annual catch data is incomplete. In 1981, approximately 700,000 pounds of American eel was harvested in Virginia and Maryland waters for European and Asian markets, but catches are significantly lower today due to stock overharvesting and changes in market demand. The European eel is the most valuable fishery in England and Wales, but stocks are seriously threatened, and these countries, along with the European Union, are attempting to develop management plans for the fishery. The Scandinavian nations have the best European eel catch data, and it indicates that the stocks have been declining since the 1960s.

Some of the options available to improve catadromous stock management are water flow manipulation, erosion control, translocation of stocks, construction of fishways that bypass dams, and artificial propagation.

See also ANADROMOUS.

Further Reading

Dadswell, Michael J., Ron J. Klauda, Christine M. Moffitt, Richard L. Saunders, Roger A. Rulifson, and John E. Cooper, eds. *Common Strategies of Anadromous and Catadromous Fishes.* Bethesda, Md.: The American Fisheries Society, 1987.

Fisheries and Oceans Canada. Available online. URL: http://www.dfo-mpo.gc.ca/home-accueil_e.htm. Accessed September 14, 2007.

National Oceanic and Atmospheric Administration: NOAA Fisheries Service. Available online. URL: http://www.nmfs.noaa.gov. Accessed March 21, 2007.

Native Fish Society. Available online. URL: http://www.nativefishsociety.org. Accessed March 20, 2007.

causeway A causeway is usually an unimproved road or a paved road constructed across marshy or periodically flooded terrain. Natural causeways may form when lava floes from volcanic eruptions near the coast comes into contact with the ocean whereas man-made causeways may also be fairly elaborate roads that cross broad water bodies or wetlands.

During medieval times, various workers such as master masons, quarrymen, woodcutters, smiths, miners, ditchers, carters, and carpenters were conscripted to build a castle or fort. These projects included the building of a causeway or a raised road over a moat, ditch, or fosse, which lead into the castle's entrance. Today, environmental and civil engineers plan and execute the construction of causeways. Such structures may entail the placing of a rock core protected on both sides by armor rock and surfacing a roadway across a fill area with a concrete or asphalt slab. Military construction personnel build pontoon causeways that provide ingress and egress across surf zones. These causeways are used for unloading amphibious ships, which facilitate the dry movement of personnel and equipment from ships to shore.

Hydrographic changes caused by causeways may negatively affect an ESTUARY and become highly publicized issues. Blockage of normal CURRENTS in a bay may have negative consequences to surrounding wetlands and the recruitment of migratory finfish that grow to maturity in the estuarine nurseries. Some environmentalists advocate the restoration of satisfactory circulation patterns and drainage by replacing the causeways with elevated roadways. Others believe that a stable ecosystem may have evolved in response to an existing causeway and ensuing changes would cause more harm than good. Prior to construction of a causeway in the United States, the present National Environmental Policy Act would require an Environmental Impact Statement to be completed. This required document for major projects or legislative proposals significantly affecting the environment is a decision-making tool that would identify positive and negative effects for the construction of a causeway.

See also BRIDGES.

cephalopods Squids, cuttlefishes, octopi, and nautili constitute a class of animals known as the cephalopods. The cephalopods, or head-footed animals as they are called (*cephalo* from Latin meaning "head," and *poda* meaning "foot"), are placed in the phylum Mollusca (soft-bodied animals), and are characterized by having a linearly chambered shell, which in some species has been greatly reduced or completely lost during the process of evolution. If an external shell is present, the animal inhabits the last (youngest) chamber with a filament of living tissue (the siphuncle) extending through the older chambers. The siphuncle secretes gases into the older, empty chambers for buoyancy control, allowing the animal to swim. The body cavity is comparatively large with a closed circulatory system with three hearts that pump oxygenated blood throughout the body and to the two-gill chambers. The head is large with complex eyes and a circle of prehensile arms or tentacles around the mouth where a parrotlike beak and radula are also found. The radula is an organ with a series of small teeth in many rows used much like a tongue bringing food into the buccal cavity for further breakdown. A muscular funnel, the siphon, is located under the mantle cavity through which water is forced by contraction of the mantle walls, providing the animal with jetlike propulsion. In certain species, one or a pair of tentacles have been modified into a copulatory organ (hectocotylus) for the deposition of sperm into the female mantle cavity for egg fertilization. There are approximately 700 living species that are exclusively marine and are either benthic (bottom-dwelling) or pelagic (open-water) inhabitants. With complex eyes, well-developed nervous systems, high metabolic rates, and other traits, these animals are well adapted for their active, predatory lifestyles. The cephalopods first appeared during the Cambrian with approximately 7,500 different forms found in the fossil record. A completely developed shell is found only in ancestral forms except for the few species of Nautilus found in the tropical PACIFIC OCEAN. These animals can be found in shallow tide pools more than 3,300 feet (1,000 m) in depth. Depending upon the species, there may be some variations in the characteristics described above.

Marine scientists and taxonomists group organisms into different categories based upon similarities in embryology, morphology, and physiology. Cephalopods are typically classified according to their shell structure or peculiarities found in soft parts such as gills. This process is important to distinguish these many organisms. For example, a clam and an octopus look nothing alike, but because the two animals share certain characteristics (they both possess a radula among other shared traits) they are placed together in the same phylum (Mollusca, one of the largest invertebrate phyla) but in different classes. In order to appropriately classify organisms, biologists rank animals as an order between the class and family. Taxonomy (the classification of organisms) uses a system called binomial nomenclature as a standard convention for naming living organisms. As the word *binomial* suggests, the scientific name of an organism is a combination of two names: the genus name and the species epithet. The real value of this system is that every described organism can be identified unambiguously with just two words. Thus, the binomial name of a species is just part of a larger classification system. The classification for the Australian Giant cuttlefish (*Sepia apama*) is provided below.

Cephalopods are usually divided into two main groups or subclasses: the *Nautiloidea,* containing a few species of nautili, and the *Coleoidea,* containing the squids, octopi, cuttlefishes, and vampire squids. The basis for this separation is that the Nautilus possesses an external shell, whereas in the other groups the shell is internal or completely lacking. In the squids and cuttlefishes, the shell is reduced and internal to facilitate swimming and aid in their predatory habits. In the octopi, the shell is completely lost and has allowed these animals to assume a more bottom-dwelling existence. A listing of some of the more common cephalopods is provided below.

Members in the subclass Nautiloidea possess a pearly, many chambered external shell coiled in one plane. Beachcombers consider these shells a good find. The head contains 38 tentacles without suckers (four modified as a spadix in males for copulation) and protected by a fleshy, lobelike hood. A radula and beak are present, and the funnel is composed of two separate folds. The eyes are small without a cornea or lens and the nervous system simple. The Nautilus is a slow swimmer and unlike its cousins it does not possess chromatophores or an ink sac. For defense, the animal simply withdraws into the shell with the hood acting as a protective covering. The animal is a bottom feeder swimming slowly over the substrate in search of food. Their fossil records are rich and represented today by two genera with five or six Indo-Pacific species. They have no commercial importance except for their shells, which are valued by collectors worldwide. The shell of the Paper Nautilus (*Argonauta nouryi*), or Argonaut as it is called, is actually formed by females as a brooding chamber and flotation device. Egg-bearing females occasionally strand themselves in vast numbers on various islands throughout the tropical Pacific, where the "shells" are then collected. Males of the species are tiny, approximately 0.6 inches (1.5 cm) long, and bear only one arm that contains the spermatophore (sperm sac). During mating, the arm of the male breaks off inside the mantle cavity of the female, providing her with sperm for egg fertilization. The tiny male may at times be found inside the brood chamber or shell of the female Argonaut.

Members in the subclass Coleoidea are the octopi, squids, cuttlefishes, and vampire squids. The shell is reduced, internal or absent, and the head and foot are combined into a common anterior structure bearing eight to 10 prehensile suckered arms and tentacles with one pair modified into copulatory organ for sperm deposition. A radula and beak are present in the mouth cavity. The funnel or siphon consists of a single closed tube where water is forced through providing propulsion. These animals possess a complex nervous system and have well-developed eyes with a cornea and a lens present. Dermal chromatophores are present, giving these animals the ability to change colors in an instant. An ink sac is also present and is used by the animals to confuse predators by ejecting a cloud of the ink while attempting escape. The ink of the cuttlefish (sepia ink) was once used in the print and photography industries.

Between the subclass and family, cephalopods are further classified by order. As an example, the subclass Coleoidea can be further classified as follows:

Order Sepioida: The cuttlefish possess a short body that is dorsoventrally flattened with lateral fins. The shell is internal, calcareous, and often chambered. The cuttlefish is able to control its buoyancy by regulating the amount of water in the shell and thus the amount of gas. There are eight short arms and two long tentacles with suckers that lack hooks borne only on spooned tips and retractable into pits. Cuttlefish are very efficient predators and are masters of camouflage. By expansion or contraction of pigment-filled chromatophores, the cuttlefish can change colors in seconds. Cuttlefish mate during the winter months with males and females coming together performing a mating dance or ritual before caressing one another with their arms and tentacles. The male deposits a sperm sac into the mantle cavity of the female; she will then retreat into a small cave or opening where she fertilizes her eggs before hanging them in clusters from the rock. Soon, after mating, the cuttlefish die, having completed their short life cycle of only 18 to 24 months. Due to its short life span, the cuttlefish fishery must be managed properly. The cuttlefish is commercially important worldwide and in some areas is threatened from overexploitation. Of the many products derived from the fishery, cuttlebone is the most familiar and is used as a source of calcium for pet birds such as parakeets.

Order Teuthoida: The squids possess an elongate, tubular body with lateral fins. The shell is internal and is reduced to a cartilaginous pen for some support of internal organs and to facilitate swimming. Squids possess eight arms and two nonretractile tentacles with suckers that often bear chitinous hooks for the retention of prey. Chromatophores and an ink sac are also present. A strong parrotlike chitinous beak and a radula are located in the buccal cavity. The genus *Architeuthis* contains the largest invertebrate animal on the planet. Better known as the Giant Squid, *Architeuthis* can reach lengths of more than 65 feet (20 m) and is a very elusive creature—first-ever photographs of this species were taken off Japan's Ogasawara Islands during September 2004. The only real knowledge concerning these animals is from dead specimens that have washed ashore or those that have been hauled up in COMMERCIAL FISHING nets. Stomach contents from sperm whales (*Physeter catodon*) have also yielded some information about these elusive creatures. Sperm whales are perhaps the only animals known to prey upon adult giant squid. Beaks of the giant squid and other hard body parts are found embedded in the ambergris of the sperm whale. Ambergris is a liquid found in the intestinal tracts of the sperm whales and was once used as an additive to perfumes. When the ambergris comes into contact with air, it becomes a very pleasant-smelling gel-like substance embedded with beaks of squids and sometimes cuttlefishes. Squids of other species are commercially important worldwide with hundreds of thousands of metric tons being landed on an annual basis. Humans use them primarily as a source of food, and as bait in both commercial and recreational fisheries. In 2002, there were 205.5 million pounds of squid landed in U.S. waters with an average vessel price of 21 cents per pound amounting to $43.5 million. Calamari or squid is a popular menu item in restaurants throughout the world.

Order Octopoda: The octopi possess a short, globular body usually without fins and the shell is completely lacking. The octopi possess eight arms joined by a web of skin (interbrachial web). Most species are bottom-dwelling inhabitants, usually associated with some type of structure such as CORAL REEFS. There are approximately 200 known and described species. They are considered by many to possess intelligence due to a suite of rather complex behaviors exhibited by some species. Octopi are commercially important worldwide and are used by humans primarily as food, especially the tentacles. They are usually caught by throwing out a line with a series of small plastic buckets attached. Octopi will retreat into the buckets, seeking shelter or a place to hide and are landed when the line is retrieved. They are also gaining popularity in the aquarium trade. Most are nocturnal predators, hiding in their lairs during daylight hours. Outside the lair's opening, octopi deposit discarded shells and other hard body parts from their prey victims. These piles of shells (midden piles) can reveal valuable information about the life history of these fascinating animals and also aid researchers in locating octopi. Most species are short-lived, from one to two years, but the Giant Pacific Octopus (*Octopus dofleini*) may live up to four years. Reproduction is similar to that of the squids, being semelparous (only one reproductive cycle) after which death occurs. The small blue-ringed octopus (*Hapalochlaena maculosa*) of the Indo-Pacific is one of the most venomous animals in the marine environment. The poison is located in the animal's saliva and is extremely toxic. It is a neurotoxin and has resulted in many human fatalities—there is no known antidote. Unlike other members of the class, octopi inject a poison into their prey victims resulting in paralysis or death.

Order Vampyromorpha: The deep-water vampire squids possess a plump, small body with one pair of lateral fins. The shell is reduced to a thin, featherlike transparent vestige. There are four pairs of arms, each with one row of distal suckers and are joined by an extensive web of skin (the interbrachial membrane). Two tendrillike, retractile filaments represent the fifth pair of arms. The radula and beak are well developed in order to seize and hold onto prey in this deep-water zone. The ink sac is vestigial as this animal employs a different strategy for escape from predators. The vampire squid is the only cephalopod that spends its entire life in the midwater pelagic zone known as the Oxygen Minimum Layer (OML), where a continuous zone of low oxygen levels persists at depths between 1,300 and 3,300 feet (400–1,000 m). The vampire squid is especially effective at removing oxygen from the water, which enables it to exist in this zone without the use of anaerobic metabolism. The blood contains hemocyanin, like other members of the class, and has a high affinity for oxygen. The vampire squid is similar in form and function to its cousins, but was placed in its own order due to the presence of sensory filaments that are used by the animal to locate food. For defense, it uses light organs on the tip of each arm and ejects a mucous cloud containing thousands of glowing spheres of blue bioluminescent (see BIOLUMINESCENCE) light in order to temporarily confuse predators allowing the vampire squid to escape. *Vampyroteuthis infernalis* is the only species in the order, and is currently of no commercial importance.

In summary, cephalopods are invertebrates found in marine (brackish and saline) environments and are very different from finfish. Like many MARINE FISH, they are active, mobile predators. Cephalopods may use fin waves and jet propulsion to capture prey or escape predators. Fish, on the other hand, propel themselves through undulations of the body or tail fin. All cephalopods are carnivorous, while some fish may forage on plant species. Many species of squid, cuttlefish, and octopus are of commercial importance. Some are caught in large traps set in LAGOONS during bottom trawling and by jigs with light attracters at night. Squids are fished with light attracters and handnets, or by hook-and-line. Octopis are occasionally caught by divers and speared by fishermen in shallow waters. They may be caught as by catch (foul-hooked) during bottom-fishing operations. Some species of squid and cuttlefish are also captured using beach seines.

See also MOLLUSKS; NETS.

Further Reading

Barnes, Robert D. *Invertebrate Zoology.* New York: Holt, Rinehart and Winston/Saunders College, 1980.

Brusca, R. C., and G. J. Brusca. *Invertebrates.* Sunderland, Mass.: Sinauer Associates, 1990.

CephBase, University of Texas Medical Branch, Galveston, Texas. Available online. URL: http://www.cephbase.utmb.edu/. Accessed September 15, 2007.

Integrated Taxonomic Information System, Washington, D.C. Available online. URL: http://www.itis.gov/. Accessed September 15, 2007.

National Resource Center for Cephalopods, University of Texas Medical Branch, Galveston, Texas. Available online. URL: http://www.utmb.edu/nrcc. Accessed September 15, 2007.

certifications Many professions use academic standing, qualifications, and certifications to ensure that only highly skilled individuals are licensed for employment in particular disciplines. Applicants for degrees and certification must meet requirements specific to the type of degree or certificate they are seeking. Basic researchers rely on degrees and publication records to assess a person's capability to advance the discipline's knowledge base. Engineering fields and marine trades utilize training, education, and certificates to qualify people to work in areas ranging from using the GLOBAL POSITIONING SYSTEM to create a digital elevation model to running a shipboard power plant. Often a particular discipline's professional associations describe the specific requirements for qualifying for certificates, which may range from carrying the title of a professional engineer to being a licensed captain of a merchant vessel.

In science fields such as biological, chemical, geological, and physical oceanography, specific certificates may not be required. However, many scientists join professional associations such as the American Academy of Underwater Sciences, the MARINE TECHNOLOGY SOCIETY (MTS), American Society of Limnology and Oceanography, or the AMERICAN GEOPHYSICAL UNION (AGU) and contribute research findings at professional conferences and, most important, in refereed journal articles. Degrees tend to categorize a person's initial role within the research community. Principal investigators manage research programs and normally have doctorates (also referred to as a doctor of philosophy or a Ph.D.). Many marine scientists holding master of science degrees are found working on industry or government research projects and depending on experience may design and manage their own programs. Scientists having bachelor's degrees in a basic science accomplish a great deal of marine science fieldwork, data processing, and reporting. The marine operations area may offer an entry-level position for someone with only a high school diploma or associate's degree. The Marine Advanced Technology and Education (MATE) Program in Monterey, California, focuses on providing certificates for operations such as operating REMOTELY OPERATED VEHICLES to awarding associate degrees that prepare students for undergraduate and graduate level marine science programs. Educational institutions such as the MATE Center also provide opportunities for at-sea and shore-based experiences in marine technology fields ranging from hydrographic surveying to AQUACULTURE.

Some marine science fields require specific education and experience requirements and then passing an examination for award of a certification or qualification. In the U.S. Navy and Air Force, specific training must be accomplished in observing weather elements such as precipitation, sky condition, visibility and obstructions to vision, TEMPERATURE, winds; plotting of weather codes must be skill demonstrated before a service member can be qualified as a weather forecaster. For another rigorous example, engineers with an "Engineer-in-Training" certificate are certified by a state as a graduate from an approved engineering program or related science curriculum and someone who has passed the National Council of Examiners for Engineering and Surveying (NCEES) eight-hour Fundamentals of Engineering (FE) examination. Additional information on training, certification, and qualification requirements can be obtained from professional and trade associations.

All mariners employed aboard U.S. merchant vessels greater than 100 Gross Register Tons are required to have a valid U.S. Merchant Mariner's Document. This certification allows mariners employed aboard U.S. merchant vessels greater than 100 Gross Register Tons to work in the deck, engineering, or steward's department of a ship. In general, the deck department is concerned with ship navigation and operation; the engineering department tasks are focused on ship propulsion and maintenance of machinery; and the steward's department handles the ship's supplies and food services. Merchant marine personnel also obtain ratings such as able seaman (*see* ABLE-BODIED SEAMEN), lifeboatman, ordinary seaman, wiper, or food handler based on their skills and experience. Deck officers such as the master, chief mate, second mate, and third mate complete many qualifications and meet criteria such as drug testing, oaths, sea service, training, practical demonstrations, and examinations.

Further Reading

American Geophysical Union. Available online. URL: http://www.AGU.org. Accessed March 19, 2007.

American Society of Limnology and Oceanography. Available online. URL: http://www.aslo.org. Accessed March 19, 2007.

Marine Advanced Technology and Education Center (MATE). Available online. URL: http://www.marine-tech.org. Accessed March 20, 2007.

The Marine Technology Society. Available online. URL: http://www.mts.org. Accessed March 19, 2007.

Nichols, Charles R., David L. Porter, and Robert G. Williams. *Recent Advances and Issues in Oceanography.* Westport, Conn.: Greenwood, 2003.

cetaceans Whales, dolphins, and porpoises make up a group of aquatic mammals known as cetaceans. They breathe through lungs, suckle their young, and have hair on their bodies at one time or another, and they are warm-blooded. Most species are exclusively

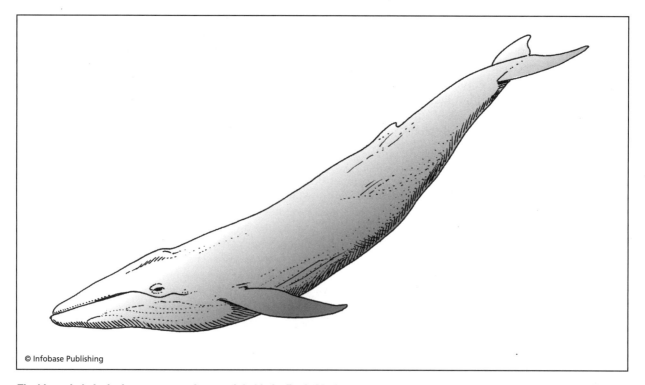

© Infobase Publishing

The blue whale is the largest mammal ever to inhabit the Earth. Marine scientists aboard the NOAA RV *McArthur II* sighted blue whales in Alaska waters for the first time in many years during July 2004.

marine, but there are some dolphins that do in fact inhabit freshwater rivers and their tributaries as well. The pink Amazon River dolphin (*Inia geoffrenis*), known as *"botos"* in Brazil, is one such species that inhabits freshwater rivers in the Amazon River basin. The common bottlenose dolphin (*Tursiops truncatus*), can be found in United States coastal ESTUARIES year round and off shore as well.

Further Reading

National Marine Mammal Laboratory, National Oceanic and Atmospheric Administration: "Whales, Dolphins, and Porpoises (Order Cetacea)." Available online. URL: http://nmml.afsc.noaa.gov/education/cetaceans/cetacea. htm. Accessed March 20, 2007.

chaos Describes complex and nonlinear systems such as turbulent fluids, weather patterns, and predator-prey relationships that are sensitive to their initial conditions. This means that a very small change in the initial conditions can become larger and larger in time. Numerical weather prediction models are started with known initial conditions, and the model is run to produce a forecast based on the initial data. If there are small errors in the data, they can propagate through the model run, producing very different results as time and space evolve over days and hundreds and thousands of miles. Since errors are typically random; vastly different long-term forecasts can be produced from only slightly different initial data. Professor Edward Norton Lorenz, a meteorologist, first noted this effect in 1961 while doing computer model experiments at the Massachusetts Institute of Technology. This effect occurs not only in computer models but also in nature. It is often referred to as the "butterfly effect." The flapping of a butterfly's wings in Kansas can influence the weather in Connecticut. This is something of an exaggeration, but it does portray the tendency of very small disturbances in wind, TEMPERATURE, humidity, and so forth, to produce a disturbance that grows with time and space. For example, the sequential introduction of dye into a fluid produces quite different patterns, demonstrating some of the properties of chaos.

The existence of chaos precludes making accurate weather forecasts of more than a few days' duration. Chaos also limits the time over which accurate ocean circulation or wave height forecasts can be made. Nevertheless, chaos is not random; discernable patterns emerge from the study of successive model outputs with slightly different initial conditions. Chaos essentially makes it possible to forecast patterns and STATISTICS but not individual case behavior.

Chaos is at present a very active branch of research in all of the Earth sciences, and indeed in all of the sciences. Marine scientists applying "chaos theory" are working toward finding the underlying order in apparently random data.

Further Reading

Gleick, James. *Chaos: Making a New Science.* London: William Heinemann, 1988.
Lorenz, Edward N. *The Essence of Chaos.* Seattle: University of Washington Press. 1993.

Charleston Bump The Charleston Bump is a geological feature rising from the ocean floor off the coast of South Carolina and Georgia. It is located 93 miles (150 km) south of Charleston, South Carolina; hence its name. The CONTINENTAL SHELF off the East Coast at this point falls to a depth of almost 2,300 feet (700 m), where it meets the Blake Plateau. Located between the shelf break with a depth of about 330 feet (100 m) and the Blake Plateau lies a ramp running from the south to the north, which rises from 2,300 feet (700 m) to 1,300 feet (400 m). This is the Charleston Bump. Then as the ramp reaches 1,300 feet (400 m), its shallowest point, it drops 330 feet (100 m) again. This clifflike wall is called a scarp. This scarp with its ledges of overhanging rocks is replete with many species of fish. North of this scarp is a deep basin that is called the scour depression. The Charleston Bump has a unique geology, and it strongly influences the flow of the GULF STREAM.

The Charleston Bump is composed of erosion resistant rocks. This rock is formed by a chemical process in which manganese-phosphorite is precipitated from the minerals that are dissolved in the seawater. This is different from rocks that are made from cooling of magma or the sediment of particles. This manganese-phosphorite formed small nodules that coated the underlying seafloor sediments. This coating is up to 3.3 feet (1 m) thick and is very resistant to erosion. It is estimated that this coating was formed during the Miocene Period (15 million years ago). Sponges and corals grow on this hard surface.

There is a large variety of fishes in the waters over the Charleston Bump. There are large numbers of squid and fishes that migrate vertically during the diurnal (daily) cycle. They rise to the surface to feed on PLANKTON during the night and then sink to the bottom during the day to avoid becoming the victim. This migration of organisms reflects sound-causing echoes in fathometers and is commonly called the DEEP SCATTERING LAYER. On the bottom of the Charleston Bump there is a large stock of wreckfish that prey on these migratory fish. The Gulf Stream continuously supplies these migratory fish to the bottom dwellers.

The Gulf Stream is one of the largest dynamic physical features on the planet. It carries over 200 Sverdrups (200×10^6 m³/s) of warm water to the north off the coast of the United States. This current transfers more heat from the southern latitudes to the northern latitudes than does the atmosphere. This mighty current reaches to the seafloor and is about 60 miles (100 km) wide with a speed of about 2.40 knots (1.20 m/s) off the coast of South Carolina. The Charleston Bump actually deflects this current seaward and generates meanders which produce a cyclonic eddy over the Charleston Bump. This eddy pumps the colder water from the deep upward. Upwelled water brings with it nutrient-rich bottom water. These nutrients cause the biological productivity of the water to increase dramatically in this area. Productivity is revealed in the chlorophyll measurements from the Coastal Zone Color Scanner, a satellite-borne sensor for measuring ocean color.

Further Reading

National Aeronautics and Space Administration. Goddard Earth Sciences Data and Information Center. "Ocean Color: Science Focus: Charleston Bump: Making Waves in the Gulf Stream." Available online. ULR: http://www.daac.gsfc.nasa.gov/oceancolor/scifocus/oceanColor/charleston_bump.shtml. Accessed September 14, 2007.

National Oceanic and Atmospheric Administration. "A Profile of the Charleston Bump." Available online. URL: http://www.oceanexplorer.noaa.gov/explorations/islands01/background/islands/sup11_bump.html. Accessed March 21, 2007.

charts, nautical Maps that have been specifically designed to meet the requirements of marine navigation are called nautical charts. These graphical portrayals of the marine environment depict elevations of selected topographic features, general configurations and characteristics of the coast, water depths obtained from hydrographic surveys, the nature or composition of the bottom from GRAB SAMPLERS, dangers and aids to navigation, the shoreline at a vertical datum such as mean high water, tide ranges from water level observations, and magnetic variation for the charted area.

Charts are available on several scales, ranging from about 1:2,500 to 1:14,000,000. Small-scale charts are used for voyage planning and offshore navigation. Charts of increasingly large scale showing more detail are used as vessels approach land. Contour lines depict points of equal depth, where the depths are the vertical distance from the chart datum to the bottom. Since 1989, chart datum has been implemented to mean lower low water for all marine waters of the United States and its territories. Mean Lower Low Water (MLLW) is the average of the lower low water tidal heights taken over a 19-year period, which is called the National Tidal Datum Epoch. This choice for a datum is particularly useful since the tide will not usually fall lower. However, water levels may fall significantly below that of the astronomically predicted tide in regions that are strongly affected by meteorological forces such as winds that set up the water or freshets. To visualize the spatial variation of water depth or what is commonly called "bathymetry" marine scientists use charts. Bathymetry is especially important in understanding motion in the ocean such as refracting, reflecting, diffracting, and shoaling waves, the generation of internal waves as currents pass over underwater sills, and the favored locations of migrating marine species. It is very useful to visualize current vector fields on top of bathymetric contours. This kind of visualization can be realized on modern electronic charts. These charts are a recent development and allow the display of digital bathymetric charts on the bridge of commercial vessels for real-time navigation. Data on tides, currents, weather, and marine hazards can be displayed on the electronic chart, using overlays. Such procedures make navigation safer and more accurate.

Most charts used in United States's waters for navigation have been developed and issued by the National Ocean Service of the NATIONAL OCEANIC AND ATMOSPHERIC ADMINISTRATION (NOAA).

Further Reading

Bowditch, Nathaniel D. *The American Practical Navigator: An Epitome of Navigation.* Bethesda, Md: Defense Mapping Agency Hydrographic/Topographic Center, 1995.

National Oceanic and Atmospheric Administration: Office of Coast Survey, National Ocean Service. Available online. URL: http://www.chartmaker.ncd.noaa.gov. Accessed June 15, 2007.

NOAA/DMA. *Chart No. 1, Nautical Chart Symbols Abbreviations and terms*, 9th ed. Joint NOAA/DMA publication, 1990.

chemical oceanographers Chemical oceanographers are scientists who study the chemistry of the sea to include the sediments below and the atmosphere above and their interrelationships. They investigate chemical processes, track the movement of chemicals in the ocean, and identify the effects of contaminants on marine ecosystems. Applied chemical oceanography involves pollution control, studies of the role of oceans in atmospheric chemistry, global climate change, and assessments of the quality of fish and fish products.

As is the case with chemistry in general, chemical oceanographers tend to be either theoretical or experimental scientists. Chemical oceanography is not an isolated discipline. It merges into physical, biological, and geological oceanography. Theoretical oceanographers develop mathematical models to explain processes such as THERMOHALINE CIRCULATION or how the variability of dissolved salts in seawater might affect different organisms. Experimental scientists use complex electronic instruments to measure the chemical properties of the oceans, such as TEMPERATURE, SALINITY, and PRESSURE by means of conductivity, temperature and depth probes (see CONDUCTIVITY-TEMPERATURE-DEPTH [CTD] PROBE), and current velocity by means of sophisticated current meters employing acoustic or electromagnetic waves, drifting buoys tracked by Earth satellites, and AUTONOMOUS UNDERWATER VEHICLE (AUV). These instruments can be deployed from well-equipped research vessels, ocean platforms, moored BUOYS, aircraft, and deep submersibles. Chemical oceanographers also deploy many oceanographic instruments, such as transmissometers, sediment corers, hydrometers, and conductivity-temperature-depth systems.

Chemical oceanographers normally have at least a master's degree. University research positions normally require a doctorate followed by postdoctoral research. With only a bachelor's degree, positions are generally limited to the technician level.

There are two general approaches to obtaining advanced degrees. Some persons attend universities that give advanced training and postgraduate degrees in physical oceanography. Others prefer to take their advanced degrees in physics and mathematics, and then participate in postdoctoral training in physical oceanography at major laboratories such as SCRIPPS INSTITUTION OF OCEANOGRAPHY, and WOODS HOLE OCEANOGRAPHIC INSTITUTION.

Chemical oceanographers are employed by major universities, government laboratories, and private industries. They perform basic research on the chemistry of the ocean and the marine atmosphere, and applied research in fields such as marine mineral exploration, pollution control and remediation, and resource management.

To keep abreast of recent developments in their field, chemical oceanographers must spend considerable time reading current technical journals, participating in scientific meetings and workshops, and taking courses in new technologies. As in almost all the sciences, they must prepare technical proposals to federal or state agencies and organizations, such as the National Science Foundation, the Office of Naval Research, and the NATIONAL OCEANIC AND ATMOSPHERIC ADMINISTRATION to obtain the funds to support their project work.

Chesapeake Bay Chesapeake Bay is a semienclosed body of water partially protected from the open sea, which is located in the mid-Atlantic along the U.S. East Coast. Marine scientists classify Chesapeake Bay as a drowned river valley ESTUARY. It is the largest estuary in the United States, spanning nearly 200 miles (321.9 km) from the Susquehanna River in the north to its entrance between the Virginia Capes. The bay's width ranges from 3.4 miles (5.5 km) near Aberdeen, Maryland, to 35 miles (56.3 km) near the mouth of the Potomac River. The bay and its tidal tributaries have around 11,684 miles (18,803.6 km) of shoreline and depths averaging 21 feet (6.4 m) with troughs that are up to 174 feet (53 m) deep. The area of land that drains into Chesapeake Bay includes parts of Delaware, Maryland, New York, Pennsylvania, Virginia, West Virginia, and Washington D.C.

The physical properties of the bay greatly influence the life cycles of SUBMERGED AQUATIC VEGETATION (SAV), marine birds, shellfish, and fish. The bay's wetlands and temperate climate offer a fertile and diverse environment for waterfowl such as mallards (*Anas platyrhynchos*), black ducks (*Anas rubripes*), canvasbacks (*Aythya valisineria*), and redhead ducks (*Aythya americana*). Winds and rains can cause high sea states and CURRENTS that uproot bay grasses. These same SAV beds that contain species such as eelgrass (*Zostera marina*) and Widgeon grass (*Ruppia maritima*) provide food and habitat for many species of waterfowl on their migration from summer breeding grounds. The beds provide a refuge from predators for finfish such as spot (*Leiostomus xanthurus*), croaker (*Micropogonias undulatus*), and young stripped bass (*Morone saxatilis*). The bay grass beds are also important for grass shrimp (*Palaemonetes pugio*) and blue crabs (*Callinectes sapidus*), especially after molting.

The majority of Chesapeake Bay has a semidiurnal tidal period. Areas that experience tides have two high and low waters and accompanying ebb and flood tidal currents per day in response to astronomical forces. The tidal day is 24 hours and 50 minutes long and is the reason why tides arrive one hour later each solar day. For some species such as horseshoe crabs (*Limulus polyphemus*), spawning activities occur in synchronization with the full moon and spring tides. Currents caused by tides, winds, rains, and differences in water levels are responsible for the transport of passive drifters such as PLANKTON, seeds, and fish larvae.

Human interactions such as overdevelopment have been blamed for declines in some of the more than 3,600 species of plants, fish, and animals that inhabit Chesapeake Bay. As an example, runoff from the watershed can pick up pollutants such as nutrients, sediment, and chemical contaminants. The ensuing siltation blocks sunlight from reaching SAV beds and smothers oyster reefs, causing problems for feeding and the settling of new larvae on old shell material. Besides high TURBIDITY from siltation, diseases and invasive species have contributed to declines in sea grasses. Other factors such as the parasites MSX (*Haplosporidium nelsoni*) and Dermo (*Perkinsus marinus*), overharvesting, and loss of habitat are contributing to declines in Eastern Oyster or *Crassostrea virginica* populations. Organizations such as the Maryland Department of Natural Resources and the Oyster Restoration Foundation are working to reestablish once flourishing SAV beds and oyster reefs. The benefit would be a natural water-cleansing system, filtering the estuary's entire water volume of excess nutrients on the order of weeks rather then today's rate of more than a year.

Many government agencies and industries are working to safeguard Chesapeake Bay. As an example, the ENVIRONMENTAL PROTECTION AGENCY (EPA), the NATIONAL OCEANIC AND ATMOSPHERIC ADMINISTRATION (NOAA) and the Department of Defense spend millions for restoration projects in the Chesapeake Bay watershed. EPA programs include air and water pollution control; toxic substances, pesticides and drinking water regulation; wetlands protection; hazardous waste management; hazardous waste site cleanup; and some regulation of radioactive materials. Activities include compliance and enforcement, inspection, engineering reviews, ambient monitoring, analysis of environmental trends, environmental planning, pollution prevention, risk assessment, and education and outreach. NOAA provides services ranging from protecting lives and property through the distribution of meteorological, hydrographic, and oceanographic information to evaluating the status and trends in stressed fisheries such as Menhaden (*Brevoortia tyrannus*). Military bases such as the United States Army's Aberdeen Proving Grounds and the United States Navy's Naval Surface Warfare Center, Dahlgren Division, have exemplified principles of stewardship through compliance with the National Environmental Policy Act. Since 1983, governors from Maryland, Virginia, and Pennsylvania, along with the mayor from the District of Columbia have sponsored a Chesapeake Bay Program that reflects a shared vision for the restoration and protection of one of our nation's most wonderful natural resources, Chesapeake Bay.

Further Reading
Curtin, Philip D., Grace S. Brush, and George W. Fisher. *Discovering the Chesapeake.* Baltimore, Md.: The Johns Hopkins University Press, 2001.
Ward, L. G., P. S. Rosen, W. J. Neil, O. H. Pilkey, Jr., O. H. Pilkey Sr., G. L. Anderson, and S. J. Howie. *Living with the Chesapeake Bay and Virginia's Ocean Shores.* Durham, N.C.: Duke University Press, 1989.
White, Christopher P. *Chesapeake Bay: Nature of the Estuary, A Field Guide.* Centreville, Md.: Cornell Maritime Press/Tidewater Publishers, 1989.

Chesapeake Bay Observing System In 1989, the University of Maryland Center for Environmental Science, Horn Point Laboratory, launched the Chesapeake Bay Observing System (CBOS) with two real-time instrumented BUOYS placed in the upper and middle reaches of the bay. Like the tides, the number of CBOS stations has increased and decreased based on user requirements and funding. The system has operated and maintained two real-time data buoys for more than 13 years and had a maximum of seven operational buoys during 2000. Sensors on the buoys acquire data on wind, TEMPERATURE, CURRENT, SALINITY, and nutrient loads. These data are then transmitted by radio and telephone links to various shore stations. Simultaneously, the data are transmitted to Horn Point Laboratories (HPL) by various means, quality controlled, and made available through the Internet. During 2004, CBOS operated six real-time buoys in the bay, arrayed from Howell Point in the north to Wolf Trap in the southern bay. This deployment included two CONTINENTAL SHELF buoys sited off Wallops Island, Virginia. Today's configuration of CBOS is in synchronization with the United States's interest in safeguarding the environment and protecting life and property through an Integrated Ocean Observing System. CBOS measurements can be linked to coastal observing systems that are located to the north and south.

The CBOS monitoring system uses different types of moored buoys. Larger long-term buoys remain on-station along the main axis of CHESAPEAKE BAY for most of the year. Smaller, portable "rover" buoys are deployed in bay or river locations requiring investigations ranging from several days to several months. Both types of buoys share similar remote sensing and data telemetry capabilities. Some of the CBOS buoys have been equipped with plug-and-play telemetry to enable real-time access via the Internet to new bio-optical, nutrient, and oxygen sensors. Buoy payloads include weather sensors such as ANEMOMETERS, underwater sensors such as ACOUSTIC DOPPLER CURRENT PROFILERS, acoustic modems to transmit data underwater, fiber optics,

The University of Maryland Center for Environmental Science at Horn Point, Maryland   *(NOAA)*

and, most important radios and antennas to transmit data to the central database. The data acquisition and dissemination systems are personal computers running software that can make the information available to cellphone and Internet users. The Smith Point and offshore CBOS stations are outfitted with Iridium Satellite Modems because other techniques, including cellphone modems have proven to be unreliable. The entire system is owned and operated by University of Maryland, and the deployment locations and durations are based on validated decisions by a consortium of academic and government partners.

Advances in observational technologies will improve applications for users. For example, researchers at Horn Point Laboratory recently demonstrated the power of combining real-time observing systems with forecast models through a National Ocean Partnership Program-funded Coastal Marine Demonstration Project. Through such investigations, Horn Point Laboratory has gained experience in being able to forecast conditions where real-time observations cannot be made. As model skill is improved through the assimilation of real data, people can be supplied with information that improves tasks that are affected by marine conditions. Whether or not observations are coupled to a model, today's CBOS information is valuable to people who have to make decisions relating to the coastal zone, that is, local government trying to assess water quality, industry leaders comparing climate patterns to energy usage, harbor pilots navigating a channel junction, scientists studying algal blooms, and recreational boaters interested in bay conditions.

Further Reading

CBOS: Chesapeake Bay Observing System, Horn Point Laboratory, University of Maryland Center for Environmental Science. Available online. URL: http://www. CBOS.org. Accessed July 22, 2008.

Chondrichthyes   Sharks, skates, rays, and chimeras all belong to the class of living fishes called chondrichthyes. Chondrichthyes comes from the Greek *chondros* for "cartilage" and *ichthys* for "fish." What these fishes all have in common is the possession of a skeleton composed of cartilage rather than true bone. There are other differences as well, but this is the primary characteristic that sets these fishes apart from all other living forms. The Agnatha or jawless fishes also possess a skeleton composed of cartilage, but, as the name suggests, these fishes lack true jaws and are placed in a separate class. The majority of living fishes possesses a skeleton composed of true bone and are placed in the class Osteichthyes with more than 21,000 different species.

The Chondrichthyes are further divided into two main groups or subclasses—the sharks, rays, and skates belong to the subclass Elasmobranchi (meaning "elastic gill") with approximately 800 different species. The chimeras are placed in the subclass Holocephali (meaning "whole head") with approximately 35 different species. These fishes are commonly referred to as "rat fishes" or "rabbit fishes."

See also AGNATHANS; FISH; OSTEICHTHYES.

climate   The average weather conditions that prevail at a location over a long period, typically many years constitutes climate. The climate of a location is generally determined by its latitude, its elevation, and its distance from the ocean. Descriptive studies of climate are concerned with the classification of climate by means of its major climate controls, such as location, TEMPERATURE, humidity, atmospheric chemistry, and atmospheric TURBIDITY. For example, one could compare a mid-latitude inland city with a city on the west coast of a continent at the same latitude and elevation. The inland city would likely have cold, dry winters and hot, wet summers; the coastal city would have cool, wet winters and cool summers. These are typical differences between continental and marine climates. Descriptive climatology makes extensive use of statistics, such as those compiled by the United States National Weather Service (*see* NATIONAL OCEANIC AND ATMOSPHERIC ADMINISTRATION).

A number of classification schemes are used to organize global climatological data; perhaps the best known is the Köppen climate classification after Wladimir Peter Köppen (1846–1940), who first grouped climates into similar types. By means of this system, it is possible to quickly determine the general characteristics of the climate of a location, even though little data from that location is available. This mass of daily weather data from locations throughout the globe is archived at the United States National Climatic Data Center in Ashville, North Carolina, and at similar facilities in Europe and Asia.

Analytical studies of climate are concerned with climate variability on all time scales. Seasonal variability occurs because the Earth is tilted 23½ degrees from the perpendicular to the ecliptic (the plane formed by Earth's orbit around the Sun, as is known by most high school and college science students). Climate variability can result from many other astronomical factors as well, such as Earth's precession (wobbling about its axis), being struck by asteroids or comets, and variations in the Sun's sunspot cycle. Earth's precession causes climate variations over time periods of tens to hundreds of thousands of years. The impacts of asteroids or comets as they struck Earth in past ages is believed to have caused mass extinctions; the most recent being that of the dinosaurs around 60 million years ago. The heavy dust layer prevailing in the atmosphere for hundreds of years would have cooled the Earth and dried out the swamps and other dinosaur-favorable habitats. A climate developed that was more suited to small but adaptable mammals.

Another climate factor is the distribution of land and water. The continental drift over millions of years produced dramatic climatic changes in the continent of Antarctica, for example, which migrated from a tropical to a polar location. Shorter-term variations appear to be related to the Sun's sunspot cycle, although no adequate physical mechanism linking climate with sunspot occurrence has yet been identified.

The effects of GLOBAL WARMING are currently under intensive study. However, it is not yet known whether such global warming will result in a warmer or cooler climate, and on what time scales. A major problem is separating natural from anthropogenic influences, such as the addition of large amounts of carbon (*see* CARBON CYCLE) into the atmosphere as a result of industrial processes. Oceanographers believe that the present trend of warming and glacier melting in the Arctic and environs could result in a major change in the circulation of the North Atlantic Ocean and a shift in the position of the GULF STREAM. This major alteration of circulation could result in a much cooler climate for Europe and eastern North America. Most climate scientists are agreed that global warming will produce more violent weather patterns than have been previously seen.

These scientists make extensive use of computer models based on the laws of motion and thermodynamics to study climate variations, and to simulate their affects on Earth's land and seas. As computer power and capacities increase, it is possible to take in more and more data, improving the accuracy and forecasts of the models. Improved models, as well as great improvements in the technology of data acquisition, such as by remote sensing satellites, have made possible the investigation of atmosphere-ocean interactions and their affects on regional and even global climate. Examples are the global affects of the occurrence of the El Niño/La Niña cycle in the tropical Pacific (*see* EL NIÑO) and the Arctic Oscillation in the North Atlantic. As a result of research, by the mid 21st century there will be sufficient data and adequate theoretical models to make climate forecasts that will answer some of today's pressing questions.

See also CLIMATOLOGY.

Further Reading
Allaby, Michael. *Encyclopedia of Weather and Climate*, rev. ed. New York: Facts On File, 2007.

climatology Climatology is the scientific study of climate that deals with describing and classifying climate in both a graphical and statistical sense. Scientific climatology studies the physical/chemical/biological controls of climate and the causes of climate variability on all time scales. Modern climatology deals with the dynamics of the atmosphere-ocean system and the affects of air-sea interaction on climate. An example of the result of these studies is the two-way feedback between the atmosphere and the ocean in the tropical Pacific resulting in El Niño and La Niña (*see* EL NIÑO).

Climatology is concerned with the heat budgets of the atmosphere-ocean system. Current topics of major interest include GLOBAL WARMING, sea level rise, and the hole in the ozone layer of the atmosphere. It involves the long-term collection of reliable data to generate mean values, variances, and the probabilities of extreme values.

Applied climatology is concerned with societal aspects of climatology to provide environmental managers with essential environmental information in a timely manner to facilitate adequate planning for the affects of climate change on human habitation. Climatologists synthesize weather conditions in a particular area and analyze meteorological elements (air TEMPERATURE, PRESSURE, winds, humidity, precipitation, visibility) to benefit farmers, aviators, mariners, scientists, and the general public.

Scientific studies make extensive use of numerical climate models implemented on super computers, which aid in the identification of important climate controls. These studies make possible the simulation of perturbations on the controls (such as temperature, humidity, and atmospheric chemistry) and the ensuing affects such as extensive flooding of island nations by rising sea level.

Microclimatology is concerned with such things as urban affects on climate, especially with regard to heating and pollution, deforestation due to logging and agriculture, river diversion, industrial plant siting, and massive vehicular traffic. Field measurements of meteorological variables have been key to advancing the use of General Circulation Models in detecting climatological impacts from wide spread deforestation in Amazonia. The models are used to evaluate changes in ALBEDO and surface roughness on heat fluxes, evaporation, and winds. Such research is aimed at understanding turbulence in the atmospheric boundary layer over land and ocean and its parameterization in numerical models.

See also CLIMATE; NUMERICAL MODELING.

Further Reading
Allaby, Michael. *Encyclopedia of Weather and Climate*, rev. ed. New York: Facts On File, 2007.

clouds Clouds are accumulations of visible water droplets or ice crystals in the atmosphere, usually the troposphere, that produce precipitation and significantly affect the thermodynamic and optical properties of the planet Earth, as well as climate. Clouds are found on all the planets of the solar system that have atmospheres, although their composition may be different from those on Earth. Clouds can reflect incoming solar radiation, in the visible electromagnetic spectral band producing a cooling affect, and they can reflect infrared radiation from the surface of Earth, producing a warming affect. They are formed when the atmosphere contains an adequate number of small particles, called hygroscopic nuclei, for water vapor or ice to condense on, and is sufficiently moist for water vapor to condense into droplets, or ice crystals, depending on the ambient temperature. The presence or absence of clouds is one of the most significant factors in the daily weather and climate. Weather satellites remotely measure cloud temperatures, moisture content, and motions, using visible, infrared, and microwave sensors.

Clouds have been used for centuries by mariners to try to forecast the weather for the next two to three days. Cloud appearance, altitude, brightness, and direction of motion were considered in the attempt to predict the weather. In the early 19th century, Englishman Luke Howard (1772–1864) introduced the modern system of cloud classification. In this system, there are three main types of clouds: cirrus, which are high, thin, wispy clouds composed of ice crystals; cumulus, which are cauliflower-shaped clouds ranging from puffy fair weather cumulus to the giant cumulus towers of severe thunderstorms; and stratus clouds, which occur in sheets of considerable horizontal extent. These clouds are subdivided into genera that are the main groups of clouds; species, relating to observed peculiarities of clouds; and varieties, relating to different arrangements of clouds with respect to one another. Cloud genera can be related to altitude as follows.

High-altitude clouds, that is, clouds above 9,842 feet (2,999.8 m) in the polar regions, 16,404 feet (4,999.9 m) in the temperate regions, and 19,685 feet (6,000 m) in the tropical regions include cirrus, cirrocumulus, and cirrostratus. Cirrus clouds are thin, wispy groupings of ice crystals. Cirrostratus are thin sheets of ice crystals through which the Sun's rays create a frosty appearance, with a ring or halo around it. Cirrocumulus are ice crystals arranged in rows of puffy clouds that resemble the organization of scales on fish, and are known as "mackerel sky."

Mid-altitude clouds are denoted by the prefix *alto,* and range from 6,562 to 19,685 feet (2,000 to 6,000 m) in height in polar regions, 6,562 to 22,966 feet (2,000 to 7,000 m) in height in temperate regions, and 6,562 to 26,247 feet (2,000 to 8,000 m) in height in tropical regions include altostratus, altocumulus, and nimbostratus clouds. Altostratus clouds are sheets of relatively uniform appearance, through which the Sun's rays often look as though shining through frosted glass. Altocumulus clouds are the cumulus cloud masses organized in rows; the cloud elements may or may not be merged into one mass. Nimbostratus clouds are gray sheets of cloud from which rain is falling (the meaning of the prefix *nimbo*). Low-altitude clouds range from the surface to 6,562 feet (2,000 m) in the polar regions, to 22,966 feet (7,000 m) in the temperate zones, and to 26,247 feet (8,000 m) in the Tropics. Stratus clouds are similar to altostratus, except at a lower altitude, and from which drizzle, ice, or snow may be falling. Stratocumulus clouds form gray, rounded masses, or rolls, of much greater thickness and density than altocumulus clouds. Moderate and extensive precipitation in the form of rain or snow is often falling from nimbostratus clouds. These are the true large area rainmakers of the atmosphere. Their thickness prevents the penetration of sunlight to the ground.

Cumulus clouds are known as clouds of vertical extent, because they can rise through all three layers of the atmosphere. These clouds form by atmospheric

Cumulus clouds like this typify the lower atmosphere above the tropical Atlantic Ocean. Observation of clouds is important because different clouds form under different circumstances.   *(NOAA/NWS)*

convection, lifting atmospheric particles until the saturation point is reached, when water vapor condenses and forms a cloud. Fair weather cumulus clouds are the puffy clouds seen in all generally fair weather environments when the atmosphere contains sufficient water vapor. Surface heating by the Sun intensifies the convection with the result that clouds that form in the morning may be converging and building into large masses by afternoon. As these masses rise into the high cloud region, strong winds may blow the tops off the clouds, forming the formidable anvil shape of cumulonimbus clouds. These are the great rainmakers of the Tropics, which also produce thunder and lightning as electric charge builds on the rising cloud droplets. Severe thunderstorms can produce cumulus mammatus clouds, consisting of great masses of cloud dropping from average cloud base. Under conditions of severe atmospheric instability, funnel clouds, known as tornadoes, can drop and extend to the surface, causing great destruction.

The movement of weather fronts determines the day-to-day weather in the temperate, or mid-latitude zones. The passage of a front is usually accompanied by a sequence of clouds, which aid the forecaster in predicting the local weather. Cirrus clouds are frequently the harbingers of an approaching warm front. Since warm fronts slope upward in the direction of movement, the sequence of clouds goes from high to low altitude. Altostratus, when the sky becomes fairly uniformly gray, and then nimbostratus, the rain clouds, frequently follow the cirrus. Cumulonimbus clouds may also accompany the front if there is unstable air above the front, bringing thundershowers. The passage of a warm front is usually followed by the arrival of a cold front, in which cold, dense, fast-moving air wedges under the warm

Lenticular alto cumulus clouds   *(NOAA/NWS)*

air of the retreating warm front. Cumulonimbus clouds are often found towering above the cold front. As the cold front passes, the temperature drops, and fair weather cumulus clouds become dominant. Severe thunderstorms are often found in squall lines moving ahead of the cold front, as well as the rotating bands of cumulonimbus clouds in HURRICANES.

The extent of cloud formation has major affects on the GLOBAL WARMING phenomenon, but the net affect of continued warming is not known. Clouds associated with convection from a warming Earth could reduce the net solar radiation, and thus produce cooling rather than heating. The eruption of Mount Pinatubo in the Philippines in the early 1990s injected vast quantities of particulate matter into the atmosphere, which resulted in greater than normal cloud development, reduced solar heating, and falling temperatures in many parts of the world.

Further Reading
World Meteorological Organization. *International Cloud Atlas-Abridged.* Geneva: World Meteorological Organization, 1956.

C-MAN stations   The National Data Buoy Center (NDBC) began installing these stations in the early 1980s for the National Weather Service (NWS), both NATIONAL OCEANIC AND ATMOSPHERIC ADMINISTRATION (NOAA) organizations. The C-MAN (Coastal-Marine Automated Network) has now grown to 60 stations around the continental United States and Alaska, as well as a few Pacific Island stations as far away as Micronesia. The C-MAN stations were developed to replace the U.S. Coast Guard Lightship stations that were superseded by automated navigational aids. A crew that took meteorological and sometimes oceanographic observations manned the Lightships.

C-MAN stations provide the NWS with a network of data-acquisition platforms in the coastal zone, filling the gap between land stations and ocean buoy stations. They provide vital data on present conditions on the CONTINENTAL SHELF to mariners, industry, and vacationers. They also provide data for input to the NWS weather forecasting models for making marine weather forecasts.

Most C-MAN stations measure barometric pressure, wind speed and direction, and air TEMPERATURE. Some stations have a greater capability and are able to measure water temperature, water level, wave parameters, CURRENT speed and direction, precipitation, and visibility. The data are processed and transmitted hourly in similar manner to the ocean data buoys. The data are reported on the NDBC Web site.

Further Reading
National Data Buoy Center: the C-MAN Program. Available online. URL: www.ndbc.noaa.gov/cman.phtml. Accessed June 14, 2004.

coastal buoy   A coastal buoy has an instrument payload and is especially designed to operate in the dynamic coastal environment. These types of moored BUOYS measure and transmit parameters such as barometric pressure, wind direction, wind speed, and gusts, air and sea TEMPERATURE; wave energy spectra from which significant wave height, dominant wave period, and average wave period are calculated. On some buoys, the direction of wave propagation is also measured and reported. Information from these buoys supports marine scientists, mariners, and recreational boaters.

The U.S. National Data Buoy Center (NDBC) employs several types of coastal buoys. Their hull diameters range from five to eight feet (1.5 to 2.4 m). Some employ line-of-sight radio communications for data telemetry. Others use satellite relays, which can send data to anywhere, in the world. The choice of hull type used usually depends on the intended deployment location and measurement requirements. There are various other types of moored buoys that are deployed in the world's oceans. At NDBC, these include 10-foot (3 m), 33-foot (10 m), and 40-foot (12 m) discus hulls and the 20-foot (6 m) Navy Oceanographic and Meteorological Automatic Device or NOMAD boat-shaped hulls. Due to their large size, 33-foot and 40-foot (10 and 12 m) buoys generally have to be towed to deployment locations behind a ship. The NOMAD buoys with their boat-shaped hulls are highly directional and have a quick rotational response.

In addition to the use of buoys in operational forecasting, warnings, and atmospheric models, moored buoy data are used for safe navigation, research programs, emergency MARINE SPILL RESPONSE, and legal proceedings. Moored buoys are a necessary technology for making sustained time series observations in the coastal ocean and are an important component of any long-term ocean observing system. Coastal buoys along with automated sensors aboard satellites and reconnaissance aircraft provide important information to help maritime nation's mitigate the effects of natural hazards.

Further Reading
National Data Buoy Center, National Oceanic and Atmospheric Administration, Stennis Space Center, Mississippi. Available online. URL: http://www.ndbc.noaa.gov. Accessed June 15, 2007.

coastal engineer A coastal engineer is a professional with a strong background in civil engineering who specializes in understanding the impacts of waves, tides, and shallow water processes on property such as BEACHES and inlets and structures such as JETTIES and seaports. Professional engineers calculate the dynamical response of coastal structures to forces ranging from subsidence to wave slamming. Specific phenomena studied include TIDES, gravity waves, wave breaking, nearshore currents, beach composition, sediment transport, SEA ICE, marine climate, and weather.

Coastal engineers tend to be experimental and commonly evaluate structures such as jetties in flumes or build physical models of entire estuarine systems to understand their complex physics. As is the case with physical oceanographers, coastal engineers develop mathematical models to explain the circulation and wave motions of the ocean, which usually involves sets of complex equations that must be solved numerically on computers. These engineers also use state-of-the-art electronic instruments or remote sensing tools to measure currents, waves, and depths or BATHYMETRY in the coastal ocean and neighboring estuaries. These instruments can be deployed from well-equipped research vessels, platforms, moored BUOYS, and aircraft. On applied projects relevant to road construction, they may use instruments such as soil test kits, digital cone penetrometers, and light weight deflectometers to determine bearing strength of road materials while building a CAUSEWAY.

One of the gateways for employment with engineering firms or federal, state, and municipal governments that work on coastal projects is a bachelor's degree in civil or coastal engineering. Course work addresses technologies associated with the design, construction, operation, and maintenance of coastal structures and facilities including BREAKWATERS, piers, wharves, channels, and pipelines. Some engineers attend universities that give advanced training and postgraduate degrees in coastal engineering.

Coastal engineering projects are usually very expensive and affect many people. In the United States, large-scale public projects are often managed by the United State Army Corps of Engineers, which operates the world famous Coastal and Hydraulics Laboratory in Vicksburg, Mississippi. A major focus of the Coastal and Hydraulics Laboratory is the effective design and maintenance of inlet navigation and shore protection projects. Coastal engineers involved in research are often called upon to evaluate the efficacy of coastal engineering projects ranging from improving seaport operations through dredging and constructing bulkheads to protecting public lands through beach nourishment and constructing jetties.

Coastal engineers may use aerial imagery to assess shoreline retreat rates or make bathymetric maps from LIDAR. They are heavily involved with storm damage prevention through physical monitoring and tracking coastal sediment budgets. They conduct geophysical studies such as sand source investigations, using equipment such as magnetometers, side-scan sonar, and coring devices. Another component of coastal engineering involves economic analyses and project planning.

See also FIELD RESEARCH FACILITY; OCEAN ENGINEER.

Further Reading

Coastal and Hydraulics Laboratory, U.S. Army Corps of Engineers. Available online. URL: http://www.chl.erdc.usace.army.mil. Accessed April 30, 2004.

Coasts, Oceans, Ports, and Rivers Institute. Available online. URL: http://www.content.coprinstitute.org. Accessed March 21, 2007.

European Union for Coastal Conservation—The Coastal Union. Available online. URL: http://www.eucc.net. Accessed March 21, 2007.

Manly Hydraulics Laboratory, Manly Vale, Sydney, Australia. Available online. URL: http://www.mhl.nsw.gov.au. Accessed March 21, 2007.

U.S. Army Corps of Engineers. Coastal Engineering Manual, 2002. Engineer Manual 1,110–2–1,100, U.S. Army Corps of Engineers, Washington, D.C. (in 6 volumes). Available online. URL: http://www.chl.erdc.usace.army.mil/chl.aspx?p=s&a=Publications;8. Accessed March 21, 2007.

coastal erosion Coastal erosion is a natural process that accounts for the movement (erosion and accretion) of BEACH materials. Atmospheric and oceanic weather, such as EL NIÑO, storms, winds, TIDES, CURRENTS, WAVES, and STORM SURGES are major factors in coastal erosion. Water movement due to the astronomical tides also produces coastal erosion. In general, water levels are rising along the North American coast due to GLOBAL WARMING, exposing more and more of the coastline to wave action. Some coastal areas are altered when resorts are built, or any other coastal structures such as groins that disrupt the fragile balance of nature. Property owners often construct stone revetments or vertical seawalls to protect their property, but these noncontinuous protection structures do not solve the erosion issue and may fail as the beach recedes. Scientists and engineers collaborate in order to measure the magnitudes and spatial patterns of coastal changes. An effective shore protection strat-

Erosion affects coasts worldwide. Climate change will accelerate sea-level rise and erosion of beaches and neccessitate building rock defenses against coastal erosion. *(NOAA)*

egy requires a complete understanding of the coastal system.

Coastal erosion usually refers to the loss of land or the removal of beach or dune materials by natural processes and also includes horizontal recession and scour that can be induced or aggravated by human activities. Accretion refers to the buildup of land by deposition of water or airborne beach material. Accretion may be attributed to human actions, such as beach buildup in response to installation of a BREAKWATER or completion of a beach nourishment project. Coastal zone managers may rely on scientists and engineers to evaluate rates of coastal erosion in order to establish eroding area boundaries. One methodology utilizes the analysis of tide synchronized aerial images to evaluate changes in the shoreline. Such changes are documented as shoreline retreat rates.

Eroding coasts lacking adequate sand and vegetation have a low probability to withstand forces generated by sea level rise, waves, tides, and other shallow water processes. Effective solutions for shoreline protection rely on a complete understanding of the physical, financial, permitting, and policy environments. Built-up areas that are subject to coastal

erosion cost tremendous amounts of money to protect. Most of the shore protection structures that are built are eventually undermined and broken apart by waves. Therefore, when possible, it makes sense to allow coastal erosion to continue and to site new structures far enough back from the shoreline, dunes, or banks to minimize the need for shore protection structures. To cite some examples, Cape Hatteras

Dune grass provides a critical stabilizing element to protect dunes on the open coast. *(NOAA)*

Light has recently been moved inshore to prevent its destruction by waves and tides. Parts of eastern Cape Cod have suffered severe erosion. Roads that were used in the 1950s were underwater by the 1970s. In 1797, President George Washington (1732–99) ordered the construction of a lighthouse at Montauk Point 300 feet in from the cliffs. He estimated that the lighthouse would last for 200 years. Today, the lighthouse is in a dangerous position very close to the edges of the cliffs.

Further Reading

Cronin, William B. *The Disappearing Islands of the Chesapeake*. Baltimore, Md.: The Johns Hopkins University Press, 2005.

Dean, Cornelia. *Against the Tide: The Battle for America's Beaches*. New York: Columbia University Press, 1999.

U.S. Geological Survey. *Landscapes, North Carolina Regional Coastal Erosion Studies*. Available online. URL: http://www.geology.usgs.gov/connections/fws/landscapes/ncarolina_erosion.htm. Accessed September 14, 2007.

Ward, Larry G., Peter S. Rosen, William J. Neal, Orrin H. Pilkey, Jr., Orrin H. Pilkey, Sr., Gary L. Andersohn, and Stephen J. Howie. *Living with Chesapeake Bay and Virginia's Ocean Shores*. Durham, N.C.: Duke University Press. 1989.

coastal inlet   A short and often narrow passage connecting a bay or LAGOON to the ocean is a coastal inlet. They are also referred to as "cuts," "inlets," "passes," and "tidal inlets." These geomorphic features are important routes for migratory fish, the exchange of nutrients between ESTUARIES and the ocean, passage of boats and ships, and marine recreational activities. Inlets are associated with complex phenomena such as strong CURRENTS and shifting channels. The scales of change within an inlet can range from seconds to centuries. For instance, current shear causes erosion by lifting certain sediment grains instantly from the bottom, while prevailing winds and their associated waves cause entire inlet systems to migrate in the direction of accretion and erosion over tens of decades. By looking at historic maps for BARRIER ISLAND systems such as the Outer Banks of North Carolina, one can see that the number of navigable coastal inlets seems to have ranged from three to 11. The main navigable passages through the Outer Banks are Oregon, Hatteras, and Ocracoke Inlets.

Aerial view of a typical East Coast inlet   *(NOAA)*

Strong hydraulic currents are often produced by local constrictions such as inlets. The flow is induced by differences in water level at either end of the inlet. The pressure head is usually caused by favorable wind events, and the resulting currents may be observed as exaggerated ebb or flood currents. The water runs from high to low water. Changes in current speed have a pronounced effect on sediment transport. Certain speeds need to be reached for erosion to occur, while slower speeds result in deposition. As waves move into shallow water, their WAVE-LENGTHS decrease and the wave heights increase. This wave shoaling process increases the chances for sediment movement. At inlets, the long-term impact of these shallow-water processes is development of a distal bar from accretion at one end of the inlet's throat and erosion at the other end. This is the basic process behind the movement of inlets, e.g., a southern migration along the Outer Banks.

Coastal inlets from Absecon Inlet in New Jersey to the Yaquina Bay Entrance in Oregon are dynamic sites that are usually well studied because of their importance to property owners and for navigation. Investigators use meteorological sensors such as ANEMOMETERS, oceanographic sensors such as current meters, pressure gauges, and drifters, side scan sonar, and aerial imagery to understand stability of the navigation channel, shoreline change, varying wave and current conditions, and sediment transport within the governing channels. Along barrier islands, which are common to the United States's East Coast, there are multiple inlet systems that have been the focus of research. Investigators are hoping to better understand processes such as rates of freshwater discharge that may be linked from inlet to inlet. At graduate schools such as North Carolina State University, students are frequently found working on leading edge coastal inlet research topics ranging from characterizing complex circulation that includes hydraulic currents to estimating the year group strength of commercially important finfish that migrate through barrier island inlets. The United States Corps of Engineers has established the Coastal Inlet Research Program in Vicksburg, Mississippi, where researchers study natural processes such as erosion and flooding in the aftermath of a HURRICANE to inlet response to stabilization efforts such as JETTY installation or dredging.

Further Reading
Coastal Inlet Research Program. U.S. Army Engineer Research and Development Center, Coastal and Hydraulics Laboratory. Available online. URL: http://www.cirp.wes.army.mil/cirp/cirp.html. Accessed March 20, 2007.

Coastal Oceanographic Line-of-Sight Buoy   The Coastal Oceanographic Line-of-Sight Buoy is an instrumented buoy customized to operate in the dynamic coastal environment. Data are telemetered by radio-frequency (RF) wireless modem to a nearby ground station. The RF modem is an external box used in a system of radio transmitters, receivers, computers, and software. It transmits a standard data stream from a device such as an onboard computer into radio signals and at the ground station converts radio signals into standard data for output by the information dissemination system. The U.S. National Data Buoy Center (NDBC) employs several types of coastal oceanographic line-of-sight or COLOS buoys with hull diameters that are either 4.9 feet (1.5 m) or 5.9 feet (1.8 m). They were originally developed with surface current meters to provide circulation information that supported sailing and yachting events during the 1996 Olympics in Atlanta, Georgia. This buoy may also provide directional wave information.

See also ANCHOR; BUOY; COASTAL BUOY; MOORING.

Further Reading
National Date Buoy Center: Center of Excellence in Marine Technology, National Weather Service, National Oceanic and Atmospheric Administration. Available online. URL: http://www.ndbc.noaa.gov. Accessed June 1, 2004.

coastline   The line that forms the boundary between land, which is called the coast, and the water is often called the coastline. Another term for this boundary is the waterline. If the waterline is referenced to a tidal datum such as mean high water, it is known as the shoreline. Drawing this interface between land and water is a complicated matter since it represents the linear location where terrestrial processes predominate and marine processes such as TIDES and breaking WAVES disappear. This location is important in determining property boundaries and in the planning and installation of coastal structures, especially coastal oil and gas activities that include field exploration, drilling, production, and well treatment.

Private agencies and governance may classify coastlines differently. For example, in Louisiana the coastline is defined as the Chapman Line, a boundary that includes certain barrier islands along Louisiana's southeast coast, thereby incorporating bays and sounds. BARRIER ISLANDS usually form parallel to and relatively close to a coastline and are separated from the mainland by a coastal LAGOON. Thus, locations using the Chapman Line must use points that are defined by latitude and longitude.

Moving away from the coastline toward the hinterland, one finds BEACHES of varying shapes and composition, dune fields that change shape over the seasons, and coastal plains comprising gently sloping strata of clastic materials. At the beach, noncohesive materials such as sand, gravel, and shell are deposited at the interface between dry land and the sea (or some other large expanse of water) and actively "worked" by present-day hydrodynamical processes such as waves, tides, and CURRENTS. Along some coasts, clastic materials that are transported and deposited by the wind build dunes that serve as a sand reservoir for beaches. Coastal plains are composed of rock fragments formed by the process of weathering, transportation, and deposition by wind, water, snow, or biological processes.

Areas located seaward from the coastline include regions such as the foreshore, surf zone, nearshore, and the offshore. The foreshore is defined as that part of the beach between mean higher high water and mean lower low water. The shore is often considered to be the strip of ground bordering any body of water that is alternately exposed or covered by tides and waves. The surf zone is the nearshore region, where incident waves break as spilling, plunging, collapsing, or surging surf on their approach to the shore. The nearshore zone extends from the swash to depths, where sediment motion is not induced by waves. Circulation in the nearshore includes LONGSHORE and RIP CURRENTS. In the offshore region, depths are such that the influence of the seafloor on wave action is small in comparison with the effect of wind.

coast type A coast type is a coupled region of wetland and deepwater approach that shares the influence of similar hydrologic, geomorphologic, chemical, biological, and physical factors. The coast is the linear strip of land that extends from the seashore inland to the first major change in terrain features. The seashore is characterized by the land area that is alternately exposed or covered by TIDES and WAVES. Depending on location on the globe, coastal characteristics are very different as a result of biological and physical forces.

Coastal classification efforts are accomplished through a specific branch of geology called "geomorphology." Geomorphology is the study of landforms and includes marine features such as margins, CONTINENTAL SHELVES, island-arcs, ESTUARIES, and BEACHES. Traditionally, coasts are classified as coasts of submergence, coasts of emergence, neutral coasts, compound coasts, leading-edge coasts, trailing-edge coasts, marginal coasts, divergent coasts, and convergent coasts. An example of coastal classification is the subdivision of Chesapeake Bay's

coastal zone into areas such as estuarine emergent wetlands, palustrine/estuarine forested wetlands, and palustrine emergent wetlands. Understanding the features of a coast is essential for coastal management. Coastal managers focus their efforts on establishing procedures to preserve sustainable marine resources. Quantifying the changes in ground conditions such as beach width, dune elevations, and density of development is useful in determining hazards, vulnerabilities, and storm impact forecasts.

During the 1970s, marine scientists and engineers such as SCRIPPS INSTITUTION OF OCEANOGRAPHY professor Francis P. Shepard (1897–1985) attempted to combine biological and physical interactions to develop descriptive coastal classification systems. Such systems characterize a coast in terms of the primary biological and physical forces that have contributed to the coastal formation process. Shepard described coasts based on energy impacting the land, where a high-energy coast is associated with strong wave action and rapid erosion and a low-energy coast is less active. In this method, coast types are classified according to similar biological and physical features. Biological interactions of importance are defined by factors such as vegetation and soils, while physical interactions are defined by climate, winds, waves, erosion, currents, and flooding. Example coast types found in the literature include drowned rivers, deltas, fiords, drumlins, volcanic, fault, wave erosion, BARRIER ISLAND, cuspate foreland, mangrove, and coral ATOLL. These classifications offer a quick picture of normal meteorological and oceanographic conditions. For example, barrier islands are associated with tidal inlets, shallow coastal lagoons, shoaling, and shifting channels.

Descriptive coastal classifications benefit the mapping and remote sensing community by providing general coast types that highlight meteorological, oceanographic, and terrain impacts. Accurate classifications are a word picture describing prevailing biological and physical conditions useful for planning operations ranging from navigation to coastal construction. A coastal classification system helps predict nearshore processes from the coast's appearance. Features can be modeled or represented using geographic data sets in a Geographic Information System (GIS). Examples of such features include shorelines, BRIDGES, marshes, maritime forests, navigation aids, and submerged aquatic vegetation beds. In the GIS, numbers, characters, images, or hydrographic surveys describe attributes for a coastal feature. The attributes of an ocean current might include depth and volumetric flow or transport.

Further Reading
Bird, Eric C. F., and Maurice L. Schwartz, eds. *The World's Coastlines.* New York: Van Nostrand Reinhold, 1985.
Shepard, Francis P. *Geological Oceanography.* New York: Crane, Rusak, 1977.

coccolithophores Coccolithophores are single-celled spherical microscopic PHYTOPLANKTON living in the upper sunlit layers of the ocean, which are surrounded by a limestone (calcite) plating that looks like fancy collar studs, soccer balls, or hub caps under an electron microscope. These platelets or "coccoliths" are produced inside the cell and extruded to the outside. The coccoliths are mineral structures consisting of calcium, carbon, and oxygen (calcium carbonate, $CaCO_3$) like the protective shells of oysters and clams. Coccolithophores are found in regions from ESTUARIES to the deep ocean, where there is enough light for photosynthesis. Coccolithophores living in the photic zone are at the bottom of the marine food chain and consumed by grazers such as copepods, which is an important step in transferring energy from producers to consumers. The other main groups of planktonic primary producers are diatoms and dinoflagelates.

When water conditions are favorable, coccolithophores have the capacity to occur in massive blooms. In the ocean, the sudden increase in the biomass of phytoplankton or ALGAE in a given area is called a bloom and are easily observed by satellites such as the Sea-viewing Wide Field-of-view Sensor or SeaWiFS. These blooms can turn the waters milky white from the dense cloud of coccoliths. Many marine scientists credit coccolithophores as being the leading calcite producers in the ocean. This calcite helps to buffer the oceans and makes up biogenous sediments such as chalk deposits. Thick deposits of coccoliths form calcerous oozes in areas above the Calcite Compensation Depth and away from continental landmasses, where terrigenous or land-derived sediments accumulate.

cold core ring A cold core ring is a rotating body of water with a relatively cool temperature. These EDDIES normally form from a meander that extends from the cooler coastal ocean into a warmer body of water. The meander, which looks like a loop in satellite imagery, breaks off from the main current and in so doing traps a pool or plug of coastal ocean as an individual eddy. As an example, satellite imagery of the GULF STREAM from NOAA's Advanced Very High Resolution Radiometer (AVHRR) depicts the warm Sargasso Sea in red and orange, and the cooler coastal waters in blue and purple, so that the rings are readily identified. In these images, the Gulf Stream is seen as a distinct linear red feature trending northeast from Florida and bending sharply seaward at Cape Hatteras located along North Carolina's Outer Banks. As the current flows northeast, meanders form and pinch off as eddies when the Western Boundary Current reaches Diamond Shoals and is deflected farther eastward. Cold core rings, which rotate in counterclockwise (cyclonic in the Northern Hemisphere) fashion because of the Gulf Stream's north to northeast flow, are seen as blue circular features in the orange-colored Sargasso Sea. Marine biologists are particularly interested in the life histories of these cold core rings since they capture whole COMMUNITIES from the biologically productive CONTINENTAL SHELF and move them into deeper less-productive waters over the ABYSSAL PLAIN. Gulf Stream meanders pinching off to trap warm Sargasso Sea water form clockwise (anticyclonic in Northern Hemisphere) rotating eddies. In AVHRR images, warm core rings are typically seen as light blue rings in the purple-colored coastal ocean. The warm core rings are less biologically productive than the cold core rings because the shelf water is more productive than the oceanic water. The process of using satellites to study phenomena such as Gulf Stream rings is called feature tracking. Marine scientists studying consecutive images of the Gulf Stream can study the ring's trajectories to better understand how these eddies that have initial diameters of several hundred kilometers dissipate their energy or decay over the course of several years, transport biota, and mix with the surrounding WATER MASS.

combatant ships Combatant ships is a classification for vessels that are used by the U.S. Navy to maintain open sea lanes and to project military power throughout the Atlantic, Pacific, Indian, and Southern Oceans; Caribbean and Mediterranean Seas; and the Persian Gulf. This classification is useful in describing the capabilities and functions of a particular warship. The classification groups vessels into types such as aircraft carriers, surface combatants, submarines, patrol combatants, amphibious warfare ships, combat logistic ships, coastal defense ships, and mine warfare ships. This functional classification benefits the navy in training personnel, planning operations, maintenance management, and in development of new systems.

Aircraft carriers conduct combat operations by fixed and rotary wing aircraft that engage in attacks against airborne, surface, subsurface and shore targets. U.S. Navy aircraft carriers are almost entirely nuclear powered in today's modernized fleet, though there are a handful of conventionally propelled (steam turbine engines powered by marine diesel fuels) carriers that remain in service. The nuclear-powered carriers include ships from the Enterprise class and Nimitz class. Conventional carriers include ships from the Forrestal, Kitty Hawk, and John F. Kennedy classes. A nuclear-powered aircraft carrier weighs about 95,000 tons and is about 1,100 feet (335 m) long, carrying about 5,500 crew and 82 aircraft, which are part of the "Carrier Air Wing" (CAW).

Surface combatants describe ships such as cruisers, destroyers, and frigates, which engage enemy forces on the high seas. Most modern U.S. Navy cruisers are part of the highly technological and capable Ticonderoga (AEGIS) class. A cruiser weighs about 10,000 tons and is about 600 feet (183 m) long and 60 feet (18.3 m) wide, carrying about 400 crew members. Modern cruisers are typically powered by gas turbine engines (similar to the jet engines found on commercial airliners), though a handful of nuclear-powered cruisers are still in service. Most modern U.S. Navy destroyers are part of the highly technological and capable Arleigh Burke class. A destroyer weighs about 10,000 tons and is about 500 feet (152 m) long and 70 feet (21 m) wide, with a crew of about 350. Most modern U.S. Navy frigates fall into the Oliver Hazard Perry class. A frigate weighs about 4,000 tons, is about 450 feet (137 m) long, 45 feet (14 m) wide, and holds a crew of about 200. During the cold war, the primary mission of destroyers and frigates was anti-submarine warfare (ASW). However, modern strategy and doctrine indicate that air warfare is the most significant threat to surface naval combatants, and current destroyer and frigate platforms reflect this, as most of their surveillance and weapons technology is now geared toward anti-air warfare (AAW) rather than ASW. For the most part, the mission of smaller combatant ships is to protect larger combatants, such as aircraft carriers and aircraft-carrying amphibious assault ships from air, surface, and submarine threats.

Submarines are self-propelled submersible vessels having a combat capability. Los Angeles, Virginia, and Seawolf class attack submarines (SSN) are designed to seek out enemy surface ships and submarines. The Ohio class ballistic missile submarines (SSBN) are used for strategic deterrence. Many submarines also employ long-range land-attack missiles, such as Tomahawk cruise missiles, to strike distant targets both on land or at sea. Submarines such as these are nuclear powered, enabling them to conduct submerged patrols for periods of 70 days or more without need of surfacing for fuel, air, or supplies. A typical nuclear attack submarine is about 350 feet (107 m) in length, 33 feet (10 m) wide, and displaces

7,000 tons or more submerged. An Ohio class SSBN is more than 500 feet (152) in length, 37 feet (11 m) wide, and displaces nearly 18,000 tons. On patrols, they can attain depths in excess of 800 feet (244 m) at speeds in excess of 25 knots (12.86 m/s). Many countries employ submarines for coastal defense. These submarines are much smaller, slower, use diesel/electric propulsion, and have a short range for missions close to their homes. Large batteries allow these submarines, once on station, to loiter silently for several days, protecting their coasts from naval attack. Unlike surface ships, submarines generally operate independently. Even when operating in support of a surface or amphibious task force, submarines must be provided safe areas in which to operate so as not to risk collision with another friendly submarine or be confused as the enemy. Submarines operate throughout the world's oceans, including under the Arctic ice cap. Successful submarine operations depend a great deal on marine science, from engineering problems such as building a ship to be stable (self-righting) both afloat and submerged, to tactical problems such as using principles of oceanography (acoustics, biologics, CURRENTS, BATHYMETRY) to give friendly submarines the advantage in hide-and-seek operations. Although not a combatant vessel, the U.S. Navy also uses nuclear submarines such as NR 1 to collect environmental data such as water depth, currents, salinity, TEMPERATURE, SEA ICE, and so forth. The data collected is used to update oceanographic conditions around the world in anticipation of future operations and for scientific studies. The NR 1 was used to search for, identify, and recover critical parts of the Space Shuttle *Challenger,* which exploded 73 seconds after liftoff on January 28, 1986.

AMPHIBIOUS WARFARE ships such as general and multipurpose amphibious assault ships are capable of long duration operations on the high seas in support of expeditionary operations by the U.S. Marine Corps. New strategies to project power in the world's littorals combine amphibious assault ships, several surface combatants, and an attack submarine to form an Expeditionary Strike Group.

Combat logistic ships provide underway replenishment, while mine warfare ships are primarily responsible for mine countermeasures. Underway transfer of matériel from a logistic ship to a combatant is made either by traditional high-line transfer between the two ships steaming side by side or by helicopter.

Coastal defense ships and patrol combatants have become increasingly important due to the global war on terror. Coastal defense ships like the Cyclone class, formerly used in support of U.S. Navy SEAL operations, focus on patrolling and interdiction. Patrol combatants navigate beyond coastal waters to provide combat operations without external support for up to 48 hours. Some of these ships have been transferred to the Coast Guard to support the homeland security effort.

The process of building and supporting the United States's combatant ships and systems rests with Naval Sea Systems (NAVSEA) Command. One of NAVSEA's functions is exercising oversight on the many specifications and standards that are involved in shipbuilding. Toward this, naval architects and marine engineers from organizations such as the AMERICAN BUREAU OF SHIPPING (ABS) generalize the fitness, safety, and environmental soundness of particular vessels through the survey and classification of marine structures from ships to floating offshore platforms. By law, the ABS provides technical standards for the construction of public ships to include naval vessels. These classifications along with other U.S. Navy acquisition rules are required to ensure that a vessel is fit for entry into naval service.

commercial fishing Commercial fishing is the business of taking marine resources for pay or purposes of sale, barter, or exchange. Fishermen use distinct craft and equipment to catch usually either demersal or pelagic species. Demersal catches include species that inhabit the seafloor, while pelagics are caught at or near the surface. The captain of a fishing boat is responsible for the fishing operation, from sailing to the fishing grounds, the method of fishing, and finally to the sale of the catch. First mates, boatswains, crew, and deckhands assist the fishing boat captain. Other members of the fishing industry include fish importers, fish receivers, fish processors, fish wholesalers, warehousemen, and AQUACULTURE farmers. Marine fisheries scientists and policy makers help regulate commercial fishing based on decisions from fish stock health and abundance, not the political or economic necessity of having a commercial fishery.

Working fishermen use traditional and modern craft that range from wooden skiffs for inshore operations like clam tonging to ocean super factory trawlers to find, harvest, and process tons of fish. Some of the primary vessels used in commercial fishing include trawlers, seiners, and long liners. Other examples include tuna clippers, drifters, salmon trollers, whale catchers, New England draggers, and lobster boats. Fishing fleets use a variety of techniques to catch fish. The four most common methods are (1) drift net fishing in which miles of nylon, curtainlike nets are suspended from BUOYS in the water to trap fish by their gills; (2) bottom trawling in which a bag-shaped net is towed over the seabed to catch bottom-living fish; (3) purse seine netting in which a bag-shaped net is pulled behind two boats until the two sides of the net meet to enclose the catch; (4) long lining in which miles of buoyed lines are set with baited hooks to catch commercially important finfish.

The men and women who go down to the sea in ships have to endure many environmental hardships, from winds and waves to pollution and harmful algal blooms. Commercial fishing is a high-risk occupation, and one owes a lot to the fishermen who face the hazards of the sea to make its natural resources available. Man-made or anthropogenic factors such as pollution can cause the closure of shellfish beds due to bacteria or toxic PLANKTON. Commercial fishermen suffer from the affects of industrial pollution, such as high concentrations of heavy metals moving up the food chain. Excessive levels of metals like mercury, lead, arsenic, cadmium, aluminum, or nickel are known to be detrimental to people. Therefore, accounts of high mercury levels in shark, swordfish, king mackerel, tilefish, and tuna affects market prices. Often high nutrient levels from nonpoint source pollution can cause increased growth rates of microscropic organisms such as dinoflagellates that leads to dangerous Red Tides or infamous *Pfiesteria piscida* outbreaks. These dinoflagellates are unicellular protists that passively drift in the sea as plankton. Some dinoflagellates have red pigmentation and can bloom or accumulate in such numbers that they form visible patches near the surface of the water called "red tides." The heavily studied toxic dinoflagellate *Pfiesteria piscicida* has been blamed for fish kills in the Albemarle-Pamlico Estuary and Chesapeake Bay.

The National Marine Fisheries Services monitors fishery landings, vessels, fishing gear and other fishery related information and makes these STATISTICS available to the public. These professionals strive to safeguard marine resources within the U.S. exclusive economic zone (EEZ) and are increasingly concerned with bycatch (unintended catch). They have developed new equipment and laws to protect marine mammals, sea turtles, seabirds, and other protected living marine resources from becoming accidentally caught by commercial fishing gear. For example, many countries require that turtle excluder devices or TEDs be installed in each net that is rigged for fishing. The TED is a metal grate fitted into the neck of a shrimp trawl, allowing certain catches such as shrimp to pass through the bars while large animals such as turtles and sharks strike the metal bars and are ejected from the net. These devices have been shown to be very effective for the shrimp industry.

Further Reading
Barker, Rodney. *And the Waters Turned to Blood.* New York: Touchstone Books, 1998.
Greenlaw, Linda. *The Hungry Ocean.* New York: Hyperion, 1999.
Junger, Sebastian. *The Perfect Storm.* New York: W. W. Norton, 1997.
Mangone, G. J. *Concise Marine Almanac.* New York: Van Nostrand Reinhold, 1986.

community   The term *community* in the field of marine science refers to naturally occurring groups of flora and fauna that inhabit and interact on or within seawater. Marine flora includes PHYTO-PLANKTON, ALGAE, and SEA GRASS. Marine fauna includes ZOOPLANKTON, invertebrates, finfish, MA-RINE MAMMALS, and MARINE BIRDS. The habitat is the immediate space where the plants and animals live. Communities may be classified in accordance to dominant species, or physical features. A "mangrove forest community" is defined by dominant species while a "deep sea vent community" is classified in accordance to the area's major physical characteristics. Species are the smallest unit of classification for flora and fauna, which are capable of breeding and producing fertile offspring. The population is the number of individuals of a particular species that live within a defined area. The ecosystem includes all the organisms in a community, the associated energy and nutrient flows, and the nonliving environment within which they interact.

Further Reading
Curtis, Helena. *Biology.* New York: Worth, 1978.

compass   A compass is a NAVIGATION DEVICE that is used aboard ships and in gauges that measure two-dimensional horizontal directions relative to Earth's magnetic North Pole. The two basic types are magnetic and nonmagnetic. Card and lensatic compasses are examples of magnetic compasses that are used for land navigation. The rectangular card compass that is found in most sporting goods stores is flat to sit on top of a map and generally made of clear plastic, with an etched ruler along one edge. The handheld lensatic compass that is commonly used by military ground units is round with a hinged cover, sighting notch, and sighting wire. Gyrocompasses and flux gate compasses are examples of directional references that leave the traditional magnetic mariner's compass as a reliable backup.

Ancient mariners used pieces of magnetic iron ore called "lodestone" to find a line oriented approximately north and south with respect to Earth's magnetic poles. Early compasses were no more complicated than a middle school student's science project. Historic accounts indicate that a serviceable compass contained a sliver of lodestone thrust through a straw that floated in a bowl of water. The technology progressed through modifications and additions such as a line marked on the inner surface of the water container parallel with ship's keel and possibly sights for taking bearings. As innovations were added, a truly reliable mariner's compass was achieved with a lubber line indicating the compass direction of the ships heading and stenciled north, south, east, west, and some intermediate direction markers that are known as the compass rose. Mariner's compasses aboard ships rest in the binnacle, a stand made of brass or nonmagnetic material.

As wooden ships gave way to steel ships, new methods to find north were necessary due to magnetic deviations caused by the vessel structure (e.g., the ship's engine and steel hull). Using principles of pendulums, gyrocompasses were developed that could navigate ships and aircraft. The mechanism behind this navigation device is an electrically spinning gyroscope within gimbals whose axle will always point toward north. Gimbals are mounting rings suspending the gyroscope for free tilt in any direction. Elmer A. Sperry (1860–1930) engineered the first marine gyrocompass in 1911, which was subsequently used by Allied naval forces during World War I. The U.S. Army Air Corps tested an artificial horizon and an aircraft directional gyro during 1929. An interesting 21st-century innovation is the flux gate compass, which directly measures the horizontal component of Earth's magnetic field with a field sensor, usually an inductor mounted to a gimbaled platform. Today's mariners use the electric flux gate compass in combination with the GLOBAL POSITIONING SYSTEM (GPS) to determine the ship's heading. Marine scientists participating in the World Ocean Circulation Experiment during 1992 were the first to develop a portable GPS and flux gate compass data logging system to obtain ship's velocity and ship's heading for the calculation of true wind. Many state-of-the-art meteorological and oceanographic instruments use a flux gate compass as a component to accurately determine directions. Current meters fixed to mooring lines might utilize an internal flux-gate compass to provide heading information, which is used to reference the current direction to magnetic north.

Even today, the proper use of shipboard magnetic compasses requires the skilled mariner to account for influences that make the compass deviate from true or geographic north. The first influence is variation, which is a measurement of how Earth's magnetic field lines vary over the entire Earth's surface (i.e., the Earth's magnetic meridians are not necessarily aligned north-south and their alignment actually varies over time). The second is deviation, the amount that the compass is deflected from the magnetic meridian due to the ship's iron. Some of the ship's magnetic properties are permanent, determined largely by the direction in which the keel was laid. Some of the ship's magnetic field is induced, varying according to the intensity of Earth's magnetic field in which it was induced. Deviation varies as the ship changes course. Also, "degaussing" equipment in the ship (electromagnetic coils designed to disguise the ship's magnetic field within the surrounding Earth magnetic field, and thus avoid magnetically detonated mines) can cause deviations in the ship's magnetic compass. The ship's effects on the magnetic compass are accounted for by careful placement of correctors. The process of correcting for deviation error is called "swinging ship." On a U.S. Navy ship, the navigator and quartermasters will swing the ship through 360 degrees, stopping each 15 degrees to measure the difference between the compass heading and a properly functioning gyrocompass. The data are recorded in a magnetic compass deviation table. The deviation to the magnetic compass resulting from the ship's degaussing currents is neutralized by a procedure called compensation. The remaining deviations caused by the degaussing coils are observed and plotted on the left side of the deviation table.

Further Reading

Bowditch, Nathaniel. *American Practical Navigator.* NIMA Publication Number 9. Bethesda, Md.: National Imagery and Mapping Agency.
Gurney, Alan. *Compass: A Story of Exploration and Innovation.* New York: W. W. Norton, 2004.

compensation depth   The depth at which the oxygen produced by photosynthesis of PHYTOPLANKTON is equal to the oxygen consumed by the respiration of phytoplankton is called the compensation depth. At the compensation depth, the net oxygen production is zero.

The light energy required for photosynthesis is only available in the surface layers of the ocean. In the sea, light intensity is rapidly reduced with increasing depth. Phytoplankton need light energy for photosynthesis and growth. Gross primary productivity is the total amount of organic matter produced as the result of photosynthesis, while net primary productivity is the amount of organic matter produced by photosynthesis, less the amount consumed by cellular respiration. In higher light intensities, phytoplankton produce more oxygen than they consume and there is a net gain in biomass, but as depth increases, photosynthesis decreases to a point where the breakdown of organic material by cellular respiration is equal to the production of organic material and there is no net primary productivity. This is the compensation depth.

If there were no mixing with waters of differing oxygen content, the compensation depth could be estimated by placing phytoplankton cells in a clear bottle. At depths shallower than the compensation depth, there would be a net gain of oxygen over time, and at depths deeper than the compensation depth, there would be a net decrease of oxygen over time.

The matter is complicated by the fact that incident solar radiation changes with the time of day and season of the year. For these reasons, the compensation depth will move up and down. Furthermore, the transparency of the water column will also influence the compensation depth. Light penetrates to a depth up to 500 feet (150 m) in the open ocean, less than 70 feet (20 m) in coastal waters, and even less in estuarine bays.

conductivity-temperature-depth probe (CTD)   A CTD is an instrument that is lowered by a cable and measures the conductivity, TEMPERATURE, and depth of the water as shown in the following figure. It has three probes. One is a ceramic probe that measures the conductivity, a thermistor that measures the temperature, and a pressure gauge that measures the ambient water PRESSURE. The temperature and conductivity are used to compute the water's SALINITY. CTDs can be either self-contained, that is, they store the information internally in memory that is downloaded to a computer when the instrument is brought back to the surface, or the results are transmitted in real-time over the armored electrical cable that lowers it back to a computer on the ship. Small CTDs can be lowered from a boat or from a pier by hand. The CTDs that connect to a cable and are much larger are capable of withstanding pressures of more than 500 atmospheres and can be lowered to the bottom of the ocean.

The sensors of CTDs have been miniaturized over the years so that chains of CTDs can be assembled. These chains can be hundreds of meters in length, and the CTD sensors can be spaced every meter along them. They are then weighted at the bottom and towed through the water. This gives a slice, instantaneously, of the salinity and temperature of

Conductivity-temperature-depth (CTD) instrument is being deployed at sea. In addition to the CTD probe, the frame supports a Niskin water sampling bottle and an acoustic "pinger" for probe location.   *(NOAA)*

CTDs speeded up the process of measuring, made the measurements more accurate, and greatly increased their vertical resolution. CTDs are capable of resolving small changes in the vertical on the order of centimeters. The CTD has revealed the microstructure of the ocean, which was unobservable using previous techniques.

constancy of composition   Constancy of composition is a general rule that the concentrations of the major dissolved ions in seawater can vary from site to site, but their relative proportions remain the same. An ion is an atom or molecule that has acquired a charge by either gaining or losing electrons. Cations are positively charged ions, while anions are negatively charged ions. These dissolved ions reach the sea through sources such as rivers, volcanism, and HYDROTHERMAL VENTS. Salts are formed by chemical reactions between a base and an acid. For example, sodium and chlorine react to form table salt (NaCl). Only six elements account more than 99 percent of these dissolved salts. They are chlorine, sodium, sulfur, magnesium, calcium, and potassium. The ions are respectively chloride (Cl-), sodium (Na+), sulfate (SO_4^{2-}), magnesium (Mg^{2+}), calcium (Ca^{2+}), and potassium (K+).

To understand movement and dilution of seawater, oceanographers need to make SALINITY mea-

the ocean. There is probably no instrument that is used more widely on oceanographic vessels than the CTD. There are also expendable CTDs that can be dropped from aircraft. The air-dropped probes deploy a float that contains a radio transmitter and antenna. When it hits the water, the float and antenna assembly deploys a probe connected by electrical cable to the float electronics that measures the conductivity and temperature profile as it descends. The data are telemetered back to the aircraft via the cable and radio transmitter.

The computed salinity and temperature as a function of depth are used to compute the density of the water column. The density differences between stations separated in distance are used to compute the geostrophic velocity (*see* GEOSTROPHY). For decades, this method of using density to estimate current shear as a function of depth has been used. Before the development of the CTD, NANSEN BOTTLES were used to measure temperature and salinity.

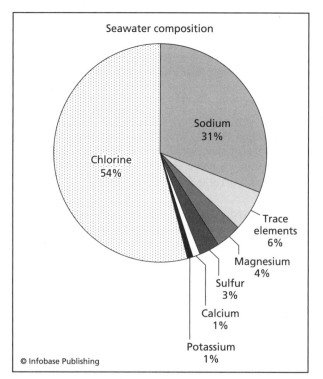

The pie chart shows the approximately constant composition of oceanic water.

The Average Concentrations of the Major Ions in Grams in One Kilogram of Seawater

Chloride	18.980
Sodium	10.556
Sulphate	2.649
Magnesium	1.272
Calcium	0.400
Potassium	0.380
Bicarbonate	0.140

surements. The information might be useful to assess impacts from increases in land runoff, the lowering of salinity locally, and the possible effects on coastal ecosystems. If a chemical oceanographer collected water samples from two different coastal stations, the total salinity might change, but the ratio of the concentrations of major ions to the total remains constant between samples. These major ions and their proportions are listed in the table above.

Salinity is a critical variable for understanding and predicting biological and physical processes. The traditional method to measure salinity is with water bottles that are analyzed in a laboratory with a salinometer. In the field, profiling Conductivity-Temperature-Depth (CTD) gauges are often used or moored *in situ* conductivity-temperature (CT) sensors. In the future, remote sensing using radar signals may provide surface salinity maps to study processes from plume dynamics to the movement of WATER MASSES.

Further Reading

Brown, Evelyn, Angela Colling, Dave Park, John Phillips, Dave Rothery, and John Wright. *Seawater: Its Composition, Properties and Behavior,* 2nd ed. London: Pergamon Press, 1995.

container Truck-sized receptacles for the storage and movement of cargo are described as containers by the transportation industry. Goods carried on a ship and covered by a bill of lading are considered to be cargo. Cargo is categorized as to whether it is roll-on/roll-off (driven on or off by means of a ramp), break-bulk (unpacked homogeneous cargo such as grain), or containerized (goods stowed in a container). Cargo containers come in many varieties in accordance to their design and characteristics. Examples include International Standards Organization (ISO) and Department of Transportation (DOT) shipping containers. More than 6 million cargo con-

tainers are offloaded from vessels onto U.S. docks by the marine transportation system, which consists of waterways, seaports, and their inter-modal connections, vessels, vehicles, and system users, as well as federal maritime navigation systems.

The international shipping community uses standardized rectangular containers for global transport by ship, rail, and highway. The standardized container lengths are 10 feet (3.1 m), 20 feet (6.1 m), 30 feet (9.1 m), and 40 feet (12.2 m), with widths fixed at eight feet (2.4 m) and heights of either eight feet (2.4 m) or eight feet six inches (2.6 m). The vast majority of containers used worldwide today comply with the ISO standard, with 20-feet (6.1m)- and 40-feet (12.2 m)-long containers predominating. These containers incorporate standard corner fittings that accept cone-shaped lugs, so that any container crane can handle almost any container. Twist-locks, devices with conelike shapes at each end, can be inserted into the corner fittings of the container in order to facilitate the loading, positioning, and lashing of containers. These devices, which can be tightened or released with the quick twist of a handle, secure the bottom container to the deck or hatch cover, or lock each container above to the one below. Containers are sometimes griped down to the decks of lighters with chains, wire lashings, and tensioning devices from the container's top corner fittings.

Containers are also manufactured to meet special needs. Tank containers contain a tank for the transport of bulk liquids or gasses surrounded by a framework with the overall dimensions of an ISO container. Dry bulk containers consist of cargo-carrying structures for the carriage of dry solids in bulk without packaging. Thermal containers are built with insulated walls, doors, floor, and roof to safeguard cargo requiring temperature control. Con-

Coast Guard boarding team members guard a container vessel suspected of carrying containers with biohazards. *(U.S. Coast Guard, PA2 Andrew Shinn)*

tainers can even be modified into equipment shelters and laboratories for scientists and engineers. A modified sea container can house the living area, office space, computers, and communications that are necessary to operate meteorological and oceanographic instruments that are deployed away from the container. At the U.S. Army Corps of Engineers Field Research Facility in Duck, North Carolina, there is a modified container housing components of a tidal reference station at the end of the 1,840-foot (560 m) pier that juts out from the dunes, across the surf zone, and into the ATLANTIC OCEAN. Investigators have used this facility to conduct intercomparison studies between various sensors, from ANEMOMETERS to directional WAVE BUOYS.

At the seaport, cranes are used to load or offload sea containers from the ship's hold or deck. The crane lifts the arriving container onto a wheeled vehicle, which carries it to the adjacent container patio, a paved yard. In the yard, machines such as wheeled straddle carriers with lifting frames remove the container from the vehicle and stack it in a specified location pending pick-up. In turn, containers can be retrieved from the stack and transported to transportation modes such as BARGES, ships, trucks, or trains. Containers are transported on public roads atop a container chassis towed by a tractor. Container chassis can either be fixed or adjustable in length.

The extremely large number of containers in use and the difficulty ascertaining what they contain gives rise to significant security concerns. Since the infamous terrorist attacks on the World Trade Center and the Pentagon in the United States on September 11, 2001, new methods have been developed to inspect containers as they enter and leave ports. These methods range from improved tags and seals, to assure that the container has not been tampered with after loading, to X-ray machines and smart sensors for tracking shipments or detecting nuclear materials. Specific rules are applied to the shipment of hazardous materials, which are capable of posing an unreasonable risk to safety and health and property during transport. Despite the large variety of security improvements, however, security controls on containers in transit remains fairly rudimentary.

Ports are maintaining deeper navigational channels to accommodate larger vessels and are upgrading cargo-handling equipment and operational procedures to increase the speed and volume of cargo throughput. Ports are striving to clear containers from the gate in minutes and offload vessels in hours. Automated container terminals are already in operations and use automated guided vehicles that can carry 20-foot (6.1 m) and 40-foot (12.2 m) containers and automated lifting vehicles that can lift and transport a loaded container without using a crane. At the Port of Rotterdam, automated stacking cranes move on rails, take containers with a spreader from an automated guided vehicle, and stack the containers six wide and two or three levels high. The port also operates a state-of-the-art vessel traffic system that can track ships on the radar screen up to 37 miles (60 km) off the coast and 25 miles (40 km) inland. These technological advancements are particularly appealing for implementation at ports with high labor costs.

See also CONTAINER SHIP.

Further Reading

Levinson, Marc. *The Box: How the Shipping Container made the World Smaller and the World Economy Bigger.* Princeton, N.J.: Princeton University Press, 2006.

The United States Mission to the European Union. *Container and Cargo Security.* Available online. URL: http://www.useu.usmission.gov/Dossiers/Cargo_Security. Accessed March 21, 2007.

container ship   Ocean-going merchant vessels designed to carry and stack thousands of containers internally or on deck are called container ships. These ships are in worldwide service, stopping at seaports for one or two days, quickly loading and offloading full or empty containers and sailing on to meet schedules at the next seaport. Commuters crossing BRIDGES over important navigation channels view these large vessels dominated by many multicolored containers in front of and sometimes behind the small superstructure containing the high technology bridge. Navigation and command functions occur on the ship's bridge.

Containers were used from the 1800s in order to reduce shipping costs, with little success. The use of specialized ships compatible with special container-handling equipment is the result of American design innovations that were demonstrated during 1956, when container service was established between Newark, New Jersey, and Houston, Texas. Specialized types of container ships have also evolved such as the Lighter Aboard Ship (also known as LASH) and Offshore Sea-Barge (also known as SeaBee), which carry floating containers or "lighters," (Roll on/Roll off) RoRo ships, which may carry containers on truck trailers, and other semi-container ships, which utilize onboard cranes and deck gear to handle containers and break bulk cargo. Naval architects are working to develop larger and faster, i.e., high-speed container ships to fulfill growing marine transportation needs.

Containerization is an efficient concept for the unitizing of cargo that is embarked aboard ships. A standard 40-foot (12.2 m) container can hold up to 20 metric tons of cargo and is designed to give maximum protection against weather, damage, and theft. One measure of container ship size is the number of 20-foot equivalent units or TEUs that can be stowed on board. A 40-foot (12.2 m) container would be equivalent to two TEUs. Most container ships carry several thousand TEUs. Another measure is to classify container ships by their capacity, where a first-generation ship would carry approximately 1,000 TEU, second-generation 2,000 TEU, third-generation 3,000 TEU, and a super container ship such as the *Regina Maersk* would be a sixth-generation capable of hauling 6,000 TEUs.

Container ships are usually built for speed, so that cargo can arrive at the proper destination fast. These ships are often fitted with bow-thrusters to permit side-to-side bow movement to shorten the time for maneuvering and docking. The sea containers have standard dimensions, and the corners are designed to be interlocking. The individual containers are usually either 20 feet (6.1 m) or 40 feet (12.2 m) in length, eight feet (2.4 m) wide and eight feet (2.4 m) high. Containers are stuffed with goods for export, moved to the seaport, and staged. During embarkation, containers are lowered, slid on special guides, and stacked in the ship's hold. They are also stowed on deck in accordance to weight, using specialized and well-maintained locking and securing equipment. Improperly stacked containers are subject to being lost overboard, especially during bad weather when twist-locks securing the containers shear. Container ships usually carry tens of thousands of deadweight tons. Shippers started using containers in the mid 1950s, following their use in World War II to speedily get material from the United States to Europe. Today, the largest container-shipping carrier in the world is M/V *Maersk Sealand*.

Container ships can be extremely costly, and the economics of container shipping centers on quick unloading and loading in order to maximize revenue operation and minimize port time. The loading and unloading work is extremely fast with the use of pier side gantry cranes and other forms of material-handling equipment. Basically, a straddle-carrier brings the container to the ship so that stacking cranes and huge forklifts can handle the containers. Since the containers have to be lowered, stacked, and the corners matched for locking, it is important to keep the ship steady during the onload or offload. As a result, state-of-the art container ships have special controls such as remotely controlled ballast pumps and valves that are manipulated by deck officers. Researchers are investigating laser applications for the semi-automated loading of containerized cargo into ship holds that are moving at the dock in response to winds, waves, and water level changes.

See also CONTAINER.

Further Reading

Mangone, G. J. *Concise Marine Almanac*. New York: Van Nostrand Reinhold, 1986.

Military Sealift Command Ship Inventory. USNS 2ND LT JOHN P. BOBO (T-AK 3008), Container & Roll-on/Roll-off Ship. Washington Navy Yard. Available online. URL: http://www.msc.navy.mil/inventory/ships.asp?ship=15&type=ContainerRollonRolloffShip. Accessed March 21, 2007.

continental margin Two major subdivisions of the seafloor include first the continental margin comprising the continental shelf and slope, and second the ocean basins, which hold the saline waters of the PACIFIC, ATLANTIC, and INDIAN OCEANS. The continental margin is a boundary between thick continental and thin oceanic crust occurring at the continental slope. The continental margin is further subdivided into Atlantic and Pacific types. The Atlantic type margins are aseismic, and have wide continental shelves and a well-developed continental rise. The Pacific type margins are seismic, have narrow continental shelves, and instead of depositing sediments on a continental rise, they are deposited in a TRENCH such as the 6.9-mile (11 km) deep Mariana Trench. At the trenches, ocean plates are being subducted into Earth's mantle. In turn, new oceanic crust is formed at the mid-ocean ridges, a major physiographic feature cutting across the ocean basins. The trenches are narrow, steep-sided troughs running parallel to the continents. They are normally found on the seaward side of island arcs, while relatively shallow seas exist on the continental side. Island arcs are seismically active, and the Japanese and Philippine Islands provide a good example. Volcanoes in the Philippines are considered to be the most deadly and costly in the world.

See also CONTINENTAL RISE; CONTINENTAL SHELF; PLATE TECTONICS; SUBDUCTION ZONE.

Further Reading

Smith, Jacqueline. *The Facts On File Dictionary of Earth Science*, rev. ed. New York: Facts On File, 2006.

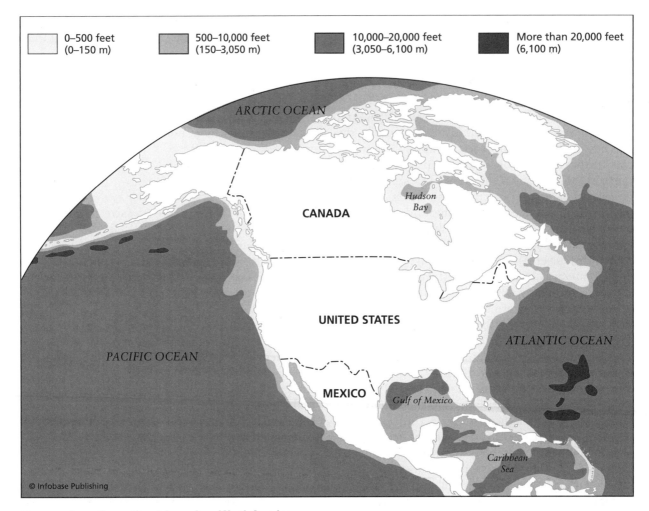

| | 0–500 feet (0–150 m) | | 500–10,000 feet (150–3,050 m) | | 10,000–20,000 feet (3,050–6,100 m) | | More than 20,000 feet (6,100 m) |

The map shows the continental margins of North America.

continental rise   The continental rise is the deepest part of the CONTINENTAL MARGIN (shoreline to the ABYSSAL PLAINS), where the continent begins. The continental rise eases the relatively steep CONTINENTAL SLOPE onto the deep ocean basins or the gently rolling ABYSSAL HILLS. The continental rise begins on the ocean side from about 2.5 miles (4 km) depth, and extends upward to about 1.2 miles (1.9 km) from the surface on the continental side (the continental slope). The continental rise is made up of thick layers of sediment that have been pulled down from the shelf and slope by the force of gravity. Much of the sediments fall downward toward the deep ocean basins through SUBMARINE CANYONS oriented generally perpendicular to the contours of the shelf and slope, often as TURBIDITY CURRENTS. Sediments such as turbidites from these gravity flows accumulate along the sloping boundary between continental and oceanic crust. These turbid or murky currents are very dense because of the large quantities of sediments that they carry and if strong enough disperse

as deep-sea fans on the abyssal plain. The thickness of the sediments may be as much as several thousand feet (km).

Turbidity currents that have sufficient momentum to run far out onto the floor of the abyssal plain create deep sea fans as the sediments spread out, sometimes extending seaward for hundreds of miles (km). They cut into the continental rise and contribute to the development of submarine canyons such as the Baltimore Canyon, a long fingerlike trough located 58 miles (93 km) from the Ocean City inlet in Maryland. The edges of these deep canyons provide good fishing locations for white (*Tetrapterus albidus*) and blue (*Makaira nigricans*) marlin. Populations of these migratory blue water (pelagic) fish are considered to be in decline due to excessive bycatch by commercial longliners targeting species such as yellowfin tuna (*Thunnus albacares*) and swordfish (*Xiphias gladius*).

In summary, the settlement of sedimentary materials builds the continental rise, the physiographic province falling between the steep continental slope and the flat abyssal plain. This sloping region may have accumulations of sediments that are thousands of feet (on the order of kilometers) thick.

Further Reading
Gates, Alexander E., and David Ritchie. *Encyclopedia of Earthquakes and Volcanoes,* 3rd ed. New York: Facts On File, 2006.
Smith, Jacqueline. *The Facts On File Dictionary of Earth Science,* rev. ed. New York: Facts On File, 2006.

continental shelf   A continental shelf is the edge of a continent covered by the sea; it lies between the shoreline and the CONTINENTAL SLOPE. Continental shelves can be wide, as in the case of Atlantic-type passive margins, such as the East Coast of North America, or narrow at Pacific-type or narrow active coasts (tectonically active with many volcanoes and earthquakes, such as the West Coast of North America). The shelf has a small slope extending seaward from the shoreline, and is thus shallow, having a global average depth of 426.5 feet (130 m). At the seaward boundary of the shelf, the continental slope increases from a tenth to a few tenths of a degree downward toward the sea to about four degrees, and then drops down to the CONTINENTAL RISE and thence to the ABYSSAL PLAINS.

Much more is known about the continental shelf because it is the closest seafloor to the continents, with abundant fisheries and rich petroleum and natural gas deposits. It is composed of sediment and sedimentary rock.

The main processes shaping the continental shelves are the erosion and consequent deposition of sediments, especially on Atlantic-type coasts. Tectonic activity is a major process on the Pacific-type coasts; scouring by ice is a major process on Arctic coasts, such as Canada and Russia.

The continental shelves of the world are characterized by high biological productivity, as evidenced by the blue-green hue of their waters, resulting from PHYTOPLANKTON. Sediments from the continents reach the shelf and constantly enrich the productivity of the waters with nitrates, phosphates, and silicates. However, proximity to the continents means that anthropogenically induced pollutant spills, as well as leakage of hydrocarbon fuels, agricultural wastes, medical wastes, and so forth, are often a major problem to environmental managers, fishermen, mariners, recreational boaters, and coastal dwellers.

TIDAL CURRENTS, weather, ice floes, seismic and volcanic activity all play roles in sculpting the floors of the continental shelves. Events such as severe winter storms, HURRICANES, earthquakes, TSUNAMIS, and landslides/avalanches can drastically reshape regions of the shelf and adjacent shores.

Further Reading
Gates, Alexander E., and David Ritchie. *Encyclopedia of Earthquakes and Volcanoes,* 3rd ed. New York: Facts On File, 2006.
Smith, Jacqueline. *The Facts On File Dictionary of Earth Science,* rev. ed. New York: Facts On File, 2006.

continental slope   The continental slope is the transition zone from the CONTINENTAL SHELF to the floor of the CONTINENTAL RISE, and thence, the deep sea. It is the transition region from the land to the ocean that most humans are familiar with; the deep sea is familiar only to mariners and researchers. While the slopes of the continental shelves are usually very small—a few tenths of degrees—the slopes of the continental slopes can range from a few degrees to steep cliffs in some regions. However, the need to compress the horizontal scale of bathymetric charts of ocean basins makes the continental slopes look much steeper than they actually are. The change in slope begins at the shelf break, usually where the depth of the shelf is about 330–660 feet (100–200 m), with 430 feet (130 m) being a global average. The continental slope provides a pathway for the continental sediments to arrive in the deep sea, especially in the regions of submarine canyons where underwater landslides called TURBIDITY CURRENTS can pour tons of sediment down into the abyssal depths.

Seismic activity and scouring during the ice ages and possibly other factors have resulted in these deep submarine canyons' cutting almost perpendicularly through the slope to the deep sea, providing very efficient pathways for sediment transport. Seismic disturbances are the usual cause for the occurrence of the turbidity currents. These currents consist of objects ranging from microscopic organic particles to sand, gravel, and even boulders that go roaring down the canyons and depositing material onto a jumbled mass at the bottom. Such turbidity currents, also known as slumps, can cut submarine telephone and telegraph cables, cause great erosion, and even result in TSUNAMIS, which can devastate nearby coastal areas.

CONTOUR CURRENTS are another cause of erosion and redistribution of sediments on the slope. Contour currents are horizontal currents moving parallel to the depth contours of the slope. They can attain high speeds and cause erosion features oriented perpendicular to the submarine canyons.

Resources on the continental slope will be the next big focus for marine industries, such as oil and gas extraction, marine mineral recovery, and deep technology development. Much more must be learned in order to perform these activities safely and efficiently, with minimal impact to the environment.

The continental slope on the East Coast of the United States is the border between the inshore shelf/slope water and the waters of the GULF STREAM, often known as the North Wall. It is a region in which warm core rings occur frequently, transporting heat and salt shoreward, and sometimes tropical plants and animals. East Coast fishermen are delighted when they can catch mahi mahi, cobia, snappers, and other southern fishes.

Further Reading

Seibold, Eugene, and Wolfgang H. Berger. *The Sea Floor: An Introduction to Marine Geology,* 3rd ed. Berlin: Springer–Verlag, 1996.

contour currents   Contour currents are strong currents that flow horizontally along steep slopes of the CONTINENTAL SHELF and the mid-ocean ridges, usually following surfaces of equal water density. These currents tend to flow around, and not over bottom topography. Physical oceanographers may also refer to contour currents as deep geostrophic currents (*see* GEOSTROPHY). These currents have traditionally been detected by photographic instruments on the ocean floor and through the study of ripple patterns, also from deep-sea photographs. Contour currents sweeping along the base of an escarpment as a result of topographic ROSSBY WAVES derived from the action of currents such as the loop current in the GULF OF MEXICO can pose a real challenge for oil drilling.

The existence of these currents was first suggested by German oceanographers Georg Adolf Otto Wüst (1890–1977) and Albert Defant (1884–1974) through study of observations of the *Meteor* expedition (1925–27). Oceanographers from the WOODS HOLE OCEANOGRAPHIC INSTITUTION measured contour current speeds of 0.3–0.8 knots (0.15–.40 m/s) at 33 feet (10 m) above the bottom during the early 1990s. They found that "abyssal storms" occurred, lasting several days. These "storms" are able to transport the upper layer of sediment. Thus, bathymetric changes such as sediment drifts, channels, erosional unconformities, and sediment waves provide strong evidence for the existence and strength of contour currents. These currents push deep WATER MASSES such as North Atlantic Deep Water and Antarctic Bottom Water along the depths of the ATLANTIC and PACIFIC OCEANS, respectively.

Further Reading

Seibold, Eugen, and Wolfgang H. Berger. *The Sea Floor: An Introduction to Marine Geology,* 3rd ed. Berlin: Springer, 1996.

convection   Convection refers to the movement of a fluid, usually predominantly vertical, that results in the transfer of properties, such as heat or salt. Convective motions are driven by buoyancy forces, which result from instabilities when part of a fluid is heated or cooled, or when salt is added. These conditions result in departures from hydrostatic equilibrium and occur in the atmosphere, in ocean waters, and in the Earth. Hydrostatic equilibrium is the stable condition in which the fluid is in equilibrium with respect to all forces.

The usual example discussed during introductory science classes is that of a pot of soup being heated on a stove. The stove heats the bottom layer of soup, which then becomes less dense than the top layer, and rises to the top. The soup on the top layer, now being denser than that on the bottom due to surface cooling sinks, where it is heated, rises, and the cycle repeats over and over again. The cooling may be improved by blowing on the surface of the soup so that it cools, becomes relatively more dense, and sinks. The circulation can be seen by observing the movement of the vegetables in the soup, which turn over repeatedly in a convection cell.

The familiar example in the atmosphere occurs when the ground is heated on a warm summer day, the heated air above it rises, and the water vapor condenses into a white puffy cumulus cloud. If the heating is strong, and the atmosphere is unstable, the cloud continues to develop, to join with other clouds in rising and thickening, until a towering

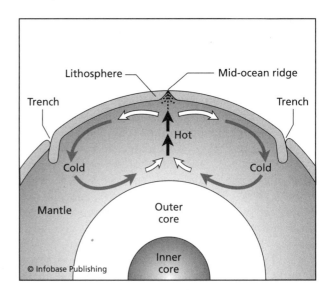

Hot material rises vertically to the top of the mantle where it spreads out and cools. The cool material sinks back through the mantle, where it is heated and starts the cycle again. The driving force is the density differences between lighter hot material and cooler denser material.

cumulonimbus cloud is formed, along with lightning and thunder—the classic summer afternoon thunderstorm.

In the ocean, convection is more complicated because the effects of dissolved salts come into play. Examples of convection in the ocean include the formation of the deepest and most widespread bottom water on the globe: the ANTARCTIC BOTTOM WATER (ABW). ABW forms when freezing and sinking water flows down the Antarctic slope of the WEDDELL SEA. As the water freezes, the freshwater separates out as ice, leaving higher salinity near-freezing water, more dense than the surrounding water, to sink to the bottom and flow outward, filling the world's ocean basins. In the Arctic, a similar process occurs, but the temperatures are not so cold. North Atlantic Deep Water (NADW) is formed in the Norwegian and Greenland Seas when freezing water sinks, again separating out the fresher water that forms ice, while the heavier high-salinity water sinks to the bottom of the basin. NADW spreads southward, flowing over the Antarctic Bottom Water. No bottom water is formed in the North Pacific Ocean/Bering Sea region because the salinity is too low for the cooled water to sink all the way to the bottom.

In mid latitudes, convective overturn occurs in the autumn, when the seasonal THERMOCLINE (strong vertical temperature gradient) breaks down due to autumnal cooling and strong winds that stir the upper layers of the ocean and cause them to mix downward into the cold lower layers, creating a more nearly homogeneous body of water. An example of this process is the forming of the so-called 64.4°F "18°C water" in the Sargasso Sea in winter. The general rule is that surface heating in spring and summer produce a strong thermocline that is strongly stratified. Strong winds and cooling in fall break down the temperature stratification and cause convective overturn and the production of a nearly uniform WATER MASS. This same process occurs in lakes. In ESTUARIES, the effects of SALINITY must always be taken into account; they sometimes override the seasonal temperature effects.

Convection is also important in studies of the geology of Earth. The movement of the lithospheric plates described in the theory of PLATE TECTONICS results from convective heating of the underside of the plates by vertical currents in the mantle. Giant convective cells rise up at spreading ridges and flow downward at subduction zones. Smaller scale convective cells are believed to be responsible for hot spots in the crust, causing formation of island chains, such as the Hawaiian Islands, and earthquakes, such as those that struck Missouri in the 19th century.

Further Reading
Aguado, Edward, and James E. Burt. *Understanding Weather & Climate,* 3rd ed. Upper Saddle River, N.J.: Pearson/Prentice Hall, 2004.
American Meteorological Society. "Glossary of Meteorology," 2nd ed. Available on CD-ROM. Boston, Mass. 2000.
Gates, Alexander E., and David Ritchie. *Encyclopedia of Earthquakes and Volcanoes,* 3rd ed. New York: Facts On File, 2006.

copepods   The copepods are small shrimplike crustaceans found in the phylum Arthropoda. Large antennae and paddlelike swimming appendages generally characterize copepods, and their name literally means "oar footed." They are ZOOPLANKTON that filter feed on PHYTOPLANKTON, such as microalgae and diatoms. However, some species are carnivorous and feed on other zooplankton. There are more than 10,000 species of copepods found throughout the world. In fact, they are the crustaceans with the largest number of species. They are found in all of the oceans, in lakes, and streams. They are very adaptable; certain species are able to adhere to the underside of ice-covered lakes and oceans. Larval copepods go through several juvenile stages called instars. These development stages occur during molting.

Copepods are a major source of protein for prey ranging from juvenile fish to baleen whales. They are the insects of the sea forming a critical link in the food chain between the phytoplankton on which they feed and the krill, fish, and whales that feed on them. Threatened fish species such as the American shad (*Alosa sapidissima*) thrive on planktonic crustaceans such as copepods, mysid shrimp, and euphausids (krill). This ANADROMOUS fish spends most of its life in large schools, entering freshwater rivers in which it was born to spawn. Other filter-feeding fish species that eat copepods include the basking (*Cetorhinus maximus*), megamouth (*Megachasma pelagios*) and whale sharks (*Rhincodon typus*), manta rays (*Manta birostris*), and about nine species of devil rays (genus *Mobula*). Free-swimming copepods are eaten by blue (*Balaenoptera musculus*), bowhead (*Balaena mysticetus*), gray (*Eschrichtius robustus*), humpback (*Megeptera novaeangliae*), minke (*Balaenoptera acutorostrata*), right (*Eubalaena glacialis, Eubalaena australis,* and *Caperea marginata*) and sei (*Balaenoptera borealis*) whales. All of these whales are filter feeders that have baleen, a sievelike device ideal for capturing zooplankton such as copepods. Besides the free-swimming copepods, there are parasitic species that attach themselves to the skin and gills of fish. Two sea lice species (*Lepeophtheirus salmonis* and *Caligus elongatus*) are responsible for commercial losses to the salmon farming industry.

coral animals Coral animals are marine organisms of the phylum Cnidaria made up of individual polyps or colonies of polyps that secrete limestone foundations. After sponges, cnidarians are the simplest form of metazoan life and the first animal phylum with cells organized as tissues. Corals are closely related to sea anemones.

A radially symmetrical polyp is the basic module of coral animals and composed of three basic tissue layers: the outer layer called the epidermis, the inner layer known as the gastrodermis, and between these two the mesoglea, a thin connective tissue made of collagen, mucopolysaccharides, and cells. The lower epidermal layer secretes a calcium carbonate skeleton, and the upper layer is in contact with the seawater, allowing general diffusion of gases and nutrients. Polyps can usually be measured in millimeters and have a mouth circled by a crown of tentacles that leads to an internal space for digestion called the gastrovascular cavity. Polyps rest in a calcium carbonate (limestone) cuplike depression, a calyx, and can be retracted into the calyx for protection from predators and environmental stress. A thin layer of tissue called the coenosarc connects all of the polyps of a colony and rests on the limestone secreted by the corals.

The polyp is the basic module of colonies that can reach meters in diameter. Colonies from different species and colonies of the same species may take on various growth forms, such as plates, bushes, trees, mounds, and pillars. Variations in water temperature, light, hydrodynamic stress, and nutrient concentrations may be causes for varying colony morphologies.

Polyps grow until they reach a certain size, and then they divide, creating two identical polyps (binary fission). Division continues throughout the life of the colony, but it may be as long as eight years before the colony of clones reaches sexual maturity at a certain minimum size (about 10 centimeters diameter). At maturity, small planktonic larvae called planulae are released and swim upward and toward light to enter surface currents. Of planulae studied, most swim to the bottom and settle within two days of release. The remaining species may swim for weeks or months for long-distance dispersal. Once the planulae settle, they metamorphose into a polyp and begin growing and dividing again. Corals can be hermaphroditic or have separate sexes, and while planulae are produced asexually by some species, cross-fertilization is most common.

Zooxanthellae are endosymbiotic dinoflagellate ALGAE that live in the cells of corals and many other reef invertebrates. They accumulate inorganic nutrients from coral waste metabolites and seawater, and the carbon fixed by the algae during photosynthesis is available to coral cells as an energy source. Although the mechanism is not clearly understood, zooxanthellae living within a tropical coral host cell provide the carbon products of photosynthesis to the host and greatly improve the skeletal growth rates of tropical corals. Given adequate conditions, such as warm water, sufficient light, and normal SALINITY levels, corals continue to build upon the calcium carbonate skeleton, and over geologic time, corals form significant limestone structures called coral reefs on Earth's surface.

Corals are noted for their high biological productivity, and this is a function of their many methods of feeding and the constant renewal of inorganic nutrients flowing over the reef. The algal-coral symbiotic relationship allows corals to function as autotrophs and maintain tight nutrient recycling in coral tissue. Corals can also function as suspension-feeding heterotrophs, using tentacles and mucus nets to capture PLANKTON. Food captured by one polyp is shared with others via the coenosarc, a gastrovascular system that circulates and digests food particles. Corals may also use extracoelenteric digestion to feed on moving prey or food particles too large for ingestion. This feeding method involves mesenterial filaments that extend from the mouth or openings in the body wall and digest food particles. Corals can also use extracoelenteric digestion to attack adjacent coral species. Some corals take advantage of dissolved organic matter in seawater by actively transporting it across the cell membrane. The ratio of heterotrophic to autotrophic feeding behavior among corals often depends on species type and local conditions, such as light availability and plankton concentrations.

Corals serve as a food source for some gastropods, crabs, polychaetes, sea urchins, and seastars. *Acanthastar planci*, the crown-of-thorns seastar, is found in the PACIFIC and INDIAN OCEANS and feeds aggressively on live corals. Butterflyfishes depend on coral as a food source, and some damselfishes destroy coral polyps to encourage growth of algal mats, their primary source of nutrition.

Corals, such as species of *Pocillopora*, *Acropora*, and *Porites*, are most commonly known as relatively shallow water organisms restricted to warm or tropical seas ranging from 65° to 97°F (18° to 36°C). These conditions foster the greatest coral diversity and productivity among scleractinian stony corals (hexacorals) corals, but corals can also survive in deep cold-water conditions with little or no light. Coral species grow in the abyssal Antarctic, and *Lophelia pertusa* is a branching coral that grows in the North Sea at depths of 825 feet (250 m) at 70° latitude. Without sufficient light, the algal-coral mechanism responsible for enhanced limestone production does not function well and reefs do not form.

See also CORAL REEF FISH; CORAL REEFS; PHYTOPLANKTON; ZOOXANTHELLAE.

Further Reading

Barnes, Richard S. K., and Roger N. Hughes. *An Introduction to Marine Ecology,* 3rd ed. Oxford: Blackwell Science, 1999.

Carpenter, Robert C. "Invertebrate Predators and Grazers." In *Life and Death of Coral Reefs,* edited by Charles Birkeland. New York: Chapman & Hall, 1997.

Hixon, Mark A. "Effects of Fishes on Corals and Algae." In *Life and Death of Coral Reefs,* edited by Charles Birkeland. New York: Chapman & Hall, 1997.

Kaandorp, Jaap A., and Janet E. Kübler. *The Algorithmic Beauty of Seaweeds, Sponges, and Corals.* Berlin: Springer-Verlag, 2001.

Muller-Parker, Gisèle, and Christopher F. D'Elia. "Interactions Between Corals and Their Symbiotic Algae." In *Life and Death of Coral Reefs,* edited by Charles Birkeland. New York: Chapman & Hall, 1997.

coral bleaching   Coral bleaching is the separation of photosynthetic endosymbiotic zooxanthellae from a coral colony, resulting in the colony turning white (bleaching). Corals are marine organisms of the phylum Cnidaria, consisting of individual polyps or colonies of polyps. Zooxanthellae are typically unicellular dinoflagellates that live in the cells of many reef organisms.

Zooxanthellae contribute to the diverse and spectacular array of coral colors. They carry chlorophyll *a* and *c* and characteristic dinoflagellate pigments diadinoxanthin and peridinin that give corals a brown, blue, or yellow-green color.

There is evidence that environmental stresses, such as pollutants, high water temperatures, air exposure, and increased ultraviolet radiation are the major causes of bleaching events. Whether coral force out the ALGAE or algae leave coral colonies voluntarily is uncertain. While the true motivation is unknown, researchers suggest that nutrient scarcity may cause coral to eject the algae or the algae exit coral cells to escape the environmental stress.

Coral bleaching events often correspond to high water TEMPERATURES during EL NIÑO events and have been recorded in virtually all coral reef ecosystems, including Central America, the Indo-Pacific, the RED SEA, and the Caribbean. The most severe observed global bleaching event to date took place in 1998 during a particularly fierce El Niño that affected almost all coral species and associated invertebrate species.

Coral bleaching does not always result in coral mortality; some coral species are more resistant to stress than others. If the host survives the stress event, zooxanthellae may repopulate the host cells and resume growth and photosynthesis.

See also CORAL REEF ECOLOGY; CORAL REEFS; ZOOXANTHELLAE.

Further Reading

Brown, Barbara E. "Disturbances to Reefs in Recent Times." In *Life and Death of Coral Reefs,* edited by Charles Birkeland. New York: Chapman & Hall, 1997.

Sorokin, Yuri I. *Coral Reef Ecology.* Berlin: Springer-Verlag, 1993.

coral reef ecology   The study of the interrelationships of biotic (living) and abiotic (nonliving) components of marine limestone skeletal structures is referred to as coral reef ecology. Coral reefs are ancient geological structures created by coral animals and other organisms that pull carbonates out of seawater to build exoskeletons, shells, and other body parts that are the basis of the coral reef ecosystem.

Reeflike structures have existed for nearly 3.5 billion years and were initially created by organisms, such as ALGAE, sponges, and bryozoans, capable of creating a rigid skeleton and anchoring to the seafloor. Not until the Ordovician (about 450 million years ago), while algae and sponges were still the primary reef-builders, did the first ancient hard corals appear (tabulates and rugoses). The Mesozoic (225 million years ago) marks the existence of the first coral reefs, and some modern reefs may be more than 2.5 million years old.

During the Neogene (23 million years ago to present) formation and melting of glaciers caused sea

Repair of a coral reef by a NOAA scientist-diver   *(NOAA)*

level fluctuations of up to 450 feet (140 m) that resulted in cyclical exposure and covering of CONTINENTAL SHELVES and reefs. About 18,000 years ago, 2-million-year-old Pleistocene glaciers began to melt, and sea levels began to rise. Sea level reached the exposed Pleistocene reef platforms about 9,000 years ago, and growth of modern reef structures began.

Sea level and tectonic changes over geologic time resulted in several geomorphological categories of modern reefs, such as ATOLLS, fringing reefs, and BARRIER REEFS. Fringing reefs are formed close to shore on rocky coastlines. If sea level is stable or carbonate production is high, the reef will grow seaward. As land subsides or sea level rises over geologic time at a high rate relative to carbonate production, a fringing reef will grow vertically to maintain proximity to the sea surface. This process results in a barrier reef and a LAGOON between the reef and the shore. As a volcanic island subsides below sea level, the barrier reef that once surrounded the island will continue to grow vertically and eventually form an annular reef called an atoll.

Water TEMPERATURE, clarity, SALINITY, and geological foundations are limiting factors in coral reef development. Coral reefs are generally limited to nutrient-poor tropical waters with year-round temperatures above 68°F (20°C). The light spectrum shifts toward blue as light intensity decreases exponentially with depth. Coral growth rate decreases with light intensity, so hermatypic (reef forming) corals are usually limited to within 230 feet (70 m) of the surface. For instance, while the GULF OF MEXICO has relatively warm waters, poor water clarity and a sand-and-sediment bottom restrict significant reef development.

Coral reef biological and physical structure is primarily a function of hydrodynamic stress on COMMUNITIES of sedentary hermatypic organisms. The most important hydrophysical factors affecting reef organisms, growth, and form are CURRENTS, WAVES, TIDES, and the THERMOHALINE structure of the water column. Tides exert regular and powerful forces that shape reef construction and affect distribution of benthic communities. Because wind and current directions are often directionally constant in the Tropics, coral reefs have many common geomorphological elements, such as reef flat, lagoon, and buttress system.

The characteristics of the reef crest and slope are a function of whether the reef is facing the wind (windward) or sheltered from the wind (leeward). The reef crest and frontal edge of windward reefs are especially exposed to wave action and usually feature a spur-and-groove system to dissipate wave energy that could be as much as 500,000 horse-power (373,000 kW). Alternating spurs (buttresses) and grooves (channels) are usually a few meters wide, and length depends on the slope of the reef's frontal edge. Grooves, where wave energy is concentrated and coral growth is limited, decrease wave stress on overall reef construction and biota. Because most energy is flowing up and down the channels, the buttresses are less turbulent and support growth that further reduces wave energy. High wave energy accompanied by storm disturbances breaks branching coral colonies, such as *Acropora palmata*, and forms a boulder zone on the reef slope. Leeward reefs are subject to significantly less wave energy. They lack the spur-and-groove system, and the slope of the frontal edge is relatively steep. Large open flats with intermittent coral cover may also characterize low wave energy regions.

A coral reef is essentially a thin layer of living tissue covering a limestone structure secreted by corals and other calcareous organisms over thousands or millions of years. Corals are animals closely related to sea anemones that produce a calcium carbonate skeleton for support and protection. Coral polyps are typically only millimeters in diameter and have a mouth circled by a crown of tentacles that leads to a simple stomach. Polyps rest in a cuplike depression, a calyx, and can be retracted into the calyx for protection from predators or environmental stress. Scleractinian corals (stony corals) are colonies of polyps, which can be meters in diameter, connected by a common gastrovascular system (the coenosarc) that covers the limestone surface between calyces.

Calcareous algae, foraminiferans, calcareous hydrozoans, molluscs, and bryozoans are examples of other construction units in the coral reef system. These reef builders may dominate reefs when conditions are not optimal for the zooxanthellae-algal limestone production mechanism.

Polyps grow until they reach a certain size, and then they divide, creating two identical polyps (binary fission). Division continues throughout the life of the colony, but it may be as long as eight years before the colony of clones reaches sexual maturity at a certain minimum size (about 3.9 inches [10 cm] diameter). At maturity, small planktonic larvae called planulae are released and swim upward and toward light to enter surface currents. Of planulae studied, most swim to the bottom and settle within two days of release, while remaining species may swim for weeks or months for long-distance dispersal. Once the planulae settle, they metamorphose into a polyp and begin growing and dividing again. Corals can be hermaphroditic or have separate sexes, and while planulae are produced asexually by some species, cross-fertilization is most common.

Corals are noted for their high biological productivity, and this is a function of their many methods of feeding and the constant renewal of inorganic nutrients flowing over the reef. The algal-coral symbiotic relationship allows corals to function as autotrophs and maintain tight nutrient recycling in coral tissue. Corals can also function as suspension feeding heterotrophs, using tentacles and mucus nets to capture PLANKTON. Food captured by one polyp is shared with others via the coenosarc, a gastrovascular system that circulates and digests food particles. Corals may also use extracoelenteric digestion to feed on moving prey or food particles too large for ingestion. This feeding method involves mesenterial filaments that extend from the mouth or openings in the body wall and digest food particles. Corals can also use extracoelenteric digestion to attack adjacent coral species. Some corals take advantage of dissolved organic matter in seawater by actively transporting it across the cell membrane. The ratio of heterotrophic to autotrophic feeding behavior among corals often depends on species type and local conditions, such as light availability and plankton concentrations.

Zooxanthellae are endosymbiotic dinoflagellate algae that live in the cells of corals and many other reef invertebrates. They accumulate inorganic nutrients from coral waste metabolites and seawater, and the carbon fixed by the algae during photosynthesis is available to coral cells as an energy source. Although the mechanism is not clearly understood, zooxanthellae living within a tropical coral host cell provide the carbon products of photosynthesis to the host and greatly improve the skeletal growth rates of tropical corals. Given adequate conditions, such as warm water, sufficient light, and normal SALINITY levels, corals continue to build upon the calcium carbonate skeleton, and over geologic time, corals form significant limestone structures called coral reefs on Earth's surface.

Of all marine systems, coral reefs are generally regarded to hold to greatest diversity of life on a per-unit-area basis. More than 60,000 reef species have been described, and this is a small percentage of the total. Nearly one-third (more than 4,000) of all modern bony fish species live on coral reefs. Reef ecologists have described 5,000 species of sponge in Australia and 6,500 species of marine mollusc in New Caledonia. Of the 1,400 known scleractinian stony corals, 800 species are associated with coral reefs.

Relatively constant abiotic conditions, such as currents, wave action, temperatures, nutrient availability, and efficient nutrient recycling within the reef system are factors that support high biodiversity and biomass levels. The structural complexity of coral reefs caused by hydrodynamic stress fosters niche diversity and leads to highly specialized organism-habitat associations and greater biodiversity.

Bioeroders are organisms that, through various activities, erode and weaken skeletons of reef-building species, creating diverse habitats for the coral reef community. They are typically small, and because they may spend their lives hiding or feeding within the reef, they are not easily observed. Bioeroder mass and numbers within the reef may be greater than reef surface biota. They shape reef growth and contribute to sediments (rubble, sand) within and surrounding the reef. Intense predation on bioeroders and competition for space has led to extensive diversification within this group of reef biota. Bioeroders can be classified into grazers (sea urchins, parrotfish, surgeonfish, filefish and other fish with chisel or beaklike teeth), etchers (bacteria, fungi, and filamentous endolithic algae), and borers (clionid sponges, bivalves, spionid polychaetes, and sipunculans (peanut worms)).

Diseases can have a major effect on coral reef health when pathogens, parasites, pollution, and habitat degradation stress corals and make them more susceptible to infection. Natural defenses against disease include mucus secretion, antibiotics, chemicals that deter parasites, a protective epidermis, and cells that will surround potential threats and produce chemicals that destroy them. Black-band disease (Caribbean and Indo-Pacific) and white-band disease (Caribbean and Red Sea and Philippines) are examples of diseases that can have devastating effects on coral reef ecosystems. Reef plants and animals are also susceptible to bacteria, viruses, parasites, and other pathogens.

Much like tropical rain forests, coral reefs are mature and biologically diverse ecosystems that are sensitive to natural and anthropogenic influences. Tidal emersion, exposure of coral reef flats and reef crests, may result in exposure to excessive light, cold, or rain that causes corals and zooxanthellae to separate. Storms, such as HURRICANES and tropical storms, are the environmental factor that causes the most significant damage to coral reefs with heavy rains that decrease salinity and powerful waves and currents that break and dislodge corals while stirring up sediments that abrade coral surfaces.

Increasing rates of population growth and consumption, especially in tropical coastal nations, are putting greater pressure on coral reef ecosystems, and coral reef ecologists are working with policy makers, economists, and social scientists to cope with the diversity and intensity of threats to one of the most ecologically valuable systems on Earth.

Ecologists assist fisheries managers to develop management plans that maximize fishing yield while maintaining sufficient stocks of breeding individuals. Marine protected areas are management tools that coordinate local communities, government, and science to protect threatened habitats and species. CITES, the Convention on International Trade in Endangered Species of Wild Flora and Fauna, is an international policy tool that regulates international illegal wildlife trafficking and depends on scientists, such as ecologists, for information on threatened reef species.

Coral reef ecology is playing a major role in the attempt to discover and understand the threats to coral reefs and their value to humans. The interrelationships within the reef system are extremely complex and dynamic, so adaptive and innovative management techniques will be necessary to sustain and improve the health of coral reef ecosystems worldwide.

See also CORAL ANIMALS; CORAL REEFS; ZOOXANTHELLAE.

Further Reading

Barnes, Richard S. K., and Roger N. Hughes. *An Introduction to Marine Ecology,* 3rd ed. Oxford: Blackwell Science, 1999.

Bryant, Dirk, Lauretta Burke, John McManus, and Mark Spalding. *Reefs at Risk.* Washington, D.C.: World Resources Institute, 1998.

Hallock, Pamela. "Reefs and Reef Limestones in Earth History." In *Life and Death of Coral Reefs,* edited by Charles Birkeland. New York: Chapman & Hall, 1997.

Hubbard, Dennis K. "Reefs as Dynamic Systems." In *Life and Death of Coral Reefs,* edited by Charles Birkeland. New York: Chapman & Hall, 1997.

Muller-Parker, Gisèle, and Christopher F. D'Elia. "Interactions Between Corals and Their Symbiotic Algae." In *Life and Death of Coral Reefs,* edited by Charles Birkeland. New York: Chapman & Hall, 1997.

Sorokin, Yuri I. *Coral Reef Ecology.* Berlin: Springer-Verlag, 1993.

coral reef fish   Coral reef fish are species that spend at least a portion of their life cycle on or near a coral reef, using reef resources. Coral reefs are highly productive tropical marine ecosystems that support high fish biodiversity and high fish biomass. Colorful fish captivate divers and snorkelers who are studying the coral reef.

High biodiversity among coral reef fish is the result of the long-term coevolution between reefs and reef organisms. As scleractinian stony corals diversified and developed more intricate reef constructions, bony fish differentiated and adapted to live in more specialized habitats and exploit an increasing variety of food sources. The Eocene (about 50 million years ago) marked a time of major reef fish diversification. Holocentridae, Gobiidae, and Chaetodontidae represent some of the oldest modern reef fish families, while Antennariidae, Bothidae, and Synodontidae have members that were pelagic until recently adapting to live in the reef environment.

Modern coral reef fish are small, with an average size of five centimeters, and classified in one of more than 100 different families. Most of these are bony fishes, but a small fraction is cartilaginous, such as the sharks and rays. A few of the major families are damselfishes and anemonefishes (Pomacentridae), wrasses (Labridae), angelfishes (Pomacanthidae), and grunts (Haemulidae). Nearly one-third (over 4,000) of all modern bony fish species live on coral reefs. The greatest species diversity is found in the Philippines, Papua New Guinea, and the Great Barrier Reef in Australia

Coral reefs are widespread throughout the Tropics, and most coral reef fish species have a broad geographic distribution compared to temperate reef species. In contrast to many nonreef fish, local communities of reef fish are commonly associated with a particular type of reef construction and reef zone and do not typically mix with neighboring COMMUNITIES. Fish associated with coral reefs are an integral link in the reef ecosystem food web and have developed many highly specialized relationships with other reef organisms, such as predator-prey, commensalisms, and symbioses.

Most reef fish have daily (diurnal) behavior patterns that consist of a period of active feeding and a period of rest. Efficient space usage is necessary on a reef, and fish often alternate resting in the same hiding places depending on if they feed during the day or at night. Vertical migrations and migrations between reef biotopes, regions of uniform environmental and biological conditions, are other examples of reef fish diurnal behavior that has developed to support feeding behavior. During migrations between biotopes, reef fish visit cleaning stations where cleaner fish (wrasses of the genus *Labroides*) feed on parasites attached to the gills and skin of migrating fish. Cleaning stations are abundant on healthy reefs, and while reef fish of all kinds line up at the cleaning stations, they are not known to feed on the cleaners.

Besides maintaining close proximity to shelter and feeding when predation is less likely, reef fish avoid predation by schooling. Light reflecting off schools of moving fish disorients predators and makes it difficult for them to isolate individual prey for attack.

Reef fish coloration is often fantastic and arguably the most beautiful and diverse in the animal kingdom. Coloration is a form of communication among reef fish. Colors and patterns enable fish to recognize their own species in turbid waters and against colorful coral reefs. Color may also serve to camouflage fish, confuse predators (false eye), or warn predators that they are toxic to eat. The colors of males are often more vibrant at breeding times, and males displaying during mating use the colors to attract the attention of females.

Symbiotic social behaviors are common among coral reef fish and other organisms on the reef. The relationship between clown fish and anemones (actinians) is one of the most well known examples. Clown fish families (genus *Amphiprion* and *Premnas*) live safely within the stinging tentacles of the anemone and eat the mucus and leftovers of the anemone. In turn, the fish defend the anemone from potential predators.

Because of high fish density and very specialized relationships between fish and respective habitats, reef space is limited and actively protected. Many species of reef fish have developed territorial behaviors to protect areas of refuge and food source. If space is invaded, fish may spread their fins and gill covers to intimidate intruders. Grunts (family Haemulidae) signal the boundaries of their territory to predators by making grunting sounds. Damselfishes that feed on ALGAE on the reef bed actively fend off other organisms that may consume the algae within their territory.

Compared to the trophic structure of fish species of temperate coasts, a very high percentage of coral reef fish are herbivores (15–50 percent of total fish biomass), and there could be as many as 10,000 herbivorous fish per 100 acres (1 ha). An abundance of herbivory is possible among reef fish because high temperatures improve the function of enzymes involved in plant material digestion, and on rocky surfaces or dead corals there is a profusion of easily digestible edible algae and seaweed. Herbivorous feeding behavior shortens the food chain and increases the efficiency of energy use at higher trophic levels. Herbivorous fish can be separated into browsers, grazers, and suckers. Browsers eat macrophytes and SEA GRASS; grazers feed on periphytonic turfs and coralline algae crusts; and suckers eat the thin organic layer off the surface of soft benthic sediments.

Plankton is constantly swept over the reef by CURRENTS and TIDES; subsequently, the abundance of planktonovores, such as damselfishes, feeding on ZOOPLANKTON is high. PLANKTON is also an important food source for cardinals, bigeyes, and soldierfishes. By feeding on passing plankton, these fish indirectly introduce energy and nutrients to the reef community. These fish are typically smaller than the average reef fish and have streamlined bodies, forked tails, and jaws that can quickly extend to catch plankton. They school during the day in LAGOONS or along the reef edge, while separating at night to feed in the water column.

The greatest percentage (40–80 percent) of reef fish species feed on zoobenthos, but total fish biomass (20–30 percent) is less than other groups because species populations are smaller and feeding specialization is greater. Goby species may dominate this category of predatory fish with more than 1,000 representatives. Examples of prey include sponges, hydroids, gastropods, crabs, polychaetes, and holothurians. Some species of butterflyfish feed primarily on corals.

The guild of fish that feed on other fish (piscivores) comprises 10 to 20 percent of reef fish species. Piscivores such as groupers, moray eels, barracudas, and reef sharks are important in controlling the density and composition of reef fish communities.

As an increasing number of fish have become specialized and adapted to living in a reef environment, various mechanisms developed to improve successful propagation. Depending on the species, days or months may separate spawning events and they may take place in the water column or on the seafloor. Permanent monogamy, as shown in some gobies, butterflyfishes, and damsels, improves breeding success and the ability to protect territory. Hermaphroditism and the ability of adult fish to change sex is common among reef fish, such as gobies, groupers, wrasses, and parrot fish, and enables fish to even out the ratio of males to females during spawning. Other ways reef fish ensure breeding success include actively protecting eggs and larvae and choosing inconspicuous spawning sites.

After larvae hatch, they remain part of the plankton for 10 to 100 days. Larvae are usually in the open ocean away from reefs and predators. They are often transparent and look very different from the adult form. Those that survive the open ocean arrive at reefs or nearby sea-grass beds as larvae or developing juveniles. Severe predation threatens the fish for the first few weeks until they become familiar with the new surroundings.

The standing stocks of coral reef fishes are estimated to be 30 to 40 times greater than stocks in most temperate region demersal fishing grounds. In the Caribbean alone, there are approximately 60,000 small-scale reef-fishing operations harvesting about 180 different species. In the western Pacific, fisheries have been estimated to collect 200 to 300 different species in quantities of 100 to 240 metric tons per year per square kilometer. The live reef fish food (primarily grouper species) business is worth nearly $1 billion annually, and the aquarium fish trade is worth more than $200 million a year.

Reef fish require healthy coral reefs for survival, and coral reefs are very sensitive to environmental changes. Destructive fishing practices, such as dynamite and cyanide fishing, are used in many countries to stun fish for easy collection, but these methods of fishing leave behind dead coral skeletons that are unable to support the reef ecosystem. Poorly regulated tourism, pollution, agricultural runoff, coastal development, and fisheries as well as changes in ocean temperatures and sea levels associated with climate change can have detrimental effects on reef health. Improvements in fisheries management, better understanding of the resource, more effective marine protected area implementation and management, and shifts in consumer behavior are necessary to ensure the survival of current reef fish biodiversity and biomass levels.

See also CORAL REEF ECOLOGY; CORAL REEFS; FISH; REEF FISH.

Further Reading

Birkeland, Charles. "Introduction." In *Life and Death of Coral Reefs,* edited by Charles Birkeland. New York: Chapman & Hall, 1997.

Choat, J. Howard, and David R. Bellwood. "Reef Fishes: Their History and Evolution." In *Ecology of Fishes on Coral Reefs,* edited by Peter F. Sale. San Diego, Calif.: Academic Press. 1991.

Hawaii Coral Reef Initiative, University of Hawaii at Manoa, Social Science Research Institute, Honolulu, Hawaii. Available online. URL: http://www.hawaii.edu/ssri/hcri/about/index.html. Accessed February 29, 2008.

Hawaii Institute of Marine Biology at Coconut Island, Kane'ohe, Hawaii. Available online. URL: http://www.hawaii.edu/HIMB/index.html. Accessed February 29, 2008.

Hixon, Mark A. "Effects of Reef Fishes on Corals and Algae." In *Life and Death of Coral Reefs,* edited by Charles Birkeland. New York: Chapman & Hall, 1997.

Pitkin, Linda. *Coral Fish.* Washington, D.C.: Smithsonian Institution Press, 2001.

Sorokin, Yuri I. *Coral Reef Ecology.* Berlin: Springer-Verlag, 1993.

Witte, Astrid, and Casey Mahaney. *Hawaiian Reef Fish.* Waipahu, Hawaii: Island Heritage, 2006.

coral reefs   Coral reefs are ancient geological structures created by coral animals and other organisms that produce a solid substrate. Coral reefs surround approximately one-sixth of the world's tropical coast-

Undamaged coral reefs, as in this photograph, are increasingly rare. Many reef resources are in need of protection due to exploitation. *(NOAA)*

lines and produce between 400 and 2,000 tons of limestone per 100 acres (1 ha) annually. While shallow living coral reefs are estimated to cover only .2 percent of the world's ocean area (about 231,661 square miles [600,000 km²]); they are one of the most biologically diverse systems on Earth.

Of all marine systems, coral reefs are generally regarded to hold the greatest diversity of life per-unit-area. Well over 60,000 reef species have been described, and this is a small percentage of the total. Relatively constant abiotic conditions, such as currents, wave action, temperatures, nutrient availability, and efficient nutrient recycling within the reef system are factors that support high biodiversity and biomass levels. The structural complexity of coral reefs caused by hydrodynamic stress fosters niche diversity and leads to highly specialized organism-habitat associations and greater biodiversity.

Reeflike structures have existed for nearly 3.5 billion years and were initially created by organisms, such as ALGAE, sponges, and bryozoans, capable of creating a rigid skeleton and anchoring to the sea-floor. Not until the Ordovician (about 450 million years ago), while algae and sponges were still the primary reef-builders, did the first ancient hard corals appear (tabulates and rugoses). The Mesozoic (225 million years ago) marks the existence of the first coral reefs, and some modern reefs may be more than 2.5 million years old.

During the Neogene (23 million years ago to present) formation and melting of glaciers caused sea level fluctuations of up to 450 feet (140 m) that resulted in cyclical exposure and covering of CONTI-NENTAL SHELVES and reefs. About 18,000 years ago, 2-million-year-old Pleistocene glaciers began to melt and sea levels began to rise. Sea level reached the exposed Pleistocene reef platforms about 9,000 years ago, and growth of modern reef structures began.

Sea level and tectonic changes over geologic time resulted in several geomorphological categories of modern reefs, such as atolls, fringing reefs, and barrier reefs. Fringing reefs are formed close to shore on rocky coastlines. If sea level is stable or carbonate production is high, the reef will grow seaward. As land subsides or sea level rises over geologic time at a high rate relative to carbonate production, a fringing reef will grow vertically to maintain proximity to the sea surface. This process results in a barrier reef and a LAGOON between the reef and the shore. As a volcanic island subsides below sea level, the barrier reef that once surrounded the island will continue to grow vertically and eventually form an annular reef called an atoll.

Living coral reefs are confined to water temperatures between 64°F (18°C) and 96°F (36°C) and perform best in the middle of that range. Growth rates differ between coral species, but most exist near their upper temperature limits and are very sensitive to sudden changes. SALINITY outside normal marine levels (33 to 36 psu), such as locations where major rivers enter the ocean, may also restrict or prevent coral reef development.

Wave energy is a major factor affecting the form and structure of coral reefs. Coral reef biological and physical structure is primarily a function of hydrodynamic stress on COMMUNITIES of sedentary hermatypic (reef forming) organisms. The most important hydrophysical factors affecting reef growth and form are CURRENTS, WAVES, TIDES, and the THERMOHALINE structure of the water column. Tides exert regular and powerful forces that shape reef construction. Because wind and current directions are often directionally constant in the Tropics, coral reefs have many common geomorphological elements, such as reef flats, lagoons, and buttress systems.

The characteristics of the reef crest and slope are a function of whether the reef is facing the wind (windward) or sheltered from the wind (leeward). The reef crest and frontal edge of windward reefs are especially exposed to wave action and usually feature a "spur-and-groove" system to dissipate wave energy that could be as much as 500,000 horsepower (373,000 kW). Alternating spurs (buttresses) and grooves (channels) are usually a few meters wide, and length depends on the slope of the reef's frontal edge. The spur and groove system serves to dissipate wave energy and encourage coral growth in less turbulent areas. Leeward reefs are subject to significantly less wave energy. They lack the spur-and-groove system, and the slope of the frontal edge is relatively steep. Large open flats with intermittent coral cover may also be associated with low wave energy regions.

As the depth of clear water increases, light penetration decreases exponentially and the spectrum shifts rapidly toward shorter wavelengths (blue). ZOOXANTHELLAE are algae that live symbiotically with coral cells and assist in limestone production. The algae require sufficient light for photosynthesis, thus limestone production by coral is indirectly related to water depth.

Coral reef ecosystems efficiently recycle nutrients in oligotrophic (nutrient-poor) waters. Currents and tides carrying PLANKTON and other particulate matter over coral reefs are a source of constant nutrient renewal. High nutrient levels may damage reef health by inhibiting the zooxanthellae-coral limestone production mechanism or encouraging resource competition from other reef organisms.

A coral reef is essentially a thin layer of living tissue covering a limestone structure that was secreted by coral animals and other calcareous organisms over thousands or millions of years. Coral animals are closely related to sea anemones and produce a calcium carbonate skeleton for support and protection. Coral polyps are typically only millimeters in diameter and have a mouth circled by a crown of tentacles that leads to a simple stomach. Polyps rest in a cuplike depression, a calyx, and can be retracted into the calyx for protection from predators or environmental stress. Scleractinian stony corals (hexacorals) are colonies of polyps, which can be meters in diameter, connected by a common gastrovascular system (the coenosarc) that covers the limestone surface between calyces.

Calcareous algae, foraminiferans, calcareous hydrozoans, molluscs, and bryozoans are examples of other construction units in the coral reef system. These reef-builders may dominate reefs when conditions are not optimal for the zooxanthellae-algal limestone production mechanism.

More than 500 million people depend on coral reefs for food or income, and nearly that same number of people live within 60 miles (100 km) of a coral reef. Recent studies estimate that reefs contribute $375 billion annually in goods and services to the world economy. Humans harvest coral reef resources for food, medicine, jewelry, and souvenirs. Researchers are considering coral reefs as a source for new medicines to treat cancers and bacterial infections. The revenue generated through reef-related tourism activities is estimated in the tens of billions.

Lack of education and lack of legal and economic incentives to conserve coastal resources in tropical developing countries are some of the problems threatening the health of coral reefs. Destructive fishing practices, such as dynamite and cyanide fishing, stun fish for easy collection but leave behind dead coral skeletons that are unable to support the reef ecosystem. Coastal development that destroys natural water filtration systems, such as mangroves and seagrass beds, increases water TURBIDITY and coral mortality. If coastal management is poor, mining, agricultural and urban runoff, and sewage can cause damage to reefs by increasing nutrient and sediment concentrations that suffocate and abrade coral animals. The aquarium trade harvests live coral and commonly uses cyanide to collect ornamental fish. Anchoring ships on coral reefs and poorly regulated tourist activities, such as diving and spear fishing, can have detrimental effects on reef communities. Coral was once mined and cut into blocks to construct massive buildings, and today coral is still extracted to make lime or mix with cement to build roads and seawalls. In Sri Lanka, coral is mined from a barrier reef that once sheltered the coast from storm waves. Without the reef, Sri Lanka is losing houses, hotels, roads, and 3,000 acres (30 ha) of land each year to erosion that was not significant when the reef was intact.

Although Earth has always been exposed to warming and cooling cycles, recent studies find that sea levels and temperatures are rising at faster rates than at any other time in Earth's history. Coral reefs are sensitive to environmental changes and have been one of the first ecosystems affected by mass extinctions in geologic history. There are signs that coral reefs are "declining" at a rapid rate, and many believe that humans are partially responsible and must take aggressive steps to ensure the persistence of this resource.

See also ATOLL; BARRIER REEF; CORAL ANIMALS; CORAL BLEACHING; CORAL REEF ECOLOGY; FRINGING REEF.

Further Reading

Barnes, Richard S. K., and Roger N. Hughes. *An Introduction to Marine Ecology*, 3rd ed. Oxford: Blackwell Science, 1999.

Birkeland, Charles. "Introduction." In *Life and Death of Coral Reefs*, edited by Charles Birkeland. New York: Chapman & Hall, 1997.

Bryant, Dirk, Lauretta Burke, John McManus, and Mark Spalding. *Reefs at Risk*. Washington, D.C.: World Resources Institute, 1998.

Hodgson, Gregor. "Resource Use: Conflicts and Management Solutions." In *Life and Death of Coral Reefs*, edited by Charles Birkeland. New York: Chapman & Hall, 1997.

Hubbard, Dennis K. "Reefs as Dynamic Systems." In *Life and Death of Coral Reefs*, edited by Charles Birkeland. New York: Chapman & Hall, 1997.

Muller-Parker, Gisèle, and Christopher F. D'Elia. "Interactions Between Corals and Their Symbiotic Algae." In *Life and Death of Coral Reefs*, edited by Charles Birkeland. New York: Chapman & Hall, 1997.

Coriolis force (effect)   The Coriolis force is a pseudoforce that must be added to the forces on a body in an accelerating frame of reference in order to satisfy Newton's second law of motion: The acceleration of a body in motion is equal to the sum of the forces acting on it divided by the mass of the body. Earth is an accelerating frame of reference, since it is rotating, so the Coriolis effect must be applied in analyzing motions in Earth's atmosphere and ocean in order for Newton's laws to be valid. Straight-line motions on Earth, seen by an observer in a fixed frame of reference in space, are perceived as curved when viewed from an observer on the surface of Earth. The phenomenon can be simply demonstrated by drawing a straight line with a piece of chalk on an old phonograph record while it is rotating on the record player. A line drawn in this manner will be curved. Creatures living on the surface of the record would perceive the motion as curvilinear.

In Earth's atmosphere, when an air or water particle begins to flow in response to the pressure gradient force due to the atmosphere trying to equalize differences in air pressure (*see* PRESSURE GRADIENT), it moves with increasing speed. As the particle accelerates, it comes increasingly under the influence of the Coriolis effect, until the magnitude of the pressure gradient force is equal and opposite to the Coriolis force, which occurs when the particle is moving at a right angle to the initial direction of motion. The deflection is to the right in Earth's Northern Hemisphere and to the left in the Earth's Southern Hemisphere. The motion achieved when the Coriolis force balances the pressure gradient

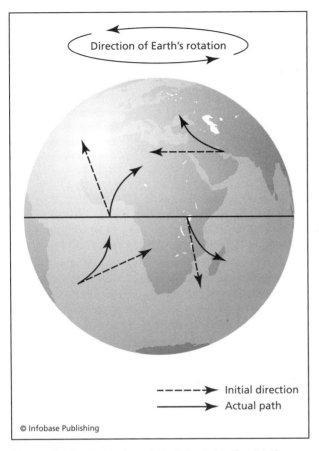

Direction of Earth's rotation

- - - - → Initial direction
———→ Actual path

© Infobase Publishing

The rotating Earth showing winds deflected to the right in the Northern Hemisphere and to the left in the Southern Hemisphere. There is no deflection at the equator.

force is called geostrophic flow in meteorology and oceanography (*see* GEOSTROPHY). The flow is then parallel to the isobars (lines of equal pressure) on a weather chart. A good approximation to geostrophic flow occurs in the upper atmosphere, and in the deeper ocean, where frictional forces are small compared to those at the air-sea, or air-ground boundaries. Pure geostrophic motion is thus not possible right at Earth's surface because of friction with the ground. The surface wind blows at a (usually) small angle inclined to the isobars, inward around a low-pressure area, and outward around a high-pressure area. The angle is larger over land than over the ocean because friction with the ground is greater than friction over a water surface.

The Coriolis effect is small in magnitude compared with the forces of gravity and the pressure gradient; therefore it takes time to act on water and air particles. It can usually be neglected for fluid flow over short time and space scales, such as the draining of water from the kitchen sink. In this case, the circular flow is caused by the shape of the drain and the rotational impulse given at the beginning of the

motion, and not by the Coriolis effect. On the other hand, in the case of the motions of winds and currents, as well as the flight of long-range ballistic missiles, the Coriolis effect must be taken into account.

The magnitude of the Coriolis effect is equal to twice the mass of the particle, times the Earth's angular velocity, times the speed of the particle, times the sine of the Earth's latitude. The sign of the motion is positive in the Northern Hemisphere and negative in the Southern Hemisphere. The Coriolis effect is zero at the equator and a maximum at the poles. The Coriolis effect results in cyclonic storms, such as mid-latitude CYCLONES as well as tropical HURRICANES, playing a large role in the circulation and heat and water balances of Earth's atmosphere and ocean.

The Coriolis effect was named after French physicist Gaspard G. Coriolis (1792–1843), who provided the first quantitative explanation of this phenomenon.

coxswain   A coxswain is an experienced sailor who has responsibility for steerage of a ship's boat such as the captain's gig and has charge of its crew. For some boats and craft, a pilot is in charge with the coxswain performing helmsman functions. The helmsman is the person who controls the rudder. Aboard a racing shell, the coxswain directs the rest of the crew. During a race, the coxswain keeps the boat moving straight, controls the pace of the boat, and motivates the crew as they row toward the finish line.

Being certified as a coxswain provides restrictions on the size of boat and how far the coxswain can navigate from the primary boat or seaport. In the United States Coast Guard, a qualified coxswain has demonstrated advanced boat handling skills as a boat team member. Once certified as a coxswain, he or she is given responsibility for directing the safe navigation of the boat, the activities of the crew, and the performance of missions. Mission essential tasks of a competent coxswain include navigation, piloting, boat handling, communication, search planning, and emergency procedures.

See also BOAT, INFLATABLE BOATS, MARINE POLICE.

craftmaster   A craftmaster is a senior enlisted sailor who directs the movement of boats such as the Landing Craft Utility, which is also known as the LCU during amphibious assault operations. The craftmaster directs docking and undocking from amphibious ships, beaching and retracting, loading and offloading, and security of equipment and supplies during the various phases of ship-to-shore movement. The craftmaster ensures that the craft employs prescribed

Rules of the Road, electronic and visual navigation, and radio and visual communications procedures. The craftmaster supervises a crew of approximately 10 sailors in all phases of craft operations and maintenance.

The U.S. Navy deploys the Mine Hunter "SWATH 1" (MHS 1) to survey coastal regions in search of mines. SWATH stands for Small Water-plane Area Twin Hull, and the craftmaster is concerned with the vessels readiness to complete missions such as executing "Q-route" surveys in which equipment is used to locate mines in different quadrants. Onboard instrumentation includes multi-beam side scan SONAR, REMOTELY OPERATED VEHI-CLES, and forward-looking sonar. These instruments are used to make detailed maps of the seafloor and to scan for moored mines ahead of the craft.

See also BOAT; COMBATANT SHIPS; HOVERCRAFT; HYDROFOIL; PATROL CRAFT; Q-ROUTES.

current   A current is a flow of fluid through a mass of similar fluid that remains stationary or is at a different velocity from the current. The wind is a current through a surrounding mass of relatively calm air. Air currents are generally horizontal but can have significant vertical components as in the updrafts of a thunderstorm. A water current is generally a horizontal movement of water, although slow vertical movements do occur, as in upwelling. Currents may be classified as tidal and nontidal. Tidal

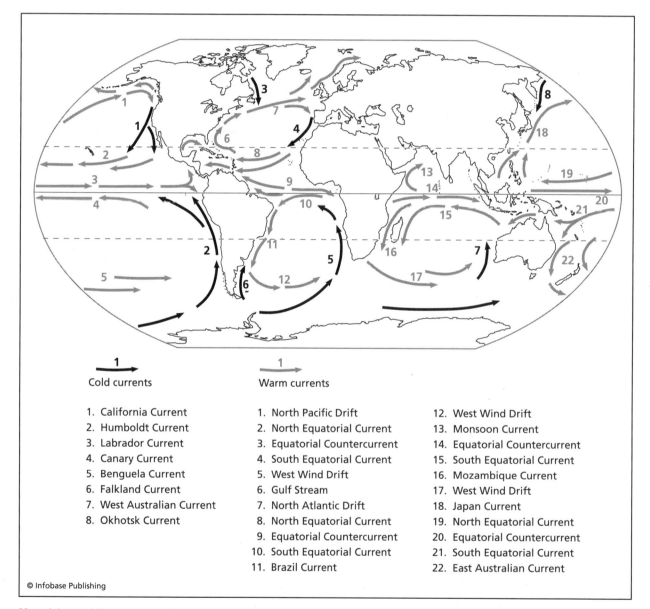

Cold currents

Warm currents

1. California Current
2. Humboldt Current
3. Labrador Current
4. Canary Current
5. Benguela Current
6. Falkland Current
7. West Australian Current
8. Okhotsk Current

1. North Pacific Drift
2. North Equatorial Current
3. Equatorial Countercurrent
4. South Equatorial Current
5. West Wind Drift
6. Gulf Stream
7. North Atlantic Drift
8. North Equatorial Current
9. Equatorial Countercurrent
10. South Equatorial Current
11. Brazil Current

12. West Wind Drift
13. Monsoon Current
14. Equatorial Countercurrent
15. South Equatorial Current
16. Mozambique Current
17. West Wind Drift
18. Japan Current
19. North Equatorial Current
20. Equatorial Countercurrent
21. South Equatorial Current
22. East Australian Current

© Infobase Publishing

Map of the world's oceans showing the approximate locations of the average positions of warm and cold currents

currents (*see* TIDE) are caused by gravitational interactions between the Sun, Moon, and Earth and are part of the same general movement of the sea that is manifested in the vertical rise and fall, called tide. Tidal currents are periodic, with a net velocity of zero over the particular tidal cycle. Tidal current is the horizontal component of the wave water particle motion, while the tide usually refers to the vertical component resulting in changes of water level. Tidal currents tend to be rectilinear, having ebbs and floods in shallow water, and rotary (continuously changing direction) in deep water. The observed tide and tidal current are the result of the combination of several tidal waves, each of which may vary from nearly pure progressive to nearly pure standing and with differing periods, heights, phase relationships, and directions.

Oceanographers have many ways to classify nontidal currents. They can be classified by geography, as in nearshore currents, offshore currents, coastal currents, and estuarine currents. Coastal currents flow generally parallel to the coast; estuarine currents are set up by large SALINITY differences in the water when salt water encounters freshwater. Nearshore currents such as LONGSHORE CURRENTS are set up inside the surf zone in response to incoming waves striking the beach at an angle. HYDRAULIC CURRENTS are set up when the winds pile water along a coast, resulting in a flow of water running downhill. This effect is fairly common along COASTAL INLETS. Such inlets have their own particular rhythm, which is generally understood by HARBOR PILOTS and FISHERMEN with knowledge on how a local inlet works.

Currents are classified by depth as surface currents, undercurrents, intermediate currents, and bottom currents. Seasonal currents exhibit changes in speed, direction, and depth as a result of seasonal change in the winds. Some currents, such as coastal currents and polar waters, and the Antarctic Circumpolar Current (ACC), which flows as a belt around Antarctica, extend all the way from the surface to the bottom. Such currents are usually winddriven in well-mixed waters of small temperature and salinity gradients.

Currents can be classified according to the physical forcing mechanism that produces them, such as wind-driven currents resulting from the overlying wind stress, THERMOHALINE (density) currents, resulting from temperature and salinity differences in the water that set up PRESSURE GRADIENTS, and riverine currents resulting from the outflow of rivers. Marine scientists, especially physical oceanographers, study how currents form in response to air and sea interactions such as the exchange of energy, momentum, and gases.

Nontidal currents include the permanent currents in the general circulatory systems of the sea, such as the GULF STREAM and the KUROSHIO CURRENT, as well as temporary currents arising from more pronounced meteorological variability, such as INERTIA CURRENTS, which are set up in the deep ocean by passing storms exerting time-varying wind stresses on the surface. Such currents can last from days to weeks, until they are damped out by fluid friction. HURRICANES drive a STORM SURGE ahead of them: mounds of water that produce significant flooding of coastal regions. Circulating rings and EDDIES permeate the major current systems of the globe. These currents are induced by shear along current boundaries and can transport water characteristics, plants, and animals great distances from their places of origin. An example is the Gulf Stream rings that transport Sargasso Sea water hundreds of miles (thousands of kilometers) northward from their places of origin. The rings can persist for months before being absorbed by surrounding waters.

Ocean currents, being vectors consisting of speeds and directions, are more difficult to measure than scalar quantities such as water level, TEMPERATURE, and salinity. For this reason, much of the knowledge of the currents of the ocean has been deduced from the distribution of temperature and salinity. For centuries, the main source of information on currents has been the reports of ship drift, which are now filed with major oceanographic data centers, such as the U.S. National Oceanographic Data Center in Silver Spring, Maryland. During the 1940s, a major breakthrough in current measurements was made by the Roberts Radio Current Meter. This current meter, named after designer Capt. Elliott B. Roberts of the Coast & Geodetic Survey, was successfully used from the 1940s through the mid-1960s. It utilized radio telemetry to send data to a receiver located away from the observation site. Recent advances in technology have made it possible to measure current fields by means of ACOUSTIC DOPPLER CURRENT PROFILERS (ADCPs), ground, aircraft, and satellite-based radars, and GLOBAL POSITIONING SYSTEM (GPS)-tracked drifting buoys.

Knowledge of ocean currents is important for accurate and safe navigation, minimizing ship transit time, search and rescue (SAR), pollutant spill cleanup, ocean structure design, fishing operations, and many other applications. The continued improvement in current measurement technology assures that safer and more efficient marine operations will be possible in future decades.

See also AGULHAS CURRENT; ALASKA CURRENT; ANTARCTIC CIRCUMPOLAR CURRENT; BEN-

GUELA CURRENT; BRAZIL CURRENT; CALIFORNIA CURRENT; CANARY CURRENT; EAST AUSTRALIA CURRENT; EASTERN BOUNDARY CURRENTS; EQUATORIAL COUNTERCURRENT; EQUATORIAL UNDERCURRENT; GUINEA CURRENT; GUYANA CURRENT; GYRE; LOOP CURRENT; NORTH EQUATORIAL CURRENT; OYASHIO CURRENT; PERU CURRENT; SOUTH EQUATORIAL CURRENT.

Further Reading

Allaby, Michael. *Encyclopedia of Weather and Climate*, rev. ed. New York: Facts On File, 2007.
Bowditch, Nathaniel. *The American Practical Navigator: An Epitome of Navigation*. Bethesda, Md.: Defense Mapping Agency Hydrographic/Topographic Center, 1995.
Pickard, George L., and William J. Emery. *Descriptive Physical Oceanography: An Introduction*. Oxford: Pergamon Press, 1990.

current meter   In marine science, a current meter is an instrument used to measure the speed and with more sophisticated sensors the direction of water flow. To measure currents during the 1800s, scientists observed the set and drift using a compass and a cord with knots tied at equal intervals attached to weighted poles from an anchored ship. An observer noted the pole's trajectory and counted the number of knots that payed out over a given time to determine the current's speed in "knots," where one knot equals 1.69 feet (0.5143 m) per second (*see* CURRENT POLE). Modern scientists still use methods to record the flow by following the current, but also measure currents passing a fixed point by using high-technology propellers that spin about a vertical axis, transducers that measure the passage of an acoustic pulse, radiometers that detect the scatter of electromagnetic energy from capillary waves, and electrodes that measure voltage differences across a magnetic field.

Drifters and sometimes environmentally friendly dyes are used to measure ocean currents by tracking the floating mass as it is moving through time and space. In the study of fluid mechanics, the process of following a "tagged water parcel" is called Lagrangian, in honor of Joseph Louis Lagrange (1736–1813), a French mathematician remembered for his study of frictionless flow problems. Drifters such as SOFAR, RAFOS, PALACE, and ARGO floats may be neutrally buoyant and are tracked by satellites, radar, radio, and sound. Dyes can be released into various parts of the water column to study how the dye patch disperses. For example, a researcher studying secondary production in the Gulf of Maine might evaluate the transport of COPEPODS by injecting dye

Deployment of a Plessey propeller-type current meter during a NOAA survey   *(NOAA)/NOS)*

into a rich patch of *Calanus finmarchicus*, an important grazer of diatoms on Georges Bank.

Another popular way to measure currents such as the KUROSHIO CURRENT in the PACIFIC OCEAN is by deploying an array of sensors connected by a mooring to determine the water's velocity at fixed locations in the water column. In the study of fluid mechanics, the process of measuring speed and direction at a fixed location is called Eulerian, in honor of Swiss mathematician Leonhard Euler (1707–83). Depending on depths, this is typically accomplished with Savonius rotor current meters (named after airfoil designs by Finnish sailor Sigurd J. Savonius (1884–1931) during the 1920s), electromagnetic current meters, or ACOUSTIC DOPPLER CURRENT PROFILERS.

Rotor current meters calculate speed by recording the rate of rotations of a propeller, and direction is determined relative to magnetic north with a compass. These instruments are oriented into the current using a vane. Electromagnetic current meters use electrodes to sense voltages that are produced across the direction of saltwater flow (seawater is a conductor of electricity; the moving conductor cutting magnetic field lines produces a voltage gradient that is recorded as a voltage function of time). The acoustic Doppler current profilers provide a profile of velocity with depth by measuring the echo of the return of reflected sound from scattering material such as plankton and other particulates that are presumed to move at the mean speed of the water. Most current meters are mounted on a mooring cable, which is deployed from a ship (*see* MOORING). Some deployments require acoustic sensors to be rigidly mounted in trawl-resistant cages on the seafloor. All these instruments provide a time series of the velocity of the ocean's water at a single latitude, longitude, and depth within the water column.

Current measurements may also be remotely sensed using sensors such as radar from shore, ships, and satellites. Ground-based radars that use the Doppler principle in a manner similar to the acoustic Doppler current profiler and the Doppler weather radars can measure ocean currents. Current measuring radars cover a range of frequencies to provide scientists and engineers with a choice of accuracies, ranges, and areas covered. The longest-range radar current meter is the over-the-horizon current meter, which can measure currents hundreds of miles away. High-frequency (HF) current meters have been the most popular, and include the CODAR (coastal ocean dynamics applications radar) in the United States, and the OSCR (ocean surface current radar) in the United Kingdom.

The principle of operation of the CODAR is that the sea surface acts like a diffraction grating to the transmitted radar waves (Bragg scattering) that are beamed at the sea surface. Since the sea surface is in motion, the moving diffraction grating effect produces side bands in the reflected radar signal spectrum that are related to the speed of the waves through the Doppler equation. Since the waves on the ocean surface that reflect the CODAR signal have a known speed, determined from the dispersion relation for ocean GRAVITY WAVES, the difference between the measured speed and the known wave speed is a measure of the current velocity. The direction of the waves can be determined by electronically steering the antenna. The OSCR operates in similar fashion.

Higher resolution, but much shorter range can be achieved with microwave radars, which use the Doppler principle, i.e. the difference frequency of two transmitted radar waves shows a Doppler shift proportional to the speed of the current. This is called the "Delta-k" radar. Delta k radars are produced in the United States and in Norway and can be mounted on BRIDGES, navigation aids, piers, and the like to mea-

sure current speeds in rivers, ESTUARIES, bays, inlets, and the like. Radar measurement of current is a subject of very active research. There are even systems for estimating current speed from aircraft and orbiting satellites, such as the synthetic aperture radar. These instruments estimate current speeds from the scattering properties of the ocean surface waves, which are superimposed on the moving sea surface.

Marine scientists and in particular physical oceanographers process and analyze speed and direction values in order to understand circulation and its spatial and temporal variability. Measuring the horizontal movement of water in the deep ocean is essential to understanding density currents that are generated at the poles and their driving forces such as pressure and gravity. Transporting heat from the equator are geostrophic currents such as the GULF STREAM and Kuroshio Current, which are controlled by a balance between PRESSURE GRADIENT forces and the CORIOLIS EFFECT. Current meters provide valuable sea truth for innovative new modeling and REMOTE SENSING efforts that are used to analyze large-scale ocean circulation and climatic changes that are leading to reductions in the extent of arctic SEA ICE.

See also MAGNETIC CURRENT METERS; OCEANOGRAPHIC INSTRUMENTS; RADAR; SAVONIUS ROTOR CURRENT METER.

Further Reading
Deacon, Margaret, Tony Rice, and Colin Summerhayes, eds. *Understanding the Oceans: A Century of Ocean Exploration.* London: UCL Press, 2001.

current pole A current pole is an outmoded device used to observe the velocity of the CURRENT. A log line attaches the current pole to the observing ship. The pole is deployed overboard and allowed to drift until the end of the line is reached. The time is recorded when the pole is put in the water, and when it reaches the end of its tether. Since velocity is equal to distance divided by time, the known length of the log line deployed divided by the time of drift gives an estimate of the speed of the current relative to the vessel. A compass bearing taken on the pole gives the relative direction of current drift, which, when corrected for the heading of the vessel, gives an estimate of the true current direction.

This method was used to gauge current speed and direction on U.S. Geodetic Survey (now National Ocean Service/ NOAA) ships. They were also deployed from U.S. Coast Guard lightships (anchored ships with light beacons in shoal areas offshore where it was not possible to build lighthouses). Current poles are still in use today, an example being the checking of the NOAA tidal current predictions at some locations in Chesapeake Bay by U.S. Power Squadron volunteer yachtsmen. They are used as teaching tools by educational institutions. Some of the tidal current predictions in the NOAA tide tables are still based on current pole measurements. Log lines to which knots are tied at regular intervals with a wood, metal, or plastic weight (the "log") are still sometimes used to determine the speed of small craft. On all vessels, the speed measuring system is called the "speed log," named for this method that was used in the days of sailing ships.

Current poles used by the U.S. Geodetic Survey were generally about 15 feet (4.57 m) long and three inches (7.62 cm) in diameter, except in very shallow areas where smaller poles were used. The poles were weighted at one end so they would float upright, with about one foot (0.3 m) above the water surface. The log line was graduated in knots and tenths of knots. These graduations were indicated by pieces of string attached by knots to the log line. The number of knots paid out in a unit of time (typically 60 seconds) indicated the speed of the current in knots, or nautical miles per hour. An extra length of line, called the stray line was an ungraduated part of the log line that allowed the current pole to drift out beyond the ship's distortion of the current.

Measurements from the light stations revealed that offshore tidal currents are rotary rather than reversing, that is, there is no slack period and the currents rotate through all points of the compass during the approximately 12 1/2 hour cycle of tidal currents off the East Coast of North America.

Current measurements from tidal current poles are Lagrangian, meaning that the observing instrument moves with the WATER MASS one is trying to measure. Tidal current poles are the forerunners of today's modern drifting BUOYS, which are tracked by radar or by GLOBAL POSITIONING SYSTEM (GPS) receivers, or other electronic means.

See also TIDE.

cutter Any U.S. Coast Guard vessel larger than 65 feet (19.8 m) is called a cutter. The term dates to the 18th and 19th centuries as a specific class of sailing vessel, especially those vessels supporting Great Britain's Royal Customs Service. Today's U.S. Coast Guard cutters include law enforcement ships (high-endurance cutters), patrol boats, buoy tenders, and harbor tugs.

See also BOAT; BUOY TENDERS; ICEBREAKING.

Further Reading

United States Coast Guard, U.S. Department of Homeland Security. Available online. URL: http://www.uscg.mil. Accessed November 18, 2006.

cyclone A cyclone is a low-pressure area in the atmosphere. Air rotation about a low-pressure area is in a counterclockwise fashion in the Northern Hemisphere and in a clockwise fashion in the Southern Hemisphere. Tropical cyclones are low-pressure weather systems in which the central core is warmer than the surrounding atmosphere. Extratropical cyclones form outside the Tropics, and the storm center is colder than the surrounding air. They are associated with fronts, that is, two colliding air masses that displace each other. In the INDIAN OCEAN and around the Coral Sea off northeastern Australia, the term *tropical cyclone* describes storms with winds that are greater than or equal to 74 miles per hour (119.09 km/hr). These storms are called typhoons in the North Pacific west of the International Date Line and hurricanes in the Atlantic basin and the Pacific Ocean east of the International Date Line.

Cyclonic storms range in size from small-scale tornadoes—funnels of violently rotating air of hundreds of feet (m) to a few miles (km) in diameter for real monsters—to hurricanes that are great systems of converged thunderstorm cells. The largest hurricanes are capable of producing winds close to 200 miles per hour (321.87 km/hr), which are called "category 5 hurricanes" in the Atlantic, and "super-typhoons" in the Pacific.

At middle and polar latitudes, there are large cyclonic winter storms that can bring days of cloudiness, high winds, and precipitation. An example of such storms are the nor'easters of the northeastern United States, which produce heavy rain or snow, severe icing, and very rough seas that cause extensive BEACH erosion and present hazards to mariners.

Large, semi-permanent cells of low pressure and cyclonic winds are found in winter in the North Pacific Ocean (the Aleutian Low) and the North Atlantic Ocean (the Icelandic Low). These are usually regions of severe winter weather.

See also HURRICANE; STORM SURGE; WATER LEVEL FLUCTUATIONS.

Further Reading

Aguado, Edward, and James E. Burt. *Understanding Weather and Climate,* 3rd ed. Upper Saddle River, N.J.: Pearson Eduction, Inc./Prentice Hall, 2004.

Bureau of Meteorology, Commonwealth of Australia. Available online. URL: http://www.bom.gov.au/index.shtml. Accessed February 29, 2008.

Hong Kong Observatory. Available online. URL: www.weather.gov.hk/contente.htm. Accessed February 29, 2008.

D

data center   A data center is a centralized storage facility for information of importance for organizations such as a corporations, universities, or government agencies. Marine information archived in a data center is especially useful for climatological studies and for decision-making purposes. The information may require permanent storage based on scientific, administrative, fiscal, legal, or historic reasons. Functions in a data center involve the storage, management, processing, and exchange of digital data and information. The utility of a data center might be measured in its efficiency to store and share quality-controlled information resources with customers.

In marine science, data centers such as the National Climatic Data Center (NCDC), the National Geophysical Data Center (NGDC), the National Oceanographic Data Center (NODC), and the National Snow and Ice Data Center (NSIDC) serve as a central repository for the storage, management, and dissemination of data and information relevant to meteorological and oceanographic data elements, parameters, and phenomena. These organizations work with other groups such as the Federal Geographic Data Committee (FGDC) to develop standards, which can be used across various data systems and facilities. Web site visitors can link to information about weather, BEACH temperatures, BUOY data, climate, severe damages to a specific area, images, maps, and atlases. These NOAA data centers respond to requests for meteorological and oceanographic data from all over the world.

In synchronization with environmental stewardship and understanding issues such as GLOBAL WARMING, a network of World and Regional Data Centers are involved in the acquisition, processing, storage, and distribution of valuable data. Data sets from institutions and international studies are stored at numerous sites according to solar, geophysical, and related environmental disciplines. These World and Regional Data Centers provide archives for ongoing investigations that offer the potential for future discoveries. The overarching goal is to establish long-term data sets that can be used by researchers to address pressing science and policy questions regarding environmental variability. Their existence serves to support scientific exchange across disciplines and governments. They ensure the preservation of valuable scientific data and the sharing of information and knowledge through international collaborations.

Further Reading

British Oceanographic Data Center. Available online. URL: http://www.bodc.ac.uk. Accessed March 24, 2007.

Data and Information Services Center (DISC): Goddard Space Flight Center, National Aeronautics and Space Administration. Available online. URL: http://daac.gsfc.nasa.gov. Accessed March 24, 2007.

Japan Agency for Marine-Earth Science and Technology: Global Oceanographic Data Center. Available online. URL: http://www.jamstec.go.jp/jamstec-e/access/okinawa/index.html. Accessed March 24, 2007.

National Climatic Data Center: National Environmental Satellite, Data, and Information Service, National Oceanic and Atmospheric Administration. Available online. URL: http://www.ncdc.noaa.gov/oa/ncdc.html. Accessed March 24, 2007.

National Geophysical Data Center: National Environmental Satellite, Data, and Information Service, National Oceanic and Atmospheric Administration, Boulder,

Colo. Available online. URL: http://www.ngdc.noaa. gov. Accessed March 24, 2007.

———. World Data Center System, National Oceanic and Atmospheric Administration. Available online. URL: http://www.ngdc.noaa.gov/wdc. Accessed March 24, 2007.

National Oceanic and Atmospheric Administration: Marine and Coastal Spatial Data Subcommittee. Available online. URL: http://www.csc.noaa.gov/mcsd. Accessed June 16, 2007.

———. National Oceanographic Data Center, National Environmental Satellite, Data, and Information Service. Available online. URL: http://www.nodc.noaa.gov. Accessed March 24, 2007.

National Snow and Ice Data Center, University of Colorado Cooperative Institute for Research in Environmental Sciences. Available online. URL: http://www-nsidc. colorado.edu. Accessed March 24, 2007.

data processing and analysis Data processing and analysis is the process of checking, preparing, organizing, and quality-controlling raw data for the purpose of producing needed information. In marine science, a combination of manual, computer assisted, and fully automated methods are used to record, classify, sort, summarize, calculate, disseminate, and store data. Data processing and analysis are fundamental skills that marine scientists use to transform raw data from sensors, instruments, and models to a form useable by customers ranging from commercial FISHERMEN and HARBOR PILOTS to COASTAL ENGINEERS and NAVAL ARCHITECTS.

Data are nothing more than values that have been collected through record keeping, observations, or measurements and organized for subsequent analysis and decision-making purposes. These data can be stored in many formats such as text, integer, or floating-point decimal. Similarly, imagery can be used to display spatial information in vector or raster formats. Vector graphics use mathematically generated points, lines, and polygons, while raster graphics are grids of individual pixels. Time series data refers to consecutive measurements that are stored in chronological order. Time series lend themselves to the development of descriptive statistics such as norms, means, and extremes. Once expensive environmental data has been quality-controlled, it should be stored in a permanent archive to establish long-term data series and to benefit other researchers.

Analysis is a key step in evaluating or making sense out of data records or converting data to information. This is done through plotting scalar and vector variables, computing descriptive statistics, and determining relationships through more complex mathematics such as time series analysis, multivariate analysis, and nonlinear signal processing. Based on the developing knowledge base, the inevitable record gaps can be interpolated by inspection or by fitting a function. As data relationships are discovered, files may be imported into graphing, mapping, or presentation software to build figures and illustrations for publications such as technical reports, professional papers, refereed journal articles, and books. Thus, a great deal of work goes into the reporting of marine science and the display of marine data.

See also CHAOS; STATISTICS; TIME SERIES ANALYSIS.

Further Reading

Global Ocean Data Analysis Project, Oak Ridge National Laboratory. Available online. URL: http://cdiac.ornl. gov/oceans/glodap/Glodap_home.htm. Accessed September 17, 2007.

IRI/LDEO Climate Data Library. International Research Institute, Lamont-Doherty Earth Observatory, Columbia University, Palisades, N.Y. Available online. URL: http://iridl.ldeo.columbia.edu. Accessed September 17, 2007.

The Math Forum @ Drexel, Drexel School of Education. Available online. URL: http://mathforum.org. Accessed March 24, 2007.

data reduction *See* DATA PROCESSING AND ANALYSIS.

datum A datum is a vertical or horizontal plane that is used to measure heights or depths and positions. Datums are critical for cartographers and hydrographic surveyors so that they can make accurate paper or electronic maps that can be used for navigation. Thus, CHARTS contain information on water depths, obstructions, and the locations of aids to navigation, which may happen to be land features. The best overall datum for mapping Earth is the World Geodetic System 1984 (WGS84). It is especially important to know the chart datum so that depth measurements or soundings can be compared to tide predictions.

The vertical datums provide a zero surface such as Mean Sea Level or MSL, to which elevations or depth soundings are referred. In the United States, the NATIONAL OCEANIC AND ATMOSPHERIC ADMINISTRATION uses its National Water Level Observation Network to determine tidal datums. For example, historic tide gauges near PORTS AND HARBORS measure the rise and fall of the tide. The raw water level data is processed and analyzed over 19-year periods to determine Mean Higher High Water,

Mean High Water, Mean Lower Low Water, Mean Low Water, and Mean Sea Level. The horizontal datum provides a reference for latitude and longitude, that is location. Examples of horizontal datums in use today include the North American Datum of 1927 (NAD 27) and the North American Datum of 1983 (NAD 83).

These datum planes are referenced to fixed points known as benchmarks. A tidal benchmark is usually sited near a tide station to which the tidal datums have been referred. Benchmarks are usually concrete blocks, brass plates, or steel plates with an inscription of elevation and location. Professional surveyors and engineers use benchmarks to find a particular elevation that can be used as a reference to determine other elevations in the survey.

For products such as the U.S. Naval Oceanographic Office's Digital Bathymetric Data Base or DBDB, the vertical datum is mean sea level and the horizontal datum is designated WGS84. Incidentally, for all practical purposes there is no difference between NAD83 and WGS84. DBDB provides ocean floor depths at various gridded resolutions to support the generation of bathymetric chart products and to provide ocean floor depth data to be integrated with geophysical parameters such as gravity and oceanographic parameters such as currents. A gridded 5 arc-minute product from these data is publicly available and widely used. An arc-minute is a unit of angular measure equal to 60 arc seconds, or 1/60 of a degree.

Further Reading

GEBCO: General Bathymetric Chart of the Oceans. Available online. URL: http://www.ngdc.noaa.gov/mgg/gebco/gebco.html. Accessed September 17, 2007.

Integrated Coastal Zone Mapping. Available online. URL: http://www.iczmap.com. Accessed March 24, 2007.

NOAA Tides and Currents, Center for Operational Oceanographic Products and Services. Available online. URL: http://tidesandcurrents.noaa.gov. Accessed March 24, 2007.

Office of Coast Survey, National Ocean Service, National Oceanic and Atmospheric Administration. Available online. URL: http://chartmaker.noaa.gov. Accessed September 17, 2007.

day shapes Special visual aids or geometric markers such as balls, cones, cylinders, and diamonds displayed on vessels during daylight hours indicating specific operations are called day shapes. To help minimize the chance of collisions, the U.S. Coast Guard's navigation rules require certain vessels restricted in their ability to maneuver to display appropriate black silhouette day shapes. For example, ma-rine scientists engaged in diving activities during the day should exhibit day shapes such as "one black ball" that indicates at anchor, the international code flag "Alpha" indicating their research vessel's restricted maneuverability, and depending on state or local law, the use of a red-and-white diver's flag to mark the diver's location. Other day shapes include "two cones apex to apex in a vertical line" to indicate trawling or fishing, "two black balls, one above the other" if a vessel is not underway, "black ball, black diamond, black ball" to signify that the vessel is restricted in maneuvering, and in the case of dredging "two round balls in a line" indicating the side in which an obstruction exists.

Further Reading

Maloney, Elbert S. *Chapman Piloting Seamanship & Small Boat Handling*. New York: Hearst Marine Books, 1994.

Navigation Center, U.S. Coast Guard. Available online. URL: http://www.navcen.uscg.gov. Accessed March 24, 2007.

dead reckoning Dead reckoning is a form of navigation developed when sailors first plied the seas and is still used today. The idea is rather intuitive—the navigator starts at a known position, for example the homeport, and sails for a specific time at a measured speed along a specific compass heading. For instance, a sailor might sail west for four hours at three knots (1.54 m/s). Thus, at the end of that time he would have traveled three nautical miles (5.56 km) along a latitude line that ran through his port. This requires a precise measure of time and a way to estimate the speed of the vessel. The direction of the vessel is measured using a COMPASS. Compasses became a standard tool in European navigation in the 12th century and were used earlier than that in the Far East. Keeping accurate time on the ship for computing longitude required centuries of technical development. However, from the 12th to the 18th century the time measurements were sufficient to measure small increments of times, for example four hours, a standard watch on board ship. These were done using hourglasses and the position of the Sun and stars in the sky. Measuring the speed of the ship was a different and more difficult task. Christopher Columbus (1451–1506), and we can assume other mariners of his age, dropped a piece of jetsam from a given forward rail on the ship and tracked it until it passed the rail near the stern of the ship. The helmsman would chant a song, that unfortunately is lost to history, and the person tracking the jetsam would calculate the speed of the vessel depending upon the syllable of the song when the jetsam passed the last rail. A more accurate method was developed later to measure distance traveled. A rope with knots evenly spaced on it was tied to the three corners of a triangular board with one tip of the board weighted. As the ship sailed along, the board was dropped over the side, and it sat in the water as a "sea anchor." As the ship moved forward, the rope was paid out and the number of knots in a measured amount of time gave the speed of the ship. Hence, the term *knots* for speed. A ship making four knots (2.05 m/s) means that four knots were paid out during the specified time. A knot is defined as a nautical mile (6,080 feet) per hour.

A famous oceanographer and naval officer, Matthew Fontaine Maury (1806–73), in the mid 19th century used the difference between the dead reckoning position of ships and the more accurate measurements of celestial position (measurements from the ship of the position based on the location of the Sun or stars in the sky), to infer the major ocean currents. Dead reckoning with celestial updates was used up until the 20th century, when electromagnetic methods began to be used. Radio signals were used in a system called LORAN (Long Range Navigation) and then satellite systems were developed in the 1960s called TRANSIT (Navy Navigation Satellite System) to be followed by the Global Positioning System, GPS, of the 1980s and beyond. To this day, dead reckoning continues to be taught at sailing schools, naval academies, and is used by commercial FISHERMEN and vacationing boaters.

Divers and autonomous vehicles use sophisticated dead reckoning procedures, which are corrected upon surfacing using the Global Positioning System. There are many autonomous vehicles and drifting BUOYS that are collecting important data to understand weather and climatic processes worldwide.

See also COMPASS; GLOBAL POSITIONING SYSTEM; NAVIGATION.

Further Reading

Maloney, Elbert S. *Chapman Piloting and Seamanship,* 64th ed. New York: Hearst Marine Books/Sterling Publishing Co., 2003.

deck gear Deck gear is the nautical term describing equipment that is used from the main deck or floor of a vessel. This gear is often divided into categories such as rigging, deck fittings, and deck machinery. Deck operations aboard many research vessels are the responsibility of the boatswain or bosun under the direct supervision of the chief mate. On research vessels, the marine technicians provide the link between the science party led by the chief scientist and crew members. The bosun is in charge of all deck hands and comes under the direct orders of the

master, chief mate, or mate. The deck department ensures a safe deck layout and oversees sampling operations that use deck gear such as winches, cranes, and gantries. Smaller WORKBOATS and BARGES might employ deck gear, pullies, and booms for dredging main navigations channels for seaports. Gear such as anchor tackle, binders, and BUOYS is "dogged down" with strong line or chain. The term *dogged down* refers to the process of holding or fastening equipment to a deck with a mechanical device. Loose equipment such as shackles, pear rings, and swivels are normally stored in bosun chests behind individually secured watertight doors. Loose or adrift gear may become a hazard to the crew in rough seas or could be lost as the vessel pitches, rolls, yaws, and heaves.

Rigging includes all the wires, ropes, and chains employed to support and operate the masts, yards, booms, and sails of a vessel. Standing rigging supports masts or king posts. Masts are vertically placed poles or spars that hoist sails, support booms, or hold electronics such as antennas, RADAR, and lights. A king or samson post is a strong vertical pole usually used to attach lines for towing or MOORING. Shrouds are the vertical ropes or wires leading from the mast to chain plates at deck level on either side of the mast, and which hold the mast from falling or bending sideways. In some rigging, turnbuckles are used to take up the slack in the shrouds and stays. These are internally threaded collars turning on two screws threaded in opposite directions. The stays lead in a forward or aft direction and are located at the mast. When they support the mast from a forward direction, they are called stays. When they support the mast from an aft direction, they are called backstays. From the era of tall ships, the term *running rigging* includes the gear used in hoisting, lowering, or trimming the sails in a vessel. On a research vessel, the running rigging would refer to moving or movable parts that are used to operate gear such as cargo runners, topping lifts, and guy tackles.

Deck fittings are permanently installed fittings on the deck of a vessel that maintain the water tightness of the deck while providing a means to attach rigging, machinery, or equipment. Traditional examples include bitts, chocks, cleats, and pad eyes. Bitts are used to secure mooring or towing lines. Lines are passed through chocks to bollards on the pier. Chocks are either closed, open, roller, or double roller. Cleats are metal fittings having two projecting horns to secure lines. Pad eyes are fixtures that are usually welded to a deck or bulkhead. They have an eye to which lines or tackle may be fastened and are used for securing or handling cargo.

Deck machinery is essential for deployment of oceanographic instrumentation and in completing maritime tasks that range from docking and offloading cargo to operations such as dredging and underway replenishment. The size and shape of the deck machinery may vary depending upon type of vessel and operations, but all are designed to lift, lower, or move cargo, equipment, and instruments. Examples of deck machinery include electric winch windlass to raise and lower anchors and cargo winches. A winch drum might contain 36,089 feet (11,000 m) of cable in order to lower survey instruments into the deep ocean TRENCHES.

In summary, marine scientists use specialized deck gear and science labs aboard research vessels in order to deploy and recover instrumented moorings, and collect profile data and water samples for analysis. Deck gear includes rigging, deck fittings such as cleats and quick releases, and deck machinery such as deep-sea mooring winches. Winches and reels are used for instrument deployment operations of various kinds including CONDUCTIVITY-TEMPERATURE-DEPTH (CTD) casts and storage and deployment of cables such as Remotely Operated Vehicle or ROV umbilicals and instrument tethers. The chief mate or bosun will assist with matters relating to deck gear that is used to handle buoy and mooring tackle. Marine technicians liaison between the ship's crew and the science party while working to maintain, repair, and assist in the operation of the onboard research equipment, sensors, water sampling systems, and over-the-side instrument handling devices.

See also BOAT; OCEANOGRAPHIC INSTRUMENTS; REMOTELY OPERATED VEHICLE; SHACKLE; SWIVEL; WINCH.

Further Reading

Research Vessel *Endeavor,* University of Rhode Island Graduate School of Oceanography, Narragansett. Available online. URL: http://techserv.gso.uri.edu. Accessed March 24, 2007.

Research Vessel *Hatteras,* Duke University Marine Laboratory, Beaufort, N.C. Available online. URL: http://www.env.duke.edu/marinelab/facilities/hatteras. Accessed March 24, 2007.

Research Vessels, WHOI Marine Operations, Woods Hole Oceanographic Institution, Woods Hole, Mass. Available online. URL: http://www.whoi.edu/marops/research_vessels. Accessed March 24, 2007.

deep chlorophyll maximum Under favorable conditions a deep chlorophyll maximum or DCM indicative of high productivity is located near the bottom of the EUPHOTIC ZONE (the zone in the upper ocean where sunlight is sufficient for photosynthesis). In order to sustain an actively growing PHYTO-PLANKTON community in a low light layer below the water's surface, conditions must provide adequate nutrients and relief from grazing by herbivores. This biomass is usually measured in terms of the organic carbon associated with phytoplankton in mg/m^3. The DCM is a common feature, which is partially or completely undetected by satellite remote sensing instruments.

The productivity of this layer is influenced by photosynthetically active radiation (PAR) intensity, upward fluxes of nutrients such as nitrate (NO_2) and phosphate (PO_4), sinking organic matter such as detritus, ZOOPLANKTON such as COCCOLITHOPHORES that feed on PHYTOPLANKTON such as diatoms, the THERMOCLINE, CURRENTS, and turbulence. PAR includes WAVELENGTHS ranging from 400 to 700 nanometers (0.4 to 0.7 micrometers) that support plant growth. The depth of penetration is dependent on suspended silt and soil particles and the amount of phytoplankton and zooplankton in the water column. Zooplankton grazers that swim down to hide in the dark during daytime diel vertical migrations may prune the DCM. This day and night or diel migration was first noticed by SONAR operators and called the DEEP SCATTERING LAYER. Physical processes such as upwelling may deepen and advect the DCM offshore. Upwelling conditions result when favorable coastal winds blow alongshore causing surface waters to move offshore that are replaced by cool nutrient-rich bottom waters. Upwelling events are associated with isotherms sloping upward toward the coast, a drop in sea-surface temperature, and increases in chlorophyll concentrations.

Investigators collect samples of the DCM by taking regular CONDUCTIVITY-TEMPERATURE-DEPTH (CTD) casts for SALINITY and temperature, and augmented with probes to collect oxygen, chlorophyll, and nutrient concentrations. Physical measurements include trajectories from the deployment of Lagrangian drifters, current maps from ship-mounted bottom-tracking ACOUSTIC DOPPLER CURRENT METERS, and meteorological data such as wind roses from shipboard weather stations. Zooplankton and nekton sampling are from towed square-shaped neuston nets designed to catch surface organisms, long and circular Reeve-type nets that can be vertically towed, and bongo nets that provide some degree of spatial replication. Neuston are planktonic organisms associated with the air-sea interface. Phytoplankton can be sampled with plankton nets, but they may also be taken from water samples with pumps or automatically closing bottles.

See NETS; PHYTOPLANKTON; REMOTE SENSING; ZOOPLANKTON.

Further Reading
Reeve, Michael R. "Large cod-end reservoirs as an aid to the live collection of delicate zooplankton." *Limnology and Oceanography*. Available online. URL: http://aslo.org/lo/toc/vol_26/issue_3/0577.pdf. Accessed April 19, 2007.

deep scattering layer (DSL) The well-defined layer of marine organisms such as ZOOPLANKTON, FISH, and squid that reflects and scatters vertically directed sound pulses has been called the deep scattering layer or DSL since the term was coined by SONAR operators during World War II. This concentration of marine organisms led by microscopic light sensitive grazers migrates daily from the MESOPELAGIC ZONE to the EPIPELAGIC ZONE. When it is daylight, zooplankton such as COPEPODS, a major consumer of PHYTOPLANKTON, rest in the dark water hundreds of meters below the surface, and at dusk they migrate to the surface to feed. Larger species such as squid, jellyfish, and vertebrate fish follow this food chain from the depths to feed on other nighttime migrators. Jellyfish and fish having gas-filled chambers cause the returns from echosounders as the DSL rises at night and descends during daytime. The DSL was intensively investigated by Marine scientists during the 1960s by means of SUBMERSIBLES making simultaneous acoustic and visual measurements.

In the PACIFIC OCEAN, species such as lanternfish (*Electrona carlsbergi*) contribute to the DSL as they feed on copepods, hyperiids, euphausiids, ostracods, and gastropods. As the small lanternfish feed near the surface, they can be eaten by squid such as the commercially important California or Market Squid (*Loligo opalescens*) and by other fishes such as the ribbon barracudina (*Notolepis rissoi*) and sea birds such as the Hawaiian dark-rumped petrel (*Pterodroma sandwichensis*). This diel migration involves a round trip that may be from 984 to 2,625 feet (300 to 800 m). Not all organisms participate in this vertical migration, especially those without swim bladders. Cartilaginous fish such as the bluespotted ribbontail stingray (*Taeniura lymma*) are "sit and wait" predators that inhabit the western Pacific and use their liver to control buoyancy.

See also CHONDRICHTHYES; SONAR.

Further Reading
Acoustic Monitoring, Vents Program, Pacific Marine Environmental Laboratory. Available online. URL: http://www.pmel.noaa.gov/vents/acoustics.html. Accessed March 24, 2007.
Bermuda Institute of Ocean Sciences, Ferry Reach. Available online. URL: http://www.bbsr.edu. Accessed April 20, 2007.
Harbor Branch Oceanographic Institution. Available online. URL: http://www.hboi.edu. Accessed April 20, 2007.
MOTE Marine Laboratory. Available online. URL: http://www.mote.org. Accessed April 20, 2007.

degree programs Since the United States is a maritime nation, there are numerous degree programs that train people for careers ranging from marine technology to basic oceanographic research. Marine scientists have basic computer skills and backgrounds in fields such as electronics, surveying, ocean mapping, basic science, or interdisciplinary fields such as environmental science. Employers in marine science generally look for people with computer and math skills and certificates from a trade school or degrees from two-year, four-year, or postgraduate schools. Entry-level positions in the marine industry might involve being a watercraft technician or an operator of a REMOTELY OPERATED VEHICLE. As marine science proliferates into other areas such as homeland security, training programs could include underwater technology for law enforcement officers and SCUBA search and rescue for first responders. Positions that currently require college degrees involve tasks such as certified hydrographic surveying, data analysis, and report writing. Senior positions might entail working as a chief scientist aboard a research vessel that is collecting data for an international oceanographic study.

Students taking courses relevant to marine science at either a junior or community college normally commute to and from school, taking classes that lead toward an associate degree in applied science. Many two-year community college programs focus on either occupational programs or transfer preparation. Marine science courses play a vital role in employee development for local business and industry that are sited in coastal states, along waterways such as the Mississippi River, and near busy port facilities. Some community college programs involve a combination of credit and noncredit courses for lifelong learning and to expose students to the positive stewardship of natural resources such as riparian wetlands, CORAL REEFS, and ESTUARIES. Two-year colleges share a philosophy and commitment to serving all segments of society through open admission, affordable tuition, and extra academic and personal support. As an example, students attending the College of Oceaneering in San Diego, California, can earn an associate of science degree in marine technology.

Four-year schools provide coursework exposing students to the basic concepts and interrelationships of physical, geological, chemical, and biological oceanographic and inshore ecosystems as well as the function and application of oceanographic equipment. Since majors, curricula, and courses may change from year to year, students interested in a bachelor of science in marine science will need to contact potential institutions for details regarding conferment of degrees. The best preparation for a marine science career would be rigorous training in calculus and applied mathematics, physics, chemistry, geology, biology, and engineering. Undergraduate degree programs in any of the above areas should provide the student with the necessary background to pursue a graduate degree in one of the subfields of marine science, biological, chemical, geological, and physical oceanography. Engineering programs such as civil, chemical, electrical, marine, mechanical, and ocean engineering as well as naval architecture often focus on design and operation of underwater vehicles, seafloor instruments, or remote-sensing systems, as well as designing structures that can withstand ocean CURRENTS, WAVES, TIDES, and severe storms. A strong background in marine technologies is very important preparation for work in the oil industry, ship building, seaport operations, and manufacturing marine equipment.

Entrance into graduate programs in marine science leading to both the master's and Ph.D. degrees requires a student to demonstrate a strong background in mathematics (through calculus), the basic sciences (e.g., biology, chemistry, geology, and physics) and some experience with computers. Once enrolled, graduate schools attempt to provide stimulating learning environments in which students study the basic principles of marine science, marine resource management, and engineering applications. At graduate school, students maintain a close interaction with faculty actively involved in marine research and management issues. Students may even test their skills by attending annual meetings of professional organizations such as the AMERICAN GEOPHYSICAL UNION, the MARINE TECHNOLOGY SOCIETY, or the OCEANOGRAPHY SOCIETY to present research posters or papers.

Advanced degrees qualify students to work in research-oriented fields such as archaeology, biochemistry, botany, chemical oceanography/chemistry, conservation, economics, geography, geoscience, mathematics, statistics and computing, microbiology, parasitology and pathology, physical oceanography, taxonomy, and zoology. More applied fields employ people holding the full spectrum of interests and academic backgrounds for work in conservation, ecotourism, engineering, environmental consulting, fisheries, food technology, geography, law, mariculture, ocean technology, pharmacology, and remote sensing. The broad range in disciplines and backgrounds is found because marine science is actually a multidisciplinary and integrated field.

Further Reading

Consortium for Ocean Leadership. Available online. URL: http://oceanleadership.org/. Accessed March 4, 2008.

Nichols, C. Reid, David L. Porter, and Robert G. Williams. *Recent Advances and Issues in Oceanography*. Westport, Conn.: Greenwood Press, 2003.

Oceanography/Marine-Related Careers: A Sea of Possible Career Options, Office of Naval Research. Available online. URL: http://www.onr.navy.mil/foci/ocean_marine/career_options.asp. Accessed March 24, 2007.

delta   The word *delta* comes from the Greek capital letter that takes the form of a triangle, (Δ). This Greek word may have its origin from the historian Herodotus (484–430 B.C.E.), who wrote of the fan-shaped mud and sand flats found at the mouth of the Nile River. Deltas are low-lying sediment covered land features that generally form at the mouths of rivers. Clastic sediment suspended by river discharge settles as a function of grain size and the force of WAVES, tides, and CURRENTS near the river mouth onto the delta bar. For example, in the absence of LONGSHORE CURRENTS, larger grain sizes are deposited near the river mouth, while finer sediments are dispersed over much greater offshore distances. Suspended sediments settle to the seafloor when currents can no longer transport the grains. Thus, deltas form when rivers transport more sediment to the coastal ocean than can be redistributed by nearshore currents. River plumes are seen from aerial and satellite imagery and identify the offshore regions where clays, silt, and sand flocculate and freshwater mixes with salt water.

Marine scientists further classify deltas in accordance to their detailed shape and the intensity of forces such as river flow, wave action, and tides. The shape of the delta is strongly influenced by the amount of sediment load and energy. However, deltas display a combination of shapes since energy levels may change over long time periods. A river-dominated delta such as the Mississippi, which projects enormous amounts of silt, usually forms into a digitate or birdfoot shape. Locations such as the Danube or Rhône delta form a more lobate structure in response to strong riverine and wave influences. Similarly, a combination of strong wave and riverine forces contributes to the formation of arcuate deltas such as the Nile delta in Egypt. The smooth rounded front that gives the Nile delta a fanlike shape is caused by waves and the ensuing longshore currents. River mouths such as the Tiber River in Italy located at a point develop into a cuspate delta as wave action pushes sediment back on both sides of the mouth, which extends the point farther out into the coastal ocean.

Deltas, the depositional features resulting from a river meeting the ocean or a lake, are usually fertile lands that support agriculture such as rice farming and are often sites for oil and natural gas exploration. The conditions contributing to their shape may change over geological time scales in synchronization with sea-level changes and sedimentation rates. The coast responds to mixed river, wave, and tidal influences through changes in the deltas shape. For example, conditions can lead to the filling in of an estuary such as the Seine River with sediment to produce the final type of delta, the estuarine delta.

See COAST TYPE; ESTUARY; TIDE.

Further Reading

Bird, Eric C. F., and Maurice L. Schwartz. *The World's Coastline.* New York: Van Nostrand Reinhold, 1985.

Shepard, Francis P. *Geological Oceanography.* New York: Crane, Rusak, 1977.

Smith, Jacqueline. *The Facts On File Dictionary of Earth Science,* rev. ed. New York: Facts On File, 2006.

density   The mass per unit volume of an object is classified as its density. Water density is of great interest to marine scientists because it determines the depth to which a water particle will sink. If a water

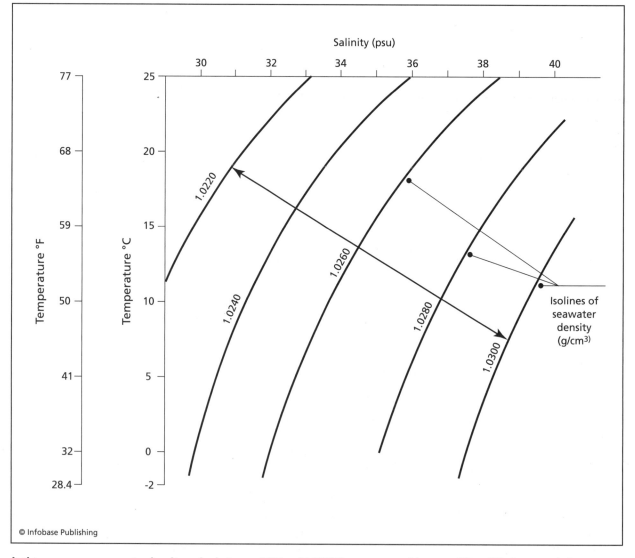

In the open ocean, seawater density varies between 1,022 and 1,030 kilograms per cubic meter (22 and 30 sigma-t units), depending on temperature and salinity.

particle is denser than the surrounding water, it will sink; if lighter, it will rise. These considerations are important to understanding how the WATER MASSES of the ocean form. Pure water is densest at a temperature of 39.2°F (4°C), which is the reason that ponds and lakes freeze from the top down. (The ice is less dense than the underlying water.) As salt is added to the water, its freezing point is lowered. Thus, water is made denser by cooling below 39.2°F (4°C), and also by increasing the SALINITY.

The average density of seawater is about 8.6 pounds per gallon (1,027 kg/m^3). Seventy-five percent of the ocean water has a density between 8.56 and 8.58 pounds per gallon (1,026.4 and 1,028.1 kg/m^3). To accurately measure the density of the water requires labor-intensive laboratory measurements. These measurements are extremely difficult to do on a moving ship. However, since density is a function of the ocean's salinity, TEMPERATURE, and PRESSURE, which are much easier to measure from a ship, the density can be calculated from the equations of state for seawater.

The equations of state for seawater, which express the density in terms of temperature, salinity, and pressure have scores of terms and are usually evaluated by means of tables or computer programs. The Joint Panel on Oceanographic Tables and Standards has developed a practical salinity scale and the accepted equations of state for seawater (Fofonoff, 1985). Professor Nicholas P. Fofonoff (1929–2003) completed considerable research relevant to circulation and thermodynamics of the GULF STREAM.

The representation of density is usually simplified by oceanographers by means of the following defining equation for the quantity σ:

$$\sigma = (\rho_{s,t,p} - 1)\, 10^3$$

where s = salinity, t = temperature, and p = pressure. In working with densities from near-surface waters (typically, 273 fathometers (500 m) or less in depth), the effect of pressure can be neglected, so it is set to zero, and the approximation for σ is σ_t, where

$$\sigma_t = (\rho_{s,t,0} - 1)\, 10^3$$

For deeper depths, the pressure effect cannot be neglected. Although seawater is only slightly compressible, compressibility cannot be neglected in the deep ocean. Indeed, if seawater were not compressible, sea level would rise about 98 feet (30 m) worldwide.

The temperature, salinity, and pressure can be measured from ships by means of a CTD (conductivity, temperature, depth) probe (*see* CONDUCTIVITY-TEMPERATURE-DEPTH [CTD] PROBE). The CTD is lowered over the side and telemeters, via a cable, the *in situ* conductivity, temperature, and depth back to the ship. Conductivity is a function of the salinity. Other methods of measuring the density include discrete measurements made using Niskin or NANSEN BOTTLES that bring water samples up to the surface as well as the measurements of the *in situ* temperature.

Density is the driving component of most of Earth's oceans, along with the wind stress. Even though the density differences are small, they can account for major currents that transport large water masses throughout the globe in the "thermohaline conveyor belt." The warm water of the Gulf Stream flows northward, where it is slowly preconditioned by evaporation, which slowly cools the water and increases its salinity. Even though these processes increase the density of the water, it is still lighter than the water it overlays. The Gulf Stream, called in this region the NORTH ATLANTIC CURRENT, separates in the waters off Newfoundland. Part of the current turns southward to rejoin the Gulf Stream system's large GYRE; however, part of it flows northward into the Norwegian Sea, where this relatively warm, yet salty water, now cools quickly and becomes much denser than the water below it. This North Atlantic water sinks to the bottom in the Norwegian Sea and begins to flow southward as a water mass called North Atlantic Deep Water. This generation of North Atlantic Deep Water is the engine that drives the thermohaline conveyor belt.

See also PERMANENT CURRENT; THERMOHALINE CIRCULATION; WATER MASS.

Further Reading

Fofonoff, Nicholas P. "Physical Properties of Seawater: a New Salinity Scale and Equation of State for Seawater." *Journal of Geophysical Research* 90 (1985): 3,332–3,342.

depth finder A navigational instrument used to measure the water depth is called a depth finder or a fathometer. Old techniques that may still be used for the sake of expediency involve the use of lead lines. In this manner, a researcher might record a water depth by lowering a weight attached to a rope over the side of the vessel. Once the rope becomes slack, it is marked and pulled back onboard. The length of rope used to reach the bottom is determined in order to estimate the depth. Fortunately, there are acoustic methods that allow navigators and researchers to sound the bottom. The use of sound to determine the depth of water became operational during World War II. The technique is called SONAR and stands for Sound Navigation and Ranging. Modern active sonar devices can be used to identify bottom structure, determine water depth, and sport-fishing variants can mark fish within the water column.

Sonar technology involves the use of a transducer that transmits a short burst of sound in the water, directed from the hull toward the bottom. The transducer is the electromechanical component of a sonar system fixed to the hull exterior, which converts electrical energy to sound energy and vice versa. Each pulse of sonar is also known as a ping. Typical frequencies for depth sounders are 50 and 200 KHz. The lower frequency has greater range but the higher frequency has better resolution. Beam widths range from 45 to 15 degrees. Following transmission, returned sound reflected from the bottom or from an object in the water is received by the transducer. The time lapsed between the transmitted pulse and received echoes from the seafloor are measured. Since the speed of sound in water is known, the simple equation

$$D = ct/2$$

where,

D = Depth,
c = speed of sound in water, and
t = time for the sound to leave and return to the transducer

is used to find the depth. For this reason, depth finders are also known as "echo sounders." The results are displayed on a strip of paper or on a display screen.

There are many reasons why marine scientists need to know the depth of the water below the surface. Most important, safe navigation requires that the research vessel or WORKBOAT does not run aground. In marine science, most scientific findings or samples are related to the immediate water depth. Water quality parameters such as TEMPERATURE, PRESSURE, and dissolved oxygen vary with depth and effect the location of certain PLANKTON and fish species. Light penetrates the water to various depths and has important effects on the productivity of PHYTOPLANKTON. The water depth also influences the propagation of WAVES and the speed of CURRENTS. Various depths and structures on the seafloor host different assemblages of benthic or bottom-dwelling organisms. Commercial FISHERMEN are dependent on accurate depth finders, water quality parameters such as temperature, and complementary navigation systems such as GLOBAL POSITIONING SYSTEM (GPS).

See also ATTENUATION; BATHYMETRY; HYDROGRAPHIC SURVEY; NAVIGATION; OCEANOGRAPHIC INSTRUMENTS.

Further Reading

NOAA Ship *Rainier*, Marine Operations Center, Pacific, National Oceanic and Atmospheric Administration. Available online. URL: http://www.moc.noaa.gov/ra/index.html. Accessed March 25, 2007.

NOAA Ship *Rude*, Atlantic Marine Center, National Oceanic and Atmospheric Administration. Available online. URL: http://www.moc.noaa.gov/ru/index.htm. Accessed March 25, 2007.

Placzek, Gary, and F. Peter Haeni. "Surface-Geophysical Techniques Used to Detect Existing and Infilled Scour Holes near Bridge Piers." In U.S. *Geological Survey Water-Resources Investigations*, Report 95-4009. Available online. URL: http://water.usgs.gov/ogw/bgas/scour/index.html. Accessed March 24, 2007.

depth recorder   *See* DEPTH FINDER.

Differential Global Positioning System   Accuracy enhancement to the United States Department of Defense GLOBAL POSITIONING SYSTEM or GPS by the United States Coast Guard (USCG) is known as Differential GPS or DGPS. The Global Positioning System is used to provide precise location based on data transmitted from a constellation of 24 satellites. GPS receivers can receive the signals from multiple satellites, and by measuring the time it took the signal to arrive they can determine one's current position on Earth with some error. By deploying a network of ground-based reference stations, that is, locations with known latitude, longitude, and elevation, the USCG is able to broadcast corrected locations in real time. The positional error of a DGPS position is 3.28 to 9.84 feet (1 to 3 m), which provides improved safety for harbor entrance and approach navigation. The USCG's DGPS system provides service for coastal coverage of the continental United States, the Great Lakes, Puerto Rico, portions of Alaska and Hawaii, and sections of the Mississippi River Basin. Differential corrections are based on the North American Datum of 1983 (NAD83) position of the reference station antennas. Positions obtained using DGPS should be referenced to North American Datum of 1983, which for all practical purposes is the same as World Geodetic System of 1984 or WGS84. Many foreign nations such as Canada have implemented standard DGPS services modeled after the U.S. Coast Guard's system to significantly enhance maritime operations in their critical waterways. Users must possess equipment that receives reference station messages, receives GPS signals, and then applies differential corrections to display enhanced location data.

Further Reading
Navigation Center, U.S. Coast Guard. Available online. URL: http://www.navcen.uscg.gov. Accessed April 19, 2007.

diffusion The random movement of ions or molecules from regions of high concentration to low concentration within a solution is known as diffusion. A familiar atmospheric illustration relates the way the smell of a freshly baked pie drifts into the living room from the kitchen. Sensory cells or olfactory nerves in the nose are stimulated by microscopic molecules released by the pie that once detected are sent to the brain, where the pie's smell is identified. Similarly, the slow motion of molecules can be illustrated by the color and taste changes in a cup of coffee once milk and sugar are added and without the aid of any turbulent mixing or stirring. Diffusion accounts for the slow motion of cream and sugar molecules as they spread through the coffee. In each of these cases, the molecules moved "down" a concentration gradient or toward areas with fewer molecules. The rate of diffusion in a fluid is usually a function of TEMPERATURE, PRESSURE, molecular weight, and molecular size.

True molecular diffusion is an ongoing process in water bodies such as ponds, lakes, and the ocean. The three kinds of diffusion that are common in the ocean are molecular diffusion that was described above, which occurs mainly in the top few millimeters or what is called the surface layer; EDDY diffusion, where external forces cause stirring; and double diffusion, which is attributed to the formation of layers of water in the ocean.

From observations and experiments in the ocean, it is seen that the rate of mixing of properties or dissolved substances occurs much more rapidly than can be explained by molecular diffusion. Motions in the ocean tend to be turbulent, which results in a much more rapid distribution of properties. Turbulent flow makes many more fluid surfaces available across which diffusion can occur. To provide a scientific explanation, oceanographers have defined turbulent diffusion coefficients analogous to molecular diffusion coefficients (the rate of diffusion is proportional to the product of the concentration gradient and the diffusion coefficient). The turbulent coefficients are much larger, allowing the dissemination of water characteristics or dissolved substances to proceed at a much greater rate than molecular diffusion. Examples of turbulent diffusion include the mixing of high-SALINITY water from the MEDITERRANEAN SEA flowing into the ATLANTIC OCEAN through the Straits of Gibraltar, and the mixing of turbid, low-salinity water from the Amazon River into the western tropical Atlantic.

The third type of diffusion is double diffusion, which describes the mixing of two different properties or substances such as temperature and salinity, which proceed at quite different rates. In the ocean, in places where laminar, low-turbulent flow is found, warm salt upper layers mix with cool, fresh lower layers, as occurs in laboratory experiments, but the mixing occurs via "tubes of mixing" a few inches (cm) in diameter and a few feet (meters) in height, called "salt fingers." Evidence of double diffusion in tranquil, strongly stratified tropical seas is determined from CONDUCTIVITY-TEMPERATURE-DEPTH or CTD instruments, which provide profiles of salinity, temperature, and density. Three of the most important data elements in describing the distribution of water in the ocean are temperature, salinity, and pressure. Salinity or the quantity of dissolved salts in ocean water is derived from measurements of the seawater's conductivity. Density in the ocean is determined by known relationships between salinity, temperature, and pressure.

In general, the magnitude of horizontal mixing is much greater than that of vertical mixing in the ocean. In the vertical, the rate of diffusion must overcome the frequently strong density stratification. For this reason, horizontal eddy diffusion coefficients are usually much larger than vertical eddy diffusion coefficients. Spilled substances in the water can make this affect visible when they spread out over areas of many square miles (km) in the horizontal, whereas they may be only a few tens of feet (m), or less, in the vertical.

Applications of turbulence theory include predicting the manner and rate in which contaminants are diffused in the ocean.

Further Reading
Bohren, Craig F. *Clouds in a Glass of Beer: Simple Experiments in Atmospheric Physics.* New York: John Wiley, 1987.
Brown, Evelyn, Angela Colling, Dave Park, John Phillips, Dave Rothery, and John Wright. *Seawater: Its Composition, Properties and Behavior,* 2nd ed. London: Pergamon Press, 1995.
Smith, Jacqueline. *The Facts On File Dictionary of Earth Science,* rev. ed. New York: Facts On File, 2006.

Digital Nautical Chart® (DNC®) The United States's National Geospatial-Intelligence Agency (NGA) produces an unclassified, vector-based digital database containing maritime significant features essential for safe marine navigation called DNC®. Vector data defines geographic features on a map as points, lines, or polygons using x, y, and z values for latitude, longitude, and elevation or depth. Features are single entities such as places and things that compose part of a map. The DNC® is built by adding layers such as cultural features, environmental features, and land cover features in a format that can be exploited by a Geographic Information System. The database is provided on a compact disk read only memory (CD-ROM) as a replication of the feature content from NGA's portfolio of hard copy harbor, approach, coastal, and topographic charts. NGA conforms to the International Hydrographic Office S-57 transfer standard for digital hydrographic data. NOAA Chart 1 describes symbology used to create the ontology of entities such as navigation aids and hazards to NAVIGATION. The ontology is a set of concept definitions that allow knowledge sharing in a geographic information system (GIS). These electronic chart products, which use a digital database to display a chart directly on a computer screen, are becoming increasingly useful and affordable. Users can build custom layers of information, declutter the navigation display, and possibly specify a minimum safety depth, and use the power of a GIS to highlight all depth curves presenting the threat of grounding.

Further Reading

International Hydrographic Organization, IHO Transfer Standard for Digital Hydrographic Data, Publication S-57, Edition 3-1—November 2000, Monaco: International Hydrographic Bureau. Available online. URL: http://www.iho.shom.fr. Accessed June 16, 2007.
National Geospatial-Intelligence Agency. Available online. URL: http://www.nga.mil. Accessed April 19, 2007.
National Oceanic and Atmospheric Administration, *Chart No. 1, Nautical Chart Symbols, Abbreviations, and Terms*. Silver Spring, Md.: National Ocean Service, 1997.

discus buoys   Discus buoys are anchored structures with round hulls used as a platform for oceanographic sensors, communication systems, and aids-to-navigation lights, shapes, sounds, and colors. In the United States, discus buoys are one of several types of buoys used by the National Data Buoy Center. The fleet of moored buoys includes discus hulls with diameters of 9.8 feet (3 m), 33 feet (10 m), and 39 feet (12 m). Other buoys include 20-foot (6 m) boat-shaped hulls, 8-foot (2.4 m) diameter coastal buoys, and saucer-shaped buoys with 6-foot (1.8 m) and 4.9-foot (1.5 m) diameters. Larger steel hulled discus buoys are reported to be more stable and rugged than smaller aluminum hulled buoys. Large discus buoys deployed in the deep ocean have sturdy MOORINGS comprising chain, nylon, and buoyant polypropylene materials designed for many years of service. The 9.8-foot (3-m) diameter buoys are ideal for deployments in the coastal ocean, while the larger buoys are ideal for long duration deep ocean deployments.

Instrumented discus buoys are outfitted with meteorological and oceanographic sensors for offshore monitoring. Meteorological sensors measure parameters such as wind, air temperature, relative humidity, and rain. Oceanographic sensors on the hull measure sea-surface TEMPERATURE and WAVES, while associated bottom- and mooring-mounted sensors are capable of providing data for CURRENTS, water levels, water temperature, SALINITY, TURBIDITY, and many other physical and biological parameters. The buoy also hosts a central computer to coordinate all the instruments and either radio or satellite communication with the shore. For deployments, these buoys may be trucked to the dock, placed in the water, towed by a research vessel to a mooring location, and anchored.

See also BUOY.

Further Reading

National Data Buoy Center, National Weather Service, National Oceanic and Atmospheric Administration. Available online. URL: http://www.ndbc.noaa.gov. Accessed April 19, 2007.

Distributed Oceanographic Data System (DODS)   A growing software capability advanced by basic oceanographic researchers to simplify all aspects of data networking has been called DODS, which stands for the Distributed Oceanographic Data System. The overarching vision involves making data accessible to remote locations regardless of local storage formats. By using DODS servers, data analyses and visualizations can be transformed for use by other oceanographers who are able to access remotely served DODS data. The server is the computer that provides the information, files, Web pages, and other services to clients that log in to exchange information. Distributed systems such as DODS are particularly useful as the United States strives to implement an Integrated Ocean Observing System, where ocean observatories can share information and track phenomena such as a GULF STREAM frontal filament that varies in time and space. Data sets made available through DODS are accessible via a Uniform Resource Locator or URL that contains the protocol type, the machine name, the directory path, and the file name. In one example, the DODS data servers provide client programs with values from the shared data sets such as latitude, longitude, and sea-surface temperatures for the filament that is being tracked. Text information might be provided about the variables in the data set and their data types. When a user logs onto the server, the word *client* refers to the user's computer and software. Standard World Wide Web browsers such as Netscape and National Center for Supercomputing Applications' Mosaic can access information that is being shared. There are hundreds of scientific datasets available online to researchers if they are using DODS servers. DODS provides relief for those collaborators who collect and store data that should be made available to other users, query archives containing multiple data formats, and manipulate data using various system interfaces. DODS resolves traditional complications to acquiring and using online data by doing the hard work of standardizing formats, managing data, and citing data that has been chosen to be shared. It allows oceanographers to open, read, sub-sample, and import directly into their data analysis applications via the Internet.

Further Reading

Federation of Earth Science Information Partners. Available online. URL: http://www.esipfed.org. Accessed April 20, 2007.

OPeNDAP, Open-source Project for a Network Data Access Protocol. Available online. URL: http://www.opendap.org. Accessed April 20, 2007.

dive equipment   Divers involved with recreational, commercial, scientific, or military diving use specially designed dive equipment to complete underwater tasks. Swimming underwater for sustained periods is usually accomplished with the aid of a face mask, swimming fins for the feet, a snorkel, and self-contained underwater breathing apparatus or scuba gear. Gear may be very different depending on depths and tasks that are being accomplished underwater.

The basic load for traditional open-circuit scuba diving consists of a mask, snorkel, fins, gas supply cylinder (air tank), regulators, buoyancy compensator, weighted belt, dive boots, and gloves. Air tanks store compressed air, and a pressure gauge provides the diver with information on how much air is in the tank. The air regulator allows the diver to breathe compressed air from the tank at low pressure. Exhaust gas is discarded in the form of bubbles with each breath. The buoyancy compensator or BC is used to help maintain position or buoyancy both in the water column and at the surface. Diving gloves and boots are often made of neoprene, a stretchy synthetic rubber made of tiny cells that restrict the flow of water. Divers may use a dry suit or a wet suit

Dive gear for deep-sea operations   *(NOAA/NURP)*

in order to conserve their body temperature in cold water. Dry suits are required for diving in the coldest water, such as through openings in Arctic ice.

Closed-circuit divers use a rebreather that injects oxygen or mixed gases into the breathing apparatus and special filters to recycle exhaled oxygen and absorb carbon dioxide (CO_2) in addition to a mask, fins, buoyancy compensator, weighted belt, dive boots and gloves. Spearfishermen and military divers were the first divers using rebreathers. Advantages from a closed-circuit rebreather include the elimination of air bubbles, more efficient use of gas, and optimized decompression characteristics. One of the dangers is low oxygen or hypoxia. In general, only highly skilled professional divers should use rebreathers.

Saturation diving relies on pressurized diving systems such as a DIVING BELL or underwater habitat, which allows divers to remain at high pressures for long periods of time (days to weeks). Gases in the diver's tissues reach equilibrium with the pressurized environment. Saturation divers have to breathe gas mixtures such as oxygen and inert gases such as nitrogen and helium. Mixed-gas diving reduces the chance for nitrogen narcosis and increases the diver's maximum operating depth. At the conclusion of the dive, divers must decompress in a recompression chamber, which can recreate the pressures at depth. Underwater robots called REMOTELY OPERATED VEHICLES and AUTONOMOUS UNDERWATER VEHICLES can accomplish some underwater tasks, reducing the need for dangerous saturation dives.

Helmets with a noncollapsible hose connected to an air supply, a protective suit, and weighted boots made up some of the first deep diving suits. Lifelines were used to control ascent and descent. Many of today's deep diving suits are armored one-atmosphere suits that contain many oil-filled joints and oxygen

Undersea Exploration: A Biographical Sketch of Jacques-Yves Cousteau

Captain Jacques-Yves Cousteau (1910–97) was a graduate of École Navale (French Naval Academy) in Brest, France, and entered military service in 1937 as a gunnery officer. During World War II, he served with the Free French Navy. In 1943, Cousteau and engineer Émile Gagnon developed the first successful self-contained underwater breathing apparatus (scuba), which freed divers of their tethered connections to support ships and permitted three-dimensional maneuverability. Scuba was first applied by military divers, who became known as frogmen, to remove enemy obstacles off beaches prior to invasion, to disarm mines, and to engage in covert activities. Following World War II, scuba was developed commercially and became known as the Aqua-Lung. The Aqua-Lung opened up whole new fields of research, such as marine archaeology, as well as making possible recreational diving for thousands of people. Scuba is widely used by police and military divers for rescue operations.

In 1947, Cousteau helped to found the Undersea Research group in Toulon, France, which became a leading center for undersea research and development. In association with MIT scientist Harold Edgerton (1903–90), he developed and applied photographic devices that have been essential to the exploration of the ocean floor. In 1957, Cousteau became the director of the Musée Océanographique de Monaco.

In 1959, Cousteau and engineer Jean Mollard designed and supervised the construction of a radically new diving vehicle, the Soucoupe (saucer), which became known as the Diving Saucer in the United States. The Diving Saucer was self-propelled and able to transport two scientists down to a depth of 1,000 feet (305 m) and perform experiments with great facility and maneuverability. The Diving Saucer was the first of the present-day deep submersibles capable of carrying a pilot and one or more observers and a scientific payload, usually including floodlights, cameras, observation chambers, and high-resolution sonars, to the depths of the ocean. The support vessel for the Diving Saucer was the R/V Calypso—a converted World War II minesweeper that Cousteau had acquired in 1950, and outfitted as a research vessel with several innovative features, including an underwater observation chamber in the bow. Cousteau also developed the first subsurface diving chamber, in which divers could live and work at the bottom of the ocean on the continental shelves without the need for frequent returns to the surface. He supervised the design of the Deepstar series of diving vehicles constructed by Westinghouse in the early 1960s. These submersibles were less maneuverable than the Diving Saucer but were capable of diving to much greater depths.

Cousteau is known worldwide for his production of many films including The Silent World in 1956 and World Without Sun, in 1966, both of which received Academy Awards for documentaries. He developed a television series, The Undersea World of Jacques Cousteau, which ran for eight years during the 1970s. He was the author of numerous popular books on the sea, including The Living Sea (1963), The Ocean World (1976), and Whales (1986). In 1974, he founded The Cousteau Society, which today has more than 300,000 members, and a popular Web site found at URL: http://www.cousteau.org/en/.

Cousteau had a long association with the National Geographic Society, which published many of the reports on his worldwide expeditions. Cousteau was associated with the U.S. Navy on many projects. He directed the Conshelf Saturation Dive Program, 1962–63. He was the co-recipient of the International Environmental Prize of the United Nations in 1977. He was awarded the Medal of Freedom in 1985 by President Reagan, and was invited to join the French Academy in 1989. He passed away on June 25, 1997.

Cousteau's work was always directed toward informing the public about the wonders and the great value of the sea and its creatures. In many households, his name is synonymous with the ocean world. He was concerned with the vital need for intelligent management of marine resources, control of pollution, cessation of whaling, the taking of marine mammals for profit, and the preservation of fostering of all marine and aquatic species.

Further Reading

Cousteau, Jacques-Yves. The Ocean World. New York: Harry N. Abrams, 1985.

———. Whales. New York: Harry N. Abrams, 1988.

The Cousteau Society. Available online. URL: http://www.cousteau.org/en/. Accessed October 17, 2007.

Rubinson, Greg. "Jacques Cousteau's The Silent Word." Available online. URL: www.salon.com/ent/masterpiece/2002/07/15/silent_world/. Accessed October 17, 2007.

—**Robert G. Williams, Ph.D.,** senior oceanographer
Marine Information Resources Company
New Bern, North Carolina

rebreather systems (*see* JIM SUIT). Cables or umbilical lines from the dive suit to the tender ship support communications. The diver may also use thruster systems to move around.

See also DIVER.

Further Reading

Ambrose, Greg. "Breathe Deep: Isle Divers Test New Gear That Recycles Air, Allowing Them to Probe Deeper and Stay Longer." *Star-Bulletin*. Available online. URL: http://starbulletin.com/96/04/03/news/story1.html. Accessed September 17, 2007.

NOAA Dive Page. Available online. URL: http://www.dive.noaa.gov. Accessed November 20, 2006.

diver In marine science, anyone equipped with breathing apparatus, weight belts, and specialized clothing who works underwater is usually called a diver. The various methods of working underwater include free diving, diving with compressed air or other gas mixtures carried by the diver, diving with compressed air from the surface, and submergence in a heavy-walled vessel such as a submersible. Scientific diving usually refers to dives conducted solely for data collection or educational activities that are under the auspices of a public or private research or educational institution or similar organization, department, or group. Scientific diving is different from commercial diving. Scientific divers are involved with observation and data gathering, while commercial divers complete tasks such as rigging heavy objects underwater, inspecting BRIDGES, tunnels, and pipelines, and cutting or welding materials underwater. For these reasons the standards and procedures for ensuring safety and risk management are different for science and commercial diving.

Free diving or what is sometimes called breath-hold diving or snorkeling involves diving with fins and a mask. This form of diving is a popular sport and has been made famous by pearl divers in Japan and Korea. Recreational free divers enjoy observing, spearfishing, and searching for elusive sea life such as conch, lobster, abalone, and finfish without air tanks, simply by diving while holding their breath. Scientists might use free diving techniques to study species

The first all-women's team of saturation divers *(NOAA/NURP)*

such as whales, dolphins, and manatees. Dives last for a minute or two, which is the approximate period that a diver can hold his or her breath or tolerate the risk of drowning from hypoxia. One danger to breath-hold diving would be sudden unconsciousness or blacking out from a lack of oxygen. A healthy and physically fit diver's body is efficient in using and transporting oxygen.

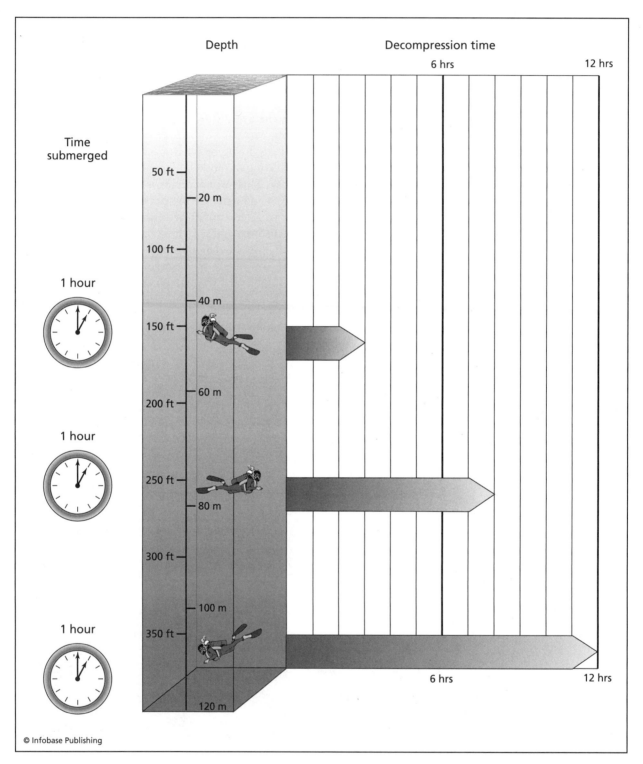

Decompression time required for divers working at the indicated depths; the deeper the dive, the greater the required decompression time.

Most divers use either open or closed circuit self-contained underwater breathing apparatus or scuba equipment. One essential piece of modern dive equipment is the buoyancy compensator or BC, an inflatable vest worn by the diver that can be automatically or orally inflated to help control buoyancy. Open circuit systems vent all expired air into the water and are the mode used by recreational divers where the diver uses only compressed air as the breathing mixture, never dives solo, and does not exceed a depth of approximately 130 feet (40 m). Military divers or frogmen often use closed circuit systems in which exhaled air is rebreathed after carbon dioxide is absorbed and oxygen is added in order to avoid being detected by showing air bubble trails. Mixed gases such as helium-oxygen can be used to go deeper than possible with compressed air. Scuba divers may be at risk for decompression problems if they ascend to the surface without proper decompression. In dive-planning tables and computer software, information is available to determine dive times for specific depths, by which the diver can avoid contracting decompression sickness. Most divers today are aided by a dive computer.

Many underwater construction and salvage projects use surface-supplied diving or hard-hat systems. Well-trained and certified commercial hard-hat divers receive their air supply while working through a long umbilical. Basically, the flexible umbilical carries air from a compressor on the dive boat or BARGE down through the water column to the diver's mask or helmet. In more complex systems, the umbilical leads into a dive suit or some larger enclosed space containing the diver. This type of diving includes underwater habitats used for saturation diving, diving bells, and rigid-helmet diving suits. Saturation divers may use mixtures, such as hydrogen-oxygen, helium-oxygen, and helium-nitrogen-oxygen. Divers breathing air at the same pressure as the surrounding water pressure are at risk for decompression problems such as the bends or air embolisms if ascent is too fast. Therefore, these commercial divers must go through long decompression periods before surfacing. Extensive tables are available for determining the length of these periods.

Heavy-walled vessels are designed to maintain an internal pressure near sea level pressure or what is called one atmosphere in order to protect occupants from surrounding water pressures that increase by one atmosphere approximately every 33 feet (10 m). As an example, during 1934, Dr. Charles William Beebe (1877–1962) and engineer Otis Barton (1899–1992) descended 3,028 feet (923 m) by squeezing into a round bathysphere consisting of a 1.5-inch-thick steel sphere lowered from a barge by steel cable. Other historical vessels include the bathy-

scaphe, which is a bathysphere with buoyancy control, and a variety of submersibles and submarines, which can travel great distances under their own power. One-atmosphere vessels may use compounds such as soda lime or lithium hydroxide to get rid of exhaled carbon dioxide (CO_2). Modern extensions of the one-atmosphere vessel are self-contained armored diving suits in which the diver becomes a small submarine. These have been featured during recent underwater investigations of marine sanctuaries during the Sustainable Seas Expeditions conducted by the NATIONAL OCEANIC AND ATMOSPHERIC ADMINISTRATION and National Geographic Society.

Diving associations such as the Professional Association of Diving Instructors (PADI) and the National Association of Underwater Instructors (NAUI) offer courses during which the student has to demonstrate a number of competencies. In addition, organizations such as the American Academy of Underwater Sciences develop, review, and revise standards for safe scientific diving certification and the safe operation of scientific diving programs. Professional organizations and adherence to approved standards help to minimize risks so that medical problems resulting from scuba diving do not go beyond middle ear squeezes that can be countered through training.

Further Reading

American Academy of Underwater Sciences. Available online. URL: http://www.aaus.org. Accessed September 17, 2007.

Ballard, Robert D. *The Eternal Darkness: A Personal History of Deep-Sea Exploration,* Princeton, N.J.: Princeton University Press, 2000.

NAUI Worldwide. Available online. URL: http://www.nauiww.org. Accessed April 20, 2007.

Richardson, Drew, ed. *PADI Open Water Diver Manual.* Rancho Santa Margarita, Calif.: Int'l PADI, 2005.

diving bell A diving bell is a watertight diving vessel, open at the bottom, that is lowered into the water, and supplied with air by hoses from the surface. Diving bells are among the oldest devices used by humans to enter and work in the water. They were used by ancient civilizations such as those of Greece and Rome. Today, the diving bell is made of high-strength, fine-grain steel and may have windows constructed from cast acrylic for pressure vessels. The American Society of Mechanical Engineers publishes codes and standards for the safe construction of Pressure Vessels for Human Occupancy, which includes diving bells (personnel transfer capsules), decompression chambers, and recompression chambers. Diving bells are lowered and raised from a mother ship or platform with a crane and serve as the

SCUBA divers deploying from a modern diving bell (NOAA/NURP)

base of operations for a dive team. Saturation divers transfer to their work locations via a diving bell. Hoses fed by pumps help maintain breathing gas in the workspace within the diving bell as it descends to the work location. Exhaled gas is evacuated from the platform. Cables attached to the mother ship also provide power and communications. Diving bells are often used in tandem with a pressurized chamber. The chamber is pressurized to the depth that the dive team will be working and provides necessary life support. Once acclimated, divers depart and enter through the diving bell. At the conclusion of underwater operations, the bell and chamber return to the surface, and the dive team decompresses in the pressure chamber. Many saturated dives, especially in support of oil production, may last up to 30 days.

See also JIM SUIT.

Further Reading

American Society of Mechanical Engineers. Available online. URL: http://www.asme.org. Accessed November 22, 2006.

NOAA's Aquarius: America's Innerspace Station. Available online. URL: http://www.uncw.edu/aquarius. Accessed November 22, 2006.

Welcome to Diving Heritage. Available online. URL: http://www.divingheritage.com. Accessed November 22, 2006.

docking pilot   A docking pilot is a mariner skilled and responsible to push, pull, and turn large ocean going vessels safely during docking and undocking maneuvers. The docking pilot is familiar with every foot of the port where he or she works. They consider factors such as vessel traffic, navigation rules, winds, CURRENTS, and water levels while guiding TUGBOATS and vessels through the docking process. Usually a port facility dispatcher will provide docking pilots with a list of ships, their dimensions, and their arrival and departure schedules. The docking pilot considers this information as well as critical data from nautical charts (*see* CHARTS, NAUTICAL), weather forecasts, traditional tide tables, local weather stations, physical oceanographic real-time systems, and vessel traffic and information systems to determine the number of tugboats needed to perform the docking process for assigned vessels.

In some ports, the HARBOR PILOT and the docking pilot are the same person. These marine pilots work together to facilitate safe navigation through coastal waters by being experts on the harbor, piers, berthing spaces, material handling equipment, and the impacts of currents, TIDES, and a host of other parameters such as wakes from nearby ships. Harbor pilots direct the ship into the port area from the seaward approaches. Then, one of the tugboats delivers the docking pilot to the ship. The docking pilot will climb up the pilot ladder from one of the tugs to con the ship into its berth. The docking pilot is responsible for the placement of tugboats that guide the ship to or away from the dock. Instructions from the docking pilot to the captain of the tugboat are communicated by radio. Once the maneuvers are completed, the docking pilots and tugs move onto their next assignment.

Mariners consider pilots to be essential since it is unlikely that the captain of a merchant vessel or warship would be familiar with the seaports that they visit. Their efforts are especially important in helping to avoid costly collision with another vessel, BRIDGE, or a pier. The docking pilot's local knowledge and the power of tugboats are applied to overcome problems associated with maneuvering a ship with a large turning radius that does not stop in a short distance. Any mishap involved in navigation can be disastrous. For this reason, the docking pilot profession requires pilots to have special training, experience, and hold U.S. Coast Guard pilot licenses. In addition, most seaports in the United States require vessels involved in foreign trade to use a pilot while navigating state waters.

See also MARINE TRADES; NAVIGATION.

Further Reading
American Pilots' Association. Available online. URL: http://www.americanpilots.org. Accessed April 20, 2007.
European Maritime Pilots' Association. Available online. URL: http://www.empa-pilots.org. Accessed April 20, 2007.
International Maritime Pilots' Association. Available online. URL: http://www.internationalpilots.org. Accessed April 20, 2007.

docking station   A docking station is a specialized MOORING providing power and control services for submersibles such as AUTONOMOUS UNDERWATER VEHICLES (AUVs) and REMOTELY OPERATED VEHICLES (ROVs). Usually, the subsea docking station is linked to the surface by an umbilical, where surface power and real-time control systems are mounted to a buoy. Docking station buoys generally use solar panels or rotors to generate power and have environmental sensors that support submersible control systems. AUVs or ROVs may receive acoustic signals from bottom-mounted hydrophones that support docking, or the submersible's navigation software provides rules that drive the AUV to successful docking. For example, as the AUV approaches the dock, it is directed to slow down and align itself with the docking axis. Once docked, communication interfaces between the mooring computer and the subsea docking station computer facilitate the exchange of information and AUV status from onboard sensors. At the mooring buoy, radio or sophisticated satellite communications such as INMARSAT or INTELSAT provide links to operations centers that may be located at offshore oil production platforms, ships, or to a laboratory ashore. Once docking occurs and the submersible is recharged, controllers specify what information he or she wishes to receive and then send the new tasks and commands to onboard computers for the AUV's next mission. Separate systems and rules at the docking station control tethers and cables, data dissemination, and command and control.

There are many systems involved in operation and maintenance of an AUV and its docking station. One of the support tasks involves monitoring the environmental conditions that are occurring from the seafloor to the surface. This is important because strong currents and wave pumping can impact navigation and docking. In the GULF OF MEXICO, the passage of EDDIES that have been shed from the Loop Current can shut down offshore oil production. The bathymetric location for deployment of a docking station should support docking operations and provide enough time for docking, recharging, uploading and downloading information, and monitoring surrounding activities. For this reason, docking stations include lights, visual sensors, acoustic sensors, and temperature gauges, pressure gauges, and current meters. The mooring itself is composed of ANCHOR, acoustic releases, chain, mooring lines of fiber optic and twisted-pair mooring cable, connection hardware, and subsurface BUOYS. Docking stations have various designs and may rotate in order to maintain alignment with the current. AUVs using a docking station can operate in an area for extended durations. Some scientists have reported that docking stations were the "garage and service station" for their AUV's mission.

Further Reading
Association for Unmanned Vehicle Systems International. Available online. URL: http://www.auvsi.org. Accessed April 20, 2007.
Indian Underwater Robotics Society. Available online. URL: http://www.iurs.org. Accessed April 20, 2007.
Monterey Bay Aquarium Research Institute. Available online. URL: http://www.mbari.org. Accessed April 20, 2007.

dredging   Dredging refers to the government regulated practice of removing sedimentary materials such as sand from the seafloor, using suction or scooping machines. Example platforms for dredging are airlift, hopper, hydraulic, and mechanical dredges. Dredging equipment includes support boats, BARGES, suction pipes, dredge pumps, discharge pipes, BUOYS, WINCHES, as well as meteorological, oceanographic, and geological instruments. On a hydraulic dredge, the rotating blade on the suction end of the pipeline is called the cutterhead. The cutterhead breaks up material on the bottom of the channel before it is sucked up through the pipe. Dredge material is deposited at a disposal site. Hydrographic surveys, data from water level gauges, and knowledge of tidal datums are key to determining when the dredge operation has achieved the correct depth.

In a drowned river ESTUARY such as CHESAPEAKE BAY, sediments enter the bay through the Susquehanna River. These sediments may have a negative effect on economics if commercial ships cannot sail into the Port of Baltimore. Therefore, for some important waterways, maintenance dredging involves keeping a navigation channel at a specific depth listed on charts to ensure safe navigation. Coastal engineers may use *in situ* sensors as well as

Dredges must operate frequently to establish and maintain clear and navigable channels. Dredging or the underwater excavation of channels is necessary to maintain a national system of waterways. This dredge is working in the Houston Ship Channel. *(USCG-PA2 James Dillard)*

physical and numerical models to study channel stability in important waterways. Research on dredging, navigation channel stability, and COASTAL INLETS is often sponsored or conducted by organizations such as the U.S. Army Corps of Engineers.

Dredge material is usually placed at overboard disposal sites, which develop into islands, or deposited into deep portions of the estuary or ocean. Research, environmental assessments, and communication with the public are essential processes in order to determine beneficial placement of dredge material in order to protect sensitive species and habitats. It is possible that dredging at certain locations would not be in the public interest. Marine scientists often conduct research to predict or assess the fate of dredge material.

In the United States, the Federal Clean Water Act protects waterways and wetlands. The U.S. Army Corps of Engineers issues permits that are required for dredging or filling in waterways and wetlands. Shifting channels and the shoaling of navigational channels at places such as the tidal entrance to harbors can pose hazards to navigation. Accumulations of sand are therefore removed from the harbor floor and moved to other locations. Coastal zone managers and policy makers depend on marine scientists to help cope with important social and environmental issues involved in dredging and the disposal of dredged materials because of ensuing physical, chemical, and biological effects. For example, if sediments contain heavy metals, the dredge material or "spoil" cannot be deposited where it will be in contact with the public. Dredge material is often used to build artificial islands and nourish

BEACHES. Coastal engineers work with mariners such as HARBOR PILOTS in order to maintain sufficient depth of water for shipping traffic to freely pass through the navigation channels and harbors.

Dredging a harbor or channel may change the water levels or circulation patterns, which may decrease the accuracy of the NATIONAL OCEANIC AND ATMOSPHERIC ADMINISTRATION's (NOAA) tide and current predictions. A new survey of tides and currents, along with deployment of a real-time wind, water level, and current measuring system such as NOAA's PORTS®, may be required to assure safe navigation.

See SEDIMENT; WATER LEVEL FLUCTUATIONS; WORKBOATS.

Further Reading

Central Dredging Association. Available online. URL: http://www.dredging.org. Accessed April 19, 2007.

Dredging, Coastal & Hydraulics Laboratory, Engineer Research & Development Center, U.S. Army Corps of Engineers. Available online. URL: http://chl.erdc.usace.army.mil/CHL.aspx?p=s&a=ResearchAreas;2. Accessed April 19, 2007.

Western Dredging Association. Available online. URL: http://www.westerndredging.org. Accessed April 19, 2007.

dyes   Water CURRENTS may be studied by releasing harmless dyes or chemicals into the flow. In the ocean, dyes are released from a vessel or aircraft, and in the laboratory dyes are injected directly into a flume. Marine scientists interested in measuring the speed and direction of surface currents tag the water parcel with either floats or dye and follow the tag. The motion of the dye is followed through aerial photography, satellite imagery, or by taking water samples. Investigators have traditionally taken multiple water samples from the water body from multiple locations as the dye plume naturally disperses. DIVERS using undersea cameras or other optical detection devices may also collect measurements to assess the spreading of the dye plume. Studies of the dispersion of dye patterns in both the horizontal and the vertical shed much light on the very small scale motions in the ocean, such as those associated with DIFFUSION of TEMPERATURE, SALINITY, oxygen, and pollutants.

This resource-intensive procedure may be improved by measuring dye dispersion with fluorometers aboard AUTONOMOUS UNDERWATER VEHICLES. The concentration of fluorescent dye in the water can be measured using a fluorometer. The fluorometer subjects a water sample to a light source and measures the resulting emitted light. Fluorescence is the process by which light is absorbed at specific wavelengths by a material and subsequently emitted at a longer wavelength. This perspective of following a tagged water parcel is called Lagrangian motion, after French mathematician Joseph Louis Lagrange (1736–1813).

The Army Corps of Engineers and ENVIRONMENTAL PROTECTION AGENCY (EPA) have determined that the nontoxic tracer dye Rhodamine WT is suitable for use in inflow studies. Upon irradiation from an external source, Rhodamine WT will fluoresce or emit light. The unit of measure is usually parts per thousand, where 1 ppt = 1 gram/liter. Other substances that have been used include chemical salts, lithium, contaminants, and tritium.

See also OCEANOGRAPHIC INSTRUMENTS; WATER LEVEL FLUCTUATIONS.

Further Reading
Army Corps of Engineers. Available online. URL: http://www.usace.army.mil. Accessed November 22, 2006.
Environmental Protection Agency. Available online. URL: http://www.epa.gov. Accessed November 22, 2006.

E

East Australia Current The East Australia Current is a warm, fast, western-boundary current flowing to the south along the east coast of Australia. Its location along the western edge of the ocean is because of Earth's rotation. The East Australia Current forms in the Coral Sea between the Great Barrier Reef and the Chesterfield Reef. It is the largest PERMANENT CURRENT close to the coast of Australia. The speed of the current varies with position and season, but is on the order of half a knot to a knot (.25–50 m/s). The current is relatively narrow, 62–124 miles (100–200 km) in width, with a volume transport on the order of 30 Sverdrups (30×10^6 m^3/s).

During the Southern Hemisphere summer, the current intensifies with the onset of the MONSOON winds. From April to December, the flow into the East Australia Current is from the Pacific subtropical water to the east. The current is the weakest during winter (June–September).

As the East Australia Current leaves the continent of Australia, it begins to form meanders and EDDIES and loses coherence. This disorganized flow generally moves across the Tasman Sea northward along the coast of New Zealand, where it turns south and into the vast eastward flow of the ANTARCTIC CIRCUMPOLAR CURRENT.

As with the BRAZIL CURRENT in the South Atlantic, the East Australia Current is less powerful than its Northern Hemisphere counterpart, the KUROSHIO CURRENT. This situation results in part from the difference in land-water distribution in the hemispheres, and from the uninterrupted global circulation of the Antarctic Circumpolar Current that allows an unimpeded return flow of water across the Pacific.

Large numbers of anticlockwise eddies have been observed between the East Australia Current and the northward flow to the east. The diameter of

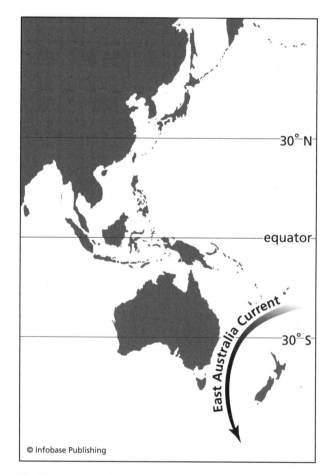

The East Australia Current is a warm current that flows parallel to the eastern coast of Australia

the eddies is about 150 miles (241 km). The eddies move at a speed on the order of 50 miles per month (80.5 km per month) toward the south. Surface speeds in these eddies can be as high as four knots (2 m/s), decreasing in depth.

Strong seasonal UPWELLING often occurs near Port Macquarie, New South Wales. The town is located on the coast, south of the mouth of the Hastings River, which empties into the Tasman Sea. Nutrient-rich waters from the upwelling induce strong PLANKTON blooms, which have been found to occur in spring in some areas of the current. Upwelling conditions may occur through both wind and current-driven processes. This type of "nutrient uplift" is observed by marine scientists using satellite imagery of sea-surface TEMPERATURE, CONDUCTIVITY-TEMPERATURE-DEPTH (CTD) sensors, and ACOUSTIC DOPPLER CURRENT PROFILERS.

See also EKMAN TRANSPORT; OCEANOGRAPHIC INSTRUMENTS; PERMANENT CURRENT.

Further Reading
Bureau of Meteorology. Available online. URL: http://www.bom.gov.au. Accessed June 18, 2007.

eastern boundary currents Broad, slow shallow currents on the east side of the Atlantic and Pacific ocean basins, in both the Northern and Southern Hemispheres, are called eastern boundary currents. They include the CALIFORNIA CURRENT, the CANARY CURRENT, the BENGUELA CURRENT, and the PERU CURRENT. These currents are on the order of 620 miles (1,000 km) wide and flow equatorward at speeds generally less than half a knot (.25 m/s) or less. They form the eastern branches of the great mid-ocean gyres (*see* GYRE), returning cooled surface waters back to the Tropics for recirculation. For example, the Canary Current returns some of the water transported to the shores of Europe by the GULF STREAM and the NORTH ATLANTIC CURRENT back to the Gulf Stream by flowing southward past Spain, Portugal and North Africa, and into the NORTH EQUATORIAL CURRENT that in turn runs westward into the Gulf Stream. The situation is not so clearly defined in the INDIAN OCEAN, where much of the eastern boundary is occupied by water, especially between Australia and Indonesia. The West Australia Current flows northward and then northwestward off the west coast of Australia. North of 20°S, the current runs westward into the Indian South Equatorial Current. The West Australia Current exhibits large seasonal variations associated with the Asian MONSOON. It is strongest in November through January, and weakest, and most variable in direction in May, June, and July.

The eastern boundary currents are not really well defined "rivers in the ocean," leading some authorities to refer to them as systems of currents. Eastern boundary currents are, however, noted for being zones of UPWELLING of nutrients from the deep layers below up to the PHOTIC ZONE. Thus, they are usually regions of high productivity. They may have a countercurrent flowing poleward inshore of the main current, or an undercurrent below the main current. For example, the Davidson Current runs inshore and sometimes underneath the California Current, and the Gunther Current flows under the Peru (Humboldt) Current. These countercurrents and/or undercurrents experience large seasonal variations, with the undercurrents sometimes surfacing. The eastern boundary currents are also characterized by slowly drifting EDDIES. Current speeds within the eddies are often the highest speeds found in the eastern boundary current systems.

Further Reading
Brown, Joan, Angela Colling, Dave Park, John Phillips, Dave Rothery, and John Wright. *Ocean Circulation.* Oxford: Pergamon Press, 1989.

ecological efficiency Ecological efficiency is the percentage or fraction of energy in a biomass produced by a trophic level that is incorporated into biomass by the next higher trophic level. Raymond Laurel Lindemann (1915–42) established the principle while working with Professor G. Evelyn Hutchinson (1903–91) at Yale University. Professor Hutchinson is considered to be the father of modern limnology. Lindemann's research was published after his death in 1942. For aquatic systems, Lindemann estimated an efficiency of at most 10 percent. For example, the efficiency of the transfer of ZOOPLANKTON energy from feeding on PHYTOPLANKTON is at most 10 percent. The other 90 percent of the energy goes into respiration, movement (mechanical energy), and heat. The Sun provides the energy input for all natural life processes, except for HYDROTHERMAL VENT communities, and is converted to stored chemical energy by the primary producers (the first trophic level). Primary producers use inorganic materials and sunlight to make cell tissue, which provides food for the second trophic level, the herbivores, and on to the third trophic level, the carnivores up the food chain to the top carnivore.

Ecological efficiency basically relates the efficiency of transfer of energy from the first trophic level, the plants or primary producers, to the highest trophic level, that is, the top carnivore. The word *trophic* comes from the Greek root "trophe" for nourishment, so a trophic level describes the "feeding" level of an organism. Trophic levels or positions that an organism occupies in a food web are important because they identify availabile energy and materials. This flow of energy and materials make their way through the web via trophic pathways. The manner in which energy and material are transferred from one level to the next is known as the ecological efficiency. As each trophic level is passed, the efficiency of the energy transfer decreases, such that the efficiency of the top carnivores is on the order of 1 percent.

Further Reading

Bush, Mark B. *Ecology of a Changing Planet.* Upper Saddle River, N.J.: Prentice Hall, 1997.
Caddy, John F., and Gary D. Sharp. "An Ecological Framework for Marine Fishery Investigations." FAO Fisheries Technical Paper 283. Rome: Food and Agriculture Organization of the United Nations, 1986. Available online. URL: http://www.fao.org/DOCREP/003/T0019E/T0019E00.HTM. Accessed June 18, 2007.
Lindemann, Raymond L. "The trophic dynamic aspect of ecology." *Ecology* 23 (1942): 399–418.

ecosystem   The plants and animals of a region, together with their chemical and physical environment, make up an ecosystem. All the plants and animals on Earth, together with their environment (the atmosphere, the oceans, and the land) make up the ecosphere. Ecosystems can be delineated in many ways. For example, an entire ocean or landmass can be considered an ecosystem. An ecosystem can also be as small as a terrarium or aquarium. Usually for working purposes, subdivisions of the large-scale ecosystems are the focus of most of the scientific investigations.

Ponds, lakes, and rivers are often studied as ecosystems. Often, the health of the ecosystem is the focus of the study. In today's industrial society, evaluation of air and water quality is a major topic, to assure breathable air and safe drinking water, and a safe environment for aquatic animals. Many regions of the industrialized countries are burdened with polluted air, acid rain, polluted water, and contaminated fish and shellfish. Contamination can occur by physical means, such as discharge of thermal effluent from nuclear power plants, chemical means, such as oil spills, and biological means (invasion of ESTUARIES by alien species).

In marine science, the global ocean, a gigantic ecosystem, can be classified in terms of the benthic (bottom) and pelagic (water) environments. The environment of the CONTINENTAL SHELF is called the neritic province; the environment of the deep ocean is called the oceanic province. The oceanic province is in turn, subdivided into the EPIPELAGIC ZONE, the upper 660 feet (200 m) of the water column (nearly the same as the PHOTIC ZONE, the volume of water of the upper ocean where significant sunlight penetrates. Below the epipelagic zone are the MESOPELAGIC ZONE, extending to 3,280 feet (1,000 m) depth, sometimes called the "twilight zone," because some faint light pervades the gloom of the region; the BATHYPELAGIC ZONE, ranging from 3,280 feet (1,000 m) to 13,123 feet (4,000 m), the average depth of the ocean basins; and the ABYSSOPELAGIC ZONE (also known as the hadal zone, the region of the deep ocean trenches). Each of these regions has its own characteristic life forms; the steadily increasing pressure of approximately one atmosphere for every 33 feet (10 m) of water depth effectively stratifies the water column into these four zones where the animals are structured to be able to cope with the crushing ambient pressures. In most cases, they cannot leave their zone without either exploding in regions of lower pressure or imploding in regions of higher pressure. Temperatures decrease with depth in the ocean; the epipelagic zone being relatively warm, the mesopelagic being a region of rapidly decreasing temperature, and the zones below being regions of near freezing temperatures.

The plants and animals with which most people are familiar inhabit either the NERITIC ZONE close to shore, or the epipelagic zone in the deep ocean, where light can penetrate to fuel photosynthesis and plant production.

The coastal oceans, including the estuaries, are the most productive regions of the seas, where runoff from the land provides a continuous supply of nutrients. Regions of UPWELLING, such as the coastal waters of California and Peru are especially productive. However, the coastal and estuarine seas are especially vulnerable to anthropogenically introduced pollution, from oil spills to wastewater discharge, and from radioactive materials to biomedical waste products.

In the deep ocean, the North Atlantic and North Pacific are very productive areas, as are the waters surrounding the Antarctic continent. In general, the waters near the Arctic and the Antarctic have a great abundance of biomass from a small number of species, whereas the tropical waters have a small abundance of biomass from a large number of species, that is, a rich species diversification.

Below the epipelagic zone, the mesopelagic or twilight zone is inhabited by such creatures as the giant squid, the principal prey of the sperm whale, dragon fish, and hatchet fish. In this zone and below are areas of primary consumers of the dead plants and animals raining down from the surface, the detritus. The zones below 3,280 feet (1,000 m) are in complete darkness, broken only by the flashes of bioluminescent animals in search of prey, trying to confuse predators or searching for mates. Strange-looking fish inhabit these regions, such as the anglerfish, gulper eels, and periphylla jellyfish.

The benthic zone, the ocean bottom, is inhabited by such creatures as sea cucumbers, crabs, clams, brittlestars, and red shrimp. The deepest zone is the hadal zone, the environment of the deep TRENCHES. Principal inhabitants in this area of total darkness and crushing pressures are sea cucumbers, probably the most numerous creatures, along with polychaete worms. Deep-sea vents, regions of the ocean floor where superheated water is boiling up through cracks in the ocean floor like geysers on land, harbor the most bizarre ecosystems on Earth. The creatures of the deep-sea HYDROTHERMAL VENTS include white clams, tube worms, crabs, shrimp, and fish. Many of these species have only recently been discovered, and many remain to be discovered. In contrast to the surface light-based COMMUNITIES, which depend ultimately on the photosynthesis of the primary producers, the primary producers of the hydrothermal vent communities are bacteria that can extract energy from the chemicals belching from the vents, such as sulfur and methane. The discovery of these incredible communities shows that life is far more adaptable and robust than had been previously believed.

The ocean contains many other ecosystems within its realm, such as CORAL REEFS, major currents, such as the GULF STREAM, and even large, warm EDDIES, rotating lenses of warm Gulf Stream water that spin off the edge of the current into the bordering waters of the CONTINENTAL SLOPE and SHELF. FISHERMEN from northern states such as New York are often surprised by drifting into a Gulf Stream eddy and catching tropical species, such as mahi-mahi. Other notable ecosystems are the estuaries, such as Chesapeake Bay and Pamlico Sound, which nurture the young of many species, the warm waters of the RED SEA, and the icy waters of the Arctic and Antarctic.

Further Reading

Byatt, Andrew, Alastair Fothergill, and Martha Holmes. *The Blue Planet: A Natural History of the Oceans.* London: DK Publishing, 2001.

Marine Ecosystems, U.S. Environmental Protection Agency. Available online. URL: http://www.epa.gov/bioindicators/aquatic/marine.html. Accessed June 18, 2007.

eddy An eddy is a rotational element of moving fluid that has a certain identity and life history. Other words relating to rotational fluid motion include VORTICITY, *whirl, vortex, whirlpool, countercurrent,* and mythical GYRES such as the Charybdis, Maelstrom, and Ixion. In nature, atmospheric and oceanic eddies occur when the current of air or water doubles back on itself, moves against the direction of main flow, or experiences large SHEAR, such as at the boundaries of the GULF STREAM. Waves in currents called meanders often increase in amplitude until they are pinched off as eddies. In the atmosphere, differences in atmospheric PRESSURE contributes to the formation of vortices from dust devils to waterspouts and from tornadoes to HURRICANES. In water bodies, eddies form when currents pass obstructions such as pier pilings or an adjacent ocean current that is flowing in the opposite direction. In the deep ocean, eddies can form when currents flow around islands or pass over topographic features. An example of the latter is the formation of "Meddies" as warm, salty water outlflows over the sill through the Strait of Gibraltar and into the ATLANTIC OCEAN at a depth of about 547 fathoms (3,282 feet [1,000 m]). As ocean currents swirl into eddies such as warm or COLD CORE RINGS, they may have different water temperatures in the center than in the surrounding ocean. Meddies retain their temperature and SALINITY characteristics for some time before eventually mixing with Atlantic Ocean waters. Large eddies have spatial scales of tens to hundreds of miles (hundreds of km) and time scales of months to years. Eddies can be both surface and subsurface. Subsurface eddies are detected by such means as SOund Fixing And Ranging or SOFAR floats, which transmit acoustic signals giving their position as functions of time.

Patches of eddies moving around in a chaotic fashion are often described as TURBULENCE. Marine scientists can quantify turbulence based on the non-dimensional Reynolds number, after British engineer Osborne Reynolds (1842–1912). Reynolds discovered that by evaluating the ratio of the fluid's velocity, density, and length of flow (inertial forces) to its thickness (viscous forces) that the flow could be classified as being laminar, transitional, or turbulent. In turbulent flow, the inertial forces dominate over viscous forces. Viscous fluids tend to damp random, turbulent motions. Turbulence is marked by small-scale chaotic behavior in the fluid motion, while a steady and organized flow field is called "laminar." Researchers are still studying the processes by which turbulent eddies cascade into larger structures, which then progress through the atmosphere and ocean as thermals or large-scale eddies. There can be a flow of energy both upward, in which small eddies merge and become larger eddies, and downward, in which large eddies break up into smaller eddies in a downward cascade of energy ending in a turbulent "inertial subrange," where the eddy's energy is dissipated as friction (heat). The British meteorologist Lewis Fry Richardson (1881–1953) described this process in the following verse:

> *Big whorls have little whorls,*
> *Which feed on their velocity,*
> *And little whorls have lesser whorls,*
> *And so on to viscosity.*
>
> —Lewis F. Richardson.
> *Weather Prediction by Numerical Processes*
> (Cambridge University Press, 1922)

Consider where a rain band that forms into a dust whirl, then the funnel cloud descends toward the ground, and wind speed increases in the dust whirl to form a mature tornado, where its intensity is as great as conditions will allow. The visible funnel inspires storm chasers and wreaks havoc for unprepared bystanders.

Maritime operations such as offshore oil production may be strongly impacted by eddies transiting operating areas. In the United States, the Minerals Management Service of the Department of the Interior is responsible for issuing, regulating, and maintaining offshore leases for exploration and exploitation in the outer continental shelf. In the leased areas of the GULF OF MEXICO, marine scientists support operators by displaying location and strength of LOOP CURRENT eddies in ocean current maps. Similar requirements exist in foreign locations since oil exploration is an international business. For example, offshore of Trinidad rings shedding from the North Brazil Current need to be tracked as they migrate across active lease blocks and disrupt safe oil production operations. In order to successfully identify, measure, and monitor eddies, oceanographers collect data from satellite images, models, current meters, and drifting data BUOYS. Oceanographic casts from a ship involve lowering a CONDUCTIVITY-TEMPERATURE-DEPTH (CTD) PROBE surrounded by water sample bottles into an identified eddy. Through analyses that include the fusion of data such as TEMPERATURE, SALINITY, oxygen, chlorophyll and depth information, marine scientists build display products that identify the location and strength of oceanographic features such as eddies.

Further Reading

"Water: The Power, Promise, and Turmoil of North America's Fresh Water." *National Geographic Special Edition* 184, no. 5A. Washington, D.C.: National Geographic Society, 1993.

edge waves   Edge waves travel along the nearshore in the surf zone. They are are actually trapped at the shoreline by the refraction of incident waves and occur under high-energy conditions that are also favorable to the development of RIP CURRENTS. Edge waves may be generated by high-energy surf zones that form in response to waves that are set up by severe and rapidly moving coastal storms such as squall lines. They may also form as a result of HURRICANES or TSUNAMI. As the hurricane piles water up on the shore, the seaward directed return flow is a rip current. Similarly, as tsunami approaching the shore change shape, they pile water over the BEACH. In fact, the first runup of a tsunami may not be the largest, emphasizing the important safety factor to not return to a beach until several hours after a tsunami hits. Similar to the STORM SURGE response, after tsunami runup, energy is reflected offshore, which may generate edge waves that travel back and forth, parallel to shore. Phase speeds associated with edge waves are much less than deep-water swells and wind waves, i.e., the speed are on the order of 1 to 10 percent of the value for shallow water swell that have wave speeds that are directly proportional to water depth.

The edge wave amplitudes are highest at the shoreline and decrease exponentially with distance. All their energy is essentially contained within the surf zone. Their amplitudes are seldom more than three feet (1 m) high. Their periods are in the one to several minutes range. In some cases, based on strength, they are important factors in shaping the beach as they impact with nearshore currents such as rip and LONGSHORE CURRENTS. Coastal engineers and oceanographers do not fully understand how these waves and currents interact with one another and with variations in shallow-water BATHYMETRY.

Further Reading

Field Research Facility, Coastal and Hydraulics Laboratory, U.S. Army Corps of Engineers. Available online. URL: http://www.frf.usace.army.mil. Accessed June 18, 2007.

Ekman spiral   The Ekman spiral is a model vector (graphical) representation of how the current due to the wind stress acting on the water surface on a rotating Earth varies with depth. The Ekman spiral was first described by oceanographer Vagn Walfrid Ekman (1874–1954). Such currents are known as pure drift currents and represent a balance between the fluid friction and the CORIOLIS FORCE. The integrated mass transport due to the drift current is 90°

Air-Sea Interactions: A Biographical Sketch of Vagn Walfrid Ekman

Vagn Walfrid Ekman was born in Stockholm, Sweden, on May 3, 1874, and died on March 9, 1954, in Gostad, Sweden. He is best known for his contributions to oceanography, especially his application of mathematics and physics to develop models of ocean and atmospheric circulation, and by inventing instruments for sampling ocean water, the ocean bottom, and for measuring currents from vessels at sea. His work is attributed to common oceanographic terms such as *Ekman spiral* and *Ekman layer.*

Ekman had early experiences with natural science as evidenced by his exposure to science by his father, Frederick Laurentz Ekman, an oceanographer, and from college lectures by Vilhelm Friman Koren Bjerknes (1862–1951), one of the founding fathers of weather forecasting. Ekman attended schools in Stockholm, and then at Uppsala University, specializing in physics, mathematics, hydrodynamics, and oceanography. His doctoral dissertation in 1902 expanded on the work of meteorologist Vilhelm Bjerknes on "wind spirals," interpreting data acquired by Bjerknes and by Fridtjof Nansen (1861–1930) on the wind drift of icebergs. Using observations by his colleague Nansen, Ekman showed that the wind-driven surface drift current moves at a 45-degree angle to the wind vector, to the right of the wind in the Northern Hemisphere, and to the left of the wind in the Southern Hemisphere (there is no deflection right at the equator). The moving surface current exerts a frictional drag on the layer of water just below it, which in turn exerts a frictional drag on the water below on down into the water column forming a current distribution known as the "Ekman spiral." The net transport of water in the Ekman layer is 90 degrees to the right of the wind direction in the Northern Hemisphere, and 90 degrees to the left of the wind direction in the Southern Hemisphere. Ekman published his thesis in 1905 as a paper entitled "On the Influence of the Earth's Rotation on Ocean Currents." Ekman did much of his work at the University of Lund, Sweden, where he was made a professor of mathematical physics in 1910, a position he held for the rest of his working life.

Ekman developed the first comprehensive and physically consistent theory of ocean circulation, first for a homogeneous ocean, and later for a stratified ocean. His equations permitted calculations of currents in depth, given the wind velocity and assumptions about the fluid friction. Ekman's theoretical models divided the ocean depths into three current regimes: a surface wind-driven current, a deep current in geostrophic balance, and a deep bottom current in which bottom friction reduces the speed of the deep current in an "Ekman spiral" in a manner similar to the surface drift current.

He was not to be confined to a blackboard or a laboratory, however. In 1925, he joined the worldwide German Meteor Expedition, which pioneered field methods of 20th-century oceanography. For his accomplishments, he was made a member of the Royal Swedish Academy of Sciences in 1935. Since 1901, the academy has awarded the Nobel Prizes in physics and chemistry on December 10, the anniversary of Alfred Bernhard Nobel's (1833–96) death. The academy also awards the Crafoord Prize for research in mathematics, astronomy, geology, and biology. This award was established by Holger Crafoord (1908–82), the Swedish inventor of the artificial kidney.

Ekman studied the internal waves of the ocean and explained a phenomenon in the Norwegian fjords called "dead water" by ship captains. When in "dead water" a ship's propeller is located at the depth of a density discontinuity in the water due to fresh glacial meltwater overlying oceanic salt water. When a ship's propeller rotates in "dead water," the power of the engine goes into generating internal waves on the density interface, rather than into propelling the ship.

In further studies through 1939, Ekman investigated ocean circulation in oceans of finite depth, stratified oceans, flow over topographic feature, such as submarine ridges and troughs, and circulation in the vicinity of various ocean basins and coastlines. His work laid foundations for the present supercomputer-based numerical hydrodynamic ocean circulation models.

Further Reading

Charton, Barbara. *A to Z of Marine Scientists.* New York: Facts On File, 2003.
The Royal Swedish Academy of Sciences. Available online. URL: http://www.kva.se/KVA_Root/index_eng.asp. Accessed July 29, 2007.

—**Robert G. Williams, Ph.D.,** senior oceanographer
Marine Information Resources Company
New Bern, North Carolina

to the right of the wind direction in the Northern Hemisphere and 90° to the left of the wind direction in the Southern Hemisphere. The spiral is a graphical means of representing the solution to the equations of motion for this particular balance of forces. Ekman's solution shows that current velocity VECTORS form a descending spiral from an angle of 45° to the right of the wind (in the Northern Hemisphere) at the water surface to an angle of 180°, or anti-parallel to the wind direction, at a depth known as the depth of frictional influence. Each vector represents the current speed and direction in a thin layer of water that exerts a frictional drag on the layer below, beginning with the surface. The drag slows each succeeding layer (vector) such that it is slightly reduced in speed and increased in angle with respect to the wind direction. The vectors decrease in length (speed) going down the spiral until they are reduced to a value of $e^{-\pi}$ at a depth called the depth of frictional influence. When the velocity vectors from the surface down to the depth of frictional influence are integrated to get the net transport, it is found to be 90° to the right of the wind direction in the Northern Hemisphere. In the Southern Hemisphere, the surface current is found to be 45 degrees to the left of the wind direction, the spiral curves down to the left until it reaches the depth of frictional influence, and the net transport is found to be 90° to the left of the wind. There is no deflection right at the equator. The water column from the surface to the base of the Ekman spiral is known as the Ekman layer.

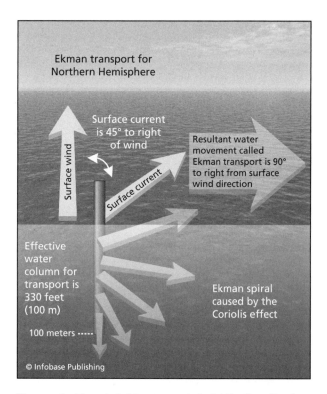

Ekman spiral for wind-driven currents in the Northern Hemisphere, showing the right (clockwise) rotation of current with depth. The resultant water movement is 90 degrees to the right of the surface wind. The water column from the surface to the bottom of the Ekman spiral is called the Ekman layer.

Ekman's spiral was motivated by Fridtjof Nansen's observations of the drift of ICEBERGS in the polar sea (*see* the essay on oceanographic innovations, page 351). Nansen observed that icebergs drifted to the right of the wind direction with the angle of drift increasing with depth. To better understand the physics, he presented the problem to the mathematical physicist Ekman, who propounded the spiral theory in a 1905 paper entitled "On the Influence of the Earth's Rotation on Ocean Currents."

It has proved difficult to directly verify Ekman's results because he assumed an ocean without boundaries and of infinite depth to simplify the analysis, and also because components of currents due to other causes are almost always present in the ocean, such as density currents, TIDAL CURRENTS, and so forth. Nevertheless, the general character of the flow predicted for the Ekman layer is known to be correct. Ekman's theory was adapted to the flow of the atmosphere over the ground and also tested in laboratory experiments, where it was verified.

Ekman extended his theory to include oceanic flows in bounded, stratified oceans, and oceans of finite depth. At the bottom of the ocean, there is also a spiral due to the friction of the ocean floor acting on the bottom current. Ekman found that the strength of the current increases from zero right at the bottom, to the magnitude of the geostrophic current right at the boundary between the interior flow of the ocean and the bottom Ekman layer, where friction is just beginning to influence the fluid flow.

Perhaps the most important result obtained from analysis of the Ekman spiral is that the net transport of wind-driven currents is 90° to the wind direction, resulting in the pileup of water in the middle of the great ocean basins, and the establishment of the great circulation gyres in the North and South Atlantic and Pacific Oceans (*see* GYRE).

Further Reading

Neumann, Gerhard, and Williard J. Pierson, Jr. *Principles of Physical Oceanography*. Englewood Cliffs, N.J.: Prentice Hall, 1966.

Ekman spiral and transport   EKMAN SPIRAL and transport theoretically explains the effect on water of wind blowing over the ocean surface, whereby the net water transport is a sum of water movement in layers down to the depth of frictional influence. The CURRENT's structure spirals with depth away from a horizontal boundary such as the surface. As the current rotates, its speed diminishes with increasing distance from the boundary. The horizontal Ekman layer caused by steadily blowing winds across the sea surface is approximately 328 feet (100 m) thick, which is shallow compared to the deep ocean. The actual depth of this layer is known as the Ekman depth and is where current velocity is opposite the velocity at the surface. The net transport of water by the wind is obtained by integrating the total flow within the Ekman layer. The movement of this water is 90° to the right of the wind in the Northern Hemisphere, and 90° to the left in the Southern Hemisphere. It is difficult to observe the Ekman spiral owing to the propagation of surface waves and the presence of turbulent mixing in the surface layer of the ocean. However, offshore Ekman transport can be observed by satellite imagery when cool ocean waters from depths of not more than 600 feet (182.9 m) move in to replace warmer water that has been transported away from favorable coasts. The cooler waters UPWELLING to the surface are rich in nutrients, which supports seaweed growth and PHYTOPLANKTON blooms. This productivity forms the energy base for populations (fish, birds, mammals) that are higher in the food chain, which usually contributes to phenomenal fishing grounds. Classic examples of upwelling are found along the California and Peru coasts.

Further Reading
Neumann, Gerhard, and Williard J. Pierson, Jr. *Principles of Physical Oceanography.* Englewood Cliffs, N.J.: Prentice Hall, 1966.

electrolysis Decomposition of a fluid by electricity is called electrolysis. Salt water conducts electricity and allows ions to move around freely, which separates compounds into simpler substances. Salt is the main material dissolved in seawater and is also widely distributed in solid deposits. As an example, the process of electrolysis with ionically bonded elements such as a solution of salt water composed of sodium chloride (NaCl) or common salt produces chlorine. Thus, for electrolysis to occur, the compound has to be ionically bonded and when in the liquid state either melted or dissolved in water. In the example, seawater is the electrical conductor or electrolyte in which current is carried by ions rather than free electrons (as in a metal such as copper).

While students perform electrolysis experiments that illustrate that water molecules are actually a compound of hydrogen and oxygen gases, marine scientists apply electrolysis to cope with galvanic corrosion. Connected pieces of metal such as CURRENT METERS and MOORING frames are often deployed into the oceans, thereby creating a battery. In the electrolytic cell where the electrolysis reaction occurs, the anode is the positive electrode, while the cathode is the negative electrode. During this process of electrolysis, the positive cations are forced to migrate toward the negative electrode (cathode). Likewise, the negative anions are forced to migrate toward the positive electrode (anode). Electrons move from the anode to the cathode. The current that flows will actually remove metal from one of the attached metal pieces such as the instrument housing through electrolysis.

To compensate for spontaneous electrolytic processes or more precisely galvanic corrosion, marine technicians attach zinc plates or collars to instrument moorings and other items such as propellers. Watermen such as lobstermen and crabbers extend the life of their lobster traps and crab pots by attaching zinc bars. Even oil drilling platforms and large ships are protected in this manner. Zincs are attached to specific items that need protection in order to induce a current flow. To complete the electrical circuit, the zincs are connected to instrument casings, propeller shafts, or other important structures. The zinc dissolves slowly while protecting the less active metal receiving the negative charge. For this reason, these zinc plates are often called "sacrificial anodes."

electronic charting display information system *See* ELECTRONIC CHARTS.

electronic charts Electronic charts contain digitized nautical information such as BATHYMETRY, locations of coastal structures, and aids to navigation for display using computers. One example developed by the National Geospatial-Intelligence Agency is called the Digital Nautical Chart or DNC®. This product is a digital database containing VECTOR information that can be used to support marine navigation using Geographic Information Systems. These vector chart systems are able to manipulate data that is stored in many files or layers. Less useful are raster chart data consisting of a digitized picture representing one layer of data. The combined navigation system that displays electronic charts is an Electronic Chart Display and Information System (ECDIS). All of these systems, which include data and computer software, are intended to allow mariners to navigate safely without paper charts in the 21st-century pilothouse. Electronic charts apply GLOBAL POSITIONING SYSTEM technologies to allow the navigator to continuously assess the position and safety of his or her vessel.

Electronic charts can be swiftly updated to display the most recent information on navigation hazards such as wrecks, toxic chemical spills, channel depth changes due to sediment movement in severe storms, and the like. They can incorporate near real-time information on winds, waves, tides, and currents from environmental measurement systems such as NOAA's PORTS® (Physical Oceanographic Real-Time System).

See also DIGITAL NAUTICAL CHART®.

Further Reading

The International Hydrographic Organization. Available online. URL: http://www.iho.shom.fr. Accessed June 18, 2007.

Office of Coast Survey, National Ocean Service, National Oceanic and Atmospheric Administration. Available online. URL: http://chartmaker.ncd.noaa.gov. Accessed June 18, 2007.

The United Kingdom Hydrographic Office. Available online. URL: http://www.ukho.gov.uk. Accessed June 18, 2007.

electron microscope   An electron microscope is a scientific instrument that utilizes a beam of electrons to produce an enlarged image. The resultant image of PHYTOPLANKTON, thin slices of rock, or other specimens illustrates shape and size, texture, composition, and atomic structure. The electron microscope facilitates basic cell, reproductive biology, and neurobiological studies and helps scientists gain insights into important topics such as microbiologically influenced corrosion in deep-sea shipwrecks.

Further Reading

National Center for Electron Microscopy, Lawrence Berkeley National Laboratory. Available online. URL: http://ncem.lbl.gov. Accessed June 18, 2007.

Physics Laboratory, U.S. National Institute of Standards and Technology. Available online. URL: http://physics.nist.gov. Accessed June 18, 2007.

"Scanning Electron Microscope." Museum of Science. Available online. URL: http://www.mos.org/sln/SEM. Accessed September 18, 2007.

"Transmission Electron Microscopy (TEM): Sample Preparation Guide." Center National de la Recherche Scientifique. Available online. URL: http://temsamprep.in2p3.fr. Accessed September 18, 2007.

El Niño   Every two to seven years, around Christmastime, the cold waters of the PERU CURRENT flowing northward off the coasts of Chile and Peru are interrupted by a flow of warm water from the northwest. The normal UPWELLING that brings cold, nutrient-rich waters to the surface off Peru is reduced and the resident fish move elsewhere, thereby cutting off the food supply of the abundant seabirds. The birds die in great numbers and foul the beaches. The unusually warm water results in heavy rainfall and catastrophic mudslides, bringing death and devastation to many of the people of Peru, Chile, and Ecuador. It was once thought that all the fish perished,

but it is now known that many schools are only moving into deeper waters. These El Niño events seem to last for one to one-and-a-half years.

The El Niño (the "boy child," which refers to the Christ child since it often comes around Christmas) has been known to the world for more than 100 years, but was originally thought to be a phenomenon local to the west coast of South America, brought on by changes in the local wind patterns. Now it is known that the El Niño is not related to local wind patterns but rather to weather and sea conditions across the entire tropical PACIFIC OCEAN, and is closely related to the Southern Oscillation (SO). The SO refers to change in atmospheric PRESSURE, winds, and ocean TEMPERATURE that occurs between the eastern and western regions of the Pacific Ocean basin. Usually, the air pressure is high in the Eastern Pacific, the winds are from the southeast (the southeast trade winds), the climate is dry, almost desertlike, and the ocean surface is cold. In the Western Pacific, around Australia and Indonesia, the pressure is low; there is abundant precipitation and warm ocean water.

The Southern Oscillation is a seesaw effect, in which these conditions reverse: The pressure in the

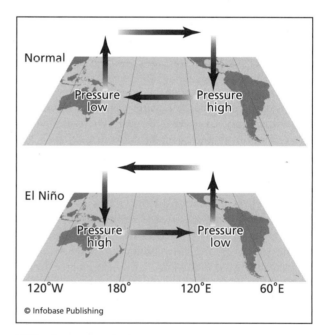

El Niño is an atmospheric and oceanic oscillation in the tropical Pacific having important consequences for weather around the globe. In the normal situation, surface pressure is high in the eastern Pacific and low in the western Pacific. This condition drives the air circulation, and the ensuing wind drives a surface sea current (the South Equatorial Current) flowing from east to west. During El Niño, the pressure is reversed, and water from the warm Equatorial Counter Current flows eastward toward Peru. This effect is called the ENSO (El Niño–Southern Oscillation).

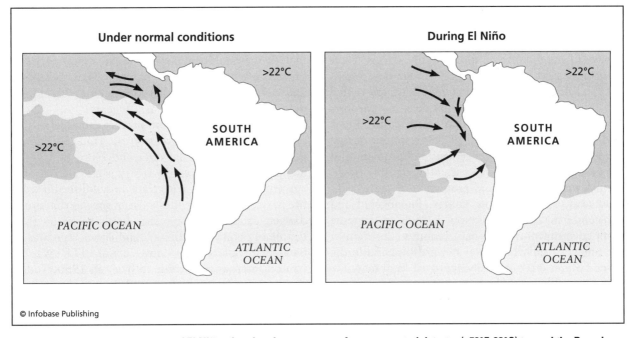

The changing temperature pattern of El Niño, showing the movement of warm equatorial water (>72°F, 22°C) toward the Peruvian coastline

Western Pacific is high, rainfall is scant, and winds are from the west. The reverse applies to the Eastern Pacific. The El Niño has been observed to develop when a huge mass of warm ocean water flows eastward toward the coast of Peru, displacing the cold waters of the Peru Current. In Australia and Indonesia, the usual summer rains are replaced by dryness and often drought, leading to failed crops and even disastrous forest and brush fires. The relation between the El Niño and the Southern Oscillation has led to the coinage of the term *ENSO*—El Niño–Southern Oscillation.

In the Eastern Pacific, changes occur in the normal weather patterns of western Central and North America in addition to South America. The reach of the El Niño is indeed far: Weather conditions in the United States, from the Pacific Northwest to the Atlantic Southeast are now known to be correlated with the onset of the El Niño–Southern Oscillation. The conditions may, in fact, be global, as weather and ocean scientists discover more and more consequences of the ENSO.

When the ENSO is over, the climate returns to "normal," but often overshoots and results in colder than normal temperatures in the Eastern Pacific and warmer than normal in the Western Pacific. This condition has been given the name La Niña (the girl child).

There is a "chicken and egg" dilemma regarding the ENSO. Does the change in atmospheric pressure and winds cause the anomalously warm sea surface, or is it the other way around; the warm sea surface changes the predominant wind patterns.

One positive result of the ENSO phenomenon is that marine scientists are now carefully studying atmosphere-ocean interactions and looking for air-sea interconnections in all of Earth's climates. Organizations such as the NATIONAL OCEANIC AND ATMOSPHERIC ADMINISTRATION (NOAA) study ENSO events such as occurrences during 1982–83 and 1997–98 in order to understand climate impacts caused by ENSO. Examples include increases in storms and droughts in the southern and eastern United States, and warmer and drier winters in the northern United States. The ENSO/La Niña has challenged climate scientists to ask whether there is such a thing as a "normal state."

Further Reading

Philander, S. George. *Our Affair with El Niño: How We Transformed an Enchanting Peruvian Current into a Global Climate Hazard*. Princeton, N.J.: Princeton University Press, 2004.

endangered species   Endangered species are attributed to populations that have been reduced to the point where they are in danger of becoming extinct. Endangerment can be caused by habitat destruction, introduction of exotic or invasive species, overexploitation, disease, and pollution. Species are plants or animals that are more or less alike in appearance, breed, and produce fertile offspring under natural conditions. A species such as the Caribbean monk seal (*Monachus tropicalis*) is extinct when living members no longer exist. The U.S. Fish and Wildlife Service likens endangered species to fire alarms. Their identification is the first step in devising plans to prevent extinction. Conservation may have prevented the extinction of the Carolina parakeet (*Conuropsis carolinensis*) that was hunted as an agricultural pest and disappeared as a result of deforestation during the early 1900s. The management and protection of the natural world has led to a scientific discipline concerned with the study and protection of the world's biodiversity: conservation biology. Endangered and threatened wildlife restoration depends on support from government and citizens, requires education of the public on environmental stewardship, and in the United States is a nationally approved legal mandate that repeatedly scores high in public opinion.

According to Paul A. Opler's 1976 article, "The Parade of Passing Species: A Survey of Extinctions in the United States," which appeared in *The Science Teacher*, more than 500 North American species had gone extinct since British colonization by the mid 1970s. Concern for the declining populations led to the Endangered Species Act (ESA), which is the federal legislation that mandates the protection of species and their habitats that are determined to be in danger of extinction. President Richard M. Nixon (1913–94) signed the ESA during 1973. The ESA specifies substantial fines for killing, trapping, uprooting (plants), or engaging in commerce in the species or its parts.

The act requires the U.S. Fish and Wildlife Service under the Department of the Interior to draft recovery plans for protected species. An endangered species list is published by the U.S. Fish & Wildlife Service to identify species protected under the ESA. The survival and recovery of a species depends on a certain minimum population base—this is called the population's "critical number." The recovery plans describe methods for the protection of a species and its habitat. Habitats must be mapped out, and programs for the preservation and management of critical habitats must be developed in order that the species can rebuild its populations.

As an example, the West Indian manatee (*Trichechus manatus latirostris*), which is found in the southeastern United States, is an example of an endangered species. Conservation biologists use photo-identification techniques to study reproductive traits and population dynamics, measure manatee response to restoration of natural hydrological cycles, and investigate the ecology of submerged aquatic vegetation and manatee diets. One of the accommodations that have been extended to mana-

tees is the establishment of "no wake zones" in certain areas of the intracoastal waterway, known to be frequented by these animals. Other accommodations include the establishment of estuarine sanctuaries to protect essential manatee habitat. Many scientists and naturalists report that the number of Florida manatees has risen in the last 20 years.

An international step to protect endangered species was the 1973 Convention on Trade in Endangered Species of Wild Fauna and Flora or CITES. Signed by 118 nations, the treaty is an international agreement that deals with trade in wildlife and wildlife parts. CITES provides a list of species that are in danger, categorizes them into levels based on their degree of "endangeredness," and either regulates or bans their trade. A well-known act of CITES was the ban on international trade in ivory in 1990. Today, there are approximately 5,000 species of animals and 28,000 species of plants protected by CITES against overexploitation through international trade. According to the convention, not one species protected by CITES has become extinct as a result of trade since the convention entered into force.

Further Reading

Czech, Brian, and Paul R. Krausman. *The Endangered Species: Act History, Conservation, Biology, and Public Policy.* Baltimore, Md.: The Johns Hopkins University Press, 2001.

Opler, Paul A. "The Parade of Passing Species: A Survey of Extinctions in the U.S." *Science Teacher* 43, no. 9 (1976): 30–34.

endpoint A biological or chemical change used as a measure of the effect(s) of a chemical or physical agent on a living organism is called an endpoint. If the anticipated change occurs, the measured or estimated dose or exposure associated with the change is recorded. Endpoints range from objective criteria such as death or deformity of the organism, to subjective criteria such as altered enzyme levels or behavior abnormalities. Researchers have distinguished between assessment endpoints (actual environmental values to be protected) and measurement endpoints (responses to a stressor with respect to the assessment endpoints).

Determining endpoints is an important part of ecological risk assessment. While the risk assessment evaluates potential adverse effects human activities have on living organisms, endpoints measure effects of a substance on an ecosystem, such as the affect of excess nitrogen from agricultural runoff on the health of a river, as measured by chlorophyll-a. During the problem formulation phase of an ecological risk assessment, researchers identify assessment endpoints that support an analysis plan. Measurement endpoints are determined during the study design and data quality phase for an ecological risk assessment. Members of the benthic community could be selected as the assessment endpoints because of their bioaccumulation capability. Numerical values given by water quality standards such as pH are good examples of measurement endpoints. Geographic Information Systems can be used to describe endpoints and to determine their spatial correlation.

Coastal zone managers might use endpoints to provide an explicit expression of an environmental value that has to be protected, for example to understand fisheries' impacts from storm water runoff near one of Chesapeake Bay's tributaries. In Chesapeake Bay, striped bass (*Morone saxatilis*) or rockfish are a valued ecological entity, where reproduction and age class structure are important attributes. Since these fish enter rivers in the spring to spawn, "rockfish reproduction and age class structure" form an assessment endpoint. Even though it might be difficult to prove causation through observations, a measurement endpoint might be that of no rockfish in the river during spring freshets. The endpoints are used to provide characteristics of CHESAPEAKE BAY rockfish that are believed to be at risk due to storm water runoff.

environmental engineer A professional who uses mathematics, basic science, and Earth science to develop solutions to environmental problems is referred to as an environmental engineer. These engineers deal with assessing or managing anthropogenic effects, that is, the consequences of human activities on the natural and man-made environments. Example projects environmental engineers work on include designing municipal water supply and wastewater treatment systems, working with health and safety experts to ensure a hazard-free working environment, managing natural resources, and mitigating pollution. Similar to ENVIRONMENTAL SCIENTISTS, environmental engineers conduct basic and applied research, analyze scientific data, and write technical reports. Efforts by environmental engineers help organizations stay in compliance with governance such as the U.S. ENVIRONMENTAL PROTECTION AGENCY, which was established in 1970 by executive order.

Environmental engineers concerned with local and worldwide environmental issues collaborate with other scientists and engineers. Multidisciplinary and integrated teams can pool skills to study and devise strategies to cope with the effects of acid rain, GLOBAL WARMING, and air pollution. They also are involved in implementing good environmental stewardship in order to protect wildlife. Experienced environmental engineers may work as consultants to help clients clean up existing hazards and follow regulations such as the U.S. National Environmental Policy Act or NEPA. Requirements and procedures for environmental assessments and environmental impact statements are described by NEPA, which is one way that federal agencies incorporate environmental considerations into planning and action. Environmental impact statements are technical reports establishing the potential impact of a construction project on the environment.

Environmental engineers often specialize in particular industry sectors such as water resources, hazardous waste, and construction where efforts focus on preventing mishaps such as oil spills rather than having to react and execute expensive clean-up plans. They apply their basic science and engineering fundamentals to evaluate complex processes that change over time and space in their business sector. Environmental engineers determine locations using the GLOBAL POSITIONING SYSTEM (GPS) and use other sophisticated sensors to measure parameters such as air and water quality. Geographic information systems, satellite imaging, and other geospatial technologies are especially useful for land reclamation, wetland mitigation, designing water distribution systems, and installing waste repositories. It is especially important for environmental engineers to stay abreast of advances in environmental science, technological innovations, and the full range of environmental issues from hazardous waste cleanup to the prevention of water pollution.

Environmental Protection Agency (EPA) A federal agency of the United States responsible for the formulation and enforcement of regulations governing the release of pollutants and other substances that may be harmful to public health and the environment, EPA is responsible for establishing national standards for environmental programs. The agency delegates to the states and Native American tribes the responsibility to issue permits and to monitor and enforce compliance. When EPA standards are not met, the agency can impose sanctions as well as provide assistance in meeting the standards. The agency funds research and monitoring programs for assuring that the nation has clean air, water, and land free of toxic chemicals, radiation, and pathogenic hazards. EPA issues research grants and graduate assistantships for developing new research and monitoring technologies. The agency performs environmental assessments, leads and coordinates partnerships with agencies of other nations, academic institutions, private sector organizations, and other U.S. federal agencies. In fact, EPA partners with more than 10,000 federal agencies, state and local governments, industries, academic institutions, and nonprofit organizations in programs to conserve water and energy, reduce greenhouse gases, and reduce toxic emissions into the air and water.

The EPA approves and monitors programs established by state and local governments. Many states have used the U.S. EPA as a template for the establishment of state EPAs, such as the California Environmental Protection Agency. The EPA performs its comprehensive and varied missions from 10 regional offices and 12 laboratories nationwide.

Further Reading

Environmental Protection Agency. "About EPA." Available online. URL: http://www.epa.gov. Accessed July 22, 2008.

environmental scientist Scientists who apply the laws, theories, concepts, and methodologies of science to issues related to the degradation of the natural environment and to the human condition are called environmental scientists. Environmental scientists are interested in all aspects of Earth and its living creatures, that is, the area often referred to as the ecosphere. Data of interest to environmental scientists includes moisture, TEMPERATURE, PRESSURE, winds, and CURRENTS as well as the locations where organisms are found. Researchers in environmental science study problems ranging from the hole in the ozone layer to the environmental problems that would be caused by burying nuclear waste products under the floor of the deep ocean basins. Investigations cover a wide range of geographies, ranging from the increase in carbon dioxide due to the clearing of tropical rainforests to the spilling of oil from pipelines laid across the Arctic tundra. Strategies are developed in order to sustain natural resources so that they will be available for future generations.

Environmental scientists work with other professionals to study and attempt to predict the environmental consequences of large-scale projects such as building military bases and airfields, marine mineral extraction, coastal recreational development, and so forth. They prepare environmental impact statements for resource managers, municipal, county, and state managers, stakeholders, and the public. Environmental scientists provide conscience in a busy and complex world by refining understanding of nature and asking questions such as, should people be allowed to hunt whales.

Environmental science really began in earnest in the mid-20th century; the consequences of industrialization were becoming painfully evident when strip-mined lands leached toxic chemicals. Polluted air and water despoiled forests. The consequences of industrialization became clear through overpopulation and the widespread occurrence of human diseases caused by inhalation, or ingestion of industrial products such as asbestos, lead paint, PCBs (polychlorinated biphenyls), and all forms of industrial waste products. The books of Rachel L. Carson (1907–64), including *Silent Spring* in 1962, along with works of other Earth science authors helped to initiate major environmental movements, subsequent legislation, and the establishment of the ENVIRONMENTAL PROTECTION AGENCY in 1970.

Perhaps the newest area of concern to environmental scientists is that of environmental terrorism: the intentional destruction of the environment for sinister purposes, such as the oil fires in the Persian Gulf set by Saddam Hussein (1937–2006) during the 1991 Gulf War, which is also known as Operation Desert Storm. This environmental sabotage was intended to despoil the atmosphere and ocean and to create environmental hazards to military forces and civilian workers.

Environmental scientists come from all fields of science; biology, chemistry, physics and the physical sciences, such as geology, meteorology, oceanography, and space science. Environmental scientists have to work as multidisciplinary and integrated teams since natural processes are linked and natural resources can be destroyed. The importance of this is seen by the extinction of species such as the passenger pigeon (*Ectopistes migratorius*) and growing lists of endangered species such as the hawksbill turtle (*Eretmochelys imbricata*). Environmental solutions to one problem will have effects on other problems. The art of environmental science might involve avoiding making some conditions worse while trying to improve others.

The training of environmental scientists usually includes a B.S. degree in a basic science and graduate study to at least the master's degree level in their chosen field of specialization, such as marine biology. Those desiring to do research should obtain the Ph.D. Some environmental scientists go on to take advanced degrees in environmental planning, policy, management, as well as environmental law.

Environmental scientists are employed by government agencies, such as the Environmental Protection Agency, the Geological Survey, and the NATIONAL OCEANIC AND ATMOSPHERIC ADMINISTRATION. State government and private sector employment opportunities may be found in academic institutions, business and industry, nonprofit organizations, stakeholder groups, and congressional lobbying groups.

To keep abreast of recent developments in their fields, environmental scientists spend considerable time reading scientific journals, attending scientific, planning, and policy meetings, and organizing public hearings and workshops. They take courses in new technologies that offer to advance the state of the art of their science. Scientists using the Internet exchange much data and information worldwide.

Further Reading

Carson, Rachel. *Silent Spring*. Boston, Mass.: Houghton-Mifflin, 1962.

Earle, Sylvia A. *Sea Change: A Message of the Oceans*. New York: Fawcett Columbine, 1995.

Environmental Protection Agency. Available online. URL: http://www.epa.gov. Accessed September 18, 2007.

epipelagic zone   The top layer of the ocean, extending from about 300 to 700 feet (91–213 m) in depth is classified as the epipelagic zone. Waters in the NERITIC ZONE, near the CONTINENTAL SHELF, are usually less clear than the open ocean waters because of higher biological productivity from land runoff and because of higher sediment content. The epipelagic zone approximately coincides with the PHOTIC ZONE; the region of significant light penetration. This is the ocean zone of primary (plant) productivity: PHYTOPLANKTON, and SEA GRASSES such as Sargassum weed. This is the home of herbivorous ZOOPLANKTON and other herbivorous marine animals. This is the region of the most familiar ocean creatures, such as sharks, tuna, mackerel, and dolphins. Since there is no production below the photic zone, except for small areas of the ocean floor where HYDROTHERMAL VENTS occur, the rain of dead organisms (detritus) forms the base of the food chain for the creatures in the deeper zones.

Because of their proximity to the surface, the plants and animals of this zone are the most vulnerable to the effects of pollution from oil spills and from overfishing by such means as long lines, which may stretch for 40 miles (67 km) in the open sea. These and many types of moored nets can become zones of death, not only for target species such as swordfish, but for all kinds of animals such as porpoises and turtles. There is scarcely any part of the world ocean that is not soiled by tar balls, plastics, torn nets, beer and soda cans, and other trash tossed overboard by people on the many commercial and recreational vessels transiting the sea. It was once believed that the oceans had a limitless capacity to absorb the waste products of human civilization. Now it is known that many ocean ecosystems are threatened with long-lasting damage, and even extinction.

Equatorial Countercurrent   The Equatorial Countercurrent is a warm CURRENT setting eastward between the North and South Equatorial Currents of the ATLANTIC, PACIFIC, and INDIAN (in the Northern Hemisphere winter) OCEANS, and is located in the region of the INTERTROPICAL CONVERGENCE ZONE (ITCZ), traditionally known by mariners as the doldrums because of the usual absence of steady winds. It is primarily a zonal current, meaning that its average direction of flow is roughly parallel to the lines of latitude. In the Atlantic and Pacific Oceans, its main axis of flow is about 7°N (where it is often known as the North Equatorial Countercurrent) and in the Indian Ocean, about 7°S. The equatorial current system is not symmetrical about the equator because the trade winds are not: the southeast trade winds reach across the equator into the Northern Hemisphere. The Atlantic Equatorial Countercurrent is the best known of all the equatorial countercurrents.

The Atlantic Equatorial Countercurrent (AECC) is located approximately between 3°N and 10°N, which forms the northern boundary of the South Equatorial Current and the southern boundary of the NORTH EQUATORIAL CURRENT. It spans nearly the distance from South America to Africa in the Northern Hemisphere summer, but is found primarily in the eastern Atlantic during the Northern Hemisphere winter. The major source water for the ECC is the retroflection (looping back on itself) of the GUYANA (north of Brazil) CURRENT, which occurs at about 5–8°N, 40°W, where part of the current turns sharply from its northwest course along the coast of Brazil to flow eastward into the countercurrent. Additional inflow comes from the North Equatorial Current.

The Atlantic Equatorial Countercurrent is on the order of 1,000 feet (300 m) deep and 250 miles (400 km) wide. The mean speed of the current is around a few tenths of a knot, increasing in speed where it flows into the GUINEA CURRENT at its eastern terminus. Volume transport in the western part of the current is on the order of 20 Sverdrups (20×10^6 m³/s), but varies greatly depending on location and season. The strength of the current is greatest in the Northern Hemisphere summer, accompanying the northward migration of the ITCZ, and least in the Northern Hemisphere winter, when the ITCZ is at its southernmost extent.

The AECC is most prominent in August and September, when it extends across the Atlantic from about 52°W to 10°W, longitude where it flows into and reinforces the Guinea Current. In October, the AECC narrows and begins to separate into a western part and an eastern part at around latitude 7°N, longitude 35°W. The separate parts contract as the Northern Hemisphere winter progresses and as the ITCZ moves southward. The greatest separation occurs during March and April, when the western branch may even disappear. It reappears in the late Northern Hemisphere spring near the equator at longitude 40°W, where the two segments begin to strengthen and merge again in August. During the time of disappearance at the surface, the current may be flowing eastward below the North and South Equatorial Currents.

The Pacific Equatorial Countercurrent flows between the westward flowing North and South Equatorial Currents, usually between latitudes 3°N and 10°N similar to its Atlantic counterpart. It spans nearly the entire Pacific, from the Philippines to the Galápagos Islands and at times (especially during El Niño years), Ecuador and Peru, where it floods the region with warm waters.

This current is more cleanly delineated and strongly developed than the Atlantic Equatorial Countercurrent. East of the Philippines, it is jointed by the Pacific North Equatorial Current. The ECC reaches its maximum speeds between 5°N and 10°N, usually below the surface.

In the Indian Ocean, the ECC is strongly influenced by the MONSOONS and by the currents in the Arabian Sea and the BAY OF BENGAL. The strength and location of the ECC varies greatly with location and season. The ECC is found mainly south of the equator in contrast to the Atlantic and Pacific Oceans, where it is north of the equator. In the Indian Ocean, during the northeast monsoon that usually occurs November through March, the ECC flows eastward from the equator to 8°S latitude. The flow north of the equator is mainly westward. During May through September, after the onset of the southwest monsoon, the flow north of the equator reverses, that is, toward the east. This flow combines with the Equatorial Countercurrent and the whole eastward flow is called the Southwest Monsoon Current.

Analysis of TEMPERATURE and SALINITY data in the tropical oceans has shown that the ECC is generally in geostrophic balance to within half a degree of the equator. The forces involved in driving the Atlantic ECC include the horizontal pressure gradient due to the pileup of water against the South American continent by the westward-flowing trade winds, a north-south pressure gradient resulting from the density distribution, and the CORIOLIS EFFECT. The current in the Pacific Ocean is similarly forced by the accumulation of water in the western Pacific due to the trade winds, the density distribution, and the Coriolis effect. In the Indian Ocean, the monsoons replace the trade winds as the major forcing function at the sea surface, and consequently, the Equatorial Countercurrent there is seasonally dependent according to the monsoon regime.

Further Reading

Knauss, John A. *Introduction to Physical Oceanography,* 2d ed. Upper Saddle River, N.J.: Prentice Hall, 1996.

Pickard, George L., and William J. Emery. *Descriptive Physical Oceanography: An Introduction,* 5th (SI) enlarged ed. Oxford: Pergamon Press, 1990.

University of Miami, Rosenstiel School of Marine and Atmospheric Science. "Ocean Surface Currents." Available online. URL: http://oceancurrents.rsmas.miami.edu. Accessed September 18, 2007.

Equatorial Undercurrent A high-speed jet of water flowing eastward at the equator at about 330 feet (100 m) depth in the ATLANTIC, INDIAN, and PACIFIC OCEANS is known as the Equatorial Undercurrent. The Equatorial Undercurrent (EUC) generally has a thickness on the order of 1,000 feet (300 m) and a width of about 180 to 300 miles (300 to 500 km). It spans almost the entire ocean basins in the Atlantic and Pacific Oceans. In the Indian Ocean, its behavior is somewhat influenced by the monsoons. The EUC has a high-velocity core, which may be as great as three knots (1.5 m/s). It is embedded in the westward-flowing South Equatorial Current. The EUC, along with the EQUATORIAL COUNTERCURRENT, are the primary routes for the water blown westward by the trade winds to return to the eastern ocean basins, in keeping with the principle of conservation of mass.

The EUC is characterized by a vertical spreading of isopleths—of TEMPERATURE, SALINITY, dissolved oxygen, nitrate, phosphate, silicate, and so forth. In the Atlantic, however, the EUC is delineated by a high-salinity core, which is fed by the high-salinity pool of water off the northeast coast of Brazil. The enhanced vertical mixing of the EUC adds to the already present equatorial UPWELLING to create "cold islands" of water along the equator, higher in nutrients than the surrounding waters. This feature is especially pronounced in the Pacific Ocean.

The Equatorial Undercurrent in the Pacific is often called the Cromwell Current after its discoverer, Townsend Cromwell, who observed the eastward set of long lines deployed in the tuna fishery during the early 1950s. Scientists from the former Soviet Union wanted to name the Equatorial Undercurrent in the Atlantic the Lomonosov Current, but this name never caught on in the international community.

The Atlantic Equatorial Undercurrent was actually discovered by John Y. Buchanan (1844–1925), a British chemist on the *Challenger* Expedition, and reported in 1886, but this discovery was forgotten until the EUC was rediscovered by oceanographers in the 1950s. The existence of an Equatorial Under-

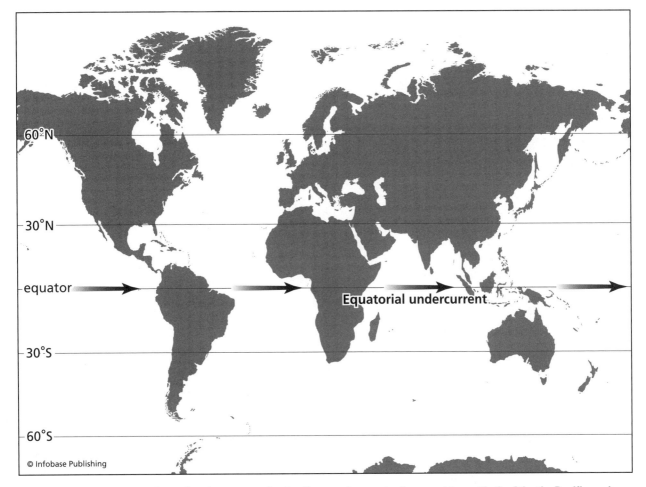

The Equatorial Undercurrent is a subsurface current jet that flows at the equator from west to east in the Atlantic, Pacific, and Indian Oceans.

current in the Indian Ocean was confirmed by oceanographers in the International Indian Ocean Expedition (1962).

The Equatorial Undercurrent is most probably a consequence of the pressure head due to the pileup of water in the western equatorial oceans caused by the stress of the westward-blowing trade winds. There is a warm, thick layer of water above the thermocline in the western ocean basins and only a thin layer of warm water above the THERMOCLINE in the eastern equatorial ocean basins. The sea surface slopes downward from west to east, while the thermocline slopes upward. This situation creates an eastward pressure gradient force along the equator. Since the CORIOLIS FORCE is zero right at the equator, the flow initiated by the pressure gradient is in a straight line. Since the effect of the Coriolis force is to cause water parcels to turn to the right in the Northern Hemisphere and to the left in the Southern Hemisphere, it acts to stabilize the flow of the undercurrent and keep it right on or very close to the equator.

KELVIN WAVES appear to be an important dynamical feature of the Equatorial Undercurrent. These are long period waves that travel along fixed boundaries. Because of the reversal of Coriolis deflection across the equator there is a fixed boundary for Kelvin waves. These waves have been observed to travel from west to east along the equator, with maximum amplitude at 0° latitude. Amplitudes decrease exponentially with distance away from the equator. Kelvin waves have also been observed to split into two waves in the eastern terminus of the EUC. One runs north while the other trends south along the western equatorial coast of Africa. Kelvin waves in the Equatorial Pacific Ocean may be an important factor in the dynamics of the EL NIÑO. These large-scale wave motions or gravity waves were discovered by Sir William Thompson (1824–1907), who later became Lord Kelvin in 1879.

Further Reading

Knauss, John A. *Introduction to Physical Oceanography*, 2nd ed. Upper Saddle River, N.J.: Prentice Hall, 1996.

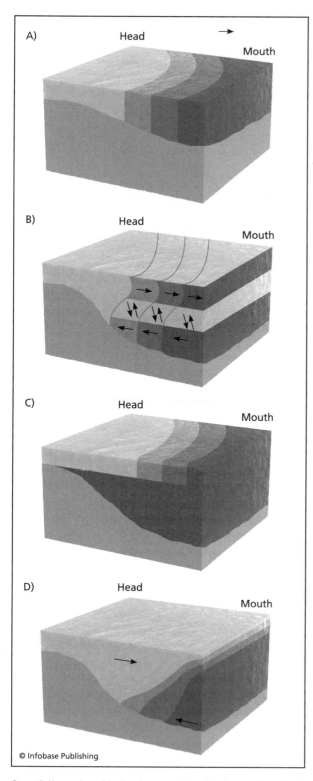

A partially enclosed body of water where freshwater from land drainage is mixed with salt water from the ocean. The four types of estuaries categorized by physical oceanographers according to salinity structure are shown in the four figures above by the isohalines (lines of equal salinity): A) the first is the vertically mixed estuary: B) the partially stratified estuary; C) the highly stratified estuary; D) the salt wedge estuary.

estuary An estuary is classified as a partially enclosed body of water, such as a bay or sound, fronting on the open sea, with a free exchange of water, so that the salt water of the ocean mixes with the freshwater of the embayment. Examples include coastal LAGOONS such as Albemarle Sound in North Carolina or Laguna Madre located in Texas and Mexico, drowned river valleys such as CHESAPEAKE BAY and Delaware Bay located along the mid-Atlantic bight, and fiords or fjords such as Grise Fiord in Canada and Lysefjorden in Norway. A bight is a slight indentation in a coast where there is not a free exchange of water. In estuaries, brackish water, which is favorable to many forms of life, is derived from the mixing of freshwater and salt water. Estuaries are generally high in nutrients (from land runoff into the rivers feeding the estuary). They serve as nurseries for many species of fish, such as mullet and striped bass (also known as rockfish). They have abundant populations of shellfish, for example, crabs, clams, and oysters, and seabirds, such as gulls and terns. The most well known estuaries are Chesapeake Bay, Galveston Bay, Puget Sound, the Thames, and the Norwegian fiords.

Estuaries are classified by their physical properties as "salt wedge," "partially mixed," and "well-mixed." Salt wedge estuaries generally occur on coastal plains where the tidal range is small, on the order of three feet (1 m) or less, leading to low TIDAL CURRENTS and little vertical mixing. Salt water flows in along the bottom while freshwater flows out to the sea on the top: a two-layer stratification. The position of the salt wedge moves inland with the incoming (flood) tide, and seaward with the outgoing (ebb) tide. The distance moved by the salt wedge is strongly affected by the weather. In the Mississippi River, for example, the mean position of the salt wedge may move dozens of miles (km) upriver during dry periods and many miles (km) seaward during periods of abundant rainfall, causing a high volume of freshwater discharge. Much of the mixing that occurs in large, shallow estuaries, such as Pamlico Sound, is due to wind and wave action rather than the tides, especially during the onset of winter storms and tropical HURRICANES.

Partially mixed estuaries occur when considerable vertical mixing takes place between the salt wedge and the freshwater above. Salt water is encountered at the surface of the estuary and increases steadily going seaward. The James River is an example of a partially mixed estuary. Partially mixed estuaries are found where the tide range is moderate (6–12 feet [2–4 m]).

Well-mixed estuaries are found in regions of large tidal excursion (greater than 12 feet [4 m]) and strong tidal currents. The lines of equal salinity (isohalines) are nearly vertical, meaning that the water is relatively homogeneous from surface to bottom, and increases in the seaward direction. The Bay of Fundy is an example of a well-mixed estuary. The vertical stratification of properties such as TEMPERATURE, SALINITY, dissolved oxygen, nitrate, phosphate and the like, is strong for a salt wedge estuary, and weak for a well-mixed estuary.

Large estuaries such as Delaware Bay have significant horizontal circulations, with freshwater flowing seaward more strongly on the right side of the estuary, and salt water flowing more strongly landward on the left side of the estuary (in the Northern Hemisphere; the reverse is found in the Southern Hemisphere). This flow pattern leads to a counterclockwise circulation that results in horizontal gradients of salinity and other properties. At the center of the circulation is a "null point," somewhat like an AMPHIDROMIC POINT in the ocean, a point of very little motion around which the circulation occurs. This flow pattern is a result of the CORIOLIS EFFECT.

The current at any particular location in an estuary results from the thermohaline (temperature-salinity) density-driven flow, known as the "residual flow," tidal flow, and the flow due to changing atmospheric and oceanic weather conditions. River discharge is high during conditions of heavy rainfall and low during drought conditions. Tides go through cycles of spring tides (large tides) and neap tides (smaller tides). The component of flow due to the weather is highly variable.

The residual flow of an estuary (the net flow after tidal and weather-induced transient flows have been averaged out over a long time period) can be measured if a large number of instruments recording temperature, salinity, current velocity, and so forth are deployed for time periods of many months. Tidal currents can be accurately predicted if the BATHYMETRY of the estuary is well known and current meter records are available from a few well-chosen locations. The flow due to weather-induced currents is difficult to predict, and requires large, complex numerical circulation models that require supercomputers to implement, and a large number of environmental sensors to provide the input data. Because of this difficulty in accurately predicting currents, many coastal harbors have installed real-time current measurement systems, such as NOAA's PORTS® (Physical Oceanographic Real-Time System) to provide essential current information to commercial users such as the shipping industry, as well as recreational boaters.

Shallow salt wedge estuaries are subject to high inputs of nutrients such as fertilizers from land runoff, which leads to excessive amounts of nutrients in the water column, resulting in blooms (prolific growths) of ALGAE and other aquatic plants, especially during warm summers. This condition is known as "eutrophication." When the algae die (the bloom "crashes"), they sink, and consume much of the dissolved oxygen in the water column, robbing the resident fish and benthic creatures of this life-sustaining gas. The absence of dissolved oxygen in the water column is a condition known as "anoxia." When anoxia occurs, many of the animals die; fish can be found floating in large numbers dead in the water. Sometimes, there is a bloom of poisonous PHYTOPLANKTON, especially dinoflagellates, which cause fish disease that results in massive fish kills. These and other planktonic-like organisms, such as *Pfiesteria piscicida* can attack the fish, leaving sores, bleeding ulcers, and fin rot, which results in their deaths. Polluted estuaries seem to be especially vulnerable to such attacks.

Estuaries are especially vulnerable to the effects of pollutant spills from agriculture, industrial plants, high-density population centers, and suburban developments. The massive spill of 11,000,000 gallons of crude oil by the tanker *Exxon Valdez* into Prince William Sound, in April 1989, is a chilling reminder of how devastating the accidental discharge of petrochemicals can be and the urgent need for environmental protection.

Estuaries are enormously high in productivity and are essential for the maintenance of coastal fisheries. They provide food, employment, and recreation to millions of people worldwide. Human activities, such as overfishing, overdeveloping, and inadequate municipal sewage treatment can destroy

Estuarine wetlands are critical to protecting the coast and its ecosystems. *(NOAA)*

these resources of incalculable value. Their productivity is also at risk because of invasive "exotic" species, such as zebra mussels (*Dreissena polymorpha*) and snakehead fishes (Channidae), which are often introduced into an estuary by discharge of ship's ballast water, or from private aquaria. When they have few natural enemies, they may outcompete the native species. A well-known historical example is the devastation of the lake trout (*Salvelinus namaycush*) fishery in the Great Lakes caused by lamprey eels (*see* AGNATHANS).

Further Reading

Australian Institute of Marine Science. Available online. URL: http://www.aims.gov.au/. Accessed June 18, 2007.

National Estuary Program, U.S. Environmental Protection Agency. Available online. URL: http://www.epa.gov/owow/estuaries/about1.htm. Accessed June 18, 2007.

Eulerian current meters   *See* CURRENT METER.

euphotic zone   The surface layer of the EPIPELAGIC (PHOTIC) ZONE, where sunlight is sufficiently strong to support photosynthesis, is known as the euphotic zone. This is the zone where marine organisms ranging in size from bacteria to kelp convert light energy into organic compounds in a rather complex set of reactions. Thus, it is in this, the very surface layer, less than 330 feet (100 m) deep, that all plants are produced. In fact, in most locations, the euphotic zone occurs within the top 100 feet (30 m) of the water column. This is the zone of primary production, which is the basis for all the food chains in the ocean, except the hydrothermal vent communities (*see* HYDROTHERMAL VENTS). The epipelagic zone is illuminated by sunlight below the euphotic zone to about 660 feet (200 m), but the light intensity is insufficient to sustain photosynthesis. Thus, creatures below the euphotic zone feed primarily on the dead material (detritus) that falls from the ocean's surface layer.

Below the euphotic zone is the disphotic zone, 660–3,280 feet (200–1,000 m) in depth, often called the "twilight zone," where a dim illumination is present, but is insufficient to support plant growth. Because of absorption of light through the water column, the only light in this zone is blue-green and violet. There is no light for photosynthesis. In order to find food in this dim region, animals tend to have big eyes, jaws, and teeth. The aphotic zone is the deepest zone, referred to as the "midnight zone," below 3,280 feet (1,000 m). There is never any light here, except for the flashes of bioluminescent predators (*see* BIOLUMINESCENCE) seeking prey, prey seeking to confuse the predators with their own bioluminescence (like countermeasures in warfare), or denizens of the deep seeking mates. In the region of hydrothermal vents, there is sometimes a glow from hot magma discharging hot gases, such as sulfur and methane into the cold ocean through cracks in the seafloor. Thus, the greatest volume of the sea is in almost complete darkness.

In the first descent of humans below 525 feet (160 m), during June 1930, William Beebe and Otis Barton reported that below 600 feet (183 m) there was a shade of blue to the ocean that no human had ever seen. Below 1,000 feet (305 m), there was darkness.

Further Reading

Ballard, Robert D. *The Eternal Darkness: A Personal History of Deep-Sea Exploration.* Princeton, N.J.: Princeton University Press, 2000.

Europa   Europa is one of the four Galilean moons of Jupiter named after Galileo Galilei (1564–1642), who discovered it in 1610. It is the second-nearest of these moons to the planet after Io. Europa is remarkable in that it may have a salt-water ocean beneath an outer crust of ice. The ocean may be as deep as 60 miles (96.6 km) and may support some form of life.

Europa is about 402,600 miles (671,000 km) from Jupiter; its orbital period is a little over nine Earth days; its radius is about 940 miles (1,570 km), which makes it a little smaller than Earth's Moon. Europa's core is believed to consist of significant amounts of iron compounds. It has a rocky mantle between the core and its hypothesized ocean, and a crust of ice above the ocean.

Europa's surface temperature is about -240°F (151°C). It is bombarded by intense radiation from Jupiter. Tidal excursions of the icy surface are on the order of 100 feet (30 m). This magnitude of excursion strongly suggests the presence of liquid water underneath the ice. Astronomers have deduced that meteoric impacts on the surface are quickly erased, implying active processes on the planet below. Features seen on the surface of Europa are similar to those seen in Earth's Arctic. The disruptions seen in the ice on the surface of Europa are also indicative of liquid water below. Thus, there is considerable evidence for an ice-covered ocean on Europa. One might wonder how liquid water could exist on the moon of a planet whose distance from the Sun is about five times that of Earth, and whose surface temperature is so low. The answer is in Jupiter's enormous gravitation, which causes tremendous tidal forces on Europa that flex the moon's surface and interior, producing characteristic patterns of cracks and ridges on the ice that would be expected

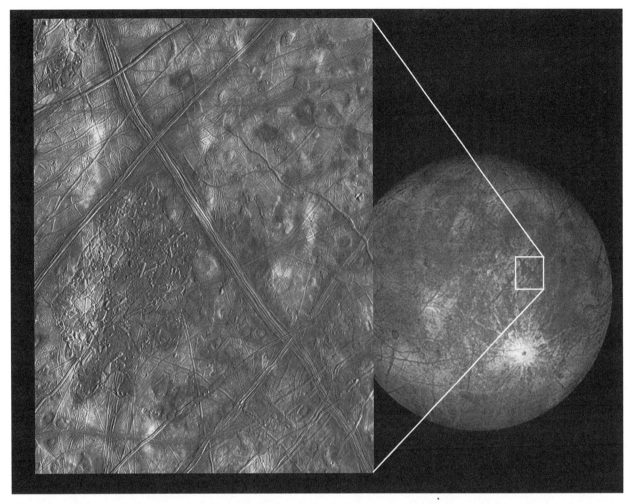

The enlarged image on the left shows a region of Europa's crust made up of blocks that planetary scientists think have broken apart and "rafted" into new positions. These features are the best geologic evidence to date that Europa (seen in a global perspective on right) may have had a subsurface ocean at some time in its past. Combined with the geologic data, the presence of a magnetic field leads scientists to believe an ocean is most likely present on Europa today (below its icy surface). Long, dark lines are ridges and fractures in the crust, some of which are more than 3,000 kilometers long. These images were obtained by NASA's *Galileo* spacecraft during September 7, 1996, December 1996, and February 1997 at a distance of 677,000 kilometers.   *(NASA/JPL)*

for an ice-covered ocean. The ice is estimated to be roughly 15 miles (25 km) thick.

This knowledge of Jupiter's remarkable moon was obtained by means of NASA's *Voyager 1* and *2* space probe flybys in the late 1970s. More recent information was obtained by NASA's *Galileo* spacecraft, which went into orbit around Jupiter in 1997 and remained in operation for four years, until it was deliberately plunged into Jupiter's thick atmosphere, where it telemetered data until friction caused it to burn up.

NASA is planning to send a spacecraft to Europa in the years ahead that will have instruments to confirm the existence of the ocean. A study is planned to launch a probe from the spacecraft to land on Europa and bore a hole through the ice and explore the

ocean below. It is possible that some form of primitive life exists similar to the chemosynthetic life forms found in the HYDROTHERMAL VENT communities on Earth. The probe would be a more advanced version of probes presently used to bore through the Antarctic ice and explore lakes covered by the ice and snow that grips the continent.

Further Reading

Chaisson, Eric, and Steve McMillan. *Astronomy: A Beginner's Guide to the Universe,* 4th ed. Upper Saddle River, N.J.: Pearson/Prentice Hall, 2004.

Mackenzie, Dana. "Is There Life Under the Ice?" *Astronomy* 29, no. 8 (August 2001): 33–37.

Pappalardo, Robert. "Jupiter's Water Worlds." *Astronomy* 32, no. 7 (July 2004): 34–41.

Exclusive Economic Zone (EEZ) The Exclusive Economic Zone or EEZ is a region of the coastal ocean adjacent to the territorial sea of the United States seaward 200 nautical miles (230 statute miles [370 km]) measured from the baseline from which the territorial sea is measured. It includes islands in the ATLANTIC and PACIFIC OCEANS claimed by the United States. These regions were formerly known as the Fishery Conservation Zone. The EEZ was established as part of the Magnuson FISHERY CONSERVATION AND MANAGEMENT ACT of 1976, whose sponsor was Senator Warren Magnuson of Washington. It states that within the EEZ, the United States has sole management authority over finfish, mollusks, crustaceans, and all animal and plant life exclusive of MARINE MAMMALS and birds. The Magnuson Act also provides for United States management authority over continental shelf fisheries resources throughout the entire migratory range of each species, unless they are within another nation's territorial waters.

The EEZ is part of a worldwide policy of the United Nation's Convention on the LAW OF THE SEA, held in 1982. This convention recognized the nations bordering on the ocean have jurisdiction over the seas within 200 nautical miles (230 statute miles [370 km]), their Exclusive Economic Zones. This convention also recognized the free access and use of the "high seas." For example, Article 56 states, "In the exclusive economic zone, the coastal State has: (a) sovereign rights for the purpose of exploring and exploiting, conserving and managing the natural resources, whether living or non-living, of the waters superjacent to the seabed and of the seabed and its subsoil, and with regard to other activities for the economic exploitation and exploration of the zone, such as the production of energy from the water, currents and winds; (b) jurisdiction as provided for in the relevant provisions of this Convention with regard to: (i) the establishment and use of artificial islands, installations and structures; (ii) marine scientific research; (iii) the protection and preservation of the marine environment; (c) other rights and duties provided for in this Convention." Much of the motiva-

tion for the EEZs came from the United States and Canada in reaction to the large numbers of fishing vessels, factory ships, deployed and drifting nets in the regions of the Grand Banks and Georges Bank. An especially large number of vessels came from such nations as the former Soviet Union and the German Democratic Republic (East Germany). This vast array of foreign fishing vessels devastated the cod and haddock fisheries, multibillion-dollar resources. United States and Canadian FISHERMEN lobbied extensively to get their governments to pass legislation limiting the entry of the foreign vessels to those with special permits. Unfortunately, this action came too late to save these fisheries from "crashing." Subsequent policies of the United States and Canada allowed overfishing by fishermen of their own nations until the Atlantic cod and haddock had almost disappeared.

In order to promote resource development in the EEZ, the NATIONAL OCEANIC AND ATMOSPHERIC ADMINISTRATION and the U.S. Geological Survey are mapping the zone floor with sophisticated SONAR equipment, including side scan and Sea Beam (multibeam at various frequencies) systems. Much of the topography of many of these regions is largely unknown in sufficient detail to perform such tasks as undersea mining, oil and gas extraction, and siting coastal ocean wastewater diffusion systems. Marine fisheries scientists are also studying overfished species and developing rebuilding plans that evaluate management benchmarks such as fishing mortality, optimal yield, and probability of recovery. Besides using the EEZ to control foreign fishing ship operations within the EEZ, efforts are under way to restore and guarantee the future of depleted fish stocks through management activities by the National Marine Fisheries Service.

Further Reading
Kunzig, Robert. *Mapping the Deep: The Extraordinary Story of Ocean Science.* New York: W. W. Norton, 2000.

extratropical cyclone *See* CYCLONE.

F

fathometer A fathometer is a device used to measure water depth by measuring the travel time of repeated acoustic pulses from a vessel to the ocean floor and return (echoes) and then applying the following formula: distance = velocity (of sound) × time × (1/2). The factor of 1/2 occurs because the travel time is the time of the pulse from the fathometer to the ocean floor and return; the depth is 1/2 of the total travel time.

This instrument, which can be used to find underwater relief such as guyots, is also referred to as a depthsounder, depth finder, depth gauge, or echo sounder. The act of measuring depth of water by using a lead line or a fathometer is called SOUNDING. Lead lines, which are still used today, are marked at intervals of fathoms and weighted at one end to manually determine the depth of water. Charting depths is an important component of marine science; basic researchers are continually improving the accuracy and precision of these measurements.

See also BATHYMETRY; DEPTH FINDER; GUYOT.

Further Reading
Bowditch, Nathaniel. *The American Practical Navigator: An Epitome of Navigation, pub. no. 9.* Bethesda, Md.: Defense Mapping Agency Hydrographic/Topographic Center, 1995.

Installation of an early fathometer on a survey vessel *(NOAA/NOS)*

feedwater Water fed to equipment with or without pretreatment for industrial applications such as power generation or desalinization is called feedwater. Pure feedwater for use in boilers is made in evaporators. Boilers are closed vessels in which water is heated and steam is generated under pressure or vacuum by the application of heat from combustible fuels, electricity, or nuclear energy. Feedwater is used to remove heat from nuclear reactor fuel rods by boiling and then generating steam. The steam becomes the driving force for the plant turbine generators, which make electricity. Feedwater is either supplied to a nuclear reactor pressure vessel in a boiling water reactor or to the steam generator in a pressurized water reactor. Feedwater in a boiling water reactor is allowed to boil in the core, while in the pressurized water reactor high-temperature water is kept under high pressure in a primary system. Most nuclear reactors producing electric power use pressurized water reactors. A nuclear reactor sustains and controls a self-supporting nuclear reaction and is the main component of a nuclear power plant. Reverse osmosis water purification involves forcing feedwater through a semipermeable membrane, where the dissolved impurities remain behind and are discharged into a waste system. Reverse osmosis equipment is used to desalinate seawater. Such desalination processes involve saline feedwater from ESTUARIES or the ocean, low-salinity product water, and by-products such as brine or very saline reject water. Wastewater (nonradioactive) discharged from nuclear power plants into lakes, rivers, bays, and the coastal ocean tends to warm the environment; it has been of some concern to environmental managers, since the resident biota may not be able to withstand the higher temperatures.

Further Reading

Department of Energy. Available online. URL: http://www.energy.gov. Accessed June 19, 2007.

Nuclear Regulatory Commission. Available online. URL: http://www.nrc.gov. Accessed June 19, 2007.

Viessman, Warren, Jr., and Mark J. Hammer. *Water Supply and Pollution Control*. New York: Harper and Row, 1985.

fetch The distance the wind has acted over the water to generate waves is called fetch. For instance, if the wind is blowing from the land out over the water, then the fetch six miles (10 km) seaward from the land. In the open ocean, where the fetch is essentially unlimited, if the wind has blown for a long time such that the energy that the wind is putting into the sea equals the energy that the WAVES lose in breaking, the waves cannot grow any more. This is called a FULLY DEVELOPED SEA. In some cases, a fully developed sea cannot occur because the body of water is fetch-limited, that is, the distance from the land to the location in question at sea is insufficient to produce an equilibrium state. Thus, for a sea to fully develop, the fetch has to be large enough and the wind has to be blowing for a critical amount of time. If the wind has been blowing for an insufficient time to produce a fully developed sea, the situation is called duration-limited.

Further Reading

Brown, Joan, Angela Colling, Dave Park, John Phillips, Dave Rothery, and John Wright. *Waves, Tides and Shallow-Water Processes*. Oxford: Pergamon Press, 1997.

Office of Naval Research. "Ocean in Motion: Waves—Characteristics." Available online. URL: http://www.onr.navy.mil/focus/ocean/motion/waves1.htm. Accessed September 18, 2007.

Field Research Facility (**FRF**) The Field Research Facility located in Duck, North Carolina, is a world-class high-technology laboratory for studying coastal processes. Investigations conducted at the Field Research Facility relate to coastal engineering, which is the study of the processes ongoing at the shoreline and construction within the coastal zone. Research thrusts involve aspects of nearshore oceanography, marine geology, and civil engineering, and projects are often directed at combating erosion or accretion at the coast. The Field Research Facility, or FRF, is part of the Engineering Research and Development Center of the U.S. Army Corps of Engineers. Researchers use state-of-the-art tools such as the 1,840-foot (560 m)-long pier that extends into the ATLANTIC OCEAN, the three-wheeled Crab for surveying, the Lighter Amphibious Resupply Cargo, or LARC, vehicle to support experiments, and the rail-mounted Sensor Insertion System, or SIS, to accurately deploy instruments anywhere within 75 feet (23 m) of the pier.

The Army Corps of Engineers hosts scientists from academic institutions and private industry to conduct their research, taking full advantage of the FRF's unique BARRIER ISLAND facilities. Some noteworthy experiments conducted at the FRF include Delilah, Duck 1994, and Sandy Duck 1997. These three experiments were used to improve the researcher's fundamental understanding and modeling of surf zone physics. Capabilities such as the Nearshore Video Imaging System are demonstrated for real-time display of surf zone parameters such as LONGSHORE CURRENTS and changing BEACH characteristics. Time series data are collected on and in the vicinity of the FRF from pressure arrays, TIDE GAUGES, WAVE BUOYS, and weather stations in order to study waves, tides, and shallow-water processes.

See also COASTAL ENGINEER; OCEAN OBSERVATORY; OCEANOGRAPHIC INSTRUMENTS.

Further Reading

Nichols, C. Reid, David L. Porter, and Robert G. Williams. *Recent Advances and Issues in Oceanography*. Westport, Conn: Greenwood, 1993.

U.S. Army Corps of Engineers: Engineer Research and Development Center. "Field Research Facility." Available online. URL: http://www.frf.usace.army.mil. Accessed September 18, 2007.

fish True fish or finfish are cold-blooded vertebrates breathing by gills and having fins for mobility. The body temperature of a fish changes with the environment. Respiration is accomplished by delicate blood vessels or gills located under the gill covers. Their fins perform specific mobility functions. They are propelled through the water by caudal and pectoral fins, while dorsal, anal, and adipose fins lend stability in swimming. The term *finfish* is sometimes used to separate true fish from species such as shellfish, crayfish, and jellyfish. In some applications such as COMMERCIAL FISHING, fish refers to aquatic animals, including whales, crustaceans, and mollusks.

There are more than 21,000 species of fish. Fish can be separated into three classes: AGNATHANS, CHONDRICHTHYES, and OSTEICHTHYES. Agnathans lack biting jaws and paired fins. They include lampreys and hagfish. The Chondrichthyes include cartilaginous fish such as sharks, rays, chimeras, and skates. Osteichthyes, or the bony fish, possess skeletons made up of true bone and include nearly all other living fish.

Fish can be distinguished by the manner in which they maintain the water balance in their bodies. The body fluids of saltwater fish such as pollock (*Pollachius virens*) have less salt than the surrounding water. These fish are always losing freshwater through osmosis. They drink seawater and excrete excess salt through their gills and urine. The body fluids of freshwater fish such as brook trout (*Salvelinus fontinalis*) have a higher concentration of salt than the surrounding water. They urinate large quantities of freshwater and use their gills to absorb salt. Fish control the intake of and expulsion of fluids by osmoregulation. Many fish have swim bladders to control their buoyancy. These are gas-filled organs that contain more or less gas as the fish needs more or less buoyancy. Some species of fish have swim bladders connected with their blood streams so that gases can pass back and forth between the bloodstream and the bladder. Most larvae and juvenile fishes have swim bladders, but may lose them after they reach adulthood. For example, bottom fishes (demersal) have no need for swim bladders. Sharks also lack swim bladders and must stay in constant motion to avoid sinking. Fishes have streamlined shapes to permit high-speed travel in the water. This is especially true of high-level predators such as tuna.

Since fish have many physical similarities, it is beneficial to distinguish among species by evaluating swimming behaviors. Marine scientists have classified active fish that are always on the hunt as cruisers. They are usually long and streamlined with red muscle tissue. A classic example would be the skipjack tuna (*Katsuwonus pelamis*). Skipjack tuna are schooling migratory fishes that feed on fishes, crustaceans, and mollusks. Other fish called lungers lurk around structures, making a short dash to capture prey swimming nearby. These fish such as the Nassau grouper (*Epinephelus striatus*) have white muscle tissue. Groupers tend to be solitary fish that feed primarily on fishes and crustaceans. There are adaptations in fish body form that affect the mechanics of swimming and moving in the water.

Many species of fish undertake migrations to either feed or breed in favorable locations in the water column. Anadromous fish such as white sturgeon (*Acipenser transmontanus*) migrate to the ocean to grow and mature, then migrate back to freshwater to spawn and reproduce. White sturgeon are demersal or bottom-seeking fish that feed on small finfish, shellfish, crayfish, and on aquatic invertebrates such as clams, amphipods, and shrimp. Catadromous fish such as the American eel (*Anguilla rostrata*) migrate from freshwater to salt water to reproduce. American eels (*Anguila rostrata*) spawn in the Sargasso Sea during the spring. The Sargasso Sea is a part of the ATLANTIC OCEAN located between the West Indies and the Azores.

Human population growth threatens fish by impairing water quality and depleting levels of dissolved oxygen. Increases in water TURBIDITY from land drainage are attributed to declines in submerged aquatic vegetation or SAV. Meadows of SAVs are critical habitat and nesting areas for juveniles of many fish species. Nutrient pollution from agricultural runoff, which elevates levels of nitrogen and phosphorus, is a significant problem leading to nitrification and fish kills. Dams and other blockages prevent migratory species such as American shad (*Alosa sapidissima*) from reaching their historic spawning grounds. In the CHESAPEAKE BAY region, removal of blockages and the construction of fish ladders at dam sites are restoration methods that are being implemented by corporations such as Baltimore Gas and Electric.

Humans also deplete fish populations by overfishing. Well-known examples are the fisheries of the Grand Banks, and Georges Bank, where decades of overfishing by fleets of vessels from many countries, especially the Soviet Union, devastated the fishery. The Magnuson Act in the United States, as well as corresponding Canadian legislation, placed restrictions on fishing by foreign vessels out to the 200-mile (321.9 km) limit, but the legislation came too late to save the cod and the haddock. The opportunity for the United States and Canada to revive the fishery was missed due to poor management, and the fishery "crashed." Thus today, populations of Atlantic cod (*Gadus morhua*) and haddock (*Melanogrammus aeglefinus*) are too small to permit commercial fishing operations.

See also MARINE FISH.

Further Reading

American Fisheries Society. Available online. URL: http://web.fisheries.org/main. Accessed November 27, 2006.

Boschung, Herbert T., Daniel W. Gotshall, David K. Caldwell, Melba C. Caldwell, and James D. Williams. *The National Audubon Society Field Guide to North American Fishes, Whales, and Dolphins.* New York: Alfred A. Knopf, 1983.

Earle, Sylvia A., ed. *Atlas of the Ocean: The Deep Frontier.* Washington, D.C.: National Geographic, 2001.

Froese, Ranier, and Daniel Pauly, eds. "FishBase: A Global Information System on Fishes." Available online. http://www.fishbase.org/home.htm. Accessed September 18, 2007.

fisheries recruitment   Fisheries recruitment is the process by which certain fish species become susceptible to fishing. Fishing activities involve the catching, taking, or harvesting of fish. The number of fish that are legally caught may be affected by variables such as the particular species growth rates and migration habits. A young fish entering the exploitable stage of its life cycle is called a recruit. Fisheries management efforts study fisheries recruitment and stock structure in order to understand the impact of fishing on a species. The stock is generally considered to be the part of a fish population that may be fished. Other important research relevant to fisheries recruitment includes investigations that assess the manner in which TURBULENCE from CURRENTS, winds, and WAVES affect PLANKTON production, distribution, migration, and the feeding ecology of larval fish. Investigators have found that turbulence may influence spawning in pelagic species such as Atlantic menhaden (*Brevoortia tyrannus*). Researchers report that maximum spawning for this species occurs off the North Carolina coast during late fall and winter in synchronization with storms. The resultant drifting eggs hatch after two to three days, and larvae are transported to ESTUARIES by ocean surface currents where they metamorphose and develop into juvenile menhaden. The spawning and hatching cycle is also dependent on parameters such as sea-surface TEMPERATURE, an important factor influencing year-class strength. For example, falling water temperatures during the egg and larval stages for largemouth bass (*Micropterus salmoides*) have been associated with small year classes. Since fish tend to spawn over a one- or two-month period annually, marine scientists refer to fish species hatched during a spawning period as a year class. Estimates of year class for species such as red drum (*Sciaenops ocellatus*) are based on the abundance of juveniles captured in bag seines. It takes several fisheries' scientists to work the fine mesh seine that has a bag midway along the length of the net. Samples from various hauls of the bag seine provide information on growth and movement.

Further Reading

Center for Quantitative Fisheries Ecology, Old Dominion University. Available online. URL: http://www.odu.edu/sci/cqfe/index.html. Accessed December 24, 2006.

International Council for the Exploration of the Sea. Available online. URL: http://www.ices.dk. Accessed June 19, 2007.

Intergovernmental Oceanographic Commission, United Nations Educational, Scientific and Cultural Organization. Available online. URL: http://ioc.unesco.org/iocweb/index.php. Accessed June 19, 2007.

fisherman   Recreational fishermen, or anglers, catch living fish, shellfish, and aquatic plants for pleasure, amusement, relaxation, or home consumption. Some anglers hire a charter boat with an experienced crew to catch sport fish such as sailfish (*Istiophorus platypterus*). Other fishermen may pay a fee to fish from a head boat (party boat) for groundfish such as red snapper (*Lutjanus campechanus*). Sport fishermen compete for trophies and cash prizes for freshwater gamefish, such as largemouth bass (*Micropterus salmoides*), and saltwater gamefish such as nearshore species like cobia (*Rachycentron canadum*) or offshore species like blue marlin (*Makaira nigricans*).

Technology has revolutionized recreational fishing. Serious fishermen use boats, with one or two powerful outboard motors and a center console, specially designed for fishing. Fish are located by sonic "fish finders," which are sophisticated echo sounders that may be integrated with digital charts and the GLOBAL POSITIONING SYSTEM. Modern fishing poles are made from carbon fiber rods, which provide good casting action and are practically indestructible. A vast array of artificial lures imitating every kind of bait, as well as the flash, noise, and texture of baits, are available to catch every kind of sport fish. Fishing techniques can be learned from sports clubs, workshops, television programs, books, magazines, and from Web sites.

Commercial fishermen earn income from the catch and sale of living and marine resources. Various types of boats, craft, and gear are used for the capture of commercially important marine species. Besides hook and line, fishermen might use trawls, gillnets, longlines, traps (or pots), and spears. Longlines are

Deployment of a large net by fishermen on several vessels yields a large catch.   *(NOAA/NMFS)*

stationary, buoyed, and anchored ground lines with attached hooks. Equipment is designed to catch pelagic migratory species or fish such as yellowfin tuna (*Thunnus albacares*) living near the surface and demersal species of fish such as Atlantic croaker (*Micropogonias undulatus*) living near the seafloor.

In coastal waters, blue crabs (*Callinectes sapidus*) are frequently caught inadvertently in shrimp trawls, and open ocean longliners fishing for tuna catch sharks accidentally.

The capture of nontargeted species is called bycatch. High-technology equipment can be used to minimize bycatch. Innovations such as Turtle Excluder Devices or TEDs are required by law to be attached to shrimp nets. By deflecting the turtles out of an escape hatch, TEDs help to protect threatened and endangered sea turtles from becoming bycatch. Bycatch is also called incidental catch, and some is kept for sale. Landings are the quantity of fish, shellfish, and other aquatic plants and animals that are brought ashore and sold. Fisheries' biologists determine fishing mortality by measuring the rate of removal of fish from a population by fishing. Sustainable rates of fishing mortality vary from species to species. Fishing is different from farming since

aquaculturists generally own organisms that will be harvested.

Further Reading

LaBonte, G. *Fishing for Sailfish*. Point Pleasant, N.J.: The Fisherman Library, 1994.

Mangone, G. J. *Concise Marine Almanac*. New York: Van Nostrand Reinhold, 1986.

National Marine Fisheries Service. National Oceanic and Atmospheric Administration. Available online. URL: http://www.nmfs.noaa.gov. Accessed June 19, 2007.

U.S. Fish & Wildlife Service. Available online. URL: http://www.fws.gov. Accessed September 18, 2007.

Fishery Conservation and Management Act

Modern fisheries programs in the United States began with the Fishery Conservation and Management Act of 1976. This legislation provided congressional mandate for a national fishery management program utilizing regional fishery management councils under the Department of Commerce's National Marine Fisheries Service. The act extended U.S. fisheries jurisdiction to 200 miles (330 km) from the coast and gave priority access to fishing grounds to vessels from the United States (*see* EXCLUSIVE ECONOMIC

Zone). Regional councils composed of members from organizations, such as the National Marine Fisheries Service, U.S. Fish and Wildlife Service, U.S. Coast Guard, and affected states work together to oversee the fisheries. The regional councils cover Caribbean, Gulf of Mexico, Atlantic, and Pacific fishing grounds. Each council is charged to produce a fishery management plan that is approved for implementation by the secretary of commerce. Management plans generally limit the number of fishermen, their gear, overall catch, and the season in order to sustain the fishery after the target harvest has been achieved. Current policies have evolved since 1976 and are now defined by the Magnuson-Stevens Fishery Conservation and Management Act.

Further Reading

National Marine Fisheries Service, National Oceanic and Atmospheric Administration. "Magnuson-Stevens Fishery Conservation and Management Act, Public Law 94-265. As amended through October 11, 1996." Available online. URL: http://www.nmfs.noaa.gov/sfa/magact. Accessed September 18, 2007.

fluid mechanics   Fluid mechanics is the study of the physics of fluids dealing with the distributions and time variations in pressure, temperature, velocity, dispersion of dissolved or suspended substances, and waves. Fluid mechanics is made up of fluid statics, the study of fluids at rest, and fluid dynamics, and the study of fluids in motion. Fluid mechanics deals with a very broad spectrum of problems, ranging from the hydraulic systems of submarines to the flow of winds in the atmosphere and currents in the ocean. The study of fluid mechanics is based on Newton's laws of motion, the laws of conservation of mass and energy, conservation of linear and angular momentum, and the equations of state of the fluid. Applications of the principles of fluid mechanics have led to the development of complex numerical circulation models of the atmosphere and the oceans. Marine scientists apply their knowledge of environmental fluid mechanics to atmospheric dynamics, physical oceanography at the smaller scales, ground-water hydrology, cloud physics and moist convection, surface-atmosphere interactions, coupled biological and physical processes, hydrologic transport processes, and sediment transport processes.

Fluid statics is the study of the behavior of liquids and gases at rest. Fluids do not resist static shear distortions, which makes them different from solids. The change in wind with height or in current with depth is defined as shear. The basic principles of fluid statics include the following: direction of force, fluid pressure, transmission of pressure, and Archimedes' principle.

Direction of force: The force exerted by a fluid on a surface is perpendicular to the surface. Pressure is force per unit area and is uniform in direction in a fluid; it is a scalar quantity.

Fluid pressure: The pressure "p" at a depth "h" in a fluid of density "ρ" is given by the hydrostatic equation

$$p = p_0 + \rho gh,$$

where "p_0" is usually atmospheric pressure at sea level, and "g" is the acceleration of gravity. High pressure in the atmosphere is usually associated with fair weather, and low pressure with storms. This parameter is especially important to determining how water levels fluctuate in response to forces such as gravitational attraction, winds, and differences in water density.

Transmission of pressure: Pascal's principle asserts that pressure applied to a fluid in a closed container is transmitted equally throughout the fluid. This principle explains the operation of the hydraulic lift, in which a small, applied force is able to lift an automobile off the ground. This concept was enunciated by the French mathematician Blaise Pascal (1623–62).

Archimedes' principle: This principle asserts that a body immersed in a fluid is buoyed up by a force equal to the weight of the fluid displaced by the body. Archimedes' principle explains why ships float, and what percentage of an iceberg is above water, the "tip" of the iceberg phenomena. For example, we know that a cubic foot of freshwater weighs 62.4 pounds (28.3 kg). Therefore, if an object placed in freshwater has a volume of a cubic foot and weighs more than 62.4 pounds (28.3 kg), it will sink. Archimedes, a famous Greek scientist, was born around 287 b.c.e. and was killed by a Roman soldier around 212 or 211 b.c.e.

Fluid dynamics is the study of fluids in motion. There are two principal approaches to this study that scientists investigate using current meters, flumes, tanks, dyes, and computer models. The Eulerian approach named after Leonhard Euler (1707–83) examines the flow parameters such as velocity, pressure, and density at discrete locations, usually by means of moored current meters in lakes and the ocean. The Lagrangian approach named after Joseph Louis Lagrange (1736–1813) examines the flow of individual parcels of fluid as they drift along. The original drifting buoys were notes in bottles. The finder was asked in the note to report the position and time at which the bottle was found.

The drift of ships due to currents has provided a means of mapping large-scale ocean circulations. Modern Lagrangian measurements are made with sophisticated drifting buoys that determine their locations by the GLOBAL POSITIONING SYSTEM (GPS). They transmit location and other ocean parameters such as pressure and temperature to satellite receiving stations.

The motion of fluids is very complex; to derive basic results, it is necessary to adopt several simplifying assumptions. A fluid thus simplified is called an "ideal fluid." The assumptions include smooth, steady flow, incompressibility, and negligible friction, no tendency to spin or form vortices. Under these assumptions, the following principles can be derived:

Continuity: The amount of fluid that enters a tube in a given time interval is equal to the amount of fluid that leaves the tube in the same time interval. If it were not so, the fluid would either accumulate in or break the tube, or material would leave the tube until it imploded due to the higher surrounding pressure. The tube need not be a real, physical tube; it may be an imaginary cylindrical surface enclosing fluid. The mathematical statement of the equation of continuity is:

$$A_1 v_1 = A_2 v_2 = \text{constant},$$

where A is the cross sectional area of the tube, and v is the velocity of flow in the tube. Subscripts 1 and 2 designate any two positions in the tube. From this principle, it is seen that fluids flow faster if the tube narrows, and slower where the tube broadens, as seen in the flow of mountain streams.

Bernoulli's equation: This equation is essentially a statement of the law of conservation of energy for fluids. It is named for the Swiss mathematician/physicist Daniel Bernoulli (1700–82). The equation is, for any two fluid volumes 1 and 2 (denoted by subscripts), as follows:

$$p_1 + \rho v_1^2/2 + \rho g y_1 = p_2 + \rho v_2^2/2 + \rho g y_2 = \text{constant}$$

p is the fluid pressure, ρ is the fluid density, v is the speed of the flow, y is the height of the fluid, and g is the acceleration of gravity. The terms in the square of the flow speed are the kinetic energies of the fluid at locations 1 and 2, and the terms in y are the potential energies of the fluid at these locations. Locations 1 and 2 were arbitrarily chosen; since they are equal to each other, they must be equal to a constant. Bernoulli's equation explains the lift force exerted by the air on an aircraft wing. The wing is designed so that the airflow is faster at the top of the wing than at the bottom, thus the air under the wing is at greater pressure than the air over the wing, providing the lift force pA, where p is the pressure, and A is the area of the wing.

Turbulence: Fluid flows can be either laminar and smooth, or turbulent and disordered. The flow of a stream is laminar (steady) flowing over a smooth bottom with smooth sides. When the flow approaches boulders and fallen timbers, it becomes turbulent (irregular), with numerous EDDIES and vortices.

Viscosity: Friction plays an important role in fluid dynamics. Fluid friction, called "viscosity," is a major factor in whether a fluid flow is laminar or turbulent. The flow of heavy oil of high viscosity tends to be laminar; the flow of light oil of low viscosity can easily become turbulent. Natural flows in the ocean are generally turbulent. The behavior of fluid friction in smooth flows is described by a "molecular viscosity coefficient," often given the symbol Greek mu "μ." Natural flows in the ocean are described by an "eddy viscosity coefficient," often indicated by a capital Roman letter, such as A. Eddy viscosity coefficients were developed when it was realized that the transfer of momentum, heat, salt, and other properties of the ocean occur at a much faster rate than would be indicated using molecular viscosity coefficients. The eddy coefficients are not on as firm a theoretical foundation as the molecular coefficients. The latter are determined by the nature of the fluid, whereas the eddy coefficients are determined largely by the nature of the flow.

Vorticity: Another important property of fluids is the vorticity, or tendency of the fluid to spin about a vertical axis. Vorticity is closely related to angular momentum. The ideal fluids studied in introductory physics courses have zero vorticity, or are "irrotational." Any fluid possessing shear, or change in velocity with location, possesses vorticity; it is not necessary for a fluid to rotate to possess this quantity. The rotating Earth itself imparts vorticity to the ocean, called "planetary vorticity," which is equal in magnitude to the CORIOLIS FORCE. Planetary vorticity is denoted by the symbol "f," which is equal to $2\Omega\sin\varphi$, where Ω is the angular velocity of Earth, and φ is the latitude. Although vorticity is a vector with three components, in the ocean, only the component of vorticity about a vertical axis is important. The planetary vorticity at the equator is zero, increasing to a value of 2Ω at the poles. Relative vorticity, usually denoted by the Greek letter zeta "ζ," is imparted to the fluid by such forces as friction due to the wind, or due to the water passing over a ridge or trough. Potential vorticity is an important concept used in explaining the circulation of the oceans.

Potential vorticity is defined as the absolute vorticity divided by z, the thickness of the layer of water with that vorticity, or

$$\text{Potential vorticity} = (\zeta + f)/z$$

From the law of conservation of angular momentum, potential vorticity must be conserved in the absence of friction. If the thickness of the layer changes, it must be accompanied by either a change in latitude or a change in relative vorticity, in order to keep the potential vorticity constant. Similarly, if the water column changes latitude, there must be either a change in thickness of the water column, or in the relative vorticity. This principle was applied by Henry Stommel (1920–92) in 1948 to explain the existence of the GULF STREAM and other western boundary currents, as well as the asymmetry of the circulation in the major ocean basins (western boundary currents are fast, deep, and narrow; eastern boundary currents are broad, shallow, and slow).

Characterizing the mechanics of fluids to include buoyancy, stability, and hydrostatics is essential to physical oceanography, coastal oceanography, and environmental sciences. As an example, oceanographers may study the pathways of migratory fish that travel from ESTUARIES to the shelf break for spawning. The fish larvae, which only swim one or two body lengths per second, must make their way across many miles (kilometers) of coastal ocean and be retained in an estuary to mature before they can migrate to ocean spawning grounds. As the young fish ride favorable currents to the shore, they become trapped in LONGSHORE CURRENTS that feed ebb tides at a COASTAL INLET. As the tide shifts to flood, the young larva are squirted into the estuaries, where they make their way to protected nursery areas such as meadows of submerged aquatic vegetation. Chemical oceanographers may study the motion of sediments, nutrients, and other chemicals that are mixed by turbulence along the front of a river plume such as the Amazon or Mississippi Rivers. The plume of fresh and sediment-laden waters is easily identified on aerial and satellite photographs.

Further Reading

Knauss, John A. *Introduction to Physical Oceanography*, 2nd ed. Upper Saddle River, N.J.: Prentice Hall/Pearson, 1996.

Stommel, H. M. *The Gulf Stream: A Physical and Dynamical Description*, 2nd ed. Berkeley: University of California Press, 1965.

food chains and webs A food chain is a graphical depiction of the various trophic levels of an ECOSYS-TEM arranged in a linear order. The food web is a representation of the trophic levels that allow for more complex feeding relationships than the simple linear representation of the food chain. Location in the food chain or trophic level identifies the number of energy-transfer steps to that level. Since a food web is made up of interconnected food chains, it requires a two-dimensional representation. The classic food chain begins with the plants, or primary producers at the first trophic level, that turn atmospheric carbon dioxide and sunlight into living cells. Examples of marine primary producers are PHYTOPLANK-TON, such as diatoms. Great blooms of diatoms occurring at mid latitudes are the pastures of the sea. Herbivores that feed on the phytoplankton form the second trophic level. These include ZOOPLANKTON, some small fishes, and other marine animals. Shrimp larvae are examples of zooplankters. The third trophic level is formed by the carnivores that eat the zooplankton. These creatures are generally small fishes, such as Atlantic herring (*Clupea harengus*). Larger carnivores, such as voracious bluefish (*Pomatomus saltatrix*), eat the herring, and top carnivores such as Mako sharks (*Isurus oxyrinchus*) eat the large carnivores. The efficiency of the energy transfer up the trophic levels becomes smaller and smaller going up the level. Thus, the efficiency of zooplankton eating PHYTOPLANKTON might be 9 percent. The efficiency of sharks eating mullet might be less than 1 percent. For example, the top carnivore, the shark, could require that many tons of phytoplankton be produced to support the zooplankton that supports the herring that support the mahi-mahi or dorado (*Coryphaena hippurus*) eaten by the shark.

In general, the feeding relationships of marine creatures are much more complex than the linear food chain. They are represented as a food web. For example, the baleen whales, even the giant blue whale (*Balaenoptera musculus*), feed on shrimplike creatures called krill (*Euphausia superba*), found in great abundance in the waters of the Southern Ocean. The largest fish in the world, the whale sharks (*Rhincodon typus*) and the basking sharks (*Cetorhinus maximus*) feed on plankton. The whales use their comblike baleen, which they have instead of teeth, to strain the krill from the seawater. Killer whales, or orcas (*Orcinus orca*), may feed on penguins or seals in the Antarctic, but in other regions, they feed on fish.

The creatures living below the epipelagic zone must live on the dead plants and animals (detritus) falling from the sunlit portion of the ocean. In this case, small fish and benthic organisms such as crabs may be feeding on larger dead fish, or even dead MARINE MAMMALS, such as whales. When a dead whale lands on the ocean floor, the creatures of the ABYSSAL ZONE have a feast, since food is very scarce. Some predators of the abyss, such as the black swallower (*Chiasmodon niger*), have expanding jaws that enable them to feed on larger creatures than themselves.

Even in the PHOTIC ZONE, there are very large creatures that feed on small plants, such as the manatees, which live in rivers and coastal regions. In the United States, the herbaceous West Indian manatees (*Trichechus manatus*) graze for food along water bottoms and on the surface. They are reported to consume 10 percent of their body weight per day in vegetation. Adult manatees can weigh up to 1,200 pounds (544 kg). In recent years, manatee populations are mostly affected by habitat loss and collisions with watercraft.

The abundance of life in the upper ocean is determined by primary productivity. In areas such as UPWELLING zones, the enrichment of the surface layers by nutrients from below increases productivity, and marine life is abundant. The Arctic and Antarctic are also areas of high productivity and abundant life. In the Tropics, deep water productivity is generally low, except for upwelling regions, such as the coast of Peru, and the "cold islands," of the equatorial Pacific. The arctic regions are characterized by a large biomass of a few species, but the Tropics are characterized by a much lower total biomass, but much greater species diversity.

In the recently discovered hydrothermal vent communities, the primary producers are the chemosynthetic bacteria, which produce cell matter not from light (of which there is none) and carbon dioxide, but rather from chemicals being discharged from the vents, such as sulfur and methane. Small creatures feed on the bacteria; larger creatures eat the smaller creatures and so on, up to the top carnivores, which are animals such as crabs and fish. Little is known about the biology of the vent communities compared to the surface communities (*see* HYDROTHERMAL VENTS).

See also ECOLOGICAL EFFICIENCY.

Further Reading
Col, Jeananda. "Enchanted Learning." Available online. URL: http://www.EnchantedLearning.com. Accessed June 20, 2007.

forced waves Forced waves require a constant input of energy to propagate through a fluid medium in contrast to free waves that travel by their acquired momentum, until they are dissipated by friction or TURBULENCE. Forced waves are found in laboratory wave tanks or in some public wave pools where a wave generator is required to maintain the wave motion. The TIDES are long WAVES that are forced by the interactions of the gravitational fields of the Earth, Sun, and Moon. Ocean surface gravity waves under a developing storm are forced by fluctuations in the PRESSURE and tangential stress of the surface winds. When the winds cease blowing, the large waves travel great distances as free "swell" waves. Waves traveling along a boundary, called "trapped waves," (sometimes referred to as EDGE WAVES or shelf waves) are forced by traveling storms of long duration. In a few days, shelf waves can transport more sediment along the coast than occurs during months of calm sea conditions.

Strong currents that flow over sills or result from pumping mechanisms can also force internal waves. Internal waves form at the interface between two layers that differ in density. A well-known example is the creation of internal waves by the subsurface outflow of Mediterranean water passing over topographic obstructions in the Strait of Gibraltar. During World War II, a German submarine was allegedly sunk in this area by breaking internal waves. A disturbing force such as the periodic astronomical forces continuously maintains true forced waves. Some internal waves propagating into the South China Sea from the Luzon Strait have been observed to fluctuate in synchronization with tidal periods. The surface manifestation of the internal waves is imaged distinctly through the use of Synthetic Aperture Radar.

Further Reading
Center for Operational Oceanographic Products and Services: National Ocean Service, National Oceanic and Atmospheric Administration. *Tide and Current Glossary.* Available online. URL: http://tidesandcurrents. noaa.gov/publications/glossary2.pdf. Accessed June 20, 2007.

forecasting In marine science, forecasting usually means predicting the future state of the atmosphere or oceans, with emphasis on precipitation, clouds, temperature, and winds in the atmosphere, and TEMPERATURE, SALINITY, ice, waves, water levels, and current velocities in the oceans. In recent years, forecasting has been broadened to include environmental conditions such as beach erosion, red tides, oil spills, local warming and chemical pollution from industrial plants, and eutrophication (greatly accelerated growth of aquatic plants due to the input of excess nitrogen into the water, such as from agricultural runoff or municipal sewage). Weather forecasting has the longest history and is the most highly developed of all the forecasting areas.

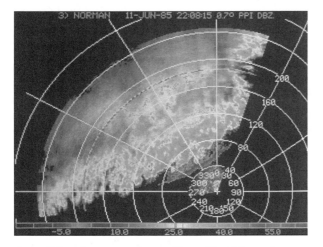

A squall line shown on radar provides critical information to severe weather forecasters. *(NOAA/NWS)*

Weather forecasting began in the late 19th century, when the invention of the telegraph made it possible to transmit weather data from stations throughout the country to a central location for analysis. Forecasts could then be prepared and issued back to the different parts of the country. The National Weather Service was organized by Congress during the 1870s to perform this task. Meteorologists using a combination of data analysis, local knowledge, and professional judgment made these early and remarkable forecasts.

Forecasting techniques during this early period included climatological forecasts based on long-term average conditions, persistence forecasts based on current conditions, extrapolating those conditions into the future, and case studies in which historical patterns are sought that are similar to current conditions. In fact, *The Farmers Almanac* was first published around 1818 and is still used by many to obtain long-range weather forecasts, gardening advice, and knowledge on lunar phases.

Weather forecasting received a great boost during the World War I era, when knowledge of the weather became critical to military operations. A group of Norwegian meteorologists led by Jakob Aall Bonnevie Bjerknes (1897–1975) developed the frontal theory of storm formation, which gave meteorologists a model that included some physics of the atmosphere. This model, including fronts (in analogy with World War I military fronts), and circulations around high- and low-pressure areas is still used today and can be seen on the daily weather maps in the newspapers and on television. Professor Bjerknes worked at the University of California at Los Angeles from 1940 to 1975 and is credited with founding the UCLA Atmospheric Sciences Department.

Prior to the advent of computers, weather forecasts were prepared by hand analysis of large amounts of data and displayed on weather maps of surface and upper air conditions, as well as maps of predicted movements of fronts and weather systems. These maps and text summaries were often transmitted by fax to stations all over the country and to ships at sea.

A critical requirement of accurate weather forecasting is to receive reports of current conditions from all over the globe. Through the offices of the World Meteorological Organization (WMO), data are funneled to weather offices in all major countries. Today's ocean observatories and networks of fully automated instrument BUOYS provide data that is essential for the development, assessment, and operation of numerical models that consider meteorological and oceanographic phenomena.

The advent of electronic digital computers in the 1950s and 1960s resulted in the development of numerical weather prediction models in which atmospheric data was fed into models based on the equations of motion and the equations of state of the atmosphere. Forecasts were made by numerically solving the equations as functions of time extending into the future. These early attempts into numerical weather prediction were less than completely successful because of limitations of data and computer power. Nevertheless, they led to the much more sophisticated models of today that are used to produce the 36-hour, seven-day, and long-term forecasts that are seen daily on television. However, meteorologists, not machines make forecasts, and so the machine output must be carefully scrutinized in light of the forecasters' knowledge.

Besides the computer revolution, modern weather forecasting has been made possible by corresponding advances in instrumentation, such as polar-orbiting and geostationary satellite platforms for visible, microwave, and infra-red (IR) measurements of cloudiness, temperature, and water vapor content of the atmosphere, Doppler weather radar, ground-based RADAR, and SONAR vertical atmospheric sounders.

The basic data analysis and weather diagnosis and forecasts are made in the United States by the National Weather Service (NWS) of the NATIONAL OCEANIC AND ATMOSPHERIC ADMINISTRATION (NOAA), U.S. Department of Commerce. The basic data on current conditions and forecasts can then be specialized and embellished with attractive graphics by meteorologists in the private sector, such as those at the Weather Channel and other network and cable television outlets, for presentation to the public. Airlines, the military, NASA, and many other industries, universities, and government agencies have their own forecasters to tailor the NWS forecast to their specific needs. In many cases, they incorporate results from their own special measurements.

Marine weather forecasts essential to the shipping industry, the Coast Guard, and recreational boaters are made for the area from the shoreline seaward to 20 miles (32.2 km). These forecasts include wave height, ocean temperature, and total water levels. The NWS uses STORM SURGE models to predict dangerous water levels, flooding conditions, and strong currents when severe storms such as tropical HURRICANES threaten the coast.

Forecasting of ocean conditions such as wave height and direction began in World War II with the development of the "Sverdrup-Munk" method during the early 1940s. This method included predictions of wave height just prior to the invasion of North Africa in 1942, and of wind and wave conditions in the Normandy beaches just prior to the landings of Allied forces. In fact, since the weather had been bad over the English Channel for many days, this prediction was critical to General Eisenhower's decision to proceed with the landings on June 6. Countless lives were saved by application of this procedure. General Dwight D. Eisenhower (1890–1969) became the 34th President of the United States.

Today, the U.S. Navy uses the most advanced models of ocean wave height and direction to provide the surface warships, embarked Marine Corps units, and other branches of the armed forces with the most accurate predictions possible prior to amphibious operations. Models have been integrated to provide coupled atmospheric and oceanographic information for deep-water, shallow-water, and surf-zone environments. Such information is critical for the successful maneuver of high-speed vessels, planing hull landing craft, and air cushion vehicles.

Forecasting of chemical and biological conditions in the oceans and lakes is at present more tenuous than weather and ocean wave forecasting. Climatological, statistical, and scientific numerical process-based models are all used to forecast environmental conditions. These process models are very complex, depending not only on modeling the physical processes, but also the chemical and biological processes, such as the consumption of PHYTOPLANKTON by ZOOPLANKTON. Frequently, not much is known of these latter processes, and bold assumptions must be made. In many cases, it is not even possible to make significant numbers of accurate measurements of the quantities being forecast. Such measurements are essential to calibrate the model and to evaluate the quality and value of the forecasts. In contrast, weather forecasts are constantly being evaluated by forecast verification, that is, comparing the predicted weather with the observed weather for the forecast period. These comparisons, which determine the forecast error, continually test weather forecasting skills and techniques.

Nevertheless, environmental managers often use models of marine chemical and biological processes in the planning of major projects, such as the siting of sewage treatment plants and the installation of waste-water diffusers in the ocean. Oil spill response models have been employed for several years to direct the efforts of the U.S. Coast Guard and other first responders in the assessment and cleanup operations necessary to minimize the damage resulting from the spill.

Accurate forecasts of weather and oceanic conditions are important for planning, whether it's to pack for a vacation or to plan energy usage for a city. The agriculture industry uses forecasts to plan for the planting and harvesting of crops. Air and cruise lines use weather conditions in order to schedule flights and plan ship tracks. Weather forecasting helps people make smart daily decisions, and thereby helps to improve safety and protect lives.

See also MARINE METEOROLOGIST.

foreshore *See* BEACH.

fossil A fossil is any remains, impression, or trace of a living organism from some past geologic age, such as a skeleton, footprint, or insect preserved in amber. Fossils are made, while remains are said to be fossilized, a process by which mineral substances by means of chemical reactions replace the organic material of the organism. Examples are the remains of an animal in petrified rock and the remains of a plant preserved in carbon. The bones of dinosaurs have been found in ancient tar pits, where the animals were trapped when feeding or seeking water.

The oldest fossils are ALGAE and bacteria. Scientists have discovered the remains of primitive bacteria in rocks of hardened silica called chert. Australian geologist Reginald C. Sprigg (1919–94) discovered the impressions of jellyfish, worms, and benthic organisms in southern Australia in 1946, a major step forward in the science of paleontology. PALEONTOLOGISTS have since been able to trace the development of life through the geologic ages.

From the study of fossils found in mountainous regions, such as the Rockies, scientists have learned when many mountain chains were underwater prior to being thrust up in epochs of mountain building. The age of the fossils is determined by radiometric dating, such as carbon 14 methods. From the study of plant fossils, paleoclimatologists can study Earth's former climates. One of the most important uses of fossils in marine science is that of dating marine SEDIMENTS. If the age of the type of fossilized organism is known, its discovery can serve as an independent means of dating the sediments in which it is found.

Today, many amateur fossil hunters derive much pleasure from hunting for and finding evidence of previous life-forms. They supplement the findings of professional paleontologists. Some locations are known for having abundant fossils; for example, there are BEACHES in which shark teeth are relatively abundant. Another application of fossil identification is the study of the possible causes of the great extinctions that have occurred in Earth's past. Scientists today have the additional powerful tool of DNA analysis to study the development and relations of past and living creatures; hence, DNA molecules are sometimes called "living fossils."

Further Reading

Bennett, Jeffrey, Seth Shostak, and Bruce Jakosky. *Life in the Universe*. San Francisco, Calif.: Addison-Wesley/ Pearson, 2003.

Prager, Ellen J., with Sylvia A. Earle. *The Oceans*. New York: McGraw-Hill, 2000.

fracture zones   Fracture zones are narrow ridges and valleys parallel to spreading ridges aligned in a staircase configuration. In fact, the spreading ridges themselves are fracture zones. The mid-ocean ridges, which circle the Earth, subdividing the ocean basins, consist of numerous fracture zones, some of which are several hundred miles (km) wide. The mid-ocean ridges contain the most rugged terrain on planet Earth.

Fracture zones consist of an active part, in which earthquakes occur frequently, and a passive part, in which the cold seawater has, over time, frozen the ridges into cold seafloor. As molten magma emerges from the spreading zones, it forms new crust that travels at a few inches (cm) per year. The farther the crust from the spreading zone, the older it is. The rate of movement of spreading sea floor has been determined by analysis of magnetic strips that record changes in Earth's magnetic field as the magnetic poles exchange places, a process for which the time history is known.

Some fracture zones form the transition from a spreading ridge to an ocean trench (*see* TRENCH). Such areas experience frequent seismic activity. Thus, these fracture zones occur along some tectonic plate boundaries (*see* PLATE TECTONICS). The Romanche trench on the floor of the equatorial

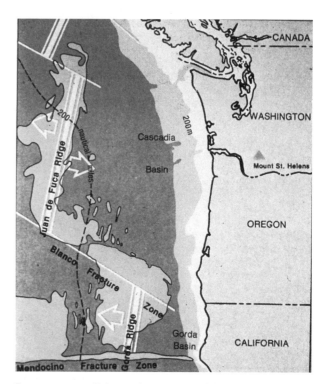

Fracture zones off the northwest coast of the United States can contain hydrothermal vents.   *(NOAA)*

ATLANTIC OCEAN is an especially large fracture zone. It displaces the axis of the Mid-Atlantic Ridge by almost 600 miles (1,000 km).

Further Reading

Erickson, Jon. *Marine Geology: Exploring the New Frontiers of the Ocean,* rev. ed. New York: Facts On File, 2003.

Kious, W. Jacquelyne, and Robert I. Tilling. *This Dynamic Earth: The Story of Plate Tectonics*. Online edition, U.S. Geological Survey. URL: http://pubs.usgs.gov/ publications/text/dynamic.html. Accessed September 18, 2007.

Public Broadcasting Service. *The Sea Floor Spread*. Available online. URL: http://www.pbs.org/wgbh/aso/tryit/ tectonics/divergent.html. Accessed September 18, 2007.

Seibold, Eugen, and Wolfgang H. Berger. *The Seafloor: An Introduction to Marine Geology,* 3rd ed. Berlin: Springer, 1996.

F-ratio   *See* STATISTICS.

free waves   Free waves are excited by a force of finite duration and then propagate freely in a fluid medium until all the energy is lost by friction or TURBULENCE. The initial input of energy may arise from winds of several days duration, as in the case of ma-

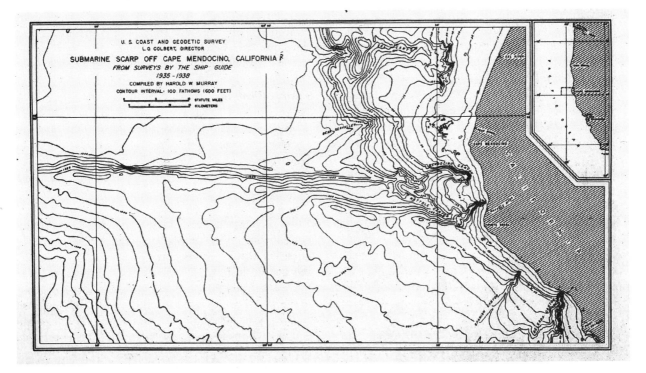

Submarine scarps off Cape Mendocino, California, delineate the fracture zone. *(NOAA)*

jor storms, or may be impulsive, of very short duration, as in the case of undersea earthquakes.

WAVES that are growing under strong winds are called "sea," while waves that continue to travel on the ocean surface after the wind stops blowing are called "swell" waves. Because ocean waves are dispersive (the speed of the waves depends on their WAVELENGTH), low frequency propagate faster away from a storm than shorter waves. Swells have a smooth, nearly sinusoidal appearance, and can travel thousands of miles (km) across Earth's surface. When reaching shore, they form the regular curling and plunging breakers that are loved by surfboarders. Examples include several world-class surfing areas such as the Banzai Pipeline on the north shore of Oahu, Hawaii, and Boca de Pascuales on the coast of the state of Colima in Mexico. TSUNAMIS are very fast surface waves created by undersea earthquakes. As a region on the ocean floor slips or rises, it produces a bump on the ocean surface that spreads out like ripples on a pond. Tsunamis can travel freely across the ocean surface at speeds approaching those of a modern jetliner. They are barely visible to mariners at sea, but as they enter shallow water, they are transformed into waves of awesome height, which wreak devastation on low-lying coastal areas. Federal, state, and local emergency management agencies promote public

awareness and tsunami hazard preparedness strategies through a tsunami warning system. Important organizations include the Pacific Tsunami Warning Center, the West Coast & Alaska Tsunami Warning Center, and the Puerto Rico Tsunami Warning and Mitigation Program.

Tidal waves, on the other hand, are not free waves, but are FORCED WAVES, which means that they are maintained by a constant input of energy from the gravitational interactions of the Earth, Sun, and Moon. Very long-period waves, such as ROSSBY (planetary) WAVES, having wavelengths hundreds to thousands of miles long, are forced by Earth's rotation. These are perhaps most familiar as the waves that develop in the jet stream that circle the globe.

Waves may be measured by simply evaluating watermarks left on a stick or processing signals from complex combinations of pressure sensors, CURRENT METERS, and accelerometers. Most wave buoys utilize accelerometers or inclinometers on board the buoys to measure the heave acceleration or the vertical displacement of the buoy hull during the sampling time period. Measuring and reporting parameters such as wave heights, periods, and directions is important in helping to protect property and save lives on the sea and near the coast.

See also OCEANOGRAPHIC INSTRUMENTS; WAVE BUOY.

Further Reading

Pacific Tsunami Warning Center, National Weather Service, National Oceanic and Atmospheric Administration. Available online. URL: http://www.prh.noaa.gov/ptwc. Accessed June 20, 2007.

West Coast & Alaska Tsunami Warning Center, National Weather Service, National Oceanic and Atmospheric Administration. Available online. URL: http://wcatwc.arh.noaa.gov. Accessed June 20, 2007.

fringing reef   A fringing reef is a geomorphological category for a modern coral reef that borders the coast of an island or continent. They are usually oriented parallel to the shoreline, require a hard substrate, adequate sunlight, tropical shallow waters (> 68°F [20°C]), and low nutrient and sediment concentrations in the surrounding waters. Corals, hydrozoans (stinging corals), alcyonarians (soft corals), and calcareous algae are examples of organisms that build fringing reefs.

If sea levels are constant or decreasing and carbonate production is high on a fringing reef, the reef will grow seaward, while vertical growth is limited by exposure to air and heat during low spring tide level. Sea level rise or land subsidence over geologic time allows fringing reefs to grow vertically and eventually form barrier reefs.

Fringing reefs develop around most islands and continents in the Indo-Pacific and Caribbean. In these regions, wave action, freshwater runoff, coastal development, and increasing nutrient and sediment concentrations due to agriculture and erosion are examples of factors that limit fringing reef development.

See also ATOLL; BARRIER REEF; CORAL REEFS.

Further Reading

Barnes, Richard S. K., and Roger N. Hughes. *An Introduction to Marine Ecology,* 3rd ed. Oxford: Blackwell Science, 1999.

Bonem, Rena, and Joe Strykowski. *Palaces Under the Sea: A Guide to Understanding the Coral Reef Environment.* Crystal River, Fla.: Star Thrower Foundation, 1993.

fully developed sea   A fully developed sea is an equilibrium point for a given wind speed where the energy that the wind is putting into the WAVES equals the energy lost by the waves through breaking and other frictional forces. A sea in this condition is fully developed. For the wind to produce a fully developed sea, the wind speed, FETCH, and duration must be adequate. This type of sea can be described theoretically by a characteristic strongly peaked spectrum. A wave spectrum shows the energy distribution as a function of wave frequency. Most of the wave's energy is found in a very narrow frequency range. The characteristic period of the wave is the frequency at the point of maximum energy. For a fully developed sea, the characteristic period increases as the wind speed increases. It is approximately eight seconds for a 20-knot (10 m/s) wind and about 16 seconds for a 40-knot (20 m/s) wind. This concept is especially important since the

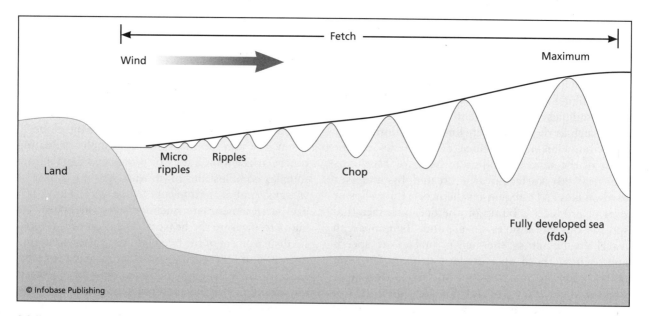

A fully developed sea is generated for a given wind speed when the fetch (length of ocean over which the wind has been blowing) and the duration (time the wind has been blowing) are sufficient.

majority of waves in the ocean are wind generated. The spectra are especially important in FORECASTING waves, designing vessels and structures, and in oceanographic research owing to the difficulty in collecting measured wave data. The spectra can be subject to geographical and seasonal limitations.

Further Reading

Chesneau, Lee S., and Capt. Michael Carr. "Waves and the Mariner," In *Mariners Weather Log* 44, no. 3. Silver Spring, Md.: National Weather Service, National Oceanic and Atmospheric Administration (December 2000): 7–19.

G

Geochemical Ocean Sections Study (GEOSECS)
The Geochemical Ocean Sections Study was a multi-investigator global ocean program that lasted from approximately 1972 to 1978 to study chemical changes in the ocean. Major institutions participating in this study were SCRIPPS INSTITUTION OF OCEANOGRAPHY, LAMONT-DOHERTY EARTH OBSERVATORY, Yale University, and the Rosenstiel School of Marine and Atmospheric Science at the University of Miami. This collaborative program promised to answer exciting questions, as it fostered cooperation among scientists of various disciplines to resolving challenging scientific questions. This effort joined CHEMICAL OCEANOGRAPHERS interested in analytical chemistry with PHYSICAL OCEANOGRAPHERS concerned with circulation to establish a deeper understanding of oceanic circulation and mixing.

Ocean scientists participating in GEOSECS planned and executed a global survey of the three-dimensional distribution of chemical, isotopic, and radiochemical tracers in the ocean. An isotope is one or more atoms of the same element. Isotopes, which are not radioactive such as oxygen 16 (^{16}O), oxygen 17 (^{17}O), and oxygen 18 (^{18}O), are called stable isotopes. Isotopes that undergo radioactive decay such as uranium-235 (^{235}U) and uranium 238 (^{238}U) are called radioactive isotopes, and the regular decay rates can be used as a clock to date materials. Principal investigators collaborated to develop a database of basic oceanographic tracers such as carbon 14 (^{14}C), helium 3 (^{3}H), radium 226 (^{226}Ra), radium 228 (^{228}Ra), radon 222 (^{222}Rn), and silicon 32 (^{32}Si). Data collection was dependent on using state-of-the-art sampling systems such as the Neil Brown CONDUCTIVITY, TEMPERATURE, DEPTH (CTD) PROBE and onboard analyses systems. This CTD is named after longtime WOODS HOLE OCEANOGRAPHIC INSTITUTION researcher Neil L. Brown (1927–2005), founder of Neil Brown Instrument Systems. Isotopes were used as chemical tracers in the oceans since WATER MASSES have distinct and unique concentrations. Thus, isotopic ratios were analyzed to indicate movement of the tracked water masses. Future researchers would be able to utilize the GEOSECS baseline to assess chemical changes in the world's oceans in order to gain insights into large-scale oceanic transport and mixing processes. Expeditions from approximately July 1972 to May 1973 were conducted in the ATLANTIC OCEAN and included approximately 121 stations. Surveys at approximately 147 stations in the PACIFIC OCEAN were conducted from August 1973 to June 1974. Approximately 141 stations recorded data from December 1977 to March 1978 in the INDIAN OCEAN. Radiotracers from cruise tracks and stations along the Atlantic, Pacific, and Indian Ocean basins made important contributions in making trace chemical profile measurements and mapping abyssal circulation.

Further Reading

International Research Institute for Climate and Society. "Oceanographic Data in the IRI Data Library." Available online. URL: http://www.iridl.ldeo.columbia.edu/docfind/databrief/cat-ocean.html. Accessed September 19, 2007.

National Research Council. *50 Years of Ocean Discovery: National Science Foundation 1950–2000*. Washington, D.C.: National Academy Press, 2000.

geodesy The United States Geological Survey, the NATIONAL OCEANIC AND ATMOSPHERIC ADMINISTRATION, and the National Geospatial-Intelligence Agency define geodesy as the science of determining the size and shape of Earth and the precise location of points on its surface. Geodesy is especially important to these agencies because they are involved in the production of maps and charts. Measurements of Earth or large areas of Earth are essential for correlating gravitational, geological, and magnetic surveys of Earth in order to make reliable maps. The famous British geodesist Colonel Alexander Ross Clarke (1828–1914) used surveying instruments such as wooden rods, steel chains, theodolites, and Bilby towers to measure meridian arcs. His base line measurements were made in Europe, Russia, India, South Africa, and Peru. The Clarke 1866 ellipsoid was one of the first acceptable models of Earth's shape since he was able to approximate both polar flattening and equatorial radius. Today, satellites provide a more accurate means to measure the size and shape of the planet.

Geodesy provides the theoretical basis for mapping and using remotely sensed images. Geometrical and astronomical measurements are used to determine the precise size and shape of Earth and to locate positions accurately on Earth's surface. Geodesy enables marine scientists to understand the ocean's surface such as the relative heights of the water and the position and extent of SEA ICE. Geodesists working at the National Oceanic and Atmospheric Administration use water level measurements or modeled water levels to determine vertical DATUMS such as mean lower low water. They also develop algorithms called datum transformations in order to convert from a horizontal datum such as North American Datum of 1927 (NAD27) to North American Datum of 1983 (NAD83). In the United States, the difference in coordinates along the y-direction from NAD27 to NAD83 can be on the order of 100s of feet (100s of m). Geodesy also includes the study of gravitational fields and their relationship to the solid structure of the planet. For example, geodesists often prepare isogonic charts showing lines on Earth's surface having equal magnetic declination. These lines are used to correct for the magnetic variation caused by changes in Earth's magnetic fields that keep a compass from pointing to the North Pole. Isogonic lines are important for cross country land, aerial, and maritime navigation, where navigators correct for magnetic variation by subtracting or adding isogonic values from their heading depending on east or west. This knowledge allows navigators to determine their position and to chart an accurate course from one point to another.

Further Reading

Burkhard, Lt. Col. Richard K. *Geodesy for the Layman.* Available online. URL: http://www.ngs.noaa.gov/PUBS_LIB/Geodesy4Layman/toc.htm. Accessed September 19, 2007.

Canadian Geodetic Service. "Canadian Spatial Reference System." Available online. URL: http://www.geod.nrcan.gc.ca. Accessed September 19, 2007.

National Geodetic Survey, National Ocean Service, National Oceanic and Atmospheric Administration. Available online. URL: http://www.ngs.noaa.gov. Accessed September 19, 2007.

geoid The geoid is a representation of Earth's surface where gravity is constant. Equipotential contours connect areas of equal gravity to create a useful model of Earth where those equipotential surfaces are perpendicular to gravity. As a result, the geoid provides a close approximation to MEAN SEA LEVEL. In response to anomalies in gravity, the geoid will tend to rise at mountains and dip at ocean basins, especially the TRENCHES. The geoid is very important to marine scientists such as geodisists and cartographers who want to map Earth's surface in a way that is useful for activities ranging from coastal construction to NAVIGATION. Observations on the geoid are transferred to a reference ellipsoid to make maps and nautical charts (*see* CHARTS, NAUTICAL). The ellipsoid provides a mathematically simple shape that describes Earth where the equatorial radius is the semi-major axis, the polar radius is the semi-minor axis, and the eccentricity is determined based on flattening at Earth's poles. The geoid, being dependent on Earth's irregular topography and BATHYMETRY cannot be calculated mathematically like the ellipsoid, but it can be measured by altimetry. For example, by using the European Space Agency's *ERS-1* satellite ALTIMETER,

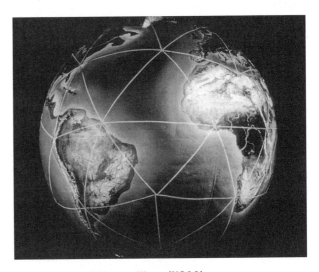

Charting the geoid by satellite *(NOAA)*

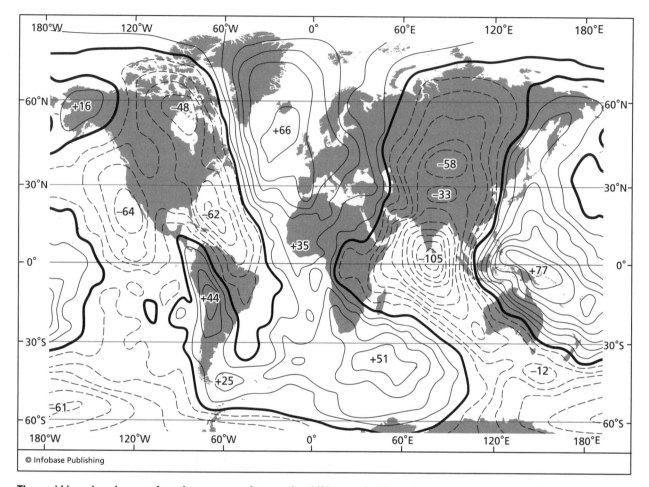

The geoid is an imaginary surface that corresponds to sea level if it extended through the continents. The geoid is perpendicular to the gravity plumb line at all locations. A plumb line is a vertical line directed to the center of the Earth.

the timing of returning microwaves that were transmitted toward Earth's surface can be precisely measured to determine elevation. It is important to remember that real sea level is not level but is in a constant state of response to astronomical tides and meteorological FORCED WAVES, which pile the water up, resulting in temporally and spatially variable WATER LEVEL FLUCTUATIONS and CURRENTS. Marine scientists are able to study variations in geoid heights over the ocean basins to locate major bathymetric features such as SEAMOUNTS.

Further Reading

Brown, Joan, Angela Colling, Dave Park, John Phillips, Dave Rothery, and John Wright. *The Ocean Basins: Their Structure and Evolution.* New York: Pergamon Press, 1989.

National Geographic Society. *National Geographic Satellite Atlas of the World.* Washington D.C.: National Geographic Society, 1998.

geological epochs Geological epochs are the shortest and most recent subdivisions of the time history of

Earth on the geologic timescale. Geological epochs begin at the end of the Mesozoic era, Cretaceous period, and the beginning of the Cenozoic era, Tertiary period. This was the time at the end of major mountain building, when the Rocky Mountains were thrust up, and when prehistoric, extinct animals passed out of existence in favor of modern animals such as the alligator and the horse, about 65,000,000 years ago.

The epochs, each of several million years duration, extend through the present, which is called the Holocene epoch, which began about 10,000 years ago. The complete geologic timescale is shown in Appendix V. This scale was developed from the principles of geology by study of the depths, configurations, and composition of the rock strata, fossil content, magnetic field orientation, and carbon 14 radioactive dating. Carbon 14 dating, however, is not possible for rocks older than the Paleozoic era because so much time has gone by that the amounts of carbon 14 in the samples would be insignificant.

The names of the timescale are not based on a system of logical classification, but have often been

Drill Ship *Glomar Challenger*

On June 24, 1966, a contract between the National Science Foundation and the University of California began Phase 1 of the Deep Sea Drilling Project (DSDP), which was based at Scripps Institution of Oceanography in San Diego, California. On March 23, 1967, a custom drill ship called *Glomar Challenger* was launched as the research vessel serving DSDP and marked a new era in scientific exploration of the world's oceanic rocks and marine sediment.

For the next 15 years, the *Glomar Challenger* crisscrossed the seas, making 96 separate voyages or "legs" and drilling at 624 sites on the seafloor. Using powerful thrusters to stay in one spot even in heavy seas, the ship could lower a drill string through 4.2 miles (7 km) of water, then drill 5,600 feet (1,700 m) or more into the sediment and rock of the seabed. Scientific operations aboard the research vessel included seismic and magnetic surveys, borehole measurements, and laboratory analysis of cores recovered from the seafloor in order to better understand the world's ocean basins. Throughout its operation from August 11, 1968, to November 11, 1983, the *Glomar Chal-* *lenger* traveled 375,632 nautical miles, recovered 19,119 cores, penetrated a total distance of 178,012 fathoms (325,548 m) below seafloor, and reached a maximum penetration depth of 5,712 feet (1,741 m) below the seafloor.

Some of the major discoveries of the DSDP through use of the *Glomar Challenger* have provided information fundamental to the understanding of the ocean floor. For example, the project confirmed theories of seafloor spreading and plate tectonics as an explanation for continental drift. Additionally, the project verified that ocean basins are relatively young in geologic terms compared to continents. The *Glomar Challenger* and DSDP were retired in November 1983.

Further Reading

Integrated Ocean Drilling Program. "DSDP Phase: *Glomar Challenger.*" Available online. URL: http://iodp.tamu.edu/publicinfo/glomar_challenger.html. Accessed October 17, 2007.

—**Bradley A. Weymer,** marine laboratory specialist
Integrated Ocean Drilling Program
Texas A & M University
College Station, Texas

derived from the locations of their discovery. For example, the geologist who discovered the Cambrian period from the study of rock strata named it after the old Roman name for Wales. Nevertheless, the naming convention has proved useful in categorizing the most significant geological events in the history of the planet Earth.

Further Reading

Erickson, Jon. *Marine Geology: Exploring the New Frontiers of the Ocean,* rev. ed. New York: Facts On File, 2003.

geological oceanographers   Scientists who study the geology of the ocean floor and surrounding coastlines are called geological oceanographers. They study the composition, age, and structure of the ocean floor and subbottom, the SEDIMENTS, minerals, topographic features, and marine fossils. They map the depths, using acoustic FATHOMETERS, and interpret the results in terms of sedimentation and crustal movement. They study underwater landslides (TURBIDITY CURRENTS), earthquakes, TSUNAMIS, undersea volcanoes, island formation, ice cover, and all other features of the solid Earth. They are interested in the formation, age, history of the ocean basins, and the processes that shaped them.

Geologists use a wide variety of tools in their studies, ranging from simple GRAB SAMPLERS to examine the surface sediments, to sophisticated drill rigs for taking long cores of the sediments and subbottom in the deep ocean. Geological oceanographers have made extensive use of underwater sound to probe the sediments and rock strata underwater. They record seismic waves to study undersea earthquake and tsunami formation, as well as the structure of Earth's mantle. They pioneered the use of underwater explosives to provide details of the structure of the rock strata under the sediments by studying the acoustic waves reflected from the bottom and rock layer and rock-sediment layer boundaries. They also use electronically controlled sound waves to probe the ocean floor. Along with GEOPHYSICISTS, geological oceanographers have been busy studying the major features and processes of the ocean basins in light of the now almost universally accepted theory of PLATE TECTONICS.

Geological oceanographers almost always have a master's degree, and those involved in research have a Ph.D. They can obtain degrees either at a marine science degree-granting university or obtain degrees in geology or geophysics and take subsequent training at major oceanographic institutions, such as SCRIPPS INSTITUTION OF OCEANOGRAPHY, WOODS HOLE OCEANOGRAPHIC INSTITUTION, or Columbia University's LAMONT-DOHERTY GEOLOGICAL OBSERVATORY.

Geological oceanographers are employed by major universities, government laboratories, and private industry (especially oil and gas companies). They perform applied research on locations of oil, gas, and other mineral deposits, national defense, earthquake prediction, and long-range underwater sound propagation. Many are involved with major field efforts that span topics from investigating seafloor processes at an ocean observatory to investigating mud flats and their occurrences in different types of tidal regimes.

Further Reading

Office of Naval Research. "Oceanography/Marine Related Careers." Available online. URL: http://www.onr.navy.mil/careers/ocean_marine/career_options.asp. Accessed June 21, 2007.

Rutgers University. "Coastal Ocean Observatory Laboratory, Rutgers Marine and Coastal Sciences." Available online. URL: http://marine.rutgers.edu/mrs. Accessed September 19, 2007.

geophysicist A scientist who specializes in the study of the composition and structure of the Earth and the forces that shaped it is called a geophysicist. They study Earth's crust and internal structure, oceans, atmosphere, adjacent space environment, gravity and magnetic fields, dynamics, and thermodynamics. Geophysicists are interested in the formation and history of the Earth, its Sun, Moon, and the inner Solar System, and their influences.

Among the best-known founding fathers of the study of geophysics were Sir Edmund Halley (1656–1742), discoverer of Halley's comet, who made important contributions to knowledge of Earth's magnetic field; Benjamin Franklin (1706–90), who proved that lightning is electricity and produced the first scientific charts of the GULF STREAM in the ATLANTIC OCEAN; and Sir William Thomson (Lord Kelvin, 1824–1907), famous for his work in thermodynamics (the Kelvin scale) gravity waves, and TIDES. During the 19th century, geophysics was known as physical geology. The field greatly expanded after 1850, with significant advances being made in the study of the solid Earth, its oceans, and atmosphere.

Many geophysicists today are working out the details and implications of plate tectonics, the motions of the crust and upper mantle brought about by convection of heat from deep within the Earth (*see* PLATE TECTONICS). Manifestations of plate tectonics include mid-ocean rifts, HYDROTHERMAL VENTS, mountain building, island formation, earthquakes, and volcanoes.

Geophysicists work with geologists in trying to predict the occurrence of earthquakes and volcanic eruptions on a time scale that would allow evacuation of threatened areas and the saving of many lives. They conduct basic investigations of the processes shaping the Earth as well as applied research on marine mineral exploration, national defense (such as mapping Earth's gravitational and magnetic fields), long-term climate prediction, disposal of radioactive waste by deep burial, and forecasting the effects of solar eruptions on communication and weather satellites.

In general, the training of geophysicists involves obtaining a bachelor's degree in a basic science (physics, mathematics), and graduate study of the desired specialty, such as space physics. Geophysics students desiring to work in research should obtain a doctorate in their chosen field of study, followed by postdoctoral work under the tutelage of established scientists who have ready access to the specialized equipment needed in this field.

Geophysicists are employed by academic institutions, government agencies, and private industry. To keep abreast of recent developments in their fields, they must spend considerable time reading current technical journals, participating in scientific meetings and workshops, and taking courses in relevant new technologies. Much new technical information and data are exchanged among scientists by means of the Internet.

Further Reading

Minerals Management Service. "Geophysicist." Available online. URL: http://www.gomr.mms.gov/homepg/lagniapp/careerpg/geophys.html. Accessed September 19, 2007.

University of Washington. "Neptune." Available online. URL: http://www.neptune.washington.edu. Accessed September 19, 2007.

geostationary satellite A geostationary satellite is an orbiting sensor payload with receivers, transmitters, and transponders that completes one oval-shaped trip along Earth's equator in 24 hours. A geostationary orbit is geosynchronous with an inclination of zero.

Geosynchronous satellites orbit at an altitude that is approximately 22,237 miles (35,787 km) above Earth's surface. This is the altitude that corresponds to an orbital period of one day. The term *inclination* describes the angle between the orbital plane of the satellite and the equatorial plane of Earth. An inclination of zero degrees corresponds to an orbit about Earth's equator that is also directed clockwise as one looks down at Earth's North Pole. A perfect geostationary orbit over the equator is hard to achieve, and therefore the satellite's actual position about the equator fluctuates slightly. These altitudes are desirable since there are few forces causing the orbit's decay. The forces involved in inserting a satellite into geostationary orbit include gravity, which accelerates the satellite toward Earth, atmospheric drag, which slows the satellite, and thrust or the force used to propel the satellite.

Sensors on the satellite are used to collect and record electromagnetic radiation that is emitted or scattered by the environment. These data are then transmitted to a receiving station where the data are processed into useful information such as satellite imagery. The stare (continuous viewing) capability of a geostationary satellite is derived from the orbital speed along the equator, which matches the Earth's rotation. These satellites appear to hover over the same spot on Earth's equator at all times. This is a particularly good attribute for communication and environmental satellites. Advantages to using geostationary satellites such as INTELSAT include fixed antennas and a reduced number of transmissions. Many communication satellites and ground stations are owned by INTELSAT, LIMITED. Communications satellites are relay stations in orbit above Earth that receive, amplify, and redirect specific electromagnetic signals. The environmental satellites provide meteorologists, oceanographers, and other Earth scientists with data on sea-surface TEMPERATURE, cloud and ice cover, smoke plumes, hydration of certain plant species to indicate climate conditions, agricultural yield, and information to help assess the health of Earth's fragile and changing ecosystems. An example of an environmental satellite is the Geostationary Operational Environmental Satellite or GOES, which is managed by the NATIONAL OCEANIC AND ATMOSPHERIC ADMINISTRATION. Television weather forecasters often display black-and-white GOES images of the Earth and clouds.

Further Reading

Jensen, John R. *Remote Sensing of the Environment: An Earth Resource Perspective.* Upper Saddle River, N.J.: Prentice Hall, 2000.

National Oceanic and Atmospheric Administration, Office of Satellite Operations. "Geostationary Satellites." Available online. URL: http://www.oso.noaa.gov/goes. Accessed September 20, 2007.

———. National Environmental Satellite, Data, and Information Service. "Satellites." Available online. URL: http://www.nesdis.noaa.gov/satellites.html. Accessed September 20, 2007.

geostrophy Geostrophy describes fluid motion where a balance of forces between the PRESSURE GRADIENT and CORIOLIS (the Coriolis effect is an apparent force that arises on rotating frames of reference, such as the rotating Earth) results in a flow parallel to the ISOBARS (lines of equal pressure). The direction of geostrophic flow is counterclockwise (cyclonic) around low pressure areas, and clockwise (anticyclonic) around high pressure areas in the Northern Hemisphere, with reversed directions in

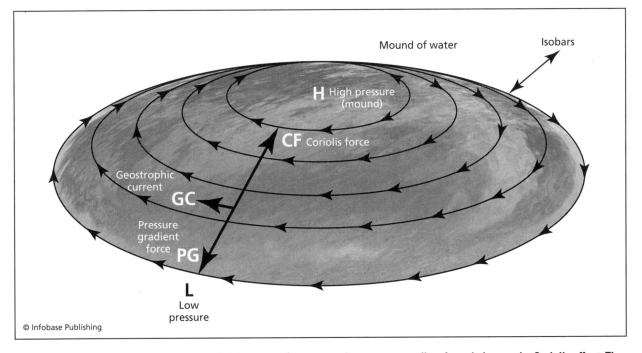

Geostrophic flow around a mound of water (high pressure) occurs as the pressure gradient force balances the Coriolis effect. The flow is clockwise in the Northern Hemisphere and counterclockwise in the Southern Hemisphere.

the Southern Hemisphere. The permanent wind and CURRENT systems on Earth are in geostrophic balance, as well as smaller-scale features such as tropical HURRICANES and mid latitude storms in the atmosphere, and meanders and EDDIES in the ocean.

Examples of geostrophic flows in the atmosphere include the jet stream, the trade winds, the mid-latitude westerlies, and the polar easterlies. In the oceans, great cyclonic rings called GYRES flow around mounds of water in the center of the ocean basins in the ATLANTIC, PACIFIC, and INDIAN OCEANS. These gyres can be divided into equatorial, subtropical, subpolar, and polar cells. The mounds are about three to six feet (1–2 m) above MEAN SEA LEVEL and are the result of convergence due to EKMAN SPIRAL AND TRANSPORT; that is, the net flow in the surface wind-driven layer is 90° to the wind direction, right in the Northern Hemisphere and left in the Southern Hemisphere. From the density and pressure fields, oceanographers can calculate the geostrophic currents by the method of dynamic computations. The dynamic method, developed by early 20th-century oceanographers, makes possible the calculation of the geostrophic current from the TEMPERATURE, SALINITY, and PRESSURE distributions of ocean areas. These quantities are much easier to measure than the ocean currents, and observed data are routinely acquired by research vessels. At-sea measurements provide vertical profiles of temperature (T) and salinity (S) as a function of pressure. These profiles are obtained either by lowering NANSEN BOTTLES and thermometers into the ocean from ships or by using modern electronic instruments such as CONDUCTIVITY-TEMPERATURE-DEPTH PROBES (CTDs) to measure these parameters. If the currents computed by the geostrophic method are to be absolute rather than relative to an arbitrary level (depth), the pressure surfaces must be referred to a known level surface, such as the GEOID.

Analysis of the geostrophically calculated currents shows that to achieve equilibrium, the fields of mass and pressure must have opposite slopes; hence, the isopycnals (lines of equal density) slope downward from left to right when facing in the direction of the current, and the isobars slope upward from left to right when facing in the direction of the current. Thus, when facing in the direction of the current, higher pressure is on the right, leading to higher water levels and an upward left-to-right slope of the sea surface in the Northern Hemisphere; the reverse is true in the Southern Hemisphere, with water piling up to the left of the current. Also for Northern Hemisphere observers facing in the direction of the geostrophic current, the higher density water is to the left, and the lower density water is to the right. Thus,

in the Northern Hemisphere, geostrophic flow is associated with a slightly elevated sea level and less dense water to the right, and water of greater density and lower sea level to the left, with the opposite conditions prevailing in the Southern Hemisphere.

The geostrophic flow is expressed mathematically as

$$V = [1/(2\Omega \sin\varphi \rho)]\Delta p/\Delta n, \text{ where}$$

V = geostrophic velocity, Ω is the angular velocity of Earth, which is 15°/hr = 7.29×10^{-5} rad/s, φ = latitude, ρ = air density, $\Delta p/\Delta n$ = pressure gradient. The term $2\Omega \sin\varphi$ is called the Coriolis parameter, often designated by the symbol f. (The symbol Δ means average change.) Using this notation, the geostrophic equation can be written as the following:

$$V = (1/\rho f) \Delta p/\Delta n.$$

Geostrophic flow occurs not only in the great wind and current systems of Earth, but also around the high and low pressure areas seen on the daily weather maps. There is clockwise (anticyclonic) flow around the high pressure areas and counterclockwise (cyclonic) flow around low pressure areas in the Northern Hemisphere. In the boundary (friction) layer nearest the ground, the wind direction is at a slight angle to the isobars, pointing inward around a low pressure

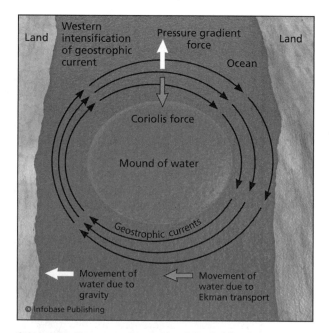

Geostrophic flow around a notational North Atlantic gyre, occuring as the pressure gradient force (PGF) balances with the Coriolis force (CF). The flow is counterclockwise in the Northern Hemisphere and clockwise in the Southern Hemisphere.

area, and outward around a high pressure area. Standing with one's back to the wind, the center of the low pressure, such as a storm, is to the left.

In the ocean, the rings and EDDIES, such as those that spin off the major permanent currents, are analogous to the high and low pressure centers of the atmosphere. Their circulations are in geostrophic balance. Ringlike circulations of diameters on the order of tens to hundreds of miles (km) break off from the GULF STREAM and are in geostrophic balance, with observed speeds as high as two knots (1 m/s). These rings can reach depths as great as 6,000 feet (2,000 m).

In summary, geostrophy continues to be an important tool. Oceanographers use this technique to map ocean circulation based on hydrographic data, while meteorologists use it to map atmospheric circulation.

Further Reading

Knauss, John A. *Introduction to Physical Oceanography*, 2nd ed. Upper Saddle River, N.J.: Prentice Hall/Pearson, 1996.

glacier ice forms Glacier ice forms result from snow and ice forming over land or water over long periods of time. Glaciers usually begin as accumulations of snowfall. As time goes by, the weight of the upper layers of the glacier compresses the lower layers into hard ice, which erodes the land below. They move on the order of several inches per day, but rates can be as high as several feet per day. Regions of major glacier formation include Alaska, Antarctica, Greenland, the Alps, the Andes, and the Himalayas. Antarctic glaciers can be more than 2.4 miles (4 km) thick in some regions. Glaciers are constantly changing; for example, accumulations may occur along the top, while ice wears away at the edges. Glaciers occur in many forms: inland ice sheets, ice shelves, ice streams, ice caps, ice piedmonts, and mountain and valley glaciers.

Physically, glaciers are basically accumulations of ice, snow, water, rock, and SEDIMENT, which can develop on land and move under the influence of gravity down topographic gradients. Rocks and sediments deposited by glaciers are called terminal moraine. The erosive actions of the glaciers scooped out most of the lake basins in Canada and the United States during the ice ages. If a glacier is accumulating ice faster than it is losing ice, it is said to be growing. Characteristics such as shape are constantly changing in response to temperatures, precipitation, and climatological processes.

By monitoring glaciers over time and around the world, glaciologists are able to construct valuable records of glacial activity and their response to climate variation. Scientists have obtained continuous records of Earth's climate by extracting ice cores from locations such as Alaska, Antarctica, and Greenland. Bubbles in the cores reveal the composition of Earth's atmosphere in past ages.

Today, there is a major concern by climatologists that glaciers are melting at an accelerating rate in response to GLOBAL WARMING. Glaciers have been retreating since the beginning of the 19th century, coinciding with the Industrial Age. Snow and ice centers study glaciers, especially their impacts on man.

Icebergs form at the edge of a glacier where it connects to the sea (*see* ICEBERG). The SEA ICE, which is permanently attached to a glacier, is called fast ice. When an iceberg breaks loose from a glacier, the process is called calving. Icebergs float in the ocean with about 10 percent above the water. That leaves about 90 percent of the iceberg hidden below water, which is why ships in polar waters keep a sharp eye out for icebergs. They are classified in accordance to their freeboard and length as growler, bergy bit, small berg, medium berg, large berg, and very large berg. Freeboard refers to the iceberg's height above water. Most of the icebergs in the North Atlantic Ocean have calved from Greenland glaciers. Those in the high latitude regions of the Southern Ocean have calved from Antarctic glaciers.

Further Reading

National Snow and Ice Data Center. "All About Glaciers." Available online. URL: http://www.nsidc.colorado.edu/glaciers/. Accessed April 20, 2005.

Smith, Orson. *Observer's Guide to Sea Ice*. Seattle, Wash.: NOAA Hazardous Materials Response Division, 2000.

Global Ocean Ecosystem Dynamics (GLOBEC) Global Ocean Ecosystem Dynamics is a multiyear international program in which marine scientists—such as oceanographers, ecologists, and fisheries biologists—are collaborating to determine the impact of global changes on populations and communities of marine animals making up marine ECOSYSTEMS. Researchers are attempting to compare and contrast the physical environment to biological processes observed in the ocean. Since there are many scales of change, physical and biological processes are evaluated at individual, population, and community levels. As an example, marine scientists from the United States working in Georges Bank are investigating finfish such as cod (*Gadus morhua*), haddock (*Melanogrammus aeglefinus*), and dominant copepod species such as *Calanus finmarchicus* and *Pseudocalanus moultoni*. Researchers evaluate the physical environment and predator and prey relationships in order to predict changes in the distribution and abundance of these species as a result of changes in their physical and biotic environment. Many scientists are also investigating how populations might respond to climate change. This work is especially important since the number of groundfish in New England's famed Georges Bank are showing signs of decline, suggesting that the impacts of overfishing are not over despite restrictions that have been placed on FISHERMEN. Anglers catch cod and haddock with rod and reel from small private boats, charter boats, and party (head) boats. Commercial fishermen employ gear using hook and line rigs, traps, and trawls.

GLOBEC is a component of the International Geosphere-Biosphere Program (IGBP), which was developed and sponsored by the Scientific Committee on Oceanic Research (SCOR), the Intergovernmental Oceanographic Commission (IOC), the International Council for the Exploration of the Sea (ICES), and the North Pacific Marine Science Organization (PICES). IGBP is a scientific research program built around a family of core projects whose mission is to deliver scientific knowledge to help human societies develop in harmony with Earth's environment. SCOR is the oldest interdisciplinary committee of the International Council of Scientific Unions, established in 1957 to coordinate international oceanographic activities. The IOC is a United Nations Educational, Scientific and Cultural Organization commission that focuses on promoting marine scientific investigations and related ocean services, with a view to learning more about the nature and resources of the oceans. ICES members exchange information and ideas on the sea and its living resources. ICES, established in Copenhagen in 1902, is the oldest intergovernmental organization in the world concerned with marine and fisheries science. PICES promotes and coordinates marine scientific research in order to advance scientific knowledge of the living resources in the North Pacific.

The goal of GLOBEC is to develop a knowledge based on the structure and functioning of the global ocean ecosystem, its major subsystems, and its response to physical forcing. Scientists are striving to forecast the marine upper trophic system response to scenarios of global change. Many GLOBEC researchers are concentrating on ZOOPLANKTON population dynamics and their responses to physical forcing. Research vessels are used to sample zooplankton and fish larvae, take CONDUCTIVITY-TEMPERATURE-DEPTH casts, and perform acoustic studies at locations such as Georges Bank. *In situ* data collected at sea may also be obtained from drifter and glide ring self-contained instruments. These data are all analyzed with remotely sensed imagery depicting sea-surface TEMPERATURE, sea surface chlorophyll-a concentration, and ocean features, such as rings, meanders, and EDDIES. Fisheries biologists measure vital physiological rates of zooplankton and fish larvae and determine fine-scale vertical and horizontal distribution of zooplankton. Often times, zooplankton egg laying, feeding, and growth rates, and larval fish feeding rates will be measured experimentally on board ship.

Further Reading

GLOBEC International Project Office. "Global Ocean Ecosystem Dynamics." Available online. URL: http://www.web.pml.ac.uk/globec/index.htm. Accessed September 20, 2007.

International Geosphere-Biosphere Programme. Available online. URL: http://www.igbp.net. Accessed June 20, 2007.

North Pacific Marine Science Organization. Available online. URL: http://www.pices.int. Accessed September 20, 2007.

Scientific Committee on Oceanic Research, International Council for Science. Available online. URL: http://www.scor-int.org. Accessed September 20, 2007.

U.S. GLOBEC (GLOBal ocean ECosystems dynamics). Available online. URL: http://www.usglobec.org. Accessed September 20, 2007.

Global Positioning System (GPS) The Global Positioning System is a United States Air Force–managed constellation of 24 satellites, allowing users to locate geographic position anywhere on Earth through continuously transmitted high-frequency radio signals, which contain time and distance data that are received and processed by a GPS receiver. Many scientists and engineers affiliated with the GPS program work at the Space and Missile Systems Center at Los Angeles Air Force Base in California. Others are assigned to the 50th Space Wing and work at Schriever Air Force Base in Colorado, which was formally known as Falcon Air Force Base. The base was renamed in 1998 to honor U.S. Air Force General Bernard A. Schriever (1910–2005). Data from monitoring stations located at Ascension Island, Cape Canaveral, Colorado Springs, Diego Garcia, Kwajalein, and Hawaii are sent to the GPS Master Control Station at Schriever Air Force Base in Colorado. Ground antenna stations at Ascension Island, Cape Canaveral, Diego Garcia, and Kwajalein receive telemetry data and transmit commands to the GPS satellites. These signals from the navigation satellites are free to any users. To access GPS for various uses, a person requires a receiver and a GPS navigation system.

The 24 satellites making up the GPS constellation are orbiting above Earth at an altitude of approximately 12,000 miles (19,312 km). Traveling at approximately 17,000 miles per hour (27,359 km/h), these satellites make two complete orbits in less than 24 hours. They are powered by solar energy and have backup batteries onboard. Rocket boosters on each satellite are fired to keep the satellites flying in the proper path, and radio signals are transmitted to locations on Earth. These radio signals travel at the speed of light or approximately 186,000 miles per second (299,338 km/s). GPS receivers take this information and use triangulation to calculate the user's exact location. Essentially, the GPS receiver compares the time a signal was transmitted by a satellite with the time the signal was received. Time-keeping is very accurate since each satellite has its own set of atomic clocks. Time differences tell the GPS receiver how far away the satellite is located. Supplementary distance measurements from additional satellites allow the receiver to determine the user's position for display on an electronic map. By using four or more satellites, the receiver can calculate altitude in addition to latitude and longitude. Sophisticated GPS software is used to calculate information such as speed, bearing, track, trip distance, distance to destination, and more.

Navigation and positioning are crucial in marine science. GPS technologies provide great advantage over positioning strategies that use landmarks, which may disappear, DEAD RECKONING, which accumulates errors quickly, and land-based radio techniques, which provide limited coverage from a network of beacons. The U.S. Coast Guard provides DIFFERENTIAL GPS through the use of ground-based receivers that monitor variations in the GPS signal. Other receivers communicate those variations so that GPS signals can be corrected for better accuracy. Differential GPS is an important resource to hydrographic surveyors that require the most accurate data on location, time, and depth. Dredge boats and fishing vessels are all applying GPS to improve their operations. Knowing the precise locations, orientations, and depths are as essential to coastal construction as knowing the location of boundaries and favorable fishing waters are to commercial fishermen. Besides using precise worldwide positioning to track ocean CURRENTS by means of drifting BUOYS, some oceanographers are even using buoys with GPS receivers to measure WATER LEVEL FLUCTUATIONS such as TIDES and WAVES. Today, GPS locations are the basis for maps and charts of Earth's features, which range from shifting channels and shoals to peaks such as Mount Everest.

Further Reading

The Aerospace Corporation. *The GPS Primer*, July 1999. Available online. URL: http://www.aero.org/education/primers/gps/. Accessed September 20, 2007.

The United States Air Force. "Air Force Link: Global Positioning System." Available online. URL: http://www.af.mil/factsheets/factsheet.asp?fsID=119. Accessed September 20, 2007.

global warming Global warming is natural and unnatural warming caused by Earth's changing astronomical positions, tectonic activity, and the amount of greenhouse gases in the atmosphere as a result of human activity. It is the observation that atmospheric temperature near Earth's surface is increasing. Earth's position and orientation relative to the Sun and even changes in energy from the Sun itself cause changes in brightness and incident solar radiation. Volcanism, continental drift, and SEA-FLOOR SPREADING influence changes in ocean circulation, one of Earth's key mechanisms for transporting heat from the equator to the poles. Certain atmospheric gases such as water vapor, carbon dioxide, and methane absorb solar radiation and reradiate this energy as back radiation. Solar energy is also reflected back into the atmosphere from clouds, dust, and ice. The progressive rise of Earth's temperature following glacial periods is caused by various factors, but the term *global warming* is often thought of in terms of warming thought to occur as a result of increased emissions of greenhouse gases. Interglacial warm periods appear to last from 15,000 to 20,000 years. Scientists show evidence that Earth has been in an interglacial period for 18,000 years.

Earth's glacial and interglacial periods are closely associated to long-term variations in solar activity and Earth's eccentricity, axial tilt, and precession. Solar flares and sunspots seem to vary on 11-year cycles. Sunspots are areas on the surface that appear to be cooler than the surrounding area. Solar flares are bursts of plasma or electromagnetic activity at the surface of the Sun. Changes in Earth's motion around the Sun are known as the MILANKO-VITCH CYCLES after Serbian astronomer Milutin Milankovitch (1879–1958). The orbital eccentricity, describing the shape of Earth's orbit around the Sun, changes every 100,000 years. Axial tilt, which is an angle from 21.5 to 24.5 degrees that Earth's axis makes with the plane of Earth's orbit changes every 41,000 years. Changes in the direction of the Earth's axis of rotation or wobble, describes the term *precession*, which has a period of 23,000 years. Combined, these three cycles alter solar radiation reaching Earth's surface, which impacts Earth's climate as evidenced by the advance and retreat of glaciers.

There are many active volcanoes such as the famed Mount Vesuvius in Italy, Mount St. Helens in the United States, and Mount Pinatubo in the Philippines, as well as areas at plate boundaries such as in the Pacific basin, where magma erupts through Earth's crust. The volcanic island Surtsey near Iceland, which rises 492 feet (150 m) above sea level and covers an area of almost two square miles (3 km^2), formed from submarine eruptions along the Mid-Atlantic Ridge that started during 1963 and subsided during 1967. Explosive volcanic eruptions inject water vapor (H_2O), carbon dioxide (CO_2), sulfur dioxide (SO_2), hydrogen chloride (HCl), hydrogen fluoride (HF) and pulverized rock and pumice 10 to 20 miles (16–32 km) into the stratosphere. These aerosols increase the reflection of radiation from the Sun back into space and thus cool the troposphere. Certain gases such as CO_2 absorb heat radiated up from Earth and warm the stratosphere.

Anthropogenic contributions to global warming include pollution from cars, factories, and power plants that adds to the greenhouse effect and heat island effects. The greenhouse effect refers to the absorption of solar radiation and reradiation in the atmosphere. Life would not be possible without the greenhouse effect. Examples of gases that absorb infrared radiation in the atmosphere are water vapor (H_2O), carbon dioxide (CO_2), methane (CH_4), nitrous oxide (N_2O), halogenated fluorocarbons (HCFCs), ozone (O_3), perfluorinated carbons (PFCs), and hydrofluorocarbons (HFCs). Heat islands are local urban areas of increased temperatures resulting from pavement and buildings that replace natural vegetation, trapped and heated air among the buildings, and waste heat. The 2007 assessment report by the United Nations Intergovernmental Panel on Climate Change indicates that anthropogenic contributions are significant.

Earth's climate system is driven by energy emitted from the Sun. Natural changes in its behavior as well as anthropogenic factors affect Earth temperatures. The forces behind global warming are a politically charged research area. Responsible scientists use the scientific method to analyze time series of Earth's temperature fields and report on their findings. Scientific results must be based on observations that lead to questions, hypotheses that attempt to answer the question, and experiments that prove or disprove the hypothesis. Marine scientists addressing global warming trends are using weather records, satellites, corals, fossils, sediment cores, and ice cores to unravel Earth's climatic patterns.

Further Reading

Collins, William, Robert Colman, James Haywood, Martin R. Manning, and Philip Mote. "The Physical Science behind Climate Change." *Scientific American*, Vol. 297, no. 2, August 2007.

Crichton, Michael. *State of Fear*. New York: HarperCollins, 2004.

Intergovernmental Panel on Climate Change, World Meteorological organization. Available online. URL: http://www.ipcc.ch/. Accessed March 8, 2008.

National Climatic Data Center. "A Paleo Perspective on Global Warming." Available online. URL: http://www.ncdc.noaa.gov/paleo/globalwarming/home.html. Accessed September 20, 2007.

National Geographic News. "Global Warming Fast Facts." Available online. URL: http://news.nationalgeographic.com/news/2004/12/1206_041206_global_warming.html. Accessed September 20, 2007.

United Nations Framework Convention on Climate Change. Available online. URL: http://unfccc.int/2860.php. Accessed June 20, 2007.

U.S. Environmental Protection Agency. Climate Change–Greenhouse Gas emission. Available online. URL: http://www.epa.gov/climatechange/emissions/index.html. Accessed March 8, 2008.

grab samplers   Grab samplers are devices for taking samples of the ocean bottom from the sediment/water interface to a depth of a few inches (cm). They are primarily used to sample the overlying SEDIMENTS, but are also used to sample the benthic flora and fauna, and to detect the presence and quantity of pollutants, such as heavy metals, in the bottom sediments. A grab sampler generally consists of a pair of jaws, which closes upon impacting the bottom, and brings a small (a few cubic inches [cm^3]) sample of the bottom material to the surface for identification and analysis on board ship, and later in a shore-based laboratory. The jaws close either by means of a mechanical device that holds the jaws apart by the tension of their weight force during descent, but releases on striking the bottom, or by a spring that snaps the jaws shut as soon as the jaws strike the bottom. A dredge-type sampler consisting of a bucket and net that is dragged along the bottom is more effective if the sediments are rocky. This sampler is generally used if information on the benthic fauna is desired.

The "snapper" type of grab sampler consists of two clamshelllike jaws that snap shut on contact with the bottom. They hold only a very small volume of sediment. The Petterson grab sampler consists of two half-cylinders held open by their own weight with supporting rods something like scissors that close the cylinders on impact. The jaws can fail to close if a stone or piece of shell is caught on their edges. The Van Veen sampler is similar, but its jaws are held tight during ascent by the lifting cable. The Van Veen sampler also has a modified version with teeth in the jaws to better penetrate closely compacted sediments. The Shipek grab sampler scoops up a bucket sample of sediment, which is then tightly closed by the action of a powerful spring.

All grab samplers disturb the sample so that it is not recovered in its natural state of layering. For this reason, samples from core samplers are greatly preferred, which not only can penetrate much deeper into the sediments and bring back a much longer sample, but the sample, although compressed, is disturbed less than in the case of the grab sampler.

Grab sampling may also be done by divers, either hardhat or SCUBA, in shallow waters. If the water is very clear, divers are sometimes employed to observe the acquisition of the grab samples to provide background data on the sampling environment. Piloted deep-research submersibles, as well as REMOTELY OPERATED VEHICLES (ROVs) often take grab samples, using specially designed equipment.

Once on board the research vessel, the samples are given a preliminary examination and then stored in labeled glass jars for analysis either in the shipboard laboratory, in the case of well-equipped, long-range vessels, or at a shore-based laboratory. The analysis that can be performed on sediment samples varies greatly, depending on the mission. A basic analysis almost always performed includes color,

Retrieval at sea of a modern grab sampler *(NOAA)*

odor, density, and presence of organisms, grain size and type, presence of foreign materials.

Grab samplers relate back to the days when mariners looked at the sediment stuck to the ship's anchor to determine bottom type, as an aid to future NAVIGATION. Different types of anchors may be required, depending on whether the bottom is sandy, rocky, or muddy. Today, the bottom type in shallow waters is indicated on navigation charts so that mariners may know in advance what to expect.

Further Reading

NOAA Coastal Services Center. "Benthic Habitat Mapping." Available online. URL: http://www.csc.noaa.gov/benthic/mapping/techniques/sensors/grab.htm. Accessed September 20, 2007.

Woods Hole Science Center, U.S. Geological Survey. "WHSC Ground-Truth Systems." Available online. URL: http://woodshole.er.usgs.gov/operations/sfmapping/samplers.htm. Accessed September 20, 2007.

gravity corer *See* PISTON CORER.

gravity waves In marine science, gravity waves are WAVES at the surface or interior of the ocean for which the restoring force is gravity. Surface waves are also referred to as "ocean wind waves" because the major generating force is the wind stress. Gravity waves include waves longer than the tiny CAPILLARY WAVES, or cat's paws, and shorter than the ocean basin-sized ROSSBY WAVES. These are waves with periods less than weeks-to-months characteristic of the Rossby waves, and greater than 0.1 s characteristic of the capillary waves. However, most marine scientists refer to gravity waves as the highly visible wind-generated waves that cause ships to pitch and roll, BEACHES to erode, and surfers to ride the tall swells onto the beach. These waves have periods ranging from about 0.1 s to 1,000 s. The short gravity waves are called ultra-gravity waves; the very long gravity waves are called infra-gravity waves. The mid-range waves, say between 1 s and 60 s are known simply as gravity or wind waves. Ocean waves become unstable and break when the ratio of wave height to WAVELENGTH exceeds 1:7. In this case, a wave 230 feet (70 m) long can only be 33 feet (10 m) high. This ratio is called wave steepness.

Although the sea surface is usually a complex assortment of waves traveling in many directions, it can be analyzed in terms of a sum of simple component sine waves of small amplitude with random phases. These component waves have wavelengths λ, periods T, and speeds c, where $c = \lambda/T$, and $k = 2\pi/\lambda$ (called the wave number). The variable "c" also stands for celerity or speed of translation. So, a wave as long as a football field (100 m) traveling

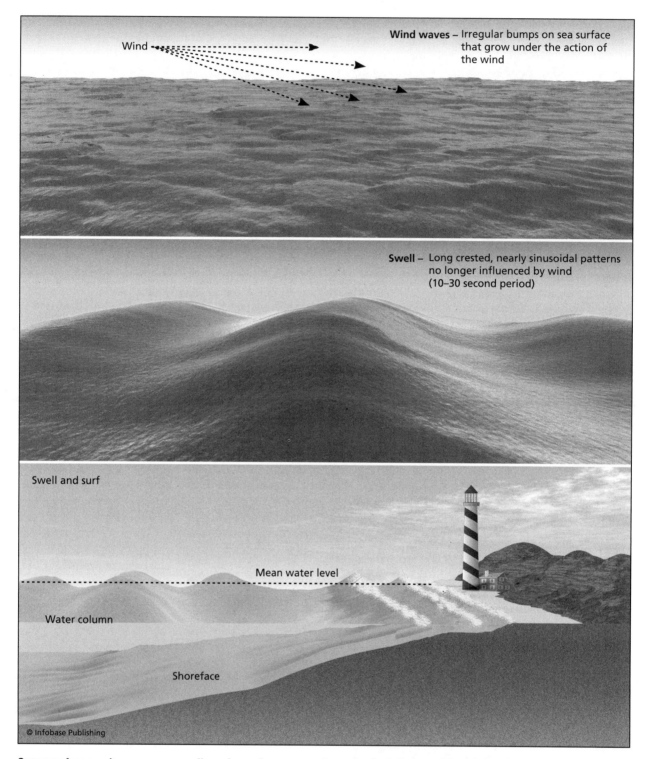

Wind waves – Irregular bumps on sea surface that grow under the action of the wind

Wind

Swell – Long crested, nearly sinusoidal patterns no longer influenced by wind (10–30 second period)

Swell and surf

Mean water level

Water column

Shoreface

© Infobase Publishing

Ocean surface gravity waves are usually made up of waves growing under the influence of the wind, and swell waves, which are waves from a distant storm. At the shoreline, both wind waves and swell break as surf.

with a period of eight seconds would have a wave speed of 24.3 knots (12.5 m/s). As an interesting side note, if the wavelength grows shorter, then the wave number becomes larger.

For water waves, it can be shown that:

$$c^2 = (g/k)\tanh(kh), \text{ where}$$

g = acceleration due to gravity, 32 ft/s² (9.8 m/s²), h = water depth, and tanh(kh) is the hyperbolic tangent of kh. Hyperbolic tangents are used to describe hyperbolas where values approach 1.0 when the argument of the tanh becomes large, and when the argument of the tanh becomes small, the value approaches the argument. The argument is the quantity in parentheses, kh. Therefore, the relationship can be simplified by noting that when kh<0.33, tanh(kh) is approximately equal to kh, and therefore,

$$c^2 = gh, \text{ and } c = (gh)^{1/2}.$$

In this case, the speed of the wave is very dependent on the water depth, a quantity that is provided on nautical charts (see CHARTS, NAUTICAL). When the depth is zero, the velocity is zero.

Waves for which this relationship holds have wavelengths (λ) greater than 20 times the water depth. Such waves are called shallow-water waves. Waves for which tanh(kh) is approximately equal to 1 have a speed approximately equal to

$$c^2 = g/k, \text{ or } c = (g/k)^{1/2}.$$

This relation is true for waves with wavelengths less than four times the water depth.

For shallow-water waves, the water depth determines the speed; for deep-water waves, the speed is a function of the wavelength, meaning that the waves are dispersive, with long waves traveling fastest, since, expressed in terms of wavelength, $c = (g\lambda/2\pi)^{1/2}$. In a wave-generating area (that is, a storm), the long waves outrun the short waves. As the waves travel over the ocean surface, the wind beats down the wave heights, and long wavelength waves run faster than short wavelength waves because of dispersion to a point where they become swell waves: smooth sinusoidal-shaped waves that travel great distances over the ocean surface (see SWELL). Swell waves are often the precursors of an approaching storm.

As these waves propagate toward shore, the wave period remains constant. The wave period is the time it takes for a wave crest to travel the distance of one wavelength. Since celerity is the ratio of wavelength to period, any decrease in speed must be accompanied by a decrease in wavelength. Therefore, as the wave shoals and slows down, the wavelength also decreases. This process groups waves together into what surfers call "sets." These waves break as surf when the wave height (H) becomes 0.78 times the water depth. or, shoaling waves break when

$$H/h = 0.78.$$

In numerical models, H is significant wave height or $H_{1/3}$. It is the average height of the highest one-third observed waves.

The energy E of a component wave can be shown to be $E = (\rho g H^2/8)$, where ρ = water density, and H is twice the amplitude (a). For deep-water waves traveling in a group, called a packet, the speed of the packet, c_{group}, as distinct from the speeds of individual deep-water waves, c_{deep}, can be shown to be $c_{group} = (1/2)c_{deep}$, where c_{deep} is the average phase speed of the individual waves. The group travels at a different speed from that of individual waves because waves are continually disappearing from the rear of the wave packet and appearing at the front of the packet. Shallow water waves are not dispersive; hence, their group velocity is the same as the velocity of the individual waves.

The passage of ocean waves is associated with particle motions of the water molecules. On the surface, deep-water waves are characterized by a circular motion, which decreases with depth, until the orbital trajectories disappear. This is why submarines can dive deep to get away from the tossing by storm seas. In shallow water, the particle motion is elliptical, becoming flatter with depth, until the motion is to-and-fro at the bottom, causing the often-observed sand ripples at the beach.

As is the case for all waves, ocean gravity waves can be refracted. For water waves, water depth plays the roll of changing the index of refraction such as glass does for light waves. Waves approaching a ridge converge toward the ridge; waves approaching a trough diverge from the trough. Water waves can be diffracted around islands and when passing through narrow inlets. The diffraction patterns can be seen when flying overhead or from satellite images. Photographic interpreters may study changing wave patterns to infer characteristics of the seafloor.

The component sine waves discussed above can be mathematically combined with the proper amplitudes, periods, and random phases to represent a real sea. The wave energy is proportional to the square of the amplitude. Thus, waves are represented as an energy spectrum. The spectrum shows the magnitude of the energy at each frequency in the wave field; the area under the spectrum graph is equal to the total energy. In wave analysis, spectra are employed that have been developed from theory, and also from measurements. Many important variables describing the waves can be derived from the spectrum, such as the total wave energy and $H_{1/3}$. Theoretical spectra are often given as parameters at different wind speeds, assuming that a FULLY DEVELOPED SEA exists. Practical wave FORECASTING models were first developed by Professors Gerhard Neumann (1911–97), and Willard Pierson (1922–2003) (*see* SEA STATE). The early wave forecasting methods used the fact that most newly generated waves travel within 45° on both sides of the predominant wind direction.

Today, many numerical wave models that incorporate the directional characteristics of the waves as well as the bottom topography are used in forecasting based on a continually improving understanding of the physics of wave generation by the wind, and increasingly accurate weather forecasts based on data from ships, BUOYS, aircraft, and satellites. Examples of wave models are SWAN (Simulating Waves Nearshore), which predicts random, short-crested wind-generated waves in shallow nearshore waters, and WAM (Wave Model), which predicts parameters such as significant wave height, mean wave direction and frequency, the swell wave height, and mean direction of swell. SWAN was developed at the Delft University of Technology in the Netherlands, and WAM was developed at the Max-Planck-Institut für Meteorologie in Hamburg, Germany, and is mostly used in deep water.

Further Reading

Knauss, John A. *Introduction to Physical Oceanography*, 2nd ed. Upper Saddle River, N.J.: Prentice Hall, 1996.

Model and Data, Max-Planck-Institut für Meteorologie, Hamburg, Germany. Available online. URL: http://www.mad.zmaw.de. Accessed January 21, 2007.

Neumann, Gerhard, and Willard J. Pierson. *Principles of Physical Oceanography*. Englewood Cliffs, N.J.: Prentice Hall, 1966.

The Official SWAN Homepage: Civil Engineering and Geosciences, Environmental Fluid Mechanics, Delft University of Technology. Available online. URL: http://www.citg.tudelft.nl/live/pagina.jsp?id=f928097d-81bb-4042-971b-e028c00e3326&lang=en. Accessed September 20, 2007.

grazing   In marine science, the process by which organisms such as ZOOPLANKTON feed on meadows of PHYTOPLANKTON, waterfowl feed on fields of submerged aquatic vegetation, and mammals such as muskrats feed on marsh grasses is called grazing. Like cattle, these grazers feed primarily on plants. The plant material or forage consumed by a grazing animal could be phytoplankton such as diatoms, dinoflagellates, and COCCOLITHOPHORES, various types and colors of ALGAE, seaweed such as eelgrass (*Zostera marina*), and marsh grasses such as *Spartina alterna flora*.

In the ocean, the microscopic plants that require sunlight, water, and nutrients to grow are called PHYTOPLANKTON. They are the primary producers and form the basis of marine food chains. Zooplankton, small fish, and some species of whales, eat phytoplankton as food. Much of Earth's oxygen supply and an entire pyramid of life, including many terrestrial plants and animals, are dependent upon marine plankton. Marine and estuarine environments are characterized by an abundance of algae. Some of the more common types include rock-weed (*Fucus gardneri*), Bladderwrack (*Fucus vesiculosus*), sea lettuce (*Ulva lactuca*), and kelp such as *Laminaria groenlandica*, which is found in Alaskan waters. Besides providing food for grazers such as sea urchins, these plants provide nursery areas for juvenile fish and fodder for waterfowl.

Further Reading

Anderson Robert F. "What Regulates the Efficiency of the Biological Pump in the Southern Ocean?" *U.S. JGOFS News* 12, no. 2, Woods Hole, Mass.: U.S. JGOFS Planning Office (March 2003): 1–4.

Microbial Observatory, University of Southern California. Available online. URL: http://www.usc.edu/dept/LAS/biosci/Caron_lab/MO/. Accessed June 21, 2007.

groin   A groin is a BEACH PROTECTION structure designed to build up the beach face by arresting LONGSHORE CURRENTS and trapping littoral materials. Longshore or littoral currents are caused by WAVES approaching the shore at an angle, and beach material moved under the influence of currents through the surf zone is called littoral drift. Longshore currents also erode headlands, fill embayments, and build SANDBARS. Groins, the narrow shore-normal structures that reduce the magnitude of the longshore currents and thereby cause the SEDIMENTS to settle, alter the natural process of erosion and accretion. Most groins are made of timber or rock and extend from the beach well into the foreshore. The region extending from the low-water mark to the upper limit of high tide is classified as the foreshore. As the groin retards erosion of the shore upstream, it tends to accelerate erosion downstream. As each groin is constructed, it deprives the down drift areas of sand nourishment. Therefore, series of groins known as a "groin field" are usually installed to protect sections of beach.

Further Reading

Barnard, Thomas. "Self-Taught Education Unit: Coastal Shoreline Defense Structures." Center for Coastal Resources Management. Available online. URL: http://ccrm.vims.edu/wetlands/techreps/coastdef.pdf. Accessed September 20, 2007.

U.S. Army Corps of Engineers. Available online. URL: http://www.usace.army.mil. Accessed June 21, 2007.

groundfish Groundfish, which are also called demersal species or bottom fish, are commercially important finfish that are found near the seafloor. There are many species that make up this group, and important fisheries exist in all the ocean basins. In the United States's Pacific Northwest, there are more than 80 species that are targeted by commercial fishermen. Along the United States's northeastern seaboard, there are more than 30 species that are of important commercial interest. In order to catch groundfish, commercial and recreational FISHERMEN use a variety of gear to include traditional baited hooks and line. Depending on conditions such as water temperatures and season, some fishermen switch gear in order to optimize their catch. In the United States, groundfish are managed through the National Marine Fisheries Service of the Department of Commerce's NATIONAL OCEANIC AND ATMOSPHERIC ADMINISTRATION (NOAA).

People are very familiar with many of the groundfish species. They are on display at seafood markets worldwide and are especially appealing for dinner at seafood restaurants. Black sea bass (*Centripristis striatus*), which are caught by many recreational fishermen from New England to Florida, are of value due to their mild taste and firm flesh. These fish tend to migrate offshore during autumn and return in spring to shallow waters where they are caught near reefs, wrecks, and other undersea structures. Alaska pollock (*Theragra chalcogramma*), which are caught by catcher/processor ships, are processed into surimi and used as imitation crab in some forms of sushi. Alaska pollock are also commonly served as fish n' chips and fish sandwiches at fast food restaurants. These fish are actually bentho-pelagic and are found at various places in the water column based on their age. Schools of larger pollock are caught near the bottom by factory trawlers. Pelagic fish spend most of their life swimming in the water column, while demersal species are found living in close relation to the bottom. Another favorite demersal species is Pacific halibut (*Hippoglossus stenolepis*), which are targeted by recreational fishermen and commercial fishermen fishing in the cold northern waters of the Bering Sea and the Gulf of Alaska. Halibut are a large flatfish with eyes on the topside of the body that swim over the sandy bottoms of the ocean floor. They are a mild-tasting, lean fish with white and firm meat. Recreational fishermen may pay large fees to catch species such as halibut on multiday fishing excursions aboard charter boats. In tropical waters, groundfish such as black grouper (*Mycteroperca bonaci*) and Warsaw grouper (*Epinephelus nigritus*) are particularly desirable as a food fish and have been overfished. In the United States, moratoriums presently exist on catching and harvesting Goliath (*Epinephelus itajara*) and Nassau (*Epinephelus striatus*) groupers. Fishermen must be careful of what fish they catch, especially since some groupers look very similar. For example, a 300-pound (136 kg) Warsaw grouper is distinguished from a Goliath grouper (also known as a Jew fish) by the number of spines on the dorsal fin. Goliath groupers have 11 spines on the dorsal fin, while Warsaw groupers have 10. In addition, the back edge of the tail of Goliath groupers is rounded, while the Warsaw's is straight.

COMMERCIAL FISHING has become a high-technology and closely monitored profession that provides numerous jobs. Depending on the vessel, groundfish fishermen utilize such fishing gears as otter trawls, traps and pots, and long lines. Many types of bottom-dwelling fish are caught when trawling open-mouthed NETS behind the boat. The mouth of a beam trawl is extended by a long spar or beam, while boards distend the opening of an otter trawl. The crew operates WINCHES to launch, tow, haul, hoist, and dump the nets.

Another harvesting method is called purse seining and involves the use of powerboats to drag a large net around a whole school of fish. Once the school is encircled, the bottom of the net is closed up and the fish are hauled aboard a mother ship.

Weighted traps and pots made of wire mesh, plastic, wood, and netting are baited and fished on the ocean floor. Traps and pots are marked by floats and may be connected together by line or fished independently. When demersal and benthic species go in they cannot get out. A sternman is usually employed to prepare pots for deployment, deploys pots once fishing grounds are reached, and quickly stages, cleans, rebaits pots, and stows pots for redeployment after they are retrieved by a hydraulic winch.

Another popular method for catching bottom fish is longlining. A long line with many hooked gangions is reeled out from a drum, anchored to the bottom at each end, and marked with floats. The gangions or leaders are fixed at regular intervals to a main ground line by a snap and hooked at the other end.

High-technology catcher/processor vessels or factory ships tend to use trawl nets to harvest fish. Workers aboard these vessels immediately process groundfish at sea, rather than delivering the catch to shore-based processing plants. Other operations utilize mother ships to receive deliveries of fish or the catcher takes the fish to shore-based processing plants.

Marine scientists such as biologists, geologists, and physicists are working together to understand

Selected List of Commercially Important Groundfish Species

Common Name	Scientific Name	Diet Consists of
Alaskan pollock	*Theragra chalcogramma*	fishes and crustaceans such as krill
American plaice	*Hippoglossoides platessoides*	invertebrates and small fishes
Atlantic cod	*Gadus morhua*	invertebrates and fish, including young cod
Atlantic pollock	*Pollachius virens*	small crustaceans (copepods, amphipods, euphausiids) and small fish
bocaccio	*Sebastes paucispinis*	fishes, including other rockfishes
Canary rockfish	*Dentex canariensis*	fish, crustaceans and cephalopods while the young feeds on plankton
copper rockfish	*Sebastes caurinus*	snails, worms, squid, octopus, crabs, shrimps, and fishes
cowcod	*Sebastes levis*	fishes, octopus, and squid
goosefish (Monkfish)	*Lophius americanus; Lophiodes kempi; Lophius piscatorius*	fishes, crustaceans, squid, waterfowl
haddock	*Melanogrammus aeglefinus*	crustaceans, mollusks, echinoderms, worms and fishes
halibut	*Hippoglossus hippoglossus; Hippoglossus stenolepis*	cod, haddock, pogge, sand-eels, herring, capelin
lingcod	*Ophiodon elongatus*	fishes, crustaceans, octopi and squid, copepods
Pacific cod	*Gadus macrocephalus*	worms, crustaceans, fishes, and octopi
Pacific ocean perch	*Sebastes alutus*	small fish and squid, octopus, shrimp, crab, and other crustaceans
quillback rockfish	*Sebastes maliger*	crustaceans (e.g., shrimps, crabs), worms, forage fish (e.g., herring, sand lance)
ped hake	*Urophycis chuss*	shrimps, amphipods and other crustaceans; squid and herring, flatfish, mackerel
rock sole	*Lepidopsetta bilineata*	mollusks, polychaete worms, crustaceans, brittle stars, and fishes
silver hake	*Merluccius bilinearis*	gadoids and herring, while smaller ones feed on crustaceans, i.e. euphausiids and pandalids
summer flounder	*Paralichthys dentatus*	small fishes, squid, seaworms, shrimp, and other crustaceans
white hake	*Urophycis tenuis*	small crustaceans, squids and small fish
windowpane	*Scophthalmus aquosus*	Arthropods, Amphipods, Caridean shrimp, Pandalid shrimp, Brachyuran crabs, Mysids, Krill, Molluscs, Sea urchins, Fish
winter flounder	*Pseudopleuronectes americanus*	shrimps, amphipods, crabs, sea urchins and snails
witch flounder	*Glyptocephalus cynoglossus*	crustaceans, polychaetes and brittle stars
wolf-fish	*Anarhichas lupus*	hard-shelled mollusks, crabs, lobsters, sea urchins and other echinoderms
yelloweye rockfish	*Sebastes ruberrimus*	fishes and crustaceans
yellowfin sole	*Limanda aspera*	hydroids, worms, mollusks, and brittle stars
yellowtail flounder	*Limanda ferruginea*	polychaete worms and amphipods, shrimps, isopods and other crustaceans and small fish such as sand lance and capelin

Information adapted from http://www.fishbase.org. Available online. Accessed September 20, 2007.

the complexity of the ecosystems and stressed fishing grounds such as Georges Banks. Fisheries management relies on decision makers to utilize science as the basis for managing natural resources such as Atlantic cod (*Gadus morhua*). Studying fish migrations may involve tagging groundfish by inserting an annotated yellow plastic tube into the back of the fish just below the dorsal fin. Anglers that catch a tagged fish should note the location and depth where it was caught, keep the fish whole, and contact fisheries personnel with information relevant to the catch. Port and fisheries biologists are experienced scientists who help train fisheries observers, conduct ground-fish sampling, study trawl logbooks and fish receiving documents, take species-composition samples, archive the data, and synthesize this data into scientific literature. Geologists, GEOPHYSICISTS, hydrographic surveyors, and CARTOGRAPHERS collaborate to map the oceans depths, using tools such as side scan SONAR, multibeam sonar, echosounders, and satellite ALTIMETERS. Seafloor sediments may be collected with corers at selected stations. Bathymetric data includes information on bottom composition or substrate type and structure, which can be used to protect critical habitat. For example, juvenile cod require gravel habitats favorable to epifauna (animals living at the ocean bottom) such as sponges, bryozoans or "moss animals," hydroids that grow in branching colonies, tunicates or "sea squirts," and tube-dwelling polychaete worms (*Filograna implexa*). PHYSICAL OCEANOGRAPHERS specialize in describing depths, WATER MASSES, CURRENTS, and WAVES that affect the life cycles of finfish. Their equipment includes *in situ* instruments and analyses that involve the investigation of spatially and temporally variable parameters such as TEMPERATURE, SALINITY, dissolved oxygen, and the speed and direction of water currents. Results range from time series plots and descriptive STATISTICS to charts and forecasts derived from numerical models. When combined, scientific datasets are used to help manage fish stocks, protect fragile seafloor, and sustain marine habitats.

Further Reading

FISHINFONetwork. Available online. URL: http://www. fishinfonet.com. Accessed June 21, 2007.

NOAA's National Marine Fisheries Service. Available online. URL: http://www.nmfs.noaa.gov. Accessed June 21, 2007.

growth of sea ice   *See* SEA ICE.

Guinea Current   The Guinea Current is a continuation of the southeastern branch of the CANARY CURRENT, flowing southward from the Cape Verde Islands along the African coast and eastward to about 3°N, where it merges with the Atlantic EQUATORIAL COUNTERCURRENT. It then flows into the Gulf of Guinea along the Nigerian coast. The Guinea Current eventually forms the northern portion of the flow into the South Equatorial Current, joining the BENGUELA CURRENT flowing up from the south.

The Guinea Current flows at a speed of about 0.5 to 1.0 knots (.25–.50 m/s); surface temperatures range from about 77°F to 82°F (25°C–28°C). It is strongest in the Northern Hemisphere summer, possibly the result of the African MONSOON, which blows from the southwest to the African continent in the Gulf of Guinea.

The Gulf of Guinea and adjacent regions north to Dakar and south to Cape Frio is characterized by complex circulations. In the Guinea Current, the ISOTHERMS slope upward toward the coast, looking in the direction of the current, showing that this current is in approximate geostrophic balance (*see* GEOSTROPHY). An upward bulge in the isotherms southwest of Dakar, Senegal, brings the THERMOCLINE (the temperature discontinuity separating the warm surface water from the cold bottom water) close to the sea surface. This feature is known as the "Guinea Dome" and is a region of UPWELLING of cool deep waters, especially in the Northern Hemisphere summer.

Recently, oceanographers theorized that the intensification of the Guinea Current in summer may be caused by internal adjustments of the density and pressure fields trying to remain in geostrophic balance, resulting in the generation of KELVIN WAVES and turbulent EDDIES. Close to the coastline of the Gulf of Guinea, the Kelvin wave splits into a northward wave traveling toward Nigeria and a southward wave traveling toward the Angolan coast. These waves enhance the coastal upwelling in that they bring deep water to the surface. Research in the Guinea Current is likely to receive greater emphasis in the future with the discovery of oil reserves off the coast of Nigeria.

Further Reading

Colling, Angela. *Ocean Circulation*, 2nd ed. Prepared for the Open University Oceanography Course Team, The Open University, Milton Keynes, U.K. Oxford: Butterworth-Heinemann. 2001.

Gulf of Maine Ocean Observing System (GoMOOS)   The Gulf of Maine Ocean Observing System or GoMOOS is a fully integrated and operational network of meteorological and oceanographic sensors as well as models that are used to make observations or forecasts of natural phenomena. GoMOOS provides users with hourly oceanographic data from the Gulf of Maine. The system is fully integrated because the many sensors transmit data to a common archive where raw data is converted into timely information products. The products report on important processes such as water TEMPERATURES, SALINITIES, CURRENTS, winds, WAVES, water levels, and other parameters that impact marine and coastal zone activities. GoMOOS is operational because the system is available 24 hours per day and seven days per week. In addition and most important, people from maritime transportation, COMMERCIAL FISHING, recreational fishing and boating, search and rescue operations, and marine spill response are using the system to make important decisions that make operations more efficient, help protect property, and save lives.

The work of controlling, processing, protecting, and facilitating access to environmental data and to provide timely information products to users is the toughest challenge for an observatory. GoMOOS attempts to make information useful by organizing data into categories such as surface marine measurements, physical ocean measurements, and chemical and bio-optical measurements. Surface marine observations such as wind speed and direction, air temperature, and significant wave height are imperative for weather FORECASTING. Physical ocean measurements such as water temperature, salinity, and current speed and direction enhance traditional weather forecasts, are essential for trajectory modeling for search and rescue (SAR), and are useful in tracking features such as EDDIES. Fisheries managers and surveyors rely on information such as chlorophyll, dissolved oxygen, and TURBIDITY on a range of activities from managing shellfish beds to sea truth commercial imagery. For sophisticated users, GoMOOS assimilates observational data automatically into an accessible geographic database on an hourly basis, which may then be queried via graphic online mapping interfaces. Internet users can retrieve up to 48 hours' worth of data from a particular station as a hypertext markup language or HTML table.

The establishment of GoMOOS during 2001 was coordinated with an effort by several federal agencies, including the U.S. Navy, the U.S. Army Corps of Engineers, the NATIONAL OCEANIC AND ATMOSPHERIC ADMINISTRATION, the U.S. Geological Survey, and the NATIONAL AERONAUTICS AND SPACE ADMINISTRATION to develop a national federation of regional observing systems. Many observing systems such as GoMOOS have been established with government funding through the National Oceanographic Partnership Program. Some economists have estimated that the potential benefit of GoMOOS is more than $30 million per year. A national network of observing systems must select and deploy sensors, maintain and operate the observational and data management component of systems, establish standards and protocols for mea-

surements, and establish data exchange and management procedures to ensure rapid access to data from geographically diverse sources. Approximately eight to 10 regions are being implemented by an intergovernmental organization known as "Ocean.US." Observing systems are envisioned with overlapping boundaries in order to share information on oceanographic, biological, and geopolitical factors. These observing systems will include data collection platforms such as moored and drifting BUOYS, instrumented offshore towers, bottom-moored instruments, stand-alone instruments, ship survey cruises, satellite imagery, and remotely and autonomously operated vehicles.

Further Reading

Global Ocean Observing System: The U.S. Global Ocean Observing System. Available online. URL: http://ocean.tamu.edu/GOOS/goos.html. Accessed September 20, 2007.

Gulf of Maine Ocean Observing System. Available online. URL: http://www.gomoos.org. Accessed June 21, 2007.

NOAA Coastal Services Center. "U.S. Coastal Observing Systems." Available online. URL: http://www.csc.noaa.gov/coos/. Accessed September 20, 2007.

Ocean.US: The National Office for Integrated and Sustained Ocean Observations. Available online. URL: http://www.ocean.us. Accessed September 20, 2007.

Gulf of Mexico The Gulf of Mexico is a semienclosed sea in the southeast United States that connects with the Caribbean Sea through the Yucatán Strait between Mexico and Cuba, and the Florida Straits between Cuba and the United States. The Gulf of Mexico basin is a large depression with shallow continental shelves west of Florida, south of Louisiana, and north of the Yucatán. The deep center of the gulf is more than 9,840 feet (3,000 m) deep. The deepest region of the Gulf of Mexico is the Sigsbee Deep, which is as deep as 14,383 feet (4,384 m). The deep flat part of the central basin is called the Sigsbee Abyssal Plain. The CONTINENTAL RISE consists of SEDIMENT transported southward from the North American continent. Salt deposits and associated salt domes are located off the Texas/Louisiana coast.

The large scale circulation in the Gulf of Mexico is dominated by the LOOP CURRENT, which has its beginning in the Yucatán Strait with the inflow of water from the North and South Equatorial Currents that flows through the Caribbean Sea. When this current enters the Gulf of Mexico, it circulates in a generally clockwise fashion around the basin and leaves by way of the Florida Straits, thus forming the beginnings of the GULF STREAM. The Loop Current experiences seasonal and shorter period variations. It often breaks into meanders and EDDIES ("eddy shedding"). Nearshore currents are determined by local winds and TIDES. On the coasts of the Florida panhandle, Alabama, Mississippi, and Louisiana, the predominant tide is diurnal, having one cycle of high and low tide each day. The coasts of southern Florida, Texas, and Mexico have mixed tides; that is, the tide oscillates between diurnal and semi-diurnal (twice daily) with the progression of the phases of the Moon. Because of the shallowness of the coastal waters, there is a strong atmospheric component to the tidal currents.

The Gulf of Mexico is the drainage basin for more than 1.4×10^6 miles^2 (3.8×10^6 km^2) of the continental United States. Significant inflows come from the Mississippi River, and the Apalachicola, Mobile, Atchafalaya, and Brazos Rivers and Rio Grande. Sediments from these rivers are deposited in the Gulf of Mexico. They become a major resource for commercial sand mining. Coastal sediment budgets that track erosion and accretion are determined by coastal engineers from organizations such as the U.S. Army Corps of Engineers.

Perhaps the Gulf of Mexico is best known as the sea where ocean drilling began in earnest for oil, tapping the vast oil and gas deposits on the seafloor. Aside from Florida, the coast is dotted with numerous oil drilling rigs supplying hydrocarbon products shoreward by means of lighters and pipelines. At present, the gulf drilling operations supply about one-quarter of U.S. natural gas and one-eighth of its oil. As the technology improves, and the demand for petro-chemicals increases, drilling operations extend into deeper and deeper water.

The Gulf of Mexico has highly productive fisheries. Gulf fisheries account for about one-fifth to one-fourth of total COMMERCIAL FISHING revenue in the United States. Especially notable are the Gulf shrimp and other shellfish. States bordering the Gulf of Mexico are major centers of recreational fishing. There are approximately 30 species of marine mammals such as Atlantic Spotted, Bottlenose, and Clymene dolphin (*Stenella frontalis, Tursiops truncatus* and *Stenella clymene*), Blainville's Beaked, Blue, and Fin whales (*Mesoplodon densirostris, Balaenoptera musculus* and *Balaenoptera physalus*), and West Indian Manatees (*Trichechus manatus*) that frequent the Gulf of Mexico.

Unfortunately, however, there are locations of extensive pollution from agricultural runoff, municipal waste, and industrial effluent. One such area is the mouth of the Mississippi River, which fans out into an area about the size of the state of New Jersey, which is called the "dead zone." Here, eutrophication from excess nitrogen in the Mississippi River has lowered the oxygen levels to the point where many species of marine life simply cannot exist. Such areas are appearing in ever-increasing numbers in all parts of the world as a result of the coastal population explosion and poor stewardship of the land and sea.

The Gulf of Mexico provides a vast supply of warm water vapor to the atmosphere. In late fall through early spring, this warm, moist air travels northward into the continental United States and meets cold, dry continental air masses from Canada and Alaska. Fronts form between these air masses, which become the loci of mid-latitude synoptic storm systems, resulting in heavy precipitation, winds, and flooding. In spring, especially, when temperature and moisture differences are large, thunderstorms often form along the fronts. These storms sometimes develop into "supercells" that spawn deadly tornadoes.

In summer and early fall, the concern is for tropical CYCLONES called HURRICANES. During the early part of the hurricane season (June–November), hurricanes often develop from tropical atmospheric waves and travel westward. Later in the season, hurricanes are more likely to form in the western Atlantic and the Caribbean. The most deadly hurricane ever to hit the United States mainland struck Galveston, Texas, in September 1900. Galveston is protected now by an extensive sea wall. Because of the hurricane threat, homes along the Gulf Coast are often built on wood piers, so as to be above the onrushing water from STORM SURGE and coastal flooding.

Further Reading

Gulf of Mexico Coastal Ocean Observing System. Available online. URL: http://ocean.tamu.edu/GCOOS. Accessed September 20, 2007.

Nipper, M., J. A. Sanchez Chavez, and J. W. Tunnell, Jr., eds. "GulfBase Resource Database." Available online. URL: http:///www.gulfbase.org. Accessed December 6, 2004.

U.S. Geological Survey. "Gulf of Mexico Integrated Science." Available online. URL: http://gulfsci.usgs.gov. Accessed September 20, 2007.

Gulf Stream The Gulf Stream is the mightiest of ocean currents, transporting warm tropical water from the western North Atlantic Ocean northward and eastward to warm the shores of northwestern Europe. The Gulf Stream is a fast, relatively narrow and deep current, which originates where the Florida Current and the Antilles Current converge, just north of the Bahamas. The Gulf Stream is the western boundary current of the North Atlantic circulation, a great GYRE spinning clockwise around the North Atlantic Basin. The Gulf Stream System includes the Florida Current,

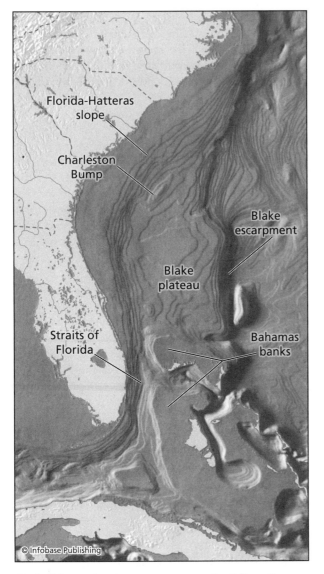

© Infobase Publishing

The Gulf Stream is a warm ocean current that follows the coast of the United States from the Straits of Florida to Cape Hatteras, where it turns to the northeast and becomes the North Atlantic Current south of the Grand Banks. The main topographic features of the bottom below the Gulf Stream off the southeastern United States (known as the Florida Current) are shown here.

which transports a large volume of water through the Straits of Florida, the Gulf Steam proper, which flows from the Florida Current to approximately the Grand Banks of Newfoundland, and into the North Atlantic Current, which flows northeastward toward the British Isles and Scandinavia. East of Newfoundland, an arm of the North Atlantic Current, sometimes called the Azores Current, runs east-southeast toward Spain and Portugal and into the CANARY CURRENT.

The Gulf Stream is the swiftest permanent current in the ocean, sometimes reaching speeds as high as six knots (3 m/s) within the Florida Current, although speeds of two to four knots (1–2 m/s) are more typical. It is a deep current reaching the bottom and scouring the ocean floor off Florida and the Carolina Capes, and extending to perhaps 3,000 feet (1,000 m) depth after it turns eastward off Cape Hatteras or Diamond Shoals. The Gulf Stream flows along the boundary between warm, indigo blue North Atlantic Central Water, and cold, green coastal water flowing southward along the east coast of Canada and the United States. The boundary between the two WATER MASSES is known as the "cold wall." In autumn through spring, cold air blowing over the warm water of the Gulf Stream from the North American continent produces intense evaporation that causes fog, a potential danger to NAVIGATION.

The Gulf Stream often meanders and may shed EDDIES, especially as it passes over the CHARLESTON BUMP and then leaves the shoreline east of Cape Hatteras. These eddies are known as "warm core rings," and "COLD CORE RINGS." Cold core rings spin off the Gulf Stream to the east into the Sargasso Sea. They may persist for up to two years before friction causes them to be absorbed by the surrounding water and lose their identity. Warm core rings spin off from the western edge of the Gulf Stream into the coastal water and drift southwestward between the Gulf Stream and the coast for as long as several months before being absorbed by the surrounding water.

Sailboats in the Newport-Bermuda race try to use these eddies to gain speed. In fact, the race is usually won by the boat that is best able to apply information on the location and strength of eddy and meander currents. Inshore of the Gulf Stream, the coastal water forms a recirculation region following the shoreline, bringing cold water as far south as Cape Hatteras, before it turns eastward and recirculates into the Gulf Stream. Knowledge on the Gulf Stream is derived from numerical models, satellite imagery, and instrumented BUOYS. These data are key to safe navigation for vessels ranging in size from fishing boats to supertankers.

The Gulf Stream has played a major role in the history of Europe and the New World. Ponce de León (1460–1521) was the first European explorer to identify the Gulf Stream, in 1519. Benjamin Franklin (1706–90) performed the first scientific study of the Gulf Stream. He collected temperature readings from ship captains and determined the general configuration of the flow by charting the pattern of temperature. He concluded that ships sailing east should follow the Gulf Stream to gain a following current, whereas ships sailing west should avoid the Gulf Stream, where the adverse current would lengthen their transit time. The shape and character of the Gulf Stream were further delineated by American hydrographer Matthew Fontaine Maury (1806–73). The first modern scientific measurements were made by the British *Challenger* expedition (1872–76) and further refined by the German *Meteor* expedition (1925–27). From the 1940s on, scientists from the WOODS HOLE OCEANOGRAPHIC INSTITUTION and other organizations have extensively investigated the Gulf Stream. Today, it is the world's most-studied current.

Physical oceanographic studies over the years have shown that the Gulf Stream is in geostrophic balance. Geostrophic flows occur when the flow is balanced by the CORIOLIS FORCE and PRESSURE GRADIENTS (*see* GEOSTROPHY). Facing in the direction of the current, the sea surface slopes upward from left to right; that is, it is slightly higher (on the order of three feet [1 m]) at its eastern edge. The thermocline slopes downward from west to east to establish a balance between the density and pressure fields. Computations of the geostrophic velocity from TEMPERATURE and SALINITY measurements agree well with direct measurements of mean velocity. In fact, the measurements made in the Gulf Stream first gave oceanographers the confidence to make worldwide applications of the geostrophic method. Transports of water volume in the Gulf Stream are very high, as great as 100 Sverdrups (100 × 10⁶ m³/s) east of Cape Hatteras. A satisfactory physical explanation of the Gulf Stream and the circulation pattern of the North Atlantic were first developed by Henry Stommel (1920–92) of the Woods Hole Oceanographic Institution in 1948. He determined that the reason for the fast, narrow Gulf Stream on the western side of the Atlantic Basin, and the slow, broad flow on the eastern side, is the result of the increase in the Coriolis force with latitude. Computer models verified his theoretical explanation.

Today, PHYSICAL OCEANOGRAPHERS are focusing their research on the variability of the Gulf Stream on time scales, ranging from millions of years to seconds, the scale of micro TURBULENCE. One of the recent advances involves the installation of instrument MOORINGS along the East Coast of the United States, which are part of the U.S. Integrated Ocean Observing System. Many of these deep-water moorings send real-time meteorological and oceanographic data on sea surface temperatures, currents, and waves to researchers located at laboratories at the University of South Florida, University of South Carolina, University of North Carolina at Wilmington, North Carolina State University, and University of North Carolina Chapel Hill. Examples of these observing systems include the South Atlantic Bight Synoptic Offshore Observation Network, Southeast Atlantic Coastal Ocean Observing System, and the Carolinas Coastal Ocean Observing and Prediction System.

The Gulf Stream contains its own ecosystems, as tropical species such as mahi-mahi or dolphinfish (*Coryphaena hippurus*), ocean sunfish (*Mola mola*), swordfish (*Xiphias gladius*), king mackerel (*Scomberomorus cavalla*), wahoo (*Acanthocybium solanderi*), and blue marlin (*Makaira nigricans*) are transported north with the current. Sargassum weed, for example *Sargassum natans* and *Sargassum fluitans* from the Sargasso Sea on the eastern edge of the Gulf Stream, provides shelter for several marine species and the larvae of many species. For example, the sargassum fish (*Histrio histrio*) swims camouflaged among the fronds, waiting for prey to swim by, while young loggerhead sea turtles (*Caretta caretta*) hide in the drifting weed mats. Even the rings breaking off from the Gulf Stream carry the ecosystems of the Gulf Stream within their circulations.

It has long been known that the transport of warm Gulf Stream water keeps western Europe warmer than regions in other parts of the world at corresponding latitudes. The Gulf Stream has many effects at smaller atmospheric scales, such as HURRICANES and mid-latitude fronts. Hurricanes moving northeast from the GULF OF MEXICO lose energy when passing over land (Florida, Georgia, South Carolina), but can reintensify when they reach the warm waters of the Gulf Stream. When winds blow southwest against the current direction, wave steepness can be increased, creating dangerous conditions for small vessels such as fishing boats. Every year, a few fishing boats are lost due to high winds and steep breaking waves in the Gulf Stream. Rough seas combined with shoal waters, such as Diamond Shoals just east of Cape Hatteras, have caused this region to be named the "graveyard of the Atlantic."

Contemporary research in the Gulf Stream makes extensive use of remote sensing to measure such quantities as temperature, height of the water surface, productivity (chlorophyll-a), wind stress, and current SHEAR. From knowledge that strong currents generally run parallel to the ISOTHERMS (lines of equal temperature), which is the case for the Gulf Stream, geologists are able to infer the ancient history of the western North Atlantic Basin.

Further Reading
Nevala, Amy E. "A Glide across the Gulf Stream: Remote-Controlled Spray's Historic Step Heralds New Era of Ocean Exploration." Available online. URL: http://www.whoi.edu/oceanus/viewArticle.do?id=3821. Accessed March 19, 2008. *Oceanus*, March 24, 2005.

Stommel, Henry M. "The Westward Intensification of Wind-driven Ocean Currents." *Transactions of the American Geophysical Union* 29 (1948): 202–206.

University of Miami Rosenstiel School of Marine and Atmospheric Science, Cooperative Institute of Marine and Atmospheric Sciences. "Ocean Surface Currents." Available online. URL: http://oceancurrents.rsmas. miami.edu/. Accessed September 20, 2007.

Guyana Current The Guyana Current is a northwest-flowing current, which parallels the approximately 280-mile (46-km) long Guyana coastline. The Guyana Current is an extension of the South Equatorial Current, which crosses the equator as it approaches the Brazil coast, then divides, with the northern branch flowing into the Guyana Current and the southern branch flowing into the BRAZIL CURRENT. The southern part of the Guyana Current, off the coast of Brazil, is also known as the "North Brazil Current." The coastal edge of the current is the "North Brazil Coastal Current."

During the course of its northward flow, the Guyana Current feeds Southern Hemisphere water into the EQUATORIAL UNDERCURRENT at the equator. Farther north, the Guyana Current feeds water into the EQUATORIAL COUNTERCURRENT. The Equatorial Undercurrent and Equatorial Countercurrent help to balance the mass transport of water from the eastern to the western Atlantic Ocean by the western flowing North and South Equatorial Currents. North of the Equatorial Countercurrent, the North Equatorial Current feeds into the Guyana Current. The Guyana Current empties into the Caribbean Sea through the Grenada Strait, located between the islands of Grenada and Trinidad and Tobago. These currents form the southern part of the North Atlantic subtropical GYRE and are set up by northeast trade winds. The North Atlantic gyre is a system of clockwise circulating currents consisting of the GULF STREAM in the west, the North Atlantic Current in the north, the CANARY CURRENT in the east, and the North Equatorial Current in the south.

Speeds in the Guyana Current as high as two knots (1 m/s) have been observed; the current speeds are more usually on the order of one knot (0.5 m/s). Oceanographers have estimated water volume transports as high as 35 Sverdrups (35×10^6 m³/s) to the northwest.

The Guyana Current is a highly complex oceanic region. The boundary between the North Equatorial Current and the Guyana Current exhibits a large amount of seasonal variation that is probably a response to freshwater discharge and to the seasonal movements of the equatorial current system. Surface salinities off northeast Brazil are low due to the rainy seasons, which occur from May to mid-August and mid-November to mid-January and the outflow of the Amazon and Orinoco Rivers. Guyana's Berbice, Demerara, and Essequibo Rivers, which are navigable for tens of kilometers for oceangoing vessels, also empty into the ATLANTIC OCEAN.

Marine scientists have determined that the Amazon River is the largest point source of freshwater entering the Atlantic Ocean. Investigators using Lagrangian drifters near the river mouths have eventually tracked their movements into the Caribbean Sea. Oceanographers using satellite tracked drifting BUOYS have determined that the rings and EDDIES of the Guyana Current are an important means for transporting South Atlantic surface and near-surface waters into the North Atlantic.

Further Reading

Bischof, Barbie, Arthur J. Mariano, and Edward H. Ryan. "The North Brazil Current." Available online. URL: http://oceancurrents.rsmas.miami.edu/atlantic/north-brazil.html. Accessed September 20, 2007.

Flagg, Charles N., R. Lee Gordon, and Scott McDowell. "Hydrographic and Current Observations on the Continental Slope and Shelf of the Western Equatorial Atlantic." *Journal of Physical Oceanography* 16 (1986): 1,412–1,429.

Johns, W. E., T. N. Lee, F. A. Schott, R. J. Zantopp, and R. H. Evans. "The North Brazil Current Retroflection: Seasonal Structure and Eddy Variability." *Journal of Geophysical Research* 95 (December 15, 1990): 22,103–122.

guyot A guyot is an undersea flat-topped mountain, or SEAMOUNT, also known as a "tabletop." Geologists believe that guyots were made when wave action eroded away the surface portions of seamounts and smoothed their sharp peaks by erosion. Their present depth is well below past changes in sea level. It is believed that this depth is a result of isostatic subsidence of their mass into the thin ocean crust. Acoustic experiments at the base of guyots show that their surfaces are tilted, and areas of depression are formed at their bases. They are often found in island-like chains with old, low mountains at one end, and steeper, younger mountains at the other end, as is the case in the Hawaiian Island group and the Fiji Island group. Their formation is related to hotspots on the seafloor, which are believed to have created these island chains. Considerable evidence exists to support this hypothesis in the Hawaiian Island group. They are most numerous in the Pacific Ocean. Some guyots are encircled by REEFS.

Guyots were named for Arnold Guyot (1807–84), a professor of geology and geography at Princeton University, by their discoverer, 20th-century American geologist Harry Hess (1906–69). Guyots were first found in the PACIFIC OCEAN by acoustic echo sounders during World War II.

Guyots are often colonized by corals and shallow water organisms, which deposit SEDIMENTS. The presence of these sediments at depths greater than coral formation supports the theory of guyot subsidence, since the coral must have been in the PHOTIC ZONE during its formation.

The surfaces of table-top guyots provide habitats for many fish and benthic organisms. Hatchet fish, lantern fish, eels, and many other species are often found in the vicinity of guyot tops. There is often a ring of dead coral rimming the guyot, providing a protective edge for the topping sediments and the guyot ecosystem.

gyre Gyres are ringlike patterns of circulation in the surface and near-surface waters of the ocean basins. Gyres move clockwise in the Northern Hemisphere and counterclockwise in the Southern Hemisphere. Gyres represent the long-term mean circulation of so-called permanent currents.

Perhaps the best-known gyre is found in the North Atlantic Ocean, where the northward-flowing GULF STREAM forms the western branch of the gyre, the eastward-flowing North Atlantic Current the northern branch, the broad, slow southward-flowing CANARY CURRENT the eastern branch, and the return

The Sargasso Sea: Sea of Legends

The Sargasso Sea is found at the center of the North Atlantic gyre; it is bounded in the west by the Gulf Stream, in the north by the North Atlantic Current, in the east by the Canary Current, and in the south by the North Equatorial Current. The sea results from Ekman transport of these great currents that encircle it with a clockwise rotation. In fact, it is one of the best examples in the ocean of the results of geostrophic circulation, initiated by the planetary wind fields in the Tropics and mid-latitudes. The Sargasso Sea is a mound of water configured as a giant thin lens floating on the deep waters below, moving with a slow anticyclonic (clockwise) circulation. Cold eddies are often spun off from the eastern side of the Gulf Stream and form cold pools in the Sargasso Sea. The waters of the Sargasso Sea are quite saline, resulting from Ekman transport of high-salinity Gulf Stream water and the limited amount of rainfall and high rate of evaporation of surface waters.

It is a unique body of water not found in any other ocean. The color of the water is a beautiful cobalt blue, the desert color of the sea, which results from an absence of the abundant phytoplankton of northern waters that color the water green to brown to yellow. However, in the Sargasso Sea, there is an abundance of Sargassum weed (*Sargassum bacciferum*), brownish floating algae that forms the habitat for a variety of creatures that drift with the weed, and, over the millennia, have adapted to it. The animals range from tiny creatures such as fish larvae that use coloration for camouflage to protect themselves from predators, but the predators also use camouflage to steal unseen to within close range of their prey. The masses of weed use tiny floats called "berries" to prevent sinking. A well-known fish found in this region is the *Histrio histrio*, an olive brown and black frogfish known as the Sargassum fish. A top predator of the Sargassum "jungle" is a fish called *Pterophryne*, which has appendages shaped like the weed and its berry floats. The Sargassum weed is known to live for a very long time, possibly for centuries. The waters around Bermuda are the breeding grounds for both European and American eels (family Anguillidae). The eels spawn and die there. The young spend two or three years growing in the Sargasso Sea and then swim back to the freshwater rivers from which their parents came.

The Sargasso Sea is a place of legends. It includes the "Bermuda Triangle," long feared by mariners, where many ships and aircraft have disappeared, including a squadron of World War II fighter planes that took off from Florida on a mission and never returned. In the early days of exploration of the North Atlantic, European seamen feared that the Sargassum weed would entrap their ships, and they would suffer doom under the hot semi-tropical sun. Naturalist Rachel Carson, in her famous book *The Sea Around Us* (New York: Oxford University Press, 1951), states that "The dense fields of weeds waiting to entrap a vessel never existed except in the imagination of sailors, and the gloomy hulks of vessels doomed to endless drifting in the clinging weed are only the ghosts of things that never were." Nevertheless, the Sargasso Sea will continue to be a place where legends are born.

Present-day problems in the Sargasso Sea include overfishing in the Bermuda Islands to satisfy the tourist demand, and accumulating tar balls and globs of oil resulting from pollutant spills that are now found in all the oceans of the world.

—**Robert G. Williams, Ph.D.**, senior oceanographer
Marine Information Resources Company
New Bern, North Carolina

flow the westward-flowing North Equatorial Current, which forms the southern boundary. The primary atmospheric driving force of this circulation is the wind stress provided by the northeast trade winds, whose steady frictional drag at the ocean surface moves water from the eastern Atlantic to the western Atlantic.

The circulation of the North Atlantic gyre exemplifies Ekman's theory that the net transport of surface water in wind driven currents is 90 degrees to the right of the wind in the Northern Hemisphere (*see* EKMAN SPIRAL). Therefore, there is a pileup of water in the center of the clockwise circulating gyre in the Sargasso Sea, which forms a high-pressure region analogous to the high-pressure cells in the atmosphere.

The corresponding gyre in the South Atlantic consists of the BRAZIL CURRENT flowing southward along the coast of Brazil, the eastward-flowing ANTARCTIC CIRCUMPOLAR CURRENT north of the Antarctic continent, the northward-flowing BENGUELA CURRENT, and the return flow of the South Equatorial Current, driven by the southeast trade winds.

Less is known about the gyres in the PACIFIC and INDIAN OCEANS than in the Atlantic Ocean. The Pacific basin, being much larger than the Atlantic, has a more complex gyre structure. In the North Pacific, the KUROSHIO CURRENT flowing northeast off the coast of Japan is the counterpart to the Gulf Stream in the Atlantic. The northern boundary of the North Pacific gyre is formed by the eastward-flowing North Pacific Current, the eastern boundary by the southward-flowing CALIFORNIA CURRENT, and the southern boundary by the North Equatorial Current, driven by the northeast trade winds as in the Atlantic. In the South Pacific, the driving force is the wind stress of the southeast trade winds transporting the surface waters of the South Equatorial Current from east to west. There is a broad generally southward flow of water east of New Zealand and between New Zealand and Australia. The southern boundary is the Antarctic Circumpolar Current, and the eastern boundary is formed by the PERU (formerly Humboldt) CURRENT, which flows into the South Equatorial Current to complete the gyre.

The exact locations and strengths of these global currents vary with season and with average climatic conditions. In the Pacific Ocean, for example, every few years there is an "El Niño" event in which warm equatorial water moves much farther eastward than usual and warms the oceanic region off the coast of Peru with consequent environmental disasters (*see* EL NIÑO). In the Indian Ocean, seasonal effects are most strongly pronounced when the major current structures change in response to the MONSOON winds.

H

hadal zone The ecological zone in the deep-ocean trenches, from depths greater than 20,000 feet (6,000 m) to the bottom of the TRENCH is called the hadal zone. There is no light, and water pressures are crushing, being 600 atmospheres and greater. TEMPERATURE and SALINITY are fairly constant, with temperatures on the order of 36°F (7°C), and salinity on the order of 36 psu (practical salinity units). The pressure increases by more than one atmosphere for every 33 feet (10 m) of depth. The bottom consists of terregenous clay and radiolarian OOZE (siliceous sediment from the skeletons of planktonic organisms called radiolarians) covering much of the ocean floor.

The dominant life forms are generally small, such as echinoderms, polychaetes, and MOLLUSKS. Echinoderms usually have tentaclelike structures called tube feet with suction pads located at their extremities. They include starfish (Asteroidea); brittle stars, basket stars, and serpent stars (Ophiuroidea); sea urchins, heart urchins and sand dollars (Echinoidea); sea cucumbers (Holothuroidea); and feather stars and sea lilies (Crinoidea). Polychaetes are segmented marine worms that possess an array of bristles on their many leglike parapodia. These animals feed on one another, or on detritus or detritus-and-mud. The holothurians-sea cucumbers are believed to be the most abundant animals in the deep trenches. Few direct observations have been made. The first were from the bathyscaphe *Trieste* in 1960. Towed observations using special equipment were made from the Danish *Galathea* expedition (1950–52) and the Soviet *Vityaz* expedition (1953–62). Scientists at SCRIPPS INSTITUTION OF OCEANOGRAPHY have also taken samples. Marine scientists are particularly interested in identifying any special adaptations that resident animals may possess, which enable them to live under this enormous pressure.

Food is fairly abundant in the hadal zone compared to what would be expected from extrapolating in depth the food abundance in the ocean basins. This is because the trenches are close to land where organic material sinks to the bottom, and the productivity of the surface waters is greater than for ocean basins, so there is a greater amount of detritus sinking to the sea floor. However, because the trenches are located along the RING OF FIRE, earthquakes are frequent. When they occur, earthquakes can cause portions of the walls of the trench to collapse, pouring down sediment and burying the communities at the bottom.

See also MARINE PROVINCES; OCEANIC ZONE; PRESSURE; SEDIMENT.

Further Reading

Ocean Explorer, National Oceanic and Atmospheric Administration. "Alvin." Available online. URL: http://www.oceanexplorer.noaa.gov/technology/subs/alvin/alvin.html. Accessed September 21, 2007.

Oceanlink: All about the ocean. Available online. URL: http://oceanlink.island.net. Accessed June 23, 2007.

hadopelagic zone The hadopelagic zone is the deepest pelagic region extending below the ABYSSOPELAGIC ZONE, which stretches out from about 20,000 feet (6,100 m) to the floor of the HADAL ZONE. This region is well below the PHOTIC ZONE, so that total darkness prevails except for the BIOLUMINESCENCE of some of the creatures. Very little is known about the hadopelagic zone. Most of the inhabitants are small, feeding on detritus and other organic matter. Samples taken from the surface during the *Galathea* (1950–52) and *Vityaz* Pacific Ocean expeditions (1953–62) revealed that animals such as bivalves, polychaetes, and sea cucumbers make up the bulk of the population, although strictly speaking they are part of the BENTHOS and not a pelagic zone. However, the observers on the bathyscaphe *Trieste*, who dove to the floor of the Challenger Deep in 1960, reported seeing a fish. It is not known whether this was a typical specimen or a rare occurrence. Humans have made no descents to the hadopelagic zone since the 1960s. Detailed knowledge of the life-forms in the hadopelagic zone will have to await improvements in technology to carry humans safely to the deepest parts of the ocean.

See also DIVING BELL; MARINE PROVINCES; OCEANIC ZONE; PELAGIC; PRESSURE; SUBMERSIBLE.

Further Reading
Ballard, Robert D. *The Eternal Darkness: A Personal History of Deep-Sea Exploration.* Princeton, N.J.: Princeton University Press, 2000.
Kunzig, Robert. *Mapping the Deep: The Extraordinary Story of Ocean Science.* New York: W. W. Norton, 2000.

halocline The halocline is a layer of ocean or estuarine water in which there is a large change in salinity relative to the layers above and below. It is the region where the rate of change, called the gradient, is a maximum compared to gradients above and below it. The halocline is important to marine scientists because these salinity changes affect the dynamics of the water circulation as well as the resident animals and plants.

Salinity changes are usually not great in the deep ocean. A change of one part of salt per thousand parts of water (1 practical salinity unit or psu) is large. However, near the coasts, especially in the regions of river mouths, salinity changes can be very large, on the order of 10 parts of salt per thousand parts of water (10 psu). Since the density of seawater is primarily determined by temperature and salinity, the halocline is often located at the region of maximum density gradient called the PYCNOCLINE. This strong vertical gradient in density is caused by temperature and salinity.

In areas such as the mouths of the Amazon, the Mississippi, the Columbia Rivers, plumes of relatively freshwater extend out into the ocean for many miles (km) before complete mixing with the surrounding salt water occurs. These plumes are easily observed by airborne sensors and satellites. The Norwegian fjords are regions of strong haloclines at the boundaries between the freshwater from glacier melt and the salty ocean water which, being denser, is found in the deep parts of the basins. Many estuaries are of the "salt wedge" type, where a front is located at the boundary between fresh or brackish water and salty ocean water. Dense seawater enters the estuary near the bottom as a wedge. Seaward of the salt wedge, the ecosystem is oceanic; landward of the salt wedge it is freshwater. Scientists look at interactions such as entrainment along the salt wedge. Entrainment is the collection and transport of objects caught by the flow of a fluid. The mixing of water particles or even larval fish at the salt wedge is an example of entrainment.

The halocline is usually represented on charts showing the distribution of temperature, salinity, and possibly oxygen and chemical parameters by means of isolines, or lines of equal values of a quantity. Lines of equal salinity are called isohalines. The isohalines crowd closely together at the halocline (like the ISOBARS showing a front on a weather map), and are widely spaced in regions of little change in salinity. On a salinity versus depth graph, the halocline is represented as the most nearly horizontal segment of the graph, showing the most rapid change of salinity with depth.

Regions of sharp haloclines, such as river mouths, can often be modeled numerically by a two-layer water flow, with fresh or brackish water being the top layer, and salty oceanic water being the bottom layer. Such models usually provide the physical oceanographer with a means of investigating the main features of the circulation. Physics-based numerical models attempt to simulate physical process such as location of the halocline by accounting for realistic basin geometry, forcing functions such as tides and winds, and boundary conditions, to reproduce existing observations or to forecast future behavior. Such a model would depict a 164–328 foot (50–100 m) layer of cold relatively freshwater beneath the deep ice-covered Arctic basin. This surface layer would be separated from the underlying warmer and saltier Atlantic water by the halocline, where the temperature is near freezing, but the salinity increases rapidly with depth. The skill of the model would be assessed using *in situ* data collected from an oceanographic MOORING, by sensors towed from a research vessel, or imagery obtained from remote sensing. From approximately 1995 through 1999, oceanographers employed a Sturgeon-class, nuclear-powered attack submarine for science cruises to the Arctic Ocean. The submarine provided a perfect platform to study ice-covered oceans since seawater samples could be extracted from regions that have never before been visited by marine scientists. This program was called Scientific Ice Expeditions or SCICEX and was instrumental in gaining a better understanding on temperature and salinity distributions in the halocline. This type of work is critical to understanding impacts from changes in the heat and freshwater content of the upper Arctic Ocean. Other important findings included the determination of SEA ICE thickness (important to the near surface salinity structure) with upward-looking sonars. These kinds of data and biological sample collections provide important sea truth for numerical modeling studies.

See also ESTUARY; SALINITY; SONAR; TEMPERATURE.

Further Reading

Arctic Submarine Laboratory. Available online. URL: http://www.csp.navy.mil/asl/index.htm. Accessed September 21, 2007.

Geophysical Fluid Dynamics Laboratory, National Oceanic and Atmospheric Administration. Available online. URL: http://www.gfdl.noaa.gov. Accessed June 23, 2007.

harbor pilot A harbor pilot takes command of a ship and guides the vessel through seaports, straits, rivers, and other confined waterways where there is a need for a mariner with local knowledge on water depths, winds, tides, CURRENTS, and natural hazards such as shifting channels and CORAL REEFS. As the ship's temporary master, the harbor pilot determines the vessel's course and speed, maneuvers to avoid hazards, and continuously monitors the vessel's position with CHARTS and navigational aides. Pilotage services entail keeping the ship safe and keeping it from running aground or interfering with other vessels navigating the waterway. The ship's crew depends on the harbor pilot to navigate their ship through the unfamiliar harbor and to a safe berth or anchorage.

Harbor pilots are generally contractors who meet the ship and control the vessel from the pilothouse as they enter or leave a port. Pilotage services are usually offered during any 24-hour period, including holidays and weekends upon request from masters of commercial vessels or even commanding officers of warships, including submarines. Services generally originate and terminate either in the seaport or nearby coastal waters. In busy seaports, harbor pilots may pilot many ships in a single day. The basic practice of piloting has to keep pace with technology. Pilothouses use a variety of manual, automated, and complicated navigation and communications systems that interface with technologies such as the U.S. Coast Guard's Vessel Traffic and Information Systems. These innovative technologies provide an increased measure of navigational safety. In fact, harbor pilots in Tampa Bay have referred to the NATIONAL OCEANIC AND ATMOSPHERIC ADMINISTRATION'S (NOAA's) physical oceanographic real-time system or PORTS® (Physical Oceanographic Real-Time System) as their "air bag and seat belt."

Marine systems that provide information that can be used as the basis for navigational decisions and from which essential and perishable information is available 24 hours a day, seven days a week, directly benefit marine operators. PORTS® are actually a network of environmental sensors disseminating fully integrated, quality controlled, real-time current, water level, and wind measurements via the Internet, telephone, or modem dial-up. Information from NOAA PORTS® is quality-controlled and archived. Archived data from PORTS® is useful to a diverse group of basic researchers at universities, engineers working at departments of transportation, coastal zone managers, and the general public, especially recreational boaters and FISHERMEN. Ultimately, harbor pilots, docking pilots, port operations' personnel such as longshoremen, and supporting systems such as weather stations directly benefit maritime commerce. These personnel and their supporting equipment are imperative to getting goods in and out of seaports.

See also DOCKING PILOT; MARINE TRADES; NAVIGATION; NOTICE TO MARINERS; PORTS AND HARBORS; TIDE.

Further Reading

North Carolina Sea Grant. "Coastwatch Early Summer 2005: Safekeepers: River, Harbour Pilots Help Ensure Safe Passage." Available online. URL: http://www.ncseagrant.org/index.cfm?fuseaction=story&pubid=135&storyid=191. Accessed September 21, 2007.

Sandy Hook Pilots. Available online. URL: http://www.sandyhookpilots.com. Accessed June 23, 2007.

Tides and Currents, National Oceanic and Atmospheric Administration. "Physical Oceanographic Real-Time System (PORTS®)." Available online. URL: http://tidesandcurrents.noaa.gov/ports.html. Accessed September 21, 2007.

U.S. Coast Guard. "Marine Safety Office, Tampa Florida." Available online. URL: http://www.uscg.mil/d7/units/mso-tampa/ports.html. Accessed September 21, 2007.

harmonic predictions Predictions of tidal heights and TIDAL CURRENTS by combining the harmonic constituents (sine and cosine waves) of the tide-producing forces of the Sun and Moon are called harmonic predictions. Usually 37 constituents are applied to represent amplitudes and phases of the tide for astronomically forced tide predictions. Predictions of the tide height and current are essential to safe navigation and optimum utilization of seaports and harbors. Most government agencies responsible for tide prediction, such as the National Ocean Service of the NATIONAL OCEANIC AND ATMOSPHERIC ADMINISTRATION (NOAA) use harmonic predictions to produce traditional tide and current tables. Source data is usually six-minute to hourly measurements of water levels and speed and direction of the current at a station that is occupied for many years.

At the present time, the tides cannot be predicted on the basis of theory alone; actual measurements of water levels and currents are needed as functions of time for durations of a year or more for accurate predictions. The time series records are then "resolved" into harmonic components by harmonic analysis, a form of Fourier analysis, which are then related to the harmonic constituents, the components at the several periods of the tidal forcing functions of the Sun and Moon (see TIDE). The constituents have been named according to their origin, and they are classified into "species"; that is, once a day (diurnal), twice a day (semi-diurnal), and mixed (a combination). For example, the largest semi-diurnal constituent with relative amplitude of 100 and a period of 12.42 solar hours is named the M_2, and is caused by the Moon. The second largest semi-diurnal constituent has relative amplitude of 47, a period of 12.00 solar hours, is named S_2, and is caused by the Sun. The largest diurnal tidal constituent has relative amplitude of 58, a period of 23.93 solar hours, and is named the K_1 for the luni-solar diurnal constituent. There is also an 18.6 year factor due to the progression of the lunar nodes; that is, the locations where the Moon's orbit crosses the ecliptic (the intersection of the plane of Earth's orbit with the celestial sphere). The 18.6-year factor is allowed for in the predictions by modifying the constituents.

The accuracy of the predictions is limited by the accuracy of the measurements on which they are based. Agencies such as the National Oceanic and Atmospheric Administration collect water level information and tidal current records at specific locations in order to generate TIME SERIES that enable harmonic prediction. Field measurements are often taken to assess the accuracy of the harmonic predictions. Marine scientists check observations for anomalies such as spikes and timing and datum shifts. Residuals are the differences between observed tides or currents and the harmonic predictions. As with any prediction, the further into the future the predication is made, the less accurate it is.

Tidal currents, being VECTOR quantities, are more difficult to predict than tidal heights, which are scalar quantities. Tidal predictions are based on astronomi-

cal forces and do not include the impact of meteorological factors such as winds and freshets (freshwater discharges from increased river flow following spring rains into saltwater bodies or basins). Tidal and non-tidal forces combine to alter coastal water levels (see WATER LEVEL FLUCTUATIONS). Nontidal forces contribute to errors in harmonic predictions.

Persistent winds can also cause water level differences at either end of a channel, which induces HYDRAULIC CURRENTS. Such fluctuations in water level and current are not in synchronization with the periodic forces of the Sun and Moon and are therefore nontidal. An actual time series of water levels or currents is composed of tidal and nontidal components. Exaggerated ebb and flood currents would be the impact of hydraulic currents on a CURRENT METER record that is being used for harmonic analysis.

The tides are greatly influenced by the shape and topography of the ocean basins. Shallow-water tides or overtides are influenced by the shape and BATHYMETRY of rivers, sounds, and embayments, and also by the sum and difference frequencies (reciprocals of the periods) of the constituents, resulting in nonlinear wave effects, such as bores. Dredging and coastal construction such as building a CAUSEWAY or BRIDGE can alter tidal predictions.

For several years, NOAA has conducted a quality assurance program in which actual water level and current measurements are compared with predictions at several key seaports around the United States. These intercomparison studies form the basis for data collection efforts that may lead to updated harmonic predictions at an important location such as the entrance to a harbor. In some cases, such as in Tampa Bay, a quality assurance study follows a maritime mishap, such as when the freighter *Summit Venture* collided with the Sunshine Skyway Bridge on May 9, 1980, following a spring squall, causing 35 fatalities. This work led to the implementation of a fully operational real-time reporting system of weather stations, water-level gauges, and current meters for parameters such as winds, water levels, and currents around Tampa Bay. The operational system provides users with access to information via telephone or computer.

The astronomically induced tides are the only values predicted in the tide and current tables; variations due to weather conditions cannot be included in the predictions. For this reason, numerical prediction models based on the fluid equations of motion and equations of state, and implemented on supercomputers, are being used to combine the astronomical tides with weather-induced water levels and currents, obtained from weather forecasts, to predict the total tide height and velocity. Such forecasts are limited by the length of time for which accurate weather forecasts are available, on the order of 6 days. At some seaports, these forecasts are supplemented by real-time tide height and current observations by electronic systems such as NOAA's PORTS® (Physical Oceanographic Real-Time System), to provide the navigators and other users of the seaport with all possible information for safe navigation.

See also TIDE GAUGE; TIME SERIES ANALYSIS; WAVE AND TIDE GAUGE.

Further Reading

Hicks, Stacy D. *Tide and Current Glossary.* Silver Spring, Md.: Center for Operational Oceanographic Products and Services, 2000.

NOAA/National Ocean Service. "Center for Operational Oceanographic Products and Services." National Ocean Service. Available online. URL: http://www.co-ops.nos.noaa.gov. Accessed September 21, 2007.

———. "Our Restless Tides: A Brief Explanation of the Basic Astronomical Factors Which Produce Tides and Tidal Currents." Available online. URL: http://www.co-ops.nos.noaa.gov/restles1.html, 1998. Accessed September 21, 2007.

heat budget The balance of incoming and outgoing radiation is called the heat budget. In meteorology and oceanography, the heat budget of the entire Earth is implied by the term *heat budget,* although heat budgets of the atmosphere and ocean, as well as smaller bodies, are also studied. Usually, the heat budget of Earth is estimated in terms of annual averages for the entire planet. The assumption is made that the average global temperature over a year is not changing; otherwise Earth would be heating up or cooling off. In the 20th century, the annual global average temperature rose; this effect is believed to be enhanced by anthropogenic activities such as the burning of fossil fuels. Heat budgets for the Northern and Southern Hemispheres are also different from each other because of the much larger percentage of land in the Northern Hemisphere. Understanding how the Sun's radiation is absorbed and reflected is very important for a wide variety of scientific and practical reasons, ranging from assessing climate to atmospherically correcting imagery that is used in marine science.

The Sun radiates most intensely in the visible band (400 to 700 nanometers) of the electromagnetic spectrum, with the peak radiation in the yellow (about 500 nanometers). For the most part, the atmosphere does not directly absorb the incoming short-wave radiation from the Sun, but rather it is absorbed from the surface of the ocean and the continents. The Earth, being much cooler than the Sun, radiates its heat in the infrared band (800 nanometers to 1 millimeter), approximately as a blackbody (a body that absorbs all the radiation falling upon it). Therefore, most of the heating of the atmosphere is from below at infrared wavelengths.

In heat budgets of Earth, the amount of incoming radiation from the Sun at the top of Earth's atmosphere is often arbitrarily assigned 100 units. Of this 100 units of the heat received, approximately 50 units are absorbed by Earth's surface, while about 30 units are scattered back into space by the atmosphere and do not act to heat Earth, while 20 units are absorbed within the atmosphere and clouds, thus accounting for the 100 units of heat. Of the 50 units received by the surface of Earth, about 20 units are lost by direct long-wave radiation, seven units are lost by conduction and convection, and 23 units are lost by evaporation as latent heat, adding up to 50 units received and 50 units radiated back. In the atmosphere, the seven units of heat lost by conduction and convection from Earth's surface are absorbed, 23 units are released by condensation as latent heat, and about eight units are absorbed from the 20 units of direct radiation from the surface. The atmosphere emits about 58 units of long-wave radiation to space, while 12 units are emitted to space directly from the surface of Earth in a WAVELENGTH band that is transparent to the atmosphere, through the "atmospheric window." Recalling that about 30 units of short-wave radiation from the Sun are scattered back into space by the atmosphere, 58 + 12 + 30 = 100, and it follows that adding the heat fluxes of long-wave and short-wave radiation, Earth receives 100 units of radiation from the Sun, and reradiates about 100 units of radiation back into space; hence, Earth as a whole is not gaining or losing heat from incoming solar radiation.

However, Earth receives more heat from solar radiation than is lost to space between latitudes 38°N and 38°S, while between these latitudes and the North and South Poles, respectively, more heat is radiated to space than is received from the Sun. This imbalance sets up temperature gradients, which are responsible for the circulations of the atmosphere and the oceans to move heat poleward and to cool the Tropics. Major ocean currents and winds, as well as mid-latitude cyclonic storms are all part of Earth's heat engine to distribute heat around the planet so that the global latitudinal heat balance is maintained.

The heat budget of the oceans is generally more difficult to estimate than that of the continents because the lower boundary is dynamic rather than static, and the physics of the air-sea boundary is not so well known as that of the air-land boundary. In addition, there are many parts of the ocean, such as the southern INDIAN OCEAN, where few direct observations are available. Innovative ocean-observing systems such as the Gulf of Maine Ocean Observing System or GoMOOS are being deployed to study heat transfer and to aid navigation. For example, GoMOOS sensors help coastal meteorologists determine when conditions are favorable for the formation of "sea smoke," a type of fog that forms when very cold air drifts across relatively warm water. Other investigators are using networks of moored instrument buoys and drifters to study processes from heat transfer during the various sea states ranging from the flat ocean to a time of precipitation and even during stormy times when the sea is fully developed. sea state codes are used to categorize the force of progressively higher seas by wave height.

Ocean water, unlike the atmosphere, is a good absorber of short-wave solar radiation, and indeed, of all electromagnetic radiation. Therefore, in contrast to the atmosphere, sunlight is readily absorbed in the surface layers of the ocean. Below 660 feet (200 m), there is little light (see photic zone). In addition, the long-wave part of the visible electromagnetic spectrum (yellow, orange, and red) is absorbed before the short-wave part of the spectrum (blue and green), so that as divers and observers on research submarines descend the ocean depths, reds and yellows quickly disappear, leaving a blue-green glow, which eventually becomes dark blue before it disappears altogether at about 660 feet (200 m) depth. Organisms having red pigmentation may be less visible to predators in deeper waters.

The ocean, then, is heated from above, and much of the heat remains in the surface layers. A cross sectional plot of the ocean's temperature along a line of longitude shows warm upper layers concentrated most strongly in the Tropics, underlain by cold bottom waters. The thermocline, the region where the vertical temperature gradient is the strongest, separates the warm, light waters from the cold, dense waters below. Studies have shown that the cold, dense water originates at the polar regions, such as the Greenland Sea in the north, and the Wedell Sea in the south. The upper few hundred feet (around 100 m) is called the "mixed layer," where weather and waves mix the surface water into a relatively homogenous mass of warm water. It fluctuates based on cycles of heating and cooling, wind, rain, and turbulent phenomena that act and interact to stir the upper ocean. The depth of the mixed layer has been found to vary from a few tens of feet (several m) to hundreds of feet (around 100 m).

The heat budget of the ocean involves the vertical exchange of heat and moisture with the atmosphere, which depends on such variables as wind speed, evaporation, and surface roughness. There are several largely empirical expressions available for calculating the fluxes of heat and water vapor over the sea, but there is not uniform agreement among researchers about the validity of these expressions. The heat budget also depends on the downward mixing of heat with the lower layers, and the flow of surface and near-surface currents, including such processes as upwelling and sinking. At the present time, this is an active area of research.

See also attenuation; climate; coral bleaching; global warming; light; marine meteorologist.

Further Reading

Atlantic Oceanographic and Meteorological Laboratory, National Oceanic and Atmospheric Administration. Available online. URL: http://www.aoml.noaa.gov. Accessed June 23, 2007.

Pacific Marine Environmental Laboratory, National Oceanic and Atmospheric Administration. Available online. URL: http://www.pmel.noaa.gov/. Accessed June 23, 2007.

SHEBA: Surface Heat Budget of the Arctic Ocean, University Corporation for Atmospheric Research. Available online. URL: http://www.eol.ucar.edu/projects/sheba. Accessed September 21, 2007.

heat flow   The term *heat flow* describes the process whereby heat flows from hot to cold, which is analogous to air flowing over an aircraft wing or water flowing through a pipe or channel. Physicists may also call this passage of thermal energy from hot to cold, heat transfer. The transfer of thermal energy generally occurs through some combination of conduction, convection, and radiation. Conduction involves the flow of heat from a region of higher temperature to lower temperature through the interaction of adjacent molecular particles. Convection describes the bulk transfer of heat by fluid motion or convection cells. Radiation describes the transfer of heat through electromagnetic waves (oscillating electric and magnetic fields) such as light. Marine scientists are particularly interested in measuring the distribution of temperatures (from the equator to the poles) in order to understand heat flow as well as the overall heat budget. At the seafloor, heat transfer is dominated by conduction, while incoming solar radiation dominates at the sea surface. Convection dominates in the mixed layer. For naval architects, ocean engineers, and marine engineers, understanding the flow of heat is especially important in the design of living spaces aboard surface and submersible vessels. For

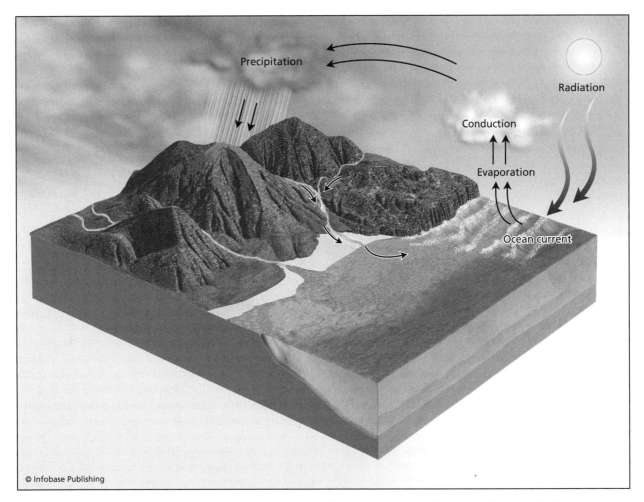

Solar insolation (incoming solar radiation) warms both the atmosphere and the oceans. The absorbed heat is distributed around the globe by the atmospheric and oceanic circulation.

oceanographers, describing heat flow from the mid-ocean ridges to the ocean interior might involve the deployment of long thermistor chains and MOORINGS with CURRENT METERS and CTD probes to quantify processes such as conduction, radiation, and advection. Observing and quantifying heat flow in the ocean is fundamental to understanding physical, chemical, and biological processes and properties near and below the seafloor. It is an essential task for exploration of natural resources such as oil and gas.

See also GEOLOGICAL OCEANOGRAPHER; MARINE ENGINEER; NAVAL ARCHITECT; OCEAN ENGINEER; OCEANOGRAPHIC INSTRUMENTS; PHYSICAL OCEANOGRAPHER.

Further Reading

U.S. Geological Survey. "An Introduction to the USGS Heat Flow Group." Available online. URL: http://earthquake. usgs.gov/heatflow/hfintroduction.html. Accessed October 27, 2007.

Woods Hole Science Center, United States Geological Survey. Available online. URL: http://woodshole.er.usgs. gov/index.html. Accessed October 27, 2007.

highly migratory species Highly migratory species—also called pelagic or oceanic species—are commercially and recreationally important finfish that move great distances in the ocean to feed or reproduce. Pelagic refers to the open ocean and does not include the seafloor or benthic habitat (*see* BENTHOS). In general, highly migratory species are those species that migrate large distances across oceans and traverse domestic and international boundaries. Examples of these highly migratory species are swordfish (*Xiphias gladius*), bluefin tuna (*Thunnus thynnus*), and blue sharks (*Prionace glauca*). Other highly migratory species, such as nurse sharks (*Ginglymostoma cirratum*), are migratory compared to other finfish but, compared to the highly migratory species listed above, are relatively sedate. Offshore fish are harvested by U.S. commercial fishermen and recreational anglers and by foreign fishing fleets. Since these fish may move through the waters of several nations, fisheries management requires international cooperation. Only a small fraction of the total harvest of highly migratory species is taken within U.S. waters.

People are familiar with many of these fish from sport fishing magazine pictures such as giant bluefin tuna being weighed at the wharf, television shows that feature the action of a jumping mako shark (*Isurus oxyrinchus*), as well as classic stories of man versus nature such as *The Old Man and the Sea*, Ernest Hemingway's Pulitzer Prize–winning novel. These fish are also on display at seafood markets worldwide and are especially appealing for dinner at seafood restaurants. For example, swordfish, which are caught worldwide by many recreational and commercial fishermen in deep temperate and tropical waters, are of value due to their appealing taste and firm flesh. These fish tend to be served baked with many flavorful seasonings and the mandatory wedge of lemon. In the 1990s, some U.S. chefs took Atlantic swordfish off their menus over concerns of the overfishing of this once plentiful species. Restaurants that support harvesting methods that do not harm the environment indicate on menus that their seafood is from sustainable sources. Thanks to domestic and international regulations, Atlantic swordfish will be almost completely rebuilt by 2010; stock assessment estimates during 2006 indicated that the biomass of North Atlantic swordfish is approximately 99 percent of the biomass necessary to support maximum SUSTAINABLE YIELD. Recreational fishermen may pay large fees to catch species such as blue marlin (*Makaira nigricans*) on offshore fishing charters. Recreational anglers and trophy hunters target blue marlin, the largest of the billfish. Blue marlin can be kept for food, and in Japan they are a popular sushi fish. Sushi refers to sweetened, pickled rice that is wrapped in raw fish fillets or "sashimi." Sushi and sashimi are wrapped together in portions and sold as sushi. There are many different types of sushi, depending on the choice of seafood.

Highly migratory species have a wide geographic distribution and are defined by the MAGNUSON-STEVENS FISHERY CONSERVATION AND MANAGEMENT ACT as tuna species including albacore tuna (*Thunnus alalunga*), bigeye tuna (*Thunnus obesus*), bluefin tuna, skipjack tuna (*Katsuwonus pelamis*), and yellowfin tuna (*Thunnus albacares*), marlin (*Tetrapturus* spp. and *Makaira* spp.), oceanic sharks, sailfishes (*Istiophorus* spp.), and swordfish. The plural abbreviation "spp." is used to refer to all the individual species within a genus. The Magnuson-Stevens Fishery Conservation and Management Act is the overarching U.S. law that dictates how U.S. fisheries, in general, are managed. Under the Magnuson-Stevens

Act, the National Marine Fisheries Service (NMFS) is responsible for helping to rebuild and maintain sustainable fisheries from 0 to 200 miles (0–321.9 km) of the United States coast. Since highly migratory species are fished by many nations, international cooperation is required to ensure that harvests remain sustainable. Thus, in the case of highly migratory species, the management authority of NMFS can also extend into the high seas for U.S. FISHERMEN.

In addition to the Magnuson-Stevens Act, other domestic laws and international laws and organizations may also govern the exploitation of highly migratory species. These include the United Nations Convention on the LAW OF THE SEA, the Atlantic Tunas Convention Act, International Commission for the Conservation of Atlantic Tunas, the Inter-America Tropical Tuna Commission, the Endangered Species Act, and the Marine Mammal Protection Act. The NMFS' Highly Migratory Species Management Division develops fisheries management plans for Atlantic tuna, swordfish, sharks, and billfish. In the Pacific, the Regional Fishery Management Councils develop these fishery management plans, which are then approved by NMFS. These plans are based on scientific information and input from interested parties, including commercial fishermen, recreational anglers, related industries, nongovernmental organizations, and governmental agencies at all levels. The plans present current information about fish stocks and the fisheries that target them, identify problems and issues to be analyzed, identify alternative means to resolve measured problems, describe analyses of the biological, ecological, economic and social effects or implications of alternative actions, and describe rationales behind the selection of proposed actions.

In order to catch highly migratory species, commercial and recreational fishermen use a variety of techniques, including trolling, gillnets, harpooning, and longlining. Trolling involves the use of fishing rods that drag hooked and baited lines or lures behind the fishing vessel. Commercial fisherman using this technique may employ electric reels. Gillnets and purse seines are also used to catch pelagic species. A gillnet is fixed to a vessel, and the panel of netting drifts behind the vessel with the current. The long net entangles unsuspecting fish; a vertical position is maintained in the water by floats that are spaced along the top and weights along the bottom. Purse-seining involves the use of powerboats to drag a large net around a whole school of fish. Once the school is encircled, the bottom of the net is closed up and the fish are hauled aboard a mother ship. Harpooning usually involves a harpooner standing on a bow platform ready to throw a 10- to 14-foot (3.05

Selected List of Commercially and Recreationally Important Highly Migratory Species

Common Name	Scientific Name	Diet Consists of
albacore tuna	*Thunnus alalunga*	fishes, crustaceans, and squids
Atlantic blue marlin	*Makaira nigricans*	fishes, octopods, and squids
Atlantic sailfish	*Istiophorus albicans*	fishes, crustaceans, and cephalopods
Atlantic white marlin	*Tetrapturus albidus*	fishes and squids
bigeye thresher	*Alopias superciliosus*	pelagic and bottom fishes and squids
bigeye tuna	*Thunnus obesus*	fishes, cephalopods, and crustaceans
blacktip shark	*Carcharhinus limbatus*	fishes, crustaceans, and squids
blue shark	*Prionace glauca*	small fishes, invertebrates, and carcasses
dolphinfish (dorado or mahi-mahi)	*Coryphaena hippurus*	small fishes and invertebrates (cephalopods, crabs, and schyphozoans)
Indo-Pacific blue marlin	*Makaira mazara*	squids, tunalike fishes, crustaceans and cephalopods
Indo-Pacific sailfish	*Istiophorus platypterus*	fishes, crustaceans, and cephalopods
longbill spearfish	*Tetrapturus pfluegeri*	pelagic fishes and squids
northern bluefin tuna	*Thunnus thynnus*	small schooling fishes (anchovies, sauries, hakes) or on squids and red crabs
pelagic thresher	*Alopias pelagicus*	small fishes and cephalopods
sandbar shark	Carcharhinus plumbeus	small fishes, mollusks and crustaceans
skipjack tuna	*Katsuwonus pelamis*	fishes, crustaceans, cephalopods, and mollusks
shortfin mako	*Isurus oxyrinchus*	tuna, herring, mackerel, swordfish, and porpoise
swordfish	*Xiphias gladius*	fishes, crustaceans, and squids
thintail thresher	*Alopias vulpinus*	schooling fishes, squid, octopi, pelagic crustaceans, and seabirds
Yellowfin tuna	*Thunnus albacares*	fishes, crustaceans, and squids

Information was adapted from http://www.fishbase.org. Available online. Accessed September 21, 2007.

to 4.27 m)- long spear that is attached to a long rope at a large pelagic fish such as a bluefin tuna or swordfish. Longlining involves the use of a mainline reeled out from a drum with many hooked gangions. The mainline is marked with floats and may be weighted in order to fish various depths, depending on the time of day. The gangions or leaders are fixed at regular intervals to the mainline by a snap and hooked at the other end. Because many gear types, including gillnets and longlines, catch species other than the targeted catch (also known as bycatch), fishermen and fishery managers work together to discover methods of fishing that reduce the number of interactions with this inadvertent catch and its associated mortalities. For example, in 2001 through 2003, U.S. fishermen worked with the National Marine Fisheries Service (NMFS) to find methods of reducing the number of sea turtles caught on pelagic longlines. The results of this three-year experiment were encouraging, and NMFS is now working with other countries to implement these methods for vessels fishing throughout the Atlantic and, if possible, in the Pacific.

Fisheries managers in both the Atlantic and Pacific Oceans are attempting to understand the complete effects of fishing on the deep-water ECO-SYSTEMS. Efforts are focused on rebuilding and sustaining populations, species, biological communities, and marine ecosystems at high levels of productivity and biological diversity. Effective methods will support the continuation of offshore activities that provide food, revenue, and recreation for humans. A list of some popular highly migratory species is provided in the table above.

Further Reading

American Sportfishing Association. Available online. URL: http://www.asafishing.org/asa. Accessed June 23, 2007.

Atlantic States Marine Fisheries Commission. Available online. URL: http://www.asmfc.org. Accessed June 23, 2007.

Greenlaw, Linda. *The Hungry Ocean: A Swordboat Captain's Journey.* New York: Hyperion, 1999.

Gulf States Marine Fisheries Commission. Available online. URL: http://www.gsmfc.org. Accessed June 23, 2007.

Highly Migratory Species, Office of Sustainable Fisheries, National Marine Fisheries Service, National Oceanic and Atmospheric Administration. Available online. URL: http://www.nmfs.noaa.gov/sfa/hms. Accessed September 21, 2007.

Pacific States Marine Fisheries Commission. Available online. URL: http://www.psmfc.org/index.php. Accessed June 23, 2007.

United Nations: Oceans and Law of the Sea, Division for Ocean Affairs and the Law of the Sea. "United Nations Convention on the Law of the Sea of 10 December 1982." Available online. URL: http://www.un.org/Depts/los/convention_agreements/texts/unclos/UNCLOS-TOC.htm. Accessed September 21, 2007.

high-speed vessels (HSVs) High-speed vessels can travel at speeds above 30 knots (15 m/s) while optimizing seakeeping, maneuvering, and stability. Seakeeping is the general term describing the performance and controllability of a vessel in the ocean. HSVs are typically less than 100 gross tons, have a shallow draft, wave resistant hulls, and travel at speeds ranging from 30 to 45 knots (15 to 23 m/s). NAVAL ARCHITECTS design HSVs to travel at high speeds under optimal cargo loads and under wave conditions such as SEA STATE three. Marine professionals use numerical descriptors to describe the sea surface with regard to wave action. For example, significant wave heights of 4.1 feet (1.25 m) characterize sea state three. Conventional vessels have a mono-hull, while HSVs tend to have innovative catamaran hulls, planing hulls, hydrofoils, or are HOVERCRAFT, which utilize an air cushion beneath the ship.

HSV operations have particular appeal to military, oceanographic, and transportation interests. To be useful, HSVs use state-of-the-art propulsion technologies to couple high speed and cargo capacity to rapidly deliver people and equipment to desired seaports. It should be noted that improving the vessel speed without paying attention to port operations would minimize the advantages that are gained at sea. To be efficient, port facilities need to facilitate fast turnaround of HSVs. Most of these vessels navigate seaways with TWO RADARS, at least two VHF radios, and a GLOBAL POSITIONING SYSTEM (GPS) in order to rapidly reach precise locations to offload people and cargo, survey ocean regions with oceanographic equipment, or to support search and rescue operations. In addition, marine scientists and engineers are evaluating the use of onboard radar to improve seakeeping and ride quality.

Military services are evaluating the utility of shallow-draft and highly maneuverable HSVs to transport combat-ready naval infantry and their equipment into littoral areas that may have austere or degraded seaports. Deck space would need to accommodate a range of military hardware including tanks, amphibious vehicles, wheeled vehicles, trailers, containers, and helicopters. These ships would be particularly useful to providing logistic support and in accomplishing noncombatant evacuation or humanitarian assistance operations. Marines from the III Marine Expeditionary Force in Okinawa, Japan, have been using an Australian-built high-speed catamaran for intratheater transport to seaports in Japan, Korea, Guam, and the Philippine Islands. Typical transits from the Naha military seaport in Okinawa, Japan, include a 400-ton load consisting of 370 combat-loaded marines, helicopters, and aviation support equipment. During Operation Iraqi Freedom, catamaran-hull HSVs were used to ferry army tanks and ammunition from Qatar to Kuwait. In the years ahead, HSVs will continue to be used to support naval expeditionary warfare concepts such as the U.S. Marine Corps's Ship-to-Objective Maneuver while saving time and money when compared to traditional military and commercial aircraft.

At-sea oceanographic research is improved through HSVs that provide flexibility in fast, highly maneuverable platforms that are capable of reconfiguring in a short time for disparate operations. The speed and stability of HSV's offer particular value in the collection of synoptic oceanographic data, where the data must be collected in short time periods. At-sea operations range from deploying arrayed instrument MOORINGS to surveying a permanent current such as the KUROSHIO CURRENT with a bottom-tracking ACOUSTIC DOPPLER CURRENT PROFILER. For this reason, fast Small-Waterplane-Area, Twin-Hull or SWATH ships are used to deliver large-ship platform steadiness and ride quality while still providing the ability to sustain normal cruising speeds in rough seas. The two submerged lower hulls, connected by struts to the superstructure, result in an extremely stable ride. These ships can conduct research in high sea states, thereby increasing their on-station availability as compared to traditional mono-hull research vessels. The U.S. Navy's first SWATH oceanographic research vessel was named R/V *Kilo Moana* and christened on November 17, 2001, at Atlantic Marine's shipyard in Jacksonville,

Florida. The R/V *Kilo Moana* is a general-purpose oceanographic research vessel, fully operational up to sea state six, and operated by the University of Hawaii Marine Center. It does not meet the true definition of a high-speed vessel since it maintains a sustained speed of only 15 knots (7.72 m/s).

High-speed ferries with service speeds above 30 knots are fast replacing conventional ferries on many existing cross-channel seaways, such as the run from Highlands, New Jersey, to Manhattan. Companies such as Incat Australia Pty. Ltd. are producing high-speed passenger and vehicular ferries. Incat HSVs utilize a "Wave Piercing Catamaran" design, which has proven particularly effective as evidenced by winning multiple Hales Trophies (a.k.a. the Blue Riband of the Atlantic) for the fastest transatlantic crossing. Other notable holders of the prestigious Atlantic Blue Riband are the *Rex, Normandie, Queen Elizabeth, Mauritarnia, S.S. United States,* and *Hoverspeed Great Britain.* Since 1933, average speeds on the best record east- and westbound crossings range from 7.31 to 41.28 knots (3.76 to 21.24 m/s). State-of-the-art equipment packages include night vision optics and communication systems ranging from digital navigation to decision-support systems; they improving safety for the burgeoning high-speed ferry service industry.

Regardless of technology, HSVs remain susceptible to the effects of high sea states. In heavy seas, HSV performance may also be reduced as planing hulls and hydrofoils loose lift, catamaran hulls respond to slamming waves against the cross structure, and hovercraft experience undesirable venting of the air cushion. HSVs are associated with elevated risks due to the reduced time that operators have to make navigational decisions when compared to conventional vessels. A 30-knot (15.43 m/s) threshold has been recognized as the point at which vessel navigation becomes less routine and the risks associated with navigational safety become more apparent. High speed on the sea may also be associated with high fuel consumptions. In heavy seas, HSVs may consume more fuel than commercial trucks with a similar cargo over land.

Further Reading

Military Sealift Command Ship Inventory. "High-Speed Vessels." Available online. URL: http://www.msc.navy. mil/inventory/inventory.asp?var=HighSpeedVesselship. Accessed September 21, 2007.

hindcasting   In marine science, hindcasting is a form of back modeling or reverse engineering to obtain objective information about natural phenomena such as WAVES. The Chinese proverb "to know the road ahead, ask those coming back" describes this methodology. Wave hindcasts are retrospective forecasts of wave height and direction based on archived wind observations or historic synoptic weather charts. This method does not use real-time wind speeds and directions to forecast wave heights and directions.

Hindcasting should not be confused with NOW-CASTING or FORECASTING. Nowcasts use current information such as wind speeds and directions to describe SEA STATES, while forecasts use real-time winds to report the forthcoming sea conditions. A hindcast is a "forecast" in the past. A major use of hindcasting is to validate numerical wave models by assuring that they predict realistic and physically realizable conditions from actual input data.

Nowcasting often involves a linear regression or other statistical correlation technique, such as between two nearby CURRENT METER sites along a navigation channel. One current meter can be removed, and the obtained regression equation can be used to nowcast currents based on the remaining current meter.

Forecasting involves estimating an actual value such as TEMPERATURE, PRESSURE, and humidity in a future time period. The terms *forecast, prediction,* and *prognosis* are often used interchangeably. By comparing prognostic information to observations, forecasters can assess their models' skill. Similarly, hindcast skill can be validated using observed data. In the case of wave hindcasts, there may be directional WAVE BUOY observations of wave heights, periods, and directions that can be compared to the hindcasts. The National Data Buoy Center maintains a robust network of wave buoys in United States waters.

Professionals from city planners to COASTAL ENGINEERS need to make informed decisions in designing structures and planning offshore operations. They need an authoritative knowledge base on climatic conditions. Data centers such as the National Climatic Data Center, or NCDC, and the National Oceanographic Data Center, or NODC, have archived wave, wind, temperature, and other weather parameters from coastal areas around the United States and its territories. The record length for observations varies from a few months at some stations to more than 50 years at other stations. These records allow meteorologists to compute hindcast data for MEAN SEA LEVEL pressure, winds, and waves. The wind is computed from the analyzed pressure fields, and the wave data are calculated from the wind fields, using numerical wave prediction models. From observations or hindcast data, or a combination of both, different kinds of STATISTICS can be produced, such as frequency tables, extreme value analysis, and duration statistics. Other environmental parameters that are hindcast include tropical CYCLONE activity from past named storms, HURRICANES, mid-latitude winter storms, and geological events, from box and gravity cores, and seismic profiles.

See also MODEL; NUMERICAL MODELING; PREDICTIONS.

Further Reading

Chen, Changsheng, Robert C. Beardsley, and Geoffrey Cowles. "An Unstructured Grid, Finite-Volume Coastal Ocean Model (FVCOM) System." *Oceanography* 19, no. 1 (March 2006): 78–89.

Coastal & Hydraulics Laboratory, U.S. Army Corps of Engineers. "Wave Information Studies." Available online. URL: http://frf.usace.army.mil/cgi-bin/wis/atl/atl_main.html. Accessed September 21, 2007.

hodograph A hodograph is an illustration of a sequence of VECTORS having a fixed origin and joined head-to-tail representing the velocity (speed and direction) of a moving particle. Dots are normally placed at the head of the vectors, and the lines are erased. This helps to show changes in CURRENT or wind directions and speed with height or depth. The technique has been traced back to the works of Sir William Rowan Hamilton (1805–65), an Irish mathematician and inventor. Hodographs are often used in oceanography and meteorology to show the vertical variation of vector fields such as acceleration, force, and momentum near boundaries, that is, near the ground, the sea surface, or the seafloor. Hodographs are figures that help the marine scientist to interpret vertical SHEAR without reviewing long tables of depths, speeds, and directions.

Perhaps the best-known hodograph in physical oceanography is the hodograph of wind-driven currents called the EKMAN SPIRAL, which is represented as a sequence of velocity vectors of decreasing length and increasing angle as water depth goes from the surface to the deepest extent of the frictional drag from the wind, the Ekman layer depth. Hodographs are used in meteorology to show the vertical variation of the velocity of the geostrophic wind with height, called the "thermal wind." Wind vectors are plotted on a polar coordinate plot relative to the same origin point, with each vector representing a different altitude.

Further Reading

NOAA's National Weather Service Weather Forecast Office, Grand Forks, N.D. "Hodographs and Soundings." Available online. URL: http://www.crh.noaa.gov/ fgf/?n=fargo57f5_hodoandsoundings. Accessed September 21, 2007.

hot spot A hot spot is a vertical convective plume of hot rock believed to originate deep within the Earth's mantle and rising to the surface, forming a volcano. However, hot spots are generally located far from plate boundaries, where mid-ocean ridge spreading and SUBDUCTION produce intense volcanic activity. Hot spots form chains of islands and SEAMOUNTS, such as the Hawaiian Ridge Emperor Seamounts, which are located on the PACIFIC OCEAN seafloor. The hot spot, which has created these islands, is located far from the boundaries of the Pacific Plate where the RING OF FIRE volcanic/earthquake activity is located. Intense localized convective motion burns through the crust, spewing lava out of such well-known volcanoes as Mauna Kea, on the Big Island of Hawaii. Hot spots are believed to remain at the same location within the mantle for tens of millions of years. The moving plates on Earth's surface pass over the hot spots, developing a volcano. As the plate moves on, the volcano is transported away from the hot spot, and a new volcano is formed. By this means, chains of islands were formed, especially in the Pacific Ocean. The volcanic rocks of which the islands are composed are found to get progressively older as distance from the hot spot increases.

By measuring the ages of the rocks, it is possible to estimate the velocity of the plate over time. In the case of the Hawaiian Island Emperor Seamount chain, it can be seen that millions of years ago the direction of the Pacific Plate changed from north to northwest, as evidenced by a bend in the "elbow" of the seamount–island chains (at about 34°N, 173°E), which extend almost to the Aleutian Islands of Alaska. Another well-known hot spot is Kerguel in the INDIAN OCEAN.

As time goes on, hot spots seem to be playing an increasingly important role in the understanding of the structure and dynamics of Earth. Significant differences in the chemical composition of the magma found at hot spots compared to that produced at plate boundaries has led to the hypothesis that hot spots arise from much deeper in the mantle than the magma found from volcanic action at plate boundaries, possibly as deep as the mantle-core boundary.

Further Reading

Monterey Bay Aquarium Research Institute. "Submarine Volcanism: Life-cycle of Hawaiian hot spot volcanoes." Available online. URL: http://www.mbari.org/volcanism/Hawaii/Default.htm. Accessed September 21, 2007.

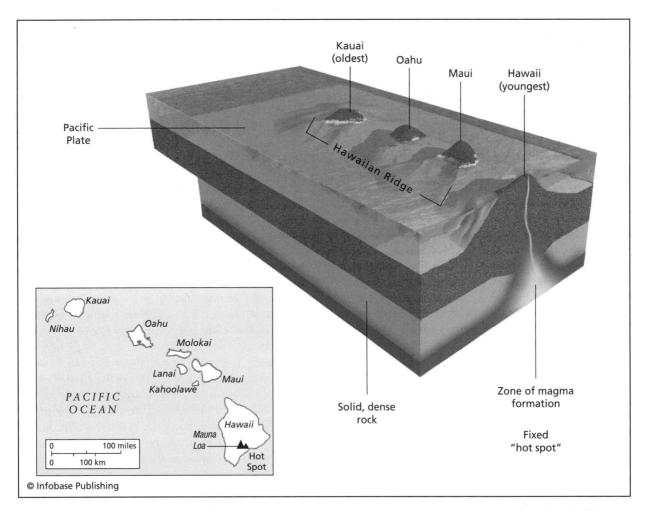

The Hawaiian Islands formed by the drifting of the Pacific tectonic plate over a hot spot. Lava continues to flow from the Kilauea volcano on the island of Hawaii.

MantlePlumes home page. Available online. URL: www.mantleplumes.org. Accessed September 21, 2007.

hot vents *See* HYDROTHERMAL VENTS.

hovercraft (air cushion vehicle, or ACV) Hovercraft are a form of amphibious vehicles capable of operation over land and water; they ride on a cushion of air provided by downward-thrusting fans and propelled forward either by a redirection of some of the downward energy or by one or more independently powered air propellers. The cushion of air is contained by a rubber skirt. The vehicle floats on the air cushion, which provides a nearly frictionless surface on water or air over which the ACV is able to attain high speeds, great maneuverability, and good fuel economy. ACVs can easily make the transition from water to land; they are used by the U.S. Marine Corps for amphibious operations and by governments and private companies for ferry transportation across such water bodies as the English Channel and between Macao and Hong Kong.

The present generation of hovercraft was developed by a British radio engineer, Sir Christopher S. Cockerell (1910–99), and a French engineer, Jean Bertin (1917–75) during the 1950s and 1960s. Their concept was to take in air through a large fan and direct the output to a large cushion on the underside of the vehicle through a system of slots. Later, the skirt was added to help build the cushion and extend the sea conditions under which the craft could operate. Today's hovercraft can attain speeds as high as 75 knots (139 kph) over water, and they operate in adverse weather and sea conditions.

Advanced hovercraft such as the *Princess Margaret* and the *Princess Anne,* readied for use in 1977, could carry 424 passengers and 54 automobiles. The craft was powered by four Rolls-Royce Marine Proteus gas turbine engines developing 3,800 horsepower. The skirt was extended to be able to cope with 13-foot (4 m) waves in Beaufort force 9 wind conditions. Severe weather conditions in the English Channel took a toll on the hovercraft ferries, which necessitated considerable down time for repairs and upgrading of components.

The former Soviet Union pioneered the military use of hovercraft. Large troop and equipment vehicles were constructed able to reach speeds as high as 70 knots (128 kph) transporting heavy tanks, troops, and artillery. The *Zubr* hovercraft, built in St. Petersburg, Russia, is an example. It is approximately 190 feet (58 m) long with a beam of 84 feet (26 m).

By the late 1980s, ACVs were in service in the U.S. Marine Corps, where they are known as Landing Craft, Air Cushioned, or LCACs. Seventeen LCACs served in the Persian Gulf region during operations Desert Shield and Desert Storm, providing much of the transport of troops and equipment from the landing ships to the BEACHES. LCACs were also used in humanitarian operations in such theaters as Bangladesh and Haiti. A major advantage of LCACs is that they can deliver payloads directly on shore. Although they are vulnerable to enemy fire, they can be outfitted with light weapons systems and can also return fire by means of tanks and artillery being transported to the beach. They are used in mine-sweeping operations on both water and the beach. The next generation U.S. Navy hovercrafts with increased lift capability are currently under development and being called Joint Maritime Assault Connectors.

LCACs used by the U.S. Marine Corps are powered by four Avco-Lycoming gas turbines developing 12,444 horsepower. Two engines drive the lift system, and the other two the propulsion system. When fully loaded with fuel and a 60-ton payload, the LCACs can develop speeds up to 50 knots (91 kph) in SEA STATE 2 over a range of 328 nautical miles (607 km). The vessels are operated by a crew of five commanded by a chief petty officer.

Further Reading

Clancy, Tom. *Marine: A Guided Tour of a Marine Expeditionary Unit.* New York: Berkley Books, 1996.

The Hovercraft Museum. "The Hovercraft in the United Kingdom Year by Year." Available online. URL: http://www.hovercraft-museum.org/years.html. Accessed April 21, 2005.

Neoteric Hovercraft, Inc. "The Hovercraft Principle." Available online. URL: http://www.neoterichovercraft.com. Accessed April 21, 2005.

hurricane   A hurricane is an extreme tropical storm with cyclonic winds equal to or greater than 74 miles per hour (120 km/h), usually with a well-developed low pressure core or "eye" in the center of the storm. The higher the wind speed, the better developed is the eye. Hurricanes almost always form over warm tropical oceans as a cluster of thunderstorms, convective cells characterized by towering cumulus CLOUDS (cumulonimbus) revealing regions of intense atmospheric

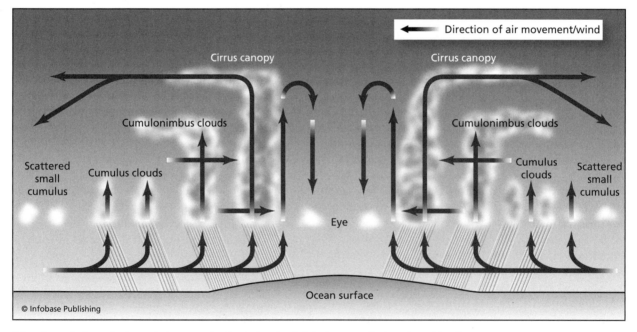

A hurricane is formed when winds exceed 64 knots (119 km/hr) during a tropical cyclone. The diagram above shows a vertical cross section of a typical hurricane showing the eye wall and the circulating bands of cumulonimbus clouds. Upper level winds determine the direction of movement of the storm.

convection rising to altitudes as great as 50,000 feet (15,240 m). The convective cells are extracting energy from the warm sea surface in the form of latent heat. Water is evaporated from the surface, and as it rises in the convection cell, it cools, eventually reaching a height where it has cooled sufficiently to condense into liquid water droplets, which build up the clouds and provide heat energy in the form of the latent heat of condensation (the heat released when water changes from the gaseous to the liquid state). It is this latent heat that is the primary engine driving the hurricane. It also explains why hurricanes develop only over very warm tropical waters.

Hurricanes in the western PACIFIC OCEAN are known as typhoons and in the INDIAN OCEAN as CYCLONES. The area of the most intense winds in a hurricane, called the "eye wall," is comprises numerous convective cells organized in a pinwheel shape, resulting in bands of very heavy precipitation and lighter precipitation. The eye of the hurricane is perhaps 12 miles (20 km) in diameter and is relatively clear. The passage of the eye often misleads observers into believing that the storm is over, when in fact it is the center of the storm, soon to return with renewed violence.

Hurricanes have been classified in terms of the destruction they cause by the Saffir-Simpson scale (see page 224). Category 1 hurricanes do minimal damage and have winds ranging from 74 to 95 miles per hour (119–154 km/h). Category 5 hurricanes, on the other hand, cause catastrophic damage from winds of greater than 155 miles per hour (250 km/h). Besides wind damage, hurricanes cause widespread flooding from intense and often prolonged precipitation. Hurricane Floyd, for example, did not cause a great deal of wind damage during its passage over North Carolina in 1999, but the heavy precipitation which followed on the heels of Hurricane Dennis, also a heavy precipitator, resulted in severe flooding and the loss of more than 50 lives.

Another destructive component of hurricanes is the STORM SURGE, the pileup of water driven ahead of the hurricane by the severe winds, which floods the shoreline as the hurricane makes landfall. The damage is magnified if the surge coincides with high water from astronomical tides.

In 1969, Hurricane Camille slammed into the Mississippi coast with a much larger than usual storm surge. The wave action caused flooding that resulted in significant loss of life. Hurricane Katrina came ashore in southeast Louisiana on August 23, 2005, and caused severe coastal flooding in Mississippi and Alabama. Flooding from levee breaks and breaches in New Orleans and surrounding areas was particularly disastrous as drainage and navigation canals allowed water to flow from Lake Pontchartrain into low areas of the city. Hurricane Katrina was the costliest and one of the deadliest hurricanes in the history of the United States.

Another source of destruction from hurricanes as they make landfall is the spawning of tornadoes, which are usually of less strength and duration than Midwest tornadoes, but can still cause considerable damage. A tornado is defined as a violently rotating column of air extending from a thunderstorm to the ground. When tornadoes accompany hurricanes, they tend to be located to the right and ahead of the path of the storm center as it comes onshore.

Regarding the climatology and meteorology of hurricanes, more is known about those in the North Atlantic than those in the Pacific and Indian Oceans. Atlantic hurricanes often strike the Gulf and East Coasts of the United States, and so have been subject to particular scrutiny.

The Atlantic hurricane season runs from June through November. During the early part of the season, hurricanes form in the western part of the North Atlantic, the Caribbean, where the warmest water is found, and at latitudes far enough north of the equator for the CORIOLIS EFFECT, which is zero right at the equator, to provide sufficient rotation to spin up a tropical hurricane. Later in the season, hurricanes are born farther to the east off northwest Africa. Hurricanes typically form from easterly waves in the

Hurricane Diana from the Tiros weather satellite, September 11, 1984 *(NOAA/NWS)*

trade winds, or convective disturbances in the INTER-TROPICAL CONVERGENCE ZONE (ITCZ). They begin as tropical depressions, then form tropical disturbances, tropical storms, and finally, full-fledged hurricanes. Fortunately, most tropical depressions do not evolve into hurricanes. Newly formed hurricanes propagate to the west, propelled either by the trade winds or upper-level winds. As the disturbances pass over the cooler water of the UPWELLING regions off the coast of northwest Africa, they weaken, but the water becomes increasingly warm moving westward, and the hurricanes strengthen and can often continue their march westward to threaten the islands of the Caribbean, the GULF OF MEXICO, and the East Coast of the United States. Late in the hurricane season, hurricanes again form primarily in the western Atlantic, as they did during the early part of the season. For the most part, the strongest storms are those that form off northwest Africa.

At some point in their paths, hurricanes turn poleward in response to the Earth's rotation. They may move into the Gulf of Mexico and strike the Texas coast, or turn northward and strike the coasts of Florida, Alabama, Mississippi, or Louisiana. Other hurricanes turn northeastward more sharply and hit the U.S. East Coast, or perhaps graze or miss it entirely and strike the Bermuda Islands.

Determining the path that a hurricane will take is an important job of the U.S. National Hurricane Center (NHC) of the National Weather Service, located in Miami, Florida. The NHC uses sophisticated computer models based on STATISTICS and the

The Saffir-Simpson Hurricane Intensity Scale

Saffir-Simpson Category	Maximum Sustained Wind Speed	Minimum Central Pressure	Storm Surge	Damage
1	74–95 mph (33–42 m/s)	>980 mb	3–5 feet (1.0–1.7 m)	*Minimal:* No major damage to most building structures. Damage primarily to unanchored mobile homes, shrubbery, and trees. Some coastal road flooding, and minor pier damage.
2	96–110 mph (43–49 m/s)	979–965	6–8 feet (1.8–2.6 m)	*Moderate:* Some roofing material, door, and window damage to buildings. Considerable damage to vegetation, mobile homes, and piers. Coastal and low-lying escape routes flood 2–4 hours before arrival of center. Small craft in unprotected anchorages break moorings.
3	111–130 mph (50–58 m/s)	964–945	9–12 feet (2.7–3.8 m)	*Extensive:* Some structural damage to small residences and utility buildings with a minor amount of curtainwall failures. Mobile homes are destroyed. Flooding near the coast destroys smaller structures with larger structures damaged by floating debris and battering waves. Low-lying escape routes inland cut by rising water 3–5 hours before hurricane center arrives. Major erosion of beaches.
4	131–155 mph (59–69 m/s)	944–920	13–18 (3.9–5.6 m)	*Extreme:* Extensive curtainwall failures with some complete roof structure failure on small residences. Major erosion of beach areas. Major damage to lower floors of structures near the shore. Terrain continuously lower than 10 feet above sea level may be flooded requiring massive evacuation inland.
5	156+ mph (70+ m/s)	<920	19+ (5.7 + m)	*Catastrophic:* Complete roof failure on many residences and industrial buildings. Some complete building failures with small utility buildings blown over or away. Major damage to lower floors of all structures located less than 15 feet above sea level and within 500 yards of the shoreline. Massive evacuation of areas on low ground within 5 to 10 miles of the shoreline may be required.

Note: The table above presents the Saffir-Simpson scale for the Atlantic and Pacific coasts to give an estimate of the expected flooding, damage to property, and requirements for evacuation in the event of an approaching hurricane of the given intensity. The Saffir-Simpson scale classifies hurricanes only on the basis of wind speed; central pressure, last used in the 1990s, is no longer used.

physics of the atmosphere and ocean to forecast hurricane tracks. The models require data obtained from many sources, including ship reports, "hurricane hunter" aircraft using dropsondes that measure air temperature, water temperature, humidity, wind velocity, pressure, and other parameters, geostationary satellites, Doppler RADAR, and other high-technology measuring systems.

Even so, there is much about hurricane dynamics that is still not known. Hurricane forecasters must weigh the consequences of errors of predicting landfall when it does not occur against errors of failing to predict landfall when it does occur. Hurricane landfall predictions require emergency management procedures, such as boarding up houses and businesses, and evacuation. If landfall does not occur at the predicted time and location, the public will have undergone great inconvenience and expense and may not respond to the next forecasted hurricane. On the other hand, failure to predict landfall when it does occur can result in great loss of life and property. The radius of uncertainty in hurricane prediction has steadily decreased from about 155 miles (250 km) in 1965, to about 100 miles (160 km) in 2000. As computer models improve along with observation technologies, it is expected that the radius of uncertainty in time and space for landfall predictions will get smaller and smaller.

The most destructive natural event ever to hit the United States was the Galveston Hurricane of September 1900, which struck and essentially destroyed the city of Galveston, Texas, with a loss of more than 6,000 lives. Increasingly accurate FORECASTING has reduced the loss of life from hurricanes, but property damage has increased enormously due to the extensive development of beachfront properties since the 1950s.

Further Reading

Landsea, Chris. "How Are Atlantic Hurricanes Ranked"? Hurricane Research Division, Atlantic Oceanographic and Meteorological Laboratory, NOAA. Available online. URL: http://www.aoml.noaa.gov/hrd/tcfaq/D1.html. Accessed March 14, 2008.

National Oceanic and Atmospheric Administration. "Hurricanes." Available online. URL: http://hurricanes.noaa.gov. Accessed September 21, 2007.

National Weather Service, NOAA. "Hurricane: A Familiarization Booklet." Bethesda: U.S. Department of Commerce, NOAA, 1993.

———. "National Hurricane Center." URL: http://www.nhc.noaa.gov. Accessed September 21, 2007.

hydraulic current A hydraulic current is water flow in a channel, resulting from a difference in water level at the two ends. The difference in water elevation creates a PRESSURE GRADIENT between the two ends of the channel, which provides a horizontal force able to move the water from the high end to the low end. Many oceanographers and coastal engineers refer to this mound of water above MEAN SEA LEVEL as the hydraulic head. These currents reach a maximum when the difference in head is greatest. As the head minimizes due to changes in winds or in reaching slack water, the hydraulic current will dissipate. The study of hydraulic currents is important since they affect safe NAVIGATION, the recruitment of fish larvae in ESTUARIES, and contribute to erosion, especially in COASTAL INLETS.

The total current in a water body is normally divided into tidal and non-tidal components (see TIDE). An oceanographer's CURRENT METER measures the total current. Tidal currents are the periodic components of the total current caused by astronomical forces. Tidal currents in shallow water tend to be reversing, while those in deep water are rotary. The term *nontidal current* describes the components of velocity that are dependent on forces such as winds, freshwater runoff, and differences in TEMPERATURE and density. The component that predominates depends on the geographic location and strength of the forcing processes.

Hydraulic currents are also found in straits in which there is a difference in time or water level of the tide wave at each end. An example is the East River in New York City, which is connected to Long Island Sound at one end and New York Harbor at the other. Since nontidal currents cannot be predicted with the accuracy of tidal currents, the NATIONAL OCEANIC AND ATMOSPHERIC ADMINISTRATION (NOAA) has installed sensors such as ACOUSTIC DOPPLER CURRENT PROFILERS (ADCPs), TIDE GAUGES, and ANEMOMETERS in various seaports to provide mariners with real-time current, water levels, and winds at difficult navigation locations. These high-technology navigation aids are called physical oceanographic real-time systems or PORTS® (Physical Oceanographic Real-Time System). They have been referred to as the "seat belts" and "air bags" for harbor pilots.

The fast flow frequently observed at BARRIER ISLAND inlets that connect the coastal ocean to bays and ESTUARIES is generally caused by water level differences between the ends of the inlet produced by the tides and often enhanced by the wind. Winds blowing in the direction opposite to the current steepen the waves and increase the difficulties for small craft trying to enter or exit the inlet. Weather systems lasting on the order of five to 10 days set up or pile the water on one side of the inlet. Oregon Inlet in North Carolina has exaggerated ebb (outgoing) and flood (incoming) flows that occur in response to favorable winds. Along the Carolina Capes nutrient-rich plumes from exaggerated ebb events attract the larval forms of commercially important finfish such as spot and croaker, which are jetted into the Albermarle-Pamlico Sound when the winds reverse. The ensuing exaggerated floods overcome the to-and-fro nature of the astronomical tides and help to retain the larval finfish in the sounds since they can only swim two or three body lengths per second.

hydrodynamics   The study of fluid motion and the movements of objects through a fluid is called hydrodynamics. Some properties of importance include DENSITY, compressibility, VISCOSITY, and nonlinear phenomena such as TURBULENCE. Compressibility is described by the condition where volume decreases as pressure increases. Most liquids are considered to be incompressible and maintain a relatively fixed volume. Viscosity describes a particular fluid's resistance to flow and generally increases as the temperature decreases. Viscous liquids such as molasses flow slowly. Irregular fluctuations in a fluid are described as turbulence. Such disturbances are measured as gusts of wind or currents that cause mixing in the ocean. Turbulence brings nutrients into the surface layer from the depths, which is important for the growth of PLANKTON.

See FLUID MECHANICS.

hydrofoil   A hydrofoil is a boat that eliminates friction between the water and the hull by rising out of the water to plane at high speed on foils. The foils are winglike surfaces attached below the front and rear of the hull to supporting legs or struts. As the boat accelerates, the hull is raised out of the water, allowing greater speeds. A hydrofoil is unique because it operates as a displacement hull, and when in planing mode, the hull is completely out of the water and only the foils are submerged. It should not be confused with a HOVERCRAFT, which raises its hull out of the water on a cushion of air.

As the hydrofoil accelerates, the foils generate lift like the wings of an airplane, but are of different dimension since the DENSITY of ocean water is approximately 1,000 times denser than air. The density of dry air at sea level is approximately 0.0807 pounds per cubic feet (1.2929 kg/m³), requiring a long foil, while the density of ocean water at sea level is approximately 64.11 pounds per cubic feet (1,027 kg/m³), allowing short foils. Scientists and engineers know that the higher density means that the foils do not have to move as fast as an airplane before they generate enough lift to push the boat out of the water. But the foils must stay submerged, or

A hydrofoil is seen here gliding on the surface of the water past fishing nets.   *(NOAA/NMFS)*

the vessel hull will abruptly return to the water, a ride-quality and sea-keeping issue. Foils are generally classified as either surface piercing, which operate at an angle to the water, or fully submerged, which operate much more like an airplane wing.

NAVAL ARCHITECTS have to cope with many challenges in the design and development of hydrofoils. Cavitation, ventilation, and planing on a bumpy ocean are just a few examples. Cavitation occurs when the water pressure is lowered to the point where water starts to boil at the foil. Lift is lost, and the boat rapidly comes off plane. Lift is also lost by ventilation when air is forced along or sucked down the lifting surface of the foil, such as when a surface-piercing foil slices through a long wave. Ladder foils or struts with series of spaced foils have been developed to improve hydrofoil operations on water bodies with WAVES. This design provides a chance for some of the foils to remain submerged to provide lift, while other foils leave the water as the trough of a wave is traversed. Most commercial hydrofoils use ladder foils, and operators understand SEA STATE limitations on planing or high-speed navigation.

Numerous experimenters such as Enrico Forlanini (1848–1930), Alexander Graham Bell (1847–1922), and Frederick W. Baldwin (1882–1948) were designing and testing hydrofoils during the early 1900s. Today, hydrofoils are used commercially as ferries and in the military as high-speed patrol boats. As an example, a hydrofoil cruise along 175 miles (282 km) of the scenic Danube River from Budapest,

Hungary, to Vienna, Austria, takes approximately six hours; a crossing of the Pearl River estuary from Hong Kong to Macao takes approximately one hour. Several navies use fast and heavily armed hydrofoils for security and even to control smugglers and drug runners. These military hydrofoils have been designed for all-weather, high sea state operations and may reach speeds from 12 to 48 knots (22.2 to 88.9 km/hr).

Further Reading
The International Hydrofoil Society. Available online. URL: http://foils.org. Accessed June 24, 2007.

hydrographer A professional who uses instruments such as echo sounders and the GLOBAL POSITIONING SYSTEM (GPS) to measure the seafloor is a hydrographer. Analysis techniques utilize mapping software such as Geographic Information Systems to describe the seafloor's physical features and adjoining coastal areas, with special reference to their use for NAVIGATION. Hydrographers throughout history have provided descriptions of the sea, lakes, and adjacent shores. Sir Francis Beaufort (1777–1857) was a hydrographer to the British Royal Navy and is most remembered for his system of estimating and reporting wind speeds. The Beaufort wind scale describes wind speed based on the appearance of the sea surface and other visible effects. Hydrographers are employed by organizations supporting operations in the open-ocean and coastal regions such as the National Geospatial Intelligence Agency (NGA), the U.S. Navy, the NATIONAL OCEANIC AND ATMOSPHERIC ADMINISTRATION (NOAA), the U.S. Army Corps of Engineers, construction companies, and dredging companies. NGA and NOAA make nautical charts that rely on hydrographic survey data that was collected by navy, army, or commercial surveyors. Local surveys are completed to support coastal construction where sub-bottom stratigraphies are identified and to ensure that controlling depths are reached in channels that have been dredged. Sub-bottom profiling identifies sediment thickness and rock structure, which is particularly important for monitoring scour around pilings that support BRIDGES and routing pipelines. For DREDGING, before-and-after surveys are used to calculate the volume of material removed from a navigation channel.

Commander Matthew Fontaine Maury (1806–73) is known for his writings and charts, which included archiving lead line soundings and other data from naval vessels and merchant marine ships. Modern hydrographic survey principles rely on the use of SONAR equipment to measure the water depth. The times it takes for sonar pings transmitted from the survey boat to reflect off the bottom and return to the boat are recorded. The time is translated into distance based on the equation:

$$V = 1/2 \, d/t$$

where

V = velocity
d = distance (in this application the distance is twice the depth), and
t = time (sec).

The coefficient $1/2$ arises because the ping travels to the seafloor and then back to the transducer.

Once depths are determined, a location is associated with the sounding from the onboard Global Positioning System (GPS). The distance between the survey vessel's GPS transceiver and satellites are calculated just as the depths were calculated by using the velocity times travel time formula. After determining the transceiver's distance from three or more satellites, the GPS software accurately determines the transceiver's latitude, longitude, and altitude. The survey team will also deploy CONDUCTIVITY-TEMPERATURE-DEPTH gauges to ensure that the echo sounder is properly calibrated to the TEMPERATURE and SALINITY of the seawater. Some sort of WAVE AND TIDE GAUGE will also be deployed to establish a vertical reference such as mean lower low water. Tidal fluctuations and survey boat motions will be removed from the survey data. This is especially important since tidal predictions do not always accurately represent actual water levels because they are often based on very old measurements and only consider astronomical forces. GRAB SAMPLERS may also be used to take SEDIMENT samples. The nature of bottom materials (sand, mud, and rock) is especially important to document on charts for purposes such as anchoring. Side scan sonars may also be deployed to provide imagery of submerged dangers to navigation.

Since hydrographic surveying is a technical discipline, there are various educational programs and certifications that may be obtained. Hydrographers may have education in fields such as Earth science, civil engineering, or cartography, and field experiences that demonstrate their ability to conduct measurements in the ocean, to perform mathematical computations associated with oceanographic measurements, and to keep complete and accurate records of the survey. They should be familiar with common terminology associated with water rights and water diversion. Surveys of the coast require hydrographers to read and use public land legal descriptions, translate written public land legal descriptions to map or chart locations, and to find those locations on the land or water. Schools such as the University of Southern Mississippi and the University of New Hampshire offer certificate programs in ocean mapping. Some of these programs meet certification requirements by the Federation Internationale des Géomètres/International Hydrographic Organization (FIG/IHO) International Advisory Board. Certification by the American Congress of Surveying and Mapping requires the surveyor to pass an examination on depth measurement, vessel positioning, horizontal and vertical controls, tides and water levels, survey planning, nautical science, and general marine science.

See also CHART; DATUM; TIDE; TIDE GAUGE; WATER LEVEL FLUCTUATION.

Further Reading

American Congress on Surveying and Mapping. Available online. URL: http://www.acsm.net. Accessed June 24, 2007.

The International Hydrographic Office. Available online. URL: http://www.iho.shom.fr. Accessed June 24, 2007.

hydrographic survey   A hydrographic survey is an orderly process of determining basic information relating to the physical characteristics of the seafloor, bodies of water, and coastal regions. Hydrographic surveyors plan and execute operations involving measurements used for determining the relative position of points on the seafloor and the adjacent coastal zone. Various electronic and acoustic instruments are used to collect data to characterize depth of water, configuration and nature of the bottom, amplitudes and directions of CURRENTS, heights and phases of TIDES, and the location of fixed objects for survey and navigation purposes. Surveys near the coast rely on control references such as benchmarks whose vertical and horizontal coordinates are regularly updated by agencies such as the National Geodetic Survey in the United States. These surveys are combined with other forms of marine information to build paper or digital nautical charts (*see* CHARTS NAUTICAL). Charts are specifically designed by CARTOGRAPHERS to meet the requirements of marine navigation, showing depths of water, nature and characteristics of the bottom, coastal elevations, man-made features, obstacles, and aids to navigation.

One of the mapping and charting authorities for the United States is the National Ocean Service (NOS) of the NATIONAL OCEANIC AND ATMOSPHERIC ADMINISTRATION (NOAA). Hydrographic surveyors associated with NOAA use survey vessels such as the NOAA ship *Fairweather,* which is outfitted with multibeam survey systems, high-resolution

side-scan SONAR, and state-of-the-art computer systems to support nautical charting. Multibeam sonar devices send to the seafloor and then receive audible signals or pings that are used to determine depths on the seafloor. This technique produces a fan-shaped swath of bathymetric data rather than a line. Side-scan sonar involves transmitting and receiving specially shaped acoustic beams out to each side of the transducer assembly or towfish in order to produce imagery that is useful for locating structure such as SEAMOUNTS, shipwrecks, pipelines, cable routes, and other features. Adjustments are made to measured depths in order to account for the manner in which the survey platform reacts to winds and waves and periodic changes in water levels from astronomical, hydrological, and meteorological conditions. In other words, multibeam sonar data is corrected for settlement and squat (drawing more water aft when in motion forward, than when at rest), heave, pitch, roll, and the heading of the vessel. Mean Lower Low Water (MLLW) is the vertical reference datum for the nautical charts produced from these data. Other important sensors used on surveys include sound velocity profilers, CONDUCTIVITY-TEMPERATURE-DEPTH (CTD) probes, water-level measurement systems, and current meters. CTDs enable surveyors to determine the velocity of sound through water from the TEMPERATURE, SALINITY, and PRESSURE, which is imperative for accurate multibeam sonar operations. WATER LEVEL FLUCTUATIONS are measured to detide the soundings and to determine water stage or the level of the water surface above the datum at a given location. CURRENT METERS, transmissometers, and GRAB SAMPLES are used to describe circulation, water clarity, and bottom types, respectively.

Data-processing routines, which identify conversion methodology from raw sounding data to the final smooth-sounding values, are used to reduce and interpret survey data. Boat sheets are used to plot details of the survey as it progresses. The final, carefully made plot from corrected data of a hydrographic survey represents the official permanent record and is called the smooth sheet. Much of this work is done at sea in the digital realm. Hydrographic survey data are compared to existing nautical charts, and the newly collected hydrographic data are used to update existing charts.

Agencies such as the International Hydrographic Organization (IHO) help to ensure that survey standards are achieved by coordinating the activities of national hydrographic offices, advocating the greatest possible uniformity in nautical charts and documents, adopting reliable and efficient methods of carrying out and exploiting hydrographic surveys, improving the knowledge base in the fields of hydrography and techniques employed in descriptive oceanography. One example is IHO S-57, the IHO Transfer Standard for Digital Hydrographic Data and its associated Electronic Navigation Chart Product Specification. IHO S-57 consists of a feature dictionary for electronic navigation charts, a data model, and an exchange format called DX-90. The U.S. Army Corps of Engineers has applied these standards to convert digital survey data of the Atchafalaya River into a River Electronic Navigational Chart. This Louisiana river, which is surrounded by swamps, provides a waterway to the GULF OF MEXICO that is approximately 172 miles (277 km) shorter than navigating the Mississippi River. Having a safely charted route saves mariners time and money and also lessens marine traffic at the Port of New Orleans.

Further Reading

Australian Hydrographic Service. Available online. URL: http://www.hydro.gov.au. Accessed September 21, 2007.

International Hydrographic Office. "IHO Transfer Standard for Digital Hydrographic Data, *IHO Special Publication No. 57 (IHO S-57)*." Edition 3.9, Monaco: International Hydrographic Bureau, November 1996.

National Oceanic and Atmospheric Administration's National Ocean Service. "Hydrographic Surveying." Available online. URL: http://oceanservice.noaa.gov/topics/navops/hydrosurvey. Accessed September 21, 2007.

———. "NOS Hydrographic Surveys Specifications and Deliverables." Available online. URL: http://chartmaker.ncd.noaa.gov/HSD/specs/Specs2007.pdf. Accessed June 24, 2007.

Navigation Services, Office of Coast Survey, National Oceanic and Atmospheric Administration. Available online. URL: http://nauticalcharts.noaa.gov/nsd/Services.htm. Accessed September 21, 2007.

Umbach, Melvin. J. *Hydrographic Manual,* 4th ed. Rockville, Md.: National Ocean Survey, 1976.

U.S. Army Corps of Engineers. *Civil Works Engineering Manual EM-1110-2-1003—Hydrographic Surveying.* Available online. URL: http://www.usace.army.mil/publications/eng-manuals/em1110-2-1003/toc.htm. Accessed September 21, 2007.

hydrological cycle   The process of water evaporating from soils, vegetation, oceans, and other bodies of water, then accumulating in the atmosphere as water vapor in clouds, and finally returning to the ocean through precipitation and runoff is called the hydrological cycle. Water is stored in the atmosphere, vegetation, lakes, ocean, and even the polar glaciers as a gas or water vapor, as a liquid, and as a solid or ice. In general terms, 97.5 percent of the water is in the ocean, 2.4 percent is in the land, and less than 1 percent is in the atmosphere. These reservoirs have a high degree of temporal and spatial variability in response to factors such as agriculture, weather, season, and climate. However, the total amount of water on Earth does not change; it is conserved. Therefore, understanding the hydrological cycle locally is very important since the distribution of water around the world is not equal. This cycle and life on planet Earth are only possible because of the unique properties of water.

Transfers of water occur for many reasons and in different locations, which are important to understand. A working knowledge of the hydrological cycle can help mitigate the effects of phenomena such as HURRICANES, flooding, and droughts. How water moves into and out of the reservoirs is complicated, but is observed though measurements of evaporation, precipitation, and the flow of streams that empty into the rivers and oceans.

The process by which liquid water at land and water body surfaces becomes water vapor in the atmosphere is called evaporation. When water is passed by vegetation back into the atmosphere, the process is called transpiration. Many scientists report that 80 percent of all evaporation is from the oceans, while the remaining 20 percent comes from inland water and vegetation. The keys to forming CLOUDS is the process of condensation, where water vapor reaches cold altitudes and, in the presence of dust particles or condensation nuclei, the molecules slow down, stick together, and form water droplets that are about 7.9×10^{-4} inches (0.02 mm) in size. Water drops on the outside of a cold metal pitcher of water provide an everyday example of condensation.

When water or ice falls back to Earth's land or water surface, it is called precipitation, which includes rainfall, snow, hail, and sleet. Rain occurs when cloud droplets grow or change form and become too heavy to remain in the cloud. The ensuing raindrops fall toward the surface; a large one may be on the order of 0.8 inches (2.0 mm). Rain falling to the ground either infiltrates into the ground as groundwater or moves across the surface as runoff. The ability of the land to absorb the water depends on its level of saturation. There is usually a level below the surface called the zone of aeration, where one finds a combination of soil, air gaps, and water. Then there is a level below Earth's surface where the ground is already fully saturated, the zone of saturation. The interface between the zones of aeration and saturation is the water table. Scientists estimate that soil water and groundwater comprise 0.5 percent of the hydrosphere's water.

Hydrologists use models of the components of the hydrological cycle to calculate the amount of runoff, the stage (height of the water) of rivers, and the potential for flooding. They may calculate the level of runoff when it rains, based on topography and land use as well as estimate river flow from elevation data or observations from stream gauges. Groundwater input into streams and rivers is generally fairly constant, increasing after precipitation and decreasing during droughts. The water table fluctuates as the amount of groundwater increases or decreases. When the entire area below the ground is saturated, flooding occurs because all subsequent precipitation is forced to remain on the surface and river stage is above the level of the riverbanks.

Further Reading

Earth Observatory, National Aeronautics and Space Administration. Available online. URL: http://earth observatory.nasa/gov/. Accessed March 8, 2008.

The Water Cycle, U.S. Geological Survey. Available online. URL: http://ga.water.usgs.gov/edu/watercycle.html. Accessed March 8, 2008.

hydrothermal vents Hydrothermal vents are cracks and fissures on the floor of the deep ocean from which geothermally heated fluid is emitted in a manner similar to that of geysers on the continents. Hydrothermal vents are typically located near regions of intense tectonic activity, such as mid-ocean SPREADING RIDGES. They are believed to be caused by cold seawater 33.8° to 39.2°F (1° to 4°C) seeping through fissures in the ocean floor that becomes heated by reservoirs of hot magma that are hotter than 650°F (343°C), and then is ejected under enormous pressure into the near freezing ocean bottom water. The hot water does not boil at these high temperatures because it is under extremely high pressure.

During this process, the heated water absorbs minerals from the ocean floor and subfloor, especially sulfur, methane, and sulfur compounds. The inclusion of these minerals causes the vented water to appear to flow out like smoke coming out of a chimney in the atmosphere. The mineral compounds precipitate or come out of suspension as the hot fluid strikes the cold water at the ocean floor and builds up deposits that are called chimneys. The fluid is often black in coloration, leading to the term *black smokers*. Sections of black smokers brought to the surface for study usually smell of sulfur and sulfur compounds. There are also "white smokers." White smokers discharge a cooler fluid than black smokers, at about temperatures between 212° and 570°F (100° and 300°C). The precipitates in white smokers are not sulfur and metals, but include silica, anhydrite, and barite. The temperature of the outflowing fluid leads to intense thermal gradients near the smokers between the geothermally heated vent fluid and the

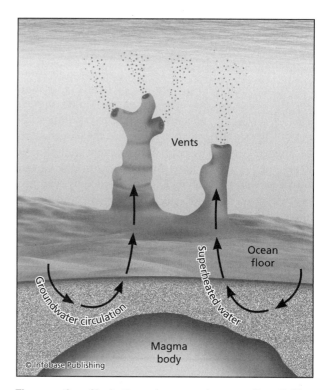

The operation of hydrothermal vents on the ocean floor. Cold seawater is cycled through cracks in the floor, heated by the hot crust, and is convected upward, along with a rich content of minerals such as sulfides.

surrounding near freezing seawater. Movement of the ocean bottom water and tectonic activities can cause the boundaries between cold and hot water to move, alternately scalding and chilling the proximate environment.

The chimneys can be tens of feet (m) high and several feet (less than 1 m) in diameter. Due to the intense tectonic activity of thermal vent regions, the vents often have an ECOSYSTEM lasting only a few tens of years before being destroyed or absorbed by newly formed crust or blasted apart by undersea volcanoes. The relatively low temperature of surrounding waters allows vent animals such as large tubeworms and clams to remain immersed in the nutrient-rich water. The diffuse vent sites develop into complex ecosystems. Large vent communities are found on the Mid-Atlantic Ridge, the Juan de Fuca Ridge in the North Pacific, and the Galápagos spreading ridge in the equatorial Pacific. They are believed to be located worldwide.

Extraordinary biological communities have been found in the vicinity of the deep-sea vents. They include giant clams (*Calyptogena magnifica*), crabs (*Austinograea williamsi*), mussels (*Bathymodiolus thermophilus*), shrimp (*Chorocaris vandoverae*), and life-forms found only in vent communities,

Cold Seeps: It Is Not All Hot Stuff

Hydrothermal vents are not the only strange locations in the ocean that harbor life without light, and that take their energy from chemicals in the water by means of chemosynthesis. Cold seeps also provide the environment for living organisms by leaking energy-rich fluids into the deep waters of the continental margins and abyssal plains. As is the case for hydrothermal vents, mats of bacteria, clams, crabs, mussels, and tube worms are found in abundance clustered around the seeps and around "brine pools" of high-salinity water that are found in the vicinity of the seeps. Chemicals such as hydrogen sulfide, methane (CH_4), ammonia (NH_4), and light hydrocarbons make up the basis of the chemosynthetic food chain. Cold-seep communities have been found in the vicinity of methane hydrates, which are of interest as a possible human fuel resource. The temperature of the water in the vicinity of many of the seeps is on the order of 40°F (4.4°C), the temperature of the ambient bottom water.

Cold seeps were first found by Barbara Hecker and Charles Paull during the 1980s in the abyssal waters of the Gulf of Mexico. Cold seeps have since been found in the waters off Oregon, Japan, Italy, and many other locations throughout the world, and at depths of hundreds of feet (tens of meters) to thousands of feet (meters) In most cases, symbiotic bacteria extract the nutrients from the water, but in some locations nutrients were extracted directly by the resident animals. Fossils have now been found of cold seep communities that existed in the distant past, especially in deposits of sulfide ores.

These findings give hope to the possibility of finding extra terrestrial life in locations such as the ocean under the ice on Jupiter's moon Europa, where sunlight would be too weak to initiate photosynthesis.

Further Reading

Department of Biology, Pennsylvania State University. "Life without Light." Available online. URL: http://www.bio.psu.edu/cold_seeps. Accessed August 15, 2007.

Monterey Bay Aquarium Research Institute (MBARI). "When is a cold seep not a cold seep?" *MBARI News,* February 10, 2005. Available online. URL: http://www.mbari.org. Accessed August 15, 2007.

—**Robert G. Williams, Ph.D.,** senior oceanographer
Marine Information Resources Company
New Bern, North Carolina

---

such as tubeworms (*Riftia pachyptila*). These animals are part of a system based not on photosynthesis, as is most life on Earth, but rather chemosynthesis, in which bacteria, called Archaea, are able to synthesize nutrients from the minerals, such as metallic sulfides, discharged by the vents. Many of the creatures, such as the tubeworms, are born with these bacteria, which provide nourishment to the worms. The tubeworms and some other vent creatures do not even have mouths or digestive systems in the usual sense of the word, but obtain their energy from the bacteria. Some scientists believe that life on Earth began in these vent communities. Astronomers are hypothesizing that if life exists elsewhere in the solar system, it may be found on Jupiter's moon EUROPA, on which an ocean is located beneath a crust of ice. The NATIONAL AERONAUTICAL AND SPACE ADMINISTRATION (NASA) is supporting the development of a probe to drill through Europa's ice and then launch a vehicle into the ocean to perform an underwater reconnaissance searching for hydrothermal vents, and possibly living organisms.

Hydrothermal vents were first seriously studied in the 1970s. The living vent communities were discovered in the late 1970s by geologists studying the vents. They were first observed by marine scientists diving in the research submarine *Alvin* in 1977. Scientists aboard the *Alvin* have observed hydrothermal vent communities in the ATLANTIC and PACIFIC OCEANS in both the Tropics and at mid-latitudes.

The vents have also been studied from research vessels on the surface that deploy special devices that act as large chain saws to saw off a portion of the black or white smokers, which is then winched to the surface, reeking of sulfur, and often with vent animals attached. The animals explode on the trip to the surface because of the great difference in pressure between their habitats thousands of feet (hundreds of m) below the ocean surface and sea level.

Hydrothermal vents are sufficiently numerous that their outflowing plumes influence the global chemistry of seawater. Research on plume composition, extent, and mixing characteristics has been in progress since their discovery.

As one example, scientists at the NATIONAL OCEANIC AND ATMOSPHERIC ADMINISTRATION's (NOAA's) Pacific Marine Environmental Laboratory (PMEL) have been studying hydrothermal vents since 1991 in the Juan de Fuca Ridge and East Pacific Rise. Hydrothermal plumes are being analyzed to determine plume dimensions, chemical composition, and mixing and diffusion processes. The plumes are being mapped from surface vessels, REMOTELY

OPERATED VEHICLES (ROVs), and research submarines. They are deploying an array of instruments that includes CONDUCTIVITY, TEMPERATURE, AND DEPTH (CTD) PROBES, optical meters to measure light attenuation, and water sampling for plume chemistry. Pacific Marine Environmental Laboratory scientists and colleagues are deploying an instrument called SUAVE, which stands for System Used to Assess Vented Emissions. SUAVE is able to analyze on site the chemical composition of vent plumes for elements such as iron, manganese, and compounds such as hydrogen sulfide.

Scientists from the U.S. Geological Survey are studying hydrothermal vents in the Gorda Ridge spreading center off the northwest coast of the United States. Because it lies within the U.S. EXCLUSIVE ECONOMIC ZONE, it is possible that metals and other minerals could be extracted for use by industry.

Further Reading

Ballard, Robert D. *The Eternal Darkness: A Personal History of Deep-Sea Exploration.* Princeton, N.J.: Princeton University Press, 2000.

National Oceanic & Atmospheric Administration, Pacific Marine Environmental Laboratory. "Vents Program." Available online. URL: http://www.pmel.noaa.gov/vents. Accessed September 21, 2007.

Nichols, C. Reid, David L. Porter, and Robert G. Williams. *Recent Advances and Issues in Oceanography.* Westport, Conn.: Greenwood, 2003.

Van Dover, Cindy Lee. *The Ecology of Deep-Sea Hydrothermal Vents.* Princeton, N.J.: Princeton University Press, 2000.

hypsometry   Hypsometry is the measurement of surface area and elevation with reference to a DATUM such as sea level. Geologists or geodesists may use

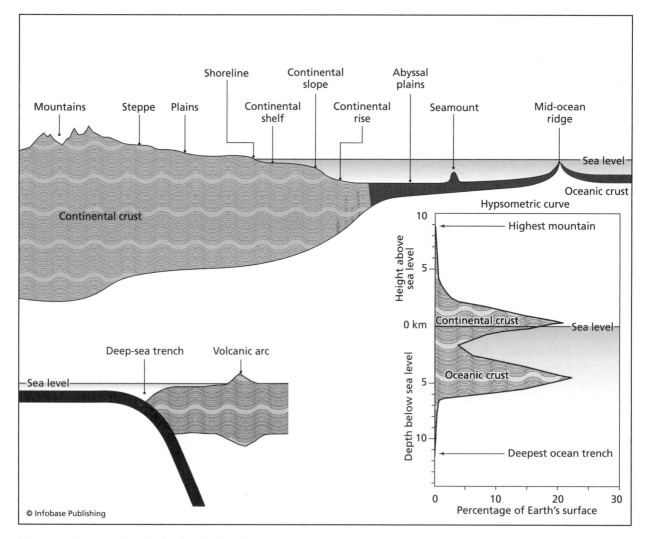

A hypsometric curve showing the distribution of land with different elevations. The inset shows the distribution of crust, indicating two fundamentally different types of crust: oceanic and continental, which have different isostatic compensation levels.

this technique to classify a planetary landscape. Elevation data is then plotted as the fraction of a planet's surface at a particular level. Hypsometry helps characterizes Earth's surface coverage, where 29 percent is land and 71 percent is ocean.

The hypsometeric curve is determined from the cumulative area (A) and the total height (H) from the lowest to the highest elevations. As one moves from the lowest elevation upward on the curve through increasing values of elevation (h), all the areas within each elevation interval are measured. With each successive increase in h, the area increases by the next new value of area. This progression is cumulative. When plotted on an x-y diagram, the relative values of h/H and a/A form the hypsometric curve. Points on the curve (a/A, h/H) are dimensionless, representing proportions of the total area and height. So, for y = 0, all values above or below the reference plane lie within the total area. The area below the curve is known as the hypsometric interval.

Surface area and elevation plots for Earth around sea level show one peak for the TRENCHES from the 36,200-foot (11,035 m)-deep Marianas Trench on the ocean floor to Mount Everest, which is 29,028 feet (8,848 m) above sea level. Two distinct levels are highlighted, such as the continent's 2,625-foot (800 m) mean elevation and the seafloor's 12,467-foot (3,800 m) mean depth. Marine scientists use these data to emphasize basic differences in continental and oceanic crust and the extent of MARINE PROVINCES such as the shelf and slope, CONTINENTAL RISE, abyssal floor, mid-ocean ridges, and trenches. On the hypsometric curve, continental boundaries extend to the CONTINENTAL SLOPE, not the shoreline.

Further Reading

National Aeronautical Space Administration. "Dr. Nicholas Short's Remote Sensing Tutorial." Available online. URL: http://rst.gsfc.nasa.gov. Accessed September 21, 2007.

I

iceberg An iceberg is a massive chunk of floating ice that varies greatly in size and shape. Icebergs usually rise from the waterline at least 16.4 feet (5 m) and have a thickness of 98 to 164 feet (30 to 50 m) and cover an area of approximately 5,382 square feet (500 m²). Smaller masses of ice are classified as either bergy bits or growlers and may be difficult to observe from vessels. Icebergs have broken away from glaciers (*see* GLACIER ICE FORMS) because of instabilities owing to cracks in the ice or strong wave action. They are not formed from SEA ICE; they are freshwater. They may be afloat, aground, or pushed up against an island or other land mass. They are described as tabular, dome-shaped, pinnacled, dry-docks, weathered, blocky, or tilted blocky. The International Ice Patrol, however, classifies them according to size as small, medium, or large. Icebergs are tracked by organizations such as the Canadian Ice Service, International Ice Patrol, and the NATIONAL ICE CENTER.

The color of icebergs, bluish white, results from the reflection and scattering of sunlight from closely spaced gas cavities throughout the ice. Snow covering the iceberg may cause a more whitish appearance. Icebergs with an aqua green appearance have rolled over to show ALGAE, which was growing on the iceberg's underside.

Most icebergs in the North Atlantic Ocean originate from Greenland glaciers, drift across the Davis Trait, and float southward in the LABRADOR CURRENT. From March to July, icebergs are a common sight along the coast of Newfoundland and generally melt a few hundred miles to the south. Icebergs in the Southern Ocean originate from glaciers on the Antarctic continent.

The sinking of the White Star liner RMS *Titanic* on April 15, 1912, resulted in the formation of the International Ice Patrol, which is internationally funded and staffed primarily by the U. S. Coast Guard. Since the establishment of this organization, only one large vessel has been lost due to collision with an iceberg. Today, most of the patrolling is done by aircraft. Icebergs and sea ice remain a natural hazard to shipping and always have an effect on NAVIGATION. Ice centers in polar regions track icebergs and sea ice conditions, using satellites to provide descriptive charts to navigators.

When an iceberg breaks loose from a glacier, the process is called calving. Icebergs float in the ocean with about 10 percent of their volume above the water. That leaves about 90 percent of the iceberg

For many years, the International Ice Patrol monitored icebergs, such as this one. *(U.S. Coast Guard)*

234

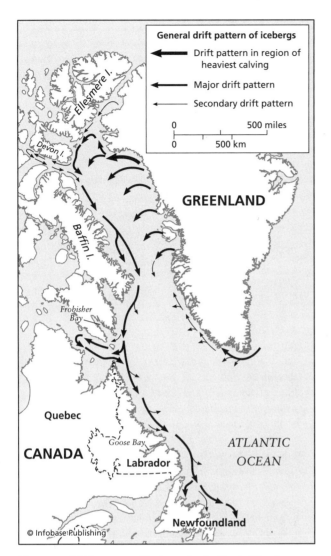

General drift pattern of icebergs

→ Drift pattern in region of heaviest calving
→ Major drift pattern
→ Secondary drift pattern

0 | 500 miles
0 | 500 km

Ellesmere I.

Devon I.

GREENLAND

Baffin I.

Frobisher Bay

Quebec

CANADA

Goose Bay

Labrador

ATLANTIC OCEAN

Newfoundland

© Infobase Publishing

The general drift pattern of icebergs in the North Atlantic Ocean *(Source: The American Practical Navigator: An Epitome of Navigation)*

Lengths and Freeboards of Major Forms of Icebergs

Form	Freeboard	Length
Growler	< 3 feet (< 1 m)	< 16 feet (< 5 m)
Bergy Bit	3–16 feet (1–5 m)	17–50 feet (5–15 m)
Small Berg	17–50 feet (5–15 m)	51–200 feet (15–60 m)
Medium Berg	51–150 feet (16–45 m)	201–400 feet (61–122 m)
Large Berg	151–240 feet (46–75 m)	401–670 feet (123–213 m)
Very Large Berg	> 240 feet (> 75 m)	> 670 feet (> 213 m)

Further Reading

Canadian Geographic. "Just the Facts." Available online. URL: http://www.canadiangeographic.ca/magazine/MA06/indepth/justthefacts.asp. Accessed September 21, 2007.

Canadian Ice Center. Available online. URL: http://ice-glaces.ec.gc.ca. Accessed June 25, 2007.

International Ice Patrol Home Page. Available online. URL: http://www.uscg.mil/lantarea/iip/home.html. Accessed April 20, 2005.

National Ice Center. Available online. URL: http://www.natice.noaa.gov/. Accessed June 25, 2007.

National Snow and Ice Data Center Home Page. Available online. URL: http://www.nsidc.colorado.edu. Accessed April 20, 2005.

icebreakers *See* ICEBREAKING.

icebreaking The process of breaking and clearing SEA ICE from seaports and shipping channels in order to maintain navigation is called icebreaking. In North American waters, Canadian and United States Coast Guard vessels perform domestic icebreaking duties in ESTUARIES, lakes, and coastal ocean regions where sea ice forms. This is especially important since many homes and workplaces in cities such as Baltimore, Philadelphia, New York, and Boston are heated by fuel oil that arrives through marine transportation. Petroleum products such as gasoline, fuel oil, kerosene, and diesel fuel are delivered by BARGE to seaports and are especially important during wintertime, when conditions are favorable for the formation of sea ice on the Great Lakes and in estuaries such as Chesapeake Bay,

hidden below the waterline, which is why ships in polar waters must maintain a vigilant watch for icebergs, especially in the frequently occurring fogs south of Newfoundland. They diminish in size due to calving, melting, and wave action. Icebergs are especially dangerous during melting. In this state, they are unstable and may fragment or overturn at any moment.

Icebergs are scientifically classified in accordance to their freeboard and length as growler, bergy bit, small berg, medium berg, large berg, and very large berg. Freeboard refers to the iceberg's height above water. The following table provides the lengths and freeboards of the major forms of icebergs.

Antarctica's Tabletops: Bergs of Note

Tabular icebergs resemble a flat tabletop and are calved from ice shelves such as the Pine Island Glacier. Glaciers cover the entire continent of Antarctica. Glaciers are large sheets of ice that form from perpetual snowfall and only moderate, short-lived melting periods, which serve only to harden and thicken these massive ice sheets. Glaciers, though made of ice, still have some liquid water properties and therefore continue to flow under their incredible weight to lower elevations, in particular toward the coastal regions of Antarctica. When these glaciers reach the ocean, they are referred to as the "ice shelf." As the glacier continues to grow and flow into the ocean, the weight of the ice shelf, coupled with the dynamic nature of the Southern Ocean (both thermally and mechanically), causes seaward pieces of the ice shelf to break free. Because its origin is glacial ice, and because of its extensive, flat surface above the waterline, the portion of the ice shelf that breaks free is referred to as a "tabular iceberg."

Tabular icebergs are frequently encountered in Antarctic waters. *(NOAA)*

By convention, scientists and imagery analysts label tabular Southern Hemisphere icebergs based on the Antarctic quadrant in which they originated. The quadrants are divided counterclockwise in the following manner:

A = 0–90W (Beliingshausen/Weddell Sea)
B = 90W–180E/W (Amundsen/Eastern Ross Sea)
C = 180E/W–90E (Western Ross Sea/Wilkes land)
D = 90E–0E/W (Amery/Eastern Weddell Sea).

When an iceberg is first sighted, the National/Naval Ice Center (NIC) in Washington, D.C. documents its original location. The letter of the quadrant and a sequential number are assigned to the iceberg for tracking purposes. For example, C-19 is the 19th Southern Ocean iceberg that had its origin between 180E/W and 90E (Quadrant C).

Since 2000, NIC has tracked more than 12 distinct Southern Ocean icebergs, including B-18, A-43, A-44, C-16, B-21, C-17, B-22, C-18, D-17, C-21, and C-19. These bergs emerged from glaciers such as the Ross Ice Shelf, the Ronne Ice Shelf, the Pine Island Glacier, the Matusevich Glacier, the Thwaites Ice Tongue, the Lazarev Ice Shelf, and the Shackleton Ice Shelf. On average, these bergs measured 15 miles (24 km) in width by 46 miles (74 km) in length. These dimensions provide some sense of the enormity of Southern Ocean tabular icebergs. The largest tabular berg to emerge from the Antarctic ice shelf since 2000 was C-19, which emerged from the Ross Ice Shelf in May 2003. When it broke free, it measured 19 miles (30.6 km) in width by 123 miles (198 km) in length, or more than 2,300 square miles (5,957 sq km) in area.

Further Reading

National Ice Center, National Oceanic and Atmospheric Administration. Available online. URL: http://www.natice.noaa.gov. Accessed October 17, 2007.

—**Kyle R. Dedrick**
Marine Information Resources Corporation
Ellicott City, Maryland

Delaware Bay, and the Hudson River. Fuel oil arrives safely to market as a result of ice patrols and continuous icebreaking activities by Coast Guard cutters such as BUOY TENDERS and powerful harbor TUGBOATS. These vessels are also used to free barges and tugs that become stuck in the ice. They also reduce the risk of flooding during the spring thaw by keeping river choke points open and the ice flowing through bottlenecks, which prevents ice dams from forming. Icebreaking ensures that waterways such as the Hudson and Delaware Rivers stay open to commercial petroleum barge traffic. This mission has a direct and positive effect on

the physical and economic well-being of millions of coastal Americans and Canadians.

The U.S. Coast Guard has been involved in open-ocean icebreaking as part of its maritime mobility mission since the late 1800s. Today, certain ships in the U.S. Coast Guard fleet are specifically designed for open-ocean icebreaking. This class of ship is called a polar icebreaker or "WAGB." Polar icebreakers are able to cut channels through approximately 6-foot (2.4 m)-thick ice at three knots (1.54 m/s), enabling tankers and bulk cargo ships to resupply remote stations in the Arctic and Antarctic. Cutters, by defini-

Icebreaking: A Marine Science Perspective

As a marine scientist, at-sea studies may be accomplished from a variety of platforms to include icebreakers. Working onboard a modern vessel such as the U.S. Coast Guard cutter *Healy* can be quite interesting as the ship breaks ice approximately five-feet (1.5 m)-thick at a speed of approximately three knots (1.54 m/s) while navigating to a research station for instrument deployments.

Many science crews in polar regions are from industry such as ROV pilots from Phoenix International, Inc., government agencies such as physical scientists from the U.S. Army's Cold Regions Research and Engineering Laboratory, or from universities such as basic researchers from the Helsinki University of Technology. These marine scientists may work aboard icebreakers, proven platforms that have even been frozen into the ice to study polar oceanography and meteorology. This feat was first accomplished from 1893 to 1896 by Norwegian explorer Fridtjof Nansen (1861–1930) in the Arctic Ocean aboard the *Fram,* a ship with a three-layered hull. Despite advances in ship design, increases in shaft horsepower, and the use of satellite technology to obtain imagery of sea ice, icebreaking seems to remain a brute force task of pushing a strong boat into ice with the hope of breaking through and cutting a channel. Riding on an icebreaker is a loud, jolting, and raucous affair. Scientists working aboard modern icebreakers may want to accomplish a myriad of tasks during icebreaking. Science plans may require acoustic profiling and bottom mapping during icebreaking, towing nets and instruments from the stern during icebreaking, and the deployment of Autonomous Underwater Vehicles and Remotely Operated Vehicles.

Ship design does affect the conduct of marine science. Most icebreaking was taken to new limits by the Union of Soviet Socialist Republics during the cold war era between 1945 and 1991. Naval architects from the Soviet Union designed and built their icebreakers with nuclear power plants. These behemoths were designed with the sole purpose of keeping northern ports open during winter, primarily for military reasons. The Russian coastline borders the Arctic Ocean, North Pacific, and the Sea of Okhotsk. Maintaining navigable waterways is important to commerce, for example in exporting oil. Icebreakers,

including the U.S. Coast Guard Cutters *Polar Star* and *Polar Stern,* employ rounded hulls to aid in the breaking of thick fast ice. Fast ice is attached to the shore or could be grounded icebergs. Round hulls do not suit themselves well to sonar transducer installations, thus most icebreakers used for marine science employ a flat bottom hull and a sharp bow that is intended to cut through ice. This design does not lend itself to breaking heavy ice, but is considered a compromise for the primary role of ocean research. One vessel employing such a design is the United States Antarctic Program Research Vessel *Nathaniel B. Palmer.*

While the largest icebreakers have the power to break straight channels for smaller vessels in convoy, it is common practice for science vessels to follow leads in the ice and to capitalize on weak ice. Leads and weak ice may be identified on ice charts provided by ice centers such as the National Ice Center in Suitland, Maryland. When the ice gets thick, vessels resort to the arduous task of "backing and ramming"—literally backing up and ramming as far into the ice as possible until the boat stops, and then doing so again, and again, and again. In estuaries such as Chesapeake Bay, buoy tenders may be employed to break ice and open shipping channels for seaports such as Baltimore Harbor.

Nearly all icebreaking vessels today also utilize satellite imagery to aid in route planning through the ice. Ice analysts use data from models, buoys, sonar, airborne reconnaissance photographs, and satellite imagery to estimate general thickness, concentration, and extent of the sea ice. Information is collected, analyzed, and made into maps at numerous cooperative ice centers. Many of these maps are created using Geographic Information Systems, which allows scientists to overlay additional geographic information as a new map layer. Ice charts are pulled from online Web sites or disseminated by facsimile to the vessel. Marine scientists involved in international ice charting collaborate and produce standard products so that mariners traveling across oceans are able to access ice information from different local authorities in the same manner and with the same look and feel.

Perhaps the most modern advancement in icebreaking technology is the use of double-acting hulls. A vessel employing a double-acting hull has a bow that is designed for open water operations and a stern that is optimized for

(continues)

tion, are 65 feet (19.8 m) in length or greater and coded with a letter *W* for U.S. Coast Guard and given an "AGB" classification, which stands for miscellaneous. These polar-class icebreakers, providing a means for search and rescue in the high latitudes as well as marine spill response in remote polar regions, are increasingly becoming sources of oil extraction.

Polar icebreakers that are presently in service are the 399-foot (122 m) *Polar Star* (WAGB-10), the 399-foot (122 m) *Polar Sea* (WABG-11), and the 420-foot (128 m) *Michael Healy* (WAGB-20). Besides clearing ice for supply ships, these Coast Guard cutters are often used as platforms for polar studies and to transport scientists to remote research stations in the Antarctic

Icebreaking: A Marine Science Perspective (continued)

U.S. Coast Guard Cutter *Polar Sea* breaking ice epitomizes the powerful engines and reinforced bows that are required to maintain channels through sea ice.   *(USCG)*

icebreaking. In open water, the vessel navigates bow first, but switches to stern first upon entering ice. This arrangement allows the propellers (Azipod® podded propulsion systems in this case) to pull the boat through the ice and pulls broken ice down and away from the breaking hull, reducing the buildup of ice in front of the vessel. As an example, the Finnish Maritime Administration's icebreaker *Botnica* employs a double-acting hull in the Baltic.

Icebreakers are important in maintaining ship channels for commerce and in providing a platform for marine science in polar waters. Icebreakers provide escort to supply ships and may serve as the resupply vessel to remote polar research stations. In some temperate climates, estuaries may also freeze, and icebreaking is required to keep seaports open. In the absence of icebreakers, operational marine scientists may compute

sea ice factors such as thickness to determine if buoy tenders may be employed to open navigation routes.

Further Reading

Cold Regions Research and Engineering Laboratory, U.S. Army Corps of Engineers. Available online. URL: http://www.crrel.usace.army.mil/. Accessed July 29, 2007.
Helsinki University of Technology. Available online. URL: http://www.tkk.fi. Accessed July 29, 2007.
Phoenix International, Inc. Available online. URL: http://www.phnx-international.com. Accessed July 29, 2007.
USCGC *Healy* (WAGB-20), U.S. Coast Guard. Available online. URL: http://www.uscg.mil/pacarea/healy/. Accessed July 29, 2007.

—**C. Reid Nichols,** president
Marine Information Resources Corporation
Ellicott City, Maryland

such as the McMurdo Station located at Hut Point Peninsula on Ross Island. The polar icebreaker/research vessel *Michael Healy* is the newest state-of-the-art cutter supporting polar research projects such as multinational climatological investigations that affect the entire world. These cutters have reinforced hulls and special icebreaking bows, a system

that allows rapid shifting of ballast to increase the effectiveness of their icebreaking, and they carry two helicopters and several inflatable boats. They support a crew, including scientists, of 110 to 180 people.

Scientists working at ice centers and the International Ice Patrol support icebreaking missions. Com-

Russian, Canadian, and U.S. icebreakers demonstrate international cooperation in icebreaking.   *(USCG, LCDR Steve Wheeler)*

features provide possible locations to efficiently break sea ice. The Ice Patrol uses U.S. Coast Guard C-130 aircraft with side and forward-looking RADAR sensors to patrol the North Atlantic and locate the boundaries of the iceberg field. This information is then provided to all mariners transiting the ice-covered regions. Such information is critical for optimal ship tracking by commercial and military ship routers.

Further Reading

Canadian Coast Guard. "Icebreaking Program." Available online. URL: http://www.ccg-gcc.gc.ca/ice-gla/main_e.htm. Accessed June 24, 2007.

Canney, Donald L. "Icebreakers and the U.S. Coast Guard." Available online. URL: http://www.uscg.mil/hq/g-cp/history/Icebreakers.html. Accessed September 21, 2007.

bined geophysical data from ice centers and geospatial data from the Ice Patrol helps analysts build products, which locate and track ICEBERGS in order to prevent disasters such as the April 15, 1912, sinking of White Star liner RMS *Titanic*. Leads or cracks in the ice may be identified in ice forecasts by the NATIONAL ICE CENTER in Suitland, Maryland. These

Indian Ocean   The Indian Ocean is the smallest of the major oceans and is surrounded by densely populated coastal regions. It is bounded on the north by the Asian landmass (north of 25°N), East Africa to the west, Indonesia and Australia to the east, and the ANTARCTIC CIRCUMPOLAR CURRENT to the south. The lack of a solid eastern boundary makes the

Scientists from the U.S. Coast Guard Cutter *Polar Sea* logging information from ice cores taken during an icebreaking operation in the Arctic Circle   *(USCG, PA3 Andy Devilbiss)*

Indian Ocean basin very different from the Atlantic and Pacific basins. It is subdivided into two basins: the Arabian Sea to the west, and the BAY OF BENGAL to the east by the Indian subcontinent. The average depth of the Indian Ocean is about 13,000 feet (3,963 m). It has adjacent or marginal seas such as the Arabian Sea, the RED SEA, the Bay of Bengal, and the Persian Gulf.

The Indian Ocean is believed to have formed at about the same time as the ATLANTIC OCEAN when, about 170–200 million years ago, the supercontinent Pangaea separated into a northern continent, Laurasia, and a southern continent, Gondwanaland. The Tethys Sea along the equator separated these two continents. As the Tethys Sea opened up and widened, it formed the Atlantic and Indian Oceans. Plate boundaries include the Indo-Australian plate, which is converging on the Eurasian Plate as evidenced by the undulating Java and Sunda TRENCH. Earthquakes result as seafloor subducts into the trenches and, depending on the force, may generate extreme TSUNAMIS.

The circulation of the Indian Ocean is significantly different from that of the Atlantic and Pacific Oceans, especially in the Northern Hemisphere summer. The wind system over the northern portion of the Indian Ocean is the Asian MONSOON, which is quite different from the trade winds that drive the equatorial Atlantic and Pacific Oceans. The Asian monsoon wind regime results in strong and persistent winds from the northeast in November through the end of March (Northern Hemisphere winter), reversing to blow from the southwest during June through September. This seasonal reversal in wind direction results in major changes to the equatorial current system and the western boundary currents.

During November through March, the NORTH EQUATORIAL CURRENT flows westward across the Indian Ocean, extending from about 8°N to the equator; the EQUATORIAL COUNTERCURRENT flows eastward, extending from the equator to about 8°S (the only major Equatorial Countercurrent south of the equator); and the South Equatorial Current flows from east to west, extending from about 8°S to about 20°S. Right at the equator, there is the eastward flowing EQUATORIAL UNDERCURRENT, extending from a few tens of feet (approximately 6 m) below the surface to a depth of several hundred feet (approximately 60 m).

The western boundary current of the Indian Ocean in the Southern Hemisphere is the AGULHAS CURRENT, which flows to the southwest along the east coast of Africa. The Agulhas Current is a narrow, swift, deep current about 60 miles (100 km) wide with speeds nearly as high as the GULF STREAM.

Water transport in the Agulhas Current has been estimated to be about 50–80 Sv (50–80 × 10⁶ m³/s). The eastern boundary current is the Leeuwin Current flowing poleward along the east coast of Australia. The Leeuwin Current is the only eastern boundary current on the planet to do so. It is possibly the result of the exchange of water with the Pacific Ocean through the straits between Indonesia and Australia. The Leeuwin Current is a weak and diffuse current with numerous EDDIES and major seasonal variations in speed and direction. At times, it appears to exist only in a statistical sense, as the resultant motion of the eddies that compose it.

During the Northern Hemisphere summer, the season of the southwest monsoon, the North Equatorial Current reverses and flows eastward, extending in width from about 8° to about 7° South, joining in eastward flow with the Equatorial Countercurrent and the Equatorial Undercurrent. Since the wind stress is from west to east at this time, the slope of the sea surface may reverse and slope upward from west to east. Some authorities believe that under these conditions, the Equatorial Undercurrent would cease to exist.

This whole eastward flowing mass of water in the north and equatorial Indian Ocean is known as the Southwest Monsoon Current. During this time, corresponding to the winter season in the Southern Hemisphere, the southeast trade winds strengthen, and the South Equatorial Current flows more strongly than in the Southern Hemisphere summer season.

During the southwest monsoon season, a strong western boundary current develops, flowing northward off the Horn of Africa—the Somali Current. This remarkable current has speeds that are sometimes greater than those of the Gulf Stream. It has a transport of about 65 Sv (65 × 10⁶ m³/s). UPWELLING in this region brings cold water along the Arabian coast. The current is very shallow compared to the Gulf Stream, however, being a few hundred feet (approximately 60 m) in depth. This is the reason it is able to develop so rapidly in response to the wind reversal, from the northeast to the southwest monsoon winds.

During the transitional seasons, April through May, and October into November, the Indian Ocean circulation is characterized by eddies and highly variable transient currents until the new wind regime is firmly established. The lack of wind forcing impacts currents. For example, northward flow of the Somali Current is either weakened or ceases all together. Circulation changes and eddy development are observed by satellites that highlight sea surface temperatures.

Understanding the Indian Ocean is important since this water body is a main transit route linking Asia, Africa, and Australia. It connects Asia, the Arabian Peninsula, Africa, with island nations such as Madagascar, Comoros, Seychelles, Maldives, Mauritius, Sri Lanka, and Indonesia. The general circulation of the Indian Ocean was known as early as the 1930s from analysis of ship drift data by (mostly) German oceanographers, but it was not until the International Indian Ocean Expedition (1963–66) that details of the circulation became known. Knowing monsoon weather patterns, circulation, and other phenomena such as severe storms and waves common to the Indian Ocean can help protect property and save lives.

WATER MASS characteristics are different from those of the Atlantic and Pacific Oceans. There is no source of cold deep water. The deep water of the Indian Ocean has its origins in the Atlantic, Pacific, and Southern Oceans. The water on the bottom is ANTARCTIC BOTTOM WATER. The water at mid depths is Antarctic Intermediate Water. Both of these water masses are flowing slowly northward, decreasing in oxygen content. Above the Antarctic Intermediate Water, above 3,281 feet (1,000 m) depth, Indian Central Water is found from the southern part of the Indian Ocean northward to about the equator, with salinities ranging from about 34.5–36.0 psu; north of the equator Indian Equatorial Water is found.

The surface water of the Indian Ocean can be divided into the high-SALINITY water of the Arabian Sea, resulting from high evaporation, especially in the Red Sea and Persian Gulf regions, and the low-salinity water of the Bay of Bengal, resulting from high-volume river discharge from the major rivers of the region, such as the Ganges and the Brahmaputra during and following the monsoon rains.

The waters of the Red Sea, bordering the northeast corner of the Gulf of Arabia, have the highest salinity of any sea on Earth, being as high as 40 psu. Temperatures are also high, with the surface temperatures on the order of 82°F (28°C) in summer and 79°F (26°C) in winter. The deep-water temperature of the Red Sea is on the order of 72°F (22°C).

The Persian Gulf is a shallow, warm body of water with summer temperatures as high as 95°F (35°C) and salinities as high as 42 psu. The western (Iraqi) shore of the Persian Gulf is noted for its rich oil deposits. This region developed some notoriety during the first Persian Gulf War when Iraqi president Saddam Hussein (1937–2006) deliberately set fire to oil rigs in the Persian Gulf in what may be the first case of waging environmental warfare.

Further Reading
Pickard, George L., and William J. Emery. *Descriptive Physical Oceanography: An Introduction,* 5th enlarged ed. Oxford: Pergamon Press, 1990.
Society for Indian Ocean Studies. Available online. URL: http://www.sios-india.org. Accessed June 25, 2007.

inertia current The transient flows of surface current initially set in motion by the wind, then continuing in motion when the wind ceases are called inertia current. This type of current results from the momentum of the moving mass of water (inertia). Momentum in this case is mass and velocity, and in ocean circulation the mass is considered to be constant. The apparent force responsible for this motion is the CORIOLIS EFFECT, the result of a rotating frame of reference, that is, the rotating Earth. The speed of the current steadily decreases until completely damped out by friction; the direction of the current rotates clockwise in the Northern Hemisphere and counterclockwise in the Southern Hemisphere. In the analysis of the physics of inertia currents, it is assumed that when the wind stops blowing, there are no PRESSURE GRADIENTS, fluid friction is neglected in the deep ocean, and the deceleration (slowing down) is balanced by the Coriolis effect. In other words, there are no fundamental forces acting on the fluid and the fluid continues moving as a result of its inertia and Earth's rotation. These assumptions lead to a pair of differential equations that can be solved to show that the radius of the inertia current r is equal to the current speed v divided by twice the angular velocity of rotation of the Earth Ω multiplied by the sine of the latitude φ. This is expressed in equation form as

$$r = v/2\Omega\sin\varphi.$$

This equation can be useful in estimating the flow of drifting flotsam in the absence of other currents. Since the Earth revolves around its axis once every 23 hours 56 minutes and 42 seconds, the angular velocity is computed by dividing 2π radians by 86,164 seconds to get the constant $\Omega = 7.29 \times 10^{-5}$ radians per second. In the absence of fundamental forces, the speed of the current v is steadily decreasing and the radius of the current's rotation is getting smaller and smaller. The equations also show that the period T of the inertial rotation is equal to π divided by the angular velocity of the Earth Ω multiplied by the sine of the latitude φ, that is,

$$T = \pi/\Omega\sin\varphi.$$

The inertia period at the poles is 12 hours, at 45 degrees latitude about 17.44 hours, at 30 degrees latitude 24 hours, and approaches infinity at the equator. The circle of inertia ranges from about .6–6 miles (1–10 km) in mid latitudes. Because of the variation of the Coriolis effect with latitude, the circles of motion are bent into irregular ellipses, being more sharply curved at high latitudes than at low latitudes. Inertia current circles migrate westward with time. CURRENT METER measurements show that inertia currents occur in all the oceans of the world in response to passing storms and the associated transient wind stresses they impose on the sea surface. Inertia currents usually have duration of several days at mid-latitudes. PHYSICAL OCEANOGRAPHERS analyzing current meter records must be careful not to confuse inertial motion with the water particle motions of INTERNAL WAVES. Internal waves propagate below the sea surface along boundaries having different densities, and in continuously stratified water. Benjamin Franklin (1706–90) first wrote about internal waves while on a sea voyage as they occurred in his stateroom oil lamp that was contaminated with water.

From a slightly different point of view, inertia currents can be viewed as resulting from the natural tendency of water and air parcels to meander and form EDDIES. For these motions, the centripetal force (mv^2/r) needed to keep the motion curvilinear is supplied by the Coriolis effect of Earth's rotation. In the absence of direct measurements of circulation, numerical models that include the physics of surface winds and inertia currents can be used to help estimate the trajectories of floating marine pollutants. The Ocean Conservancy maintains information on marine debris for the entire United States. The center's program encourages people to collect shoreline litter, keep count of the types of litter collected, and to report their collection statistics. Armed with information on permanent currents, storms, and inertia currents, researchers can estimate the trajectories of materials deliberately cast overboard.

Further Reading

Ocean Conservancy magazine. Available online. URL: http://www.oceanconservancy.org. Accessed September 21, 2007.

inflatable boats Inflatable boats have air-filled components such as floors and hulls and are specially designed by NAVAL ARCHITECTS to meet a variety of needs, ranging from rowing out to a moored cabin cruiser, fishing and water skiing, to lifesaving and military operations. The burgeoning civil and military inflatable boat industry stems from U-shaped designs by Zodiac engineer Pierre Debroutelle (1886–1960). His 1934 creation contained two lateral buoyancy chambers connected by a wooden transom. To-

day's inflatable boats usually have hard transoms, inflatable keels, and high-pressure inflatable floors. These improvements decrease the weights of these boats and makes them easier to handle if they have to be quickly inflated or deflated. Rigid Inflatable Boats or RIBs have air-filled chambers with fill and relief valves that make up the hull, rigid floor material made from fiberglass or aluminum, and stringers that lock the floorboard elements in place. Most Ribs have multiple air chambers that need to be protected from being punctured and deflated. To anchor the boats, some operators use special grapnel-style ANCHORS since they are easy to stow and have no sharp points or rough edges.

The main benefits for RIBs are buoyancy, storage, and easy assembly. Depending on the size of inflatable boats, a davit system may be required to launch and retrieve the inflatable from a mother vessel. Davits are a pair of small cranes, usually at the transom, which are used to lift a dinghy and suspend it over the water while under way.

There are many other categories of inflatable boats. The smallest inflatables or dinghies have soft transoms and can be maneuvered with oars, a paddle, or a low-horsepower motor when a motor mount is attached. Yachtsmen conveniently employ inflatables as tenders that transport passengers and provisions back and forth from the moored vessel. There are also inflatable kayaks or "duckies" that require the use of double-bladed paddles. Sport boats have walk-around tubes that must be fully inflated, a hard transom, and a sectional floor made of wood, fiberglass, composite, or aluminum. They also have inflatable or wood keels. These boats can be folded up once the floor is removed. Rollups have a hard transom that can be rolled up with the floor remain-

The *Zodiac,* used to ferry personnel and for many other purposes, with its inventor, Pierre Debroutelle   *(Zodiac, France)*

ing in the boat. Inflatable boats tend to cause a lot of spray and are very slippery when wet, especially around the inflatable tube or collar.

Inflatable boats are very important platforms for marine science. The research vessel *Thomas G. Thompson,* which is operated by the University of Washington, has one 26-foot (7.9 m) RIB workboat, one 15-foot (4.6 m) inflatable workboat, and a 19-foot (5.8 m) RIB rescue boat. They can close in on instrumented BUOYS for service and maintenance, can be used to support divers, and are used frequently to complete hydrographic surveys in shallow waters. RIBs outfitted with differential GLOBAL POSITIONING SYSTEMS, echosounders, side-scan SONAR towfish, CONDUCTIVITY-TEMPERATURE-DEPTH (CTD) PROBES, and computers used for data acquisition and NAVIGATION are ideal for hydrographic surveys to assess pre- and post-dredge depths, map channel cross sections, and to support oceanographic studies. Oceanographers use these data to display processes such as WAVES, CURRENTS, and water levels as overlays on top of bathymetric (*see* BATHYMETRY) data.

Military and law enforcement units use inflatable boats since they are able to provide fast response in unfavorable sea conditions as well as handle shallow water. Military reconnaissance units launch inflatable rollups from aircraft, landing craft, and even from submerged submarines. Coxswains are able to navigate heavily reinforced rubber inflatable boats from over the horizon and transit across the surf zone or maneuver into riverine environments. For extreme weather, RIBs provide a high-speed and seaworthy craft for coastal resupply, surveillance, and security missions. One of the most important features for coastal patrol ships used for search and recovery is the ability to efficiently launch and recover RIBs, even in heavy seas such as up to eight-feet (2.5 m) wave heights. RIBs have become a mainstay for rescue services and law enforcement around the world.

infragravity waves Infragravity waves are waves of lower frequency than wind-generated ocean surface gravity waves, but higher in frequency than the semi-diurnal (twice daily) tides. This range embraces a period band between half a minute and 12 hours. This is a usually quiet band in the surface water wave energy spectrum. Infragravity waves include surf beat, EDGE WAVES, SEICHES, and TSUNAMIS. These phenomena, especially tsunamis, may contain large amounts of energy when they occur, but they are very infrequent, leading to low average values.

Surf beat describes the fluctuating heights of sea and swell waves breaking as they reach the shoreline. Surf beat appears to be a nonlinear effect of the superposition of shoreward moving wave trains in shallow water, as the wave trains combine (modulate). Measurements of surf beat require extensive filtering of wind wave records to extract the surfbeat signals from the much stronger wind wave signals. These infragravity motions have periods on the order of several minutes and dissipate through the surf zone.

Edge waves are generated by severe and rapidly moving storms such as squall lines. They propagate along a beach in the surf zone, contributing to water level changes. They are also present when strong winds pile up water against the shore, which must flow seaward as RIP CURRENTS. These resurgent phenomena have phase speeds that are about 1 to 10 percent that of deep-water swell and wind waves. Their amplitudes are highest at the shoreline and decrease exponentially seaward with distance. All their energy is essentially contained within the surf zone. Their amplitudes are seldom more than three feet (1 m) high. Their periods are in the one-to-several-minutes range. They can be important factors in forming the configuration of BEACHES.

Seiches are wind-induced oscillations in ports, harbors, embayments, and lakes. They occur typically when winds have been blowing strongly in one direction for hours to days, and then stop, or change direction. The pile of water then sloshes back and forth much as the water in a bathtub. COASTAL ENGINEERS designing wharfs along a seaport may consider potential flooding impacts by observing seiching with water level gauges or estimate the WATER LEVEL FLUCTUATIONS from the surface wind data, using well-accepted algorithms.

Tsunamis are generated by underwater earthquakes or landslides, or could be generated by a large body such as the chunk of a volcanic island sliding into the sea during eruption, or a meteor landing in the ocean. These long waves can travel across oceans with small amplitudes (wave heights), but speeds close to that of a jetliner. When they reach a shoreline, they undergo a transformation into terrifying and destructive waves of enormous height. Through natural shoaling processes, the WAVELENGTHS decrease and the wave heights increase. Extensive networks such as the Pacific Tsunami Warning Center consist of seismographs, TIDE GAUGES, numerical models, sirens, and communications systems. Injury and loss of life is minimized when coastal populations are warned that a tsunami may be approaching the shore. The warnings help to reinforce evacuation orders and procedures that should have been practiced by local residents.

Some universities, government agencies, and laboratories such as the U.S. Army Corps of Engineers'

FIELD RESEARCH FACILITY focus intense research on waves, to include infragravity waves. As an example, during October 1990, the U.S. Army Corps of Engineers, the Office of Naval Research, and the Naval Research Laboratory sponsored the "Duck Experiment on Low-frequency and Incident-band Longshore and Across-shore Hydrodynamics," or DELILAH to describe directional wave information from approximately 90 instruments from the shoreline seaward out to the 42.7-foot (13 m)-depth contour.

Further Reading

Birkemeier, William A., Cindy Donoghue, Charles E. Long, Kent K. Hathaway, and Clifford F. Baron. "The DELILAH Nearshore Experiment: Summary Data Report, Technical Report CHL-97-24." U.S. Army Corps of Engineers, September 1997.

O. H. Hinsdale Wave Research Laboratory, Oregon State University. Available online. URL: http://wave.oregon state.edu. Accessed September 21, 2007.

infrasonic waves Sound waves below the average frequency of human hearing, that is, below the range 20–20,000 Hz, are called infrasonic waves. Infrasonic waves can be generated by such natural phenomena as ocean WAVES, winds, earthquakes, volcanoes, cosmic impacts, and HYDROTHERMAL VENTS. They can also be generated by BRIDGES, machines, such as an aircraft flying through turbulence, and nuclear weapons detonation.

Infrasonic waves can be detected by means of seismographs and low-frequency microphones called "geophones." These instruments are routinely deployed by engineers, geologists, and geophysicists. Innovations in meteorology could involve the detection of tornadoes that generate infrasonic pulses.

Low-frequency microphones placed under the water (hydrophones) can detect underwater earthquakes, which may generate TSUNAMI. They are part of the Pacific Tsunami Warning System. Research is ongoing to determine if infrasonic waves can be useful in the prediction of major, destructive earthquakes, volcanic eruptions, and tsunamis.

Infrasonics plays a large part to ensure that arms control treaties are not violated. Hydrophones are employed in the detection of violations of underwater nuclear weapons tests. Underground nuclear explosions are monitored by detecting the seismic waves generated by the explosion. Difficulties arise from background noise and naturally occurring seismic disturbances. Atmospheric nuclear explosion detection methods involve measuring radionuclides, electro-magnetic pulse (EMP), optical flash, and shock waves.

Geologists use instruments called air guns in seismic reflection profiling from ships to obtain information on the SEDIMENTS and sub-bottom structure. They can operate at frequencies at the fringe of the lower bound of human hearing/upper range of infrasonic waves. In general, the lower the operating frequency of the instrument, the greater the range. It is anticipated that many new applications of infrasound to the geosciences will be made in the 21st century as the technology improves.

Further Reading

Earth System Research Laboratory, National Oceanic and Atmospheric Administration. "Infrasonics Program." Available online. URL: http://www.etl.noaa. gov/about/review/as/bedard. Accessed September 21, 2007.

Institute of Electrical and Electronic Engineers (IEEE) The Institute of Electrical and Electronic Engineers is a nonprofit U.S. engineering organization that was originally founded in 1884 by two predecessors, the American Institute of Electrical Engineers and the Institute of Radio Engineers. As a professional society, IEEE has been organized to advance the theory and application of electrotechnology and allied sciences such as ocean engineering. IEEE has more than 365,000 members in over 150 countries and serves as a catalyst for technological innovation by developing, defining, and reviewing standards within the electronics and computer science industries. The institute supports the needs of its members through a wide variety of programs such as conferences and services and an online database for IEEE published information since 1999. IEEE's longstanding mission involves the creation, development, integration, sharing, and application of knowledge about electronic, computer, and information technologies for the benefit of humanity and the profession.

Since marine science includes applications of electrical and electronic engineering, IEEE has an Oceanic Engineering Society as well as a Geoscience and Remote Sensing Society. The Oceanic Engineering Society of IEEE holds meetings for the presentation of papers and their discussion, and also has committees focused on issues ranging from submarine cables and connectors to BUOY MOORING and communications. IEEE through the Oceanic and Engineering Society and the Marine Technology Society has combined resources to host an annual professional meeting known as The Oceans Conference. IEEE's Geoscience and Remote Sensing Society seeks to advance geoscience and remote sensing science and has committees focused on issues ranging

Selected Technical Committees

Ocean Engineering Society

Air/Space Remote Ocean Sensing

Autonomous Underwater Vehicles

Current Measurements

Environmental Technology

Information Processing

Marine Communication, Navigation, and Positioning

Modeling, Simulation, and Visualization

Non-Acoustic Image Processing

Oceanographic Instrumentation

Sonar Signal and Image Processing

Submarine Cable

Underwater Acoustics

Geoscience and Remote Sensing Society

Data Fusion

Data Standardization and Distribution

Instrumentation and Future Technologies

from sensor innovations and frequency allocation to data collection and processing. The Geoscience and Remote Sensing Society annually sponsors the International Geoscience and Remote Sensing Symposium or IGARSS, an international conference for the exchange of ideas that are published in a formal proceedings document.

Scientists and engineers are thought of as professionals regardless of whether they work for federal, state, or local government, large or small business, or as independent consultants. Many marine scientists have obtained formal degrees, most perform internships, and some, such as engineers and surveyors, obtain licenses or certifications to practice their profession. Membership in professional societies such as IEEE provides important forums for the exchange of ideas, continuing education opportunities, technical achievement awards for selected members, and advancement of the profession. The spectrum of participation for an active member might involve participating in stimulating discussions at committee meetings, writing a technical paper and presenting findings at the joint Marine Technology Society/IEEE Oceans Conference or at IGARSS, publishing a paper in the *Journal of Oceanic Engineering,* or chairing a session at either the Oceans Conference or IGARSS. Example technical committees are listed in the table above. It is important to realize that marine science is a growing field because of the contributions that have been made through many multidisciplinary scientists and professional organizations.

Further Reading

IEEE Geoscience and Remote Sensing Society. Available online. URL: http://www.grss-ieee.org/. Accessed September 21, 2007.

IEEE Ocean Engineering Society. Available online. URL: http://www.oceanicengineering.org. Accessed September 21, 2007.

Institute of Electrical and Electronics Engineers, Inc. Homepage. Available online. URL: http://www.ieee.org. Accessed January 28, 2005.

interferometry Interferometry refers to the reinforcing or canceling of wave trains of the same frequency from spatially separated sources observed at a point location. If the wave trains are in phase, or nearly so, their amplitudes add, producing constructive interference. If they are 180° out of phase, they produce destructive interference; if their amplitudes are identical, they will cancel. These effects, true for small amplitude sine and cosine waves, follow from the superposition principle, which shows that the wave amplitudes add linearly. Thus, if the waves are of the same frequency and amplitude, and are in phase, the combined amplitude will be double that of the individual waves; if they are 180° out of phase, they will cancel. This latter effect is used in aircraft to reduce the engine noise level in the cabin.

An important attribute of interfering waves is that they are coherent, or have a fixed phase relationship. In general, light waves from sources such as candles, fluorescent, or incandescent bulbs are incoherent, because they output different frequencies, their phases are random, and interference effects are not observed. In the case of the laser, however, laser light has the remarkable property of generating light waves of a single frequency and fixed phase relationship; thus, interference effects between identical lasers are readily observed and are often used to study interference in the physics laboratory. Interference between sound waves occurs when an observer is listening to the output of two loudspeakers driven electronically by a single source such as a signal generator in different parts of a room. This is so because as the distances from the listener to the loudspeakers vary as the listener changes location, the phase differences between each of the loudspeakers vary from 0° to 360° (or multiples thereof), causing interference effects ranging from total reinforcement to total cancellation.

At optical frequencies, interference patterns consist of a series of light and dark bands, called bright and dark fringes. Interference in sound waves causes the sound intensity to fluctuate as location is varied. Interference is also observed when incident light reflects from the top and bottom surfaces of thin films.

Instruments making use of the interference phenomenon are called interferometers. They may use sound, visible light, or any electromagnetic wave, such as radio, radar, infrared, X-rays, and so forth. The ordinary interferometer used in the college physics laboratory uses a beam splitter to separate a light beam into two paths, whose path lengths can be varied by adjusting the instrument, resulting in a pattern of bright and dark fringes. The output of the interferometer can be used to produce an interferogram, or plot of the interference pattern of light and dark fringes, often superimposed on an observation grid, such as a terrain or bathymetric map. In the case of sound, the acoustic intensity is plotted on a graph of intensity versus location, usually in terms of distance from a line, which bisects the speakers and is perpendicular to the plane of the speakers.

Interferometry is used in marine and Earth sciences to make precise measurements of distance, such as movements of topographic or bathymetric features, following the occurrence of an earthquake or volcano. Very long baseline interferometry (VLBI) refers to interferometers with a great distance between the receiving elements; in some cases the diameter of Earth. Even greater distances such as those between celestial bodies are used by CARTOGRAPHERS and HYDROGRAPHERS to make precise measurements of the shape of the GEOID. Oceanographers use VLBI to accurately determine long-term changes in water levels, and to make fine adjustments to numerical supercomputer ocean tide models. Geophysicists make extremely accurate measurements of Earth's rotation period by means of VLBI, using celestial sources of electromagnetic (EM) waves, such as the very distant proto-galaxies known as quasars. Interference effects are used to electronically "steer" the beams of multi-element SONAR and RADAR systems (arrays), obviating the need for rotating antennas.

Further Reading

Goddard Space Flight Center, Very Long Baseline Interferometry, National Aeronautics and Space Administration. Available online. URL: http://lupus.gsfc.nasa.gov/index.htm. Accessed September 21, 2007.

internal wave   An internal wave is a gravity wave propagating in the interior of the ocean where there are strong DENSITY, TEMPERATURE, and SALINITY gradients. Investigations of internal waves in the oceans, MARGINAL SEAS, and lakes indicate that disturbances at the density interface, called the PYCNOCLINE, generate internal waves. Surface expressions of these phenomena are seen in satellite imagery where internal waves are propagating into the coastal ocean north of Delaware Bay, from the Strait of Gibraltar into the Gulf of Cadiz, and into the South China Sea from the Luzon Strait. The pycnocline can be displaced up or down in the coastal ocean near a shelf break or adjacent to a sill in a strait or fjord when lighter, warmer, and fresher water overlays denser, colder, and saltier water. Other mechanisms for generating internal waves include SHEAR stresses between fluid layers, water displacements by ship wakes, and strong TIDAL CURRENTS. The intrusion of permanent currents into a pass can cause the upward or downward oscillations of the pycnocline that have been attributed to the generation of internal waves. Internal waves can also be generated by the pressure and stress fluctuations of fast-moving storm systems.

Internal waves behave somewhat like surface gravity waves where energy propagation is across a density discontinuity marked by the familiar air-sea

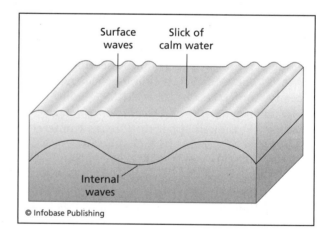

© Infobase Publishing

Internal waves are waves on density boundaries within the fluid that can be progressive or standing. There is a slight disturbance on the surface in the form of small waves above the internal wave crests and slicks (calm water) between the crests.

boundary. The main difference between surface and internal waves is that the density difference between air and water is much greater than that between fluid layers within the ocean. This lesser difference results in slower speeds, larger amplitudes, and lower energies for internal waves compared to surface waves. Internal waves in shallow water cause small displacements of surface water particles that result in lines of convergence and divergence. These processes result in the creation of slicks of oily material on the surface that can be used to track internal waves beneath. The surface expressions of internal waves, although small, are often sufficient to produce scattering patterns for electromagnetic waves that are detectable from spacecraft.

The spatial and temporal scales for internal waves are typically on the order of kiloyards (km) for WAVELENGTH and hours for wave period. Wave amplitudes are on the order of tens of feet (greater than 3 m), which is much larger than surface gravity waves. As internal waves propagate into shallow water, they slow down, wavelengths decrease, amplitudes increase, and as they grow unstable, they break, causing a high degree of TURBULENCE and mixing.

The upper limit to the frequency of internal waves is determined by the buoyancy frequency, also called the Brunt-Väisälä frequency. This is the frequency that a water parcel exhibits when displaced from its equilibrium position (determined by its density). Like a mass on the end of a spring, the displaced fluid parcel will oscillate about its equilibrium position where the restoring force, proportional to the density gradient, acts in a manner analogous to the restoring force of a spring attached to a mass. The greater the displacement, the greater the restoring forces. The buoyancy frequency ranges from minutes in shallow water to hours in deep water, depending on the density stratification. Buoyancy frequency is named after esteemed meteorologists David Brunt (1886–1965) and Vilho Väisälä (1889–1969).

Long wave motions such as TIDES, KELVIN WAVES, and ROSSBY WAVES can have internal components to their motions. For example, as a shoreward-propagating tide wave in a two-layered ocean enters shallow water, topographic features on the bottom, such as reefs or ridges at the depth of the pycnocline, can cause vertical oscillations that propagate away as internal waves.

Internal waves occur not only in two-layer water bodies, but in the generally continuously stratified ocean where density increases steadily with depth. The deep ocean can be regarded as a series of layers of gradually increasing density. Thus, internal waves are possible along boundaries between layers; in fact, internal waves can occur over all or a considerable part of the water column.

Oceanographers measure internal waves by observing the fluctuations of density with depth and time. In many cases, the salinity changes with depth are not large, or can be estimated, and time series of temperature profiles can be used to obtain the amplitudes and frequencies of internal waves. During the 1950s and 1960s, Dr. Eugene C. LaFond (1909–2002) of the U.S. Naval Electronics Laboratory, San Diego, California, developed towed THERMISTOR (resistance temperature sensors) chains and used them to make extensive measurements of internal waves. German oceanographer Albert Defant (1884–1974) developed much of the early theory of internal waves, substantiated by measurements in the Baltic Sea and the North Atlantic.

Researchers are working to develop MODELS and PREDICTIONS for the generation, propagation, and fate of internal waves. Improving the knowledge base by being able to understand and predict these phenomena has many important societal benefits. MARINE BIOLOGISTS are particularly interested in the mixing of nutrient-rich deep waters with nutrient-poor surface waters that occurs as the internal waves trap water and break. Scientists and engineers working in offshore oil exploration are concerned with the impact of internal waves on survey operations with REMOTELY OPERATED VEHICLES and AUTONOMOUS UNDERWATER VEHICLES, and drilling operations with submersible, jackup, drill ship, or semi-submersible platforms. Similarly, personnel working aboard submarines are concerned with understanding impacts on buoyancy and sound travel as they navigate through the water column. During World War II, a German submarine (U-boat) is believed to have been lost while transiting the Strait of Gibraltar on the density interface between

Internal waves in the atmosphere are often made visible by cloud formations. *(NOAA/NWS)*

the upper and lower water layers. A breaking internal wave may have caused the vessel to lose stability and sink. Prior knowledge on locations, generating mechanisms, magnitudes, and propagation of internal waves are particularly helpful to increase navigation safety and in understanding underwater sound transmission.

Further Reading

Knauss, John A. *Introduction to Physical Oceanography*, 2nd ed. Upper Saddle River, N.J.: Pearson/Prentice Hall, 1996.

Welcome to the Ocean Internal Wave Online Atlas, Center for Remote Sensing, College of Marine and Earth Studies, University of Delaware. Available online. URL: http://atlas.cms.udel.edu/. Accessed March 9, 2008.

International Council for the Exploration of the Sea (ICES)

The International Council for the Exploration of the Sea is the world's oldest intergovernmental organization devoted to marine and fisheries science. ICES was founded in Copenhagen, Denmark, in 1902 by eight northern European nations for the purpose of promoting and encouraging the exploration of the seas, primarily the North Atlantic Ocean, and the development of international fisheries research. There are now 20 member nations of ICES: Belgium, Canada, Denmark (including Greenland and Faroe Islands), Estonia, Finland, France, Germany, Iceland, Ireland, Latvia, Lithuania, the Netherlands, Norway, Poland, Portugal, Russia, Spain, Sweden, the United Kingdom, and the United States.

The leadership of ICES is provided by the ICES Council, made up of two delegates from each of the member states. A bureau serves as the executive committee for the council, and consists of a president, a first vice president, and five vice presidents, all elected by the delegates. This group provides direction to consultative, advisory, and science committees, as well as more than 100 working and study groups. ICES provides advice to national governments on regional fisheries management and control of pollution.

Present tasks for the ICES members include describing the state and variability of the marine environment, determining the role of climate variability, evaluating sea water contamination and eutrophication, developing fisheries assessment tools, and coordinating international monitoring and data management programs.

Examples of ICES programs include studies of the invasion of North Atlantic waters by alien species, the near collapse of eel stocks, the possibilities for recovery of cod stocks, and the importance of SEAMOUNTS as oases for marine life.

Further Reading

Hedley, Chris. "Internet Guide to International Fisheries Law." Available online. URL:http://www.oceanlaw.net/texts/summaries/ices/htm. Accessed April 23, 2005.

International Council for the Exploration of the Seas. Available online. URL: http://www.ices.dk/. Accessed July 23, 2008.

Intertropical Convergence Zone (ITCZ)

A region of light and variable winds close to the equator, found in the ATLANTIC, PACIFIC, and INDIAN OCEANS, and the intervening landmasses is known as the Intertropical Convergence Zone. The ITCZ is actually a latitude belt that girdles the globe. Its mean position is generally north of the equator, but is located south of the equator in the Indian Ocean and the western Pacific Ocean during the Northern Hemisphere winter. At this time, it is also south of the equator across the African and South American continents and just north of Australia. Its position is closely related to that of the thermal equator: the narrow belt around Earth of maximum temperature. At the vernal equinox (March 21), the ITCZ begins to move north, and by the time of Northern Hemisphere summer, it is generally north of the equator at all locations, except for central Brazil. Nowhere is this northward movement more evident than in central Asia, where the ITCZ is located across the Tibetan Plateau. At the autumnal equinox (September 21), the ITCZ begins its move southward as the Northern Hemisphere begins to cool.

The ITCZ owes its existence to the convergence of the trade winds. In the oceans, it is generally found in the region of the eastward flowing EQUATORIAL COUNTERCURRENT. The absence of steady westward wind stress in the ITCZ allows a return flow to the east of the WATER MASSES piled up in the western ocean basins by the trade winds.

The ITCZ was formerly known as the "doldrums," a region of calms that was avoided by sailing ships. Today, the ITCZ is also known as the equatorial convergence, because it is the region of convergence of the trade winds.

As a result of high sea surface temperatures that heat the air from below, the ITCZ is an area of extensive, deep atmospheric convection, leading to numerous thunderstorms. These thunderstorms are often so numerous that they cluster together in regions of extensive convection. For this reason, the pilots of small aircraft prefer to avoid flying through the ITCZ, with its sudden winds, rain squalls, and powerful updrafts and downdrafts. Precipitation under the ITCZ is very high, resulting in the vast equatorial rain forests of Africa, South America (Brazil), and Indonesia. Tropical depressions and disturbances occur frequently along the ITCZ. A small percentage of these storms develop into HURRICANES, when there is extremely strong atmospheric convection over water and favorable upper level winds.

Marine scientists study the convection in the ITCZ by means of weather balloons, RADAR, satellite imagery, instrument-supporting tethered balloons, moored and drifting meteorological and oceanographic BUOYS, and aircraft using radar and laser remote sensing systems. These studies contribute essential information to further understanding the interaction between tropical and mid-latitude weather systems, and the relations between the ITCZ and global climate.

invasive species Invasive species—also called a non-native, exotic, nonindigenous, or alien species—are plants or animals that are introduced into a new habitat, lack indigenous predators, and have an aggressive growth pattern. These non-native plants or animals can upset an ECOSYSTEM, causing native species to suffer, decline, or even become extinct. Ecosystem management programs are needed to combat invading non-native species. Strategies to manage invasive species include surveillance, control measures, and the reestablishment of native species. Control measures include eradicating, suppressing, reducing, or managing invasive species populations. Effective management prevents the spread of invasive species such as kudzu (*Pueraria montana*) from areas where they are present and restores native species to the ecosystem. Kudzu is native to Japan and is spreading along the southeastern United States, where it climbs over and kills native plants and trees. It is controlled by grazing over many years, burning, mowing or cutting, and through the application of herbicides.

In the United States, environmentalists and legislators have worked together to identify problems associated with invasive species. Prime examples to mitigate in aquatic and coastal regions include controlling outbreaks of zebra mussels (*Dreissena polymorpha*) that clog water supply pipes, slowing the spread of phragmites (*Phragmites australis*) that reduces the diversity of mixed wetland communities, and elimination of the voracious northern snakehead (*Channa argus*) that decimates populations of native fish. Local community volunteers have helped collect data on the extent of invasive species through surveys and remediation activities. In some states, for example Maryland, grade school students have worked hard on service-learning projects to remove nuisance water chestnuts from Chesapeake Bay tributaries such as the Bird River. After laws such as the National Invasive Species Act of 1996, coastal zone managers at the federal, state, and local government levels build inclusive organizations that work together at preventing the introduction of invasive species and coping with their impacts.

Invasive species such as water chestnuts (*Trapa natans*), a submerged aquatic vegetation, may also interfere with commercial, agricultural, or recreational activities. These species are introduced in many ways such as in discharged ballast water, through intentional stocking to control another problem, or as discarded aquarium specimens. The control of invasive species is complex. Methods include controlled burns, flooding, or other natural habitat processes. Some submerged aquatic vegetation becomes so thick that mechanical harvesters are utilized to unclog waterways. Biocontrols such as tiny aquatic insects like milfoil weevils (*Euhrychiopsis lecontei*) have also been used with success to control some forms of submerged aquatic vegetation that have become an overwhelming nuisance.

Marine scientists study factors such as the arrival of exotic species, their establishment, and ultimate integration into the ecosystem. Establishment occurs when new arrivals such as the European green crab (*Carcinus maenas*) can sustain themselves through local reproduction and recruitment. It feeds on many types of organisms, particularly bivalve MOLLUSKS (e.g., clams, oysters, and mussels), polychaetes (marine worms), and small crustaceans. Recent information suggests that European green crabs can outcompete other crustaceans for food and habitat. FISHERMEN commonly use green crabs as bait for tautog (*Tautoga onitis*) fishing at breakwaters and JETTIES along the northeast Atlantic coast. Green crabs are credited with causing significant losses to Manila clam (*Ruditapes philippinarum*) stocks in Washington and California. The Manila clam is a long-established exotic species from Japan and is not immune to foraging by the European green crab.

Invasive species are often able to displace indigenous species because they have no natural enemies. *(NOAA/NMFS)*

A Selected List of Invasive Vegetation and Animals Found in the United States

Name		Description
COMMON	SCIENTIFIC	
Alewife	*Alosa pseudoharengus*	Native to the Atlantic coast, these anadromous fish migrate up rivers and small streams to spawn and then return to sea shortly after spawning. They have been introduced to lakes through stocking and in places such as the Great Lakes have a hard time surviving temperature extremes, causing large Alewife kills.
Asian swamp eel	*Monopterus albus*	Native to Asia, this species has become established in Hawaii, Florida, and Georgia through aquarium releases or escapes from fish farms. This hardy species normally inhabits muddy ponds, swamps and rice fields. It can burrow into moist earth during the dry season to survive long periods without water. Its voracious appetite, generalized feeding habits, and ecological flexibility make it a threat to native species.
Brazilian water-weed	*Egeria densa*	A Brazilian submerged aquatic plant that has escaped from aquariums. This plant has grown uncontrolled in Washington lakes and in California's San Francisco Bay and the Sacramento-San Joaquin Delta.
Common reed	*Phragmites australis*	An aggressive invader of marshes, which displaces species that provide valuable forage for wildlife, especially waterfowl. Phragmites are still used for thatch roofs in places such as Ireland.
Eurasian ruffe	*Gymnocephalus cernuus*	This freshwater fish is common to the Black and Caspian Seas of eastern Europe and was probably introduced into the Great Lakes from ships' ballast waters. It competes with native fish such as yellow perch (*Perca flavescens*) for food and habitat.
Eurasian water-milfoil	*Myrophyllum spicatum*	An African and Eurasian submerged aquatic plant that has escaped from aquariums and spreads by vegetative propagation through stem fragmentation. It forms dense mats on lakes, ponds, and reservoirs throughout the United States.
European green crab	*Carcinus maenas*	This crustacean has made its way from Europe's North and Baltic Seas to the United States in ballast water. This crab poses a threat to mollusk populations if it establishes itself.
Giant reed	*Arundo donax*	Plants native to Mediterranean regions such as North Africa, which thrive along waterways in southern United States. This plant can tolerate saline environments.
Giant Salvinia	*Salvinia molesta*	Native to Brazil, this fast-growing, mat-forming and floating fern is thought to have arisen from water gardening. It thrives in tropical, sub-tropical, and warm temperate areas.
Hydrilla	*Hydrilla verticillata*	A submerged aquatic plant native to Asia, Africa, and Australia that is found throughout the coastal states. Hydrilla form dense canopies that often shade out native vegetation.
Lionfish	*Pterois volitans*	Native to the sub-tropical and tropical regions of the Pacific and Indian Oceans and the Red Sea, lionfish have been found along the coast from Florida to Cape Hatteras, North Carolina. They are a popular saltwater aquarium fish that have venomous spines.
Mute swan	*Cygnus olor*	An exotic waterfowl species from Eurasia that threatens native waterfowl and has a negative impact on submerged aquatic vegetation beds along the East Coast (*see* picture on page 249).

Name		
COMMON	SCIENTIFIC	Description
Northern snakehead	*Channa argus*	Native to Asia and Africa, this obligate air-breathing freshwater fish released into the wild by aquarists and people trying to establish a local food source. Reproducing populations have been found in Florida and Maryland.
Nutria	*Myocastor coypus*	A semi-aquatic rodent native to South America that is found in and around fresh and saltwater ponds and marshes. Nutrias were initially introduced into North America and farmed for their fur. Since their introduction, some animals have escaped these farms and established localized breeding populations from Texas to Virginia and in the Great Lakes area.
Round goby	*Neogobius melanostomus*	A brackish to freshwater fish native to the Black, Azov, and Caspian Sea basins transported and released into the Great Lakes and Mississippi River accidentally from the ballast water of ships traveling from Eurasia.
Sea lamprey	*Petromyzon marinus*	A parasitic anadromous fish common to the Atlantic that may have entered the Great Lakes via the artificially created Erie and Welland Canals.
Veined rapa whelk	*Rapana venosa*	A large predatory Asian mollusk common to the Sea of Japan that feeds on bivalves such as oysters, clams, and mussels. This invasive species was introduced through ships' ballast water and possibly the transport of egg masses with marine farming products.
Water chestnut	*Trapa natans*	Native to Europe, Asia, and Africa, water chestnuts form dense floating mats, severely limiting light, and four sharp spines on the fruits can cause painful wounds. Water chestnuts have infested mid-Atlantic states such as Pennsylvania, Maryland, and Delaware.
Zebra mussel	*Dreissena polymorpha*	Native to the Balkans, Poland, and the former Soviet Union, Zebra mussels have spread through introduction by ballast water of oceangoing commercial vessels and waterway discharge.

The table above highlights some invasive species that are common in the United States.

Further Reading

Florida Integrated Science Center, U.S. Geological Survey. "The Nonindigenous Aquatic Species Program." Available online. URL: http://cars.er.usgs.gov/Nonindigenous_Species/nonindigenous_species.html. Accessed September 21, 2007.

Marine Life Information Network for Britain and Ireland. Available online. URL: http://www.marlin.ac.uk. Accessed June 25, 2007.

National Institute of Invasive Species Science. Available online. URL: http://www.niiss.org/cwis438/websites/niiss/home.php?WebSiteID=1. Accessed September 21, 2007.

National Invasive Species Information Center. Available online. URL: http://www.invasivespeciesinfo.gov. Accessed June 25, 2007.

isobar   An isobar is a line of constant pressure on a scientific chart, such as a weather map. Moving along an isobar, there is no change in barometric pressure. Isobars are displayed on a number of different reference surfaces, such as the surface chart, constant altitude charts, and constant density charts. The most familiar of these is the daily surface weather map, printed in newspapers, magazines, and seen on television and the Web, which shows barometric pressure of a part of Earth's surface. Isobaric charts help meteorologists to visualize the pressure patterns and describe and forecast the weather. For many years, areas of low pressure have been associated with stormy weather, whereas areas of high pressure have been associated with fair weather. Mariners carried barometers aboard ships as early as the 16th century.

Wind directions, especially upper-level winds, are roughly parallel to the isobars. At the surface, friction exerts a drag force on the moving air that makes the surface winds blow at a slight angle to the isobars toward centers of low pressure and away from centers of high pressure. This is how the wind "fills" surface low-pressure areas.

Isobars are a special case of isopleths, which are lines of constant value in an analysis, often presented in map form. Isobars are similar to the contour lines on a topographic map, which show the hills and valleys of terrain. These features are roughly concentric circles on a map, with hills having values increasing toward the center of the circle, and depressions having values decreasing toward the center of the circle. In a similar fashion, an isobaric weather chart shows the "hills" and "depressions" of air, corresponding to high and low pressure, respectively.

Isobars are usually given in units of millibars, where a millibar is equal to 100 Pascals (N/m^2). When the isobars crowd together, there is a "steep pressure gradient" (change of pressure with distance), and a resultant strong wind. When the isobars are far apart, there is only a weak pressure gradient, and the winds are light.

Synoptically gathered, that is, at the same time, barometric pressure data from meteorological stations is used to construct isobaric contours. Expert meteorologists once did the analysis by hand, but now computer programs perform this task, using numerical algorithms for interpolation of data between points of observation. The use of computers greatly speeds up the analysis and FORECASTING of the weather and lets meteorologists focus on forecasting rather than plotting data.

Isobaric analysis is also used in oceanography, where subsurface pressure is determined from analysis of temperature and salinity data for a region. The

units of pressure in oceanography are usually decibars, which are tenths of bars, or 100 times as great as the units of atmospheric pressure. In shallow water, the depth in decibars is approximately equivalent to the depth in meters. Pressure gradients (*see* PRESSURE GRADIENT) are large in regions bounding fast currents, such as the GULF STREAM. The pattern of isobars also reveals the presence of warm and cold core EDDIES, corresponding to high- and low-pressure regions of the atmosphere.

isostatic equilibrium (isostacy) Isostatic equilibrium is the means by which Earth's crust floats on the mantle. The crust, being composed of less dense material than the mantle, is supported by a buoyant force so that it has achieved vertical equilibrium with the downward gravitational force. As stated in Archimedes's principle, a floating body is supported by a force equal to the weight of the fluid it displaces. This is why steel ships can float and how ice cubes float in a glass of water. The larger the volume of the floating ice cube, the greater is its volume under water. This same principle explains how continents float in the soft, plastic material of the upper layer of the mantle, called the "asthenosphere." The asthenosphere is not a true liquid, but it can flow under the ambient conditions of extremely high heat and pressure, which occur deep within the Earth.

Since the crust under the oceans is denser than continental crust, oceanic crust sinks deeply into the mantle. Continental crust under mountain ranges,

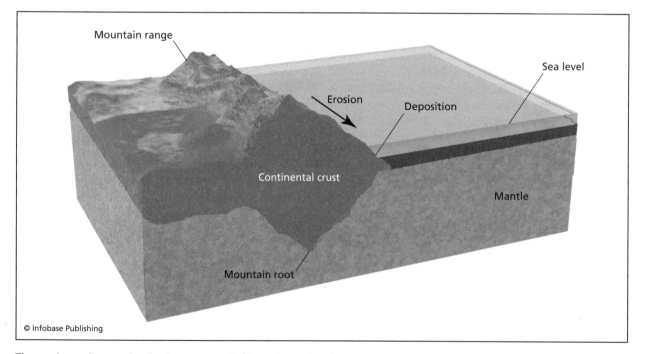

© Infobase Publishing

The continents float on the plastic upper mantle (the asthenosphere) in a manner similar to ice floating in water.

supporting the weight of the mountains, also sinks deeply into the mantle. The crust under the oceans is thin, on the order of three to six miles (5 to 10 km) in thickness, whereas the crust under mountain ranges is thick, on the order of 48 miles (80 km). In other words, a large volume of light continental crust is required to support the mountains. The crust, along with the topmost layer of the mantle to which it adheres (called the "lithosphere"), forms the tectonic plate (*see* PLATE TECTONICS) that move in response to convective forces resulting from the formation of new crust rising up along the mid-ocean ridges and sinking (subducting, *see* SUBDUCTION ZONES) back into the mantle on the opposite sides of the plates.

Isostacy can also be seen in the geologic processes associated with the last Ice Age. During that time, glaciers covered the North American continent south to the latitude of New York. The weight of the glaciers depressed the continents farther down into the asthenosphere. When the glaciers retreated, isostacy caused the continental crust to "rebound" or rise in response to the reduction of the downward weight force.

isotherm An isotherm is a line of constant temperature on a scientific chart, such as a weather map, or a thermodynamic diagram. Moving along an isotherm, there is no change in temperature. Isotherms are displayed on a number of different reference surfaces, such as the surface chart, constant altitude charts, and constant density charts, as well as charts showing the distribution of temperature with height in the atmosphere. The most familiar of these is the isothermal map that often appears along with the daily surface weather map in today's newspaper. Isothermal charts, along with isobaric charts, help meteorologists to visualize temperature and pressure patterns, and thus describe and forecast the weather.

Isotherms are a special case of isopleths, which are lines of constant value in an analysis, often presented in map form. Isotherms are similar to the contour lines on a topographic map, which show the hills and valleys of terrain. Isotherms are not always represented as lines; they are often shown as changes in color or in shades of the same color. This format is frequently used in national newspapers and magazines to enhance appearance and readability. Climatologists often represent average global temperatures by color changes at the isotherms.

Isotherms are usually given in units of degrees Celsius, or degrees Fahrenheit in the United States. When the isotherms crowd together, there is a "steep temperature gradient" (large change of temperature with distance), which often indicates a weather "front," that is, a line of collision between cold and warm air masses. When the isotherms are far apart, there is only a weak temperature gradient, and the presence of a front is not likely.

Synoptically gathered, that is, at the same time, temperature data from meteorological stations is used to construct isothermal charts. Expert meteorologists once did the analysis by hand, but now computer programs perform this task, using numerical algorithms for interpolation of data between points of observation. The use of computers greatly speeds up the analysis and forecasting of the weather.

Isothermal charts are also used by oceanographers, where subsurface temperature is determined from analysis of physical oceanographic data for a region. Temperature gradients are large in regions bounding fast currents, such as the GULF STREAM. Temperature increases dramatically in moving from the U.S. East Coast to the axis of the Gulf Stream, shown on charts of polar-orbiting satellite-obtained temperature data as a crowding of the isotherms. Also often seen on the charts in the boundaries of the Gulf Stream are thermal EDDIES. The circulation in the eddy can be inferred from the temperature. The circulation around warm core eddies in the Northern Hemisphere is clockwise; the circulation around cold core eddies is counterclockwise. Isothermal charts of the sea surface are widely used by the commercial fishing industry to locate fish such as tuna and swordfish.

Isotherms are also often displayed on vertical temperature sections obtained from research vessels equipped with CTDs (*see* CONDUCTIVITY-TEMPERATURE-DEPTH [CTD] PROBE) and/or from lines of moored oceanographic BUOYS equipped with temperature sensors. When the section is taken in tropical or sub-tropical waters to great depths, a major feature on these charts is the crowding of isotherms at the depth of the THERMOCLINE, where temperature falls very rapidly with increasing depth from high values in the tropical surface water, to low values in the cold deep and bottom water.

J

jet stream   The jet stream is an intense core of high speed winds that encircle the Earth at latitudes where strong temperature gradients called semi-permanent *fronts* are found, or regions where strong temperature contrasts exist. The jet stream was discovered during World War II by U.S. Army Air Force *B-29* bomber crews on their way to Japan from Pacific island bases. It is often described as a river of wind flowing high in the atmosphere. It migrates principally as a result of seasonal changes in the Sun's rays or insolation. Insolation is the solar radiation received above the Earth's surface.

The best-known jet stream is the Midlatitude Jet Stream, which occurs above the Polar Front (the intense horizontal temperature contrast between polar air masses to the north, and tropical air masses to the south). The jet stream is found at altitudes of 4.4 to 7.2 miles (4.0 to 12.0 km). Its speeds range from about 120 to 240 miles per hour (200 to 400 km/h), but can occasionally be even higher; its width is 60 to 300 miles (100 to 500 km), and it is several miles (km) in vertical thickness. The jet stream is approximately in geostrophic balance (*see* GEOSTROPHY), which means that the CORIOLIS FORCE is equal and opposite to the pressure gradient force, due to atmospheric pressure differences (*see* PRESSURE GRADIENT), and that the flow is parallel to the ISOBARS, the lines of equal pressure.

In the Northern Hemisphere winter, when temperature gradients are large, there is a second jet stream located around latitude 25°N. During this season, the Midlatitude Jet Stream can drop southward as far as 30°N latitude. In the summer, the Midlatitude Jet Stream remains at roughly lat 50°N latitude.

The jet stream is an important factor in Earth's weather. Severe mid-latitude storms usually follow the most intense temperature gradients, which means that the jet stream is associated with stormy weather. This pattern is found in the United States and Canada. In winter, severe storms can occur as far south as Florida, but in the summer they are generally found only in the northernmost states and in Canada. Upper-level winds are important factors in "steering" major weather systems, such as winter storms and tropical HURRICANES. Thus, knowledge of the positions of the jet stream(s) is important to weather FORECASTING. Mid-latitude winter storm tracks usually follow perturbations in the jet stream. Marine scientists interested in the spawning behavior of pelagic fish such as menhaden (*Brevoortia tyrannus*) are particularly interested in knowing how the jet stream steers storms and determines locations of high and low air pressure. The landmasses contribute to the zonal pattern of semi-permanent high- and low-pressure cells. Some important examples are known as the Siberian and Azores High and the Aleutian and Icelandic Low.

As the jet stream flows, instabilities occur, which result in the formation of large (near-horizontal) waves called ROSSBY WAVES, after Carl-Gustaf Rossby (1898–1957), who was the first to understand them. Rossby waves have long WAVELENGTHS on the order of 2,400 to 3,600 miles (4,000 to 6,000 km). When a Rossby wave develops in the jet stream, cold polar air can be transported to the south, while warm tropical air is transported to the north. As the wave deepens, it becomes unstable and eventually pinches off, leaving cold EDDIES to the south and warm eddies to the north. The formation of these eddies provides one mechanism to distribute the accumulating heat of the Tropics, helping to balance Earth's HEAT BUDGET. Eventually, the eddies decay, the jet stream smoothes out, and the Rossby wave cycle begins all over again.

Further Reading

National Weather Service, Jet Stream—Online School for Weather, National Oceanic and Atmospheric Administration. "The Jet Stream." Available online. URL: http://www.srh.noaa.gov/srh/jetstream/global/jet.htm. Accessed September 24, 2007.

jetty A jetty is a BEACH PROTECTION structure extending into the ocean, lake, or river to influence WAVES, TIDES, and CURRENTS in order to stabilize a channel and protect a harbor. Jetties are designed and built by coastal engineers at the entrance of COASTAL INLETS, harbors, and rivers to direct and confine littoral currents and to prevent silting of the entrance channels. Most jetties are narrow, elongated features made of rock or riprap. Riprap and armor stone are interlocking man-made rocks or large quarry stones used for shoreline protection. Dual jetties at an inlet may extend into the ocean for up to a mile (1.7 km) in order to prevent longshore drift from filling channels and to provide protection for navigation.

Jetties cause the same problems of downdrift erosion as GROINS. In some cases, the rates of erosion and accretion at the jettied channel entrances decrease, and centers of deposition along adjacent coasts move away from the harbor entrances. Some jettied harbor entrances require accumulations of sand deposits to be pumped around the jetty structures to adjacent beaches. Changes in BATHYMETRY caused by jetties also modify the manner in which incident waves break. The wave reflection, convergence, and shoaling near a jetty may generate large waves that peel or break unevenly along the crest. In contrast to spilling waves common to flat and straight shorelines, peeling and plunging surf near jetties becomes a favorite surfing location for surfers. RIP CURRENTS tend to form as jetties divert LONGSHORE CURRENTS offshore, especially during storms or high winds.

Further Reading

Bureau of Beaches and Coastal Systems, Florida Department of Environmental Protection. Available online. URL: http://www.dep.state.fl.us/beaches. Accessed September 24, 2007.
COPRI: The Coasts, Oceans, Ports, and Rivers Institute of the American Society of Civil Engineers. Available online. URL: http://content.coprinstitute.org/index.html. Accessed September 24, 2007.
U.S. Army Corps of Engineers. "CIRP: Coastal Inlets Research Program." Available online. URL: http://cirp.wes.army.mil/cirp/cirp.html. Accessed September 24, 2007.

JIM suit The JIM suit was one of the first commercial atmospheric diving suits using a rebreathing system. They were named after diver Aubrey Jesse "Jim" Jarrett (1905–64), who explored the wreck of the RMS *Lusitania* during October 1935, using a Tritonia diving suit. The passenger liner was sunk off the coast of Ireland by a torpedo from a German U-boat on May 7, 1915. The one-atmosphere Tritonia dive suit, designed by Joseph Salim Peress (1896–1978), was the precursor to the JIM suit. Oceaneering, a company supporting offshore oil exploration, has used the JIM suit with its oil-filled arm and leg joints in harsh North Sea environments. Some of its earliest test dives were in support of oil platforms located in the cold and deep Canadian Arctic. The JIM suit received further recognition during 1985, when

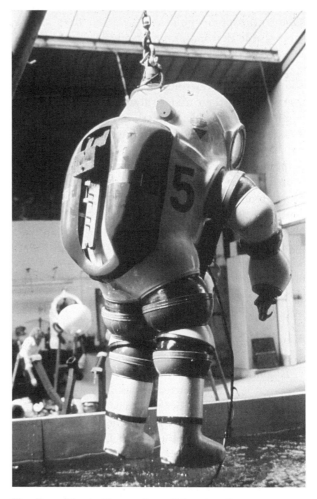

The Jim suit is used in very deep diving operations. (NOAA/NURP)

divers in the JIM suit helped recover a Wellington bomber that was discovered in the 754-foot (230-m) deep Loch Ness in Scotland (the largest body of freshwater in the United Kingdom). Dr. Robert H. Rines and his group from the Academy of Applied Science in Maine have been credited with finding the bomber. This artifact was especially important since this wreck was the second to last of the 11,461 Wellington bombers manufactured during World War II. The other was a trainer on display in a museum. Divers attached and then inflated lifting bags to parts of the aircraft to carefully raise sections to the surface.

Development of the JIM suit can be traced back to British inventor John Lethbridge (1675–1759), who is remembered for his underwater diving engine that was used by the East India Company to salvage cannons and chests of silver from the seafloor. This early diving apparatus consisted of a 6-foot (1.8 m)-long oak cylinder with a window and watertight greased leather gloves, which allowed the diver's arms to manipulate ropes and shackles underwater. By contrast, the JIM suit is an articulated anthropomorphic single-person deep diving submersible that maintains the internal pressure at or very near one atmosphere. Divers wearing the magnesium JIM suit weighed approximately 1,100 pounds (4,893 Newtons) in air and were winched overboard for their underwater work. The JIM suit eliminated the physiological hazards of ambient pressure and the need for decompression. The first JIM suit divers were able to ascend and descend without any consequences or delay while benefiting from eight universal joints and hand pinchers to complete underwater tasks for about three hours at depths of 984 feet (300 m). Later versions or series of the JIM suit allowed divers to descend with improved visibility, joints, and hands to approximately 1,968 feet (600 m). Twenty-first-century high-technology atmospheric diving suits are used for many operations from search and rescue to salvage and underwater construction.

Further Reading

NOAA Dive Page, National Oceanic and Atmospheric Administration. Available online. URL: http http://www.dive.noaa.gov. Accessed June 26, 2007.

NOAA's Undersea Research Program, Office of Oceanic and Atmospheric Research, National Oceanic and Atmospheric Administration. Available online. URL: http://www.nurp.noaa.gov/index.htm. Accessed June 26, 2007.

Joint Global Ocean Flux Study (JGOFS) The Joint Global Ocean Flux Study was an international

program to understand biogenic elements in the ocean, such as carbon. The program began during the late 1980s with a primary focus to survey CO_2 and the bio-optical properties of the sea surface, to develop long-term time series, to observe biogeochemical processes, to describe and predict time-varying fluxes of carbon, and to build a biogeochemical database. It was a multidisciplinary program involving scientists from more than 30 countries.

JGOFS was a multidisciplinary and integrated effort focused on biogeochemistry. Under the supervision of more than 250 principal investigators, approximately 343,000 nautical miles (635,236 km) of ocean were explored culminating in approximately 650 scientific publications. A final conference was held on May 5–8, 2003, in Washington, D.C. The conference title was "A Sea of Change: JGOFS Accomplishments and the Future of Ocean Biogeochemistry."

Researchers showed that CO_2 is one of the major greenhouse gases, which allows energy from the Sun to be trapped in the atmosphere, thereby causing it to heat up and induce the GLOBAL WARMING that many scientists have been trying to monitor. The ocean is now viewed as a reservoir for CO_2, which actually stores up to 50 times more CO_2 than does the atmosphere. Investigators using numerical models reported that the oceans take up at least a third of the anthropogenic carbon dioxide by dissolving it into water that then loses contact with the atmosphere because of sinking or vertical mixing in the water column. For these reasons, JGOFS research thrusts were aimed at answering questions such as:

- What is the background level of CO_2?

- How much CO_2 can the ocean hold?

- How does CO_2 get from the atmosphere into the ocean?

- How long does CO_2 reside in the ocean?

- How does the biology of the ocean affect the CO_2 levels?

- How can we monitor it globally in a cost-effective way?

Further Reading

Integrated Marine Biogeochemistry and Ecosystem Research, Institut Universitaire Européen de la Mer

The Joint Global Ocean Flux Study (JGOFS)

The Joint Global Ocean Flux Study (JGOFS) is a group of international scientists, managers, government officials, and technicians that is laying the groundwork to understand how the carbon dioxide in the atmosphere moves from the atmosphere into the ocean. Understanding the carbon cycle in the environment has taken on great importance in determining the policies of nations and their economies. It is clearly evident from the available data that humankind has been adding measurable amounts of CO_2, carbon dioxide, to the atmosphere since the beginning of the Industrial Revolution. As more nations are becoming major industrial centers, such as India, South Korea, China, and Brazil, more carbon dioxide will be entering the atmosphere.

It is also known that CO_2 is one of the major greenhouse gases that allows more of the energy from the Sun to be trapped in the atmosphere, thereby causing it to heat up. This induces the global warming that many scientists have been trying to monitor. The ocean is a reservoir of carbon dioxide and actually stores up to 50 times more CO_2 than does the atmosphere. Numerical models indicate that the oceans are taking up at least a third of the anthropogenic carbon dioxide by dissolving it into water that then loses contact with the atmosphere because of sinking or vertical mixing in the water column. What is the background level of CO_2? How much CO_2 can the ocean hold? How does it get from the atmosphere into the ocean? How long does it reside in the ocean? How does the biology of the ocean affect the CO_2 levels? How can we monitor it globally in a cost-effective way? These are some of the questions that JGOFS has tried to answer.

The JGOFS emerged under the sponsorship of the Scientific Committee on Oceanic Research (SCOR) and the International Council of Scientific Unions (ICSU). In January 1988, the international planning committee for JGOFS met for the first time and later in the year JGOFS assumed responsibility for the carbon dioxide measurements program. The committee set up a science plan that consisted of five JGOFS program elements.

The following are its program elements:

Time-series   Two time-series stations were started, one at Hawaii and the other at Bermuda. The objective of these time-series is to provide well-sampled seasonal resolution of biogeochemical variability. Data records for more than 10 years now exist at these two sites. There are also permanent moorings to assess these waters that are relatively low in nutrients and cannot support plant life.

Process Studies   The objective of the process studies is to understand the causal effects in ocean models on the oceanic biogeochemical systems. The four major U.S. JGOFS Process Studies were (1) North Atlantic Bloom Experiment, (2) Equatorial Pacific Process Study, (3) Arabian Sea Process Study; and (4) Antarctic Environment and Southern Ocean Process Study.

CO_2 Study   A large-scale survey to provide a composite, basin to global scale, biogeochemical view of the ocean. This includes satellite observations of ocean color that employs the ocean color sensor Sea-viewing Wide Field Sensor (SeaWiFS) to measure surface chlorophyll distribution over the world's oceans.

Synthesis and Modeling   Synthesize knowledge gained from JGOFS and related studies into models that reproduce current understanding of the ocean carbon cycle

Data Management   The JGOFS's Data Management Office manages a data system to provide access to JGOFS data sets comprising information generated from the program's activity elements.

More than 20 countries have participated in reaching these goals, and papers are being published on the results. On May 4, 2003, the JGOFS era officially ended. The groundwork has been laid for the baseline understanding of the global carbon cycle.

—**David L. Porter, Ph.D.**
The Johns Hopkins University
Applied Physics Laboratory
Laurel, Maryland

(IUEM). Available online. URL: http://www.imber.info. Accessed June 27, 2007.

International Geosphere-Biosphere Programme, Royal Swedish Academy of Sciences. Available online. URL: http://www.igbp.net. Accessed June 27, 2007.

International JGOFS. Available online. URL: http://ijgofs. whoi.edu/jgofs.html. Accessed June 27, 2007.

U.S. Joint Global Ocean Flux Study. URL: http://usjgofs. whoi.edu. Available online. Accessed June 27, 2007.

Joint Oceanographic Institutions for Deep Earth Sampling (JOIDES)   Formation of the Joint Oceanographic Institutions for Deep Earth Sampling marked the beginning of the Deep Sea Drilling Project (DSDP), the largest ocean drilling project of all time, and possibly the most significant marine geological project in history. Starting in 1963 as the JOIDES program, it had the objective of investigating the history of the ocean basins of the world. The original partners included LAMONT-DOHERTY EARTH OB-SERVATORY (LDEO) of Columbia University, the Rosenstiel School of Marine and Atmospheric Science at the University of Miami, the SCRIPPS INSTITUTION OF OCEANOGRAPHY, and the WOODS HOLE OCEANOGRAPHIC INSTITUTION. Now this group includes Oregon State University, Texas A&M University, the University of Hawaii, the University of Rhode Island, the University of Texas, and the University of Washington.

The JOIDES program expanded into the DSDP when a specialized research drilling ship capable of drilling into the ocean floor thousands of feet (hundreds of m) below the sea surface, the *Glomar Challenger,* was acquired by the program. Most of the theory and observations confirming the validity of PLATE TECTONICS theory were developed and implemented as part of this program, which included a series of two-month cruises that produced vast quantities of data on the ocean basins, such as BATHYMETRY, structure, and composition. The *Glomar Challenger* has been replaced by a newer vessel of even higher capability, the *JOIDES Resolution.*

The oceanographic institutions of the United States were joined in this program by institutions and vessels from Canada, France, Germany, Japan, Russia (the former Soviet Union), and United Kingdom.

Further Reading

Davis, Richard A., Jr. *Oceanography: An Introduction to the Marine Environment,* 2nd ed. Dubuque, Iowa: Wm. C. Brown, 1991.
Integrated Ocean Drilling Program. Available online. URL: http://www.iodp.org/home. Accessed June 26, 2007.
Joint Oceanographic Institutions. Available online. URL: http://www.joiscience.org/joi. Accessed June 26, 2007.

K

katabatic winds Katabatic winds are caused by a layer of cold air forming near the ground over elevated, sloping terrain through radiative cooling. Typically, these winds form on a clear night with a negligible geostrophic wind (caused by atmospheric pressure gradients; *see* GEOSTROPHY). Thus, the air close to the ground over elevated terrain can become colder than air at the same atmospheric level that is not near the ground due to terrain elevation differences. The result is cold, dense air near the ground over elevated terrain, flowing downhill due to gravitational forces. Katabatic winds are most pronounced and cover the largest areas in Greenland and Antarctica.

Antarctica experiences great extremes in wind speeds. Temperature inversions in the atmosphere over the high interior can cause some of the strongest winds. Inversions serve to trap the coldest, most dense air close to the ground. Gravity pulls the cold dense air downhill from the interior of the continent to the coastal regions as if the cold air were a liquid. The interior winds are called "inversion winds." The CORIOLIS EFFECT then diverts the winds to the west, creating coastal easterly winds. Thus, the polar plateau is a persistent source of cold, dense air and strong winds along the margins of the continent. In the interior of Antarctica, surface winds flow over mostly gentle slopes in the terrain. In some areas, the terrain funnels the winds in accordance with the Bernoulli effect, well known in fluid dynamics, causing convergence in the airflow and therefore causing an increase in wind velocity. Daniel Bernoulli (1700–82) described the behavior of fluids under varying conditions of flow and height. These intensified winds, once they reach the coast, are considered katabatic winds. Thus, Antarctic katabatic winds begin as inversion winds.

Like inversion winds, gravity drives them downhill, but steeper slopes and narrow valleys serve to significantly intensify katabatic winds. Katabatic winds can accelerate from near zero to as much as 50 to 66 feet per second (15–20 m/s) very quickly. There are two types of katabatic winds, ordinary and extraordinary. Ordinary katabatic winds have a constant direction but a highly variable speed. They form from inversion cooling in a small drainage basin. Thus, the katabatic wind event first depends on the drainage basin "filling up" with cold, dense air. After an event, it takes some time to "refill" the basin which, in turn, leads to another katabatic wind event. Extraordinary katabatic winds can persist for days or weeks. They usually occur between Cape Denison and Port Martin. Convergent katabatic flow from the East Antarctic Ice Sheet makes the Cape Denison-Commonwealth Bay region of Adelie Land the windiest spot on Earth. The mean annual wind speed is 50 miles per hour (80 km/hr), and maximum measured wind velocities exceed 200 miles per hour (330 km/hr).

See also FLUID MECHANICS.

Further Reading
Center for Ocean–Atmospheric Prediction Studies Library. "Katabatic Winds Bibliography." Available online. URL: http://www.coaps.fsu.edu/lib/biblio/katabatic-wind.html. Accessed September 24, 2007.
National Snow and Ice Data Center. Available online. URL: http://nsidc.org. Accessed June 27, 2007.

kelp The members of the laminariales, collectively known as the "kelps," include the largest and structurally most complex of all the algae. The laminariales make up an order of the phylum, Phaeophycophyta, commonly known as the brown ALGAE or dusky seaweeds. Their color is attributed to an accessory pigment called fucoxanthin, which masks the presence of the green pigment chlorophyll. The sporophyte (meiospore-producing generation) of the kelp is typically constructed of three parts: the blade (lamina), the stalk (stipe), and the holdfast (hapteron).

In sexual reproduction, the diploid sporophytes become fertile and develop unilocular sporangia, within which meiosis occurs, producing meiospores (zoospores). After successive mitotic divisions, 32–64 haploid, laterally biflagellate zoospores are released that grow into microscopic filamentous male or female gametophytes. Laterally biflagellate sperm, produced by the male gametophytes, fertilize eggs produced by the female gametophytes. The zygote grows into a new sporophyte, which eventually develops the macroscopic form characteristic of the species.

As part of the marine flora, the kelp provides food, shelter, spawning areas, and a substrate for the attachment of numerous marine animals. In many places along the coast, certain species belonging to the order laminariales, form dense growths known as "kelp forests." Larger forests grow in the waters colder than 68°F (20°C). The forests range from beds a few feet (m) in depth, as in the Atlantic laminaria beds of Nova Scotia, to the giant plants of the Pacific Coast that grow up from holdfasts as deep as 100–150 feet (30–45 m) below the surface. In the latter habitat, one finds species belonging to the genera, *macrocystis* (the giant kelp), *nereocystis* (bull whip kelp), and *pelagophycus* (the elk kelp).

Many commercial applications have been found for kelp. The algin and alginate industry is based on the extraction of alginic acid from the larger brown seaweeds and some of the smaller ones. The most common form of algin is a soluble sodium salt of alginic acid. The American Kelco Co. on the Pacific Coast is the world's major producer. On the U.S. Pacific coast, *macrocystis,* and to a lesser extent, *nereocystis* and *pelagophycus* form the raw material for the algin industry. Algin's water-holding and gelling properties make it useful as an emulsifier, thickener, and smoother. It is found in numerous products, including baked goods, syrups, puddings, salad dressings, beer, milkshakes, paper coatings, and explosives. Kelps are also used as fertilizer (potash sources) and may be a possible energy source in the future. The kelps can be converted to methanol, which can then be converted economically to gasoline.

Further Reading

Boney, Arthur D. "Aspects of the Biology of the Seaweeds of Economic Importance." *Advances in Marine Biology* 3 (1965): 105–253.
Chapman, Valentine J., and David J. Chapman. *Seaweeds and Their Uses.* London: Chapman and Hall, 1980.
Hillson, Charles J. *Seaweeds.* University Park, Pa.: The Pennsylvania University Press, 1977.

Kelvin waves Kelvin waves are long-period, low-frequency shallow water (WAVELENGTH 25 or more times water depth) gravity WAVES that are trapped by vertical boundaries, such as COASTLINES, and propagate along those coastlines counterclockwise in the Northern Hemisphere and clockwise in the Southern Hemisphere. Kelvin waves are sufficiently long so that they are influenced by the CORIOLIS EFFECT, which causes the surface of the wave to slope exponentially downward away from the boundary. The slope results in the waves being in geostrophic balance (*see* GEOSTROPHY) in the direction perpendicular to their directions of propagation. The extent of the slope (the e-folding distance) is given by the ratio of the phase speed of the wave v to the Coriolis effect f, or v/f, where $f = 2\Omega\sin\varphi$, Ω being Earth's angular velocity, and φ being the geographic latitude. The ratio v/f is known as the Rossby radius of deformation, a significant parameter in many types of atmospheric and oceanic wave motions. The shallow-water wave phase speed v is given by the shallow-water wave equation (*see* WAVES),

$$v = \sqrt{gH}$$

where g is the acceleration of gravity, and H is the water depth. Kelvin waves can be from tens to hundreds of miles (km) long, and can occur both as surface waves, or internal waves propagating along the THERMOCLINE, the region of rapid temperature change between the surface mixed layer and cold, deep water below. When one swirls a mug of liquid such as tea or coffee in a circular motion to the left or right, a trapped wave develops that is analogous to a Kelvin wave. These long-period waves are named after British mathematician and physicist Lord William Thomas Kelvin (1824–1907).

The following equation describes the height Y of a propagating Kelvin wave along a boundary:

$$Y = Y_0 \exp(-x/R_d)$$

where R_d is the Rossby radius of deformation, equal to $(\sqrt{gH})/f$, Y_0 is the wave amplitude at the boundary, and x is the distance along the boundary from some reference point. (Interested readers may consult a mathematics text to learn more about exponential functions of the form e^x, where e is the constant 2.718.)

Examples of Kelvin waves in the ocean include the deep ocean TIDES, such as those in the North

Atlantic Ocean, in which the wave crests propagate counterclockwise around the North Atlantic basin, and slope downward from the coastlines to a low point called the AMPHIDROMIC POINT in the central basin around which the waves rotate. The disturbances initiating the EL NIÑO conditions in the eastern Pacific Ocean propagate across the Pacific along the equator as internal Kelvin waves. In this case, the equator serves as a boundary because the algebraic sign of the Coriolis parameter f is positive in the Northern Hemisphere, and negative in the Southern Hemisphere, trapping a double Kelvin wave that slopes away from the boundary (the equator) in both hemispheres. When the equatorial waves strike the boundary, the coastline of South America, they split into two Kelvin waves, one traveling northward along the coasts of Peru and Ecuador and the other traveling southward along the coastlines of Peru and Chile. The rapid changes in the upper INDIAN OCEAN in response to the shifting MONSOON winds can be explained by equatorial Kelvin waves set up by the changing pressure and wind stress. Kelvin waves are thus seen as a mechanism for the rapid transfer of upper ocean conditions from one location to another on time scales of days to weeks and are an important link in the transfer of energy from the atmosphere to the ocean and from the ocean to the atmosphere.

Theoretical properties of Kelvin waves are related to Poincaré waves, which are shallow water waves that have varying velocities in their cross sections and have several modes of oscillation. Kelvin waves are impacted by boundaries such as the coast or the mug as described in the earlier analogy. Poincaré waves have water particle orbits that are more elliptical. These types of inertia gravity wave (restoring forces are inertia and gravity) are named after French mathematician Henri Poincaré (1854–1912).

Further Reading

Colling, Angela, ed. *Ocean Circulation,* 2nd ed. Milton Keynes, U.K.: Butterworth Heinemann, 2001.

Knauss, John A. *Introduction to Physical Oceanography,* 2nd ed. Upper Saddle River, N.J.: Pearson/Prentice Hall, 1996.

knot   A knot is a unit of speed that has been traditionally used for maritime and aviation purposes. One knot is equal to one international nautical mile per hour (6,076 feet per hour or 1,852 mph). The international nautical mile is a unit of distance equal to the length of an angle of one minute of any great circle of the Earth, usually the meridian. These values were standardized in 1929 by the International Hydrographic Organization, then adopted in 1954 by the U.S. Departments of Commerce and Defense.

The knot is permissible to be used with the International System of Units.

The term *knot* originated from the log line used to determine the speeds of sailing ships. A log was attached to a line with a number of equally spaced knots and thrown overboard. The number of knots let out in a specific time interval gave the speed in knots (*see* TIDE POLE). The time interval was standardized at 28 seconds, and the distance between the knots at 47.25 feet (14.397 m).

The knot is approximately equal to 1.6878 feet / second (.5144 m/s). Since the nautical mile is slightly larger than the terrestrial mile, by a factor of about 1.15, the knot is correspondingly larger than the terrestrial mile per hour, that is, a knot is approximately 1.15 terrestrial miles per hour. In addition to being the traditional marine unit of speed, aviators and weather forecasters often give wind speeds in knots.

Further Reading

Taylor, Barry N. "A Guide for the Use of International System of Units (SI)." NIST Special Publication 811, 1995 edition. National Institute of Standards and Technology. Available online. URL: http://www.physics.nist. gov/cuu/pdf/sp811.pdf. Accessed June 27, 2007.

krill   Krill are small pelagic shrimplike invertebrates known as euphausiids that are an important source of food for many marine species. There are approximately 85 species of krill, and they are generally small in size, ranging from less then 0.5 inch (1 cm) to six inches (15 cm) long. These ZOOPLANKTON tend to spend days at depths of approximately 320 feet (100 m) and swim toward the surface at night to feed on phytoplankton.

See also MARINE CRUSTACEANS.

Kurile Current   *See* OYASHIO CURRENT.

Kuroshio Current   Also known as the Black Current or the Japan Current, the Kuroshio Current is one of the world's most significant permanent currents. It is a western boundary current of the North Pacific Ocean, which flows past the coast of Japan, and by means of the Kuroshio Extension, runs into the eastward-flowing North Pacific Current. The North Pacific Current (also known as North Pacific Drift) is reinforced along its northern boundary by cold waters from the Bering Sea transported southward by the OYASHIO CURRENT. The name Kuroshio means "black stream" in Japanese, referring to the dark color of its waters. The Kuroshio is defined to begin at the western terminus of the Pacific North Equatorial Current east of the Philippines, and then

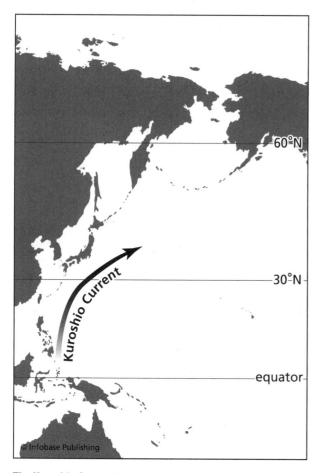

The Kuroshio Current is a warm current that flows parallel to the eastern coast of Japan. It is the western branch of the North Pacific gyre and is similar in many ways to the Gulf Stream.

clonic gyre of the North Pacific. The other branches are the eastward-flowing North Pacific Current, the southward-flowing CALIFORNIA CURRENT, and the westward-flowing Pacific North Equatorial Current, which flows back into the Kuroshio at its source region, thus completing the GYRE. As the eastern North Pacific Current flows into the southward-flowing California Current, a branch splits off to flow northward along the North American continent into the Gulf of Alaska, called the Alaska Current. The currents rotate about the Gulf of Alaska, forming a secondary gyre to the great North Pacific gyre.

In a manner similar to the Gulf Stream, the Kuroshio exerts a significant influence on the climate of the North Pacific region, bringing warm tropical water northward to warm otherwise subpolar regions. The Kuroshio Current flows on the Pacific side of Japan and warms areas as far north as Tokyo. A branch of the Kuroshio known as the Tsushima Current enters the Sea of Japan through the Korea Strait.

Although variability in the Kuroshio Current is not so well known as that of the Gulf Stream, it has been found to have significant seasonal variations, being strongest in summer and weakest in spring and fall, with strengthening by the southwest Asian monsoon, and weakening with the northeast winds predominant in other seasons. Strong winds have been observed to accelerate the current in speed but not in direction. The Kuroshio has one feature that sets it apart from the Gulf Stream. Although both are characterized by meanders and eddies, the Kuroshio has a large meander just south of the subsurface Izu-Ogasa Ridge with deep passes at around 3,280 feet (1,000 m) depth. This meander can persist for several years. Understanding its cause and variability is an active area of research by physical oceanographers.

Further Reading

Goddard Earth Sciences Data and Information Services Center, National Aeronautical and Space Administration. Available online. URL: http://disc.sci.gsfc.nasa. gov. Accessed September 24, 2007.

Japan Agency for Marine-Earth Science and Technology. Available online. URL: http://www.jamstec.go.jp/e/. Accessed June 27, 2007.

National Marine Fisheries Service, National Oceanic and Atmospheric Administration. "LME #49: Kuroshio Current. Large Marine Ecosystems of the World." Available online. URL: http://www.edc.uri.edu/lme/text/kuroshio-current.htm. Accessed September 24, 2007.

runs northward past Taiwan to the Japanese islands. The Kuroshio splits at Yakushima, with a western branch flowing northward through the Korea Strait, the Tsushima Current, and a more powerful eastern branch flowing through Tokara Kaikyo and then along the coast of Shikoku.

The Kuroshio is the North Pacific counterpart of the GULF STREAM, transporting warm tropical water northward to high latitudes. Like the Gulf Stream, it is fast, deep, and narrow. Speeds as high as four knots (2 m/s) occur. The width of the current east of Japan is perhaps 60 nautical miles (110 km); the depth of the current is as deep as 13,000 feet (4,000 m). Volume transports of the Kuroshio are on the order of 40 Sv (40×10^6 m³/s), and of the Kuroshio Extension on the order of 65 Sv (65×10^6 m³/s).

Like the Gulf Stream in the Atlantic, the Kuroshio is the western branch of the great anticy-

Labrador Current The Labrador Current is a cold, low-SALINITY current flowing southeastward along the Labrador Coast, where some of the cold water enters the Gulf of St. Lawrence and some becomes the Nova Scotia Current. The Nova Scotia Current passes around the southern tip of Nova Scotia, where it enters the Gulf of Maine. The Labrador Current originates in the cold arctic water flowing southeast through the Davis Strait, and from the warmer waters of the West Greenland Current that retroflects (loops around) in the northern part of the Labrador Sea. The West Greenland Current retroflection joins the Labrador Current off the northern coast of Newfoundland, where it is augmented by freshwater outflow from the fjords of the Labrador Coast.

The speeds found in the Labrador Current average around 0.5 knots (0.25 m/s) but can be as high as 1.5–2.0 knots (0.75–1.0 m/s). There is uncertainty in the seasonal variations in current speed and direction. This current is noted for transporting ICEBERGS southward from the west coast of Greenland into the northwestern Atlantic. It was one of these icebergs that caused the demise of the RMS *Titanic* on April 14, 1912. Today, the International Ice Patrol, the U.S. and Canadian Coast Guards, and ice centers such as the U.S. NATIONAL ICE CENTER collaborate to warn mariners of ice conditions. Arctic ice BUOYS transmit data to meteorologists and oceanographers working at forecast offices in the United States, Canada, and Europe.

The Labrador Sea, just to the east of the Labrador Current, is an important source of intermediate and deep water for the ATLANTIC OCEAN. During the winter, cold, dry arctic air cools the surface water and causes it to freeze, rendering out freshwater and increasing the salinity and DENSITY. This combination of low TEMPERATURE and higher salinity results in the sinking of the cold, dense surface

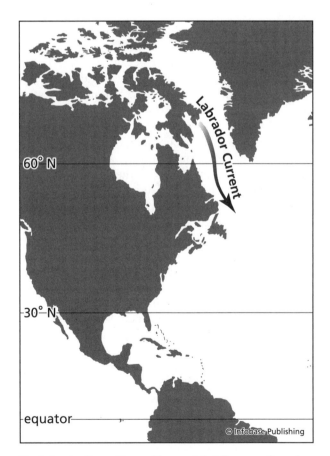

The Labrador Current is a cold current that flows southward parallel to the northeastern coast of Canada.

water to great depths, spreading south and east. South of Greenland, this water, called western North Atlantic Deep Water, combines with eastern North Atlantic Deep Water from the Norwegian and Greenland Seas (*see* WATER MASS). Large variations in winter weather conditions result in large variations in the amounts of deep water formed. Climatologists are concerned that GLOBAL WARMING (*see* CLIMATOLOGY), resulting in the melting of glaciers and SEA ICE, could lower the surface salinity and consequently the density of the surface water sufficiently to prevent it from sinking. This change in surface salinity could play a major role in altering the climate of the North Atlantic region, since oceanographers believe that the cold, sinking Labrador Sea water is one of the driving forces of the GULF STREAM. Should this sinking not occur, the stream could be diverted south of its present position, resulting in much harsher winters for Europe and northeastern North America.

The Labrador Current has two branches that flow parallel to each other, one in the shallow coastal water of the CONTINENTAL SHELF, the other in the deep water of the CONTINENTAL SLOPE. Oceanographers have estimated the volume transport of these components to be around three Sverdrups (Sv) (3×10^6 m³/s) for the shelf component, and 16 Sv (3×10^6 m³/s) for the slope component. The Sverdrup is a unit of measure named after oceanographer Harald U. Sverdrup (1888–1957) and is equivalent to 264 million U.S. gallons per second. Volume transports computed from direct measurements of current from moored BUOYS are larger than corresponding transports computed from temperature and salinity data using the geostrophic method, which gives the density-driven component of the circulation (*see* GEOSTROPHY). This result suggests that externally imposed pressure fields, such as that of the wind stress, are significant contributors to the flow of the Labrador Current.

As it continues flowing southward, the current converges with the Gulf Stream system off Newfoundland in the Grand Banks region and adds its cold water to the warm EDDIES of the stream. The boundary between the warm and cold waters is called the "cold wall." The Gulf Stream system continues east of the Grand Banks, where it is known as the North Atlantic Current. The Grand Banks is a region of high biological productivity and has been one of the world's great fisheries for hundreds of years.

The infamous fogs of the Grand Banks are created when the cold waters of the Labrador Current meet the warm waters of the Gulf Stream and the water vapor in the cold air is rapidly evaporated.

Water color changes dramatically from the dull green of the Labrador Current to the cobalt blue of the Gulf Stream. The Labrador Current bifurcates in this region, with part flowing eastward along the northern edge of the North Atlantic Current, and part flowing southwestward around the southeastern tip of Newfoundland.

This latter current is known as the Labrador Current Extension. It flows inshore of the Gulf Stream and transports cold arctic water from the Labrador Current to the northeast coast of the United States. It flows southwestward over the continental shelf around Cape Cod as far south as the region north of Cape Hatteras. Here the current retroflects and flows northward along the western boundary of the Gulf Stream, forming a counterclockwise GYRE of coastal water between Newfoundland and Cape Hatteras. Its average speed is about 0.6 knots (.3 m/s).

Further Reading

Bowditch, Nathaniel. *The American Practical Navigator: An Epitome of Navigation.* Bethesda, Md.: Defense Mapping Agency Hydrographic/Topographic Center, 1995.

Rosenstiel School of Marine and Atmospheric Science, University of Miami. "Ocean Surface Currents." Available online. URL: http://oceancurrents.rsmas.miami.edu/index.html. Accessed September 25, 2007.

lagoon   A lagoon is a shallow body of water separated from the ocean by a sandspit or a BARRIER ISLAND, generally along land margins or surrounded by a CORAL REEF. It contains salt or brackish water and has few if any inflows from rivers and estuaries. Coral reef lagoons occur only in waters warm enough for coral to thrive. The connection with the ocean is made by means of one or more inlets.

Coastal lagoons are formed as part of the barrier island process, with sand and or other SEDIMENTS being transported inland to fill the embayment. Coastal lagoons are found on the southeast Atlantic and Gulf Coasts of the United States. The water temperature of the lagoon is usually warmer than the adjacent ocean in the summer and colder in the winter. SALINITY variations in lagoons are often quite large, ranging from below that of coastal ocean water during times of heavy rain, to far above the salinity of coastal ocean water when dry conditions prevail. Laguna Madre, Texas, is an example of a coastal lagoon that becomes hypersaline, having up to 100 practical salinity units (psu) during the dry season, but becomes highly diluted compared with the adjacent ocean water during rainy seasons. The tendency of such lagoons to be deficient in oxygen is, in this case, offset by the abundance of plants growing on the bottom. Laguna Madre supports a large population of benthic (bottom-dwelling) organisms and fish. Pamlico Sound, in coastal North Carolina, supports a large nursery for Atlantic fish, such as mullet and red drum. The health of coastal lagoon ECOSYSTEMS is vulnerable to major atmospheric events, such as droughts and severe storms, as well as pollution from human industrial and agricultural activities.

Coral lagoons are found in abundance in the Pacific Ocean, particularly within ATOLLS, which formed as coral reefs develop. They are generally circular in shape. These lagoons often provide homes for many species of fish and shellfish, which provide a good source of food for native populations and visiting mariners. Many coral lagoons are threatened by destruction of coral due to GLOBAL WARMING, as well as human activities such as blasting reefs to harvest fish and reef-supported animals.

Further Reading

Center for Coastal Studies, Scripps Institution of Oceanography, official Web page of the University of California, San Diego. Available online. URL: http://www-ccs.ucsd.edu. Accessed September 25, 2007.

Virginia Coastal Reserve, U.S. Long-Term Ecological Research Network. Available online. URL: http://www.vcrlter.virginia.edu/. Accessed June 27, 2007.

Lagrangian current meters   *See* CURRENT METERS.

laminar flow   Laminar flow refers to the motion of a viscous fluid (one with internal friction) in which particles of the fluid flow in parallel layers (sheets called lamiae), each of which has a constant velocity but is in motion with respect to its adjacent layers, so that the layers slide over one another. An extreme example of laminar flow is seen in the flow of heavy motor oil. The flow of a viscous fluid in a circular pipe is a type of flow that has been extensively studied by mechanical engineers.

In the ocean, a region of laminar flow occurs near the bottom for shallow-water GRAVITY WAVES. Very strong SHEARS exist at the surface and bottom boundaries of this flow. As a general rule in both the atmosphere and the ocean, laminar flows may be found close to boundaries. The motion of the air and sea is, for the most part, turbulent rather than laminar, which means that complex and usually turbulent (*see* TURBULENCE) multidirectional EDDIES or whirls are superimposed on the average flow direction.

If laminar flow is found adjacent to a fluid boundary such as a stream bed, it is known as a viscous sublayer and is often no more than a few milli-

meters thick. The flow over smooth, fine mud and sand is often laminar and is described as hydrodynamically smooth flow. Flow over bumpy sand with large pebbles and rocks, or over rocky bottoms, is called hydrodynamically rough flow, and it generally results in the development of turbulence. Meteorologists and oceanographers have performed a considerable amount of research on the study of current and wind velocity profiles near solid and liquid boundaries.

See also FLUID DYNAMICS; OCEANOGRAPHY.

Further Reading

Wright, John, Angela Colling, and Dave Park. *Waves, Tides and Shallow-Water Processes,* 2nd ed. Milton Keynes, U.K. Oxford: Butterworth-Heinemann, 2001.

Lamont-Doherty Earth Observatory (LDEO)   A leading research institution for the Earth sciences located about 15 miles (24.1 km) north of New York City, on the west bank of the Hudson River, LDEO employs about 200 research scientists and numerous student interns.

LDEO was established in 1949 as a gift to Columbia University from Florence Corliss Lamont (1873–1952), an alumna, following the death of her husband, Thomas W. Lamont. Columbia decided to use the Lamont Estate as a center for geological research. Following a major contribution in 1969 by the Henry L. and Grace Doherty Charitable Foundation (a major supporter of oceanographic institutions), the observatory was renamed the Lamont-Doherty Geological Observatory. Its present name, Lamont-Doherty Earth Observatory was acquired in 1993 in view of the increasing scope and breadth of its scientific observations.

LDEO began its scientific work with a world-ranging research schooner, the *Vema,* which obtained detailed data on the geology of the world's ocean basins and the circulation of its oceans. The present primary research vessel is the 239-foot (72.9 m) *Maurice Ewing.* Typical missions for this vessel include seismic imaging, deep coring, and acquiring data from deep ocean boreholes. The observatory maintains seismic networks and keeps seismometers in readiness for rapid deployment anywhere on Earth. LDEO has collected and maintains the world's largest collection of deep-sea and SEDIMENT cores, consisting of more than 13,000 samples.

Available equipment for visiting and resident scientists includes all types of state-of-the-art instrumentation for rock, soil, and water sampling and analysis. Specific instrumentation includes mass spectrometers, scanning electron microscopes, scanning tunneling microscopes, and X-ray diffraction equipment. The massive amount of data collected is analyzed on resident supercomputers and computer networks.

LDEO is perhaps best known worldwide for its contributions to the discoveries of mid-ocean ridges, subduction zones, submarine volcanic zones, and the ensuing mapping of these features by renowned geologists Bruce Heezen (1924–77) and Marie Tharp (1920–2006). LDEO has been a major center for research in the development and application of the theory of PLATE TECTONICS in the latter years of the 20th century and the early years of the 21st century.

Further Reading

Lamont-Doherty Earth Observatory (LDEO), Earth Institute at Columbia University. Available online. URL: www.ldeo.columbia.edu. Accessed September 25, 2007.

land and sea breeze   Land and sea breeze refers to a coastal atmospheric circulation that usually begins

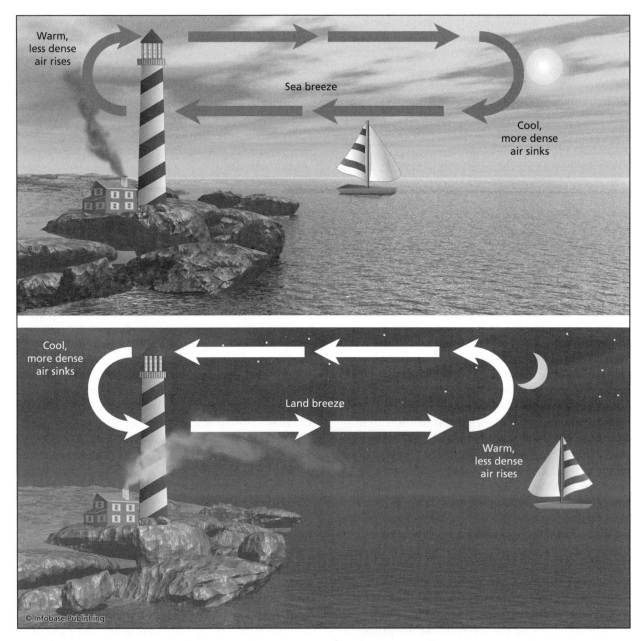

Local breezes are generated by unequal heating of the Earth's surface due to local conditions. In the daytime, the land gets much hotter than the water—hot air rises over land, drawing cooler air in from over the water. At nighttime, the land cools off much more rapidly than the water—warmer air over the water rises, drawing off cooler air from over the land.

on warm, sunny summer days with the coastal land heating faster than the ocean (water has a higher specific heat than soil), causing the air overlying the land to rise, creating a low-pressure area. The pressure over the water is then higher, and a horizontal PRESSURE GRADIENT exists between the sea and the coast. A wind develops in response to the pressure gradient that is known as the "sea breeze." As the heated air over the land continues to rise, the pressure aloft is increased and becomes larger than that at the corresponding altitude over the sea, resulting in a horizontal land-to-sea pressure gradient and a consequent seaward flow aloft that cools and sinks over the ocean to close the cycle. The boundary between the maritime air and the warmed air over the land is known as the "sea-breeze front." Over lakes, it is called the "lake-breeze front."

At night, the land cools off faster than the sea (due to the lower specific heat of soil compared to water), and so the flow reverses, producing a "land breeze" from land to sea. The land breeze is usually not so strong as the sea breeze.

Large, deep lakes experience "lake breezes," which are similar to land and sea breezes. The deeper the lake, the more heat energy it can store, and so the lake breezes are more strongly developed over deep lakes than over shallow lakes. The Great Lakes, for example, are noted for their lake breezes that cause a welcome cooling of the surrounding lands in the summertime.

Meteorologists study land-and-sea breeze circulations to determine their effects on other weather phenomena such as thunderstorms. They use MODELS and weather stations to locate the land-and-sea breeze fronts. There are at present numerous investigations on the manner in which the land-and-sea breeze fronts transport pollutants.

Further Reading

Hsu, Shih-Ang. *Coastal Meteorology.* San Diego, Calif.: Academic Press, 1988

La Niña A condition of colder than normal surface and near surface water in the equatorial central and eastern Pacific Ocean is classified as La Niña. UPWELLING along the coasts of Peru and Ecuador is enhanced, as well as the atmospheric high pressure cell that generally prevails over the eastern Pacific. Dry, in some cases near-desert conditions prevail along the coast of Peru. The upwelling of nutrients enhances the productivity of the PERU CURRENT, building up the anchovetta fishery and providing food for multitudes of seabirds whose guano yields the raw material for Peru's fertilizer industry, a major source of income for the country.

La Niña (Spanish for "the girl child") is the opposite of the EL NIÑO (Spanish for "the boy child") condition, in which warm water moves from the western to the eastern Pacific Ocean and cuts off the upwelling that usually prevails along the Peruvian coast in the region of the Peru Current. Both El Niño and La Niña are linked to the Southern Oscillation (SO), in which the normally high atmospheric pressure over the central and eastern Pacific lowers, while the normally low atmospheric pressure over the western Pacific rises. The changing pattern of air pressure in the equatorial Pacific results in a decrease in the east-to-west PRESSURE GRADIENT, and an ensuing weakening or even a secession of the westward-blowing trade winds. When the westward-directed wind stress diminishes, or ceases altogether, the huge pool of warm water in the western Pacific northeast of Australia and Indonesia moves eastward into the central and eastern Pacific, cuts off the upwelling in the Peru Current, and an El Niño event begins. When the El Niño event is over, the decreasing temperatures of the ocean CURRENTS of the region often drop below normal and into a cold state, which is known as La Niña. The trade winds return to their normal pattern, but with greater intensity. The combined air-sea phenomenon is known as the ENSO (El Niño Southern Oscillation). La Niña events frequently, but not always, follow El Niño events. They sometimes occur independently.

Possible consequences of La Niña events now being studied by climate scientists include dry conditions along the California coast, greater precipitation in the Pacific Northwest, and possibly a greater number of Atlantic HURRICANES. In general, they are the opposite conditions of those believed to be related to the El Niño events. However, conditions vary greatly for each succeeding La Niña. La Niña's effects on the atmosphere are therefore expected to differ somewhat for each occurrence.

Further Reading

The Green Lane, Environment Canada. "What Is La Niña?" Available online. URL: http://www.msc-smc.ec.gc.ca/education/lanina/what_is/index_e.cfm. Accessed September 25, 2007.

NOAA La Niña Page, National Oceanic and Atmospheric Administration. Available online. URL: http://www.elnino.noaa.gov/lanina.html. Accessed June 27, 2007.

laser A laser is a device that produces a nearly parallel, nearly monochromatic beam of light by exciting atoms to higher energy levels and causing them to radiate their energy as photons of light coherently, that is, in phase. Laser is an acronym for *light a*mplification by *s*timulated *e*mission of *r*adiation. The study of the interaction of light or radiation with matter is called spectroscopy. Lasers and spectrometers are especially important in remote sensing for the identification of substances through the spectrum of emitted or absorbed radiation.

There are two general types of lasers: pulsed lasers, such as the ruby laser, which are energized by pulses of electrical energy and transmit discreet pulses of laser light at the excitation frequency; and continuous lasers, such as the helium-neon laser, in which the body of the laser contains helium and neon gases that are stimulated by an input of electrons that emit a continuous beam. Pulsed lasers can produce bursts of energy as short as nanoseconds with power levels as high as 10^9 watts. Liquid lasers can be tuned over a range of wavelengths. Most lasers emit light of a single frequency, which may be red, green, blue, or yellow, depending on the application. Gas lasers are now available that transmit in the near infrared and in the ultraviolet spectral bands.

The principle of operation is that an intense electric current or auxiliary radiation source bombards the atoms in the laser tube, "pumping up" their energy levels. The pumped-up atoms then decay not to their ground states but to their normal excited states, providing more excited atoms than ground state atoms, a condition known as "population inversion." The atoms in their excited state emit photons, which stimulate other excited atoms to emit photons of the same energy and wavelength. By means of totally and partially silvered mirrors at each end of the laser tube, STANDING WAVES are set up, and an intense beam of coherent radiation emerges from the partially silvered end of the tube. During each reflection through the tube, more and more atoms are stimulated to emit photons so that the emitted light rapidly builds to a high intensity.

In marine science, lasers are often used as the source of radiation in remote sensing systems such as LIDAR. Laser ALTIMETERS are pulsed lasers that make possible the very precise determination of range by measuring the travel times of the laser pulses to and from the target back to the receiver. They have been used to measure the precise distance from the surface of Earth to a reflector placed by the Apollo astronauts on the Moon. They are used in Earth satellites to map the topography of the land and ocean surfaces, to measure wave height, movements of BEACH sand after severe storms, and BRIDGE clearance in navigation channels. Lasers are used in precise coastal navigation and surveying systems. Pulsed lasers in the blue-green range are used to determine water depth in clear, shallow water bodies and coastal regions, and to make biomass assess-

ments of schooling fish. Continuously emitting lasers are used to excite fluorescence in ALGAE to measure primary productivity (chlorophyll-a) from aircraft and earth satellites, to determine the extent or characteristics of slicks from oil and other pollutant spills. They are used in automated weather observing stations of the National Weather Service to estimate the TURBIDITY of the atmosphere by measuring the amount of attenuation in a laser beam projected over a known path. The same technique is used in the ocean to determine the turbidity of ocean waters. Laser systems exploiting Raman scattering (the inelastic scattering of light from molecules) are under development to study the concentration of carbon dioxide in the upper ocean. The number of marine science applications grows with the rapidly expanding laser technology base.

Further Reading

Bell Labs History. "What Is a Laser?" Available online. URL: http://www.bell-labs.com/history/laser/laser_def. html. Accessed September 25, 2007.

International Laser Ranging Service. Available online. URL: http://ilrs.gsfc.nasa.gov. Accessed June 28, 2007.

Law of the Sea   The Law of the Sea is a body of international rules and principles developed by many countries to regulate ocean space. For centuries, maritime countries and even some without a coast have realized that international agreements about the sea were necessary, especially to define marine boundaries and to manage natural resources such as fisheries and minerals. Integrating rules across many countries, which govern the use of the oceans and their resources, leads to management practices that are effective and sustainable. Today, the United Nation's Convention on the Law of the Sea provides a collaborative forum for the development of a comprehensive regime of law and order in the world's oceans and seas. United Nations conferences on the Law of the Sea have been conducted since 1958. It is an especially important activity of the United Nations since all problems of ocean space are closely interrelated and need to be addressed as a whole.

One of the foundational provisions within the Law of the Sea relates to agreements over definitions for territorial seas, contiguous zones, the CONTINENTAL SHELF, the EXCLUSIVE ECONOMIC ZONE, and the high seas (the high seas are international waters where all countries enjoy the traditional freedoms of navigation), overflight, scientific research, and fishing. For example, HIGHLY MIGRATORY SPECIES are referred to in Article 64 of the Law of the Sea, which provides for the rights and obligations of coastal and other states whose nationals fish for these species.

Annex 1 provides an agreed list of species considered highly migratory and supports the notion that management of resources such as highly migratory species requires cooperation among the many coastal states.

The Law of the Sea establishes a 12-nautical mile (22.2 km) distance from the COASTLINE (a minute of arc along a great circle of the Earth) as a territorial sea where coastal countries may enforce their laws and regulate the usage and exploitation of natural resources. Naval and merchant ships are provided the right of "innocent passage" through territorial seas of a coastal country. For the purpose of law enforcement, this contiguous zone extends for 24 nautical miles (44.4 km) from a maritime country's shores. For example, the U.S. Coast Guard can arrest and detain suspected drug smugglers, illegal immigrants, and customs or tax evaders who violate laws of the United States within its territory or the territorial sea. For island countries such as the Philippines and Indonesia, "archipelagic waters" are defined as a 12-nautical mile (22.2 km) zone extending from a line drawn joining the outermost points of the outermost islands of the group that are in close proximity to one another.

Supplementing the 12- and 24- mile (22.2- and 44.4-km) zones of resource utilization and law enforcement, where countries enjoy special rights over the exploration and use of marine resources, the Exclusive Economic Zone (EEZ), consisting of the seabed and its subsoil, extends from a coastal states' territorial sea to the outer edge of the CONTINENTAL MARGIN, or to a distance of 200 nautical miles (370.4 km). When there is an overlap between coastal states, the involved states must delineate the actual boundary. The three countries with the largest area of EEZ are the United States, France due to its many territories, and Australia.

In conclusion, the Law of the Sea encompasses several United Nations conventions on the topic and one international treaty. Key events are the First United Nations Convention on Law of the Sea, the Second United Nations Convention on Law of the Sea, and the Third United Nations Convention on Law of the Sea. The treaty, which has been signed by 148 countries, resulted from the Third United Nations Convention on Law of the Sea. The Secretariat for the Law of the Sea resides within the United Nations Division for Ocean Affairs and the Law of the Sea.

Further Reading
Churchill, Robin, and Vaughan Lowe. *The Law of the Sea.* Manchester, U.K.: Manchester University Press, 1999.
Oceans and Law of the Sea, U.N. Division for Ocean Affairs and Law of the Sea. Available online. URL: http://www.un.org/Depts/los/. Accessed September 25, 2007.

lidar   A lidar is a remote sensing instrument consisting of a LASER transmitter and an optical receiver, which is generally a telescope, along with associated detection and signal processing equipment. Lidar is an acronym for *light detection and ranging*. It can be thought of as laser RADAR, which uses the same principles as radar, but transmits in the visible, infrared, or ultraviolet range of the electromagnetic spectrum.

Usually the transmitter and receiver of the lidar are aligned and placed close together in order to measure the energy and signal characteristics of the signal backscattered (*see* BACKSCATTER) by the target, that is, the object(s) of study. The laser light beam is coherent, highly focused, and almost purely monochromatic in wavelength, so that the echoes can be readily interpreted by means of scattering theory, much of which assumes a single wavelength. The range of the beam to successive targets can be determined from the time of the laser pulse to travel out and back, since the speed of light in air is nearly constant. By range-gating the signal, a matrix of range bins is recorded from which a profile of the quantity of interest can be obtained, in a manner similar to that of the ACOUSTIC DOPPLER CURRENT METER. The laser sensors of lidars have excellent resolution (the ability to distinguish between two closely spaced objects), but suffer high absorption from clouds and water vapor.

Lidar systems are usually deployed from aircraft, but are now being installed on weather satellites with greater frequency. They can also be deployed on the ground in the upward-scanning mode. Airborne lidar may be used for topographic and bathymetric surveying. In this case, the lidar is coupled with a GLOBAL POSITIONING SYSTEM (GPS) to determine aircraft position and an Inertial Navigation System (INS) to determine the aircraft attitude. Meteorologists may use lidar from the ground to measure winds. Satellite-borne lidar systems have been developed to measure geophysical parameters such as changes in elevation of the Greenland and Antarctic ice sheets.

Lidar systems can be classified as laser ceilometers to measure cloud-base heights and internal cloud structures, Doppler lidars to measure the velocity of aerosol or cloud targets, polarization lidars to measure water vapor phase in clouds, and Raman lidars to measure the concentrations of atmospheric constituents and pollutants. DIAL lidars (an acronym for Differential Absorption

Lidar) are used to measure the ozone concentration. The DIAL works by measuring the difference in absorption of light at different wavelengths. Lidars have been used to measure upper-level winds in the atmosphere, atmospheric temperature, and the amount of Saharan dust in the tropical atmosphere.

Further Reading

EAARL: Experimental Advanced Airborne Research LIDAR, Hydrospheric and Biospheric Sciences Laboratory, NASA GSFC/Wallops Flight Facility. Available online. URL: http://inst.wff.nasa.gov/eaarl/. Accessed March 9, 2008.

Earth Observing Laboratory, The National Center for Atmospheric Research. Available online. URL: http://www.eol.ucar.edu/lidar. Accessed June 28, 2007.

The International LIDAR Mapping Forum. Available online. URL: http://www.lidarmap.org. Accessed June 28, 2007.

Joint Airborne Lidar Bathymetry Technical Center of Expertise. Available online. URL: http://shoals.sam.usace.army.mil. Accessed June 28, 2007.

light   Light is defined as energy traveling as waves or as vibrations of electric and magnetic fields through the universe. The electromagnetic nature of light was first proved in 1864 by James Clerk Maxwell (1831–79), an English theoretical physicist who solved the set of electromagnetic equations now named in his honor. Heinrich Hertz (1857–94), a German physicist, experimentally verified the existence of electromagnetic waves. The frequency of electromagnetic waves, the Hertz, or Hz, was named for him.

Light travels at 186,282.397 miles per second (299,792,458 m/s) in free space and was measured accurately by Nobel laureate Albert Abraham Michelson (1852–1931) along the banks of the Severn River on Chesapeake Bay during 1875. Visible light is the narrow band of electromagnetic radiation that lies in the wavelength range between 400 and 700 nanometers (4,000–7,000 angstroms), which may also be called the visible light spectrum. The colors of visible light are seen in a rainbow and are red, orange, yellow, green, blue, and violet, along with various combinations and shades of these colors. Within the visible part of the spectrum, red light has a longer wavelength than violet light. When combined, all of the wavelengths present in visible light form colorless white light which can be refracted and dispersed into its component colors by means of a prism or water vapor. A prism is a transparent material with two or more straight faces at an angle to one another and was used in old sailing vessels (deck prisms) to help spread light throughout a cabin.

The colors red, green, and blue are classically considered the primary colors because they are fundamental to human vision. Sensors called photometers are used to measure light.

See also LIGHT AND ATTENUATION; PHOTOGRAPHER; REMOTE SENSING.

light and attenuation   Light in the sea is attenuated much more severely than in the atmosphere. Attenuation occurs because light is both absorbed and scattered by water molecules and by suspended matter. The greatest distance that a diver can see in clear oceanic water is about 100 feet (30 m). The decrease in light intensity with depth is expressed by the exponential equation

$$I = I_0 \exp(-kz)$$

where I_0 is the solar radiation that penetrates the surface of the ocean, I is the light intensity at depth z, and k is the attenuation coefficient of light in the ocean. (Interested readers may consult a mathematics text to learn about exponential functions.)

As a rule of thumb, there is insufficient light for a diver to work at depths much below 330 feet (100 m) in the deep ocean. In shallow coastal waters, which tend to be turbid from suspended matter, visibility may be almost zero at 33 feet (10 m) depth. Divers working in coastal waters must learn to work by feel, or by using instruments that use sound, which has no trouble penetrating even turbid coastal water.

Light in the sea is differentially attenuated with depth, depending on its wavelength. The average human eye can perceive light in the spectral range 350 nanometers (deep violet) to 750 nanometers (deep red), which includes all the colors of the rainbow, violets, blues, greens, yellows, oranges, and deep reds. The wavelength of the light increases continuously from violet to red. The shorter the wavelength, the greater the amount of light energy. Red light is absorbed in the upper few meters of the water column, then orange, yellow, and so forth, down to blue. Below 330 feet (100 m) depth in clear seawater, all light appears blue to violet. Below 6,560 feet (2,000 m) depth, there is virtually no light, save that of bioluminescent (*see* BIOLUMINESCENCE) creatures and lava flows at underwater volcanoes and spreading zones.

Objects underwater appear out of focus to swimmers because of refraction, or bending of the light rays, that occurs at the water-eye boundary. FISHERMEN at the surface of the water perceive swimming fish as being displaced from their actual positions. Spear fishermen in Polynesia learned long ago to compensate for this effect. Wearing goggles or a div-

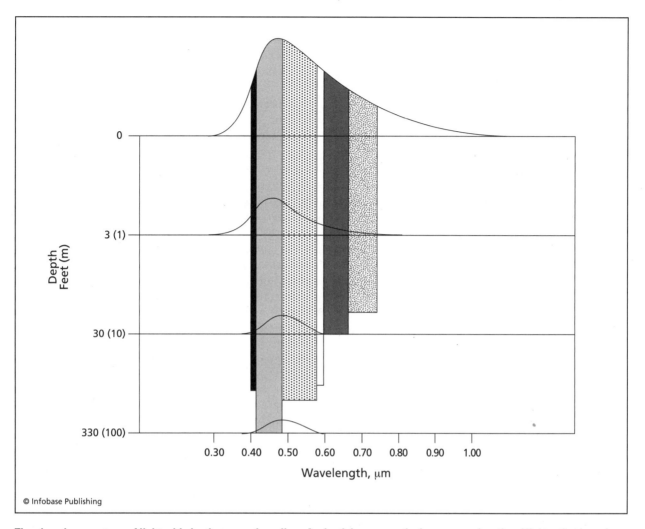

The changing spectrum of light with depth as seen by a diver. As depth increases, the longer wavelengths of light—that is, red, orange, yellow—are attenuated, and at depths of 330 feet (100 m) the scene appears deep blue.

ing mask puts a layer of air between water and the eye, correcting the vision. However, even with goggles or a mask, objects appear magnified, and peripheral vision is reduced.

Even though light is rapidly attenuated with depth, in exceptionally clear shallow waters blue-green high-powered LASERS deployed in aircraft are used to map the bottoms of shoal areas. Such instruments can document changes to the shoreline and the bottom topography produced by winter gales and HURRICANES, and by human activities such as DREDGING.

Aircraft-mounted lasers are also used to excite fluorescence in the chlorophyll of the upper layers of the ocean and thus provide a means of remote sensing to measure chlorophyll-a content, from which charts of primary production can be compiled.

The first instrument devised to measure optical transparency of the water is the SECCHI DISK, named for its inventor, Father Pietro Angelo Secchi (1814–1900), a Catholic priest and scientist. The Secchi disk is lowered from a ship or pier until the white disk just disappears from view. The depth is measured by the amount of line paid out, which is called the "Secchi depth." Charts of equal Secchi depth are available for some parts of the ocean and are used for rough determinations of water clarity. Today, modern measurements of optical transparency, called optical beam transmittance, are made with solid-state sensors such as lasers transmitting over a fixed path length. Such devices can be lowered from the deck of a ship to great depths, providing a vertical profile of water clarity. Scientific divers may also use underwater spectrometers to measure optical properties in the ocean.

Further Reading

Pickard, George L., and William J. Emery. *Descriptive Physical Oceanography: An Introduction,* 5th ed. Oxford: Pergamon Press, 1990.

limiting factors Environmental factors that regulate the growth, reproduction, or distribution of organisms are called limiting factors. Success of a population is dependent on those limiting factors that affect its environment. For example, the limiting factor in the lack of biodiversity in the tundra is low temperature. Since limiting factors are closely related to the tolerance levels species have to physical and chemical factors, they focus on abiotic (nonliving) rather than biotic (living) factors. Abiotic factors include seawater properties, CURRENTS, and even the type of bottom or substrate. Limiting factors prevent an organism from reaching its full biotic potential by restricting development and productivity. On land, limiting factors may include soil nutrients, water, light, and TEMPERATURE. Limiting factors in marine ECOSYSTEMS include light, nutrients such as nitrogen, SALINITY, temperature, pH, and even dissolved oxygen.

Whether on land or in the ocean, levels of light or carbon dioxide are limiting factors for primary producers that use photosynthesis for the production of carbohydrates. Solar radiation is a limiting factor during certain seasons and is limiting below the COMPENSATION DEPTH, where photosynthetic oxygen production is balanced by respiratory uptake. In the ocean, the compensation depth is at the lower limit of the PHOTIC ZONE or approximately located where light is 1 percent of its incident intensity. Below this depth there will be little, if any, PHYTOPLANKTON growth. Phytoplankton such as diatoms, dinoflagellates, and macroscopic seaweeds such as eelgrass (*Zostera marina*) are important primary producers, converting inorganic materials into organic compounds by photosynthesis.

Nitrogen (N) is a large component of animal waste and takes the form of nitrogenous compounds such as urea and uric acid. Without good controls such as waste stabilization ponds, untreated materials can enter coastal regions. ESTUARIES often respond with increased growth of phytoplankton ALGAE in the process known as eutrophication, which may result in gross degradation of the habitat, including anoxia, objectionable odors, and ultimately fish kills.

The effect of SALINITY on the survival and development of fish and plants is extremely important. Plants and fish can live only within their levels of tolerance. Salinity is defined as the amount of various salt minerals in a given volume of water. Mangrove species show specific salinity tolerance ranges. For example, black mangroves (*Avicennia germinans*) tend to live in coastal tidal areas throughout the Tropics and subtropics. These communal plants thrive in this brackish water environment where salt water mixes with fresh. During the 1980s, marine scientists studying Ronco croaker (*Bairdiella icistia*), orangemouth corvina (*Cynoscion xanthulus*), and sargo (*Anisotremus davidsoni*) in the Salton Sea (a lake located in the southeastern corner of California) found that the optimal range of salinity for these fish was between 33–37 parts per thousand (ppt) salt. Extreme salinities up to 45 parts per thousand (ppt) were found to adversely affect growth, food consumption, food assimilation, and respiration. Today, salinities in the Salton Sea have reached 44 parts per thousand (ppt), about 25 percent higher than ocean water. This elevated salinity explains the decline in orangemouth corvina, one of the regions popular sport fish.

Light and temperature changes are major limiting factors, which can greatly alter environmental conditions in marine ecosystems and are especially important with respect to biological impact in coastal areas. For example, CORAL REEFS are found in tropical regions between latitude 30°N-and-S, and do not survive if water temperatures are not in the range between 73.4° and 77°F (23°–25°C). These same coral reefs are limited by light and cannot survive depths of greater than 164 feet (50 m). Runoff and SEDIMENT load kills coral reefs since it increases TURBIDITY and does not permit the sunlight to reach the coral. The coral needs sunlight so that the ZOOXANTHELLAE contained in their polyps can perform photosynthesis.

The normal "parts Hydrogen" or "pH" of seawater is slightly alkaline and in the range of 8–8.3 on the pH scale. The term *pH* describes the relative concentration of positive hydrogen ions (H+) and negative hydroxyl ions (OH-) in a solution. It is a logarithmic scale describing the acidity of a substance. A pH value of 7.0 means that equal concentrations of the above ions are present and the solution is said to be neutral. A pH value below 7.0 means that more hydrogen ions (H+) are present than hydroxyl (OH-) ions and the solution is more acidic; a pH of above 7.0 means the solution contains more hydroxyl ions (OH-) than hydrogen ions (H+) and is more alkaline. Seawater that is acidic could dissolve the calcium carbonate shells or exoskeletons of certain corals, plankton, and algae.

Most forms of marine life depend on oxygen (O). The amount of dissolved oxygen in a body of water is an indication of its purity. When water is polluted by organic material such as sewage, its levels of dissolved oxygen generally drop. The organic material serves as food for microorganisms. As the microorganisms multiply, oxygen is depleted, and fish along with other forms of aquatic life that depend on the oxygen to live will die. If there are no oxygen-consuming pollutants in water, the concentration of dissolved oxygen will largely be determined by water temperature and salinity. The level of oxygen required by different species varies, but most fish require at least 4.0 milligrams per liter to survive for long periods.

longshore current   The longshore current—also called a littoral or alongshore current—flows close and almost parallel to the shoreline as a result of waves striking the shore at an oblique angle. The approaching or incident waves change shape or transform as they enter increasingly shallow water. Beachgoers can actually observe wavelengths shortening as wave heights increase until the incident waves break as SURF. These same waves turn or bend to become parallel to depth contours in a process that marine scientists call "wave refraction" (*see* WAVES). Some of the wave energy is reflected seaward, and, if objects such as BREAKWATERS or islands are in the propagation path, the waves will be diffracted (bend) around

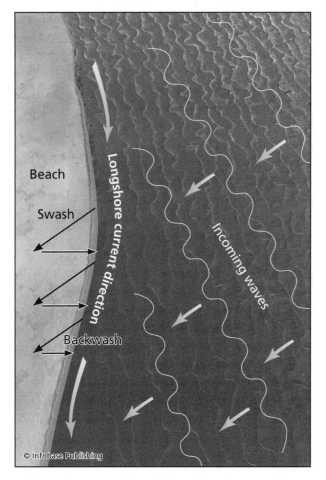

The longshore current results from the component of wave energy moving parallel to the beach; it appears whenever the incoming waves approach the beach at an angle.   *(Source: NOAA)*

the obstacle. The direction of the wave-induced longshore current can be determined from the direction of incident waves.

The two primary types of currents in the nearshore zone are longshore and RIP CURRENTS. This nearshore region generally extends from the swash (upsurge of water) zone through the breaker zone. The swash zone moves with WATER LEVEL FLUCTUATIONS and includes the swash mark or the limit where water runs up and runs down the beach. The swash mark is usually a wavy line of fine sand, foam, seaweed, and possibly driftwood. The run-up is a mass of water from a breaking wave that returns to the sea. During the summer at popular flat sandy beaches, adventurers can be seen jumping on skimboards for brief rides in ankle-deep uprush. Return water in synchronization with favorable longshore currents can cause rip currents. For example, longshore currents may converge to form the neck of a rip current or they may collect water against obstacles such as a JETTY or pier, which is diverted seaward as a rip current. This flow may cut deep holes in seemingly shallow water. The deeper nearshore water often appears darker in color and wave breaking is greatly reduced. The rip currents generally span the entire water column until they pass through the surf zone.

Surf zone processes such as longshore currents contribute to beach erosion and accretion. The surf zone is the stretch of shallow water where waves break as they approach the shore. Waves striking the shore at an angle as opposed to straight on will cause the swash to move up the beach at an angle. The swash moves the SEDIMENT particles (usually sand and shells) up the beach at this angle. Backwash brings the suspended material straight down the beach. The resultant effect is a current and slow movement of particles along the shore. The movement of suspended material through the surf zone is called longshore drift and is one of the principal processes in the construction of beach deposits such as spits and SANDBARS. Beach buildup or the accumulation of beach material due to natural action of waves, currents, and wind is called accretion. Coastal engineers and oceanographers are interested in measuring and predicting longshore currents. Marine scientists are advancing marine science at research facilities such as the Naval Research Laboratory at Stennis Space Center in Mississippi by investigating longshore currents. They are using video imagery to better understand wave processes. Video feeds of the nearshore report parameters such as breaker heights, wave periods, wave directions, longshore current speeds, and surf zone width in real time. One such system is called the Video Imaging System for Surf Zone Environmental Reconnaissance or

VISSER, which makes possible the rapid assessment of wave and current conditions and facilitates the synthesis of individual point measurements.

Further Reading

Beaches, Office of Water, U.S. Environmental Protection Agency. Available online. URL: http://www.epa.gov/beaches. Accessed June 28, 2007.

Coastal Data Information Program, Scripps Institution of Oceanography. Available online. URL: http://cdip.ucsd.edu. Accessed June 28, 2007.

Field Research Facility, Coastal Hydraulics Laboratory, Engineer Research and Development Center. Available online. URL: http://www.frf.usace.army.mil. Accessed June 28, 2007.

long-term environmental observatory A long-term environmental observatory is a nonstandard term describing locations where instruments are deployed to measure meteorological, oceanographic, and geophysical phenomenon, such as LAND AND SEA BREEZE circulation, WAVES and CURRENTS, and tectonic activity. Observatories often utilize expensive and state-of-the-art weather stations, fully instrumented BUOYS, seismic sensors on the seafloor, and research vessels that collect *in situ* measurements in nearby regions. Since the Latin phrase *in situ* means "in place," environmental scientists call instruments that collect data either in the atmosphere, in the water column, or in the sediments, "*in situ* sensors."

The frequency of making observations or sampling is related to the characteristic time "signal" and location of the phenomenon being studied. Typically, very long time series are desired to study such major signals as climate change. Some phenomena may occur over both small and large horizontal distances as well as over a range of depths. Data that are collected over extensive time periods can be merged together, integrated, and otherwise analyzed to determine status and trends on important processes such as coastal erosion. However, there are many challenges to maintaining data over long time periods such as maintaining and calibrating instruments, overcoming technological obsolescence, updating data fields, and data sharing. Such archives and the ensuing analyses are critical to building a baseline of information from which scientists can assess causes for environmental changes. These data may be used to improve knowledge in a variety of concerns from air and water quality to GLOBAL WARMING and SEA LEVEL rise.

Marine scientists integrate and analyze the data sets from long-term observatories so as to identify environmental change and improve understanding of the causes of change; the observatories make these long-term data sets available as a basis for research and prediction. The temporal and spatial scales of change for particular phenomena vary according to abiotic and biotic forces. Abiotic forces are parameters such as TEMPERATURE, moisture, sunlight, pH, dissolved oxygen, and SALINITY; biotic forces involve interactions with other organisms, competition, predation, symbiosis, nest sites, and habitat modification. A viable long-term environmental observatory would provide information to support operations such as merchant ships trying to enter a busy seaport as well as basic researchers who are trying to advance the knowledge base on INTERNAL WAVES, harmful algal blooms, and the status of endangered species.

Rutgers University has implemented the Long-term Ecosystem Observatory at a depth of 49 feet (15 m). It is called LEO-15 and collects oceanographic data with high temporal resolution, which can be used to answer natural process questions across various scientific disciplines. LEO-15 data are collected from *in situ* instruments. Oceanographic instruments include ocean CURRENT METERS, THERMISTORS, plankton samplers, and optical back scatter sensors, which are connected via a fiber optic link back to shoreside computers. Sea surface temperature is derived from the Advanced Very High Resolution Radiometer mounted on Earth satellites. Data for the imagery are collected from satellite receiving stations. Meteorological data such as wind speed and direction, air temperature, barometric pressure, and humidity are collected from sensors attached to a 230-foot (70 m) tower. The actual design of LEO-15 was implemented by scientists, engineers, and technicians from the Rutgers Institute of Marine and Coastal Sciences and the WOODS HOLE OCEANOGRAPHIC INSTITUTION.

Further Reading

Center for Coastal Monitoring and Assessment, National Ocean Service, National Oceanic and Atmospheric Administration. Available online. URL: http://ccma.nos.noaa.gov. Accessed June 28, 2007.

Coastal Ocean Observation Lab, Institute of Marine and Coastal Sciences. Available online. URL: http://marine.rutgers.edu/mrs/LEO/LEO15.html. Accessed June 28, 2007.

Intergovernmental Oceanographic Commission, Global Ocean Observing System. Available online. URL: http://www.ioc-goos.org. Accessed June 28, 2007.

Joint Oceanographic Institution, Ocean Observatories Initiative. Available online. URL: http://www.joiscience.org/ocean_observing/initiative. Accessed June 28, 2007.

Long-Term Ecological Research Network. Available online. URL: http://www.lternet.edu. Accessed June 28, 2007.

Ocean.US. Available online. URL: http://www.ocean.us. Accessed September 25, 2007.

Loop Current   The Loop Current is a clockwise circulation in the GULF OF MEXICO extending from the Yucatán Current flowing northward through the Yucatán Strait into the Gulf of Mexico, to the Florida Current, which flows out through the Florida Straits into the Atlantic Ocean to begin the GULF STREAM. The Loop Current is a western boundary current of the Gulf of Mexico; it is therefore a fast current, extending to great depths, and tending to be unstable so that meanders and EDDIES are frequently observed features. Eddies cast off by the Loop Current tend to drift slowly toward the west into the central and western parts of the Gulf of Mexico. The CONTINENTAL SHELF borders the Loop Current to the west (toward Mexico) and the north (toward Texas, Louisiana, Mississippi, Alabama, and Florida).

Speeds in the Loop Current are on the order of two knots (1.0 m/s) at the surface in the Yucatán Strait, decreasing to about .8 knot (0.4 m/s) at 3,280 feet (1,000 m) depth. Sometimes there is southward flow under the Loop Current about 657 feet (200 m) above the CONTINENTAL SLOPE. These currents are important for migratory fish species. They can affect the speed of vessels and the extensive drilling operations being conducted in the Gulf of Mexico. There is great variability in the position of the Loop Current. Sometimes it flows almost directly from the Yucatán Strait to the Straits of Florida; while at other times, it bows westward and northward into the inner Gulf of Mexico, as far as the Mississippi delta region. Although there appear to be seasonal variations in the Loop Current, there does not appear to be a well-defined seasonal pattern. Volume transports in the current have been estimated to be from 24 to 30 Sverdrups (24×10^6 m³/s to 30×10^6 m³/s) in the vicinity of the Yucatán Strait.

Knowledge of changes in the Loop Current circulation are important to marine operations in the Gulf of Mexico, such as oil and gas extraction, fishing, and pollution control. Marine scientists can use satellite imagery, CURRENT METERS, and instruments deployed from oil production platforms to determine the speed of eddies and their associated currents. High-current speeds are forceful enough to halt drilling operations.

Further Reading

Rosenstiel School of Marine and Atmospheric Science, University of Miami. "Ocean Surface Currents." Available online. URL: http://oceancurrents.rsmas.miami.edu/index.html. Accessed September 25, 2007.

lunar day   A unit of time determined by one rotation of the Earth with respect to the Moon. In astronomical terms, it is the interval between two successive passages of the Moon over the local meridian (imaginary line of longitude passing directly overhead). The mean lunar day is approximately 24.84 solar hours in duration (about 24 hours, 50 minutes), that is, 1.035 times as long as a mean solar day. The solar day, the unit of time determined by one rotation of the Earth with respect to the Sun, is the unit used in daily timekeeping. The "extra" 50 minutes in the lunar day occurs because the Moon's position, as seen from Earth, appears to shift eastward against the background stars. The Earth must rotate eastward a little extra each day to bring the Moon over the local meridian; hence the Moon rises about 50 minutes later each night.

The lunar day, like the solar day, is divided into 24 hours, with each hour, in turn, divided into 60 minutes and each minute into 60 seconds. The durations of these lunar and solar time units differ by small amounts.

The mean lunar day, that is, the mean duration of all lunar days over a year, differs by a small amount from the actual lunar day because the speed of the Moon varies slightly in its orbit around the Earth, just as the mean solar day differs from the actual solar day because the speed of the Earth varies in its orbit around the Sun. Three other small deviations between the mean lunar day and the actual lunar day occur because (1) there are small perturbations in the rotational velocity of the Earth about its axis, (2) the Earth precesses (wobbles slightly, like a slowing top) about its rotation axis, and (3) the plane of the Moon's orbit precesses slightly about the Earth.

The lunar day defines lunar time, which is useful in studies to predict the TIDES, since the influence of the Moon in producing Earth's tides is about twice as great as the influence of the Sun. Furthermore, the phases of the Moon have served humankind for thousands of years as a clock for periods greater than a day and less than a year. The word *month* is derived from *moon* since the Moon passes through one complete cycle of phases in approximately 30 days.

Further Reading

The U.S. Naval Observatory. Available online. URL: http://www.usno.navy.mil. Accessed June 28, 2007.

lysocline   A region in the deep ocean where the rate of dissolution of calcite increases dramatically is called the lysocline. Dissolution occurs when a solute in contact with a solvent dissolves to form a solution. The lysocline is below the upper layers of the ocean that are supersaturated with calcium carbonate ($CaCO_3$) and located at greater depths where $CaCO_3$ starts to dissolve due to increasing PRESSURE and decreasing TEMPERATURE. This region is a transition zone between the dissolution and burial of $CaCO_3$ SEDIMENTS. It is the depth that separates well-preserved (high calcite) from poorly preserved (low calcite) carbonate shells (sediments).

Calcite represents the stable form of $CaCO_3$ and is the main component of seashells. The solubility of $CaCO_3$ increases with increasing pressure and decreasing temperature, so that the cold and deep waters tend to be undersaturated with respect to $CaCO_3$. The depth at which seawater becomes undersaturated and shells first start to show signs of dissolution is called the lysocline. Above the lysocline, dissolution occurs at a slower rate than the rate at which the shells sink, which causes $CaCO_3$ to accumulate in sediments. At greater depths, the rate of sinking is not high enough to keep up with dissolution, and no carbonate accumulates in the sediments. Thus, in response to temperature and pressure, the lysocline, or the depth at which calcite starts to dissolve, tends to be located in shallower water in the higher latitudes and becomes deeper toward the equator as the temperature of the surface layers increases. The balanced depth at which $CaCO_3$ entirely disappears from deep-sea sediments is called the carbonate compensation depth or CCD. This depth occurs around 12,138 feet (3,700 m).

See also CHEMICAL OCEANOGRAPHERS; GEOLOGICAL OCEANOGRAPHERS.

Further Reading

Brookhaven National Laboratory. Available online. URL: http://www.bnl.gov/world. Accessed June 28, 2007.

Carbon Dioxide Information Analysis Center, Oak Ridge National Laboratory. Available online. URL: http://cdiac.ornl.gov. Accessed June 28, 2007.

Lamont-Doherty Earth Observatory, Columbia University. Available online. URL: http://www.ldeo.columbia.edu. Accessed June 28, 2007.

Lawrence Livermore National Laboratory. Available online. URL: http://www.llnl.gov. Accessed June 28, 2007.

M

magnetic current meter (electromagnetic current meter)   An electromagnetic current meter is an instrument used in marine science to measure the speed and direction of water movement through the use of electromagnetic induction. Electromagnetic induction involves the generation of an electromotive force and current by a changing magnetic field, and was first described by Michael Farady (1791–1867) in 1831. In Farady's famous experiment, he used two coils of wire wound around opposite sides of an iron ring. He magnetized the iron by passing a current through the first coil. He discovered that an electromotive force was induced when current flowed through the second coil. When the current was turned off in the first coil, a current flowed through the second coil, but in the opposite direction. A current meter applies Faraday's Law by measuring voltages resulting from seawater flow (a moving conductor) through a magnetic field. The meter uses alternating current to generate a magnetic field through internal coils, in which the electrons move back and forth at regular cycles as the applied voltage alternates between positive and negative. Sensing electrodes are used to measure voltage changes as water flows around the outside of the instrument. Directions may be determined by measuring the rectangular Cartesian current components with orthogonal pairs of electrodes; the orientation of the instrument with respect to magnetic north is measured by an internal instrument compass.

Today's current meters are precision instruments greatly improving observations of ocean currents from a CURRENT POLE. Modern CURRENT METERS are automated to provide stable velocity readings that can be digitally displayed. Hydrologists, meteorologists, civil engineers, and oceanographers frequently use them. Accurate *in situ* observations require current meters to be properly assembled, adjusted, calibrated, and maintained. Measurements in the field or natural environment are called *in situ*, as opposed to well-controlled laboratory measurements. Magnetic current meters can be sited in a range of locations from flumes and wave tanks to irrigation channels and the deep ocean. For many applications, they offer a capability that is superior to rotor and acoustic sensors.

Current meters may be mounted on a mooring cable, which is deployed from a ship (*see* MOORING) or fixed to the bottom on a sturdy frame. Electromagnetic current meters provide excellent linearity, directional response, and do not have excessive power requirements. They are susceptible to biofouling, especially on the sensing electrodes, and only measure velocities from a point. Biofouling is the overgrowth of ALGAE, marine invertebrates, and other organisms on sensors, intake pipes, vessel hulls, and other marine structures. It can clog rotors or restrict water flow around an instrument. Rotor-type current meters are somewhat outdated and susceptible to data degradation by wave pumping in shallow water, giving false high readings. Rotating units can be configured to measure flow with vertical-shaft or horizontal-shaft axes. Electromagnetic current meters are usually less expensive than acoustic Doppler current meters (*see* ACOUSTIC DOPPLER CURRENT PROFILER). Some acoustic current meters can provide a profile of velocity with depth. They use more power and have increased quantities of data to manipulate and display. Electromagnetic current meters are in wide use owing to their portability and ease of use.

Magnuson-Stevens Fishery Conservation and Management Act The United States's public law governing the conservation and management of ocean fishing is known as the Magnuson-Stevens Fishery Conservation and Management Act. Also called "Public Law 94-265," this act controls heavy foreign fishing in U.S. waters, promotes sustainable fisheries, and links the fishing community more directly to management processes. It establishes exclusive U.S. management authority over all fishing within the EXCLUSIVE ECONOMIC ZONE (EEZ), all ANADROMOUS fish throughout their migratory range, except when in a foreign nation's waters and all fish on the CONTINENTAL SHELF. Anadromous species such as American shad (*Alosa sapidissima*) and Atlantic sturgeon (*Acipenser oxyrhynchus*) spend most of their life in salt water but migrate into freshwater tributaries to spawn. Foreign fishing within the United States's EEZ is prohibited unless conducted pursuant to a governing international fishery agreement and permit, and only if the foreign nation extends reciprocity to U.S. fishing vessels. The act also establishes eight Regional Fishery Management Councils responsible for the preparation of Fishery Management Plans (FMPs). These plans help achieve optimum yield from fisheries in the council's region. A fishery includes one or more stocks of fish based on geographical or scientific characteristics and any fishing for such stocks. Membership is designed to include commercial fishermen, recreational fishermen, and representatives of the public interest in marine conservation. The councils provide an ECOSYSTEM approach for responsible fisheries, whereby all people involved in fishing accept responsibility to provide fisheries information, embrace collaborative research, participate in the fishery management process, comply with regulations, avoid waste, and develop training to instill a responsible fishing ethic. Today's eight councils are as follows:

- *Caribbean Fishery Management Council*: Headquartered in San Juan, Puerto Rico, this is the only council that does not include one of the 50 states. Members are responsible for the creation of management plans for fishery resources common to the waters off Puerto Rico and the U.S. Virgin Islands.

- *Gulf of Mexico Fishery Management Council*: Headquartered in Tampa, Florida, this 17-member council manages fisheries in the EEZ of the GULF OF MEXICO. States with voting representation on the council include Texas, Louisiana, Florida, Alabama, and Mississippi.

- *Mid-Atlantic Fishery Management Council*: Headquartered in Dover, Delaware, this 25-member council is responsible for management of fisheries in federal waters, which occur predominantly off the Mid-Atlantic coast. States with voting representation on the council include New York, New Jersey, Pennsylvania, Delaware, Maryland, Virginia, and North Carolina. North Carolina is on two councils, the Mid-Atlantic and South Atlantic Fishery Management Councils.

- *New England Fishery Management Council*: Headquartered in Newburyport, Massachusetts, this 18-member council manages fishery resources within the federal 200-mile (321.9 km) limit off the coasts of Maine, New Hampshire, Massachusetts, Rhode Island, and Connecticut.

- *North Pacific Fishery Management Council*: Headquartered in Anchorage, Alaska, this 15-member council has voting members from Alaska, Washington, and Oregon. They help manage resources over a 900,000-square-mile (2.3×10^6 km^2) Exclusive Economic Zone, which includes the Gulf of Alaska, Bering Sea, and Aleutian Islands.

- *Pacific Fishery Management Council*: Headquartered in Portland, Oregon, this 19-member council has voting representatives from Oregon, Washington, California, and Idaho. Some members represent state or tribal fish and wildlife agencies, and some are private citizens who are knowledgeable about recreational and COMMERCIAL FISHING or marine conservation. They help manage fisheries off the coasts of California, Oregon, and Washington.

- *South Atlantic Fishery Management Council*: Headquartered in Charleston, South Carolina, this 17-member council is responsible for the conservation and management of fish stocks within the three to 200-nautical-mile (5 to 370 km) limit off the Atlantic coasts of North Carolina, South Carolina, Georgia, and east Florida to Key West.

- *Western Pacific Fishery Management Council*: Headquartered in Honolulu, Hawaii, this 16-member council is responsible for management of the Exclusive Economic Zone around the Territory of American Samoa, Territory of Guam, State of Hawaii, the Commonwealth of the Northern Mariana Islands, and U.S.

Pacific island possessions, an area of nearly 1.5 million square miles (3,884,982 km²).

A formal FMP is written by the Fishery Management Council to describe a fishery and all governing laws. The plan documents essential fish habitats and when completed is submitted to the U.S. secretary of commerce for implementation. Essential fish habitats are necessary waters and seafloors for spawning, breeding, feeding, and growing to maturity. Formal changes to an FMP are also submitted to the secretary of commerce as an amendment for review and approval. The approved FMP is a federal plan. These efforts help to maintain sustainable fisheries by eliminating overfishing and rebuilding stocks.

See also CORAL REEF FISH; FISH; FISHERMAN.

Further Reading
Caribbean Fishery Management Council. Available online. URL: http://www.caribbeanfmc.com/. Accessed May 1, 2006.
Gulf of Mexico Fishery Management Council. Available online. URL: http://www.gulfcouncil.org/. Accessed May 1, 2006.
Mid-Atlantic Fishery Management Council. Available online. URL: http://www.mafmc.org/mid-atlantic/about/about.htm. Accessed May 1, 2006.
National Oceanic and Atmospheric Administration. *Bycatch: A National Concern.* National Marine Fisheries Service, 1997.
National Research Council. Ocean Science Board. *Sharing the Fish: Toward a National Policy on Individual Fishing Quotas.* Washington D.C.: National Academy Press, 1999.
New England Fishery Management Council. Available online. URL: http://www.nefmc.org/about/index.html. Accessed May 1, 2006.
North Pacific Fishery Management Council. Available online. URL: http://www.fakr.noaa.gov/npfmc/. Accessed May 1, 2006.
Pacific Fishery Management Council. Available online. URL: http://www.pcouncil.org/. Accessed May 1, 2006.
South Atlantic Fishery Management Council. Available online. URL: http://www.safmc.net/. Accessed May 1, 2006.
Wallace, R. K., and K. M. Fletcher. *Understanding Fisheries Management: A Manual for Understanding the Federal Fisheries Management Process, Including Analysis of the 1996 Sustainable Fisheries Act,* 2nd ed. Mississippi-Alabama Sea Grant, 2001.
Western Pacific Fishery Management Council. Available online. URL: http://www.wpcouncil.org/. Accessed May 1, 2006.

major plates The blocks of the crust and uppermost layer of the mantle that interact with other plates in tectonic (forces or conditions within Earth's crust that cause movement) activity are called major plates. The rigid plates are essentially "floating" on the plastic (gumlike) upper mantle called the asthenosphere, and interacting in response to heating by CONVECTION currents UPWELLING from deep within the mantle.

The major plates of the Earth are the African Plate, the Antarctic Plate, the Eurasian Plate, the Indian-Australian Plate, the North American Plate, the Pacific Plate, and the South American Plate. There are six MINOR PLATES.

Lithospheric plates can be convergent, divergent, or transform. The plates diverge at mid-ocean ridges, where new crust is being formed. They converge at subduction zones, where one of the plates is sliding under the other and plunging back down into the mantle; transform plates have a common boundary in which they are sliding past each other.

Because of their lower density, continental plates tend to remain on the surface of the Earth; oceanic plates on collision dive back into the mantle because of their greater density. In the mantle, they melt and over millions of years are "recycled" back as new ocean floor.

The thickness of continental crust is on the order of 25 to 60 miles (40–100 km); the oceanic crust is much thinner, three to five miles (5–8 km). Because continental crust remains on the surface of the Earth, it is much older than oceanic crust (which gets subducted back into the mantle many times) and is reformed as new rock each time it returns to the surface.

Continental crust is thickest under mountain ranges and thinnest on the CONTINENTAL MARGINS, which are underwater and form the edge of the continents. Therefore, continental plates consist of both continental crust and oceanic crust.

As a specific example, the crust of the North American Plate is formed at the Mid-Atlantic Ridge and moves steadily westward at a speed on the order of a few inches (cm) per year. At the North American continental margin, the crust changes from oceanic to continental and considerably thickens. With time, SEDIMENTS build up on the continental margin. Magma (molten rock under the surface) from the asthenosphere cools and solidifies under the plate, making it thicker and heavier in going from east to west. Along the North American West Coast, the North American Plate converges with the Pacific Plate, which plunges beneath it.

MID-OCEAN SPREADING RIDGES and SUBDUCTION ZONES are noted for their volcanoes and earthquakes. In the West Coast of the United States, the San Andreas Fault region has seen several major earthquakes in the past hundred years.

Northwest of North America, two oceanic plates are converging: the Pacific Plate and the Eurasian Plate. The Alaska Trench is the surface expression of this subduction zone; the Aleutian Islands are being thrust upward above it. The Aleutians are known for frequent and severe volcanic eruptions and earthquakes, sometimes generating catastrophic TSUNAMIS.

See also MINOR PLATES; PLATE TECTONICS; SUBDUCTION.

Further Reading

Erickson, Jon. *Marine Geology: Exploring the New Frontiers of the Ocean*, rev. ed. New York: Facts On File, 2003.

manganese nodules   Manganese nodules are lumpy accumulations of heavy minerals such as manganese, zinc, and copper precipitated onto the seafloor and first analyzed more than 130 years ago during the Challenger Expedition (1873–76). HMS *Challenger* was a wooden corvette commanded by Captain George Strong Nares (1831–1915); Sir Charles Wyville Thomson (1830–82) served as the chief scientist. Manganese nodules may form as hydrogenous sediment when warm water from HYDROTHERMAL VENTS mixes with cold, deep waters over the ABYSSAL PLAIN. Metal elements in the hot volcanic waters precipitate or crystallize as slightly flattened black spheres. Most manganese nodules tend to be the size of a pea, but occasionally are found as large as softballs. They are collected from the seafloor by researchers using BOX CORERS and chain dredges. Box corers collect undisturbed samples of the seafloor's surface and are best utilized on soft bottoms. Dredges are dragged behind a vessel and scrape materials from the seafloor. They are best utilized on hard bottoms.

The specific composition of manganese nodules varies with location in the OCEAN BASINS. The major

Manganese nodules obtained from box cores of the deep ocean floor   *(NOAA)*

elements are manganese and iron with lesser amounts of copper, nickel, and cobalt. The nodules are approximately 25 to 30 percent manganese and form around a core that may be a shark's tooth, bone fragment, or meteorite. They grow very slowly from .04 to 8 inches (1 to 200 mm)/million years. Due to their position outside the EXCLUSIVE ECONOMIC ZONE in dense fields at depths from 2.5 to 3.7 miles (4 to 6 km), deep-sea mining operations are too costly. Sufficient manganese oxide such as pyrolusite and rhodochrosite is easily mined from the Earth's crust in places such as Russia, Brazil, Australia, South Africa, Gabon, and India.

Manganese is an important chemical element, it is Mn in the periodic table, has an atomic number of 25, and is recognized as an essential trace nutrient. Manganese is a constituent found in all steels or iron alloys. (Alloys are mixtures of two or more metals.) Other elements in steel include chromium, nickel, molybdenum, copper, tungsten, cobalt, and silicon. Steel is an important structural material for vehicles, vessels, BRIDGES, machinery, tools, and fasteners. Nutrients such as manganese are essential dietary elements required in small quantities. Manganese in excess is toxic-causing neurodegeneration.

mangrove swamp   Mangrove swamps or forests are coastal wetlands found along the world's tropical and subtropical low-energy shorelines that are dominated by perennial mangrove trees. These evergreen trees are tolerant of the brackish waters common to ESTUARIES and saltier waters found along the shores of the coastal ocean. Mangrove swamps have a dense canopy, and the trees thrive in saturated low-oxygen mud flats that would be a hazardous environment for any other tropical plant. Mangroves may be associated with CORAL REEF coasts, and they colonize areas from the equator to 28 degrees north and south latitude. These swamps produce organic matter, cycle nutrients, stabilize shorelines, and help establish food webs.

Out of approximately 100 species of mangroves, there are four typical species of mangrove trees. The black mangrove (*Avicennia germinans*) grows to heights of 24 feet (8 m) and has a root system composed of long, horizontal underground cable roots that produce hundreds of thin, vertical pneumatophores in the water around the tree. Pneumatophores or breathing roots are the respiratory organs in wetland plants and characteristic of both black and white (*Laguncularia racemosa*) mangroves. The black mangrove does not grow on stilt roots and is one of the most salt-tolerant of the mangrove species. The white mangrove has prominent stilted roots and is found farther from the

Mangrove roots grow through the water into the river or estuarine bed. They serve as a habitat for many types of creatures. *(NOAA)*

water than red and black species. They can grow to heights of 49 feet (15 m) and provide a source for tannin. Bark from the white and red mangrove (*Rhizophora mangle*) trees may be harvested for the leather tanning industry. To harvest the tannins, trees are cut down and debarked. The bark may be ground into a powder and then mixed with water to make a solution which rawhide can be soaked in prior to being pressed. Red mangroves

grow from 16 to 66 feet (5–20 m) tall and are supported above the water by stilt roots. They can tolerate more salt than the black mangroves and are found at lower intertidal positions than other species. This evergreen is an invasive species in Hawaii, where it was introduced during the early 1900s to stabilize the shoreline. Found farthest from the waters edge are the shorter sweet mangroves (*Maytenus phyllanthoides*), which usually grow from six to 12 feet (2 to 4 m) high. These trees have been used in reforestation efforts following HURRICANES in southwest Florida.

Most mangrove seedlings are released into the water by adult plants. The long, cigar-shaped seedlings float like a fishing bobber. As the growing seedling drifts, leaves sprout from the upper end and roots from the lower. Once the seedling reaches shoal water with a sand, silt, mud, or clay substrate, it takes root. Unfavorable locations prevent the stranding of seeds or propagules due to factors such as high wave conditions that prevent the accumulation of fine sediments or breaks up root structures. Preferred locations are deltaic and estuarine coasts. After several years, the young tree produces prop roots that spread, aiding the tree in oxygen uptake. Prop roots also help support the mangrove from above ground. The resultant mangrove swamps

Mangrove swamps are critical to protecting the Florida coastal environment. *(NOAA)*

become dense and create many habitats for marine life.

Mangrove swamps serve as feeding, breeding, and nursery grounds for a wide range of marine life. The entanglement of trees and roots common to a mangrove swamp are full of bacteria and ALGAE that break down plant litter into a sticky mud that provides nutrients and further habitat for successive species. The muds provide a refuge for MARINE CRUSTACEANS such as the spotted mangrove crab (*Goniosis cruentata*). These crabs eat mangrove leaves and are preyed upon by wading birds such as the Roseate Spoonbill (*Ajaja ajaja*). The mangrove rivulus (*Rivulus marmoratus*) is a killifish that fertilizes its own eggs, hides in crab burrows, and feeds on invertebrates such as mosquito larvae. Cuban naturalist Felipe Poey (1799–1891) described this hermaphrodite and mangrove swamp inhabitant in 1880. Juvenile Caribbean spiny lobster (*Panulirus argus*), striped mullet (*mugil cephalus*), and schoolmaster or mangrove snapper (*Lutjanus apodus*) are examples of species that mature in mangrove swamps. Sport fishermen also catch gamefish such as tarpon (*Megalops atlanticus*), common snook (*Centropomus undecimalis*), and spotted sea trout or weakfish (*Cynoscion nebulosus*) in mangrove swamps.

Mangrove swamps are important ECOSYSTEMS that are threatened by human encroachment. Conservationists are combating habitat loss by increasing public awareness and education about mangrove swamps, establishing conservation areas and integrating mangrove restoration into current coastal planning.

Further Reading

GLOMIS: GLObal Mangrove Database & Information Service, University of the Ryukys, Okinawa, Japan. Available online. URL: http://www.glomis.com/. Accessed March 9, 2008.

World Atlas of Mangroves, Food and Agriculture Organization of the United Nations (FAO). Available online. URL: www.fao.org/forestry/site/mangroveatlas/en. Accessed March 9, 2008.

marginal sea A marginal sea is a water body connected to one of the major oceans and partly enclosed by land. The adjacent coasts are strips of land extending inland to where marine vegetation is no longer found. The boundary between the coast and the shore is the COASTLINE. The basin is filled with saline water and extends from the coastline; it includes the CONTINENTAL SHELF, CONTINENTAL SLOPE, CONTINENTAL RISE, ABYSSAL PLAIN, and MID-OCEAN SPREADING RIDGES. In some cases, especially in the PACIFIC OCEAN, marginal seas form behind island arcs such as in the Aleutians, Japan, the Philippines, the

Marginal Sea	Description
Andaman Sea	Indian Ocean—bounded to the north by Myanmar, to the east by Thailand and Malaysia, to the west by the Andaman and Nicobar Islands, and to the south by Sumatra and the Strait of Malacca.
Arabian Sea	Indian Ocean—merges with the Gulf of Oman to the northwest, which flows through the Strait of Hormuz into the Persian Gulf, the Gulf of Aden in the southwest, and the Indus delta to the northeast.
Baltic Sea	Atlantic Ocean—connects to the North Sea via the Skagerrak, Kattegat, and Danish Straits to the northwest and includes the Gulf of Bothnia, Finland, and Riga to the east.
Banda Sea	Pacific Ocean—centered between the Indonesian Archipelago, New Guinea, and Australia. It connects to the South China Sea through the shallow Java Sea.
Barents Sea	Arctic Ocean—located north of the Scandanavian Peninsula and bounded by the Norwegian Sea in the west and the Arctic Ocean to the north and east.
Bay of Bengal	Indian Ocean—bordered to the north by the Ganges Delta, to the east by the Malay Peninsula, and on the west by India.
Beaufort Sea	Arctic Ocean—sited northeast of Alaska, north of Canada's Yukon Territory, Northwest Territories, and west of Canada's arctic islands.
Bering Sea	Pacific Ocean—outlined by the Kamchatka Peninsula to the west, the Aleutian Islands to the south and southeast, Alaska to the east, and the Chukchi Peninsula to the north. Connects to the Chukchi Sea through the Bering Strait. Includes Gulf of Anadyr, Norton Sound, and Bristol Bay.
Black Sea	Atlantic Ocean—lies between southeastern Europe and Asia Minor. It is connected to the Sea of Azov to the north. To the south it connects through the Bosporus to the Sea of Marmara, the Dardanelles, and the Aegean Sea.

Marginal Sea	Description
Caribbean Sea	Atlantic Ocean—bounded to the north by North America, Cuba, Hispaniola, Puerto Rico, Jamaica, and the Cayman Islands, to the east by the Leeward and Windward Islands, to the south by South America, and to the west by Central America. It connects to the Gulf of Mexico through the Yucatán Channel.
Chukchi Sea	Arctic Ocean—located off the east coast of Siberia and the northwestern coast of Alaska. Pacific waters enter via the Bering Strait.
Coral Sea	Pacific Ocean—situated in the South Pacific between Australia on the west, Vanuatu and New Caledonia on the east, between New Guinea and the Solomons on the north, and New Zealand to the south. It merges into the Tasman Sea.
East China Sea	Pacific Ocean—bordered by China, South Korea, and Japan, it is bounded to the north by the Yellow Sea and connects to the South China Sea through the Taiwan Strait. It connects with the Sea of Japan through the Korea Strait, which is divided by Tsushima Island. The Ryukyu Islands separate the East China Sea from the North Pacific Ocean.
East Siberian Sea	Arctic Ocean—situated on the eastern section of the Siberian shelf. It is separated from the Laptev Sea by the New Siberian Islands and connects to the Chukchi Sea through Long Strait.
Greenland Sea	Arctic Ocean—extends along Greenland's east coast and Norway's west coast. Bounded by Spitsbergen to the northeast and Iceland to the southwest. Adjoins the Barents Sea and connects to the Atlantic Ocean via the Denmark Strait.
Gulf of Maine	Atlantic Ocean—trends north along the coast from Cape Cod in Massachusetts to Cape Sable in Nova Scotia, Canada. The Bay of Fundy is located to the north and to the east are Georges Bank, Brown Bank, and the Atlantic Ocean.
Gulf of Mexico	Atlantic Ocean—bordered to north by the shoreline of U.S. states Texas, Louisiana, Mississippi, Alabama, and Florida and to the south by the shoreline of Mexican states Tamaulipas, Veracruz, Tabasco, Campeche, and Yucatán. At its eastern end it connects to the Caribbean Sea through the Straits of Florida and the Yucatán Channel.
Iceland Sea	Atlantic Ocean—located on the Iceland shelf and bounded by the Kolbeinsey Ridge to the west and the Jan Mayen Ridge to the east. Adjoins the waters of the Norwegian Sea to the east, the Greenland Sea to the north, the Denmark Strait to the west, and the North Atlantic Ocean to the south.
Irish Sea	Atlantic Ocean—adjoins the Celtic Sea at St. George's Channel to the south and the Atlantic Ocean through the narrow North Channel.
Java Sea	Pacific Ocean—connects the Banda Sea with the South China Sea through the Karimata Strait. It is connected to the Andaman Sea through the Strait of Malacca. Adjoins the Flores Sea and is connected to the Sulawesi (Celebes) Sea through the Makassar Strait.
Kara Sea	Arctic Ocean—located on the Siberian shelf and bounded by Novaya Zemlya to the north and the Taymyr and Yamal Peninsulas to the south. Connects to the Barents Sea on the west and the Laptev Sea to the east.
Labrador Sea	Atlantic Ocean—joins Baffin Bay through the Davis Strait to the north and Hudson Bay through the Hudson Strait to the west. Adjoins the Atlantic Ocean between Greenland and Canada to the southeast.
Laptev Sea	Arctic Ocean—sited off the coast of northern Siberia to the west of the East Siberian Sea and to the east of the Kara Sea.
Lincoln Sea	Arctic Ocean—adjoining the Arctic Ocean north of Greenland and Canada. It is situated northeastward of Ellesmere Island and northwest of northern Greenland. It connects to Baffin Bay through Nares Strait, which includes Robeson Channel, Hall Basin, Kennedy Channel, Kane Basin, and Smith Sound.
Mediterranean Sea	Atlantic Ocean—bounded by Europe to the north, by Africa on the south, and by Asia to the east. Connected to the Atlantic Ocean through the Strait of Gibraltar to the west and to the Black Sea on the northeast through the Dardanelles and the Bosporus. Numerous islands and peninsulas create the Balearic Sea, Ligurian Sea, Tyrrhenian Sea, Ionian Sea, Adraitic Sea, Aegean Sea, Sea of Crete, and Sea of Marmara.
North Sea	Atlantic Ocean—separates the British Isles from continental Europe. Linked to the Celtic Sea through the Strait of Dover and English Channel. Adjoins the Norwegian Sea to the north, where it separates Scotland from Norway.

(continues)

Marginal Sea	Description
Norwegian Sea	Arctic Ocean—consists of the waters between the continental shelves of Norway to the east and Iceland to the west. It adjoins the Barents Sea to the northeast, the Greenland Sea to the northwest, the Iceland Sea to the west, and the North Sea to the southeast. The Shetland and Faroe Islands are located toward its southern extent.
Persian Gulf	Indian Ocean—surrounded by Iran and Iraq to the north, Kuwait, Saudi Arabia, Bahrain, Qatar, United Arab Emirates, and Oman to the east and south. It connects with the Arabian Sea and the Gulf of Oman through the Strait of Hormuz.
Philippine Sea	Pacific Ocean—situated in the western Pacific Ocean. Located between the Philippine archipelago to the west and the Mariana and Caroline Islands to the east. To the north are the Ryukyu Islands and Japan. Connects to the South China Sea through the Luzon Strait, which is divided by the Bataan and Babuyan Islands.
Red Sea	Indian Ocean—separates the African and Asian continents. The Sinai Peninsula divides the northern part into the shallow Gulf of Suez to the west and the deep Gulf of Aquaba to the east. The Mediterranean Sea is linked to the Gulf of Suez by the man-made Suez Canal. The Red Sea connects to the Indian Ocean through the Gulf of Aden.
Ross Sea	Pacific Ocean—adjacent to the Ross Ice Shelf and flanked by Marie Byrd Land and the southern extent of the Transantarctic Mountains. It connects to McMurdo Sound, which is free of pack ice in late summer and therefore an important staging point for scientific investigations.
Sea of Cortez	Pacific Ocean—bordered by Baja California and mainland Mexico. The Sea of Cortez or Gulf of California trends southeastward into the Pacific Ocean. The Colorado River delta is located in the northern end of the gulf near Puerto Penasco, Mexico.
Sea of Japan	Pacific Ocean—enclosed on the east by the Japanese islands, the west by the Korean Peninsula, and the north and northwest by Russia. It is connected to the East China Sea in the south, the Sea of Okhotsk in the north, and the Pacific Ocean in the east via the Korean Strait and the Tatar strait.
Sea of Okhotsk	Pacific Ocean—bounded by the Siberian coast to the west and north, the Kamchatka Peninsula to the east, and the Kurile Islands to the south and southeast.
South China Sea	Pacific Ocean—Bordered to the west by Vietnam, Thailand, and the Malay Peninsula. Includes the Gulf of Thailand and the Gulf of Tonkin. It is connected to the East China Sea via the Taiwan Strait, the Andaman Sea via the Strait of Malacca, the Java Sea via the Karimata Strait, the Sulu Sea via the Balabac Strait, and to the Philippine Sea via Luzon Strait.
Sulawesi Sea	Pacific Ocean—encircled by the Philippines, Indonesia, and Malaysia. Separated from the Sulu Sea to the north by the Sulu Archipelago. Joins the Java Sea to the south through the Makassar Strait. Adjoins the north part of the Moluccan Sea to the east. Also known as Celebes Sea.
Tasman Sea	Pacific Ocean—situated off the southwest coast of Australia. It is surrounded by New Zealand to the east, Tasmania to the southwest, and the Coral Sea to the north.
Weddell Sea	Atlantic Ocean—Located at the southern most tip of the Atlantic Ocean. Sited west of Antarctica, southeast of South America and bordering the Antarctic (Palmer) Peninsula. It is adjacent to the Ronne Ice Shelf. The Bransfield Strait connects the Drake Passage with the waters of the Weddell Sea. The Drake Passage is the open ocean between South America and Antarctica.
Yellow Sea	Pacific Ocean—surrounded on the north and west by China and the Korean Peninsula to the east. It connects to the Bohai Sea and Korea Bay in the north and in the south merges into the East China Sea.

Ryukyus, and the Solomons. These dividers form as a result of volcanism caused by the subduction of oceanic plates into trenches (see PLATE TECTONICS). The Kuril Islands form an arc that extends from the southern tip of the Kamchatka Peninsula in Russia to Hokkaido, the second largest island in Japan. Straits connecting larger bodies of water are common to marginal seas. For example, the Bering Sea is associated with dividers such as the Aleutian Islands along the northern part of the Pacific Rim and connectors such as the Bering Strait, which links the Bering Sea with the Chukchi Sea. *Pacific Rim* is a frequently used term referring to countries in East Asia, North America, South America, and Australia that border the Pacific Ocean. Marginal seas may also be connected regionally such as the Nordic Seas, which refers to the Norwegian, Greenland and Iceland Seas. Marginal seas along with the Arctic, Atlantic, Indian and Pacific Oceans make up the global ocean. Water flowing out of a marginal sea may flow across a sill into the deep sea. Sills are depositional features that separate one deep-water area from another. Forceful flow across the sills can induce EDDIES and INTERNAL WAVES. The surface manifestations of these phenomena have been observed with satellite imagery, especially by international research programs investigating the South China Sea.

Marginal seas are semi-enclosed basins connected to the ocean. The table that begins on page 282 lists the name of a marginal sea, the associated ocean, and a description of the important boundaries, dividers, and connectors. Large landlocked of inland water bodies such as the Caspian Sea (the largest salt lake in the world), Aral Sea, Dead Sea, and the Sea of Galilee are not marginal seas because they do not connect to the ocean.

Further Reading

International Hydrographic Organization. *Limits of Oceans and Seas, Special Publication 23,* 3rd ed. Monte-Carlo: IHO, 1953.

National Geographic Society. *National Geographic Atlas of the World,* 7th ed. Washington, DC: National Geographic Society, 1999.

mariculture The brackish or saltwater farming and husbandry of marine animals and plants in tanks, ponds, cages, pens, or net-enclosed areas in the ocean is known as mariculture, marine aquaculture, or seafarming. AQUACULTURE refers to the cultivation of freshwater animals and plants. The terms *aquaculture* and *mariculture* are sometimes used interchangeably. Mariculture operations are primarily used to produce seafood such as red drum (*Sciaenops ocellatus*), red abalone (*Haliotis rufescens*), chemicals such as algin, carrageenin, iodine, mannitol (marine MACROPHYTES), and cultured pearls from silver-lip pearl oysters (*Pinctada maxima*). Mariculture operations are also established for population enhancement of threatened and endangered species. Restoration efforts in Delaware, Maryland, and Virginia for species such as American shad (*Alosa Sapidissima*) have been successful along the East Coast of the United States. Research and development in mariculture is especially important as scientists solve problems such as waste disposal and disease prevention in the cultivation of fish, shellfish, and seaweeds in natural and controlled environments. Some of the most common species raised by mariculture are listed in the following table.

Common Name	Scientific Name
Atlantic salmon	*Salmo salar*
banana prawn	*Fenneropenaeus indicus* and *F. merguiensis*
blood cockle	*Anadara granosa*
blue mussel	*Mytilus edulis*
Chilean blue mussel	*Mytilus chilensis*
cobia	*Rachycentron canadum*
coho salmon	*Oncorhynchus kisutch*
eastern oyster	*Crassostrea virginica*
European seabass	*Dicentrarchus labrax*
fleshy prawn	*Penaeus chinensis*
giant kelp	*Macrocystis pyrifera*
gilthead seabream	*Sparus aurata*
gracilaria seaweeds	*Gracilaria asiatica, G. lemaneiformis, G. tenuistipitata,* and *G. verrucosa,*
green mussel	*Perna viridis*
Japanese carpet shell	*Ruditapes philippinarum*
Japanese flounder	*Paralichthys olivaceus*
Japanese kelp	*Laminaria japonica*
Japanese scallop	*Patinopecten yessoensis*
jumbo tiger prawn	*Penaeus monodon*
Mediterranean mussel	*Mytilus galloprovincialis*
milkfish	*Chanos chanos*
nori	*Porphyra yezoensis, P. tenera, P. haitanensis, P. pseudolinearis, P. kunideai, P. arasaki,* and *P. seriata*
northern quahog	*Mercenaria mercenaria*
Pacific oyster	*Crassostrea gigas*
Peruvian scallop	*Argopecten purpuratus*
rainbow (steelhead) trout	*Oncorhynchus mykiss*
Red Sea bream	*Pagrus major*
red seaweeds	*Eucheuma denticulatum* and *Kappaphycus alvarezii*
red snapper	*Lutjanus campechanus*
seagrapes	*Caulerpa racemosa*
striped bass	*Morona saxatilis*
striped mullet	*Mugil cephalus*
southern flounder	*Paralichthys lethostigma*
wakame	*Undaria pinnatifida*
whiteleg shrimp	*Penaeus vannamei*
yellowtail	*Seriola quinqueradiata*
Zanzibar weed	*Eucheuma cottonii*

Further Reading

Center for Tropical and Subtropical Aquaculture, Waimanalo, HI. Available online. URL: http://www.ctsa.org/. Accessed March 9, 2008.

Food and Agriculture Organization of the United Nations (FAO). *FAO Technical Report, No. 500*. Rome: Food and Agriculture Organization of the United Nations. 2006. Also available online. URL: http://www.fao.org/docrep/009/a0874e00.htm. Accessed March 9, 2008.

Mariculture Committee, International Council for the Exploration of the Sea (ICES), Copenhagen, Denmark. Available online. URL: http://www.ices.dk/iceswork/mcc.asp. Accessed March 9, 2008.

NOAA Aquaculture Program, National Ocean Service, National Ocean Service, National Oceanic and Atmospheric Administration, U.S. Department of Commerce. Available online. URL: http://aquaculture.noaa.gov. Accessed March 9, 2008.

Marine Advanced Technology Education (MATE) Center   Founded in 1997, the Marine Advanced Technology Education (MATE) Center creates an education system for marine technicians by offering technical programs, curriculum development, best-practice demonstrations, student involvement and outreach activities necessary to develop workforce capacity, and cutting-edge skills for the marine industry. Programs range from summer institutes focused on topics such as building and operating Remotely Operated Vehicles (ROVs) (*see* REMOTELY OPERATED VEHICLES; ROV's) to curricula relevant to Geographic Information Systems (GIS), and information on marine technology certificate and degree programs.

The MATE Center received initial funding from the National Science Foundation as an outcrop from legislation during 1992 that created a number of Advanced Technological Education (ATE) programs. These ATE centers are aimed at increasing the number of skilled technicians to work in "strategic advanced-technology fields." Within the environmental science field, there is an Advanced Technology Environmental Education Center located on the campus of Scott Community College in Bettendorf, Iowa, a Marine Advanced Technology Education Center located on the campus of Monterey Peninsula College in Monterey, California, and a Northwest Center for Sustainable Resources located on the campus of Chemeketa Community College in Salem, Oregon.

The MATE Center is especially important since marine science jobs employ a wide range of technical people such as biologists, chemists, geologists, physicists, mathematicians, oceanographers, meteorologists, hydrologists, engineers, CARTOGRAPHERS, computer scientists, and marine technicians. Marine scientists are sought out by operators to collect data, process data, analyze data, and distribute beneficial marine information. Marine information is essential to support the managers of ocean resources, the builders of offshore oil production platforms, nuclear power plants, and forecasters of natural hazards such as HURRICANES and TSUNAMIS.

The MATE Center is best known for its action-packed summer institutes and involvement in the National Ocean Sciences Bowl. During the summer of 2005, the MATE Center conducted their 4th Annual National Student ROV Competition at the NASA Johnson Space Center's Neutral Buoyancy Lab in Houston, Texas. Teams learned about ROV technologies and then competed to assemble an ROV capable of meeting navigation and recovery challenges set forth by the competition. Since 1998, the National Ocean Sciences Bowl or NOSB® has engaged students and teachers at the regional and national level to assess their knowledge on the world's oceans and awareness of topics from marine resources and ocean exploration to global climate change and sea level change. On this front, the MATE Center in cooperation with the Monterey Bay National Marine Sanctuary hosts a regional competition called the Otter Bowl. Regional events are timed so that winning teams from around the country can compete at the NOSB® during the spring. These forums in concert with organizations such as the MATE Center are able to improve science education as they link curricula to the National Science Education standards. Such programs have enabled the MATE Center to bring the exciting world of marine science and technology into high school and community college classrooms.

Further Reading

Advanced Technology Environmental Education Center, home page. Available online. URL: www.ateec.org. Accessed November 1, 2005.

Marine Advanced Technology Education Center home page. Available online. URL: www.marinetech.org. Accessed November 1, 2005.

National Ocean Sciences Bowl. Available online. URL: http://www.nosb.org/. Accessed September 25, 2007.

Northwest Center for Sustainable Resources home page. Available online. URL: www.ncsr.org. Accessed November 1, 2005.

marine archaeologist   A marine archaeologist is a scientist involved in historical maritime studies, the exploration of relics, past sea life, and nautical anthropology through discovery, analysis, and conservation of marine artifacts and monuments.

A Selected List of Deepwater, Shallow-Water, and Coastal Archaeological Sites

Description	Location	Comments
Brenton Reef Lightship *LV 39*	Light Vessel *LV 39* was discovered in Massachusetts Bay in 180 feet (55 m) of water during July 2004. The bay has been a center of commercial shipping for more than 300 years and covers approximately 800 miles (1,287 km) of coastline from the tip of Cape Cod Bay to the New Hampshire border.	Light Vessels (LV) served to guide mariners where there was a need for warnings but in a spot where lighthouses could not be built. *LV 39* was in service at various locations from 1875 to 1935. Its final service involved marking dangerous shoals near Brenton Point. *LV 39* sank during 1975 while being towed to a shipyard in Beverly, Massachusetts, the oldest landing in the United States.
Emanuel Point Shipwreck	A colonial Spanish ship and all of it contents. were located in 12 feet (3.7 m) of water in Pensacola Bay during 1992	Discovered in 1992 by the Florida Bureau of Archaeological Research. Thought to be part of the expedition of Tristán de Luna, who led the first attempt by Europeans to colonize Florida in 1559.
HMS *DeBraak*	After sinking off of Cape Henlopen during 1798, the Brigantine *DeBraak* was discovered in 1984 and raised during 1986.	The Zwaanendael Museum in Lewes, Delaware, displays approximately 20,000 artifacts salvaged from the Brigantine *DeBraak*
HMS *Sussex*	Rests in deep water in the western Mediterranean sea.	The 80-gun British warship sank during a storm in 1694 with a valuable cargo of coin.
Japanese Midget Submarine	University of Hawaii researchers using deep-diving submersibles discovered a Japanese midget submarine about three to four miles off Pearl Harbor during 2002.	The submarine was sunk by a U.S. Navy destroyer on December 7, 1941, and is now considered a war grave and an important monument that bears witness to the outbreak of World War II in the Pacific.
Monitor National Marine Sanctuary	The *Monitor* sank on December 31, 1862, while being towed approximately 16 miles (25.7 km) in 240 feet (73 m) of water off Cape Hatteras, North Carolina. It was discovered in August 1973 by Duke University researchers.	The Monitor National Marine Sanctuary protects the wreck of the Civil War ironclad USS *Monitor*. There is major deterioration of the hull. A Monitor Center is located at the Mariners' Museum in Newport News, Virginia.
North American B-25C Mitchell	Recovered during September 2005 at depth of 145 feet (44.2 m) in Lake Murray, South Carolina.	The B-25C was the first version of the Mitchell to be mass-produced for service in World War II. This particular bomber was ditched during a training mission on April 4, 1943, and the crew escaped safely.
Nuestra Señora de Atocha	One of several Spanish galleons that sank during September 1622 off the Florida Keys during a hurricane and discovered in 1985.	The *Nuestra Señora de Atocha* was bound for Spain and heavily laden with treasure. Relicts from such shipwrecks yield important clues into colonial America.
Port Royal, Jamaica	Port Royal in Jamaica was destroyed on June 7, 1692, by an earthquake. Underwater archaeological investigations of the drowned 17th-century city began during 1981.	One of the largest towns in the English colonies and a haven for pirates.
Steamship *Great Britain*	Launched during 1843 by the Great Western Steamship Company. The vessel ran aground in shoal water off Ireland. It made many voyages between England and Australia. The vessel was restored in dry dock in Bristol, the same seaport where it was built.	The SS *Great Britain* was the world's first iron ocean liner driven by a screw. This ship was designed to provide luxury transatlantic travel and set new standards in engineering, reliability, and speed. Today, she is a popular museum ship in Bristol, United Kingdom.
Tel Nami Project	Haifa University archaeologists have excavated a Bronze Age seaport on Israel's west coast near the Mount Carmel region, approximately 16 miles (25 km) south of the modern seaport of Haifa.	The Carmel coast provides a natural harbor along the Mediterranean Sea and numerous archaeological sites that are key to understanding maritime history as well as the history of biblical times.

Archeological sites usually contain the physical remains of past human activities and records documenting the scientific analysis of these artifacts. Research and interpretation of submerged cultural resources such as shipwrecks and seaports may involve REMOTE SENSING surveys and nonintrusive mapping investigations to full-scale excavations.

The table on page 287 provides a selected listing of deep-water, shallow-water, and coastal archaeological sites. The preservation of shipwrecks is very important since a finite number of ships are lost in each period, by each culture.

See also ARCHAEOLOGIST.

marine biologist Marine biology is a discipline under the broader field of biological oceanography. Marine biologists study or work on projects involving the study of marine plants, animals, and protests (one-celled organisms, such as protozoans and eukaryotic algae) in riverine, estuarine, and oceanic environments. Species may range in size from microscopic bacteria and ALGAE that drift about in the water column to migrating whales such as the blue whale (*Balaenoptera musculus*), the largest mammal. Marine biology is a popular undergraduate program for students entering biological oceanography graduate programs. Biological oceanographers study organisms and their biological processes within the marine environment. Some people use the terms *marine biology* and *biological oceanography* interchangeably.

See also BIOLOGICAL OCEANOGRAPHER.

Further Reading
Thurman, H. V., and E. A. Burton. *Introductory Oceanography,* 9th ed. Upper Saddle River, N.J.: Prentice Hall, 2001.

marine birds Marine birds are warm-blooded vertebrate animals with wings, feathers, and a beak or bill that depend on marine habitats for survival. All birds have a four-chambered heart and lay eggs. Marine birds swim, wade, or feed in the water and are often categorized by their behaviors. They have salt glands, which concentrate salt from blood near the sinuses. Excess salt is "sneezed" out. Many marine birds have webbed feet to paddle through the water and walk on mud. There are diving birds such as the brown pelican (*Pelecanus occidentalis*), pelagic birds such as the wandering albatross (*Diomedea exulans*) with an 11-foot (3.4 m) wingspan that can soar above the waves for long time periods, penguins such as the emperor penguin (*Aptenodytes Forsteri*), which inhabit the frozen Antarctic and "fly" underwater, shorebirds of the order Charadriiformes, such as

bristle-thighed curlew (*Numenius tahitiensis*), which scramble to and fro in the swash zone in search of Pacific mole crabs (*Emerita analoga*), and waders such as the long-legged Egyptian plover (*Pluvianus aegyptius*), which may clean meat fragments from crocodile teeth. The swash zone is the area between wave runup and rundown on a BEACH.

Some marine birds undergo long migrations and find coastal lands such as marshes to breed. Arctic terns (*Sterna paradisaea*) fly approximately 20,000 miles (32,000 km) each year as they migrate from the Arctic to the Antarctic and back. Marshes provide a suitable habitat for predators such as osprey (*Pandion haliaetus*), belted kingfishers (*Ceryle alcyon*), anhinga (*Anhinga anhinga*), and limpkin (*Aramus guarauna*). Marine birds take advantage of low tide to forage on mudflats. Some such as gulls are scavengers eating crippled fish that have been wounded by schools of predators such as bluefin tuna (*Thunnus thynnus*), king mackerel (*Scomberomorous cavalla*), or Spanish mackerel (*Scomberomorous maculatus*), bait scraps from fishing vessels, or may even be found inland feeding on refuse near Dumpsters and landfills. Herbivorous species such as dabbler ducks feed on SUBMERGED AQUATIC VEGETATION, seeds, grasses, small insects and animals with their broad and short bills. Marine birds compete more for nesting space than for food.

Diving birds (pelecaniformes) such as cormorants, pelicans, and boobies are predators. Pelecaniformes all have webbed feet and a gular sac (throat sac). In pelicans, the throat sac is attached to the lower mandible of the bill and traps fish and water as the bird plummets into the sea after schooling finfish. Pelicans such as the brown pelican (*Pelecanus occidentalis*) are amazing fliers and are commonly observed cruising in single file along the nearshore zone on feeding trips. American white pelicans (*Pelecanus erythrorhynchos*) scoop fish with their enormous bill while swimming. The bill is pointed downward in order to drain water and then raised to swallow captured fish. Cormorants are fish eaters that dive from the air or from surface into shoal waters, where they swim with webbed feet to capture prey, using their hooked bills. Boobies such as the northern gannet (*Morus bassanus*), a pelagic bird ranging from Canada to the GULF OF MEXICO, are recognized for their incredible and aerodynamic plunges for fish from the air. Northern gannets are the largest of the boobies and head for shore to nest and breed.

Pelagic (seagoing) birds such as the ashy storm-petrel (*Oceanodroma homochroa*), northern fulmar (*Fulmarus glacialis*), and the sooty shearwater (*Puffinus griseus*) live much of their lives offshore. They are

Emperor penguin
Aptenodytes forsteri

Wandering albatross
Diomedea exulans

Little shearwater
Puffinus assimilis

Great black-backed gull
Larus marinus

Arctic tern
Sterna paradisea

Brown pelican
Pelecanus occidentalis

Great cormorant
Phalacrocorax carbo

Common seabirds

especially adapted for the ocean as evidenced by salt glands that empty into their nose and the excretion of nitrogenous wastes as insoluble uric acid to conserve water. Species often have an external tube on the horny covering of the bill. Skimming the surface of permanent currents such as the GULF STREAM are pelagic birds such as the black-capped petrel (*Pterodroma hasitata*), Cory's shearwater (*Calonectris*

Selected List of Marine Birds

Common Name	Scientific Name	Characteristics
common loon	*Gavia immer*	diving bird
coot	*Fulica americana*	diving bird
double-crested cormorant	*Phalacrocorax auritus*	diving bird
red-throated loon	*Gavia stellata*	diving bird
ancient murrelet	*Synthliboramphus antiquus*	pelagic bird
black-footed albatross	*Phoebastria nigripes*	pelagic bird
Leach's storm-petrel	*Oceanodroma leucorhoa*	pelagic bird
Laysan albatross	*Phoebastria immutabilis*	pelagic bird
red-billed tropicbird	*Phaeton aethereus*	pelagic bird
red-tailed tropicbird	*Pheaton rubricauda*	pelagic bird
tufted puffin	*Fratercula cirrhata*	pelagic bird
white-faced storm-petrel	*Pelagodroma marina*	pelagic bird
white-tailed tropicbird	*Pheaton lepturus*	pelagic bird
American avocet	*Recurvirostra americana*	shorebird
Atlantic least tern	*Sterna antillarum*	shorebird
black-bellied plover	*Pluvialis squatarola*	shorebird
black-necked stilt	*Himantopus mexicanus*	shorebird
Bonaparte's gull	*Larus philadelphia*	shorebird
Dunlin	*Calidris alpina*	shorebird
Caspian tern	*Sterna caspia*	shorebird
Forster's tern	*Sterna forsteri*	shorebird
laughing gull	*Larus atricilla*	shorebird
marbled godwit	*Limosa fedoa beringiae*	shorebird
mew gull	*Larus canus*	shorebird
ring-billed gull	*Larus delawarensis*	shorebird
western gull	*Larus occidentalis*	shorebird
American bittern	*Botaurus lentiginosus*	wading bird
black-crowned night heron	*Nycticorax nycticorax*	wading bird
glossy ibis	*Plegadis falcinellus*	wading bird
great egret	*Egretta alba*	wading bird
least bittern	*Ixobrychus exilis*	wading bird
wood stork	*Mycteria americana*	wading bird
Bufflehead	*Bucephala albeola*	waterfowl
canvasback duck	*Aythya vallisneria*	waterfowl
greater scaup	*Aythya marila*	waterfowl
green-winged teal	*Anas crecca*	waterfowl
mallard	*Anas platyrhynchos*	waterfowl
mottled duck	*Anas platyrhynchos*	waterfowl
pied-billed grebe	*Podilymbus podiceps*	waterfowl
pintail duck	*Anus acuta*	waterfowl
ruddy duck	*Oxyura jamaicensis*	waterfowl
wood duck	*Aix sponsa*	waterfowl

diomedia), Audubon's shearwater (*Puffinus ihermin-ieri*), Wilson's storm-petrel (*Oceanites oceanicus*), band-rumped storm-petrel (*Oceanodroma castro*), bridled tern (*Sterna anaethetus*) and sooty tern (*Sterna fuscata*). They feed on fish larvae, flying fish, and other small fish, squid, and crustaceans such as COPE-PODS, which are often captured on the ocean surface at night. These birds only return to land to breed.

Penguins are equipped to handle the extreme cold of the Antarctic and sub-Antarctic islands. They cannot fly in the air, but use their wings like paddles to dart around in polar waters to feed on Antarctic krill (*Euphausia superba*), squid, and small finfish. Penguins may dive to great depths to get their meals. Penguins have little surface area compared to their actual volume owing to small feet, wings, and heads. They have shiny, waterproof feathers to keep their skin dry. They also have a thick insulating layer of fat under the skin. Several example species are the African penguin (*Spheniscus demersus*), royal penguin (*Eudyptes schlegeli*), rock hopper penguin (*Eudyptes chrysocome*), yellow-eyed penguin (*Megadyptes antipodes*), macaroni penguin (*Eudyptes chrysolophus*), and the erect-crested penguin (*Eudyptes atratus*).

Shorebirds include plovers, sandpipers, gulls, and terns. The principal shorebirds consist of plovers and sandpipers. Plovers such as the piping plover (*Charadrius melodus*) are fast fliers with sharp pointed wings. They have round heads, short bills, plump bodies, long legs, and short tails. Sandpipers such as the purple sandpiper (*Calidris maritima*) and the rock sandpiper (*Calidris ptilocnemis*) have elongated bodies and slender bills. Plovers and sandpipers feed by probing in the mud for insects, crustaceans, MOL-LUSKS, and worms, or picking food items off the surface of the water. Gulls are normally found along the coast, typically nest on the ground, and feed on crabs and small fish. Terns, which are in the same family as gulls (Laridae), appear to hover over the water until they site a small fish, whereupon they plunge in after their prey.

Wading birds such as the great blue heron (*Ardea herodias*), the wood stork (*Mycteria americana*), and the sandhill crane (*Grus canadensis*) all have long legs, long bills, and long necks to wade into the water to catch their food. They are at the top of the food chain in wetlands such as salt marshes, swamps, and ESTUARIES, which they use for foraging and breeding. While many catch their prey with a quick jab of the bill, the colorful roseate spoonbill (*Ajaja ajaja*) lowers its open bill into the water and sweeps from side to side, closing its mandibles when it touches small fish. The Florida Everglades, one of the largest wetlands in the world, is a major breeding ground for wading birds in the eastern United States.

Seeing marine birds in flight, admiring their colors and crests, or observing a perilous dive to capture prey fascinates many beachgoers. Birdwatchers and FISHERMEN may use eight, nine or 10-power handheld binoculars for long-distance bird watching. The number indicates the magnification, or how much larger, or closer, the object will appear versus what is seen with normal vision. Since many commercial fish swim in schools and prey on smaller schools of fish, fishermen watch to see where marine birds such as gulls, cormorants, and terns are feeding. By monitoring where the birds hunt, they are able to find large schools of fish and increase their catch. Marine birds have inspired many countries, states, and organizations to name a particular species as a mascot or state bird. Canada chooses the common loon (*Gavia immer*), which breeds across Canada as the national bird. The California gull (*Larus californicus*) is the Utah state bird, and the brown pelican (*Pelecanus occidentalis*) is the Louisiana state bird.

Seabirds face many threats from man such as being oiled by marine spills, loss of habitat, entanglement in fishing gear, ocean pollution from dumping, and possibly by increases in global temperature. The table on page 290 is a selected listing of marine birds; there is no taxonomic or evolutionary group of marine birds.

Further Reading

Cornell Lab of Ornithology. "All About Birds." Available online. URL: http://www.birds.cornell.edu/AllAbout-Birds/BirdGuide. Accessed September 25, 2007.

Grand Manan Whale and Seabird Research Station. Available online. URL: http://www.gmwsrs.org/main.htm. Accessed August 16, 2006.

Monterey Bay National Marine Sanctuary Site home page. Available online. URL: http://bonita.mbnms.nos.noaa.gov/sitechar/bird.html. Accessed September 25, 2007.

USGS, Patuxent Wildlife Research Center. "Patuxent—Bird Population Studies." Available online. URL: http://www.mbr-pwrc.usgs.gov/. Accessed September 25, 2007.

Western Atlantic Shorebird Association. Available online. URL: http: www.vex.net/~hopscotc/shorebirds/. Accessed September 25, 2007.

marine crustaceans Marine crustaceans are a group of free-living and sessile (attached) invertebrates living in salt or brackish water, best known for their hard outer shell such as American lobsters (*Homarus americanus*), blue crabs (*Callinectes sapidus*), and white shrimp (*Panaeus setiferus*). Brackish water is a mix of freshwater and salt water common to ESTUARIES. Marine crustaceans might be parasitic such as Griffen's isopod (*Orthione griffenis*), which

A Selection of Interesting and Commercially Important Crustaceans

Common Name	Scientific Name	Regional Locations
Alaskan prawn, California prawn, or spot prawn	*Pandalus platyceros*	Spot prawn are large benthic shrimp found in subtidal rocky and sandy north Pacific habitats.
American lobster, Maine lobster	*Homarus americanus*	American lobsters are found in the western North Atlantic, from Labrador to Cape Hatteras in North Carolina.
blue crab	*Callinectes sapidus*	Blue crabs are native to the western Atlantic from Nova Scotia, Canada, to northern Argentina, including Bermuda and the Caribbean.
blue king crab	*Paralithodes platypus*	Blue king crabs occur most often between 147–246 feet (45–75 m) depths on mud-sand substrate adjacent to gravel rocky bottom. They range the North Pacific, from British Columbia to the Bering Sea and from the Aleutian Islands to the Sea of Japan.
blue shrimp	*Litopenaeus stylirostris*	Blue shrimp favor water depths less than 90 feet (27 m) and are native to the eastern Pacific, from Mexico to Peru.
brown shrimp	*Farfantepenaeus aztecus*	Brown shrimp are found offshore in depths from 90 to 180 feet (27–55 m) from Massachusetts, around the Florida peninsula, and to the lower Yucatán Peninsula. They are the most abundant around the Bay of Campeche along the Texas coast.
California spiny lobster	*Panulirus interruptus*	California spiny lobsters favor rocky coastlines in the eastern North Pacific, from San Luis Obispo in California to Rosalia Bay in Baja California, Mexico.
Caribbean spiny lobster, rock lobster, Florida lobster, West Indian langouste	*Panulirus argus*	Caribbean spiny lobsters are found in shallow reefs or eel grass beds along the western Atlantic, from North Carolina to Rio de Janeiro in Brazil, including the Gulf of Mexico, Bermuda, and the Caribbean.
Chinese white shrimp, fleshy prawn, Oriental shrimp	*Fenneropenaeus chinensis*	Chinese white shrimp are native to the eastern North Pacific, off the coasts of China and western coast of Korea.
common shrimp, brown Shrimp	*Crangon crangon*	Common shrimp favor shallow sand and mud bottoms. They are found in coastal regions along the eastern Atlantic, Atlantic coast of Europe, and in the Mediterranean, Baltic, and Black Seas.
Dungeness crab, Pacific edible crab, market crab	*Cancer magister*	Dungeness crabs prefer waters less than 295-feet (90 m) deep with sandy bottoms. They are found in the western North Pacific, from Alaska to central California.
European green crab, shore crab, green crab, Joe Rocker	*Carcinus maenas*	The green crab lives predominantly in the intertidal area and favors estuaries and salt marshes. It is indigenous to the northeastern Atlantic, from Norway to North Africa.
European lobster, common lobster, hummer	*Homarus gammarus*	European lobsters favor continental shelves to depths of 492 feet (150 m). They are found on hard substrates in the eastern Atlantic from northwestern Norway (Lofoten Islands) south to the Azores and the Atlantic coast of Morocco. They are also found in the Mediterranean and the northwest coast of the Black Sea.
Florida stone crab	*Menippe mercenaria*	Florida stone crabs are found in the western North Atlantic, from North Carolina to Texas, including the Gulf of Mexico, Cuba, and the Bahamas.
galatheid crab, pinchbug, squat lobster	*Euminida picta*	Galatheid crabs are found from 656 to 2,624 feet (200 to 800 m) from Norway and Greenland in the north to the Bay of Biscay in the south. They prefer rock and cobble slopes, seamounts, and have been observed around hydrothermal vents, cold seeps, and whale falls.

Common Name	Scientific Name	Regional Locations
giant crab, Tasmanian crab, Australian giant crab	*Pseudocarcinus gigas*	Giant crabs are found in high latitudes in waters that are from 59 to 1,640 feet (18–500 m) deep. They occur along the edge of the continental shelf around south and western Australia.
giant tiger prawn, black tiger shrimp	*Penaeus monodon*	Giant tiger prawns are often found over muddy sand or sandy bottoms at depths from 65 to 164 feet (20–50 m). They are native to the Indian and western Pacific Oceans near the coasts of East Africa, South East Asia, South Asia, and Australia.
golden crab	*Chaceon fenneri*	Golden crabs live at depths greater than 1,000 feet (305 m) and favor rock ledges along the upper continental slope. They occur along the U.S. east coast, in waters off Bermuda, throughout the Charleston Bump area, and into the Gulf of Mexico.
golden king crab, Adak brown king crab	*Lithodes aequispina*	The golden king crab prefers rocky substrates and waters from 590 feet (180 m) to 3,000 feet (914 m). They are found along the continental slopes of the North Pacific Ocean and the Bering Sea near Alaska and Russia.
Gulf stone crab	*Menippe adina*	The Gulf stone crab is an estuarine and coastal species found in waters less than 167 feet (51 m) from North Carolina to northwest Florida and the Caribbean Sea.
hair crab, Korean hair crab	*Erimacrus isenbeckii*	High-latitude crabs found north of the Alaskan peninsula and around the Pribilof Islands.
hairy crab	*Haplogaster mertensii*	The hairy crab ranges from Atka in the Aleutian Islands to Puget Sound, Washington.
helmet crab	*Telmessus cheiragonus*	Helmet crabs are found among submerged aquatic vegetation and range the intertidal to 360 feet (110 m) from the Chukchi Sea and Norton Sound, Alaska, to Monterey, California. They also are found from Siberia to Japan in the western Pacific.
Jonah crab	*Cancer borealis*	Jonah crabs favor soft bottoms and are found in deep waters of the continental slope. They range the western North Atlantic, from Nova Scotia in Canada to the Tortugas at the southernmost tip of the Florida Keys.
northern stone crab	*Lithodes maja*	Northern stone crabs favor hard and soft substrates at depths from 213 feet (65 m) to 2,650 feet (800 m). They are found in the northwest Atlantic from Greenland to New Jersey and off western Europe in locations such as the Koster-Väderö Trough and the Gullmarfjord near Sweden and along the coast of Norway.
northern shrimp, northern prawn, deep-water prawn	*Pandalus borealis*	Northern shrimp favor mud and clay bottom at depths from 66 to 4,364 feet (20 to 1,330 m). They are found throughout the North Atlantic, North Pacific, and Arctic Oceans.
Norway lobster	*Nephrops norvegicus*	Norway lobsters favor muddy bottoms that are from 66 to 2,650 feet (20 to 800 m) deep. They are found in the eastern Atlantic region from Iceland, the Faeroes and northwestern Norway (Lofoten Islands), south to the Atlantic coast of Morocco. They are also found along the western and central basin of the Mediterranean.
Pacific pink shrimp, ocean shrimp, Oregon shrimp or pink shrimp	*Pandalus jordani*	Pacific pink shrimp are found from California to Alaska over muddy bottoms.
pink shrimp	*Farfantepenaeus duorarum*	Pink shrimp are found in the western Atlantic, from Maryland to Quintana Roo in Mexico.

(continues)

A Selection of Interesting and Commercially Important Crustaceans (continued)

Common Name	Scientific Name	Regional Locations
red king crab, Alaskan king crab	*Paralithodes camtschaticus*	Red king crabs are found in waters less than 590 feet (180 m) deep in the North Pacific Ocean. In Asia, they occur from the Sea of Okhotsk and the Shelf of Kamchatka in eastern Russia to the Aleutian Islands.
ridged slipper lobster, shovelnose lobster	*Scyllarides nodifer*	Slipper lobsters prefer mud, sand, shell and coral bottoms from depths of six to 298 feet (2 to 91 m). The ridged slipper lobster is found in the western Atlantic Ocean off North Carolina through the Gulf of Mexico to Yucatán.
scaly slipper lobster, common slipper lobster	*Scyllarides squammosus*	Slipper lobsters are found in the Indian Ocean and Pacific, including Hawaii. They are found in shallow waters around coral reefs.
Snow crab, tanner crab, queen crab	*Chionoecetes opilio*	Snow crabs favor mud, sand, and gravel bottoms that range from 229 to 918 feet (70 to 280 m) deep. They have a circum-arctic distribution occurring from the eastern Bering Sea northward to the Beaufort Sea and in the western Atlantic Ocean south to Casco Bay, Maine.
southern tanner crab	*Chionoecetes bairdi*	Southern tanner crabs prefer green-and-black mud, fine gray-and-back sand, and shell bottoms. They have been found from shoal water to 1,552 feet (473 m) deep. They range the North Pacific, from the Shelf of Kamchatka in eastern Russia to the Bering Sea and from the Alaska Peninsula down to Oregon State.
white shrimp, northern white shrimp	*Panaeus setiferus*	White shrimp favor shallow mud, peat, clay, and sand bottoms. They are found in estuaries and along the coast in the western North Atlantic, from New Jersey to Campeche in Mexico.
whiteleg shrimp, white shrimp	*Penaeus vannamei*	Whiteleg shrimp prefer shoal waters to 236 feet (72 m) and are native to the eastern Pacific, from Sonora in Mexico to northern Peru.

Some species such as pinchbugs and squat lobsters have common names that are not accepted by the American Fisheries Society.

attacks burrowing mud shrimp (*Upogebia puget-tensis*), filter feeders such as Antarctic krill (*Euphausia superba*), which catches tiny PHYTO-PLANKTON drifting in polar waters, predators like the Peacock mantis shrimp (*Odontactylus scyllarus*) with cocked claws, or scavengers such as the common or Atlantic mud crab (*Panopeus herbstii*), which feed off detritus and baby clams in a salt marsh. They range in size from small branchiopods to the Japanese spider crab (*Macrocheira kaempferi*), the largest of the arthropods. Branchiopoda include brine shrimp (*Artemia salina*) or sea monkeys, the endangered San Diego fairy shrimp (*Branchinecta sandiegoensis*), which inhabits vernal pools, and marine water fleas (e.g., *Podon polyphemoides, Evadne Nordmanni,* and *Daphnia magna*). Vernal pools are small, temporary bodies of water that form in depressions. They are common along coastal terraces of Southern California, where geologic forces have lifted alluvial landscape surfaces above sea level.

Crustaceans have a head, thorax, and abdomen. Their bodies are covered with a chitinous exoskeleton, which may be thick and calcareous (as in the lobsters), or thin and transparent (as in water fleas). Chitin is a polysaccharide that is insoluble in water and somewhat resistant to acids. In some crustaceans, the shell may be impregnated with calcium carbonate for extra strength. Since the exoskeletons or carapace do not grow, they are periodically shed when the crustacean outgrows its shell. A new soft shell lies just beneath the old hard shell. The crustaceans excrete hormones, which cause the hard shell to split so that the animal can back out of its old shell. Once this takes place, the new soft shell takes time to harden. Soft-shelled crustaceans are especially vulnerable to attack from fish, other crustaceans, MARINE BIRDS, reptiles, and mammals.

Molted or shed shells are food for marine bacteria that break down chitin's fibrous structure to simple sugars and amino acid. Researchers at North Carolina State University have used chitin to filter pollutants from industrial waste water.

Crustaceans have two pairs of antennae, stalked compound eyes, and two mandibles on the head. Antennae are jointed, extend forward from the head, and are used for sensing. The compound eyes contain tens to thousands of ommatidia or sensors, which detect brightness and sometimes colors. Researchers have studied the compound eyes of the rock mantis shrimp (*Neogonodactylus oerstedii*) to develop REMOTE SENSING and imagery analysis systems. This species is found throughout the Caribbean reef and grass flat habitats, living in coral rubble, cavities in coralline ALGAE, and among shells. The first pair of thoracic appendages in most crustaceans has claws or pincers for crushing. Crustaceans breathe via gills that are located where the legs attach to the thorax. The most primitive species (lobsters and shrimp) have well-developed abdomens, whereas the swimming crabs have reduced, almost vestigial, abdomens. The crab body is short and wide from tip to tip, and flat. The tips are actually lateral spines. Pairs of legs on each body segment are especially adapted for swimming and walking along the seafloor. The blue crab uses the last pair of flattened and paddlelike legs to swim sideways through the water. Some crustaceans may detach or discard legs or claws to escape a predator's grasp in a process called autotonomy. Fortunately, if the crustacean loses an appendage, it can grow another. The tail fan, tail stem, or telson on a crustacean is a terminal segment that does not have any appendages. The American lobster uses the tail fan like a paddle to dart backward. Flippers that propel shrimp through the water are located just below the sharp tail stem. These tails are totally absent in the swimming crabs.

Most crustaceans are separate male and female individuals or dioecious. The shape of the pleopod may help distinguish male and female crustaceans. On a male crab, this abdominal appendage or apron is narrow and long like the Washington monument (or an inverted "T") and on a female crab it resembles the shape of a pyramid (immature female) or the dome of a capitol building (mature female). Female crabs carry eggs under their apron, which are eventually fertilized and released into the water to hatch. Once the adult female hard crab has fertilized and extruded her eggs on the abdomen or abdominal flap, she is sometimes called a "sponge crab." Some sedentary crustaceans such as acorn (*Semibalanus balanoides* and *Chthamalus antennatus*) and goose-neck (*Pollicipes polymerus*) barnacles are hermaphrodites. They possess both male and female reproductive systems, producing both eggs and sperm. Barnacles are filter feeders using their feathery legs or cirri to trap PLANKTON. British naturalist Charles Darwin (1809–82) was first to study and classify barnacles, which helped in his later descriptions of natural selection.

The actual mechanisms by which crustaceans spawn are complex and varied. Mating is linked to molting. Reproduction is usually sexual. In mating, a male impregnates a newly molted female. Fertilization of the egg does not occur at the time of mating. Sperm is stored in the female's body until her eggs are fully developed. The eggs are fertilized when the female extrudes them under her abdomen. During spawning, eggs flow from an opening in the female over the sperm and are then attached to her swimmerets or pleopods by a natural adhesive. Barnacles cross-fertilize by using a retractable tube containing sperm to reach outside the shell as far as several inches to a neighboring barnacle to copulate. Barnacles store fertilized eggs inside their shell. Upon hatching, planktonic larvae develop through different life stages in the water column. Larval forms of the crustaceans are planktonic, and the animal gradually develops into the juvenile and adult forms through successive molts.

Crustaceans are an important food source for predators. Small planktonic crustaceans, such as COPEPODS and krill, are a major link in the marine food chain between photosynthetic phytoplankton and finfish and marine mammals. Crustaceans such as crabs are crucial in recycling nutrients trapped in the bodies of dead organisms. Branchiopoda are sometimes cultured as live food for aquarium fish and are also important bio-indicators to monitor water quality. Fisheries biologists usually refer to commercially important MOLLUSKS and crustaceans as shellfish. Species vary based on geographic regions. For example, smaller species of shrimp inhabit cold waters in the northern Atlantic and Pacific. Shrimp are therefore classified as cold water or northern shrimp, warm water, tropical or southern shrimp, and freshwater shrimp. The majority of shrimp landings in the United States are brown (*Penaeus aztecus*), white (*Penaeus setiferus*), and pink (*Penaeus duorarum*) shrimp caught along the GULF OF MEXICO. In the United States, shellfish are considered a delicacy. Lobsters are associated with New England and the Gulf of Maine, Atlantic blue crabs with Maryland and Chesapeake Bay, and Dungeness crabs (*Cancer magister*) with Washington and Puget Sound.

Crustaceans are the insects of the sea and important in many food webs. As an example, krill consume phytoplankton near the surface during the day and retreat to deeper waters at night. Krill consumers include penguins, seals, and baleen whales. Krill means "whale food" in Norwegian, and the baleen whales swim into schools of krill with their mouths open. They take in hundreds of gallons of water, close their mouths and squirt the water through the baleen. Tough flexible baleen bristles trap the krill and the whale swallows them. A single whale can eat several tons of krill a day. Scientists report that the population of Antarctic krill has declined in concert with the loss of SEA ICE in Antarctica and overfishing. This loss in sea ice is attributed to GLOBAL WARMING. Such a crisis in krill populations would challenge the top of the food chain, which belongs to leopard seals (*Hydrurga leptonyx*), killer whales (*Orcinus orca*), and polar bears (*Ursus maritimus*). There are tens of thousands of crustacean species living in the ocean. Some of the more interesting and commercially important crustaceans are listed in the table that begins on page 292.

Further Reading
Corson, Trevor. *The Secret Life of Lobsters*. New York: HarperCollins, 2004.
Warner, William W. *Beautiful Swimmers*. Boston: Little, Brown, 1976.

marine engineer A marine engineer is a professional mariner with a degree in a nautical science, general engineering, or marine engineering who specializes in understanding shipboard systems. Students working toward becoming a marine engineer take mathematics and science courses to learn the principles behind ship design and power plants and how to solve engineering problems related to vessels. Typical issues include operating technical systems, understanding corrosion and other forms of wear and tear, environmental hazards, and safety. Marine engineers operate and maintain propulsion engines, boilers, generators, pumps, and other marine machinery. They may work aboard ferries, dredges (*see* DREDGING), TUGBOATS, towboats, research vessels, and deep-sea merchant ships. The crew may have a chief engineer and a first, second, and third assistant engineer. Assistant engineers stand periodic watches and oversee the operation of engines and machinery. Marine oilers work belowdecks under the direction of the ship's engineers. They lubricate gears, shafts, bearings, and other moving parts of engines and motors, read pressure and temperature gauges and record data, and may repair and adjust machinery.

Professional mariners ply canals, rivers, harbors, the Great Lakes, and ocean navigation channels in order to safely move people and cargo from one seaport to another. A typical merchant ship has a captain, deck officers or mates, a chief engineer, and assistant engineers, plus cooks, stewards, seamen, oilers, electricians, machinery mechanics, and communicators. In the United States, licensed deck and engineering officers have either graduated from the U.S. Merchant Marine Academy or one of a half dozen state academies, (e.g., Maine Maritime Academy, Massachusetts Maritime Academy, State University of New York Maritime College, Texas Maritime College, California Maritime Academy, and Great Lakes Maritime Academy) and passed a written examination. Academy graduates are commissioned as ensigns in the U.S. Naval Reserve, and some go on active duty in the U.S. Navy. Some U.S. marine engineers work aboard foreign-flag vessels, tugboats, or take relevant land-based jobs with organizations such as the Maritime Administration, U.S. Geological Survey, Department of Commerce, shipping companies, marine insurance companies, and manufacturers of marine-related machinery. Some marine engineers serve on oceanographic research vessels with universities, or institutions, such as the WOODS HOLE OCEANOGRAPHIC INSTITUTION.

No training or experience is required to become an ABLE-BODIED SEAMAN or deckhand on vessels operating in harbors or on rivers or other waterways. These workers generally learn skills on the job. Experienced seamen and deckhands are eligible to take Coast Guard examinations to qualify as a mate, pilot, or captain. Persons with at least three years of at-sea experience may also become a licensed marine engineer if they meet requirements and pass the appropriate U.S. Coast Guard examination.

See MARINE TRADES.

Further Reading
Taylor, David A. *Introduction to Marine Engineering*. Boston: Butterworth-Heinemann, 1996.

marine fishes Marine fishes are limbless aquatic vertebrates with fins for mobility and internal gills for breathing in saltwater environments. Their habitat is a complex solution of mineral salts eroded from Earth's crust and physical-chemical-biologic interactions associated with sea life ranging in size from microbes such as *Pyrolobus fumarii* (Archaea) discovered near HYDROTHERMAL VENTS at the bottom of the ATLANTIC OCEAN to the blue whale (*Balaenoptera musculus*), the largest animal on Earth. Marine fish species have adapted to different SALINITY regimes, ranging from the relatively fresh Baltic Sea to the saline RED SEA and PERSIAN GULF. Marine fish must live in water with favorable salinities to avoid dehydration because of osmosis—the movement of water through a semi-permeable membrane such as skin and gills. When concentrations of dissolved minerals on one side of a cell membrane are not equal to concentrations on the other side, water flows from the direction of higher concentration to lower until equilibrium is reached. Temperature and pH are two other important parameters to fish health. Aquarium enthusiasts are particularly interested in matching water quality to a fish's known tolerances to create an optimal habitat for expensive marine fish species.

Salinity, TEMPERATURE, and dissolved oxygen are important factors that determine the abundance and distribution for marine fish. A significant variation in salinity is caused by physical factors such as ice melt, river inflow, evaporation, precipitation, winds, wave motion, and ocean CURRENTS that cause horizontal and vertical mixing of the seawater. An average expression for the salinity of the oceans indicates that there are approximately 35 pounds (15.9 kg) of salt per 1,000 pounds (453.6 kg) of seawater. Large salinity gradients occur in locations where freshwater from rivers meets and mixes with coastal ocean water, in locations where freshwater is floating on top of dense saline water, such as ESTUARIES, and in locations where evaporation usually exceed precipitation, such as the eastern Mediterranean Sea. Daily and seasonal air temperatures also influence water temperatures, where light, warm water floats on top of heavy, colder water. During extreme conditions such as droughts, temperatures in water bodies such as Chesapeake Bay can rise as a result of ensuing lower water levels. Reduced freshwater inflow from the drought results in higher salinity in the estuary. Salinity and temperature directly affect the amount of dissolved oxygen in water. As water temperature increases, the amount of dissolved oxygen in the water decreases. Similarly, the more saline the water, the less oxygen the water can hold.

Some fish species only live in freshwater, some thrive in salt water, and some have adapted to live in either fresh or salt environments. Stenohaline fish, which are found usually in either fresh or salt water, can survive only in a restricted range of salinities, while euryhaline fish, which may inhabit estuaries, tolerate a wide range of salinities. Fish having blood with lower salt concentrations than the surrounding water are well suited for the marine environment. Saltwater fish have to work against osmosis by drinking large amounts of water and excreting salt ions through their gills. They have concentrated urine. Freshwater fish have higher concentrations of salt in their blood and constantly have to excrete less-concentrated water absorbed by osmosis. Salin-

ity plays a major role in the lives of ANADROMOUS fish such as sea lamprey (*Petromyzon marinus*), Atlantic sturgeon (*Acipenser oxyrinchus oxyrinchus*), Atlantic shad (*Alosa sapidissima*), Atlantic herring (*Clupea harengus harengus*), salmon (fishes of the family Salmonidae), and striped bass (*Morone saxatilis*) that spawn in freshwater and then spend their adult lives in the sea. During late summer, when salinity and temperatures are at their highest levels in Chesapeake Bay, oceanic fish such as Spanish mackerel (*Scomberomorous maculatus*) enter the bay to feed on schools of Atlantic menhaden (*Brevoortia tyrannus*). Atlantic menhaden are one of the most abundant species of finfish in estuarine and coastal Atlantic waters. The Spanish mackerel return to more southerly ocean waters when water temperatures drop below 70°F (21.1°C). During summertime in the Chesapeake Bay region, water temperature and PHYTOPLANKTON levels increase in relation to excess nitrogen and phosphorous that is washed into the bay from tributaries. A combination of environmental stress, algal blooms, and disease-causing organisms are the most likely reasons behind many fish kills in the bay. As algal blooms die, they drop to the seafloor and decompose, using up the oxygen of the deeper water. The depleted levels of oxygen in the bay's water, a state known as hypoxia, may kill oxygen-dependent sea life. Most marine life cannot survive when oxygen levels fall below two parts per million.

Other critical parameters for saltwater fish are pH and temperature. Rapid changes can stress and even kill marine fish. The solubility of gases such as carbon dioxide (CO_2) increases with decreasing temperature or salinity. The average pH of the ocean is 8.1, and marine fish tend to live in seawater that is slightly alkaline (pure distilled water is neutral with a pH of 7). The range of temperature in which fish are most comfortable may be quite large because most fish are poikilothermic or cold-blooded. Fish metabolisms slow when the water is cold and speed up when the water warms. Although temperature requirements of fish are flexible, fish do have their limits. If the water is too cold, metabolisms slow to life-threatening rates. If it is too warm, fish may overheat and die, or their immunity to disease is decreased. These changes in temperature may also impact the water's pH and ability to hold dissolved oxygen. Most dissolved CO_2 in the oceans is converted to carbonic acid (H_2CO_3), which decreases the pH, putting more stress on marine fishes. Fisheries managers and scientists are therefore concerned over the increasing level of CO_2 emissions.

See also CORAL REEF FISH; FISH.

Further Reading

Fisheries and Aquaculture Department, Food and Agriculture Organization of the United Nations. Available online. URL: http://www.fao.org/fishery/. Accessed March 8, 2008.

Froese, Ranier, and Daniel Pauly, eds. 2007.FishBase. Available online. URL: http://www.fishbase.org. Accessed March 9, 2008.

marine macrophytes Marine macrophytes are rooted, freely floating, or attached vascular plants and ALGAE-inhabiting coastal and oceanic regions large enough to be seen by the naked eye. The term *macrophyte* is formed from two Greek words, *macro* meaning large or great and *phyte* referring to plants.

In general, marine macrophytes may either be vascular plants such as marsh grasses or algae such as seaweeds. Their shortest dimensions (leaf blades) are generally greater than 0.02 inches (0.5 mm). Vascular plants have conducting tissues such as the xylem (for water) and phloem (for food). Examples include emergent plants such as cattails (*Typha latifolia*) and

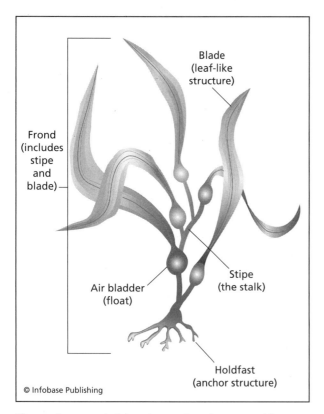

Macro-algae attach themselves to the substratum with holdfasts. Air bladders of pneumatocysts along the stipe provide floatation and keep the plant upright in the water column. The stipe is flexible and very different from the stem of land plants; it does not carry water or nutrients. The leaflike structures are called blades and collect sunlight for photosynthesis.

Long's bulrush (*Scirpus longii*) with root structures. The algae secure themselves to the seafloor with a holdfast and have floats that help the frond, stipe, and blades sway about in the water. The leafy appendages or shoots of the algae are called fronds. The stalklike portion of the algae is called the stipe. Marine macrophytes may form vast forests (for example, *Macrocystis pyrifera*) or expansive floating blooms (for example, *Sargassum natans*) that may be washed ashore as tangled clumps and mats. Macrophytes are sensitive to light and nutrient levels. They may improve water quality, minimizing SEDIMENT suspension by dampening WAVE action and slowing CURRENTS. They stabilize the sediments.

Macrophytes composing a salt marsh form the basis of a food web, where microbes and ZOOPLANKTON consume fallen plant material or litter called detritus and small organic particles. The grazers may also help prevent algal blooms resulting from excess nutrients such as phosphorus in marine ecosystems. Predators such as gastropods, crustaceans, MOLLUSKS, and fish rely on these drifting organisms for food energy. As primary producers, macrophytes such as *Amansia glomerata* are an important food source for grazers such as sea turtles and widgeon grass (*Ruppia maritima*) for mammals such as the West Indian Manatee (*Trichechus manatus*), muskrats (*Ondatra zibethicus*), and nutria (*Myocastor coypu*). The importance of macrophytes such as SEA GRASSES makes them an indicator species to assess the health of a particular water body.

Macrophytes provide important habitat for all types of marine animals. In places such as Florida, popular sport fish including the tarpon (*Megalops atlanticus*), common snook (*Centropomus undecimalis*), crevalle jack (*Caranx hippos*), and jewfish grouper (*Epinephelus itajara*) forage in sea-grass habitats. In the open ocean and in coastal LAGOONS, assemblages of macrophytes provide food and cover and are an especially important nursery area for many marketable or commercially important finfish. Macrophytes may also become nuisance species. In the many estuaries in the United States, the Asian water chestnut (*Trapa natans*) outcompetes other SUBMERGED AQUATIC VEGETATION and clogs waterways. These plants are poor forage for waterfowl, and the fruit has four sharp spines that hamper swimming and other water activities.

Macrophytes have many commercial applications. They are harvested and processed as thickeners for food and beauty products. For example, Irish moss (*Chondrus crispus*), a red alga, is raked from intertidal waters of the North Atlantic in Europe and North America and processed into the thickening agent carrageenan. Food scientists might use carrageenan to help thicken ice cream or pudding, to keep mixtures together, to make compounds stickier, and to prevent sugars from crystallizing. It is very much like cornstarch or gelatin. Kelp harvesting is accomplished by hand and also with cutters fitted to vessels having conveyor belts that help move harvested material to bins. They are processed into the valuable thickening agent called algin. Algin is found in the cell wells of kelp and has many food science and pharmaceutical applications. Like carrageenan, algin is used as a thickener and binding agent. The pharmaceutical industry uses algin to make tablets, dental impressions, facial creams, and lotions. Iodine and mannitol may also be extracted from some kelp plants. Iodine is added to salt and other foods to prevent thyroid gland disorders and goiters, while mannitol is used as an osmotic diuretic. Kelp plants are also used as livestock fodder. Potash and acetone extracted from kelp have been used to make explosives. The table at left is a list of commonly studied macrophytes.

Commonly Studied Macrophytes

Scientific Name	Common Name	Description
Amansia glomerata	No common name	red algae
Caulerpa taxifolia	killer algae	green algae
Chondrus crispus	Irish moss	red algae
Egeria densa	Brazilian water-weed	red algae
Egregia menziesii	feather boa	brown algae
Halodule beaudettei	shoal grass	sea grass
Halophila decipiens	paddle grass	sea grass
Halophila engelmannii	star grass	sea grass
Halophila johnsonii	Johnson's sea grass	sea grass
Hypnea musciformis	hooked weed	red algae
Lyngbya majuscula	tangled-hair seaweed	blue green algae
Macrocystis pyrifera	giant bladder kelp	brown algae
Nereocystis luetkeana	bull kelp	brown algae
Pelagophycus porra	elk kelp	brown algae
Posidonia oceanica	neptunegrass	sea grass
Potamogeton pectinatus	Sago pondweed	floating leaved plant
Scirpus longii	Long's bulrush	emergent plant
Trapa natans	water chestnut	sea grass
Ruppia maritima	widgeon grass	sea grass
Sargassum natans	gulfweed	brown algae
Syringodium filiforme	manatee grass	sea grass
Thalassia testudinum	turtle grass	sea grass
Tolypiocladia glomerulata	red seaweed	red algae
Typha latifolia	cattail	emergent plant
Ulva lactuca	sea lettuce	green algae
Zostera marina	eel grass	sea grass

marine mammals   Marine mammals are warm-blooded animals with hair on their bodies; they produce milk for their young such as cetaceans (whales, dolphins, and porpoises) and pinnipeds (seals, sea lions, and walruses). These mammals spend most of their lives in the marine environment. Marine mammals, such as the sperm whale (*Physeter macrocephalus*) show incredible adaptation to the marine environment. These whales dive to depths of thousands of feet (m) in search of and to do battle with their prey, the giant squid (*Architeuthis; see* CEPHALOPODS), and can remain under water for periods longer than an hour. Whales and other cetaceans have sophisticated SONARS for location of prey (*see* CETACEANS), and for communication. Many marine mammals, such as bottlenose dolphins (*Tursiops truncates*) and killer whales (*Orcinus orca*), are among the most intelligent in the animal kingdom. Polar bears (*Ursus maritimus*), which have webbed feet and whose nostrils close when they are underwater, are also

In the United States, manatees live in the rivers of Florida, where landfill, draining of wetlands, and careless recreational boaters threaten their extinction.   *(NOAA)*

A Selected List of Marine Mammals

Common Name	Scientific Name	Status
Amazonian manatee	*Trichechus inunguis*	endangered
Beluga whale	*Delphinapterus leucas*	depleted
blue whale	*Balaenoptera musculus*	endangered, depleted
bottlenose dolphin	*Tursiops truncatus*	depleted
bowhead whale	*Balaena mysticetus*	endangered, depleted
Chinese River dolphin	*Lipotes vexillifer*	endangered, depleted
Dugong	*Dugong dugon*	endangered
finback whale	*Balaenoptera physalus*	endangered, depleted
Guadalupe fur seal	*Arctocephalus townsendi*	threatened, depleted
gray whale	*Eschrichtius robustus*	endangered, depleted
Hawaiin monk seal	*Monachus schauinslandi*	endangered, depleted
humpback whale	*Megaptera novaeangliae*	endangered, depleted
Indus River dolphin	*Platanista minor*	endangered, depleted
killer whale	*Orcinus orca*	endangered, depleted
Mediterranean monk seal	*Monachus schauinslandi*	endangered, depleted
northern fur seal	*Callorhinus ursinus*	depleted
northern right whale	*Eubalaena glacialis*	endangered, depleted
northern sea otter	*Enhydra lutris kenyoni*	threatened
polar bear	*Ursus maritimus*	potentially threatened
right whale	*Balaena glacialis*	endangered
Saimaa seal	*Phoca hispida saimensis*	endangered, depleted
sei whale	*Balaenoptera borealis*	endangered, depleted
southern right whale	*Eubalaena australis*	endangered, depleted
southern sea otter	*Enhydra lutris nereis*	threatened
sperm whale	*Physeter macrocephalus*	endangered, depleted
spinner dolphin	*Stenella longirostris*	depleted
spotted dolphin	*Stenella attenuata*	depleted
Steller sea lion	*Eumetopias jubatus*	endangered, threatened, depleted
Vaquita	*Phocoena sinus*	endangered, depleted
West African manatee	*Trichechus senegalensis*	threatened
West Indian manatee	*Trichechus manatus*	endangered

Note: Some entries are from the Federal List of Endangered and Threatened Wildlife and Plants; they are reported as endangered or threatened. An "endangered" species is one that is in danger of extinction throughout all or a significant portion of its range. A "threatened" species is one that is likely to become endangered in the foreseeable future. A "depleted" species or population stock is below its optimum sustainable population.

considered to be marine mammals. Polar bears inhabit SEA ICE adjacent to shorelines and prey primarily on seals such as ringed seals (*Phoca hispida hispida*). Their habitats in the Arctic are currently threatened by the affects of ice melting due to GLOBAL WARMING. Kelp beds are a favorite location for thickly furred sea otters (*Enhydra lutris kenyoni, Enhydra lutris lutris,* and *Enhydra lutris nereis*) that are found in northern Pacific waters off of Washington, Alaska, Japan, Russia, and California. They eat abalone, clams,

This harbor seal is a familiar face to many mariners and recreational boaters. *(NOAA)*

octopi, sea urchins, sea stars, snails, and squid. Sea otters are the smallest of the marine mammals. In tropical coastal locations, cyclindrically-shaped manatees (*Trichechus inunguis, Trichechus manatus,* and *Trichechus senegalensis*) swim in shallow fresh and saltwater areas with flipperlike forelimbs and horizontal paddlelike tails. These mammals are primarily herbivorous, feeding on benthic algae, MANGROVE, SUBMERGED AQUATIC VEGETATION, and overhanging plants.

Many populations of marine mammals have declined due to human activities. Fishing practices such as tuna purse-seine fishing (as they are frequently caught in the nets) has been considered a major threat to dolphin species, and other practices such as clubbing, poisoning, and uncontrolled hunting activities have threatened various species of pinnipeds with extinction (*see* NETS). Manatees are especially susceptible to collisions with powerboats. For these reasons, the 1972 Marine Mammals Protection Act or MMPA protects marine mammals in U.S. waters. This act also established a federal Marine Mammal Commission, whose members are appointed by the president of the United States. The importation of marine mammals and marine mammal products into the U.S. is strictly prohibited. Some marine mammals may be designated as "depleted" under the MMPA. Endangered and threatened marine mammals are further protected under the Endangered Species Act (ESA). Endangered species are in danger of extinction throughout all or most of their range. The ESA was established in 1973 by President Richard M. Nixon (1913–94) and provides for the conservation of endangered and threatened species of mammals as well as birds, fish, invertebrates, and plants and their essential habitats.

Marine biologists may focus their studies on marine mammals, their life processes, and the rela-

tionship of marine mammals to other ECOSYSTEM inhabitants. Besides biology, these scientists have a strong background in chemistry, physics, mathematics, and computer sciences. They may also have skills in acoustical analysis, biostatistics, genetic analysis, and biomolecular analyses. The table on page 300 lists the common and scientific names for some important marine mammals.

Further Reading

Gulf of Maine Research Institute. "Marine Mammals." Available online. URL: http://octopus.gma.org/marine mammals. Accessed September 26, 2007.

International Marine Mammal. Available online. URL: http://www.imma.org/index.html. Accessed August 10, 2006.

The Marine Mammal Center. Available online. URL: http://www.marinemammalcenter.org/. Accessed September 26, 2007.

Marine Mammal Commission. Available online. URL: http://www.mmc.gov/. Accessed August 10, 2006.

National Marine Mammal Laboratory, Alaska Fisheries Science Center, National Oceanic and Atmospheric Administration. Available online. URL: http://www.afsc.noaa.gov/nmml. Accessed March 9, 2008.

NOAA Fisheries, Office of Protected Resources. "Marine Mammals." Available online. URL: http://www.nmfs.noaa.gov/pr/species/mammals/. Accessed September 26, 2007.

U.S. Fish and Wildlife Service. "The Endangered Species Program." Available online. URL: http://www.fws.gov/endangered/. Accessed September 26, 2007.

marine meteorologist A marine meteorologist is a scientist who studies the physics of the marine atmosphere and often the ocean below it, as well as the relations between the two. His or her biggest job is to observe and forecast (*see* FORECASTING) the weather. Meteorologists study the distribution and motions of mass and energy in the atmosphere. Marine meteorologists study air-sea interactions, especially with regard to tropical storm formation. They provide predictions of hurricane paths and intensities, and the specific effects of hurricanes, such as maximum winds, water levels (STORM SURGE), and flooding (*see* HURRICANE). Elected officials use these forecasts as well as disaster management teams to decide upon, plan, and implement evacuations of coastal areas in danger from hurricanes and other tropical and midlatitude storms.

Specific factors studied include the distribution of temperature, PRESSURE, moisture content, incoming solar radiation, transports of heat and water vapor, clouds, low-level and upper-level winds, and in the ocean TEMPERATURE, SALINITY, DENSITY,

CURRENTS, WAVES, and TIDES. By understanding and describing environmental factors, a meteorologist might determine favorable flight paths for aircraft or seaways for ships.

Phenomena that are described by research papers include distributions of sea surface temperatures, coastal flooding, severe weather, tropical warm water pools, land and sea breeze circulation, SEA ICE distributions and thickness, EL NIÑO Southern Oscillation, and climate change. Investigations involve other scientists, the collection of *in situ* data, remotely sensed imagery, and numerical modeling. Findings may be presented to other meteorologists and interested personnel at conferences.

Meteorologists develop mathematical models to explain the circulation of the atmosphere that involve sets of complex equations that must be solved numerically on computers. Models used for understanding the physics of the atmosphere are called diagnostic models; models that are used for weather prediction are called prognostic models (*see* MODELS). To assess the skill of these models, researchers compare results to ship observations, data recorded in the International Comprehensive Ocean-Atmosphere Dataset, real-time information or archived data from OCEAN OBSERVATORIES, imagery from satellites, or specifically designed field measurement programs. Organizations such as the National Oceanographic Data Center and the National Climatic Data Center provide archived data for scientific and commercial application.

Meteorologists use a great variety of instruments to measure the physical and chemical properties of the atmosphere, such as electronic thermometers to measure temperature, barometers to measure pressure, ANEMOMETERS to measure wind velocity, and hygristers for measurement of humidity. These instruments can be incorporated in balloon-borne instrument packages to measure profiles of the lower atmosphere. The positions of the balloons are tracked either visually or by RADAR to give additional information about wind speed and direction used by weather forecasters (*see* METEOROLOGICAL INSTRUMENTS). Marine meteorologists rely on networks of moored buoys and coastal stations to provide weather and ocean surface information to make their forecasts. An example is the developing network of ocean observatories in the United States (*see* OCEAN OBSERVATORY).

At-sea measurements of meteorological variables are very important to these weather forecasts since the oceans, seas, and lakes cover about 70 percent of Earth's surface. Mariners may access local forecasts, warnings, and current weather information from weather radio transmitters for safety at sea and for optimizing their travel routes. Mariners use weather information and experience when planning passages through a plethora of potential weather hazards such as hurricanes or typhoons and tropical cyclones. Recreational boaters may employ electronic barometers, handheld anemometers, and portable weather stations, while professional mariners may rely on precision barometers, certified hygrometers, radar, and sophisticated weather stations. The National Weather Service publishes the *Mariners Weather Log,* which contains articles, news, and information about marine weather events and phenomena, storms at sea, and weather forecasting,

In recent years, significant amounts of meteorological data have been acquired by weather satellites using visual and infrared imaging to provide storm locations, hurricane tracks at sea, cloud cover, wave height and direction, and wind speed and direction. Wind speed can be measured by radar scatterometers that measure the roughness of the sea surface and the corresponding wind speed that would cause this roughness. An example of a basic research result that is now used operationally is WindSat, a satellite-based multifrequency polarimetric microwave radiometer. The Naval Research Laboratory Remote Sensing Division, the Naval Center for Space Technology, and the National Polar-Orbiting Operational Environmental Satellite System Integrated Program Office developed this satellite capability. On large islands, such as the Hawaiian Islands, weather forecasters use Doppler weather radar to obtain wind velocity and precipitation for local mariners, aviators, farmers, and city dwellers. These instruments can be deployed from well-equipped research vessels, ocean platforms, moored BUOYS, and aircraft. Marine meteorologists also work closely with PHYSICAL OCEANOGRAPHERS and climatologists in deploying instruments such as radar altimeters and synthetic aperture radars on Earth satellites for studies of ocean topography and its relation to the global wind patterns and ocean circulation.

Hurricane forecasters have several new instruments for improving their forecasts, including dropsondes (instrument packages dropped from aircraft), ocean surface probes, high-resolution aircraft radars, and "aerosondes," small unpiloted aircraft that can fly through the hurricanes at very low altitudes. The large volumes of data are incorporated into numerical models that predict the hurricane track and intensity. Research meteorologists are continually improving the resolution of these models; the current model has a resolution of 7.5 miles (12 km), and experimental models are under development with resolutions of less than one mile (1.61 km). These new models will better predict where the most intense winds and storm surge will occur (and hence the most potential for damage).

Marine meteorologists normally have as a minimum a master's degree. University research positions normally require a doctorate, followed by postdoctoral research. With only a bachelor's degree, positions are generally limited to the technician level. In addition to education, the American Meteorological Society has several certification programs such as the Certified Broadcast Meteorologist program, the Seal of Approval, and the Certified Consulting Meteorologist program. These certifications help recognize broadcast and consulting meteorologists who have achieved a certain level of competence and foster high standards of professionalism among meteorologists.

There are two general approaches to obtaining advanced degrees in marine meteorology. Some persons attend universities that give advanced training and postgraduate degrees in meteorology. Others prefer to take their advanced degrees in physical oceanography, physics, and mathematics, and then participate in postdoctoral training in marine meteorology at major laboratories such as the Woods Hole Oceanographic Institution, which has a cooperative postdoctoral program with the Massachusetts Institute of Technology. The graduate student would take several core courses in fluid mechanics, thermodynamics, air-sea dynamics, dynamical meteorology, coastal oceanography, and geophysical fluid mechanics.

Meteorologists are employed by the National Weather Service (NWS) of the National Oceanic and Atmospheric Administration (NOAA), major universities, government laboratories, television stations, and private industries. Research meteorologists perform basic research on the physics of the atmosphere and applied research in fields such as marine weather forecasting, military science, climatology, air pollution control and remediation, resource management. Marine meteorologists will contribute greatly to society by assessing threats posed by global warming through the collection and analysis of past and present data on worldwide temperature trends. Numerical modelers using supercomputers apply state-of-the-art physical principles to study changes in the atmosphere and ocean on scales ranging in time from seconds to decades and in space from a specific sensor location to the globe. Meteorologists from local weather observers to professors are supporting society by providing objective information that is used in ways that range from planning daily activities to forecasting energy usage.

To keep abreast of recent developments in their field, meteorologists must spend considerable time reading current technical journals, participating in scientific meetings and workshops, and in taking courses in new technologies. As in almost all the sciences, they must prepare technical proposals to federal or state agencies and organizations, such as the National Science Foundation, the Office of Naval Research, and the National Oceanic and Atmospheric Administration (NOAA). to obtain the funds to support their project work. They can expand their resources by partnering with scientists from other organizations and by making use of existing databases available on the World Wide Web. In democratic countries, researchers may testify before elected officials such as the U.S. Congress in order to explain phenomena such as global climate change, which affects the way people are currently carrying out their lives.

Further Reading

American Meteorological Society. Available online. URL: http://www.ametsoc.org/. Accessed September 9, 2006.

Hsu, S. A. *Coastal Meteorology*. San Diego, Calif.: Academic Press, 1988.

National Oceanic & Atmospheric Administration's National Data Buoy Center. "United States Voluntary Ship Observing Program." Available online. URL http://vos.noaa.gov/. Accessed September 26, 2007.

NOAA Satellite and Information Service. "ICOADS: The International Comprehensive Ocean-Atmosphere Data Set Project." Available Online, URL: http://www.ncdc.noaa.gov/oa/climate/coads/. Accessed September 26, 2007.

marine operations Marine operations include the duties and tasks associated with ensuring that equipment, vessels, and technologies enable marine scientists and engineers to accomplish their maritime missions. For many organizations, the term *marine operations* refers to those people directly involved with the support of vessels and their associated equipment. Equipment would include pallets, cargo nets, forklifts, and cranes that are used to onload and offload vessels. Access to storage lots, wharfs, and berthing spots may be essential and have to be scheduled at busy seaports. Vessels might range from watercraft, boats, barges, and lighters to tenders, fishing vessels, and container ships.

Since 70 percent of the Earth is covered by water, the full spectrum of professional activities occurring on waterways, in estuaries, and in the ocean has a rich history and has evolved along with technology. Today's routine commercial marine activities range from the docking, loading, and undocking of vessels in order to transport cargo to distant seaports to the operation of dredges and the maneuver of support craft to maintain safe controlling depths in navigation channels. Government organizations such as the U.S. Coast Guard (USCG) support marine operations by deploying and operating aids to navigation such as markers and BUOYS and enforcing maritime laws. State-of-the art navigation support systems include the USCG's Vessel Traffic and Information System or VTIS and the NATIONAL OCEANIC AND ATMOSPHERIC ADMINISTRATION's Physical Oceanographic Real-Time Systems or PORTS™. In marine science, there are a myriad of traditional operations such as hydrographic surveying, ocean drilling, oceanographic and fisheries research, scientific diving, and servicing deployed instruments attached to MOORING lines, tethers, and buoys, or those deployed at the seafloor. Research platforms include buoys, piers, BRIDGES, watercraft, boats, research vessels, AUTONOMOUS UNDERWATER VEHICLES, and aircraft. Equipment includes NETS, *in situ* electronic instruments, DYES and drifters, and remote sensing instruments (*see* OCEANOGRAPHIC INSTRUMENTS).

Planning, which includes administration, logistic support, scheduling, and control, is essential to marine operations. Planners acquire the necessary sailing instructions, nautical charts (*see* CHARTS, NAUTICAL), atlases, and tide and current tables to study all the navigation areas common to their mission. Various documents are written such as cruise plans, field plans, and data quality-assurance plans to identify goals and risks to the project. These reports are critical to ensuring the success of coastal engineering projects, data collection efforts, and research cruises. The data quality-assurance plan describes the study area, required measurements, types of instruments, documents operator checks of equipment, the levels of training for field personnel, and error budgets. At the conclusion of marine science cruises, data are described in data reports and archived at repositories such as the National Oceanographic Data Center. Scientists use quality-assured data to study processes and phenomena, which may be described in technical reports, trade magazines, World Wide Web sites, conference proceeding articles, and refereed journal articles. Ensuring operational success rests in the ongoing maintenance and outfitting of marine vessels, equipment, and gear. Besides ships, small boats facilitate marine research and sampling near the mother ship or in local nearshore and estuarine environments. Control of complex marine operations may be the responsibility of a coxswain for the maneuver of a small boat that could be involved in search and rescue, the boatswain for the deployment of buoys off the stern of a buoy tender, captain or master of a fishing boat involved in trolling for shellfish, the staff civil engineer in the port operations division of a bustling seaport, a harbor pilot for the docking (*see* DOCKING PILOT) of a large ship, or a principal investigator or chief scientist for an oceanographic expedition.

Personnel participating in marine operations include scientists, licensed masters, mates and MARINE ENGINEERS, and unlicensed members of the engine, steward, and deck departments. The senior person in a science party responsible for the outcome of a scientific expedition is called the chief scientist. Science parties are generally multidisciplinary, including MARINE BIOLOGISTS, PHYSICAL OCEANOGRAPHERS, CHEMICAL OCEANOGRAPHERS, and GEOLOGICAL OCEANOGRAPHERS. There are many other important participants such as survey, marine, and electronic technicians who operate mission, communication, and navigation equipment. On a scientific cruise, the research vessel's officers and crew provide mission support and assistance to embarked scientists who may be from various government laboratories, the academic community, and industry.

Since cargo in coastal nations such as the United States is carried by water-borne transportation, maritime commerce benefits from accurate navigation, charting, real-time weather observations and coastal forecasts, and search and rescue services. Technical changes in the marine industry are improving safety, safeguarding natural resources, and increasing efficiency. For example, emerging OCEAN OBSERVATORIES such as LEO-15 may actually provide real-time displays of parameters such as WATER LEVEL FLUCTUATIONS, CURRENTS, TEMPERATURES, SALINITIES, dissolved oxygen, and wind. Satellite imagery might be provided to highlight oceanographic features such as algal blooms or the extent of SEA ICE, while hydrographic surveys provide BATHYMETRY that supports the display of wave modeling output such as wave height, wave period, and wave direction. This information may be used to plan the movement of ships, which is called OPTIMAL SHIP TRACK PLANNING. The data can be used to make information products that help responders to mitigate the effects of marine mishaps quickly. Dangerous events that require immediate and planned response include ship fires, coastal flooding, collisions, and sinkings.

Further Reading

Bishop, Joseph M. *Applied Oceanography.* New York: John Wiley, 1984.

National Oceanic & Atmospheric Administration. "NOAA Marine Operations." Available online. URL: http://www.moc.noaa.gov. Accessed September 26, 2007.

Woods Hole Oceanographic Institution home page. Available online. URL: http://www.whoi.edu/marops. Accessed September 26, 2007.

marine police Marine police, or harbor police, are usually representatives from government organizations charged with the responsibility of maintaining law and order on waterways. Efforts by marine police improve boater safety and coastal security through the enforcement of laws relating to water traffic, immigration, and conservation. They provide search and rescue services and allow other law enforcement personnel to reach locations not easily accessible from land. Their vigilance and response helps prevent and mitigate crime on vessels and in the coastal zone. Marine police are often seen patrolling coastal waters, canals, rivers, harbors, and ESTUARIES in a variety of vessels such as personal watercraft, INFLATABLE BOATS, and small- to medium-size motorboats. Like the U.S. Coast Guard, marine police have the authority to stop and board vessels in order to check for compliance with state and federal laws.

Marine police officers are responsible for a multitude of unique police duties. They are usually responsible for the enforcement of commercial and recreational fishery laws and regulations. Tasks involve checking recreational fishing licenses, monitoring COMMERCIAL FISHING catch and gear for compliance, and providing educational programs. By helping to protect fishery stocks and habitat, marine police are helping to ensure the sustainability of today's marine resources for future generations.

Marine police have advanced training to provide effective counterterrorism patrols to military installations, shipyards and nuclear power plants, and other high-value maritime facilities. Many officers have completed the Marine Law Enforcement Training Program held at the Federal Law Enforcement Training Center in Glynco, Georgia. They are also specialists in life saving, SCUBA diving, small arms, and explosives.

There are many other personnel and organizations that help make marine police work successful. MARINE ENGINEERS and mechanics ensure that watercraft and boats are kept in top running condition. Communications personnel answer calls ranging in nature from a simple question pertaining to a regulation to dispatching marine police personnel to boating accidents, marine spills, and other mishaps. Personnel from agencies such as the National Marine Fishery Service and the U.S. Fish & Wildlife Service train and deputize marine police to enforce federal fish and wildlife laws.

A U.S. Coast Guard inspection team prepares to board a foreign fishing trawler. *(USCG)*

marine pollution Marine pollution refers to the introduction of substances by humans into the marine environment, resulting in harm to living resources, presenting hazards to human health, downgrading the quality of seawater, and hindering marine activities such as fishing and swimming. Some major types of pollutants that have been the focus of recent research are chemicals and oil, sewage, garbage, radioactive waste, thermal pollution, and eutrophication (an overaccumulation of nutrients that support a dense growth of ALGAE and other organisms that decay and deplete the shallow waters of oxygen). Even though 70 percent of Earth is covered by water, the oceans are still vulnerable to water pollution. Pollutants are a real problem since they spread naturally by TURBULENCE, are transported by CURRENTS, eventually accumulate up the food chain, and can be consumed by humans who may become ill. The Great Pacific Garbage Patch is an accumulation of marine debris resulting from circular flow in the North Pacific GYRE. The size of this patch is unknown, but it has been estimated to be greater than twice the size of the continental United States.

Chemicals and oil get into the ocean in a variety of ways from ballast water dumping by ships to runoff from land drainage. With large increases in coastal populations, the main source of pollution may be storm water runoff and sewage from municipalities. This also includes fertilizers from lawns, pesticides such as chlordane, and hydrocarbons that reach the ocean from cars, powerboats, and gas docks. Toxic chemicals such as mercury, dioxin, dichlorodiphenyltrichloroethane or DDT, and polychlorinated biphenyl or PCB may also enter ocean systems through food chains and therefore affect organisms at places that differ from where the toxins were released. These toxic pollutants poison marine animals and plants, as well as humans who eat contaminated sea life. Some oil entering the ocean is natural and occurs as seepage from the seafloor and eroding sedimentary rocks, which release oil. Catastrophic pollution sources, which quickly grab media attention, are marine spills from tankers, offshore oil wells, and pipelines that cause extensive, immediate, and long-term damage to coastal and marine habitats. The resultant ECOSYSTEM damage (such as

Radioactive Waste Disposal in the Sea

According to the United States Department of Energy (DoE), radioactive waste commonly contains highly radioactive elements such as cesium, strontium, technetium, and neptunium. Governments, environmental, and community organizations have expressed great concern over the storage of radioactive waste because some of these materials remain radioactive, and therefore potentially dangerous to humans, for millions of years. Most scientists agree that the safest means of radioactive waste disposal is by burying it deep under the ground in what is called a "geologic repository." Some oceanographers have examined the possibility of locating a geologic repository offshore, which would require burial of the radioactive waste materials under the ocean floor. The United States operates nuclear reactors at commercial power plants, at government research facilities, and on about 40 percent of the U.S. Navy's submarines and ships. About 20 percent of electricity in the United States is generated by nuclear power.

Nuclear power facilities use fuel made of solid ceramic pellets of enriched uranium to produce electricity. One pellet can produce almost as much electricity as can one ton of coal. The energy released by the nuclear fuel produces heat, which is used to convert liquid water into steam. The resulting pressurized steam then turns giant turbines that generate electricity. After three or four years of use in a nuclear reactor, uranium pellets are no longer useful for producing electricity and so are considered "spent fuel". Workers then remove the spent fuel from the reactor, despite its continued high radioactivity. After workers remove spent fuel from a reactor, they place it in a pool of water. One can compare this pool to a steel-lined concrete swimming pool. The water in the pool removes heat from the spent fuel and protects workers and the general population from radiation. After the spent fuel has cooled, some facilities place the fuel in dry containers made of steel and concrete to impede the remaining radiation.

After the cold war, the United States began closing nuclear weapons plants and disposing of nuclear weapons materials. Weapons production results in "high-level radioactive waste." Until the late 1970s, the United States acquired materials for nuclear weapons by reprocessing spent nuclear fuel from nuclear reactors. Nuclear scientists reprocess fuel by chemically treating it to separate uranium from plutonium. The result of reprocessing is a highly radioactive "sludge." The U.S. Department of Energy will not permit shipment of this material to a repository until it is solidified and stored in stainless steel canisters.

Currently, radioactive waste is stored in temporary facilities located in 131 sites in 39 states. Most storage the marsh lands surrounding Prince William Sound, Alaska, which includes seabirds, mammals, fisheries, and coastal residents) evokes international concern through photographs and reports by scientists and journalists who document such disasters.

Sewage is another pollutant that reaches the ocean and can cause serious health problems, especially near swimming BEACHES and shellfish beds. Following storms, sewage systems can overflow, causing rainwater along with raw sewage to flow directly into coastal waters. Sewage increases nutrient-loading and particulate matter that is suspended in the water. Consequences include the degradation of benthic invertebrate communities and disappearance of SUBMERGED AQUATIC VEGETATION beds due to the ensuing high TURBIDITY. In sewage-laden waters, the presence of diseased fish and the occurrence of fish kills become more prevalent. High nutrient concentrations can contribute to red tides, massive blooms of microscopic algae, which in some cases cause illness through skin contact, inhalation, or ingestion. Agricultural waste, such as that from hog and poultry farms, is similar to sewage in its effects on the environment. For example, waste from hog farms is often contained in hog lagoons, and sprayed on neighboring fields as fertilizer. However, under heavy rain and strong winds, the lagoons may overflow, saturating the neighboring water bodies with excess nutrients and dangerous pathogens that may result in fish kills and human disease.

The world's oceans have historically been a dumping ground for garbage, which includes old fishing NETS, plastics, general household garbage, medical waste, and trash from oceangoing vessels and large cities. Some municipalities routinely use BARGES to dump garbage at sea. Ships routinely dump their garbage at sea, and some aircraft dump waste from lavatories over the oceans. Plastics, poisons, and toxins may be ingested by sea life, which can negatively affect the entire food chain. Litter such as plastic six-pack holders can be especially dangerous to ocean life that gets entangled in it, becoming injured and drowning. Wastes can block out sunlight and will ultimately settle on the seafloor, possibly smothering benthic sea life.

facilities are located near large bodies of water. In the United States, more than 161 million people live within 75 miles (125 km) of nuclear waste storage sites. These storage facilities are considered safe, but none is considered permanent, as they will not withstand rain and wind for thousands of years over which the waste will remain hazardous to humans.

Experts worldwide continue to study the following four primary options for permanently disposing of nuclear waste:

1 permanently storing the material at temporary storage sites;
2 burying it in the ocean floor or polar ice sheets;
3 sending it into outer space; and
4 placing it deep underground in a geologic repository.

Most scientists agree that disposal in an underground repository is the best long-term solution. The theory behind deep geologic disposal is to keep waste dry and isolated for as long as possible, so that its radiation can slowly diminish to safe levels. Underground storage also protects waste from weather phenomena that can disperse radioactive materials into the atmosphere, freshwater bodies, and oceans. The U.S. Department of Energy (DoE) has laid out plans for an underground repository at Yucca Mountain, Nevada. In response to public and congressional concerns, the DoE plans to store only nonexplosive, nonflammable solid radioactive waste materials inside Yucca Mountain. Meanwhile, some ocean scientists continue to study the idea of permanently storing radioactive waste beneath the ocean floor. Some of the ocean floor is highly dynamic in terms of volcanism and PLATE TECTONICS. As a result, ocean scientists must carefully select a reasonable location for deep-sea burial of radioactive waste such that waste materials cannot be released and dispersed into the ocean environment itself. Additionally, ocean scientists must carefully consider the inherent problems of ocean floor burial such as the difficulty of building an undersea repository. Oceanic transportation to the burial sites would also involve considerable risk and cost.

Further Reading

Nadis, S. "The Sub-Seabed Solution." *Atlantic Monthly* 278, no. 4 (199): 28–39.
U.S. Department of Energy Office of Civilian Radioactive Waste Management. "Fact Sheets: What Are Spent Nuclear Fuel and High-level Radioactive Waste?" Available online. URL: http://www.ocrwm.doe.gov/factsheets/doeymp0338.shtml. Accessed October 17, 2007.

—Kyle R. Dedrick
Marine Information Resources Company
Ellicott City, Maryland

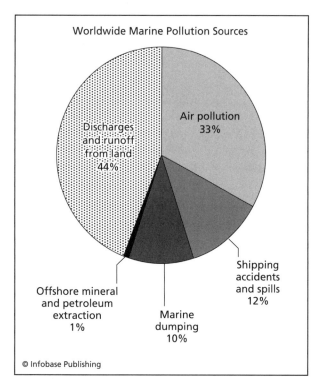

Worldwide Marine Pollution Sources

Air pollution 33%

Discharges and runoff from land: 44%

Offshore mineral and petroleum extraction 1%

Marine dumping 10%

Shipping accidents and spills 12%

© Infobase Publishing

The marine environment is contaminated by potentially harmful substances. The pie diagram presents the major sources of marine pollution. Most sources of ocean pollution are seen to originate on land.

Since the emergence of nuclear weapons in 1945, followed by nuclear power plants, radioactive waste has found its way into the ocean. It enters the ocean from nuclear weapons testing, the dumping of wastes from nuclear fuel cycle systems, and nuclear accidents. Some scientists and engineers have proposed plans to dispose of radioactive waste on the ocean floor. Methods include encasing it in concrete and setting it on passive regions of the seafloor, dropping it into the depths in torpedoes, and depositing containers in shafts drilled into the ocean floor. The disposal of radioactive wastes onto the ocean bottom was made illegal by the 1972 International Convention on the Prevention of Marine Pollution by Dumping of Wastes and Other Matter, which is commonly called the "London Convention."

Power plants located along the coast use seawater for cooling purposes, which is returned to the marine environment as warm-water discharge. Hot water discharged by factories and power plants causes thermal pollution by increasing ambient water temperatures. The Calvert Cliffs Nuclear Power Plant in Maryland has cooling water outflow into Chesapeake Bay that has become a favorite winter fishing location for striped bass (*Morone saxatilis*). Anglers carefully fish this jet of warm water gushing

up from the bay's bottom that forms a half-mile (0.81 km) RIP CURRENT flowing away from the power plant. Thermal discharge may be a serious problem where organisms are near their thermal maximum; for example, tropical marine vegetation in a heated bay may be in stress. Temperature increases change the level of oxygen dissolved in a body of water, thereby disrupting the water's ecological balance. These changes will kill some plant and animal species while encouraging the spread of others. The effect of thermal pollution on the marine environment is an active research topic in which investigators study the effects on communities immediately adjacent to the discharge.

Natural and man-made processes may combine to form a water body that becomes rich in dissolved nutrients. This process is called eutrophication and may result in a deficiency of dissolved oxygen, producing an environment that favors plant over animal life. Fertilizers such as nitrogen and phosphorus used on land reach the ocean via rivers and streams. High nutrient concentrations often result in PHYTOPLANKTON blooms such as red tides. Red tides are actually a population explosion of dinoflagellates that are toxic and fatal to certain fish species. Since these blooms contain toxic phytoplankton, they are also called harmful algal blooms. The Gulf Coast of Florida has in recent years experienced extensive red tides that have killed large numbers of sport fish and fouled beaches. There is a "dead zone" at the mouth of the Mississippi River the size of the state of New Jersey in which agricultural waste from the Mississippi River and tributaries has caused severe eutrophication and environmental stress. The ensuing lack of oxygen makes this region unfavorable to sea life. This hypoxic zone tends to form in the summer and negatively affects COMMERCIAL FISHING and recreational boating.

Further Reading

Center for Marine Conservation. "Ocean Conservancy." Available online. URL: http://www.cmc-ocean.org. Accessed September 26, 2007.

Gattuso, Jean-Pierre, editor. "Marine Debris." In *Encyclopedia of Earth*, Environmental Protection Agency and National Oceanic and Atmospheric Administration (Content Sources). Available online. URL: http://www.eoearth.org/article/Marine_debris. Accessed March 9, 2008.

Valdez Science home page. Available online. URL: http://www.valdezscience.com. Accessed September 26, 2007.

Virginia Institute of Marine Sciences. "Marine Pollution." Available online. URL: http://www.vims.edu/bridge/pollution.html. Accessed September 26, 2007.

A containment boom helps to confine oil spilled from a drilling rig.   *(NOAA)*

marine protected area   A region that is afforded legal protection against development and uncontrolled extractive and consumptive activities, such as fishing, is a marine protected area. Such areas tend to have different shapes, range in size from one to hundreds of square miles (thousands of km^2), have varying management characteristics, and may be established for specific purposes. One of the largest marine protected areas is managed by the Australian government and includes Heard Island and the McDonald Islands. It is located in the Southern Ocean about 2,485 miles (4,000 km) south of Australia.

Protection of marine areas is being accomplished internationally. There are many different types of marine protected areas, and they are known by various names. These protected areas include conservation areas, critical habitats, fishery management zones, monuments, parks, research reserves, sanctuaries, and wildlife refuges. A viable marine protection area provides society with a tool to safeguard biodiversity, manage fisheries, and restore depleted animal and plant populations. Their boundaries should be determined based on the input of stakeholders, such as fishermen, divers, scientists, mariners, and community members. Designated protected areas provide locations where scientists can study rich and complex marine habitats. These areas have been established to protect many different habitats. They can include rocky reefs and shorelines, coral reefs, sand and mud seafloors, estuaries, and deep-ocean shipwrecks.

In order to establish a successful marine protected area, managers need to know the organisms that exist in the region, understand habitat associations and predator and prey relationships, and the physical environment extending from the seafloor to the surface. Research within the protected area is important in developing a baseline from which changes can be measured. Changes to the region need to be documented, as practices such as anchoring on coral reefs and fishing are stopped and tourism increases. Some protected areas use geographic information systems to track and display the many features and attributes that are important for management. A management goal might involve boosting the abundance, diversity, and size of marine species living within protected areas.

A Sample of Marine Protected Areas

Marine Reserve	Location	Protected Habitat
Abalone Cove Ecological Reserve	California	A 1.17-mile (1.9 km) shoreline consisting of a rocky intertidal zone, with rocky outcrops and sand in the subtidal. Surrounding this site are sandy coves, kelp forests, and reef habitat.
Achang Reef Flat	Guam	Coral reefs, submerged aquatic vegetation, and mangroves. Offshore islets are composed of limestone, and deepwater channels provide habitat for predatory species.
Ahihi-Kinau	Hawaii	One of several areas on Maui where fishing is restricted. The land section of the reserve is composed of the last lava flow on Maui and brackish water ponds.
Al Yasat	United Arab Emirates	A group of islands and their surrounding waters covering approximately 186 square miles (482 sq km). The islands are surrounded by coral reefs and provide habitat for sea turtles and dugong.
Bajo de Sico Bank	Puerto Rico	A restricted offshore fishing area off the west coast of Puerto Rico near Mayagüez Bay.
Bodega Marine Life Refuge	California	Rocky shore, salt marsh, and tidal flats on Pacific coast near the San Andreas Fault.
Buxton Woods Coastal Reserve	North Carolina	Barrier island coast with maritime evergreen forest, maritime shrub swamp, maritime shrub thicket, freshwater sedge, and salt marsh communities.
Caye Caulker	Belize	Belize Barrier Reef and Submerged Aquatic Vegetation.
Channel Islands	California	Waters surrounding Anacapa, Santa Cruz, Santa Rosa, San Miguel, and Santa Barbara islands hosts sandy beach to deep rock habitats, including rocky intertidal, shallow eelgrass, surf grass, rock and kelp forests, and mid-depth rocks.
Dar es Salaam	Tanzania	Islands surrounded by coral reefs, mangroves, and sea grass beds.
Dry Tortugas National Park	United States	Shallow and deep coral reefs and sand in an elliptical, atoll-like formation approximately 70 miles (113 km) west of Florida Keys.
Egg Rock Wildlife Sanctuary	Massachusetts	A rock island located at the entrance to Nahant Bay. The island is about 80 feet (24 m) high and three acres in area. The seafloor near Egg Rock is composed of broken rocks and boulders with some sandy areas.
Folkestone Marine Reserve and Park	Barbados	Fringing reefs, patch reefs, and offshore bank reefs in the eastern Caribbean. The park includes a museum, aquarium, science zone, recreational area, and several locations for water sports.
Galápagos	Ecuador	Waters surrounding the Galápagos Islands, which include submarine volcanoes or seamounts and coral reefs. Important open-water habitats called *bajos* are the shallow waters that are found over seamounts.
Glover's Reef	Belize	Sand and mangrove, cayes, cocal, littoral thicket, seagrass beds, and reef.
Goat Island	New Zealand	Rocky shores, deep reefs, underwater cliffs, canyons, and sand flats.
Great Barrier Reef	Australia	The largest marine reserve, with coral reefs, mangrove estuaries, sandy and coral cays, continental islands, seagrass beds, algal and sponge "gardens", sandy or muddy seabed communities, continental slopes, and deep-ocean trenches.
Heard and McDonald Islands	Australia	Sub-Antarctic islands with benthic areas associated with cobbles, basaltic sand, mud, and siliceous diatom oozes and territorial seas. Heard Island—80 percent ice-covered, mountainous, and dominated by a large massif (Big Ben) and an active volcano (Mawson Peak). McDonald Islands—small and rocky.
Hol Chan	Belize	Coral reefs, submerged aquatic vegetation beds, coastal lagoon, and mangrove swamp near the island of Ambergris Caye in the Caribbean.
Island Beach State Park	New Jersey	An approximate nine-mile (14.5-km) stretch of undeveloped barrier island coast located at the southern end of Barnegat Peninsula. The sand dunes, maritime forest, and salt marshes are prime habitat for migratory birds and seabirds such as osprey.

Marine Reserve	Location	Protected Habitat
James V. Fitzgerald	California	Rocky intertidal and subtidal Monterey shale habitat located in San Mateo County near San Francisco, California.
Kermadec Islands	New Zealand	Eleven small volcanic islands located 466 miles (750 km) northeast of Aukland. The shallow seafloor is dominated by corals, and protected areas extend over a 12 nautical mile (22 km) radius, centered around every outcrop.
Little Tinicum Island Natural Area	Pennsylvania	A two-mile long (3.2-km-long) forested island located in the Delaware River containing tidal mudflats.
Malindi-Watamu, Mombasa, Kiunga, and Mpumhuti	Kenya	Coral reefs, sea grass beds, mangroves, and lagoons along the Indian Ocean.
Matuku	New Zealand	Salt meadows, saltmarshes and mangrove forests along with intertidal mudflats and shell-spits.
Narrow River	Rhode Island	A six-mile-long (9.5-km-long) estuary paralleling the west passage of Narragansett Bay. Habitats include tidal waters, fresh water, mud flats, salt marsh, streams and seeps, red maple swamps and oak-maple woodlands.
Octopus Hole Conservation Area	Washington	Octopus Hole is a rocky habitat located on southern Hood Canal, a natural fjord. Depths are from 15 ft. (5 m) to 65 ft. (20 m). Fissures, boulders, and ridges provide crevices for a variety of marine life.
Padilla Bay National Estuarine Research Reserve	Washington	A shallow bay filled by the Skagit River. Sea grass meadows, tidal flats and sloughs, salt marshes, and upland forests and meadows.
Port Honduras Marine Reserve	Belize	Mangrove swamps, cayes, and barrier reef form an estuarine complex that is connected to the Caribbean Sea.
Queen Anne's Revenge	North Carolina	The alleged shipwreck of the infamous pirate Edward Teach, or Blackbeard, located in a 20- to 25-foot (6- to 7.6-m) deep channel between Shackleford and Bogue Banks near Morehead City and Beaufort seaports.
Quillayute Needles Refuge	Washington	Sited along the Pacific northwest coast of the United States from Flattery Rocks south to Copalis Beach. Habitat consists of islands, rocks, and reefs that are vital sanctuary for seabirds, sea lions, harbor seals, sea otters, and whales.
Round Island Coastal Preserve	Mississippi	A barrier island on the Gulf of Mexico covered by pine forests with interior marshes. This habitat is an important feeding, resting, and wintering ground for migratory birds and is the site of a great blue heron rookery.
Seasonal Area Management (SAM) East	Massachusetts	This area includes some of the deep waters in the Gulf of Maine and shallow waters of George's Bank. This critical habitat is used by North Atlantic right whales between May and August.
Soufrière Scott's Head Marine Reserve	Dominica	Reef nursery areas for pelagic and reef fish; hot springs. The reserve is divided into a fish nursery, recreation, fishing priority, and scuba diving areas.
Sumilon Marine Reserve	Philippines	Established as a fish sanctuary in 1974, Sumilon Island is a coralline island surrounded by fringing reefs. The protected area covers approximately 0.15 square miles (0.39 sq km).
Togiak National Wildlife Refuge	Alaska	The Togiak National Wildlife Refuge's coastline provides coastal cliffs, estuaries, mudflats, river outlets, wetlands, and coastal tundra.
Tonga Marine Reserve	New Zealand	Coastal forest, sheltered granite shores, and rocky sub-tidal reefs.
Urca de Lima Underwater Archaeological Preserve	Florida	The 1715 Spanish shipwreck Urca de Lima is located in the Atlantic Ocean, north of Ft. Pierce Inlet State Park and Pepper Park. It lies in 10-15 ft (3-4.6 m) of water, approximately 600 ft (183 m) offshore.
Vatia	American Samoa	South Pacific coral reef habitat.

(continues)

A Sample of Marine Protected Areas *(continued)*

Marine Reserve	Location	Protected Habitat
Wallops Island National Wildlife Refuge	Virginia	Barrier island, woodlands, and salt marsh.
Wassaw National Wildlife Refuge	Georgia	A barrier island with salt marshes, mud flats, maritime forests, dunes, and beaches.
Yachats Marine Garden	Oregon	A large, rocky bench, broken into sections by large fissures and small, coarse-grained sandy beaches. It characterizes the coast near the mouth of the Yachats River and Bay. The intertidal habitat has extensive mussel beds.
Zekes Island Estuarine Reserve Dedicated Nature Preserve	North Carolina	This site is a complex of islands, salt marshes, tidal flats, dunes, and shrub thickets.

Note: These locations provide a natural classroom for marine education and help preserve national history and culture.

Marine protected areas save marine resources from many forms of exploitation, while providing opportunities for recreation and study. Countries such as the United States have also established MARINE SANCTUARIES in the open ocean that are analogous to federal parks such as Yellowstone National Park. Today's researchers are applying scientific methods to determine optimal locations to establish fully protected marine areas as an oceans-management tool. The table that starts on page 310 lists a selection of marine protected areas.

Further Reading
Marine Protected Areas of the United States. Available online. URL: http://www.mpa.gov. Accessed June 17, 2006.

marine provinces Marine scientists have partitioned the ocean into natural regions called marine provinces, based on physical forces, availability of light and nutrients, complexity of the marine food web, and bio-optical properties. These science-based zones differentiate unique features, organisms, and processes within the ocean realm.

The seafloor is important as a habitat for flora and fauna (benthos) and for burrowing species living in the substrate. The seafloor has major features such as a flat or gently sloping CONTINENTAL SHELF, a steep CONTINENTAL SLOPE cut by submarine canyons, a SEDIMENT-laden CONTINENTAL RISE, a flat ABYSSAL PLAIN, and where plates are subducting, a plunging TRENCH. BATHYMETRY changes as one moves from deep ocean regions, across the continental shelf break into the coastal ocean, and into shallow nearshore regions. Changes in depth affect the propagation of surface WAVES such as swell. WAVELENGTHS for these storm-generated waves will decrease as the wave heights increase during the wave's transit from deep to shallow water. These waves will break as SURF along offshore reefs and BEACHES. Other physiographic regions such as the mid-oceanic ridges are associated with specific processes such as hydrothermal circulation and concomitant HYDROTHERMAL VENTS. SEAFLOOR SPREADING also explains the location of undersea mountains called SEAMOUNTS. The peaks of these seamounts that once extended above sea level are flattened by wave action. As these weathered features move into deeper water by plate motions they are classified as table mounts or GUYOTS. They are often near HOT SPOTS and, depending on climate and depth, promote the growth of CORAL REEFS as they move away from the hot spot.

Biological environments over the continental shelf including the water above it are called the neritic province (*see* NERITIC ZONE). This "brown water" region extends from mean lower low water to the seaward edge of the shelf (the shelf break), where the depth is usually less than 656 feet (200 m). The continental shelf varies in width; for example, it is very narrow along the California coast and wide north of Siberia. The deep "blue water" region away from the continental shelf break is called the pelagic province (*see* PELAGIC ZONE). It includes the water column (an idealized column of water from the bottom to the surface) and leaves out the seafloor and coastal waters. The pelagic province is subdivided into the epipelagic or PHOTIC ZONE, the mesopelagic or twilight zone, the dark bathypelagic where temperatures range from 35.6 to 39.2°F (2 to 4°C), the

ABYSSOPELAGIC ZONE, where many colorless creatures are adapted to temperatures ranging from 32 to 39.2°F (0 to 2°C), and the cold high-pressure hadalpelagic layers (*see* HADOPELAGIC ZONE).

Zones extending from the shoreline to depths of 19,685 feet (6,000 m) include the littoral, bathyal, and ABYSSAL ZONES. The littorals trend offshore from the shoreline to a depth of approximately 600 feet (183 m) and are subdivided into three regions known as the supralittoral, the intertidal, and the sublittoral zone. Littoral bottoms in MARGINAL SEAS such as the GULF OF MEXICO are usually composed of sands, while benthic regions such as the abyssal and hadal zones consist of soft, fine-grained mud. In the deep ocean, species diversity decreases with depth. Researchers report that some of the dominant abyssal species are black fish with large heads such as abyssal rattails (*Coryphaenoides murrayi*), crustaceans such as vent shrimp (*Rimicaris exoculata*) that can sense dim light emanating from hydrothermal vents, suspension and deposit feeders, gelatinous sea cucumbers, bivalves, and wriggling brittle stars.

Large marine ECOSYSTEMS are biologically productive coastal areas ranging from ESTUARIES seaward to the region of fronts, EDDIES, and JETS, which delineate inshore from open-ocean waters. These regions have distinct hydrography and are considered to be the area where most pollution, overexploitation, and coastal alteration occurs. They may be on the order of 77,220 miles^2 (200,000 km^2), or more, and hold significant marine resources, making them the logical setting for coastal zone management. Scientists from organizations such as the World Conservation Union, the Intergovernmental Oceanographic Commission of United Nations Educational, Scientific and Cultural Organization or UNESCO, and the U.S. NATIONAL OCEANIC AND ATMOSPHERIC ADMINISTRATION report that 64 large marine ecosystems produce 95 percent of the world's annual marine fishery biomass yields.

Physical and chemical changes common to a particular marine region have led scientists to partition the water column into biogeochemical provinces. These provinces are defined by independent geographical and oceanographic boundaries such as distances to the coast, current systems, river inputs, or location of oxygen minimum zones. Such regions are high-productive areas and have varying input from land drainage. Adjacent biogeochemical provinces may be distinguished through oceanographic measurement programs. For example, differences may be observed after analyzing vertical distributions or profiles of TEMPERATURE, SALINITY and DENSITY. One tool to highlight differences, especially in the upper 656 feet (200 m) of the water column, is a temperature and salinity (T-S) diagram. Biogeochemical provinces help researchers focus on a location to understand the distribution of chemicals such as mercury (Hg). Mercury can enter the atmosphere through various methods such as coal combustion and the incineration of wastes. It reaches the oceans through atmospheric deposition. Therefore, knowing the air-sea exchange of mercury is important since the bioaccumulation of this chemical is a health concern to wildlife and humans. Satellites provide an important source of data to quantify oceanographic processes such as the upper mixed-layer structure, heat storage, and sea surface temperature, which play important roles in the evolution of atmospheric events. Scientists work with satellite images from sensors such as the multispectral ocean color sensor developed by NASA, called Sea-Viewing Wide Field-of-View Sensor or SeaWiFS, and the synthetic aperture radar developed by the Canadian Space Agency, called RADARSAT 1 and 2. This imagery is useful to study biogeochemical provinces such as the marginal ice zone. The marginal ice zone is located near the edge of the sea ice pack where there are complex interactions among ocean waves and ice cover as well as atmospheric and oceanic gradients. Important gradients include rapidly changing wind speeds, temperatures, and salinities.

Science-based criteria that define sections of the environment facilitate a better understanding of temporal and spatial changes occurring over scales that range from moments to epochs. In the same way as meteorologists have classified climate, marine scientists have developed provinces to reflect differences among ocean environments. These provinces, zones, and partitions help scientists to comprehend the whole system by understanding its similarities. Such knowledge is beneficial by equipping governments to better manage

A Selection of Notable Ocean Basin Depths

Ocean	Mean Depth	Greatest Depth
Arctic	4,019 feet (1,225 m)	Molloy Deep 18,399 feet (5,608 m)
Atlantic	11,801 feet (3,597 m)	Puerto Rico Trench 28,681 feet (8,742 m)
Indian	12,175 feet (3,711 m)	Sunda Trench 25,358 feet (7,729 m)
Pacific	13,042 feet (3,976 m)	Mariana Trench 36,161 feet (11,022 m)

An Alphabetical Selection of the World's Large Marine Ecosystems (LMEs)

Agulhas Current	East-Central Australian Shelf	Kara Sea	Red Sea
Antarctica	East China Sea	Kuroshio Current	Scotian Shelf
Arabian Sea	East Greenland Shelf	Laptev Sea	Sea of Japan
Arctic Ocean	East Siberian Sea	Mediterranean Sea	Sea of Okhotsk
Baltic Sea	Faroe Plateau	Newfoundland-Labrador Shelf	Somali Coastal Current
Bay of Bengal	Guinea Current	New Zealand Shelf	South Brazil Shelf
Barents Sea	Gulf of Alaska	North Australian Shelf	South China Sea
Beaufort Sea	Gulf of California	North Brazil Shelf	Southeast Australian Shelf
Benguela Current	Gulf of Mexico	Northeast Australian Shelf/Great Barrier Reef	Southeast U.S. Continental Shelf
Black Sea	Gulf of Thailand	Northeast U.S. Continental Shelf	
California Current	Hudson Bay	North Sea	Southwest Australian Shelf
Canary Current	Humboldt Current	Norwegian Shelf	Sulu-Celebes Sea
Caribbean Sea	Iberian Coastal	Northwest Australian Shelf	West Bering Sea
Celtic-Biscay Shelf	Iceland Shelf	Oyashio Current	West-Central Australian Shelf
Chukchi Sea	Indonesian Sea	Pacific Central-American Coastal	West Greenland Shelf
East Bering Sea	Insular Pacific-Hawaiian	Patagonian Shelf	Yellow Sea
East Brazil Shelf			

Note: These ocean realms encompass coastal areas from river basins and estuaries to seaward boundaries of continental shelves and the outer margins of permanent currents.

their exclusive economic zones. A nation's EXCLUSIVE ECONOMIC ZONE extends 200 nautical miles (371 km) offshore and represents the boundary where it has jurisdiction over natural resources to include fishing. Establishing marine provinces provides an important tool to understanding how a system reacts to changes such as increasing eutrophication of coastal areas or to a mishap such as an oil spill. Scientists are developing a baseline of information on interactions within marine provinces. Sudden changes from the baseline can be indicative of global climate change. Tables on pages 313 and 314 give examples of ocean basin depths and large marine ecosystems, respectively.

Further Reading

The Environmental Data Center. "Large Marine Ecosystems of the World." Available online. URL: http://www.edc.uri.edu/lme/intro.htm. Accessed September 26, 2007.

marine reptiles Marine reptiles are mostly cold-blooded vertebrates, comprising the sea turtles, salt-water crocodiles, marine iguana, and sea snakes that inhabit salt water and in some cases travel great distances to lay eggs. Reptiles have lungs, and their hearts are three-chambered, except in crocodilians,

which have a four-chambered heart. Their position in the animal kingdom is between amphibians and birds. Marine reptiles generally live in coastal waters common to tropical or subtropical regions. They are restricted to warm waters since they are ectothermic or cold-blooded, depending on external temperatures to control metabolic rates. Marine reptiles have body fluids that are much less salty than seawater, which will lead to dehydration unless there is a mechanism to get rid of salt. Marine reptiles drink seawater and excrete excess salts through specialized salivary, lingual, lachrymal, or nasal salt glands. These secretory cells actually concentrate sodium (Na^+) and chloride (Cl^-) ions from the blood. Marine reptiles tend to excrete nitrogenous wastes as ammonia, while terrestrial reptiles excrete waste as uric acid.

Sea turtles are air-breathing reptiles found in most tropical and subtropical marine habitats. They migrate hundreds of miles between nesting and feeding grounds. Sea turtles are widely distributed in warm oceans, and some have been listed as endangered or threatened species. They are different from terrestrial tortoises and fresh or brackish water terrapins because their limbs have evolved into flippers. The sea turtle's head and limbs are nonretractable.

While swimming, streamlined sea turtles are observed using front flippers to paddle and their hind flippers for steering. Some sea turtles such as the green turtle (*Chelonia mydas*) feed on submerged aquatic vegetation. Other carnivorous species such as the leatherback (*Dermochelys coriacea*) feed on jellyfish, crabs, shrimp, snails, sea urchins, and MOLLUSKS. The leatherback is the largest of the sea turtles, weighing up to 1,400 pounds (635 kg). They use salt-secreting tear (lachrymal) glands to cope with the osmotic challenges of the ocean. After reaching sexual maturity, male and female sea turtles mate in shallow offshore waters. Females swim ashore through the SURF zone, dig a hole above the high water mark with their hind flippers, lay their golf ball–size eggs in the hole, cover it up, and return to the ocean. After approximately two months, the eggs hatch, the baby turtles dig their way out of the nest, and scurry toward the ocean, using light cues. Many predators such as sea gulls and crabs are waiting to eat the baby turtles. Only a very small proportion of sea turtles survive to maturity. Scientists report that most sea turtles migrate long distances to nest at the same locations where they hatched. Other reasons contributing to the endangered or threatened status of sea turtles involve accidental deaths as bycatch, pollution, and being harvested for meat, eggs, shells, and leather. Bycatch refers to the unwanted capture of species taken by fishing strategies such as trawling or longlining. The application of technologies such as turtle excluder devices or TEDs on trawl NETS and the use of circle hooks by longliners are helping to make measurable progress in restoring sea turtle populations. Light from cities found near turtle nesting beaches may also contribute to problems. The artificial light tends to lead hatchlings toward the hinterlands and certain death instead of toward the lighter horizon and challenges of the sea.

Many species of marine turtles are currently threatened with extinction. (NOAA)

Saltwater crocodiles are large carnivorous reptiles that inhabit ESTUARIES, river mouths, and tidal inlets located in tropical and subtropical coastal regions. They have eyes, ears, and nostrils located near the top of the head, flattened bodies and tails, short legs, and powerful jaws. When the crocodile is submerged, their ears and nostrils have valves that close. In the eastern Indian Ocean, Australia, and some western Pacific Islands, saltwater crocodile (*Crocodylus porosus*) favor MANGROVE SWAMPS or riverbanks, grow to 23 feet (7 m), and may weigh 2,500 pounds (1,134 kg). Depending on size, these Indo-Pacific saltwater crocodiles eat a large range of animals from crustaceans and vertebrates (such as fish, turtles, snakes, and birds) to large prey such as monkeys, kangaroo, wild boar, and buffalo. In southern Florida, the Caribbean, southern Mexico, and along the Central American coast south to northwest South America, American crocodiles (*Crocodylus acutus*) inhabit brackish areas where fresh and salt waters mix, such as coastal wetlands and canals. They may grow to 15 feet (4.6 m) and weigh up to 450 pounds (204.5 kg). The American crocodile feeds on insects, crabs, fish, waterfowl, and small mammals and is less aggressive than the infamous *Crocodylus porosus*. Crocodiles are more aggressive than alligators, and in crocodiles the long lower fourth tooth protrudes on the side of the head when the mouth is closed. In alligators the teeth of the lower jaw fits into pits in the upper jaw. Male and female crocodiles mate while submerged, and the female will lay her eggs several weeks later. She will find a suitable site to build a mound of vegetation and mud from which she digs a hole in the center, lays 40 to 80 eggs, and covers them up. The eggs hatch within three months and the baby crocodiles head to the shore. Some female crocodiles have been observed cracking ready-to-hatch eggs in their mouth and carefully transporting the young to the shore. Crocodiles are the largest reptiles and may be the closest living animal to the dinosaur.

The marine iguana (*Amblyrhynchus cristatus*), which lives in large colonies on the Galápagos Islands, is the only marine lizard. The Galápagos are an archipelago of approximately 25,000 square miles (64,749.7 km²) located 600 to 700 miles (966 to 1,126.5 km) west of Ecuador in equatorial waters. Charles Darwin (1809–82) described the behavior of the different species of iguana during his studies of the Galápagos Islands sometime during September and October 1835. Like other lizards, iguanas are cold-blooded, egg-laying animals with an excellent ability to adapt to their environment. Sizes seem to relate to their distribution across the islands, where the largest are found on Fernandina and Isabela

Table of Common Sea Turtles

Scientific Name	Common Name
Caretta caretta	loggerhead sea turtle
Chelonia agassizi	East Pacific Green sea turtle
Chelonia mydas	green sea turtle
Dermochelys coriacea	leatherback sea turtle
Eretmochelys imbricata	hawksbill sea turtle
Lepidochelys kempii	Kemp's Ridley sea turtle
Lepidochelys olivacea	Olive Ridley sea turtle
Natator depressus	Australian flatback sea turtle

Commonly Studied Sea Snakes

Scientific Name	Common Name
Acalyptophis peronii	horned sea snake
Aipysurus apraefrontalis	short-nosed sea snake
Aipysurus duboisii	reef shallows or Dubois's sea snake
Aipysurus eydouxii	stagger-banded or spine-tailed sea snake
Aipysurus foliosquama	leaf-scaled sea snake
Aipysurus fuscus	dusky sea snake
Aipysurus laevis	golden or olive sea snake
Aipysurus pooleorum	shark bay sea snake
Aipysurus tenuis	brown-lined sea snake
Astrotia stokesii	Stokes's sea snake
Disteira kingii	spectacled sea snake
Disteira major	olive-headed sea snake
Disteira walli	Wall's sea snake
Emydocephalus annulatus	annulated or turtle-headed sea snake
Enhydrina schistose	beaked sea snake
Ephalophis greyi	north-western mangrove sea snake
Hydrelaps darwiniensis	black-ringed mangrove sea snake
Hydrophis atriceps	black-headed sea snake
Hydrophis brooki	Brooke's sea snake
Hydrophis belcheri	Belcher's hydrophis sea snake
Hydrophis caerulescens	blue or dwarf sea snake
Hydrophis coggeri	slender-necked sea snake
Hydrophis cyanocinctus	blue-banded sea snake
Hydrophis czeblukovi	fine-spined sea snake
Hydrophis elegans	elegant sea snake
Hydrophis fasciatus	barred sea snake
Hydrophis geometricus	geometrical sea snake
Hydrophis gracilis	slender sea snake
Hydrophis inornatus	plain sea snake
Hydrophis mcdowelli	small-headed sea snake
Hydrophis melanocephalus	black-headed sea snake
Hydrophis melanosoma	black-banded sea snake
Hydrophis ornatus	reef sea snake or Ornate sea snake
Hydrophis pacificus	large-headed sea snake

Islands and smaller marine iguanas are found on Genovesa. Their black coloration is used to absorb solar radiation after swimming in the cold ocean to eat macrophytic marine ALGAE. Clumps of dark red-brown or green algae growing on intertidal rocks form the bulk of the marine iguana's diet. Marine iguana are skillful swimmers, using a flattened tail and webbing on all four feet for swimming. Claws help anchor the iguana in the breaking waves, surf, and currents. Like other marine reptiles, the marine iguana must excrete salt, which it does by sneezing through salt glands located above the eye and connected via a duct to the nostrils. In fact, colonies of tens to thousands of marine iguanas are observed basking in the sun on warm lava rocks along the shore. Females and males grow from about two feet (0.6 m) to 4.3 feet (1.3 m) in length, respectively.

Snakes inhabiting the sea are long, slender, and limbless reptiles of order Serpentes with flattened rudders like tails. They are found only in the warm waters of the Indian and Pacific oceans and use lateral undulation for locomotion. They are mostly found in shallow coastal waters. A few species such as the brown-lipped sea krait (*Laticauda laticaudata*) swim to shore to lay their eggs, but most like deadly yellow-bellied sea snakes (*Pelamis platurus*) live and breed completely at sea. Sea snakes are air breathers with valved nostrils set on the top of the snout. They have one lung that extends the length of their body, which is lined with blood vessels to aid in oxygen (O_2) absorption, and at the end is a sac to store air. Sea snakes can also absorb oxygen from the water through their skin and stay submerged for about two hours if necessary. They have a gland under the tongue called the sublingual gland that helps get rid of salt from the seawater they drink. Like its land cousins, cobras, kraits, and mambas, a sea snake can unhinge its jaw to feed on fish eggs,

Scientific Name	Common Name
Hydrophis schistosus	hook-nosed sea snake
Hydrophis vorisi	estuarine sea snake
Lapemis curtus	Shaw's sea snake
Lapemis hardwickii	Hardwicke's or spine-bellied sea snake
Parahydrophis mertoni	northern mangrove sea snake
Pelamis platurus	yellow-bellied sea snake

Note: True sea snakes are pelagic air-breathing reptiles found in tropical Pacific seas. They are the only marine reptiles reproducing at sea.

small fish, crabs, and squids. All sea snake species are venomous. Their fangs tend to lie at the back of the mouth rather than at the front, and the venom delivery system is poor.

In summary, of the four classes of reptiles (testudines, rhynchocephalia, crocodilians, and squamates), only testudines (turtles), crocodiles, and squamates (lizards and snakes) live in the marine environment. Approximately 80 of the 6,800 reptile species inhabit marine regions. The largest number of marine reptilia comes from the squamates, which includes approximately 71 species of sea snakes. There are no true marine amphibians because of their thin skin. Reptiles (owing to migrating sea turtles) inhabit all the oceans except those of polar regions. The Disney film *Finding Nemo* popularized marine reptiles with the transport of sea turtle characters Crush and Squirt on the East Australian Current. Sea turtles such as the leatherback sea turtle are remarkable; they are the most ancient of living reptiles, migrate long distances, and are capable of diving well below the EUPHOTIC ZONE to chase prey, avoid ships, and escape predators. (Depending on water clarity, the euphotic zone extends to about the 164 feet (50 m) depth, where there is still enough light in the water for photosynthesis.) The marine reptiles have adapted to a life at sea and need to be protected due to habit loss from human development, trade in turtle shell curios and jewelry, overfishing, and illegal trade. International laws such as the Convention on International Trade in Endangered Species of Wild Fauna and Flora and the U.S. Endangered Species Act help to protect many species of marine reptiles.

Further Reading

Busch Entertainment Corporation. "Sea Turtles." Available online. URL: http://www.seaworld.org/animal%2Dinfo/ info%2Dbooks/sea%2Dturtle. Accessed September 26, 2007.

Dunson, M.A. *The Biology of the Sea Snakes*. Baltimore, Md.: University Park Press, 1975.

Keuper-Bennett, Ursala, and Peter Bennett. "Turtle Trax: A Page Devoted to Marine Turtles." Available online. URL: http://www.turtles.org. Accessed September 26, 2007.

marine reserve *See* MARINE PROTECTED AREA.

marine salvage Marine salvage is property saved from a wrecked vessel, or the recovery of the vessel and its property. Rescues and marine salvage play a large part in ocean lore. From the tall ship days to the present, marine salvors (seafarers who help salvage a vessel) have been in the business of reclaiming damaged, discarded or abandoned material, ships, craft and floating equipment for reuse, repair, refabrication, or scrapping. Salvors who come to the assistance of vessels in distress in order to save persons and property from destruction are entitled to present a claim for a salvage reward. A passing boater, also known as a "chance salvor," may also impose this charge. A professional salvor may charge some percentage of the value of the salvaged vessel. In order for the salvor to exact a reward, there must have been a marine peril, and the salvage services must have been successful and rendered voluntarily. Marine perils include fire, collisions, leaking, sinking, grounding, and distress brought upon by mechanical breakdowns and equipment failures. Typical acts of salvage in ESTUARIES and the coastal ocean involve the rescue and tow of boats that have run aground, run out of fuel, or have mechanical problems.

The 1989 International Convention on Salvage in London, United Kingdom, which was signed by the United States and ratified by the U.S. Senate, established the general principles of marine salvage. Maritime lawyers specialize in cases associated with maritime commerce and navigation, which includes lawsuits relevant to marine salvage matters.

Further Reading

Admiralty and Maritime Law Guide. "International Convention on Salvage, 1989." Available online. URL: http://www.admiraltylawguide.com/conven/salvage1989.html. Accessed September 27, 2007.

marine sanctuary Governments may establish a MARINE PROTECTED AREA with specific purposes and zones and classify them as a marine sanctuary. Sanctuaries have varying shapes and sizes, but are generally developed to provide a safe habitat for threatened species or to protect historically significant locations such as shipwrecks and lighthouses. Sanctuaries are designed to protect natural and cultural features while allowing people to use and enjoy the marine environments in a sustainable way. Usage of the protected habitat is regulated in order to restore depleted fisheries and promote environmentally sound marine recreation. The government of Australia has been a leader in conservation and enacted the Great Barrier Reef Marine Park Act on June 16, 1975. The Great Barrier Reef Marine Park Authority provides zoning, management plans, permits, and education in an effort to conserve the Great Barrier Reef, which consists of hundreds of islands and thousands of coral reefs.

In the United States certain MARINE PROTECTED AREAS, designated under the National Marine Sanctuaries Act of 1972, are managed as National Marine Sanctuaries. These areas contain the remains of important shipwrecks, are breeding and feeding grounds of whales, sea lions, sharks, and sea turtles, and critical benthic habitats. Some of these sanctuaries are zoned for specific purposes such as "no-take" areas (wildlife must not be killed or captured). They are analogous to Yellowstone National Park,

Table of U.S. Marine Sanctuaries

Name	Location	Habitat
Channel Islands National Marine Sanctuary	Pacific Ocean	An approximate 1,252-nautical-mile^2-area extending from mean high water to six nautical miles (nm) offshore from San Miguel, Santa Rosa, Santa Cruz, Anacapa, and Santa Barbara Islands.
Cordell Bank National Marine Sanctuary	Pacific Ocean	Located on the continental shelf, about 43 nautical miles northwest of the Golden Gate Bridge and 18 nautical miles west of the Point Reyes lighthouse. The protected area includes the 4.5-mile (7.5-km)-wide by 9.5-mile, (15.8-km)-long granitic bank.
Fagatele Bay National Marine Sanctuary	Pacific Ocean	A fringing coral reef ecosystem on the island of Tutuila in American Samoa.
Florida Keys National Marine Sanctuary	Atlantic Ocean	Barrier reef, sea grass meadows, and mangrove swamps make up this marine protected area.
Flower Garden Banks Marine Sanctuary	Gulf of Mexico	The United States's northernmost assemblage of coral reefs, located about 105 miles (169 km) directly south of the Texas-Louisiana border.
Gerry E. Studds Stellwagen Bank National Marine Sanctuary	Atlantic Ocean	Sited between Cape Ann and Cape Cod, in the southwest corner of the Gulf of Maine. Stellwagen Bank is a shallow, primarily sandy feature, curving in a southeast to northwest direction for 19 miles (30.6 km).
Gray's Reef National Marine Sanctuary	Atlantic Ocean	17 nautical miles^2 (58 km^2) of live-bottom reefs on the continental shelf.
Gulf of the Farallones National Marine Sanctuary	Pacific Ocean	An area of 948 nautical miles^2 (3,252 km^2) off the northern and central California coast.
Hawaiian Islands Humpback Whales National Marine Sanctuary	Pacific Ocean	A series of five marine protected areas distributed across the main Hawaiian Islands.
Monitor National Marine Sanctuary	Atlantic Ocean	The *Monitor* shipwreck is in approximately 235 feet (71.6 m) of water 16.1 miles (25.9 km) off Cape Hatteras, North Carolina.
Monterey Bay National Marine Sanctuary	Pacific Ocean	Consisting of 276 miles (444 km) of shoreline and 5,322 miles^2 (13,734 km^2) of ocean, habitats include kelp forests and a submarine canyon.
Northern Hawaiian Islands Marine National Monument	Pacific Ocean	4,500 miles^2 (11,655km^2) of coral reefs in the Hawaiian archipelago.
Olympic Coast National Marine Sanctuary	Pacific Ocean	This Pacific Northwest marine protected area includes the Nitinat, Juan de Fuca, and Quinault submarine canyons.
Thunder Bay National Marine Sanctuary	Great Lakes	An underwater preserve, encompassing 448 miles^2 (1,244 km^2) of northwest Lake Huron.

which was established in 1872 by President Ulysses S. Grant (1822–85).

One of the first marine sanctuaries in the United States was established 16.1 miles (25.9 km) off the coast of North Carolina during 1975 to protect the wreckage of the Civil War ironclad USS *Monitor,* which sank on December 31, 1862, in rough seas. The Channel Islands, Grays Reef, Looe Key, and Gulf of the Farallones Marine Sanctuaries were established during 1980 and 1981. Fagatele Bay in American Samoa was designated as a National Marine Sanctuary during 1986. Cordell Bank off the California coast was designated in 1989, followed by the Florida Keys in 1990, and the Flower Garden Banks, Gerry E. Studds Stellwagen Bank, Hawaiian Islands Humpback Whales, and Monterey Bay National Marine Sanctuaries during 1992. The Olympic Coast National Marine Sanctuary was designated during 1994 and the Thunder Bay National Marine Sanctuary during 2000. Thunder Bay is the first sanctuary in the Great Lakes. On June 15, 2006, President George W. Bush designated the largest single area in United States history for protection as the Northwestern Hawaiian Island Marine Monument. A table of marine sanctuaries in the United States is provided on page 318.

Further Reading

Channel Islands National Marine Sanctuary. Available online. URL: http://channelislands.noaa.gov/. Accessed September 27, 2007.

Cordell Bank National Marine Sanctuary. Available online. URL: http://cordellbank.noaa.gov/. Accessed June 19, 2006.

Fagatele Bay National Marine Sanctuary. Available online. URL: http://fagatelebay.noaa.gov/. Accessed June 19, 2006.

Florida Keys National Marine Sanctuary. Available online. URL: http://floridakeys.noaa.gov/. Accessed June 19, 2006.

Flower Garden Banks Marine Sanctuary. Available online. URL: http://flowergarden.noaa.gov/. Accessed June 19, 2006.

Gerry E. Studds Stellwagen Bank National Marine Sanctuary. Available online. URL: http://stellwagen.noaa.gov/welcome.html. Accessed June 19, 2006.

Gray's Reef National Marine Sanctuary. Available online. URL: http://graysreef.noaa.gov/. Accessed June 19, 2006.

Gulf of the Farallones National Marine Sanctuary. Available online. URL: http://farallones.noaa.gov/. Accessed June 19, 2006.

Hawaiian Islands Humpback Whales National Marine Sanctuary. Available online. URL: http://hawaiihumpbackwhale.noaa.gov/. Accessed June 19, 2006.

Monitor National Marine Sanctuary. Available online. URL: http://monitor.noaa.gov/. Accessed June 19, 2006.

Monterey Bay National Marine Sanctuary. Available online. URL: http://montereybay.noaa.gov/. Accessed June 19, 2006.

Northwestern Hawaiian Islands Marine National Monument. Available online, URL: http://hawaiireef.noaa.gov/. Accessed June 19, 2006.

Olympic Coast National Marine Sanctuary. Available online. URL: http://olympiccoast.noaa.gov/. Accessed June 19, 2006.

Thunder Bay National Marine Sanctuary. Available online. URL: http://thunderbay.noaa.gov/. Accessed June 19, 2006.

marine spill response Immediate actions taken to mitigate the effects of the inadvertent discharge of contaminants into estuarine and oceanic regions to include BEACHES and adjoining coastal lands are referred to as marine spill response.

There are many contaminants such as biological agents, chemicals, and fuels that are shipped in specially designed vessels every day. Common examples include ammonia and oil that are transported by vessels and transloaded at seaports. Ammonia is considered a high health hazard because it is corrosive to the skin, eyes, and lungs. Oil generally refers to all oil products and their by-products to include gasoline, petroleum, crude (mineral) oils, lube oil, sludge, oil refuse, oil mixed with other wastes, and other liquefied hydrocarbons. Oil is toxic and physically and chemically alters coastal habitats. Spilled oil damps wave action and effectively smothers flora and fauna.

Vessels transporting oil and terminals handling these products are required by the U.S. Oil Pollution Act of 1990 to develop oil spill contingency plans. This act was signed into law approximately one-and-a-half years after the U.S. flag tank ship *Exxon Valdez* ran aground March 24, 1989, on Bligh Reef in Prince William Sound, Alaska, spilling more than 11 million gallons of crude oil (this oil spill was the largest in U.S. history). Besides strengthening planning and prevention activities, the Oil Pollution Act established liability standards and requires spill removal equipment to be on hand. It expands the federal government's ability to react to a mishap and helps provide necessary and sufficient resources through tools such as the Oil Spill Liability Trust Fund and designates response organizations such as the U.S. Coast Guard, the agency entrusted with the enforcement of maritime pollution laws. Other agencies such as the National Weather Service (NWS) from the NATIONAL OCEANIC AND ATMOSPHERIC

Chronology of Significant Oil Spills since March 1967

Name	Year and Location	General Description of Oil Spill
supertanker *Torrey Canyon*	1967, Celtic Sea off Land's End, the most southwesterly point of Great Britain.	Ran aground and wrecked on Pollard's Rock in the Seven Stones reef between the Isles of Scilly and Land's End, England. In excess of 38 million gallons (143,845,648 liters) were spilled and spread along the south coast of England and the Normandy coast of France.
oil tanker *SS Witwater*	1968, Caribbean coast off Panama, near Galeta Island.	Ruptured hull in heavy seas, 588,000 gallons (2,225,822 liters) spilled.
Liberian tanker *Arrow*	1970, Chedabucto Bay in Nova Scotia, Canada.	Struck Cerberus Rock and sank in heavy weather; approximately 4.5 million gallons (17,034,353 liters) were spilled.
Sweden freighter *Othello*	1970, Tralhavet Bay east of the island of Vaxö in the Baltic Sea.	A tanker collision spilling approximately 18 million gallons (68,137,412 liters).
South Korean tanker *Sea Star*	1972, Gulf of Oman.	Collision with the Brazilian tanker *Horta Barbosa,* spilling about 35.3 million gallons (133,625,036 liters).
Liberian tanker *Argo Merchant*	1976, southeast of Nantucket Island near the Massachusetts coast.	Ran aground on Middle Rip Shoal during heavy weather, approximately 7.7 million gallons (29,147,671 liters) spilled.
tanker *Hawaiian Patriot*	1977, 300 nautical miles (556 km) off Honolulu, Hawaii, in the Pacific Ocean.	Ship developed a leak and caught fire, releasing approximately 30.4 million gallons (115,076,518 liters) into the northern Pacific.
very large crude carrier *Amoco Cadiz*	1978, English Channel, three miles (4.8 km) off the coast of Brittany, France.	Grounded on Portsall Rocks while outside of shipping lanes owing to failure of the steering mechanism; more than 65 million gallons (246,051,766 liters) were spilled.
Petroleos Mexicanos exploratory well *IXTOC I*	1979, Campeche Bay, 43.2 nm (80 km) northwest of Ciudad del Carmen, in the Gulf of Mexico.	Offshore wellhead blowout; 147 million gallons (556,455,532 liters) released.
supertanker *Atlantic Empress*	1979, Caribbean Sea off the coast of Tobago; 243 nm (450 km) east of Barbados.	Collision with the supertanker *Aegean Captain;* 43 million gallons (162,772,707 liters) of oil were spilled and burned during ensuing explosions and fire.
tanker *Burma Agate*	1979, eight miles (13 km) offshore of Galveston Island in the Gulf of Mexico.	Collision with the freighter *Mimosa* near Galveston, Texas, resulting in spillage and burning of 10 million gallons (37,854,118 liters).
Norwuz oil field	1983, Persian Gulf.	Military destruction during Iran-Iraq War, approximately 42 million gallons (158,987,295 liters) spilled.
British oil tanker *Alvenus*	1984, Gulf of Mexico and Galveston Bay.	Ruptured hull from grounding on the Calcasieu River Bar; 2.8 million gallons (10,599,153 liters) spilled.
oil tanker *Arco Anchorage*	1985, Strait Juan de Fuca, Port Angeles harbor on Washington's Olympic Peninsula.	Ran aground and spilled approximately 239,000 gallons (904,713 liters) of oil.
tanker *Nova*	1985, Arabian Gulf, 75.5 nm (140 km) south of Kharg Island.	Tanker collision, approximately 21 million gallons (79,493,647 liters) spilled.
Occidental Petroleum's oil production platform *Piper Alpha*	1988, North Sea off Scotland.	Explosions and gas fire killed 167 workers, unknown amount of oil spilled.
oil tanker *Exxon Valdez*	1989, Prince William Sound, Alaska	Ran aground on Bligh Reef and spilled approximately 11 million gallons (41,639,510 liters) of oil.
Argentine tour ship *Bahia Paraiso*	1991, Palmer Station, Antarctica.	Ran aground and spilled 200,000 gallons (757,082 liters) of oil.

Name	Year and Location	General Description of Oil Spill
Persian Gulf War	1991, Persian Gulf (Kuwait and Saudi Arabia).	Environmental sabotage by Iraqi forces during the Gulf War; estimates range from 240 to 335 million gallons (908.5 million to 1.268 billion liters) released.
barge *Vista Bella*	1991, Caribbean Sea, 10.4 nm (19.3 km) northeast of Nevis Island.	Sinking in 328 fathoms (600 m) of water after towing cable snapped, released up to 560,000 gallons (2,119,831 liters).
barge *Bouchard B155*	1993, Egmont Channel in lower Tampa Bay, Florida.	Collision with another vessel, and release of 336,000 gallons (1,271,898 liters).
barge *Morris J. Berman*	1994, Caribbean Sea, 15 nm (274 m) off San Juan, Puerto Rico.	Grounding on coral reef off Punta Escrambron after tow cable failure; released 800,000 gallons (3,028,329 liters).
Norwegian-owned tanker *Sea Empress*	1996, Irish Sea just outside the Milford Haven estuary in South Wales.	Spilled 37.6 million gallons (142,331,483 liters) after running aground on rocks below the cliffs of St. Anne's Head during a storm; polluted the Pembroke and Gower coastlines.
single-hull tanker *Erica*	1999, off the northwest coast of France, inundating 216 nm (400 km) of Brittany's beaches with oil.	Spilled 3 million gallons (11,356,235 liters).
Ecuadorean-registered tanker *Jessica*	2001, near San Cristóbal Island in the Galápagos and 600 miles (966 km) from mainland Ecuador in the Pacific Ocean.	Ran aground on Schiavoni Reef, spilling 175,000 gallons (662,447 liters).
Bahamese tanker *Prestige*	2002, near Galicia Island in northwest Spain.	The *Prestige,* carrying about 20 million gallons (75,708,236 liters) of heavy fuel oil, split in two and sank after sustaining damages in heavy weather.
Norwegian vehicle carrier *Tricolor*	2002, English channel impacting the Belgian and Dutch coasts.	The *Tricolor* collided with the French container ship *Kariba* in heavy fog, spilling more than 25,000 gallons (94,635 liters).

Note: Details are provided to aid online searches that describe response, recovery, and ongoing effects from these disasters.

ADMINISTRATION (NOAA) provide meteorological and oceanographic support by adapting existing NWS products to the spill site. Large spills may also require support from NOAA's Office of Response and Restoration. NOAA scientists utilize MODELS to forecast the spill's movement or trajectory and predict weathering effects of the oil.

Rapid response is required to mitigate the impact of marine spills, which are often caused by ship groundings, collisions, and fires, or from pipeline and storage tank leaks. In the advent of a spill, practiced and informed emergency personnel are needed to help save lives, protect property, remove oil, and rescue and rehabilitate distressed and oiled animals. Minimizing the elapsed time between the incident and the beginning of response operations is crucial. It is especially important to contain spills in shallow water before the forces of winds, WAVES, and CURRENTS can work the pollutant into the SEDIMENTS where they may be resident for many years. Some of the overarching spill response functions include discovery and notification, preliminary assessment, and determining the extent of damage to fish, birds, and mammals due to the spill. These response personnel also plan and implement a wildlife rehabilitation plan.

Accurate knowledge on the ambient conditions such as winds, TEMPERATURE, waves, and SALINITY at the time of a spill are foundational to planning and executing response operations. The slick is spread and transported by winds, waves, and currents. Depending on its DENSITY, the oil may be distributed throughout the water column (water from the surface to the bottom) and into the bottom sediments by these forces. Weathering by evaporation, dissolution, oxidation, emulsification, and microbial degradation alters the oil's physical and chemical properties. By knowing the type of oil spilled, the magnitude of the spill, and physical conditions at the spill site, responders can develop an optimal response for spill cleanup. Techniques include using booms (floating barriers with a skirt that extends below the surface and chain ballast for stability) to contain spills or to minimize the release of contamination that would otherwise spread or migrate to other areas. Response options also include mechanical recovery with skimmers, chemical treatment, and *in situ* burning. Various types of skimmers (free-floating, side-mounted, built in to a vessel, handheld) are used to remove oil from the sea surface. Weir skimmers float on the surface with an opening adjusted to allow the oil to flow into the device and into temporary containers. Drum, disc, and brush skimmers pick up the oil, which is collected and pumped into temporary containers. The recovered contaminants are sometimes stored on container BARGES. Chemical treatment or dispersants are sprayed over the affected area when the oil is thin on the surface and covers a large expanse. For large spills, aircraft may be employed to apply dispersants. In extreme cases over large areas, *in situ* burning is a response option. This technique involves the use of fireproof containment booms. It can be implemented only when the spill is located at a safe distance from populated areas and the winds are blowing away from those areas. Marine spill responders must utilize the best combination of response options to combat a marine spill while taking the surrounding weather conditions into account.

Marine spills will occur due to various mishaps in the future. In each event, there will undoubtedly be similarities with past spills and a unique set of circumstances that contributed to the incident. Natural spreading of the spill in response to waves, tides, and currents is a major factor in determining the response options. Mitigation will require careful assessment by experienced personnel to ensure effective response performance. These calamities can be minimized through preparedness, adequate resources, new technology, increased maritime safety, and through analysis of past spill response operations. A chronological list of significant oil spills since March 1967 is provided on page 320.

Further Reading

Exxon Valdez Oil Spill Trustee Council home page. Available online. URL: http://www.evostc.state.ak.us. Accessed October 10, 2007.

U.S. Environmental Protection Agency. "Oil Program." Available online. URL: http://www.epa.gov/oilspill. Accessed September 27, 2007.

Marine Technology Society (MTS) A nonprofit, international, professional organization established in 1963 to promote the exchange of information in ocean and marine engineering, science, and policy. The society supports marine professionals from academia, industry, and government. MTS has more than 2,600 members in the United States and abroad.

MTS's formal mission involves dissemination of marine science and technical knowledge promoting the education of ocean engineers, technicians, and those involved in ocean-related policy issues; advancing the development of tools and procedures to explore, study, preserve, and further the responsible and sustainable use of the ocean.

MTS combines resources with the Oceanic Engineering Society of the INSTITUTE OF ELECTRICAL AND ELECTRONIC ENGINEERS to host an annual professional meeting known as MTS/IEEE Oceans. This conference specializes in the fields of marine technology and ocean engineering. MTS/IEEE Oceans 2005 was held in Washington, D.C., with the theme of "One Ocean." The technical program committee for these oceans conferences selects technical papers for presentation covering many marine technology and engineering topics and themes. Papers are written in a standard format and published in a conference proceedings document.

Scientists and engineers are professionals regardless of whether they work for federal, state, or local government, large or small business, or as independent consultants. Many marine technologists and engineers have obtained formal degrees, most perform internships, and some, such as engineers and surveyors, obtain licenses or certifications to practice their profession. Membership in professional societies such as MTS provides important forums for the exchange of ideas, continuing education opportunities, technical achievement awards for selected members, and advancement of the profession. The spectrum of participation for an active member might involve participating in stimulating discussions at committee meetings, writing a technical paper and presenting findings at the joint Marine Technology Society/IEEE Oceans Conference, or publishing a paper in the *Marine Technology Society Journal* or participating in section meetings or a technical committee.

It is important to realize that marine technology is a growing field because of the contributions that have been made through many multidisciplinary scientists and professional organizations. MTS presents a large number of awards to professionals and students who have made outstanding contributions to marine science. For example, the society formally recognizes the accomplishments of individuals who have demonstrated the highest degree of technical accomplishment in marine science, engineering, or technology with the prestigious Lockheed Martin Award for Ocean Science and Engineering. The Society presents an MTS Outstanding Student Section Award to any MTS student section demonstrating superior performance in furthering the society's

MTS Technical Committees

Industry and Technology

Buoy technology

Cables and connectors

Deepwater field
   development

Diving

Manned underwater
   vehicles

Moorings

Oceanographic
   instrumentation

Offshore structures

Remotely operated
   vehicles

Renewable energy

Ropes and tension
   members

Seafloor engineering

Underwater imaging

Unmanned underwater
   vehicles

Government and Public Affairs

Marine law and policy

Marine minerals
   resources

Marine security

Ocean economic potential

Ocean observing systems

Ocean pollution

Education and Research

Marine archaeology

Marine education

Marine geodetic
   information system

Marine materials

Ocean exploration

Physical oceanography/
   meteorology

Remote sensing

objectives. The table above lists the MTS technical committees.

Further Reading

Marine Technology Society. Available online. URL: http://www.mtsociety.org. Accessed May 4, 2006.

marine trades The term *marine trades* describes many ocean-related businesses and for-profit technical jobs that require special skills. These jobs are diverse, spanning the spectrum of MARINE OPERATIONS from building and sailing vessels to repairing and maintaining ships, their components, and support structures. Marine trades people apply their skills in the harsh and unforgiving marine environment. Workplace challenges include motions such as pitch, the fore-and-aft motion of a vessel, and roll, the vessel's side-to-side motion, which make almost any task more difficult at sea than ashore, as well as inducing sea sickness in many of the ship's company. Other environmental factors include intense solar radiation, winds, high humidity, extreme temperatures, SEA ICE, depth changes, and CURRENTS.

Marine trades support marine science, maritime transportation, navigation safety, recreation, and national security.

United States ship and boat builders provide commercial and recreational vessels for statewide, national, and international users. Major ship and boat builders are found in coastal states such as Maine, New Jersey, Virginia, North Carolina, Mississippi, Louisiana, Texas, California, Oregon, Washington, and Michigan. Shipyards produce BARGES, boats, floating dry docks, submarines, offshore platforms, and ships. Shipyards are fixed facilities with dry docks and fabrication equipment to build, convert, or alter ships. Dry docks are enclosed structures that allow a vessel to be raised so that water can be pumped out and repairs made to the hull. Boats and watercraft are built in factories or plants that specialize in working with wood, fiberglass, steel, and composite materials. Most boats are built to meet or exceed National Marine Manufacturers Association and U.S. Coast Guard standards. The boat builders design and build hulls, decks and superstructures, and install fuel and water tanks, power plants, marine electronics, cleats, and rigging. Boats are generally sold at dealerships and boatyards. Jobs associated with recreational boating are found in marketing and sales, customer service, finance, parts and accessories, engine repair, and general marina or boatyard labor. Skilled technicians to service boats are in high demand.

Many marine trade jobs stem from the fisheries industry. In places such as the Bering Sea, billions of pounds of groundfish are harvested each season. FISHERMEN work aboard fishing boats such as crabbers, gillnetters, longliners, purse seiners, trollers, and trawlers. Deckhands work with fishing gear, pick and sort fish from NETS or traps, clean and maintain the vessel, and assist with daily chores such as cooking and cleaning. Factory trawlers utilize crews of approximately 100 people to process catches of species such as walleye pollock (*Theragra chalcogramma*) and Pacific cod (*Gadus macrocephalus*). Fishing jobs require long hours in cold and wet environments, and pay is usually a percentage of the catch. Crew members on factory trawler butcher lines may work repetitive tasks 18 hours per day, seven days per week, processing catch such as lingcod (*Ophiodon elong-atus*), Tanner crabs (*Chionoecetes bairdi* and *Chionoecetes opilio*), and Pacific halibut (*Hippoglossus stenolepis*). Factory trawler jobs are physically demanding and require employees to operate machinery, sort catch, and load and unload supplies and products. Processing tasks include hand filleting, inspecting, freezing, and

Selected Marine Trade Jobs

Descriptive Title	Short Job Description
ABLE-BODIED SEAMAN	Stands bridge watch or lookout when under way. Maintains interior and exterior spaces and deck gear, and operates machinery during loading of stores and scientific operations.
boat mechanic	Repairs gas and diesel engines, electrical systems, plumbing systems, and performs general maintenance on marine systems.
bulk terminal specialist	Handles bulk cargo. Operates and maintains bulk terminal machinery and equipment. (*see* CONTAINER)
cartographic technician	Develops and modifies engineering, surveys, photogrammetric and geotechnical drawings, maps, and charts. (*see* CARTOGRAPHERS)
coastal resources program specialist	Supports environmental programs that create, restore, enhance, protect, and preserve coastal zones and other natural resources.
conservation specialist	Inspects and performs enforcement, emergency response, data collection and evaluation, and investigation tasks to ensure compliance with regulations.
crane operator	Loads/unloads cargo to/from vessels, using shipboard and seaport cranes.
damage control person	Performs emergency repairs to decks, structures, and hulls by emergency pipe patching, plugging, and shoring.
deckhand	Handles lines to moor vessel to wharves, ties one vessel to another, or rigs towing lines.
diver	Commercial divers perform underwater work such as welding as directed from a topside supervisor or lead diver. Scientific divers perform research tasks or educational activities that do not involve ship husbandry or construction.
dock foreman	Transfers cargoes from/to BARGES and provides supervision to complete job.
draftsperson	Draws structural and mechanical features of ships, docks, and other marine structures and equipment.
dredge mates	Operates anchor winch and monitors the condition of wire cables, WINCHES, pulleys, blocks, pipeline, barges, and MOORING parts.
electrician	Maintains and installs onboard electrical systems.
electronics technician	Maintains and repairs all makes and types of shipboard machines, electronic communications, television systems, and mobile communication equipment.
engineering technician	Performs basic math computations and measurements associated with marine power systems. Becomes familiar with procedures, rules and regulations.
engine person	Operates, services and repairs internal combustion engines aboard ships.
equipment specialist	Designs and tests new or modified marine equipment.
evaporator	Controls evaporators to concentrate chemical solutions to specified density by evaporation.
firetug deckhand/ fire fighter	Performs manual deck and firefighting duties aboard a fireboat.
fisheries technician	Assists in fishery research and management projects by installing and taking out fish weirs and/or wheels, collecting biological and sociological data, tagging fish, and maintaining equipment.
fisherman	Performs crew member tasks to catch fish and transport products to markets.
fishery vessel observer	Records fishing efforts, locations, and rates. Determines composition of catch and bycatch. Monitors regulatory compliance.
fork lift operator	Operates industrial trucks equipped with forklift, stand up reach trucks, sit down forklift, and pallet jacks, in an effort to lift, stack, tier, or move pallets of product in loading, unloading, moving and checking materials.
glass blower	Fabricates, modifies, and repairs experimental and marine science laboratory glass products, using variety of machines and tools.
machinist	Fabricates, manufactures, and repairs metal parts, using various types of machine tools and shop equipment.

Descriptive Title	Short Job Description
marina manager	Manages, coordinates, and directs the overall operations of marina properties.
marine carpenter	Constructs, repairs, and remodels in accordance with the standard practices aboard a vessel.
marine mechanic	Repairs marine diesel engines, pumps, and related marine equipment.
marine machinery mechanic	Removes, installs, overhauls, repairs, tests, and maintains steam turbines, gas engines, and diesel engines.
marine machinery repairer	Removes, installs, overhauls, repairs, tests, and maintains marine machinery and equipment.
marine mammal trainer	Prepares food for animals, draws blood, monitors water quality, hoses off decks, cleans pools, sets up audio-visual equipment, checks microphones, washes windows surrounding pools, and conducts underwater sessions with animals to maintain learned behaviors and teach new behaviors.
marine operator	Operates boats and watercraft. (*see* COXSWAIN)
marine police officer	Preserves life and property, enforcing state, municipal, and federal laws as related to coastal and seaport activities.
meteorological technician	Observes and records weather conditions for use in forecasting. (*see* MARINE METEOROLOGIST)
oiler	Performs routine engine room and mechanical duties
painter	Paints and finishes the surfaces of vessels and related marine equipment.
photographer	Takes and processes still photographs in the marine environment.
piledriver	Hoists, sets, designs, and assists in the driving of steel and wooden piles, erecting struts and braces, and making other wharf and bridge repairs.
pump operator	Operates liquid cargo transfer systems.
safety technician	Inspects safety equipment, properly disposes of hazardous materials, and serves as member of the spill response team.
seaport security guard	Maintains security of seaport, docks, and grounds by preventing crime and vandalism.
sheet metal worker	Fabricates, alters, and repairs sheet metal.
ship and pipe fitter	Assembles and maintains parts of a ship. Performs various welding processes to repair, modify and install various pipes, structures, foundations and fittings associated with shipboard hydraulic, pneumatic and water systems.
shipwright	Maintains, fabricates, repairs, and installs wood parts, structures, and equipment aboard vessels.
small engine mechanic	Repairs and adjusts the electrical and mechanical equipment of inboard and outboard boat engines.
stevedore	Carries or moves cargo by handtruck to wharf and stacks cargo on pallets or cargo boards to facilitate transfer to and from ship.
steward	Oversees general stateroom and passageway cleanliness and all food service operations
survey technician	Operates, adjusts, and maintains survey instruments including water level gages, tide stations, levels, theodolites, electronic hydrographic and distance measuring equipment, GPS/RTK receivers, and other precision equipment used in conducting a variety of survey work.
welder	Performs wide variety of welding, brazing, and metal cutting jobs.
wildlife and fisheries technician	Assists biologists and technicians of higher level in various natural resource management or research activities, including use, maintenance, and calibration of scientific and technical equipment.
wildlife educator	Provides wildlife education programs and services.
wildlife enforcement agent	Enforces all fish and game laws relative to protection of wildlife and fisheries resources.
wiper	Maintains, cleans, paints, and preserves the ship.

packaging. Primary products from the pollock fishery include fillets, minced fish called "surimi," fishmeal, oil, and roe (eggs of a female fish). Crab sticks are the cylindrical sticks made of surimi, to which crab meat coloring and flavoring is added. Dockworkers and fish pitchers are also hired to unload fish from boats. The most famous fish pitchers can be found slinging Chinook salmon (*Oncorhynchus tshawytscha*) at Pike Place Fish Market in Seattle, Washington.

Marine technician jobs vary in accordance with employers and missions. Positions with industry may relate to vessel operations, hydrographic surveying, cable-laying operations, and oil exploration. Marine techs may hold dive certifications, a U.S. Coast Guard captain's license, and operate survey boats. They need to be fully familiar with hydrographic survey equipment such as differential global positioning systems, sub-bottom SONAR, side-scan sonar, ACOUSTIC DOPPLER CURRENT PROFILERS, sediment corers, FATHOMETERS, and REMOTELY OPERATED VEHICLES. They complete logs and file reports. Positions with academia involve work with multidisciplinary science teams and research vessels. The marine technician may inspect, maintain, and rebuild research equipment such as AUTONOMOUS UNDERWATER VEHICLES or manage the daily operations of a university oyster hatchery. The marine tech aboard a research vessel is usually a professional with an engineering or science degree. He or she operates and deploys the ship's oceanographic equipment and instruments. Responsibilities often include data collection, processing, and report preparation. Marine technicians also work for the government, designing, building, and operating marine systems, weather and oceanographic data collection systems, and satellite systems. They may also manage real-time reporting systems such as NOAA Physical Oceanographic Real-Time Systems or PORTS® to help provide citizens with marine information that can help protect lives and save property.

The table on pages 324 and 325 lists selected marine trade jobs.

Further Reading

Marine Advanced Technology Education (MATE) Center. Available online. URL: http://www.marinetech.org. Accessed September 27, 2007.

maritime signals　Maritime signals are flags used to communicate visually between ships. Early Phoenician, Athenian, and Roman sailors with their ramming warships are reported to have communicated at sea by hoisting various pennants. Today's recreational boaters fly burgees or pennants that are generally triangular in shape from the top of the forward-most mast. Burgees represent boat manufacturers, yacht clubs, or local U.S. Power Squadron affiliation. The U.S. Power Squadron is a nonprofit educational organization promoting safe boating as well as participating in various important projects to validate nautical charts (*see* CHARTS, NAUTICAL) and traditional tide and tidal current tables. Serious sailors use approximately 40 flags representing 26 letters of the alphabet, 10 numerals, four special flags, three repeaters or substituters, and an answering pennant to communicate with other knowledgeable seafarers. Signal flags are also used to decorate or dress the vessel for ceremonial and festive occasions such as fleet parades. During holidays such as Independence Day, vessels anchored in city harbors such as Annapolis, home of the U.S. Naval Academy, have signal flags strung end to end and hung bow to stern from their halyards.

The International Code of Signals provides a means of communication in situations related to maritime safety, especially because of language barriers. The prescribed signal flags and pennants are easily observed from great distances and have specific meanings. Individually or in combination, they spell out important messages. Individual letters signify important and common messages such as the letter flag "ALPHA" flown not less than one meter in height means "diver below" and is flown to protect the vessel from collision. Similarly, the letter flag "DELTA" when stationary signals "keep clear of me, I am maneuvering with difficulty." Signal flags can also be grouped in pairs to display important information. For example, "DELTA" and "VICTOR" mean, "I am drifting." Besides the code pennants, there are other systems for communicating the International Code of Signals. The codes can be spelled verbally over a radio, using the full names of the letters, such as "ALPHA" for "a." They can also be transmitted by flashes of light, sound, or radio clicks by using Morse code. Samuel Finley Breese Morse (1791–1872) developed Morse code, a means of communication using dots and/or dashes. A dot is a signal of short duration, and a dash is a signal of longer duration. Series of dots and/or dashes correspond to letters of the alphabet.

Another method used by seafarers and beach patrol lifeguards is semaphore. The shipboard signal person holds flags in various positions to send messages letter by letter to another ship. Semaphore signals can be sent and received much more quickly than flaghoists or Morse code with flashing lights. A Navy signal person can communicate in semaphore if a ship turns off its telephones and electronics. Semaphore is best suited for short distances and clear daylight conditions. Some ocean lifeguards still learn basic semaphore signals, as well as how to operate rescue BUOYS, the dangers of RIP CURRENTS, and lifesaving techniques. However, the principal means of communication by most beach patrols is by radio.

Further Reading
National Imagery and Mapping Agency. *International Code of Signals for Visual, Sound and Radio Communications,* Publication 102, U.S. ed., 1969 ed., revised 2003.

Martha's Vineyard Coastal Observatory Martha's Vineyard Coastal Observatory is an oceanographic and meteorological real-time observing system. During 2001, the WOODS HOLE OCEANOGRAPHIC INSTITUTION launched the Martha's Vineyard Coastal Observatory (MVCO). The observatory consists of a sheltered shore station, a nearby 33-foot (10 m)-tall mast outfitted with meteorological sensors, and oceanographic stations located approximately one mile (1.6 km) offshore in 40 feet (12 m) and 50 feet (15 m) of water. The meteorological mast is sited at the Katama Air Park, a large sandplain, grassland and conservation area behind the dune field; it collects wind speed and direction, humidity, precipitation, solar and infrared radiation, momentum, heat, and moisture fluxes. Separating the shore and weather station from the oceanographic sensors is the southern barrier island shore of Edgartown, Massachusetts, facing the high swells of the ATLANTIC OCEAN. Oceanographic measurements from the two offshore stations include WAVES, CURRENTS, seawater TEMPERATURE, SALINITY, and light transmission. The site at the 50-foot (15 m) depth includes a unique tower with carefully installed sensors that help researchers understand processes such as how the faster-moving air imparts energy into the slower-moving water. Air-sea interactions occur as a consequence of air being in contact with the sea surface. Raw data and power are transmitted between the shore station and the meteorological and oceanographic sensors via fiber-optic and copper cables permanently embedded beneath the dunes, shore, and seafloor.

Researchers from Woods Hole Oceanographic Institution, also known as WHOI (pronounced "hoo-ee"), have established MVCO to facilitate long-term, multidisciplinary research in the nearshore region. As the data from the station are analyzed, researchers can begin to identify patterns in the nearshore movement of sand that causes shoaling and shifting channels and the impact of winds on beaches. Data can be used to provide improved forecasts of severe weather and to assess coastal erosion rates. Operational meteorologists can build products such as daily weather average graphs and wave charts that highlight information on coastal storms. Updated information is made available 24 hours per day, seven days per week, to anyone with access to the Internet. Recreational boaters can query the MVCO to evaluate wave heights, wave periods, and direction. Over time, the growing long-term database will allow researchers to better study NATURAL HAZARDS such as nor'easters and sea-level rise.

The installation of MVCO with funding from the National Science Foundation, Office of Naval Research, and WHOI is in synchronization with the United States's interest in safeguarding the environment and protecting life and property through an Integrated Ocean Observing System. MVCO is just one of several observatories located along the U.S. East Coast and will become an asset to the National Science Foundation Ocean Observatory Initiative. Other examples include the South East U.S. Atlantic Coastal Ocean Observing System, FIELD RESEARCH FACILITY, Chesapeake Bay Observing System, and the LONG-TERM ECOSYSTEM OBSERVATORY at 15 meters depth or LEO-15. Marine scientists able to access observatories located from Cape Canaveral to Cape Cod can track oceanographic processes such as coastal water level changes during a HURRICANE passage. Data from these observatories provide a basis for the study of hydrodynamics, sediment transport and shoreline evolution, benthic biological processes, gas transfer, aerosol physics, and coastal meteorology.

Further Reading
Martha's Vineyard Coastal Observatory. Available online. URL: http://www.whoi.edu/mvco. Accessed September 27, 2007.
NOAA Coastal Services Center. "U.S. Coastal Observing Systems." Available online. URL: http://www.csc.noaa.gov/coos. Accessed September 27, 2007.
Ocean.US: The National Office for Integrated and Sustained Ocean Observations. Available online. URL: http://www.ocean.us. Accessed September 27, 2007.

mass spectrometer The mass spectrometer, or "mass spec," is a laboratory instrument for identifying a substance and the kinds of particles present in the given substance. In the Earth sciences, the most frequent use for the mass spectrometer is for the analysis of organic compounds in the environment. Advanced mass spectrometers make possible the analysis of samples ranging from large and complex molecules to tiny viruses.

The analysis is performed by first preparing an extract from a sample of interest, such as water, SEDIMENT, plant or animal organism, or pollutant, which is then inserted into the mass spectrometer. The output is a mass spectrum in which the particles are sorted by mass, in a manner analogous to the spectrum of light, where the light photons are sorted according to their wavelength. The particles are first ionized and then beamed through an electromagnetic field. The amount of their deflection in the spectrometer is proportional to their mass. The masses of the particles making up the sample provide information about the structure of the substance, and thus, its identity. The results of the analysis can be recorded on a strip chart, or in a computer file. Today's state-of-the-art mass spectrometers display the data, along with the results of the analysis.

Although the appearance of mass spectrometers has changed greatly since their development, the basic principle of operation is the same. The sample of interest is vaporized and ionized to produce positive ions by bombardment with electrons from an electron beam. The ions are then injected into the mass spectrometer, maintained at a high vacuum to prevent scattering by air molecules, and accelerated through an electric potential difference. The ions then enter a magnetic field that is perpendicular to their direction of motion. The magnetic field bends their trajectories by an amount that depends on the ions' masses and the (known) magnetic field strength. This result follows from the equality between the magnetic and centripetal forces on the ions, or

$$qvB = mv^2/r, \text{ so that } r = mv/qB$$

where B is the magnetic field strength, v is the ion's velocity, m is the mass of the ion, q is the charge on the ion, and r is the radius of curvature of the ion's path through the instrument. As an example in a demonstration instrument, the radius followed by the positive ion (the nucleus) of ordinary hydrogen might be 0.1 m; the radius of a heavy isotope of hydrogen; deuterium might be 0.2 m.

The magnetic field can then be adjusted so that only ions of a particular mass strike a fixed target, or detector. This is possible because the electric and magnetic fields cause all the ions of a particular mass to travel in a specific curved trajectory that corresponds to a pathway through the instrument. Ions of other than the selected mass will be scattered against the walls and not detected. The detector produces a signal proportional to the intensity of the ion beam. By adjusting the magnetic field strength and comparing it with the signal from the detector, ions of different masses are sent through the pathway. This procedure is repeated until the entire range of masses has been scanned, producing a spectrum of the masses. Mass spectrometers in common use today include time-of-flight analyzers, quadrupole, and quadrupole ion trap analyzers.

Further Reading
Bell, Suzanne. *Encyclopedia of Forensic Science*, rev. ed. New York: Facts On File, 2008.
Rosen, Joe. *Encyclopedia of Physics*. New York: Facts On File, 2004.
Scripps Center for Mass Spectrometry. Available online. URL: http://masspec.scripps.edu/. Accessed December 11, 2005.
————. Available online. URL: http://masspec.scripps.edu. Accessed March 9, 2008.

maximum depth of distribution   The maximum depth of distribution is an objective biological element used to assess the ecological quality of a habitat. For endangered SUBMERGED AQUATIC VEGETATION species such as Johnson's sea grass (*Halophila johnsonii*), this element may be defined as the depth at which the deepest shoot is found within a 164 feet (50 m) transect perpendicular to the shore of the Indian River Lagoon in Florida. A transect is usually a narrow strip along which investigators measure or count organisms to estimate populations and variability. Depths are measured with an echo sounder, locations recorded by a GLOBAL POSITIONING SYSTEM receiver, and other notes such as time are transcribed to a field log. The maximum depth of distribution could also relate to the location of other phytobenthos (plants of the benthic zone) or demersal finfish such as the generally abundant blue hake (*Antimora Rostrata*). In the case of submerged aquatic vegetation, the maximum depth of distribution is important since TURBIDITY is one of the major factors attributed to losses in sea grass coverage. Similar biological elements used to assess status and trends are historical presence, species, percent cover, canopy height, and flowering. Other factors might relate to hydrodynamics such as retention time, sediment features such as organic matter content, and climatic constraints such as meteorological normals. Meteorologists classify standard values of a data element that have been averaged at a given location over a fixed number of years as a normal.

maximum economic yield   Maximum economic yield is a statistical term used to define harvest that gives the highest profit. Like terrestrial farmers, fishermen (recreational, subsistence, and commercial) and aquaculturists need to consider net revenues with respect to the costs of labor and capital equipment. Thus, the maximum economic yield would be the positive difference of total revenues minus total costs from fishing effort. The total cost associated with fishing effort is a combination of fixed vessel costs and variable expenses relevant to crew, gear, bait, fuel, and days at sea. Maximum economic yield is one of several management targets assessed by fishery scientists. Other targets are total allowable catch and maximum sustainable yield. Total allowable catch limits may be determined from sample trawling to determine the relative abundance of each species, the number of reproductively mature individuals, and the probable size of the spawning stock for that year. Maximum sustainable yield involves finding a balance where the current harvest will not affect future harvests. In determining the maximum catch that can be removed from a stock over an indefinite period, one should also consider environmental variability. These targets are especially difficult to achieve since there are complex relationships among population size, population growth rates, abundance, and variables such as sea surface TEMPERATURE. These targets are normally estimated by evaluating historical catches or by using sophisticated mathematical MODELS.

See also AQUACULTURE; FISHERIES RECRUITMENT; FISHERMAN; MARICULTURE; YEAR CLASS STRENGTH.

Further Reading

Froese, Ranier, and Daniel Pauly, eds. FishBase. Available online. URL: www.fishbase.org. Accessed September 27, 2007.

National Marine Fisheries Service. Available online. URL: http://www.nmfs.noaa.gov/. Accessed August 28, 2006.

mean sea level (MSL)   Mean sea level is a statistical representation of the sea surface that is usually computed by averaging all water levels over a 19-year period. This period relates to the approximate 19-year lunar cycle (235 lunar months) between recurrences of the full Moon on the same date. This discovery is credited to the Athenian astronomer Meton (5th century B.C.) and is sometimes called the Metonic cycle. Observing WATER LEVEL FLUCTUATIONS requires the placement of a recording instrument at a tide station. These facilities or tide houses usually include weather stations, water level gauges, THERMISTORS to measure water temperatures, and CURRENT METERS. The data is saved and transmitted to a repository for long-term archival. Shorter averaging periods for series of consecutive water level observations (time series) should be specified as monthly mean sea level or yearly mean sea level. Changes in

water levels do occur over long and short time scales and in response primarily to astronomical (gravitational attraction of the Moon and Sun) and meteorological (set up by winds, runoff from precipitation, freeze-up during winter, and melt-out during spring) forces.

The computation of mean sea level is not trivial. There are many natural processes and human (anthropogenic) causes that affect the characterization of water level fluctuations. Locally, coastal lands may be subsiding in an ESTUARY where expanding marshes are sinking or rising due to elastic rebound of expanding lands from the last ice age. Changes in elevations may also occur due to tectonic actions such as earthquakes. GLOBAL WARMING may also cause rising sea levels due to processes such as thermal expansion of ocean waters and the melting of glaciers. For these reasons, average water levels from a single location should be referred to as relative sea level or RSL because the measurement specifies the height of the sea in relation to local landmarks where the land may be rising or sinking. Geodetic surveys of the tide station are also made to account for any settling at the site. A robust network of water level gauges is essential in analyzing sea level. For example, tide stations making up the National Water Level Observation Network are used to measure sea level variations around the United States.

Captain Sir James Clark Ross (1800–62), a British Antarctic explorer, called mean sea level, the "Zero Point of the Sea." MSL is often represented by a triangle on scientific graphs, which depict sea level. A MOORING diagram for an oceanographic research project might depict MSL at the waterline near the drawing of a BUOY with an upside down triangle (∇). Important vertical datums (base elevations used as references for heights or depths) such as the North American Vertical Datum of 1988 are related to local tidal datums such as mean lower low water and mean higher high water. These vertical datums are used in many REMOTE SENSING and mapping applications. For example, CHART datum is often referenced to mean lower low water. Datums and datum transformations are established by the National Geodetic Survey of the NATIONAL OCEANIC AND ATMOSPHERIC ADMINISTRATION (NOAA); the organization was first established in 1807 by President Thomas Jefferson (1743–1826) as the Survey of the Coast.

The National Ocean Service of the NOAA computes MSL from primary or reference tide stations that are tied to geodetic controls such as benchmarks or monuments. Benchmarks are permanently fixed markers (usually 3.5 inches (8.9 cm) diameter brass or bronze disks) with an inscription of location and elevation. They can represent a surveyed point in the

National Geodetic Vertical Network. The complicated determinations of MSL are important for many applications requiring altitude such as meteorologists launching weather balloons so that they can make upper air measurements, aviators planning and executing flight plans, and parachutists making free fall high-altitude, low-opening parachute jumps.

Further Reading

American Congress on Surveying and Mapping. Available online. URL: http://www.survmap.org/. Accessed September 28, 2007.

Kusky, Timothy. *Encyclopedia of Earth Science.* New York: Facts On File, 2005.

National Geodetic Survey. Available online. URL: http://www.ngs.noaa.gov/. Accessed September 28, 2007.

National Ocean Service. "Our Restless Tides." Available online. URL: http://tidesandcurrents.noaa.gov/restles1.html. Accessed October 10, 2007.

———. "Tide and Current Glossary." Available online. URL: http://tidesandcurrents.noaa.gov/publications/glossary2.pdf. Accessed August 12, 2006.

NOAA Tides and Currents. Available online. URL: http://tidesandcurrents.noaa.gov/index.shtml. Accessed September 28, 2007.

Pugh, D. T. *Tides, Surges and Mean Sea Level.* New York: John Wiley, 1987.

U.S. Geological Survey. Available online. URL: http://www.usgs.gov/. Accessed August 12, 2006.

measure of angle Measure of angle refers to the determination of the magnitude of the angle, usually in degrees, between two intersecting lines. Euclid (~330 B.C.E.– 260 B.C.E.) defined a plane angle as "the inclination to one another of two lines in a plane which meet one another and do not lie in a straight line." An angle can be thought of as the amount of opening between two straight lines meeting at a point. The point where the two lines meet is called the vertex of the angle. Angles can be labeled by a letter inside the angle near the vertex, or by three letters, with the second letter at the vertex and the first and third letters on the two sides of the angle. The degree is defined by constructing 360 equally spaced radii of a circle; 1/360th part of the circle is thus one degree. Degrees are further subdivided into 60 minutes in a degree, and 60 seconds in a minute. In scientific work, angles are also expressed in radians; a radian is defined by the angle subtended by an arc of the circle equal in length to the radius. There are 2π radians in a circle; a radian is thus equal to 57.2957795131 degrees.

The measurement of angles is a frequent requirement in piloting and NAVIGATION at sea. The angles between navigation aids, between an aid and a point on shore, to a vessel at sea, and between vessels at sea are all frequently required to safely navigate a ship to its port of destination. Celestial navigation requires the precise measurement of angles between the Sun and Moon and the horizon, and between stars and the horizon. Angles can be conveniently plotted on paper, using a protractor, a half circle with its rim subdivided into 180 equal parts. Angles can be plotted on navigation charts, using the compass rose imprinted on the chart and a pair of parallel rules, which are two rulers joined together by a linkage so that they can move, but always remain parallel. Angles can be measured on deck by taking the difference between compass bearings to desired reference points. At sea, a pelorus, (a fixed frame without a directive compass element but provided with vanes) is used in conjunction with a compass to give either magnetic or true bearings. Small craft typically use magnetic compasses (*see* COMPASS), which depends on the direct force of Earth's magnetic field to turn the compass card in the appropriate horizontal direction. Large commercial vessels are usually equipped with gyrocompasses, which have one or more gyroscopes as the directing elements. Gyrocompasses have smaller errors caused by the rolling motion of the vessel in a seaway.

The marine SEXTANT is an instrument specifically designed to measure angles at sea very precisely, especially the altitudes (angles above the horizon) of celestial bodies. It can also be used, however, to measure angles between locations of interest, such as ships or BUOYS on the horizon, or between shoreline navigation aids such as lighthouses.

measure of time The measure of time is a procedure that informs people when events take place in their daily lives and makes possible accurate NAVIGATION, astronomical prediction, and the conduct of scientific experiments. Time is a difficult quantity to define, except in terms of itself. Physicists, beginning with Sir Isaac Newton (1642–1727) have struggled with understanding time. Newton's suggestion that time is like an ever-flowing stream with events moving past us has been generally rejected by the physicists of today, beginning with Albert Einstein (1879–1955), who, in the early 20th century, showed that the passage of time is relative and depends on the velocity and gravity of the bodies involved. Until recently, such questions were of little concern to most humans, who understand the passage of time quite well from the point of view of getting to work on time, of setting new athletic records. However, recent advancements in the accuracy of time measurements make possible the measurement of relativistic effects in daily life, such as the perception of the time required for a certain satellite to complete its orbit around the Earth.

Early civilizations based measurement of time on the movement of celestial bodies, mainly the Sun and the Moon. The day was determined to be one rotation of the Earth, the month, one completion of the Moon's orbit, and the year, one completion of the Earth in its orbit around the Sun. The Babylonians are believed to have developed the presently used units of time, that is, of 12 months in a year, about 30 days in a month, 24 hours in a day, 60 minutes in an hour, and 60 seconds in a minute. The year was defined to be 365 ¼ days, which presented the problem of defining the number of days in the month for each month, and introducing the need for leap years. Today, the movement of astronomical bodies is still an important basis for the measure of time.

Two definitions of the day are in common use: the sidereal day and the solar day. The sidereal day is the time required for the Earth to make one complete revolution with respect to the stars. Since the stars are so far away (the nearest star, Proxima Centauri, being about 4.23 light-years distant) the time required for a given star to rise in the east essentially does not change. Astronomers like sidereal time because a given star always rises from the same location at the same time. However, time based on the solar day is the time in common use. The solar day is the time required for the Earth to make one complete revolution with respect to the Sun, for example, from noon to the following noon. Since the Earth is orbiting the Sun, each day carries it past the previous location, and the Earth must turn a little more (about one degree) until the Sun is directly over the meridian, defining noon. Thus, the solar day is slightly longer than the sidereal day (about four minutes). The motions of the Earth also cause the length of the solar day to vary slightly over the year. Because the planet speeds up when closest to the Sun (Northern Hemisphere winter), the length of the day is slightly longer than when farthest from the Sun (Northern Hemisphere summer). In order to avoid the need for clocks with variable hours, ordinary time is based on the "mean solar day," the mean (average) length of day, which is defined to be exactly 24 hours. The reading of a sundial, which reads the time based on the actual position of the Sun, must be corrected to obtain a reading that agrees with local clock time. The correction, the difference between mean solar time (clock time) and actual solar time is called the "equation of time." The equation of time must also be applied in computing a ship's or aircraft's position from SEXTANT readings.

In order to allow the local time of sunrise and sunset to be approximately the same at a given latitude, 24 time zones have been defined, with the meridian of Greenwich, U.K., being the "zero" or "prime meridian." Other time zones are approximately spaced at intervals of 15 degrees of longitude east and west of the prime meridian, with some bending in consideration of national boundaries. The east and west longitudes meet on the other side of the Earth at the international dateline, where there is a jump of one day across the line. Westward travelers over the dateline must add a day to their calendars, while eastward travelers must subtract a day.

To keep time, early civilizations relied on sundials, and, when the weather was cloudy and at night, they used hourglasses or water clocks, which counted time as the amount of water flowing from one chamber of the clock to another. Mechanical clocks were developed in the later Middle Ages. Dutch physicist Christian Huyghens (1629–95) made major improvements to the mechanical clock, such as the use of the pendulum and the spiral balance spring in the early 17th century. In the 18th century, an English carpenter, John Harrison (1693–1776) made a clock sufficiently accurate and rugged to provide the measurements from which the longitude at sea could be calculated, a great breakthrough in navigation. Very accurate mechanical clocks called ship's chronometers are still in use on many vessels for this purpose.

Over the years, the requirements for accurate time measurements steadily increased. In the early 20th century, accurate clocks were temperature and humidity compensated, and operated in a vacuum to overcome affects of atmospheric pressure variations. In 1927, Warren Marrison (1896–1980), a Bell Telecommunications Laboratories' engineer discovered that when a quartz crystal is excited by an electric current, it resonates at a very precise frequency. By the late 1940s, precision quartz crystal clocks had accuracies of about one second every 30 years. Quartz crystal oscillators have since been combined with transistors and integrated circuits to provide precision time bases for everything from computers to wristwatches. Most modern ships now use a quartz crystal–based ship's chronometer for their timekeeping and navigation needs.

However, the present day standard is the cesium beam (atomic) clock, which is accurate to more than a nanosecond (10^{-9}s) per day. This standard is based on findings by Harold Lyons (1913–98) and collaborators at the then National Bureau of Standards in Washington, D.C. (now called the National Institute for Standards and Technology, or NIST) that cesium atoms changing energy states did so with great stability in their resonant frequencies. These instruments are so accurate that the second has been redefined in terms of the frequency of oscillation of the cesium atom. This technology was an essential component in the development of the GLOBAL POSITIONING SYSTEM (GPS). Clock technology is continuing to improve today to the point where the relativistic effects referred to at the beginning of this article can actually be measured.

Further Reading

U.S. Naval Observatory. Available online. URL: www. usno.navy.mil. Accessed June 23, 2008.

Mediterranean Sea   The Mediterranean Sea is a MARGINAL SEA of the ATLANTIC OCEAN connected by the Strait of Gibraltar separating Spain and Morocco. The Mediterranean is bounded by Europe to the north, North Africa to the south, and west Asia to the east. The Black Sea is a marginal sea of the Mediterranean, connected by the Dardanelles, the Sea of Marmara, and the Bosporus, which separate Greece and Turkey. The term *Mediterranean* is also used to describe seas with some similar characteristics, such as the Caribbean Sea, which is also known as the American Mediterranean.

The Mediterranean Sea is centered at about 38°N latitude, between roughly 5°W and 35°E longitude. The length of the Mediterranean Sea is about 2,400

miles (3,865 km); the area is about 1.15×10^6 square miles (2.88×10^6 km^2). The average depth of the Mediterranean Sea is about 4,600 feet (1,500 m). The greatest depth, more than 16,000 feet (5,000 m) is in the Hellenic Trough, which lies between Greece and the island of Crete. Three peninsulas emerge from Europe into the Mediterranean—the Iberian Peninsula, Italy, and the Balkan Peninsula. The Mediterranean Sea also separates Europe from Africa.

In the life cycle of ocean basins, the Mediterranean Sea is considered to be in a terminal stage of development as opposed to mature and declining examples such as the Atlantic and Pacific Oceans, respectively. The Mediterranean is almost landlocked and is a remnant of the vast ancient sea called Tethys, which was squeezed almost shut during the Oligocene Epoch, 30 million years ago. Crustal plates carrying Africa and Eurasia are still grinding together causing earthquakes and the potential eruption of Mounts Etna, Vesuvius, and Stromboli, all volcanoes in Italy. The Mediterranean is divided into an eastern basin with a mean depth of about 13,780 feet (4,200 m), and a western basin with a mean depth of about 11,150 feet (3,400 m). The basins are separated by a sill of average depth of about 1,310 feet (400 m) extending from Sicily to the Tunisian coast. The eastern basin of the Mediterranean Sea consists of the Adriatic Basin east of Italy, the Ionian

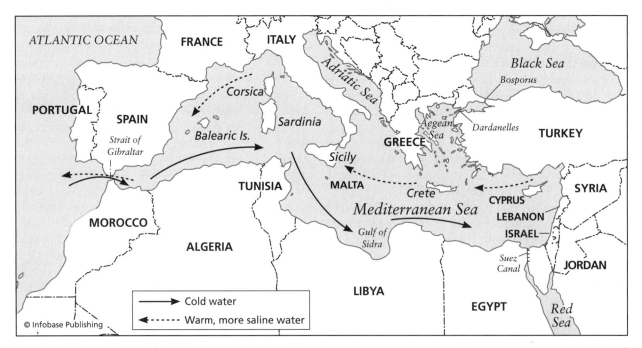

The Mediterranean Sea connects to the Atlantic Ocean by the Strait of Gibraltar, to the Black Sea by the Dardanelles and the Bosporus, and to the Red Sea by the Suez Canal. The water in the eastern Mediterranean is warmer and more saline than in the western Mediterranean. Cold water from the Atlantic flows in and along the coast of North Africa. Warm, more saline water returns to the Atlantic at subsurface depths through the Strait of Gibraltar.

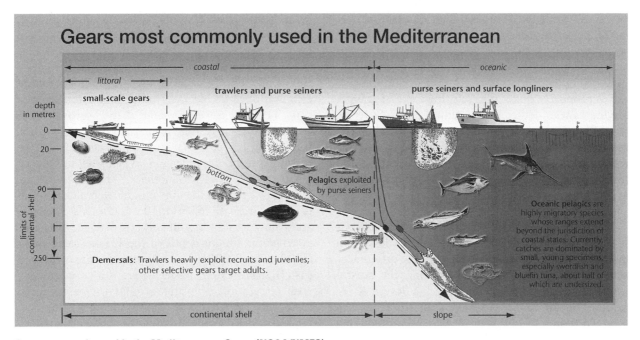

Gears most commonly used in the Mediterranean

Gears commonly used in the Mediterranean Sea *(NOAA/NMFS)*

Basin, from the Strait of Sicily to Greece and Crete, and the Aegean Basin north of Crete. The Levantine Basin lies to the east of Greece between Greece and the Levant (Lebanon). In the west, the Alboran Basin lies between Gibraltar and the Gulf of Lions, to the south of France. The Balearic Basin is to the east of the Alboran basin; the Tyrrhenian Basin is east of Corsica and Sardinia, and to the west of the Italian peninsula.

A high rate of evaporation, about 40 inches per year (100 cm/year), which exceeds river runoff and precipitation, results in high values of surface SALINITY (and hence, DENSITY) in the eastern Mediterranean Sea, which sinks and produces a subsurface flow of dense, highly saline (about 37.8 practical salinity units, or psu) water (called the Levantine Intermediate Water, or LIW) flowing out into the Atlantic through the Strait of Gibraltar, with a compensating inflow of less saline Atlantic water above, which is on the order of 36.3 psu. The loss of outflowing water induces a westward-flowing CURRENT along the southern shore of Europe from the source of the water south of eastern Turkey to the Strait of Gibraltar.

The dense MEDITERRANEAN WATER flowing outward through the Strait of Gibraltar forms a stable tongue of intermediate density that can be identified across the North Atlantic Ocean, and even into the South Atlantic. The temperature and salinity of the Mediterranean water is about 55.4°F (13°C) and 37.3 psu as it begins to flow down the CONTINENTAL SLOPE from the Strait of Gibraltar into the Atlantic

This highly saline flow produces a layer of water of maximum salinity in the eastern North Atlantic Ocean, which gradually decreases with distance from the strait as the water is mixed and thus diluted with colder and fresher Atlantic water. The average depth of this flow is about 3,280 feet (1,000 m); it can be readily identified by the temperature and salinity distribution, which results in a characteristic signature in the temperature-salinity (T-S) diagram (a thermodynamic diagram often used in oceanography to identify specific water masses and is used to infer the flow of water from one region to another).

The high salinity Mediterranean water at its source region in the eastern basin is known as the Levantine Intermediate Water (LIW), formed off the south coast of Turkey in the winter, when dry winds from the northwest and abundant sunshine result in evaporation greatly exceeding precipitation and river runoff. At its source region, the LIW has a temperature of about 59°F (15°C), and a salinity of 39.1 psu. It has been observed flowing west from about 655 feet (200 m) to 1,970 feet (600 m) deep along the south shore of Europe, and thence into the Atlantic Ocean.

The sinking and replenishing of water from the surface results in relatively high levels of dissolved oxygen, about 4.8 milliliters per liter. This process sets up vertical CONVECTION cells in which the dense surface water sinks, and the less dense water from near the bottom rises to the surface to be again warmed and made more salty in a continuing cycle. Other physical processes of note are the setup of

INTERNAL WAVES as Mediterranean waters flow over the sill and the formation of month- to year-long rotational flows called "meddies." Internal waves propagate along density interfaces such as the THERMOCLINE. Like surface waves on the ocean, internal waves propagate under the restoring force of gravity. Mediterranean eddies or "Meddies" are subsurface anticyclonic whirlpools with a warm and salty core caused by the interaction between North Atlantic water flowing in through the Strait of Gibraltar and the outgoing high salinity flow of MEDITERRANEAN WATER, sometimes known as the Mediterranean Undercurrent.

The tides of the Mediterranean are semi-diurnal and of small magnitude, almost always less than three feet (1 m) high. Tide ranges are the differences between high and low waters. Along the Bristol Channel, a major inlet connecting the Severn River in Great Britain to the Atlantic Ocean, the difference between high and low water level is approximately 49 feet (15 m), while tidal ranges along the Mediterranean Sea are typically four inches (10 cm). Consequently, the ancient Phoenicians, Greeks, and Romans saw little need to study the tides until they began sailing beyond the Strait of Gibraltar to such destinations as Great Britain.

The Mediterranean Sea is relatively low in nutrients such as nitrates and phosphates since very few rivers flow into it, transporting suspended nutrient-rich SEDIMENTS. Two exceptions are the Rhône River in France and the Nile River in Egypt. These rivers have built up significant delta cones seaward of their mouths, with the Rhône contributing the greater amount, producing a larger cone. These land farms are often fertile and important for agriculture.

The climate of the Mediterranean region is generally warm and arid. The dominant winds are the Mistral, a northwesterly cool wind blowing down from the French Alps that brings clear weather; the Bora, a cold dry winter wind; the dry and hot Sirocco, which blows from North Africa into the Mediterranean Sea; and the humid Khamsin wind, also from Africa. Since climate is a primary factor impacting marine organisms, it helps to determine distinctive characteristics. Mediterranean climate resembles those of the lands bordering the Mediterranean Sea, and some of the important agricultural crops come from the olive trees (*Olea europaea*), European wine grapes (*Vitis vinifera*), citron trees (*Citrus medica*), pomegranate shrubs (*Punica granatum*), and cork oak trees (*Quercus suber*).

The BLACK SEA, which is marginal to the Mediterranean, connected by high-speed currents through the shallow (130–300 feet [40–90 m]) Dardanelles and the Bosporus, is notable in oceanography for being a two-layered system, with an upper layer of mostly freshwater and a lower layer of very little to no dissolved oxygen, called an anoxic basin. This condition results from the fact that, unlike the Mediterranean, the river runoff and precipitation exceed evaporation, making the surface water too light to sink. The lack of refreshment of the bottom waters from the surface results in their stagnation. The dissolved oxygen in the water has been replaced by hydrogen sulfide, resulting in an evil-smelling WATER MASS of very low productivity. The depth of the boundary between the upper and lower water masses is on the order of 165 feet (50 m); it seems to fluctuate from year to year and perhaps with the seasons.

The maximum depth of the Black Sea is about 7,200 feet (2,200 m). The bottom consists of an eastern and western basin. There is a counterclockwise circulation of water around its western and eastern basins. Owing to the sinking of organic matter and density gradients, there is a sharp boundary between oxic and anoxic Black Sea environments. This has contributed to findings by MARINE ARCHAEOLOGISTS of well-preserved late Roman or early Byzantine vessels from about C.E. 410–520. Ancient Greeks and Romans carried bulk food products in amphorae, two-handled transport jars that have been salvaged and are on display in museums.

Further Reading

Pickard, George L., and William J. Emery. *Descriptive Physical Oceanography: An Introduction,* 5th ed. Oxford: Pergamon Press. 1990.

Mediterranean water   Mediterranean water is a subtropical WATER MASS of the Atlantic Ocean that is characterized by high TEMPERATURE and SALINITY. Dense Mediterranean water flows outward through the Strait of Gibraltar, where it forms a stable tongue of intermediate density that can be identified across the North Atlantic Ocean and even into the South Atlantic. The temperature and salinity of the Mediterranean water are about 55.4°F (13°C) and 37.3 practical salinity units (psu), respectively, as it begins to flow down the CONTINENTAL SLOPE from the Strait of Gibraltar into the Atlantic. Salinity units are from the Practical Salinity Scale of 1978, which uses equations and conductivity ratios carefully fit to the real salinity of diluted North Atlantic seawater. Mediterranean water is highly saline and produces a water layer of maximum salinity in the eastern North Atlantic Ocean, which gradually decreases with distance from the Strait of Gibraltar as the water is mixed and thus diluted with colder and fresher Atlantic water. The average depth of this flow is about 3,280 feet (1,000 m). It can be readily identified within the water column from temperature/ salinity profiles by research vessels using CONDUCTIVITY-TEMPERATURE-DEPTH (CTD) profilers. Mediterranean water produces a characteristic signature in the Temperature-salinity (T-S) diagram (a thermodynamic diagram used in oceanography to identify specific water masses and to infer the flow of water from one region to another) that shows anomalously high salinity at around 50°F (10°C) compared to the norm for its depth range.

Mediterranean water is formed in the eastern MEDITERRANEAN SEA off the south coast of Turkey in the winter, when dry winds from the northwest along with abundant sunshine result in evaporation (loss of water) greatly exceeding precipitation and river runoff (gain of water). German oceanographer Georg Wüst (1890–1977) referred to this water as the Levantine Intermediate Water (LIW). At its source region, the LIW has a temperature of about 59°F (15°C), and a salinity of 39.1 psu, where the high rate of evaporation at the sea surface, about 40 inches per year (100 cm/year), results in high values of surface salinity (and hence, density), that causes the water to sink and produce a subsurface flow of dense, highly saline (about 37.8 practical salinity units, or psu) water. It has been observed flowing west from about 655 to 1,970 feet (200–600 m) deep along the south shore of Europe, and thence flowing out into the Atlantic Ocean through the Strait of Gibraltar, with a compensating inflow of less saline Atlantic water above, which is on the order of 36.3 psu.

There is another source of the dense high salinity Mediterranean water to the south of France, offshore the Riviera, called the Western Mediterranean Deep Water (WMDW), which combines with the LIW before flowing out into the Atlantic. Compensating incoming Atlantic water, required by the condition for continuity of mass (that is, the Mediterranean Sea cannot, on the average, be gaining or losing water), flows eastward as a current along the north shore of Africa, running into the Levantine Basin to replace the outflowing LIW to complete a counterclockwise circulation around the entire Mediterranean Sea.

The sinking and replenishing of water from the surface results in relatively high levels of dissolved oxygen, about 0.16 ounces (4.8 ml) per 2.1 pints (l). This process sets up vertical CONVECTION cells in which the dense surface water sinks and the less dense water from near the bottom rises to the surface to be again warmed and made more salty in a seasonal cycle. Mediterranean water is also observed to be low in nutrients, since river flow into the Mediterranean Sea is not great.

Further Reading
Pickard, George L., and William J. Emery. *Descriptive Physical Oceanography: An Introduction,* 5th ed. Oxford, U.K.: Pergamon Press, 1990.

mesopelagic zone The oceanic region from 656 to 3,281 feet (200–1,000 m) lying below the EPIPELAGIC and above the BATHYPELAGIC, ABYSSOPELAGIC, and HADOPELAGIC ZONES is called the mesopelagic zone. Red, orange, and yellow light is mostly absorbed in the upper layers of the ocean, and the remaining blue wavelengths are absorbed in the mesopelagic zone, which is also called the "twilight zone." This disphotic zone is bordered by the PHOTIC ZONE above and the APHOTIC ZONE below. Fish inhabiting this midwater region are unique and may have lurelike appendages (luminous structures used to attract prey), long, sharp teeth, expandable jaws and stomachs, and bioluminescent (*see* BIOLUMINESENCE) features. Predators such as the anglerfish (*Lophius piscatorius*), Pacific viperfish (*Chauliodus macouni*), and black dragonfish (*Idiacanthus atlanticus*) may be captured in the mesopelagic zone by midwater trawls or observed by REMOTELY OPERATED VEHICLES at depths ranging from approximately 66 to 660 feet (20–2,000 m). Many creatures in the mesopelagic zone migrate toward the surface at night and return to depth during the day as evidenced by acoustic reverberations called the deep-scattering layer or DSL that have been detected by echo sounders. Sound is scattered by organisms with gas-filled chambers or air bladders such as jellyfish and fish. Lantern fish such as the blue lanternfish (*Tarletonbeania crenularis*), northern lanternfish (*Stenobrachius leucopsarus*), and the California headlightfish (*Diaphus theta*) having photophores on the head and body are abundant in the DSL. Photophores are light-producing organs. These fish provide an important food sources for a variety of other MARINE FISHES and MAMMALS.

See also ATTENUATION; MARINE PROVINCES; NETS.

meteorological measuring instruments Meteorological measuring instruments are devices for measuring the important quantities of the atmosphere, such as PRESSURE, TEMPERATURE, humidity, precipitation, and wind speed and direction. For the most part, these are the quantities needed to make weather forecasts (*see* FORECASTING). Weather occurs in the troposphere, the lowest layer of atmosphere extending above Earth's surface from five to 10 miles (8–16 km). Atmospheric measurements must be made frequently, periodically, and at locations within and extending well beyond the forecast area. Meteorological instruments used in marine meteorology are, for the most part, the same as those used over land. However, they must be fabricated with greater ruggedness to withstand ship and buoy motions, salt spray, or even immersion in salt water. Instruments deployed on moored buoys must be able to survive for long periods of time under rough sea conditions (*see* BUOYS). They may feed their data into a radio transmitter for transmitting in real-time via satellite relay, or observations may be saved on solid-state recorders.

.   Typical weather instruments include aneroid barometers to measure pressure, cup ANEMOMETERS, weather vanes, and hygristers for measurement of humidity. One of the earliest instruments used to measure the change in atmospheric pressure was the barometer, which dates back to Evangelista Torricelli (1608–47), who devised the principle during 1643. Research meteorologists use more sophisticated instruments, such as sonic anemometers (working on the Doppler principle), quartz crystal pressure sensors, and acoustic or Doppler RADAR wind profilers that work on the same physical principles as the ACOUSTIC DOPPLER CURRENT PROFILER (ADCP). Weather forecasters also require profiles of atmospheric temperature, humidity, pressure, wind speed and direction at regular (typically six-hourly) intervals obtained from weather balloons carrying rawinsondes, or instrument loggers that transmit the readings of the sensors to stations on the ground or at sea.

At-sea measurements of meteorological variables are very important to weather forecasts since the oceans, seas, and lakes cover about 70 percent of Earth's surface. Mariners may access local forecasts, warnings, and current weather information from weather radio transmitters for safety at sea and for optimizing their travel routes. Mariners use their weather knowledge and experience when planning passages through a plethora of potential weather hazards such as HURRICANES or typhoons and tropical CYCLONES. Recreational boaters may employ electronic barometers, handheld anemometers, and portable weather stations, while professional mariners may rely on precision barometers, certified hygrometers, radar, and sophisticated weather stations.

In recent years, significant amounts of meteorological data are acquired by weather satellites

Meteorological measuring instruments including anemometers, wind vanes, temperature and humidity sensors, rain gauges, sounding balloons, and Doppler radar   *(NOAA/NWS)*

using visual and infrared imaging to provide storm locations, hurricane tracks at sea, cloud cover, wave height and direction, and wind speed and direction. Wind speed can be measured by radar scatterometers that measure the roughness of the sea surface and the corresponding wind speed that would cause the roughness. A large amount of research has gone into designing the mathematical algorithms, the hardware and software needed to perform these measurements and analyses. Satellite measurements are very important because large sections of the ocean, such as in the South Pacific, are frequented by few ships. The area involved is too great to instrument with island stations, and the required number of moored buoys would be prohibitive. On large islands, such as the Hawaiian Islands, weather forecasters use Doppler weather radar to obtain wind velocity and precipitation for local mariners, aviators, farmers, and city dwellers. However, detailed weather radar information is limited to a few tens of miles (tens of kms) from the transmitter.

Changing meteorological parameters that are measured by instruments, identified by weather satellites, and described by almost instant communications provide valuable information on weather. However, sudden weather changes can and do occur; people are caught at sea and have to use good seamanship skills to prevent mishaps. For recreational boaters, there is no reason to venture away from the dock when information about impending storms and high winds is being continuously broadcast on both radio and television. Weather routing helps commercial shipping companies avoid heavy weather, which saves time and prevents damage to vessels and cargo. In heavy weather, vessel speed should be reduced; all persons should don a life jacket; all loose objects and rigging should be secured; and all hatches and openings should be covered or closed.

Further Reading
Aguado, Edward, and James E. Burt. *Understanding Weather and Climate*, 4th ed. Upper Saddle River, N.J.: Pearson/Prentice Hall, 2007.

meteorologist   *See* MARINE METEOROLOGIST.

METOC officers   METOC officers are United States sailors and marines familiar with processes associated with meteorology and oceanography (METOC) in order to support fleet units. Scientifically oriented servicemen designated as naval METOC Officers provide for the readiness and safety of fleet and shore establishments. They support naval strategy and tactics by applying the sci-

ences of physical oceanography, meteorology, and geospatial information and services to naval operations. They assess and predict the effect of the environment on naval platforms, weapon systems, and sensors from design, through development, test and evaluation, to operational implementation.

Navy METOC officers may trace their lineage to Lieutenant Matthew Fontaine Maury (1806–73), who is best known for his synthesis of thousands of logbooks and charts that considered winds and CURRENTS to determine optimal navigational routes. Under the direction of the Oceanographer of the Navy, today's METOC officers are serving at the Naval Meteorology and Oceanography Command and the Naval Oceanographic Office at the John C. Stennis Space Center near Bay St. Louis, Mississippi, the historic Naval Observatory in Washington, D.C., the Fleet Numerical Meteorology and Oceanography Center in Monterey, California, at several field activities located in foreign countries such as Japan, and aboard ships. They serve in a variety of billets from Operations Aerology Division (OA) officers aboard aircraft carriers and amphibious attack ships to Mobile Environmental Team leaders aboard destroyers, cruisers, and frigates. Many METOC officers earn advanced degrees in meteorology and oceanography by the time that they are promoted to lieutenant commander.

Marine Corps METOC officers have the challenging responsibility to provide meteorology support to Marine Expeditionary Forces that traditionally maneuver from sea to inland coastal areas. They advise and assist commanding officers assigned to at-sea or deployed Marine Expeditionary Units that are often first responders to natural disasters such as the Indian Ocean earthquake and TSUNAMI that ravaged southern Asia during December 2004. Marines from 15th Marine Expeditionary Unit (MEU) delivered more than 98,000 pounds (44,452 kg) of disaster relief aid to hard-hit areas throughout the island of Sumatra during Operation UNIFIED ASSISTANCE. Marine Corps METOC officers plan and execute support services that are provided by Marine Expeditionary Force METOC support teams (MSTs) and coordinate within the deployed forces as appropriate to meet operational requirements. Marines analyzing weather and oceanographic phenomena may also be located aboard command and control ships or at the Naval Oceanographic Office.

Navy and Marine METOC officers apply their training to analyze environmental conditions and provide forecasts to various warfare commanders, warfighters, and planners. They utilize sensors such as weather stations, numerical MODELS that pinpoint features such as the THERMOCLINE, or forecast phenomena such as permanent CURRENTS and waves, imagery of the sea surface and seafloor, nautical charts, and information from the U.S. Naval Observatory such as the *Nautical Almanac*. This important reference is an annual book containing astronomical and tidal information. Aboard ship, the Navy OA Division and Marine Intelligence Section spaces are packed with computer monitors, satellite systems, data feeds, and METOC software suites. These highly trained men and women present environmental information as displays generated on command and control systems, PowerPoint presentations, or in technical presentations or briefs prior to military operations. These briefs may support COXSWAINS maneuvering small boats into previously uncharted rivers or helicopter pilots transporting heavy externally lifted cargoes to recently established landing zones. As part of a joint task force, the senior METOC officer on the staff provides or arranges METOC support services and products to the staff and coordinates METOC support services and products for the joint force. The joint force involves two or more services applying air, land, sea, and space forces in a unified and synergistic effort. A METOC officer may also be the officer-in-charge of the joint METOC forecast unit.

Throughout the ages, understanding the effect of environmental parameters on weapons system and platform performance have been important components of command. Applied meteorology and oceanography are essential skills for understanding and characterizing the naval environment. In the military service, METOC officers, aerographer mates, weather forecasters, and instrument technicians collect and analyze data to produce environmental intelligence, which reduces uncertainty about the sea and coastal conditions that naval forces face. These men and women produce weather maps, updated nautical charts, and tide and current predictions essential to support amphibious operations and missions such as humanitarian relief, noncombatant evacuation operations or NEOs, and direct actions against enemy forces or compounds. The development and implementation of complex weapons systems such as planing hulls like the Expeditionary Fighting Vehicle and multiengine tiltrotor aircraft like the MV or CV-22 Osprey increase the need for precise environmental information.

Further Reading

Commander, Naval Meteorology and Oceanography Command Stennis Space Center. Available online. URL: http://www.navmetoccom.navy.mil. Accessed March 9, 2008.

U.S. Naval Observatory, Department of the Navy. Available online. URL: http://www.usno.navy.mil. Accessed March 9, 2008.

microbial loop   The trophic link between microbes such as bacteria and viruses and ZOOPLANKTON is called a microbial loop. This trophic level represents a level of consumption in a micro food chain, which complements the classic marine food chain where drifting plants called PHYTOPLANKTON are primary producers. The food chain simply diagrams who eats whom, while the microbial loop depicts energy losses as dissolved organic matter or DOM that is looped back to PLANKTON such as COPEPODS. DOM can be likened to a soup of molecules, which sustains microbial food webs in the ocean. Larger organisms such as zooplankton cannot directly ingest it. Microbes such as bacteria and viruses, which may be as small as 0.000787mil (.02 micrometers), use DOM as their carbon and energy sources.

This microbial loop is present wherever larger marine organisms exist and is a significant research area. Microorganisms can be seen only under the most powerful microscopes. Researchers estimate that a teaspoon of seawater has on the order of a million bacteria, while virus counts are on the order of tens of millions. Microscopic marine grazers eat bacteria and release organic material into the water. Other microbes convert this DOM into raw nutrients, which are available to other plankton, completing the microbial loop. Marine scientists such as Dr. Frede Thingstad from the Institute of Microbiology at the University of Bergen have shown that the ecological role of microbes is much more important than simply recycling dead organisms and organic waste.

Further Reading

CYCLOPS, Cycling of Phosphorus in the Mediterranean. Available online. URL: http://earth.leeds.ac.uk/cyclops/. Accessed May 18, 2006.

Kirchman, David L., ed. *Microbial Ecology of the Oceans.* New York: John Wiley, 2000.

Rassoulzadegan, Fereidoun, lead author; Jean-Pierre Gattuso, topic editor. "Marine Microbial Loop." In *The Encyclopedia of Earth.* Ed. Cutler J. Cleveland (Washington, D.C.: Environmental Information Coalition, National Council for Science and the Environment). Available online. URL: http://www.eoearth.org/article/Marine_microbial_loop. Accessed March 10, 2008.

mid-ocean spreading ridges   Mid-ocean spreading ridges are the major topographic feature of the global ocean. The mid-ocean ridges run continuously around the Earth for a total distance of about

100,000 miles (60,000 km) and divide the oceans into basins. Transform faults (fractures in the crust along which lateral movements take place) normal to the spreading axis segment the ridge into "spreading centers" about 18 to 30 miles (30 to 50 km) in width. This feature has prompted scientists to liken the pattern of the mid-ocean ridges on global topographic maps to the stitching on a baseball.

The spreading ridges are regions where hot magma is UPWELLING from magma chambers below the seafloor and is flowing out of the rift valleys to form new crust. In so doing, the continents and the tectonic plates on which they are riding are being forced apart (see PLATE TECTONICS). The magma cools as it flows away from the rift at the center of the ridge, forming igneous rock. In some areas such as Iceland, the spreading ridge for the Atlantic, the Mid-Atlantic Ridge, reaches to the surface where new islands, such as Surtsey, are born as ejections of hot lava spewing out of the cold ocean into the even colder atmosphere. As the lava cools, it forms the new islands. Eventually, wind and waves will wear the igneous rock down into soil and a new habitat for plants and animals will begin.

The processes active in the mid-ocean ridges have been paramount in the development of the theory of plate tectonics. Hydrothermal activity, earthquakes, SEAFLOOR SPREADING, and magnetic striping are all common to the ridges. Scientists such as Harry Hammond Hess (1906–69), a Princeton University geologist and a Naval Reserve rear admiral; Robert Sinclair Dietz (1914–95), a geologist with the NATIONAL OCEANIC AND ATMOSPHERIC ADMINISTRATION and an Air Force Reserve lieutenant colonel; Frederick John Vine (1939–88), a marine geophysicist at Cambridge University, and Drummond Hoyle Matthews (1931–97), a professor in the department of geodesy and geophysics at Cambridge University, were instrumental in discovering these seafloor processes. Today's investigators use *in situ* OCEANOGRAPHIC INSTRUMENTS such as networks of deep-ocean hydrophone arrays to monitor low-level seismicity.

In the context of the major OCEAN BASINS, the mid-ocean ridges are bounded by ABYSSAL HILLS that fade into ABYSSAL PLAINS as the SEDIMENTS thicken to conceal the small-scale roughness of the terrain below. The plates may have continents riding on them, which have changed their configurations many times since the formation of Earth. At times in the past, they have formed a single supercontinent (the most recent being Pangaea), which eventually split apart into the continents of modern times. The force that powers this motion is the convection of hot magma in the Earth's mantle, resulting from heat left over from the formation of the Earth and the decay of radioactive rocks deep within the Earth.

On the ocean floor, the horizontal movement of the magma forces tectonic plates apart at speeds on the order of a few inches (cm) per year (see MAJOR PLATES). New crust is being formed at a faster rate in the Pacific than in the Atlantic. At the other end of the plate, the crust has become cold and heavy, and sinks back into the mantle in a process called "subduction" (see SUBDUCTION ZONES).

The Mid-Atlantic Ridge is divided into "spreading centers" 186 to 311 miles (300–500 km) long. A V-shaped valley delineates the center of the ridge about 19 to 31 miles (30–50 km) in width. The small-scale topographic features of the ridge become smoother with distance normal to the spreading ridge axis. The greater the distance, the older the seafloor, and the greater amount of time for which sediments can build up from the continuing "snow" of detritus from the upper sunlit layers of the ocean that fill in the valleys and depressions in the rock floor. The crust of the continents is lighter than that under the oceans, and therefore more buoyant. Since continental plates do not subduct unless they converge, continental crust is much older than oceanic crust; oceanic crust renews itself every few million years as the subducted plates are "recycled" through the mantle.

Geologic features frequently formed in mid-ocean spreading ridges include sheet and pillow lavas that freeze in contact with the cold, deep waters of the ocean floor. The large-scale features of the ridge have been mapped by side-scan SONAR; the smaller scale features such as ravines and pillow lavas have been observed from deep-sea submersible craft such as R/V *Alvin,* from the WOODS HOLE OCEANOGRAPHIC INSTITUTION. These observations along with magnetic measurements helped to confirm the theory of plate tectonics.

Further Reading

Erickson, Jon. *Marine Geology: Exploring the New Frontiers of the Ocean.* New York: Facts On File, 2003.

Seibold, E., and W. H. Berger. *The Sea Floor: An Introduction to Marine Geology,* 3rd ed. Berlin: Springer, 1996.

Volcano Hazards Program, U.S. Geological Survey. Available online. URL: http://volcanoes.usgs.gov. Accessed March 10, 2008.

Volcano World. Available online. URL: http://volcano.und.nodak.edu/. Accessed May 19, 2006.

Milankovitch cycles   The 20th-century Serbian mathematician Milutin Milankovitch (1879–1958) recognized that minor changes in orbital eccentricity, changes in the tilt of Earth's axis, and precession of the equinoxes cause slight but important variations (now known as Milankovitch cycles) in the amount of solar energy reaching Earth. Milankovitch estimated that these three conditions contribute to changes in Earth's climate responsible for the advance and retreat of glaciers. Paleontologists investigating ice ages occurring during the Pleistocene Epoch, which began about 1.8 million years ago, have applied these cycles to understand cold (glacial) and warm (interglacial) periods, shifting climate, and changes in sea level.

The path that the Earth takes about the Sun takes approximately one year and is not fixed. Milankovitch showed that over the course of 100,000 years

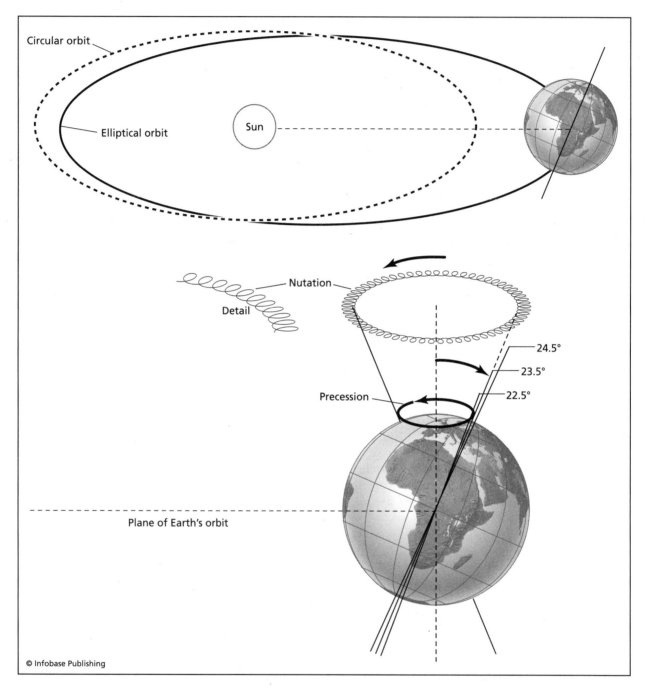

© Infobase Publishing

Orbital variations of the Earth, called Milankovitch cycles, cause changes in the amount of incoming solar radiation. Shown here are changes in the eccentricity (departure from circular) of the orbit, irregularity in the precession (nutation), and precession (wobble) of the equinoxes.

Earth's orbit changes from a circular to an elliptical track. When the orbit around the Sun is most elliptical, there are larger differences in the distance between the Earth and Sun at perihelion than during aphelion. Perihelion is the point during Earth's orbit when it is closest to the Sun, and during aphelion it is farthest from the Sun. Earth's current orbital path is nearly circular, but as the orbit becomes more elliptical, a significant change in incident solar radiation will result. The amount of solar energy received makes perihelion much warmer than aphelion.

Earth is normally tilted at an angle of 23.45° as it orbits the Sun, which explains why there are seasons. The equator is at an angle of 23.45° to the ecliptic plane. This celestial plane is defined by Earth's circular orbit about the Sun. Solar radiation strikes each of Earth's hemispheres obliquely during the winter (winter solstice occurs around December 21 for the Northern Hemisphere and June 21 for the Southern Hemisphere) and more directly during the summer (summer solstice occurs around June 21 for the Northern Hemisphere and December 21 for the Southern Hemisphere). Thus, more solar radiation is received during the summer than in the winter in accordance with the general formula energy flux = power/area (watt/m^2). The ground receives more energy during the summer (high temperatures) and less energy during the winter (low temperatures). Earth has intermediate positions during its revolution about the Sun, which are called vernal (spring) and autumnal (fall) equinoxes. At the equinoxes, each hemisphere receives exactly the same amount of light from the Sun. Milankovitch showed that over the course of 41,000 years, the axis Earth makes with the plane of Earth's orbit changes from 22.5° to 24.5°. The ensuing changes in obliquity correspond to variation in amounts of incoming solar energy.

While Earth, an oblate spheroid, spins counterclockwise from east to west on its revolution around the Sun, the axis wobbles like a top or precesses. The discovery of precession is credited to the Greek astronomer Hipparchus (190 B.C.–120 B.C.). Over the course of 26,000 years, this conical motion traces a circle on the celestial sphere. The celestial sphere is an imaginary sphere of gigantic radius with Earth at the center. Earth's axis is currently pointed toward the star Polaris, and it will slowly get closer until around the year 2100. The change in direction of Earth's axis of rotation determines the timing of the seasons relative to the position of Earth along its orbit around the Sun. Seasons will gradually shift over thousands of years.

The variation of Earth's exposure to the Sun's rays, or insolation, changes as a result of astronomical motions. Variance in insolation contributes to global climate patterns. Variability in the amount of incoming solar radiation also affects global heat transport, which involves changes in permanent CURRENTS and wind systems. Published scientific studies of deep-sea sediment cores confirm that periods of climate change correspond with Milankovitch's Earth-Sun cycles. Quantifying the effect that humans have on climate remains a topic of research.

Further Reading

Adams, J., M. Maslin, E. Thomas. "Sudden climate transitions during the Quaternary." Environmental Sciences Division, Oak Ridge National Laboratory. Available online. URL: http://www.esdornl.gov/projects/qen/transit.html. Accessed May 18, 2006.

Anderson, Edwin J. "The Croll-Milankovitch Hypothesis." Available online. URL: http://astro.ocis.temple.edu/~andy/Contents/Research/orbitalforcing.htm. Accessed September 28, 2007.

Berger, A. "Where Astronomy Meets Geology: From Ice Ages to Global Warming." Available online. URL: http://www.igpp.ucla.edu/colloquia/lectures/berger/. Accessed May 18, 2006.

Crowley, T., and K. Kim. "Milankovitch Forcing of the Last Interglacial Sea Level." *Science* 265, no. 5178 (1994): 1,566–1,568.

Davis, O. K. "The astronomical theory of climatic change." Available online. URL: http://www.geo.arizona.edu/palynology/geos462/21climastro.html. Accessed May 18, 2006.

Hays, James D., John Imbrie, and Nicholas J. Shackleton. "Variations in the Earth's Orbit: Pacemaker of the Ice Ages." *Science* 194, no. 4270 (1976): 1,121–1,132.

Imbrie, John, and Katherine Palmer Imbrie. *Ice Ages: Solving the Mystery.* Berkeley Heights, N.J.: Enslow, 1979.

Kaufman, Y., and J. Yoram. "On the Shoulders of Giants." Milutin Milankovitch (Earth Observatory, NASA). Available online. URL: http://earthobservatory.nasa.gov/Library/Giants/Milankovitch/milankovitch.html. Accessed May 18, 2006.

MacDonald, Gordon J., and Richard A. Muller. "Origin of the 100 kyr Glacial Cycle: Eccentricity or Orbital Inclination?" Available online. URL: http://www.muller.lbl.gov/papers/nature.html. Accessed November 4, 2002.

Rial, J. A. "Pacemaking the Ice Ages by Frequency Modulation of Earth's Orbital Eccentricity." *Science* 5427 (1999): 564–568.

Seldomridge, Leish A. "Resources on the Milankovitch Cycle." Available online. URL: http://www.geology.fsu.edu/~kish/dynamic/global/LAS.htm. Accessed May 18, 2006.

Stern, David P. "Precession." Goddard Space Flight Center, NASA. Available online. URL: http://www-istp.gsfc.nasa.gov/stargaze/Sprecess.htm. Accessed May 18, 2006.

Thomas, Ellen. "The pacemaker of the Ice Ages: Milankovitch Cycles in Climate." Wesleyan University, Middletown, Conn. Available online. URL: http://www.ethomas.web.wesleyan.edu/ees123/milank.htm. Accessed May 18, 2006.

mining   Mining at sea refers to the extraction of minerals from ores on or under the seabed. The most economically valuable mineral is petroleum, which is now being extracted from the ocean floor in large quantities (*see* OFFSHORE DRILLING). Other minerals that are mined by industry include sulfur extracted from salt domes off the coast of Louisiana, coal and iron extracted from mines bored horizontally seaward off the coast of Japan, and diamonds from mines bored horizontally under the ocean floor off the coast of South Africa.

The possibilities and prospects of deep-sea mining have excited the marine science and engineering communities since the day MANGANESE NODULES were discovered by the HMS *Challenger* Expedition (1872–76) more than a century ago. Manganese nodules are potato-sized aggregates of minerals, especially manganese, iron, cobalt, copper, and nickel that litter the seafloor in some areas of the ABYSSAL PLAINS in the deep ocean, at depths on the order of 12,000 feet (4,000 m), especially the North Pacific north of the equator. They form in deep, quiet waters by the oxidation and subsequent precipitation of minerals out of solution, and accumulate on "seeds" such as particles of OOZE, or even shark's teeth, somewhat in the manner of a pearl forming in an oyster. The geographic distribution is limited to relatively calm areas of the deep ocean because in regions of heavy sedimentation, such as the CONTINENTAL RISE, the SEDIMENTS in a short time would bury the nodules.

The metallic elements that form the nodules may have originated in HYDROTHERMAL VENTS, especially those known as "black smokers," where sulfur and large quantities of minerals are ejected from sub-bottom reservoirs of hot magma, when nearly freezing sea water under high pressure seeps through cracks under the seafloor. It is heated by the hot magma, and then ejected violently through the vents back into the sea, along with a load of suspended minerals from the molten rock below. Many of these minerals precipitate out immediately as sulfides and form black "chimneys" that build up around the vents.

Various mining methods have been proposed for retrieving the nodules, including DREDGING by ocean-bottom "crawlers," (bottom-traveling devices) or retrieval of the rock by ROVs (REMOTELY OPERATED VEHICLES). The rock and sedimentary materials are then pulverized into "slurry," a mixture of seawater, minerals, and sediment, by means of a hydraulic system and pumped to the surface by means of a pipeline to a support vessel. The support ship separates the nodule fragments and sedimentary material on board, and discharges the sediment and deep water back into the ocean. Such a procedure necessitates a high level of technology to control vehicles on the bottom from surface ships in sometimes rough and dangerous sea while maintaining the flow of slurry up the pipe. The mining operation generates plumes of sediment that could alter the composition of the water column and bury the benthic animals below.

Thus far, the costs and risks of deep-sea mining for manganese nodules have been too high to implement an actual operation. However, as the consumption of these metals increases, depleting terrestrial sources of these minerals, their value may rise to the point where it becomes economically feasible to conduct seabed mining operations.

Today, actual mining operations at sea are being conducted off the coast of Namibia for diamonds by the DeBeers Marine Corporation. Two methods are used. In one, a vertical tunnel cutter drill bit digs into the seafloor to separate the diamondiferous gravel from the seafloor material and the sediments. In the other, a remotely operated bottom crawler sucks up the gravel, which is then processed to remove the sedimentary material. The corporation plans to expand this operation to include the west coast of South Africa. This expansion involves expensive geophysical surveying, using the latest technology such as AUTONOMOUS UNDERWATER VEHICLES (AUVs) equipped with side-scanning SONARS and swath bathymetric mapping systems for locating and characterizing ore bodies on the seafloor.

There is also active commercial interest in mining minerals that have accumulated in brine pools deep in the RED SEA at depths greater than 6,000 feet (2,000 m). In these deep-sea pools, iron is concentrated to more than 8,000 times its average in sea water, copper 100 times, and so on for many metals. The mineral deposits seem to have originated in hydrothermal vents and from mineral deposits in newly forming sea floor associated with the rift at the bottom of the Red Sea (*see* PLATE TECTONICS). It is likely that this rifting of continental margins has resulted in the release of polymetallic sulfides, which may prove to have high concentrations of extractable minerals.

Further Reading

Antrim, Caitlyn L. "Deep Seabed Mining the Second Time Around." *Sea Technology* 47, no. 8 (2006): 17–24.

Bishop, Joseph M. *Applied Oceanography.* New York: Wiley-Interscience, 1984.

Erickson, Jon. *Marine Geology: Exploring the New Frontiers of the Ocean,* rev ed. New York: Facts On File, 2003.

Seibold, Eugene, and Wolfgang H. Berger. *The Sea Floor: An Introduction to Marine Geology,* 3rd ed. Berlin: Springer, 1996.

Stevenson, Ian, Paul Nicholson, and Thys Heyns. "High-Resolution 3D Geophysics in Marine Diamond Mining." *Sea Technology* 47, no. 8 (2006): 10–16.

minor plates   The minor plates are the small lithospheric blocks of Earth's crust and upper mantle involved in the interaction with other plates in tectonic (forces or processes within the crust resulting in movement) activity (*see* PLATE TECTONICS). The rigid plates are essentially "floating" on the asthenosphere, the plastic uppermost part of the mantle. Movement of the plates occurs in response to heating by convection currents from below. The minor plates are the Arabian Plate, the Caribbean Plate, the Cocos Plate, the Nazca Plate, and the Philippine Plate.

   See also MAJOR PLATES.

model basin   A model basin is a physical basin or tank used to carry out hydrodynamic tests with scale models of ships, submarines, boats, craft, deep-ocean platforms, coastal structures, and OCEANOGRAPHIC INSTRUMENTS. Models for testing might include scaled-down versions of wave-piercing catamaran hulls for a high-speed vessel, mooring and riser systems for an offshore oil production platform, or twin JETTIES for a COASTAL INLET. The performance of oceanographic instruments such as an ACOUSTIC DOPPLER CURRENT PROFILER or a WAVE BUOY can be assessed against CURRENTS and waves that have been set up in the tank. The sea-keeping abilities of the scale-model hull or the stability of an offshore platform can be studied prior to building full-scale prototypes. Sea-keeping refers to a vessel's performance, controllability, and dynamic response in a seaway. Studies in wave tanks help ocean engineers determine the type of riser support a spar buoy might need for oil exploration in the GULF OF MEXICO. A spar is a deep-floating cylindrical hull that supports a deck. Rigid or flexible risers connect the offshore installation to the seafloor. Experiments in a model basin provide data and information that is used for the purpose of designing full-size ships, boats, craft, and structures or assessing the skill of innovative instruments to measure currents or waves.

   The United States Navy and Army both maintain premier model basins and wave tanks that are used for a variety of projects directly relevant to naval architecture, ocean engineering, and coastal engineering. The Naval Sea Systems Command's David Taylor Model Basin began operations in Carderock, Maryland, during 1939 and is considered a landmark in mechanical engineering by the American Society of Mechanical Engineers. It comprises three independent towing basins for shallow water, deep water, and high-speed applications. The water level in the shallow water basin can be varied to simulate rivers, canals, and restricted channels. These U.S. Navy towing basins are used by other government agencies, academia, and industry for projects such as assessing capabilities of trawl NETS, calibration of CURRENT METERS, and testing of AUTONOMOUS UNDERWATER VEHICLES. The U.S. Army Engineer Research and Development Center, Coastal and Hydraulics Laboratory, in Vicksburg, Mississippi, models the physical environment, using wave tanks and flumes. These design tools are used for many coastal projects, from assessing erosion of coastal lands to determining the fate of dredge material.

   Physical models are actually miniature reproductions of a physical system such as an embayment, harbor, or river mouth. Scale models provide insights into a system's behavior by observing results not readily examined in the real world. These studies help engineers optimize the design of BREAKWATERS and jetties that are intended to reduce wave heights and minimize sedimentation of a navigation channel. Wave tanks enable investigators to estimate the impact coastal structures will have on wave heights and directions with emphasis on wave diffraction and refraction (*see* WAVES).

Further Reading

Naval Sea Systems Command, the Carderock Division. Available online. URL: http://www.dt.navy.mil/. Accessed October 10, 2007.

Oregon State University, O.H. Hinsdale Wave Research Laboratory. Available online. URL: http://wave.oregonstate.edu/. Accessed October 10, 2007.

University of Michigan, Marine Hydrodynamics Laboratory. Available online. URL: http://www.engin.umich.edu/dept/name/facilities/mhl. Accessed October 10, 2007.

U.S. Army Corps of Engineers, Coastal and Hydraulics Laboratory. Available online. URL: http://chl.erdc.usace.army.mil/. Accessed October 10, 2007.

models   Models are representations of reality developed to understand complex processes, causes, and effects and to aid in predictions. In marine science, physical models are tangible representations of a system. Diagnostic models state conditions at a specific time, while prognostic models consider conditions as a function of time. Prognostic models highlight the evolution of phenomena over time. Models may be used to simplify the natural world and to analyze environmental processes that occur across cascading scales from seconds (TURBULENCE) to eons (climate change). Model studies allow marine scientists to test their hypothesis and to compare observations with assumptions.

Physical scale models can be built to study circulation in ESTUARIES or the response of SEDIMENT transport in a COASTAL INLET with JETTIES. Three-dimensional physical models may be built with various sorts of construction materials such as clay, sands, pebbles, and concrete and incorporate flumes to simulate water dynamics. It may fit on a table or fill a warehouse. In December 2003, the Louisiana Department of Natural Resources and Louisiana State University obtained a physical model of the Mississippi River Delta to study how freshwater and sediment diversions can help restore wetlands. The Sogreah Laboratory in Grenoble, France, built this model.

Early work to understand the dynamics of ocean circulation, especially wind-driven circulation, was done with laboratory models, tanks, and flumes that focused on such problems as the effects of Earth's rotation on fluid flow, especially the affects of vorticity. Correct modeling of these phenomena required an analysis of characteristic length and time-scale differences between the model and the ocean, by means of the dimensionless numbers of fluid dynamics, such as the Reynolds number, the Froude number, the Rossby number, and many others. The Reynolds number, after Osborne Reynolds (1842–1912), is the ratio of the inertial force to the viscous force. The Froude number, after the NAVAL ARCHITECT William Froude (1810–79), is the ratio of the inertial force to the force of gravity. The Rossby number, after the meteorologist Carl-Gustaf Arvid Rossby (1898–1957), is the ratio of inertia to the CORIOLIS force. These ratios demonstrate the importance of friction, gravity, and the Coriolis effect, respectively. The Coriolis effect, an apparent deflection, is caused by the Earth's rotation.

Numerical models that utilize computers to solve differential equations that describe physical phenomena have largely replaced physical models. Terms in the equation each account for specific processes such as gravity, pressure, density, friction, and rotation of the Earth. Solutions to these equations of motion provide forecasts relevant to circulation (feet/second, knots, or m/s; direction) or some other change of process per unit of time (flux). Other numerical models that account for momentum transfer of the faster-moving air into the slower-moving water,

BATHYMETRY, and shoreline structures are used to study phenomena such as waves. Some models are coupled so that the output from one serves as input for the other. In this case, the model might simulate time-dependent distributions of water levels, currents, TEMPERATURES, SALINITIES, and WAVES. The skill of predictive models is assessed or verified using observed information or sea truth data. Numerical models may run on personal computers or Cray supercomputers, depending on their complexity.

See also MODEL BASIN; NUMERICAL MODELING.

Further Reading

Louisiana Department of Natural Resources. Available online. URL: http://dnr.louisiana.gov/. Accessed August 29, 2006.

Princeton Ocean Model. Available online. URL: http://www.aos.princeton.edu/WWWPUBLIC/htdocs.pom/index.html. Accessed August 30, 2006.

U.S. Army Corps of Engineers, Engineer Research and Development Center, Waterways Experiment Station. Available online. URL: http://www.wes.army.mil/. Accessed August 30, 2006.

Von Arx, William S. *An Introduction to Physical Oceanography.* Reading, Mass.: Addison-Wesley, 1962.

Mohorovicic discontinuity (Moho)   The Moho is the bounding zone between Earth's crust and mantle. The Moho was detected by the change in velocity of SEISMIC WAVES, that is, earthquake waves, from about three miles per second (6–7 km/s) in the crust to about five miles per second (8 km/s) in the mantle. It is named after Andrija Mohorovičić (1857–1936), a Croatian physicist who discovered the discontinuity in 1909. Physically, it is a layer of rapid density transition, in which the density of Earth increases downward from about 2.9 to 3.0 gm/cm^3 in Earth's crust under the oceans to about 3.3 gm/cm^3 in the upper layer of Earth's mantle. The change in density is greater in going from Earth's continental crust to the mantle, since continental crust is lighter than oceanic crust with an average density of about 2.7 to 2.8 gm/cm^3. The crust is much thinner under the oceans than under the continents, being as shallow as about three miles (5 km) under the oceans, and as deep as 37 miles (60 km) under continental mountain ranges. The Moho extends from about 22 to 50 miles (35–80 km) in depth.

The change in Earth's density results from both a change in chemical composition of the rock layers and the temperature and pressure of the environment. The crust under the oceans is composed of basaltic silicate minerals, which have high iron and magnesium concentrations. The crust under the continents is generally granite, whereas the material of the mantle is mostly olivine and pyroxene. Since in general, the TEMPERATURE and PRESSURE increase with depth, the mantle is at a significantly higher temperature than the crust. Temperature and pressure are such that the upper mantle flows like lava. The crust and the very top of the mantle have combined and broken apart to form the tectonic plates that "float" on the near-plastic layer of the upper mantle. The plates are called the lithosphere, and the upper mantle, on which the lithosphere floats, is called the asthenosphere. The asthenosphere heats the crust from below and produces mid-ocean spreading, island building, volcanic activity, and earthquakes. All these phenomena can be shown in the theory of PLATE TECTONICS to be related to convection within the mantle.

In the 1960s, SCRIPPS INSTITUTION OF OCEANOGRAPHY professor Walter Munk (1917–) and colleagues proposed to drill a hole through the crust under the PACIFIC OCEAN to the mantle. The project was named Mohole, and drilling was performed in the vicinity of the Hawaiian Islands. The technology of the time, however, was not up to the task, and in 1966, the U.S. Congress cancelled funding for the project. Even so, Mohole was the forerunner of the OCEAN DRILLING PROGRAM, which has yielded significant improvements in deep-sea drilling technology. The project resulted in greatly improved understanding of the solid earth under the oceans.

Some might wonder if humans have ever seen a piece of the mantle. In fact, they have in a few surface outcrops on land where continents collided in past geologic eras, where deep folding occurred and the mantle was stripped off the crust in a process called delamination.

Further Reading

Davis, Richard A. *Oceanography: An Introduction to the Marine Environment*, 2nd ed. Dubuque, Iowa: 1991.

mollusks Mollusks are invertebrate organisms of the phylum Mollusca, generally characterized as slow-moving, bilaterally symmetrical, and lacking an internal skeleton and metameric (internal) segmentation.

The first mollusks appeared in the Cambrian Period (about 500 million years ago). Today, seven extant classes comprise over 110,000 species. The classes are: Monoplacophora, Aplacophora, Polyplacophora (chitons), Gastropoda (nudibranchs, snails, and limpets), Scaphopoda, Bivalvia (mussels, oysters, and clams), and Cephalopoda (octopi and squids). Gastropoda and Bivalvia contain 99 percent of living mollusk species.

Mollusks range in size from snails a millimeter long to giant clams that have shells weighing more than 500 pounds (230 kg) and giant squids more than 55 feet (17 m) long. Mollusks have a wide geographic distribution and inhabit rivers, lakes, oceans, mudflats, mountains, forests, CORAL REEFS, and deserts. Pulmonate land snails are found at elevations of 19,700 feet (6,000 m), and Protobranchiate bivalves live at ocean depths of 29,500 feet (9,000 m). Mollusks are known to burrow, bore, crawl, and swim. They feed on a wide range of food types. For example, almost all cephalopods are active pelagic carnivores that feed on fish, crustacea, and other cephalopods; however, bivalves are generally sedentary suspension feeders or deposit feeders.

The basic molluskan body plan can be divided into three regions. First, the ventrally located, usually well-differentiated head-foot is responsible for locomotion and contains most sensory organs and some nerve concentrations. Second, the visceral mass contains organs of reproduction, digestion, and excretion. Third, the mantle (pallium) is a fold of skin that hangs over the visceral mass and covers the space containing the gills. The mantle uses calcium supplied by food and seawater to excrete a calcium carbonate shell that covers the soft body of many mollusks.

Each class of Mollusca has a unique body and shell form. Species of Bivalvia have two hinged calcareous valves, and Cephalopoda may have an internal structure (squids), an external coiled shell (*Nautilus*), or lack a calcareous structure altogether (octopi). The Gastropoda usually have a single-piece shell that may be coiled (snails), conical (limpets), or a flattened spiral. Slugs are gastropods that lack shells. Chitons (Polyplacophora) have eight plates that form a shell, and because of the shape of their shell, Scaphopoda are known as "elephant's-tusk-shells."

Primitive mollusks have separate sexes, fertilize externally, and release large numbers of eggs; however, more advanced mollusks are commonly hermaphroditic, may fertilize internally, and release fewer and larger eggs. Larvae may be a free-swimming or crawling-life stage and can last from only a few days to months.

See also BIVALVES; CEPHALOPODS.

Further Reading

The Mollusca, University of California Museum of Paleontology, Berkeley, California. Available online. URL: http://www.ucmp.berkeley.edu/taxa/inverts/mollusca/mollusca.php. Accessed March 10, 2008.

Morton, J. E. *Molluscs*. London: Hutchinson University Press, 1968.

monsoon A seasonal reversal in wind direction occurring twice a year over much of the tropical ocean, producing distinctly different weather, is called a monsoon. Usually, one monsoon is very dry; the other brings heavy rain. The best-known monsoon wind system is the Asian monsoon blowing over the INDIAN OCEAN and South and Southeast Asia, bringing dry weather in the Northern Hemisphere in winter and heavy rains in summer. The monsoons are the result of the change in pressure distribution from winter to summer and the ensuing changes in atmospheric conditions. During winter, the land cools more rapidly, and to a much greater extent than the sea (the heat capacity of water is much greater than that of soil). The air above the land becomes cold and dry, cold dry air being heavier than warm, moist air, leading to a high pressure cell over land and relatively low pressure over the sea. This pressure distribution produces a PRESSURE GRADIENT from land to sea with a consequent development of land- to-sea winds. In the summer, the land heats more quickly and to a greater extent than the sea, resulting in low atmospheric pressure over land,

in a manner similar to the LAND AND SEA BREEZE, but on a much larger scale.

The Asian monsoon produces dramatic climate changes over the Indian subcontinent and surrounding lands. In winter, cooling air over the great landmass of central Asia develops into a strong high-pressure cell over Siberia that unleashes cold, northeasterly winds over South and Southeast Asia. In the summertime, hot air over India can lead to surface temperatures that are often greater than 100°F (38°C). The ensuing low pressure draws in the cooler, moist air over the Indian Ocean as southwesterly winds over the Indian subcontinent bring welcome relief from the heat and dryness, with much-needed rain to irrigate the crops. There are times when a delay in the onset of the summer monsoon can be catastrophic for farmers; crop failure may result in severe famine. Usually in India, the winter monsoon lasts from November through April; the summer monsoon lasts from late May and ends in September. There is a transition period between the monsoons of several weeks when wind directions are quite variable.

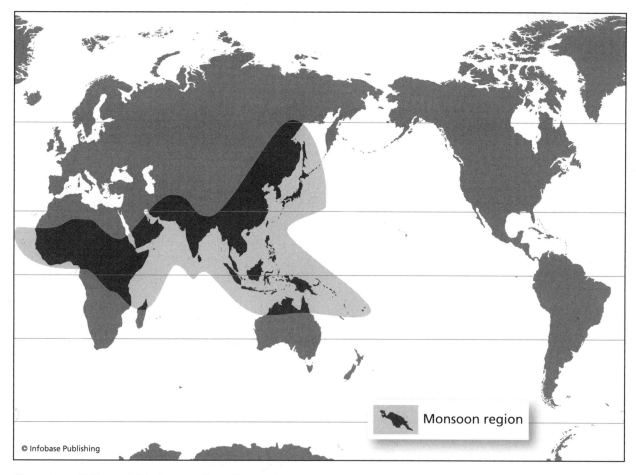

© Infobase Publishing

Monsoon region

The regions of Africa and Asia that are affected by monsoons

In South Asia, the east-to-west-running Himalaya Mountains create a barrier to the southwest monsoon that confines the summer monsoon rains to the Indian subcontinent. However, the general northward movement of the INTER-TROPICAL CONVERGENCE ZONE (ITCZ) helps to create a westward-flowing (easterly) jet stream that increases rainfall over Southeast Asia, Indonesia, the Arabian Sea, and east Africa. As summer fades into winter, the ITCZ retreats southward toward the equator, the rain subsides, and the easterly jet stream is replaced by the polar jet stream over the Himalayas, bringing cold winter winds.

The physics of the basic airflow of the monsoon is similar to the land and sea breeze, but on a much larger scale. During the winter monsoon, the sinking air over the Tibetan high-pressure area is dried by adiabatic (without gain or loss of heat) compression, whereas in summer, the low pressure air over the continent produces convergence and rising air, which condenses at altitude giving off latent heat (similar to what happens in a HURRICANE), pumping energy into developing rain clouds. Thus, the effects of dry winter winds and summer rainfall are enhanced.

Rainfall is further augmented by orographic lifting of the southwest monsoon winds up the southern slopes of the Himalayas. The water vapor in the uplifted air condenses and forms clouds and rain. The result of these moisture-inducing processes is a rainfall of intensity and duration unknown in temperatelands.

Smaller monsoons of much less consequence to the residents of the regions are found in the southwestern United States and northern Mexico, where most of the rainfall over the desert falls in the summer with the southwest winds bringing moist air from the Gulf of California over Mexico and Arizona, New Mexico, and southwest Texas. Because there is no east-to-west mountain range to provide orographic lifting, which is very important to developing the intense rain clouds in the Asian monsoon, the North American monsoon is only a weak counterpart. However, it does provide moisture to the parched desert and is a welcome relief from the heat.

The monsoon system over the Indian Ocean exerts a strong and persistent northeast wind stress during November through March, reversing direction to exert a stress from the southwest during June through September. During November through March, the NORTH EQUATORIAL CURRENT flows westward across the Indian Ocean, extending in width approximately from the equator to about 8°N; and the South Equatorial Current flows from east to west, extending in width from about 8°S to about 20°S. At the equator, the EQUATORIAL UNDERCURRENT, extends from a few tens of feet (m) below the sea surface to a depth of several hundred feet (m).

During the Northern Hemisphere summer, when the southwest monsoon is in full force, the North Equatorial Current reverses direction and flows eastward to join the Equatorial Countercurrent and the Equatorial Undercurrent, forming an eastward-flowing mass of water known as the Southwest Monsoon Current. During this season, a strong northward-flowing current develops off East Africa, known as the Somali Current because of its proximity to the Somali coast. The Somali Current can be nearly as swift as the GULF STREAM, but it is much shallower, permitting it to reverse rapidly with the onset of the monsoon wind regimes.

During the transitional seasons, late April through May, and October into November, the winds are quite variable, with consequent transient currents and EDDIES prevailing in the equatorial Indian Ocean until the new monsoon becomes established.

Further Reading

Allaby, Michael. *Encyclopedia of Weather and Climate*, rev. ed. New York: Facts On File, 2007.

mooring A mooring is a marine structure using cables, lines, chains, and anchors to secure ships, boats, platforms, or buoys in a fixed place. Buoys attached to moorings might be used to mark positions on the water, to alert mariners of hazards, to indicate changes in bottom contours or navigation channels, or to provide a platform for marine science instruments and communications equipment.

Moorings for boats are usually made up of a pick-up float, pennant, mooring buoy, sets of swivels and shackles, light chain, heavy chain, and deadweight anchors. Some moorings use other components such as subsurface floats, which help prevent mooring tackle from scouring the bottom. Mooring buoys are usually painted white and are marked with identification numbers. Pennant lines should be twice the distance from the bow chock to the water's surface. The boat ties up to the pennant line, which is attached to the mooring line or chain. The length of mooring chain is dependent on water depth and the size of the boat. In an anchorage, it is usually less than 2.5 times the maximum depth of the water. Anchors are chosen based on the bottom type and the size or weight of the vessel (*see* ANCHOR). Vessels larger than 65 feet (19.8 m) are usually moored under the supervision of a harbormaster. Moorings for ships may be composed of multiple anchors that are connected to a riser chain by ground legs. Steel

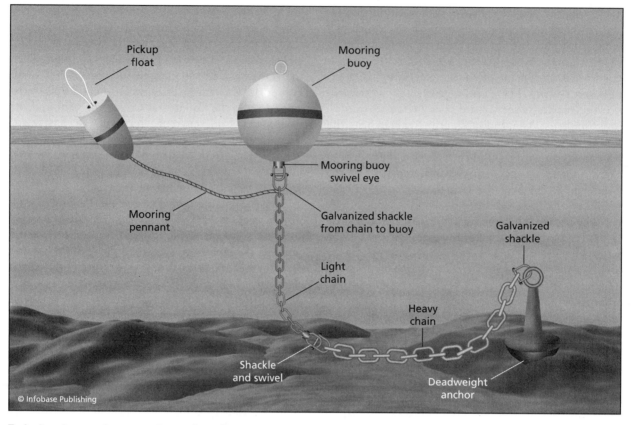

Typical marine mooring system in a yacht anchorage

ground rings are used to join riser chain to ground legs. Depending on the anchorage and why the ship is mooring, multiple mooring points may be required. A typical example might involve a bow and stern mooring. The mooring line for a ship is called a hawser. In marine parks, reserves, and sanctuaries recreational moorings are often provided to help protect benthic organisms from the ravages of anchoring. Damages caused by anchors in sea-grass beds and CORAL REEFS can take decades for recovery.

Navigation buoys of varying color, shapes, sizes, and light and sound characteristics are moored in rivers, ESTUARIES, intercoastal waterways, and the ocean as aids to navigation (*see* BUOYS and BUOY TENDER). They are made of materials such as steel, foam, or plastic, and have counterweights below the waterline. Lifting or mooring bails are attached to the hull. Waterlines are painted on the buoy at the prescribed distance below the top horizontal surface or at the widest diameter. Bails are semicircular handles that are fixed to the buoy hull and possibly to a cement anchor. The counterweight helps provide buoy stability in wind and current. Electricians, welders, and other technical personnel outfit buoys with all necessary signal equipment and hardware

and apply specific procedures for the deployment and maintenance of aids to navigation.

Most navigation buoy moorings consist of a buoy, bridle, shackle and swivel combinations, riser chain, chafe section, bottom chain, and a sinker. Shackles are used to connect buoys to bridles, to connect lengths of chain, and to attach chain to sinker bails. Swivels are used in the mooring to allow the buoy to twist without causing the chain to kink. The chafe section is the length of chain that rubs against the ocean floor as the buoy swings, rises, and falls with CURRENTS, TIDES, and storm waves. Sinkers of different weights and styles are used to hold the buoy in position. They are deadweight anchors and are usually made of concrete clumps. Depending on bottom conditions, expensive cast iron mushroom anchors may be used to hold the mooring. Bent steel rods are cast into the concrete sinkers to make a bail to connect mooring chain. Deployment depths have to keep buoy counterweights free of the bottom and are dependent on the size of chain.

Marine scientists utilize instrumented moorings when experiments or investigations require observations over extended periods of time. The basic elements of an oceanographic mooring are similar to moorings for vessels or navigation buoys. They require a buoy platform, connectors, a mooring line or cable, and an anchor. Sensors such as weather stations are fixed to the buoy platform, and CURRENT METERS or sediment traps may be installed inline or onto the mooring line or cable. Additional floatation may be fixed to the mooring line or cable to compensate for the weight of instruments or to enable recovery of instruments from a damaged mooring. Oceanographic moorings may also include acoustic releases so that the research vessel's crew can signal and retrieve the buoy, mooring, and sensors. Oceanographic moorings are designed to be as vertical as possible in order to obtain quality measurements. Mooring lengths are chosen to minimize swing, and lines or cable diameters are small in order to minimize drag and reduce turbulence. Depending on where measurements are required, marine scientists use either surface or subsurface moorings. Surface moorings are designed to allow measurement of the marine boundary layer, mixed layer, and bottom waters. In deep-water regions where surface measurements are not required, subsurface moorings may be employed. In this scenario, the mooring is not affected by winds and surface waves. Oceanographic moorings usually utilize deadweight anchors such as concrete sinkers or stacks of railroad wheels.

In designing a successful mooring, the marine scientist must first select a study location and determine the appropriate instruments that are going to be positioned in the water column or the marine boundary layer. A marine tech usually determines a suitable mooring configuration that is appropriate for the associated bottom soil conditions, water depth, winds, currents, and waves. Engineering tasks involve evaluation of environmental loads on mooring elements from winds, WATER LEVEL FLUCTUATIONS, and currents. Once this is done, mooring components need to be designed and built. This includes selecting appropriate chain sizes, obtaining necessary chain fittings such as shackles, pear rings, swivels, and mooring rings, determining mooring line geometry, casting concrete sinkers, and obtaining and outfitting the appropriate buoy. Once the capacity of the mooring is fully understood, use limitations should be specified. Certain buoy components may be built in the marina, while others are completed aboard the research vessel.

Moorings are critical to safe and efficient marine operations. They need to be designed and installed utilizing great care and well-planned efforts. Mooring components must sustain anticipated loads without failure. Calamities may result from anchors dragging or the breakage of mooring chain, cable, connectors, or lines. For permanent moorings, engineers design the moorings to survive in winter and tropical storms. Designers consider the current, wind, and wave characteristics for a 100-year return event such as a major HURRICANE or typhoon. The consequences of mooring failure are severe and include collisions, oil spills, grave environmental damage, and loss of property and life.

See SHACKLE; SWIVEL.

Further Reading

Berteaux, Henri O. *Buoy Engineering.* New York: John Wiley, 1976.

N

Nansen bottle The Nansen bottle is a device, designed by Fridtjof Nansen (1861–1930) and Otto Pettersson (1848–1941) in 1910 for taking water samples at prescribed depths in the ocean from a research vessel. Temperature is usually also measured when the water samples are taken. The body of the bottle is a brass cylinder of 2.5 pints (1,200 ml) capacity. The interior of the cylinder is coated with tin or silver, or, more recently, polytetrafluoroethylene (Teflon®). At each end of the cylinder is a plug valve. A rod that is attached to the wire cable used to deploy the bottles at their proper depths connects the plug valves. The bottles are lowered in the open position so that water can flow through them freely. When a "messenger," or metal cylinder is attached to the wire cable and released, it slides down the wire at about 656 feet per minute (200 m/min.) and trips the first (most shallow) bottle, which falls through a 180-degree angle and closes tightly. The force of the rotation of the bottle causes a second messenger attached to the bottle cable clamp to be released, and it in turn trips the bottle below it, which releases its messenger, which trips the third bottle and so on down to the end of the cable, where the lowest bottle is deployed. After all bottles have tripped, the cast is retrieved, and each bottle is removed from the wire cable and placed in an on-deck laboratory by a marine technician or graduate student. Sampling the water column with Nansen bottles is called taking a "Nansen cast." To make measurements comparable among research vessels, the International Association of Physical Oceanography has specified standard depths, in meters, of 10, 20, 30, 50, 75, 100, 150, 200, 300, 400, 500 meters (33, 66, 99, 165, 248, 328, 493, 657, 990, 1,320, 1,650 feet) for the "shallow cast," and 600, 700, 800, 1,000, 1,200, 1,500, 2,000, 2,500, 3,000, 4,000 meters (1,980, 2,310, 2,640, 3,281, 3,937, 4,922, 6,562, 8,203, 9,843, 13,124 feet) for the "deep cast." Usually, one bottle is set to trip just below the surface. As a side note, remember one foot is equal to .3048 meters, and standard atmospheric pressure is about 14.7 pounds per square inch (one atmosphere) of pressure. Therefore, at a depth of 33 feet (10 m) beneath the sea surface, the total absolute pressure is two atmospheres, where one atmosphere is caused by the weight of the air in Earth's atmosphere, plus one atmosphere for the weight of 33 feet (10 m) of seawater. Neglecting higher order effects, at a depth of 13,123.4 feet (4,000 m), the pressure is 400 atmospheres plus the weight of Earth's atmosphere.

In general, 12 bottles is the maximum used in any single cast. The bottles are rigged to a frame sometimes called a rosette. A powerful WINCH and strong cable are needed to haul the bottles back from depths as great as several miles (thousands of meters). The instrument package is usually lowered to within several feet (1 m) of the seafloor. Once the instrument is raised and gear secure, the ship can sail to a new station.

Each bottle has a hand-operated valve, so that the water sample may be drained into a glass or plastic bottle for analysis on board ship. Usually, water samples are analyzed for SALINITY by means of a "salinometer," which determines salinity from the electrical conductivity of the sample (*see* OCEANOGRAPHIC INSTRUMENTS). In the early 20th century, salinity was determined by means of chemical titration. Often the samples are also analyzed for dissolved oxygen, either by chemical titration or by electronic sensors.

Oceanographic Innovations: A Biographical Sketch of Fridtjof Nansen

Fridtjof Nansen (1861–1930) was a Norwegian oceanographer, explorer, statesman, and humanitarian. Considered by many to be the first modern polar oceanographer, he was also a marine zoologist and a leading citizen of his native Norway.

Nansen studied zoology at the University of Oslo, beginning in 1881. His first polar expedition, an exploration of Greenland, took place in 1888. He and Otto Pettersson (1848–1941) designed the Nansen bottle, first reported in 1925, for taking subsurface water samples. When outfitted with reversing thermometers, the bottles became the principal instruments used for decades in taking measurements at hydrographic stations. Nansen pioneered oceanographic field methods on board the research vessel *Fram,* during the Norwegian North Polar Expedition (1893–96), which was intentionally frozen into the Arctic ice in order to provide a solid measurement platform. In 1895, his polar expedition came closer to the North Pole, within 272 miles (438 km), than had any predecessor. During this expedition, Nansen made careful observations of drifting pack ice, from which he deduced the nature of the motion of the water responding to the wind. He observed that the ice drifted at an angle of 20 to 40 degrees to the right of the wind direction. He hypothesized that the wind-driven velocity structure of the upper ocean is divided into moving layers, each layer of which imparts a shearing stress to move the layer below. He provided his observations to his colleague Vagn Walfrid Ekman (1874–1954), a theoretical oceanographer, who derived the now-famous relationship for wind-driven currents, called the Ekman spiral. Ekman published his theory in 1902, which relates the balance between frictional effects in the ocean and the fictitious forces arising from planetary rotation.

As a delegate to the International Council for the Exploration of the Sea (ICES) during the Stockholm Conference of 1899, Nansen spoke with King Oskar II of Sweden and Norway (1829–1907) relevant to a "bottle post" received from polar explorer Salomon August Andrée

(1854–97), who had disappeared along with photographer Nils Strindberg (1872–97) and civil engineer Knut Hjalmar Ferdinand Fräkel during an arctic balloon expedition of 1897 to the North Pole. Andrée determined that he could beat Nansen's approach to the North Pole by riding the winds. The remains of Andrée and his crew were discovered during 1930 on the island of Kvitoya in the Arctic Ocean by the Norwegian Bratvaag Expedition while investigating glaciers and seas in the Svalbard Arctic region. In 1901, Nansen was chosen to be director of the prestigious International Council for the Exploration of the Sea (ICES), an organization that coordinates and promotes marine research in the North Atlantic. He was appointed professor of oceanography at the University of Oslo in 1908. He led major oceanographic expeditions and published his findings in several books.

A statesman as well as a scientist, in 1905 Nansen helped to separate Norway from Sweden. He was appointed minister to Great Britain in 1906. Following World War I, he was very active in providing assistance to refugees, directing the return of Russian and German prisoners of war. He is credited with saving millions of Russian lives during the famine of 1921. He was Norway's delegate to the League of Nations. For his humanitarian work, Nansen was awarded the Nobel Peace Prize in 1922. In 1995, he was celebrated in Russia on a 50-ruble coin. In 1999, readers of one of Norway's serious newspapers, *Aftenposten,* voted Nansen as the most prominent Norwegian of the century.

Further Reading

International Council for the Exploration of the Sea. Available online. URL: http://www.ices.dk/indexfla.asp. Accessed July 29, 2007.

International Polar Year. Available online. URL: http://www.ipy.org/. Accessed July 24, 2008.

Norway, the official site in the United States. Available online. URL: http://www.norway.org/. Accessed July 24, 2008

—**Robert G. Williams, Ph.D.,** senior oceanographer
Marine Information Resources Company
New Bern, North Carolina

Sometimes the samples are examined for biological content, but biological samples are usually taken by means of bottles or NETS (e.g., bongo nets) specially designed for the purpose. Water samples, which can be collected at exact depths, are critical in examining PLANKTON. Bongo nets are two fine-meshed plankton nets towed together on a single frame to catch tiny plants, PHYTOPLANKTON, as well as animals, ZOOPLANKTON.

Marine scientists try to understand variability in plankton populations at various depths. Marine flora and fauna are affected by temperature, salinity, and VISCOSITY of the water. Viscosity is a measure of internal friction or resistance to fluid flow and in the ocean is a function of temperature and salinity. Plankton can exploit viscosity by increasing their contact with the surrounding water with spines and moving appendages.

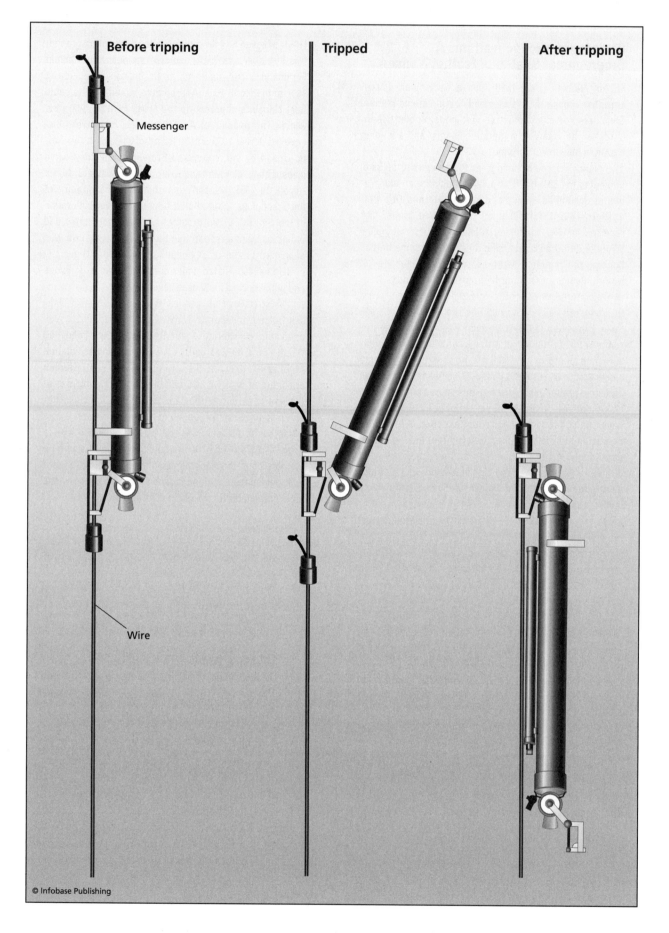

Before tripping

Messenger

Wire

Tripped

After tripping

Nansen bottles are usually supplied with frames supporting reversing thermometers to enable determination of the temperature at the depth of the water sample. This is possible by means of reversing thermometers, in which the mercury column in the thermometer splits when the bottle is reversed (inverting the thermometers). The height of the inverted column on the temperature scale etched in glass on the instrument then gives the temperature at the depth of the sample. Since the deployed cable may be at an angle to the vertical due to subsurface currents or drift of the ship, the true depth may not be equal to the amount of wire paid out, as indicated on deck by a meter. The depth is determined accurately from the difference in reading between conventional or "unprotected" thermometers, in which the ambient pressure affects the height of the mercury column, and "protected" thermometers, in which the mercury column is protected from the effects of water pressure by a glass sheath. When the bottle is retrieved, the protected and unprotected temperatures are recorded and the sampling depth determined from special tables, or by computer programs that perform the same calculations as the tables. The temperature of good-quality reversing thermometers can be read to 0.02°F (0.01°C). This is a measure of precision. Similarly, the salinity of the water sample can be determined by chemical titration to 0.02 parts per thousand, now known as practical salinity units, or psu. However, salinity measurements determined from the electrical conductivity of the water sample are about an order of magnitude more accurate than those determined from chemical titration; hence, the conductivity method is now almost exclusively used on research vessels.

By this means, the general features of the temperature and salinity distributions of the global ocean have been determined from thousands of archived Nansen casts taken since the early 20th century. Numerous charts and atlases have been produced to show distribution of properties either during specific cruises, or by inclusion of data from many expeditions. Such data can be obtained from the National Oceanographic Data Center in the United States, or from World Data Center A in Washington, D.C., or World Data Center B, in Moscow, Russia. The temperature and salinity data can be used to determine water density and geopotential, from which the density-driven or "baroclinic" component of current in the direction perpendicular to the ship's track line can be determined.

Nansen bottles today have been largely replaced by plastic "Niskin" bottles, which have rubber plungers at both ends to close the bottle. The plungers are activated, and thermometers reversed by the release of tension in rubber tubing, accomplished by the downward momentum of the "messengers." Niskin bottles come in a variety of sizes to facilitate sampling for chemical or biological properties of the water. In physical oceanography, they are frequently deployed concentrically around in situ electronic CONDUCTIVITY-TEMPERATURE-DEPTH (CTD) PROBES, and can be activated from the surface by sending a signal down an electrical cable. Such a configuration of Niskin bottles is called a "Rosette Sampler." The temperature and salinity from the water samples can then be compared with the electronic readout of the CTD as a means of data quality control, checking for such problems as electronic drift, biofouling of the sensors, and damaged components.

National Aeronautics and Space Administration (NASA)

NASA is the independent federal agency that develops and implements the civilian programs of the U.S. government that deal with aeronautical and astronautical research and the development of launch vehicles, spacecraft, and the technology to support ventures in the upper atmosphere and space. NASA was established as an independent agency in 1958, primarily from the National Advisory Committee for Aeronautics (NACA), which had been organized by President Woodrow Wilson (1856–1924) to supervise the scientific study of the problems of flight. Although NASA is most famous for its space missions, it has a considerable effort in the Earth sciences, including the launching and maintaining of satellites for study of the Earth, and in support of other agencies such as the NATIONAL OCEANIC AND ATMOSPHERIC ADMINISTRATION (NOAA), the U.S. Geological Survey (USGS), and the National Reconnaissance Office (NRO).

NASA's facilities include a headquarters in Washington, D.C., and laboratories and flight centers throughout the United States, including the Kennedy Space Center at Cape Canaveral, the Johnson Space Center in Houston, and the Jet propulsion Laboratory in Southern California. Most of Earth science, including study of Earth's oceans and atmosphere, is conducted from the Goddard Space Flight

(opposite page) Diagram of a Nansen bottle, a container used to obtain subsurface samples of seawater, as it trips when hit by a downward-falling messenger weight that releases the upper drawing clamp and causes the bottle to swing through nearly 180 degrees, filling and sealing the water bottle, and breaking the mercury columns of the reversing thermometers.

Center in Beltsville, Maryland, formed in 1959, and named for Dr. Robert H. Goddard (1882–1945), the American pioneer in liquid-fueled rocket research. Goddard today includes a workforce of 12,000 civil servants and contractors studying problems ranging from the origin of the universe to the composition of the solar system to Earth's climate. Earth scientists at Goddard study global climate, including such variables as wind distribution, temperature, precipitation and chemical composition. Earth sciences are administered and conducted by the Earth Science Directorate.

One of the major programs of the Earth Science Directorate is the Earth Observing System (EOS), which consists of a science component and a data system component supporting a coordinated series of polar-orbiting and low inclination satellites for long-term global observations of the land surface, biosphere, solid Earth, atmosphere, and oceans. This research is directed toward understanding Earth as an environmental system by determining how its components have developed, how they function, how they interact with one another, and how they evolve over long and short time scales. This program has the goal of helping to quantify the practical impacts that natural and human activities have on Earth's resources during the next century. Goddard scientists have made possible advances in weather observing and FORECASTING systems, terrain and oceanic topographic mapping, and improvement of information systems technology.

Among the recent satellite systems of interest to Earth scientists is the Aqua, a NASA Earth science mission satellite focused on the measurement of the constituents of Earth's water cycle, including CLOUDS, precipitation, SEA ICE, and snow cover. NASA Goddard has developed and is supporting environmental Earth satellites, such as the Geostationary Operational Environmental Satellite (GOES), which is a key element of U.S. weather observing and forecasting, the QuikSCAT, which is a satellite supporting a microwave RADAR system that measures near-surface winds under all weather and cloud conditions over Earth's oceans, and the TERRA, which observes changes in Earth's solar radiation budget, along with changes in the land/ocean surface and interactions with the atmosphere. The TERRA project is a partnership with the aerospace agencies of Canada and Japan. Other satellites of interest to Earth scientists include the Activity Cavity Radiometer Irradiance Monitor spacecraft to measure total solar irradiance for the U.S. Global Climate Research Program, the Earth Observing 1, successor to the highly successful "Landsat" series of satellites that were of great value to farmers, foresters, geologists, economists, city planners, and

many others, and the Stratospheric Aerosol and Gas Experiment (SAGE), designed to enhance understanding of long-term variations of key components of Earth's atmosphere.

Other well-known NASA Earth environmental satellites include the TOPEX/Poseidon, an American-French system that reportedly measured the ocean surface topography to an accuracy of 1.65 inches (4.2 cm), which has been succeeded by the *Jason-1,* also an American-French satellite managed by the Jet Propulsion Laboratory. The TOPEX-Poseidon mission enabled climate scientists to forecast the 1997–98 EL NIÑO. It also contributed to an improved understanding of ocean circulation and global climate.

See also GEOSTATIONARY SATELLITE; REMOTE SENSING.

Further Reading
"Goddard Space Flight Center." Available online. URL: http://www.gsfc.nasa.gov/gsfc/about/about_gsfc.html. Accessed October 11, 2007.
The Jason Education Project, Texas A & M University. "OceanWorld." Available online. URL: http://ocean-world.tamu.edu/print/resources/bookmarks/satellites.htm. Accessed October 11, 2007.
NASA History Division home page. Available online. URL: http://history.nasa.gov/. Accessed June 3, 2006.

National Ice Center The National Ice Center is an operational organization represented by the United States Navy, the NATIONAL OCEANIC AND ATMOSPHERIC ADMINISTRATION, and the U.S. Coast Guard; it focuses on providing worldwide operational ice analyses, SEA ICE atlases, outlooks, and forecasts for the armed forces of the United States and allied nations, U.S. government agencies, and the private sector. Typical products are produced by METOC OFFICERS, AEROGRAPHER MATES, and civilian scientists that cover Arctic, Antarctic, Great Lakes, and Chesapeake Bay ice conditions. Charts identifying the location of ICEBERGS, ice edge, ice thickness, polynyas, and ice leads support ice-breaking operations and the navigation of vessels. Polynyas are areas of water surrounded by sea ice that are often opened by KATABATIC WIND systems (the down slope flow of cold air from a glacier). They can accelerate rapidly along the edge of the Greenland and Antarctic ice sheets. Extensive breaks or cracks in the sea ice are called leads, and depending on factors such as weather and CURRENTS can close quickly. Organizations such as the International Ice Patrol utilize ice analyses and forecasts to monitor and respond to dangers near the Grand Banks of Newfoundland, where the infamous sinking of the White Star liner *Titanic* occurred on April 15, 1912, after it struck an iceberg and approximately 1,517 passengers and crew perished. Headquarters for the International Ice Patrol are located at the Coast Guard Research and Development Center in Groton, Connecticut.

Researchers are taking a close look at the rate at which Arctic sea ice is melting for several important reasons. A smaller solid ice field would have significant impacts on global climate, such as rising sea level, modified ocean circulations, and loss of arctic animal habitats. Changes in sea ice conditions may affect maritime commerce by altering marine transportation routes. Changes in the extent of polar ice will affect surface reflectivity. Physical responses to changes in heat transfer from the atmosphere to the ocean at the poles include changes in cloudiness, humidity, and ocean currents.

Further Reading

Alfred Wegener Institute for Polar and Marine Research. Available online. URL: http:www.awi-bremerhaven.de/. Accessed October 11, 2007.
Canadian Ice Service. Available online. URL: http://ice-glaces.ec.gc.ca/. Accessed October 11, 2007.
National Ice Center. Available online. URL: http://www.natice.noaa.gov. Accessed October 11, 2007.
U.S. Coast Guard. "International Ice Patrol." Available online. URL: http://www.uscg.mil/lantarea/iip/home.html. Accessed October 11, 2007.

National Oceanic and Atmospheric Administration (NOAA) This federal agency within the Department of Commerce is responsible for mapping and charting of the oceans, environmental data collection, monitoring and prediction of conditions in the atmosphere and the ocean, and management of marine resources and habitats. NOAA's mission is "to understand and predict changes in the Earth's environment and conserve and manage coastal and marine resources to meet our Nation's economic, social, and environmental needs." The agency conducts research and gathers data about the global oceans, atmosphere, space, and the Sun and applies this knowledge to science and services, such as weather and ocean tide predictions. NOAA was formed in 1970 from several existing federal agencies, several of which are among the oldest government science organizations in the United States. NOAA is dedicated to enhancing the national safety and economic security through the prediction and research of weather and climate-related events and providing stewardship of the nation's coastal, marine, and Great Lake resources.

NOAA's strategy consists of seven interrelated goals that include: (1) advance short-term warnings and forecast services, (2) implement seasonal to interannual climate forecasts, (3) assess and predict decadal to centennial change, (4) promote safe navigation, (5) build sustainable fisheries, (6) recover protected species, and (7) sustain healthy coastal ECOSYSTEMS. NOAA meets its strategic goals through its civilian and uniformed workforce. NOAA employees work in the following major organizations:

1. National Weather Service (NWS)

2. National Ocean Service (NOS)

3. National Marine Fisheries Service (NMFS)

4. National Environmental Satellite, Data, and Information Service (NESDIS)

5. Oceanic and Atmospheric Research (OAR), and

6. Marine and Aviation Operations (NOAA Corps)

The Commissioned Corps of NOAA consists of approximately 250 officers and is the smallest of the United States's seven uniformed services. These multidisciplinary line offices are essential to collect and provide the nation with appropriate environmental information in a timely fashion and in usable formats.

The National Weather Service is the descendant of the Weather Bureau, which was organized in 1870 within the U.S. Army. In 1890, the Weather Bureau was transferred to the Department of Agriculture as a civilian agency. Its first weather map for the Washington, D.C., area was published in 1895; it established a HURRICANE warning service in 1896 and began regular weather observations in 1898.

Today, the NWS issues thousands of weather, river stage and flood forecasts, as well as severe storm warnings, based on a vast observational network of weather stations incorporating such new technologies as weather satellites (designed and deployed by NASA), Doppler RADAR, automated surface-observing systems, sophisticated computer models, and high-speed communications systems. The observation network includes land-based stations, satellite observations, BUOY observations at sea, and aircraft observations, including the hurricane hunters. In 2005, the NWS deployed the Advanced Weather Interactive processing System (AWIPS) to improve climate, water, and weather products and services that help protect life, property, and the economy. It is estimated that the NWS's long-range predictions for the 1997–98 EL NIÑO episode helped the state of California avoid about $1 billion in losses.

The National Oceanic and Atmospheric Administration's (NOAA's) National Marine Sanctuary Program

In 1972, exactly 100 years after the first national park was created, the United States made a similar commitment to protecting its marine treasures by establishing the NOAA National Marine Sanctuary Program. Since then, 13 national marine sanctuaries, representing a wide variety of marine environments, have been designated that protect over 18,000 square miles (61,740 sq km) of the nation's maritime treasures. From historical treasures such as the wreckage of the civil war ironclad the USS *Monitor* to the pristine coral reefs of Fagatele Bay in American Samoa, the program is mandated under the National Marine Sanctuaries Act, previously known as Title III of the Marine Protection, Research, and Sanctuaries Act of 1972, which authorizes the secretary of commerce to identify and designate certain areas of the marine environment that are of special national significance due to their ecological, historical, educational, recreational, and aesthetic qualities. The act also provides authority for the comprehensive and coordinated conservation and management of these marine and freshwater environments, and activities affecting them, in a manner that complements existing regulatory authorities. One of those sites, the Florida Keys National Marine Sanctuary, has taken the protection of its marine resources one step further by establishing the Tortugas Ecological Reserve. After years of planning, and the signature of the Florida governor and cabinet on April 24, 2001, regulations designed to protect the diverse marine life and pristine coral reefs of the Tortugas in a no-take Ecological Reserve took effect on July 1, 2001, in the westernmost waters of the Florida Keys National Marine Sanctuary. The Tortugas Ecological Reserve is broken into two sections: Tortugas North and Tortugas South. Tortugas North is approximately 151 square nautical miles (518 km²) and covers some of the most pristine reef environments found in North America. Tortugas South is approximately 60 square nautical miles (206 km²) and encompasses Riley's Hump and deep-water spawning grounds for groupers, snappers, and other commercially and recreationally important species. The implementation of this reserve provided an excellent opportunity for NOAA to investigate and determine the effects of human disturbance (such as, elimination of consumptive sampling and physical impacts) on the functioning of coral reef and deep water algal and sea grass ecosystems. In order to determine the efficacy of this management action, several long-term monitoring actions were proposed,

A flagship of the NOAA fleet—the NOAA ship *Ron Brown*   (NOAA)

including evaluation of the local and regional areas in terms of larval fish transport mechanisms utilizing satellite drifters placed on the surface to determine circulation patterns, pathways, and possible transport vectors into the Florida Keys reef tract; changes in adult fish biomass utilizing divers performing point and transect counts of fishes recording species, numbers, and size classes at 30 permanent stations located in and out of the reserve for comparative analysis; changes in ecosystem structure and complexity utilizing side-scan and multibeam sonar for habitat characterization, and video transects of coral reefs and associated habitats such as sea grass and algal beds in order to determine species abundance, diversity, and percent cover and the use of rugosity (extent wrinkled) measurements of coral reefs to determine habitat complexity.

The National Centers for Coastal Ocean Science (NCCOS) headed by the Center for Coastal Fisheries and Habitat Research Laboratory in Beaufort, North Carolina, have been conducting this important research in the Tortugas before the actual implementation of the Reserve to establish baseline data up to the present day. It is hoped that these findings can be applied to management issues not only in the Dry Tortugas but to other NOAA trust resources in the future.

Further Reading

National Oceanic and Atmospheric Administration. "Florida Keys National Marine Sanctuary." Available online. URL: http://florida.keys.noaa.gov. Accessed August 13, 2007.

—Craig Bonn
Beaufort Marine Lab
Beaufort, North Carolina

The National Ocean Service began at the beginning of the 19th century, when President Thomas Jefferson established the Survey of the Coast. In 1878, the name was changed to the Coast and Geodetic Survey to reflect the inclusion of geodesy in the mission of the organization. Today, NOS prepares nautical and aeronautical charts, tide and current predictions, and manages the National Geodetic Survey, which specifies latitude, longitude, height, scale, gravity, and orientation of locations throughout the nation. The NOS deploys real-time state-of-the-art current and wind measurement systems in critical areas in the nation's harbors, and transmits the data to mariners navigating the channels (*see* PORTS). NOAA is the trustee for 12 marine PROTECTED AREAS, working through NOS, along with a growing number of partners and volunteers to protect the National Marine Sanctuaries, the marine equivalent to the National Parks.

NOAA's National Marine Fisheries Service is the descendant of the U.S. Commission of Fish and Fisheries, the first conservation agency, which began its work in 1871 to protect, study, manage, and restore fish and fisheries. Today, the NMFS uses a scientific approach to fisheries management to sustain living marine resources, seeking to balance public needs and interests in the use and enjoyment of these resources while preserving their integrity. Under NMFS management, U.S. fisheries have begun to recover the devastation wrought by foreign fishing fleets, and U.S. FISHERMEN have been taking a larger share of the total catch within the 200-mile (321.9 km) zone instituted by the MAGNUSON-STEVENS FISHERY CONSERVATION AND MANAGEMENT ACT.

Two recent examples include rebuilding swordfish stocks and developing a long-term strategy for restoring threatened and endangered salmon in the Pacific Northwest. NOAA's NMFS has been a strong advocate for reduction of worldwide commercial whaling through its efforts in providing guidance to the International Whaling Commission. NMFS has instituted a program of installation of turtle exclusion devices on commercial fishing equipment to protect threatened sea turtle populations.

The tens of thousands of environmental observations made by civilian and military government agencies worldwide are archived by NOAA's National Environmental Satellite, Data, and Information Service (NESDIS). Critical data and imagery that is derived from satellites is used for a range of operations from weather FORECASTING to fighting wild fires. Weather and climate observations are quality controlled and archived in readily accessible standardized formats at the National Climatic Data Center in Asheville, North Carolina. Geodetic and space weather data (primarily the activity level of the Sun and its discharge of highly energetic ionized particles in the "solar wind") are archived at the National Geophysical Data Center in Boulder, Colorado, while oceanographic data are stored at the National Oceanographic Data Center in Silver Spring, Maryland.

These data are available to scientists and environmental planners and managers worldwide.

NOAA's Oceanic and Atmospheric Research (OAR) laboratories conduct applied research in air-sea interaction, severe storm formation, weather and climate prediction, ocean circulation, HYDROTHERMAL VENT chemistry and biology, and many other topics relating to providing a safe environment and a good quality of life.

The NOAA Corps' main purpose is to operate NOAA's ships and aircraft, and to provide planning and operational services to all NOAA programs and activities. Ships and boats that make up the fleet include hydrographic survey, oceanographic research, and fisheries research vessels. The airplanes and helicopters are essential platforms for many aerial surveys and scientific investigations. NOAA's aircraft operate globally from the Tropics to polar regions on relevant projects, from improving hurricane prediction MODELS to global climate change. NOAA aircraft include the Lockheed Orion, Gulfstream IV-SP, Cessna Citation II, DeHavilland Twin Otter, Gulfstream Jet Prop Commander 1000, Rockwell Aero Commander, and the Lake Seawolf. NOAA is headquartered in Silver Spring, Maryland, and has facilities and laboratories in various locations throughout the nation.

Further Reading

Kusky, Timothy. *Encyclopedia of Earth Science*. New York: Facts On File, 2005.
National Hurricane Center home page. Available online. URL: http://www.nhc.noaa.gov. Accessed October 12, 2007.
National Oceanographic and Atmospheric Administration home page. Available online. URL: http://www.noaa.gov. Accessed October 12, 2007.
Shea, Eileen L. "A History of NOAA." Available online. URL: http://www.history.noaa.gov/legacy/noaahistory_1.html. Accessed October 12, 2007.

natural hazards Natural hazards are dangerous phenomena that destroy life and property as a result of meteorological, oceanographic, and geological processes that span varying temporal and spatial scales from microscale events such as water spouts to global-scale events such as climate change. Microscale processes occur over lengths of a mile (< 2 km) and only last for a few minutes. WATERSPOUTS like tornadoes have a twisting funnel-shaped cloud, but they form over water from updrafts of warm, moist air. Such phenomena may impact a river, a boat yard, or an ocean beach. Mesoscale events such as squall lines may last an hour or less and are so small they may only affect a beach resort. They have length scales on the order of a few miles to hundreds of miles (5 km ≤

length ≤ 2,000 km). The next scale in the spectrum is the synoptic scale, which includes air masses, fronts, and high- and low-pressure systems that span hundreds to thousands of miles (100s to 1,000s of km) with changes generally occurring over five- to ten-day periods. Synoptic scale storms along the East Coast of the United States bring abundant moisture from the Atlantic Ocean and Gulf of Mexico and generate high rainfall, which may lead to coastal flooding and erosion. Global or planetary scale phenomena occur on the time scales of months or years and cover continents and general circulation patterns such as the trade winds, prevailing westerlies, and doldrums. Marine scientists and other professionals try to understand the cause and effects of natural hazards. Basic and applied science efforts are focused on predicting where and when natural hazards occur and helping mitigate their effects.

As cool and warm air masses collide, unequal electrical charges build up in the atmosphere because of friction. In some cases the cloud base becomes so highly charged that an opposite charge is induced on the ground. Lightning between clouds and between clouds and the ground neutralizes these unequal charges similar to a giant spark. Since salt water is a good conductor, lightning in coastal areas is particularly dangerous. People standing on the beach provide the shortest path from the cloud base to the ground and could be struck and killed by the lightning. A lightning strike lasts for a few milliseconds and delivers approximately 300 kilovolts. Consequently, as thunderstorms approach the coast, fishing and other beach-related activities become extremely dangerous. High winds associated with squalls may generate high waves and rip currents. Squalls are severe local storms with sustained winds of 16 knots (8.23 m/s) or higher and are usually associated with thunder and lightning. As storms approach the coast, people should get out of the water, leave beach areas, and seek shelter in a building or an automobile. In the absence of shelter, they should find the lowest spot away from umbrellas and avoid open spaces. Thunderstorms may also produce large hail, which can damage cars, boats, and crops. Beware of flash flooding caused by slow-moving thunderstorms, especially in coastal built-up areas where much of the ground is covered by impervious surfaces.

Tropical depressions, tropical storms, hurricanes, and typhoons are low-pressure weather systems where the central core is warmer than the surrounding atmosphere. Extratropical cyclones form outside the Tropics, and the storm center is colder than the surrounding air. In the Indian Ocean and around the Coral Sea off northeastern Australia, the term *tropical cyclone* describes storms with winds

that are greater than or equal to 74 MPH (119.1 km/hr). These storms are called typhoons in the North Pacific west of the International Date Line and hurricanes in the Atlantic basin and the Pacific Ocean east of the International Date Line. The primary damaging forces associated with these storms are high-level sustained winds, heavy precipitation, tornadoes, storm surge, and wind-driven waves. Based on central pressure, maximum sustained winds, and storm surge, the Saffir-Simpson Hurricane Intensity Scale rates severity of a hurricane on a scale from one to five. Category three and four are considered extreme hurricanes, and a category five is considered catastrophic. All hurricanes are capable of inflicting great damage and loss of life. When a hurricane makes landfall, it may totally change the landscape. Erosion and flooding may produce new coastal inlets along a barrier island. It may take a city years to rebuild in the wake of a catastrophic hurricane such as Hurricane Katrina. Hurricane Katrina was a category three storm at landfall in Grand Isle, Louisiana, on August 29, 2005. Surge waters were forced into Lake Pontchartrain, and water levels rose approximately three feet (1 m). Earthen levees separating Lake Pontchartrain, Lake Borgne, and the Mississippi River from New Orleans were breached at several locations by the surge, causing massive flooding in New Orleans, a city that is below sea level. Loss of life and property were exacerbated from this flooding. When hurricane watches and warnings are issued, people should closely monitor radio, television, or National Oceanic and Atmospheric Administration (NOAA) Weather Radio for official bulletins of the storm's progress and instructions from civil defense authorities. Preparation includes planning travel routes ahead, properly shielding windows, having adequate on-hand supplies, and following the guidance of city officials in order to evacuate from designated locations in the face of hurricanes, typhoons, and cyclones.

At middle and polar latitudes, there are large cyclonic late fall, winter, and early spring storms called Nor'easters that can bring days of cloudiness, high winds, and precipitation. Nor'easters are named for the winds that blow in from the northeast and move these ocean storms up the East Coast from the Carolina Capes along the Gulf Stream, a narrow and fast permanent current of warm water that lies off the East Coast of the United States. These ocean storms produce heavy rain or snow, severe icing, very rough seas, and high surf that combine to cause extensive beach erosion. As these storms move northeast and the low pressure system deepens, heavy seas are produced that present navigational hazards to mariners. An example is the so-called perfect storm or Halloween storm of late October 1991, which sank the fishing vessel *Andrea Gail*. Best-selling author Sebastian Junger describes these events in *The Perfect Storm*.

Many WAVES develop from storms in regions called generating areas. In these generating areas, energy from the wind is transferred into the ocean. Waves propagating away from the generating area have a smooth form and are called SWELLS. The wave's speed depends on the amount of energy it received from the ocean storm. The energy that is imparted into the wave is derived from the wind velocity (speed and direction), how long the wind blew (time or duration), and the distance over which the wind blew (FETCH). Since the storm winds tend to blow over various distances at variable speeds for from five to 10 days, the generated waves have different heights, lengths, and periods. Waves leaving a generating area may have originated from different locations and will therefore arrive at a particular beach from different angles. As the waves travel across the ocean and approach the shore, they feel the bottom and change shape. This process is called shoaling and involves the decrease in WAVELENGTHS from friction and a corresponding increase in wave heights. Usually when wave heights are approximately 78 percent of the water depth, they become unstable and unleash stored energy upon the beach as SURF. The manner in which the energy is dissipated determines the nature of the breaking and is characterized as some combination of plunging, spilling, surging, and collapsing surf. Breaking waves may end up piling water upon the shore, which is returned offshore through LONGSHORE CURRENTS, which feed RIP CURRENTS. High-energy surf zones can be very dangerous to beachcombers, swimmers, surfers, and people using body boards and sea kayaks. One of the most spectacular and dangerous surf breaks in the world is the Banzai Pipeline at the northern shore of Oahu, Hawaii. Unpredictable shore breaks can knock people down, causing unconsciousness and serious neck and spinal injuries. Strong currents can carry people into sharp CORAL REEFS and large waves. Some waves are so large that even professional surfers have been killed when they fell off their surfboards. When visiting the beach, one should never turn away from the ocean, swim at beaches with lifeguards, observe posted warnings and lifeguard instructions, avoid walking on wet, rocky ledges, stay away from the water's edge during big surf, and never surf or bodyboard in big waves unless an expert.

The Earth is divided into seven MAJOR and six MINOR PLATES that are each moving relative to one another. According to PLATE TECTONICS, the plates interact along their margins as evidenced by geological processes such as the formation of mountain belts and volcanoes and the occurrence of earthquakes. Volcanic eruptions such as Tambora in Indonesia during 1815 and the 1991 Mount Pinatubo eruption in the Philippines have caused many fatalities, major damage, and severe social and economic impact. Some researchers conclude that cold weather during 1861, also known as the "year without summer," was caused by the Tambora eruption, owing to increased reflection of solar radiation from volcanic dust trapped high in the atmosphere. Earthquakes occur at mid-ocean ridges where the plates are being pulled apart, at margins where plates are sliding past one another, and at the deep TRENCHES where one plate is thrust under the other. Submarine earthquakes can displace large volumes of water, which force sea waves called TSUNAMIS. *Tsunami* is a Japanese term meaning harbor, "tsu," and wave "nami." Persons caught in the path of a tsunami are at extreme risk from being struck or crushed by debris, or drowning. On December 26, 2004, the second-largest earthquake ever reported on the Richter scale off the west coast of Northern Sumatra, Indonesia, generated tsunamis that killed more than 230,000 people and inundated coastal communities. The Richter scale assigns a single number from 1 (micro) to 9 (rare or great) to quantify the magnitude of an earthquake. This scale is credited to Charles Richter (1900–85). The Sumatra-Andaman earthquake was reported as a 9. Tsunamis may also be forced by submarine landslides, submarine volcanic eruptions, and by a large meteorite impacts in the ocean. The Tsunami Warning System in the Pacific Basin comprises participating international member states. Professionals such as oceanographers, geologists, and seismologists monitor seismological and tidal stations throughout the Pacific Basin to evaluate earthquakes that may cause a tsunami. The Pacific Tsunami Warning Center, located near Honolulu, Hawaii, is the operational center of the Pacific Tsunami Warning System and provides tsunami warning information to national authorities in the Pacific Basin. Scientists predict tsunami arrival times at selected coastal communities within the geographic area defined by the maximum distance the tsunami could travel in a few hours. A tsunami watch with additional predicted tsunami arrival times is issued for a geographic area defined by the distance the tsunami could travel in a subsequent time period. Local authorities and emergency managers are responsible for formulating and executing evacuation plans for areas under a tsunami warning. During a tsunami warning, people should avoid low-lying coastal areas.

See also CYCLONE; RIP CURRENT; ROGUE WAVE; TSUNAMI.

Further Reading

Junger, Sebastian. *The Perfect Storm: A True Story of Men Against the Sea.* New York: Harper Paperbacks, 1998.

Knauer, Kelly, ed. *Nature's Extremes: Inside the Great Natural Disasters that Shape Life on Earth.* New York: Time Books, 2006.

NASA, Earth Observatory. "Natural Hazards." Available online. URL: http: //earthobservatory.nasa.gov/Natural Hazards/natural_hazards_v2.php3. Accessed October 12, 2007.

The Smithsonian Institution. "Global Volcanism Program." Available online. URL: http://www.volcano.si.edu/. Accessed July 24, 2008.

nautical charts   *See* CHARTS, NAUTICAL.

naval architect   A naval architect is a professional who designs and engineers vessels for the recreational boating and maritime industry. Naval architects consider hull form development, strength, stability, and operational characteristics of vessels. Vessels to be built include small craft, BOATS, and ships. BARGES may be designed to store raw materials that have to be transported up the Mississippi River from Baton Rouge, Louisiana, to Minneapolis, Minnesota. Yachts might be designed for sport fishing so that fishermen can cruise from the Ocean City Inlet, Maryland, to the GULF STREAM to participate in the annual white marlin (*Tetrapterus albidus*) fishing tournament. A seagoing TUGBOAT would be designed with a special propulsion system to push, pull, and maneuver a disabled vessel into an anchorage. A fireboat is designed with space for tanks, pumps, hoses, and nozzles to fight fires onboard vessels, at docks, and in warehouses. Cruise ships have to be designed to transport many passengers to exotic ports of call. A research vessel must be designed with wet and dry laboratories, WINCHES, and ample deck space for at-sea scientific studies. Warships are designed to carry weapon systems, store ordnance, and to patrol shallow coastal oceans. Naval architects may also design anchor chains or computer-controlled propellers that hold floating rigs in place.

Naval architects are employed by ship and boat builders, offshore constructors, design consultants, departments of defense, and regulatory and ship survey societies such as the AMERICAN BUREAU OF SHIPPING. Those involved in building ships specialize in engineering requirements, design testing, and operational assessments. Requirements deal with the vessels function, capacity, and speed. Naval architects, like COASTAL and OCEAN ENGINEERS, may work hard to reconcile requirements with cost and actual performance. They complete detailed drawings, plans, build scale models, and prototypes that are tested in a MODEL BASIN, lake, ESTUARY, or ocean to ensure that the vessel meets design objectives. Scale models of the hull could be evaluated in a model basin. DYES and video cameras help naval architects analyze the flow around the hull, the structural body of the boat that rests in the water. The hull influences the amount of resistance that a boat generates as speed is increased. Prototypes may be made of materials such as wood, steel, aluminum, and composites. Many small craft hulls are still constructed of wood. Steel is a strong and cost effective material for boat hulls that is used globally. Aluminum is slightly more difficult to work with, but provides a strong, lightweight, durable, and easy to maintain hull. Composites are two or more materials having different properties that are combined. Fiberglass is a composite consisting of glass strands that are usually woven into a cloth and coated in a resin. The primary advantage of a composite is that the combined material takes advantage of the constituent material's good characteristics.

Naval architects working for the Naval Surface Warfare Center are involved with designing and testing planing hull amphibious vehicles for the U.S. Marine Corps and high-speed vessels for the U.S. Navy. Measures such as ride quality and sea-keeping are assessed in various sea states (wind-generated wave heights are characterized in sea state codes; *see* SEA STATE). This type of work requires extensive technical writing, verbal reporting, and ultimately the preparation of vessel operation manuals.

Students interested in earning a bachelor of science in naval architecture must complete college courses covering both the arts and sciences, including the essential subjects mathematics, physics, and English. The naval architecture curriculum provides a broad education in engineering mechanics and their application to the design and construction of vessels, vehicles, platforms, and marine systems. Individual courses focus on structural mechanics, hydrodynamics, marine power systems, and marine dynamics. Structural mechanics relates to the design and analysis of vessels and appendages such as keels, rudders, and propellers, including static strength, fatigue, dynamic response, safety, and production. Marine hydrodynamics covers resistance, maneuvering, and sea-keeping characteristics of submerged and floating bodies. Marine power systems are the mechanical systems onboard a vessel to include the main propulsion system and propeller design. Topics in marine dynamics relate to structural vibrations and rigid body responses of vessels to winds, WAVES, and CURRENTS. Naval architecture is often combined with marine engineering. A MARINE ENGINEER deals primarily with power plants. Many naval architects have advanced degrees such as a master of science or, a Ph.D. in naval architecture and marine engineering. Naval architects may need a professional engineer license for some jobs. In order to be licensed, after earning a college degree, the naval architect must pass the open-book Fundamentals of Engineering Exam, and then with about four years of experience the naval architect takes the open book Principles and Practices Exam.

Further Reading

Royal Institution of Naval Architects. Available online. URL: http://www.rina.org.uk. Accessed July 22, 2006.

Society of Naval Architects and Marine Engineers. Available online. URL: http://www.sname.org. Accessed July 22, 2006.

Naval Oceanographic Office (NAVOCEANO)

The U.S. Naval Oceanographic Office is a U.S. Department of Defense organization located at Stennis Space Center, Mississippi, consisting of about 1,000 military and civilian oceanographers and other scientists, engineers, and technicians responsible for collecting and analyzing geophysical data pertaining to the world's oceans, in the interest of supporting naval operations conducted by the United States and its allies. NAVOCEANO operates eight vessels, which accommodate scientists conducting oceanographic research and surveys to support Department of Defense interests. NAVOCEANO also hosts one of the world's largest major "shared resource" computing centers, which enables scientists from NAVOCEANO and other government organizations to analyze large volumes of data and run complex NUMERICAL MODELS, as are often required in oceanographic studies.

The traditional primary mission of the Naval Oceanographic Office (NAVOCEANO) has been to conduct hydrographic, bathymetric, and magnetic surveys of the world's oceans. NAVOCEANO analyzes data collected by these surveys and delivers results to the National Geospatial Intelligence Agency or NGA (formally National Imagery and Mapping Agency or NIMA and prior to that was named Defense Mapping Agency or DMA) so that NGA may produce nautical charts (see CHARTS, NAUTICAL), used by naval vessel captains to ensure safe navigation. Products created at NAVOCEANO include digital bathymetric databases, current maps, ice forecasts, Special Annotated Imagery Littoral (SAIL) studies of the littoral battlefield, and Special Tactical Oceanographic Information Charts (STO-ICs). NAVOCEANO relies heavily on a small fleet of hydrographic and oceanographic survey ships and state-of-the-art sound navigation and ranging (SONAR) technology to collect most of its high-fidelity data.

Satellites and aircraft, and the sensors integrated into these platforms, have also become important means of collecting large amounts of data on the world's oceans very quickly and efficiently. However, many of NAVOCEANO's civilian and military employees take most pride in the fact that they have ventured into some of the world's most exotic ocean-covered places to conduct surveys long before any significant maritime traffic could even attempt to do so safely.

During the cold war (approximately 1945–91) and during the period of heightened tensions in the Middle East and the Arab world following the cold war, NAVOCEANO took on many additional responsibilities related to monitoring the world's oceans. For example, NAVOCEANO is responsible for collecting and analyzing oceanographic data such as ocean TEMPERATURE, SALINITY, and DENSITY, and any other geophysical parameter that affects sound propagation in the ocean. This work is to ensure that sonar systems function properly and to provide a catalogue of ocean sound sources so that enemy submarines may be detected and friendly submarines may be disguised. During the first Gulf War (Operations Desert Shield / Desert Storm, 1990–91) NAVOCEANO took on the complex task of monitoring and predicting ocean and near-shore currents in order to predict the drift of explosive nuisance mines. They used complex computer-based models of the ocean that required very detailed representations of parameters such as BATHYMETRY, ocean currents, depths, waves, and water densities. These models require extensive computing power in order to produce useful results in a reasonable amount of time. In fact, NAVOCEANO boasts one of the world's largest supercomputing facilities and the world's largest oceanographic library, the Matthew Fontaine Maury Oceanographic Library.

NAVOCEANO is responsible for communicating the information it gathers and analyzes to ocean modeling and prediction organizations and directly to naval personnel, ships' captains, Marine Corps units, and other organizations that may be affected by various ocean conditions. NAVOCEANO also tests and implements numerical techniques designed to assess and forecast ocean conditions. Such techniques range in focus from SEA ICE prediction to identifying ocean conditions indicative of EL NIÑO.

In summary, NAVOCEANO collects, analyzes, and archives geophysical data and runs models pertaining to ocean conditions to ensure safe navigation for Department of Defense seagoing vessels and to enhance the at-sea performance of naval weapon and sensor systems. At NAVOCEANO, the environmental sciences provide a foundation for U.S. national security.

Further Reading
Naval Oceanographic office, Stennis Space Center. Available online. URL: https://oceanography.navy.mil. Accessed March 10, 2008.
Nichols, C. R., D. L. Porter, and R. G. Williams. *Recent Advances and Issues in Oceanography.* Westport, Conn.: Greenwood, 2003.

Naval Research Laboratory   The Naval Research Laboratory (NRL) was founded on July 2, 1923, and has kept pace with basic research issues surrounding the Department of Navy's operational environment of air, land, sea, and space. The Naval Research Laboratory receives direction from its parent organization, the OFFICE OF NAVAL RESEARCH. NRL is the corporate research laboratory for the U.S. Navy and Marine Corps and conducts a broad program of scientific research, technology, and advanced development. Research at NRL is diverse and ranges from studies of the deep-ocean TRENCHES to monitoring the Sun's behavior. Two Nobel laureates, Jerome Karle (1918–   ) and Herbert A. Hauptman (1917–   ) from NRL, received the Nobel Prize in chemistry in 1985 for devising direct methods employing X-ray diffraction analysis in the determination of crystal structures. Many research programs are described in the annual document entitled *The NRL Review,* which may be accessed from the NRL Web site.

Further Reading
Nobelprize.org. Available online. URL: http://nobelprize. org. Accessed November 27, 2006.
United States Naval Research Laboratory. Available online. URL: http://www.nrl.navy.mil. Accessed November 27, 2006.

navigation   Navigation is the art and science of determining position and direction on the Earth's surface. Mariners practice navigation to steer a course and control the position of a vessel in order to travel safely and accurately from one preset location to another, usually from the port of departure to the destination. Nautical charts, sailing directions, the *Nautical Almanac,* and tide tables are traditional products used for navigational purposes by vessels. Wise navigators compile all available background information and use geographical or celestial references over the surface of the Earth to determine the vessels' position, called a "fix." Predictions of the locations of the vessel at future times can be made from the navigator's plot of recent fixes on a nautical chart, along with present speed and heading (direc-

tion) of the vessel and estimates of offset by winds and CURRENTS. This type of navigation is called DEAD RECKONING. Methods of navigation used on today's commercial and military vessels include:

1. Dead reckoning involves advancing the vessel's position by extrapolating the time history of the vessel's speed, distance, and heading since the most recent fix. Navigators include information on water depths, currents, and, in coastal waters, navigation aids.

2. Piloting involves navigation in coastal waters, usually within sight of land, where frequent fixes can be taken by means of ranges and bearings to prominent land features, islands, navigation aids, and depths. A DEPTH FINDER or FATHOMETER is used to measure water depth. The fathometer contains a transceiver mounted under the vessel that transmits pulses of high-frequency sound, which reflects off the sea bottom and is received by the transceiver. Electronic circuitry in the fathometer measures the time between the transmission and reception of each pulse, from which distance to the bottom is automatically calculated, and displayed on a chart, digitally, or on a screen.

3. Celestial navigation involves using the positions of the Sun, Moon, stars, and planets to obtain lines of position, using spherical trigonometry, tables, and almanacs. A sextant is used to measure the angle of celestial objects above the horizon; an accurate clock called a ship's chronometer is used to read the hours, minutes, and seconds from the observation; and the *Nautical Almanac* allows calculation of the exact location of the Sun, Moon, planets, or stars. Today, celestial navigation is mainly used as a backup to satellite and electronic navigation.

4. Radio navigation involves finding position by means of radio direction finding and hyperbolic navigation systems, such as loran or (*lo*ng-range *ra*dio *n*avigation). Loran involves establishing position based on the time displacement between signals from two or more fixed shore based antennas.

5. Satellite navigation involves determining position by means of signals transmitted by instruments placed in orbit on artificial satellites. The GLOBAL POSITIONING SYSTEM (GPS) is a much more accurate radionavigation system then Loran. GPS fixes requires receiving signals from at least three satellites from the constellation of 24 satellites.

A Brief History and Current and Future Programs at the Johns Hopkins University Applied Physics Laboratory

The Applied Physics Laboratory (APL) is the largest division of the Johns Hopkins University. It is a not-for-profit research laboratory dedicated to solving complex critical U.S. challenges, located halfway between Washington, D.C., and Baltimore, Maryland. There are more than 3,300 employees, two-thirds of the staff being engineers or scientists. APL was founded in 1942 to address problems U.S. forces were facing in World War II. Today APL's 130 specialized research and test facilities work on problems ranging from sending probes to Mercury and Pluto to enhancing a naval battle group's war-fighting capability. APL's funding level is about $500 million each year. In addition, APL houses part of the Whiting School of Engineering's part-time program. It is one of the largest part-time programs in the world. A large majority of the faculty is made up of the researchers and engineers at APL. They transfer their hands-on practical knowledge of engineering, computer science, biomedicine, and so forth, to their students, which number around 3,500.

The following documents some of the unclassified history of APL, delineates some of the laboratories present programs, and describes the direction that APL will lead in the future.

HISTORY
World War II
It was 1944 that the U.S. naval fleet in the Pacific was being faced with a new threat that was almost undefendable—suicidal air plane pilots who would fly their bomb-laden aircraft into naval ships. These pilots were the equivalent in accuracy to today's smart bombs. There was little way to protect the fleet from this threat. APL joining with industry and the government developed the proximity fuse. This was a fuse that went into antiaircraft shells that had an embedded radar that would cause the warhead to exploded in the proximity of aircraft. The accuracy of these shells brought to the fleet an unprecedented protection. So much so, that along with the atom bomb and radar, the proximity fuse was judged to be one of the three most valuable technology developments of World War II.

Navigation
On October 4, 1957, the former Soviet Union rocked the world with the launch of the world's first artificial satellite. Sputnik broadcast a signal as it circled the globe. Scientists from the APL brought some antennas out on the front lawn of the laboratory to listen to the signal that the satellite was broadcasting. They heard the unmistak-

able Doppler shift of the satellite as it flew overhead. They realized that they knew they could locate the satellite by listening for the maximum of the Doppler shift and marking the exact time. Doppler shift is the apparent change in frequency that is observed when an object moves toward or away from an observer. This is true whether the signal is the horn of a train, light of a star, or the electromagnetic signal from a satellite. As the source approaches the observer the frequency increases as a function of the speed of the source. Christian Johann Doppler (1803–53) first noticed this phenomenon, which Edwin Hubble (1898–1953) used to explain the red shift apparent in the stars that were moving throughout the universe.

Since they knew their location, they could calculate the location of the satellite. Then they figured they could do the reverse. If they knew the orbital location of the satellite and measured the Doppler shift from the ground, they could compute the location on the ground. This led to the design, development, testing, launch, and operation of the Transit Satellite System. This revolutionized navigation on the sea and in the air. It was the precursor to what is today the Global Positioning System. APL scientists and engineers were responsible for the Transit Program.

PRESENT PROGRAMS
Space
The APL also has a presence in space. More than 50 satellites have been built at APL. These include the old Transit Satellites, which were the precursors to today's GPS system, and the spectacular NEAR mission, which visited and landed on the asteroid Eros in 2001. Missions are presently being developed to visit the planets Mercury, Mars, and Pluto and to put twin satellites in orbit around the Sun to take three-dimensional images of it. APL continues to push the limits of science in space by developing techniques to search for water on the Moon and measure the ice of Europa.

Weapons
APL is developing future hypersonic weapons. Supersonic flight refers to flight that travels between 1.2 to five times the speed of sound. Hypersonic refers to flight that travels at speeds greater than five times the speed of sound, about 3,300 miles per hour. Hypersonic weapons are needed to strike heavily defended or deeply buried targets with a very short time between launch and arrival. To develop weapons that can travel this fast requires exotic fuels that are inherently unstable and therefore very unsafe for storage aboard ships. APL is developing a dualcombustor ramjet (DCR) engine,

(continues)

A Brief History, and Current and Future Programs at the Johns Hopkins University Applied Physics Laboratory
(continued)

which uses a fuel that is safe to handle on board ships and planes and still meets the desired performance. Jet aircraft use thrust from jet engines for propulsion. Turbojets, which are found on commercial aircraft, use a compressor to bring the fuel and air into the high-pressure combustor. Ramjet engines also have a combustion chamber, but it is the air entering the jet that is "rammed" into the combustion chamber that produces the high pressure needed for combustion. Ramjets only work once they are moving as it is the forward motion of the engine that rams in the air. APL has demonstrated a high-efficient combustion using cold liquid jet fuel—a much safer fuel. Tests have demonstrated speeds of up to Mach 6, that is more than 4,000 miles per hours. APL, the Defense Advanced Research Projects Agency, and the Office of Naval Research are working on developing hypersonic flight for these new strike missiles.

Fleet

APL has developed the Cooperative Engagement Capability for the U.S. Navy. This is an operational system of systems. It allows ships in a battle fleet to share their sensor data and weapons. Thus, instead of a ship acting autonomously, it can tie into the radars of the entire fleet. This enables them to have the "big picture," and to fire weapons that are on another ship. This helps the CEC to counter the cruise missile threat when they are the maximum distance from the fleet. It is a true example of the whole being greater than the sum of its parts. Benefits include the use of another ship's radar should the one with the weapons lose the use of their radar. This capability is being presented to other NATO nations with the idea that entire battle fleets from many countries could work as one unit in meeting the air defense needs of the fleet. APL is transferring this system to the U.S. Navy's contractor for production. APL is also the technical development agent for the U.S. Navy's Stan-

dard Missile program. This is an air defense missile used to intercept and destroy missiles fired at surface ships.

Land Mines

Hidden land mines are an international threat to all of humanity. These kill and cripple thousands of innocent adults and children every year. APL is developing a low-cost mine detector system that can be carried in a backpack to a site, set up, and then used autonomously or with remote control to detect land mines. A prototype of this system is now being built.

Homeland Security

APL has responded to the new requirements of a nation under attack. The tragic events of September 11, 2001, led the staff to expedite many projects in the area of biological/chemical detection and surveillance, mail inspection, system security, and decontamination. Systems for monitoring over-the-counter drug use, doctor's visits, and emergency room patients have led to models for alerting authorities to biological and chemical attacks. APL has led local authorities in mock attacks, which resulted in refining how they respond to attacks and other emergencies.

FUTURE

The effects on society by the research scientists of APL will be seen in the newly designed currency, in digital images imprinted with signature keys to preserve the integrity of those images, and in testing and developing better passenger restraints for automobiles. These are just a few of the thousands of projects ongoing at APL. All are designed to serve the nation's interests.

Further Reading
John Hopkins Applied Physics Laboratory. Available online. URL: http://www.jhuapl.edu/. Accessed July 29, 2007.

—**David L. Porter, Ph.D.**
Applied Physics Laboratory
The Johns Hopkins University
Laurel, Maryland

GPS has become a key system for the measurement of time, frequency, and position.

A navigator must be skilled at analysis and interpretation of information from many sources. Modern navigators are often aided by electronic charting and other navigation systems that provide electronic chart displays of all available position information, as well as hazards to navigation, weather, and oceanographic information. A navigator should never rely

completely on electronic and satellite systems, which may fail. Recreational and commercial navigators must always be prepared to fall back on the use of paper charts, magnetic compasses, celestial navigation, and piloting.

Large ship navigation can be subdivided into the inland waterway, harbor approach, coastal, and open ocean phases. For the inland waterway, harbor approach, and coastal phases of navigation, the methods of piloting are used to move arriving and departing vessels safely to and from their ports of destination. Piloting involves obtaining the pertinent nautical charts and required publications, such as NOTICE TO MARINERS, *Coast Pilots, Sailing Directions, Tide and Tidal Current Tables.*

The nautical chart (*see* CHART, NAUTICAL) is the center of the navigator's work station. The ship's position, course made good, and estimated future positions are all plotted on this chart. Nautical charts provide depth information, locations of navigation aids such as lighthouses and BUOYS. The navigator takes bearings and ranges on prominent landmarks and navigation aids by means of sextants (*see* SEXTANT) when visibility is good, or by radar or radio direction finder when visibility is poor. Positions are also determined by means of satellite-based Global Positioning System (GPS). The navigator studies the tide and current tables to determine the set and drift of the vessel, weather forecasts to allow for the effects of tides, currents, winds and waves, as well as storm conditions, small craft warnings and the like. Large vessels are often required by law to take a pilot on board to conduct the vessel's passage to or from the harbor, or through difficult or constricted passages. Many vessel masters choose to take a pilot on board even if he or she is not required to do so by law. Pilots have detailed knowledge of local hazards to navigation such as sunken wrecks, inlets, shoal areas, and adverse currents and weather. They can safely conduct the vessel through harbor traffic lanes. The volume of marine traffic in a busy port often makes safe navigation difficult. Very large vessels may require DOCKING PILOTS to safely maneuver the vessel into the pier or terminal and secure it properly to the dock. The open ocean phase involves collision avoidance, avoiding shoal, optimal ship tracking, avoiding storms, and any other hazard or situation that would prevent or delay the ship from scheduled arrival at its destination port.

The nautical chart is the focal point of the navigator's activity. (NOAA/NOS)

During conditions of bad weather and poor visibility, radio direction finding proves its value. Vessels are able to take ranges and bearings to radio beacons, aeronautical radio beacons, and even some radio stations. Radio beacons identify themselves by a series of dots and dashes. Radio beacons generally operate continuously under all weather conditions. Loran-C is a radio navigation system for use in open coastal waters and the open ocean. Loran-C is called a hyperbolic system because its receivers measure the time difference between signals from pairs of transmitting stations. The time difference corresponds to a difference in distance (distance = velocity × time). The locus of points having the same time difference from a station pair forms a hyperbolic line of position or LOP. The intersection of two or more LOPs gives the vessel's location. Loran-C systems are accurate to about a quarter of a nautical mile (1.9 km).

Radar navigation depends on taking ranges and bearings to landmarks as well as radar navigation beacons. Radar beacons emit signals that are triggered by a ship's radar. The signal giving the range and bearing information to the transmitter is displayed on the planned position indicator (PPI) on the vessel's bridge. Ranges and bearings to two or more known points fix the position of the ship. The methods used for plotting radar information are very similar to those used for visual piloting.

The most popular and accurate navigation system available today is the Global Positioning System (GPS), consisting of 25 Earth satellites. GPS measures the distances between satellites in orbit and a receiver on or above the Earth and computes a sphere of position from these distances (the position of the ship or aircraft is somewhere on the sphere of position). In general, four satellites are used to provide the intersection of four spheres of position, thus fixing the vessel's position to an accuracy adequate for most commercial vessels. Military and research vessels and aircraft that require greater accuracy can use differential GPS (*see* DIFFERENTIAL GLOBAL POSITIONING SYSTEM (DGPS) that relies on difference in distance computations to provide greater accuracy. The positional error of a DGPS position is three to 10 feet (1–3 m). GPS provides vertical as well as horizontal position information. The U.S. Coast Guard's DGPS system provides service for coastal coverage of the continental United States, the Great Lakes, Puerto Rico, portions of Alaska and Hawaii, and a greater part of the Mississippi River basin.

With the success of electronic navigation systems, especially GPS, many recreational and small commercial boat operators believe that it is no longer necessary to learn and practice celestial navigation, which for many years was the primary system of position location on the open sea, out of sight of land. This assumption, however, is incorrect. In the event of the loss of the GPS and other electronic receivers overboard, it is essential to be able to navigate by the position of celestial bodies. Electronic receivers can also be broken during heavy weather, or experience technical problems, which degrade accuracy. Atmospheric conditions such as severe thunderstorms can make the acquisition of electronic navigation signals difficult. Finally, it has always been a maritime tradition that navigators use all available position information along with their analysis and professional judgment to ascertain the vessel's location with a minimum of error. Even airline navigators check their positions periodically with star sights.

Celestial navigation or navigational astronomy is concerned with locating one's position using the celestial coordinates of the Sun, Moon, stars, and planets; their apparent motions; and time, measured by a precision ship's chronometer. Celestial coordinates refer to the positions of these bodies on the celestial sphere, an imaginary sphere concentric with the Earth, on which all celestial bodies are imagined to be projected. Ancient astronomers pictured the heavenly bodies as being located on the surface of this sphere. The celestial sphere appears to rotate because the Earth is rotating with it. Celestial coordinates of the stars are fixed on the sphere, unlike their altitudes and azimuths that change with time.

The navigator can determine how far he or she is from the projection of the star's position on the surface of the Earth by using a SEXTANT to observe the star's bearing, or azimuth, and its angle above the horizon, or altitude. These measurements, combined with accurate time are used to enter tables, such as the *Nautical Almanac* to provide a line of position (LOP). Two additional observations give circles of position. The point of intersection of these circles is the fix. Navigators take additional star observations whenever possible to confirm the accuracy of the fix.

Having some or all of the navigation information described above, the navigator enters it on the traditional paper nautical chart or on an electronic chart to display the ship's current position and cruise trackline through the water, called the course made good. On the bridge of today's commercial and military vessels, the information from electronic navigation systems can be fed directly into a computer; visual sightings are entered by hand. The computer then displays the ship's position, trackline, water depth, future course, hazards to navigation, tides, weather, and the like on the electronic chart. An electronic chart consists of four compo-

nents: (1) real-time positioning system, usually GPS, (2) computer hardware, (3) electronic chart data, and (4) software that displays and manipulates real-time sensor input and chart data. The electronic chart system, known as ECDIS (Electronic Chart Display and Information System), has freed the navigator from computational drudgery so that his or her attention can be given to optimizing the ship's route and assuring safe passage by avoidance of shoals, obstacles, and bad weather. Although there are several types of electronic charts, ECDIS is the only internationally standardized system.

Further Reading

Bowditch, Nathaniel. *The American Practical Navigator: An Epitome of Navigation, 2002 Bicentennial Edition.* Publication No. 9, National Imagery and Mapping Agency, 2002.

Gurney, A. *Compass: A Story of Exploration and Innovation.* New York: W. W. Norton, 2004.

Maloney, Elbert S. *Chapman: Piloting, Seamanship and Boat Handling,* 64th ed. New York: Sterling, 2003.

Perugini, Nick. "Behind the Accuracy of Electronic Charts." *Sea Technology* 42, no. 3 (2001): 33–37.

navigation aids   Navigation aids are visible, audible, and electronic symbols, equipment, and devices such as lights, horns, bells, markers, BUOYS, and beacons placed along coasts and navigable waters as guides to mark safe water. These markers are also called "aids to navigation," which is abbreviated as ATON by the U.S. Coast Guard. Navigation aids assist mariners in determining their position in relation to land and marine hazards and are established by government and private authorities for piloting purposes. On board a vessel, navigation lights prevent

This navigation aid is typical of those deployed in the shallow coastal waters of the southeastern United States. (Note the solar power cells.) Some also serve as a habitat for sea birds, such as ospreys. *(NOAA)*

Maintaining navigation aids is a big job for the U.S. Coast Guard. (NOAA)

collisions by indicating location and the relative heading of one vessel as seen from another. Lighthouses and lightships of distinctive shape and color, have lights, sound signals, and radio beacons to mark shoals and channel entrances. Lightships are specially equipped vessels anchored at specific locations. Sound signals are used when vessels are in sight of each other and are meeting or crossing at specific distances. Radio beacons are transmitters capable of broadcasting a characteristic signal specifically to aid navigation at night or in other situations of reduced visibility. Audible signals sounded from lights, buoys, and markers to assist mariners during periods of low visibility are called fog signals. Markers or ranges show the navigator that he or she is maneuvering along the centerline of the channel when the markers are observed in-line, one directly behind the other. Floating objects anchored to the seafloor are called buoys. Aids to navigation have distinctive shapes and colors indicating their purpose and procedures for navigation. Navigation aids are the road signs and traffic lights for navigable waters. These aids to navigation assist mariners in following channels, avoiding dangers such as shoals, and piloting safely until making safe landfall.

See also CHART, NAUTICAL; COMPASS; DEPTH FINDER; DIGITAL NAUTICAL CHART; GLOBAL POSITIONING SYSTEM; NAVIGATION; NAVIGATION DEVICES.

Further Reading

United States Coast Guard. "Navigation Center." Available online. URL: http://www.navcen.uscg.gov. Accessed October 12, 2007.

———. *Navigation Rules (International-Inland),* COMDTINST 16672.2D, U.S. Department of Homeland Security. Washington, D.C.: U.S. Government Printing Office, 1999.

navigation devices   Navigation devices are instruments used to determine a vessel's position, velocity, and predicted course. Perhaps the most familiar navigation device is the SEXTANT, used in celestial navigation on the ocean to measure the angle of celestial objects above the horizon; an accurate clock called a ship's chronometer is used to read the hours, minutes, and seconds at the time of the observation. The *Nautical Almanac* then allows calculation of the exact locations of the Sun, Moon, planets, and stars, from which the vessel's position can be determined, using tables and spherical trigonometry. The sextant is also employed in piloting to take bearings and ranges on prominent landmarks and NAVIGATION AIDS when sailing in coastal waters whenever visibility permits. When this information is entered on a paper or electronic navigation chart, the vessel's position, course made good, and predicted course can be determined. Another familiar navigation aid, the DEPTH FINDER, also called a FATHOMETER, is used to measure water depth. The fathometer contains a transceiver mounted under the vessel that transmits pulses of high-frequency sound that reflect off the sea bottom and are received by the transceiver. Electronic circuitry in the fathometer measures the time between the transmission and reception of each pulse, from which distance to the bottom is automatically calculated and displayed on a chart or digitally on a screen. In addition to alerting the ship's captain of dangerous shoal (shallow) waters, the depth readings are used on the navigation chart to confirm the ship's position.

During conditions of bad weather and poor visibility, radio direction finding is used to determine the ship's position. Vessels are able to take ranges and bearings to radio beacons, aeronautical radio beacons, and even some radio stations. Radio beacons identify themselves by a series of dots and dashes. Radio beacons generally operate continuously under all weather conditions. Loran-C is a radio navigation system for use in open coastal waters and the open sea. It is called a hyperbolic systems because Loran-C receivers measure the time difference between pairs of transmitting stations. The time difference corresponds to a difference in distance (distance = velocity × time). The locus of points having the same time difference from a station pair forms a hyperbolic line of position, or LOP. The intersection of two or more LOPs gives the vessel's location. Loran-C systems are accurate to about a quarter of a nautical mile (1.9 km).

Radio navigation systems are being adapted to technological advances. Loran-C provides coverage for maritime navigation in U.S. coastal areas. It provides navigation, location, and timing services for both civil and military air, land, and marine users.

Loran-C is approved as an en route supplemental air navigation system for both Instrument Flight Rule (IFR) and Visual Flight Rule (VFR) operations. The Loran-C system serves the 48 continental states, their coastal areas, and parts of Alaska. Another radio navigation system, OMEGA, was terminated on September 30, 1997, after 26 years of continuous service. OMEGA measured position relative to transmitting station pairs by making phase comparisons of the signals from these pairs. OMEGA was a worldwide radio navigation system of medium accuracy that was suitable for general commerce. There were eight transmitters widely spaced throughout the world. It was a very low frequency (VLF) system, which operated between frequencies of 10 and 14 kHz.

RADAR navigation depends on taking ranges and bearings to landmarks as well as radar navigation beacons. Radar beacons emit signals that are triggered by a ship's radar. The signal giving the range and bearing information to the transmitter is displayed on the planned position indicator (PPI) on the vessel's bridge. Ranges and bearings to two or more known points fix the position of the ship. The methods used for plotting radar information are very similar to those used for visual piloting with a sextant.

The premier navigation system available today is the Global Positioning System (GPS), consisting of 25 Earth satellites (*see* GLOBAL POSITIONING SYSTEM). GPS measures the distances between satellites in orbit and a receiver on or above the Earth, and computes a sphere of position from these distances (the position of the ship or aircraft is somewhere on the sphere of position). In general, four satellites are used to provide the intersection of four spheres of position, thus fixing the vessel's position to accuracy adequate for most commercial vessels. Military and research vessels and aircraft that require greater accuracy can use differential GPS (*see* DIFFERENTIAL GLOBAL POSITIONING SYSTEM [DGPS]) that relies on difference in distance computations to provide greater accuracy. The positional error of a DGPS position is three to 10 feet (1–3 m). GPS provides vertical as well as horizontal position information. The U.S. Coast Guard's DGPS system provides service for coastal coverage of the continental United States, the Great Lakes, Puerto Rico, portions of Alaska and Hawaii, and a greater part of the Mississippi River basin.

The ship's navigator must analyze, interpret, and integrate all of this information to ensure safety of the vessel and crew. Modern navigators are often aided by electronic charting and other computer-based navigation systems that provide electronic chart displays of all available position information, as well as hazards to navigation, weather, and ocean-ographic information, such as TIDES, CURRENTS, WAVES, shoaling areas, and the like. The U.S. Coast Guard provides many systems, devices, and personnel to operate and maintain Aids To Navigation or ATON (*see* NAVIGATION AIDS). The U.S. Coast Guard uses specialized BUOY TENDERS and other platforms such as a BARGE to tend aids to navigation buoys.

On the bridge of today's commercial and military vessels, the information from electronic navigation systems is input directly into a computer; visual sightings are entered by hand. The computer then displays the ship's position, trackline, water depth, future course, hazards to navigation, tides weather, and the like on the electronic chart. The electronic chart system, such as ECDIS (Electronic Chart Display and Information System; the only internationally standardized system) consists of four components: (1) real-time positioning system, usually GPS, (2) computer hardware, (3) electronic chart data, and (4) software that displays and manipulates real-time sensor input and chart data. Numerous manual backup systems (paper charts, compass, barometer, sextant, binoculars, and horns) are used in the case of power outages or the failure of automated systems. Modern sailors are familiar with age-old navigation technologies as well as highly accurate state-of-the-art systems dependent on satellite technologies.

Specialized navigation devices are used for the subsurface navigation of submarines, AUTONOMOUS UNDERWATER VEHICLES (AUVs), and divers involved in activities such as oil drilling, salvage operations, marine archaeology (*see* MARINE ARCHAEOLOGIST), and the deployment and retrieval of OCEANOGRAPHIC INSTRUMENTS. Acoustic beacons are often deployed at the beginning of a project and used as reference points for relative position triangulation for the duration of the project. Absolute positions are determined from the location of an acoustic beacon with respect to a known position determined by GPS navigation. Underwater microphones, also known as hydrophones, receive signals from the beacons. Distances and directions can be determined relative to the beacons from the time delays of the acoustic pulses from each beacon; each individual beacons is identified by a signature signal, or sequence of dots and dashes. The methods used are similar to those in radio navigation.

Further Reading

Bowditch, Nathaniel. *The American Practical Navigator: An Epitome of Navigation, 2002 Bicentennial Edition.* Publication No. 9. National Imagery and Mapping Agency, 2002.
Gurney, A. *Compass: A Story of Exploration and Innovation,* New York: W. W. Norton, 2004.

Perugini, Nick. "Behind the Accuracy of Electronic Charts." *Sea Technology* 42, no. 3 (2001): 33–37.

U.S. Coast Guard. "Navigation Center." Available online. URL: http://www.navcen.uscg.gov/. Accessed October 12, 2007.

nearshore currents Waves approaching the shore, TIDES, and processes such as shoaling cause CURRENTS in the nearshore region known as nearshore currents. The nearshore extends seaward from the low tide line past the point where waves are breaking. Shoaling refers to the natural reduction in WAVELENGTH and concomitant increase in wave height as waves propagate shoreward. The incident waves become unstable and break as surf along the BEACH. The manner in which the waves break or release energy is dependent on factors such as beach gradients and winds. These breaking waves are classified as spilling, plunging, collapsing, and surging surf. Spilling breakers gradually dissipate energy over a long distance, while plunging breakers dissipate energy over a shorter distance. Collapsing describes plunging breakers that have been arrested by the wind or are in transition from plunging to surging. Surging breakers predominate on steep slopes. The waves are associated with a shoreward mass movement of water. The ensuing increase in water level is called wave setup. Waves striking the shore at an oblique angle

result in water flowing parallel to the shore, which is called the longshore current. Longshore currents can converge and form the neck of a rip current. As waves pile on the beach, waters move offshore from the beach through the surf zone as a rip current. These combined processes account for the nearshore currents. These nearshore currents are important in carrying materials along the beach as longshore drift and flushing the water outside the breaker zone through rip currents.

See also LONGSHORE CURRENTS; RIP CURRENTS.

Further Reading

University of Wisconsin Sea Grant Institute. "Coastal Natural Hazards." Available online. URL: http://www. seagrant.wisc.edu/coastalhazards. Accessed October 12, 2007.

Wright, J., A. Colling, and D. Park. *Waves, Tides, and Shallow-Water Processes*, 2nd ed. Oxford: Butterworth-Heinemann, 2000.

neritic zone Ocean waters above the CONTINENTAL SHELF are known as the neritic zone. They extend from the low tide line to the shelf break. The shelf slopes gently seaward from the shoreline to the shelf break, where the depth is approximately 71 fathoms (130 m). The neritic zone can also be defined as the shallow PELAGIC zone that includes regions

Selected Neritic Finfish from the International Union for Conservation of Nature and Natural Resources Red List

Common Name	Scientific Name	Comments
albacore tuna	*Thunnus alalunga*	Commercial pelagic highly migratory species. This species is used for canned white tuna and is a popular, widespread commercial fish.
Baltic sturgeon	*Acipenser sturio*	Demersal ANADROMOUS species. Besides the Baltic Sea, it was widespread during the last century in the Black, Northern, and Mediterranean Seas and the northern part of Atlantic Ocean.
black grouper, Warsaw grouper	*Epinephelus nigritus*	Demersal species of reef fish usually found in very deep rocky bottoms that grows to very large proportions over its long lifespan.
black sea bass, giant sea bass	*Stereolepis gigas*	Demersal species of sea bass usually found in very deep water that grows to very large proportions over its long lifespan.
blue skate	*Dipturus batis*	Demersal species tolerating a wide range in depth and temperatures. Caught as bycatch in bottom trawls.
bocaccio rockfish	*Sebastes paucispinus*	Demersal reef fish that feed on a variety of fishes, crabs, and squids.
calico grouper, speckled hind	*Epinephelus drummondhayi*	Demersal reef fish species inhabiting offshore rocky bottoms.
Chinese bahaba, giant yellow croaker, yellow-lipped croaker	*Bahaba taipingensis*	Demersal species found in East and South China Sea coastal waters. The swim bladder of this species is valued for its medicinal properties and as a general tonic for health.

(continues)

Selected Neritic Finfish from the International Union for Conservation of Nature and Natural Resources Red List (continued)

Common Name	Scientific Name	Comments
coelacanth	*Latimeria chalumnae*	A large lobed-finned demersal species that inhabits the steep slopes of volcanic islands in the Indian Ocean. Considered a living fossil and was thought to have died out with the dinosaurs.
grey nurse shark	*Carcharias taurus*	Inhabits shallow coastal waters from the surf zone down to 197 feet (60 m). Population declines along the east coast of Australia.
jewfish, goliath grouper	*Epinephelus itajara*	Found in shallow, inshore waters over rock, coral, and mud bottoms to depths of 151 feet (46 m) and feeding on crustaceans such as spiny lobsters, shrimps, and crabs.
knifetooth sawfish	*Anoxypristis cuspidata*	A benthic species found in the estuarine and coastal reaches of the Indian and Pacific Oceans. These fish are easily captured in net gear fished in shallow sandy or muddy bottom areas.
largetooth sawfish	*Pristis perotteti*	A demersal species found in estuaries, lagoons, and shallow coastal waters less than 33 feet (10 m) deep. They feed on benthic invertebrates that are stirred up from the substrate by the saw. Sawfish are especially vulnerable to being captured in fishing nets.
Maltese ray	*Leucoraja melitensis*	A demersal species inhabiting the Mediterranean Sea.
northern bluefin tuna	*Thunnus thynnus*	A large pelagic and highly migratory species preferring subtropical and temperate waters that occasionally comes close to shore. This species is one of the largest of the tunas and is being used in farming industries in the Mediterranean, North America, and Japan.
porbeagle	*Lamna nasus*	A pelagic shark found in temperate and cold waters on the continental shelf as well as in deep ocean basins worldwide. Considered a big gamefish.
smalltooth sawfish	*Pristis pectinata*	Demersal species inhabiting tropical estuaries and found close to shore over muddy and sandy bottoms. They are easily entangled in net gear.
southern bluefin tuna	*Thunnus maccoyii*	A large pelagic and highly migratory species preferring temperate waters. This species is one of the largest of the tunas. In Australia, fishermen keep some in MARICULTURE pens or tuna farms.
spotted handfish	*Brachionichthys hirsutus*	A demersal species found on coarse to fine sand and silt, in coastal waters of the continental shelf of southwest Pacific. It is being outcompeted by an invasive starfish, which preys on its egg clusters.
striped dogfish, striped smooth-hound	*Mustelus fasciatus*	A small schooling demersal shark that feeds on benthic invertebrates.
wreckfish	*Polyprion americanus*	A large, slow-growing, deep-water gamefish that is found near caves, shipwrecks, and other types of bottom structure. Large adult fish are often found beyond the neritc zone in regions rising above the abyssal plane such as the Charleston Bump.

Note: The Red List also includes MARINE MAMMALS, birds, reptiles, and crustaceans. Red List species are considered to be at risk of global extinction.

such as the surf zone, nearshore zone, PHOTIC ZONE, EXCLUSIVE ECONOMIC ZONE, and large marine ecosystems (*see* MARINE PROVINCES). Nutrients are delivered into the neritic zone by land runoff, outflow from rivers and ESTUARIES, and through resuspension from the bottom by WAVES and CURRENTS. There is usually sufficient sunlight for photosynthesis and high levels of primary productivity as evidenced by the amount of chlorophyll that is detected in satellite images from SeaWiFS, a satellite sensor launched in 1997. Chlorophyll, resident in floating plants called PHYTOPLANKTON, reflects green and yellow-green light. SeaWiFS stands for Sea-viewing Wide Field-of-view Sensor and measures the amount of reflected light of different colors from the sea surface. The neritic fishery includes both demersal and pelagic stocks. The fishery encompasses all the species caught in the neritic zone as well as the fishing gear and methods. Changes in species abundance and distribution are caused by variability in the oceanic environment, long-term global climate change due to both natural processes and human impacts, pollution from industrialization, damages to coastal habitat through construction of dams, and unsustainable fishing operations, such as overfishing and bycatch. The table that begins on page 369 lists selected neritic finfish from the International Union for Conservation of Nature and Natural Resources Red List.

See also CONTINENTAL SLOPE.

Further Reading

The IUCN Red List of Threatened Species™. Available online. URL: http://www.iucnredlist.org. Accessed July 23, 2006.

nets Nets are open mesh bags or devices made from various natural or synthetic materials to capture, separate, or hold marine plants and animals. Scientists working principally in the fields of biological oceanography (*see* BIOLOGICAL OCEANOGRAPHER), limnology, fisheries, AQUACULTURE, and environmental studies use nets and their associated hardware such as bridles, swivels, headropes and footropes, messengers, release mechanisms, otter doors, collecting buckets, and ground gear such as rollers. Various types of nets are deployed to capture specimens from specific locations such as murky salt marshes, turbulent COASTAL INLETS, wave-dominated nearshore regions, or deep offshore blue water. Net sizes, mesh sizes, and net shapes are very important to obtaining the correct samples and determining the correct number of specimens. Small specimens will swim, crawl, or drift through large mesh. Researchers must understand the impact of oceanographic conditions on their gear. For example, ocean CURRENTS can offset a trawl from the planned tow path.

Researchers working from shore, platforms such as docks, and from small watercraft may use beach seines, cast nets, dip nets, or fish traps to sample shallow-water organisms. Migration or life cycle studies may require the capture of commercially important finfish such as southern flounder (*Paralichthys lethostigma*) to include assessing the numbers, sizes, and catch locations for larvae, juvenile, and adult forms. A seine is a long rectangular net used close to shore or on tidal flats that have a favorable snag-free and firm seabed. The end of the net is

Nets (mesh fabrics used to capture marine specimens) being hoisted aloft for maintenance *(NOAA/NMFS)*

tied to the upper and lower extents of a pole. Wading researchers grasp the poles and move the net vertically through the water. The upper edge of the seine may contain floats and the lower edge, which drags along the bottom and contains lead weights. Cast nets are circular monofilament or cotton mesh nets with lead weights strung around the perimeter that spread the net open when thrown with a spinning motion. They are predominately used by skilled FISHERMEN to catch baitfish such as South American pilchard (*Sardina sagax*), bay anchovy (*Anchoa mitchilli*), and ballyhoo (*Hemiramphus brasiliensis*), which feed on PLANKTON. Dipnets are bag-shaped and held open by a square or rounded frame on the end of a long pole. They are used to scoop organisms from the water, either from the bank of an ESTUARY or at sea from the deck of a floating platform or ship. Fisheries biologists may use hooplike fish traps called fyke nets in shallow estuaries to sample fish communities. Fyke nets are typically 40 feet (12 m) long; 3.28 feet (90 cm) wide by 2.5 feet (75 cm) high with a 0.75-inch (1.9 cm) diamond-shaped mesh nylon netting that is smaller than the intended catch. Wings or leads direct swimming fish into the trap. Segments of the net may contain funnel-shaped throats, which prevent fish from leaving as they enter a particular section. The net is anchored in place using posts. Managed fish such as Atlantic croaker (*Micropogonias undulatus*), bluefish (*Pomatomus saltatrix*), butterfish (*Peprilus triacanthus*), speckled trout (*Cynoscion nebulosus*), striped bass (*Morone saxatilis*), and weakfish (*Cynoscion regalis*) are funneled into the end of the fyke trap into a bag. The bag has a small opening at the end, which can be untied to release the fish.

Onboard a research vessel, standard collecting gear might include several types of plankton nets and bulky trawls. Plankton nets come in various shapes and sizes from simple dipnets to complicated towed net systems. They are designed to catch neuston, PHYTOPLANKTON, and ZOOPLANKTON. Neuston nets may be made of silk with the finest of mesh to catch neuston (plankton floating at the air-sea interface to include certain types of fish eggs and fish larvae). Phytoplankton, zooplankton and some free-swimming organisms are collected using ring nets, closing plankton nets, bongo nets, and multinet midwater or pelagic trawls. Towed nets typically have a round or rectangular opening and look like a funnel that leads into a collection bucket at the end of the net. When two ring nets are attached to a single frame and towed together, they are called a bongo net. Trawls are actually towed nets, closed by a bag and extended at the opening by wings. Strong steel cables (wire rope, warp) that are payed out by a WINCH connect the net to the research vessel or trawler. Low-opening trawls are designed to roll or drag nets along or near the bottom to capture demersal and benthic species. High-opening trawls are designed to catch semi-demersal or pelagic species. Demersal species such as summer flounder or fluke (*Paralichthys dentatus*), Atlantic cod (*Gadus morhua*), whiting (*Merlangius merlangus*), and red grouper (*Epinephelus morio*) live near the seafloor. Pelagic species such as blue shark (*Prionace glauca*), northern anchovy (*Engraulis mordax*), Jack mackerel (*Trachurus symmetricus*), and California market squid (*Loligo opalescens*) are found living in the ocean above the bottom. A Tucker trawl is used to sample mid water regions with a multiple opening and closing net system. The mesh might be 0.01 inches (150 micrometers) to collect small zooplankton. Once the net is at the correct depth, a mechanical messenger is sent down the towline. A messenger is a metal weight that slips down the line to trigger the opening of the net. The first messenger is sent down the wire to open the first net. A second messenger intended to close the first net and open a second net is sent down the wire when the Tucker trawl is at the desired depth. Tucker trawls are commonly rigged with either two or four nets. Some scientists also fix OCEANOGRAPHIC INSTRUMENTS such as temperature and depth recorders to their trawls to confirm actual depths fished. Careful attention must be paid to finely meshed nets such as plankton nets, which may become clogged by ALGAE or trash and require high-pressure spraying of the net to clear the nylon or silk mesh. John Dove Isaacs, III (1913–80) and Lewis W. Kidd (1923–82) from SCRIPPS INSTITUTION OF OCEANOGRAPHY devised what is known today as the Isaacs-Kidd Mid-Water trawl or IKMT to collect zooplankton and freely swimming animals at high speed. The IKMT has been adapted for use by fisheries biologists in Pacific waters near Hawaii to catch and study the larvae of highly migratory species such as blue marlin (*Makaira mazara*), shortbill spearfish (*Tetrapturus angustirostris*), striped marlin (*Tetrapturus audax*), swordfish (*Xiphias gladius*), and wahoo (*Acanthocybium solanderi*). The mouth of this long net is held open by a frame with a V-shaped aluminum diving vane. A series of hoops of decreasing size maintain the round shape of the net. The large bag-shaped end of the net is called the codend and may be finely meshed to contain a variety of midwater zooplankton. *Codend* is the nautical term describing the back of the net where catch collects as the net drags through the water.

There are many other applications where mesh nets are used in coastal regions. A pound net is a long net strung like a fence on wooden piles or poles that have been driven into the seafloor toward a heart-shaped enclosure, which acts as a funnel. The longer leader net guides fish trap the fish in the crib

Net fishermen deploying a purse seine from a small boat (NOAA/NMFS)

Selected Net Gear Used by Marine Scientists and Commercial Fishermen

Equipment	Description
beach seine	Generally a long net used from shore to about 15 feet (4.6 m) to catch organisms living on a sandy bottom. The beach seine is set from and returned to the beach. It may be deployed by a small boat a short distance from shore and pulled in to the beach by several persons. This method is highly effective for sampling the nearshore and intertidal zones that typically support the highest diversity of fishes in coastal marine areas.
beam trawl	A beam made of wood or metal rigidly maintains the horizontal opening of the trawl net. The beam may be 39 feet (12 m) long. Examples include the sole beam trawl and English and the arch-shaped Irish trawls.
bongo net	Two conical mesh plankton nets with circular openings, fastened together side by side and towed behind a research vessel.
Boris Goshawk bottom trawl	The trawl design originated in the 1950s for the capture of groundfish, the general term referring to demersal species. (*See* bottom trawl.)
Boris Goshawk rockhopper trawl	Rockhopper trawls have large rubber wheels designed to roll over seafloor obstructions such as rocky reefs, corals, and other structures on the seafloor. Designed to catch groundfish. (*See* bottom trawl.)
bottom trawl	A trawl net that is dragged along the seafloor to catch fish such as flounder and haddock. Examples include Baca trawl, Bering Sea combination trawl, Aleutian cod trawl, Aleutian cod combination trawl, bobbin trawl, eastern otter trawl, Nor'eastern bottom trawl, otter trawl, roller trawl, Sand Point cod trawl, seguam 6-seam hard bottom trawl, selective flatfish trawl, shrimp trawl, Stuart trawl, and Yankee trawls.
cast net	A circular hand-thrown net designed and constructed to spread out and strike the water surface in a circular shape and manner. After it hits the water, weights drag it down and the attached drawstring pulls the net into a purse.
Clark-Bumpus plankton net	A net design originating in 1939 by George L. Clarke (1905–87) and Dean Franklin Bumpus (1912–2002) to retrieve samples of microscopic surface phyto and zooplankton.
Cobb (Stauffer) trawl	Net with a square mouth designed to catch juvenile fish, similar to the Marinovich and Methot trawls.
Danish seine	A seine or cone-shaped otter trawl that is hauled over an area of about .78 square miles (2 km^2) to a stationary vessel. The very long towing ropes disturb clouds of mud, which help herd the fish into the net.
dipnet	D-shaped, circular, or triangular shaped nets fixed to a long handle. Net frames and handles may be made of lightweight and strong materials such as extruded aluminum. Also landing nets.
driftnet	Gillnets that are not anchored and drift freely with the current, separately or, more often with the boat to which they are attached. They are deployed across the path of migrating fish schools. Marine animals strike the net and become entangled in its meshes.
fish traps	A system of nets that are hung on poles. Fish swim naturally along the shore or channel and hit the lead, which directs the fish through into an enclosure (heart) and through a tunnel into the pot. Examples include Alaskan trapnet, creels, eel pots, Fyke nets, hoop nets, minnow traps, Pennsylvania trapnet, pots, pound nets, and semi-oval fish traps.
gillnet	Meshed fishing gear designed to catch the fish by the gill cover, entanglement, or wedging and thereby preventing escape. There are many types of gillnets such as sink gillnets, anchored gillnets, and driftnets. Most are hung with corks on the topside and have a lead line installed along the bottom. The ends of the gillnet are marked with a buoy.
hoop net	A fish trap consisting of a series of staked interconnected conical pockets narrowing at equal distances to a point. It may use leads or wings to channel fish into the hoops.
Isaacs-Kidd Mid-Water trawl	A small trawl net designed to capture a wide range of pelagic species, although it is generally most effective for larval and juvenile fishes or small forage species.
lift nets	Netting panels or mesh bags with the opening facing upward, which are submerged, left for the time necessary for light or bait to attract fish over the opening, then lifted out of the water.

(continues)

Selected Net Gear Used by Marine Scientists and Commercial Fishermen *(continued)*

Equipment	Description
manta net tow	A modified neuston net with a rectangular mouth attached to a frame that supports square aquaplanes. These extensions stabilize the net when it is towed and keep the top of the net at the sea surface. The towing bridle may be rigged so that one side is longer than the other in order to force the mouth away from the ship at a slight angle.
metro net	Fine meshed nets attached to fixed square frames with bridle attachments in the middle of the side-pieces of the frame. They are designed to capture zooplankton and micronekton.
mid-water trawl	Trawl nets towed above the bottom. Examples include Aleutian wing trawls, Bering Sea pelagic trawls, high-speed mid-water rope trawl, IC whitefish mid-water trawl, Marinovich trawl, Methot trawl, Patagonian semi-pelagic trawls, radial trawls, Stauffer trawls, and wonder trawls.
Multiple Opening/ Closing Net and Environmental Sampling System or MOCNESS	A system of nine nets that can be opened and closed on command to capture plankton at specific depths. It may also carry sensors to measure conductivity, temperature, pressure, fluorescence, optical transmission, dissolved oxygen, and light levels while being towed.
neuston net	Rectangular nets designed to collect surface plankton samples. Floats may be attached to the top of the frame. Frames may be rigged together to make a double neuston net. The net is towed through the water so that half the net is in the water and half out.
otter trawl	A large net with floats strung along the headrope and chains dangling from the footrope. The net is held open by the force of water moving against two massive otter doors or boards. The chains drag along the bottom and cause fish to swim up into the net.
Paranzella net	A seine-like net towed between two vessels.
plankton net	Various sized (from relatively small hand nets to nets with a mouth more than 10 feet (3.05 m) across). The size of the mesh (apertures) varies and is used to describe the net. Nets are hauled in vertical and horizontal directions. Example nets include the bongo net, conical ring net (NORPAC net), Juday net, Nansen net, plummet net, WP2 net, and WP3 net.
pull net	Any net where fish are caught by horizontal dragging. Examples include seine, trawl, and scoop nets.
purse seine	Nets with corks on the top and a lead line along the bottom. They are closed at the bottom by means of a free-running line through rings located in the body of the net above the lead line. Examples include herring seine and round haul seine.
pushnet	A device with handles, yoke, runners, net frame, and bumper bar used to sample juvenile fish and shrimp in estuaries. It may be bow-mounted to small workboats.
ring net	A simple net consisting of a fine-meshed bag attached at its mouth, or opening, to a metallic ring. The net itself is terminated in a bottle or jar where unfiltered plankton and other particulate matter collect.
seine	A rectangular net used to encircle fishes, usually weighted at the bottom and with floats at the top. For example, beach seine.
set net	Gillnets that are anchored in one place.
skimmer trawl	Used in shallow water where there is a large amount of floating debris.
surface trawls	Trawls designed to sample juvenile finfish in their offshore epipelagic habitat. Examples include the Kodiak pair trawl and the Nordic 264 rope trawl.
tangle net	Very small meshed net that catches fish by teeth rather than gills.
Thomas drop-net system	A helicopter deployed drop net, which is lowered around a selected school of targeted species, pursed, and left floating.
trammel net	Gillnet made with two or more walls joined to a common float line. Often used at or near the bottom.
try net	A trawl where bridles are typically joined at a shackle that is attached to a single towing warp. The warp is normally strung over a davit or block mounted centrally on the stern of the vessel. This arrangement allows for vessels other than trawlers to successfully fish the net.

Equipment	Description
Tucker trawl	A mechanical open and close net with a fixed frame and weighted bottom bar that is designed to catch plankton.
turtle nets	Dipnets for the safe handling of sea turtles have handles that are at least six feet (1.82 m) long and can support a minimum of 100 pounds (34.1 kg) without breaking or significant bending or distortion. The net hoop has an inside diameter of at least 31 inches (78.74 cm) and a bag depth of at least 38 inches (96.52 cm). The mesh opening may be no more than three inches × three inches (7.62 cm × 7.62 cm). Estuarine turtles are often captured using hoop nets.
umbrella net	The umbrella net is most effectively operated from sites affording an elevated vantage point. The extended net is simply sunk to the bottom and then lifted as rapidly as possible when the fish swim over it. (*See* lift nets.)
universal trawl	A trawl that can be used for mid-water and bottom trawling.

Note: Some terms such as *bottom trawl* are general and others such as *fyke net* describe specific designs.

or holding pen. Fishermen come alongside the pound with a small power boat, pull up the net and dip fish such as striped bass, bluefish, catfish (*Ictalurus punctatus and Ameiurus catus*), croaker, flounder, white perch (*Morone americana*), spot (*Leiostomus xanthurus*), weakfish, and river herring (*Alosa pseudoharengus* and *Alosa aestivalis*) into a larger fishing boat. In estuaries such as the lower Chesapeake Bay and in the coastal ocean, purse seines are used to catch species such as menhaden (*Brevoortia tyrannus*). Two boats circle a long net around a large school of menhaden and then draw the weighted net closed at the bottom. Hydraulic winches are used to haul the catch into a larger vessel that can process the catch. Gillnets are used by fisheries biologists and fishermen to catch fish by their gill covers. Swimming fish encounter the net and may become snared while retreating. Depending on the mesh size, gillnets can catch larger fish while allowing smaller fish to pass through. Long gillnets may be stored on a reel on the stern of a boat. The reel helps the watermen unwind the net into the water without tangling. Nets are also used to protect beachgoers from dangerous marine life. Along some ocean beaches such as in Australia, straight, rectangular nets are suspended in the water column between BUOYS to ward off sharks. Lines of floats fixed to the top and sinkers at the bottom keep the nets upright in the water. MARICULTURE applications involve the deployment of nets to hold farmed species in a confined area.

Marine scientists and concerned citizens are helping to establish guidelines for environmental responsibility and ecological awareness that are needed for a sustainable ocean. It is especially important that nets are employed using knowledge, techniques, and procedures that do not diminish the ability of wild species to survive. Trawling has become heavily regulated owing to overfishing and the physical damage done to the seafloor by scraping trawl doors and rollers. One recent scientific advance involves the application of numerical model output and satellite images to locate and then clean up ocean areas where flotsam and jetsam such as ghost nets concentrate. A ghost net is an abandoned net or one that has been lost during a storm. This net will continue fishing without anyone to retrieve its catch. Efforts are under way to reduce the numbers of abandoned fishing lines, traps, and nets, which are a leading cause of entanglement of marine wildlife. Another area that relates to stewardship involves developing innovative fishing gear that reduces the incidental catching of marine life along with the targeted species. Some researchers report that bycatch with gillnets is the single greatest threat to porpoise and dolphin populations. State-of-the-art nets may be outfitted with sound devices such as acoustic pingers that warn MARINE MAMMALS away from the net, guards that preclude entry for some species, and excluder devices allowing inadvertently captured species an escape path. Organizations such as the Marine Stewardship Council are working toward preserving fish stocks for the future. They reward environmentally responsible fisheries management and practices with their distinctive blue (or sometimes black) product label. The table starting on page 373 provides a listing and description of selected net gear used by marine scientists and commercial fisherman.

Further Reading

Isaacs, John D., and Lewis W. Kidd. "A Midwater Trawl." Scripps Institution of Oceanography. Ref. No. 51–57 (November 15, 1951).

National Oceanic and Atmospheric Administration. "Fisheries." Available online. URL: http://www.noaa.gov/fisheries.html. Accessed August 17, 2006.

Northeast Fisheries Science Center. "Fisheries Observer." Available online. URL: http://www.nefsc.noaa.gov/femad/fsb. Accessed October 12, 2007.

nitrogen cycle The continuous sequence of events by which nitrogen is converted into chemicals that can be used by plants and animals for sustaining life and building tissue is called the nitrogen cycle. The cycle is driven by uptake of nitrogen by living organisms from green plants such as PHYTOPLANKTON, conversion to tissue by means of photosynthesis, release back into the environment by excretion, or death and decay of the organism that fertilizes the growth of new green plants and begins the cycle again.

In the ocean, the greatest uptake of nitrogen occurs near the surface, in the PHOTIC ZONE, where photosynthesis is occurring. As organisms die and sink to the ocean floor, the nitrogen is removed from the near surface layers and transferred to the deeper ocean below the photic zone, and cannot be used again by the near-surface plants and animals until brought back into solution by decomposition and recirculated back to the surface. The vertical circulation of nitrogen from the deep layers to the surface is strong only in the Arctic and Antarctic regions of the ocean and in areas of strong UPWELLING, which occur for the most part close to shore along the west coasts of the Americas and Africa and in areas of divergent flow in the deep ocean, such as occur north of the equator in the eastern PACIFIC OCEAN (the so-called cold water islands).

Although nitrogen is the most abundant element in the atmosphere (occurring as N_2) and is extremely abundant in the lithosphere, where it occurs as chemical compounds in the rocks, it cannot be used in these forms by marine organisms. Nitrogen occurs in a number of different states in the ocean, including molecular nitrogen, nitrites (NO_2), nitrates (NO_3), and ammonia (NH_4). It is the latter three forms that are essential to the marine ECOSYSTEM. Nitrogen is a necessary component in the production of amino acids from which proteins are made. Plants take up nitrogen, which is then consumed by herbivorous animals that are consumed by carnivorous animals, and so on up the food chain.

The PHYTOPLANKTON are the most important nutrient-producing plants in the ocean. Although there are plant-eating fish, most of the oceanic herbivores are the ZOOPLANKTON. The higher trophic levels then build on successions of larger and larger carnivores. The expiration and subsequent decay of the plants and animals by microbacteria results in the production of inorganic nitrogen, which is used by the living plants and animals in an endless cycle of plant growth and photosynthesis, beginning with the grazing of the plants by herbivores, predation by carnivores, production of nitrogen by their fecal matter, and by the subsequent death and decomposition of plants and animals, leading to renewed growth.

Nitrogen compounds differ in their abundance in seawater. Nitrates are most abundant in the upper layers, ammonia is approximately evenly distributed with depth, and nitrite is most abundant in the thermocline region. The net vertical distribution of inorganic nitrogen shows relatively low values near the surface because of uptake by marine plants, mostly phytoplankton, increasing rapidly at the THERMOCLINE and below. Since the bulk of the marine plants live near the surface, this distribution may limit the primary productivity in the oceans. There is a seasonal variability in the nitrogen of the upper ocean (the photic zone). There are low values in the summer with strongly developed thermoclines and high uptake by marine plants, and high values in winter and spring, with strong wind-induced mixing and relatively low plant abundance due to low levels of incident sunlight. In regions of the ocean where there is strong upwelling, such as off the coasts of Peru and California, nitrogen is brought to the surface, enabling a continuing high level of productivity. Typical values of nitrogen concentration below the thermocline are about 35 microgram atoms per liter in the Pacific and INDIAN OCEANS, and 20 in the ATLANTIC OCEAN.

In riverine and estuarine waters, the natural nitrogen cycle has been altered by human activities, such as application of organic and chemical fertilizers to boost agricultural production, removing wetlands to build subdivisions, and discharging organic wastes (sewage) into the water. The excess nitrogen causes enormous blooms of phytoplankton, which eventually die and sink, cutting off light to the fish and benthic organisms below. Much of the dissolved oxygen in the water column is used to decompose the vegetation, and little is left for living animals, a condition called eutrophication.

The excess nitrogen also reacts with oxygen in the atmosphere, forming nitrous oxides, which remove ozone from the upper atmosphere, and add "greenhouse" gases to the lower atmosphere, adding to GLOBAL WARMING. As is the case with fossil fuels, massive addition of nitrogen compounds into the air and water by humans can result in catastrophic upset of the often delicate balances of nature.

See NUTRIENT CYCLES.

Further Reading

Armstrong, F. A. J. "Inorganic Nitrogen in Sea Water." In *Chemical Oceanography*, edited by J. P. Riley and G. Skirrow. London: Academic Press, 1965.

———. "Phosphorus." In *Chemical Oceanography*, edited by J. P. Riley and G. Skirrow. London: Academic Press, 1965.

Davis, Richard A., Jr. *Oceanography: An Introduction to the Marine Environment*. Dubuque, Iowa: Wm. C. Brown, 1991.

Flinders University of South Australia. "The Nutrient Cycle." Available online. URL: http://www.es.flinders.edu.au/~mattom/IntroOc/notes/intro2.html. Accessed October 12, 2007.

Smil, Vaclav. "Global Population and the Nitrogen Cycle." *Scientific American* 277, no. 1 (July 1997): 76–81.

NOMAD buoy The NOMAD buoy is a moored deep-sea weather BUOY of the NATIONAL OCEANIC AND ATMOSPHERIC ADMINISTRATION's (NOAA's) National Data Buoy Center (NDBC). The official name of this buoy is the "6-Meter NOMAD," referring to its length, about 20 feet (6 m). NDBC's moored buoys are deployed in coastal and deep waters off the coasts of the continental United States, around Hawaii, in the Bering Sea, and in the mid-Pacific. The acronym NOMAD stands for Navy Oceanographic and Meteorological Automatic Device. These buoys transmit regular coastal and offshore routine weather observations used in official weather forecasts. They are instrumented to measure such quantities as air and sea temperature, barometric pressure, surface wind speed and direction, surface humidity, wave height, and, in some cases, wave direction. ANEMOMETERS fixed to these buoys are usually 17 feet (5 m) above the sea surface. These buoys may also be equipped with THERMISTOR chains and CURRENT METERS.

The NOMAD has an aluminum boat-shaped hull that ensures survivability in heavy seas, while being relatively cost effective. The boat hull has a highly directional response to incoming waves, as well as a quick rotational response. There have been no known capsizings of NOMAD buoys. They are corrosion resistant and have minimal effects on magnetic COMPASSES used in instruments for buoy orientation, wind, current and wave direction. They are easy to transport by flatbed truck, rail, or aircraft.

NOMAD buoys have been in widespread use since the 1960s and are very important in helping to protect life and property on the oceans. As an example, a 6-meter NOMAD buoy is moored at station 41001, approximately 150 nautical miles east of Cape Hatteras, in a water depth of 14,525 feet (4,427 m). It is owned and operated by the National Data Buoy Center. It measures air temperature, air pressure, sea temperature 3.3 feet (1 m) below the sea surface, wind velocity, wave height and period. It reports hourly observations that are available in tabular form with some graphics in near-real time on the Internet. Recreational boaters and professional mariners may access these data in developing float plans or ship's tracks (*see* OPTIMAL SHIP TRACK PLANNING). They are essential starting points in planning search and rescue operations or responding to marine spills.

Further Reading

"National Data Buoy Center: Moored Buoy Program." Available online. URL: http://www.ndbc.noaa.gov/mooredbuoy.shtml. Accessed October 12, 2007.

North Atlantic Current (NAC) The North Atlantic Current also known as the North Atlantic Drift is generally conceded to be the continuation of the GULF STREAM, forming south of the Grand Banks and west of the Mid-Atlantic Ridge. The NAC is the northern branch of the North Atlantic GYRE: a clockwise circulation of currents driven by the prevailing winds. The currents include the Gulf Stream, the North Atlantic Current, the CANARY CURRENT, and the NORTH EQUATORIAL CURRENT. Water from slope currents along the CONTINENTAL SLOPE of North America adds volume to the current that can have a transport as large as 35 Sverdrups (35×10^6 m³/s) depending on location and season. As part of the Gulf Stream system, the North Atlantic Current transports large amounts of heat and salt from the tropical Americas to the shores of northern Europe, making the northwestern part of the continent much warmer than it otherwise would be.

Some water from the LABRADOR CURRENT is believed to recirculate into the North Atlantic Current in the Newfoundland Basin, an area where current meanders and EDDIES frequently occur, especially to the south of the Grand Banks. These features are frequently seen in thermal infrared images recorded by weather and research satellites. Often the flow in this current is more nearly a succession of loops, meanders, and eddies rather than a continuous river in the ocean.

The North Atlantic Current splits west of the British Isles, with a branch going northward into the Irminger Current to feed the NORWEGIAN CURRENT and a branch going southward to flow into the Canary Current, thus completing the northeastern portion of the North Atlantic gyre.

Speeds in the North Atlantic Current can be as high as two knots (1 m/s), although the average speed of the current is much less than the Gulf Stream

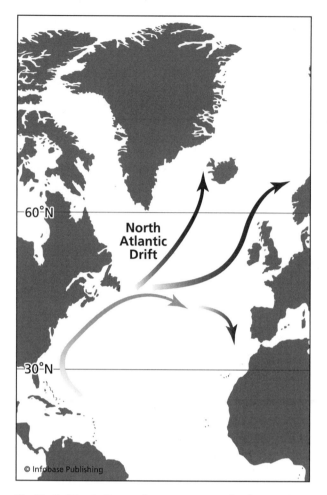

The North Atlantic Current is a warm current that is a continuation of the Gulf Stream. It breaks into branches, one flowing to the east of Iceland past the coasts of northwestern Europe, the other turning southward toward the coast of Portugal and the Canary Islands.

proper. Current speeds are measured with CURRENT METERS (*see* OCEANOGRAPHIC INSTRUMENTS), radar, and by the geostrophic method.

Weather conditions in the region of the NAC tend to be severe in the Northern Hemisphere winter, with strong interactions between the Bermuda-Azores high-pressure cell, and the Icelandic low-pressure cell, leading to cyclone intensification. The frequent passage of fronts coming off the North American continent and the associated storm systems result in high winds and seas. FISHERMEN of the Grand Banks know that between October and April sea conditions are often too severe for the operation of small fishing boats and will make fishing activity extremely hazardous, even for large ships. In the spring, ICEBERGS drifting south from Labrador and Greenland make conditions even more hazardous. Fog often prevails at the northern edge of the NAC, where

warm waters coming originally from the Gulf Stream are overlain by cold air masses moving eastward off the Maritime Provinces of Canada. Conditions such as these resulted in the sinking of the White Star liner *Titanic* in April 1912, and the subsequent institution of the International Ice Patrol. During World War II, Allied merchant ships transporting war materials to Britain had to contend with these severe weather conditions as well as with German U-boats. More recently, the area south of the Grand Banks was the region of the perfect storm, a coincidence of a hurricane and two mid-latitude storms that resulted in the sinking of the sword fishing boat *Andrea Gail*, with a loss of all hands.

Further Reading

Clarke, R. A., H. Hill, R. F. Reiniger, B. A. Warren. "Current System South and East of the Grand Banks of Newfoundland." *Journal of Physical Oceanography* 10, no. 1 (1980): 25–65.

The Cooperative Institute for Marine and Atmospheric Studies, University of Miami Rosenstiel School of Marine and Atmospheric Science. "Ocean Currents." Available online. URL: http://oceancurrents.rsmas. miami.edu. Accessed October 12, 2007.

Junger, Sebastian. *The Perfect Storm: A True Story of Men against the Sea*. New York: Harper Perennial, 1999.

North Equatorial Current (NEC) The North Equatorial Current (NEC) is a warm, wide, slow, shallow current spanning the ocean basins of the world between approximately latitudes 10°N and 25°N. The North Equatorial Current rolls westward, dragged by the northeast trade winds such as those that propelled the sailing ships of Columbus and the Spanish and Portuguese explorers of the 15th and 16th centuries to the New World.

The North Equatorial Current is the southern branch of the Northern Hemisphere subtropical anticyclonic (clockwise) GYRE circulating the waters of the Atlantic, Pacific, and Indian Oceans. The NEC forms the northern boundary of the North Equatorial Countercurrent (*see* EQUATORIAL COUNTERCURRENT), which returns some of the westward-transported water to the eastern part of the ocean basins.

The Atlantic NEC begins around the Canary Islands off the northwest coast of Africa and terminates in the region of the Antilles (Windward Islands), where it feeds into the Antilles and Caribbean Currents, which are source waters for the GULF STREAM system. The Pacific North Equatorial Current begins off the west coast of Panama and terminates in the region off the east coast of the Philippines,

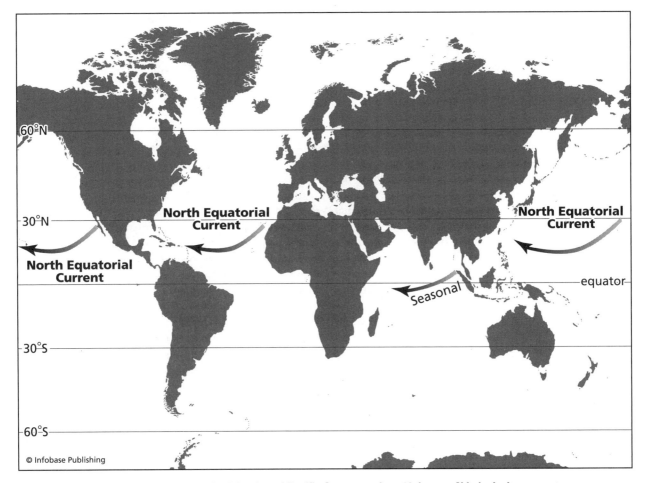

The North Equatorial Current is found in the Atlantic and Pacific Oceans at about 10 degrees N latitude; it transports ocean surface water westward in response to the trade winds.

the Philippine Sea, and flows into the source waters of the KUROSHIO CURRENT (the Pacific counterpart of the Gulf Stream). The NEC in the Indian Ocean is seasonal, being present only during the time of the northeast MONSOON. The Indian Ocean North Equatorial Current begins in the region west of the island of Sumatra and flows westward to East Africa, where the flow branches north and south. The southward flow eventually feeds into the AGULHAS CURRENT. The current in the Pacific is rather similar to that in the Atlantic, but it is stronger and more clearly delineated. The current in the Indian Ocean is present only in Northern Hemisphere winter, during the northeast monsoon. More is known about the Atlantic North Equatorial Current than those in the Pacific and Indian Oceans.

The Atlantic North Equatorial Current has a mean westward speed of 0.2 to 0.3 knots (0.10–0.15 m/s), with peak values of about 0.3 knots (0.15 m/s) in the Northern Hemisphere. It follows the northeast coast of Brazil and the Guianas into the Caribbean and Antilles Currents. The annual mean transport

of the NEC has been estimated to be in the order of 8.5 Sverdrups (8.5×10^6 m³/s), with the lowest transports occurring in the late Northern Hemisphere winter and the highest transports in the Northern Hemisphere summer. The annual variation in transports is about 2.3 Sverdrups. According to scientists of the PIRATA (Pilot Research Moored Array in the Tropical Atlantic) program (1998 et esq.), near surface temperatures in the NEC range from 86°F to 90°F (30°C–32°C) in late summer and early fall; in winter they range from 75.2°F to 82.4°F (24°C–28°C). Salinities range from the order 35.2 psu in winter to 36.4 psu in summer.

Since EKMAN TRANSPORT due to the stress of the trade winds at the sea surface transports water northward away from the southern edge of the NEC there is upwelling of colder deep water to replace the surface water.

Seasonal variations are most extreme in the Indian Ocean, where the northeast and southeast monsoons are found rather than the northeast trade winds of the Atlantic and Pacific. In the Northern

Hemisphere winter, the northeast monsoon blows from central Asia toward Africa and drags the surface waters of the Indian Ocean North Equatorial Current westward in similar fashion to the northeast trade winds in the Atlantic and Pacific Oceans. However, in the Northern Hemisphere summer, the southwest monsoon blows from Africa to South Asia. Under these conditions, the eastward flowing Southwest Monsoon Current replaces the westward flowing NEC. The southwest monsoon also sets up a strong northward flowing current off the Horn of Africa, the Somali Current.

Seasonal variations in the equatorial wind fields are also responsible in the Atlantic and Pacific Oceans for a displacement of the INTERTROPICAL CONVERGENCE ZONE (ITCZ), from a position around the equator during the Northern Hemisphere winter northward to around 10°N during the Northern Hemisphere summer. The ITCZ is a region of frequently light, variable winds between the Northeast Trades and the Southeast Trades. In the days of the great sailing ships it was known as the Doldrums, dreaded by sea captains because their ships could be calmed for days or even weeks without the steady force of the tradewinds. The axis of the ITCZ generally delineates the thermal equator. There is a corresponding displacement of the equatorial system of surface currents by as much as 10 degrees of latitude, with the axis of the North Equatorial Countercurrent approximately parallel to the ITCZ. At the approach of the Southern Hemisphere summer, the surface equatorial currents return southward.

The stress of the northeast trade winds acting on the surface of the water results in an Ekman transport of water 90° to the right of the wind direction facing in the direction of the current, generally from south to north, which piles up water on the northern edge of the current. This pileup results in a north-to-south downward slope of the sea surface, which keeps the current in geostrophic balance. As a consequence of this motion, the THERMOCLINE (zone of maximum vertical change in temperature separating the light, warm surface waters from the cold, bottom waters) deepens from south to north to balance the pressure gradient force due to the sloping sea surface (see GEOSTROPHY). Since surface water is being transported northward away from the southern edge of the NEC, there is upwelling of colder deep water to replace the surface water, with a consequent enhancement of biological productivity, since tropical surface waters are generally deficient in nutrients.

Major international oceanographic expeditions, which deployed large numbers of current meters in the Atlantic NEC, include the Soviet tropical hydrophysical test range experiment in 1970, the joint U.S.-Soviet *Polymode* in 1974–75, the GARP (Global Atmospheric Research Program) Atlantic Tropical Experiment (GATE) in 1974, *Eastropac* (in the eastern tropical Pacific Ocean, 1967–68) and the *Indian Ocean Expedition* in the 1970s. *WOCE* in the 1990s was an experiment to investigate all of the oceans. These expeditions investigated the medium- to small-scale features in these oceans such as meanders, eddy, and rings, the current structures in depth, as well as the temperature and salinity structure (see CURRENTS). For example, during the Soviet tropical hydrophysical test range experiment, oscillations in the current having a timescale on the order of a month were found at depths from 656 to 4,922 feet (200–1500 in), in addition to shorter wavelength oscillations of inertial and tidal periods.

Much knowledge of these currents was obtained from satellite altimeters measuring the topography of the sea surface, as well as surface temperature during the TOPEX/Poseidon satellite missions in the 1990s (TOPEX/Poseidon was launched on August 10, 1992).

Further Reading

Bischof, Barbie, Elizabeth Rowe, Arthur J. Mariano, and Edward H. Ryan. "The North Equatorial Current." Available online. URL: http//oceancurrents.rsmas.miami.edu/atlantic/north-equatorial.html. Accessed March 11, 2008.

Bowditch, Nathaniel. *The American Practical Navigator. An Epitome of Navigation. 1995 ed.* Bethesda, Md.: Defense Mapping Agency Hydrographic/Topographic Center, 1995.

Coiling, Angela. *Ocean Circulation,* 2nd ed. Prepared for the Oceanography Course Team, the Open University, Walton Hall, Milton Keynes, England. Oxford: Butterworth-Heinemann, 2001.

Knauss, John A. *Introduction to Physical Oceanography,* 2nd ed. Upper Saddle River, N.J.: Prentice Hall, 1996.

Monin, A. S., V. M. Kamenkovich, and V. G. Kort. (John J. Lumley, ed. of the English translation) *Variability of the Oceans.* New York: Wiley, 1977.

Pickard, George L., and William J. Emery. *Descriptive Physical Oceanography: An Introduction,* 5th (SI) enlarged edition. Oxford: Pergamon Press, 1990.

Richardson, P. L., and D. Walsh. "Mapping climatological seasonal variations of surface currents in the tropical Atlantic using ship drifts." *Journal of Geophysical Research* 91 (September 15, 1986): 10,537–10,550.

University of Miami, Rosenstiel School of Marine and Atmospheric Science. "A Web-based ocean current reference site." Each current has links, summary text, plots, and numerical model (HYCOM) simulations. Available online. URL: http://oceancurrents.rsmas. miami.edu/index.html. Accessed February 7, 2004.

North Pacific Current The North Pacific Current flows eastward across the North Pacific Ocean from the eastern end of the KUROSHIO CURRENT Extension (about 180° east longitude) to the North American coast between latitudes 30°N and 40°N. The Oyashio Current contributes cold subarctic waters flowing south from the Bering Sea and the Sea of Okhotsk. The eastern end of the current bifurcates into the northward flowing ALASKA CURRENT and the southward flowing CALIFORNIA CURRENT. The California Current is a slow, broad eastern boundary current that flows between California and the Hawaiian Islands. The Alaska Current flows northward along the coast of Canada and Alaska and then turns southwest to form the Aleutian Current. The North Pacific Current forms the northern branch of the North Pacific gyre; the eastern branch is the California Current; the southern branch is the Pacific North Equatorial Current; the western branch is the Kuroshio Current and Current Extension. The westerly winds that drive the North Pacific Current result in a maritime climate for the west coast of the United States and Canada, with major weather systems, such as severe winter storms, frequently pounding the coasts of Oregon and Washington and then moving eastward across the North American continent.

Further Reading

Bowditch, Nathaniel. *The American Practical Navigator: Publication No. 9.* Bethesda: Defense Mapping Agency Hydrographic/Topographic Center, 1995.

Pickard, George, and William Emery. *Descriptive Physical Oceanography: An Introduction,* 5th ed. Oxford: Pergamon Press, 1990.

Norwegian Current The Norwegian Current is an extended northward flow of the NORTH ATLANTIC CURRENT after it divides southeast of the Grand Banks and flows north of Scotland toward the Norwegian coast. The Norwegian Current flows northeastward along the northwest coast of Norway. The Norwegian Current is greatly influenced by the bottom topography; it tends to follow the Norwegian TRENCH, the deepest part of the Norwegian Sea. The depth of the current has been observed to be on the order of 164 to 328 feet (50–100 m). The Norwegian Current has an average speed on the order of 0.5 knots (0.26 m/s) and a volume transport on the order of three to six Sverdrups ($3–6 \times 10^6$ m³).

The three main Atlantic currents that flow poleward are the North Atlantic, Irminger, and Norwegian Currents. As the Norwegian Current splits between Iceland and Norway, its temperature is lowered by cold Arctic air, and the intrusion of brackish or nearly freshwater from the rivers and fjords of Norway as well as from the Baltic Sea lower its salinity. The western branch of the warm Norwegian Current enters the Barents Sea from the south, while the eastern branch continues poleward into the Arctic Ocean, flowing past Spitsbergen, the largest of the Svalbard Islands.

Since the Norwegian Current is warmer than the surrounding Arctic waters, it can affect the amount of ice cover in the Barents Sea; hence, it has a significant effect on regional climate. The southern extent of the Barents Sea is ice-free even in winter and in late summer as a result of the Norwegian Current's transport of warm water. Geologists have found regions surrounding the Svalbard archipelago and western parts of the Barents Sea to be rich in natural resources. Petroleum companies have discovered fossil fuel reserves from exploratory wells that are drilled by oil rigs such as the *Deepsea Bergen*. These vessels are specially designed and classified for operations in harsh environments and deep waters. They include a floating platform on pontoons or a platform perched on long legs, anchor cables, derrick, cranes, helipad, crew boats, antennas, and oil-processing areas.

Atlantic currents such as the Norwegian Current can transport and redistribute pollution in the Arctic. Understanding the fate of water contaminates is important because data on contaminants in the deep sea are sparse, and because of consequences to the marine food chain. Oceanographers have found that pollutants in the Arctic Ocean are highest in the surface waters and decrease with depth. This would indicate that pollutants are entering the Arctic Ocean through a combination of river runoff, atmospheric input, and release from SEA ICE. Synthetic Aperture Radar (SAR; *see* RADAR) is used by environmental protection organizations to monitor and detect oil on water surfaces. The oil slicks appear as dark patches on images because of the damping effect of the oil on waves that is measured by backscattered signals from the radar transceiver.

Further Reading

The Cooperative Institute for Marine and Atmospheric Studies, University of Miami Rosenstiel School of Marine and Atmospheric Science. "Ocean Currents." Available online. URL: http://oceancurrents.rsmas. miami.edu. Accessed May 19, 2006.

Notice to Mariners A Notice to Mariners is a written or broadcast report providing timely marine safety information for the correction of navigation charts, light lists, and sailing directions from a wide variety of sources both foreign and domestic. Light lists provide detailed information on all lights, BUOYS, day beacons, ranges, fog signals, radio beacons, and RADAR beacons. Sailing directions such as the *U.S. Coast Pilot* describes channels, hazards, winds and currents, restricted areas, port facilities, pilotage service, and many other types of information needed by a navigator for safe and efficient navigation. In waters surrounding the United States, Notice to Mariners corrections are published by the U.S. Coast Guard, the National Geospatial Intelligence Agency (NGA), and the Canadian Hydrographic Service. Cartographers from organizations such as the Maritime Division at NGA collect the various Notice to Mariners documents, interpret the documents for changes, and apply the changes to existing nautical chart editions. Notice to Mariners are essential to all navigators for the purpose of keeping charts, light lists, and sailing directions current.

See also CHARTS, NAUTICAL.

Further Reading
Maloney, Elbert S. *Chapman: Piloting, Seamanship and Boat Handling*, 64th ed. Sterling: New York, 2003.
Navigation Center, U.S. Coast Guard. Available online. URL: http://www.navcen.uscg.gov. Accessed on September 18, 2006.

nowcasting Nowcasting refers to short-term FORECASTING techniques used in the natural sciences, such as meteorology, oceanography, and space science. Nowcasts usually span a time of a few hours, whereas routine weather forecasts span a time of 36 hours. Nowcasts and forecasts are present and future science-based predictions on environmental parameters or conditions. In meteorology, nowcasts often refer to severe weather forecasts used to issue tornado watches and warnings, especially when circulation "signatures" (characteristic RADAR echoes of a hook shape) are observed in the Doppler radar images. Other events for which nowcasts are issued include severe thunderstorms, hailstorms, microbursts (strong straight-line downbursts of wind), and flash floods. Beside the Doppler radar, weather satellite images and very fast interactive computers are employed to make the nowcast. Nowcasting was employed at recent Olympic sailing races to provide the participants with environmental information to optimize their racing strategy and tactics.

In oceanography, nowcasting is used by the U.S. Navy to forecast ocean thermal conditions for submarine and antisubmarine warfare operations, wind and wave conditions for amphibious landings, and aircraft operations. Government agencies such as NOAA's National Ocean Service (NOS) use nowcasts to forecast the CURRENT at several locations in ship channels from current measurements at a single location. Statistical techniques may be employed or other numerical methods to describe physical processes such as WATER LEVEL FLUCTUATIONS or reversing tidal currents. Relationships may be used to nowcast events at one location from *in situ* measurements at another site. For example, the currents at Port Manatee in Tampa Bay, Florida, are predicted from measured currents in the vicinity of the Sunshine Skyway Bridge, as part of that estuary's physical oceanographic real time system or PORTS® (Physical Oceanographic Real-time System) program.

Nowcasts are made for space weather events such as solar flares or prominences that are expected to affect communication satellite operations or even damage or destroy the satellites' electronic equipment. Solar flares can cause high levels of radiation in space. When the warning time is sufficient, the space agencies are able to rotate the satellite away from the solar disturbance, thus preventing damage or destruction to sensitive transmitters and receivers.

Further Reading
Coastal Data Information Program. Available online. URL: http://cdip.ucsd.edu. Accessed September 30, 2006.
National Weather Service, National Centers for Environmental Prediction. Available online. URL: http://www.ncep.noaa.gov. Accessed September 30, 2006.

numerical modeling Numerical modeling is a method that uses the power of computers to solve differential equations that describe physical phenomena. The Navier-Stokes equations are a set of equations that describe the time rate of change of the momentum of a fluid. Basically, they express Newton's Second Law—the acceleration of a body of fluid is caused by the sum of the forces acting on it. The forces are the PRESSURE GRADIENT, frictional effects, gravity, and the effects of Earth's rotation. The Navier-Stokes equations which are named after Claude-Louis Navier (1785–1836) and George G. Stokes (1819–1903) describe all the motions that a fluid can have from capillary waves on the surface to tidal flows that encircle the globe. These equations cannot be solved analytically; they must be solved numerically; that is the field of numerical modeling.

For example the gradient of the velocity term is $\partial u/\partial x$, where u is the velocity of the fluid in the

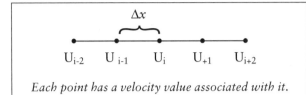

Each point has a velocity value associated with it.

x-direction, and x is the distance. This differential term can be represented as the finite difference between two adjacent terms. Suppose the velocity gradient needs to be calculated at a number of equally spaced points in a channel each separated by a distance Δx. Each point has associated with it a velocity and a pressure. At a point designated as position i, the value of the velocity is U_i. To compute the velocity gradient at the point i, the term, $\dfrac{\partial u}{\partial x}$ becomes $\dfrac{(U_i - U_{i-1})}{\Delta x}$. This form of representing the velocity gradient by this difference formula is called the finite difference method. Using this equation the velocity gradient can be calculated at each point. Each point then has a set of numerical estimates of the terms that make up the Navier-Stokes equations. These equations are then solved on a computer one step at a time. The time evolution of the flow is defined by the governing equations.

This method, called the finite difference method, can be used to solve the Navier-Stokes equations that describe the tides, or surface waves, or whatever the desired phenomena are that is being studied. This is the method that is used to solve the weather predictions shown on TV. The forecast models that show tomorrow's weather are just numerical estimates of what the atmosphere is really doing. The same is true of ocean models. Because fluids have time scales and length scales that are small, on the order of hours and kilometers respectively, it would take a computer years to accurately model a month of ocean motion.

There are many types of models that exist. Some forecast the physical processes, such as the speed of the current or height of the wave for the ocean, and some forecast the growth of a species of fish. The latter models are much more complicated in that they rely on accurate models of the physical processes and then must model the growth, prey rates,

and predator rates for a specific species. Any model that leads to a prediction must then be verified by comparing it against the measurements that are taken. This is called "ground truthing." For example, the height of a STORM SURGE is forecast from a combined meteorology and oceanography model, and compared with actual heights reached by the water. The more accurate the model is when compared to "ground truth," the more reliability the users have in their models. The results of the model predictions versus the actual water levels are incorporated into future versions of the model to provide continual improvement of the forecasts.

Some models, called global circulation models (GCMs), will encompass the entire globe. A GCM of the ocean would also include as input the heating of the surface and the wind stress. These values would be derived from an atmospheric model. Of course, when doing modeling, there are two competing forces on the modeler. The modeler desires to have the smallest time steps and smallest spatial resolution possible. This procedure creates accurate requirements for the measurements. If the currents in Chesapeake Bay are modeled, one would desire a length scales on the order of meters. However, as the length scale is decreased, the amount of time it takes to compute an answer goes up as the square of the number of grid points in one direction! So there is strong pressure to decrease the number of grid points, which results in increased length scales. A length scales of one kilometer would give too course a resolution that would miss many estuary currents that swirl around the bay.

There are many ways to solve problems numerically. These include finite deference methods, finite element models, spectral models, and analog models. Each has a benefit and a cost associated with it. Analog models are small-scale physical models that are used to reveal the underlying physics of a large-scale phenomenon. Small airplanes are tested in a wind tunnel before full-scale airplanes are built. Similarly, a rotating tank can mimic the flow in the North Atlantic and develop flows similar to the GULF STREAM.

Models will continue to grow and be a critical tool for understanding the currents of the ocean and air masses of the planet. As computer resources increase in storage and speed, models will approach more closely their natural analogs. All progress must be tempered by the statement "Just because we know the answer to a problem, does not mean we know anything about the processes that gives us that answer." The main purpose of these models is to reveal the underlying physical processes leading to better understanding of the aquatic environment, which in turn leads to improved predictions.

Further Reading

Pond, Stephen, and G. L. Pickard. *Introductory Dynamical Oceanography,* 2nd ed. Oxford: Pergamon Press, 1983.

nutrient cycles   Nutrient cycles refer to the continuous sequence of events by which nutrients are converted into forms that can be used by plants and animals for sustaining life and building tissue. In the oceans, the primary nutrients are usually considered to be nitrates, phosphates, and silicates. While silica is not a nutrient per se, it is an essential constituent to build the shells of diatoms and radiolarians (*see* PLANKTON). The cycle is driven by uptake of nutrient chemicals by living organisms, conversion to tissue by means of photosynthesis or chemosynthesis, release back into the environment by excretion or death and decay of the organism to begin the cycle again, and distribution by the circulation within the oceans. In the ocean, the greatest uptake of nutrients occurs near the surface, in the PHOTIC ZONE, where photosynthesis is occurring. As organisms die and sink to the ocean floor, the nutrients are removed from the near surface layers, and transferred to the deeper ocean below the photic zone, and cannot be used again by the near-surface plants and animals until brought back into solution by decomposition and recirculated back to the surface. The vertical circulation of nutrients from the deep layers to the surface is strong only in the Arctic and Antarctic regions of the ocean, and in areas of strong UPWELLING, which occur for the most part close to shore along the west coasts of continents and in areas of divergent flow, such as occurs north of the equator in the eastern PACIFIC OCEAN.

One of the most important nutrient cycles involves the flow and recirculation of the many forms of nitrogen. Nitrate (NO_3), as an example, moves with water in the soil to enter rivers that run into the sea. Some bacteria convert nitrogen into ammonium (NH_4), called nitrogen fixation. The expiration and subsequent decay of plants and animals by bacteria result in the production of inorganic nitrogen, which is then used by living marine plants and animals in an endless cycle. An early investigator into marine nutrient concentrations was Alfred Redfield (1890–1983). He measured amounts of nitrogen and phosphorus in marine particulate matter and the surrounding water.

The cycle of phosphorus bears similar characteristics to the cycle of nitrogen. The most abundant forms of phosphates are the orthophosphates, which are inorganic compounds produced by bacteria breaking down organic phosphates from decaying tissues. This chemical conversion takes place rather rapidly and thus remains high in the water column, making phosphorus available to upper layer plants and animals. Unlike nitrogen, phosphorus availability is generally not a limiting factor in biological productivity. Typical values of the mean concentration of phosphorus below the thermocline are about 30 microgram-atoms per liter in the Pacific and Indian Oceans, and about 18 in the Atlantic Ocean. The vertical distribution of phosphorus shows minimum values in the surface layers, resulting from uptake (mostly by phytoplankton), and increasing values in depths of a few hundred meters, then rapidly increasing in the thermocline, down to a maximum at around 3,280 feet (1,000 m). The Pacific and Indian Oceans are higher in phosphate concentration than the Atlantic. This is likely the result of the more vigorous circulation in the middle and deep layers of the Atlantic Ocean compared with the other oceans. As with nitrogen, the seasonal maximum in phosphate is in the late winter, when productivity is low, and is a minimum in summer, when productivity is highest corre-

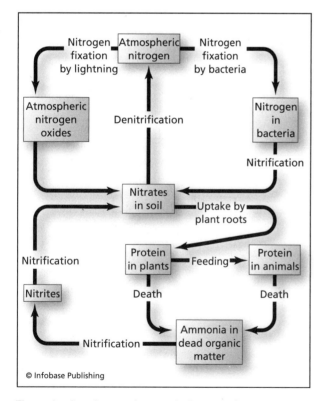

The cycle of nutrients and energy in the ocean (*see* NITROGEN CYCLE)

sponding to high levels of sunlight. Areas of upwelling of phosphorus are generally very high in biological productivity.

Silicon follows a cycle much like that of nitrogen and phosphorus. The source of silicon in the ocean is the weathering of rocks, occurring as silicon dioxide (SiO_2), and in clay minerals. The vertical distribution of silicon shows values near zero at the sea surface, increasing to values on the order of 180 microgram-atoms per liter above the seafloor. This distribution is probably explained by the uptake of silicon by diatoms and radiolarians, as well as their precipitation downward to the seafloor after expiration. As with the other nutrients, silicon has relatively low values in the Atlantic Ocean, and high values in the Pacific, resulting from the ring of volcanoes (see RING OF FIRE). Typical values of silicon concentration below the thermocline are about 180 microgram atoms per liter in the Pacific Ocean, about 40 in the Atlantic Ocean, and about 60 in the Indian Ocean.

Sulfur is an important nutrient for organisms surviving on chemosynthesis, rather than photosynthesis, such as the animals populating the HYDROTHERMAL VENT regions of the ocean floor, where there is perpetual darkness except for the phosphorescent glow of deep-sea animals and the glow of high temperature fluids in the vents. These animals acquire their nutrients from bacteria, which are able to synthesize hydrogen compounds from the chemicals discharged by the vents.

Sulfur compounds, however, such as hydrogen sulfide (H_2S) are toxic to the marine plants and animals of the photic zone. In the Black Sea, where an anaerobic (without oxygen) environment is found below the thermocline, the abundance of hydrogen sulfide results in a lifeless, evil-smelling volume of water.

Further Reading

Armstrong, F. A. J. "Inorganic Nitrogen in Sea Water." In *Chemical Oceanography*, edited by J. P. Riley and G. Skirrow. London: Academic Press, 1965.

———. "Phosphorus." In *Chemical Oceanography*, edited by J. P. Riley and G. Skirrow. London: Academic Press, 1965.

Davis, Richard A. Jr. *Oceanography: An Introduction to the Marine Environment*. Dubuque, Iowa: Wm. C. Brown, 1991.

"The Nutrient Cycle." Earth Sciences 1. The Flinders University of South Australia. Available online. URL: http://www.es.flinders.edu.au/~mattom/IntroOc/notes/intro2.html. Accessed July 24, 2008.

nutrient regeneration Nutrient regeneration refers to the replenishment of nutrients in the marine environment by marine animals through discharge of urine and fecal matter. The animals in the cycle include herbivores, such as ZOOPLANKTON, and carnivores, such as finfish. After the expiration of the animal, the organic material is decomposed by bacteria, returning inorganic nitrates, phosphates, and silicates to the environment. Thus, the nutrients are again made available to the living plants. Compounds such as ammonia (NH_4) from urine are taken up almost immediately by the primary producers, the PHYTOPLANKTON. The majority of the heavier materials, feces, and dead organisms such as whales, sink to the bottom and may be lost to the animals of the PHOTIC ZONE, especially in the deep ocean basins that are on the order of 12,000 feet (4,000 m) deep.

For this reason, shallow waters are usually more productive than deep waters because the depth of sinking is limited to a few hundred feet (tens of m), leaving the material subject to transport back to the surface layers by vertical circulations brought about by tidal circulations, TURBULENCE, Langmuir cells (convection cells in the upper ocean that set up lines of convergence along which the plankton can be concentrated), wind waves, INTERNAL WAVES, and the like. Irving Langmuir (1881–1957) was the first person to study bubbles marking vertical rotating cells of water seen in response to wind over water in rivers, lakes, seas, and oceans. The most productive areas in the ocean are the zones of UPWELLING, in which nutrients are recycled back to the surface by strong vertical water motions. Upwelling zones are generally found on the west coasts of continents, such as California and Peru.

The strong, permanent THERMOCLINES found in the subtropics and Tropics tend to restrict vertical circulations that would renew the nutrients in the surface layer. These regions tend to have low levels of primary production, and consequently spare food chains. As an example, the waters of the Sargasso Sea are low in primary producers, which enhance the clarity of the water, leading to the beautiful cobalt blue color known as the "desert color" of the sea. Areas of high productivity tend to blue-green, green, or even yellow-green, indicating high amounts of chlorophyll produced by the phytoplankton. Polar waters, where strong vertical circulations are found, are generally rich in regenerated nutrients. However, in the Northern Hemisphere productivity is limited by low light levels during winter. Summertime conditions are the opposite in "the land of the midnight Sun." Recently, marine scientists have suggested that productivity in some parts of the ocean could be increased by putting nutrients into select areas of the sea surface, much as farmers fertilize their fields on land.

Further Reading

Sumich, James L. *An Introduction to the Biology of Marine Life.* Dubuque, Iowa: Wm. C. Brown, 1980.

Thorne-Miller, Boyce. *The Living Ocean: Understanding and Protecting Marine Biodiversity,* 2nd ed. Washington, D.C.: Island Press, 1999.

O

ocean basins Ocean basins constitute the physical marine environment between the continental rises of the outer CONTINENTAL MARGINS. For example, the North Atlantic Basin represents the deep part of the ocean between Europe and North America. The ocean basins are submarine topographic depressions underlain by oceanic crust. Oceanic crust is the thin crust found under the world's oceans, which is gener-ally on the order of three to four miles (5–6 km) in thickness, in contrast to the continental crust, which can be tens of miles (km) thick. The oceanic crust is generally dense basalt, including olivine and sima, whereas the continental crust contains large amounts of light quartz and sial.

The major ocean basins of the world are the PACIFIC, ATLANTIC, INDIAN, ARCTIC, and SOUTHERN

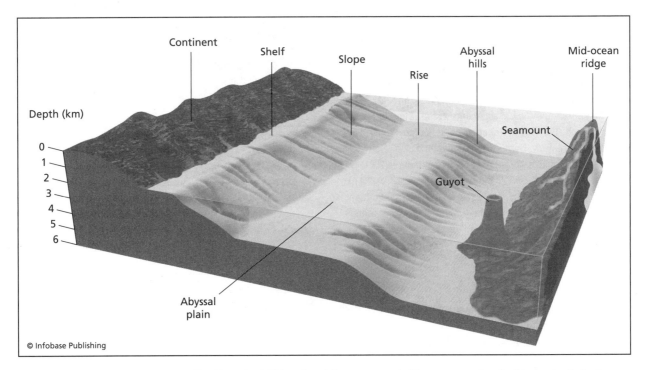

Ocean basins are depressions on the Earth's surface (lithosphere) that are occupied by seawater. A typical ocean basin is shown conceptually in block form with the continental margin merging smoothly into the abyssal plain and the rise of the abyssal hills into the mid-ocean ridge.

Ocean Explorations: A Biographical Sketch of Dr. Robert D. Ballard

Dr. Robert D. Ballard, born on June 30, 1942, in Wichita, Kansas, and raised in San Diego, California, is best known as an ocean explorer who found the sunken RMS *Titanic* on September 1, 1985. The wreckage was found on the seafloor 350 miles (560 km) off Newfoundland. Explorations of the ocean basins, which cover almost 70 percent of Earth's surface and 96 percent of the planet's water, require tremendous planning and resources. Dr. Ballard's feats, application of technologies, and scientific reporting rival Charles Darwin's observations during the voyage of the HMS *Beagle* (1831–36), bathymetric soundings charted by U.S. Navy Lt. Matthew Maury during the mid 1800s, and the HMS *Challenger* expedition from 1872 to 1876.

Ballard was trained as a scientist. He earned a bachelor's degree in physical science from the University of California, Santa Barbara, in 1965, and then attended graduate schools at the University of Southern California and at the University of Hawaii, studying oceanography. He ultimately received a Ph.D. in marine geology and geophysics from the University of Rhode Island in 1974. He spent more than three decades as a researcher at Woods Hole Oceanographic Institute (WHOI), where he led and participated in deep-sea explorations. At WHOI, he also helped to develop manned submersibles and Remotely Operated Vehicles for marine research. He has been a visiting scholar at Stanford University and holds 16 honorary degrees.

Ballard served in the military on analytical assignments. In the U.S. Army, he worked as an intelligence officer from 1965 to 1967; later, he transferred his commission to the United States Navy, where he was assigned as a meteorology and oceanography (METOC) officer. He served on assignments at the Office of Naval Research, the Naval Oceanographic Office, the Office of the Chief of Naval Operations, and the National Research Council. He retired as a commander in 2001.

Ballard has led or participated in more than 110 deep-sea expeditions that have focused on mid-ocean ridges and on hydrothermal features such as warm-water vents, black smokers, white smokers, and their associated animal communities. He has conducted research aboard the Deep Submergence Vehicle *Jason*, which deployed the tethered remotely operated vehicle *Jason Jr.* His innovations include using video cameras controlled from *Alvin* to transmit images to sites on land. His *Jason* Project provided allowed hundreds of thousands of schoolchildren to view undersea explorations as they were unfolding. Some of Ballard's recent discoveries include ancient Phoenician ships off the coast of Israel during 1999, 1,500-year-old wooden ships in the Black Sea during 2000, and during 2002, the remains of President John F. Kennedy's boat, PT *109,* which was sunk in the Solomon Sea during World War II. He is the author of more than 50 refereed scientific articles and 20 popular scientific books, including *Return to Midway, The Water's Edge, The Eternal Darkness: A Personal History of Deep-Sea Exploration,* and the children's book *Ghost Liners.* Ballard also has been featured in a dozen *National Geographic* television specials, including *In Search of Noah's Flood, Pearl Harbor: Legacy of an Attack, The Battle for Midway,* and the record-breaking *Secrets of the Titanic.*

He is the president of the Institute for Exploration, a division of the Sea Research Foundation, in Connecticut. Ballard's many achievements have resulted in numerous awards and honors, including the Robert Dexter Conrad Award for Scientific Achievement, the Lone Sailor Award, the Explorer's Club Medal, the Marine Technology Society Compass Distinguished Achievement Award, the Lindbergh Award, and the National Geographic Society's most prestigious award, the Hubbard Medal.

Further Reading

Ballard, Robert D. *Eternal Darkness: A Personal History of Deep-Sea Exploration.* Princeton, N.J. Princeton University Press, 2000.

Charton, Barbara. *A to Z of Marine Scientists.* New York: Facts On File, 2003.

Nichols, Charles R., David L. Porter, and Robert G. Williams. *Recent Advances and Issues in Oceanography,* Westport, Conn.: Greenwood, 2003.

Mystic Aquarium & Institute for Exploration. Available online. URL: http://www.mysticaquarium.org. Accessed July 29, 2007.

—**C. Reid Nichols,** president
Marine Information Resources Corporation
Ellicott City, Maryland

OCEANS. The ocean basins also contain distinguishing features like large MARGINAL SEAS, such as the MEDITERRANEAN and the GULF OF MEXICO in the Atlantic. Basins are separated by continents, except in the Antarctic, where the Southern Ocean is joined by the Atlantic, Indian, and Pacific Oceans. The average depth of the world ocean is 12,238 feet (3,730 m). The many marginal seas along with the ocean basins make up the global ocean.

There are three predominant physiographic provinces within the ocean basins: the abyssal floor, the mid-ocean ridge, and the deep ocean TRENCHES. The abyssal floor consists of the ABYSSAL PLAIN and the ABYSSAL HILLS. The abyssal plains are the large, relatively flat (slopes of less than 1:1,000) areas of water depths from about 10,000 to 20,000 feet (3,000–6,000 m). The abyssal hills can be thought of as the foothills of the mid-ocean ridges. They are on the order of 165 to 820 feet (50–250 m) in height above the surrounding plain, and on the order of a few miles (km) in horizontal dimension. Abyssal hills are particularly abundant in the Pacific Ocean.

The Mid-Ocean Ridge, sometimes known as the Ocean Ridge, (*see* BATHYMETRY) is the longest chain of mountains on Earth, extending around the globe like stitching on a baseball for about 50,000 miles (80,000 km). It rises from the abyssal hills about 3,280 to 9,630 feet (1,000–3,000 m) and is found at water depths on the order of 8,200 feet (2,500 m) below MEAN SEA LEVEL. It is known as the Mid-Ocean Ridge because the first such feature was discovered in the Atlantic Ocean, where the ridge does indeed run through the middle of the North and South Atlantic Oceans. In the Indian and Pacific Oceans, however, the ridge is closer to one continent or the other.

The Mid-Ocean Ridge is characterized by a high level of seismic activity, including volcanic eruptions and earthquakes. The ridge, in fact, is a spreading zone of hot magma welling up from the interior of the Earth and slowly pushing apart the great tectonic plates that cover the surface of the Earth. The spreading ridges are cut by fracture zones in which may be miles (tens of km) wide and extend for up to 1,860 feet (3,000 km) in length transverse to the ridges. These zones are characterized by faults, which can displace the ridges by hundreds of miles (km). The Mid-Atlantic Ridge is perhaps the most clearly defined, with a rift valley that runs along the top of the ridge, which is tens of miles (km) wide and up to 1.2 miles (2 km) deep. UPWELLING material moves up through this valley and fills the ever-expanding rift zone as the oceanic crust moves farther apart at speeds on the order of a few inches (cm) per year. The East Pacific Rise is moving apart at higher speeds than are found in the Mid-Atlantic Ridge.

The third province, the deep-sea trenches, are the regions where the old oceanic crust plunges back down into the Earth at the edge of continents, such as the Peru-Chile Trench along the west coast of South America. Trenches are characterized by a V-shaped profile, with the steepest side being the landward side, and a much lesser slope along the seaward

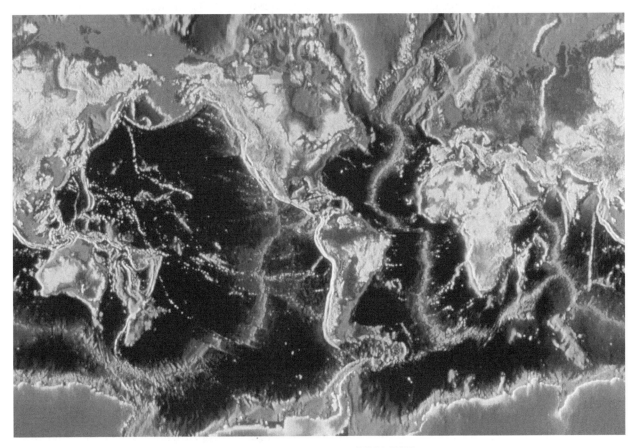

This topographic map shows the ocean basins and mid-ocean ridges. *(NOAA-NOS)*

side. The deep ocean trenches are the deepest parts of the world ocean, having depths on the order of 20,000 to 36,000 feet (6,700–11,000 m) below the sea surface. They may be thousands of miles (km) long and hundreds of miles (km) wide. Trenches are frequently accompanied by island arcs, which are often volcanic zones. In these regions, the material of the crust goes back down into the mantle, where it is heated and "recycled," that is, it is available to be pushed up long the Mid-Ocean Ridges to again make new crust. The regions where this down plunging occurs are called SUBDUCTION ZONES. Most of the subduction zones are found around the rim of the Pacific Ocean, where the Pacific Plate is thrusting downward below the surrounding continental plates, such as the North American Plate. Subduction zones are characterized by extensive volcanic eruptions and earthquakes; hence the name RING OF FIRE for the subduction zones of the Pacific.

The ocean basins are covered by SEDIMENTS from the land (lithogenous) and from the continuous rain of dead marine organisms (biogenous) that may be as deep as several miles (km). Lithogenic sediments include the beach deposits that beachcombers stroll along while searching for intricate shells and other treasures. There are also cosmogenous sediments, such as meteors raining in from outer space, and hydogenous sediments, chemical precipitates. Glacial marine sediments are derived from glaciers. As one moves from the continents to the center of ocean basins, the amount of lithogenous sediments decreases and the amount of biogenous sediment increases.

There are only five major types of sediment in all of the world's ocean basins: calcareous OOZE, siliceous ooze, deep-ocean clay, turbidites, and glacial marine sediments. The preponderance of sediments are biogenic and lithogenic. Sediments derived from chemical precipitation such as metal sulfides and MANGANESE NODULES are hydrogenic. Sediments act to smooth the terrain by filling in canyons and basins and forming the smooth abyssal plains of the world. Much of the world's ocean basin areas are covered with pelagic sediments, that is, fine-grained sediments that settle from the open sea. Among common types of pelagic deep-sea sediments are clays, diatomaceous oozes (from skeletons of the diatom plankters), and radiolarian ooze, from the skeletons of radiolarians.

Further Reading

Davis, Richard A. Jr. *Oceanography: An Introduction to the Marine Environment,* 2nd ed. Dubuque, Iowa: Wm. C. Brown, 1991.

Erikson, Jon. *Marine Geology: Exploring the New Frontiers of the Ocean,* rev. ed. Facts On File, 2003.

Kusky, Timothy. *Encyclopedia of Earth Science.* New York: Facts On File, 2005.

Seibold, Eugene, and Wolfgang H. Berger. *The Sea Floor: An Introduction to Marine Geology,* 3rd ed. Berlin: Springer, 1996.

Ocean Drilling Program   The Ocean Drilling Program is a large-scale international oceanographic research program that focused on investigations of the history and structure of Earth as recorded in sea-floor sediments and rocks. Cores were obtained from the scientific Drill Vessel *JOIDES Resolution.* The Ocean Drilling Program succeeded the 1966 to 1983 Deep Sea Drilling Project, which utilized the D/V *Glomar Challenger* to drill and core in the ATLANTIC, PACIFIC, and INDIAN OCEANS as well as the MEDITERRANEAN and RED SEAS. This early geological research helped confirm the theory of continental drift that was advanced by Alfred Lothar Wegener (1880–1930). The D/V *Glomar Challenger* recovered more than 19,000 cores and drilled to a depth of 952 fathoms (1,741 m).

The Ocean Drilling Program (ODP) was initiated with funding from the U.S. National Science Foundation (NSF) and the JOINT OCEANOGRAPHIC INSTITUTIONS FOR DEEP EARTH SAMPLING (JOIDES), a consortium of 22 international partners. Joint Oceanographic Institutions, Inc. (JOI), a group of 18 U.S. institutions, managed the program. Texas A&M University, College of Geosciences, was the science operator, and LAMONT-DOHERTY EARTH OBSERVATORY at Columbia University provided logging services and administered the Site Survey Data Bank. ODP's purpose was to conduct basic research into the history of the ocean basins and the overall nature of the crust beneath the ocean floor. For this purpose, they converted a more advanced drill vessel from industrial use and rechristened it *JOIDES Resolution.* From 1985 to 2003, this vessel traveled almost nonstop around the world, drilling 650 holes on 110 expeditions (or "legs"). At the conclusion of ODP, the program included more than 20 nations. *JOIDES Resolution* demobilization activities were completed on September 30, 2003, the last day of the Ocean Drilling Program.

The Integrated Ocean Drilling Program commenced in October 2003 as a followup to the Deep Sea Drilling Project and the highly successful ODP. Principal partners in this effort are the United States, Japan, and the European Union. The United States will operate the D/V *JOIDES Resolution,* which uses seawater as the primary drill fluid. Japan provides the D/V *Chikyu,* which uses mud as a drill fluid, allowing deeper drilling. The European Union is the framework for economic and political cooper-

Ocean Drilling Program Summary

Operations days	Distance traveled	Sites visited	Holes drilled	Cores recovered
6,591	355,781 nmi	669	1,797	35,772

	Latitude	Longitude	Geographic area and expedition number
Northernmost site	80.5°N	8.2°E	Arctic Ocean, Leg 151, Site 911
Southernmost site	70.8°S	14.6°W	Weddell Sea, Leg 113, Site 693

	Meters	Feet	Miles	Geographic area and expedition number
Deepest hole penetrated	2,111	6,926	1.3	East Pacific Ocean, Leg 148, Hole 504B
Shallowest water depth	38	123	0.02	Northwest Pacific Ocean, Leg 143
Deepest water depth	5,980	19,620	3.7	West Pacific Ocean, Leg 129
Most core recovered in a single expedition	8,003	26,258	5	Southeast Atlantic Ocean, Leg 175
Total core cored	321,482	1,070,535	202.8	
Total core recovered	222,704	741,704	140.5	

Data courtesy of IODP-USIO (JOI Alliance). For more information, refer to URL: iodp.tamu.edu. Accessed July 25, 2008.

ation between 15 European countries. They provide mission specific platforms for ocean drilling. One of the Integrated Ocean Drilling Program goals involves drilling into the MOHOROVICIC DISCONTINUITY, the boundary between the crust and mantle.

For ODP drilling accomplishments, the review shown above of the statistics garnered by the *JOIDES Resolution* from January 1985 to September 2003 (ODP Legs 100–210) reveals the intensity with which scientific expeditions were conducted.

ocean engineer An ocean engineer is a professional who designs and engineers offshore structures. Ocean engineering is a broad discipline sharing skills with several other marine science–oriented disciplines such as coastal engineering, naval architecture, and physical oceanography. Ocean engineers may find themselves designing BUOYS, platforms, submarine pipelines, underwater vehicles, and complex MOORINGS. Buoys have hulls, instrument payloads, satellite communication equipment, electronic systems, and batteries. They may operate at the entrance of a seaport to provide vital real-time information on weather,

SEA STATE, and CURRENTS. The buoy mooring is designed based on mission, hull type, location, and water depth. Some directional WAVE BUOYS may have elastic moorings that minimize the effect of mooring cable on buoy motions. There are various types of offshore platforms, but all have to be designed to accommodate workers for the safe operation of machinery for drilling and producing oil and natural gas. They may rise thousands of feet (100s of m) above the seafloor. Ocean engineers design anchor chains to moor large ships or floating piers in anchorages. Successful engineering projects require the ocean engineer to be an expert on the environment, one who knows biological and physical forces that affect operational structures and vehicles.

Design consultants, offshore constructors, ports, departments of defense, and government agencies employ ocean engineers. They are involved in the exploration and utilization of the ocean's resources. Their work is usually tied to oil or gas exploration or the design of offshore drilling platforms, harbor facilities, and underwater machines. Those involved in constructing or maintaining structures in the ocean or in the supervision of geotechnical or geophysical offshore work must understand the impacts of wind, waves, currents, and sea level on the operation. The ocean engineer completes engineering requirements, design tests, and makes operational assessments. Requirements might deal with the function, capacity, and speed of an AUTONOMOUS UNDERWATER VEHICLE or AUV. Scale models and prototypes of the AUV might be tested in a MODEL BASIN, lake, ESTUARY, or ocean to ensure that the vehicle meets design objectives. Ocean engineers can process and statistically analyze long series of ocean observations (time series) such as wave heights, wave periods, and wave directions to determine the probability of extreme waves or evaluate consecutive satellite images of an EDDY that is moving toward an offshore platform to determine the impact on important diving operations. Ocean engineering, like coastal engineering, marine engineering, and naval architecture, requires extensive technical writing, verbal reporting, and ultimately the preparation of operation manuals.

Students interested in earning a bachelor of science in ocean engineering must complete college courses covering both the arts and sciences, including the essential subjects mathematics, physics, and English. The ocean engineering curriculum provides a broad education in engineering mechanics and their application to the design and construction of vessels, vehicles, platforms, and marine systems. Individual courses focus on structural mechanics and marine dynamics. Structural mechanics relates to the design and analysis of vessels and platforms, including static strength, fatigue, dynamic response, safety, and production. Marine hydrodynamics covers resistance, buoy motion, and mooring characteristics of submerged and floating bodies. Topics in marine dynamics relate to structural vibrations and rigid body responses of structures to winds, waves, and currents. Ocean engineers may need a professional engineer license for some jobs. In order to be licensed, after earning a college degree, the ocean engineer must pass the open book Fundamentals of Engineering Exam, and then after about four years of experience, take the open book Principles and Practices Exam.

Further Reading

American Association of Port Authorities. Available online. URL: http://www.aapa-ports.org/. Accessed July 27, 2006.

American Society of Naval Engineers. Available online. URL: www.navalengineers.org/. Accessed July 27, 2006.

Marine Technology Society. Available online. URL: www.mtsociety.org/. Accessed July 27, 2006.

Oceanic Engineering Society. Available online. URL: www.oceanicengineering.org/. Accessed July 27, 2006.

Society of Naval Architects and Marine Engineers. Available online. URL: www.sname.org/. Accessed July 27, 2006.

oceanic zone An oceanic zone is a region of the ocean seaward of the CONTINENTAL SHELF break, which includes the bathyal, ABYSSAL, and HADAL ZONES. The oceanic zone is the sub-region of the PELAGIC zone farthest from land. Therefore, nutrients are sparse in this zone, which is usually deeper than 656 feet (200 m). PHYTOPLANKTON are found in the euphotic layer, where there is sufficient light for photosynthesis. Primary production is absent below the euphotic layer, where no light arrives. Some marine animals inhabit specific layers in the water column such as the EPIPELAGIC, MESOPELAGIC, and BATHYPELAGIC ZONES. Fisheries found in the mesopelagic and bathypelagic regions are often classified as deep-water species. Many HIGHLY MIGRATORY SPECIES such as yellowfin tuna (*Thunnus albacares*), wahoo (*Acanthocybium solandri*), and dolphinfish (*Coryphaena hippurus*) are epipelagic, while deep-water examples include blue whiting (*Micromesistius poutassou*), sablefish (*Anoplopoma fimbria*), and orange roughy (*Hoplostethus atlanticus*). Organisms in the aphotic zone tend to be carnivores, suspension, or detritus feeders. Dead plant and animal material or detritus and other matter such as dust drifting down into the depths from the PHOTIC ZONE are called "marine snow." The oceanic zone comprises the largest ocean region from shallow wave induced surface currents to cold slow moving deep-water currents.

See also MARINE PROVINCES; NERITIC ZONE.

ocean observatory An ocean observatory is a fully integrated and operational suite of instruments permanently deployed to measure meteorological, oceanographic, and geophysical phenomena, such as LAND AND SEA BREEZE circulation, WAVES and CURRENTS, water quality, and tectonic activity. Observatories are facilities that often utilize expensive and state-of-the-art weather stations, OCEANOGRAPHIC INSTRUMENTS, seismic sensors on the seafloor, numerical models (*see* NUMERICAL MODELING), remotely sensed images, and research vessels that collect in situ measurements in nearby regions. Since the Latin term *in situ* means "in place," environmental scientists call instruments that collect data either in the atmosphere, in the water column, or in the SEDIMENTS in situ sensors. Remotely sensed data are collected from a distance, usually from sensors on board aircraft or satellites. A fully integrated observatory utilizes necessary and sufficient in situ and remotely sensed capabilities to meet information requirements. They provide real-time information and may be enhanced by NUMERICAL MODELING. This combination is especially useful for trajectory modeling.

An ocean observatory provides information to support operations such as currents for a merchant ship trying to enter a busy seaport, as well as long-term datasets for basic researchers who are trying to advance the marine science knowledge base. Multiple data elements and parameters are collected in order to build informative products. Data elements include scalars such as TEMPERATURE, PRESSURE, dissolved oxygen, wave height, and WAVELENGTH, and vectors such as winds and currents. Winds and currents have a magnitude (speed) and direction. It is also helpful to display phenomena such as currents over BATHYMETRY. Information from a viable ocean observatory should be available 24 hours per day, seven days a week, so that operators can make important decisions. The ideal system helps to improve maritime

commerce, protect property, and save lives. Quality-controlled data and information from an ocean observatory can be used to assess trends and as initial information to mitigate marine spills.

In the United States, numerous ocean observatories have been implemented through the National Oceanographic Partnership Program (NOPP), a collaboration of approximately 15 federal agencies focused on providing leadership and coordination to national ocean research efforts. NOPP created the National Office for Integrated and Sustained Ocean Observations, an interagency office called "OCEAN. US." Since being established in 2000, Ocean.US has catalyzed the coordinated development of the Integrated Ocean Observing System (IOOS). Agencies contributing funds and support to Ocean.US include the NATIONAL OCEANIC AND ATMOSPHERIC ADMINISTRATION, U.S. Navy, National Science Foundation, NATIONAL AERONAUTICS AND SPACE ADMINISTRATION, Minerals Management Service, ENVIRONMEN-

Regional Associations

Regional Association	Location	Participating States & Territories
Alaska Ocean Observing System	Gulf of Alaska, Bering Sea, Arctic Ocean	Alaska
Caribbean Regional Association for an Integrated Coastal Ocean Observing System	Caribbean Sea	Puerto Rico, U.S. Virgin Islands
Central and Northern California Ocean Observing System	Pacific Ocean, off California coast	California
Great Lakes Observing System	Lake Erie, Huron, Michigan, Ontario, Superior, and St. Clair.	Michigan, Ohio, Minnesota
Gulf of Mexico Coastal Ocean Observing System	Gulf of Mexico	Florida, Alabama, Mississippi, Louisiana, and Texas
Mid-Atlantic Coastal Ocean Observing Regional Association	Atlantic Ocean, Chesapeake Bay	Massachusetts, Rhode Island, New York, New Jersey, Delaware, Maryland, and Virginia
Northeastern Regional Association of Coastal Ocean Observing Systems	Gulf of Maine	Maine
Northwest Association of Networked Ocean Observing Systems	Pacific Ocean	Washington, Oregon, and California
Pacific Islands Ocean Observing System	Pacific Ocean	American Samoa, the Commonwealth of the Northern Mariana Islands, the Federated States of Micronesia, Guam, Hawaii, the Republic of the Marshall Islands, and the Republic of Palau
Southeast Coastal Ocean Observations Regional Association	Atlantic Ocean	North Carolina, South Carolina, Georgia, Alabama, and Florida
Southern California Coastal Ocean Observing System	Santa Monica Bay, Pacific Ocean	California

TAL PROTECTION AGENCY, U.S. Army Corps of Engineers, U.S. Coast Guard, and U.S. Geological Survey. The developers, managers, and scientists involved with observatories have many goals such as improving predictions of climate change, supporting marine operations, mitigating the effects of NATURAL HAZARDS such as HURRICANES and nor'easters, complementing homeland security measures, reducing public health risks, and protecting and restoring healthy marine ECOSYSTEMS by facilitating ecosystem-based management of natural resources.

Regional associations have been established around the United States to facilitate integration of observing efforts within geographical areas. Regions include ESTUARIES and the coastal ocean around Alaska, the Pacific Northwest, Northern, Central, and Southern California, Hawaiian Islands, Great Lakes, Northeast, Mid-Atlantic, Southeast, GULF OF MEXICO, and the CARIBBEAN SEA. The coastal ocean extends from the upstream limit of where waters are affected by tides (head of tide) to the seaward boundary of the EXCLUSIVE ECONOMIC ZONE. These regional associations manage the design, implementation, operation, and development of their ocean-observing systems to benefit local users. All observatories meet local information requirements. They include sensors to measure phenomena, technologies to transmit data, and standard procedures to archive and manage data. Science-based analyses are used to produce products that can be disseminated to disparate users, ranging from university and government researchers to commercial and recreational mariners. A National Federation of Regional Associations is planned to coordinate the network of regional associations.

The IOOS is a major step toward development of a permanent Global Ocean Observing System or GOOS, which aims to distribute data and derived products, including analyses, forecasts, and assessments from worldwide ocean observations. GOOS, like the Integrated Ocean Observing System, would consist of networked sensors, models, and remotely sensed imagery. Regional associations such as the Gulf of Mexico Coastal Ocean Observing System provide linkages with coastal observing programs of other nations in the Gulf of Mexico such as Belize or Mexico. Observatories are also workbenches providing sea truth data to test and track innovations such as autonomous gliders, which sample the ocean with onboard sensors while automatically cruising in a sinusoidal pattern from the sea surface to the ocean floor and back. Gliders are approximately 6.5 feet (2 m) long with wingspans of approximately four feet (1.2 m). They change their vertical and horizontal position in the water by pumping mineral oil between bladders located inside and outside of the glider hull. Ocean-observing systems foster the development and sharing of technologies and the transfer of important data on phenomena such as WATER LEVEL FLUCTUATIONS, sea surface temperatures, and SALINITIES across boundaries. Countries may collaborate to share resources to better monitor and understand severe weather and climate changes along with improved descriptions and forecasts about the state of the ocean. The table on page 393 lists several of these associations that have established links with one another in order to track phenomena such as algal blooms and to collaborate with other countries, such as Canada, on a myriad of topics ranging from fisheries to climate change.

Further Reading

Global Ocean Observing System. Available online. URL: http://www.ioc-goos.org/. Accessed October 15, 2007.

IOSS Web Portal. Available online. URL: http://www.tide-sandcurrents.noaa.gov/opendap.html. Accessed March 10, 2008.

Marine Technology Society. "Ocean Observing Systems." Marine Technology Society Journal 37, no. 3 (Fall 2003).

National Federation of Regional Associations. Available online. URL: http://www.ocean.us.nfra. Accessed March 13, 2008.

NOAA Coastal Services Center. "Coastal Observing Systems." Available online. URL: http://www.csc.noaa.gov/bins/c-obs.html. Accessed October 15, 2007.

The Oceanography Society, *Oceanography* Special Issue: Ocean Biogeographical Information System 13, no. 3 (2000). Available online. URL: http://www.tos.org/oceanography/issue-archive/13-3.html. Accessed August 24, 2006.

Ocean.US. The National Office for Integrated and Sustained Ocean Observations. Available online. URL: http://www.ocean.us/. Accessed October 15, 2007.

oceanographic buoys   *See* BUOYS.

oceanographic instruments   Oceanographic instruments are devices for the measurement of quantities essential to understand the biological, chemical, geological, and physical characteristics of the ocean and the marine atmosphere. Instruments have been designed to collect and analyze samples, to make direct measurements *in situ*, and to take measurements remotely from satellites, aircraft, ships, BUOYS, and underwater platforms and observatories. Quantities of interest include TEMPERATURE, PRESSURE, speed, direction, distance, intensity, color, grain size, texture, and weight. Instruments can be classified as those common to the parent sciences of biology, chemistry, and physics and those specially designed for use in the ocean. Almost all measurements made at sea must be associated with a specific location and time; hence, oceanographers of all disciplines must employ accurate navigation systems such as the GLOBAL POSITIONING SYSTEM (GPS). They are concerned about making the measurement platforms as stable as possible so that the data are not corrupted by the motion of the instruments. Marine biologists, for example, want to know the precise depth at which a species of squid is found; excessive vertical platform (ship) motion due to the pitch, roll, and heave of the ship distort this information. PHYSICAL OCEANOGRAPHERS need to know the depth at which current measurements are made. Vertical MOORING motions however, can distort this information. The exact positions of data samples acquired from high-speed towed vehicles, or AUTONOMOUS UNDERWATER VEHICLES (AUVs) must be known.

Many instruments should be routinely serviced, tested, and calibrated to maintain accuracy and measurement repeatability. Calibration involves comparing and adjusting outputs to a known scale. As an example, the International Association for the Physical Sciences of the Oceans, or IAPSO, provides filtered natural seawater as a standard for calibration purposes. Ampules of standard water are ideal for calibrating salinometers and other marine instruments. Temperature and pressure sensors on CONDUCTIVITY-TEMPERATURE-DEPTH PROBES (CTDs) are routinely calibrated at facilities such as the WOODS HOLE OCEANOGRAPHIC INSTITUTION's Calibration Laboratory. Preoperational checks ensure that instruments are working properly prior to final setups on the deck of a ship. WAVE BUOYS and CURRENT METERS may be evaluated in a towing basin such as large tanks found at the Waterways Experiment Station in Vicksburg, Mississippi, or at the Naval Surface Warfare Center in Carderock, Maryland. Some instruments such as a SECCHI DISK are also deployed by hand along the shore or from a pier. A Secchi disk is lowered into the water to measure water transparency. Aircraft may deploy specially manufactured instruments that measure atmospheric and oceanic profiles. Maintenance and tests on instruments and sensor payloads are necessary tasks to collecting high-quality data.

All branches of marine science make extensive use of computers for data acquisition, data quality control, data processing, and analysis. Computers have greatly extended the use of real-time systems that make the data available to the investigators as soon as, or very shortly after, they are acquired. Many marine scientists write their own computer programs to analyze oceanographic data. In the sections following, examples of the important instruments of the major disciplines of oceanography are discussed.

Instruments to Study Oceanic Life-forms

MARINE BIOLOGISTS use a wide variety of gear, including plankton nets, sleds, seines, gill nets, trawls, underwater low-light TV cameras, and water quality measuring equipment such as bottles, THERMISTORS (temperature sensitive resistors), and dissolved oxygen meters. NETS of varying fabric and mesh sizes are deployed at various depths to sample the nekton and PLANKTON, and bottom trawls to sample the benthos. Water samples are collected and analyzed by microbiologists to find and study bacteria, viruses, and plankton responsible for phenomena such as BIOLUMINESCENCE, algal blooms such as red and white tides, and fish kills. NANSEN BOTTLES with plug valves at either end have been traditionally used to collect seawater; they were invented by Norwegian oceanographer, polar explorer, and Nobel Peace Prize winner Fridtjof Nansen (1861–1930). To study finfish, ichthyologists use special equipment to sort, examine, and store adult and larval fish. Dry laboratories are used for examining larval fish and for other analyses such as the study of otoliths. In the skull of bony fish, crystalline structures of the inner ear composed of calcium carbonate and protein are called otoliths. They aid the fish in balance, orientation, and sound detection. Marine fisheries labs are equipped with stainless steel processing areas with precision balances to sort, measure, and examine adult fish for age, diet, diseases, and parasites. Stereomicroscopes and cameras are used to process, photograph, and draw fish larvae. Tissue grinders are used for compositing fish samples for studies of bioenergetics and contaminant concentration determinations.

The bathyscaphe, developed in the 1930s, afforded marine biologists a chance to observe the ocean's plants and animals *in situ* at great depths, but since descent and ascent were governed by very strong cable systems, horizontal mobility was severely limited. In the 1960s, a new generation of self-propelled deep-sea submersibles emerged, including the *Aluminaut*, the *Star 3*, the *Deep Star* series, and the *Alvin*, which is still in use today, although it is completely rebuilt (*see* SUBMERSIBLES). These vehicles enable biologists to study the ocean's plants and animals in their natural environment. By the late 20th century, marine biologists were using Autonomous Underwater Vehicles such as MIT's *Odyssey* equipped with cameras and other sensing equipment to observe biological communities in action.

Instruments to Study Seawater

CHEMICAL OCEANOGRAPHERS have traditionally used water samples for chemical titration and analysis of oxygen, salts, and nutrients. Surface samples may be taken either from Niskin bottles on a CTD or by lowering a dip bucket. Shale Niskin (1926–88) the founder of General Oceanics, Inc., developed the Niskin bottle, which has replaced Nansen bottles. Chemical analyses are performed in the ship's laboratory spaces or at onshore research facilities. For example, marine technicians measure the temperature, SALINITY, dissolved oxygen, nutrients, and concentrations of chlorofluorocarbons (CFCs) from water samples that may be collected with Niskin bottles. Recent electronic techniques, common to shore-based chemical laboratories, including gas chromatography, mass spectrometry, infrared spectrometry, and colorimetric chemical analysis using spectrographs coupled to computers, have made these investigations more efficient and accurate. Autonomous vehicles have been designed that have a continuous intake of water for automatically analyzing biological productivity, measuring traces of pollutants, radioactive substances, and other seawater constituents.

Conductivity-temperature-depth (CTD) profilers have been developed to measure the salinity and temperature profiles of the ocean. The CTDs are installed on moorings, dropped from aircraft, lowered from ships, and installed on AUVs that survey programmed ocean routes. Niskin bottles of varying sizes are used to obtain large samples of seawater from various depths at specific stations. In some cases, moorings or autonomous drifting collection devices take water samples at discrete times and store them for later retrieval. These samples are analyzed using lab-based measurement methods. When the sample is on board, automated chemical analyzers are capable of measuring carbon dioxide, nitrate, phosphate, silicate, ammonium, and dissolved oxygen. These pumping systems can be used at depths from the surface down to about 33 feet (10 m); they can also be installed on board research submarines or on ROVs and AUVs.

Instruments to Study the Seafloor

GEOLOGICAL OCEANOGRAPHERS have greatly improved the techniques for acquiring samples of the ocean bottom. Early in the 20th century, simple GRAB SAMPLERS were used, consisting of a pair of jaws lowered on a cable to the seafloor that snapped shut, acquiring a sample of the bottom SEDIMENT. The PISTON CORER, lowered by ship cable to the ocean floor, where a suspended weight was dropped, driving a steel cylinder into the sea bottom extracting a core of sediment some 10 feet (3 m) in length largely replaced the grab sampler for deep-sea work. The corer was then retrieved and reset for another drop. The cores could be processed and analyzed on board ship or stored for later processing on shore. In general, the

youngest sediment is at the top of the core, and the oldest sediment is at the bottom. Shell or skeletal remains of organisms contained in sediment cores are useful in interpreting past environmental conditions, especially by paleo-oceanographers. Marine geologists conduct high-resolution seismic-reflection surveys in offshore regions for purposes ranging from seafloor exploration for discovering oil to investigating hazards posed by landslides, TSUNAMIS, and potential earthquake faults. Analyses of the sub-bottom sediments and structure below coring depth were made by recording the bottom-reflected/refracted sound waves on a hydrophone (underwater microphone) from explosive charges set off in the water column by another ship some distance away. Explosive and electromechanical sound sources such as sparkers and boomers are still used today to send sound waves into the ocean floor, to be recorded by arrays of hydrophones, suspended either from ships or buoys. By this means, the layers of sediment and rock under the seafloor are mapped and charts of oceanic regions constructed.

GEOPHYSICISTS routinely collect data from satellites, conduct geological surveys, construct maps, and use instruments such as gravimeters, tilt meters, and magnetometers to measure Earth's gravity and magnetic fields. Imagery of the bottom can also be obtained from side-scan SONARS towed behind various-size vessels. Many geophysicists use these instruments to search for oil, natural gas, minerals, and groundwater. Geophysicists also use computer modeling to portray water layers and the flow of water or other fluids through porous materials such as deep-ocean vents. Data from these many geophysical instruments may be processed and saved in Geographic Information Systems for the purpose of making maps and charts.

Among the laboratory instruments used by geologists and geophysicists are X-ray diffractometers to determine the crystal structure of minerals, petrographic microscopes to study rock and sediment samples, and seismographs to measure SEISMIC (earthquake) WAVES generated by movements in the Earth's crust. Most of what is known about the interior of the Earth has been learned from these measurements. The Earth's interior has a solid metal inner core of nickel and iron that is approximately 746 miles (1,200 km) in diameter, a molten nickel and iron outer core, a dense and mostly solid rock mantle, and a thin silicate rock crust. Heat from the inner core causes molten materials in the outer core and mantle to move around.

Geophysicists make extensive use of numerical models to analyze complex seismic data to develop insights into the structure of the Earth's mantle and core, especially the convective motions that are believed to drive tectonic plate motions (*see* PLATE TECTONICS). The asthenosphere is about 112 miles (180 km) thick and includes the lower mantle. The lithosphere includes the crust and uppermost mantle and is about 62 miles (100 km) thick. Mechanical coupling between the asthenosphere and the lithosphere is one of the forces that drives plate tectonics. The asthenosphere acts as a conveyor belt for the overlying lithosphere. These plate motions are complicated by variations in lithospheric thickness.

Instruments to Study the Physical Properties of the Ocean

PHYSICAL OCEANOGRAPHERS deploy arrays of current meters, water quality (temperature, salinity, oxygen, turbidity, nutrients) sensors, pressure sensors, wave and TIDE GAUGES, and drifting buoys. Some drifting instruments such as wave buoys are fitted into standardized air deployment packages for airdrop. Many recent observation programs have also been undertaken to improve computer modeling and simulation capabilities.

Physical oceanographers deploy large numbers of expendable instruments from both research and military or commercial vessels, such as the expendable bathythermograph, or "XBT," which is a streamlined weight with a temperature sensor and a long coil of conducting copper wire attached to a recorder on deck. The probe can be launched from a ship steaming at full speed. The instrument recorder on deck records temperature versus depth as the wire pays out until it breaks, thus ending the cast. The ensuing XBT profile documents how temperature varies with depth. Nansen bottles with reversing thermometers, of great historical importance in the exploration of the oceans, have been largely replaced by CTDs, which measure temperature with platinum resistance thermometers or thermistors, salinity by temperature and electrical conductivity, and depth by pressure transducers. The data are transmitted up an electrically conducting cable to a recorder on deck. By this means, continuous profiles of temperature and salinity versus depth (pressure) can be obtained. These data are generally used to compute the density and to estimate the currents from the density distribution by means of geostrophic computations (*see* GEOSTROPHY), but they also open a window on the small-scale structure and fluctuations of the ocean.

In order to investigate these short-period temperature fluctuations in the ocean, internally recording thermographs were deployed on moored buoys. These often initially consisted of a camera that took pictures of a mercury thermometer at constant time intervals. When the buoy was retrieved, the thermometer readings could be read from the film. More recent versions use thermistors, platinum resistance, or quartz crystal thermometers to record data on solid-state recorders. These sensors can be constructed with very short time constants, and, using rapid sampling, record the high-frequency temperature fluctuations in the sea. In the late 20th century, data could be sent in real-time from moored buoys to shore stations using packet radios, or across the world using earth Satellites. One of the first examples of real-time reporting of water levels, currents, and winds was demonstrated in Tampa Bay by NOAA's National Ocean Service (*see* NATIONAL OCEANIC AND ATMOSPHERIC ADMINISTRATION (NOAA). This system, according to one harbor pilot is the "seat belt and airbag" for the shipping industry.

"Thermistor chains," that is, long chains of hard rubber blocks, in which THERMISTORS are imbedded, measured high-frequency temperature fluctuations in depth. The data were transmitted up a wire cable to a recorder on deck. The chains were deployed on buoys or towed from ships. By this means, INTERNAL WAVES in vertically stratified ocean waters were recorded and the data analyzed. The results revealed a spectrum of temperature (really density) fluctuations with periods from months to minutes, and wavelengths from many hundreds of miles (km) to a few inches (cm).

A REMOTE SENSING method to obtain sea surface temperatures over large areas involves the use of radiometers on satellites. Satellites such as the National Oceanic and Atmospheric Administration's Advanced Very High Resolution Radiometer or AVHRR measure the amount of thermal infrared radiation given off by the surface of the ocean. Besides bodies of water, radiometers measure electromagnetic radiation from snow, ice, and CLOUDS. Sensors on satellites can only measure the surface of the ocean, and information about the ocean interior must be derived. Marine scientists use a variety of instruments to collect in situ data simultaneously with remote sensors. The in situ data helps in the analysis of imagery and is often called sea-truth data. Thermistors, underwater spectrometers, sediment traps, transmissometers, and sonar are some examples to measure water temperature, reflectance, suspended solids, turbidity, visibility, and water depth.

In the past, current speed and direction were measured by observing the drift of an object such as a TIDE POLE from a moored ship. Some of these measurements are still the basis for predictions in the traditional tidal current tables. Currents can be directly measured either by Eulerian current meters, which record the current velocity at a fixed location, or by Lagrangian current meters, which drift with the current and telemeter reports of change of position with time. Mathematicians Leonhard Euler (1707–83) and Joseph-Louis Lagrange (1736–1813) made major advances in fluid dynamics. The tide pole is an early form of Lagrangian current meter.

Another example of Lagrangian current measurements is found in the reports of "ship drift," in which the average current velocity is estimated from the deviation of a ship's time and position compared to its DEAD RECKONING time and position. It is assumed that the deviation in position is the result of offset by winds and currents. In the ship drift method, if the wind effect is neglected, the current can be estimated. These reports provide a major source of information on long-term currents; in fact, the first atlases of ocean currents were prepared by this means. Since the mid 20th century, ship drift data from both research vessels and ships of opportunity have been sent to oceanographic data centers, such as the U.S. National Oceanographic Data Center, in Silver Spring, Maryland (*see* NATIONAL OCEANIC AND ATMOSPHERIC ADMINISTRATION), where archives of ship drift current velocity estimates are maintained. Such data are also available as the International Comprehensive Ocean and Atmospheric Data Set or ICOADS, which can be obtained from the NOAA National Climatic Data Center.

The ship drift method of current estimation is only as good as the navigation of the ship, that is, the estimated ship's position. In recent years, navigation methods have improved steadily. The state-of-the-art method used today is the GLOBAL POSITIONING SYSTEM (GPS). Using GPS and modern, highly accurate timepieces (chronometers), the accuracy of ship drift current measurements has improved enormously from the early reports filed in the 19th century.

Lagrangian measurements can also be made by determining the trajectories of drifting buoys. Early drifting buoys consisted of floating logs, wooden crosses, parachute drogues, and DYE patches. In waters near shore, optical range and bearing systems (such as those left over from World War II) can be used to determine the position of the drifting buoy. The speed of the current can be computed from the change in position with time. As in the case of ship drift, modern drifting buoys use GPS receivers to transmit their positions to ships, land stations, or even Earth satellites. The data can be sent to investigators worldwide over the Internet.

One of the earliest Eulerian current meters was the Ekman current meter, a mechanical instrument that recorded the number of turns of a propeller in a unit time to get current speed, and dropped small metal balls in a container graduated in 10 degree increments as current samples were taken to determine direction. Electrical versions of the instrument were developed, such as the Roberts Radio Current Meter, which was deployed on moored buoys and radioed the data back to a servicing vessel. These instruments were basically underwater versions of the cup ANEMOMETER with wind vane, used for decades to measure wind speed and direction in the atmosphere.

Savonius rotor current meters were developed in the mid 20th century to minimize the considerable affect of vertical motions of the current meters resulting from ship and buoy motions. The Savonius rotor consists of two oppositely facing half cylinders with flat endcaps. The number of turns per unit time of the rotor is proportional to the speed of the current, which is often recorded by means of the closure of a magnetic switch activating an electrical current upon each rotation. Many early Savonius rotor current meters used photographic recording. The addition of a vane provided the direction of the current. Current meters such as the Richardson current meter, developed for deep-ocean use, incorporated a Savonius rotor for current speed, a vane for current direction, a thermistor for water temperature, a battery for the power supply, and a film recording system. In an early application of these instruments, physical oceanographers at the Woods Hole Oceanographic Institution deployed a set of Richardson current meter-equipped buoys along a line from Woods Hole, Massachusetts, to Bermuda during the 1950s.

Later models of Savonius rotor current meters incorporated better design and signal processing to minimize errors, magnetic tape recording, and long-life batteries to permit deployment times of many months in the deep ocean. In shallow water, rotor current meters might erroneously record high current velocities as they respond to oscillatory wave motions. They are biased by vertical flow.

More recent current meters have used the electromagnetic principle, that a moving current of salt water, being a conductor, creates a magnetic field, which in turn creates an electric current that can be readily measured. Early electromagnetic (EM) current meters used electrodes, whereas more recent EM current meters are inductively coupled. MAGNETIC CURRENT METERS have been crafted from telephone cables under the ocean, using the ocean current moving through Earth's magnetic field to generate the electric current through the cable. By this means, the current and resulting water transport were measured for the Florida Current as it passes through the Straits of Florida.

In the late 20th century, acoustic Doppler "speed logs" used to measure ship's velocity relative to the water were modified to provide the velocity of the water relative to the ship. These instruments were rapidly improved and configured for deployment on research vessels, moored buoys, bottom platforms, ocean drilling rigs, and even ships of opportunity. In their present form, they are called acoustic Doppler current profilers (ADCPS), and represent the present state-of-the-art in current measurements. Most major oceanographic expeditions today equip one or more vessels with ADCPs for current profiling while the vessels are under way.

Ocean current-measuring radars (*see* RADAR) employing the Doppler principle were also developed in the late 20th century, but these were not deployed operationally very often, in part because of the high cost of the instruments. Perhaps the most well known land-based radar developed in the United States was (and still is) the CODAR (Coastal Ocean Dynamics Applications Radar). CODAR is a readily transportable, remote sensing, and high-frequency (HF) radar that measures the dynamic properties of the ocean surface, such as surface wind waves, wind stress, surface currents, and movements of sea ice. CODAR can also be used to track transponder-equipped drifting buoys to provide a "ground truth" for Lagrangian current measurements.

Another current measuring radar system of the 20th century is the OSCR (Ocean Surface Current Radar), HF radar requiring a large land-based array of antenna elements developed in the United Kingdom. Higher-frequency (microwave) radars were designed to study the currents at shorter range, but providing greater resolution than HF radars. Pioneering work in this field was done by researchers such as Albert Hoyt Taylor (1879–1961) at the U.S. NAVAL RESEARCH LABORATORY. More recent innovations include the MIROS Direction Wave and Surface Current Radar and the ProSensing multifrequency delta-k radar. MIROS AS, a company focused on water monitoring, provides meteorological and oceanographic sensors for MARINE OPERATIONS and is located in Asker, Norway. ProSensing, a systems engineering firm, develops radar and radiometer sensing systems and is located in Amherst, Massachusetts. They developed coastal delta-k radar under sponsorship of the National Oceanic and Atmospheric Administration.

At the close of the 20th century, these radar systems were still being improved. Wide application of this method of measuring surface currents can be expected in the 21st century, from land-based systems, shipboard systems, and Earth satellites. In fact an operational CODAR system is located on the coast of New Jersey and is operated by Rutgers University. Currents can also be mapped by visual and infrared imaging systems aboard environmental satellites by several methods. One relatively simple method, at least in principle, is measurement of the duration of the displacement of thermal patterns on the surface of the ocean from one location to another. This procedure is basically the "velocity equals distance divided by time" relationship. The method is complicated by the temporal change in the patterns, which occurs along with the spatial change; hence, a mathematical pattern recognition procedure must often be used. Other procedures make use of the change in optical and thermal properties of currents compared to their surroundings (such as the warm waters of the GULF STREAM) to provide nearly synoptic maps of the variable features of the current, such as meanders and EDDIES. Satellite-based systems are limited, however, to measuring the properties of the sea surface. It is often possible to infer characteristics of the subsurface circulation from the surface data by means of numerical models based on ocean physics.

A method under development that can measure directly the subsurface currents of the oceans is called ACOUSTIC TOMOGRAPHY. The name "tomography" was chosen because of the similarity of the technique to X-ray tomography used to provide visual X-ray sections of the human body, as in the CAT scan. In the case of ocean acoustic tomography, vertical arrays of sound sources and receivers are installed on the ocean floor many miles (km) apart, and acoustic signals are transmitted over different paths to show the three-dimensional structure of the ocean volume. When a sound pulse is beamed over a known distance from the source array to a receiving array, the travel time is slightly less if it is traveling in the direction of the current, and slightly greater if it is moving in the direction opposed to the current. By measuring changes in arrival times over many different paths, the velocity field within the volume of the acoustic array may be mapped to within tenths of a knot (cm/s). These technologies are developing rapidly, and it is to be expected that they will be improved and made more accessible to researchers throughout the world in the 21st century.

Further Reading

Barrick, Donald E., Michael W. Evans, and Bob L. Weber. "Ocean Surface Currents Mapped by Radar," *Science* 198, 4313, (October 1977): 138–144.

Bowditch, Nathaniel. *The American Practical Navigator.* Bethesda, Md.: Defense Mapping Agency Hydrographic/Topographic Center, 1995.

Davis, Richard A., Jr. *Oceanography: An Introduction to the Marine Environment.* Dubuque, Iowa: Wm. Brown, 1991.

Lipa, Belinda J., and Donald E. Barrick. "Least-Squares Methods for the Extraction of Surface Currents from Crossed-Loop Data: Application at ARSLOE." *IEEE Journal of Ocean Engineering*, vol. OE-8, 4 (October 1983): 226–253.

Pickard, George L., and William J. Emery. *Descriptive Physical Oceanography: An Introduction*, 5th ed. Oxford: Pergamon Press, 1990.

Prandle, David, and D. K. Ryder. "Measurement of Surface Currents in Liverpool Bay by High-frequency Radar." *Nature* 315 (May 1985): 128–131.

U.S. Department of Commerce, Coast and Geodetic Survey. *Manual of Current Observations*. Washington, D.C.: U.S. Government Printing Office, 1950.

oceanography Oceanography is the study of the oceans and the atmosphere above the oceans. Oceanography has its basis in the scientific disciplines of biology, chemistry, geology, and physics. Therefore, the field of oceanography has been divided into four main branches, biological oceanography, chemical oceanography, geological oceanography, and physical oceanography. Many oceanograpers have graduate degrees and strong backgrounds in the basic sciences. They form mulidisciplinary and integrated teams to conduct research on the oceans.

Oceanography is a relatively young science; a systematic, scientific study of the oceans is only about 200 years old. Benjamin Franklin (1706–90) studied the GULF STREAM using temperature measurements provided by ship captains in the late 18th century. The U.S. Coast Survey (now the National Ocean Service, U.S. NATIONAL OCEANIC AND ATMOSPHERIC ADMINISTRATION) was established in 1807 to study and chart coastal waters for navigation. The U.S. Navy's Depot of Charts and Instruments (now the U.S. Naval Oceanographic Office) was founded in 1830. Alexander Dallas Bache (1806–67), grandson of Benjamin Franklin, took over the leadership of the Coast Survey in 1843 and directed much research to define the boundaries of the Gulf Stream. Commodore Mathew Fontaine Maury (1806–73; called the "Pathfinder of the Seas"), was director of Departments of Charts and instruments from 1842–61 and compiled vast amounts of information on ocean CURRENTS, and climatic conditions (*see* METOC OFFICER). A British expedition, the voyage of the HMS *Challenger* (1873–76) was government-funded to perform the study of the seas for scientific purposes and produced a 50-volume report in 1895. The German *Meteor* Expedition of the late 1920s obtained the first comprehensive sets of data and analyses on the WATER MASSES of the Atlantic Ocean. In the United States, the Scripps Laboratory (now the SCRIPPS INSTITUTION OF OCEANOGRAPHY) was established in 1925, and the WOODS HOLE OCEANOGRAPHIC INSTITUTION was established in 1930. During World War II, oceanographers such as Harald Ulrik Sverdrup (1888–1957) used their skills to forecast wave and weather conditions for Allied troops preparing to invade the beaches of Africa, Europe, and the Pacific. Following World War II, with the advent of the cold war, oceanography was given a major funding boost to provide environmental information for planning and operations in antisubmarine warfare. Technologies developed in World War II, especially SONAR, were used to map the floor of the global ocean and to reveal the mechanism by which the continents are spreading apart or crashing together (*see* PLATE TECTONICS).

Today, oceanographers of all backgrounds use technological developments of the late 20th and early 21st centuries to maximum advantage. Biological oceanographers can now directly observe marine animals in action from small manned SUBMERSIBLES, REMOTELY OPERATED VEHICLES (ROVs), and AUTONOMOUS UNDERWATER VEHICLES (AUVs). They are using sonar and LIDAR to locate and make bio-assessments of schools of fish. Geological oceanographers and geophysicists are drilling into the ocean floor to deeper and deeper depths. They are studying landforms, sedimentation, earthquakes, and volcanoes from satellites equipped with visual and microwave remote imaging systems. Chemical oceanographers are using towed subsurface vehicles to sample and analyze the water near HYDROTHERMAL VENTS and to track the vent plumes. Physical oceanographers are measuring currents remotely with radar and acoustic scanning systems, and studying super typhoons with satellite visual and infrared images (*see* OCEANOGRAPHIC INSTRUMENTS).

The advent of the use of digital systems and computers in the late 20th century has revolutionized oceanography as it has the other sciences. Instruments can be deployed on the floor of the ocean and record data on solid-state devices for months at a time. Vast amounts of data on ocean circulation can now be analyzed and used to help develop and calibrate numerical computer models predicting such things as the affects of GLOBAL WARMING on the Gulf Stream. Laser-based systems in aircraft can make possible the assessment of BEACH erosion following the passage of HURRICANES and major storms. Real-time systems deployed in ports and harbors can provide information on winds, waves, currents, and tides to harbor pilots, commercial shippers, the military, and recreational boaters to provide for the safe and efficient operation of port activities (*see* PORTS AND HARBORS).

Biological Oceanography

Biological oceanography is the study of the plants and animals that make the ocean and its environs their home. It encompasses the location and classification of the plants and animals of the marine environment, including bacteria, PHYTOPLANKTON, ZOOPLANKTON, the fishes, and MARINE MAMMALS such as otters, dolphins, and whales, and marine reptiles such as turtles, snakes, and crocodiles. Biological oceanographers study the basic productivity of the ocean, the food chain, and the ECOSYSTEMS of estuarine, coastal, shallow, and deep waters. They categorize the reproduction and life cycles of marine animals and plants. They study the uptake of nutrients by marine plants, the interrelations of plants and animals, the effects of light, temperature, salinity, pH, and nutrients on marine creatures.

Marine pathologists investigate invertebrates, fishes, marine mammals, and seabirds. They study the affects of parasites, diseases, and anthropogenically introduced pollutants on animals and plants of the ecosystems of rivers, ESTUARIES, and coastal waters. Fisheries biology is an important division of biological oceanography. Many marine biologists work in the area of development, maintenance, and management of fisheries. They try to restore fisheries where stocks have been depleted by overfishing. This is the situation now extant for the Atantic cod (*Gadus morhua*) fishery in the Georges Bank area, which was severely overfished by vessels from Europe and the Soviet Union during the latter part of the 20th century.

Marine biologists study the effects of pollutants such as petrochemicals on living creatures and develop methods of toxic spill remediation. They are presently engaged in a battle to prevent alien species, often transported by ballast water in the holds of ships, or by release from aquaria, from taking over local environments. Examples of such invasive species include water chestnuts (*Trapa natans*), zebra mussels (*Dreissena polymorpha*), common lionfish (*Pterois volitans*), several species of catfish (e.g., *Pylodictus olivaris* and *Clarias batrachus*), and South American nutria (*Myocastor coypus*). Microbiologists study bacteria and viruses in the ocean and seek to develop new drugs for combating human ailments.

Chemical Oceanography

Chemical oceanography is the study of the composition of seawater, the elements found in the sea, and their exchange with the atmosphere, the bottom sediments, and the marine plants and animals. Chemical oceanographers study the geochemistry of seawater, the interrelations among and distribution of nutrients. They are concerned about the cycles of salinity, oxygen, carbon, and nutrients, and are particularly interested in the absorption of carbon dioxide from the atmosphere, and the pH balance of the ocean. If the ocean temperature becomes too high, or the water is too acidic, for example, CORAL BLEACHING is likely to occur with the eventual death of the CORAL REEF. Chemical oceanographers are concerned about the flux of sulfides and metallic compounds from hydrothermal vents into the deep ocean, and whether this flux has a major effect on the composition of seawater. They are concerned about the rate of absorption of carbon from the atmosphere, and whether the oceans can absorb most of the excess carbon dioxide being pumped into the atmosphere by industrialized nations.

Geological Oceanography

Geological oceanography is the study of the configuration and composition of landforms bordering the seas, as well as the floor and the subfloor of the ocean. Geological oceanographers are interested in the depth and composition of the sediments above the ocean bottom, the deposition of the sediments from lands bordering the sea, and the deposition of organic material and calcium carbonate resulting from the sinking of dead marine plants and animals over millions of years. Geological oceanographers study the formation of marine minerals, and they map regions of the ocean floor that may contain oil and gas deposits suitable for drilling. They are involved in developing new resources such as extraction of fuel from methane hydrates, which are found on the floor of the ocean in great abundance in areas such as the CONTINENTAL SHELF of the Mid-Atlantic coast of the United States. Geological oceanography is closely related to marine geophysics in areas such as the study of earthquakes, volcanoes, and landslides under the ocean, and the consequent TSUNAMI waves that sometimes accompany such events. Geological oceanographers are extensively involved in efforts to delineate regions of the ocean floor where rifts and fault zones are likely to cause earthquakes leading to devastating tsunami waves such as the great tsunami wave that struck Indonesia and South Asia in December 2006.

The structures of the ocean borders of continents, island chains, and the subfloor of the ocean are included in the study of geophysics. Earth's crust under the ocean is much thinner and younger than that under the continents. This crust has been created by UPWELLING of hot magma from deep within the mantle and is extruded into the cold ABYSSAL ZONE of the ocean in regions such as the mid-ocean rift. The force of this new crust is opening the Atlantic basin and pushing apart the continents of Europe and North

America, as well as Africa and South America. Around the Pacific Rim, however, crust is sinking in areas such as the west coast of North and South America, the coast of Japan, and Indonesia. This rising and sinking of hot magma from the mantle is part of the mass balance of the solid Earth. It results in the convective forces that drive the great tectonic plates on which the continents ride (*see* PLATE TECTONICS).

Marine geophysics also includes the study of Earth's geopotential (gravity) and magnetic fields under the ocean. The study of the alignment of magnetic materials in the rock of the ocean floor provided considerable evidence for the drift of continents and the subsequent development of the modern theory of plate tectonics. The geopotential and magnetic fields within the ocean and its environs are of considerable interest to the scientific and commercial communities.

Physical Oceanography

Physical oceanography is the study of the physics of the ocean and its environs, including the lower marine atmosphere. It includes the study of waves, TIDES, CURRENTS, TURBULENCE, and the distribution of physical properties of seawater, such as TEMPERATURE, SALINITY, and DENSITY. Oceanographers study the propagation of sound in the sea and use sound waves as REMOTE SENSING instruments to provide profiles and sections of ocean properties. They are interested in the heat and salt budgets of the oceans and the lower atmosphere; they also study the exchanges of heat with the atmosphere on time scales from minutes to eras. They perform long-term studies of climatology and the affects of global warming. They study the formation of water masses of the ocean, and the transport of water, heat and salt from one part of the ocean to another, such as in the sinking of surface water in the seas around Greenland and Labrador to form the North Atlantic Deep Water (*see* WATER MASSES), which then spreads southward and eastward at depths close to the ocean floor. Physical oceanographers work with MARINE METEOROLOGISTS to understand the exchanges of heat and salt between the polar and subtropical regions. They study the effects of major oceanic/atmospheric events such as the EL NIÑO on the climate of the ocean and the bordering continents, as well as other long-term interactions between the ocean and the atmosphere.

Physical oceanography also includes the study of the physics of estuaries, beaches, and coastal zones. Physical oceanographers work with engineers and meteorologists to predict the STORM SURGE caused by hurricanes, as well as storm tides due to winter storms at mid-and polar latitudes. They work with weather forecasters to provide mariners with information on winds, waves, locations and intensities of storms. This information not only greatly enhances maritime safety, but enables global shipping companies to optimize their sea routes (*see* OPTIMAL SHIP TRACK PLANNING). Physical oceanographers along with other marine scientists, engineers, and policy makers have been instrumental in the deployment and operation of ocean observation systems such as the FIELD RESEARCH FACILITY in Duck, North Carolina, and the GULF OF MAINE OCEAN OBSERVING SYSTEM in Portland, Maine.

Further Reading

Charton, Barbara. *The Facts On File Dictionary of Marine Science,* new ed. New York: Facts On File, 2008.

Davis, Richard A. *Oceanography: An Introduction to the Marine Environment,* 2nd ed. Dubuque, Iowa: Wm. C. Brown, 1991.

Nichols, Cied R., David L. Porter, and Robert G. Williams. *Recent Advances and Issues in Oceanography.* Westport, Conn.: Greenwood, 2003.

Oceanography Society, The   A professional society with the overarching goals to (1) disseminate knowledge about oceanography and its application through research and education, (2) promote communication among oceanographers, and (3) provide a constituency for consensus-building across the many disciplines that make up oceanography. The Oceanography Society was founded in 1998 and hosts periodic professional meetings, publishes *Oceanography,* a magazine that promotes and chronicles all aspects of ocean science and its applications, and recognizes exemplary accomplishments in oceanography through several awards. The society has approximately 1,500 members who generally work in the fields of biological, chemical, geological, physical oceanography, engineering, education, or public policy. The Jerlov Award is provided to oceanographers making significant advances in the nature and consequences of light in the ocean. Nils Gunnar Jerlov made many advances in ocean optics and applied his work to the classification of WATER MASSES. The Walter Munk Award is intended to recognize distinguished acoustics research. Walter Heinrich Munk (1917–   ) is an esteemed contributor to the fields of physical oceanography and geophysics. At present, he is emeritus professor of geophysics at the SCRIPPS INSTITUTION OF OCEANOGRAPHY in La Jolla, California. The Oceanography Society Fellows Program has been established to recognize individuals who have attained eminence in oceanography though their outstanding contributions to the field of oceanography or its applications during a substantial period of years.

See also AMERICAN GEOPHYSICAL UNION; AMERICAN METEOROLOGICAL SOCIETY; AMERICAN SOCIETY OF LIMNOLOGY AND OCEANOGRAPHY; MARINE TECHNOLOGY SOCIETY; OCEANOGRAPHY.

Further Reading

Arnone, Robert A., Michelle Wood, and Richard W. Gould Jr. "The Evolution of Optical Water Mass Classification." *Oceanography* 17, no. 2 (2004): 14–15.

International Council for the Exploration of the Sea. Available online. URL: http://www.ices.dk/indexfla.asp. Accessed February 25, 2007.

Nichols, C. R., D. L. Porter, and R. G. Williams. *Recent Advances and Issues in Oceanography.* Westport, Conn.: Greenwood, 2003.

The Oceanography Society. Available online. URL: http://www.tos.org. Accessed February 25, 2007.

Ocean.US   Ocean.US is the Web site of the national office for planning and coordination of the Integrated Ocean Observing System or IOOS in the United States. During 2000, it was established as an intergovernmental agency to develop IOOS, the United States's contribution to a Global Ocean Observing System.

See also OCEAN OBSERVATORY.

Further Reading

Ocean.US: The National Office for Integrated and Sustained Ocean Observations. Available online. URL: http://www.ocean.us/. Accessed August 15, 2006.

Office of Naval Research (ONR)   The Office of Naval Research or ONR is responsible for all aspects of science and technology programs that support the U.S. Navy and the U.S. Marine Corps. To ensure the success of scientific research programs and to ensure their relevancy to operational Naval and Marine Corps forces, ONR develops research thrusts and ongoing collaborative relationships with universities, private and public research institutions, and other government agencies, as well as industry. The various departments or "Codes" within ONR have specific areas of expertise and focus. Codes include 31: Information, Electronics, and Surveillance; 32: Ocean, Atmosphere, and Space; 33: Engineering, Materials, and Physical Science; 34: Human Systems; 35: Naval Expeditionary Warfare, and 36: Industrial and Corporate Programs. Oceanographic studies fall largely under Code 32. This department also includes the Naval Space, Science & Technology (S&T) Program Office. ONR examines various capabilities for observing, modeling, and predicting oceanographic processes. ONR's oceanographic efforts attempt to better understand and exploit environmental pro-

cesses, sensors, and the data they provide, physical models of the ocean, assimilation of observed data into modeled results and vice versa, and model and algorithm validation studies.

The U.S. Navy has a great interest in antisubmarine warfare (ASW). ASW includes the science of detecting, locating, and classifying submarines with acoustic (SONAR) and non-acoustic technologies. The Navy and Marine Corps are both very interested in the concept of mine warfare (MIW). MIW includes the science of detecting, localizing, identifying, and neutralizing mines in both the ocean and littoral (coastal) environment, and improving offensive mining capabilities. ONR also focuses on maritime intelligence, surveillance, and reconnaissance (ISR) and space exploitation. ISR includes the science of providing maritime "situational awareness" through development and exploitation of REMOTE SENSING and space communications capabilities.

ONR's research interests also include high-resolution (both spatial and temporal) characterization of littoral and open ocean processes, real-time environmentally adaptive sensors and systems, development and use of distributed and autonomous ocean systems through data assimilation, exploitation of multispectral and broad-band sensors, assimilation of multiple data streams into predictive models, innovative techniques for modeling and validation of MODELS, broadband acoustic sources and volumetric arrays, space technology for ocean and atmosphere remote sensing, new classification algorithms, and automated target recognition for cluttered undersea environments. ONR recently funded the University of North Carolina to install the South East Atlantic Coastal Ocean Observation System. This system, which is also called SEACOOS, may become an important component of the U.S. Integrated Ocean Observing System.

ONR's undersea laboratory, SEALAB II, permits many types of manned undersea experiments. *(U.S. Navy)*

Further Reading
Office of Naval Research. Available online. URL: www.
onr.navy.mil. Accessed October 17, 2003.

offshore drilling Offshore drilling refers to the exploration and exploitation of a marine mineral resource, primarily petroleum and natural gas on the CONTINENTAL SHELF, but extending onto the CONTINENTAL SLOPE and CONTINENTAL RISE as new technologies are developed to support deep drilling. Petroleum is a complex mixture of hydrocarbons, and the primary fuel of modern civilization.

Initially, petroleum deposits under the sea were extracted from wells in shallow water of a few tens of feet (m) in depth on the inner continental shelves, beginning in 1896 from a wooden pier in Summerland, California, but over the years, improving technology has made possible the extension of drilling of exploratory wells to 10,000 feet (3,048 m) or more. The Transocean drillship *Discover Deep Seas* started

Drill Ship *JOIDES Resolution*

The scientific research vessel *JOIDES Resolution* began operations in 1978 as the *Sedco/BP 471*, originally an oil exploration vessel. In January 1985, after being converted for scientific research, the vessel began working for the Ocean Drilling Program (ODP). The vessel was contracted again in June 2004 to serve the Integrated Ocean Drilling Program (IODP). The *JOIDES Resolution* is owned by Overseas Drilling Limited, which is a joint venture company owned 50 percent by Transocean and 50 percent by DSND Shipping AS. The vessel is named for the *HMS Resolution*, which explored the Pacific Ocean, its islands, and the Antarctic region under the command of Captain James Cook more than 200 years ago. Like its namesake, the purpose of the current *Resolution* is to sail for scientific exploration. But this time those discoveries lie deep beneath the oceans.

The physical dimensions of the ship are remarkable. Fitted with a drilling derrick standing 202 feet (61.5 m) above the waterline, the 469-foot (143 m)-long ship is 68.9 feet (21 m) wide. During a leg, the crew positions the ship over the drill site, using 12 computer-controlled thrusters as well as the main propulsion system. The rig can suspend as much as 30,020 feet (9,150 m) of drill pipe to an ocean depth as great as 27,018 feet (8,235 m).

Near the center of the ship is the "moon pool," which is a 23-foot (7 m)-wide hole through which the drill string is lowered. Each pipe joint is about 93.5 feet (28.5 m) and weighs about 1,925 pounds (874 kg). The drill crew uses the draw works to thread each joint to the drill string. The process of lowering the drill bit, which is affixed to the end of the drill string, takes about 12 hours in 18,045 feet (5,500 m) of water. To core through the seafloor, the entire drill string is rotated. The thrusters mounted underneath and facing perpendicular to the long axis of the ship keep the massive vessel from rotating during drilling operations.

The ship has a "lab stack" where several laboratories are organized on seven floors with more than 12,000 square feet (1,115 sq m) of space. On Level 1 (Hold Deck) at the bottom of the vessel is the general cold storage. Level 2 (Lower 'Tween Deck) has the refrigerated core storage and most of the gym. (The 'tween deck is the space between two continuous decks.) The gym is continued on Level 3 (Upper 'Tween Deck), which also houses the electronics shop and photography lab. On Level 4 (Main Deck) are the computer user room, computer center, and science lounge. Level 5 (Forecastle Deck) (The upper deck near bow and crew berthing) contains chemistry, microbiology, paleontology, and X-ray laboratories. Level 6 (Bridge Deck) is where core handling, sampling, and description are done. Also on this deck are the physical properties and paleomagnetism laboratories, and the core photo table. Level 7 (Lab Stack Top) contains the downhole measurements and thin section laboratories. At the fantail of the ship (Poop Deck), the underway geophysics lab contains the equipment that gathers ship position, water depth (bathymetry), and magnetic/seismic information useful in studying the topography of the seafloor. Living quarters, the galley, the hospital, and library are located in the forward section of the ship. The *JOIDES Resolution* can maintain operations with a crew of 112 persons for up to 90 days.

Work aboard the ship never ceases; operations continue 24 hours a day, seven days a week. Even during port call, work continues around the clock. On any given expedition, the ship's complement consists of as many as 50 scientists/technicians and 65 crew members. Data and figure courtesy of IODP-USIO (JOI Alliance).

Further Reading
Integrated Ocean Drilling Program. "U.S. Drillship Tour & History." Available online. URL: http://iodp.tarnu.edu/publicinfo/drillship.html. Accessed October 17, 2007.

—**Bradley A. Weymer,** marine laboratory specialist
Integrated Ocean Drilling Program
Texas A & M University
College Station, Texas

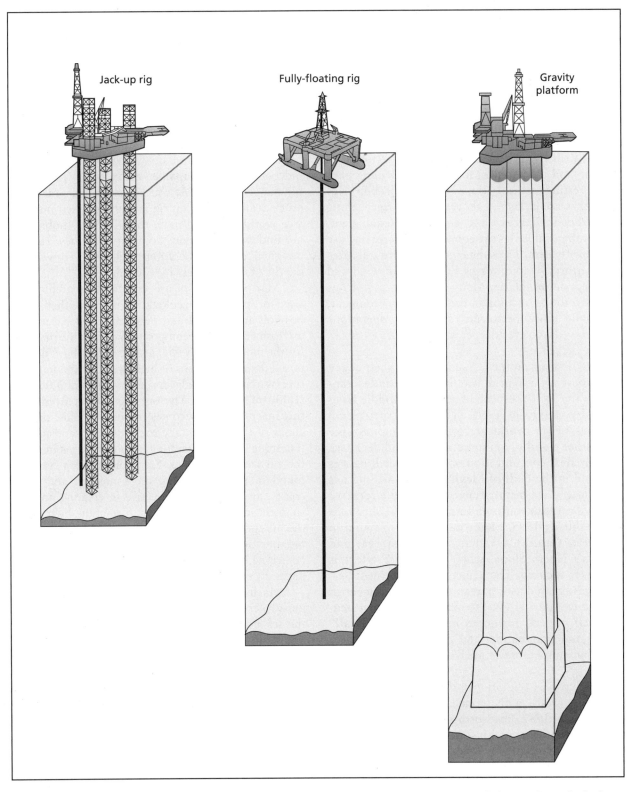

Jack-up rig

Fully-floating rig

Gravity platform

Types of drilling rigs include the jack-up rig, a free-floating rig that extends three or four telescopic legs to the seabed when ready to drill. The fully floating rig is mounted on a free-floating, self-powered hull. Once in position, several motors keep it on station. The vessel can move from one drill site to the next under its own power. The gravity platform, used in relatively shallow sites, is towed to station and then lowered into position when ready to drill. Once production is started, this rig is not moved.

a well in 10,011 feet (3,051 m) of water while constructing ChevronTexaco's Toledo well in the Alaminos Canyon of the GULF OF MEXICO. Offshore wells supply approximately 20 percent of the world's supply of petroleum and 5 percent of its natural gas. Petroleum is by far one of the most valuable mineral resources mined from the sea.

Drilling rigs include floating ships operating in water less than 100 feet (30 m). In deeper waters, mobile drill rigs are often used, which are towed to the site and then sunk. Most of these drilling platforms resting on the seafloor are called "jackup" rigs since they are jacked up above the sea surface, along with the drilling rig and supporting equipment. Semi-submersible rigs are coming into increasing use in deeper waters. The submerged structure stabilizes the platform by minimizing the effects of wind and wave action on the rig. Many different types of rigs are used for oil drilling in the marine environment, especially since petroleum extraction operations have moved into ever-deeper water (*see* OIL PRODUCTION PLATFORMS).

Oil drilling on the U.S. continental shelf began in earnest after World War II and expanded rapidly in the 1970s, especially after the Middle Eastern oil embargo in 1983. (The largest petroleum reserves in the world are found in Saudi Arabia and other nearby countries of the Middle East). The greatest concentration of offshore drilling rigs is found in the Gulf of Mexico, where drilling has continued since installation of a fixed platform 12 miles (19.3 km) south of Terrebonne Parish, Louisiana, during 1947. Major fields are also found in Southern California waters, the North Sea, and Alaska's North Slope, where an intense effort is currently focused. The Minerals Management Service was created as a bureau of the Department of the Interior during 1982 to manage outer continental shelf mineral resources in an environmentally sound and safe manner and to collect, verify, and distribute mineral revenues from federal and Indian lands.

In Europe, exploration of the North Sea began in the early 1960s, with production on a small scale beginning in 1967. The North Sea is an extremely difficult operating area with severe winter storms, high winds, and dangerous waves as high as 100 feet (30 m). Drilling for petroleum and natural gas is an inherently dangerous activity. In March 1980, the Norwegian platform Alexander L. Kielland collapsed in a storm with the loss of more than 100 people. The North Sea is the leading source of Europe's oil reserves and is one of the largest non Organization of the Petroleum Exporting Countries (OPEC) regions in the world. The OPEC countries include Algeria, Indonesia, Iran, Iraq, Kuwait, Libya, Nigeria, Qatar, Saudia Arabia, and the United Arab Emirates.

Offshore drilling has increased greatly worldwide due to the increasing oil prices and the desire to become more independent of Middle Eastern reserves. The high price of oil in 2005–06 has resulted in a veritable explosion of interest in exploiting all possible oil fields. Although oil prospecting has many risks, including financial risks due to the high development and operating costs, major and independent oil companies are pressing the search for major reserves. At the present time, practically every available drilling rig is being pressed into service worldwide, and more than 30 new mobile rigs are under construction. Some of these new rigs are designed to drill in 12,000 feet (4,000 m) of water to depths of 50,000 feet (15,000 m).

Marine geologists are constantly searching the seas for promising geologic formations that might bear oil and natural gas. Petroleum and natural gas are formed by the decomposition of organic material, for the most part PHYTOPLANKTON, that have decomposed under conditions of high pressure and temperature overlain by thick layers of SEDIMENT of 3,000 feet (1,000 m) or more. The organic carbon from these tiny marine plants becomes trapped in folds, or anticlines of the sedimentary rock, usually sandstone or limestone. Geologic faulting provides spaces in which the oil can accumulate. Salt domes, such as those found in profusion along the Louisiana and Texas coasts, are particularly effective in creating an environment suitable for petroleum formation, by trapping the oil and gas. This compressing and heating of the organic material to form petroleum is known as "cracking." About 1.5 million barrels of oil are released naturally every year from the seabed into the ocean.

Marine geologists use explosive charges and air guns to create powerful sound waves that reflect off the sea bottom and reflect or refract from the rock and sediment layers below to provide a picture of the sedimentary structure and its likelihood of bearing oil. If the area looks sufficiently promising, a test boring is made by a drilling platform. As the well is drilled, it must be lined with steel casings to prevent cave-ins and to provide a conduit for the oil to flow to the platform on the surface. A blowout preventer is installed on the steel casing to prevent a geyserlike explosion of the oil when the drill bit penetrates the cap rock. If the well produces a significant amount of oil, additional wells are drilled, and a field may be developed. The offshore region of the Mid-Atlantic coast of the United States appeared geologically promising as a source of petroleum, especially in the region of the Baltimore Canyon Trough, but has not proved to be productive, with many dry wells illustrating how even very promising geological structures do not always yield the expected significant quantities of petroleum. Companies such as Exxon Mobil have been prohibited from oil exploration offshore of the Carolina capes by moratoria since 1982.

Recent petroleum explorations have focused on deeper water as shallow reserves are used up. The technological challenges in drilling in water thousands of feet (m) deep are severe. Floating platforms must be used that are held in place by ANCHORS, ballast, and computer-controlled stabilizing systems. Controlling a drill bit almost two miles below the surface of a rough sea, along with the associated pipes and risers, is a remarkable technical achievement possible only very recently.

New technologies are also advancing the capabilities of diving services (see DIVER), service vessels (workboats), personnel transport vessels, pipelines and terminals to transport petroleum from shallow water fields, and tankers to transport petroleum and natural gas from deep-water drilling platforms to shoreline terminals for subsequent processing and distribution. Exploration and development of deep-water drilling sites requires the extensive utilization of REMOTELY OPERATED VEHICLES (ROVs) for such activities as drill bit insertion and repair, and AUTONOMOUS UNDERWATER VEHICLES (AUVs) for performing site surveys. As the search for new petroleum reserves continues at a frantic pace, the capabilities of these vehicles are being pushed to the limit.

Present areas of active exploration and development include the coastal and continental slope regions of Brazil, Angola, all of West Africa, and Southeast Asia, with special interest focusing on the deep-water (greater than 1,000 feet (300 m) areas. Shallow-water areas of great interest include the Caspian Sea and western Australia.

The high costs and great difficulty of offshore oil production and transportation can be seen in the August and September 2005 paths of destruction created by hurricanes Katrina and Rita in the Gulf of Mexico, where more than 100 oil platforms were destroyed and more than 50 damaged, with associated damage to pipelines and other infrastructure. A year later, the Alaska Pipeline from the Alaskan North Slope to the U.S. West Coast was required to be shut down due to rust and other maintenance problems, which had resulted in relatively small oil spills. Nevertheless, the turbulent conditions in the Middle East provide a constant driving force in the search for undersea oil and gas.

Offshore drilling can potentially yield sources of energy other than petroleum and natural gas, including geothermal energy from pressurized deposits of natural gas and from methane hydrates (crystalline enclosures around methane molecules formed under low temperatures and high pressures on several areas of the seafloor, such as off Cape Hatteras). However, extracting the methane fuel has not been feasible due to the danger of explosions. Giant plumes of explosive methane have been observed rising from the sea floor. Nevertheless, Japan is conducting a major study to determine safe methods for separating out the methane fuel. Oil shale is another possible source of hydrocarbon fuels, but at present the cost of extraction is too high to be economically feasible.

In general, offshore drilling is conducted in blocks leased by governments. In the United States, the Minerals Management Service (MMS) leases blocks within the EXCLUSIVE ECONOMIC ZONE to drilling companies in areas such as the Gulf of Mexico and Southern California. Prior to leasing, an environmental impact statement must be prepared, based on all available historical environmental data, to show that the environment would not be threatened by the drilling activity. Nevertheless, oil spills do occur, producing significant environmental damage (see MARINE POLLUTION). The technology to prevent and limit oil spills is growing steadily; for example, new tankers are required to have double hulls to minimize the effects of a collision.

As long as the present turmoil continues in the Middle East, industrialized nations will devote great resources to the exploration and development of new petroleum and natural gas reserves as well as alternative sources of energy. In addition, some countries store petroleum for use in emergencies. The U.S. Strategic Petroleum Reserve (SPR) is the world's largest supply of emergency crude oil stored in huge underground salt caverns along the Gulf of Mexico coastline. A decision to withdraw crude oil from the SPR is made by the president under authority of the Energy Policy and Conservation Act of December 1975.

Further Reading

Erickson, Jon. Marine Geology: Exploring the New Frontiers of the Ocean, rev. ed. New York: Facts On File, 2003.

Mangone, Gerard F. Concise Marine Almanac. New York: Van Nostrand Reinhold, 1986.

Minerals Management Service. Available online. URL: http://www.mms.gov/. Accessed August 29, 2006.

OPEC: Organization of the Petroleum Exporting Countries. Available online. URL: http://www.opec.org/home/. Accessed August 29, 2006.

Pagano, Susanne. "Strong Oil, Gas Demand Drives Drilling, Production Activities." Sea Technology 47, no. 4 (2006): 28–36.

———. "Worldwide Expansion of Oil Drilling Activities." Sea Technology 42, no. 4 (2001): 10–16.

Seibold, Eugene, and Wolfgang H. Berger. *The Sea Floor: An Introduction to Marine Geology*, 3rd ed. Berlin: Springer, 1996.

oil production platforms Oil production platforms are specially designed steel, composite, and concrete structures deployed in the ocean to drill for and produce oil and gas. Composites are materials such as fiberglass, which when combined have properties superior to the individual materials. Giant oil production platforms are built in dockyards, then towed to exploration fields, and modules are assembled on the high seas. They are generally installed on the CONTINENTAL SHELF and are built to survive the 100-year storm, a storm and its associated waves, surge, and winds that would occur once in a century. For example, the distance from the lowest deck to MEAN SEA LEVEL is designed to exceed large storm waves. This distance is called airgap. During Hurricane Katrina, wave heights in the GULF OF MEXICO were 55.5 feet (16.91 m) at 5:00 A.M. central time on August 29, 2005. This wave information was reported from a 3-m National Data Buoy Center DISCUS BUOY deployed 64 nautical miles (118.5 km) south of Dauphin Island (*see* BUOYS). Marine scientists approximate maximum wave heights as 1.9 times the significant wave height (*see* WAVES). Thus, the maximum wave height was approximately 105 feet (32 m). The combination of Hurricanes Katrina and Rita during August 2005 destroyed approximately 100 production platforms in the Gulf of Mexico and associated pipelines. Some were salvaged, and others were sunk to make artificial reefs.

Oil is a finite resource that continues to be discovered and mined in more complex areas, such as polar climates where ICEBERGS complicate drilling operations, in regions near the CONTINENTAL RISE associated with strong EDDIES such as GULF STREAM rings, or off the continental shelf in waters deeper than 10,000 feet (3,048 m). Oil fields are the result of dead animal and plant materials that have been buried under layers of SEDIMENTS. The associated PRESSURE and TEMPERATURE are sufficient to convert the biogenic material into hydrocarbons. Geophysicists study the propagation of shear waves (S-waves) and primary waves (P-waves) to obtain initial clues on the composition and structure of sediments on the seafloor. They are looking for economically favorable reservoirs of hydrocarbons that are trapped under sedimentary rocks. These reservoirs appear as anomalies or bright spots on seismic surveys of the seafloor. A favorable location may become the site for the installation of an exploratory platform to drill a test well. After analysis, the hydrocarbons may be obtained and processed via a system of production platforms and connecting pipelines. Some oil and gas fields are associated with natural seeps. These hydrocarbons bubble up to the surface. Specially designed tents or gathering systems are used to capture the natural gas, which is then transported to onshore facilities by pipeline. Once most of the oil has been recovered from an oil field, the offshore platforms may be dismantled and refurbished or used as an artificial reef. Programs such as Rigs to Reefs in Louisiana and Texas have been instrumental in using decommissioned rigs as habitats for commercially important finfish and MOLLUSKS.

Offshore platforms are like ships, and there are many designs and layout solutions derived by multidisciplinary teams of scientists, engineers, and technicians (*see* OCEAN ENGINEER). They may be unique structures and are sometimes classified as fixed platforms, semi-submersible platforms, submersible platforms, jackup platforms, tension-leg platforms, spar platforms, and drill ships. Fixed platforms have negligible motions and are built on legs that are anchored directly onto the seafloor. Submersible platforms usually have two hulls, the mat and the main hull. Once the platform is towed to its shallow-water site, the lower hull or mat is flooded, lowering the entire unit to its operating draft, which rests on the seafloor. During redeployment, the lower hull is deballasted, raising the entire unit to its towing draft. A submersible drilling rig is a bottom-supported unit. Jackup platforms are self-supported, resting on hydraulic legs that can be raised and lowered. Semi-submersible platforms are partially submerged when the rig reaches the drill site and buoyant legs allow the platform to float. The legs act as a keel to keep the structure upright, and MOORING lines are used to keep the platform in place. Tension leg platforms are floating rigs where vertical motions are reduced through moorings that tether the rig to the seafloor. Tethers and risers may be made of large chains and pipes, respectively. Spar platforms are tall vertical cylinders held in place by mooring lines or tethers. Innovations include telescoping spar platforms that utilize slip joint mechanisms to extend buoyancy chambers to form a stable, wave-independent platform. Many derricks have similar capabilities where upper sections are nested inside the lower section of the structure. Derricks are the tower structures that support the drillstring. Once oil production is complete or the spar platform needs to be moved, the telescoping mechanism is retracted and the spar platform is prepared for movement by a BARGE. Drill ships are vessels that are outfitted as floating production, storage, and offloading platforms. They can store 100,000 barrels of crude oil and utilize seafloor and satellite positioning systems to guide

Ocean Platforms for Research: The Harvest Platform

The Harvest Platform is one of three oil and gas production platforms owned by Plains Resources Inc. in the Point Arguello Field that is offshore of Point Conception in Santa Barbara Co., California. Harvest Platform was originally developed by Texaco in 1985 and is located about 6.7 miles (10.8 km) from the coast in about 675 feet (205 m) of water. Approximately 3.2 miles (5.14 km) of 12-inch (304.8 mm) diameter wet oil and eight-inch (203.2 mm) diameter sour-gas pipes connect Harvest Platform to Hermosa Platform. Wet oil is a mixture of crude oil and water. Sour gas contains toxic compounds such as hydrogen sulfide (H_2S). Hydrogen sulfide is a colorless gas with an offensive smell like rotten eggs. Subsea pipelines are used to pump processed oil and gas approximately 10 miles (16 km) from the Platform Hermosa to the inland Gaviota oil and gas processing facility.

Offshore oil and gas production platforms are usually instrumented with a combination of meteorological and oceanographic instruments that support operations that range from deploying divers and Remotely Operated Vehicles to maneuvering rig equipment and docking vessels. Data are used to characterize the physical environment and forecast oceanographic conditions such as sea state. Operators need to understand effects on drilling from prevailing winds that develop waves and cause upwellings and circulations such as the southerly slow-moving California Current, the northward flowing Davidson Countercurrent, and turbulent eddies. These data, especially the wind at the ocean surface, developing waves, and the flow of the ocean's surface layer, are also important to help mitigate a marine spill and are provided to clean up crews that are working from Oil Spill Response Vessels such as the *Mr. Clean III,* which is usually moored near Harvest Platform.

Harvest Platform is well known for ongoing marine-based research, test, and development. It provides sea truth data or in situ verification for National Aeronautics and Space Administration's (NASA's) and the centre National d'Etudes Spatiales' (CNES') TOPEX/Poseidon mission. CNES is the French space agency, which is located in Toulouse, a city in southern France near the Garonne River. TOPEX is NASA's radar altimeter, and Poseidon is CNES's radar altimeter. The imagery is used to study ocean surface topography or wave height by allowing investigators to observe the ocean's response to wind and thermal forcing by the atmosphere. Harvest Platform supports a spatial array of approximately six pressure gauges that provide parameters such as wave height, peak wave frequency, and peak wave direction. Other interesting measurements include air temperature, sea surface temperature, water vapor that is measured from a water vapor radiometer, and complementary wave information from a directional wave buoy managed by the National Data Buoy Center (NDBC). If the Harvest Platform sensors are down for maintenance, wave height, wave period, and wave direction information can be obtained from the NDBC buoy. These data are transferred to the Coastal Data Information Program approximately eight times per day. This location near the entrance to Santa Barbara Channel is particularly important for measuring swells that are approaching the coast and for making surf forecasts.

Offshore structures have a certain lifespan and may serve a variety of uses. With the development of ocean observatories, some platform owners have already partnered with marine scientists to use portions of the structure as a research station. On the East Coast of the United States the South Atlantic Bight Synoptic Offshore Observational Network or SABSOON utilizes navy-owned towers offshore of Georgia as a platform for meteorological and oceanographic sensors. In other cases, no longer needed platforms are sunk to form artificial reefs that benefit the marine environment, commercial fishermen, recreational fishermen, and divers.

Further Reading

Coastal Marine Institute. Available online. URL: http://www.coastalresearchcenter.ucsb.edu/cmi/Main.html. Accessed July 29, 2007.

Scripps Institution of Oceanography, Coastal Data Information Program. Available online. URL: http://cdip.ucsd.edu. Accessed November 29, 2004.

South Atlantic Bight Synoptic Offshore Observational Network, Skidaway Institute of Oceanography. Available online. URL: http://www.skio.peachnet.edu/research/sabsoon. Accessed July 29, 2007.

—**C. Reid Nichols,** president
Marine Information Resources Corporation
Ellicott City, Maryland

---

computers that control high-powered thrusters, which keep the ship on station. Some of the important components of a rig are mud tanks and pumps, the derrick, drawworks, power plants, and topdrive. Drilling fluids are called mud, and drawworks reel the drilling line out and in as the bit penetrates into the designated sedimentary layers or strata. Topdrives turn the drillstring, which includes components such as drill pipe. Drilling fluids are pumped to the drill bit through this pipe. The key component

A Selected List of Submersible and Semi-Submersible Platforms and Drill Ships by Location

Location	Platform Name	Description
Andaman Sea	Doo Sung	A semi-submersible rated at 1,500 feet (457 m) drilling in the Gulf of Martaban offshore of Myanmar.
Arabian Sea and Bay of Bengal	Aban II, III, IV, V, & VI	Refurbished jackup rigs rated from 250 to 300 feet (76 to 91m) drilling in oilfields such as the Bombay High and offshore of the Coromandel Coast.
Arctic Ocean	Frontier Discoverer	A 525-foot (160 m) drill ship being modified for operations in the Beaufort and Chukchi Seas.
Arctic Ocean, Barents Sea	Murmanskaya	A jackup rig rated at 300 feet (91 m) and capable of drilling to 20,000 feet (6,096 m), which has operated in the Barents Sea. It was transported to Murmansk by the heavy-lift vessel *Transshelf*.
Atlantic Ocean	Deep Venture	A dynamically positioned drill ship outfitted for operations in water depths up to 4,200 feet (1,280 m). Moored near Buenos Aires, Argentina.
Atlantic Ocean	GSF Aleutian Key	A semi-submersible rated at 2,300 feet (701 m) that was moved from the Gulf of Guinea near the Cote d'Ivoire to Trinidad in the Caribbean Sea.
Atlantic Ocean	GSF Grand Banks	A semi-submersible rated at 1,500 feet (457 m) that is drilling in the Jeanne d'Arc Basin near Newfoundland, Canada.
Atlantic Ocean	Ocean Alliance	A deep-water semi-submersible rated at 5,000 feet (1,524 m) and capable of drilling to 25,000 feet (7,620 m). It is presently off the coast of Brazil at a depth of 4,731 feet (1,442 m).
Barents Sea	Eirik Raude	A semi-submersible designed to drill 30,000 feet (9,144 m) below the Earth's surface in water depths up to 7,500 feet (2,286 m) and to keep drilling during gale force winds and 40-foot (12-m) waves.
Black Sea	Prometeu	jackup
Caspian Sea	Dada Gorgud	An upgraded semi-submersible capable of drilling in water depths of 1,557 feet (475 m). It is currently drilling in the Azeri oilfield.
Caspian Sea	Istiglal	A semi-submersible rated at 2,297 feet (700 m) drilled a 10,000 foot (3,048 m) hole into the Caspian Sea in 32 days and drilled to a total depth of 21,943 feet (6,668 m).
Caspian Sea	Lider	A semi-submersible with accommodations for 130 people. It can drill in water depths from 246 feet (75 m) to 3,281 feet (1,000 m).
Galveston Bay	Ocean Star	A jackup rig and museum at Galveston Pier 19.
Gulf of Mexico	Atwood Richmond	A submersible capable of working at depths of 70 feet (21 m) that was upgraded in 2000 and has operated in the Gulf of Mexico for more than 18 years.
Gulf of Mexico	Cajun Express	A modern semi-submersible with a 30,000-square-foot (9,150 m²) unobstructed main deck, reduced rig-structure vibrations, and double-skinned pontoons and columns. Drilling operations are conducted in the deep Mississippi Canyon.
Gulf of Mexico	Deepwater Nautilus	A conventionally moored propulsion-assisted semi-submersible drilling rig rated for water depths up to 8,000 feet (2,438 m). Hurricane Katrina caused the rig to drift off course by approximately 80 miles (128 km) from its moored position.
Gulf of Mexico	Gunnison	The ninth spar platform moored in deep Gulf of Mexico waters at a depth of 3,150 feet (960 m).
Gulf of Mexico	Horn Mountain	A truss spar deployed 85 miles (137 km) offshore of Venice, Louisiana, at a depth of 5,423 feet (1,653 m) in the Mississippi Canyon. The 555-foot (169 m)-long cylinder that floats upright was not damaged by hurricanes during 2005.
Gulf of Mexico	Magnolia	A tension leg platform (TLP), which set the world record for TLP water depth at 4,674 feet (1,425 m) during 2004.
Gulf of Mexico	Marco Polo	A deep-water tension leg platform rated at 4,300 feet (1,310 m) located about 150 miles (240 km) offshore of Louisiana.

Location	Platform Name	Description
Gulf of Mexico	Mars	A tension leg platform operating at a depth of 2,940 feet (896 m) in the Mississippi Canyon that lost its derrick during Hurricane Katrina.
Indian Ocean	Aban Abraham	A 487.7-foot (149 m) dynamically positioned drillship operating off the west coast of Africa.
Indian Ocean	Atwood Eagle	A semi-submersible platform capable of operating in waters that are 5,000 feet (1,524 m) deep. It is currently drilling in the Carnarvon Basin offshore of Australia.
Indian Ocean	Bombay High North	A semi-submersible was destroyed in a fire after colliding with a multipurpose support vessel that ruptured a riser pipeline in the Arabian Sea during July 2005. The platform was lost in two hours.
Mediterranean Sea	Atwood Hunter	A 2001 upgraded semi-submersible operating offshore of Egypt.
Mediterranean Sea, Black Sea	Atwood Southern Cross	A highly utilized semi-submersible rated for 2,000 feet (610 m) that has been transported from the Mediterranean Sea to Southeast Asia and back to the Mediterranean.
Mediterranean	Labin	A cantilever jackup rig with living space for 100 people, rated at 347.77 feet (106 m), and able to drill to 19,685 feet (6,000 m). The rig is deployed in the Mediterranean near the city of Ravenna in Italy.
North Atlantic	Hibernia	A platform rig operating in the Jeanne d'Arc Basin, 196 miles (315 km) east of St. John's, Newfoundland, Canada.
North Sea	Ekofisk 2/4 C and 2/4 X	Platform rigs located in approximately 250 feet (76 m) of water operating in the aging Ekofisk oilfield that was first discovered in 1969. The Ekofisk oilfield complex located approximately 200 miles (321.9 km) offshore of Norway is being modernized.
Pacific Ocean	Gail	A fixed platform located in 739 feet (225 m) of water approximately 10 miles (16 km) offshore in the Santa Barbara Channel. Its value as an artificial reef has been studied, especially for populations of cowcod (*Sebastes levis*) and bocaccio (*Sebastes paucispinis*).
Pacific Ocean, Santa Barbara Channel	Hermosa	An eight leg, five-deck platform located approximately 10 miles (16 km) offshore of Point Argüelles, California, in 603 feet (184 m) of water. It is 4.8 miles (7.7 km) from Platform Hidalgo and 3.2 miles (5.1 km) from Platform Harvest.
Persian Gulf	AD22	A 100-foot (30.5 m) rated jackup rig replacing AD19, an independent leg jackup that collapsed and sank offshore of Khafji, Saudi Arabia, during 2002.
Persian Gulf	Al Mariyah	A 110-foot (33.5 m) rated three-leg jackup rig that collapsed during 2000.
Persian Gulf	Al Yasat	A 180-foot (55 m) rated jackup rig working the large Umm Shaif oilfield, which was first discovered in 1958.
Persian Gulf	Junana	A jackup rig being modernized to a cantilever beam rig with new engines, pumps, and drilling machinery. Living spaces will be replaced with a new 100-person unit. The helideck will be replaced with an aluminum version. The rework period is approximately 260 days, and two weeks of sea trails are scheduled to take place offshore of Abu Dhabi.
Persian Gulf	Noble Charles Copeland	A jackup rig with living space for 100 people, rated at 280 feet (85 m), and able to drill to 20,000 feet (6,096 m). The rig is deployed off the coast of Qatar.
Persian Gulf	WilPower	A new jackup rig capable of drilling to 30,000 feet (9,144 m) in 375 feet (114 m) of water, destined for Saudi Arabia.
Red Sea	Bennevis	A self-propelled jackup rig rated at 250 feet (76 m) operating offshore of Egypt in the Gulf of Suez.
Red Sea	Kamose	A jackup rig with living space for 80 people, rated at 300 feet (91 m), and able to drill to 20,000 feet (6,096 m). The rig is deployed in the Gulf of Suez (leading to the Suez Canal) offshore of Egypt in 210 feet (64 m) of water.

(continues)

A Selected List of Submersible and Semi-Submersible Platforms and Drill Ships by Location
(continued)

Location	Platform Name	Description
South Atlantic	Petrobas 36	Large semi-submersible, which sank during 2001 while operating in the Campos Basin.
South China Sea	COSL 941	The first jackup rig in the Peoples Republic of China using automated drilling machine control technologies. It commenced drilling operations in the Gulf of Tonkin during June 2006.
South China Sea, Natuna Sea	Nanhai II	A semi-submersible rated at 1000 feet (305 m) and is capable of drilling to 25,000 feet (7,620 m). It is presently located offshore of the Natuna Islands of Indonesia.

of the rig is the drill bit, which cuts, grinds, crushes, and otherwise pulverizes the rock. The floating platforms are anchored to the seafloor by complicated hardware such as mooring lines and riser pipes.

The crew reaches the platform by helicopter that lands on a helipad or at docking areas by boat. Offshore platforms also provide the living quarters, offices, and working areas for the crew. The platform rig Hibernia in the North Atlantic is one of the largest platforms and has accommodations for 185 people.

Working aboard an offshore platform is hard and dangerous work since the objective is the extraction of hydrocarbons, a volatile substance under extreme pressure, from beneath the hostile and unforgiving ocean. Offshore drilling rigs operate 24 hours per day, seven days per week. There are usually two crews that work 12-hour work shifts or tours. The crew includes derrickmen, engineers, geophysicists, mechanics, motormen, welders, roughnecks, and roustabouts (*see* MARINE TRADES). The supervisor of the rig crew is called the driller, and he or she controls the major rig systems such as the pumps and drawworks. Control rooms are packed full with computer screens, control levers, and electronic instrumentation. Many professionals support the operation and maintenance of offshore oil platforms from saturation divers burying pipeline at great depths to pilots involved in aerial surveys to detect oil pollution. The table on pages 410–412 provides a selected listing of submersible and semi-submersible platforms and drill ships by location. Thousands of oil production platforms have drilled through layers of sand, silt, and rock to reach trapped oil and gas deposits.

Further Reading

Mangone, Gerard F. *Concise Marine Almanac.* New York: Van Nostrand Reinhold, 1986.
Minerals Management Service. Available online. URL: http://www.mms.gov/. Accessed August 29, 2006.
OilOnline. Available online. URL: http://www.oilonline.com/key/rig_locs.asp. Accessed September 1, 2006.
Rigs to Reefs Information. Available online. URL: http://www.gomr.mms.gov/homepg/regulate/environ/rigs-to-reefs/information.html. Accessed September 2, 2006.
World Meteorological Organization. *Guide to Wave Analysis and Forecasting,* 2nd ed. WMO-No.702. Geneva, Switzerland: WMO, 1998.

ooze　Ooze is soft sediment on the ocean floor composed largely of the remains of planktonic organisms, such as diatoms, foraminifera, and radiolarians (*see* PLANKTON). Oozes must contain at least 30 percent of biogenic material. They make up the bulk of the sediments in the deep ocean, especially in the Pacific Ocean.

Oozes can be classified as calcareous ooze or siliceous ooze. Calcareous ooze is made up of skeletal material consisting mainly of calcium carbonate. Three types of organisms form calcareous ooze: COCCOLITHOPHORES, foraminifera, and pteropods. Coccolithophores are microscopic plants that live in shallow, open waters. Their structure consists of many calcareous plates that disintegrate when they die and drift to the bottom. Foraminifera are microscopic organisms that live in shallow water and fall to the bottom upon death. Pteropods are small snails (gastropods) that are on the order of a millimeter in length, and float in shallow, open ocean waters.

Siliceous sediments, resulting from planktonic organisms secreting silica, are not as abundant as calcareous sediments, but do constitute a majority of the sediments at some locations. In high latitudes, diatoms (microscopic plants) are abundant and so make a major contribution to the sediments in the form of diatomaceous ooze. In fact, these are the dominant sediments in many sub-arctic and arctic

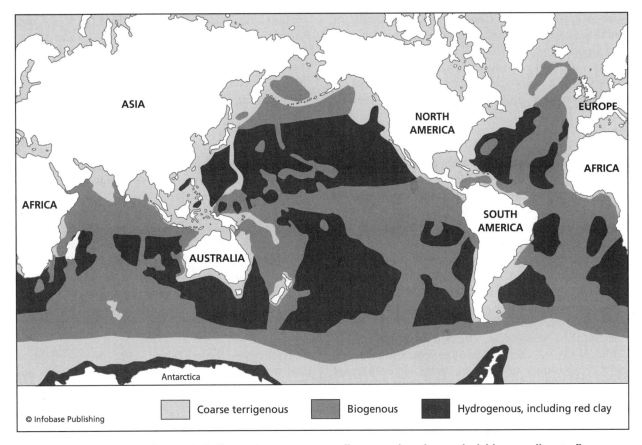

The map legend reads: Coarse terrigenous | Biogenous | Hydrogenous, including red clay

© Infobase Publishing

The distribution of ocean sediments, including terrigenous coarse sediments such as river or glacial-borne sediments, fine sediments such as red clay, biogenous sediments such as calcareous and siliceous oozes

regions. In equatorial waters, there are large populations of radiolarians, protozoa whose skeletal remains are prominent in marine sediments in the form of radiolarian ooze.

See also SEDIMENT.

open coast beach   An open coast beach is a beach that is found along the coastal ocean that is not sheltered by inland topography. Inland water areas may be protected or enclosed by headlands within narrow straits. Headlands or promontories may have considerable elevation or may just be a stretch along the shore where the direction of the coastline changes. A pocket beach may be located between two headlands.

See also BEACH.

optimal ship track planning   Optimal ship track planning is a procedure for minimizing the sailing time from the port of embarkation to the port of destination by scientific analysis of recent information on environmental factors such as winds, WAVES, and CURRENTS, and applying these factors to vessel sea-keeping characteristics, such as dimensions, engine power, fuel capacity, cargo, general condition of the vessel, and speed in various sea states. Under perfect sailing conditions of calm seas and flawless ship performance, the minimum sailing distance between two ports on Earth is the great circle route, the minimum distance between two points on a sphere. The ocean, however, is seldom calm, and vessels will surge, heave, sway, pitch, roll, and yaw in response to the seas. Because of weather and sea conditions, ocean vessels could reduce sailing time by planning a route that would avoid storms and adverse sea conditions, even though the absolute distance is longer than the great circle route. This is so because vessels lose speed when sailing into heavy seas, which results in severe rotational motions such as pitch and roll, which may also threaten the stability and structural integrity of the ship, damage or loss of cargo, and the safety of the crew. Planners work toward optimizing transit by factoring in sea-keeping and ride-quality issues in making the most efficient course. Optimal ship track planning may also be referred to as weather routing or meteorological navigation. For many years, ship captains have consulted climatological information on winds, tides, and currents, such as is found in the *Coast Pilot,* a publication of the U.S. National Ocean Service (formerly U.S. Coast and Geodetic Survey), as well as the tide and current tables, sailing directions, bathymetric charts, current atlases, and the like. However, this information is based on average conditions, often over many years, and may be quite different from conditions encountered by the ship on its voyage. More recently, weather forecasts received by ships' facsimile machines, and expendable bathythermographs (XBTs; *see* OCEANOGRAPHIC INSTRUMENTS) could be used to amend and update the climatological information. For example, it was found that the region of highest velocity of the GULF STREAM could be located from satellite images and a few XBT drops along the CONTINENTAL SLOPE. However, it was up to the ship's bridge officers to integrate and interpret all this information, a difficult task at best, and especially difficult considering the collateral duties of managing and operating a major vessel.

For these reasons, Dr. Richard W. James (1925–2007) of the U.S. Naval Oceanographic Office developed a procedure in 1957 for using environmental data, forecasts, and nowcasts (*see* NOWCASTING) to plan the optimum ship route to minimize sailing time and maximize crew, vessel, and cargo safety. This procedure was tested by commercial and naval vessels and found to save time and money. It became formalized in subsequent years as "optimum ship track planning" and incorporated satellite infrared airborne radiation thermometer (ART), and infrared and microwave imaging measurements from satellites and aircraft.

In the beginning, U.S. Navy planners and other oceanographers did the analysis, but this procedure involved subjective analysis and could be applied only to limited number of ships. As computer technology improved, much of the analysis could be done on digital computers using, for example, wind and wave FORECASTING models. Computerization enabled a much wider application of optimum ship track planning, as it could be done quickly and for several ships at once.

Today, the environmental data, forecast model data, position data from the GLOBAL POSITIONING SYSTEM (GPS), and ship response characteristics from computer models can be fed into computer programs that integrate all this information and input to the electronic navigation system on the bridge of the vessel for electronic chart display. Ocean observatories provide real-time environmental information that may be accessed via the Internet or by cellular telephones that can improve the routing of vessels through coastal oceans and into seaports.

A number of private firms now provide this service to shipping companies and the military, in partnership with federal agencies such as the U.S. Coast Guard and the NATIONAL OCEANIC AND ATMOSPHERIC ADMINISTRATION (NOAA), and with scientific laboratories. Recreational boaters planning cruises or races or just putting together a float plan often engage the services of these firms to provide optimum route planning. Similar procedures are used for both military and civilian aircraft.

Further Reading

Bishop, Joseph M. *Applied Oceanography.* New York: Wiley-Interscience, 1984.

Cook, Bob. "Ocean-Pro Weather Services: Advanced Offshore Weather, Consulting, Instruction, and Resources." Available online. URL: http://www.ocean-pro.com/htmfiles/routing2.htm. Accessed July 28, 2006.

James, Richard W. *Application of Wave Forecasts to Marine Navigation,* Special Publication 1. U.S. Naval Oceanographic Office, NSTL, Station Mississippi, 1957.

Osteichthyes (bony fishes)   The largest class of cold-blooded vertebrates that breathe through one pair of gill openings, and use paired fins and a gas bladder for swimming are called Osteichthyes (bony fishes). The bony flap that covers the gills is called the operculum. The gas bladder controls the fish's buoyancy and in some species is important for hearing. These finfish have a skeleton of bone, scales, jaws, and paired nostrils. They are called bony fish and are distinguishable from sharks and their relatives (Chondrichthyes), hagfish (Myxini), and lampreys (Cephalaspidomorphi). There are more than 23,500 species of bony fish, and more than half live in the marine environment. Bony fish are found from the Tropics to the poles, in fresh, salt, and brackish waters, and from the deepest TRENCHES to the highest mountain lakes. Many species migrate. Those swimming between fresh and saline environments are called diadromous. Catadromous species live in freshwater and spawn in the ocean. ANADROMOUS species live in the ocean but spawn in freshwater. One of the largest species of bony fish is the ocean sunfish (*Mola mola*), which can grow more than 14 feet (4.2 m) long from dorsal fin tip to anal fin tip and 10 feet (3 m) in horizontal length. Large sports fish include the black marlin (*Makaira indica*), blue marlin (*Makaira nigricans*) and bluefin tuna (*Thunnus thynnus*).

See also FISH.

oxygen minimum layer   The oxygen minimum layer is a minimum in the vertical distribution of dissolved oxygen in the deep ocean at roughly 274 fathoms (500 m) depth, occurring in the disphotic zone.

The middle region of the ocean is often called the twilight or disphotic zone and extends from the PHOTIC (surface to 100 fathoms [183 m]) to the APHOTIC (550 fathoms (1,006 m) to seafloor) ZONES. The associated photosynthesis present in surface waters does not occur in the twilight zone. In addition, the utilization of oxygen by living animals, and the oxidation of sinking dead organisms (detritus) result in consumption of dissolved oxygen, without replacement. Much of the sinking dead material comes from PHYTOPLANKTON. Dissolved oxygen in the ocean generally has a value of between one and six milliliters per liter of water (the SI unit for dissolved oxygen is the µmol/kg, but most of the oceanographic literature uses ml/l). The location of the oxygen minimum is in the region of the main PYCNOCLINE, where a rapid increase of DENSITY with depth inhibits the vertical motions that could bring dissolved oxygen from surface layers. The pycnocline is generally considered to be the transition region between the surface and the deep waters below. Above the oxygen minimum in the photic, or sunlit zone, surface waters are generally high in dissolved oxygen, reaching a maximum at the surface, especially in the storm-tossed waters of the Arctic and Antarctic. Organisms adapted for life in the oxygen minimum layer are usually sedentary species with a slow respiration rate such as the shortspine thornyhead (*Sebastolobus alascanus*), inhabiting depths from 55 to 465 fathoms (100–850 m). Another example is the vampire squid (*Vampyroteuthis infernalis*), which has been observed floating motionless in the water column by scientists using REMOTELY OPERATED VEHICLES. Below the oxygen minimum layer, ocean water recently at the surface is transported by ocean currents, especially in the South Atlantic and South Pacific Oceans, where cold, dense, oxygen-rich water from the Antarctic sinks and spreads northward, and in the North Atlantic Ocean, where North Atlantic Deep Water is formed and sinks in the region of the Greenland Sea. The North Pacific, and northern Indian Oceans which have few sources of deep water, are relatively low in oxygen compared to the North Atlantic. Anoxic conditions in deep water occur in barred basins such as the Black Sea, where there is no connection with deep, ventilated high-latitude water. The Bosporus sill separates the Black Sea basin from the Sea of Marmara, which opens into the Aegean and the Mediterranean Seas. In the Black Sea, the bottom waters are cut off from surface layers by a very strong pycnocline and is almost devoid of dissolved oxygen. Instead of oxygen, hydrogen sulfide, produced by the reduction of sulfate by bacteria, is present.

See also MARINE PROVINCES.

Oyashio Current (Kamchatka Current)   The Oyashio Current is a subarctic current flowing southwestward from the Bering Sea, through the Bering Strait, along the Kurile Islands, and into the KUROSHIO CURRENT off the coast of Honshu, Japan, with some additional flow from the Sea of Okhotsk. Oyoshio means parent (*oya*) stream (*shio*) in Japanese. The Oyashio is the southern branch of a counterclockwise circulation in the Bering Sea. The Oyashio Current is the Pacific counterpart of the LABRADOR CURRENT in the Atlantic. Differences occur in part because the North Pacific does not have open access to the Arctic Ocean, as does the North Atlantic, and because the Bering Sea is quite shallow.

The Oyashio and Kuroshio are western boundary currents in the northern Pacific Ocean. Western boundary currents are strong permanent currents and a major component of ocean GYRES. The region of convergence of the Oyashio and Kuroshio is believed to be the region of formation of the North

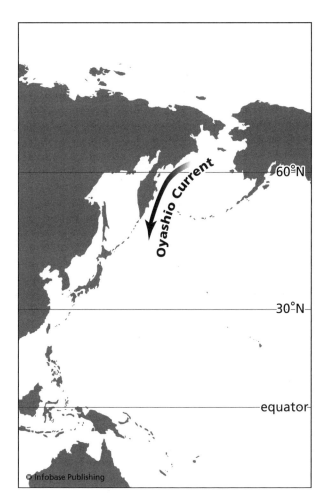

The Oyashio Current is a cold current that flows southward along the coast of Russia toward Japan and in many respects is the Pacific counterpart to the Labrador Current.

Pacific Intermediate Water (NPIW), an important Pacific Ocean WATER MASS found beneath the surface water mass (the North Pacific Central Water). The North Pacific Intermediate Water is of considerable interest to oceanographers because it may be one of the very few water masses that are not generated at the sea surface. The NPIW may be formed as cold arctic surface water of relatively low SALINITY and high oxygen content sinks to the level of its own DENSITY.

Temperatures in the Oyashio Current are generally on the order of 39 to 41°F (4–5°C), and salinities on the order of 33.7 to 34.0 psu. Average velocities are generally less than 1.3 knots (2.4 km/hr) toward the southwest. The combined flow of the Kuroshio and Oyashio Currents flows eastward toward the coasts of British Columbia and Alaska as the North Pacific Current. Warm core rings spinning off the Kuroshio Current often become embedded in the Oyashio Current, leading to regions of large velocity SHEAR, vertical motion, and high biological productivity. In fact, primary productivity is very high throughout the whole region of confluence of the Oyashio with the Kuroshio, creating one of the world's riches fisheries.

Further Reading

Kuroshio/Oyashio Currents Zone (KR/OY). Available online. URL: http://www.pmel.noaa.gov/np/pages/sub regions/kroy.html. Accessed June 3, 2006.

NOAA Fisheries Northeast Fisheries Science Center. "Oyashio Current LME (Large Marine Ecosystem)." Available online. URL: http://na.nefsc.noaa.gov/lme/text/lme51.htm. Accessed October 15, 2007.

Pickard, George L., and William J. Emery. *Descriptive Physical Oceanography: An Introduction,* 5th ed. Oxford: Pergamon Press, 1990.

P

Pacific Ocean The Pacific Ocean is a body of water covering approximately 30 percent of the Earth's surface, lying between the Southern Ocean, Asia, Australia, and the Americas. It is the largest of the world's five oceans and includes the Arafura Sea, Banda Sea, Bering Sea, Celebes Sea, Coral Sea, East China Sea, Gulf of Alaska, Gulf of Tonkin, Java Sea, Philippine Sea, Sea of Japan, Sea of Okhotsk, Solomon Sea, South China Sea, Tasman Sea, and the Yellow Sea. Strategic straits or canals (chokepoints) include the Bering Strait, Panama Canal, Luzon Strait, and the Singapore Strait.

The equator divides the Pacific Ocean into the North and the South Pacific Oceans. The Pacific has an average depth of 2,154 fathoms (3,940 m), which includes depths associated with the mid-ocean ridges, deep basins such as the Northwest Pacific Basin, and TRENCHES such as the Mariana Trench. New ocean floor is formed at spreading centers such as the East Pacific Rise. Trenches are formed where the Pacific plate is subducting or sliding under other plates. The Challenger Deep in the Mariana Trench near Guam is the deepest location at a depth of approximately 5,966 fathoms (11 km). The process of subduction contributes to seismic activity such as earthquakes and volcanism, contributing the Pacific Ocean's RING OF FIRE, which stretches across the Aleutian Islands of Alaska, across the western coasts of North and South America (North and South America Plates) and along New Zealand and the eastern edge of Asia (Indo-Australian Plate and Eurasian Plate). Significant portions (approximately 75 percent) of the world's volcanoes lie within the Ring of Fire. Shallower regions are found along the CONTINENTAL SHELVES and the marginal seas. The southwestern

Pacific Ocean includes many low coral islands and rugged volcanic islands (many are unnamed). They form archipelagos such as the Louisiade and Solomon Island Archipelagos, which contain Guadalcanal, New Georgia, and Bougainville Islands. The Japanese occupied many of these islands during World War II, and MARINE ARCHAEOLOGISTS and historians are studying wreckage from the Battle of the Coral Sea. The Coral Sea's western boundary includes Australia's Great Barrier Reef.

Unequal heating of the atmosphere from season and latitude (variable insulation) occurring across the large Pacific Ocean contributes to air pressure differences that drive global wind patterns that contribute to the establishment of PERMANENT CURRENTS. Globally, polar easterlies tend to occur between 60° and 90° latitude, and prevailing westerlies occur between 30° and 60° latitude. In the Northern Hemisphere, persistent winds called northeast trade winds blow from northeast to southwest between approximately 9° and 30° north latitude. The southeast trade winds blow persistently between 0° and 25° south latitude. Islands only interrupt these trade winds as they blow across the Pacific. High pressure forms about 30° degrees latitude in response to the divergence (cool air sinking) of the westerlies. This region of little wind is called the horse latitudes by mariners. Low pressure forms between approximately 0° and 10° latitude in response to convergence of the trade winds (warm air rising), producing the INTERTROPICAL CONVERGENCE ZONE (doldrums).

In response to these winds and the Earth's rotation, surface currents in the northern Pacific are dominated by a clockwise, warm-water gyre (the North

Pacific Drift to the north, the CALIFORNIA CURRENT to the east, the NORTH EQUATORIAL CURRENT to the south, and the KUROSHIO CURRENT to the west) and in the southern Pacific by a counterclockwise, cool-water gyre (the South Equatorial Current to the north, the PERU CURRENT to the east, the ANTARCTIC CIRCUMPOLAR CURRENT to the south, and the EAST AUSTRALIA CURRENT to the west). Researchers using satellites are actively studying temperature variability and the extent of what is called the Pacific Warm Water Pool, which is found along the equatorial Pacific. Models, satellite imagery, and instrumented BUOYS are used in the northern Pacific to investigate the formation of SEA ICE in areas such as the Bering Sea and Sea of Okhotsk, especially during winter. A temperature contrast between the Eurasia landmass and the Pacific Ocean sets up MONSOON winds. During summer, Eurasia heats up more than the surrounding Pacific Ocean, creating onshore winds that bring moisture-laden maritime air from over the Pacific onto land. During winter, when Eurasia cools, drier, cool air blows offshore.

As heat is transported toward the poles by winds and currents, the Pacific Ocean provides the energy for severe storms. For example, tropical CYCLONES, called typhoons in the western North Pacific (west of the international dateline), may be generated from May to December; tropical cyclones called HURRICANES may form in the eastern and central North Pacific (east of the dateline) from June through October. In the southwest Pacific Ocean (west of 160° east longitude), these storms are called severe tropical cyclones. Similarly, over larger cyclical time scales, EL NIÑO/LA NIÑA phenomena (temperature fluctuations) occur in the equatorial Pacific, influencing weather in the Western Hemisphere and the western Pacific. El Niño conditions occur when the easterly trade winds weaken, allowing warmer waters of the western Pacific (the warm water pool) to travel eastward. As these waters reach South America, cool, nutrient-rich seawater that is upwelled along the coast of Peru is replaced by warmer water. In synchronization with surface temperature changes in the tropical eastern Pacific Ocean is the Southern Oscillation, which results from the air pressure difference between Tahiti, the largest island in French Polynesia, and Darwin, Australia. La Niña phenomena are characterized by unusually cold sea surface temperatures found in the eastern tropical Pacific. El Niño occurs much more frequently than La Niña, but both affect global climate and marine ecosystems.

Meteorological patterns such as wind, evaporation, and rainfall also influence surface SALINITIES in the Pacific. Surface salinities may reach 37 practical salinity units (psu) where there is excess evaporation in subtropical regions as evidenced by dry air or regions of high pressure such as in the southeastern Pacific. Evaporation rates depend on factors such as solar radiation, humidity, and wind, so salinity is highest in mid latitudes. Surface salinities are low where precipitation is high. Along the equator, surface salinities may be around 34 psu, especially in the western Pacific. There is considerable rainfall associated with the Asian monsoons. Salinities around 32 psu are found in the extreme northern zone of the Pacific where there is influence of rivers, especially from Canada. But river flow is not as significant in the Pacific compared to the other oceans.

Surface salinity is lower in the North Pacific than in the South Pacific; salinity is generally lower in the entire Pacific than in the Atlantic. Surface temperatures are highest in the equatorial region, and generally higher in the west than in the east. Low temperatures are found in regions of UPWELLING, such as the Peru Current in the South Pacific, and the California Current in the North Pacific, as well is in areas of upwelling in the eastern Equatorial Pacific known as cold islands.

The WATER MASS structure of the Pacific is more complex than the Atlantic. Below the surface mixed layer, Pacific Equatorial Water spans almost the entire breadth of the ocean, from west to east. This water mass extends from below the THERMOCLINE to about 2,600 feet (800 m) in depth and is the most saline in the Pacific Ocean. The thermocline is deep in the western Pacific, about 820 feet (250 m) in depth, but rises to about 165 feet (50 m) in the east. North of the Equatorial Water mass, the North Pacific Central Water ranges to about 40° north latitude. ("Central Water" usually refers to near-surface water masses in the central region of ocean basins.) In the North Pacific, the Pacific Subarctic Water mass extends across the ocean, with a temperature on the order of 36–39°F (2–4°C), and a salinity about 33.5–34.5 psu. South of the Equatorial Water mass, South Pacific Central Water extends to the Subtropical Convergence at about 40° south latitude. South of the Central Water, is the Subantarctic Water, which is formed between the Subtropical Convergence, and the Antarctic Polar Front; south of the Equatorial Water mass, the South Pacific Central Water extends to the Subtropical Convergence at about 40° south latitude.

The Antarctic Intermediate Water (AIW) (occurring also in the Atlantic and Indian Oceans) is found below the Central Water masses; but, in contrast to the Atlantic, it penetrates only as far north as the equator. The AIW has a temperature of about 36°F (2.2°C) and a salinity of about 33.8 psu. The corresponding water mass to the north is the North Pacific

Intermediate Water. The deep water of the Pacific Ocean, below about 1,094 fathoms (2,000 m) to the bottom, is quite uniform, with temperatures on the order of 34–36°F (1.1–2.2°C), and salinities from 34.65 to 34.75 psu. In contrast to the Atlantic, there is no local source for Pacific deep-water masses; the salinity in the northern regions is too low to provide sufficient density for the surface water to sink when cooled, as it does in the Atlantic south of Greenland and Labrador. Also, the Bering Strait is too shallow for formation of deep water. The Pacific Deep Water mass has several sources, including North Atlantic Deep Water, Antarctic Bottom Water, and water from the Indian Ocean. The deep waters of the Pacific have lower dissolved oxygen values than the Atlantic. This is believed due to a slower deep circulation.

In an effort to sustain resources, Large Marine Ecosystems have been designated for coastal areas extending to the seaward boundaries of continental shelves and the outer margins of the major current systems. The Large Marine Ecosystems may help to protect endangered marine species, for example certain MARINE MAMMALS and MARINE REPTILES and reduce oil pollution, especially in the Philippine and South China Seas. Examples of Large Marine Ecosystems in the Pacific Ocean include the East and West Bering Sea, Gulf of Alaska, California Current, Pacific Central American Coastal, Humboldt Current, Insular-Pacific Hawaiian, Sea of Okhotsk, Oyashio Current, Sea of Japan, Kuroshio Current, East China Sea, South China Sea, Sulu-Celebes Sea, Indonesian Sea, and the shelf around eastern Australia and New Zealand. Large Marine Ecosystems are highly productive and are being assessed by international scientists to ensure a sustainment of natural resources. One of the roles of TREATIES AND GOVERNANCE involves helping to maintain natural resources at renewable levels.

See also ARCTIC OCEAN; ATLANTIC OCEAN; GYRE; INDIAN OCEAN; MARGINAL SEAS; PLATE TECTONICS; SOUTHERN OCEAN; SUBDUCTION ZONE.

Further Reading

Archaeological Institute of America. Available online. URL: http://www.archaeological.org. Accessed February 28, 2007.

Central Intelligence Agency. "The World Factbook." Washington, D.C. Available online. URL: https://www.cia.gov/cia/publications/factbook. Accessed December 24, 2006.

International Hydrographic Organization. Available online. URL: http://www.iho.shom.fr. Accessed December 24, 2006.

Large Marine Ecosytems: Information Portal. Available online. URL: http://woodsmoke.edc.uri.edu/Portal. Accessed October 15, 2007.

National Oceanic and Atmospheric Administration. "Fagatele Bay National Marine Sanctuary." Available online. URL: http://fagatelebay.noaa.gov/html/intro.html. Accessed February 28, 2007.

Pacific Marine Environmental Laboratory, National Oceanic and Atmospheric Administration. "Tropical Ocean Atmosphere Project." Available online. URL: http://www.pmel.noaa.gov/tao. Accessed March 1, 2007.

Pickard, George L., and William J. Emery. *Descriptive Physical Oceanography: An Introduction*, 5th ed. Oxford: Pergamon Press, 1990.

United States Geological Survey. "Active Volcanoes and Plate Tectonics, 'Hot Spots' and the 'Ring of Fire'." Available online. URL: http://vulcan.wr.usgs.gov/Glossary/PlateTectonics/Maps/map_plate_tectonics_world.html. Accessed October 15, 2007.

paleontologist A paleontologist is a scientist who studies paleontology, or extinct and formerly living organisms. Paleontology comes from the Greek word roots *paleo* meaning "ancient," *onto* meaning "being or existence," and *logy* meaning "the study of." Paleontologists generally study life-forms that existed during past geological periods through analysis of fossil records (*see* GEOLOGICAL EPOCHS). Paleontology is a broad field overlapping with disciplines such as biology, climatology, geology, and physics. Paleontologists have been instrumental in helping to identify how long-term physical changes on Earth have affected living creatures such as the dinosaurs. For example, Luis Alvarez (1911–88) and his son, Walter Alvarez, from the University of California discovered evidence that an asteroid may have caused the mass extinction at the Cretaceous-Tertiary (K-T) boundary while investigating tectonic paleo-magnetism in Italy during 1980. The Cretaceous period lasted 81 million years from about 146 to 65 million years ago and was followed by the Tertiary period. The Alvarezes hypothesized that clay layers enriched in iridium (Ir), a metal of the platinum family found near the K-T boundary, were a direct result of fallout following a giant asteroid impact on Earth.

Paleontologists attempt to discover how biotic and physical factors have shaped past life and develop rules relevant to those biotic responses. They attempt to apply their knowledge to better understand changes in today's global environment. Paleontologists may find fossils and then carefully excavate the site where the remains, impressions, or traces of organisms preserved in rock are found. This work relies on a wide variety of tools, ranging from heavy equipment such as bulldozers to hand tools such as shovels, drills, hammers, chisels, and brushes. Careful records of measurements such as lengths and weights, as well as images such as drawings and photographs are maintained of artifacts and fossils. The fossils and fragments may be further protected with glue sprays or castes of plaster. Some artifacts may be packaged and sent to laboratories for high-technology testing. In the laboratory, scientists use techniques such as deoxyribonucleic acid (DNA) testing or stable isotope analyses to unlock secrets from organisms that lived long ago. Researchers have shown that DNA extracted from cave bears (*Ursus spelaeus*) is related to DNA from modern brown (*Ursus arctos*) and polar (*Ursus Maritimus*) bears. Scientists may also consider stable isotopes of oxygen to determine paleo-climate and paleo-ocean information. The three types of oxygen atoms are called ^{16}O, ^{17}O, and ^{18}O. Oxygen atoms in rocks do not have identical proportions of ^{16}O, ^{17}O, and ^{18}O atoms, but the ratios of these isotopes are controlled by their masses. For example, ratios of oxygen isotopes such as $^{18}O/^{16}O$ from bone phosphates from *Tyrannosaurus rex* have been used to study this animal's thermophysiology.

Some paleontologists focus research on determining the climatic and ocean patterns existing before humans began measuring weather parameters such as temperatures from a thermometer, precipitation from a rain gauge, sea-level PRESSURE from a barometer, and CURRENTS by measuring the drift of ships. Paleontologists interested in paleo-climates and the paleo-ocean use proxy records to infer past climate conditions. Fossilized coral or other warm water species can be used to determine locations and temperatures of water bodies. For example, most coral species require water temperatures between 64.4° and 86°F (18° and 30°C). Eduard Suess (1831–1914) using fossil records from Europe and Africa located the "Tethys Sea" between the paleo-continents Laurasia and Gondwana. Paleo-oceanographers speculate that today's remnants of the Tethys Sea are the Black, Caspian, and Aral Seas. The Caspian Sea, which separated from the Black Sea approximately 5 million years ago, is the world's largest lake.

Paleontologists have almost always earned a master's degree, and those involved in basic research a Ph.D. The courses that are most pertinent to paleontology include mineralogy, stratigraphy, sedimentary petrology, invertebrate paleontology, ecology, invertebrate and vertebrate zoology, evolutionary biology, and genetics. Paleontologists must have a firm grasp on statistical analysis and solid computer skills. They are employed as college and university professors where they are often faculty members in geology departments. Museums employ professional paleontologists, sometimes as curators, who might also carry out their own research and teach. Cura-

tors manage and oversee museums, historic properties, and archaeological sites. Besides preserving, cataloging, and utilizing collections, they may develop and produce exhibits or educational materials. A much smaller number of paleontologists work for governments, usually in geological mapping or in managing ruins.

Further Reading

NOAA Satellite and Information Service. "NOAA Paleo-climatology." Available online. URL: http://www.ncdc. noaa.gov/paleo/paleo.html. Accessed September 11, 2006.

Paleontological Research Institution. Available online. URL: http://www.priweb.org. Accessed September 12, 2006.

The Paleontological Society. Available online. URL: http:// www.paleosoc.org. Accessed September 12, 2006.

The Society of Vertebrate Paleontology. Available online. URL: http://www.vertpaleo.org/. Accessed September 12, 2006.

parasitism The condition of living at the expense of, on, or within a host organism in order to obtain sustenance or protection without particular compensation is called parasitism. In many cases, the parasite benefits while the host is harmed. For example, diadromous sea lamprey (*Petromyzon marinus*) are parasites on Atlantic salmon (*Salmo salar*) and lake trout (*Salvelinus namaycush*). Sea lamprey spend the early stages of their life in freshwater streams and rivers. They migrate during their middle stage to salt water and return during their final stage to freshwater as breeding adults. In water bodies such as the Great Lakes, sea lamprey benefit by drinking the body fluids of fish, but the fish is debilitated through loss of scales, blood, and acquiring disease and discomfort. The lamprey uses a suction-disk mouth filled with small, sharp teeth and a filelike tongue to attach to the host fish, punctures the skin, and drains the fish's body fluids. In the ocean, many other biological interactions occur in which one species is favored and the other is inhibited. For example, small parasitic crustaceans such as sea lice include *Argulus canadensis*, *Caligus elongatus*, and *Lepeophtheirus salmonis*. Sea lice are saltwater CO-PEPODS that hold onto their host with clawed antennae and feed on their host's mucus, skin, and blood. The skin of most bony fish is covered with waterproof scales. Glands in the skin where the scales are embedded secrete a layer of mucus that covers the fish's entire body. Breaks in this mucus layer may be harmful to the fish. Marine mammals such as whales suffer from such parasites as barnacles and whale lice.

Further Reading
Department of Fish and Game Headquarters. "Common Parasites of California Marine Fishes." Available online. URL: http://www.dfg.ca.gov/mrd/parasites.html. Accessed June 29, 2007.
MarineBio.org. Available online. URL: http://www.marinebio.com. Accessed June 29, 2007.

patrol craft   Patrol craft are small vessels operated by navy, coast guard, and marine police for coastal defense and homeland security missions. They are used for patrolling and interdiction in blue (open ocean) and brown (coastal ocean, ESTUARIES, and rivers) water environments. During the United States's involvement in Vietnam from approximately 1965 to 1973, Fast Patrol Craft or Swift boats were used in the Mekong River. This river runs from the Tibetan Plateau through China, Burma, Thailand, Laos, Cambodia, and Vietnam before emptying into the South China Sea. The very flat Mekong delta area is susceptible to flooding, especially during intense summer MONSOON rains. Operators had to consider impacts of river flow, changing water levels, waves, obstacles, and vegetation as these 51-foot (15.5-m) long boats maneuvered at speeds up to 23 knots (42.6 km/hr) with a draft of approximately four feet (1.2 m). Modern versions are used to support naval special warfare missions within the ships' specified capabilities. Example ships are approximately 170 feet (52 m) long with a maximum navigational draft of eight feet (2.4 m). U.S. Coast Guard patrol boats are cutters that range in length from 110 feet (33.5 m) to 123 feet (37.5 m) long and travel at speeds of approximately 26 knots (48 km/hr). Crews on board U.S. Coast Guard cutters participate in SEARCH AND RESCUE missions, prevent illegal immigration, and respond to those who are trying to float into U.S. waters, interdict go-fast drug smuggling vessels, and mitigate MARINE SPILLS. Police may have specially configured patrol craft to fight crime and to respond to marine mishaps. These vessels are usually boats with a V-hull or a cathedral hull and outfitted with state-of-the-art electronics, SCUBA equipment, and attachments such as a tow bar. Some maritime nations, such as the Netherlands, utilize police patrol craft that are approximately 26 feet (8 m) long and can maneuver at speeds up to 25 knots (46.3 km/hr). They have work spaces in the wheelhouse for the helmsman, navigator, and a communications officer and are designed to handle heavy seas (Beaufort Seven). Such vessels have wide decks and fenders to facilitate boarding operations. They may also be able to deploy rubber inflatable boats to operate in very shallow water. The Beaufort Scale was devised in 1805 by Admiral Francis Beaufort (1774–1857) and equates wind force to the corresponding appearance of the sea surface. For example, Beaufort Seven indicates that when winds are near gale force, the sea heaps up, waves are from 13 to 20 feet (4 to 6.1 m) high, and white foam streaks off the breakers. Marine scientists may support the navigation of patrol craft with imagery from satellites, output from wave and circulation models, and the display of data from instrumented buoys.

See also BOAT; COMBATANT SHIPS; HYDROFOIL.

Further Reading
Cutler, Thomas J. *Brown Water, Black Berets: Coastal and Riverine Warfare in Vietnam.* Annapolis, Md.: Naval Institute Press, 1988.
NAVY.mil. Available online. URL: http://www.navy.mil. Accessed October 3, 2006.
United States Coast Guard. Available online. URL: http://www.uscg.mil/uscg.shtm. Accessed October 3, 2006.

pear ring   A nearly round rigging device commonly used as an attachment point for moorings or as a sling link for the suspension of objects is called a pear ring. This device has a length (L), diameter (D), and two different sized widths (W$_1$ and W$_2$). The pear ring is an ideal link for towing where the towline connects to the narrow (tapered) end of the pear ring and several lines of a bridle are attached to the wider end of the pear ring. The pear ring can also be attached to straps and belts. The wide end may accommodate webbing, multiple ends of chain, or multiple rings. The narrow end is ideal for attaching to a clip, single ring, or chain link. The pear ring is often attached to swivels by shackles in a mooring and is also frequently attached to hooks. There are many pear ring sizes and depending on their use are made from materials such as stainless steel, galvanized steel, bronze, or painted steel. Small pear rings may be fitted with a threaded gate, for example, to secure keys. Other types of rings are perfectly round, oval, dee-shaped, and triangular or delta shaped.

See also MOORING; SHACKLE.

pelagic   Pelagic refers to inhabitants of open water marine realm found above the CONTINENTAL SHELF, SLOPE, RISE, ABYSSAL, and HADAL ZONES that are free of direct influence of the shore or ocean bottom. Neritic waters extend landward from the shelf-slope break, while oceanic waters extend seaward. Pelagic organisms are generally free-swimming planktonic or nektonic organisms. PLANKTON are tiny organisms that drift through the ocean's layers and serve as food for many marine species. MARINE BIOLOGISTS collect pelagic invertebrates with NETS, pumps,

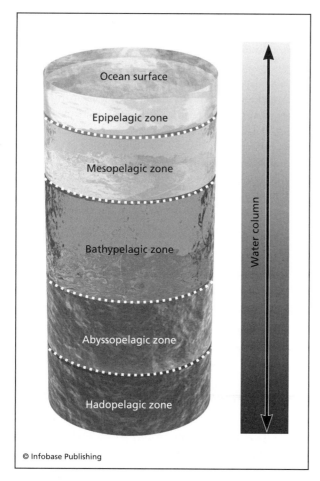

The relative locations of the pelagic zones (open-ocean regions away from land or inland waters in the oceanic water column)

and specialized collecting devices, over depths ranging from the bathypelagic to the neustonic layer. Eggs that drift about in the water column may also be considered as pelagic. Nektonic organisms such as squid, fish, and whales are active swimmers. Pelagic fish species tend to aggregate on structure such as SEAMOUNTS or near seaweed mats. Fisheries biologists sample fast-moving fish species with trawls. A behavior of the pelagic species such as schooling contributes to their dominance in overall fish landings in the ATLANTIC and PACIFIC OCEANS and the MEDITERRANEAN SEA. The pelagic cormorant (*Phalacrocorax pelagicus*) eats fish and crustaceans and can dive to depths of approximately 100 feet (30 m) to catch its prey. Animals living on or under the substrate are called benthic. Some organisms such as the loggerhead turtle (*Caretta caretta*) live in various ECOSYSTEMS (terrestrial, neritic, and oceanic zones) over the course of their lives.

Further Reading

Pelagic Shark Research Foundation. Available online. URL: http://www.pelagic.org. Accessed June 29, 2007.

UN Atlas of the Oceans. Available online. URL: http://www.oceansatlas.org/. Accessed September 17, 2006.

permanent current The permanent current is the average background current that flows regardless of season, local weather conditions, or stage of the TIDES. Permanent currents include the major oceanic currents of the global circulation system, such as the GULF STREAM, the KUROSHIO CURRENT, and the

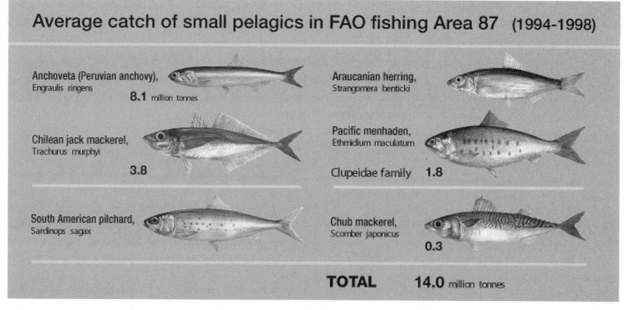

Pelagic fish caught off the coasts of Chile and Peru *(NOAA/NMFS)*

NORTH EQUATORIAL CURRENT. These currents are readily identified regardless of local conditions, although their local flows are modified by seasonal variations, which may produce such phenomena as meanders, EDDIES, and rings. Permanent currents include lesser-known currents such as the coastal current flowing southwest along the East Coast of the United States inshore of the Gulf Stream, sometimes known as the Labrador Current Extension (see LABRADOR CURRENT). It varies in strength and direction with seasonal conditions, but the basic flow is always present.

The permanent current at a given location (if there is one) can be calculated from CURRENT METER measurements by taking averages over many tidal cycles and over the passage of many weather systems. Hydrodynamic models consider permanent currents for applications such as predicting the trajectory of marine spills. Other factors to consider for spills include drift from winds, waves, and tides.

The driving mechanisms of the permanent currents include the wind stress, as in the case of the North Equatorial Current driven by the trade winds, and the density distribution, as in the case of the Gulf Stream.

Further Reading

Mangone, Gerard, J. *Concise Marine Almanac.* New York: Van Nostrand Reinhold, 1986.

Persian Gulf A MARGINAL SEA of the INDIAN OCEAN, the Persian Gulf is separated from the Gulf of Oman by the Strait of Hormuz, a strategic chokepoint. It is surrounded by Iran and Iraq to the north, Kuwait, Saudi Arabia, Bahrain, Qatar, United Arab Emirates (UAE), and Oman to the east and south. The Shatt al-Arab River, which forms part of the boundary between Iraq and Iran, is found along the western end. It is a shallow, well-mixed ESTUARY formed at the confluence of the Tigris, Euphrates, and Karun Rivers and ends as a well-developed delta. SEDIMENTS transported by the Shatt al-Arab River settle to the bottom as the sediment-laden water meets the Persian Gulf. The deltaic plain is noticeably fan-shaped, with the deepest water occurring along the Iranian coast, and a broad shallow area along the Arabian coast. Water depths near the Iraqi coast are usually several feet (a few meters) deep. Some of the islands in the Persian Gulf are Abu Musa, Al Jirab, Auhah, Bahrain, Bubiyan, Farur, Greater Tunb, Hendorabi, Hormuz, Kish, Larak, Lavan, Lesser Tunb, Nabi Salah, Qeshm, and Sirri Islands. Qeshm Island borders the north side of the Strait of Hormuz. The island state, Bahrain, lies between Qatar and the Saudi Arabian coast.

The Persian Gulf is about 615 miles (990 km) long with an area of approximately 90,000 square miles (233,099 km²). The narrowest location is 35 miles (56 km) in the Strait of Hormuz, and the widest location is 210 miles (338 km) between the UAE to the south and Iran to the north. It is a shallow water body, with an average depth of 115 feet (35 m). Depths deeper than 300 feet (90 m) are rare and occur in current-scoured channels around islands. The deepest sounding is 328 feet (100 m). Tides in the Persian Gulf are mainly diurnal, having one high and low tide per day. Some areas such as those near the Strait of Hormuz are characterized by semi-diurnal (two cycles per day), and other areas have mixed tidal signals. Near the mouth of the Shatt al-Arab mean higher high water is 10 feet (3 m), while regions near the Strait of Hormuz have mean high water spring tides of 12 feet (3.7 m). Mean higher high water is the mean of a series of higher high waters collected during a tidal day. The height of mean high water springs is the annual average of the heights of two successive high waters during those periods of 24 hours (during the new and full moons) when the range of the tide is greatest. The circulation in the Persian Gulf has a component called anti-estuarine, which means that the high-density (highly saline) water caused by the intense evaporation sinks and flows through the Gulf of Oman into the Indian Ocean and is replaced by surface water. This circulation produces an accumulation of carbonates on the seafloor of the gulf. The outflow of the deep water helps to remove pollutants from the Persian Gulf. The circulation in the southern part, close to the Gulf of Oman, also has a seasonal component related to the MONSOON wind regime of the Indian Ocean region. In the summer, the winds from the southwest monsoon are to the north, tending to drive surface flow northward, while in the winter, the northeast monsoon tends to drive surface water southward. In the northern and central parts of the gulf, the winds are primarily from the north-to-northwest, which drive surface water southward. The actual currents at a particular location depend on the relative strengths of the tidal, wind-driven, and anti-estuarine flows.

Water levels along the coast can be affected by local winds, which depending on direction can push water away from the coast (offshore winds) or cause flooding of the low sebkhas (onshore winds). Sebkhas or salt flats are flat areas where natural brines prevent the growth of vegetation. Dunes, mounds of sand deposited by the wind, surround some sebkhas. The sebkhas result from shallow lagoons lying behind BARRIER ISLANDS. They form from the accumulation of sediments that push up algal mats, leaving them exposed to the intense solar radiation and evaporation, especially after they have been flooded by especially high tides or storms. The high rate of evaporation makes the shores of the Persian Gulf one of the few areas of the world's oceans where evaporate minerals can form. Minerals enriched by this process include magnesium and calcium in the form of gypsum.

Persian Gulf water temperatures usually range from 75° to 90°F (24° to 32°C), but may reach as high as 96°F (36°C), crediting it with the warmest seawater on the globe. These temperatures can pose a challenge for some vessels having water-cooled engines and have little cooling effect for those working or living along the coast. Salinities are reported to range from 37 to 40 psu and the average for the world's ocean is 35 psu. High-SALINITY water flows out of the Persian Gulf and into the Gulf of Oman. There is more evaporation occurring than precipitation; the greatest amount of freshwater comes from the nutrient rich Shatt al-Arab estuary. These water temperatures (> 68°F [20°C]) and salinities (> 38 psu) make up the saline and dense WATER MASS known as Persian Gulf Water, which is found at a depth of 273 to 328 fathoms (500 to 600 m) in the Indian Ocean south of Madagascar.

The Persian Gulf has vast natural resources from marshes and CORAL REEFS to sand, pearls, and oil reserves. Marshlands such as the Hawr Al-Hawizeh in Iraq and the Hawr Al-Azim in Iran are transitional features between land and water. Fish, crabs, and shrimps live in the marshes where stems, leaves, and roots provide food and shelter from predators. They provide important habitat for MARINE MAMMALS such as the smooth-coated otter (*Lutrogale perspicillata*), the bandicoot rat (*Erythronesokia bunnii*), and the black-tufted gerbil (*Gerbillus mesopotamiae*). Migratory birds such as the pygmy cormorant (*Phalacrocorax pygmaeus*), marbled teal (*Marmaronetta angustirostris*), white-tailed eagle (*Haliaeetus albicilla*), imperial eagle (*Aquila heliaca*) and slender-billed curlew (*Numenius tenuirostris*) winter over in these marshes. In addition, MANGROVES found along the coast of Iran, Saudi Arabia, Bahrain, and the UAE are important habitat for species such as the white-collared kingfisher (*Halcyon chloris*). In general, light penetrates the clear Persian Gulf water column from top to bottom, providing the basic setting for SUBMERGED AQUATIC VEGETATION (SAV) and fringing reefs around the islands. The UAE has utilized approximately 2.8 billion cubic feet (80,000,000 m³) of dredged sand resources to build a palm island extending into the Persian Gulf, large enough to be seen with satellites. This complex has been built in the shape of a palm tree surrounded by BREAKWATERS. Hotels and land-scaped gardens are now being constructed on this man-made island, which is connected to the mainland by a bridge. Natural high-quality pearls coming from the Persian Gulf are from bivalves such as the gulf pearl oyster (*Pinctada radiata*), black-lip oyster (*Pinctada margaritifera*) and the winged oyster (*Pteria macroptera*). Oyster reefs in the Persian Gulf are mainly located along the Saudi Arabian, Bahrain, UAE, and Qatar coasts. The oysters coat irritants such as a grain of sand inside their shells with layers of nacre (mother-of-pearl) to isolate the foreign object.

The greatest economic resource coming from the Persian Gulf is oil, which is extracted from regions such as the Median trough. Such areas are generally found at low latitudes within 30° of the equator, where biologic productivity is high. This trough is also the axis of the Arabian plate and Iranian subplate and holds significant petroleum reserves. Once oil is extracted from the Persian Gulf oil fields, it is piped from an OIL PRODUCTION PLATFORM to oil terminals such as the Ra's at-Tannurah Oil Terminal and tank farm in Saudi Arabia.

Commercial fishing provides an important economic contribution as evidenced by fleets fishing for species such as black pomfret (*Parastromateus niger*), deep flounder (*Pseudorhombus elevatus*), frigate tuna (*Auxis thazard thazard*), grunt (*Pomadasys kaakan*), Hilsa shad (*Tenualosa ilisha*), humphead snapper (*Lutjanus coccineus*), Klunzinger's mullet (*Liza klunzingeri*), longtail tuna (*Thunnus tonggol*), narrow-barred Spanish mackerel (*Scomberomorus commerson*), orange-spotted grouper (*Epinephelus coioides*), yellowfin seabream (*Acanthopagrus latus*), giant seacatfish (*Arius thalassinus*), silver pomfret (*Pampus argenteus*), tigertooth croaker (*Otolithes ruber*), and wahoo (*Acanthocybium solandri*). Important shrimp species include the green tiger prawn (*Penaeus semisulcatus*), which is caught in shrimp trawls, and black tiger prawns (*Penaeus monodon*), which may be harvested from shrimp farms. Considerable fishing effort is made from small boats, including many wooden dhows (traditional Arab sailing vessels). The main fishing gear used by Persian Gulf countries includes trawls, gillnets, wire traps, and hook-and-line (*see* NETS).

Anthropogenic factors such as major wars in the past two decades, oil spills, and development threatens the Persian Gulf ECOSYSTEM. Rapid development reduces the extent of marshes and mangroves stands. The extent of coral reefs and their associated biota have been decimated through a combination of pollution, overfishing, and runoff and natural bleaching events. CORAL BLEACHING is caused when corals lose the symbiotic inter-cellular algae (*see* ZOOXANTHEL-

LAE) from which corals derive their color, most of their food, and their ability to grow their skeletons that forms a reef structure. Temperature changes and TURBIDITY are examples of stresses causing the loss of zooxanthellae, which photosynthesize during the daylight. Marine fisheries in the Persian Gulf are reported to be under stress by oil pollution, continuous land reclamation, and habitat loss. This is especially pronounced when critical habitat such as spawning grounds for fish and rookeries for birds are destroyed by construction or pollution. A prime nursery area and spawning ground for many species is the Shatt-al Arab.

Further Reading

Central Intelligence Agency. "The World Factbook." Available online. URL: https://www.cia.gov/cia/publications/factbook/geos/xo.html. Accessed October 7, 2006.

Hutchinson, Stephen, and Lawrence E. Hawkins. *Oceans: A Visual Guide.* Sydney, Australia: Firefly Books, Ltd., 2005.

Mangone, Gerard J. *Concise Marine Almanac.* New York: Van Nostrand Reinhold Company, 1986.

National Geographic Society. *National Geographic Atlas of the World,* seventh ed. Washington, D.C.: National Geographic Society, 1999.

Yao, Fengchao. "Water Mass Formation and Circulation in the Persian Gulf, and Water Exchange with the Indian Ocean." Available online. URL: rsmas.miami.edu/divs/mpo/About_MPO/Seminars/2004/Yao_18Feb04.pdf. Accessed October 15, 2007.

Peru (Humboldt) Current A cold current flowing northward along the western coast of South America, the Peru Current begins off the west coast of Chile at about latitude 40°S as an offshoot of the ANTARCTIC CIRCUMPOLAR CURRENT just before it reaches the Drake Passage. The current flows close inshore past Chile, Peru, and Ecuador, to around the southwest corner of Colombia, and then into the westward flow of the Pacific South Equatorial Current.

Although this current is sometimes called the Humboldt Current, after Alexander von Humboldt (1769–1859) who explored it in 1805, the name Peru Current (Corriente del Peru) was officially established by the Ibero-American Oceanographic Conference in April 1935. This cold current contributes to dry conditions along the arid and rocky Peruvian coast. It runs toward the equator along the eastern edge of the PACIFIC OCEAN.

The Peru Current is wide and slow moving in common with the eastern boundary currents of other oceans (*see* EASTERN BOUNDARY CURRENT), such as the CALIFORNIA CURRENT in the North Pacific and the CANARY CURRENT in the North

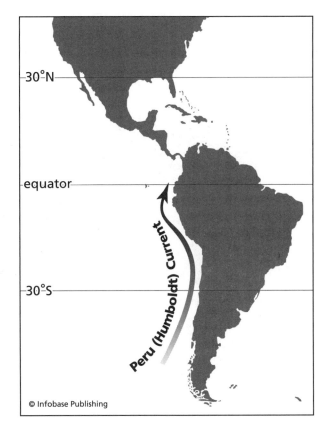

The Peru (Humboldt) Current is a cold current that flows northward from the Southern Ocean along the western coast of South America

Atlantic. It forms the eastern limb of the South Pacific GYRE. There is a poleward-flowing current beneath the Peru Current, the Gunther Current, similar to the Davidson Current flowing poleward beneath the California Current in the North Pacific.

The Peru Current is best known for the UPWELLING of cold nutrient-rich water that sustains a rich anchovetta and tuna fishery and a very large bird population. The upwelling results from EKMAN TRANSPORT away from the coast by the prevailing southeast winds. The cold upwelled surface water is responsible for the frequent occurrences of fog along the Peruvian coast.

When the Peru Current reaches tropical latitudes off the coast of Peru, its cool temperatures contrast sharply with the surrounding equatorial waters. In fact, "La Costa" along the Pacific Ocean has a dry, mild climate.

Every few years (two to seven) generally in December, the northward flow of the Peru Current off northern Peru and Ecuador is interrupted, and the cool Peru surface water is replaced by the warm equatorial water of the South Pacific Equatorial Countercurrent, which moves southward as the

Purse seiners operating in the Peru Current off the coast of Chile *(NOAA/NMFS)*

(Southern Hemisphere) summer season progresses. Under these conditions, the upwelling is cut off, the anchovetta move to other waters, or move to greater depths, and the tens of thousands of birds that depend on them for food perish, fouling the waters and the coastline. In addition, the warm water results in greater evaporation and rainfall in the normally arid regions of western Peru, with consequent floods and mudslides.

This condition, known as the "El Niño," once thought to be a result of local changes in the winds is now understood to result from conditions ranging across the vast South Pacific Ocean, with global consequences for the weather (*see* EL NIÑO). Its occurrence is closely related to the atmospheric phenomenon called the "southern oscillation" (SO); the terms have been combined into the acronym ENSO. These events coincide with Christmas, hence the name El Niño, meaning the Christ child, and local FISHERMEN know that the associated precipitation is the harbinger of a bad fishing season.

Further Reading

Narragansett Laboratory, National Marine Fisheries Service, National Oceanic and Atmospheric Administration. "Humboldt Current: LME #13, Large Marine Ecosystems if the World, Large Marine Ecosystem Program." Available online. URL: http://www.edc.uri.edu/lme/text/humboldt-current.htm. Accessed June 27, 2007.
World Wildlife Fund. "Humboldt Current." Available online. URL: http://www.worldwildlife.org/wildworld/profiles/g200/g210.html. Accessed June 29, 2007.

pH   The abbreviation pH stands for parts hydrogen, a scale with values ranging from 0 to 14, which indicates the level of acidity or alkalinity. The pH scale represents the concentration of hydrogen ions in a solution. Lower pH values correspond to acidic solutions, pH values of 7 are neutral, and higher pH values represent alkaline (basic) solutions. The ocean is slightly alkaline with an average pH of 8.1. pH is an important water measurement, which may be measured with an electronic pH meter in the environment or with litmus paper in the lab. Litmus paper contains special chemicals called acid-base indicators that are made from lichens. After the litmus strip is dipped into a solution, it will change colors; red means the substance is acidic and blue means it is basic.

photic zone   The photic zone is the layer of the ocean that is penetrated by sunlight to a depth of approximately 109 fathoms (200 m). In this region light is scattered and absorbed by the ocean. Seawater contains water molecules, which reflect blue wavelengths of light, particulate matter that scatters light (which may be seen in satellite images as sediment plumes), and reflections of light from the shallow seafloor (which highlights channels and shoals on aerial photographs). The amount of light attenuating material found in water or water clarity may be measured with a TURBIDITY meter or nephelometer in Nephelometric Turbidity Units. Many researchers still use a SECCHI DISK to measure turbidity. The black-and-white disk is lowered into the water until it can no longer be seen, and this result is known as the Secchi depth, a measure of turbidity. In the ocean, absorption is strong in the red wavelengths and weak in the blue wavelengths, as water molecules absorb the longer wavelengths first. This absorption of the red, orange, yellow, and green wavelengths (optical or visible spectrum) of light is especially noticeable to coral reef SCUBA divers who cannot make out the bright orange colors of the clown fish (*Amphiprion ocellaris*), which lives in symbiosis with anemones such as the giant sea anemone (*Stichodactyla gigantea*) or the magnificent anemone (*Heteractis magnifica*). The shorter wavelengths that are reflected by water particles are the blues and violets. PHYTO-PLANKTON in the ocean absorb red and blue wavelengths and reflect green wavelengths owing to chlorophyll. Chlorophyll is a green pigment vital to photosynthesis.

The photic zone is most important to support the primary producers of almost all marine food webs, phytoplankton. The photic zone provides solar energy that is used by phytoplankton to make organic compounds such as glucose ($C_6H_{12}O_6$) from inorganic compounds such as carbon dioxide (CO_2) and water (H_2O). Organic compounds include a chain of carbon atoms onto which hydrogen atoms are bonded. Inorganic compounds do not contain carbon chemically bound to hydrogen. Some important inorganic compounds in marine science are sodium chloride (NaCl), which makes up approximately 80 percent of the dissolved material in seawater, and calcium carbonate ($CaCO_3$), which is formed by corals (reefs), MOLLUSKS (oyster shells), and algae (coccoliths). It is interesting to note that shells in deep water dissolve due to the higher concentration of carbon dioxide (*see* CALCITE COMPENSATION DEPTH). The photic zone is also referred to as the epipelagic, euphotic, or sunlight zone.

See also DEEP CHLOROPHYLL MAXIMUM; EPIPE-LAGIC ZONE; EUPHOTIC ZONE; MARINE PROVINCES; PHOTOGRAPHER.

Further Reading

NASA Goddard Space Flight Center, Goddard Earth Sciences Data and Information Services Center. "Oceans: The Blue, the Bluer, and the Bluest Ocean." Available online. URL: http://daac.gsfc.nasa.gov/oceancolor/scifocus/oceanColor/oceanblue.shtml. Accessed October 15, 2007.

photographer Photographers record and manipulate still images, using film and/or digital methods. In marine science, a photographer's roles and responsibilities might extend to making a video or involvement with visual journalism. The photographer may be a member of a science party who is focused on documenting important field procedures, organisms, and features such as BEACH materials, gradients, and bank heights. Today's computer and optical technologies provide means for the description of shallow-

and deep-water organisms, phenomena, or processes as a combination of text and audio, images in the form of still and moving messages, informational graphics, and graphical designs for print and interactive multimedia presentations. Visual products can be disseminated via compact disk read-only-memory or CDROM, digital versatile discs or DVDs, the World Wide Web, and in traditional hard-copy formats such as film.

Photographers may perform their trade for the marine industry such as an oceanographic instrument manufacturer or a commercial diving company, news media or public affairs organizations concerned with current topics such as MARINE POLLUTION or maritime mishaps, or scientific organizations involved in basic or applied research. Some will practice the art of underwater photography and may support ocean exploration or the work of MARINE ARCHAE-OLOGISTS. Marine cameras, camcorders, and underwater lights are essential equipment. Some cameras, strobe lights, and video equipment may also be mounted to underwater structures or REMOTELY OPERATED VEHICLES. Aerial photographs may be taken to evaluate changes along a coast following severe weather. These images may be evaluated to determine rates of accretion or erosion along the coast. The U.S. Geological Survey's Earth Resources Observation Systems Data Center archives aerial and satellite images that are especially useful for mapping and studying physical features of the coast such as BEACHES, shoreline, shifting channels, and vegetationlike CORAL REEFS and MANGROVE SWAMPS. Researchers may routinely use a digital portable field microscope combined with a digital camera and a microscopic module to image microscopic objects. Marine science discoveries may also be highlighted through scientific photographs such as light microscopy images and electron microscopy images. To encourage the art of marine photography, the Rosenstiel School of Marine and Atmospheric Science at the University of Miami sponsors an annual underwater photography contest. This competition is open to any amateur photographer.

To ensure the collection of useful images, photographers must assess environmental conditions, utilize the optimal camera for a specific shoot, determine correct angles to make the composition most interesting, and know how to best capture their subject matter, especially if the animals are minute, skittish, dangerous, or exceedingly large. Underwater photographers must understand the transmission and absorption of light through the water column. They compensate by restoring colors with mixed-light photography and the use of filters. In low-light conditions, the photographer may use multiple

Modern Physical Oceanographer: A Biographical Sketch of Walter Heinrich Munk

Walter Heinrich Munk was born in Vienna, Austria, on October 19, 1917, and was sent by his family to the United States at the age of 14 to become a banker. Munk instead developed a distinguished career as a physical oceanographer, educator, and geophysicist who has made major advances in almost all areas of the physics of the ocean and the solid Earth. His long career spans pre–World War II to the present. He was fascinated with physics, which he first studied at Columbia University, and subsequently earned bachelor's and master's degrees from the California Institute of Technology. At the outbreak of World War II, and prior to his involvement with Scripps, Munk served in the U.S. Army Ski Battalion. However, he was shortly reassigned to assist Professor Harald Sverdrup (1888–1957) at the Scripps Institution of Oceanography, University of California, in war-related research. He was both an oceanographer with the University of California Division of War Research and a meteorologist for the Army Air Force. During the war, Munk, along with Professors Sverdrup, Jakob Bjerknes (1887–1975), and Jorgen Holmbroe trained U.S. military personnel in military meteorology and oceanography; this training program later became the prototype for university courses in meteorology and oceanography in the United States during the cold war years.

During World War II, he, along with Professor Sverdrup, developed the "Sverdrup-Munk" method of forecasting ocean waves and swell. This procedure made it possible for military planners to forecast ocean wave conditions prior to Allied assaults on enemy-held territory, resulting in saving the lives of many Allied soldiers in landings from North Africa to Normandy, and in the island hopping campaigns of the Pacific theater of the war. After World War II, he continued his education at Scripps and was awarded the Ph.D. in 1947. At that time, he joined the faculty of the University of California at San Diego as assistant professor.

In 1954, he became professor of geophysics. In 1955, he founded the University of California Institute of Geophysics and Planetary Physics and directed the institute until 1984. He is currently professor emeritus at the University of California.

In the 1950s, he did research on long waves in the surf zone, the "quiet band" in the ocean wave spectrum, and developed concepts of surf beat, edge waves, tsunami propagation, and long-distance swell propagation. He studied and improved tidal prediction. He pioneered the use of SCUBA gear in physical oceanography. Professor Munk was one of the earliest modelers of the ocean, and oceanography students have studied his models of ocean circulation for the last half century. He has a wide range of interests in science and has published major works on topics ranging from tide gauges to the physics of the rotation of the Earth, described in a book co-authored with Gordon MacDonald, *The Rotation of the Earth* (Cambridge University Press, 1960).

---

strobes. In the PHOTIC ZONE with ample light and clear water, the photographer may still require fill light and correction filters in order to achieve good color saturation. A fill light provides artificial light to illuminate the shadow areas of a scene. Filters may be colored glass or plastic, which are used to modify light, especially color content. Some photographers may also develop and print their own photographs or technically manipulate and enhance their photographic images to create a desired effect.

Further Reading

San Diego Underwater Photographic Society, Inc. Available online. URL: http://www.sdups.com. Accessed August 9, 2006.

U.S. Geological Survey. "Earth Resources Observation and Science." Available online. URL: http://edc.usgs.gov. Accessed August 9, 2006.

University of Miami, Rosenstiel School of Marine and Atmospheric Science. Available online. URL: http://www.rsmas.miami.edu. Accessed August 9, 2006.

physical oceanographer   Scientists who study the physics of the ocean and often the atmosphere above it, as well as the relations between the two are physical oceanographers. They study the distribution and motions of mass and energy. Specific phenomena studied include the distribution of TEMPERATURE, SALINITY, and DENSITY in the ocean, temperature and water vapor in the atmosphere, ocean wind waves, TIDES, CURRENTS, INTERNAL WAVES, SEA ICE, the propagation of light and sound in the sea, marine climate, and weather.

As is the case with physics in general, physical oceanographers tend to be either theoretical or experimental scientists. Theoretical oceanographers develop mathematical models to explain the circulation and wave motions of the ocean, which usually involve sets of complex equations that must be solved numerically on supercomputers. Experimental scientists use complex electronic instruments to measure the physical properties of the oceans, such as temperature, salinity, and pressure by means of CON-

Munk advocated a proposal, along with Harry Hess (1906–69), in the early 1960s to drill through the Earth's crust and the Mohorovicic discontinuity (Moho) under the ocean to the mantle. While this project was too ambitious for the technology of the time (and still is), it helped to start the Deep Sea Drilling Project, funded primarily by the U.S. National Science Foundation. This program has continued for decades and has revolutionized understanding of the Earth's geophysics. During this time, Munk developed many new oceanographic instruments in partnership with Scripps Institution of Oceanography engineer Frank Snodgrass.

In 1979, Professors Munk at the Scripps Institution of Oceanography and Carl Wunsch (1941–) at the Massachusetts Institute of Technology proposed using an acoustical analog of X-ray tomography to determine the temperature and heat content of ocean waters over the vast reaches of the Pacific Ocean. That concept has been implemented in the Acoustic Thermometry of Ocean Climate (ATOC) project. Munk organized the Heard Island Experiment in which underwater acoustic signals transmitted at Heard Island in the South Indian Ocean could be heard around the world. However, groups of environmentalists opposed the experiment because of possible adverse effects on marine mammal hearing and behavior. Experiments were conducted with the participation of marine mammal biologists. Along with oceanographers Peter Worcester and Carl Wunsch, Munk published *Ocean Acoustic Tomography* in 1995.

Munk showed that a major factor in ocean tidal friction is the generation of internal waves at seafloor topographic features, such as mountain ranges, seamounts, and guyots. Tidal friction is important because it is gradually slowing the Earth's rate of rotation; some day in the distant future, the Earth will keep the same face to the Moon as the latter does now to the Earth.

Munk also served on numerous important research committees, such as the Naval Research Advisory Committee. In 1984, Munk became one of four chairmen of Navy Research in Oceanography, under the secretary of the navy. He was elected to membership in the National Academy of Sciences in the United States, the Russian Academy of Sciences, and the Royal Society. He was awarded the National Medal for Science by President Ronald Reagan (1911–2004) in February 1985.

Further Reading

Charton, Barbara. *A to Z of Marine Scientists*. New York: Facts On File, 2003.

Munk, Walter Heinrich, Peter Worcester, and Carl Wunsch. *Ocean Acoustic Tomography*. Oxford: Cambridge University Press, 1995.

Scripps Institution of Oceanography Archives. "Walter Heinrich Munk Biography." Available online. URL: http://scilib.ucsd.edu/sio/archives/siohstry/munk-biog.html. Accessed October 17, 2007.

—**Robert G. Williams, Ph.D.**, senior oceanographer
Marine Information Resources Corporation
New Bern, North Carolina

DUCTIVITY-TEMPERATURE-DEPTH (CTD) PROBES, and current velocity by means of sophisticated CURRENT METERS employing acoustic or electromagnetic waves, drifting BUOYS tracked by Earth satellites, and AUTONOMOUS UNDERWATER VEHICLES (AUVs). These instruments can be deployed from well-equipped research vessels, ocean platforms, moored buoys, aircraft, and deep SUBMERSIBLES. Physical oceanographers also deploy remote sensors, such as radar ALTIMETERS, and synthetic aperture RADARS on Earth satellites.

Physical oceanographers normally have at least a master's degree. University research positions normally require a doctorate followed by postdoctoral research. With only a bachelor's degree, positions are generally limited to the technician level.

There are two general approaches to obtaining advanced degrees. Some people choose to attend universities that give advanced training and postgraduate degrees in physical oceanography. Others prefer to take their advanced degrees in physics and mathematics, and then participate in postdoctoral training in physical oceanography at major laboratories such as the SCRIPPS INSTITUTION OF OCEANOGRAPHY, and the WOODS HOLE OCEANOGRAPHIC INSTITUTION.

Major universities, government laboratories, and private industries employ physical oceanographers. They perform basic research on the physics of the ocean and the marine atmosphere, as well as applied research in fields such as marine weather FORECASTING, military science, climatology, marine mineral exploration, pollution control and remediation, resource management, ship routing, underwater acoustics, and marine safety.

To keep abreast of recent developments in their field, physical oceanographers must spend considerable time reading current technical journals, participating in scientific meetings and workshops, and in taking courses in new technologies. As in almost all the sciences, they must prepare technical proposals to federal or state agencies and organizations, such as the National Science Foundation, the Office of Naval

Research, and NOAA, to obtain the funds to support their project work. They can expand their resources by partnering with scientists from other organizations and by making use of existing databases available on the World Wide Web.

Further Reading

AMS-UCAR. 2000 *Curricula in the Atmospheric, Oceanic, Hydrologic, and Related Sciences, American Meteorological Society and the University Corporation for Atmospheric Research.* Available online. URL: http://www.ametsoc.org/curricula/index.html. Accessed June 29, 2007.

University of Delaware Sea Grant College Program. "Marine Science Careers." Available online. URL: http://www.ocean.udel.edu/seagrant/education/marinecareers.html. Accessed October 15, 2007.

physical scientist Physicists or other scientists specializing in the study of nonliving matter and energy are physical scientists. Examples of physical scientists are physical chemists, astronomers, meteorologists, physical oceanographers, and geophysicists. Physical scientists often apply the findings of physicists, chemists, and mathematicians working on the frontiers of basic science to their areas of specialization. For example, PHYSICAL OCEANOGRAPHERS studying nonlinear interactions of ocean wind waves have applied recently developed techniques of particle physics to understand the transfers of energy among ocean wave components. Physical scientists often work closely with engineers in the development of new technology measurement systems, such as underwater acoustic tomography.

In general, the training of physical scientists involves obtaining a bachelor's degree in a basic science (physics, chemistry, mathematics), and graduate study of the desired specialty, such as underwater acoustics. Physical science students desiring to work in research should obtain a doctorate in their chosen field of study, followed by post-doctoral work under the tutelage of established scientists who have ready access to the specialized equipment needed in this field.

Academic institutions, government agencies, and private industry employ physical scientists. To keep abreast of recent developments in their fields, physical scientists must spend considerable time reading current technical journals, participating in scientific meetings and workshops, and in taking courses in relevant new technologies. Much new technical information and data are exchanged among scientists by means of the Internet.

Further Reading

Physics Central, American Physical Society. Available online. URL: www.physicscentral.com. Accessed June 30, 2007.

phytoplankton Phytoplankton are microscopic organisms that drift about in the upper layers of lakes, ESTUARIES, and the ocean. They use chlorophyll for photosynthesis by fusing water molecules and carbon dioxide (CO_2) into carbohydrates in the presence of sunlight. Phytoplankton are one of the two major divisions of plankton, the other being zooplankton. The division is made not according to plants or animals, as is the case with terrestrial species, but is based on whether or not the organisms can photosynthesize. Some photosynthesizers ingest other organisms when light levels are low, exhibiting animal-like characteristics. For example, the euglenoids can hunt for food under conditions of reduced light. They have chloroplasts, a pigmented organelle (eyespot) that helps the organism orient toward light and a flagellum for mobility. Organelles are distinct structures within a cell that carry out specific functions.

Phytoplankton are autotrophic organisms, that is, they are the primary producers in the marine food chain. Phytoplankton use carbohydrates as "building blocks" to grow. Marine animals then consume the phytoplankton to get these same carbohydrates. In this way, energy is transferred through feeding to primary consumers such as zooplankton and then to secondary and tertiary consumers such as fish.

Some of the most common phytoplankton are diatoms, dinoflagellates, and COCCOLITHOPHORES. Diatoms are unicellular ALGAE with cell walls composed of silica. These rigid cell wells comprise two valves or shells and are called tests. Diatoms inhabit the oceans, freshwater, and even damp soils. Their tests are preserved in the fossil record. Dinoflagellates are protists with two flagella and a cellulose covering. They may be primary producers or predators on other protozoa. They have characteristics of both plants and animals and are best known from summer blooms that are called "red tides." Coccolithophores are single-celled algae that surround themselves with calcium carbonate tests or coccoliths. Blooms of coccolithophores cause what is sometimes called a "white tide." The microscopic tests (skeletons) of phytoplankton such as coccolithophores comprise biogenous sediments such as chalk, a soft, white, and powdery rock.

See also PLANKTON; ZOOPLANKTON; ZOOXANTHELLAE.

Further Reading

American Society of Limnology and Oceanography. Available online. URL: http://aslo.org/index.html. Accessed November 27, 2006.

The Euglenoids Project home page. Available online. URL: euglena.msu.edu/. Accessed October 15, 2007.

Guiry, Michael D., and G. M. Guiry. *AlgaeBase.* Available online. URL: http://www.algaebase.org. Accessed October 15, 2007.

National Oceanographic Data Center, National Oceanic and Atmospheric Administration. "Plankton Data." Available online. URL: www.nodc.noaa.gov/General/plankton.html. Accessed November 27, 2006.

Smithsonian Environmental Research Center. "Phytoplankton Guide." Available online. URL: http://www.serc.si.edu/labs/phytoplankton/guide/index.jsp. Accessed November 28, 2006.

piston corer   A piston corer is an oceanographic instrument used to take a vertical sample of the sediment at the bottom of a body of water. Cores are usually taken when a simple GRAB SAMPLE of the bottom is inadequate. Piston corers, also known as gravity corers, are lowered from a research vessel to the vicinity of the bottom, where the corer is allowed to free-fall. When it strikes the bottom, gravitational force drives it into the SEDIMENT, thus obtaining a relatively long cylindrical sample of the material. In fine-grain sediments, such as clay and mud, cores as long as 66 feet (20 m) can be obtained; in coarse-grain, compacted sediments, the core may be only 1.6 feet (0.5 m) long. The diameter of the core sample is on the order of a few inches (cm).

The corer consists of a cylindrical piston, usually made of stainless steel, with a plastic liner, which holds the core. A sharp nose helps the corer to penetrate the bottom. Weights are added, as much as 2,200 pounds (1,000 kg) for deep-sea corers, to the piston to increase the force obtained at the water/sediment interface, and drive the corer farther into the sediments. Sometimes an explosive is detonated on impact with the bottom to drive the corer deeper into the sediment. Most modern corers have a probe attached that strikes the bottom first, releasing the corer from the ship's cable into free fall at a preset and adjustable distance. One refinement has been to create a partial vacuum in the core liner by means of a piston released when the corer strikes the bottom, thus reducing the friction of the sediment passing into the liner. The nose of the corer is tapered and sharp to penetrate the sediment. Some corers have a "core catcher" attached to prevent the core from falling out of the corer back to the bottom.

Once the sample has been taken, the corer and core are winched back aboard the research vessel for study aboard ship, then preserved for more complete analysis at a shore-based laboratory. Some cores are prepared for archiving for long-term storage at a core "library." When the core has been taken, the load on the WINCH is at a maximum to break the corer loose from the sediment. In the early days of oceanography, very often the cable would break, the corer would be bent, or the nose damaged, resulting in lost data. Getting a large corer back aboard ship during rough sea conditions is always hazardous to the scientists and crew. However, modern materials and careful design have minimized some of these difficulties.

Piston corers come in many sizes, from less than a meter for use on small vessels doing estuarine research, to many meters for use in the deep sea aboard large research vessels. They have been used since 1947 to collect cores of soft sediment such as clay without disturbing the sediments. Once a core is retrieved to the ship, the sample is cut into manageable lengths or segments. The segments are split in half lengthwise so that the collected samples can be subsampled and studied. Piston cores are used aboard the 420-foot (128 m) U.S. Coast Guard cutter *Healy*. Research vessels like the *Healy* include deck cranes, A-frames, and winches.

A major problem with the analysis and interpretation of the data is that when the core is taken, the sediments are compressed inside the corer, so that the core does not really represent the true vertical environment under the sea bottom. This is particularly vexing to biologists who want to study the spatial distribution of benthic organisms. Geologists wanting to study the stratigraphy of the ground under the water must make corrections for the compaction of the sediment in order to get a true picture of the distribution of quantities like sediment density and porosity. This distortion in the samples from piston/gravity corers provided the motivation to develop new types of corers such as box corers and vibrocorers. The need for deeper cores to characterize the ocean bottom has led to the use of geophysical drill rigs, which rely on mechanical drilling to produce cores of much greater length than those obtainable by piston/gravity corers.

See also BOX CORER; OCEAN DRILLING PROGRAM; VIBROCORER.

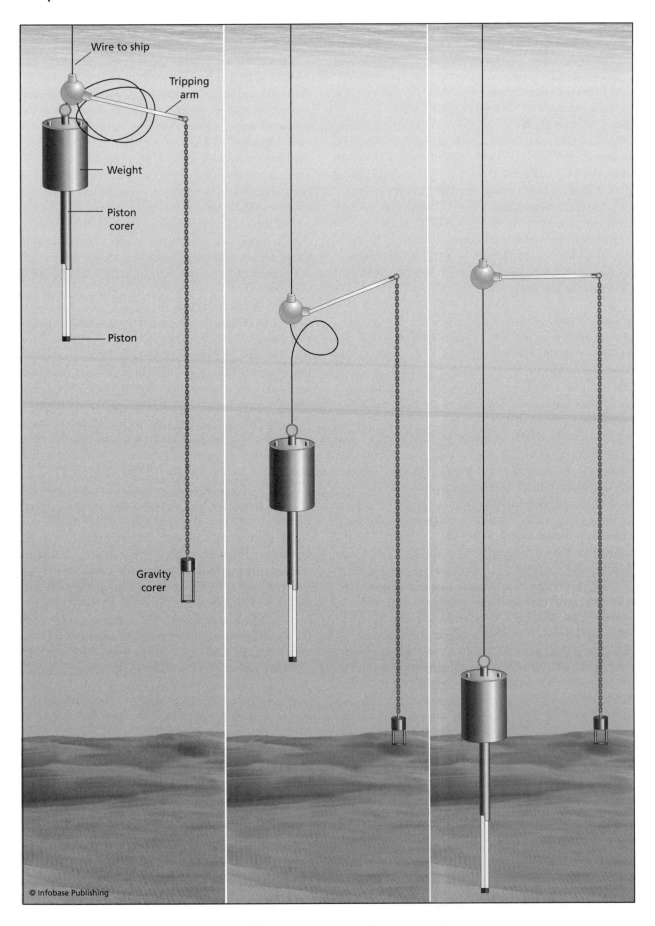

Wire to ship

Tripping arm

Weight

Piston corer

Piston

Gravity corer

Further Reading
National Geophysical Data Center. Available online. URL: www.ngdc.noaa.gov. Accessed November 22, 2006.
United States Coast Guard. Available online. URL: www.uscg.mil. Accessed November 22, 2006.

plankton Plankton are small organisms that passively drift with the CURRENTS. Professor Viktor Hensen (1835–1924) is credited with coming up with the name from the Greek *planktos* meaning "wandering." Floating microscopic plants that obtain their energetic requirements through photosynthesis are classified as phytoplankton. Animal plankton such as copepods and jellyfish are classified as zooplankton. Plankton may be able to swim by means of flagella or other appendages, but they are not strong enough to swim against a current. A flagellum is a long, whiplike organelle protruding from a cell, which is used for motility. Larval forms of some vertebrates and invertebrates are called meroplankton because they spend part of their life cycle in the plankton prior to becoming part of the nekton or benthos as adults. Organisms spending their entire life cycle as plankton are called holoplankton.

Plankton may be sampled using either a fine mesh plankton net towed from a research vessel or a water sampling system (*see* NETS and NISKIN BOTTLE). They may be categorized according to size. Plankton measuring less than two micrometers in size are called picoplankton, between two and 20 micrometers nanoplankton, between 20 and 200 micrometers microplankton, between 0.2 and 20 millimeters mesoplankton, between 20 and 200 millimeters macroplankton, and greater than 200 millimeters megaplankton.

Regardless of size, planktonic organisms are the most abundant life forms on Earth and play a crucial role in the marine food chain. They are usually the base of the food web. Phytoplankton are the dominant primary producers in the ocean. They are strongly affected by TEMPERATURE, light, and nutrient availability. For this reason, they dominate the PHOTIC ZONE, or the depth to which sunlight penetrates the water. Given excess nutrients, the populations of phytoplankton will bloom, causing fluctuations in the amount of available oxygen. Zooplankton are grazers of phytoplankton and forage for larger animals. Zooplankton such as COPEPODS are also consumed by fish such as the whale shark (*Rhincodon typus*) and baleen whales such as the pygmy right whale (*Caperea marginata*).

See also CETACEANS; PHYTOPLANKTON; ZOOPLANKTON.

Further Reading
American Society of Limnology and Oceanography. Available online. URL: http://aslo.org/index.html. Accessed November 27, 2006.
Census of Marine Life. Available online. URL: http://www.coml.org. Accessed November 27, 2006.
National Oceanographic Data Center, National Oceanic and Atmospheric Administration. "Plankton Data." Available online. URL: http://www.nodc.noaa.gov/General/plankton.html. Accessed November 27, 2006.
State University of New York at Stony Brook. "Plankton." Available online. URL: http://life.bio.sunysb.edu/marinebio/plankton.html. Accessed November 27, 2006.

plankton bloom A rapid increase of phytoplankton occurring during periods of favorable light conditions and nutrient buildups is called a plankton bloom. Some physical factors contributing to blooms are reduced flushing rates, elevated SALINITIES, and delivery of micronutrients from land drainage, especially iron. These factors may follow the transit of severe storms such as a HURRICANE, or long periods of summer drought, when the water column becomes highly stratified and vertical mixing is inhibited. This environment leads to eutrophic conditions that result in low or anoxic water near the bottom that is toxic to benthic organisms and bottom-feeding fishes. Blooms or high concentrations of plankton may discolor the water with pigments and produce what may be called red, white, and brown tides. Dinoflagellates such as *Alexandrium tamarense, Noctiluca scintillans,* and *Pfiesteria piscicida* usually cause red tides, which may be harmful to fish and people owing to toxins produced by these plankton. Filter feeders such as some zooplankton and MOLLUSKS may ingest and become contaminated by harmful dinoflagellates, causing paralytic shellfish poisoning. Paralytic shellfish poisoning is a rare form of food poisoning from eating contaminated shellfish such as mussels, cockles, clams, scallops, oysters, crabs, and lobsters. Harmful algal blooms may also kill MARINE MAMMALS. Large quantities of COCCOLITHOPHORES such as *Emiliania huxleyi* are reported to turn waters the color of milk. Marine scientists have used satellite imagery as well as photography from the space shuttle and International Space Station to study the extent of white tides. Brown tides result from a bloom of microalgae (picoplankton) such as *Aureococcus*

(opposite page) The piston corer relies on gravity aided by the piston to obtain core samples of up to several feet in length.

The Deep Sea Drilling Project

On June 24, 1966, a contract between the National Science Foundation and the University of California began Phase 1 of the Deep Sea Drilling Project (DSDP), which was based out of Scripps Institution of Oceanography in San Diego, California. DSDP was founded in order to investigate the evolution of ocean basins by drilling cores of oceanic sediments and underlying oceanic crust. In 1967, the Joint Oceanographic Institutions for Deep Earth Sampling (JOIDES), funded by NSF, leased a drill ship named the *Glomar Challenger* to serve as the research vessel for DSDP, and launched a new era in scientific exploration of Earth's oceanic rocks and marine sediment (*see* the "Drill Ship: The *Glomar Challenger*" essay).

Scientific operations aboard the *Glomar Challenger* included seismic and magnetic surveys, borehole measurements, and laboratory analysis of cores recovered from the seafloor in order to better understand the world's ocean basins. Some of the major discoveries of the DSDP have provided many fundamental understandings of the ocean floor. During the first DSDP expedition (or "leg"), core samples were taken from what are now known as salt domes. This discovery has since become extremely important for oil companies because many oil

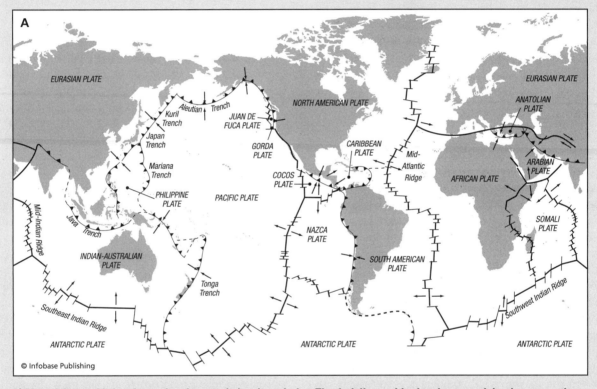

A) Map of the world showing major plates and plate boundaries. The dark lines with triangles are subduction zones, the double lines are mid-ocean ridges, and the single lines are transform faults.

anophagefferens. This species lacks flagella and body scales and has contributed to ecological damage and destruction of commercial shellfisheries. Vessels are considered to be potential vectors for long-distance transport and local-scale dispersal of plankton, which may bloom when transported to rich feeding grounds, possibly in ship's ballast and bilge tanks.

See also PHYTOPLANKTON; PLANKTON.

plate tectonics   Plate tectonics is the theory of movement of the Earth's crust due to convection currents in the upper mantle, which accounts for the distribution of the continents, mountain building, earthquakes, and volcanoes. Plate tectonics was preceded by the theory of continental drift, much of which was developed by Alfred Wegener (1880–1930), a German meteorologist. Although other scientists had suggested the movement of continents over time, Wegener was the first to put together a complete description of how this might have occurred. His ideas were based on the similarity in outline of the continents on both sides of the ATLANTIC

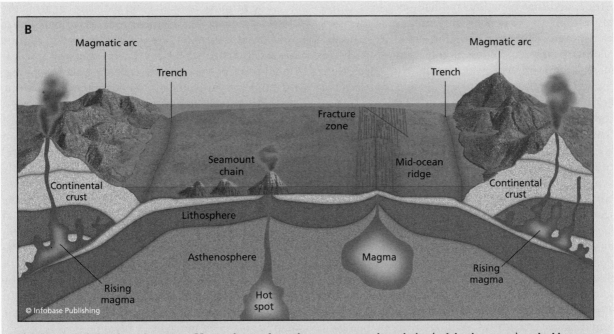

B) A diagram of plate boundary types. Magmatic arcs form above convergent boundaries (subduction zones), and mid-ocean ridges form along active, divergent boundaries. Fracture zones extend past transform boundaries (plates sliding in opposite directions) and are associated with plate segments moving in the same direction. A hot spot is also shown that supplies magma from the upper mantle.

reservoirs are found in close proximity to salt domes. However, the purpose of DSDP and the *Glomar Challenger* was scientific exploration.

One of the most important discoveries occurred during DSDP Leg 3 along the Mid-Atlantic Ridge. The information scientists gathered from drilled cores gave definitive proof in support of Alfred Wegener's theory of continental drift. The cores provided evidence that oceanic crust is created at spreading centers and migrates outward in either direction to ultimately be subducted under less dense continental crust. Many geologists consider this finding to be the "smoking gun" for the plate tectonics theory and seafloor spreading.

Another important discovery was that oceanic crust is geologically much younger than continental crust. DSDP scientists found that oceanic crust reaches a maximum age of 200 million years, as opposed to roughly 4 billion years preserved in continental crust. This means that oceanic crust continually progresses through a life cycle, from spreading centers to subduc-

tion zones where oceanic rocks (basalt) are consumed back into Earth's mantle.

The DSDP drilled approximately 624 holes into the seafloor throughout the world ocean. For 15 years, the *Challenger* sailed 96 legs, which averaged to about one hole every 308,880 square miles (800,000 sq km). During its existence, DSDP involved the participation of five nations and provided the foundation for successive programs: the Ocean Drilling Program and the Integrated Ocean Drilling Program (*see* OCEAN DRILLING PROGRAM). The *Glomar Challenger* and DSDP were retired in November 1983.

Further Reading

The Deep Sea Drilling Project, Reports and Publications home page. Available online. URL: http://www.deepseadrilling. org/. Accessed on October 17, 2007.

—**Bradley A. Weymer,** marine laboratory specialist
Integrated Ocean Drilling Program
Texas A & M University
College Station, Texas

OCEAN, which had been noticed as early as 1596 by German geographer Abraham Ortelius (1527–98), as well as the similarity of fossils and geologic strata at corresponding locations in South America and Africa, and North America and Europe. The explanation for these observations during his time was that the continents had been connected by land bridges, which then sunk into the oceans. Wegener hypothesized that all the continents were at one time connected as the supercontinent that he named Pangaea, which broke apart 200 to 250 million years ago. Wegener found that glaciers had occupied parts of North

Africa, and that New England had once been tropical, leading to the conclusion that these landmasses had moved. Mountain building could then be explained by the collisions of continents that thrust material upward. For example, the Himalayas were formed when India crashed into Asia. At first, he had no suitable theory to explain how things as large as continents could drift, but he developed a hypothesis that the centrifugal force of the Earth's rotation and the tidal gravitational forces combined to move the continents over the upper mantle. In seeking evidence, he tried to estimate the speed of movement of landmasses, but greatly overestimated it. His research was discredited by the leading geologists of his day when they showed that the forces generated by the centrifugal force of the Earth and the gravitational forces were much too weak to move continents through solid rock, like ships sailing on the ocean.

Interest in Wegener's work was revived in the 1950s and 1960s, when new technology, especially sounding FATHOMETERS facilitated the mapping of the topography of the ocean floor (BATHYMETRY), revealing rift zones and a mid-ocean ridge system that circled the Earth like the stitching on a baseball. Other evidence included the pattern of magnetic signatures in undersea rocks and the increasing depth of SEDIMENTS with increasing distance from the mid-ocean ridges. The discovery of seafloor spreading in the 1960s was the beginning of the acceptance of the idea of continental drift, as well as providing the explanation of how this motion takes place in the developing theory of plate tectonics. In plate tectonics theory, the rigid plates on a sphere (somewhat like an eggshell with cracks) rotate around a "pole of spreading," through which passes an "axis of spreading," not necessarily coincident with a planet's axis of rotation, as shown by Swiss mathematician Leonard Euler (1707–83).

Bathymetric and geophysical data collected during the 20th century showed that blocks of continental crust and adjacent oceanic crust correspond to the "plates" on a sphere, and appear to be "floating" on the upper mantle, the asthenosphere. The lithosphere is solid and includes the continental and oceanic crust. The asthenosphere is semi-molten (plastic) and lies between 62 to 155 miles (100 to 250 km) to a depth of approximately 373 miles (600 km) below the lithosphere. The lithosphere is broken up into a number of plates such as African, Antarctic, Australian, Eurasian, North American, South American, Pacific, Cocos, Nazca, and the Indian plates. The plates are in motion, and the edges are either moving away from one another (divergent boundaries), pushing against one another (convergent boundaries), or sliding past one another (transform boundaries).

The primary force driving the plates apart is the CONVECTION of hot magma: rock that flows under great heat and pressure from the upper mantle toward the surface of the Earth; and the gravitational force of the opposite ends of the plates plunging under continents at the locations of oceanic TRENCHES (deep depressions in the ocean floor); a process called subduction (*see* SUBDUCTION ZONE). The energy driving the convection is the heat from deep within Earth's lower mantle and upper core. This is residual heat acquired and subsequently retained when, during the time of planet formation, planetesimals smashed into one another to create the hot, young Earth, as well as from the heat produced by radioactivity in the rocks. Major milestones in the evolution of plate tectonics theory include the 1962 publication of Harry Hess's *History of Ocean Basins,* and a map of the *World Ocean Floor* by Bruce C. Heezen (1924–77) and Marie Tharp (1920–2006), in 1977 (versions of portions of their map having been published earlier).

Plate tectonics theory shows that the production of new oceanic crust is accomplished by the separation of lithospheric plates along the mid-ocean ridges, which permits magma (molten rock under the surface) to well up out of the rift found at the ridgeline and spread out in opposite directions perpendicular to the ridge axis. The process functions something like two conveyor belts, with the belts carrying new crust moving slowly (on the order of inches (cm) per year) away from the axis of the ridge. Evidence for this process includes the increasing depth of oceanic sediment moving away from the rift toward the continents, the age of the rock floor increasing with distance from the rift, and magnetic polarity of the rock base in reversing strips corresponding precisely to reversals of Earth's magnetic poles throughout the geologic ages. Recently, using sophisticated techniques of very long baseline INTERFEROMETRY, geophysicists have confirmed that the seafloor and the associated tectonic plates are moving with velocities on the order of those predicted by plate tectonics theory, that is, on the order of inches per year.

Subduction zones, the area where an oceanic tectonic plate plunges down into Earth's mantle, occur at the opposite ends of the tectonic plates from the spreading ridges where they are formed. The deep-ocean TRENCHES are found directly above the subducting plate. Volcanic island chains are often located on the continental side of the trench, for example, the Japan Trench. The volcanoes on these islands have been formed by magma being forced upward by the downward-moving plate, and are the most powerful on Earth. Subduction zones are also regions of strong earthquakes, which often generate large TSUNAMIS. The entire Pacific Ocean is bounded by subduction zones, as can be seen from the distribution of ocean trenches around its perimeter. The large number of volcanoes associated with this process is collectively known as the *Ring of Fire*.

Subduction is necessary to preserve the mass balance of material on the surface of the Earth. The way it occurs is that the newly formed plates move away from the ridges; they acquire sediment on top and cooling magma underneath from the mantle, both of which thicken the plate and make it heavier as time goes on. Eventually, the plate becomes too heavy to remain on the surface; gravity pulls it down into the mantle, where, over millions of years, it melts and is "recycled" as a new plate.

Subduction zones may be found at converging oceanic plates or by an oceanic plate and a continental plate. In the latter case, the oceanic plate, being denser, always plunges under the continental plate. Because the oceanic plates are continually being "recycled," they are very young compared to the age of the continental plates, which have been on the surface of the Earth for billions of years.

An example of a subduction zone is found in the west coast of the United States, where the Pacific plate is diving under the North American plate, and in the process, forming mountain chains such as the Cascades. The San Andreas Fault is found in this region; major earthquakes have occurred recently in both San Francisco and Los Angeles.

When oceanic plates collide, island arcs are formed, as in the case of the Aleutian chain. The curvature of the arcs is a feature of the spherical Earth (the intersection between a sphere and a plane), and can be seen quite distinctly in following the chain on a map from Alaska to Russia.

Along the margins of the tectonic plates, transform faults form where edges of the plates are sliding past one another in strike-slip (lateral) movements. They are believed to result from lateral strain. Almost all the transform faults on Earth are located on mid-ocean spreading ridges where they run perpendicular to the ridgeline and offset the centerline of the ridge. In the case of the Mid-Atlantic Ridge, transform faults segment the ridge into independent sections up to about 40 miles (64.4 km) wide and 20 to 30 miles (32.2 to 48.3 km) long. These sections are known as spreading centers. They can be a few miles to hundreds of miles long. This feature causes numerous scientists to compare the mid-ocean ridges to the stitching on a baseball. The transform faults result in steep cliffs along the fracture zone between segments of the ridges.

The theory of plate tectonics has revolutionized the science of geophysics and continues to provide a wellspring of insight into the outer and inner workings of planet Earth.

See also GEOLOGICAL OCEANOGRAPHERS; HYDROTHERMAL VENTS; MAJOR PLATES; MID-OCEAN SPREADING RIDGES; MINOR PLATES.

Further Reading
Charton, Barbara. *A to Z of Marine Scientists.* New York: Facts On File, 2003.
Davis, Richard A. *Oceanography: An Introduction to the Marine Environment.* Dubuque, Iowa: Wm. C. Brown, 1991.
Erickson, Jon. *Marine Geology: Exploring the New Frontiers of the Ocean.* New York: Facts On File, 2003.
Gates, Alexander E., Ph.D., and David Ritchie. *Encyclopedia of Earthquakes and Volcanoes,* 3rd ed. New York: Facts On File, 2007.
Heezen, Bruce C., Marie Tharp, and Maurice Ewing. "The Floors of the Oceans, I. The North Atlantic." *Geological Society of America Special Paper 65,* 1959.
Hess, Harry H. "History of Ocean Basins." In *Petrologic Studies: A Volume in Honor of A.F. Buddington,* edited by A. E. J. Engel. Boulder, Colo.: Geological Society of America, 1962.
Kious, W. Jacquelyne, and Robert I. Tilling. *This Dynamic Earth: The Story of Plate Tectonics.* Washington, D.C.: U.S. Government Printing Office, 1996. Available online. URL: http://pubs.usgs.gov/gip/dynamic/dynamic.html. Accessed March 10, 2007.
Kusky, Timothy. *Encyclopedia of Earth Science.* New York: Facts On File, 2005.
Menard, Henry W. *Ocean Science: Readings from Scientific American,* introduction by H. W. Menard. San Francisco, Calif.: W. H. Freeman, 1977.
Rothery, Dave. *The Ocean Basins: Their Structure and Evolution,* 2nd ed. Prepared for the Open University Oceanography Course Team, The Open University, Milton-Keynes, U.K. Oxford: Butterworth-Heinemann, 1998.

polar orbiting satellite A polar orbiting satellite travels in an orbit that passes over the Earth's poles. These satellites are launched into nearly circular Sun-synchronous, low-Earth orbits, with altitudes that range from 378 nautical miles to 458 nautical miles (700 to 850 km). Since the satellite maintains a fixed position relative to the Sun, it will not enter the Earth's shadow and obtains twice daily coverage of every portion of the globe. The near-circular orbits at these altitudes have orbital periods of approximately 100 minutes, giving the satellite approximately 14 orbits per day. The Earth rotates beneath the polar orbiting satellite, allowing sensors to obtain global coverage from a single satellite. By contrast, the GEOSTATIONARY SATELLITE is positioned above the equator at an altitude of 19,323 nautical miles (35,787 km) above the Earth's surface and rotates around the Earth at the same rate as the Earth spins.

Sensors on polar orbiting satellites are used to collect and record electromagnetic radiation that is emitted or scattered by the environment. Polar orbiting satellites often collect information during an entire Earth orbit and then download data directly to the receiving station when they are within radio line of sight. The combination of antenna azimuth and elevation required to point an antenna at a polar orbiting satellite varies continuously. The receiving station is also called the tracking, ground, or Earth station. Data received at the ground station are processed into useful information such as satellite imagery.

Polar-orbiting satellites are commonly used for meteorological, oceanographic, and geophysical applications that range from weather FORECASTING to tracking ocean features such as EDDIES to analyzing crop yields. Sensors include the Advanced Very High Resolution Radiometer that is onboard the NOAA Polar-Orbiting Operational Environmental Satellites to detect cloud cover and the surface temperature. It is critical for weather forecasting by the National Weather Service. Defense Meteorological Satellite Program satellites carry numerous sensors such as the Special Sensor Microwave Imager, Special Sensor Gamma/X-Ray Detector, and the Special Sensor Magnetometer, which are used primarily to collect terrestrial and space weather data. Space weather describes changes in the interplanetary magnetic field, solar flares, sunspots, and disturbances in the Earth's magnetic field. The effects of space weather can disrupt communications and damage satellites. Multispectral imagery is derived from Landsat and SPOT and is very useful for agencies such as the Department of Agriculture for vegetation analysis. These satellites collect data from various portions of the electromagnetic spectrum such as ultraviolet, visible, infrared, and microwave energy coming from the Earth's surface. The Landsat Program is a series of Earth-observing satellite missions jointly managed by the NATIONAL AERONAUTICS AND SPACE ADMINISTRATION (NASA) and the U.S. Geological Survey. This resource has been instrumental in observing CORAL REEFS and marine protected areas. The French government Satellite Probatoire d'Observation de la Terre or SPOT satellite imaging system collects green, red, and near infrared spectral data.

In the United States, the National Polar Orbiting Environmental Satellite System (NPOESS) combines Department of Commerce, NASA, and Department of Defense military polar orbiting satellite capabilities. This tri-agency Integrated Program Office will be instrumental in collecting and disseminating imagery relevant to Earth's weather, atmosphere, oceans, land, and near-space environment.

Further Reading

Jensen, John R. *Remote Sensing of the Environment: An Earth Resource Perspective.* Upper Saddle River, N.J.: Prentice Hall, 2000.

National Geophysical Data Center, Earth Observation Group, Defense Meteorological Satellite Program. Available online. URL: http://www.ngdc.noaa.gov/dmsp/dmsp.html. Accessed August 8, 2006.

NOAA Satellite and Information Service, National Environmental Satellite, Data, and Information Service. "National Polar-orbiting Operational Environmental Satellite System." Available online. URL: http://www.ipo.noaa.gov/about_NPOESS.html. Accessed October 15, 2007.

SPOT IMAGE. Available online. URL: http://www.spotimage.fr/html/_167_.php. Accessed August 8, 2006.

USGS Landsat Project. Available online. URL: http://landsat7.usgs.gov/index.php. Accessed August 8, 2006.

population Population refers to all the members of a single species that inhabit a specified geographic region. In fisheries research, populations are considered to be groups of animals genetically related due to interbreeding. In practice, it is often difficult to identify which members of a group, such as fishes, make up a population. Differences in individuals can result from geographic separation, differing environmental circumstances, and also variability to stresses, such as overfishing and pollution. Marine scientists may focus research on population structure and dynamics of exploited populations, effects of fisheries on benthic or pelagic communities, conservation, and the effects of environmental contaminants on marine organisms.

The term *stock* can be used as an alternative, where *stock* refers to a group of individual fishes that is genetically or behaviorally identifiable. NOAA's National Marine Fisheries Service uses the term STOCK in the context of regional fisheries management. Fisheries biologists use fishing logbooks, observer data, well-planned fishery surveys, and population models to make stock assessments for species such as sharks, which are often caught by commercial bottom longline gear or as gamefish in fishing derbies or rodeos. Some example coastal species are the Atlantic sharpnose sharks (*Rhizoprionodon terraenovae*), finetooth shark (*Carcharhinus isodon*), sandbar shark (*Carcharhinus plumbeus*), and scalloped hammerhead shark (*Sphyrna lewini*). Research by the Shark Population Assessment Group at the National Marine Fisheries Service's Panama City Laboratory is helping to assess shark populations in U.S. ATLANTIC OCEAN and GULF OF MEXICO waters. This type of work is especially important for species that reproduce at very slow rates and bear very few offspring.

The term *population* is also used in ecology to refer to the second of five hierarchical levels, which are: (1) the individual organism that includes any form of life including plants and animals; (2) the population (individuals of the same species); (3) the COMMUNITY (a number of populations); (4) the ECOSYSTEM,

NOAA's Physical Oceanographic Real-Time System (PORTS®)

The National Oceanic and Atmospheric Administration's Physical Oceanographic Real-Time System (PORTS®) is a system developed and implemented by NOAA's National Ocean Service (NOS) that measures, integrates, and disseminates observations and predictions from a variety of state-of-the-art instruments measuring real-time currents, water levels, winds, air and water temperatures, salinities, and other variables at multiple locations in the United States, with a data dissemination system that incorporates telephone voice response as well as computer modem dial-up. Instruments used include acoustic Doppler current profilers, acoustic water level gauges, and electronic conductivity-temperature-depth (CTD) gauges. PORTS provide much more accurate information than can be extracted from the traditional tide and current prediction tables that are available either online, or through private publications. This is so because traditional predictions include only the astronomical tidal component of water levels and currents computed from the positions of the Earth, Sun, and Moon, and historical time series of water height and velocity data. PORTS information, however, includes real-time current and water level data and may include predictions based on statistical or numerical hydrodynamic models.

The objectives of PORTS are to promote navigation safety, improve the efficiency of U.S. ports and harbors, and ensure the protection of coastal marine resources. This information is critical for safe and cost-effective navigation, search and rescue, hazardous material and oil spill prevention and response, and scientific research. PORTS are operating in many of the Nation's busiest ports and harbors, including Narragansett Bay, Rhode Island; New Haven, Connecticut; New York/New Jersey; Delaware Bay and River; Chesapeake Bay; Houston/Galveston; Los Angeles/Long Beach; San Francisco Bay; the Columbia River; Tacoma, Washington; Anchorage, Alaska; and Soo Locks, Minnesota. The first PORTS was installed and is operating in Tampa Bay, Florida. Information from PORTS, when combined with up-to-date nautical charts and precise positioning information, such as from the Global Positioning System (GPS) provides the mariner with a clear picture of potential dangers that threaten navigation safety. PORTS also enables shipping companies to maximize their cargo by reducing uncertainties in under-keel clearances. In the new global economy, ports are extremely busy; commercial vessels are getting larger and increasingly difficult to maneuver in close quarters, both with respect to bottom clearance and collision prevention. In many ports, clearances in areas such as turning basins are only a few feet for very large vessels. Knowledge of the currents, water levels, winds, waves, and water density can increase the amount of cargo moved through a port by enabling mariners to make use of every bit of dredged channel depth with greater safety.

PORTS can be accessed by mariners by means of telephone voice response and the Internet. For many locations, PORTS provides forecasts of water levels and currents that enable vessel owners and operators to make good decisions regarding loading of the vessel and minimizing transit time with greater safety.

PORTS are tailored to local needs and can be configured with a large variety of sensors and data transmission procedures to fit operational requirements and budgets. The improved efficiency of the port will more than offset the cost of equipment, installation, and operation. NOAA has estimated that for each additional foot of vessel draft, there is an increased profit per transit of between $36,000 and $288,000. Hence, PORTS improves the safety, efficiency, and profit of the nation's ports and harbors.

Further Reading

Nichols, C. Reid. "Operational Characteristics of the Tampa Bay Physical Oceanographic Real-Time System." In *Advances in Hydro-Sciences and Engineering.* Vol. 1, part B. Edited by Sam S. Y. Wang. University: The University of Mississippi Press, 1993.

NOAA Tides and Currents. "Physical Oceanographic Real-Time System (PORTS)." Available online. URL: http://www.tidesandcurrents.noaa.gov/ports.html. Accessed August 15, 2007.

Williams, R. G., Geoffrey W. French, and C. Reid Nichols. "Nowcasting of Currents in Tampa Bay Using a Physical Oceanographic Real-Time System." In *Advances in Hydro-Science and Engineering,* vol. 1, part B. Edited by Sam S. Y. Wang. University, Mississippi: The University of Mississippi Press, 1993.

—**Robert G. Williams, Ph.D.,** senior oceanographer
Marine Information Resources Company
New Bern, North Carolina

where environmental factors interact with the populations; and (5) the biome level, where ecosystems are grouped according to the general categories of organisms. BIOLOGICAL OCEANOGRAPHERS are especially concerned with understanding how habitat structure affects demographic processes. They consider physical, chemical, and geological characteristics of the ocean environment in order to understand the ecology of marine organisms.

Populations can be subject to both explosions and crashes. A population explosion refers to a rapid increase in the number of individuals of a species in a specified location. If the number of individuals exceeds the capacity of the environment to provide the needed resources such as oxygen and food, a population crash often follows. A population crash is a sudden and catastrophic decline in the numbers of a given species within a study area. The introduction of excess nutrients into an ESTUARY from agricultural runoff or storm drain overflow can result in a PHYTOPLANKTON bloom in which the excess nitrogen acts as a fertilizer, greatly increasing the number of plants, until the light is cut off below the surface, and the plants die; their decomposition consumes the available oxygen, and the fishes and benthic organisms die from anoxia. Such a condition is called eutrophication, a serious problem for many coastal regions of the United States. Some population crashes are the consequence of climatic patterns such as EL NIÑO, which may last for three to seven years. One observed effect of El Niño is an increase in ocean temperatures when tropical PACIFIC OCEAN trade winds die out. This phenomena affects UPWELLING along the coast of South America and commercially valuable fisheries such as Pacific sardines (*Sardinops sagax*) and Peruvian anchovetta (*Engraulis ringens*) that school in cool, nutrient-rich upwelled waters.

See also YEAR CLASS STRENGTH.

Further Reading

Census of Marine Life, Consortium for Oceanographic Research and Education. Available online. URL: http://www.coml.org. Accessed June 30, 2007.

Pacific Marine Environmental Laboratory, National Oceanic and Atmospheric Administration. "What is an El Niño?" Available online. URL: http://www.pmel.noaa.gov/tao/elnino/el-nino-story.html. Accessed October 15, 2007.

Panama City Lab, National Marine Fisheries Service. Available online. URL: http://www.sefscpanamalab.noaa.gov. Accessed November 16, 2006.

Thorne-Miller, Boyce. *The Living Ocean: Understanding and Protecting Marine Biodiversity*, 2nd ed. Washington, D.C.: Island Press, 1999.

Wyman, Bruce, and L. Harold Stevenson. *The Facts On File Dictionary of Environmental Science*, 3rd ed. New York: Facts On File, 2007.

ports and harbors   Ports and harbors are facilities where piers and docks support operations such as docking, undocking, discharge and boarding of passengers, and cargo handling. Ports are an important component of a maritime transportation system along with intercoastal waterways and shipping. In the United States, the maritime transportation system consists of more than 361 public and private deep-water and intercoastal waterway ports and more than 24,000 miles (40,000 km) of inland and coastal navigable waterways. Piers located at the port are the structures specifically designed to secure a vessel for the loading and unloading of cargo. The term *dock* is also used to indicate that a cargo-handling area is located near the berthing areas, the structures where vessels normally tie up. The structure used to onload and offload cargo, especially when attached to land, may also be called a wharf or quay. Important commodities that are transloaded at ports are natural resources such as iron ore, coal, and limestone, which are used in the production of steel. Other commodities include crude petroleum, refined petroleum products, and residual fuel, which are used for energy. Containerized cargoes such as textiles, manufactured goods, household goods, and groceries are handled at container ports and terminals (*see* CONTAINER AND CONTAINER SHIP). Busan Port located at the southern end of the Korean peninsula in the Republic of Korea (South Korea) is an extensive, expanding, and modern containerport. By 2011, it will have 0.93 miles (1.49 km) of breakwaters and 31 berthing spaces.

Harbors are culturally and economically important natural or man-made geographic features that provide shelter and anchorages for vessels. Inland harbors may be associated with canals, such as the six-mile (9.7 km) canal in Zeebrugge, Belgium, which connects to the North Sea, with basins, such as in Bremen, Germany, and with tide gates, such as those in Bremerhaven, Germany. Bremen and Bremerhaven are both on the Weser River, which flows about 300 miles (483 km) through central and northwest Germany to the North Sea. The tide-gates type harbor utilizes locks to control water levels within the harbor. River harbors such as Jacksonville Harbor in Florida may consist of quays that are either cut into the banks or run parallel to the banks and piers, which extend into the flowing river. Approximately 23 miles (37 km) of the St. Johns River is dredged to maintain a governing depth of 40 feet (12 m) in the main shipping channel for users of the Jacksonville

Harbor. Indentations along a coastline such as in Kingston, Jamaica, are classified as natural harbors since they afford shelter from wind and waves without any jetties or breakwaters to provide shore protection. Kingston became the major port facility in Jamaica after an earthquake on June 7, 1692, destroyed the infamous 17th-century city Port Royal, a haven for pirates and buccaneers. A harbor lacking shelter from the wind, sea, and swell is called an open roadstead, such as the oil pipeline terminal Ra's at-Tannurah on the PERSIAN GULF in Saudi Arabia.

In locations such as Cherbourg in Normandy, France, a coastal harbor may be reinforced with breakwaters to improve protection provided by natural sources. Cherbourg is located at the end of a peninsula, which separates the Gulf of St. Malo from the Bay of Seine. In November 1984, the wreck of the Confederate steamship *Alabama* was discovered in about 200 feet (61 m) of water off Cherbourg (*see* MARINE ARCHAEOLOGIST). The USS *Kearsarge* sank the CSS *Alabama* during a naval battle in 1864. Cherbourg Harbor was the second to last stop for the White Star liner *Titanic*'s cruise to New York City. The *Titanic* hit an ICEBERG and sank approximately 400 miles (644 km) from Newfoundland, Canada, on April 15, 1912. (Cherbourg was liberated from Nazi Germany by American forces around June 27, 1944, during World War II. Today, the city is known as Cherbourg-Octeville.)

Another type of coastal harbor is referred to as a tide-gates harbor because its water levels are controlled by locks. An example is the Port of Mumbai in India, which opens to the Arabian Sea and is lined by MANGROVE SWAMPS along the northwestern and eastern shores of the harbor.

The terms *ports, harbors,* and *seaports* are often used synonymously. Harbor refers to the principal water area of the port. Since mariners may be unfamiliar with local conditions, harbor pilots are licensed to direct ships in and out of the harbor. The navigational area may be associated with peculiar shoals, currents, and obstructions that are not adequately reflected on charts and sailing directions. Port facilities must keep pace with world trade, changes in ship sizes and ensuing drafts, and inland transportation networks. Draft is the number of feet or meters that the hull of a ship is submerged beneath the surface of the water. Draft changes due to factors such as SALINITY and the weight of cargo onboard. Channel depths may change in response to tides and winds. Accurate information on water depths, winds, and currents is necessary for the safe operation of vessels within the port. This information may be obtained from real-time observation systems such as NOAA's PORTS, or from charts and tables that are continually updated (*see* NOTICE TO MARINERS). In most major ports, incoming and departing vessels are required to take a HARBOR PILOT onboard, who has the necessary knowledge of local conditions.

Efficient ports are designed with optimal waterside and landside connections. Ports can be viewed as the hub for ship, truck, and rail transportation systems. At the port, cargo is separated into individual shipments and routed to different destinations. Factors affecting port efficiency include cargo-handling capability, the amount of labor required to move cargo, available storage area, waterside access limitations, the capacity of road and rail connections, and inland transportation availability. With security threats on the rise as evidenced by criminal terrorism, new security imperatives provide an everyday challenge for transportation providers worldwide. High-technology solutions such as the utilization of mass spectrometers (*see* MASS SPECTROMETER) to detect unusual signals (chemical, biological, and radiation) are being integrated into everyday port operations. The instrument analyzes ions according to their mass-to-charge ratio, and the number of ions at a given mass value (intensity) is determined electrically. It can be used by security personnel to determine the relative abundances of various kinds of particles within a given sample.

See also ANCHOR; ARCHAEOLOGIST; BREAKWATERS; JETTY; MOORING.

Further Reading

Bremen KEY PORTS Portal. Available online. URL: http://www.keyports.de. Accessed October 9, 2006.

JAXPORT, The Jacksonville Port Authority. Available online. URL: http://www.jaxport.com. Accessed October 9, 2006.

Maritime Administration, United States Department of Transportation. Available online. URL: http://www.marad.dot.gov. Accessed October 9, 2006.

Mumbai Port Trust. Available online. URL: http://mumbaiport.gov.in. Accessed October 9, 2006.

National Geospatial Intelligence Agency. *The World Port Index,* 18th ed., Publication 150. Bethesda, Md.: National Geospatial Intelligence Agency, 2005.

Port Authority of Jamaica. Available online. URL: http://www.portjam.com. Accessed October 9, 2006.

Port de Cherbourg. Available online. URL: http://www.port-cherbourg.com. Accessed October 9, 2006.

Port of Busan. Available online. URL: http://www.busanport.com. Accessed October 9, 2006.

Port of Zeebrugg. Available online. URL: http://www.zeebruggeport.be. Accessed October 9, 2006.

Saudi Port Authority. Available online. URL: http://www.ports.gov.sa. Accessed October 9, 2006.

predictions Predictions refer to estimations of future values of the state of the ocean and atmosphere, by interpreting statistics of past and present values, or by process-based numerical models, such as ocean circulation models based on the equations of motion, or both. In common usage, the word *predictions* has much the same meaning as *forecast,* but in marine science, *forecast* is usually applied to phenomena not easily predicted, such as the weather, whereas *predictions* refers to the future times and heights of the tide and associated current velocities resulting from the gravitational influences of the Moon and Sun on the oceans of the world, which can be determined years in advance. However, astronomical factors alone cannot account for the observed tides. Because of the influence of varying shorelines, water depths, and ocean basin configurations, time series of actual tidal times, water levels, and CURRENTS must be used. Maximum accuracy can be obtained by taking equally spaced samples that may occur on the order of minutes for as long as 19 years. Thus, the record is in synchronization with the longest lunar period, the lunar node that corresponds with eclipses of the Sun that are repeated at different places on Earth every 18.6 years. This record length would include WATER LEVEL FLUCTUATIONS induced by the precession of the Moon's orbital plane around the Earth in 18.6 years and the concomitant effect of the Moon on the precession of the equinoxes (*see* TIDES).

Predictions for every year are generally made available in hard copy publications as tables, on CD ROMS, and on the World Wide Web. Detailed predictions of high and low tides for every day of the year are made in the United States by NOAA's National Ocean Service (NOS) for major ports such as Baltimore, Boston, New York, New Orleans, and San Francisco, which are known as reference stations, with predictions for more than 3,000 less frequented subordinate locations made with reference to the observations for the major ports. Although NOS no longer publishes the tide tables, the data are furnished to commercial contractors to prepare products for distribution to the public. Predictions for major ports are usually made by means of long-time series of water levels of several years' duration; predictions for adjacent or subordinate locations are made by comparing water levels with the reference station Current measurements are more difficult and costly to make than water level measurements, because currents are vectors with both speed and direction, whereas water levels are scalars with only height above a reference level. Many of the older current measurements required that a research vessel be standing by to deploy the CURRENT METERS and record and retrieve the data. Current measurements

for the reference stations may be made over periods of months to more than a year; subordinate station current predictions are based on comparison of relatively short periods of data acquisition by current meters with simultaneous measurements at the major port. Current predictions are available from NOAA/NOS for more than 2,700 subordinate stations.

Traditional methods of tidal predictions for reference stations include combining astronomical data on the positions of the Moon and Sun with longtime series of observations of water levels, and often currents as well. A type of Fourier series analysis, called harmonic analysis is performed on the data to yield the amplitudes and phases (times) of the tidal constituents. Each constituent represents a periodic change in the relative positions of the Earth, Moon, and Sun. A single constituent is given in the form of the following:

$$y = A \cos (at + \varphi), \text{ where}$$

y is a function of time as expressed by the
  symbol t and is reckoned from a specific origin,
A is the amplitude of the constituent,
a is the constituent speed in degrees per hour,
t is time, and
φ is the phase in degrees.

The period of the constituent is the time required for the phase to change through 360 degrees, and is the cycle of the astronomical conditions represented by the constituent (*see* TIDES). Predictions for subordinate stations involve comparing water level and current amplitudes and phases with those of the reference station (*see* TIDE TABLES). Predicted tides and currents at the subordinate or secondary station are reduced to height and time differences or velocity ratios and time differences with respect to the reference station. These tabular predictions, however, do not include the effects of the weather, which can be considerable. In many regions, atmospheric events such as the passage of strong fronts and storms can dominate the water levels and currents. In a long channel, winds may set up water levels higher at one end than the other, which causes hydraulic currents to form. Similarly, freshets my cause unusual drainage patterns that overcome the reversing nature of tidal currents. Freshets are the increased discharge from heavy rain or a spring thaw. Changes made to the BATHYMETRY of the port such as DREDGING can degrade the accuracy of published tide and current predictions. For this reason, it is highly desirable for the port authorities to provide the resources to install real-time wind, water level, and current measuring systems to provide accurate and up-to-the-minute information to mariners. One example of such a system is NOAA's Physical Oceanographic Real-Time System (PORTS®), installed in several ports in the United States (*see* essay "Physical Oceanographic Real-Time Systems"). This information greatly enhances the safety, efficiency, and economies of the ports in which real-time systems are installed.

Predictions, especially current predictions, can also be affected by changes to navigation channels due to natural processes, as well as dredging by the port authorities. If such major changes are made to a port, it is necessary to reacquire data on which to base new predictions. Researchers may deploy current meters or pressure gauges over several tidal cycles to check the accuracy of predicted currents or water levels. The residual is usually the observed curve minus the predicted curve. Fluctuations in the residual are compared to known weather events or watershed discharge. Unexplained differences may be due to human error in making measurements or changes within in the basin.

The accuracy and regional applicability of predictions have increased greatly over the years with improvements being made in measurement technologies, such as the ACOUSTIC DOPPLER CURRENT PROFILER (ADCP), numerical circulation models of PORTS AND HARBORS, analytical techniques, and real-time computer and communications technologies. The most recent predictions of winds, water levels, currents, and waves in a port equipped with a PORTS system can be transmitted to the navigation bridge of an incoming cargo or passenger vessel and displayed on electronic charts to provide the master of the vessel with all the environmental information he or she needs to safely and efficiently navigate to the dock.

See also FORECASTING, WAVE PREDICTION.

Further Reading
National Ocean Service, National Oceanic and Atmospheric Administration. "Tides and Currents Products." Available online. URL: http://www.tidesandcurrents. noaa.gov/products.html. Accessed November 10, 2006.

pressure Pressure refers to the force per unit area exerted by the molecules of a substance such as air or water. The pressure is proportional to the density and the kinetic energy of the substance. Increased kinetic energy means that the molecules of the substance have a higher average velocity, and thus a greater effect in impinging on a surface. The pressure at a point in the atmosphere or the oceans is the same in any direction; hence, it is a scalar quantity. In general, as the TEMPERATURE of a substance increases, the average molecular speed increases, and hence the pressure. In the case of the atmosphere, which can be

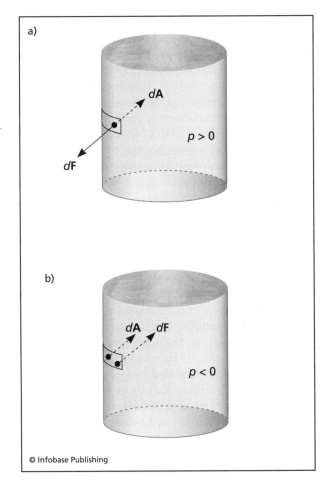

The pressure *p* in a fluid causes a force on all contact surfaces. If *d*F is an infinitesimal force resulting from pressure *p*, then *d*F = *pd*A, where *d*A is the vector of magnitude *d*A that is perpendicular to the area element and pointing toward the fluid. The situation is depicted for fluid in a container, showing the direction of force on an element of inner wall surface for positive fluid pressure (a) and for negative pressure (b).

approximated as an ideal gas, the pressure is directly proportional to temperature, everything else being held constant. The pressure that the atmosphere exerts on a square meter of the surface of the Earth is simply the weight of the overlying column of air from the ground to the limits of the atmosphere. In the case of a water body, the pressure at a subsurface point in the water is equal to the weight of the overlying water column plus the weight of the overlying atmosphere. In the ocean, measured pressure is reported as either absolute pressure, which is the total weight of the water and air above, or gauge pressure, which is only the weight of the water, the pressure at the sea surface being defined as equal to zero.

Atmospheric pressure can be measured by the height of a column of mercury in a glass tube sitting in an open reservoir of mercury, or, by a metal box in which the air has been evacuated and a pointer attached to the box so that as the box expands or contracts in response to changes in air pressure, the pointer moves over the face of a dial. The former is called a mercury barometer, and the latter is called an aneroid barometer. Barometers are used as ALTIMETERS in aircraft, where the barometer is calibrated to read the height of the aircraft in feet or meters. Mariners know that rising pressure is indicative of good weather, while falling pressure signals the approach of a storm.

The Earth's surface exhibits day-to-day variations in pressure, as the temperature and water vapor content change. Air pressure decreases exponentially with altitude as less and less of the atmosphere lies above the measurement point. For every 3.5 miles (5.6 km) increase in height, the mass of air above decreases by one half. Dry air is denser than moist air, and is thus heavier, causing a higher pressure. By recording the pressure changes at numerous stations, meteorologists build up a pattern of highs and lows, which are presented as isobars (*see* ISOBAR) on synoptic weather charts, such as seen in daily newspapers and on the televised weather report. Weather forecasters use this information to determine wind speeds and direction. Similarly, ocean currents flow around highs and lows of oceanic pressure that can be determined from the height of sea surface, called ocean surface topography. In the United States, atmospheric pressure is reported in units of millibars, where one standard atmosphere is equal to 1,013 millibars or 14.7 pounds (6.7 kg) per square inch. In other countries, pressures are usually reported in SI units, the unit of pressure being the Pascal or Newton per square meter. Since 10 millibars is equal to 1 kiloPascal, the standard atmosphere is 101.3 kiloPascals.

If there is a pressure difference between two locations on the surface of the Earth, there will be a force created called the pressure gradient force. This force is represented by the equation

Pressure Gradient force = $\Delta p / \Delta x$, where

Δp = pressure at station two (mb) - pressure at station one (mb)

Δx = location of station two (m) - location of station one (m).

In other words, the pressure gradient force is equal to the change in pressure Δp divided by the distance between the stations Δx. Since the pressure gradient is different for different directions, it is a vector quantity. When an air mass starts to move from a high-pressure region to a low-pressure region, the rotation of the Earth comes into play as the CORIOLIS FORCE, causing it to turn to the right in the

Northern Hemisphere and to the left in the Southern Hemisphere. If friction is negligible, as it is in the upper atmosphere, the parcel will continue turning until the Coriolis force is equal and opposite to the pressure gradient force, which occurs when the flow is parallel to the isobars. The flow moves clockwise around high-pressure areas and counterclockwise around low-pressure areas in the Northern Hemisphere, and the reverse in the Southern Hemisphere. Students of meteorology often use the saying, "A high clock over a low counter" in order to remember this concept. Motion resulting from this balance between the horizontal Coriolis forces and the horizontal pressure forces is called geostrophic flow.

Air moves with much greater speed over greater distances in the horizontal than in the vertical. When the air is not in motion in the vertical direction, it must be in equilibrium, which means that the gravitational (downward) force due to the weight of a parcel of air exactly balances the buoyancy force (Archimedes' force). If Δp is the pressure, Δz is the change in height above the surface, the pressure is given by $\Delta p = p_0 - g\rho\Delta z$, which is called the hydrostatic equation, where p_0 is the surface pressure, and g is gravity, the force acting toward the center of the Earth. This equation is essentially the vertical component of Newton's Second Law of Motion (Force = mass × acceleration) for the atmosphere, and is an essential element in the equations describing the state of the atmosphere in numerical weather prediction models. Archimedes of Syracuse (ca. 287–ca. 212 B.C.) was a Greek mathematician and engineer. Sir Isaac Newton (1642–1727) was an English physicist and mathematician who determined the three laws of motion, gravity, and invented calculus.

In the ocean, the same general behavior is observed as for the atmosphere, since air and water are both fluids in motion on the rotating Earth. However, seawater is about 1,000 times denser than air, having a density on the order of 10.3 pounds per gallon (1,025 kg/m^3) compared to about 0.14 ounces per gallon (1.03 kg/m^3) for air. Hence, a small change in vertical position in the sea produces a much greater change in pressure than in the atmosphere. For example, at a depth of 32.8 feet (10 m), the pressure change from the surface is equal to

$$g\rho\Delta z = 100,450.0 \text{ Pa, where}$$

g = gravity = 9.8 m/s^2,
ρ = density of seawater = 1,025 kg/m^3, and
Δz = (depth - surface) = 10 m.

A similar calculation for air shows that the change in pressure due to a decrease of 32.8 feet (10 m) in height at the surface is only 126.4 Pa. A pressure equal to the weight of the atmosphere at the surface is called one atmosphere of pressure. In the sea, the pressure increases by approximately one atmosphere for every 32.8 feet (10 m) increase in depth. At 3,281.0 feet (1,000 m) depth, the pressure is about 100 atmospheres; at 13,124 feet (4,000 m) depth, the average depth of the ocean basins, the pressure is about 400 atmospheres.

Oceanographers have traditionally used the decibar, or .1 bar, as the unit of pressure, since the pressure in decibars is approximately equal to the depth in meters in the upper layers of the ocean. A pressure of 2,500 dbar is roughly equal to a depth of 1,367 fathoms (2,500 m). At deeper depths, the approximation is not as accurate because seawater is slightly compressible. Oceanographers use a CONDUCTIVITY-TEMPERATURE-DEPTH (CTD) PROBE to measure SALINITY, temperature, and depth in the water. Depth is inferred from the pressure sensor on the instrument. Temperature is measured with a THERMISTOR. Salinity is determined from the water sample's induction, which is measured by the conductivity probe.

Since pressure is a function of density, which in turn depends on the temperature, salinity, and depth, the pressure at a given depth varies slightly with temperature and salinity, leading to differences in pressure at the same depth at different locations, resulting in horizontal pressure gradients. These differences give rise to an internal pressure field depending on the THERMOHALINE (temperature and salinity) characteristics of the water. There are also externally imposed pressure differences at horizontal surfaces, resulting from the pileup of water due to winds. For example, the Atlantic trade winds cause the water to pile up against the South American continent, leading to a generally upward slope in the sea surface between equatorial Africa and Brazil. Water will run down this upward slope due to the pressure gradient force.

The total pressure in the ocean is the sum of the externally (atmospherically) imposed pressure, called the barotropic component, and the density or internally imposed pressure, called the baroclinic component. In general, the baroclinic situation is common in the surface layers, while barotropic conditions are manifested at great depth, where density is determined by pressure rather than temperature and salinity. These components are often treated separately by numerical ocean circulation models (*see* MODELS).

The externally imposed water levels can now be measured from Earth satellites by means of RADAR, and from these measurements maps of the sea surface topography can be drawn, showing the regions of high and low pressure in a manner similar to the highs and lows of pressure on synoptic weather maps. From these charts of ocean surface topography, the barotropic component of the ocean circulation can be determined.

See also PRESSURE GRADIENT.

Further Reading

Aguado, Edward, and James E. Burt. *Understanding Weather and Climate*, 4th ed. Upper Saddle River, N.J.: Pearson/Prentice Hall, 2007.

pressure gradient   A pressure gradient is a change in force per unit area in space over a fixed time. The spatial area might be the distance as measured from two pressure gauges, where one is deployed on the seafloor in the coastal ocean approach to a tidal inlet while the other gauge is at the end of the main navigation channel on the estuarine side of the coastal inlet. In meteorology, the pressure gradient would be calculated as the difference in pressure measured along a line perpendicular to the ISOBARS on a weather chart. When isobars are very close together, the pressure gradient is steeper and winds are stronger, and when they are far apart, the pressure gradient is weaker and so are the winds. Meteorologists still use weather balloons that measure atmospheric variables such as pressure from an attached radiosonde (instrument package). Data such as atmospheric pressure from multiple balloons is transmitted to radio receivers on the ground from which the pressure gradient may be determined. The first use

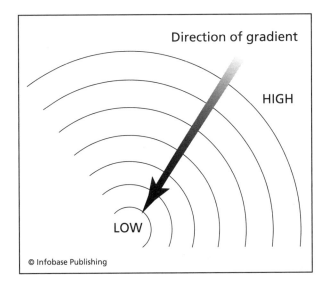

The pressure gradient force is the rate at which pressure changes with distance in air or water between centers of high and low pressure.

of balloon-borne radiosondes is credited to Russian meteorologist Pavel A. Molchanov (1893–1941) during 1930.

In the absence of other forces, fluids such as water and air move from high to low pressure. This gradient in pressure results in a net force that extends from regions of high pressure to regions of low pressure. The pressure gradient force (*see* PRESSURE), $\Delta p/\Delta x$, is equal to the change in pressure (Δp) divided by the distance between the stations (Δx). In the atmosphere and ocean, this force results in winds and CURRENTS. Another force of interest, the CORIOLIS FORCE, results from the rotation of the Earth. The Coriolis force, named after Gaspard-Gustave de Coriolis (1792–1843), causes the parcel to turn to the right in the Northern Hemisphere and to the left in the Southern Hemisphere. In the case of upper-atmosphere winds and deep-ocean currents, when friction is negligible, the fluid parcel will continue turning until the Coriolis force is equal and opposite to the pressure gradient force. This equality occurs when the flow is parallel to the isobars. The flow moves clockwise around high-pressure areas and counterclockwise around low-pressure areas in the Northern Hemisphere, and vice versa in the Southern Hemisphere. This circulation is called geostrophic flow (*see* GEOSTROPHY).

Since pressure is a function of DENSITY, which in turn depends on the TEMPERATURE, SALINITY, and depth, the pressure at a given depth varies slightly with temperature and salinity, leading to differences in pressure at the same depth at different locations, resulting in horizontal pressure gradients. These differences give rise to an internal pressure field depending on the thermohaline (temperature and salinity) characteristics of the water (*see* THERMOHALINE CIRCULATION). Horizontal pressure gradients are much smaller in magnitude than the vertical pressure gradients, but are important in maintaining the ocean circulation. The horizontal pressure gradient drives the ocean water toward regions of low pressure. This gradient arises from spatial gradients in the pressure at the ocean surface and pressure interior to the ocean. For example, externally imposed pressure differences at horizontal surfaces result from the pileup of water by winds. The Atlantic trade winds cause the water to pile up against the South American continent, leading to a generally upward slope in the sea surface between equatorial Africa and the coast of Brazil.

The total pressure in the ocean is the sum of the externally (atmospherically) imposed pressure, called the barotropic component, and the density or internally imposed pressure, called the baroclinic component. These components are often treated separately in numerical ocean circulation MODELS. For example, baroclinic pressure gradients arise from density gradients, which can be estimated through the hydrostatic relation devised by Blaise Pascal (1623–62):

$$P = \rho\, g\, z, \text{ where}$$

P = pressure,
ρ = density of seawater 64 pounds/foot^3
   (about 1,027 kg/m^3),
g = the acceleration of gravity 32 feet/second^2
   (9.81 m/s^2), and
z = the depth.

This equation is used since ocean pressures will remain constant except for changes in fluid density such as higher-salinity water. Marine scientists may measure pressure in the ocean with a pressure gauge relative to atmospheric pressure. Measured pressures are often called gauge pressures since they are relative to another pressure, atmospheric pressure. The absolute pressure at a site in the ocean can be found by adding the reference pressure to the gage pressure. Thus,

$$P_{abs} = P_{atm} + P_{gage}, \text{ where}$$

P_{abs} = Absolute pressure,
P_{atm} = Atmospheric pressure, and
P_{gage} = Pressure measured from a submerged
    pressure gauge.

Absolute pressure is the total pressure, and gauge pressure is simply the pressure read or recorded on a gauge.

The externally imposed water levels can now be measured from Earth satellites by means of highly accurate RADAR ALTIMETERS, and from these measurements, maps of the sea surface topography can be drawn, showing the regions of high and low pressure in a manner similar to the highs and lows of pressure on synoptic weather maps. From these charts of ocean surface topography, the barotropic component of the ocean circulation can be inferred from the geostrophic relations. In regions of the ocean that are nearly homogenous, such as the polar seas, the barotropic component alone gives a reasonable approximation to the total circulation.

The baroclinic component of the circulation is determined by the distribution of density with depth; hence, it must be calculated from temperature and salinity measurements from ships using CONDUCTIVITY-TEMPERATURE-DEPTH (CTD) PROBES, or from instruments distributed over depth attached to moored BUOYS. This component of the circulation is important in highly stratified parts of the ocean, such as in the Tropics and subtropics, where large temperature gradients such as the THERMOCLINE cause large gradients in density and thus in pressure.

Further Reading

Ocean-Modeling.org. Available online. URL: http://www.ocean-modeling.org. Accessed June 30, 2007.

Rosmorduc, Vinca, Jérôme Benveniste, Olivier Lauret, Maria-Pilar Milagro Perez, and Nicolas Picot. J. Benveniste and N. Picot (Ed.). *Radar Altimetry Tutorial*. Collecte, Localisation, Satellites, France. Available online. URL: http://www.altimetry.info. Accessed June 30, 2007.

primary production   The synthesis of organic matter by plants (*see* ALGAE, MARINE MACROPHYTES, PHYTOPLANKTON), which is the main source of energy and nutrition for primary consumers, is called primary production. Chains or rings of carbon atoms (C) onto which hydrogen atoms (H) are bonded are considered to be organic compounds. Primary producers are able to fix carbon through photosynthesis, which can be described by the following simple formula:

$$6CO_2 + 6H_2O \xrightarrow{\text{sunlight}} C_6H_{12}O_6 + 6O_2$$

The product of photosynthesis is a carbohydrate, such as sugar glucose ($C_6H_{12}O_6$), and oxygen (O), which is released into the atmosphere. This production in marine ECOSYSTEMS forms the base or primary component of the food web and produces food for primary consumers such as herbivores. Echinoderms such as purple-spined sea urchin (*Arbacia punctulata*), red sea urchin (*Strongylocentratus franciscanus*), purple sea urchin (*Strongleocentrotus purpuratus*), and northern sea urchin (*Strongylocentrotus droebachiensis*) are primary consumers as they feed on brown algae (*see* KELP).

Primary producers form the basis of the food chain, which expands from herbivores to carnivores, with humans very often at the top of the food chain. If primary production can be boosted, the entire food chain benefits. Because of this, suggestions have been made in recent times to add fertilizer or nutrients to areas of the ocean that are very low in primary productivity.

In general, plants are the primary producers, but some types of plankton can photosynthesize and are also capable of hunting for their food if light levels are low (*see* PHYTOPLANKTON). Since photosynthesis occurs primarily in the visible portion of the spectrum (*see* LIGHT), marine scientists may estimate primary production using REMOTE SENSING. One example is ocean color data from the Sea-viewing Wide Field-of-view Sensor or SeaWiFS.

Further Reading

Goddard Space Flight Center, National Aeronautical and Space Administration. "Ocean Color Web." Available online. URL: http://oceancolor.gsfc.nasa.gov. Accessed June 30, 2007.

The Long Term Ecological Research (LTER) Network. Available online. URL: http://www.lternet.edu. Accessed December 3, 2006.

pycnocline   The pycnocline is a water layer exhibiting a rapid increase of DENSITY with depth. In the deep ocean, the pycnocline results from the combined effects of TEMPERATURE and SALINITY, which have a zone of maximum change in depths between approximately 328 feet (100 m) to approximately 3,280 feet (1,000 m) below the surface. In shallow coastal areas, the pycnocline often results from a large salinity gradient called the HALOCLINE located at the border of relatively fresh or brackish estuarine water and open ocean water. As temperature decreases with depth, there is usually a corresponding increase in density. The densest layers in the ocean are at the bottom, and the least dense layers are near the surface. Cold salty water sinks below warm freshwater.

In open ocean waters, temperature is usually the dominant effect over salinity in determining the density and coincides with the THERMOCLINE, a layer of rapidly changing temperature, normally decreasing from a warm surface layer to cool or cold mid-depth and bottom layers. In tropical and subtropical waters, the density structure usually consists of a mixed layer that is 165 to 660 feet (50 to 200 m) deep, a layer below this extending from 660 to 3,280 feet (200 to 1,000 m) depth in which density increases rapidly with depth (the pycnocline), and a deep layer where density increases slowly to the seafloor. This characterization is often called the three-layer ocean. In the ocean, the pycnocline is found in the same region as the thermocline and the halocline. The increasing pressure of the deep water also increases the density.

In polar waters, there is generally no thermocline and no pycnocline, except in regions where melting ice distributes a layer of relatively low-density freshwater overlying the heavier ocean water. The density increases with increasing salinity as freshwater freezes and saline waters sink. Density is less sensitive to change in the cold waters and is more sensitive to change in the deeper waters. Thus, variations in density due to temperature and salinity are small but very important. The temperature and salinity values define WATER MASSES.

The density distribution is a major factor in determining the stability of the water column. As a general rule, density increases with depth, so that the heavier water is underneath the lighter water in a stratified configuration, and vertical stability is maintained. Except in the polar regions, the increase of density is not uniform, but increases rapidly in the pycnocline layer. The strong temperature decrease with depth that is characteristic of the global ocean results in a strongly stratified water column that inhibits vertical mixing. In addition, density (weight of water divided by the volume) hardly changes and is approximately 62.4 pounds per cubic foot (1,000 kg/m^3) for pure water and about 64 pounds per cubic foot (1,027 kg/m^3) for seawater.

The rate of change of density with depth determines a quantity called the static stability of the water column. The static stability (E) is defined as:

$$E = -(1/\rho)(\Delta\rho/\Delta z), \text{ where } \rho \text{ is density (kg/m}^3\text{) and } z \text{ is depth (m)}.$$

If E > 0, the water column is stable; if E = 0, it is neutral; and if E < 0, the water column is unstable. A stable configuration means that if a water parcel undergoes a small displacement, it will return to its original position; if it has neutral stability, it will remain where it is; if it is unstable, it will accelerate away from its original position. A cone is stable on its base, neutral on its side, and unstable on its apex. In the upper thousand or so meters of water in the open ocean, values of stability (E) are about 10^{-5} per meter to 10^{-6} per meter, with higher values in the pycnocline. Below 3,280 feet (1,000 m) depth, stability decreases and may be as low as 10^{-8} per meter in the deep TRENCHES.

The pycnocline often acts as a barrier to the vertical movement, or mixing of water, especially in warm tropical regions where there is a strongly developed thermocline, although it does not prevent dead organisms from sinking through it to the bottom, as is sometimes asserted in folklore. Because of this effect, the ocean is sometimes approximated as a two-layer system, with well-mixed water of low density in the upper layer of around 328 to 1,640 feet (100 to 500 m) in thickness, and a lower layer of high-density water. Such approximations are often used in numerical circulation models and in determining sound propagation. Sound travels very fast near the surface of the ocean in warm water and decreases with depth to a minimum at a depth of approximately 547 fathoms (1,000 m). As pressure increases with depth, so does the speed of sound.

In mid-latitudes, a seasonal pycnocline develops in response to the seasonal thermocline in summer, when heated near-surface water forms a layer of low-density water overlying heavier, colder water. The seasonal pycnocline is especially strong in shallow coastal regions such as the Mid-Atlantic Bight off the East Coast of the United States on the CONTINENTAL SHELF. (A bight is an indentation or concave stretch of coastline.) The pycnocline persists throughout late spring, summer, and into early autumn. In October, strong winter storms usually disrupt the pycnocline and mix the water from the surface to the bottom, as well as cool the surface water, making it heavier and causing it to sink. In subtropical waters such as the Sargasso Sea in the region of Bermuda, there is also a seasonal pycnocline, which dissipates in winter, resulting in a relatively uniform water mass "the 64.4°F (18°C) water" in the upper ocean.

Further Reading

National Oceanographic Data Center. "The Global Temperature-Salinity Profile Program." Available online. URL: http://www.nodc.noaa.gov/GTSPP/gtspp-home.html. Accessed October 30, 2006.

Pickard, George L., and William J. Emery. *Descriptive Physical Oceanography: An Introduction,* 5th ed. Oxford: Pergamon Press, 1990.

Q

Q-routes Q-routes are predefined navigable paths established near major maritime port facilities for the purpose of clearing naval mines. Naval mines or "sea mines" are self-contained explosive devices laid on the seafloor, attached to MOORINGS, released to drift about, or deployed as smart, self-propelled torpedoes waiting for a ship signature. These weapons are manufactured or purchased for offensive and defensive naval warfare. To counter these threats, especially by terrorists, Navy and Coast Guard organizations worldwide are often assigned a mission of Mine Countermeasures and Clearing (MCM). These missions prevent or reduce the damage that may be caused to civil, merchant, and military shipping by underwater mines. One key component to MCM is to identify a safe route, known in the mine warfare lexicon as a "Q-route," into seaports.

One of the baseline tasks of MCM involves understanding the environment, which requires high-resolution hydrographic surveys. These surveys generally extend along established routes from the entrance of major seaports and well out onto the CONTINENTAL SHELF where there are shipping lanes. The crew operates specialized equipment aboard the minesweepers such as side-scan SONAR and REMOTELY OPERATED VEHICLES. They may collect water and bottom samples to overcome challenges to the detection and neutralization of mines posed by the water column and complex seafloors. Side-scan sonar is a towed system where the "towfish" or transducer assembly transmits and receives sound from each side of the instrument. Some sound energy is reflected back to the sonar, where the signal is processed to produce images that delineate surfaces and objects. Factors such as bottom composition and BATHYMETRY, water clarity, SALINITY, TEMPERATURE gradients, and ambient noise levels affect the sonar. One trick to finding mines that may have been planted in a navigation channel or seaport is knowing what is already there, especially since most waterways contain a fair amount of minelike objects such as cars, refrigerators, and 55-gallon (208.2 l) drums that all appear on the sonar operator's screen. If an unknown object is identified by the side-scan sonar survey, a remotely operated vehicle can be deployed with onboard lights and video camera to take a closer look and send back high-resolution images. The Q-route surveys make the process of detecting mines easier since the MCM units can focus on changes from one survey to the next. The survey would cover established routes into, out of, and between seaports. High-technology capabilities such as AUTONOMOUS UNDERWATER VEHICLES may also be deployed to investigate new minelike contacts for further analysis and neutralization. MCM ships and aircraft are specially designed and equipped to conduct these surveys. Some MCM ships use Small Waterplane Area Twin Hull (SWATH) technology to gain added stability as well as forward-looking sonar to detect mines ahead of the ship. Helicopters are used in MCM operations with scanning lasers and to tow side-scan sonar or hydrofoil mine-clearing sleds.

Q-routes also minimizes the area the MCM commander is responsible to keep clear of mines. Their existence enhances safety of passage by friendly shipping and may actually delineate the best route through a marine minefield. Completing Q-route surveys has the secondary benefit of collecting data necessary to update nautical charts or to support

DREDGING operations. All surveys require depth information, and the work done by MCM personnel also includes determining bottom composition (sand, mud, rock). MCM personnel also have the ability to lay tripwire sensors to determine activity of potential mine layers.

Many maritime nations are vulnerable to mining of their harbors and shipping lanes because mine laying is a cheap and relatively unsophisticated means of terrorism and economic warfare. Experts estimate it would take years to redevelop a database of minelike objects along Q-routes if existing survey operations were allowed to lapse. Some experts in MCM technology recommend the use of shore-based or moored sensing systems to detect and alert first responders to potential mine-laying events. Indicators might include unusual surface water splash, seismic spikes, and direct visual observation, which could allow for quick response to a mine-laying event. Routine hydrographic surveys remain the foundation for establishing and maintaining safe maritime Q-routes.

Further Reading

Chappell, Gordon. California State Military Department, the California State Military Museum. "Historic California Posts—Forts Under the Sea: Submarine Mine Defense of San Francisco Bay." Available online. URL: http://www.militarymuseum.org/Mines.html. Accessed October 15, 2007.

Demine Web Site. Available online. URL: http://www.demine.org. Accessed January 14, 2007.

NATO Undersea Research Center. Available online. URL: http://www.nurc.nato.int. Accessed January 14, 2007.

quadrat   A quadrat is a spatial unit or rectangular plot used to divide a study area into cells for detailed analyses. MARINE ARCHAEOLOGISTS, BIOLOGISTS, and coastal zone mangers may build a frame out of polyvinyl chloride (PVC) pipe of known area as a sampling unit for population studies. Quadrats used in marine science are usually either 10.7 square feet (1 m²) or 16 square feet (1.5 m²). The quadrat is used to quantify the number or percent cover of a given species within a given area. This objective measure allows researchers and coastal zone managers to sample several parts of an area and draw conclusions about the whole area. For example, a marine biologist working in Chesapeake Bay may don SCUBA gear and use quadrats to estimate the extent of invasive species such as water chestnuts (*Trapa natans*) in a tributary. Instead of counting each and every plant within a SUBMERGED AQUATIC VEGETATION (SAV) bed, the investigators sink their quadrats down in

random places to the seafloor and count all the plants within the square. The divers may have laminated information cards that help them in the identification of specific SAV species. The sampling is usually repeated numerous times so that survey results will have reliable information on what types of organisms and how many of each species are found within the SAV beds, without having to count all the SAV within a particular river. This type of work may also be accomplished following aerial surveys that help classify the location and extent of invasive species within SAV beds.

Quadrats may be used in other habitats such as CORAL REEFS, oyster reefs, mud flats, BEACHES, and salt marshes. Researchers make biomass measurements in a salt marsh by placing quadrats at various intervals along a transect and recording the percent cover by species. The biomass is actually the total living biological material in a given area or of a biological community. It is measured by dry weight per given area. Therefore, all the plants within the quadrat are harvested (cut) at the surface of the sediment and sorted into living and dead aboveground biomass. Living plants are sorted by species. Plant material is then dried in an oven and weighed on an electronic balance.

Further Reading

Chesapeake Bay Program. "Water Chestnut." Available online. URL: http://chesapeakebay.net/water_chestnut.htm. Accessed October 29, 2006.

National Renewable Energy Laboratory. "Biomass Research." Available online. URL: http://www.nrel.gov/biomass. Accessed June 30, 2007.

Quality (Q) factor   A Q-factor is a dimensionless number describing the damping of a slightly damped oscillator, such as a plunger in liquid attached to a spring. A damping force such as friction will eventually stop the oscillations. The Q-factor describes the characteristics of mechanical, electrical, and acoustic systems operating near resonance. Resonance occurs when the frequency of a driving force (the stretch of a spring or the gravitational pull of the Moon) equals the natural frequency of an oscillating system. The natural frequency describes how a system normally oscillates. Most systems have several resonant frequencies or will oscillate at a maximum amplitude at several frequencies. Examples are a child on a playground swing (only one resonant frequency), automotive shock absorbers, AC circuits, and acoustic transducers. In marine science, the Q-factor is particularly of interest in designing the transmitting and receiving characteristics of acoustic sources and receivers (hydrophones).

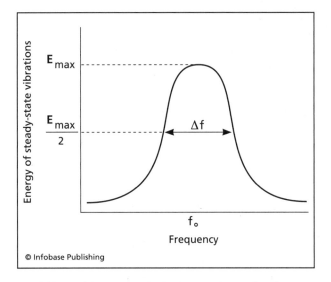

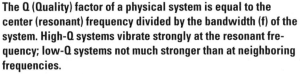

The Q (Quality) factor of a physical system is equal to the center (resonant) frequency divided by the bandwidth (f) of the system. High-Q systems vibrate strongly at the resonant frequency; low-Q systems not much stronger than at neighboring frequencies.

The Q-factor highlights the sharpness of the resonance of the system. A large "Q" means the damping to the system is small and the resonance is sharp; a small "Q" means the damping is large and the resonance is broad. For a radio, a high Q-factor receiver requires fine tuning to a precise central frequency, while a low Q-factor receiver will work over a range of frequencies around the central one.

The Q-factor is usually represented by a graph of the average power delivered to an oscillator by a sinusoidal driving force versus the driving frequency. The Q-factor is expressed mathematically as $Q = f_0/\Delta f$, where f_0 is the resonant frequency of the system, and Δf is the frequency bandwidth of the resonance.

Engineers must be careful to design a building, a structure, or a ship so that it does not resonate at a frequency of one of the environmental forcing functions, such as the wind or the waves. A famous example of faulty structural design is the Tacoma Narrows Bridge, which collapsed in 1941, when the natural frequency of the bridge resonated with the frequency of the wind vortices passing through the Narrows. Modern skyscrapers must be designed so that they do not resonate at the frequencies of earthquake (seismic) waves.

Radio and television sets are designed so that their tuning circuits resonate sharply with the desired station frequencies so that they respond only to signals broadcast at the desired transmission frequency and discriminate against signals outside that frequency band.

The Q-factor also applies to systems in nature. In the case of underwater sound transmission, bubbles in the water, air-filled swim bladders in fish, air sacs in drifting sea weed will resonate and send back a large reflected signal when insonified by acoustic waves corresponding to their resonant frequencies. The larger the Q-factor of, for example, a fish's swim bladder, the stronger the reflected signal when the fish is struck by a sound wave of its resonant frequency. A layer of swimming fish and plankton found at various depths in many seas of the world known as the DEEP SCATTERING LAYER caused considerable difficulty in the operation of SONAR systems.

In the case of ocean tides, the Q-factor for an ocean basin can be determined from the frequencies present in the incoming tide wave compared to the natural frequencies of oscillation of the basin. Studies of the Q-factor for the tides have revealed that the diurnal (once daily) tides are more strongly affected by friction with the bottom than the semidiurnal (twice daily) tides (see TIDES).

Tides in the Bay of Fundy on the North American Atlantic coast between Maine and Nova Scotia are extremely high because the dimensions of the bay are such that many of the natural frequencies of oscillation of the water in the bay closely match the frequencies of the incoming tide waves. The Q-factor for the Bay of Fundy is therefore large compared to most embayments.

Further Reading

Time and Frequency Division, Physics Laboratory, National Institute of Standards and National Institute of Standards and Technology. Available online. URL: http://physics.nist.gov/TechAct.2000/Div847/div847.html. Accessed June 30, 2007.

quartermaster A quartermaster is a military occupational specialty. Soldiers assigned as quartermasters plan and execute logistical matters. A sailor who is a quartermaster attends to the vessel's helm, binnacle, and signals. The helm is the area on a vessel where operational controls are located. The binnacle is the stand for a magnetic compass. Navy quartermasters are trained to solve complex navigational issues with traditional and state-of-the-art tools. The quartermaster or helmsman may use a combination of visual reference, magnetic COMPASS, and a rudder angle indicator to follow courses directed by officers onboard the bridge. Logistical matters relate to planning and controlling the flow and storage of goods and services from point of origin (factory) to point of consumption (customer). Army quartermasters are trained to receipt for, store, and issue supplies.

Sailors who deal with supplies are normally called storekeepers.

See also NAVIGATION; NAVIGATION AIDS; NAVIGATION DEVICES.

quay A structure of solid construction along the edge of a port or harbor where boats are docked for loading and unloading is called a quay. A similar, but open structure providing berthing spaces and cargo-handling facilities is called a wharf. Quays are generally made from concrete, while wharfs are fixed platforms composed of a deck that is supported by pilings. Piles are long structural elements composed of timber, steel, or reinforced concrete, which are embedded into the ground to support vertical loads and to resist lateral forces. Another coastal structure called a mole provides berthing for ships. It is adapted from a BREAKWATER, where the vessel is moored on the sheltered side. ABLE-BODIED SEAMEN on the deck crew handle lines used to moor the ship to the quay.

Many quays have to be modernized to keep pace with DREDGING operations required to bring deeper draft ships into important commercial harbors. In designing the quay wall, engineers consider how weights or loads are distributed throughout the quay. They consider loads caused by wind, earthquakes, and gravity. The wall consists of retaining and bearing concrete structures as well as a deck slab and depending on WATER LEVEL FLUCTUATIONS (tides) may be on the order of a hundred feet (tens of meters) high from foundation to the top of the quay. During construction, soft sediments below the quay foundation are removed down to the rock. Some quay walls consist of precast concrete blocks that are floated to their location and sunk on top of a gravel bed and rock foundation. The back of the walls may be backfilled by quarried stone and covered with geo-textile fabric, a soil stabilizer helping to prevent scour and uneven settlement.

Piers or wharfs may also extend from the quay into navigable waters to facilitate additional berthing spaces for the loading and unloading of cargo. Wharf maintenance might involve removal and replacement of pilings, joisting, and decking. The lifting, setting, and driving of piles is done with a BARGE that contains a pile driver. Wharves may also be outfitted with street lights, navigation lights, fire alarms, electrical services, plumbing, tide houses, and buildings. A tide house is the facility used to make tidal observations and contains oceanographic and meteorological sensors. Modern wharves are constructed from concrete-filled steel tubular bearing piles that support a reinforced concrete suspended deck structure. Attached to the wharf are fittings such as bollards, cleats, fenders, and ladders. Bollards and cleats are fixed to the quay or wharf deck to moor vessels. Bollards are usually short-cast iron posts with a larger diameter near the top that secures mooring warps (docklines) or hawsers. The larger diameter end helps prevent docklines from slipping off the bollard. Wooden or rubber fenders (rub strips) may be fixed to the quay or wharf to absorb the shock or impact created by rubbing of the hull of a vessel with the quay. Some vessels or boats will hang pieces of rope, wood, or floats over the side of a vessel or boat to protect it from chafing.

Quays, wharves, piers, and docks are key components to PORTS AND HARBORS and the history of a maritime nation. Maritime archaeologists may study these structures, and their technological development, and renovate key facilities for educational purposes. In Maryland's Eastern Shore on Chesapeake Bay, the Joppa Steamboat Wharf Project involves replication of a once thriving wharf and warehouse complex on the shores of the Choptank River. The project also includes building of a visitor's center, a transportation museum, and the restoration of several Chesapeake Bay Skipjacks. Watermen used to dredge oysters (*Crassostrea virginica*) on the Chesapeake, using wide-bodied single-mast skipjack sailboats. (Skipjacks are sloops having one mast and two sails, a mainsail and a foresail.) The wharf project replicates a passenger terminal during the steamboat era, the period from approximately 1882 to 1912. This renovated wharf facility is based on historical photographs and insurance maps that document early 20th-century wharf structures. Modern wharfs such as Fisherman's Wharf in San Francisco, California, blend the old with the new. Fisherman's Wharf is known for its historic waterfront and maritime museum, seafood restaurants, cable cars, San Francisco Bay and the Golden Gate Bridge. It remains a key center for waterborne commerce, and at the end of a pier near Crissy Field, which borders San Francisco Marina to the east and the Golden Gate Bridge to the west, is the oldest continuously operating TIDE GAUGE in the Western Hemisphere.

Further Reading

Center for Operational Oceanographic Products and Services, National Oceanic and Atmospheric Administration. *Tide and Current Glossary.* Available online. URL: http://tidesandcurrents.noaa.gov/pub.html. Accessed October 23, 2006.

Choptank River Heritage Center. Available online. URL: http://www.riverheritage.org. Accessed October 23, 2006.

R

radar A device for determining the presence and location of an object by measuring the time for the echo of a radio wave to return from it and the direction from which it returns is called radar. The term *radar* was originally an acronym standing for *r*adio *d*etection *and r*anging. It was developed between the world wars in Britain, France, and Germany. A radar system sends out a train of pulses of electromagnetic waves in the radio or microwave frequency band. If the waves strike an object (target), they are reflected and return as an echo to a receiver, usually co-located with the transmitter. By watching blips on a

The NOAA ship *Ron Brown* sails toward a thunderstorm under study by the ship's Doppler radar. *(NOAA)*

screen, the radar operator can determine the range and direction of the target. Doppler radars employ the principle that the frequency of the reflected signal is shifted up or down depending on the motion of the target. By this means, the velocity of moving objects can be measured. Police radars employ this principle to catch speeders. Weather Doppler radars are used to detect rain and wind shear, which often indicate the presence of severe weather systems. By means of Doppler weather radar, forecasters can determine the size, shape, speed, and direction of travel of major storms, such as thunderstorms, tornadoes, and HURRICANES. PHYSICAL OCEANOGRAPHERS can use radar to track drifting BUOYS to measure ocean CURRENTS; BIOLOGICAL OCEANOGRAPHERS can track marine birds and whales on the surface; MARINE METEOROLOGISTS use radar to identify and study the development of hurricanes at sea.

Radar systems consist of a transmitter for generating pulses of radio waves, usually by means of a magnetron or klystron tube, a modulator that controls the transmitter, an antenna to send out the pulses of waves and to receive the echoes, and a receiver to collect the echoes and amplify them for display on an indicator, usually a fluorescent screen of a cathode ray tube called a planned position indicator, or PPI. Radio waves travel at the speed of light in a vacuum, about 186,000 miles per second (300,000 km/s), but slightly slower in a medium such as Earth's atmosphere. Radio waves are severely attenuated in seawater, and so are not useful within the volume of the ocean.

Radar has been a very useful tool in marine science since World War II in such tasks as ship and aircraft navigation, shoreline delineation, sea ice mapping, tracking of drifting buoys to make Lagrangian maps of surface currents, locating equipment on the surface, such as buoys, surface instrument platforms, drilling rigs, and so forth. In the 1970s, scientists began developing radar as a REMOTE SENSING tool, to make measurements of wind speed, wave height, current velocity, spilled pollutant trajectories, and the like. For the most part, marine scientists and engineers have used high frequency (HF) radars in the three to 30 MHz band for measurements of coastal waves and currents, with some more recent systems using the very high frequency (50 to 1,000 MHz) band. HF radars typically have long-range (tens of miles (km) and moderate resolution. Recently, microwave radars in the general band eight to 40 GHz (X, K_u, K, and K_a bands) are being developed to measure waves and winds at short ranges (tens of feet (m)) but much higher resolution than the HF radars. HF and VHF radars are cur-

rently in use to provide monitoring of sea surface winds, waves, and currents, coastal pollution, SEDIMENT transport, and BEACH erosion.

A popular coastal radar developed in the United States called the CODAR (coastal ocean dynamics applications radar), has been used worldwide to determine the dynamic properties of the ocean surface, such as winds, waves, currents, movements of pollutants, and movements of sea ice. The principle of operation of the CODAR relies on the analogy of sea surface waves to the Bragg scattering of X-rays due to diffraction by the planes of atoms in a crystal lattice. The radar extracts from the sea surface the waves that are exactly half the radar wavelength. Since the waves are in motion, waves traveling toward or away from the radar antenna produce a Doppler shift in the received echo. The Doppler shift appears in the spectrum of the echo as two sharp peaks, symmetrically located about the transmitted frequency. The Doppler shift is directly proportional to the velocity of the ocean waves. In the absence of current, the location of the peaks (sidebands) is known precisely, because the ocean-wave dispersion relation specifies the phase speed of the waves as a function of their wavelength. In the presence of an ocean current, an additional Doppler shift occurs, since the waves are transported by the current. This additional Doppler shift is proportional to U/λ, where U is the speed of the mean current in the direction of the radar beam, and λ is the radar wavelength. Thus, the location of the Bragg peaks in the radar echo spectrum provides a measure of current velocity in the direction of the radar transmission. This technique is similar to determining the speeds of stars and galaxies by the displacement of spectral lines from their known locations at rest velocity (the famous "red shift" in astronomy) that shows that the universe is expanding. By deploying two or more radars, the current vectors in each radar direction can be determined, and the net current vectors within the radar operating area computed to show both the speed and direction of the current at regular grid points.

Several other current measuring radar systems were developed in the 20th century. One is known as the ocean surface current radar, or OSCR. It is high-frequency radar requiring a land-based array of antenna elements that are steered electronically to receive signals from all required directions. OSCR has been deployed operationally at the South Florida Ocean Measuring Center near Fort Lauderdale. Higher frequency, that is, microwave radar systems, such as the MIROS in Norway, and the multifrequency delta-k radar in the United States, are now being used to measure currents and waves at shorter range but higher resolution. These radars can be deployed on such platforms as OIL PRODUCTION PLATFORMS, buoys, and BRIDGES.

Weather and special purpose satellites have proved to be very useful platforms on which to mount remote-sensing radar systems. The most straightforward application is that of radar altimeters (*see* ALTIMETER), that can measure sea surface height from the travel time of the radar pulses to the sea surface, and back to the transmitter in space. From these measurements, the slopes of the sea surface can be derived over distances of tens of miles (km), providing information on the geostrophic currents at the sea surface (*see* GEOSTROPHY). These measurements are especially useful in remote areas of the globe, such as the SOUTHERN OCEAN, where few ship and buoy measurements are available. Another type of measurement uses a "radar scatterometer" to measure the scattering of the reflected radar beam from the sea surface. The scattering is related to the spectrum of the short-wavelength waves on the surface (capillary waves and ripples), which depend on the wind speed in a well-defined way. By this means, the radar scatterometers can measure wind speed, as well as sea surface roughness. The first scatterometer was deployed on the first satellite dedicated to ocean remote sensing, the *Seasat*. Unfortunately, the *Seasat* failed after operating for only a month, but the data so impressed the U.S. Navy that new systems were planned and deployed.

Today, satellites in orbit using DIFFERENTIAL GLOBAL POSITIONING SYSTEM (DGPS) positioning information can provide sea surface height at some areas to within a few centimeters. The possibilities of radar remote sensing of the sea surface, whether from the ground or from space are very great; significant advances in this technology can be expected within the next few years.

Still another type of radar in use today is the synthetic aperture radar, or SAR, which uses the high velocity of the satellite platform to simulate a very long antenna, by storing the echoes in memory, and processing a large number of echoes at one time. This procedure greatly enhances the system resolution. SAR systems have proven very effective in mapping drifting ice.

See also AERIAL SURVEY; REMOTE SENSING; SEA ICE.

Further Reading

Allaby, Michael. *Encyclopedia of Weather and Climate,* rev. ed. New York: Facts On File, 2007.

Coastal Ocean Observation Lab, Rutgers University. Available online. URL: http://www.marine.rutgers.edu/cool. Accessed June 30, 2007.

Nichols, C. Reid, David Larsen Porter, and Robert G. Williams. *Recent Advances and Issues in Oceanography.* Westport, Conn.: Greenwood Press, 2003.

Pickard, George L., and William J. Emery. *Descriptive Physical Oceanography: An Introduction,* 5th ed. Oxford: Pergamon Press, 1990.

Radar Ocean Sensing Laboratory, Rosenstiel School of Marine and Atmospheric Science, University of Miami. Available online. URL: http://www.rsmas.miami.edu/groups/rosl. Accessed July 1, 2007.

SuperDARN, Super Dual Auroral Radar Network, Johns Hopkins University Applied Physics Laboratory. Available online. URL: http://superdarn.jhuapl.edu/. Accessed July 1, 2007.

radiometric dating Radiometric dating involves estimating the age of natural substances such as rocks or fossils by determining the amount of a long–lived radioactive element and its decay products. Using the decay process as a clock is called radioactive dating and is an application of Nobel Prize winner Antoine-Henri Becquerel's (1852–1908) work with uranium (U). For example, the radioactive isotope uranium 238 decays slowly into the stable isotope lead 206 and is used to estimate the age of the mineral zircon ($ZrSiO_4$). It takes 4.5 billion years for one half of the uranium 238 to turn into lead 206. Dating materials based on radioactive decay is useful since the process is not affected by weathering, chemical state, TEMPERATURE, or PRESSURE. Some radionuclides are found naturally, such as potassium and radiocarbon (carbon 14). Potassium is a vital nutrient, and carbon 14 is found as trace amounts on Earth.

Radioactive potassium 40 is common in micas, feldspars, and hornblendes and can be used to date sediment cores. A radioactive isotope of potassium (K), potassium 40 (^{40}K), decays to argon-40 (^{40}Ar) and calcium (Ca) very slowly. The procedure makes the assumption that any argon (a rare gas found in Earth's atmosphere) that was originally present in the sample to be dated was lost and is now resident in the air. Any ^{40}Ar found in the sample from the core must be the result of the decay of ^{40}K. Argon is chosen for analysis rather than calcium because it is very unlikely that any argon is present in the sample except through the decay of potassium.

Radioactive potassium has a half-life (the time required for half the atoms in a radioactive isotope to decay) of 1.28 billion years, which means that 50 percent of it will have decayed into calcium and argon in this time. In another 1.28 billion years, another 50 percent will have decayed into calcium and argon, and so on down through the ages. At the end of its half-life, 90 percent of the ^{40}K decay product is calcium, and 10 percent is argon. By calculating the ratio of argon to potassium, the age of the sample can be determined. Thus, if the ratio is 0.1, the sample is 1.28 billion years old. If the ratio is 0.3, three quarters of the material ($\frac{1}{2}$ in the first 1.28 billion years, plus $\frac{1}{4}$, or $\frac{3}{4}$) in the second 1.28 billion years will have decayed so that the sample is 2.56 billion years old, and so forth. Since some of the argon does in fact escape into the atmosphere, a correction must be applied. Other materials used for long-term dating include rubidium 87, which decays into strontium 87, and uranium 238, which decays into protactinium (Pa). The latter is useful for dating sedimentary materials less than 60,000 years old. For example, since uranium is much more soluble than its decay products thorium (Th) and protactinium, these latter elements precipitate to the ocean floor when formed by their parent (uranium) material.

Carbon-14 (^{14}C) dating is perhaps the most familiar radiometric dating for organic materials. In marine science, this often involves estimating the dates of layers of SEDIMENT from BOX CORERS and PISTON CORERS, and from samples obtained by drilling. By this means, scientists can find the rate of mixing of the top layers of sediment by marine animals, sedimentation rates, and the occurrence of anomalous events such as bombardment by meteors or expulsion by volcanoes. The half-life of carbon-14 is 5,700 years. Each year, 1.2 percent of ^{14}C decays into nitrogen 14 (^{14}N). Since ^{14}C is found in the atmosphere as a result of the bombardment of atmospheric nitrogen 14 by cosmic rays, all living things have ^{14}C in their makeup. Since lost ^{14}C in the bodies of living organisms is replaced, carbon-14 dating is used to estimate the elapsed time since the death of the organism. The procedure of dating involves determining the ratio of radioactive carbon to total carbon present, which is about one part in 10^{12} in living bodies. The maximum age that can be deter-

mined in marine sediments from deep cores is about 40,000 years.

The use of radioactive isotopes in marine science may also be applied to study sediments and WATER MASSES. A list of isotopes that are often used in radiometric dating is provided in the table on this page. Ratios of lead 210 (^{210}Pb): lead 206 (^{206}Pb) in core samples are determined by gamma spectrometry and used to estimate sedimentation rates. Lead 210 has a half-life of 22.3 years and decays to lead 206. Tritium (^{3}H) is used as a tracer to study transport and to estimate the time since a water mass was last at the ocean surface in direct contact with the atmosphere. Tritium is most effective in studies that involve age calculations less than 10 years and in shallow waters where there is the least amount of naturally occurring helium (He). Tritium has a half-life of 12.45 years and decays to helium 3. Radioactive dating is also an important tool in the study of paleontology (*see* PALEONTOLOGIST), paleoclimatology, and GLOBAL WARMING.

Further Reading

Hewitt, Paul G., John Suchocki, and Leslie A. Hewitt. *Conceptual Physical Science-Explorations*. San Francisco: Addison-Wesley, 2003.

Newman, William L. *Geological Time*. USGS Information Services, U.S. Geological Survey. Available online. URL: http://pubs.usgs.gov/gip/geotime. Accessed July 1, 2007.

Seibold, Eugene, and Wolfgang H. Berger. *The Sea Floor: An Introduction to Marine Geology*, 3rd ed. Berlin, Germany: Springer, 1996.

Rapidly Installed Breakwater (RIB) Rapidly Installed Breakwater refers to a V-shaped floating coastal structure designed to deflect incident waves from the apex of the V. The BREAKWATER is made of booms composed of high-strength marine fabrics that are linked together and moored. The result is a sheltered area that can be used to support operations such as lightering. The pool of calmer water is located inside the V and extends into the lee of the structure. The nautical term *lee* refers to the area that is sheltered from the wind. Incident waves are those landward-moving waves that would disrupt marine operations in the coastal ocean, for example cargo offloading. Lightering operations include loading or unloading a vessel by means of BARGES alongside.

In the coastal ocean, a RIB would support rescue and recovery operations and marine operations such as DREDGING, BRIDGE repair, and JETTY construction, especially when the operations must be conducted in mildly energetic seas, such as SEA STATE

Most Commonly Used Isotopes for Radiometric Dating

Radioactive Parent Material	Ultimate Daughter Product	Half-Life
Carbon-14 (^{14}C)	Nitrogen 14 (^{14}N)	5, 700 years
Potassium 40 (^{40}K)	Argon 40 (^{40}Ar)	1.3×10^{9} years
Rubidium 87 (^{87}Rb)	Strontium 87 (^{87}Sr)	4.9×10^{10} years
Uranium 235 (^{235}U)	Lead 207 (^{207}Pb)	7.0×10^{8} years
Uranium 238 (^{238}U)	Lead 206 (^{206}Pb)	4.5×10^{9} years

3 when wave heights range from three to five feet (.9 to 1.5 m). A scale model of the RIB was tested during May and June 1996 in Currituck Sound, adjacent to the FIELD RESEARCH FACILITY in Duck, North Carolina.

Further Reading

Engineering Research and Development Center, U.S. Army Corps of Engineers. Available online. URL: http://www.erdc.usace.army.mil. Accessed December 8, 2006.

Field Research Facility, Coastal and Hydraulics Laboratory, U.S. Army Corps of Engineers. Available online. URL: http://www.frf.usace.army.mil. Accessed December 8, 2006.

recruitment Recruitment is the process by which new members enter a population by either reproduction or immigration. In the ocean, successful recruitment may be related to factors such as spawning density, spawning period and length, and fecundity.

See also FISHERIES RECRUITMENT.

Further Reading

American Fisheries Society. Available online. URL: http://www.fisheries.org. Accessed December 13, 2006.

Biological Resources, United States Geological Survey. Available online. URL: http://biology.usgs.gov. Accessed December 13, 2006.

Center for Coastal Fisheries and Habitat Restoration. Available online. URL: http://www.ccfhr.noaa.gov. Accessed December 13, 2006.

Pacific Fisheries Environmental Laboratory. Available online. URL: http://www.pfeg.noaa.gov. Accessed December 13, 2006.

Redfield ratio An empirical relationship based on the chemical composition of PLANKTON and seawater is called the Redfield ratio. Alfred Clarence Redfield (1890–1983) studied molecular nitrate and phosphate ratios in PHYTOPLANKTON and found them to have an elemental ratio of carbon to nitrogen to phosphorus (C:N:P) of approximately 106:16:1. This ratio closely tracks the ratio of nutrients in the ocean. Significant deviations from the Redfield ratio provide information on the potential for one nutrient to be used up by phytoplankton while leaving a surplus of other nutrients. Elemental ratio information provides coastal zone managers with objective information to base decisions regarding nutrient management strategies that help control phytoplankton growth.

See NUTRIENT CYCLES.

Red Sea The Red Sea is a narrow MARGINAL SEA of the INDIAN OCEAN, located between Africa and the Arabian Peninsula. The Red Sea is separated from the Gulf of Oman by the strategic Strait of Bab el-Mandab and is connected to the MEDITERRANEAN SEA through the Gulf of Suez and the Suez Canal. The strait is approximately 48 miles (77 km) long with western and eastern channels. The canal is an engineering feat, completed in 1869, that stretches for approximately 120 miles (193 km) from Port Said, Egypt, to Suez, Egypt. The Red Sea is surrounded by Egypt, Israel, and Jordan to the north, Sudan and Egypt to the west, Saudi Arabia and Yemen to the east, and Somalia and Eritrea to the south. To the north, the Red Sea splits into the Gulf of Suez and the Gulf of Aqaba to form the Sinai Peninsula. It is one of the most heavily traveled waterways in the world, supporting maritime commerce between Europe and Asia.

The Red Sea basin was created through the separation of the Arabian plate from the African plate (*see* PLATE TECTONICS). This separation takes the form of a geological depression called the Great Rift Valley, which starts in Syria, crosses the Red Sea, and stretches into east Africa. In the Red Sea, SEAFLOOR SPREADING is occurring at HYDROTHERMAL VENTS such as the Atlantis II Deep, Kebrit Deep, Zabargad Fracture Zone, and Shaban Deep. Seawater circulates through sub-bottom Miocene (23.8 to 5.3 million years ago) evaporite deposits, obtaining geothermal heat and dissolving solids before finally surfacing in the depression as a deep-sea brine pool. Evaporites are sedimentary deposits formed from the evaporation of seawater. These deep hyper-saline waters provide habitat for specialized microbial communities classified as extremophiles or extreme halophiles. Archaea, one of the simplest forms of life, fit into this category. They do not require sunlight for photosynthesis and absorb carbon dioxide (CO_2), nitrogen (N_2), or hydrogen sulfide (H_2S) and give off methane gas (CH_4) as a waste product. Archaea that inhabit highly saline environments have adapted to be the most salt-tolerant organisms on Earth.

The Red Sea is about 1,200 miles (1,930 km) long and up to 225 miles (362 km) wide with an area of approximately 169,100 square miles (438,000 km²). Compared to other seas, it is a relatively young basin, with an average depth of 1,611 feet (491 m). The deepest sounding is 9,974 feet (3,040 m) located off Port Sudan. TIDES are predominately semidiurnal (two cycles per day) and ranges are around 1.6 feet (.5 m) at Port Said near the Mediterranean Sea and 1.6 feet (.5m) at the northern part of the Strait of Bab el-Mandab. The Strait of Bab el-Mandab is a transition point between Red Sea tides and mixed tides with a range of 6.6 feet (2m) in the Gulf of Aden. Circulation in the Red Sea varies according to meteorology (northeasterly and southwesterly MONSOONS), TEMPERATURE and SALINITY (THERMOHALINE) variations, and tidal forces. Currents generated by the wind have speeds that are approximately 3 percent of the wind speed. Wind-driven surface currents can generate UPWELLING of deep waters or downwelling of surface waters depending on the wind direction along a barrier such as the coast. In general, Red Sea circulation follows an anti-estuarine pattern where relatively freshwater enters the Red Sea above outflowing dense saline waters.

The Red Sea's climate, which holds records for the most severe heat and humidity combinations on Earth, helps to form a range of water temperatures and salinities that make up Red Sea water, which flows into the Gulf of Aden in a bottom layer over the shallow Hanish sill. Red Sea water temperatures usually range from 70° to 77°F (21° to 25°C) with an occasional reading near 96°F (36°C). The hot climate and lack of rivers flowing into the Red Sea contribute to salinities ranging from 36 to 42 psu—a consequence of evaporation exceeding freshwater inputs such as precipitation, few rivers or streams, and limited connection with the less saline Indian Ocean. Therefore, seawater temperatures of 71.6°F (22°C) and salinities (> 39 psu) make up the saline and dense WATER MASS known as Red Sea Water, which is found at a depth of 547 to 602 fathoms (1,000 to 1,100 m) in the Indian Ocean south of Madagascar.

The Red Sea is known for its natural resources such as CORAL REEFS, sea-grass beds, and mangroves and is listed as a Large Marine Ecosystem (see MARINE PROVINCES). Many Large Marine Ecosystems have been established to apply ecosystem-based management, approaches that consider physical factors such as BATHYMETRY, currents, and water mass, as well as biological factors such as productivity and food webs to assess status and trends. Close to shore, fringing reefs border the Red Sea coastline, while farther offshore corals grow into large and complex barrier reef structures. SUBMERGED AQUATIC VEGETATION (SAV) such as shoal grass (Halodule wrightii) and serrated-leaf seagrass (Cymodocea serrulata) are common on or near coral reefs. Most inhabit the sandy bottoms of LAGOONS, and some grow in sand at the base of the outer reef crest. They provide forage for many fish, dugongs, and sea turtles. Mangrove species such as gray mangrove (Avicennia marina) are commonly found as fragmented strands along with occasional red mangroves (Rhizophora mucronata) in quiet waters behind barrier coral reefs. Mangrove stands are typically thin, ranging from 164 to 984.3 feet (50 to 300 m) in width and extending from 328.1 feet (100 m) to more than 12.4

miles (20 km). The mangroves range in size from small bushes to trees reaching up to 13 to 20 feet (4 to 6 m) in height. Mangroves provide food (decomposed leaves or detritus) and shelter for fish, shrimp and crabs. Mangroves are grazed by camels and are being planted along some coastlines such as in Eritrea. Endangered species that depend on sea-grass beds and MANGROVE SWAMPS along the Red Sea coast include sea turtles and dugongs. The hawksbill (Eretmochelys imbricata) and the green sea turtle (Chelonia mydas) are the most common species of sea turtles. Green sea turtles forage on the bases of sea grasses, and hawksbill sea turtles eat encrusting organisms such as sponges, tunicates, and MOLLUSKS. Green sea turtles are the largest of the hard-shelled sea turtles. Dugongs (Dugong dugon) are herbivorous marine mammals that inhabit shallow bays and lagoons having ample sea-grass beds. Unlike Florida manatees (Trichechus manatus latirostris), dugongs rarely enter riverine environments. As a conservation measure, marine protected areas such as the Ras Mohammed National Park in Egypt, the Red Sea Marine Peace Park in Jordan, and the Sanganeb National Park in Sudan have been established to study populations such as declining longtail butterfly rays (Gymnura poecilura) and giant guitarfish (Rhynchobatus djiddensis).

COMMERCIAL FISHING provides an important economic contribution as evidenced by fleets composed of factory made fiberglass boats and traditional boats such as sambuqs and houris using trawls, gillnets, and hook-and-line gear (see NETS). Today's sambuqs are multiday 49 to 82-foot (15 to 25 m)-long wooden boats outfitted with inboard diesel engines, while houris are single-day 23 to 30-foot (7 to 9 m)-long fishing boats with outboard engines. Many of the commercial species are harvested near fringing reefs that are usually from 0.6 to 1.9 miles (1 to 3 km) wide. Some of the important native and highly migratory fish are blacktip reef sharks (Carcharhinus melanopterus), dogtooth tuna (Gynmosarda unicolor), greasy grouper (Epinephelus tauvina), Indian mackerel (Rastrelliger kanagurta), king seerfish (Scomberomorus commerson), longtail tuna (Thunnus tonggol), orangespotted trevally (Carangoides bajad), redmouth grouper (Aethaloperca rogaa), spangled emperor (Lethrinus nebulosus), twobar seabream (Acanthopagrus bifasciatus), twinspot red snapper (Lutjanus bohar), and yellowspotted trevally (Carangoides fulvoguttatus). Commercially important shellfish include crustaceans such as green tiger prawn (Penaeus semisulcatus), Indian white shrimp (Penaeus indicus), and spiny lobsters (Panulirus homarus) and mollusks such as trochus (Trochus dentatus and T. vergatus), giant

A Selected List of Sharks Common to the Red Sea

Common Name	Scientific Name	Comment
Arabian smooth hound	Mustelus mosis	Bottom dwelling in inshore and offshore waters.
bigeye houndshark	Iago omanensis	Inhabits deep waters from 55 to 547 fathoms (100 to 1,000 m).
bigeye thresher	Alopias superciliosus	Observed near coral reef drop-offs or seamounts.
blacktip shark	Carcharhinus limbatus	Rarely found in water deeper than 98 feet (30 m).
blacktip reef shark	Carcharhinus melanopterus	Prefers shallow waters over coral reefs.
grey reef shark	Carcharhinus amblyrhnchos	Swims near the surface to depths of 153 fathoms (280 m).
milk shark	Rhizoprionodon acutus	An abundant inshore and offshore shark.
oceanic white tip shark	Carcharinus longimanus	A large deep-water shark that sometimes travels in groups.
pelagic thresher	Alopias pelagicus	Inhabits the surface to at least 83 fathoms (152 m) deep.
Scalloped hammerhead shark	Sphyrna lewini	Found in coastal waters.
short fin mako	Isurus oxyrinchus	Inhabits neritic waters from the surface down to 82 fathoms (150 m).
silky shark	Carcharhinus falciformis	Coastal and oceanic, but more common near shelves and slopes, from the surface to depths of 273 fathoms (500 m).
silver tip shark	Carcharinus albimarginatus	Coastal and pelagic, from the surface to a depth of 437 fathoms (800 m).
sicklefin lemon shark	Negaprion acutidens	Lives in shallow inshore and offshore waters to 50 fathoms (92 m).
sliteye shark	Loxodon macrorhinus	Swims near the surface and bottom; at depths from four to 44 fathoms (7 to 80 m).
snaggletooth shark	Hemipristis elongatus	Inshore and offshore down to 16 fathoms (30 m).
spot tail shark	Carcharhinus sorrah	Associated with reefs and is found in shallow water to 76 fathoms (140 m).
tawny nurse shark	Nebrius ferrugineus	Found from intertidal to depth of 38 fathoms (70 m).
tiger shark	Galeocerdo cuvier	Inhabits coastal and pelagic waters, from the surface to about 76 fathoms (140 m) depth.

clams (*Tridacna maxima*), and pearl oysters (*Pinctada margaritifera*). Echinoderms of commercial importance are certain sea cucumbers, benthic scavengers that are known internationally as teat fish. Species that are harvested are black sea cucumber (*Holothuria nobilis* and *H. vagabunda*), red sea cucumber (*Holothuria vatiensis*), brown sea cucumber (*Holothuria marmorata*), and the sand sea cucumber (*Holothuria scabra*).

Anthropogenic factors such as overfishing, oil spills, and development threatens the Red Sea ECOSYSTEM. The Red Sea is known for its many species of sharks, which have declined as demand for their meat, liver oil, fins, and skins has increased. Shark meat is traditionally sun-dried, salted, or hot-smoked by local fishermen. Increasing demand for dried shark fins has contributed to excessive fishing effort by foreign vessels upon many shark species. Some of the more common species of Red Sea sharks are listed in the table above. The Red Sea has remained healthy due to a lack of runoff from precipitation and rivers and comparatively small population centers. However, coral reefs and mangroves located along the coast are threatened by pollution from offshore oil production and the deballasting of merchant ships transiting the heavily trafficked Red Sea. MARICULTURE activities such as shrimp farms pose a serious danger to mangrove strands owing to degradation of water quality and consequent spread of diseases caused from shrimp farm effluents. Development along the Red Sea coast should be done in a manner that protects coral reef ecosystems. Organizations such as the Egyptian Environmental Affairs Agency have published regulations that prohibit the destruction of the natural coastline, tidal flats, and coral reefs.

Further Reading

Central Intelligence Agency. *The World Factbook.* Available online. URL: www.cia.gov/cia/publications/factbook/geos/xo.html. Accessed October 7, 2006.

Hutchinson, Stephen, and Lawrence E. Hawkins. *Oceans: A Visual Guide.* Sydney, Australia: Firefly Books, 2005.

Large Marine Ecosystems of the World. Available online. URL: http://woodsmoke.edc.uri.edu/Portal/. Accessed October 16, 2006.

Mangone, Gerard J. *Concise Marine Almanac.* New York: Van Nostrand Reinhold, 1986.

Marine Science Station, The University of Jordan. Available online. URL: http://www.ju.edu.jo/resources/new-MSS/index.html. Accessed October 17, 2006.

Ministry of State for Environmental Affairs, Egyptian Environmental Affairs Agency. Available online. URL: http://www.eeaa.gov.eg/. Accessed October 17, 2006.

National Geographic Society. *National Geographic Atlas of the World,* seventh edition. Washington, D.C.: National Geographic Society, 1999.

South Sinai Sector, National Parks of Egypt. Available online. URL: http://www.touregypt.net/parks/parks.htm. Accessed October 16, 2006.

reef fish   Reef fish are species that live on or around reefs. The reefs are associated with deposits of skeletal carbonate materials that form a rich habitat for a variety of other plants and animals. Some fish inhabit the reef only as juveniles, while other species use nearby MANGROVE SWAMPS and sea-grass beds as a nursery and then migrate to the reef later in life. Some commercially important reef species include groupers, grunts, porgies, and snappers. Colorful species such as blue damselfish (*Chrysiptera cyanea*), lionfish (*Pterois volitans*), and clown anemonefish (*Amphiprion ocellaris*) are often captured and displayed in reef aquariums and fish tanks with live corals and associated invertebrates.

See also FISH; MARINE FISHES; OSTEICHTHYES.

Further Reading

American Fisheries Society. Available online. URL: http://web.fisheries.org/main. Accessed November 27, 2006.

Froese, Rainer, and Daniel Pauly, eds. "FishBase." Available online. http://www.fishbase.org/home.htm. Accessed November 27, 2006.

Noordeloos, Marco, Mark Tupper, Yusri Yusuf, Moi K. Tan, See L. Tan, Shwu J. Teoh, and Shabeen Ikbal. "ReefBase: A Global Information System on Coral Reefs." Available online. URL: http://www.reefbase.org. Accessed December 13, 2006.

A reef ecology diver monitors the health of the reef. (NOAA/NMFS)

The Reef Environmental Education Program, The Reef Fish Survey Project. Available online. URL: http://www.reef.org/data/surveyproject.htm. Accessed December 13, 2006.

refractometer   A refractometer is an instrument to measure the SALINITY of seawater samples, usually on deck of a research vessel or in a shore-based laboratory. The refractometer, by incorporating an INTERFEROMETER, is able to measure the very small differences in the optical index of refraction (a measure of the change of direction of a light ray in passing obliquely from one medium to another) as a function of salinity and TEMPERATURE with accuracy acceptable for many purposes (the refractive index of seawater increases with increasing salinity and decreasing temperature). At a specific temperature, the refractive index of pure seawater depends only on the salinity. A seawater standard sample is used to convert the instrument readings to international standard values. A monochromatic light source to illuminate the sample is required because the refractive index of seawater varies with WAVE-LENGTH. A LASER is particularly useful for this purpose. This method of determining salinity is not widely employed; for a time it was commonly used on board research vessels of the Soviet Union. Saltwater aquarium enthusiasts often use refractometers since maintaining the proper salinity is crucial for the survival and health of marine fish and corals. This methodology may be used in laboratories to assess the purity of other liquid samples by comparing the sample's refractive index to the value for the pure substance.

See also SALINOMETER.

relative dating Relative dating involves determining the age of components of the Earth and the remains of its living organisms, such as sedimentary rock layers, lava flows, and fossils in relation to the age of known components. For example, the relative age of a newly discovered fossil is determined by comparing it with similar fossils whose ages are known. A major assumption in relative dating is that of uniformitarianism, which asserts that processes that operated in the remote past are not different from those today, such as the rate of erosion of sandstone under the action of flowing water. Thus, the effects of sediment deposition, mountain building, erosion, and fossil characteristics are taken into account. The absolute age of a sample however, cannot be determined by relative dating; for that, RADIO-METRIC DATING (estimating the age of an object by means of the nuclear decay of naturally occurring radioactive isotopes) must be employed.

Several of the principles used by Earth scientists to determine relative ages are given in the table below. Direct evidence from observations of the rock layers may be used to estimate the relative age of rock layers. The environment that existed when the rock was being formed might be inferred from the texture. For example, sandstones with ripple marks might indicate a coastal or riparian habitat. Geologists and PALEONTOLOGISTS have correlated fossils from various parts of the world in order to develop relative ages to particular strata. This is called relative dating and identifies whether layers are older or younger than another. Earth scientists use relative dating to sequence the age of artifacts, fossils, and rocks.

Perhaps the most familiar example of relative dating in the United States is the Grand Canyon, in which the layered structures of the rock are exposed, and the lowermost layers are found to be older than the uppermost layers. By this means, it has been determined that many areas of the United States such as the Permian Basin of west Texas were overlain by ancient seas. The Rocky Mountains are relatively young; the Appalachians relatively old.

Principles of Relative Dating

Principle	Method
Original horizontality	Layers of sediment are deposited horizontally on the undisturbed surface parallel to the gravitationally level surfaces
Superposition	Each layer is older than the one above and younger than the one below. Layers at an angle to the local horizontal have been moved by crustal disturbances.
Cross-cutting	An intrusion or fault that cuts through preexisting rock is younger than the rock through which it cuts.
Inclusion	Pieces of one rock type enclosed by another are older than the rock enclosing them.
Fossils	Fossils contained within rock are compared with similar fossils of known age to determine the age of the rock containing them.
Lithological correlation	The mineralogy of a newly discovered rock is compared with rocks of similar mineralogy and known age.
Geophysical correlation	The physical/chemical properties of the sample is compared with materials of similar composition and known age.
Standard geologic column	A standard composed of rock sequences from different areas matched to a timescale based on computed absolute ages of the rocks, that is, the radiometric dates.
Magnetic reversals	The pattern of magnetic stripes in a sample is compared with the known times of reversals in the Earth's magnetic field. This principle provided considerable evidence for the theory of plate tectonics.

Oceanographers use relative dating of core and drill samples to map the distribution of sediments, bottom types, and fauna over during past ages.

The geologic timescale giving the eras, periods, and durations of conditions of the Earth and its inhabitants (*see* Appendix V) was derived by means of relative dating. A list of principles of relative dating can be found on page 461.

Further Reading

Hewitt, Paul G., John Suchocki, and Leslie A. Hewitt. *Conceptual Physical Science-Explorations*. San Francisco: Addison Wesley, 2003.

Kusky, Timothy. *Encyclopedia of Earth Science*. New York: Facts On File, 2005.

Remotely Operated Vehicle (ROV) An unmanned tethered submersible is called a remotely operated vehicle. It is controlled by an ROV pilot or operator. The tether, or umbilical cable, provides power and communications (video and data signals). Today's ROV technologies can be traced back to submersible systems such as the Cable-controlled Undersea Recovery Vehicle (CURV), which was developed in the early 1960s by the U.S. Navy's Naval Ordnance Test Station. The ROV *CURV* is best remembered for recovering a lost hydrogen bomb in 467 fathoms (853 m) of water offshore Spain during 1966. ROVs vary in size and capabilities from micro class ROVs that may be operated from a small boat to working class ROVs that are launched and recovered using a ship's crane. Besides flotation, motors, thrusters, and powerful manipulator arms, payloads include video cameras, lights, and OCEANOGRAPHIC INSTRUMENTS such as hydrophones, magnetometers, pressure gauges, SONARS, THERMISTORS, transmissometers, and water samplers. Equipment is com-

The ROV NURP1 is used for a great variety of research purposes. *(NOAA/NURP)*

monly modular and can be uninstalled and installed depending on mission requirements. The ROV operator or pilot uses a video console and text information (depth, heading, and geographic position) along with a joystick to control thrusters that provide for horizontal, lateral, or vertical movements. Today's state-of-the-art ROVs are used in a variety of applications such as mine countermeasures, pipeline and telecommunication inspections, salvage, SEARCH AND RESCUE, trenching and cable burying, and underwater exploration. Some well-known ROVs are the two-body *Argus/Little Herc* system operated by the Institute for Archaeological Oceanography at University of Rhode Island, the two-body *Jason/Medea* system operated by the WOODS HOLE OCEANOGRAPHIC INSTITUTION, the *RCV-150* operated by the Hawaii Undersea Research Laboratory, *ROPOS* operated by the Canadian Scientific Submersible Facility, the Spectrum Series *Phantom* S2 operated by the National Geographic Society, and *Tiburon*, which is operated by the Monterey Bay Aquarium.

See also ARCHAEOLOGIST; AUTONOMOUS UNDERWATER VEHICLE; MARINE SALVAGE; ROV OPERATOR; SUBMERSIBLE.

A Remotely Operated Vehicle, a tethered underwater robot, breaks the surface after extensive underwater operations. *(NOAA/NURP)*

Further Reading

Deep Submergence Laboratory, Woods Hole Oceanographic Institution. Available online. URL: http://www.dsl.whoi.edu. Accessed February 6, 2007.

Hawaii Undersea Research Laboratory, University of Hawaii. Available online. URL: http://www.soest.hawaii.edu/HURL. Accessed February 7, 2007.

Institute for Archaeological Oceanography, University of Rhode Island. Available online. URL: http://iao.gso.uri.edu. Accessed February 7, 2007.

Monterey Bay Aquarium Research Institute. Available online. URL: http://www.mbari.org. Accessed February 6, 2007.

National Geographic Society. Available online. URL: http://www.nationalgeographic.com. Accessed February 7, 2007.

National Undersea Research Center for the North Atlantic and Great Lakes, University of Connecticut. Available online. URL: http://www.nurc.uconn.edu. Accessed February 7, 2007.

NOAA's Undersea Research Program, National Oceanic and Atmospheric Administration. Available online. URL: http://www.nurp.noaa.gov. Accessed February 7, 2007.

ROPOS.COM: Canadian Scientific Submersible Facility. Available online. URL: http://www.ropos.com. Accessed February 7, 2007.

remote sensing The process of measuring (or sensing) the physical parameters of a system without making direct contact with that system. Examples of remote sensing in everyday life are a medical doctor using X-rays to sense bone structure, or a police officer using a RADAR gun to measure vehicle speeds on the road. There are two broad categories of remote sensing. One system is passive and the other active. A passive system is one that "listens" for what is going on. A cat's ears are a passive detection system. An active system is one that sends out a signal and "listens" for its echo. A bat's SONAR system is an active detection system.

There are acoustical remote sensing systems and electromagnetic sensing systems, both active and passive. An ACOUSTIC DOPPLER CURRENT PROFILER (ADCP) is an example of an active system. This instrument can sit on the bottom of the ocean or be moored above it looking vertically up or down, sending out sound waves that reflect off particles in the water and are received back at the instrument. By range-gating the measurements, that is, dividing the received signals into short time intervals, a profile in the vertical return signal is obtained. Thus, there are a number of vertical bins, typically one meter in height, where the Doppler shift of the sound waves is measured. The water moving in these bins and advecting the small particles that the sound is reflected off causes the Doppler shift of the sound waves. From the Doppler shift the three-dimensional velocity can be calculated. The ADCP measures at a sampling rate of about one Hertz and can estimate the three-dimensional velocity from the instrument to a distance of about 984.3 feet (300 m).

Another use of sound to remotely measure the ocean is the application of acoustic tomography. Some medical doctors study images of human internal organs created using a technique called CAT Scan. CAT stands for computer aided tomography.

The same technique used to create these images is used in the ocean, except that instead of an X-ray device inches from the area being scanned, moorings with acoustic transmitters and receivers are placed hundreds to thousands of kilometers from one another. Applying tomographic techniques to the arrival times of the acoustic signals remotely senses the ocean temperatures and currents (see ACOUSTIC TOMOGRAPHY).

Remote sensing scientists are concerned with sensing energy in the electromagnetic spectrum and interpreting the results. The example often seen is the TEMPERATURE of the ocean as observed from space. A constellation of NOAA satellites circles the globe, providing the sea surface temperature (SST). This is a passive system called the Advanced Very High Resolution Radiometer (AVHRR), which measure at a number of frequencies the radiation being emitted from the ocean. AVHRR observations are then used to calculate the SST. A number of active systems are orbiting Earth that use radar waves to measure the distance from the satellite to the ocean surface below; they are called altimeters (see ALTIMETER). The first altimeter in space was on board *Skylab* in the 1970s. This altimeter measured the sea surface height (SSH) to an accuracy of about 19.7 inches (50 cm). The latest altimeter is on *Jason-2* and is able to measure the SSH of the ocean to an accuracy of about 0.8 inches (2 cm). Data from this instrument is capable of measuring the location of the Gulf Stream as well as individual mesocale EDDIES that meander about the ocean. There are many other types of remote sensing systems. Some, like those mentioned above, are space-based, but there are also land-based remote sensing systems, such as the Coastal Ocean Dynamics Radar. CODAR is a set of instruments that sit on the coast and send out radar waves that scatter off the surface waves, which allow estimates to be made of the surface velocity. There are many other examples of remote sensing of the color of the ocean as well as of its salinity.

Remote sensing is an important tool for the marine scientist. It allows the researcher to make measurements globally or in areas that would be too expensive using conventional in situ measurements. Marine scientists may use special sensors to collect important ground truth, such as sunphotometers to measure aerosols and underwater spectrometers to measure bottom reflectance. Remote sensing when integrated with technologies such as geographic information systems is especially useful to marine scientists focused on mapping the oceans. As evidence of this, *The World Atlas of Coral Reefs* by Mark D. Spalding, Corinna Ravilious, and Edmund P. Green was selected as the GIS/Mary B. Ansari Best Reference Work Award for 2002.

See also AERIAL SURVEY; HYDROGRAPHIC SURVEY.

Further Reading

Geoscience Information Society. Available online. URL: http://www.geoinfo.org. Accessed March 17, 2007.

Jensen, John R. *Remote Sensing of the Environment: An Earth Resource Perspective.* Upper Saddle River, N.J.: Prentice Hall, 2000.

Ocean Surface Topography, Jet Propulsion Laboratory, California Institute of Technology, National Aeronautics and Space Administration. Available online. URL: http://sealevel.jpl.nasa.gov/technology/instrument-altimeter.html. Accessed July 1, 2007.

Spalding, Mark D., Corrina Ravilious, and Edwin P. Green. *The World Atlas of Coral Reefs.* Berkeley: University of California Press, 2001.

residence time Residence time refers to the average time spent by a mass or substance in a reservoir such as an aquifer, ESTUARY, the ocean, or the atmosphere. In some reservoirs, residence time may be confused with flushing time, or the time it takes for the freshwater inflow to equal the total amount in the system (freshwater replacement time). Determining residence times is complex and requires an understanding of biogeochemical processes as well as tidal flushing, pollutant dispersion, current patterns, sedimentation, and erosion. Biogeochemists determine the form, fate, and movement of elements through biological, geological, and chemical materials. Substances of concern may be nonpoint-source pollution that enters the reservoir as runoff from land rather than from a pipe, which is a source-of-point pollution. In the case of nutrients such as nitrogen (N) and phosphorus (P), an investigator may determine groundwater discharge to streams that provides a large amount of flow that eventually enters estuaries. Nitrogen and phosphorus transported by surface runoff to estuaries may accelerate eutrophication, thus affecting the usage of water resources for drinking, fishing, and recreation. Quantifying the residence time of groundwater in a watershed assists in developing an understanding of the movement of nutrients such as nitrogen and phosphorous. Estimation of residence time is very complex, especially in the varied geologic settings of most estuaries. Researchers often use tracers to estimate the age of nutrient-laden groundwater and associated residence time. Typical tracers are chlorofluorocarbons (CFCs) and tritium (^3H), which in the discharge example above are transported to groundwater from the atmosphere as precipitation infiltrates to the water table. Anthropogenic CFCs such as trichlorofloromethane (CFC11), dichlorodifluoromethane (CF2C12), trichlorofluoromethane (CFC13), and trichlorotrifluoroethane (C2F3C13) were first developed in the early 1930s as coolants in air-conditioning and refrigeration. Tritium input to groundwater results from atmospheric testing of nuclear devices that began in 1952. Radioactive decay of ^3H produces the noble gas helium 3 (^3He). Investigators assess concentrations of CFCs and ratios of ^3H to ^3He to estimate an "apparent" age of the mass or substance. For the groundwater, the age is determined as the time elapsed since the parcel of water was isolated from the atmosphere following the time when rainwater seeps into the ground and is added to an aquifer (recharge). Groundwater dating with CFC11, CFC12, and CFC113 is possible because the atmospheric mixing ratios of these compounds are known, Henry's law of solubility, and concentrations in air and young water (< 50 years since recharge) are relatively high and can be measured. By Henry's law, the weight of a gas dissolved by a liquid is proportional to the PRESSURE of the gas upon the liquid (provided that there is no chemical reaction taking place). The solubility of a gas (tracer) in a liquid (seawater) depends on TEMPERATURE, the partial pressure of the gas (tracer) over the liquid (seawater), the nature of the solvent (seawater), and the nature of the gas (tracer). This solubility relationship is named after English chemist William Henry (1774–1836). Dissolved CFCs are widely employed by marine scientists as transient tracers for water dating and ocean mixing and circulation studies. Water samples for CFC analysis may be collected using Niskin bottles mounted on a rosette frame. Understanding the residence time of a mass or substance is critical for resource managers attempting to develop actions to mitigate pollution or reduce nutrient loads entering an estuary. With respect to issues such as GLOBAL WARMING and climate change, the residence time for greenhouse gases such as chlorofluorocarbons, hydrofluorocarbons, perfluorocarbons, and sulfur hexafluoride in the atmosphere is particularly important.

Further Reading

Kusky, Timothy. *Encyclopedia of Earth Science.* New York: Facts On File, 2005.

United States Environmental Protection Agency. Available online. URL: http://www.epa.gov. Accessed December 24, 2006.

United States Geological Survey. "Water Resources of the United States." Available online. URL: http://water.usgs.gov. Accessed December 24, 2006.

Vents Program, Pacific Marine Environmental Laboratory, National Oceanic and Atmospheric Administration. Available online. URL: http://www.pmel.noaa.gov/vents/index.html. Accessed December 24, 2006.

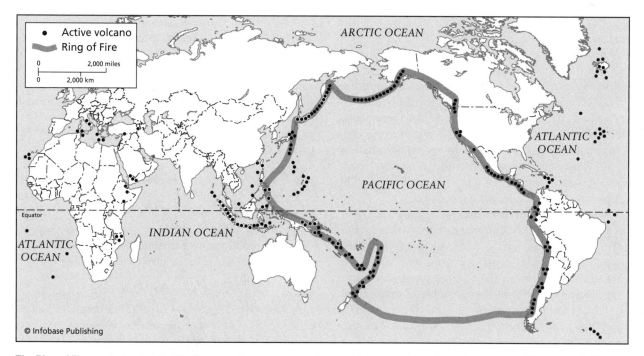

The Ring of Fire encircles the Pacific Ocean and encompasses the most intense earthquake and volcano activity on Earth. The distribution of volcanoes in the ring and other major volcanic areas of the world are also shown.

Ring of Fire A popular term describing the large number of volcanoes bordering the PACIFIC OCEAN is the Ring of Fire. In fact, the Ring of Fire contains most of the volcanoes on planet Earth. The volcanoes have been thrust up, many in an island arc-building process, as a consequence of edges of the Pacific seafloor plunging under, or subducting, the continental plates that surround them (*see* SUBDUC-TION ZONES). Deep ocean TRENCHES delineate subduction zones, where the plate of dense oceanic crust is moving below the surface under the neighboring plate. For example, in the case of western North America, the Pacific plate is diving underneath the North American plate. This process is accompanied by strong earthquakes and volcanic eruptions, and by the thrusting up of the Cascade Mountains. The earthquakes and volcanic landslides sometimes create giant TSUNAMIS. The Ring of Fire extends a distance of about 30,000 linear miles (50,000 km), surrounding the Pacific Ocean.

This deformation of the Earth and the study of faulting and folding that leads to the circum-Pacific "Ring of Fire" is called tectonics (*see* PLATE TECTON-ICS). The geological model that describes the Earth's lithosphere, which is comprised of crust and the uppermost section of mantle, is a series of approximately 13 more or less rigid plates that move in relation to one another. These plates change size and position over time as new seabed is created at the mid-ocean ridges and old seabed is destroyed at the trenches. Conse-

quently, these seismically active destructive margins are volcanic. The archetype Ring of Fire Indonesian volcano Krakatoa, which erupted in 1883 with consequent loss of more than 83,000 lives, provides the classic example of volcanism along the Pacific margins.

Further Reading

Gates, Alexander, Ph.D., and David Ritchie. *Encyclopedia of Earthquakes and Volcanoes,* 3rd ed. New York: Facts On File, 2007.

Cascades Volcano Observatory, United States Geological Survey. "'Hot Spots' and the 'Ring of Fire.'" Vancouver, Washington. Available online. URL: http://vulcan. wr.usgs.gov/Glossary/PlateTectonics/Maps/map_plate_ tectonics_world.html. Accessed March 2, 2007.

rip current A rip current—also called a rip tide—is a narrow, strong CURRENT that flows seaward from the surf zone. The surf zone is the stretch of shallow water where waves break as they approach the shore. Waves approaching the shore pile water up on the beach, which dissipate as a return flow through the surf zone. This return flow or rip current is on the order of 10 meters wide with speeds of one to two meters per second through the surf zone. Observers recognize rip currents as discolored water extending offshore, or from foam lines or seaweed streaks that mark the current's seaward edge. The discoloration is caused by the transport of material such as eroded sand. The plumes of dirty water, often mushroom-shaped, extend from the beach outward beyond the surf and are easily identified on aerial imagery of the coastline.

Coastal structures such as GROINS, JETTIES, and piers may cause rip currents by interrupting movements of water that run parallel to the beach. As waves and currents collect water against these obstacles, the alongshore currents are diverted as rip currents that can cut deep holes in seemingly shallow water. The rip currents generally span the entire water column until they pass the surf zone. The deeper water often appears darker in color.

Because rip currents are hazardous to swimmers, beach patrols recommend that swimmers stay away from groins, jetties, piers, or other structures extending from the beach into the surf zone. Rip currents can pull even the strongest swimmers into deeper water. Most deaths occur when people caught in the current try to swim against the current to reach the shore. The panicked swimmer becomes totally exhausted and drowns. People caught in a rip current are usually advised to wade or swim parallel to the beach until the current weakens. Otherwise, people are advised to float with the current beyond the breakers, then swim shoreward at an angle away from the current. At beaches with lifeguards who have been trained to identify rip currents from their elevated towers, warnings are posted when they are observed. A yellow flag indicates "caution, dangerous conditions," while a red flag indicates "stop, don't enter the water."

Surf fishermen scout the location of breakers and look at beach composition in pursuit of rip currents that carry small baitfish and mole crabs beyond the breakers. Wave heights are usually less over the rip current that is cutting loose SEDIMENT from the bottom. Large fish forage along these sloughs and channels for menhaden, crabs, and minnows. Some fishermen place a mushy pulp of sea life such as cut fish or minced mollusks called chum into a rip current from the beach. This chum creates a slick that is intended to attract sport fish such as bluefish, striped bass, and channel bass to the surf caster's fish-finder rig.

COASTAL ENGINEERS and oceanographers are interested in measuring and predicting rip currents since they are dangerous to people and cause erosion. These professionals are advancing understanding of nearshore currents by investigating beach profiles and wave processes at coastal laboratories such as the United States Army Corps of Engineers' FIELD RESEARCH FACILITY located in Duck, North Carolina. They also study the impact of rip currents on offshore sediment transport, shoreline evolution, and pollutant transport. Rip currents are attributed

to beach erosion and changes in the to-and-fro movement of SANDBARS. Public interest in rip currents is high due to safety issues and concerns to protect beachfront property. The science behind rip currents can be applied to define the conditions and situations when weather offices and beach patrols should issue local advisories.

See also BEACH; CURRENT METER; LONGSHORE CURRENT; SURF.

rogue waves   Rogue waves are gigantic ocean waves characterized by extreme height and very steep-sided troughs that run ahead or just behind the waves, often described as "holes in the sea." They are level-crested, that is, extending a long distance perpendicular to the direction of travel. They are also called "killer waves" or "freak waves," but are officially known as extreme storm waves (ESW). In 1933, the USS *Ramapo* accurately measured a wave 112 feet (34 m) high in the Pacific Ocean. These monster waves, although rare, are being encountered with greater frequency as vessel traffic increases worldwide.

Rogue waves, no matter how large, begin "life" as a "dance" between sea and air, in which the atmosphere transfers energy to the sea. The energy transfer occurs by means of both pressure fluctuations and wind stress. Waves are formed wherever wind acts on water, but very large and intense storms, especially HURRICANES and severe winter CYCLONES are necessary to produce rogue waves. The average height of the waves is governed by the time the wind has acted over the sea surface, the FETCH, or distance over which the wind has blown, and the speed of the wind. The ocean surface can be modeled as a random collection of waves of various shapes, periods, and heights, generated both by the local wind and by distant storms. If the random phases of very large waves combine in just the right way, their crests grow higher and their troughs grow deeper, generating rogue waves of truly awesome dimensions.

When opposing currents, or shoaling bottoms, are found near areas of frequent severe storms, very large storm waves are often transformed into rogue waves. One such area is off the east coast of South Africa, where strong winds from the south polar region generate huge waves, which run into the southward-flowing AGULHAS CURRENT (opposing current). Supertankers routinely transit this region en route to the PERSIAN GULF. These ships can be especially vulnerable to rogue waves because the waves catch the vessel with water under bow and stern, and the deep trough under the mid-section; that is, the mid-section has no water under it for a moment. This lack of support produces enormous stresses called sagging and hogging, which can cause

A giant wave smashes into a NOAA research vessel operating in the Bering Sea. *(NOAA)*

the vessel to split in two. Several supertankers have been lost in this way. Because of the sagging and hogging danger, smaller vessels are often safer than the large supertankers.

Small vessels are still at great risk, however. Rogue waves are believed to be the cause of the much-publicized sinking of the fishing vessel *Andrea Gail* off Sable Island, Nova Scotia, Canada, during the so-called perfect storm, when in October 1991 a nor'easter combined with a tropical hurricane with winds up to 120 miles per hour (200 km/hr) to produce huge seas. If a vessel encounters the trough of a rogue wave, the bow drops down and the superstructure is swept away due to the enormous force of the wave breaking over the deck. Many vessels, which disappeared without a trace, have probably been lost in this way.

In shallow water, very large waves in otherwise calm seas that seem to appear from nowhere are sometimes called rogue waves. In Daytona Beach, Florida, two or three 10 to 18 foot (3–5 m) waves smashed onto the beach on the night of July 3, 1993. Fortunately, there were no fatalities from these waves, although there were several parked vehicles on the beach. The waves arose from a relatively calm sea; no great storms were in the area. After examin-

ing a number of possible causes, including underwater landslides, scientists concluded that the waves were produced in a region of squalls, thunderstorms, and downbursts acting in concert to generate a few gigantic waves. A "wall of white water," a rogue wave, was seen between the hypothesized generating area and the shore. Apparently, the great wave broke before hitting the beach and the parked vehicles, but the remnant was frightening enough.

Giant waves, called killer waves, or sneaker waves, have caused fatalities on the rocky coasts of Oregon and Washington. These waves have suddenly appeared, often from relatively calm seas, and swept unwary victims away. The physics of the formation of these waves must be better understood, and FORECASTING methods developed to avoid infrequent, but continuing tragedies.

See also SOLITARY WAVES; WAVES.

Further Reading

The GKSS Research Center, Geesthacht, Germany. Available online. URL: http://www.gkss.de/. Accessed March 2, 2007.

Junger, Sebastian. *The Perfect Storm: A True Story of Men Against the Sea.* New York: HarperCollins, 1998.

Nickerson, Jerome W. "Freak Waves." *Mariners Weather Log* 37, no. 4 (1993): 14–19.

Shillington, F. A., and E. H. Schumann. "High Waves in the Agulhas Current." *Mariners Weather Log* 37, no. 4 (1993): 24–28.

Rossby waves   Rossby waves are very long waves in the atmosphere and the ocean that generally propagate zonally, that is, parallel to lines of latitude, and in which the restoring force is provided by the rotation of the Earth in the form of the CORIOLIS FORCE. Rossby waves, named for their discoverer, Carl-Gustav Rossby (1898–1957), are also known as planetary waves. In the atmosphere, they have WAVE-LENGTHS thousands of miles (km), and in the ocean, hundreds of miles (km). Rossby wave amplitudes are primarily horizontal rather than vertical, as is the case for wind-driven water waves.

All waves propagate by means of a restoring force that increases with increasing displacement from the equilibrium position. In the cases of a freely swinging pendulum as well as surface wind waves on water, the restoring force is gravity. In the case of Rossby waves, the restoring force is physically equal to the change in the Coriolis force with latitude, β, described mathematically as $\beta = (\Delta f/\Delta y)$, where f is known as the "Coriolis parameter," and is equal to twice the angular velocity of the rotating Earth (2Ω) multiplied by the sine of the latitude (φ), and y is the north-south coordinate. The mathematical Δ notation refers to change in the Coriolis parameter with latitude. For example, a Rossby wave might be propagating along the eastward-flowing North Equatorial Countercurrent. As a water particle in the wave moves away from the equilibrium position northward, the increasing value of β provides the restoring force to bring it back to equilibrium, and similarly, particles moving southward are brought back to equilibrium. This process continues as the wave crosses the ocean basin. The phase velocities (velocities of points of fixed phase) of Rossby waves are always westward, although the group velocities (the velocities at which groups, or packets of waves, travel) can be eastward with respect to the mean flow. The energy of the waves travels at the group velocities (*see* WAVES).

The dynamics of Rossby waves can be explained in terms of the conservation of potential vorticity—the oceanic and atmospheric equivalent of the principle of conservation of angular momentum in physics. In very basic terms, the vorticity of a spinning column of water, such as found in a Rossby wave, is the sum of the planetary vorticity provided by the Earth, and the relative vorticity provided by the local rotation of the water. In the Northern Hemisphere, for example, counterclockwise motion is regarded as positive; clockwise motion negative. As the water in the crest of the Rossby wave moves northward, the planetary vorticity increases or becomes more positive; the local vorticity must decrease or become more negative to keep the sum constant and conserve potential vorticity. This process retards the northward motion, eventually sending the water in the wave southward. This time, the relative vorticity must increase to offset the decrease in planetary vorticity and thus keep the sum of the two constant. The details of this analysis show that the phase speed of a Rossby wave in the ocean is equal to the following:

$$C = -\beta / (k^2 + f^2/gh)$$

where $\beta = \Delta f/\Delta y$, $k = 2\pi/\lambda$, where λ is the wavelength, $f = 2\Omega\sin\varphi$, that is, twice the sine of the latitude φ times the Earth's angular velocity of rotation Ω. Rossby waves travel slowly in the ocean; at mid-latitudes, they can take months to cross an ocean basin; in equatorial latitudes, they move much faster, taking perhaps weeks to cross a basin. Rossby waves, as well as KELVIN WAVES are believed to be important in the onset of the EL NIÑO phenomenon. The equator, for which the Coriolis parameter is equal to zero and increases uniformly in both the north and south directions, appears to serve as a waveguide for these waves. Kelvin waves, when reflected from a continental boundary such as occurs in west equatorial Afri-

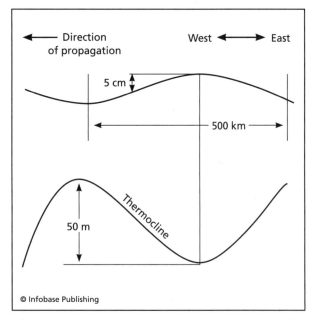

Direction of propagation

West ← → East

5 cm

500 km

50 m

Thermocline

© Infobase Publishing

Rossby waves in the ocean have very long horizontal scales, on the order of hundreds of miles (km), but displace only a few inches (cm) in the vertical at the surface, and a few feet (m) to hundreds of feet (m) at the thermocline.

ca's Gulf of Guinea can split and propagate northward and southward along the African coast, and initiate a Rossby wave along the equator at the same time! This process is believed to occur in the Pacific Ocean along the coast of South America as well, and would be a major factor in the onset of the El Niño.

In the atmosphere, upper-level flow is usually marked by a series of ridges and troughs in atmospheric pressure; the largest of these are the Rossby waves. They exert great impact on the daily weather. They can transport cold arctic air to low latitudes and warm tropical air to high latitudes in a few days, sometimes producing record-breaking temperatures. They also cause divergence and convergence in the upper atmosphere. When air at high altitudes diverges, it can draw up air from below creating a convergence at the surface with associated cloud development and precipitation. Converging air at high altitudes produces sinking, and high-pressure fair weather systems at near surface altitudes.

The phase velocity of Rossby waves in the atmosphere is given by

$$C = U - bL^2$$

where U is the westerly component of the wind speed in meters per second within the wave, $b = (5.3 \times 10^{-3})$ $\cos(\varphi)$ where φ is the latitude, in SI units, and L is the wavelength in meters. Although Rossby waves propagate westward, there is a critical speed of the wind at which they remain stationary. By setting C = 0 in the above equation, the critical wavelength is found to be $L = \sqrt{U/b}$. Very long Rossby waves in the atmosphere occasionally move upwind, a condition called retrograde motion.

The importance of understanding and modeling the role of Rossby waves in global climate studies is just beginning to be fully appreciated. Rossby waves have been observed to produce major variability in western boundary currents such as the GULF STREAM by imparting changes in velocity and direction (meanders) and by spinning off warm and cold core EDDIES.

Further Reading

Aguado, Edward, and James E. Burt. *Understanding Weather and Climate*, 4th ed. Upper Saddle River, N.J.: Pearson/Prentice Hall, 2007.

Colling, Angela, ed. *Ocean Circulation*, 2nd ed. Oxford: Butterworth-Heinemann, 2001.

Knauss, John A. *Introduction to Physical Oceanography*, 2nd ed. Upper Saddle River, N.J.: Pearson/Prentice Hall, 1996.

ROV operator   The pilot of a remotely operated vehicle is often called an ROV operator. He or she is involved with the operation and maintenance of these SUBMERSIBLES to include installing various sensors or items of equipment such as CURRENT METERS, pressure gauges, cameras, lights, acoustic positioning systems, hydrophones, motors, and manipulators to the ROV. Planned subsea missions, which may involve SEARCH AND RESCUE, oceanographic studies, marine archaeology, hydrographic surveys, tank and pipe repair, cable maintenance and burial, and drilling rig support, determine actual payloads. The ROV operator has to be familiar with at-sea operations, especially launch and recovery systems, as well as payloads.

Large ROVs are deployed and retrieved from the deck, using an A-frame or crane. The A-frame is located on the stern and is a special type of crane that helps to lower or raise the ROV with the help of a WINCH. Small or micro ROVs may be deployed by hand from a small boat. The operator pilots the ROV with a joystick, based on real-time video feeds and navigation data received via a tether (umbilical cable) that connects the unmanned submersible to the mother ship. The tether provides the power and data uplink between the ROV and shipboard control center. Controlling the ROV's thrusters allows the robot to make horizontal, lateral, and vertical movements in the water. The operator may have to pull information from various computer screens or windows in order to complete the mission successfully. Tasks that the ROV operator might complete involve planning the mission by evaluating environmental conditions such as tidal currents, operating a launch and recovery system, undocking and docking using a tether management system, navigation, and operating manipulator or cutting arms. Onboard lights, cameras, and video allow the operator to record imagery below the ocean surface. Before and after each mission, all electronic and hydraulic systems must be maintained, calibrated, and checked for serviceability.

Further Reading

Marine Advanced Technology Education Center. Available online. URL: www.marinetech.org. Accessed October 21, 2006.

Marine Technology Society. Available online. URL: www.MTSociety.org. Accessed October 21, 2006.

NOAA's Undersea Research Center, University of North Carolina at Wilmington. Available online. URL: www.uncwil.edu/nurc. Accessed October 21, 2006.

Remotely Operated Vehicle Committee of the Marine Technology Society. Available online. URL: www.rov.org. Accessed on October 21, 2006.

S

salinity   Salinity is a measure of the dissolved solids in sea water. The traditional definition used throughout most of the 20th century defines salinity as the total amount of dissolved solids per thousand by weight when all the carbonate is converted to oxide, the bromide and iodide are converted to chloride, and organic matter is oxidized. The dissolved solids are derived from chemical erosion of rocks of the Earth's crust, and, also, from the discharge from HYDROTHERMAL VENTS in the oceanic ridges. All natural waters have some dissolved materials in them. For this reason, water is often referred to as the universal solvent.

A general rule is that the concentrations of the major dissolved ions in seawater can vary from site to site, but their relative proportions remain the same (*see* CONSTANCY OF COMPOSITION). An ion is an atom or molecule that has acquired a charge by either gaining or losing electrons. Cations are positively charged ions, while anions are negatively charged ions. These dissolved ions reach the sea through sources such as rivers, volcanism, and hydrothermal vents. Salts are formed by replacing the hydrogen ions of an acid with another positively charged chemical, often a metal such as sodium in sodium chloride (NaCl). Only six elements account for more than 99 percent of these dissolved salts. They are chlorine, sodium, sulfur, magnesium, calcium, and potassium. The ions are respectively chloride (Cl^-), sodium (Na^+), sulfate (SO_4^{2-}), magnesium (Mg^{2+}), calcium (Ca^{2+}), and potassium (K^+). In summary, the total value of salinity can change from region to region, but the ratio of the concentrations of major ions to the total remains constant for all samples. These major ions and their proportions are listed in the table on page 472.

Today, the preferred definition among oceanographers expresses salinity as a function of electrical conductivity, given below. In the early to mid 20th century, salinity was determined by chemical titration with silver nitrate, a slow process with an accuracy of about .02 parts per thousand. Today's electronic SALINOMETERS are accurate to about .003 parts per thousand. The Joint Panel on Oceanographic Tables and Standards has developed a practical salinity scale and the accepted equations of state for seawater. Professor Nicholas P. Fofonoff (1929–2003) from the WOODS HOLE OCEANOGRAPHIC INSTITUTION is best known as a leader in the development and introduction of the now standard Practical Salinity Scale. His research focused on the dynamics of ocean circulation, physical properties and thermodynamics of seawater, and the application of BUOY systems to measurements of ocean currents. The Buoy Group under the direction of Fofonoff formulated a new definition of salinity based on electrical conductivity, in which the practical salinity of a seawater sample is given by the ratio of the electrical conductivity at a temperature of 59°F (15°C) and pressure of one standard atmosphere (14.7 pounds (6.8 kg) per square inch) to that of a potassium chloride (KCl) solution of a given mass fraction at the same temperature and pressure. The "parts per thousand," or ppt designation of salinity has given way to salinity values as pure numbers, or followed by "psu," standing for practical salinity units.

Salinity is a critical variable for understanding and predicting biological and physical processes. The traditional method to measure salinity is to collect samples at depth with water bottles that are then

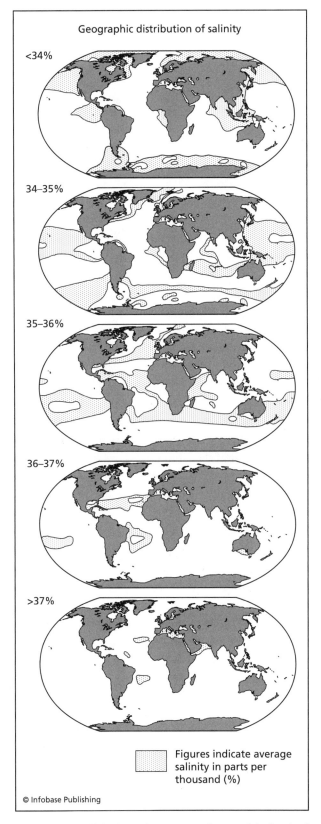

Geographic distribution of salinity

<34%

34–35%

35–36%

36–37%

>37%

Figures indicate average salinity in parts per thousand (%)

© Infobase Publishing

Ocean surface salinity depends on evaporation, precipitation, land runoff, and upwelling. In general, the polar oceans have relatively low salinity (<34 psu), the mid-latitude regions moderate salinity (34-35 psu), the equatorial regions, fairly high salinity (35-36 psu), and the subtropical regions, the highest salinity (>37 psu).

analyzed in a laboratory on shipboard or ashore with an electronic salinometer. Some marine scientists surround CONDUCTIVITY-TEMPERATURE-DEPTH (CTD) gauges with a rack of water bottles called a rosette. The salinity measurements are standardized worldwide by means of filtered natural seawater ampules of standard seawater provided by the International Association for the Physical Sciences of the Oceans (IAPSO) for calibrating salinometers and other marine instruments. Interestingly, saline water bodies such as the RED SEA and the Dead Sea are thought to have a therapeutic effect, especially for the skin, due to their high salt and mineral content.

In the field, profiling CTD gauges are often used or moored in situ conductivity-temperature (CT) sensors. Temperature and pressure sensors on CTDs are routinely calibrated at facilities such as the Woods Hole Oceanographic Institution's Calibration Laboratory. Preoperational checks ensure that instruments are working properly prior to final setups on the deck of a ship. The use of CTDs has greatly increased the process of measuring water properties, made the measurements more accurate, and greatly increased their vertical resolution. CTDs are capable of resolving small changes in the vertical on the order of centimeters in depth, and thousandths of a part per thousand in salinity. The CTD has revealed the microstructure of the ocean, which was unobservable using previous techniques. In the future, REMOTE SENSING using RADAR signals may provide surface salinity maps to study processes from plume dynamics to the movement of WATER MASSES.

The concentration of salt in deep ocean water is about 35 parts per thousand. Rivers have a concentration of about 1/10th of a part per thousand. Shallow seas, where evaporation is high, have higher salinities. Salt concentrations increase as water evaporates. The Red Sea has the highest salinity of any sea in the ocean, often above 40 parts per thousand. Salinities in the open ocean range from about 33 parts per thousand to 37 parts per thousand. Salinity shows general distribution characteristics with latitude, such that the highest values of open ocean salinity occur at the subtropical high regions of the great mid-ocean GYRES where precipitation is low and evaporation is high; a low value occurs near the equator where tropical precipitation is high, and minimum values in the polar seas where precipitation is very high, and evaporation is low. The vertical distribution of salinity shows variation at the surface, and in coastal regions, of several parts per thousand, but uniformity in the deep layers of the world oceans.

The waters of MARGINAL SEAS often have low salinities due to runoff (*see* HYDROLOGICAL CYCLE).

Numerous rivers may feed these arms of the ocean. For example, salinities in the Baltic Sea are on the order of 10 parts per thousand or less. The Baltic Sea in northeastern Europe is one of the world's largest bodies of brackish water, receiving freshwater input from rivers, rainfall, and infiltration. In other geographic locations, evaporation may greatly exceed precipitation. Salinity in coastal LAGOONS such as the Laguna Madre on the south Texas coast can be as high as 50 parts per thousand, owing to high evaporation.

The freezing point of water decreases as the salinity increases, which explains why polar ocean waters can have temperatures lower than the freezing point of pure water, 32°F, (0°C). Salinity decreases the temperature at which water is most dense. Freshwater is most dense at 39°F (4°C). Salinities increase as the seawater freezes because salts are left behind as the ice crystals take form as frazil. The cold water becomes denser and sinks until it reaches surrounding waters with a greater density. One of the most important applications of salinity and temperature observations as a function of depth is the computation of the density of the water column. Once the temperature and salinity are known, the density can be calculated from the equations of state for seawater. These equations, which express the density in terms of temperature, salinity, and pressure, have scores of terms, and are usually evaluated by means of tables or computer programs.

The density differences between stations separated in horizontal distance are used to compute the geostrophic velocity (see GEOSTROPHY). For decades, this method of using density to estimate current SHEAR as a function of depth has been used. The vertical density distribution also determines the stability of the water column and its susceptibility to INTERNAL WAVES. In general, the density always increases with depth, because of the overriding effect of temperature on density. However, in polar waters, and even in coastal waters off New England in winter, temperature may increase slightly with depth. However, the salinity of the deep water is almost always higher than that at the surface and prevents the density from decreasing with depth.

Density is the driving component of most of the Earth's oceans, along with the wind stress. Even though the density differences are small, they can account for major currents that transport large water masses throughout the globe in the so-called thermohaline conveyor belt. The warm water of the Gulf Stream flows northward where it is slowly preconditioned by evaporation, which slowly cools the water and increases its salinity. Even though these processes increase the density of the water, it is still lighter than

the water it overlays. The Gulf Stream, called in this region the NORTH ATLANTIC CURRENT, separates in the waters off Newfoundland. Part of the current turns southward to rejoin the Gulf Stream system's large gyre (see GULF STREAM); however, part of it flows northward into the Norwegian Sea and part toward Iceland, where this relatively warm, yet salty water, now cools quickly and becomes much denser than the water below it. This North Atlantic water sinks to the bottom in the Norwegian Sea and begins to flow southward as a water mass called North Atlantic Deep Water. This generation of North Atlantic Deep Water is the engine that drives a deep global circulation known as the Thermohaline Conveyor Belt (see THERMOHALINE CIRCULATION).

Salinity changes are usually not great in the deep ocean. A change of one part of salt per thousand parts of water (1 permanent salinity unit or psu) is large. However, near the coasts, especially in the regions of river mouths, salinity changes can be very large, on the order of 10 parts of salt per thousand parts of water (10 psu), and constitute a salinity front, or HALOCLINE. Since the density of seawater is primarily determined by temperature and salinity, the halocline is often located at the region of maximum density gradient or the PYCNOCLINE.

In areas such as the mouths of the Amazon, the Mississippi, and the Columbia Rivers, plumes of relatively freshwater extend out into the ocean for many miles (km) before complete mixing with the surrounding salt water occurs. The Norwegian fjords are regions of strong haloclines at the boundaries between the freshwater from glacier melt and the salty ocean water, which, being denser, is found in the deep parts of the basins. Many ESTUARIES are of the "salt wedge" type, where a front is located at the boundary between fresh or brackish water and salty ocean water. Dense seawater enters the estuary near

The Average Concentrations of the Major Ions in Grams in One Kilogram of Seawater

Chloride	18.980
Sodium	10.556
Sulphate	2.649
Magnesium	1.272
Calcium	0.400
Potassium	0.380
Bicarbonate	0.140

the bottom as a wedge. Seaward of the salt wedge, the ECOSYSTEM is oceanic; landward of the salt wedge it is a freshwater ecosystem. Scientists look at interactions such as entrainment along the salt wedge. Entrainment is the collection and transport of objects caught by the flow of a fluid, for example, the mixing of water particles or even larval fish along the boundary of the salt wedge.

Regions of sharp haloclines, such as river mouths, can often be modeled numerically by a two-layer water flow, with fresh or brackish water being the top layer, and salty oceanic water being the bottom layer. Such models usually provide marine scientists and engineers with a means to investigate the main features of the circulation, as well as to plan for remedial action in the event of pollutant spills.

See also DENSITY; OCEANOGRAPHIC INSTRUMENTS; SEA ICE; TEMPERATURE; WATER MASS.

Further Reading

Fofonoff, Nicholas P. "Physical Properties of Seawater: A New Salinity Scale and Equation of State for Seawater." *Journal of Geophysical Research* 90 (1985): 3,332–3,342.

Large Marine Ecosytems of the World, LME#23: Baltic Sea. Available online. URL: http://www.edc.uri.edu/lme/text/baltic-sea.htm. Accessed February 19, 2007.

Open University Oceanography Course Team. *Seawater: Its Composition, Properties and Behavior,* 2nd ed. Oxford: Pergamon Press, 1995.

Pickard, George L., and William J. Emery. *Descriptive Physical Oceanography: An Introduction,* 5th ed. Oxford: Pergamon Press, 1990.

salinometer A salinometer is an instrument for measuring salinity, now usually based on electrical conductivity. In the 19th and nearly the first half of the 20th century, salinity measurements were based on chemical titration with a silver nitrate solution that determined the chlorinity, and then calculated the salinity from the formula:

$$salinity = 1.80655 \times chlorinity$$

This procedure is generally accurate to + and -.02 practical salinity units (psu), and is slow and difficult to perform well at sea. In the 20th century, it was discovered that it is both faster and more accurate to determine the salinity of seawater from measurements of electrical conductivity. Today, almost all measurements of salinity use the electrical conductivity method. The precision of most shipboard salinometers is better than + and -.003 psu.

The U.S. Coast Guard led the conversion from titration to conductivity. During the 1930s, the Coast Guard used an alternating current bridge to determine the salinity of water samples while on International Ice Patrol. By 1956, several oceanographic laboratories were using conductivity bridges with improved circuits to obtain + and -.003 psu accuracy. In 1957, Gerald L. Esterson of the Chesapeake Bay Institute developed an improved salinometer that did not require the troublesome electrodes of most existing salinometers. In 1961, Neil L. Brown (1927–2005) and Bruce V. Hamon in Australia developed an inductive salinometer, which has now come into general use. In this instrument, the sample is pumped into a conductivity cell that consists of two wire coils that induce a closed-loop electric current through the core and the surrounding water. The second coil measures the induced current that is directly proportional to the electrical conductivity of the surrounding water. In this procedure, the temperature of the sample is measured, and its affect compensated for in the instrument. The first *in-situ* salinometers, the S of the STD (salinity-temperature-depth) probes, used analog circuits to compensate for the temperature affect. This permitted direct reading of the physical properties of the water column, which enormously enhanced the vertical resolution and the amount of data that could be obtained during an oceanographic cruise. However, one problem that surfaced rather quickly was that errors in the salinity profile were being introduced because the time constants on the salinity and temperature probes were different, yielding erroneous results in parts of the water column where gradients are large, such as the THERMOCLINE and HALOCLINE. These errors limited the use of these instruments in making detailed studies of the ocean's fine structure in regions of large gradients.

The solution appeared with the extensive adoption of digital systems and computers for use in acquisition of data in real-time. The procedure is to use not an STD, but a CTD (conductivity-temperature-depth probe) that measures conductivity, temperature and pressure, and the temperature compensation to determine salinity is calculated on a digital computer and applied to the conductivity, rather than calculating and applying correction factors within the instrument with analog circuits. The modern CTD has enabled great advances in understanding of the fine structure (step structure) of the ocean in regions of large gradients, and made possible the development of theoretical models of such processes as double diffusion. Salinometers are standard equipment aboard oceanographic research vessels and are often used to calibrate and verify results from a CTD. Salinometers compare the conductance of a sample of seawater directly and simultaneously against the conductance of a sample of standard seawater. Results are verified against International Association for the Physical Science of the Ocean (IAPSO) Standard Seawater. The IAPSO Standard Seawater Service provides seawater standards of precisely known electrical conductivity ratio for the calibration of salinometers and other salinity measurement devices. IAPSO Standard Seawater is the only salinity standard approved by the oceanographic community.

See also CONDUCTIVITY-TEMPERATURE-DEPTH PROBE; REFRACTOMETER; SALINITY; TEMPERATURE.

Further Reading

Kjerfve, Bjorn, J. "Measurement and Analysis of Water Current, Temperature, Salinity, and Density." In *Estuarine Hydrography and Sedimentation,* edited by K. R. Dyer. Cambridge: Cambridge University Press, 1979.

Pickard, George L., and William J. Emery. *Descriptive Physical Oceanography,* 5th ed. Oxford: Pergamon Press, 1990.

sandbars A sandbar is a linear hummock of sand in the surf zone, oriented generally parallel to the BEACH, and always underwater. Sandbars are on the order of a few feet in width and height, but tens of feet to hundreds of feet in length. They can occur singly or in small groups. If the sandbar extends above water at low tide, it is called a ridge; if there are several ridges, the areas between the ridges are called runnels.

Sandbars play an important role in maintaining the integrity of the beach. During severe storm conditions, when waves are attacking the beach, typically in winter, the waves pounding the shore transport sand off the beach into the surf zone. This sand accumulates to form one or more sandbars. The sandbar or bars help to protect the beach from further attack by the waves. Very severe storms may develop two or three sandbars. The largest waves moving shoreward then break on the outer bar, thus expending most of their energy before striking the beach.

During periods of calm water, typically in summer (excepting HURRICANES), the long, low swells, often called groundswell, transports the sand from the sandbars back to the beach, thus restoring the beach to its former state. Sandbars thus serve the purpose of protective storage of sand during storm conditions, and the beach is able to maintain itself if it has access to an adequate supply of SEDIMENT.

Complex sandbars of curvilinear shape are found at mouths of rivers and inlets. Tidal currents deposit these sandbars. The current is a maximum through the channel or throat of an inlet, so that suspended sediments are carried through the inlet and deposited at the mouth, where current speed is insufficient to retain the sediment in suspension. If the sediments build up on the ocean side of the inlet, they forms an "ebb-tide delta"; if on the LAGOON or bay side, a "flood-tide delta." In regions of small tidal range, called micro tidal regions, the sand builds up and the inlet must be dredged frequently to maintain sufficient water depth for navigation. A well-known example is Oregon Inlet in the Outer Banks of North Carolina. Deltas form at river mouths by the same type of action. A large delta is found at the mouth of the Mississippi River. Sandbars in rivers and inlets are often hazards to navigation. Vessel masters must be constantly alert to avoid running aground on them.

Unfortunately, much of the beach development of the past century has acted counter to nature's way that beaches protect themselves. Condominiums, JETTIES, GROINS and the like prevent the natural movement of sand, and thus the beaches are denied the nourishment they need to maintain themselves.

There is still much about the formation and movement of sandbars and other features of the beach that COASTAL ENGINEERS do not understand. The U.S. Army Corps of Engineers have supported extensive research at their FIELD RESEARCH FACILITY at Duck, North Carolina, in order to quantify the behavior of moving sand, the development of beach features, and the often ensuing of erosion and loss of much of the beach. Researchers from academia, industry, and government have been deploying state-of-the-art instruments in the surf zone and beyond in a series of intense experiments. These experiments have shown the importance of EDGE WAVES in modulating LONGSHORE CURRENTS and on forming many features of the beach, such as the cusp-shaped beaches found worldwide. Edge waves travel parallel to the beach, and are very long in period, on the order of minutes. They are generally not observable by swimmers. Walter Munk of the SCRIPPS INSTITUTION OF OCEANOGRAPHY first measured them in the 1950s. Only recently has sufficient data been available to begin to realize the importance of these waves in shaping beach features. Edge waves also appear to be related to the formation of the dreaded RIP CURRENTS, by interacting with swell waves.

Fishermen casting in the surf for bluefish, striped bass (rockfish), red drum, and the many other game fish, have long been aware of sandbars, ridges, and runnels. Food tends to accumulate between the bars, and thus to attract game fish. Bars are detected from the breaking waves far offshore and the lighter color of the water over the bars. Dark strips between the bars are the runnels, which provide important clues as to where the fish are located.

Further Reading
Dean, Cornelia. *Against the Tide: The Battle for America's Beaches.* New York: Columbia University Press, 1999.

Savonius rotor current meter A Savonius rotor current meter is designed to minimize the degradation of measurements of horizontal water current velocity due to vertical motion by means of two oppositely-faced half cylinders rotating about an axis (usually vertical) in a hooded enclosure. The vertical motions are often due to wind, WAVES, CURRENTS, and TURBULENCE. The rotors are made of light plastic and suspended in carefully machined bearings to minimize friction. The Savonius rotor current meter was designed to measure velocities as low as .04 knots (2 cm/s). Recording of data can be accomplished by means of small magnets embedded in the rotor and housing assembly that actuate electrical impulses each time the rotor is turned. The impulses provide a time-varying voltage proportional to water current. Photographic means of data recording have also been employed.

Savonius rotors are integral to many types of current meters that are deployed from ships, BUOYS, and fixed platforms. A vane with magnetic COMPASS is usually incorporated into the assembly to measure the direction of the water current. They are also commonly used in ANEMOMETERS. The Savonius rotor is named after Sigurd J. Savonius (1884–1931) of Finland, who designed a vertical shaft wind machine that turns when the wind is blowing from any direction. Savonius windmills are often two half 55-gallon drums fixed to a central shaft. The drums face in opposing directions, and the shaft turns as one drum catches the wind bringing the opposing drum into the flow of the wind. The turning shaft is used to provide ventilation, drive pumps, or grind grain.

These current meters were extremely popular in the 1960s, '70s, and '80s, but are being replaced by solid-state acoustic Doppler current profilers (ADCPs) and electromagnetic current meters.

See also ACOUSTIC DOPPLER CURRENT PROFILER; CURRENT METER; OCEANOGRAPHIC INSTRUMENTS.

scattering Scattering of energy in multiple directions as it passes through a medium such as water or the atmosphere was first described by John W. Strutt (Lord Rayleigh, 1842–1919). Scattering is the unpredictable manner in which energy, such as light, propagates after striking particulates such as dust in the atmosphere or suspended silts in the ocean. Scattering is a function of the WAVELENGTH of electromagnetic energy and the scatterer's size in the medium. The more particulate matter, the more the energy is scattered. Some common scatterers include bubbles, droplets, density interfaces in fluids, defects in crystalline solids, surface roughness, cells in organisms, and even clothing fibers. Today, scientists refer to molecular scattering (gas molecules) as Rayleigh scattering, scattering from larger diameter particulates (smoke and dust) as Mie (non-molecular) scattering, and scattering from water vapor as nonselective scattering. Water tends to appear blue owing to the scattering taking place in the violet, dark blue, and light blue portions of the spectrum. Scattering from particulates that are larger than the wavelength results in Mie scattering, which results in the white light from mist and fog, especially when using bright headlights.

See also APHOTIC ZONE; ATTENUATION; LIGHT; PHOTIC ZONE.

scope Scope describes the ratio of the running length of line and chain that attaches to an anchor (rode) to the depth of water where a buoy or vessel is moored. The scope determines the sweep of a moored vessel or buoy and its holding power. To properly compute the scope for anchoring, first determine the vertical distance from the eyelet on the buoy hull that connects to the mooring line or the bow roller (pulpit) on the bow of the vessel to the seafloor. This distance includes the bow's height above the water (freeboard), and the water depth, which may fluctuate owing to tides, winds, and waves. As a simple example, consider a 29-foot (8.8 m) research boat supporting dive operations to assess the health of Olympia oysters (*Ostreola conchaphila*) in Puget Sound during a calm day. To anchor safely, the captain will utilize a five to one scope. The distance from the bow roller to the water for this boat is four feet (1.2 m), and at high tide the planned anchorage is 20 feet (6.1 m) deep. In picking the anchorage, the captain also considered the vessel's swing in order not to affect dive operations and sampling near the oyster reef. This calculation relates to vessel length and rode length where as follows:

swinging radius (SR) = length of vessel + [(rode)^2 - (depth + freeboard)^2]$^{1/2}$, so

SR = 29 feet + [(165 feet)^2 - (20 feet + 4 feet)^2]$^{1/2}$ = 192 feet (58.5 m)

The swinging radius will always be less than the length of the rode and the overall length of the boat. The rode was determined by adding the freeboard and depth to get 33 feet (10.1 m) and multiplying by five to get a total length of 165 feet (50.3 m). If the winds started to increase, the captain could improve

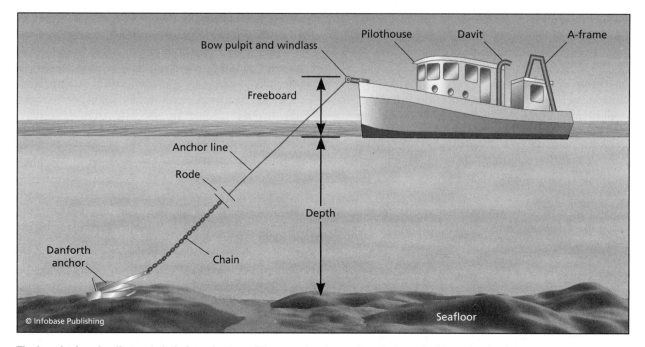

The length of anchor line and chain from the bow of the vessel to the seafloor is the rode. The ratio of rode to depth is the scope.

holding power by increasing the scope to seven to one. The correct scope will compensate for horizontal and vertical forces and reduce loading on the anchor. Scopes for marine equipment such as traps, pots, and instrumented buoys vary in accordance to WAVES, TIDES, and CURRENTS. For example, Tropical Atmosphere Ocean (TAO) surface buoys extending approximately 16 feet (4.9 m) above the sea surface have been moored under a great deal of tension with heavy sinkers, steel cables, and subsurface floats at depths between 820 to 3,281 fathoms (1,500 to 6,000 m) with a short scope (ratio of mooring length to water depth) of 0.985 in order to keep the donut-shaped (toroid) buoy nearly vertical. Data from these buoys have been used to study climate variability and air-sea interactions.

See also ANCHOR; BUOYS; MOORING.

Further Reading
National Data Buoy Center, National Oceanic and Atmospheric Administration. Available online. URL: http://www.ndbc.noaa.gov. Accessed October 31, 2006.

Scripps Institution of Oceanography (SIO)
Scripps Institution of Oceanography (SIO) is one of the oldest and most prestigious centers of oceanographic research in the world. SIO embraces all the marine sciences and engineering to study the biology, chemistry, geology, physics, and climatology of the oceans of the Earth. It provides leadership in education, with training for students in the ocean sci-

ences from the technician to postdoctoral level, as well as maintaining many public service ocean education programs.

Scripps was founded in San Diego, California, in 1903 by University of California zoology professor William Emerson Ritter (1856–1944), his students, and a farsighted group of West Coast community leaders. The institution was endowed by philanthropist Ellen B. Scripps (1836–1932) and businessman Edward W. Scripps (1854–1926). In 1912, Scripps became a part of the University of California and expanded its focus and research to become the nation's first multidisciplinary marine institute. It was given its present name in 1925. Today, more than 300 research programs are in progress.

The mission of Scripps is "To seek, teach, and communicate scientific understanding of the oceans, atmosphere, Earth, and other planets for the benefit of society and the environment." Scripps vision is "To be an international leader in originating basic research, in developing scientists, and in advancing the science needed in the search for a sustainable balance between the natural environment and human activity." To this end, Scripps is staffed with an impressive group of leading marine scientists and engineers of the world.

Scripps maintains a fleet of four first-class oceanographic research vessels and the platform FLIP, which is able to maintain a vertical orientation at the sea surface to enable studies of air-sea interaction and near surface ocean dynamics. The fleet roams the world's oceans in the solution of the many varied problems undertaken by the institution. They are home-ported in San Diego at Point Loma.

In addition to its research vessels, Scripps maintains a research pier, a hydraulics laboratory, with wind and wave channels, an analytical laboratory, a visualization chamber, a library that is the University of California's principal collection of ocean specimens, and world-class supercomputer facilities.

Over the years, Scripps's scientists and engineers have made major contributions to marine science and technology that have revolutionized the study of the oceans, the overlying atmosphere, air-sea interactions, and global climatology.

Further Reading
Scripps Institution of Oceanography. Available online. URL: sio.ucsd.edu/. Accessed October 16, 2007.

scuba The acronym scuba stands for *s*elf-*c*ontained *u*nderwater *b*reathing *a*pparatus and has become an ordinary word describing underwater activities requiring gear such as masks, snorkels, fins, high-pressure air tanks, demand regulators, buoyancy compensators, weighted belts, dive boots, knives, and gloves. The buoyancy compensator or BC is often built onto the tank harness like a vest and is connected to the air supply. The BC is a bladder that is used to keep the diver neutrally buoyant. Air is released or expelled from the bladder to control the overall buoyancy of the diver and his or her dive equipment. This important piece of safety equipment also allows the diver to control ascent and descent. Air tanks are made of steel or aluminum and are designed to operate safely at pressures ranging from 2,250 to 3,500 pounds per square inch (15.51 to 24.13 Mega N/m^2). As a means of comparison, air pressure at sea level is only about 14.7 pounds per square inch (101,352.93 N/m^2). The demand regulator provides oxygen at ambient pressure to the diver when he or she inhales. Pressure from the air tank is reduced through pressure regulating valves and a hose. One valve (first stage) is attached at the top of the tank, and another valve is part of the mouthpiece (second stage), which is at the free end of the air hose. Marine scientists often study shallow-water environments, using scuba diving equipment and may employ more technical means such as mixed gas, helmeted, or saturation diving when conducting research in deep waters. Most scientific diving programs require current certifications, membership in a diving association, continuing education, and regular equipment inspections.

See also DIVE EQUIPMENT; DIVER.

seafloor spreading Seafloor spreading involves the production of new oceanic crust by the separation of lithospheric plates along the mid-ocean ridges. Magma (molten rock under the surface) wells up out of the rift found at the ridgeline and spreads out in two directions normal to the ridge axis. The process functions something like two conveyor belts, with the belts describing new crust moving slowly (on the order of inches per year) away from the axis of the ridge. Evidence for this process includes the increasing depth of oceanic SEDIMENT moving away from the rift toward the continents, the age of the rock floor increasing with distance from the rift, and magnetic polarity of the rock base in reversing strips corresponding with reversals of the Earth's magnetic poles throughout the geologic ages. Recently, using sophisticated techniques of very long baseline INTERFEROMETRY, geophysicists have confirmed that the seafloor and the associated tectonic plates are moving with velocities on the order of those predicted by plate tectonics theory, that is, on the order of inches per year.

The discovery of seafloor spreading in the 1960s was the beginning of the acceptance of the idea of continental drift, as well as providing the explanation of how this motion takes place in the developing theory of PLATE TECTONICS.

sea grass Flowering plants (angiosperms) that use their leaves, underground stems (rhizomes), and roots to obtain nutrients from sediment and water are commonly called sea grass. A sea-grass ECOSYSTEM is an important environment (usually tidal) that includes all the other interrelated nonliving and living things. Sea grasses help stabilize fine SEDIMENTS with their root systems, which branch off from the rhizomes and are important in helping to maintain water quality. They support a considerable biomass and diversity of associated species. Sea grasses provide food for many marine animals (mammals, birds,

Sea grass is an important factor in estuarine health. *(NOAA)*

Sea Grass Species Found Worldwide

Amphibolis Antarctica	Phyllospadix serrulatus
Amphibolis griffithii	Phyllospadix torreyi
Cymodocea angustata	Posidonia angustifolia
Cymodocea nodosa	Posidonia australis
Cymodocea rotundata	Posidonia coriacea
Cymodocea serrulata	Posidonia denhartogii
Enhalus acoroides	Posidonia kirkmanii
Halodule beaudettei	Posidonia oceanica
Halodule. Bermudensis	Posidonia sinuosa
Halodule ciliate	Posidonia ostenfeldi
Halodule emarginata	Potamogeton pectinatus
Halodule pinifolia	Posidonia robertsoniae
Halodule uninervis	Ruppia cirrhosa (spiralis)
Halodule wrightii	Ruppia maritime
Halophila australis	Ruppia megacarpa
Halophila baillonis	Ruppia tuberose
Halophila beccarii	Syringodium filiforme
Halophila capricorni	Syringodium isoetifolium
Halophila decipiens	Thalassia hemprichii
Halophila engelmanni	Thalassia testudinum
Halophila johnsonii	Thalassodendron ciliatum
Halophila hawaiiana	Thalassodendron pachyrhizum
Halophila minor	Zostera asiatica
Halophila ovalis	Zostera caespitosa
Halophila ovata	Zostera capensis
Halophila spinulosa	Zostera capricorni
Halophila stipulacea	Zostera caulescens
Halophila tricostata	Zostera japonica
Heterozostera tasmanica	Zostera marina
Lepilaena cylindrocarpa	Zostera mucronata
Lepilaena marina	Zostera muelleri
Phyllospadix iwatensis	Zostera noltii
Phyllospadix japonicus	Zostera novazelandica
Phyllospadix scouleri	

Note: Sea grasses are submerged vascular plants. There are approximately 60 species of sea grasses found worldwide. Sea-grass beds composed of a handful of species may cover many miles of coastal waters and provide habitat for thousands of species of flora and fauna.

fish, crustaceans, echinoderms) and shelter for others (ZOOPLANKTON, juvenile fish, and soft crabs). Sea grasses are typically found in shallow coastal marine waters in small patchy beds where there is adequate LIGHT, favorable salinities, and sufficient nutrient availability. Light is one of the primary determinants of the maximum depth at which these plants can grow, since a lack of light interferes with their ability to photosynthesize. Other factors affecting sea grass distribution include TURBIDITY, SALINITY, TEMPERATURE, CURRENT and wave action, pollution, and destruction of beds from DREDGING, boat propellers, and anchoring. A list of sea grass species is provided in the table at left.

See also SUBMERGED AQUATIC VEGETATION.

Further Reading

Friends of the San Juans, Friday Harbor. Available online. URL: www.sanjuans.org. Accessed December 23, 2006.

Green, Edmond P., and Frederick T. Short, eds. *World Atlas of Seagrasses.* Berkeley: University of California Press, 2003.

Lippson, Alice Jane, and Robert L. Lippson. *Life in the Chesapeake Bay,* 2nd ed. Baltimore: The Johns Hopkins University Press, 1997.

NOAA Coastal Services Center. *Benthic Habitat Mapping: Online Data.* Available online. URL: www.csc.noaa.gov/crs/bhm/online.html. Accessed December 23, 2006.

NOAA Fisheries, Office of Protected Resources. "Johnson's seagrass (*Halophila johnsonii*)." Available online. URL: www.nmfs.noaa.gov/pr/species/plants/johnsonsseagrass.htm. Accessed December 23, 2006.

United Nations Environment Program, World Conservation Monitoring Center. Available online. URL: www.unep-wcmc.org. Accessed December 23, 2006.

World Seagrass Association. Available online. URL: www.worldseagrass.org. Accessed December 23, 2006.

sea ice Frozen seawater forming a floating boundary between flowing waters and the cooler atmosphere is called sea ice. Sea ice takes different forms as the water begins to freeze and melt with the polar seasons. Even though the ice is formed in saline water, salt molecules are rejected back into the liquid as the ice forms. The sinking of this dense saline water sets up the deep ocean currents.

Some of the many types of sea ice include frazil, brash ice, fast ice, pancake ice, and pack ice. Frazil represents the first stage in sea ice growth and consists of small needlelike ice crystals, which under calm conditions form thin sheets of ice. Accumulations of floating ice not more than 6.6 feet (2 m) across are called brash ice. In shallow water, when the sea ice becomes anchored to the land and does

Sea ice (ice forming on the surface of the ocean when seawater freezes) is a frequent navigation hazard in Alaska. *(NOAA/NOAA Corps)*

that drifts due to winds and currents is pack ice. Some of the more permanent forms of sea ice are found in polar regions such as the Arctic and Antarctic. For example, multiyear ice has survived at least one melt season and is usually at least 6.6 to 13.1 feet (2 to 4 m) thick. New ice forms on the underside of the multiyear ice.

Scientists do not consider icebergs to be sea ice because the icebergs are composed of ice originating from glaciers. For example, icebergs that are tracked in the North Atlantic originate from the snow that formed glaciers in Greenland. Scientists tend to use aerial and satellite imagery, BUOYs with meteorological and oceanographic sensors, and weather stations to study the formation and melting of sea ice. Some features of concern include sea ice thickness, compactness and type, motion, ice-margin phenomena such as EDDIES, and the impact waves on the ice pack. Marine scientists working at ice centers may support ice breaking operations by finding cracks or leads in the sea ice. If located in a favorable position, the ice breaker can open a shipping channel along the lead.

See also GLACIER ICE FORMS; ICEBERG; ICEBREAKER.

not move with winds or currents, it is classified as fast ice. Circular pieces of ice approximately four inches (10 cm) thick and 0.1 to 9.8 feet (0.03 to 3 m) in diameter with raised edges that form as ice rubs together is called pancake ice. Ice in the open ocean

The edge of the ice pack   *(NOAA/NOAA Corps)*

Further Reading

Arctic and Antarctic Research Institute (Russian Federation). Available online. URL: www.aari.nw.ru. Accessed July 25, 2008.

Canadian Ice Service. Available online. URL: www.ice-glaces.ec.gc.ca. Accessed December 15, 2006.

Danish Meteorological Institute. Available online. URL: www.dmi.dk/dmi/index/danmark.htm. Accessed December 15, 2006.

The Finnish Institute of Marine Research. Available online. URL: www.fimr.fi. Accessed July 25, 2008.

National Ice Center. Available online. URL: www.natice.noaa.gov. Accessed December 15, 2006.

National Snow and Ice Data Center. "State of the Cryosphere." Available online. URL: nsidc.org. Accessed December 15, 2006.

Norwegian Meteorological Institute. Available online. URL: met.no. Accessed July 25, 2008.

Swedish Meteorological and Hydrological Institute. Available online. URL: http://www.smhi.se. Accessed December 15, 2006.

sea level Sea level is the vertical level of the ocean after short-term water level fluctuations are averaged out. Sea level does not consider variations in water level due to wind waves. When this elevation is referred to as MEAN SEA LEVEL, it is the average height of the sea surface, based upon at least 19 years of hourly observations of tidal heights on the open coast or in adjacent waters having free access to the ocean. Mean sea level, commonly abbreviated as MSL, and referred to simply as "sea level," serves as a reference surface for all altitudes in upper atmospheric studies. Long-term processes such as changes in the size of polar ice caps, PLATE TECTONICS, climatic variability, and waterway discharge affect sea level by changing the amount of water that is stored in the oceans.

See also DATUM; TIDES; WATER LEVEL FLUCTUATIONS; WAVES.

seamounts Seamounts are isolated underwater mountains, usually volcanoes, generally circular in plan view, that rise more than one mile (2 km) above the floor of the ocean. Small seamounts tend to be rather sharply peaked; large seamounts have gently sloping sides. Seamounts that rise above the ocean surface form islands, or ATOLLS. An atoll is a ring-shaped island protected by CORAL REEFS, and containing a LAGOON.

Although found worldwide, seamounts are most abundant in the Pacific Ocean, especially near transform fault zones, such as mid-ocean ridge spreading zones and SUBDUCTION ZONES. There are some linear groupings of seamounts that were probably formed by hot spots under the ocean floor. Hot spots are zones of vertical convection from the mantle to the crust and may be located anywhere under a tectonic plate. Most geologists believe that the Hawaiian Islands and the Emperor Seamounts, extending northwest from Hawaii, were formed by the motion of the Pacific Plate over a hot spot. They note that the older extreme northwest islands, such as Midway, are much less geologically active than the main Hawaiian Islands, which are now passing over the hot spot. Evidence for this is found in the form of Kilauea Volcano, which is at present very active, with lava pouring into the Pacific Ocean. The Loihi Seamount, now below sea level, appears to be in the process of island formation and is being extensively studied by geologists and geophysicists.

Seamounts are a deep-ocean feature; islands near continents, such as Cuba and the United Kingdom are usually made of continental material, mostly granite. Although hundreds of seamounts have been located, most are probably still undiscovered.

Some seamounts have been found with coral tops that are below the level of coral growth. These seamounts are believed to have subsided into the ocean floor from their own weight.

Atolls can provide convenient places for anchoring small vessels. However, during storms, waves break on the windward side of the surrounding reefs, since their walls rise abruptly from the seafloor, making the transition from deep to shallow water in a very short horizontal distance. Negotiating passages through the reefs can be very dangerous, due to rapid shoaling and focusing of wave trains by the topography. Very large waves can also occur on the leeward side of an atoll from recombining wave trains that have been bent around the reefs.

Oceanic islands also affect the local atmospheric circulation, since convective cells develop during the day, due to the more rapid warming of the land compared to the water, producing cumulus clouds that have often been used by mariners to locate Pacific islands. The islands furnish homes for birds, reptiles, and small mammals, and are the centers of marine ECOSYSTEMS.

See also CONVECTION; HOT SPOT; PLATE TECTONICS.

Sea of Japan The Sea of Japan is a marginal sea of the Pacific Ocean; it is surrounded by North and South Korea and Russia to the west and Japan (Hokkaido, Honshu, and Kyushu Islands) to the east. It is primarily connected to the Sea of Okhotsk through the Strait of Tartary (also Mamiya Strait and Strait of Nevelskoi) and La Pérouse Strait (also Soya Strait)

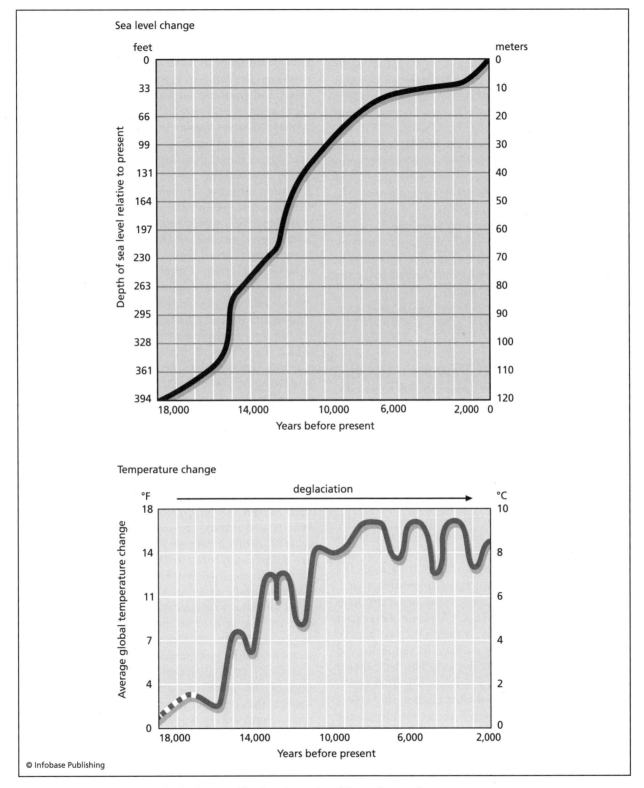

Sea-level change in recent geologic time resulting from ice melt and thermal expansion

to the north and the East China Sea through the Korea Strait to the south. Straits are narrow channels of water connecting two larger bodies of water. The warm Tsushima Current flows north into the Sea of Japan through the Korea Strait along the west coast of Kyushu and Honshu Islands. The Tsushima Current enters the Sea of Okhotsk as the Soya Warm Current. The Sea of Japan has a surface area of about 377,600 mile^2 (978,000 km^2) and a mean depth of 5,748 feet (1,752 m). The Sea of Japan has complicated BATHYMETRY that includes the Japan Basin in the north, the Yamato Basin in the southeast, and the Tsushima Basin in the southwest. The maximum depth in the Japan Basin is 12,276 feet (3,742 m). The Sea of Japan is located on the Eurasian plate with seismically active miner plates such as the Amurian and Okhotsk plates. Tsunamis generated in the Sea of Japan can be especially devastating since the earthquake source lies so close to the coast. A tsunami is a series of very long wavelength ocean waves generated by the sudden displacement of water by earthquakes, landslides, or submarine slumps. On some charts, the Sea of Japan is referred to as the "East Sea" or "East Sea of Korea."

See also KUROSHIO CURRENT; MARGINAL SEA; PACIFIC OCEAN; TSUNAMI.

Further Reading

Japan Meteorological Agency. Available online. URL: http://www.jma.go.jp/jma. Accessed December 15, 2006.

The National Institute of Advanced Industrial Science and Technology, Geological Survey of Japan. Available online. URL: http://www.gsj.jp. Accessed December 15, 2006.

Search and Rescue (SAR)   In marine science, the application of oceanography and marine meteorology to locate those lost and rescue those in distress supports emergency operations called Search and Rescue. Maritime law and tradition require the master of a vessel to assist those in trouble at sea. First responders in marine environments are usually Coast Guardsmen, the Air National Guard, or marine police who train to minimize the loss of life, injury, and property damage by rendering aid to persons in distress on the sea. The U.S. Coast Guard maintains Search and Rescue or SAR facilities on the East, West and Gulf Coasts; in Alaska, Hawaii, Guam, and Puerto Rico, as well as on the Great Lakes and inland U.S. waterways. The U.S. Coast Guard has had the mission of Search and Rescue since 1831, and is the world leader in the development and improvement of SAR methodologies. Other, on-the-scene vessels such as naval ships may assist rescues. Civilian boaters who are members of the U.S. Coast Guard Auxiliary assist both the Coast Guard and local police, often in inland waters where there are no major Coast Guard vessels.

Satellites play a large role in Search and Rescue. Satellites in low-Earth and geostationary orbits are used to detect and locate aviators, mariners, and land-based users in distress. The satellites relay distress signals from emergency beacons to a network of ground stations and ultimately to the NATIONAL OCEANIC AND ATMOSPHERIC ADMINISTRATION's Mission Control Center in Suitland, Maryland. The Mission Control Center processes the distress signal and passes initial data from a registration database to the appropriate Search and Rescue authorities. Example information includes caller identification, the nature of the distress, information on winds, sea state, currents, water and air temperature, the distress signal, emergency contact information, vessel identifying characteristics, and location. Signals are transmitted from digital 406 MHz Emergency Position-Indicating Radio Beacons (EPIRBs), which are automatically activated radios designed to aid in Search and Rescue efforts.

An important aspect in SAR is rapid response, which is essential to those in distress. Whatever the methods used in SAR, the first step in the operation is developing the search plan. The plan must have the following elements: determination of the most probable location of the search object, determination of the size of the search area, selection of a search pattern, and coordination of available units. The first responders immediately begin determining atmospheric conditions such as wind speed, wind direction, wind chill, and oceanographic conditions such as SEA STATE, CURRENTS, and depths. Atmospheric conditions are determined from airborne sensors, dropwindsondes, and instrumented buoys. Sea surface currents are determined with numerical models, air-deployed satellite-tracked drifting BUOYS, and potentially other in situ measurements such as from an OCEAN OBSERVATORY. Computer technologies are also utilized in the planning and execution of rescue operations. Example technologies are trajectory models, geographical information systems, and the GLOBAL POSITIONING SYSTEM. If the location of the vessel or survivors in distress is not known, the computer programs output probabilities of locations, taking into account the uncertainties in the input information. The initial locations for search are updated using wind and current data, as well as the results of previous searches in the area. The computer procedures optimize the search consistent with conducting a thorough rescue effort while at the same time providing for the most rapid search possible.

When over the search area, thermal infrared cameras may be used to find survivors in the water during both day and evening searches. Infrared is particularly valuable since humans radiate more heat than the surrounding water. Improved understanding and prediction of winds, seas, and currents, along with improving location, communication and rescue technologies will significantly improve the safety and reliability of maritime transport and business at sea.

Further Reading

Association for Rescue at Sea. Available online. URL: http://www.afras.org. Accessed January 22, 2007.

Bishop, Joseph M. *Applied Oceanography*. New York: John Wiley, 1984.

International Satellite System for Search and Rescue. Available online. URL: www.cospas-sarsat.org. Accessed October 16, 2007.

National Association for Search and Rescue. Available online. URL: www.nasar.org/nasar. Accessed January 22, 2007.

National Search and Rescue Committee, U.S. Coast Guard. Available online. URL: www.uscg.mil/hq/g-o/g-opr/nsarc/nsarc.htm. Accessed January 22, 2007.

Search and Rescue Satellite Aided Tracking. NOAA Satellite Operations Facility, National Oceanic and Atmospheric Administration. Available online. URL: www.sarsat.noaa.gov. Accessed January 22, 2007.

U.S. Coast Guard Headquarters. "Rescue 21." Available online. URL: www.uscg.mil/rescue21. Accessed January 22, 2007.

U.S. Coast Guard Search and Rescue home page. Available online. URL: www.uscg.mil/hq. Accessed October 16, 2007.

sea state   Dominant sea surface characteristics such as wave height, wave direction, and wave period are called sea state. Scales have been developed to describe sea states or the properties of wind-generated waves on the surface of the ocean. The term *sea* actually refers to locally generated waves produced by the wind and may be described by wave periods and heights. Sea waves are short-crested and irregular waves. Surface waves and currents are generated by winds (*see* WAVES). Under marine conditions with breezes and gales, energy is lost by the waves as a result of white-capping. Swells are waves that were generated in a storm area and have propagated out of the generating area. They have longer WAVELENGTHS than the local wind-driven waves.

There are various methods to measure waves, including wave staffs, free-floating BUOYS, pressure gauges, acoustic probes, ground-, aircraft-, and satellite-based RADARS and LIDARS. The computation of sea state is a derived parameter extremely important to marine forecasters, mariners, OCEAN ENGINEERS, and scientists.

Waves are the dominant forcing mechanism in the ocean and must be understood in order to design ships, coastal structures, and to assess processes such as BEACH erosion. Observers on the deck of a ship, standing on a beach, or on a dock may estimate wave heights, directions, and periods. The observer focuses on well-formed waves and may use references such as the length of a ship to estimate wavelengths. In many operational settings, wave buoys outfitted with accelerometers measure wave parameters. The directional wave buoy records the rise and fall of waves as well as wave directions. Bottom-mounted pressure gauges are also used and may be incorporated into TSUNAMI warning systems. They detect the frequency and size of propagating waves. Waves are also measured from

WMO Code Table 3700 Sea State Codes

Code Figure	Descriptive Terms	Height (in feet)	Height (in meters)
0	Calm (glassy)	0	0
1	Calm (rippled)	0 to .33	0–0.1
2	Smooth (wavelets)	.33 to 1.64	0.1–0.5
3	Slight	1.64 to 4.1	0.5–1.25
4	Moderate	4.1 to 8.2	1.25–2.5
5	Rough	8.2 to 13.12	2.5–4
6	Very rough	13.12 to 19.69	4–6
7	High	19.69 to 29.53	6–9
8	Very high	29.53 to 45.93	9–14
9	Phenomenal	>45.93	>14

space, and wavelengths and directions may be determined by analyzing satellite radar imagery. Imagery analysis systems process electromagnetic signals, which often identify wavelike patterns. For active systems, transmit times of radar and lidar signals from sensor to sea surface and back can be used to estimate waves and currents.

Oceanographers and scientists often characterize wave heights by averaging the highest one-third of all waves occurring during a particular time period and calling this measure significant wave height ($H_{1/3}$). The World Meteorological Organization (WMO) has prescribed the use of Table 3700 (see page 483) for sea state as a standard for the quantification of well-developed wind waves by observers.

The Beaufort Scale was originally developed in 1805 by Sir Francis Beaufort (1774–1857) as a system to estimate wind strengths without the use of instruments. With each Beaufort number there is a description of the state of the sea surface. For example, Beaufort Number 1 is, "Sea like a mirror" and 12 is, "The air filled with foam and spray. Sea completely white with driving spray; visibility very seriously affected." The Beaufort Scale is still widely used and helps to integrate various components of marine weather such as wind strength, sea state, and observable effects into a word picture.

Some scales of wave motion, such as gravity waves, are mathematically correlated to spectra such as the standard Pierson-Moskowitz spectra accounting for the relationship between winds and waves. This wave spectrum represents a fully developed sea, the time when waves reach equilibrium with the wind. Wave energy spectra describe the energy in the sea surface as a function of the frequencies of the component waves that make up the surface. Since the energy in each component wave is determined by the square of the amplitude ($^1/_2$ the wave height), the spectrum gives the distribution of wave heights with frequency. This spectral form for fully developed wind seas was first described by New York University professor Willard J. Pierson (1922–2003) and Lionel Moskowitz in 1964. Their findings were based on measurements of winds and waves from two weather ships in the North Atlantic Ocean. The Neumann spectrum, after Gerhard Neumann (1911–96), enabled the relation between wind speeds and wave energy. This result, along with work by Pierson that modeled the sea surface as a Gaussian random process, made a practical wave forecasting system possible. Both of these results were published in 1952, and, in 1955 compiled as "Observing and Forecasting Ocean Waves by Means of Wave Spectra and Statistics," U.S. Navy Hydrographic Office Pub. No. 603, by Willard J. Pierson, Gerhard Neumann, and Richard W. James of the Hydrographic Office (a predecessor of the U.S. Naval Oceanographic Office). Since that time, many new spectral forms have been developed using both theory

Beaufort Wind Scale

No.	Mean Wind Speed		Descriptive WMO Terms	Probable Wave Height	
	Knots	M/S		Feet	M
0	<1	0.0-0.2	Calm	0	0
1	1-3	0.3-1.5	Light Air	.25	0.1-0.2
2	4-6	1.6-3.3	Light Breeze	.5	0.3-0.5
3	7-10	3.4-5.4	Gentle Breeze	2	0.6-1.0
4	11-16	5.5-7.9	Moderate Breeze	4.92	1.5
5	17-21	8.0-10.7	Fresh Breeze	6.56	2.0
6	22-27	10.8-13.8	Strong Breeze	11.48	3.5
7	28-33	13.9-17.1	Moderate Gale	16.4	5.0
8	34-40	17.2-20.7	Gale	24.6	7.5
9	41-47	20.8-24.4	Strong Gale	31	9.5
10	48-55	24.5-28.4	Whole Gale	39	12.0
11	56-64	28.5-32.7	Storm	49	15.0
12	>64	above 32.7	Hurricane	>49	>15

and measurements and are central to the art and science of forecasting sea state.

See also GRAVITY WAVES; OCEANOGRAPHIC INSTRUMENTS; WAVE AND TIDE GAUGE; WAVE PREDICTION.

Further Reading

Bowditch, Nathaniel. *The American Practical Navigator.* Available online. URL: http://164.214.12.145/pubs/. Bethesda, Md.: National Geospatial-Intelligence Agency, 2002. Accessed December 23, 2006.

Coastal and Hydraulics Laboratory, Engineer Research and Development Center, Waterways Experiment Station, U.S. Army Corps of Engineers. Available online. URL: http://chl.erdc.usace.army.mil. Accessed December 23, 2006.

Coastal Data Information Program. Available online. URL: http://cdip.ucsd.edu. Accessed November 8, 2006.

National Data Buoy Center, National Oceanic and Atmospheric Administration. Available online. URL: http://www.ndbc.noaa.gov. Accessed November 9, 2006.

Neumann, Gerhard, and Willard J. Pierson, Jr. *Principles of Physical Oceanography.* Englewood Cliffs, N.J.: Prentice Hall, 1966.

U.S. Army Corps of Engineers. *Coastal Engineering Manual.* Engineer Manual 1110-2-1100, U.S. Army Corps of Engineers, Washington, D.C. (in 6 volumes), 2002.

Secchi disk The Secchi disk is a simple device used to estimate the diffuse attenuation of light in water. The Secchi disk is a weighted circular disk divided in quadrants and painted alternating white and black. An observer lowers the disk into the ocean or lake with the disk parallel to the water's surface. The depth at which the Secchi disk disappears is called the Secchi depth, or depth of visibility, and a measure of water transparency. This measurement is important for estimating the depth of the EUPHOTIC ZONE (light zone), which is of great importance to photosynthetic primary production. Since the Secchi depth depends on the size of the disk, the diameter has been standardized at 11.8 inches (30 cm).

The Secchi disk is probably the simplest of all oceanographic measuring devices and has been used for more than 100 years. Father Pietro Angelo Secchi (1818–78), who was a scientific adviser to the pope in 1865, developed this instrument to determine water clarity in the MEDITERRANEAN SEA. Even though it is still a widely used device, the quality of the data is always in question. An individual measurement is not very meaningful, but large numbers of measurements yield useful statistics. How deep it is when it can no longer be seen is a function not only of the clarity of the water but the sky, the observer, and the state of the ocean surface.

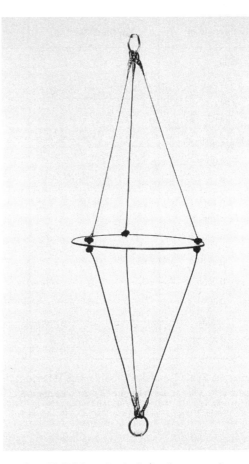

The Secchi disk is a simple device for measuring water clarity. It is lowered from the surface until it is not visible and the vanishing depth is recorded as the Secchi depth. *(NOAA)*

There are many applications that depend on knowing how much light is penetrating the water column. Scientists use Secchi depth datasets in combination with space-based irradiance data to develop algorithms that allow the Secchi depth to be calculated over spatially extensive areas from satellites such as SeaWiFS. Because of their low cost and easy application, Secchi disks are widely employed by volunteer workers in water-quality studies.

sediment Unconsolidated grains of inorganic and organic material that is suspended in and being transported by air, water, or ice and is deposited onto the seafloor is called sediment. After storms or snow melt, lithogenous sediments (topsoil, sand, and minerals) may be washed from the land into rivers, ESTUARIES, and the coastal ocean. Sedimentary rocks form from the accumulation and lithification of sediment. As sediment particles accumulate, they compact, become cemented together, and lithify or harden. The cementing agents that form solid rock conglomerate are silica, calcite, or iron oxides. Sedimentary rocks include

Millimeters (mm)	Micrometers (μm)	Wentworth size class		Rock type
4096		Boulder		Conglomerate/Breccia
256		Cobble	Gravel	
64		Pebble		
4		Granule		
2.00				
1.00		Very coarse sand		
1/2 0.50	500	Coarse sand		
1/4 0.25	250	Medium sand	Sand	Sandstone
1/8 0.125	125	Fine sand		
1/16 0.0625	63	Very fine sand		
1/32 0.031	31	Coarse silt		
1/64 0.0156	15.6	Medium silt	Silt	Siltsone
1/128 0.0078	7.8	Fine silt		
1/256 0.0039	3.9	Very fine silt		
0.00006	0.06	Clay	Mud	Claystone

Udden-Wentworth Scale grain-size classification scheme. This scale was first proposed in 1898 by Udden and then adopted in 1922 by Wentworth. This scale is frequently used by marine scientists and sedimentologists.

conglomerates such as sandstones, siltstones, and shales. Sandstone is composed of silicate grains such as quartz, feldspar, chert, and mica. Siltstone is composed of silt-sized grains that are finer than sand but coarser than clay. Shale is derived from mud or clay minerals. Rough, angular pieces of broken-up rocks may also consolidate without being smoothed by the action of CURRENTS, WAVES, and glaciers into what is called breccia. The cementing agents that fill the spaces to form the solid rock are silica, calcite ($CaCO_3$), and iron oxides. Hydrogenous sediments may precipitate from the overlying water or form chemically, such as manganese nodules. Other forms are phosphates, carbonates, and metal sulfides. Salts such as halite, gypsum, and calcite, which precipitate out of the water, are called evaporites. These sediments form very slowly from chemical reactions that take place in the ocean. Cosmogenous sediments such as iron spherules or silicate condrules are derived from outer space. Organic or biogenous sediments are composed of calcium carbonate ($CaCO_3$) tests secreted by PLANKTON such as foraminifers and COCCOLITHOPHORES, as well as pteropods and silica (SiO_2) tests secreted by diatoms and radiolarians. Pelagic sediments consisting of more than 30 percent shell fragments of microscopic organisms are called oozes such as calcareous or siliceous ooze.

Sediments may be classified by the Udden-Wentworth scale, which categorizes material according to grain size such as clay, silt, sand, gravel, cobble, and boulders. The geometric scale of grain sizes was named after geologists Johan August Udden (1859–1932) and Chester K. Wentworth (1891–1969). Grains are often analyzed with sieves and hydrometers. Sieves are mesh screens with openings of known sizes used to measure the maximum diameter of a sediment grain. Hydrometers measure the suspension density at a specific depth at different times to determine the amount of sand, silt, and clay in a sample.

The process of sedimentation is very important to be controlled and monitored. Too much sedimentation near the coast may keep sunlight from reaching SUBMERGED AQUATIC VEGETATION and smother reefs composed of shellfish or corals. Sediment plumes are easily tracked with aerial photographs and satellite imagery.

See also BEACH COMPOSITION; MANGANESE NODULES; OOZE.

Further Reading

Coastal and Hydraulics Laboratory, Engineer Research and Development Center, U.S. Army Corps of Engineers. Available online. URL: http://chl.erdc.usace.army.mil. Accessed December 18, 2006.

Divins, D. L. "NGDC Total Sediment Thickness of the World's Oceans & Marginal Seas." Available online. URL: http://www.ngdc.noaa.gov/mgg/sedthick/sedthick.html. Accessed October 16, 2007.

Global Environment Monitoring System, United Nations Environment Program. Available online. URL: http://www.gemswater.org. Accessed December 18, 2006.

International Hydrological Program, United Nations Educational, Scientific, and Cultural Organization. Available online. URL: http://typo38.unesco.org/index.php?id=240. Accessed December 18, 2006.

U.S. Environmental Protection Agency. "Water Science." Available online. URL: http://www.epa.gov/waterscience. Accessed December 18, 2006.

U.S. Geological Survey. "Suspended Sediment Database." Available online. URL: http://co.water.usgs.gov/sediment. Accessed December 18, 2006.

———. "Water Resources of the United States." Available online. URL: http://water.usgs.gov. Accessed December 18, 2006.

sediment transport Bottom-lying materials may be slid, rolled, hopped, or suspended off the seafloor by forces such as ice floes, gravity, winds, waves, and running water (streams, rivers, and currents) in a process called sediment transport. Sediment transport begins in the benthic boundary layer, where one can observe both laminar and turbulent flows. The boundary layer is the fluid layer near the seafloor. At the seafloor, fluid velocity must be zero, and the boundary layer is a thin film that depends on surface texture, the fluid's velocity profile, and fluid mass properties such as SALINITY. Laminar flows are smooth and orderly where adjacent layers or laminas slip past each other inducing little mixing. In turbulent flows, fluid particles move along irregular paths, which induce mixing and the transfer of momentum from one portion of the fluid to another. Turbulent flow is responsible for erosion and entrainment of the sediment from the seafloor and sustained movement of sediment above the seafloor. The process of lifting grains from rest from the seafloor and putting them into motion is called entrainment. Grains are deposited or eroded in accordance with the magnitude of the flow. In general, stronger flow moves larger grains.

The classic Hjulström diagram depicts the relationship of fluid flow, transport, deposition, erosion, and grain size. Filip Hjulström (1902–82) published his research findings on the morphological activity of rivers during 1935. His investigations

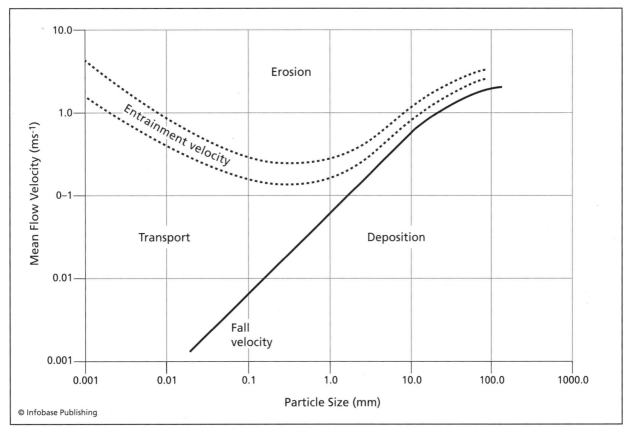

Hjulström diagram. The graph takes sediment size and channel velocity into account. The x-axis shows the size of particles in millimeters. The y-axis shows the velocity of the river in centimeters per second. The three curves on the diagram show when different-size particles will be deposited, transported, or eroded.

at Uppsala University in Sweden focused on the river Fyris. This river, which runs through Uppsala, was a port prior to silting over and remains a natural laboratory for hydrological studies. The Hjulström diagram illustrates how grain size and flow strength determine how grains are transported. Grains around 0.039 inches (1 mm) in diameter require the least energy to erode, as they are sands that do not coagulate. Particles smaller than these fine sands are often clays that require a higher velocity to produce the energy required to separate small, clay particles that have coagulated. Larger particles such as pebbles are eroded at higher velocities, and very large objects such as cobble and boulders require the highest velocities to erode. When the velocity drops below the critical velocity, particles will be deposited or transported, instead of being eroded (depending on the current's energy and particle size).

Sediment transport is an ongoing process in the marine environment. As waves move into shallow water and begin to feel bottom, they impact the seafloor. They also contribute to the formation of rip currents, EDGE WAVES, and shear waves. Longshore currents cause the alongshore transport of sand known as longshore drift. Rip currents are responsible for the cross-shore sediment transport. Weather conditions contribute to erosion and accretion of sand and the to-and-fro motion of seasonal sandbars. Further offshore near the shelf break, sand deposits may become unstable and slide off the shelf, cutting submarine canyons. The ensuing TURBIDITY CURRENTS (underwater landslides) are an example of sediment transport under the influence of gravity. These powerful, short-lived, currents consist of dilute mixtures of sediment and water having a density greater than the surrounding water. They were first described after investigations of submarine cable breaks following an earthquake at Grand Banks, off Nova Scotia, Canada, on November 19, 1929.

Knowledge of sediment erosion, transportation, and deposition is very important for economic reasons, such as safeguarding navigation, maintaining channel depths, pollution mitigation, and planning ocean construction projects. Sedimentary features are spectacular along erosional coasts such as the White Cliffs of Dover, which are composed of coccoliths, and sea arches and stacks along the coast of Iceland. Such features form in response to the composition of coastal bedrock and waves, tides, and currents. Depositional features include spits, sandbars, and sand dunes. One classic example is the extensive barrier island system in North Carolina known as the Outer Banks. Barrier islands are long

offshore deposits of sand lying parallel to the coast. Sediments may also be commercially important such as sands mined from the GULF OF MEXICO or diatomaceous earth composed of silica tests (from the diatoms) that are commonly used in filters, abrasives, and absorbents.

See also BARRIER ISLAND; CURRENT; LONGSHORE CURRENT; RIP CURRENT; SEDIMENT.

Further Reading
Coastal and Hydraulics Laboratory, Engineer Research and Development Center, U.S. Army Corps of Engineers. Available online. URL: www.chl.erdc.usace.army.mil. Accessed December 18, 2006.

Minerals Management Service. Available online. URL: www.mms.gov. Accessed December 18, 2006.

Sediment Transport Instrumentation Facility, Woods Hole Science Center, U.S. Geological Survey. Available online. URL: www.woodshole.er.usgs.gov/operations/stg. Accessed December 18, 2006.

seiche A seiche is a standing wave in an enclosed or semi-enclosed body of water, such as a lake, ESTUARY, bay, or seaport, which results in the rise and fall of water. In a standing wave, there is no net propagation of energy. Seiches can be set up by a steady wind that piles up water at one end of the basin, and then stops blowing, allowing the water to slosh back and forth much as disturbed water in a bathtub, until the energy of the disturbance is dissipated by friction. Coastal seiches can also be induced by long-period waves coming in from the open ocean.

The seiche in a closed basin can be explained physically as the result of the interference of two progressive waves, yielding a standing wave of period as follows:

$$T = \lambda/v = 2\,l\,/(gh)^{1/2}$$

where T is the period of the seiche, λ is the wavelength, v is the wave phase speed, l is the length of the basin, g is the acceleration due to gravity (32 feet/s^2 or 9.8 m/s^2), and h is the water depth, since by wave theory (*see* WAVES), the WAVELENGTH of the fundamental (first normal mode) oscillation is twice the length of the basin, or $2 \cdot l$, and, if the wave is in shallow water, then $v = (gh)^{1/2}$ (the speed of shallow water waves). The first normal mode is an oscillation in which all particles are moving with the same frequency and phase. In the case of water waves in the seiche, the nodes correspond to horizontal motion, with zero vertical motion; the antinodes correspond to vertical motion with zero horizontal motion. If the embayment is open to the sea, however, the corresponding period is

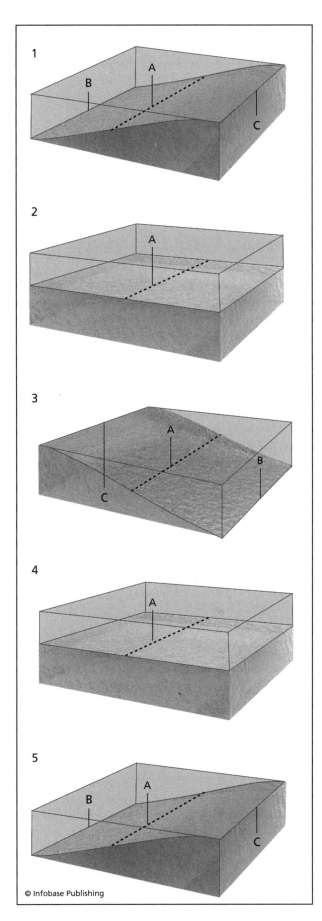

Seiches are standing waves in a lake, harbor, or estuary that oscillate about a point called the node (point A in the diagram). At the node, the seiche moves horizontally but not vertically. At the antinode, a crest or trough of a seiche (B and C in the diagram), the water moves vertically but not horizontally.

$4l/(gh)^{1/2}$, where the speed is the same as for the basin closed at both ends, but the wavelength is four times the length of the basin. The seiche has a node at the mouth of the basin and an antinode at the head. This situation corresponds to vibrating air in an organ pipe open at one end, with a node at the closed end, and an antinode at the open end. The water level oscillations will be large if the incoming waves oscillate at nearly the resonant frequency of the basin.

In order to predict times and water levels of incoming seiches, this simple theory is not sufficiently accurate, and it is necessary to take into account the exact topography and BATHYMETRY of the basin as input to a numerical model, as well as the speed and duration of the wind, the incoming wave characteristics, and so forth. A number of successful seiche models were developed during the latter part of the 20th century for application to the Atlantic, Pacific, and Gulf Coasts of North America, as well as the Great Lakes.

Seiches can have a positive affect on the environment by aerating and mixing the water, and providing transport to the larvae of fish or shellfish. However, powerful seiches with large wave heights can also cause destruction of boats anchored in restricted harbors. The wharves of a port facility must be engineered to withstand flooding caused by seiches and waves from severe weather.

Further Reading

Wright, John, Angela Colling, and Dave Park. *Waves, Tides, and Shallow-Water Processes.* Oxford: Butterworth/Heinemann, 2002.

seismic waves Seismic waves are waves in the solid Earth that originate from movements of the crust along fault lines, SUBDUCTION ZONES, and in all locations in which stresses can build up and then be suddenly released. The sudden release of the stress often results in an earthquake in which seismic waves propagate outward from a focus (the locus of the beginning of the quake) along the surface of the ground, and also within the volume of the Earth. The location on the surface of the Earth above the focus is called the epicenter.

Seismic waves can be detected by seismometers or seismographs, instruments that are very sensitive to small displacements of the Earth and associated forces. Seismometers are deployed in great numbers

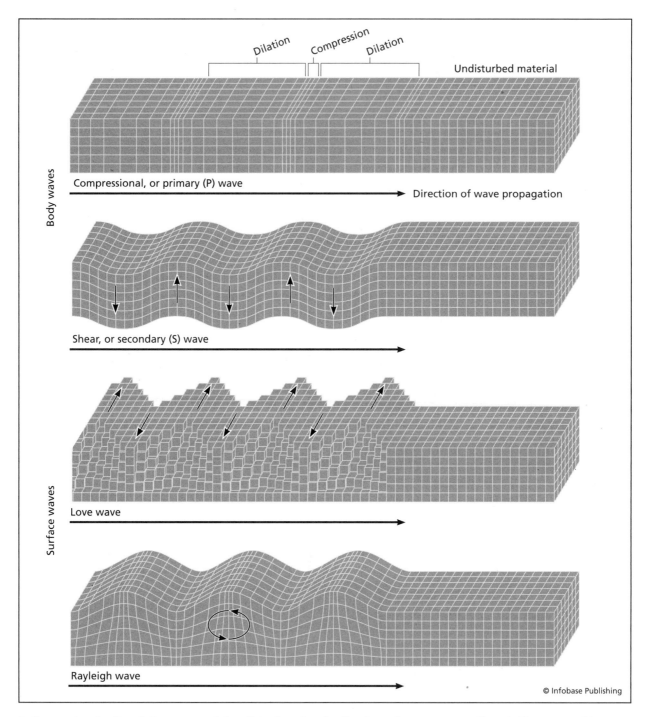

Dilation Compression Dilation

Undisturbed material

Compressional, or primary (P) wave

Direction of wave propagation

Shear, or secondary (S) wave

Love wave

Rayleigh wave

Body waves

Surface waves

© Infobase Publishing

A diagram showing the relative movement of particles in rock and soil as the various waves pass through. They are roughly divided into body waves, which are fast and travel throughout the volume of the Earth, and surface waves, which are slower and travel only along the Earth's surface.

in the industrialized countries of the world, and in increasing numbers at locations in which earthquakes are frequent. The first waves to reach the seismometer are the "P," or primary waves. These are the fast compressional waves (waves in which the displacement is longitudinal, *see* WAVES) that travel at speeds on the order of about four miles per second

(6 km/s). The next group of waves to reach the seismometer are called "S" waves, for secondary waves. These are shear waves, in which the particle motions associated with the wave are at right angles to the propagation of the wave, much as waves on a rope. These waves travel more slowly than the P waves, at about two miles per second (3.5 km/s). Earthquakes

emit P and S waves; the P waves damage structures by suddenly changing their volumes; S waves damage structures by a back-and-forth shaking. If the seismic waves resonate with the normal modes (natural frequencies of vibration) of the structure, great damage can be done.

Oceanographers are generally most interested in earthquakes that occur under the sea, which can produce seismic sea waves called TSUNAMIS. These terrifying waves are produced when a fault line breaks, causing blocks of oceanic crust to rise or sink, thus suddenly displacing a huge amount of water and generating a seismic sea wave. Tsunamis can also be generated by volcano eruptions and TURBIDITY CURRENTS, in which a great volume of solid material suddenly breaks loose and displaces a similar volume of water, creating a seismic wave.

Seismic waves are not always destructive. Many thousands of small earthquakes occur that do little or no damage. The signals from these events are studied by geophysicists who use them to chart the structure of the subterranean Earth. By this means, the present model of the Earth's interior has been developed, which shows a thin crust overlying a plastic mantle, with a liquid core of iron-nickel, and in the very center, a solid core of iron nickel (solid because the enormous pressure squeezes the atoms together). The development of this model was made possible because S waves do not travel through liquid, whereas P waves do. Seismometers on the side of the Earth opposite the earthquake detect P waves but no S waves, indicating that there is liquid near the center of the earth.

Not all undersea earthquakes produce tsunamis. A system of pressure recorders is deployed on the floor of the PACIFIC OCEAN to detect earthquakes and consequent tsunamis, as part of the international Tsunami Warning System (TWS), with 26 participating member states. A similar system would have warned Southeast Asian coastal dwellers of the onset of the 2004 INDIAN OCEAN tsunami (known officially as the Sumatra-Andaman earthquake) that resulted in hundreds of thousands of casualties.

Further Reading

Department of Earth and Space Science, University of Washington. "The Physics of Tsunamis." Available online. URL: www.geophys.washington.edu/tsunami/general. Accessed October 16, 2007.

Erickson, Jon. *Marine Geology: Exploring the New Frontiers of the Ocean.* New York: Facts On File, 2003.

Gates, Alexander E., Ph.D., and David Ritchie. *Encyclopedia of Earthquakes and Volcanoes,* 3rd ed. New York: Facts On File, 2007.

Kusky, Timothy. *Encyclopedia of Earth Science.* New York: Facts On File, 2005.

National Oceanic and Atmospheric Administration. "Tsunamis." Available online. URL: www.noaa.gov/tsunamis. Accessed July 15, 2007.

settlement The process by which organisms and suspended materials reach resting places in or near the seafloor is called settlement. Larval forms of many species (especially marine invertebrates) may fix themselves to hard substrates such as pilings, shell material, or the seafloor itself prior to metamorphosing into an adult. The settlement of inorganic matter such as sediments is also important to pelagic and benthic habitats. For example, sediments settle to the seafloor when the energy of flowing water is unable to support the load of suspended material.

Biotic settlement is an area of active marine research since the process may follow complex biological triggers, or settlement may occur as larvae become attracted to members of their own species. Tracking the movement of marine organisms from where they were born to their ultimate destination covers different MARINE PROVINCES such as the CONTINENTAL SHELF and coastal ESTUARIES or surface waters to the seafloor. For reef-building corals, biologists have shown that species such as *Acropora millepora, Pocillopora meandrina,* and *Montipora capitata* are inhibited from settling by algal tufts and sediments. These social, sessile hard corals spawn approximately once per year. After mass spawning, bundles of eggs and sperm are released into the water column, where millions of eggs are fertilized near the surface. The fertilized eggs enter a brief larval stage where researchers are trying to identify cues associated with settlement and metamorphosis of the coral larvae. The coral larvae sink and settle in favorable habitats such as rocky areas or hard substrates to begin the creation of new coral colonies. Survival for coral larvae is low due to physical factors such as poor water quality and unfavorable currents and biological factors such as predator abundance. The larval stages of many REEF FISH and scores of invertebrates are known to involve a dispersal phase followed by the selection of a suitable reef environment in which to settle. Larval finfish such as the panda clownfish (*Amphiprion polymnus*) are known to settle less than 600 feet (182.9 m) away from their parents, while other marine species may undergo very long active or passive migrations until settling in to their permanent habitats. Organisms may use swimming abilities and sensory mechanisms (chemical, sound, and visual cues) to locate a suitable settlement. Successful settlement is fundamentally important for the replenishment of species (*see*

RECRUITMENT) and the sustainment of fragile ECO-SYSTEMS such as coral reef communities.

Sediments are commonly analyzed and are an important water quality indicator. Sediments reach estuaries and the ocean from land drainage (debris flows and bank failures). Essentially, TURBULENCE is important in the transport of suspended particles such as silt and sand. Grains that are lying on the seafloor are entrained into the water when a critical shear stress is reached. The grain is transported by favorable currents and settles when currents can no longer hold the grain in suspension. Settling rates can be determined with sediment traps that are placed on the seafloor and collected after a period of time to provide the net settlement rate. Excess sedimentation and TURBIDITY can smother marine flora and fauna. Analysis of aerial and satellite imagery can also assist in determining the extent and concentration of suspended sediments. Studies utilizing images may require coincident field sampling to collect sea truth data (in-situ measurements collected to verify and calibrate remotely sensed images) and hydrologic and sediment transport models to predict sediment concentrations.

See also SEDIMENT; SEDIMENT TRANSPORT.

Further Reading

American Oyster, Chesapeake Bay Program Office. Available online. URL: http://www.chesapeakebay.net/american_oyster.htm. Accessed July 15, 2007.

North Inlet-Winyah Bay National Estuarine Research Reserve. Baruch Marine Field Laboratory. Available online. URL: http://www.northinlet.sc.edu. Accessed July 15, 2007.

Pacific Coral Reefs Website, United States Geological Survey. "Coral Reefs." Available online. URL: http://coral-reefs.wr.usgs.gov. Accessed December 26, 2006.

sextant The sextant is an instrument that measures vertical angles at sea very precisely, especially the altitudes, or angles above the horizon, of celestial bodies to determine the position of a ship or aircraft. It can also be used in piloting to measure horizontal angles between objects of interest, such as ships or BUOYS on the horizon, or between shoreline navigation aids, such as lighthouses. Its name comes from a part of the instrument called the arc, which is one-sixth of a circle. The arc is supported by a frame, which also supports an index arm that is the radius of the circle, a telescope, two parallel mirrors, shades of varying darkness to observe the Sun, and a micrometer or vernier scale to read angles precisely. One of the mirrors is called the index glass, and the other is called the horizon glass. Adjusting two screws at each end of the mirrors aligns the instru-

ment. The horizon glass is only half-silvered to allow a direct view of the horizon.

In operation, the navigator brings the direct ray from one object (usually, the horizon) into coincidence with the double reflected ray from the other object (usually, the Sun or star) by moving the index arm. This method is based on two laws of optics, which say that (1) for a light ray incident on a plane mirror, the angle of reflection is equal to the angle of incidence, and (2) the angle between the first and final directions of a ray of light that has undergone double reflection from mirrors in the same plane is twice the angle that the two reflecting mirrors make with each other.

The modern sextant was invented independently by John Hadley (1682–1744) in Britain, and by Thomas Godfrey (1704–49) in the United States, in the early 1730s. The sextant replaced navigational instruments such as octants and astrolabes. Very few astrolabes have survived the ages, and some are on display at the Adler Planetarium and Astronomy Museum in Chicago, Illinois. Originally, only instruments having an arc of 60 degrees were called sextants (60 degrees is one-sixth of a circle), but today any altitude-measuring instrument is called a sextant.

Modern sextants use a rotatable micrometer to measure angles to a theoretical accuracy of $1/10$th of a minute of arc. The sextant makes use of a telescope to view the horizon directly, and shade glasses to protect the eye while viewing the image of the Sun. In practice, the navigator sights on the horizon in the direction of the Sun with the telescope, then brings the double-reflected image of the Sun or star into coincidence, that is, so that the lower limb of the Sun appears exactly on the horizon. From this angle, and nautical almanacs, the latitude of the vessel can be determined. Navigators use this sighting and an accurate ship's chronometer and tables to determine the longitude.

The simplest means of measuring latitude using the sextant is to sight on the North Star, Polaris; the altitude of Polaris is just equal to the latitude. Small corrections are required from tables, since Polaris is not exactly over the Earth's North Pole. During the day, latitude can be most simply determined by using the sextant, a chronometer, and tables to determine the altitude of the Sun at local noon, that is, the passing of the Sun through the local meridian. This technique is widely used by small craft operators and does not require knowledge of spherical trigonometry, which is required when navigating by the Sun and stars in arbitrary positions.

Marine sextants for commercial use are carefully crafted of very fine materials, which allow precision machining, while preserving ruggedness and resistance to corrosion. Inexpensive instru-

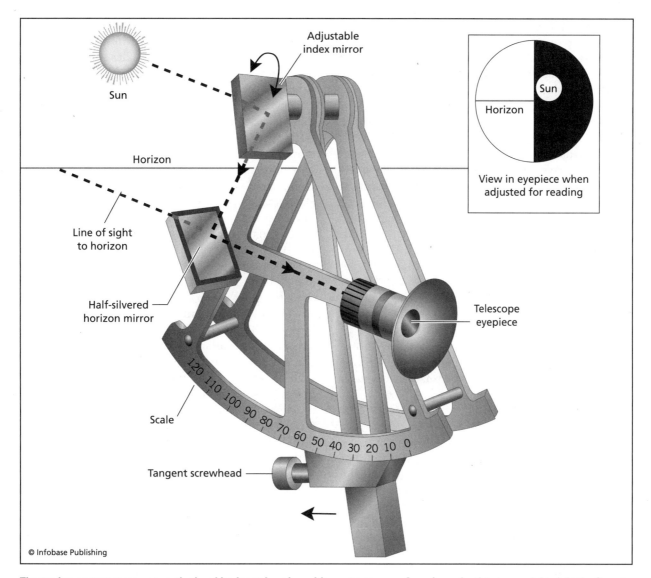

Adjustable
index mirror

Sun

Horizon

Line of sight
to horizon

Half-silvered
horizon mirror

Scale

120 110 100 90 80 70 60 50 40 30 20 10 0

Tangent screwhead

Telescope
eyepiece

Sun

Horizon

View in eyepiece when
adjusted for reading

© Infobase Publishing

The marine sextant measures vertical and horizontal angles with great accuracy. A marine scientist at sea might sight the Sun at noon to find the research vessel's latitude.

ments made of plastic are available for students to learn the operation of the sextant. Although the principles are straightforward, it takes a great deal of study and practice to accurately determine a vessel's position.

See COMPASS; MEASURE OF ANGLE; CHARTS, NAUTICAL; NAVIGATION.

Further Reading

Adler Planetarium and Astronomy Museum. Available online. URL: http://www.adlerplanetarium.org. Accessed July 15, 2007.

Ifland, Peter. "The History of the Sextant." Lecture, October 30, 2000, University of Coimbra. Available online. URL: http://www.mat.uc.pt/~helios/Mestre/Novemb00/H61iflan.htm. Accessed July 15, 2007.

Physical Sciences Collection, Smithsonian, National Museum of American History, Kenneth E. Behring Center. "Navigation." Available online. URL: http://american history.si.edu/collections/navigation. Accessed July 15, 2007.

shackle A shackle is generally a U- or D-shaped connector secured by a bolt and/or pin that links hardware such as the ends of turnbuckles, hooks, padeyes, mooring lines, and lengths of chain. Turnbuckles are often used to adjust the length and tension of a shroud or stay. Hooks may be used to lift cargo from ship holds. The hook may also serve as a quick-release device that is engaged while under strain of an oceanographic mooring's deadweight ANCHOR. The anchor is released and the mooring is set by the pull of a lanyard or tag line. Usually the BUOY is deployed first off the fantail and the mooring line is payed out while arrays of sensors are eased into the water column by marine technicians as the ship moves slowly toward the deployment station. Padeyes are metal devices with a hole for a shackle or pin. On a vertical padeye, the axis of the hole is parallel to the deck. On a horizontal padeye, the axis is perpendicular to the deck. Shackles are important for moorings where mooring pennants may be shackled to mooring chain or in weight handling where shackles connect slings to a load.

Shackles come in many shapes and types and depending on their use are made from materials such as stainless steel, galvanized steel, bronze, or painted steel. There are screw pin, round pin, or bolt and nut type shackles. Shackles with proper load limits must be used in marine rigging. Usually the same size shackle is used as the chain that it is connecting. The size of a shackle is determined by the diameter of the beam (A). It is not determined by the diameter of the pin. In general, the working load in tons is three times the shackle diameter squared as follows:

$$\text{Working load (tons)} = 3\ A^2$$

Suitable shackle pins or bolts are forged from alloy steel, heat treated and tempered to provide greater strength. Steel is an iron-based mixture or alloy that also contains manganese, silicon, copper, and possibly chromium, nickel, molybdenum, or tungsten. Shackles are usually marked with raised or indented letters with size (inches and millimeters) and working load limits in tons. Working load limits should never be exceeded.

Safety shackles utilize a bolt that contains a hole to accommodate a cotter pin for locking the nut on the bolt. The pin is installed to prevent nuts from rotating on bolts. An added measure of safety may be achieved by passing wire through drilled holes of adjacent bolts or capscrew heads to prevent rotation. Anchor shackles have exaggerated bows and chain-type shackles have parallel sides. Prior to deployment, all shackles and their associated pins should be visually inspected for corrosion, wear, deformation, surface cracks, or any other condition that may lead to failure.

See also MOORING; OCEANOGRAPHIC INSTRUMENTS.

shear Shear is the change in velocity of a fluid with position. One common example in the atmosphere, which is observed by kite fliers, is wind shear or the change in wind with height. Any angler going fishing with a cane pole and "bobber" can also observe vertical shear. Once the line is cast into the water, the "bobber" moves much faster at the surface, leaving the leader and baited hook slightly behind. This happens because of any wind blowing across the water; drag on the leader and hook that are rigged below the bobber, and the fact that the deeper layers of the water are moving slower until there is virtually no motion at the bottom. If the hook and leader are moving ahead of the "bobber" and if the "bobber" is going underwater, it is time to pull in the catch. Shears are important in the study of fluids because they can initiate VORTICITY, the tendency to spin, which can produce vortices such as whirlpools, ocean EDDIES, or tornadoes in the atmosphere. The change in velocity imparts shearing stresses to the surface and floor of the ocean, which generate waves and drive the great ocean currents.

Shear is important at all space (vertical and horizontal) and timescales of motion. Marine scientists study the small-scale shears of winds and currents over various types of ground surfaces, such as water, grass, forest, and over and around obstacles such as islands. In general, if the surface is smooth and if the viscosity (fluid friction) is large, the flow is LAMINAR, or can be described by layers of smoothly flowing plates of fluid (laminae). If the surface is rough, and the viscosity is low, the shear can initiate turbulent (*see* TURBULENCE) or highly disorganized flow. In fluids such as the atmosphere and ocean, the layers interact where layers on top shear past layers below. A simple picture described by many professors is a book glued to a table. If a shearing force is applied to the cover of the book, the pages will slide and deform the book. This shearing force also causes a stress between layers. This, in part, is how the faster-moving air transfers energy into the slower-moving water. Through empirical investigations, current speeds generated by winds over the water are approximately 3 percent of the wind speed.

When studying the strength of materials, marine scientists may attempt to determine the ability of materials such as SEDIMENTS to resist deformation. The term that is used is *shear strength*. One important application involves determining the shear stress that initiates erosion and deposition of sediments. Some of the important fluid properties in determining thresholds for erosion and deposition are specific weight, viscosity, and speed. The investigator must also know sediment properties such as particle size, shape, specific gravity, and the grains' position in the seabed. The shearing force gives rise to shear stresses, which ultimately move a grain of sand (*see* SEDIMENT TRANSPORT) from the seabed.

Changes in winds or currents over a horizontal distance are called shear vorticity. When looking in the flow direction, if the velocity decreases to the right, a clockwise torque will be set up producing negative vorticity. In contrast, if the flow velocity decreases to the left a counterclockwise torque will be generated producing positive vorticity. However, vorticity is a three-dimensional vector, and depending on scale requires consideration of the CORIOLIS FORCE, shear force, and curvature (*see* VORTICITY). Studying horizontal shear is often accomplished in flumes, by deploying oceanographic instruments, or through physics-based time domain numerical models (*see* MODEL BASIN; MODELS; and NUMERICAL MODELING).

Wind shear at small scales is an extremely important factor in generating ocean wind waves, since it imparts stresses to the surface that, along with small-scale pressure fluctuations induce wave motion. At larger scales, wind shear at the surface produces the stresses that drive the great permanent currents, such as the NORTH EQUATORIAL CURRENT driven by the northeast trade winds (*see* EKMAN SPIRAL AND TRANSPORT). Friction with coastal boundaries is important to producing shear that initiates rotational flow, such as around SANDBARS and headlands.

Detection of strong wind shears is important to the safe operation of airports. Many airports have installed systems such as Doppler RADAR to monitor wind shear at critical points along the flight paths. Knowing the current velocity profile is important for navigating in shallow water. Measurements of these profiles can be made with a bottom-mounted ACOUSTIC DOPPLER CURRENT PROFILER or arrays of CURRENT METERS fixed to a mooring (*see* OCEANOGRAPHIC INSTRUMENTS). Detection of current shear is important to safe navigation in areas such as inlets, river bends, channel junctions, and narrow ship channels. For example, large supertankers could be unexpectedly turned in a narrow seaway, such as the Houston Ship Channel, and present a hazard to other vessels, which navigate the channel in close order. Physical Oceanographic Real-Time Systems (PORTS®) have been installed to improve the safety and efficiency of selected seaports (*see* essay on PORTS, page 439).

Shear is an important kinematic parameter affecting other processes such as the propagation of sound. Being able to quantify the forces of layers of water moving over one another is paramount to determining factors such as mixing, turbulence, erosion, and wave development. In fact, researchers affiliated with the WOODS HOLE OCEANOGRAPHIC INSTITUTION developed and tested a shear meter during the late 1990s. Their instrument was a 32.8-foot (10 m)-long SPAR BUOY that floated horizontally in the water column and measured the flow difference from rotors at either end of the buoy. It measures revolutions per hour at the density layer where the buoy has been configured to float. The instrument was tested in Seneca Lake, New York, in 1996 and then deployed for approximately 100 days in the ATLANTIC OCEAN during spring 1998.

Shear is also important in geological processes. Stresses tangential to a surface such as the boundary between two rock masses or two tectonic plates (*see* PLATE TECTONICS) are known as shearing stresses that tend to cause one plate to slide past another. This is often not a smooth process, but rather a buildup of forces followed by a sudden release in which a deformation occurs. If this happens near the surface, an earthquake is the usual result. The San Andreas Fault in California is one result of such shearing stresses. Continental margins in North Brazil and the Guinea Coast of Africa are parallel to the fracture zones there and are called shear margins.

Further Reading

Center for Applied Coastal Research, University of Delaware. Available online. URL: http://www.coastal.udel.edu. Accessed January 28, 2007.

Coastal and Hydraulics Laboratory, Engineer Research and Development Center, U.S. Army Corps of Engineers. Available online. URL: http://chl.erdc.usace.army.mil. Accessed January 28, 2007.

Coastal & Ocean Fluid Dynamics Laboratory, Applied Ocean Physics & Engineering Department, Woods Hole Oceanographic Institution. Available online. URL: http://www.whoi.edu/science/AOPE/cofdl. Accessed January 28, 2007.

Earth and Space Research: A Nonprofit Research Institute, Seattle, Washington. Available online. URL: http://www.esr.org. Accessed January 28, 2007.

Integrated Ocean Drilling Program. Available online. URL: http://www.oceandrilling.org. Accessed January 28, 2007.

Physical Oceanographic Real-Time System (PORTS®), Center for Operational Oceanographic Products and Services (CO-OPS), National Ocean Service, National Oceanic and Atmospheric Administration. Available online. URL: http://tidesandcurrents.noaa.gov/ports.html. Accessed January 28, 2007.

ship's routing The analytical procedure of planning a route that minimizes sailing time between the ports of embarkation and destination is called ship's routing. It is also an important publication of the International Maritime Organization (IMO), describing the general guidelines for ship routing, identifying deep water routes and regions to be avoided, and describing traffic control that has been adopted by the IMO. All of this information is distributed by means of NOTICES TO MARINERS and Sailing Directions, and is incorporated on the most recent nautical charts (*see* CHARTS, NAUTICAL). Notices to Mariners are periodicals that advise mariners on matters affecting navigational safety, including new hydrographic discoveries, shifting channels, and the location of navigational aids. Sailing Directions are guides with information relevant to a particular coastal region to assist mariners in planning ocean passages. Nautical charts support navigation by depicting depths of water, nature of bottom, elevations, configuration and characteristics of the coast, dangers, and aids to navigation.

As a general term, ship's routing refers to planning a route to minimize sailing time between the ports of embarkation and destination, avoiding hazardous areas such as shoals and underwater obstructions, as well as severe weather conditions, minimizing the possibility of collisions between ships, and providing for safe navigation in ports and harbors (*see* OPTIMAL SHIP TRACK PLANNING). It is a component of vessel traffic management that is based on principles such as good order and predictability. Coast Guardsmen work tirelessly to reduce the rate of collisions, rammings, and groundings, while aiming to protect life and property. Supporting systems include Vessel Traffic Information Systems (VTIS) and Physical Oceanographic Real-Time Systems. These information systems provide real-time information that aids decision makers. For example, the VTIS might include a precision navigation system and the Coast Guard's DIFFERENTIAL GLOBAL POSITIONING SYSTEM (DGPS) to provide a transiting vessel with precise information regarding its own position as well as the location and maneuvering data of other vessels in the harbor.

Further Reading

Office of Coast Survey, National Ocean Service, National Oceanic and Atmospheric Administration. Available online. URL: http://chartmaker.ncd.noaa.gov. Accessed December 11, 2006.

U.S. Coast Guard. Available online. URL: http://www.coastguard.mil. Accessed December 11, 2006.

U.S. Maritime Administration. Available online. URL: http://www.marad.dot.gov. Accessed December 11, 2006.

solitary wave A solitary wave is a stable wave that propagates across large distances without disbursing or transforming. The solitary wave, also called soliton, consists of a single crest that is not necessarily small compared to the depth of the water. In the ocean, they may be generated by flow over deep sills, which disrupts the THERMOCLINE, or by a standing harmonic wave in shallow water. The pumping of the thermocline sets up the wave. Solitary waves generally propagate along density interfaces such as the PYCNOCLINE, and play a significant role in mixing and may affect acoustic propagation. Researchers can generate one solitary wave or a soliton in a flume. In fact, they were first observed and described by British civil engineer John Scott Russell (1808–82) as he studied the propagation of a wave forced by the sudden displacement of water from a BARGE that stopped abruptly along the Union Canal at Hermiston near Edinburgh, United Kingdom. His findings and continued work with solitary waves in a wave tank were published in 1845. Russell, who designed ship hulls and conducted research in ship design, is credited with the discovery of solitons.

The soliton can be regarded as an extreme case of a finite amplitude wave known as the Stokes' wave in which the WAVELENGTH greatly exceeds the water depth (as distinct from a small amplitude wave in which the ratio of wave amplitude to wavelength must be very small, such as the simple harmonic wave in introductory physics courses). In contrast to the simple harmonic wave in which the water particles move in complete circles as the wave passes, the Stokes' wave involves a circular motion slightly open at the top that allows a small forward transport of the surface water. (For wind-driven waves, this transport is on the order of 3 percent of the wind speed, a rough figure often used in tracking the movement of oil spills.) The extreme form of Russell's soliton has a sharp crest of 120°, and the profile is represented mathematically as a hyperbolic secant function.

Solitons are used to model the propagation of TIDAL BORES that occur when the rising tide forces the tidal wave front to move faster than the natural speed of the waves. When this occurs, a shock wave develops called a tidal bore that propagates with a very steep front, moving upstream as a roaring wall of water. Tidal bores on the Amazon River, called *pororocas* (rock crushers), can be greater than 25 feet (7.6 m) in height. The highest tidal bores on record reach a height of approximately 29 feet (9 m) on the Qiantang River near Hangchow, China. Other locations with tidal bores include the entrance to the Daly River in Northern Territory, Australia; the Bay of Fundy in Nova Scotia, Canada; the Severn River near Gloucester, England; the Garonne River between Cambes and Cadillac and the Dordogne River near St. Pardon, France; the Hooghly (Hugli) River in northeastern India; and the Turnagain Arm near Anchorage, Alaska, in the United States. On occasion, kayakers and surfers have been known to ride tidal bores, which can be especially dangerous.

Solitary waves may also be involved in the formation of the gigantic ship-killing rogue waves. They may form suddenly as an outcrop of wind, wave, and CURRENT interactions. The right superposition of waves may generate the rogue wave in an otherwise chaotic sea. These large waves have been reported from ships at sea and offshore oil production platforms. They are extremely dangerous and a topic of research for marine scientists. There is little in situ data, and investigations are conducted with numerical models or in wave tanks.

See also INTERNAL WAVES; ROGUE WAVE.

Further Reading

Colling, Angela, and Dave Park (for the Open University Course Team). *Waves, Tides, and Shallow Water Processes.* Milton Keynes, U.K.: Butterworth/Heinemann, 2002.

Eilbeck, Chris. "Solitons home page." Department of Mathematics, Heriot-Watt University. Available online. URL: http://www.ma.hw.ac.uk/solitons. Accessed October 16, 2007.

Irving, Robert. "Solitary Waves." *Mariners Weather Log* 37, no. 4 (1993): 20–23.

Lamb, Sir Horace. *Hydrodynamics,* 6th ed. New York: Dover, 1945.

sonar   Sonar is actually an acronym that stands for *s*ound *n*avigation *a*nd *r*anging. Leonardo da Vinci (1452–1519) was reportedly one of the first scientists to imagine this technique. The first patent for a sonar type device is credited to Lewis Fry Richardson (1881–1953), a meteorologist from Great Britain who is best remembered for his work identifying fluid turbulence. A major milestone in the development of sonar came in 1912 with Professor Reginald A. Fessenden's (1866–1932) design and construction of a moving coil transducer for the projection and reception of sound underwater. The Fessenden transducer allowed vessels to communicate with one another underwater using Morse Code, as well as to detect echoes from underwater objects (echo ranging). Morse code, after Samuel F. B. Morse (1791–1872), uses sequences of dots and dashes to transmit information. In 1914, a Fessenden oscillator made possible the detection of an ICEBERG at a distance of two miles (3.2 km) from the transmitting ship. This oscillator was installed on all U.S. submarines during World War I. During this time, the German Navy inflicted heavy losses on Allied shipping. Thus the location of enemy submarines became a task of major importance for the British Navy, and later the U.S. Navy.

The term *sonar* is commonplace as the technique has become routine in marine operations ranging from bathymetric surveying to underwater navigation. Sonar in the ocean is analogous to radar in the atmosphere. Sonars basically determine the time it takes for an echo to bounce off an underwater object and return to the sound source. Since sonar involves the transmission and reception of sound in the water, the technique depends on the environment (for example, TEMPERATURE, SALINITY, and PRESSURE) and the transmission and receiving equipment. A transducer is the electromechanical component of a sonar system that transmits short bursts of sound into the water. Following transmission, returned sound reflected from the bottom or from an object in the water is received by the transducer. Modern active sonar devices can be used to identify bottom structure, determine water depth, and locate wrecks, specific bottom types, well heads, pipelines, and a myriad of other industrial applications. Certain sport fishing and commercial fishing variants mark fish within the water column, as well as locate holes in the bottom and temperature discontinuities such as the thermocline, under which large fish may be found.

Sound is transmitted much more efficiently through water than through the air. The average speed of sound through a standard ocean and atmosphere is 3,403.5 miles per hour (1,521.5 m/s) and 769.5 miles per hour (344 m/s), respectively. Sounds propagate even faster through the ocean with increasing temperature, salinity, DENSITY, and pressure. Therefore, scientists may be particularly interested in the structure of the THERMOCLINE, HALOCLINE, and PYCNOCLINE, especially as these features relate to sound transmission in ocean water. In fact, a marine scientist could estimate changes in ocean temperature by observing the resultant change in speed of sound over long distances. Changes in the water column affect the time lapsed between transmitted pulses and received echoes from the seafloor. But, given the parameters that affect the speed of sound, the simple equation is as follows

$$D = ct/2$$

where,

D = Distance,

c = speed of sound in water, and

t = time for the sound to leave and return to the transducer and is used to find a target in the water or on the bottom.

Sonars are classified as passive, when underwater microphones (hydrophones) are used to detect underwater noises (ambient noise) such as submarines and whales, or active, when a sound pulse is transmitted to reflect off a target and return to the sending vessel (echo ranging). Modern sonar systems generally consist of arrays of transmitters and receivers to form directive beams that can accurately locate both the distance and direction to a target, even under conditions of high ambient noise. State-of-the-art sonars employ lenses and mirrors just as optical systems do, but they are made of piezoelectric materials such as barium titanate.

Some marine scientists with a strong background in physics specialize in the discipline of underwater acoustics; they are also called acousticians. Their research is generally focused on the generation, control, transmission, reception, and effects of sound. Physically, sound is a quickly varying pressure wave within the sea. Sound waves are longitudinal waves, so that the pressure variations occur in the direction of propagation, as in waves on the child's toy known as the "slinky." Traditional tasks involve designing transducers, acoustical measurement techniques, noise and vibration control, and, most important, the effects of underwater sound in response to phenomena such as INTERNAL WAVES. Today, acousticians working in the ocean are required to ensure that the intensity of transmitted sound does not damage protected species such as MARINE MAMMALS. In addition to passive listening to animal sounds, tiny transducers can be embedded in an animal's flesh (without harm) and used to track the animal. Such techniques have been used to track whales off the Hawaiian Islands.

Important applications of sonar include wreck location and detailed structural mapping of the seafloor by means of side-scanning sonar, charting the CONTINENTAL SHELF and SLOPE by multibeam sonar, mapping ocean currents and eddies by ACOUSTIC TOMOGRAPHY, characterizing the sediments and subsediment basements by geophysical sonars, and identification and determination of behavioral characteristics of species of whales and other marine animals from the sounds they make. Acoustic navigation can be conducted by precise location of acoustic TRANSPONDERS on the ocean floor. Such techniques are very important to installing underwater structures and pipelines.

See also AMBIENT NOISE; ATTENUATION; DEPTH FINDER.

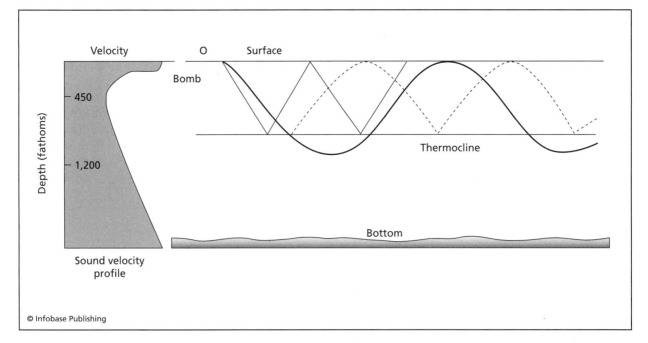

Sonar systems can obtain long ranges of many miles by making use of the "sound channel," a region of the water column of minimum sound velocity that acts to refract and thus focus sound rays within the channel; that is, the sound channel acts as a wave guide.

Further Reading

Clay, Clarence S., and Herman Medwin. *Acoustical Oceanography: Principles and Applications.* New York: Wiley-Interscience, 1977.

MacLeish, William, ed. "Sound in the Sea." *Oceanus* 20, no. 2 (1977), 1–75.

Museo Leonardiano di Vinci. Available online. URL: http://www.leonet.it/comuni/vinci/. Accessed March 3, 2007.

Urick, Robert J. *Principles of Underwater Sound.* New York: McGraw Hill, 1983.

Woods Hole Science Center, US Geological Survey. "A Brief History of the Use of Sound in Ocean Exploration." Available online. URL: http://woodshole.er.usgs.gov/operations/sfmapping/soundhist.htm. Accessed January 7, 2007.

sounding Sounding is the charted or measured depth of water, or the activity of measuring the depth. For centuries, mariners used lead lines, or lines marked at intervals, such as every three feet (1 m), attached to a lead weight that was thrown or lowered overboard, and the amount of line paid out determined when the weight hit bottom, at which time the depth was measured. Nineteenth-century oceanographic expeditions used sounding machines for determining the depth in the deep ocean. The sounding machine consisted of a reel of wire to one end of which was attached a weight that carried a device for recording the depth. The machine was equipped with a crank or motor for reeling in the wire. Such measurements are subject to error in that the line may not be vertical. Given the average depth of the deep oceans as about 12,000 feet (4,000 m), it required a great deal of time to make just a few measurements. The true shape of the ocean floor was not learned until after World War II, when great numbers of measurements were made by means of acoustic echo sounding to determine water depths, in which the water depth is determined by measuring the time interval between emission of a sonic or ultrasonic pulse and the return of its echo from the ocean bottom. The instrument used for this purpose is called an echo sounder or FATHOMETER.

See also BATHYMETRY.

sound in the sea Sound in the sea is a compressional (longitudinal) wave moving through the ocean. Particles in the wave move parallel to the direction on which the wave is moving. Surface waves (transverse waves) are different because they move perpendicular to the direction in which the wave is moving. Sound waves have a source such as the vibrations of a boat propeller that disturbs the water particles. The sound wave is transported from the source to a receiver such as a SCUBA diver's ears by means of water particle interactions. In the open ocean, passive underwater acoustic monitoring allows the detection of some important phenomena. Passive monitoring involves the use of instruments to detect natural sound energy. Some researchers use an active approach, which entails transmitting acoustic energy and then receiving the reflected or backscattered energy. Marine scientists my use hydrophones to study sound sources such as earthquakes, volcanic eruptions, landslides, marine mammal and fish vocalizations, weather, and ship traffic. Hydrophones convert acoustic energy into electrical voltage, which is then read on a meter and saved as an important element in a monitoring network.

See also ATTENUATION; SONAR.

Further Reading

NOAA PMEL Vents Program, Hatfield Marine Science Center. "Acoustic Monitoring." Available online. URL: http://www.pmel.noaa.gov/vents/acoustics.html. Accessed July 17, 2007.

South Atlantic Current (SAC) The South Atlantic Current is the southern branch of the South Atlantic GYRE, which flows from the Falkland (Malvinas) Islands to the southern tip of the Union of South Africa. The other branches are the BENGUELA CURRENT, which runs northward along the west coast of Africa; the South Equatorial Current, which flows from equatorial Africa to the northeast coast of Brazil; and the BRAZIL CURRENT, which flows southward off the coast of Brazil and completes the gyre by joining the northward flow of the Falkland (Malvinas) Current, and turns eastward into the South Atlantic Current. The SAC is smoothly joined to the ANTARCTIC CIRCUMPOLAR CURRENT (ACC) forming a continuous eastward flow. The southern boundary of the current approximately coincides with the subtropical convergence, a zone of rapid TEMPERATURE and SALINITY change separating relatively warm subtropical South Atlantic surface water from the cold sub-Antarctic surface water.

Mean speeds in the SAC are on the order of 0.5 to .7 knots (0.25–.35 m/s). The speed and constancy of the current increase steadily from its northern boundary to about 40°S (the general locus of the subtropical convergence), where it merges smoothly with the Antarctic Circumpolar Current. Temperatures are on the order of 15°C along the northern edge of the current, decreasing to temperatures on the order of 7–8°C south of the subtropical convergence. Salinities in the South Atlantic Current are on the order of 35 psu in the northern section. As the subtropical convergence is crossed, salinities decrease to around 34 psu as more Antarctic water is encountered. After it crosses the Atlantic, the SAC feeds subtropical South Atlantic water into the western edge of the Benguela Current as it flows northward off the coast of Angola. The water to the north of the SAC is warm and salty, forming the center of the South Atlantic gyre, in a manner similar to the Sargasso Sea water in the North Atlantic, which is at the center of the North Atlantic gyre. The Southern Hemisphere mid-latitude westerly winds, part of the great anticlockwise wind system circulating over the ocean current gyre, are a primary driving force for this current. The SAC is in approximate geostrophic balance (*see* GEOSTROPHY), with the isopycnal surfaces (surfaces of constant density) sloping downward from south to north, and the isobaric surfaces (surfaces of constant pressure) sloping upward from south to north. Hence, the sea surface, being approximately isobaric, slopes upward toward the north into the subtropical convergence. The resultant distribution of surface heights can be used to estimate the geostrophic currents from satellite ALTIMETER data.

The German Meteor Expedition (planned and organized by A. Merz) determined the general features of the South Atlantic Current and of the South Atlantic Ocean in 1925–27, when it occupied more than 310 oceanographic stations, mostly in the South Atlantic. The Meteor Expedition is best known for its mapping of deep-ocean relief that included documenting the mid-ocean ridge. The ridges are rugged mountainous features rising 0.6 to two miles (1–3 km) above the adjacent ABYSSAL PLAINS.

The counterparts of the South Atlantic Current in both the PACIFIC and INDIAN OCEANS are both eastward-flowing currents with a continuous border with the northern edge of the Antarctic Circumpolar Current as is the Atlantic SAC. The South Pacific Current spans the distance from New Zealand to the coast of Peru; the South Indian Current spans the distance from the east coast of South Africa to the west coast of Australia. These two currents lie in traditionally data-poor areas far removed from the main shipping lanes. Hence, remotely sensed observations from satellite altimeters are particularly valuable here. Recently, observations are being acquired from ships and aircraft supplying bases on the Antarctic continent.

Further Reading

Bowditch, Nathaniel. *The American Practical Navigator An Epitome of Navigation*. Bethesda, Md.: Defense Mapping Agency Hydrographic/Topographic Center, 1995.

Physical Oceanography Program, Office of Earth Science, National Aeronautics and Space Administration. "Ocean Motion and Surface Currents." Available online. URL: http://oceanmotion.org/. Accessed October 16, 2007.

Pickard, George L., and William J. Emery. *Descriptive Physical Oceanography: An Introduction*, 5th ed. Oxford: Pergamon Press, 1990.

South Equatorial Current (SEC)   The South Equatorial Current (SEC) is a warm, wide, and generally shallow current spanning the great ocean basins of the world approximately between the latitudes of 4°N and 15°S–25°S in the Atlantic and Pacific Oceans, and between 10°S and 25°S in the Indian Ocean. The major part of the Atlantic and Pacific branches of the SEC are found south of the Equator, but extend northward a few degrees into the Northern Hemisphere. The Indian Ocean SEC is found entirely south of the Equator. The South Equatorial Current flows westward, driven by the stress of the southeast trade winds.

The South Equatorial Current is the northern branch of the Southern Hemisphere subtropical anticyclonic (counterclockwise) GYRE circulating the waters of the Atlantic, Pacific, and Indian Oceans. The Atlantic SEC forms the northern branch of the gyre, the Brazil Current the western branch, the South Atlantic Current the southern branch, and the Benguela Current the eastern branch. The SEC has as its source region the flow of the Benguela Current into the Gulf of Guinea in West Africa, and extends westward to the coast of Brazil. At the eastern extent of the Brazilian coast, Cape Sao Roque, it splits with the northern branch flowing into the Guyana Current (North Brazil Coastal Current), and the southern branch turning below Natal and flowing into the Brazil Current. The northern branch also feeds the Equatorial Countercurrent and Equatorial Undercurrent. The Atlantic South Equatorial Current is stronger and more extensive than the North Equatorial Current.

The Pacific South Equatorial Current flows westward, generally between latitudes 4°N and 10°S. It constitutes the northern branch of the South Pacific gyre flowing westward from South America across the Pacific Ocean to Australia and New Guinea; The western branch of the gyre is the East

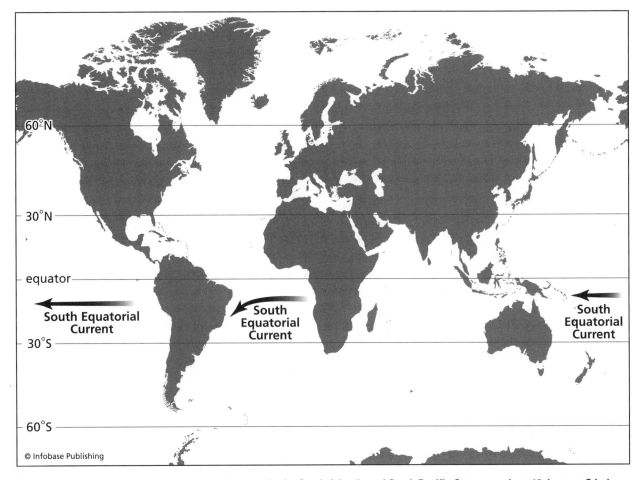

The South Equatorial Current flows from east to west in the South Atlantic and South Pacific Oceans at about 10 degrees S latitude in response to the southeast trade winds. In the Northern Hemisphere summer, the current may cross over the equator into the Northern Hemisphere in the western part of the ocean basins.

Australia Current, the southern branch is the South Pacific Current (bordering the West Wind Drift, or Antarctic Circumpolar Current), the eastern branch being the Peru Current. In the mid Pacific region, between longitudes 150°W and 180°W, a considerable portion of the current turns south to form the center of the South Pacific gyre. The part continuing westward splits as it approaches the Coral Sea and the coast of Australia, with part flowing north toward New Guinea and part turning south to flow into the East Australia Current.

The Indian South Equatorial Current is driven by the monsoon regime. The northern boundary of the current tends to vary between 9°S and 11°S, being farther north during the season of the southwest monsoon, and farther south during the season of the northeast monsoon. The Indian SEC is the northern boundary of the ocean wide counterclockwise gyre; the Agulhas Current is the western boundary, the West Wind Drift the southern boundary, and the West Australia Current the eastern

boundary. The current flows from the west coast of Australia westward across the Indian Ocean to the African coast. As the SEC approaches the coast, part flows through the Mozambique Channel between Madagascar and the mainland, and part flows along the east coast of Madagascar. The two branches join at the southern tip of Madagascar to form the Agulhas Current, a very strong current notorious for giant waves that have sent supertankers to the bottom of the ocean. The northernmost part of the South Equatorial Current flows northward along the coast of East Africa and contributes to the Somali Current in Northern Hemisphere summer, a very strong shallow current. During the Northern Hemisphere summer, the Indian North Equatorial Current and Equatorial Countercurrent are replaced by the Southwest Monsoon Current, resulting in almost continuous eastward flow north of the SEC.

In the Atlantic and Pacific Oceans, where the South Equatorial Current flows on both sides of the

Equator, the sea surface slopes downward north and south away from the equator, while the thermocline slopes upward to the north and south away from the Equator to maintain geostrophic balance (*see* GEO-STROPHY). This dynamic behavior results from the fact that the Coriolis parameter, proportional to the sine of the latitude, changes algebraic sign in going across the equator, and is thus zero right at the equator (*see* CORIOLIS EFFECT).

Recent expeditions and studies have shown that the South Equatorial Current is a complex set of current branches, with many regions of frequent meanders and eddies. For example, scientists from the University of Miami, NOAA's Atlantic Oceanographic and Meteorological Laboratory, and many other institutions in South America, Europe, and Africa have shown that the Atlantic SEC is comprised of three branches: a northern SEC; a central SEC; and a southern SEC, with the branches sometimes being separated by countercurrents and undercurrents. They have begun to determine seasonal behavior and prepare estimates of water volume, heat, and salt transports.

Further Reading

Bonhoure, Dorothee, Elizabeth Rowe, Arthur J. Mariano, and Edward H. Ryan. "The South Equatorial Current System." *Ocean Surface Currents*, 2004. Available online. URL: http://oceancurrents.rsmas.miami.edu/atlantic/south-equatorial.html. Accessed March 11, 2008.

Bowditch, Nathaniel. *The American Practical Navigator: Publication No. 9*. Bethesda: Defense Mapping Agency Hydrographic/Topographic Center, 1995.

Southern Ocean   The Southern Ocean is the circumpolar body of water lying between 60° south latitude and the South Pole. The International Hydrographic Organization named this region the fifth ocean during the spring of 2000. The other four are the Arctic, ATLANTIC, INDIAN, and PACIFIC OCEANS. The Southern Ocean includes the southern extents of the Atlantic, Indian, and Pacific Oceans, covering about 7,846,000 square miles (20,327,000 km²) and surrounding Antarctica. MARGINAL SEAS include the Amundsen Sea, Bellingshausen Sea, Ross Sea, a portion of the Scotia Sea, and the WEDDELL SEA.

Antarctica is credited with being the coldest place on Earth and is almost entirely covered by glacial ice. CONTINENTAL SHELF areas are approximately 160 miles (257 km) wide with maximum widths in the vicinity of the Weddell and Ross Seas of more than 1,600 miles (2,570 km). The deep continental shelf is covered by glacial deposits varying widely over short distances. Complex BATHYMETRY includes oceanic rises, ABYSSAL HILLS, glacial troughs, submarine valleys, moraine ridges, and drumlins. The South Sandwich Trench lies on the eastern side of the South Sandwich Islands. These volcanic island arcs result from subduction of the South American Plate beneath the South Sandwich Plate. Moraines are hill-like piles of glacial sediments parallel to the former ice margin. Drumlins are teardrop-shaped piles of glacial till that are parallel to the direction of glacial ice motion. These features delineate the drainage of glaciers and are telltale signs of erosion and deposition on the shelf by glaciers and ice streams. The seafloor of the Southern Ocean ranges from 13,100 to 16,400 feet (4,000 to 5,000 m) below sea level over most of the area. Bathymetric surveys have been completed by research vessels such as the *Academik Boris Petrov, Bio-Hesperides, Maurice Ewing, Nathaniel B. Palmer,* and *James Clark Ross* in support of the United States Southern Ocean Global Ocean Ecosystems Dynamics (SO GLOBEC) program. The GLOBEC program is a multidisciplinary and integrated research effort among physicists, biologists, and chemists focused on the interplay of physics, biology, and chemistry in the Southern Ocean.

One of the notable features of the Southern Ocean is the ANTARCTIC CIRCUMPOLAR CURRENT, which circulates east around Antarctica within a band that ranges from 45° to 60° south latitude. As this current approaches the South American continent, part of the WATER MASS is deviated northward as the PERU (HUMBOLDT) CURRENT, which brings cold Antarctic water to the coast of Chile and Peru. The Antarctic current also flows northward as a subsurface current between Australia and New Zealand. Scientists are investigating several degree temperature changes measured across the current and variance in the composition of water samples extracted from the flow. ICEBERGS and SEA ICE frequent the Southern Ocean throughout the year and are a problem for shipping. Typical temperatures in the Southern Ocean range from 28.4° to 37.4°F (-2 to 3°C) and SALINITIES 33 to 34 psu. Huge tabular icebergs are calved from the ice shelf in the Southern Ocean's Weddell Sea. An iceberg is classified in accordance with its freeboard (height above water) and length as a growler, bergy bit, small berg, medium berg, large berg, or very large berg (*see* GLACIAL ICE FORMS). Sailboats competing in races such as the Volvo Ocean Race use RADAR and lookouts with night vision goggles to avoid hazards such as growlers as they approach Cape Horn, the tip of South America. The approximately 500-mile (805 km)-wide Drake Passage occurs where the Antarctic Circumpolar Current is squeezed between the continental landmasses of South America and Antarctica. This region is particularly interesting to marine scientists since seawater circulating around the Southern Ocean has to pass through the Drake Passage, thereby providing an interesting cross section of the Antarctic Circumpolar Current.

Because of the high wind speeds in the Southern Ocean and the unrestricted FETCH, seas are often very high and ROGUE WAVES are comparatively frequent. Ships navigating Antarctic waters must be armored against collisions with ice. Historically, the passage around Cape Horn or through the Drake Passage was dreaded by sailing ship masters because of the cold, frequent severe weather, and shoal areas. Prior to the completion of the Panama Canal, in 1914, East Coast shipping bound for California and points north had to round the cape.

There is a great abundance of krill, small shrimplike creatures, due to the vigorous vertical water circulations and high productivity adjacent to the Antarctic continent, and great numbers of whales gather to feed. Prior to the formation of the International Whaling Commission in 1946 and subsequent restrictions on whaling, the slaughter of whales by maritime nations such as Norway, Iceland, Japan, Russia, and the United States drove several species to near extinction.

See also ANTARCTICA; ANTARCTIC CIRCUMPOLAR CURRENT.

Further Reading

Antarctic and Southern Ocean Coalition. Available online. URL: http://www.asoc.org. Accessed December 24, 2006.

Byrd Polar Research Center. Available online. URL: http://www-bprc.mps.ohio-state.edu. Accessed December 24, 2006.

Central Intelligence Agency. *The World Factbook.* Available online. URL: https://www.cia.gov/cia/publications/factbook. Accessed December 24, 2006.

International Hydrographic Organization, Monaco. Available online. URL: http://www.iho.shom.fr. Accessed December 24, 2006.

Palmer Long-Term Ecological Research. Available online. URL: http://pal.lternet.edu. Accessed December 24, 2006.

Scientific Committee on Antarctic Research. Available online. URL: http://www.scar.org. Accessed December 24, 2006.

Southern Ocean Regional Data Center, British Oceanographic Data Center. Available online. URL: http://www.bodc.ac.uk/projects/international/argo/southern_ocean. Accessed December 24, 2006.

U.S. Antarctic Resource Center. Available online. URL: http://usarc.usgs.gov. Accessed December 24, 2006.

Volvo Ocean Race. Available online. URL: http://www.volvooceanrace.org. Accessed December 26, 2006.

spar buoy A spar buoy is a long and thin ballasted structure that floats in a vertical position. In inland waters, small spar buoys are often used as regulatory buoys. They may be painted white with orange reflective tape and black lettering that provides information such as speed limits. They may be attached to deadweight anchors to mark obstructions or reefs. Spar buoys are also used as navigation aids, especially in rivers, ESTUARIES, and intercoastal waterways. Large manned spar buoys are used for research and offshore exploration. Marine scientists use large spar buoys as a stable research platform where sensors are deployed to measure the atmosphere and the ocean. Studies from a spar buoy include experiments in acoustics, optics, and air-sea intereactions. Spar buoys used in offshore oil production are designed to meet the rigors of deep-sea operations. They include an elongated submerged hull with oil storage chambers and variable ballast chambers.

See also BUOYS; OIL PRODUCTION PLATFORMS.

Further Reading

Minerals Management Service, U.S. Department of the Interior. Available online. URL: http://www.mms.gov. Accessed January 31, 2007.

specific gravity Specific gravity is defined as the ratio of the DENSITY of a material to the density of pure water. It is a useful descriptive quantity because it is dimensionless and does not depend on the units used, and it indicates whether a body made of a certain material will float or sink in pure water. It can be measured by means of Archimedes' principle; that is, a body is buoyed up by a force equal to the weight of water displaced by the body. Seawater is slightly denser than pure water, and so would have a specific gravity greater than one. It exerts a greater buoyancy force than pure water, to which any swimmer who lives near the ocean can attest. Aquarium enthusiasts often use a hydrometer to estimate SALINITY. The hydrometer will sink until it displaces its own weight. In denser fluids, it displaces less fluid and floats higher, while in less dense fluids it floats lower. Hydrometers are calibrated based upon the specific gravity of water at 60°F (15.6°C) being 1.000.

specific heat Specific heat is defined as the amount of energy (heat) needed to raise the TEMPERATURE of a unit mass of a substance by 1°. If a quantity of energy, Q, is transferred to a sample of mass, m, and it raises the temperature by an amount ΔT, then the specific heat, c, is equal to:

$$c = Q/m\Delta T.$$

The SI unit of specific heat is the joule per kelvin per kilogram [J/kg·K]. The measure of how well a substance resists changing its temperature when it absorbs or releases heat is another way of defining specific heat.

The greater the specific heat of a substance, the more energy must be added per unit mass to cause a given temperature change. When the temperature increases, Q and ΔT are taken to be positive, and energy is transferred into the system. When the temperature decreases, Q and ΔT are taken to be negative, and energy transfers out of the system. The ability of water to moderate or stabilize temperatures depends on its relatively high specific heat. Water resists changing its temperature as evidenced by the relatively large quantity of heat for each degree of change that is lost or absorbed when there is a temperature change.

Specific heats are different at constant pressure (c_p) and at constant volume (c_v). For seawater, the specific heat is also a function of the SALINITY. Water has the highest specific heat of ordinary substances. This means that the very large lakes and oceans have high thermal inertia and are slow to warm in summer and slow to cool in winter. This characteristic moderates the climate of locations close to the sea, or very large lakes. During winter, energy is transferred from the water to the air, thus warming the adjacent communities; in summer, energy is transferred from the air to the water, thus cooling the adjacent communities. West Coast cities in the path of the prevailing westerly winds off the PACIFIC OCEAN, such as San Francisco, have more moderate climates than East Coast cities, such as New York, where the prevailing westerly winds have blown over the North American landmass.

Further Reading

Michon, Scott. "Earth's Big Heat Bucket." Available online. URL: http://earthobservatory.nasa.gov/Study/HeatBucket. Accessed December 4, 2006.

Serway, Raymond A., and John W. Jewett, Jr. *Physics for Scientists and Engineers*, 6th ed. Belmont, Calif.: Brooks/Cole-Thomson Learning, 2004.

sport fish Sport fish are fish that are caught for sport and usually subject to fishing regulations. They may also be called game fish. People involved in sport fishing are often referred to as anglers. They catch or attempt to catch for personal use (not for sale or barter) freshwater, marine, or ANADROMOUS fish species by hook and line. Freshwater species such as bluegill (*Lepomis macrochirus*), largemouth bass (*Micropterus salmoides*), and northern pike (*Esox lucius*) are caught in streams, ponds, lakes, and rivers. Marine species, which include cobia (*Rachycentron canadum*), weakfish (*Cynoscion regalis*), and tarpon (*Megalops atlanticus*), are caught in ESTUARIES and the ocean. Anadromous species such as Atlantic salmon (*Salmo salar*), striped bass (*Morone saxatilis*), and white sturgeon (*Acipenser transmontanus*), are born in freshwater, migrate to the ocean to grow into adults, and then return to freshwater to spawn. Nongame species such as oyster toadfish (*Opsanus tau*) and clearnose skate (*Raja eglanteria*) are usually not regulated.

Humankind has a long history of fishing for consumption and sport. Anglers use various types of

A record tuna catch of the early 1900s *(NOAA/NMFS)*

gear to catch fish for recreational purposes. Fishing rods are long tubular poles made of fiberglass, graphite, or a combination of fiberglass and graphite. Lengths and diameter differ depending on whether or not the fishing style involves casting or trolling baits. Handles on rods are designed to seat various types of reels, which are designed to accommodate fishing line. Reels have many parts and come in various styles such as flyreels, spinning reels, closed-face reels, revolving-spool reels, round reels, and big-game reels. The line is made of twisted silk and linen threads or synthetic materials such as braided nylon (multiple small fibers), monofilament (single strand nylon), woven polyester threads, and stainless steel wire. Various types and sizes of hooks are used with live baits, dead baits, and casting or trolling lures. Anglers may fish from the banks of a river, along an ocean beach, and in various types of vessels from skiffs in inland waters to large yachts in the open ocean.

Billfish are the best-recognized sport fish and are popularized on various television shows and in sport fishing magazines. They are large and fast predators with a rounded spearlike upper jaw, and they display jumping behaviors when caught. Fisherman cruise out onto the CONTINENTAL SHELF in sport fishing boats to catch these pelagic fish, which are generally found in the upper 300 to 600 feet (100 to 200 m) near submarine canyons, fronts, and weed lines. Billfish are highly migratory species, moving thousands of miles annually throughout the world's tropical, subtropical, and temperate oceans and MARGINAL SEAS. Billfish species include blue marlin (*Makaira nigricans*), white marlin (*Tetrapturus albidus*), sailfish (*Istiophorus albicans and Istiophorus platypterus*), and longbill spearfish (*Tetrapturus pfluegeri*). Fishing strategies include trolling with dead baits or drifting with live baits. Commonly used baits are ballyhoo (*Hemiramphus brasiliensis*), striped mullet (*Mugil cephalus*), and bonito (*Sarda chiliensis and Sarda sarda*). Typical trolling lures imitate forage fish such as Scombridae (the family of mackerels, tunas, and bonitos) and Cephalopoda such as squid and cuttlefish.

The numbers of fishermen and anglers across the globe are very large, and the numbers of species caught may rise or decline due to natural causes or mismanagement of a particular fishery. Overfishing can cause stocks such as Atlantic cod (*Gadus morhua*), found on both sides of the North Atlantic, to rapidly decline. In response, fisheries' biologists and policy makers are working together with fishing communities to reduce fishing mortality and promote rebuilding of spawning stock biomass. Fish management involves marine scientists who are working to maintain healthy, sustainable marine ECOSYSTEMS that provide future fishing opportunities for the next generation of anglers. Fish management involves activities such as monitoring, research, stocking, habitat improvement, evaluation, information and education, investigation, and controlling access to certain areas.

See also AQUACULTURE; FISH; FISHERMAN; HIGHLY MIGRATORY SPECIES; MARICULTURE; MARINE FISHES.

Further Reading

American Fisheries Society. Available online. URL: http://web.fisheries.org/main. Accessed November 27, 2006.

Angler Information, the Staff Office for Intergovernmental and Recreational Fisheries, National Marine Fisheries Service, National Oceanic and Atmospheric Administration. Available online. URL: http://www.nmfs.noaa.gov/irf/angler.html. Accessed January 28, 2007.

The Billfish Foundation. Available online. URL: http://www.billfish.org/new. Accessed January 28, 2007.

Froese, R., and D. Pauly, eds. *FishBase*. Available online. http://www.fishbase.org/home.htm. Accessed November 27, 2006.

LaBonte, G. *Fishing for Sailfish*. Point Pleasant, N.J.: The Fisherman Library, 1994.

World Wildlife Fund. Available online. URL: http://www.worldwildlife.org. Accessed January 28, 2007.

standing wave A standing wave is a wave (*see* WAVES) that oscillates without progressing, also known as a stationary wave. A standing wave results when a wave in a system interferes with itself after reflecting from the system's boundaries. Everyday examples include plucked or bowed strings, water sloshing in a bathtub, and the vibrating air columns in organ pipes. In the case of a stretched string on a mandolin, for example, when a string is plucked, transverse waves travel along the string and reflect from both ends, forming a standing wave. The locations along the standing wave where the string medium is not moving are called nodes, while locations of maximum vibrational amplitude are called antinodes. The lowest frequency of vibration, in which the particles of the string oscillate in unison, is called the fundamental or first harmonic. The string can also oscillate in two loops, called the second harmonic, in three loops, the third harmonic, and so forth. This theory has application in marine science to the study of seiches and tides. In the case of a SEICHE, a standing wave is generally the result of strong winds or changes in barometric pressure that occur in lakes and semi-enclosed water bodies of the ocean. However, the horizontal water current is a maximum at the node, and a minimum at the antinodes, where the motions parallel to the boundaries of the wave are entirely vertical. The period of a standing water wave depends on the length and depth of the water body.

Stationary wave theory is applied to the study of the TIDES and assumes that the basic tidal movement in the open ocean is a system of standing wave oscillations; progressive waves become important only as the tide advances into tributary waters such as MARGINAL SEAS, bays, ESTUARIES, and rivers.

Further Reading

Knauss, John A. *Introduction to Physical Oceanography,* 2nd ed. Upper Saddle River, N.J.: Pearson/Prentice Hall, 1996.

statistics Statistics refers to an inductive science that attempts to draw conclusions about nature from observations and the mathematics of probability theory. Statistics is concerned about the errors of observations, and whether the observations from a limited sample of data really represent the population of data as a whole. Statistical methods are used in marine science to analyze past data and to make inferences about the future. Statistical methods are widely used in meteorological and oceanographic FORECASTING. Statistical methods indicate the best procedures for obtaining data required as input to numerical prediction models. Statistical information may also be used to assess the skill or to calibrate numerical models (*see* NUMERICAL MODELING). Statisticians design sampling schemes to produce information that is representative of the phenomenon under study. For example, marine scientists measure deep-water wave heights, periods, and directions at a given location as a function of wind speed, duration, and FETCH, and seek to infer characteristics of deepwater waves under similar generating conditions at any other location or time. They apply the theory of probabilities to forecast extreme events such as 100-year floods and wave heights.

Many processes in marine science appear to be drawn from a normal population (governed by the normal probability distribution) that can be characterized by a "mean" μ, the average or expected value of the population, and a "standard deviation," σ. When the mean and standard deviation are known for a normal distribution, confidence intervals, or statements of uncertainty, can be made about specific samples. In practice, the mean and standard deviation of a population are not known, but can be estimated from limited amounts of data. The mean

of x_i observations of, say wave height, can be estimated by the following simple average

$$x_{avg} = (1/n)\Sigma x_i, \text{ where } i = 1,2,3,...,n.$$

and the standard deviation can be estimated by

$$s = [1/(n-1)\Sigma(x_i - x_{avg})^2]^{1/2},$$

where the sum is over the total number of observations. The estimated standard deviation provides a rough measure of the intensity of the data as fluctuations about the mean. The variance, the square of the standard deviation s^2, is proportional to the mechanical energy in a process such as water wave generation (*see* TIME SERIES ANALYSIS). For samples drawn from a normal population, it is possible to determine confidence intervals around the mean and standard deviation to determine their probability of being correct.

To determine confidence intervals, one starts with the theoretical "t-distribution," given by

$$t = (x_{avg} - \mu)/[\Sigma(x_i - x_{avg}^2)]^{1/2}/n(n-1)$$

If the value of t is obtained from tables or a computer program, then an interval can be constructed around the mean with a given probability of enclosing the true value of the mean. For example, if the limits of t are chosen as $-t_{.05}$ to $t_{.05}$, then the probability Pr of the mean being between two specified limits is 90 percent, that is, there is a 90 percent chance of enclosing the true value of the mean μ, or, in equation form:

$$Pr[x_{avg} - t_{.05}[\Sigma(x_i - x_{avg})^2]^{1/2}/n(n-1) < \mu < x_{avg} + t_{.05}[\Sigma(x_i - x_{avg})^2]^{1/2}/n(n-1)] = .90.$$

Similar methods can be applied to determine the confidence limits bounding the variance, except that the chi-square distribution is used instead of the t-distribution.

One of the most widely employed tools in statistics is the frequency distribution, which organizes and summarizes the data in either a table or a graph called a histogram. The histogram is a plot of the number of occurrences (frequencies) within each data class interval. The shape of the histogram provides clues as to the theoretical probability distribution function (such as the normal distribution), that governs the process, and which enables probability statements to be made about the process. The class intervals must be carefully chosen, because if they are too large, important details will be lost; if they are too small, the histogram does not do much organizing of the data.

In marine science, one often deals with vector processes, which are characterized by a magnitude and a direction, such as current velocity (see VECTOR). Vector data can be represented as two-way frequency distributions, such as a table of current speed versus direction. The table defines data intervals, in which the number of occurrences (frequencies) of current for a given speed and direction are entered into the table in class intervals. Contour lines can then be drawn through the frequency data (isoplething) to provide a visual display; for example, at intervals of every 100 observations. Another procedure is to develop a "current rose," in which vectors showing current direction are superimposed over the compass directions, the length of the vectors corresponds to the speeds, and the line thickness of the vectors corresponds to the frequencies of occurrence of currents of that speed and direction.

Current charts based on ship's observations have been developed that show current roses in each 1° square of an ocean area, such as the North Atlantic. Meteorologists and oceanographers have also used 10° x 10° squares or Marsden squares to identify and archive the geographic position of meteorological data.

A quick measure of the tendency of the current to flow in one direction, called the "stability," or "persistence," is to divide the resultant vector average speeds by the simple mean current speed. If the current always flows from the same direction, the stability is 1; if it is equally likely from any direction, it is 0. A statement about the current speed without a statement of the stability would not have much meaning.

A major use of statistics is to fit equations to data. For example, it may be hypothesized that sound speed is a linear function of temperature. By means of a mathematical procedure called least squares analysis, data can be fit to a straight line, and error, or confidence intervals, determined. If the confidence intervals are large, a higher order curve fitting can be performed and new confidence intervals determined. By this means, experimenters can assess whether their data can be described theoretically by an equation or frequency distribution. For data on several variables, such as sound speed as a function of TEMPERATURE, SALINITY, and PRESSURE, multivariate least squares techniques can be used to fit equations to the data.

The association between two variables, say sound speed and temperature, can be determined by calculation of a "correlation coefficient," that can vary from -1 to +1. A value of 0 would mean no relationship between the variables, a value of +1 would mean that the temperature change is always accompanied by sound speed change, and a value of -1 would mean that a positive temperature change would always be accompanied by a negative sound speed change. Statistical procedures are used to place confidence limits on correlation coefficients. For time series data, the concept of correlation coefficient can be extended to correlation functions that describe the behavior of the time series. Time series can be Fourier-transformed to obtain the energy spectrum. For data of several variables, a "correlation matrix" is calculated that gives the correlation between any two variables. Many multivariate statistical techniques such as analysis of variance, canonical correlation, and principle component analysis are available to the data analyst.

For details on these and many other procedures, the reader is referred to books on probability and statistics. Contributions by scientists such as Leon E. Borgman (1928–2007) have been instrumental in applying ocean wave statistics to the risk analysis of ocean structures. Professor Borgman was inducted into the National Academy of Engineering during 1999.

There are many innovations in characterizing the environment. Some recent advances have been made in evaluating natural systems as fractals that describe nonlinear relationships. Fractals are self-similar geometric figures that may share the same statistics at different spatial locations. The determination of order within nonlinear systems is the subject of CHAOS theory.

Further Reading

Gleick, James. *Chaos: Making a New Science.* New York: Penguin, 1987.

Kendall, Sir Maurice. *Multivariate Analysis.* New York: Hafner Press, 1975.

Neumann, Gerhard, and Willard J. Pierson, Jr. *Principles of Physical Oceanography.* Englewood Cliffs, N.J.: Prentice Hall, 1966.

Panofsky, Hans A., and Glenn W. Brier. *Some Applications of Statistics to Meteorology.* University Park, Pa: Mineral Industries Extension Service, 1958.

Walpole, Ronald E., and Raymond H. Myers. *Probability and Statistics for Engineers and Scientists.* New York: MacMillan, 1972.

stenothermal Stenothermal is a biological term describing an organism that is capable of living or growing within a limited temperature range. For example, marine biologists classify fish such as burbot

(*Lota lota*) which are typically found in waters below 53° Fahrenheit (12°C), as cold water stenothermal species. Burbot, a freshwater cod, inhabit glacial rivers such as the Yukon and Tanana Rivers in Alaska and are reported to migrate away from warmer waters. CORAL REEF communities, where water temperatures are greater than 65°F (18°C), are inhabited by many stenothermal species of vertebrates and invertebrates. Damselfishes such as the Panamanian sergeant major (*Abudefduf troschelii*) are warm-water stenothermal species. This fish has a compressed ovoid body and distinctive black stripes on a silvery white body tinged with yellow.

Temperature affects the physiological system of stenothermal species more significantly than eurythermal species, which grow over a wider temperature range. Stenothermal marine species favor different zones in the water column, even as one extends offshore from the shallow neritic and out into the deep oceanic zones. Eurythermal species tend to be found above the THERMOCLINE or in coastal waters where temperatures fluctuate, while stenothermal species, requiring constant temperatures, are found below the thermocline or in the Tropics. Some species migrate with the temperature gradient in order to maintain a constant body temperature; others slow down in the cold; and specialized species make physiological adjustments.

Marine scientists are interested in how organisms adapt to the environment, especially extreme environments such as the cold waters of deep sea and polar marine regions or the hot temperatures associated with HYDROTHERMAL VENTS. Of particular interest are thermal habitats such as coral reefs, which may be severely stressed by cool waters. Scientists studying the effect of temperature on physiology may assess heart function and metabolism in varying species under different temperatures. Understanding physical factors such as temperature, which affects the success of an organism, is important to forecasting distribution patterns of species as a response to climate change.

See also REEF FISH.

Further Reading

Arctic Studies Center, Department of Anthropology, National Museum of Natural History, Smithsonian Institution. Available online. URL: http://www.mnh.si.edu/arctic. Accessed February 1, 2007.

NOAA's Coral Reef Information System (CoRIS). Available online. URL: http://coris.noaa.gov. Accessed February 1, 2007.

stock assessment   Stock assessment is the application of statistical and mathematical tools to fisheries data in order to obtain a quantitative understanding of the stock's status and to assess whether changes are due to natural or human-related causes. A stock is one species forming a group of similar ecological characteristics that is subject to assessment and management. Stock assessment is the process of studying the effects of fishing on fish populations. Statistical and mathematical tools are used to describe the numbers of selected fish species that share common ecological and genetic features. Fisheries surveys collect data that is controlled by marine scientists, for example by collecting fish samples from research vessels or by observing catches aboard a COMMERCIAL FISHING vessel. Data not only includes estimates of the total removals due to human activities (for example, fishery landings and discarded bycatch), but biological data such as growth rates, age of sexual maturity, maximum longevity, and the proportion of each age group dying each year due to natural causes, and other factors that affect stock productivity. These data are analyzed using models to estimate historical and recent trends in the stock, the number of small fish entering the fishery each year (recruitment level), and the fraction of the stock alive at the beginning of the year that are killed by fishing (exploitation rate). Sophisticated models consider parameters such as age and size as well as physical oceanographic effects such as TEMPERATURE, SALINITY, circulation, and SEA STATE. The goal for this analysis is determining whether or not the stock is within biologically acceptable yields for healthy stocks, or within the expected rates of rebuilding for depleted stocks.

Fisheries biologists that understand the life histories of the target species, catch statistics, abundance, and complex ecosystem dynamics use the stock assessments as a basis to establish biologically acceptable yield for healthy stocks, as well as the expected rate of rebuilding for depleted stocks. A biological threshold may be established that indicates that a limit has been reached. If the stock is below a minimum threshold, the stock is in an "overfished condition." Status determinations are often made using these kinds of biological reference points. Once stocks are assessed to be in an "overfished condition" or if overfishing is occurring, fisheries managers must implement measures to rebuild the stock and eliminate overfishing. Such actions synchronize with international efforts to develop sustainable harvest policies that eliminate overfishing of both target and associated species. Continued research by marine scientists enables fisheries managers to optimize fishing mortality targets while still maintaining a low risk to target and associated species and the marine ecosystem.

There is much research to be done since marine scientists are still identifying new species and unlocking the secrets about the life history of known species. Research vessels are required to complete surveys that indicate abundance, spatial and temporal distributions (stock structure), associated species, habitat, and oceanographic variables. Observers on fishing vessels provide important information on species composition, amounts of each species kept and discarded, and fishing effort. The application of marine science to fisheries management is critical for maintaining sustainable marine fisheries. This is especially true since a contemporary fishing fleet applies technological advances to improve the harvest of marine fisheries. Collaboration among fishermen and marine scientists is critical even if it simply involves sharing information on the movement, growth, and capture of marine species (for example tagging studies). Fisheries management in the 21st century must involve adequate data collection, science-based stock assessment, and harvest control.

See also BIOLOGICAL REFERENCE POINT; FISHERIES RECRUITMENT; FISHERY CONSERVATION AND MANAGEMENT ACT; YEAR CLASS STRENGTH.

Further Reading

Mace, Pamela M., Norman W. Bartoo, Anne B. Hollowed, Pierre Kleiber, Richard D. Methot, Steven A. Murawski, Joseph E. Powers, and Gerald P. Scott. "Marine Fisheries Stock Assessment Improvement Plan: Report of the National Marine Fisheries Service National Task Force for Improving Fish Stock Assessments, NOAA Technical Memorandum NMFS-F/SPO-56, Silver Spring, Maryland: National Marine Fisheries Service, National Oceanic and Atmospheric Administration, U.S. Department of Commerce." Available online. URL: http://www.st.nmfs.noaa.gov/StockAssessment/tm_spo56/2001ImprovePlan.doc. Accessed February 3, 2007.

storm surge Storm surge describes water fluctuations in the coastal ocean or lakes, resulting from water pushed toward the shore by the force of the winds. Storm surge may also be called wind setup. It is a rise in water levels attributed to interactions among wind speed, wind direction, FETCH, atmospheric pressure, tides, waves, BATHYMETRY, and nearshore slopes or BEACH profiles. Fetch is the distance over which the wind blows and generates waves upon the water. Extreme storm surges are associated with severe tropical CYCLONES OR HURRICANES and have caused repeated damages in places such as the GULF OF MEXICO, where there is a broad and shallow CONTINENTAL SHELF. Storm surges may also increase

The storm surge of Hurricane Eloise rushes toward shore. *(NOAA/National Weather Service)*

along ESTUARIES, owing to the funneling effect of the converging shoreline.

In the Northern Hemisphere, winds circulate in a counterclockwise fashion around a low-pressure system. When winds swirl in a counterclockwise fashion in the Northern Hemisphere or clockwise in the Southern Hemisphere, the flow is called "cyclonic." For a system that is moving shoreward, waters are raised (set up) to the right of the storm and lowered (set down) to the left. An advancing surge may also combine with high tides and superimposed with SURF to create severe flooding. For low-lying areas that are densely populated, storm surges will be devastating as evidenced by Hurricane Katrina, which made landfall on August 29, 2005, near Buras, Louisiana. The storm surge caused catastrophic damage along the Gulf of Mexico coast, devastating many coastal townships such as Venice, Louisiana, in Plaquemines Parish and cities such as New Orleans and Slidell in Louisiana and Bay St. Louis and Waveland in Mississippi. Gentle gradients off the coast allow a greater surge to inundate the coastal zone. Storm surges propagating into seaports, which are usually located in estuaries, may cause severe damage to ships, marinas, and boats. Vessels tend to break free from their moorings, especially if the SCOPE and mooring lines have not been adjusted for the sudden rise in water level.

Along BARRIER ISLAND coasts, storm surge may overflow the dune system, causing overwash fans of beach material. Overwashes may not only destroy dune structures and vegetation but may contribute to the flooding of beach properties. Storm surge and erosion are a natural recurring process. Dunes and beaches tend to rebuild themselves during quieter conditions as low-energy waves transport sand from offshore bars and the surf zone to the beach, causing accretion. Thus, the call by researchers and coastal zone managers to protect the beach and dune system from real estate development at the expense of major geological features such as the foredune ridge. The foredune ridge is the highest of a series of dunes and is somewhat stabilized by vegetation such as American beachgrass (*Ammophila breviligulata*), beach pea (*Lathryrus japonicus*), sea oats (*Uniola paniculata*), and seashore lupine (*Lupinus littoralis*). The dune system helps to block storm surge and provides a bank of sand for natural beach nourishment. Extreme storm surge during winter nor'easters along the East Coast of the United States or during severe tropical cyclones may even erode enough beach material along a barrier island to create a new COASTAL INLET. For example, along the Outer Banks of North Carolina there have been between three and 11 major coastal inlets. Oregon Inlet, which separates Bodie Island from Pea Island, was formed in the wake of a hurricane during 1846.

A storm surge is a water fluctuation that may last from a few minutes to a few days. Moving cyclonic atmospheric pressure systems and the ensuing wind stress are the principal causes. As the wind blows, it creates surface currents that are impeded by friction in shallow water, which causes the water level to rise. The water along the downwind portion of the moving low-pressure system is set up, while the water along the upwind portion of the moving low-pressure system is set down. The lowered water levels may be particular important in situations such as when Hurricane Andrew skirted the west Florida coast during August 1992. The cyclonic winds drastically lowered and then raised water levels in Tampa Bay. Measurements of the storm surge were recorded in water level gauges along the coast where the surge was estimated by subtracting the predicted tides from the actual recorded water levels. Water levels were recorded every six minutes on the National Ocean Service's water level gauges. Structures must contend with not only the surge, but also, storm waves and surf which propagate on top of the surge. Other factors in the water level fluctuation are the tidal signal and the inverse barometer effect. Tides are the periodic rise and fall of waters from astronomical forces, while the inverse barometer effect is the change in sea level from changes in atmospheric pressure.

See also SURF; TIDE; WATER LEVEL FLUCTUATION; WAVES.

Further Reading

Center for Operational Oceanographic Services and Products, National Ocean Service, National Oceanic and Atmospheric Administration. Available online. URL: http://tidesandcurrents.noaa.gov. Accessed March 10, 2007.

National Oceanic and Atmospheric Administration. "Hurricane Andrew–10 Years Later." Available online. URL: http://www.noaa.gov/hurricaneandrew.html. Accessed March 10, 2007.

———. "Hurricane Katrina–Most Destructive Hurricane to Ever to Strike the U.S." Available online. URL: http://www.katrina.noaa.gov/. Accessed March 10, 2007.

stratigraphy   Stratigraphy is the study of the formation, composition, sequence, and correlation of rock strata or layers. Investigators assume that the youngest rock layers lie on top of the older rocks (the law of superposition). Historical events may be determined through analysis of strata found along cliff faces, through excavations and deep core samples. Marine scientists focus on the relative and numerical age relationships in the sediments. Sedimentary rocks form from the accumulation and lithification of sediment. As sediment particles accumulate, they compact, become cemented together, and lithify or harden.

Stratigraphy can be broken down into a number of different branches. Geologists, interested in lithostratigraphy, determine which types of rock dominate certain layers based on factors such as grain size, color, and mineralogic composition. Fragments of rock included within a sample (another rock) are older than the surrounding rock. Geologists may also use fossils to determine the order and age of strata. For example, trilobites are associated with the Paleozoic era (542–251 million years ago). These ancient arthropods went extinct before the dinosaurs. Extinct fossils do not reappear in younger strata. Evaluating the fossilized remains of flora and fauna in the layers is called biostratigraphy. Scientists with knowledge of the Earth's magnetic field can study orientations of magnetites in the strata. The hot liquid outer core, along with axial rotation, causes the Earth's magnetic fields, and there is daily wondering and episodic reversals in polarity. The reversals are supported by geological evidence that dates back to the mid-Jurassic, such as when lava flows solidified (*see* PLATE TECTONICS and SEAFLOOR SPREADING). Large 20th-century research programs such as the Integrated Ocean Drilling Program have shown that the seafloor is older with increasing distance from spreading centers (oceanic ridges). This science is called magnetic stratigraphy and uses the polarity record to date sedimentary features. Determining ages based on chemical composition is called chemostratigraphy; it usually involves analysis of isotopes of carbon (C), hydrogen (H), nitrogen (N), oxygen (O), silicon (Si) and sulfur (S). Elements of these chemicals have the same atomic number but different atomic mass. Abundances of these isotopes are sufficiently high to assure precise measurement of isotope ratios by mass spectrometry. Basically, geochemists look at different ratios of isotopes, for example oxygen (oxygen 16, oxygen 17, and oxygen 18) obtained from ice cores, CORAL REEFS, and sediments to evaluate global temperatures and sea level. The numbers 16, 17, and 18 refer to the atomic mass. Thus, oxygen 16 is called "light oxygen" and oxygen 18 is called "heavy oxygen." When seawater evaporates, the concentration of water vapor is enriched in oxygen 16, as evidenced by ratios of oxygen 18 (^{18}O) to oxygen 16 (^{16}O). Sea level also responds to climatic warming (rise) and cooling (fall), which can be correlated to ratios such as ^{18}O/^{16}O or δ^{18}O.

Marine scientists can learn a great deal about the Earth's history from the sediments, especially since sediments accumulate continuously. Researches may drill a stratigraphic test hole hundreds of feet deep to obtain core samples for lithostratigraphic, biostratigraphic, magnetostratigraphic, chemostratigraphic, and paleontologic data. Cores may contain a wide range of sediment types from biogenic OOZES and chalk to lithogenic materials. The location of calcareous nannofossils tells researchers something about the depth of the ocean based on the calcite COMPENSATION DEPTH. A cryogenic magnetometer may be used to determine the magnetization of sediment samples at regular intervals along the core. Stable isotopes may help in assessing temperature variations in the deep ocean or in the growth of ice sheets. A mass spectrometer may be used to provide the isotopic ratio of the sample relative to a given standard. Other techniques such as radioisotope dating can be used alongside stratigraphy in order to determine average rates of deposition, rock composition, the width and extent of the strata, the fossils contained, and ultimately the geological history. Marine paleontologists are often involved in dating and correlating stratified rocks by means of stratigraphy, especially biostratigraphy. Some high-technology methodologies employed by marine geologists may include using side-scan SONAR to characterize surface sediments, ground-penetrating RADAR to image bedrock, and VIBRACORERS to collect sediments.

See also OCEAN DRILLING PROGRAM; PALEONTOLOGIST; RADIOMETRIC DATING; RELATIVE DATING; SEDIMENT.

Further Reading

Commission for the Geological Map of the World. Available online. URL: http://ccgm.free.fr/index_gb.html. Accessed February 9, 2007.

Gon, Sam, III. "A Guide to the Order of Trilobites home page." Available online. URL: http://www.trilobites.info. Accessed October 16, 2007.

Haeni, F. Peter, Marc L. Buursink, John E. Costa, Nick B. Melcher, Ralph T. Cheng, and William J. Plant. "Ground-Penetrating RADAR Methods Used in Surface-Water Discharge Measurements." Available online. URL: http://water.usgs.gov/ogw/bgas/publications/gpr_surfwater/index.html. Accessed October 16, 2007.

International Commission on Stratigraphy home page. Available online. URL: http://www.stratigraphy.org. Accessed February 9, 2007.

Kusky, Timothy. *Encyclopedia of Earth Science*. New York: Facts On File, 2005.

National Museum of Natural History, Smithsonian Institution. "Geological Time." Available online. URL: http://www.nmnh.si.edu/paleo/geotime/main/index.html. Accessed February 9, 2007.

subduction zone   A subduction zone is an area where an oceanic tectonic plate plunges down into Earth's mantle. The deep-ocean TRENCHES are found directly above the subducting plate. Volcanic island chains are often found on the continental side of the trench, for example, the Japan Trench. The volcanoes on these islands have been formed by magma being forced upward by the downward-moving plate and are the most powerful on Earth. Subduction zones are also regions of powerful earthquakes, which often generate large TSUNAMIS. Subduction zones bound the Pacific Ocean, as can be seen from the ocean trenches on its perimeter. The large number of volcanoes associated with this process is known as the RING OF FIRE.

Subduction zones are at the opposite ends of the tectonic plates (*see* PLATE TECTONICS) from the spreading ridges where they are formed. As the newly formed plates move away from the ridges, they acquire sediment on top and cooling magma underneath from the mantle, both of which thicken the plate and make it heavier. Eventually, the plate becomes too heavy to remain on the surface, and it sinks down into the mantle, where, over millions of years, it melts and is "recycled" as a new plate.

Subduction zones may be found at converging oceanic plates, or by an oceanic plate and a continental plate. In the latter case, the oceanic plate, being denser, always plunges under the continental plate. Because the oceanic plates are continually being "recycled," they are very young compared to the age of the continental plates, which have been on the surface of the Earth for billions of years.

An example of a subduction zone is found in the West Coast of the United States, where the Pacific

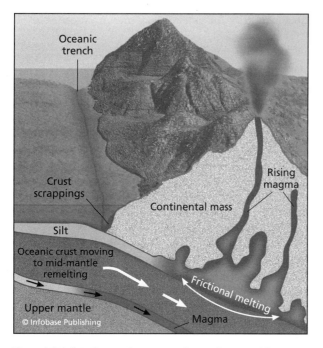

Oceanic trench

Crust scrappings

Silt

Oceanic crust moving to mid-mantle remelting

Upper mantle

© Infobase Publishing

Rising magma

Continental mass

Frictional melting

Magma

The subduction of oceanic crust on the seafloor provides new molten magma for volcanoes that fringe the deep-ocean trenches and provide material for new crust.

plate is diving under the North American plate, and in the process, forming mountain chains such as the Cascades. The San Andreas Fault is found in this region; major earthquakes have occurred recently in both San Francisco and Los Angeles.

When oceanic plates collide, island arcs are formed, as in the case of the Aleutian chain. The curvature of the arcs is a feature of the spherical Earth (the intersection between a sphere and a plane), and can be seen quite distinctly in following on a map the chain from Alaska to Russia.

submarine canyons   Submarine canyons are undersea canyons cutting through and generally perpendicular to the continental margin. They typically have V-shaped steep sides, emanating from river valleys of present and past major rivers, such as the Hudson. The Hudson canyon is a continuation of the Hudson River Valley. Many of the canyons were cut during the ice ages when sea level was lower and the rivers extended almost to the present CONTINENTAL SLOPE. The SEDIMENT from these rivers could then have traveled down the canyons like waterfalls, providing the continents with major pathways to discharge sediment onto the CONTINENTAL RISE and the ABYSSAL PLAIN. Catastrophic events such as earthquakes and severe storms can cause TURBIDITY CURRENTS (underwater mudslides) that scour the canyon walls and dump great quantities of sediment onto the rise. Again, this

transport would have been greater during the ice ages, when severe storms were more frequent and water levels were low, so that that wave action extended very deep into the water column and close to the bottom. For northern canyons, the material deposited by turbidity currents includes terminal moraine from the glaciers, deposited during the ice ages.

These bypass zones, which transport sediments from continents to the oceans, can be sized in different manners. The largest cross section is the Zhemchug Canyon, named for the former Soviet Union expeditionary vessel *Zhemchug*. The second-largest in area and in volume of incision is the Navarin Canyon. The longest is the Bering Canyon. These large canyons are all found on the North American margin of the Bering Sea.

Further Reading

SIMoN: Sanctuary Integrated Monitoring Network, Monterey Bay National Marine Sanctuary Offices. Available online. URL: http://www.mbnms-simon.org. Accessed August 6, 2007.

submerged aquatic vegetation (SAV)   The term used to describe vascular seed plants that are anchored to the sea bottom by true roots is submerged aquatic vegetation (SAV). Normally these plants remain completely submerged, but their uppermost shoots may be exposed during an extremely low tide (such as spring tide). SAVs have true leaves, stems, and roots (that is, these organs are supplied with vascular tissue such as the xylem and phloem). They produce rather inconspicuous flowers and fruits that contain seeds. The plants may spread by seed dispersal (sexual reproduction) or by the growth of rhizomes (horizontal stems that grow slightly beneath the bottom) that send up new shoots at the edges of existing grass beds (asexual reproduction).

SAVs are seasonal and die back with the approach of cold-water temperatures. Growth resumes with the return of warm-water temperatures. During the growing season, plants may be cropped off, and they will grow back. Crab scrapers (watermen who employ a device called a crab scrape—a V-shaped steel frame with a smooth metal bar and trailing net that glides along the bottom—drag their "scrapes" through the grass beds in search of peelers and soft crabs. The "scrapes" lop off the SAV plants and a bale of grass comes up with the scrape when the watermen retrieve it. The plants grow back from the roots, rhizomes, and shoots left behind. One could think of the practice as mowing an underwater lawn. The "scrapes" lack teeth and therefore, do not dig up the roots and rhizomes that anchor the plants to the bottom. Soft-shelled clammers, on the other hand, have been criticized for using their hydroescalators in and around grass beds because they dig up the bottom in pursuit of the soft-shell clam (*Mya arenaria*) that lives a foot or more beneath the bottom.

There are both freshwater and saltwater species of SAVs. An ESTUARY such as CHESAPEAKE BAY provides a habitat for many different species because of its wide range of salinities. Up rivers and in the upper bay, one may find freshwater species such as Wild Celery (*Vallisneria Americana*) and Redhead grass (*Potamogeton perfoliatus*), while in the Tangier Sound area, one finds a mix of Widgeon grass (*Ruppia maritima*) and eelgrass (*Zostera monina*). SAVs can grow only when sufficient light penetrates the water. They normally grow to a depth of nine feet, but too many nutrients in the water may spawn excess PHYTOPLANKTON growth that clouds the water. Sediments from runoff may suspend, also blocking light, or settle, smothering the plants.

Submerged aquatic vegetation is ecologically important for several reasons. It provides oxygen through photosynthesis, excellent habitat for numerous organisms, food for waterfowl, stabilization of bottom sediments, and slow uptake and release of nutrients. In addition, SAVs buffer against shoreline erosion by reducing wave energy and filter out sediments carried by runoff by slowing water causing the suspended particles to settle to the bottom or stick to the plants.

In the last few decades, SAVs have declined due primarily to an increase in TURBIDITY and a decrease in water clarity. Runoff from coastal construction may stress or kill SAV beds. SAVs favor clear water that does not contain so many nutrients that PLANKTON blooms are triggered.

See also SEA GRASS.

Further Reading

Bay Grass Key, Maryland Department of Natural Resources. URL: http://www.dnr.state.md.us/bay/sav/key/. Accessed March 14, 2008.

Lippson, Alice Jane, and Robert L. *Life in the Chesapeake Bay,* 2nd ed. Baltimore and London: The Johns Hopkins University Press, 1997.

submersible The term *submersible* describes various types of autonomous, remotely operated, and manned underwater vessels used for marine science research and exploration, especially in the deep sea. Submersibles are much smaller and have a shorter range than submarines. Autonomous Underwater Vehicles, or AUVs, are untethered underwater robots capable of navigating and carrying out prescribed missions such as exploration, pipeline monitoring, and hydrographic surveying without an onboard pilot. They can operate near obstructions where tethered systems may become entangled. The AUV known as ABE, which stands for Autonomous Benthic Explorer, is best known for imaging HYDROTHERMAL VENTS on the seafloor. Remotely Operated Vehicles or ROVs are connected by a communications cable to a mother ship. An ROV operator controls the submersible from a platform (ship, dock, BARGE) at the surface. Like AUVs, ROVs can be effectively used in treacherous conditions without endangering human lives. ROVs are used extensively by the oil and salvage industry. They are loaded with cameras, SONAR, lights, and robotic arms. The British Royal Navy's Submarine Rescue Team uses the ROV *Scorpio* and the manned submersible *LR5* for deep-sea rescue. Some other famous manned submersibles include the deep submergence vessel *Alvin* operated by the WOODS HOLE OCEANOGRAPHIC INSTITUTION in the United States; the *Mir-1* and *Mir-2* operated by the P. P. Shirshov Institute of Oceanology in Russia; the French submersible *Nautile,* which is operated by the French Research Institute for Exploitation of the Sea (Ifremer); and the Japanese research submersible *Shinkai,* which is operated by the Japan Agency for Marine-Earth and Science Technology.

Besides academic research and exploration, some commercial companies have operational submersibles that take tourists to depths of approximately 120 feet (36.6 m) to see wrecks, reefs, and marine life. Most of these submersibles are electric powered, outfitted with a full suite of bow, vertical, and stern thrusters, underwater communications, lights, viewing portals, and ballast tanks. These vessels meet requirements specified by classification organizations such as the AMERICAN BUREAU OF SHIPPING. A dive may include approximately 60 passengers and last for approximately 45 minutes.

The U.S. Navy's rescue submarine DSRV1   *(U.S. Navy)*

The Perry submarine diving through the ice   *(NOAA/NURP)*

Several surface vessels or tenders support the submersible. The crew is involved with vessel operations, ensuring that all safety requirements are met, and highlighting to the passengers interesting finds such as several green sea turtles swimming gracefully near a wreck, sharks napping in a cave, or a school of barracuda drifting in the water column. Locations where opportunities exist to explore the undersea world in a submersible include locations near the Caribbean and Hawaiian Islands.

See also AUTONOMOUS UNDERWATER VEHICLE; REMOTELY OPERATED VEHICLE.

A Westinghouse submersible of the "Deep Star" series
(NOAA/NUR)

Further Reading

Association for Unmanned Vehicle Systems International. Available online. URL: http://www.auvsi.org. Accessed February 7, 2007.

Atlantis Adventures, Vancouver, B.C., Canada. Available online. URL: www.atlantissubmarines.com/. Accessed March 14, 2008.

The Autonomous Systems and Controls Laboratory, Virginia Polytechnic Institute and State University. Available online. URL: http://www.ascl.ece.vt.edu. Accessed February 6, 2007.

Center for Autonomous Underwater Vehicle (AUV) Research, Naval Postgraduate School. Available online. URL: http://www.cs.nps.navy.mil/research/auv/auvframes. html. Accessed February 6, 2007.

Center for Ocean Technology, University of South Florida. Available online. URL: http://cot.marine.usf.edu. Accessed February 6, 2007.

Deep Submergence Laboratory, Woods Hole Oceanographic Institution. Available online. URL: http://www. dsl.whoi.edu. Accessed February 6, 2007.

Institut français de recherche pour l'exploitation de la mer, Issy-les-Moulineaux, France. Available online. URL: http://www.Ifremer.fr. Accessed February 6, 2007.

Japan Agency for Marine-Earth and Science Technology. Available online. URL: http://www.jamstec.go.jp. Accessed February 6, 2007.

Monterey Bay Aquarium Research Institute. Available online. URL: http://www.mbari.org. Accessed February 6, 2007.

P. P. Shirshov Institute of Oceanology of the Russian Academy of Sciences. Available online. URL: http://www.ocean.ru/index_en.htm. Accessed February 6, 2007.

surf Sea waves that break on a shore or on shoals, usually accompanied by lines of foamy water caused by entrapped air as the waves break, are called surf. As SWELL propagates from distant storms into shallow water, the wave crest may outrun the slower-moving base, causing the formation of breakers or surf. This occurs because of friction as the waves are shoaling and WAVELENGTHS are decreasing as wave heights are increasing. Waves in deep water are limited by wave steepness. They break when they become unstable, which occurs as the ratio of wave height to wavelength increases beyond 1/7. In shallow water, the incident wave's stability is a function of depth and the beach gradient. Incident waves break as surf when depth becomes less than 1.3 times the wave height. The wave height is two times the amplitude. Other factors influencing how waves break near the shore include TIDES, CURRENTS, winds, and interactions with other waves.

Breaking waves in the surf zone may dissipate their energy as heat, sound, weathering of rock, movement of bottom material, and the exchange of gases. Interestingly, smaller waves may build up a BEACH, while larger, more energetic waves may tear down a beach by moving beach materials offshore—a major cause of beach erosion. Grains are deposited on the shore by the smaller waves (see SEDIMENT TRANSPORT). Shoreline communities spend many millions of dollars rebuilding beaches. The basic types of breaking-water waves are spilling, plunging, collapsing, and surging surf. Spilling surf occurs on gently sloping beaches and dissipates its energy gradually as evidenced by the frothy crests, which spill down the leading slope of the breaker. Plunging surf is found on gentle beach slopes and dissipates its energy dramatically as the crest of the wave curls over and crashes just forward of the wave's base. Plunging surf is the preferred type of wave for surfing with its characteristic arched back and concave front. Plunging waves usually result from long period swell or beaches that have a slightly steeper gradient than those where spilling surf dominates. Surging

surf is caused by steep beach gradients where the wave's base rushes (surges) up the beach before the crest can break. When surf is in transition (plunging to surging) owing to rapid bathymetric changes or from being arrested by offshore winds, the crest collapses, resulting in foam. Such waves are called collapsing surf.

Surfing is a popular sport in which a person attempts to paddle a surfboard just ahead of an incident wave that is going to break. The surfer slides down the forward slope of the wave from the crest toward the trough as the wave begins to shoal. Surfers study weather reports and beaches to find favorable swell (periods between 13 to 15 seconds, or longer) that will break along a gently sloping beach or reef. Surfers may try to ride the breaking wave all the way to shore or to where the surf passes from shallow water into deeper water closer to shore. Depending on the BATHYMETRY, the wave may reform and break again in even shallower water close to shore. Surfers call the forward slope of a breaking wave "the face," and the crest "the lip." Face height is the length of the steepening shoreward front of the wave from the top of the crest to the low part of the trough. It is the region where surfers ride the waves. In plunging surf, the space from the lip and along the face is called the barrel. Catching a barrel ride or riding the tube is considered one of the most difficult maneuvers in surfing. On the north shore of the Hawaiian island of Oahu is the Pipe, a

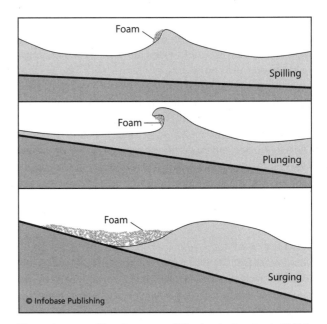

The main types of breakers are spilling (top), plunging (middle), and surging (bottom). The surging breakers are most constructive in replenishing beach sand.

favorite place for surfers because of its very large and stable swell waves.

See also GRAVITY WAVES; WAVES.

Further Reading
Los Angeles/Oxnard, National Weather Service Forecast Office. Available online. URL: http://www.wrh.noaa.gov/lox/. Accessed October 16, 2007.
National Weather Service Forecast Office. Available online. URL: http://www.prh.noaa.gov/hnl. Accessed January 4, 2007.
New Orleans/Baton Rouge, Louisiana, National Weather Service Forecast Office. Available online. URL: http://www.srh.noaa.gov/lix/. Accessed January 4, 2007.
Newport/Morehead City, North Carolina, National Weather Service Forecast Office. Available online. URL: http://www.erh.noaa.gov/er/mhx/. Accessed January 4, 2007.
Open University Course Team. *Waves, Tides, and Shallow-Water Processes,* 2nd ed. Oxford: Butterworth-Heinemann, 1999.
Storm Surf, Surf Forecast-Surf Report-Buoy Report-Buoy Forecast Wave Models-Weather Models home page. Available online. URL: http://stormsurf.com. Accessed October 16, 2007.

A modern surface combat vessel: the U.S. Navy's destroyer *Fife*   *(U.S. Navy)*

surface roughness In marine science, surface roughness is a vertical length scale referring to the presence of buildings, trees, dunes, sand waves, surface waves, and other irregular topographic, bathymetric, or sea surface features. The coastal zone is usually associated with micro-relief that affects land drainage, flood control, and critical habitats. At the seabed, sand ripples form in response to the effect of CURRENTS and waves on different grain sizes. On the sea surface, roughness is influenced by winds, precipitation, and in the case of marine spills, surface oil coverage. As oil films spread over the sea surface, they damp or reduce the height of surface waves. Being able to quantify surface roughness is especially important for NUMERICAL MODELING and REMOTE SENSING. Sea surface roughness generated by wind, waves, tides, and currents is also important to acousticians since it has a major effect on how sound travels underwater.

See also BEACH; SEDIMENT TRANSPORT; WAVES.

surface warfare The traditional definition of surface warfare involves activities that are conducted to destroy or neutralize enemy naval surface forces and merchant vessels. Surface warfare is also considered to be the community of naval personnel that are assigned to surface ships and involved with missions described as antisubmarine warfare, antiair warfare, antisurface warfare, land attack, and theater air missile defense. In general, surface warfare officers or SWOs are the men and women who lead sailors within the many specialized divisions or departments of a surface ship, such as operations and navigation, gunnery, and damage control. Example surface combatants are cruisers, destroyers, and frigates. Surface combatants provide protection of sea and air routes, ports, coastal airfields, and facilities and substantial command, control, and communications capabilities. Personnel aboard some surface combatants called Meteorology and Oceanography (METOC) officers and aerographers' mates prepare meteorological and oceanographic analyses and forecasts, including acoustic prediction products, for use by appropriate operational personnel. Most important, METOC information and forecasts are applied to determine environmental effects on ship sensors and weapons systems.

See also AEROGRAPHER'S MATE; AMPHIBIOUS WARFARE; METOC OFFICER.

Further Reading
Committee on Environmental Information for Naval Use, Ocean Studies Board, Division on Earth and Life Sciences, National Research Council. *Environmental Information for Naval Warfare.* Washington, D.C.: The National Academies Press, 2003.

sustainable yield The sustainable yield is the amount of a renewable resource that may be harvested on a continuous basis without compromising the resource's ability to renew itself. Since resources obtained from the ocean are not infinite, it is important to quantify the amount of fish, shellfish, plants, and minerals that are being harvested or extracted. Therefore, it is important to determine the maximum sustainable yield so that adequate populations or quantities of harvested or extracted resources remain available for future generations. Finfish are one of the most important resources that humans derive from the sea. People consume approximately two-thirds of the fish caught worldwide, and the remaining third is used commercially for feed or fertilizer. Fisheries and AQUACULTURE are economically important for many coastal communities. Fish is an especially important animal food in the diets of people from developing countries having access to fishing grounds.

Marine fisheries scientists and marine geologists determine maximum sustainable yields for marine resources. Fisheries scientists balance catch limits against high-technology fishing gear such as factory trawlers capable of harvesting threatening amounts of fish. In the United States, scientists from the Minerals Management Service (MMS) of the U.S. Department of the Interior manage the nation's rich offshore mineral resources located on the outer CONTINENTAL SHELF. Initiatives in the energy sector are studied by the MMS since development and use of fossil fuels depletes nonrenewable natural resources and is not without a cost to society. Scientists are developing alternative fuel sources such as ethanol and biodiesel, which are not presently viable replacements for hydrocarbons. Ethanol is an alcohol-based fuel produced by fermenting and distilling starch crops (sugar beets, sugar cane, and corn) that have been converted into simple sugars. Biodiesel is a fuel that can be manufactured from vegetable oils, animal fats, or recycled restaurant greases. Many renewable fuels are not yet produced in sufficient quantities for general use or require engine modifications.

Uncontrolled harvesting of marine resources leads to population declines, while moratoriums on overharvested species such as striped bass (*Morone saxatilis*) in Chesapeake Bay has led to population rebounds. Uncontrolled fishing and mining results in poor stewardship of natural resources as evidenced by trawls dragged across the seafloor with devastating effects on benthic communities. Trawling is a traditional commercial fishing method of catching marine fishes and invertebrates by pulling funnel-shaped nets through the ocean (*see* NETS). Destruction by bottom trawlers is observed through side-scan SONAR imagery and underwater photography or videography and is analogous to clear-cutting in forests. By killing benthic invertebrates as well as key predators, fishing affects food webs and fundamentally alters the ECOSYSTEM.

Overharvesting, pollution, and natural disasters have led to a decline in populations of marine resources such as finfish. A once booming cod fishery is now trying to recover from years of mismanagement. Fisheries managers normally set annual total allowable catches based on STOCK ASSESSMENTS. Unfortunately, commercial fishermen respond to such measures by catching as many fish as possible in the shortest amount of time. These policies lead to intensive fishing in all kinds of weather that results in loss of gear and dangerous working conditions. Modern management strategies call for an individual quota system where a given fishing vessel and crew are allotted a percentage of the total annual catch. In this way, a particular crew can fish when it is most convenient and safe. With fishing pressure spread out over a longer period of time, fish stocks are showing some improvement, especially in the Canadian Grand Banks and the Georges Banks off the East Coast of the United States, where the fisheries had collapsed.

There are numerous programs such as Seafood Watch and the Seafood Choices Alliance that aim toward raising consumer awareness about the importance of buying seafood from sustainable sources. These types of organizations help support environmentally friendly fisheries and aquaculture operations. Such efforts are key to reducing bycatch, overfishing, and pollution and exposing illegal fishing methods. With respect to reducing dependence on hydrocarbons that are being mined from the seafloor, alternative fuel vehicles and hybrid electric technologies are being used by low-mileage but high-use operations such as fork lifts in warehouses, airplane tugs and baggage carts at airports, and shuttle buses having a centralized fueling capability. Community interest in conservation ensures that marine resources will exist over the long term without endangering species or the health of the surrounding ecosystem.

Further Reading

Blue Ocean Institute. Available online. URL: http://www.blueocean.org. Accessed October 25, 2006.

FishWise: Healthy Seafood for You and the Oceans. Available online. URL: http://www.sustainablefishery.org. Accessed October 25, 2006.

Food and Agricultural Organization of the United Nations, Fisheries and Food Security. Available online. URL: http://www.fao.org/focus/e/fisheries/intro.htm. Accessed October 25, 2006.

Mineral Management Service, Securing Ocean Energy and Economic Value for America. Available online. URL: http://www.mms.gov. Accessed October 25, 2006.

Monterey Bay Aquarium, Seafood Watch. Available online. URL: http://www.mbayaq.org/cr/seafoodwatch.asp. Accessed October 25, 2006.

Seafood Choices Alliance. Available online. URL: http://www.seafoodchoices.org. Accessed October 25, 2006.

swell Swell refers to surface WAVES that have been generated at a remote storm location and have traveled a considerable distance from their place of origin. At the generating area, winds blowing across the water build waves. At some point, the waves are fully developed and are no longer growing by the force of the wind. Such waves leave the generating area and are called swell. Swell waves are gravity waves of long period (10–15 s), a WAVELENGTH in deep water of several hundred feet (meters), and generally low height (a few feet, or fractions of a meter to 3–4 m. Marine scientists refer to the short-period waves still being created by winds as seas. Seas are short-crested and irregular, while swell are smooth, long waves with well-defined crests.

Understanding the development and propagation of swell is important to maritime safety and coastal zone management. Swell waves originate in major storms and propagate outward within a cone of 30°–45° angle from the predominant wind direction. Because of the dispersive characteristics of water waves (*see* GRAVITY WAVES), long waves travel fastest and move ahead of the storm. As the swell waves travel, their height decreases as their length increases because of three factors.

1 They propagate over an increasingly wide front such that the wave energy is spread thinner and thinner.

2 Adverse winds beat down the swell waves.

3 Energy is lost due to friction with the atmosphere and the ocean.

On calm days, the swell height may be only a few inches. Swell can pass through active wave–generating areas with very little modification. Optimal ship routes are planned to avoid storms and heavy seas in order to identify the best possible SEA STATES and CURRENTS to enhance seakeeping and ride quality (*see* OPTIMAL SHIP TRACK PLANNING).

When swell reach a distant shore, they become modified as shallow water waves moving in toward the beach in regular lines that break as the combers favored by surfing enthusiasts (*see* SURF). Swell is often the harbinger of distant storms, such as HURRICANES or nor'easters. Storms in the northern Pacific may generate swell in excess of 20 feet (6 m) high that grow into even larger surf such as those documented on the Bonzai Pipeline on the North Shore of Oahu Island in Hawaii or at the Mavericks surf break, just north of San Francisco, California. In the littoral zone, beaches may be closed owing to dangerous swell reports and the potential for bone-crushing surf, strong RIP CURRENTS, and overwash on coastal roads. The littoral zone is subdivided into the supralittoral zone, which is underwater during storms, the intertidal zone, which includes the foreshore, and the sublittoral zone extending offshore from the low tide line to a depth of approximately 200 feet (60 m). For this reason, marine scientists, city managers, meteorologists, fishermen, and surfers may track conditions along the coast by assessing information from weather charts, numerical wave models such as WAVEWATCH III, satellite images, tide tables, buoy reports, and even local webcams of the surf zone. One of the most reliable tools for measuring swell height and periods are networks of WAVE BUOYS that can warn boaters and beachgoers of impending dangerous marine and surf conditions.

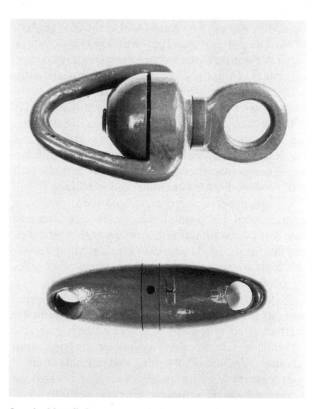

A swivel is a fitting composed of rotating elements (eye-bolt, nut, and box). Swivels normally connect mooring lines and chain and prevent twisting. Heavy swivels or swivel anchor connectors are designed to slide over anchor rollers and prevent snagging. *(NOAA, Y. Berard)*

Further Reading

Coastal Data Information Program, Ocean Engineering Research Group, Integrative Oceanography Division, Scripps Institution of Oceanography. Available online. URL: http://cdip.ucsd.edu. Accessed January 21, 2007.

FNMOC Wave Watch 3 (WW3), Fleet Numerical Meteorology and Oceanography Center. Available online. URL: https://www.fnmoc.navy.mil/PUBLIC/WW3/index.html. Accessed January 21, 2007.

National Centers for Environmental Prediction, National Weather Service, National Oceanic and Atmospheric Administration. "NOAA/NCEP Operational Wave Models." Available online. URL: http://polar.ncep.noaa.gov/waves. Accessed October 16, 2007.

National Data Buoy Center, National Weather Service, National Oceanic and Atmospheric Administration. Available online. URL: http://www.ndbc.noaa.gov. Accessed January 21, 2007.

swivel A swivel generally refers to a connector or coupler in which one end freely turns about a headed pin. The rings or eyes on the swivel may be attached to a mooring to rotate in order to keep lines from tangling or chain from twisting. Swivels come in many shapes and types, and depending on their use are made from materials such as stainless steel, galvanized steel, bronze, or painted steel. Some common examples include eye and eye swivel shackles, swivel lifting hooks, and mooring swivels. Swivels may be used to attach a permanent fixed piece of hardware to a removable fitting, where a swiveling action is required. They are often used in fishing, commercial boating, and marine science applications in conjunction with blocks and rigging to provide articulation or rotation. They are used in moorings, and extreme load swivels are manufactured to handle working loads that may exceed hundreds of metric tons.

See also ANCHOR; MOORING; SHACKLE.

T

temperature Temperature is a physical quantity characterizing the molecular kinetic energy in a physical body. The range of temperature in the global ocean is from less than 33.8°F (1°C) to more than 91.4°F (33°C); seawater freezes at about 28.6°F (-1.9°C) at the surface. The exact freezing point depends on the SALINITY. This property can have a dramatic effect on the state of the water. For example, if the salinity in shallow seas such as the Baltic is lower than 24.7 practical salinity units (psu), the whole body of water cools before the freezing point is reached. Further cooling at the surface results in the formation of ice; the surface water becomes lighter than the surrounding water and does not sink. If the salinity is greater than 24.7 psu, the entire water body turns over until the freezing point is attained, resulting in much greater cooling for the entire water column when the salinity is higher. This effect is of concern in regard to GLOBAL WARMING. The extensive melting of Arctic ice may result in increased precipitation and decreased surface and near-surface salinity in regions such as south of Greenland. Since density depends in part on salinity, the surface water may not be dense enough to sink and replenish the North Atlantic Deep Water, which might interrupt the global THERMOHALINE circulation and perhaps cause changes such as the GULF STREAM flowing farther to the south than at present. Such an occurrence would result in colder winters for eastern North America and Europe.

The rate of chemical reactions usually increases with temperature. Other properties of seawater are greatly affected by the temperature, including density and solubility. Seawater of 30.0 psu has its maximum density at 27.5°F (-2.5°C). As the temperature rises above that, the density decreases (with constant salinity). Dissolving power increases with increasing temperature. Water temperature affects the amount of dissolved oxygen water can hold. For example, cool water holds more oxygen than warm water. Aquarists are very familiar with these properties in maintaining tanks that sustain marine flora and fauna.

Water has a very high heat capacity; that of pure water at 77°F (25°C) and atmospheric pressure is 4.186 joules per kilogram per degree Celsius (1 cal/g°C), greater than any other common substance. The exact value of the heat capacity of seawater depends on the salinity. It retains heat and can act as a giant heat pump in distributing the heat from tropical regions to high latitudes. Because of this, many coastal areas such as Nova Scotia and northwestern Europe are warmer than they otherwise would be at their latitudes. Such locations have what is called a marine climate in contrast to inland areas such as Kansas that have continental climates. The annual temperature range of marine climates is less than that of continental climates, being cooler and moister in summer, and warmer and moist in winter. The air temperatures are moderated by proximity to the ocean.

The vast bulk of the depths of the ocean are very cold. Less than 10 percent of the ocean is warmer than 50°F (10°C), and more than 75 percent is below 39.2°F (4°C). The primary reason for this is that sunlight is heavily absorbed by seawater; most of the radiant energy of the Sun is absorbed above 328.1 feet (100 m) depth. Thus, in general, the oceans are characterized by a warm surface layer, a zone of rapid temperature decrease called the THERMO-

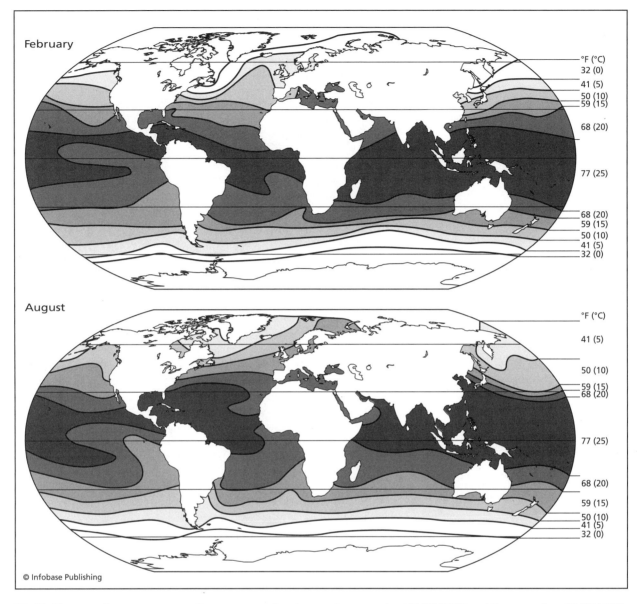

Worldwide sea surface temperatures. Average sea surface temperatures are about 62.6°F (17°C), and they range from 28.4°F (-2°C) in high latitudes (polar seas) to 96.8°F (36°C) at low latitudes (Persian Gulf). The temperature differences between February and August show that there is considerable seasonal variation, even in mid-ocean.

CLINE, and very cold water from below the thermocline to the ocean floor. The overlying surface layer of well-mixed water is called the mixed layer. At high latitudes, strong winds and cold surface waters result in strong convective mixing, in which the water column is very cold from surface to bottom, a nearly homogeneous vertical distribution of temperature. In tropical latitudes, a warm, light surface layer overlies the cold, dense water that has flowed down from Arctic and Antarctic regions to fill the world's ocean basins. The thermocline in the tropical regions is shallow. The net result is that a contour graph of temperature and depth of an ocean basin from north to south looks like a cross section of an onion, with the layers concave side upward, with warm waters at the top in the Tropics and equatorial regions, and cold waters in the polar and deep regions. Temperature anomalies occur as a result of the ocean circulation; the Gulf Stream transports warm surface water northward along the East Coast of North America, and eastward to northwest Europe; the same pattern applies to the KUROSHIO CURRENT in the Pacific, where warm water is transported to northwest North America. On the other hand, cold Arctic waters are transported southward inshore of the Gulf Stream by the LABRADOR CUR-

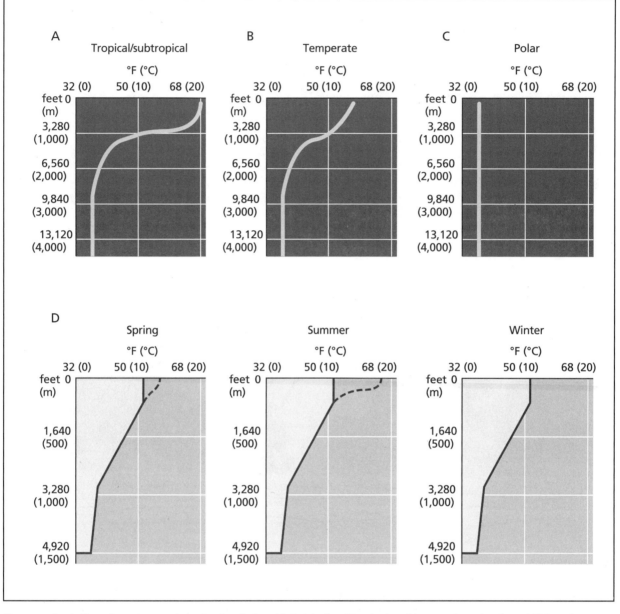

An approximate three-layer temperature structure is found in tropical, subtropical, and temperate waters (a and b). In polar waters (c), there is no thermocline, and the temperature difference between surface and bottom waters is small. In subtropical and temperate latitudes, a seasonal thermocline (d) is formed in the near surface layers by solar heating.

RENT, and inshore of the Kuroshio by the OYASHIO CURRENT.

Like the seasons and climate, the temperature boundary between surface and deep waters varies due to a combination of incoming solar radiation (geography) and water depths. The shallow waters of mid-latitudes, such as New York Bight experience a seasonal thermocline in summer and autumn, which disappears when strong winter winds and cooling increase the density and the mixing of surface layers

into the deep layers. (A bight is a curving shoreline, which forms a wide bay.) The thermocline is still the boundary between the mixed layer above and the deep layer below.

The very deepest waters of the ocean basins, the waters of the deep TRENCHes that are under great pressure, experience a slight increase of temperature due to compress ional heating. To allow for this effect, oceanographers use the concept of potential temperature: the temperature a parcel of water would have if brought to the surface without gain or loss of heat from the surrounding water. As an example, a parcel of water at a depth of 16,404.2 feet (5,000 m) and salinity of 35.0 psu with an in situ temperature of 33.8°F (1°C), if brought to the surface without gain or loss of heat (that is, adiabatically), would have a temperature of 30°F (0.58°C). Potential temperature is usually designated by the Greek letter theta (θ).

Temperature is seen to be one of the key parameters measured by any weather station, ocean station, ship, or BUOY. Temperatures are measured in situ by THERMISTORS or thermometers. Marine scientists are especially concerned with the difference between air and sea temperature, which determines whether the energy flow is from the ocean to the atmosphere, or from atmosphere to ocean. In earlier times, meteorological offices had voluntary observing ships obtain samples of water for measuring temperature with canvas buckets. The at-sea meteorologists took temperature measurements on deck immediately after the water was drawn. They were careful to keep the water stirred and took readings after the thermometer was in the water for 30 seconds. Benjamin Franklin (1706–90) prepared the first map of the Gulf Stream from temperature readings taken by cooperating ship captains. State-of-the-art techniques to measure sea surface temperatures remotely include the use of radiometers such as the Advanced Very High Resolution Radiometer, which is on board certain NOAA satellites.

The temperature of the seawater is also measured by reading the temperature of the engine-room intake temperature, where water is usually drawn off from a tap or valve. A thermograph may also be installed to automatically record sea temperatures at the intake. This water is not at the surface, and the temperature may be contaminated by the ambient ship temperature. A vast archive of historical temperature measurements are collected and made available to marine scientists and engineers through the International Comprehensive Ocean and Atmospheric Dataset (ICOADS).

Today, most marine scientists use conductivity-temperature-depth (CTD) probes to measure the profile of ocean temperature. Temperature profiles are taken from dedicated research vessels, moored and drifting oceanographic buoys, SUBMERSIBLES, REMOTELY OPERATED VEHICLES (ROVs), and AUTONOMOUS UNDERWATER VEHICLES (AUVs). Temperature measurements are often taken by volunteer mariners on ships of opportunity. Weather and ocean satellites measure ocean temperature remotely from space to provide global coverage, and, by this means, greatly improved weather FORECASTING and the knowledge of global climate.

How water temperature influences fishing for small pelagics

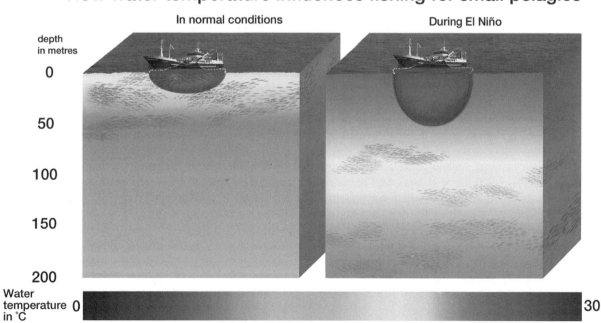

Temperature conditions have a great effect on fishing. During El Niño, fish sink below maximum net depth, and the catch is poor. *(NOAA/NMFS)*

The temperature structure of the ocean is the major factor, at least in the deep ocean, determining the sound velocity structure, which is of great importance to the operation of FATHOMETERS and SONAR systems. Hence, the navies of the world have gone to great effort and expense to chart the average temperature distribution of the oceans. Knowledge of the temperature structure is also essential to the correct interpretation of geophysical data taken at sea by means of explosive devices, air guns, and other prospecting tools.

Water temperature also largely determines the location of major fisheries, such as tuna and swordfish. Fishermen and fisherwomen carefully follow the patterns of surface temperature available from maps of satellite and aircraft data, as well as taking their own subsurface measurements. Some species of fish are known to locate along certain isotherms. NOAA's National Marine Fisheries Service provides detailed temperature and other information to U.S. fishing fleets.

See also ALBEDO; CONDUCTIVITY-TEMPERATURE-DEPTH PROBE; DENSITY; FEEDWATER; ISOTHERM; STENOTHERMAL.

Further Reading

Knauss, J. A. *Introduction to Physical Oceanography*, 2nd ed. Upper Saddle River, N.J.: Prentice Hall, 1996.

National Environmental Satellite, Data, and Information Service, National Oceanic and Atmospheric Administration. "Advanced Very High Resolution Radiometer–AVHRR." Available online. URL: http://noaasis.noaa.gov/NOAASIS/ml/avhrr.html. Accessed October 16, 2007.

Neumann, G., and Willard J. Pierson, Jr. *Principles of Physical Oceanography*. Englewood Cliffs, N.J.: Prentice Hall, 1966.

territorial sea   A territorial sea is the defined extent of inland and coastal waters that are under the jurisdiction of a nation or state. The term is also known as "territorial waters." The notion of territorial sea becomes complex when water bodies are borders, water resources must be shared, and the region includes sea-lanes that are essential for maritime commerce. Countries abiding by the LAW OF THE SEA have agreed, "the boundaries of a coastal state extends, beyond its land territory and internal waters and, in the case of an archipelagic State, its archipelagic waters, to an adjacent belt of sea, described as the territorial sea." The sovereignty includes the entire air space, water column, seafloor, and strata. The third convention on the Law of the Sea establishes a 12-nautical mile (22 km) distance from the coastline where coastal countries may enforce their laws and regulate the usage and exploitation of natural resources. In addition, naval and merchant ships are provided the right of "innocent passage" through the territorial seas of a coastal country. There are other zones that complicate relationships among countries such as demarcation lines relevant to inland and international (high seas) navigation rules, the 24-nautical mile (44 km) "contiguous zone" that supplements the 12-nautical mile boundary, and the 200-nautical mile (370.4 km) EXCLUSIVE ECONOMIC ZONE, which is aimed at preserving a country's natural resources. Boundaries are drawn on nautical charts that are developed by hydrographic offices that collaborate by sharing important maritime information such as datums, tidal constituents, and surveys.

See also MARINE PROVINCES.

Further Reading

Division for Ocean Affairs and Law of the Sea, the United Nations. Available online. URL: http://www.un.org/Depts/los. Accessed May 22, 2005.

International Hydrographic Office. Available online. URL: http://www.iho.shom.fr. Accessed February 11, 2007.

thermistor   A thermistor is a temperature-sensitive electrical resistor made of semiconducting materials, such as manganese oxide or nickel oxide. Thermistors have large, negative temperature coefficients, that is, increasing temperature decreases their resistance. In addition, the change of resistance with temperature is nonlinear, unlike metals, in which the resistance is directly proportional to the temperature—the higher the temperature, the higher the resistance. For thermistors, the electrical resistance decreases at a significantly higher than linear rate with increasing temperature, thus making the thermistor an ideal means of measuring temperature electrically. Thermistors can be inserted into Wheatstone bridge circuits and then calibrated over a desired temperature range to make laboratory measurements of temperature.

In oceanographic applications, thermistors can be installed singly, in groups in instrument pods, or in vertical or horizontal strings. They can be mounted on buoys, platforms, SUBMERSIBLES, AUTONOMOUS UNDERWATER VEHICLES (AUVs), or towed in chains behind research vessels. Thermistor chains are a key instrument in the study of INTERNAL WAVES that manifest themselves as subsurface temperature fluctuations.

A particularly interesting historical application was the isotherm follower, designed and used in the 1950s by E. C. LaFond (1909–2002) of the U.S. Navy Electronics Laboratory in San Diego, California. This instrument consisted of from one to three booms installed on an offshore oceanographic platform, which supported underwater sensors, mainly thermistors, incorporated into a feedback circuit driving a servo motor/mechanical WINCH system on the deck of the platform. The instrument package was designed to float up and down on a preset ISOTHERM, thus providing a time record of vertical temperature fluctuations, and by inference, internal waves. When a thermistor recorded temperatures below that of the preset isotherm, the feedback circuit raised the instrument to a shallower depth where the temperature was higher. When a thermistor recorded a temperature above that of the preset isotherm, the feedback circuit lowered the instrument to a cooler temperature. The instrument package would thus seek and follow the proper isotherm, measuring its depth by a pressure sensor.

Today, thermistors are widely used by the National Weather Service and the National Data Buoy Center to measure subsurface ocean temperature on moored weather buoys, for example, the NOMAD buoy, C-MAN stations, and on drifting buoys. The data are transmitted via satellite to shore stations for inclusion in the synoptic observations of the national weather observation network.

See also C-MAN STATIONS; NOMAD BUOY; PRESSURE.

thermocline The thermocline is a water layer in which the temperature decreases rapidly with depth, generally separating warm surface waters from cold bottom waters. The thermocline, being usually the principal factor in governing the DENSITY of the water, is usually colocated with the pycnocline, the maximum vertical density gradient. As temperature decreases with depth, there is usually a corresponding increase in density.

In tropical and subtropical waters, the temperature structure usually consists of a mixed layer of 165–660 feet (50–200 m) depth, a layer below this extending to 660–3,280 feet (200–1,000 m) depth in which temperature decreases rapidly with depth (the thermocline), and a deep layer where temperature decreases slowly to the bottom. The increasing pressure of the deep water slightly increases the temperature. In polar waters, there is generally no thermocline, and the waters are well mixed, except in regions where melting ice distributes a layer of relatively fresh low-density water overlying the heavier ocean water.

Dynamical and climatological factors also play a major role in thermocline depth. For example, in the eastern equatorial PACIFIC OCEAN, the thermocline is usually rather shallow, at roughly 164 feet (50 m) in depth or less in regions of coastal UPWELL-

ING, but in the western part of the Pacific, the thermocline is about 820 feet (250 m) in depth, underlying a thick warm water layer. The same type of distribution is also found in the equatorial ATLANTIC OCEAN; the thermocline is very shallow, sometimes only a few fathoms (m) deep off the coast of Guinea, but tens of fathoms (hundreds of m) deep off the coast of Brazil.

Basically, the thermocline develops because the major source of heat for the oceans, solar radiation, extends only to the top 330 feet (100 m) or so of the water column in the clearest ocean water. All green plants, such as seaweeds and PHYTOPLANKTON, are restricted to this depth. The depth of light extinction is much shallower in turbid coastal waters. Below this depth, the sunlight is completely absorbed and darkness prevails. Hence the upper sunlit layers of the global ocean tend to be warm, the lower layers very cold. In the Arctic and Antarctic regions, cold air and strong winds mix the water column down to deep depths, and a thermocline is absent.

In temperate latitudes, a seasonal thermocline usually develops in late spring, summer, and early autumn that may be a few feet (m) to tens of feet (m) deep. For example, in New York Bay and on Georges Bank, surface waters are warm in summer, while bottom waters are cold, from intermittent upwelling of deeper waters offshore, and from the Labrador Extension Current that flows along the coast inshore of the GULF STREAM from Labrador nearly to Cape Hatteras. These seasonal fluctuations control the growth of phytoplankton and hence the ZOOPLANKTON and all animals above in the food chain. In the Sargasso Sea near Bermuda, there is a shallow seasonal thermocline above the main thermocline that dissipates in winter, resulting in a nearly homogeneous mass of water called the 64.4°F (18°C) water.

See HALOCLINE; LIGHT; PYCNOCLINE; TEMPERATURE.

Further Reading
Knauss, John A. *Introduction to Physical Oceanography,* 2nd ed. Upper Saddle River, N.J.: Pearson/Prentice Hall, 1996.

thermohaline circulation Motion in the ocean that is driven by differences in density within the water column is called thermohaline circulation. The density of seawater is a function of temperature (Greek *thermo* = heat) and salinity (Greek *haline* = salinity). The density in the ocean changes with depth, and the ocean is usually stably stratified as seawater density increases with depth and is neutral when there is no change in density with depth. If density decreases with depth, then the water column is unstable and seawater will respond with lighter water rising while heavier water sinks. This movement from density differences in the water column is called thermohaline circulation. Other forces causing circulation are the astronomical tide-generating forces (*see* TIDE) and wind stress (*see* STORM SURGE).

Globally, wind-driven surface currents move heat from the equator toward the poles. As these currents reach the poles, they cool and become much more dense, especially as surface waters freeze, leaving behind salt. The dense waters sink and flow out into the ocean basins, where mixing takes place. Sinking water in the North Atlantic Ocean near the Norwegian and Greenland Seas forms North Atlantic Deep Water (NADW). As the fresher water forms ice, the heavier high-salinity water sinks to the bottom of the basin. NADW spreads southward, flowing over denser bottom waters such as Antarctic Bottom Water (AABW). Similarly, AABW forms when freezing and sinking water flows down the Antarctic slope of the WEDDELL SEA. Bottom waters are not formed in the North Pacific Ocean/Bering Sea region because the salinity is too low for the cooled water to sink all the way to the bottom. The bottom waters move cold water toward the equator. This circulation on a global scale is often referred to as global thermohaline circulation, a process helping to regulate climate.

PALEONTOLOGISTS or paleoceanographers use principles of physical oceanography along with data on the ancient positions of oceans and continents to make estimates of what ocean circulation would have looked like during prehistoric times. By understanding past climates and current rates of climate change, marine scientists are predicting a reduced-circulation state owing to increased rainfall and warming over the North Atlantic. Presently, the ATLANTIC OCEAN is generally more saline than the PACIFIC OCEAN because there is more evaporation of freshwater, which increases the salinity. The bottom AABW and NADW are overlain by intermediate (for example, Antarctic Intermediate Water, Subarctic Intermediate Water, and Arctic Intermediate Water) and surface waters. The waters near the surface are well mixed by the wind and replace the sinking water at the poles. The conveyor system is affected by locations of the continents and in the movement of heat from the equator. Numerical models are used to assess circulation changes given changes in many important and complex variables such as solar radiation, SEA ICE, and rainfall. Researchers have compared simulated transport (volumetric flows) of water in the Atlantic from the late 1800s through the year 2100 to estimated GLOBAL WARMING scenarios. In reality, global climate is a complex balance among processes occurring within the linked atmosphere, biosphere, cryosphere, geosphere, and hydrosphere.

Tidal bores occur on several rivers in Alaska. *(NOAA Corps)*

See CONVECTION; DENSITY; SALINITY; TEMPERATURE; WATER MASS.

Further Reading

Allaby, Michael. *Encyclopedia of Weather and Climate*, rev. ed. New York: Facts On File, 2007.

Ocean and Climate Change Institute, Woods Hole Oceanographic Institution. Available online. URL: http://www.whoi.edu/institutes/occi/viewTopic.do?o=read&id=501. Accessed March 11, 2007.

Thermohaline Circulation, Climate Timeline Information Tool, National Geophysical Data Center, National Oceanic and Atmospheric Administration. Available online. URL: http://www.ngdc.noaa.gov/paleo/ctl/thc.html. Accessed March 11, 2007.

UNEP/GRID-Arendal, United Nations Environment Program. Available online. URL: http://www.grida.no. Accessed March 11, 2007.

tidal bores

Tidal bores are tide waves propagating up a shallow, sloping ESTUARY or river as a SOLITARY WAVE. The leading edge is an abrupt rise in water level, breaking continuously and presenting a roaring, turbulent surface. Undulations in the water surface may follow behind the bore front.

Tidal bores are found in rivers and estuaries having a large range in tide, on the order of 20 feet or more. Seen in profile, the tidal bore would resemble a mathematical step function.

The height of tidal bores is typically two to three feet (.5m), but can be much larger, as on the Amazon River, where the height of the bore can be as high as 20 feet (5 m) with speeds on the order of 12 knots (6 m/s).

Bores occur because the rising tide forces the wave front to move faster than it can propagate in water of that depth (as calculated by the shallow water wave equation $c = \sqrt{gh}$), creating a shock wave in the water.

Beside the Amazon, famous tidal bores occur in the Yangtze River in China, the Trent River in the United Kingdom, the Elbe in Germany, the Petitcodiac (a tributary of the Bay of Fundy in Canada), and Cook Inlet in Alaska.

Further Reading

Hicks, Steacy. *Tide and Current Glossary*. Silver Spring, Md.: U.S. Dept. of Commerce, NOAA, 1989.

Wright, John, Angela Colling, and Dave Park, eds. *Waves, Tides and Shallow-Water Processes*. Milton Keynes, U.K.: Butterworth/Heinemann, 1999.

tidal currents

Tidal currents are the horizontal movements of the water caused by astronomical forces. In coastal areas, incoming tidal currents are known as flood tides, while the outgoing tides are called ebb tides. In offshore regions, tidal currents change direction clockwise in the Northern Hemisphere and counterclockwise in the Southern. They are sometimes called rotary currents with speeds varying throughout the tidal cycle, passing through the two maxima in approximately opposite directions and the two minima with the direction of the current at approximately 90° from the directions of the maxima.

See also HARMONIC PREDICTION; TIDE; TIME SERIES.

tidal power plant

A tidal power plant is a coastal or offshore structure that generates electricity from the power of TIDAL CURRENTS, which are the to-and-fro vertical movements of water caused by astronomical forces. The basic mechanism involves submersible water turbines, where kinetic energy from the flow is converted to mechanical energy by turning a bladed rotor. Locations favorable for the generation of electricity by tidal power should be classified as macrotidal with a tidal range above 13 feet (4 m). The La Rance Tidal Power Station near St. Malo, France, has been in operation since the mid-1960s. It was built like a dam, which is called a barrage (a large weir), where sluice gates allow the tidal basin to fill on the flood and exit through the turbine system on the ebb. When the gates are closed, water levels are raised to facilitate the diversion of water through the turbines. Similar structures began operations in Kislogubsk, near Murmansk, Russia, on the Barents Sea during 1968; in Zhejiang Province, China, at the Jiangxia tidal power station near the East China Sea during 1980; and in Annapolis Royal, Nova Scotia, Canada, by the Bay of Fundy during 1982. Maximum tidal ranges in the Bay of Fundy are reported to be 56 feet (17 m), the largest in the world. There are other tidal plants in the world and locations where tidal flow or permanent currents are sufficient to turn turbine blades in synchronization with the tides.

New designs apply axial flow rotors to drive a generator via a gearbox similar to the wind turbines (wind farms) that are seen on mountain passes. These rotors, which are approximately 36 feet (11 m) across, may be mounted on collars and slide on a tubular steel column driven into the seafloor. They operate at depths from 65 to 98 feet (20 to 30 m) deep. The rotor and collar assembly is raised to the surface for maintenance purposes. These tidal turbines have been installed in the Bristol Channel near Lynmouth in North Devon, United Kingdom, Loch Linnhe on the west coast of Scotland, and Kvalsundet, a narrow strait off Norway's coast, south of Hammerfest.

Selected Tide-Dominated Coastal Locations with a Tidal Range More Than 20 Feet (6.1 m)

Country	Site of Tide Station	Mean High Water Spring	
		Feet	Meters
Argentina	Punta Loyola, Patagonia	39.0	12.0
Australia	Derby, Kimberley	36.7	11.2
Canada	Saint John, New Brunswick	26.0	7.9
China	Chuanshi Island, Fujian Province	20.3	6.2
India	Bhavnagar, Gujarat State	33.5	10.2
Republic of Korea	Pyongtaek, Gyeonggi Province	28.2	8.6
Russia	Mys Astronomicheski, Kamchatka	30.8	9.4
Mozambique	Beira	21.3	6.5
United Kingdom	Port of Bristol, Avonmouth	43.3	13.2
United States	Anchorage, Alaska	29.5	9.0

Note: These regions have large tides in part because of the funneling effect of the basin.

Power from the tides is a renewable source of energy and helps to reduce mankind's dependence on fossil fuels. Renewable energy comes from a resource that is either not depleted by its use (such as in solar, wave, or wind energy) or can be replenished at a rate comparable to the rate at which it is consumed (such as biofuels). A major advantage for tidal power plants is the lack of pollution or greenhouse gases. A disadvantage, especially for a gate-structure dam (barrage), is the resulting change in SEDIMENT TRANSPORT, which may cause the ESTUARY to fill with sediments. Challenges to building, maintaining, and operating tidal power plants include cost, saltwater corrosion, marine biofouling, access to maintain submerged components, and damage from severe weather. Marine science research will involve the analysis of environmental impacts (sound, collision, migration) and the development of viable designs that provide a societal benefit. The following table is a selection of tide-dominated coastal locations with a tidal range that exceeds 20 feet (6.1 m). Macrotidal regimes are associated with strong ebb and flood currents. Tide ranges were determined from the nearest tide station.

Further Reading

Electric Power Research Institute. Available online. URL: http://my.epri.com. Accessed February 3, 2007.

House of Commons. "Science and Technology: Seventh Report." Available online. URL: http://www.parliament.the-stationery-office.co.uk/pa/cm200001/cmselect/cmsctech/291/29102.htm. Accessed October 16, 2007.

Nova Scotia Power, Inc. "Tidal Technology." Available online. URL: http://www.nspower.ca. Accessed February 4, 2007.

Ocean Renewable Energy Coalition. Available online. URL: http://www.oceanrenewable.com. Accessed February 3, 2007.

The Rance Tidal Power Plant, Electricité de France. Available online. URL: http://www.edf.fr/html/en/decouvertes/voyage/usine/retour-usine.html. Accessed February 3, 2007.

Rudkin, E. *Survey of Energy: Marine Current Energy.* London: World Energy Council. Available online. URL: http://www.worldenergy.org. Accessed February 3, 2007.

United Nations Atlas of the Oceans. Available online. URL: http://www.oceansatlas.com. Accessed February 4, 2007.

tide Vertical and horizontal movements of water caused by the gravitational forces of the Moon and the Sun acting on the Earth make up the tide. The first written records on tides are credited to Herodotus of Halicarnassus (484 B.C.–425 B.C.), the father of history. Sir Isaac Newton (1642–1727) established the foundation for understanding the tides when he developed the universal law of gravitation. Newton's amazing insight was that every object in the universe attracts every other object with a force directed along the line of centers for the two objects. The gravitational force is proportional to the product of their

Acoustic Tide Gauges: A Sound Way to Measure Water Levels

Traditional tide gauges have used a float in a large cylindrical tube called a stilling well to record the water level. The level of the float in the well can be recorded by eye and paper, by photographic, or electrical systems, the latter being the most common. The purpose of the stilling well is to "still" the high-frequency water level oscillations due to surface wind waves, or waves produced by passing ships, and thus make the measurements more representative of the tide height.

The present state-of-the-art system for measuring tide height uses an acoustic pulse traveling in air down a sound tube of small diameter (order .5 inches (13 mm)) to the water surface and then reflects the pulse back to a receiver which is often colocated with the transmitter. The water level is then determined using the distance-time formula

distance = velocity × time / 2

to determine the distance to the water surface, and thus the water level. The principle is the same as that used in ship's fathometers that send sound pulses from a transducer below the ship to the ocean floor and back to determine water depth. The measurements must be referred to a standard level, such as that provided by carefully surveyed bench marks, in order to be comparable to all other tide gauge readings.

A difficulty with this measurement is that the sound speed in air is affected by temperature and humidity, with temperature being the greater effect. If the temperature is the same in the tube, the measurement will be accurate, but if the sound speed varies due to changing temperature within the tube, the measurement will be less accurate. To overcome this problem, two or more thermistors (electrical resistors with a negative temperature coefficient) are often installed in the tube to provide a means of correcting the measurements. In addition, an acoustic reflector is placed in the sound tube to provide an acoustic reflection at a constant distance that can be used as an in-field calibration of the sensor.

The sound tube is usually installed inside a larger plastic tube (diameters on the order of six inches (15 cm) that serves as a stilling well, to reduce the effects of the short period wind waves.

Acoustic tide gauges are sometimes used with the transmitter in open air, but this method does not allow for careful temperature compensation.

Data sampling rates are on often on the order of every six minutes, with each measurement consisting of between 150 and 200 acoustic pings at one-second intervals. This sampling rate is high compared to the traditional float gauges, both because acoustic echoes are inherently noisy and must be averaged for mean values, and also to resolve the short period surface motions that are not damped out by the enshrouding stilling well.

Acoustic tide systems often include other sensors, such as pressure gauges, anemometers, salinometers, current meters, and the like. They are usually a part of real-time oceanographic measurement systems that provide the navigator and harbor pilot with detailed tide, current, wind and wave information to bring the vessel safely to port, and to the dock.

Two examples of acoustic tide gauge systems in use are NOAA's Next Generation Water Level Measurement System (NGWLMS) in the United States, and Australia's Sea Level Fine Resolution Acoustic Measuring Equipment (SEAFRAME) system in Australia and Pacific islands. These systems are in widespread use and can resolve water level fluctuations on the order of three millimeters. They transmit their data to central facilities via land line or satellite link for near-real data analysis, processing, and distribution.

Further Reading

Intergovernmental Committee on Surveying and Mapping. "Australian Tides Manual." Special Publication No. 9. Available online. URL: http://www.icsm.gov.au/tides/SP9/links/IOCVIII_acoustic.html. Accessed October 17, 2007.

—**Robert G. Williams, Ph.D.,** senior oceanographer
Marine Information Resources Company
New Bern, North Carolina

masses and inversely proportional to the square of the separation between the two objects. This law is easier to understand by considering the following formula:

$$F_g = G (m_1 * m_2) / r^2, \text{ where}$$

F_g = gravitational force,
G = universal gravitation constant (6.67×10^{-11} N m²/kg²),

m_1 = mass of object one,
m_2 = mass of object two, and
r = the separation between the objects' center.

This equation allows one to calculate the force of gravitational attraction between any two objects of known mass and known separation distance. Newton applied such findings to establish the notion of an equilibrium tide that assumes that the entire

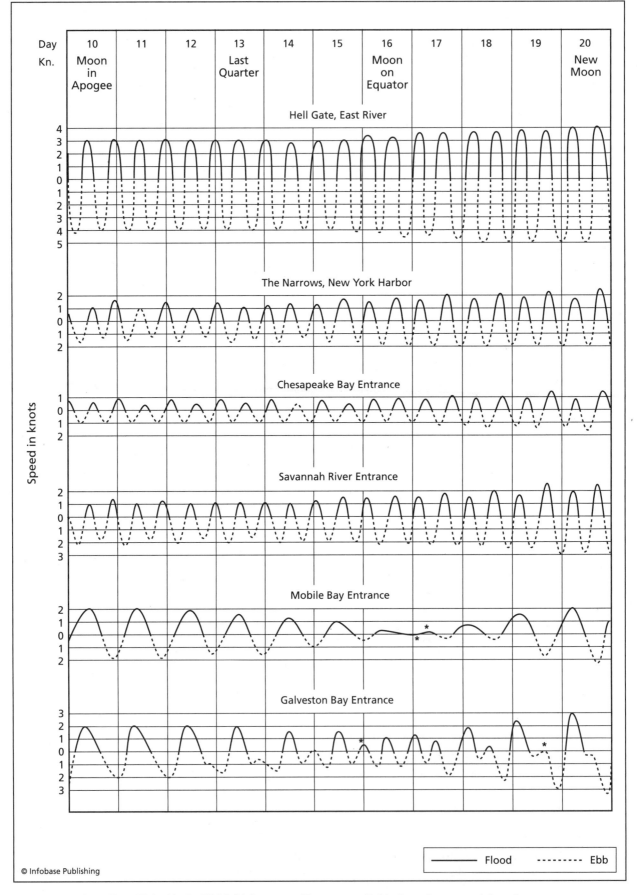

Tidal currents formally published in the NOAA tidal current tables, now available through commercial vendors.

Earth is covered in water and explains spring tides occurring at new and full Moons and neap tides occurring at the first and third quarters (quadrature). French mathematician and astronomer Pierre-Simon, marquis de Laplace (1749–1827) suggested a dynamic theory where the tides are driven by astronomical forces. Sir William Thomson (1824–1907), also known as Lord Kelvin, advanced the science even further by devising a method of harmonic analysis to explain the periodic rise and fall of the tides. He is credited with inventing one of the first tide-predicting machines. The prediction of tides was complicated for early tidalists owing to the irregular distribution of land and water on the Earth and the corresponding effects of friction and inertia. Ultimately, collections of longtime series of data and the reduction of tidal data by harmonic analysis resulted in being able to reproduce tidal curves as a sum of sinusoidal curves that are directly related to astronomical conditions. Observations of WATER LEVEL FLUCTUATIONS may be studied at a particular reference station from which up to 19 years of at least hourly data may be collected to predict tides. Such data may be compared to secondary or subordinate stations.

Tides are actually the periodic rise and fall of the water (or atmosphere) in synchronization with changes in position of the Sun and Moon relative to points on the rotating Earth. They are, after all, a forced shallow water wave. In a tidal cycle, there are high waters and low waters. In some locations, which are called semidiurnal, there are two high and low waters per tidal day; in other locations, which are called diurnal, there is one high and low tide per

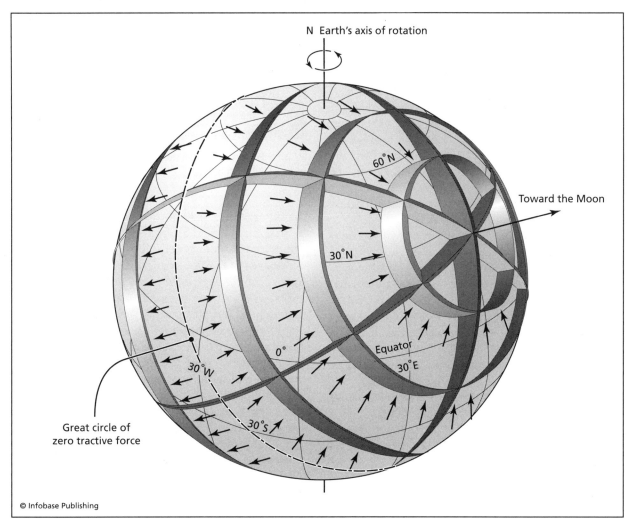

The equilibrium tide model showing the horizontal component of tide-producing forces in small arrows for when the Moon is over 30 degrees N latitude. The large fences represent the height of the tidal bulge at any point on the Earth. The Earth rotates through these tidal bulges that remain generally stationary.

tidal day; and still others, which are called mixed, have a combination of the two. Mixed areas may be mostly semidiurnal or diurnal.

Tides will propagate upstream and lose energy due to friction. Their shape not only changes due to friction, but in direct response to the changing tidal forces. The limit or "head of tide" is found somewhere upstream where the mean range is less than 0.2 feet (6 cm). The maximum height reached by the rising tide is called high water, and the minimum height reached by the falling tide is called low water. Depending on astronomical forces and the geographic location, other phases may be observed such as lower high water or the lowest of the high waters due to the declinational effects of the Moon and Sun and similarly lower low water or the lowest of the low waters during a tidal day.

The horizontal movement of water resulting from the tide-producing force is called TIDAL CURRENT. Tidal currents in shallow water tend to be rectilinear in which the approaching landward current is called the flood and the seaward current is called the ebb. The transition from flood to ebb, or vice versa, is called slack water. Like the vertical tides, tidal currents may be semidiurnal, diurnal, or mixed. In deep water, tidal currents are rotary and shift direction around the compass rose, that is the water follows the path of an ellipse. Rotation as a consequence of the CORIOLIS FORCE is clockwise in the Northern Hemisphere and counterclockwise in the Southern Hemisphere. In general, tidal current speeds increase as the water depth decreases. The net velocity over a tidal cycle should be zero.

Water level fluctuations are also caused by atmospheric pressure, wind, precipitation, and temperature. The magnitude of these nontidal forces can be observed by subtracting the astronomically predicted tides from the observed tides. The residual is a direct measure of the contribution by meteorological type events to the total water level record or currents, and possibly errors in measurement or prediction.

The mathematical principles used to describe the periodic tides are called harmonic analysis. Marine scientists consider that the tide is a sum of sinusoidal waves. The general equation is

$$h(t) = H_0 + A \cos (at + \alpha) + B \cos (bt + \beta) + C \cos (ct + \gamma) + \dots$$

where h is the height of the tide at any time (t). The height of the mean water level above the datum (mean lower low water) is H_0. Each cosine term is known as a harmonic constituent and the coefficients A, B, and C are the amplitudes derived from

NOAA tide houses contain tide gauges, recording instruments, and possibly satellite radios. *(NOAA/NOS)*

the observed water level time series. The expression in parentheses is the phase, which has maximum values at 180° and is zero at 90° and 270°. Periods and speeds of the tide are derived from astronomical data and are independent of the location for the time series. The variables α, β, and γ are initial phases that depend on the time series. Determining the amplitudes and phases of the constituents of the actual tide from time series is done by Fourier analysis and results in the computation of harmonic constants (amplitudes and phases). The method is harmonic since the harmonic constituents (cosine terms from above) are combined to determine the composite tide. In the real world, shallow water effects and the interaction among the constituents distorts the sinusoidal shape of the tidal curve. Many marine scientists attempt to account for 37 constituents in a longtime series of water levels or currents. The constituents are added together as a sum of cosine terms to harmonically predict the tides.

Tidal records form the basis to study important scientific issues such as sea level rise. Even though the techniques of the 19th century provide an adequate model to assess the periodic tides and tidal currents, there are many new and exciting initiatives to predict tides using shorter length time series or information such as waterlines that are derived from images. A waterline is the interface between land and water at the time of imaging. With increases in computational power of the 21st century and innovations in sampling theory such as chaotic signal processing, marine scientists may be at the cusp of creating some new techniques to predict the periodic rise and fall of the tide. In some seaports or important channel junctions, navigators can receive real-time water level and current information that supports navigational decisions. Improvements in reporting water levels provides a direct benefit to society, especially for those people who navigate in shallow channels, manage seaport operations, mine natural resources along the coast such as diamonds, or live by the sea.

See also HARMONIC PREDICTION; OCEANO-GRAPHIC INSTRUMENTS; PREDICTION; PRESSURE; TIME SERIES ANALYSIS; WAVE AND TIDE GAUGE; WAVE PREDICTION.

Further Reading

Hydrographic Office, Kowloon, Hong Kong. Available online. URL: http://www.hydro.gov.hk/main.htm. Accessed March 5, 2007.

Hydrographic Service, Royal Netherlands Navy, The Hague, Netherlands. Available online. URL: http://www.hydro.nl. Accessed March 5, 2007.

The International Hydrographic Organization, Monaco. Available online. URL: http://www.iho.shom.fr. Accessed March 6, 2007.

National Oceanic and Atmospheric Administration. "Our Restless Tides: A Brief Explanation of the Basic Astronomical Factors Which Produce Tides and Tidal Currents." Available online. URL: http://tidesandcurrents.noaa.gov/pub.html. Accessed March 4, 2007.

———. "Tides and Currents, Center for Operational Oceanographic Products and Services." Available online. URL: http://tidesandcurrents.noaa.gov. Accessed January 21, 2007.

Parker, Bruce B. *Tide Analysis and Prediction.* NOAA Special Publication NOS COOPS 3. Silver Spring, Md.: National Ocean Service, National Oceanic and Atmospheric Administration, 2007.

Swedish Maritime Administration, Norrköping, Sweden. Available online. URL: http://www.sjofartsverket.se/. Accessed March 5, 2007.

The United Kingdom Hydrographic Office, Somerset, United Kingdom. Available online. URL: http://www.ukho.gov.uk. Accessed January 21, 2007.

tide gauge A tide gauge is an in situ oceanographic instrument that measures changes in water levels as a function of time. These instruments range in sophistication from a simple graduated staff where visual observations can be recorded manually to state-of-the-art electronic sensors and communication equipment that telemeter water level records to a data acquisition system. Older tide gauges relied on a float in a stilling pipe communicating with the sea through a small hole that filtered out shorter waves. The tide level in early gauges was read by eye, while more recent gauges used changing electrical resistance with submersion to activate chart or magnetic tape recorders. Modern tide gauges use pressure sensors, especially those in the deep sea or in surf zones, and acoustic pulses transmitted down a vertical plastic tube The tube serves the same basic function as the stilling pipe by damping out high-frequency wave motions. The travel time of the acoustic pulse transmitted down the tube and back depends on the water level. Corrections are made for varying temperature and humidity to provide for the most accurate measurement possible. The data are recorded digitally on solid-state recorders, or transmitted via satellite in real time. Usually several tide gauges are installed in a sheltered location, and associated electronic equipment is housed in what is called a tide house. They are often sited on a pier near an important seaport, at research stations, or along a stretch of coastline. Many stations in the United States are part of the National Water Level Observation Network. This network is composed of approximately 175 long-term, continuously operating water-level stations including the Great Lakes. Data from these stations provide the basis for determining tidal datums and factors such as sea level rise. In tidal locations, the time series of high and low tides are used to determine by harmonic analysis the amplitudes and phases of the tide with respect to the astronomical forces. Similar measurements are made in nontidal rivers and streams to determine the water level by stream or river gauges. Water levels in streams are a function of hydrometeorological forces.

See also OCEANOGRAPHIC INSTRUMENTS; PREDICTION; PRESSURE; WAVE AND TIDE GAUGE; WAVE PREDICTION.

Further Reading

National Oceanic and Atmospheric Administration. "NOAA Tides and Currents." Available online. URL: http://tidesandcurrents.noaa.gov. Accessed October 16, 2007.

The United Kingdom Hydrographic Office, Somerset, United Kingdom. Available online. URL: http://www.ukho.gov.uk. Accessed January 21, 2007.

tide staff A tide staff is a simple instrument to measure the height of the tide consisting of a vertical staff supporting a length scale from which the water level can be read visually. It is called a fixed staff when mounted so that it cannot easily be removed. A portable staff is designed for removal from the water when not in use. For this type of staff, a fixed mounting is used so that the staff will always have the same elevation when installed for use. Some tide staves use an electric monel metal tape, voltmeter, and electrical power source, such as a battery. When contact is made with the water, the circuit is completed, and the voltage provides a quantitative measure of the tide height, using a calibration graph.

See also TIDE.

tide tables Tide tables provide predictions, usually a year in advance, of the times and amplitude of the tide at a reference or primary station, usually located in a major port. Tides are the vertical rise and fall of the water, while TIDAL CURRENTS are the horizontal floods and ebbs. Tide table provide information on heights and times of maximum and minimum water levels; tidal current tables provide information on maximum and minimum current speeds, directions, and times of slack water. These predictions at reference stations are usually supplemented by tidal height differences, velocity ratios, and time differences by means of which additional predictions can be obtained for secondary locations.

Tide tables in the United States were formerly provided in hard copy by NOAA's National Ocean Service, but annual predictions are now available from NOAA only in electronic form. The predictions are made available to private companies, many of whom publish printed tables.

Tide tables provide only the astronomical component of the tide and tidal current. Observed tide heights, slack times, and current speeds will vary from the predicted values due to weather conditions, especially strong winds and storms. For this reason, many major ports provide near real-time information on tide, current and wind conditions based on actual measurements and short-term forecasts.

Tide tables have been provided by the National Ocean Service (formerly the U.S. Coast and Geodetic Survey) since 1890.

See also PREDICTION; MODELS; TIDE; TIME SERIES ANALYSIS.

Further Reading
National Oceanic and Atmospheric Administration. "Center for Operational Oceanographic Products and Services." Available online. URL: http://co-ops.nos.noaa.gov/. Accessed October 16, 2007.

time series analysis Mathematical and statistical procedures aimed at quantifying temporal and spatial variability of time-varying data are called time series analysis. Data sets of observations that are sequential in time (nonrandom) or are spaced at equal intervals are called time series. Common examples of a time series are long records of six-minute water levels; hourly WATER TEMPERATURE, and SALINITY; daily barometric pressure measurements; monthly air temperature measurements; and yearly fish population estimates. These records may contain trends, periodicities, cycles, and noise. Generally scientists manipulate the data to learn what processes are occurring in time or in frequency. Various time series analysis techniques also serve as possible starting points to identify environmental mechanisms and interrelationships.

Traditional methods of time series analysis include Fourier transforms, where a signal is converted to a series of sine and cosine functions; wavelet methods, where signals are of finite duration; and singular spectrum analysis, where a signal is broken down into unique components. This understanding of the system's patterns is used to predict future behavior or to define limits of the system. Phenomena of interest may be water level changes inside and outside a COASTAL INLET, ebb and flood currents, the frequency of coastal storms, accretion and erosion, and the recruitment of commercially important finfish.

Harmonic analysis describes mathematical principles used to describe recurring phenomena. In marine science, these mathematics are especially useful in studying harmonics such as tidal elevations in water level records. For tidal processes (*see* TIDES), the periods at which the major forcing functions occur, such as the gravitational forces resulting from the positions of the Moon and Sun relative to the Earth, are precisely known; it remains to compute the amplitudes and phases of the responding water levels. The general equation is

$$h(t) = H_0 + A \cos(at + \alpha) + B \cos(bt + \beta) + C \cos(ct + \gamma) + ...$$

where h is the height of the tide at any time (t). The height of the mean water level above the datum (mean lower low water) is H_0. Each cosine term is known as a harmonic constituent and the coefficients A, B, and C are the amplitudes derived from the observed water level time series. The expression in parentheses is the phase, which has maximum values at 180° and is zero at 90° and 270°. Periods and speeds of the tide are derived from astronomical data and are independent of the location for the time series. The variables α, β, and γ are initial phases that depend on the time series. Determining the amplitudes and phases of the constituents of the actual tide from time series is done by Fourier analysis and results in the computation of harmonic constants (amplitudes and phases). The method is harmonic since the harmonic constituents (cosine terms from above) are combined to determine the composite tide.

Jean-Baptiste-Joseph Fourier (1768–1830), a French mathematician, is credited for providing the theoretical basis and developing the techniques to define functions as a series or expansion of sines and cosines of various amplitudes and frequencies. The Fourier transform decomposes a time-domain signal (that is, a function of time) into sums of sinusoidal components. The resulting function provides a complete picture of frequency space and may be used to discuss the harmonic structure of phenomena such as tides. For example, the Fourier transform of a water level signal can be analyzed for its frequency content, but will not provide the time that a high or low water will occur. The inverse Fourier transform provides the method to go from the frequency domain into the time domain. The time series of sun spot observations going back to the 17th century was one of the first time series to be analyzed by Fourier theory.

For oscillatory signals such as current speed and water level, marine scientists often compute the power spectrum that identifies a signal's power (energy per unit time) falling within given frequency intervals. The power spectrum is a plot of the magnitude or relative value of some parameter against frequency. It is usually represented on a logarithmic scale since the power can vary by orders of magnitude across the frequency spectrum. It is computed by taking the Fourier transform of the signal (for example, water level as a function of time, or tide record) and turning it into a frequency spectrum. The power spectrum highlights periodic processes by showing energy peaks (for example water level squared) at important frequencies (the inverse of the period). Slopes on the curve are indicative of the rate at which fluctuations are occurring. Total energy in a signal (for example h(t) from the second paragraph) is equal to the area under the square of the magnitude of its Fourier transform. Two important considerations in spectral estimation are resolution and stability. Resolution determines the capability to see fine detail in the spectrum, whereas stability is a measure of the statistical confidence that can be placed in the estimation at a particular frequency. Resolution and stability determine the length of record required to answer, with a given level of confidence, the ques-

tions being asked of the spectrum. It is always an advantage to obtain as long a data record as possible to obtain the maximum resolution (since time is the inverse of frequency, a long record in time is necessary to distinguish between closely spaced frequency components). However, the process being observed must be reasonably stationary over the length of the time of the observations. Hence, a trade-off must be made between resolution and statistical confidence in the estimated spectrum.

Frequencies of interest relate to processes such as gravity wave action, tides, synoptic scale storms, sun spots, climate, and the MILANKOVITCH CYCLES. The time series data is collected from bottom pressure gauges, tide gauges, current meters, WAVE BUOYS, and weather stations. In the case of wave buoys, a Fourier transform is applied to time series of vertical buoy hull displacements to compute the power spectrum where significant wave height, peak spectral period, and peak spectral direction are determined.

The spectrum of a single time series is called the auto spectrum; spectra of multiple time series are described by cross spectra, phase spectra, and magnitude-squared coherence. The time domain analysis yields the autocorrelation function for a single time series, and the cross correlation functions between time series. The two procedures are related in that the auto spectrum is the Fourier transform of the autocorrelation function, and the cross spectrum is the Fourier transform of the cross correlation. It is necessary to compute the statistical confidence of the spectral estimates, as spurious peaks without any physical significance may occur in the computed spectra due to random statistical fluctuations. The correlations among several time series such as wave height in orthogonal directions, pressure, and velocity can be computed by means of the correlation matrix, which provides the correlations between any pairs of variables as functions of the time lags between the time series. The Fourier transform of the correlation matrix can be taken to yield the spectral components as a function, for example, of frequency and direction, or as the x- and y-components of the wavenumber spectrum. These later products will enable the determination and prediction of wave height as a function of direction.

The technique of wavelets (small waves) involves the representation of a time-domain signal in terms of a fast decaying oscillating waveform (a function that integrates to zero as it fluctuates about the x-axis). This technique aims toward determining the frequency components of a signal at a certain moment in time. In other words, the signal is cut or filtered into several parts and then the parts are analyzed separately in accordance with resolution and scale. Wavelets are applicable in situations where the signal contains discontinuities and sharp spikes or the signal has a trend (nonstationary). In the case of wavelets, time series are analyzed based on time-scale representations, while Fourier transforms are based on time-frequency representations. High scales (low-frequency) correspond to dilated signals, and small scales (high-frequency) correspond to contracted signals. The wavelet plot is three-dimensional along axes of frequency (scale), time, and amplitude.

Another analysis method is called singular spectrum analysis or SSA. Basically the single time series is rotated so that the maximum variability is visible or that the most important gradients are observed. This diagnostic signal-processing tool involves the decomposition of a signal that is represented by a short noisy time series. SSA tries to overcome the problems of finite sample length and the noisiness of sampled time series not by fitting an assumed model to the available series, but by using a covariance matrix, instead of a fixed sum of sines and cosines (Fourier theory). The covariance matrix is a group of points that describes how the data varies about the mean of the dataset. The method is used to extract the major signals in oceanic and atmospheric datasets. For examples, SSA has been used in analyzing the Southern Oscillation Index, which may be computed from the monthly fluctuations in sea level pressure differences in the South Pacific between Tahiti in French Polynesia and Darwin in Australia. MEAN SEA LEVEL pressure at Tahiti is usually higher than at Darwin. If the pressure at Darwin rises relative to Tahiti, the SOI becomes negative indicative of an EL NIÑO event.

See also GRAVITY WAVES; WAVE AND TIDE GAUGE; WAVES.

Further Reading

American Mathematical Society. Available online. URL: http://www.ams.org. Accessed January 23, 2007.

Graps, A. "An Introduction to Wavelets." *IEEE Computational Sciences and Engineering* 2, no. 2 (Summer 1995): 50–61.

Schureman, P. *Manual of Harmonic Analysis and Prediction of Tides*, Special Publication No. 98, U.S. Department of Commerce, Coast and Geodetic Survey. Washington, D.C.: U.S. Government Printing Office, 1958 (reprinted May 1988).

trade certifications Trade certifications are awarded by many education and certification programs for marine technologists who are engaged in activities such as diving, hydrographic surveying, navigation, operating REMOTELY OPERATED VEHICLES (ROVs), and

safety management. While academic degrees establish a knowledge base for entry into a marine science occupation, there are numerous certifications required to support marine science. In addition, certain activities require continuing education or recertification to ensure that tasks are completed within known working and safety standards.

Participation in scientific and commercial diving programs requires numerous certifications and management procedures. A DIVER might hold certifications such as Open Water Diver, Advanced Open Water Diver, Master Diver, Surface Supplied Air Diver, Mixed-Gas Diver, and Bell/Saturation Diver. A certification card or "C-card" verifies that a diver is qualified for recreational SCUBA diving, and professional divers may hold a commercial diver certification card. In the United States, the Occupational Safety and Health Administration (OSHA) exempt certain types of scientific diving from rules governing commercial diving. Marine science organizations with a diving program should have a dive safety manual and a diving control board, which ensures that all divers are properly trained to complete underwater research. Scientific divers are observers and data collectors, while commercial divers are involved in inspection and construction. A scientific diver might be trained to operate an OCEANOGRAPHIC INSTRUMENT, while a commercial diver is trained in welding and demolitions in order to place and remove heavy objects underwater.

Organizations such as the International Hydrographic Organization have established standards and certifications to ensure that highly skilled individuals complete hydrographic surveys. Class A and B certifications are internationally recognized. A surveyor with a Class A certification has a comprehensive and broad-based knowledge in all aspects of the theory and practice of hydrography (see HYDROGRAPHER) and allied disciplines. This certification is for surveyors who practice analytical reasoning, make decisions, and develop solutions to nonroutine problems. A surveyor with a Class B certification has demonstrated a practical comprehension of hydrographic surveying. He or she has demonstrated the skill to carry out the various hydrographic surveying tasks. There are many other programs that do not meet requirements for international recognition but adequately train support personnel employed in hydrographic operations.

Navigation is a personal skill that is very helpful to marine scientists, seamen, and anyone else plying coastal and open ocean waters. Schools and certifications in map reading, coastal navigation, GLOBAL POSITIONING SYSTEM, electronic charting, and celestial navigation increase safety and are beneficial on projects that require a detailed understanding of positioning and seamanship. In the United States, a captain's license is required to carry up to six paying passengers on vessels up to 100 gross tons. A master's license is required for any inspected vessel certified to carry more than six paying passengers. A captain's license and a master's license are used in jobs such as charter boat fishing and those providing commercial assistance to vessels that may be aground, disabled or out of fuel, or experiencing some other malfunction where towing services have been requested.

Marine scientists often apply engineering know-how, mechanical tools, and standardized procedures to operations in the marine environment. Marine scientists may also receive certifications that they have met certain standards for material-handling equipment such as cranes and forklifts, commercial diving, and underwater operations. It should be noted that military occupational skill training from a Department of Defense school, experiences gained at an academic institution, or certifications by a recreational agency are not necessarily accepted as a certification to complete work for certain marine operations. For example, an ocean engineering student working on a SUBMERSIBLE project at a university is not necessarily qualified to operate underwater equipment at an offshore rig. An ROV OPERATOR might carry a ROV Pilot/Technician Certification Card, and these certifications are often tied to the specific demands of offshore operations. Similarly, marine technicians aboard research vessels may work toward obtaining many different certifications to operate specific types of shipboard electronic and mechanical equipment. Even aquarists and marine spill response technicians might need to receive specific safety training and certification on particular items of equipment, operation techniques, and procedures.

Safety is an important issue in marine science and covers a wide variety of workplaces from subsurface environments to airborne laboratories. Regulations and inspection efforts by professionals involved in marine safety are aimed at improving safety for all those who work at or travel on the sea. Safety management involves every level of the organization in order to maintain a culture that reduces accidents, especially in the often harsh and unforgiving marine environment. Physical safety may be tied to legal obligations to employers such as ensuring that employees have appropriate certifications. Organizations such as OSHA require that every employer establish and maintain an effective injury and illness prevention program. Supervisors are then required to provide appropriate safety training covering all safety hazards related to the job. In the case of mis-

haps, all evidence is gathered and investigators review "black boxes," logbooks, charts, and other documents. They conduct interviews to consider human factors and try to re-create the physical conditions that lead to the calamity. Usually, results from safety investigations by organizations such as the U.S. Coast Guard are made available to the general public.

See also CERTIFICATIONS, MARINE TRADES.

Further Reading

Association of Commercial Diving Educators. Available online. URL: http://www.acde.us. Accessed February 12, 2007.

Association of Diving Contractors International. Available online. URL: http://www.adc-int.org. Accessed February 12, 2007.

Institute of Navigation. Available online. URL: http://www.ion.org. Accessed February 11, 2007.

International Cartographic Association. Available online. URL: http://www.icaci.org. Accessed February 11, 2007.

International Federation of Surveyors. Available online. URL: http://www.fig.net. Accessed February 11, 2007.

International Hydrographic Office, Monaco. Available online. URL: http://www.iho.shom.fr. Accessed February 11, 2007.

Marine Safety Laboratory, United States Coast Guard. Available online. URL: http://www.rdc.uscg.gov/msl/Home/tabid/202/Default.aspx. Accessed February 12, 2007.

The Marine Technology Society. Available online. URL: www.mts.org. Accessed February 11, 2007.

transect　A transect in marine science refers to a straight line placed within a study area where environmental measurements are taken. Marine scientists might study the diversity of intertidal organisms found along a BARRIER ISLAND by placing a number of transects perpendicular to the shore and taking samples at predetermined intervals along the transect's length. Transect mapping allows the investigator to describe the physical location and distribution of resources, ECOSYSTEMS, and possibly determine an inventory of species common to a particular habitat. High-technology surveys utilizing the GLOBAL POSITIONING SYSTEM along a transect might employ state-of-the-art acoustic, mid-water trawling, and underwater video technologies to determine STOCK ASSESSMENTS for certain species. The College of Charleston in South Carolina is studying the coastal ocean by sampling and analyzing physical and biological data collected from 20 stations along several transects that extend from the inner CONTINENTAL SHELF offshore Fort Sumter, the location where the

U.S. Civil War commenced after a Confederate artillery attack on April 12, 1861, and out to the shelf break, where the GULF STREAM transports warm waters flowing northward. The Gulf Stream begins near Cape Canaveral, Florida, and heads north until it hits the CHARLESTON BUMP off the South Carolina coast and is directed northeast along the rest of the Carolina Capes. Near Diamond Shoals off Cape Hatteras, the Gulf Stream begins to meander and shed Gulf Stream rings (*see* EDDY). The Charleston Bump is a deep-water bottom feature rising from the Blake Plateau, which attracts HIGHLY MIGRATORY SPECIES such as marlin (*Makaira nigricans, Tetrapturus albidus*), sailfish (*Istiophorus platypterus*), and swordfish (*Xiphias gladius*). Transects are linear features used in various habitats such as BEACHES, CORAL REEFS, deep ocean, mud flats, oyster beds, and salt marshes to provide a reference area for detailed scientific analyses.

Transects are of importance in determining the geostrophic velocity of currents; the usual procedure is to plan a cruise track perpendicular to the CURRENT, so that the line of oceanographic stations is normal to the current, providing the maximum information on the current's velocity distribution (*see* GEOSTROPHY).

See also QUADRAT.

Further Reading

Caro-COOPS, Carolinas Coastal Ocean Observing and Prediction System. Available online. URL: http://nautilus.baruch.sc.edu/carocoops_website. Accessed October 29, 2006.

The Cooperative Institute for Marine and Atmospheric Studies, Ocean Surface Currents. Available online. URL: http://oceancurrents.rsmas.miami.edu. Accessed October 29, 2006.

Projects Oceanica, College of Charleston. "Transects." Available online. URL: http://oceanica.cofc.edu/Transects/home.htm. Accessed October 29, 2006.

transform fault A transform fault is a tectonic plate margin where edges of the plates are sliding past one another in strike-slip (lateral) movements. They are believed to result from lateral strain. Almost all the transform faults on the Earth are located on MID-OCEAN SPREADING RIDGES where they run perpendicular to the ridgeline and offset the centerline of the ridge. Transform faults are the active part of oceanic fracture zones where they transfer motion between the spreading segments that they offset and are thus seismically active only between the spreading segments. Lithosphere is neither created nor destroyed at the transform fault. Along the mid-ocean ridge, molten mantle material comes to the surface at the fracture zones to create new crust. In the case of the Mid-Atlantic Ridge, transform faults segment the ridge into independent sections up to about 40 miles (64.37km) wide and 20 to 30 miles (32.19 to 48.28 km) long. These sections are known as spreading centers. They can be a few miles to hundreds of miles long. It is this feature that causes numerous scientists to compare the mid-ocean ridges to the stitching on a baseball. The transform faults result in steep cliffs along the fracture zone between segments of the ridges.

The San Andreas Fault between the North America and Pacific plates in California is a continental example of a transform fault. Other examples are found between Vanuatu and Fiji in the South Pacific, Israel, South Island of New Zealand, and in Sumatra.

See also PLATE TECTONICS.

Further Reading

Erickson, Jon. *Marine Geology: Exploring the New Frontiers of the Ocean.* New York: Facts On File, 2003.

Gates, Alexander E., and David Ritchie. *Encyclopedia of Earthquakes and Volcanoes,* 3rd ed. New York: Facts On File, 2007.

Rothery, Dave. *The Ocean Basins: Their Structure and Evolution,* 2d ed. Prepared for the Open University Oceanography Course Team, The Open University, Milton-Keynes, U.K. Oxford: Butterworth-Heinemann, 1998.

transponder A transponder is a transmitter that emits a signal in response to an interrogating signal that enables the location of the transponder to be identified. Atmospheric transponders transmit electromagnetic waves, whereas underwater transponders transmit acoustic waves. A transponder differs from a beacon, such as the familiar lighthouse or navigation buoy, in that it transmits only when interrogated.

Transponders are used in navigation, harbor and coastal surveying, location of scientific instruments and other equipment, and military missions. An example of an electromagnetic transponder is the RACON, called a RADAR beacon that is really a transponder that transmits a signal in response to ships' navigation radar emissions. RACONs enable suitably equipped vessels to determine their direction and position relative to the transponder.

Acoustic transponders are used in marking the deployment sites of receiver/transmitters with scientific instruments moored on the ocean floor for long durations. Servicing and recovery vessels use SONAR to interrogate the transponders in order to locate the instruments on return voyages.

Modern navigation system transponders usually send digital signals that are received by digital shipboard navigation systems and fed into computers that can incorporate the transponder signal with other navigation data to provide an accurate, absolute location of the ship as well as the deployed instruments (*see* NAVIGATION). Transponders are extensively deployed by geologists and engineers in at-sea oil and gas exploration operations to locate test wells, wellheads, underwater pipelines and nodes, as well as by MARINE ARCHAEOLOGISTS to locate shipwrecks.

Often the first task of manned undersea SUBMERSIBLES is to deploy transponders on the ocean floor to be used for navigation of the study area. In shallow water, small transponders can be positioned by DIVERS to locate deployed equipment, or to perform search patterns for everything from downed aircraft to buried treasure. On return trips, divers use small handheld acoustic interrogators containing embedded computers to provide them with ranges and bearings to the objects to be located.

Further Reading
Bowditch, N. *The American Practical Navigator: An Epitome of Navigation.* Bethesda, Md.: Defense Mapping Agency Hydrographic/Topographic Center, 1995.

trawler A trawler is a displacement hull fishing vessel that drags open-mouthed nets called trawls. These vessels may be classified differently depending on size or how the cone-shaped nets are rigged and towed. The smallest trawlers are seaworthy yachts that are a copy of the trawler design; these vessels do not generally tow nets. North Carolina State University uses a 48-foot (14.6 m) trawler, the research vessel *Humphries,* as a platform for SEDIMENT and fisheries sampling in ESTUARIES such as the Neuse River. Lobster boats range in size from 45 to 65 feet (14 to 20 m); the New England draggers are usually less than 100 feet (30 m) in length. Lobster boats and deck gear such as hydraulic WINCHES are used to deploy and retrieve lobster pots in a string or trawl. A lobsterman and sternman might set 25 trawls per day. There are usually 10 connected pots per trawl. Shrimp boats are around 100 feet (30 m) long and are dwarfed by the factory trawlers, which are well over 145 feet (44 m) long. Shrimp trawlers (shrimpers) deploy warps, the lines used to tow nets, through blocks mounted on horizontal outriggers in order to tow otter trawls. A third net may also be towed directly from the stern. Side-trawlers are recognized by their aft superstructure and wheelhouse. The fish hold is situated amidships, and the trawl winch is located in front of the superstructure. Warps

pass through forward and aft gallows. On a vessel, gallows are a deck-mounted frame. Modern stern trawlers have mostly replaced side-trawlers. Stern-trawlers have a working deck toward the rear where nets are set and hauled. Pair trawlers involve the use of two trawling vessels that fish along a parallel track about 0.62 miles (1 km) apart with trawl nets deployed from the stern. As the trawl nets are pulled through the water, they catch all sorts of marine life. Targeted species are finfish such as Alaska pollock (*Theragra chalcogramma*), black cod (*Notothenia microlepidota*), haddock (*Melanogrammus aeglefinus*), and saithe (*Pollachius virens*), and crustaceans such as shrimp and crabs. The catch collects at the cod-end of the net, which can contain many tons of fish. The fish are sorted, processed, and frozen. Fisheries scientists may use trawls aboard research vessels or study the catch aboard fishing vessels using trawl gear.

See also BIOLOGICAL OCEANOGRAPHER; BOAT; FISHERMAN; MARINE BIOLOGIST; NET.

Further Reading
Fisheries Global Information System, Food and Agriculture Organization of the United Nations. Available online. URL: http://www.fao.org/fi/default.asp. Accessed February 10, 2007.

treaties and governance Treaties and governance includes international, national, state, and local agreements, as well as laws and policies aimed at protecting and sustaining ocean resources. Treaties are instruments (legal agreements) between two or more states through which the states establish rights and obligations among themselves. They are critical in establishing oceanic jurisdictions, especially in defining a maritime nation's coastal boundaries. Policy makers establish organizations, departments, and people to administer ocean policies. In the United States, governance is facilitated by federal organizations such as the NATIONAL OCEANIC AND ATMOSPHERIC ADMINISTRATION and the U.S. Coast Guard, State Departments of Natural Resources, and local marine police.

The establishment of ocean jurisdictions is exemplified by international agreements (conventions) such as the LAW OF THE SEA, which has been instrumental in establishing the EXCLUSIVE ECONOMIC ZONE (EEZ) and territorial seas. Countries may gather and make agreements, which establish obligations of all states to protect and preserve the marine environment. Such conventions urge all countries to cooperate on a global and regional basis in formulating rules and standards. Of utmost importance is the granting of coastal states the right to protect and preserve their

EEZ. Organizations such as the International Maritime Organization help to enforce the agreements and are involved in litigation regarding liability and compensation issues and the facilitation of international maritime traffic.

Ocean resources are protected through laws and law enforcement. In the United States, foundational legislation includes the Outer Continental Shelf Lands Act of 1953, Coastal Zone Management Act of 1972, Marine Mammal Protection Act of 1972, Ocean Dumping Act of 1972, Port and Waterway Safety Act of 1972, Deepwater Ports Act of 1974, Magnuson-Stevens Fishery Conservation and Management Act of 1976, and the Estuaries and Clean Waters Act of 2000. The Ocean Act of 2000 also established the United States Commission on Ocean Policy. This commission was instrumental in highlighting and integrating national efforts. Legislation may be improved from time to time owing to scientific advances or new knowledge through amendments.

Effective coastal management safeguards watersheds, protects people and property from natural hazards, conserves and restores ecosystems, manages the shore face, and supports maritime commerce. Modern approaches apply management decisions that relate to natural biological and physical boundaries or ecosystems. Successful legislation and governance is often based on having champions from government, such as state governors, and is developed with input from interested parties from federal, state, local, nonprofit, and public interest groups. A good example is the Chesapeake Bay Agreement that was initiated in 1983 and modified in 1987 and 2000. Its purpose is to focus partners such as the State of Maryland, the Commonwealths of Pennsylvania and Virginia, and the District of Columbia on protecting and restoring living resources and estuarine habitats, improving water quality, managing coastal lands, and engaging the community. CHESAPEAKE BAY is one of the largest and most productive ESTUARIES in the world.

See also ECOSYSTEM; FISHERY CONSERVATION AND MANAGEMENT ACT.

Further Reading

Chesapeake Bay Program. Available online. URL: http://www.chesapeakebay.net. Accessed February 14, 2007.

Council of Great Lake Governors. Available online. URL: http://www.cglg.org. Accessed February 14, 2007.

Gulf of Maine Council on the Marine Environment. Available online. URL: http://www.gulfofmaine.org. Accessed February 14, 2007.

International Maritime Organization. Available online. URL: http://www.imo.org. Accessed February 14, 2007.

International Program Office, National Ocean Service, National Oceanic and Atmospheric Administration.
Available online. URL: http://international.nos.noaa.gov/welcome.html. Accessed February 14, 2007.

Nature Conservancy. Available online. URL: http://www.nature.org. Accessed February 14, 2007.

Oceans, Coasts, and Estuaries, Oceans and Coastal Protection Office, U.S. Environmental Protection Agency. Available online. URL: http://www.epa.gov/owow/oceans. Accessed February 14, 2007.

United States Commission on Ocean Policy. Available online. URL: http://www.oceancommission.gov. Accessed February 14, 2007.

———. *An Ocean Blueprint for the 21st Century.* Available online. URL: http://www.oceancommission.gov/documents/full_color_rpt/000_ocean_full_report.pdf. Accessed February 14, 2007.

trench In marine science, a trench refers to a very deep depression in the ocean floor that is narrow, with asymmetrical V-shaped cross sections, having very steep sides. Shallower trenches with flat bottoms are known as troughs.

Trenches reveal the presence of SUBDUCTION ZONES, where the oceanic crust of a tectonic plate is being overridden by another plate and is being forced down into the asthenosphere (the upper plastic part of the mantle). The deep ocean trenches have curved or "arcuate" shapes denoting these subduction zones (*see* BENIOFF ZONE). They are the regions of the most powerful earthquakes and volcanoes on the planet.

Trenches generally begin at depths greater than 20,000 feet (7,000m) and extend as deep as 36,000 feet (11,000 m). The crushing pressures found in trenches go from 600 to 1,100 atmospheres. Temperatures are on the order of 36°F (2.2°C); salinities are on the order of 35 psu (practical salinity units). The habitat of the deep trenches is called the HADAL ZONE. The bottom may be composed of terregenous clay and oozes such as radiolarian OOZE, the skeletal remains of radiolarians.

Volcanic island arcs border the trenches on their landward sides. The island arc sides of the trenches are steeper than the seaward sides. Trenches have a characteristic asymmetrical V-shape in cross section of perhaps a few miles across the trench. The largest trench, the Peru-Chile trench, is about 2,500 miles (4,167 km) long.

Most trenches are found around the rim of the Pacific Ocean called the RING OF FIRE because of the large number of volcanoes. The deepest trench marking the deepest spot on the surface of the Earth is the Challenger Deep in the Mariana Trench (discovered during the *Challenger* expedition of 1872–76, which is about 36,000 feet (10,973 m) deep. A few trenches are found in the INDIAN OCEAN and the ATLANTIC OCEAN, such as the Puerto Rican

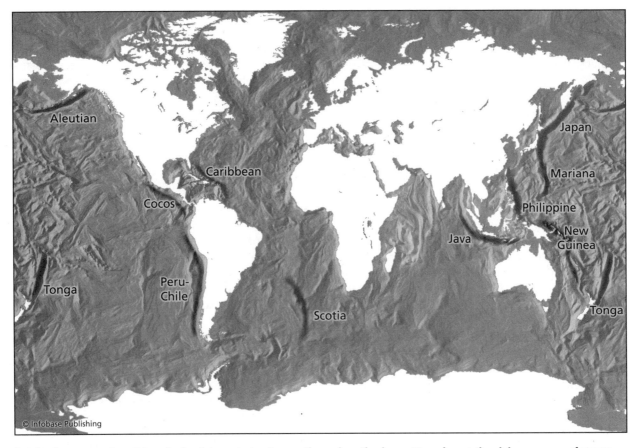

Subduction zones where lithospheric plates entering the mantle produce the deepest trenches on the globe are areas of strong seismic and volcanic activity.

Trench just seaward of Puerto Rico. The Mariana Trench was visited by French explorer Jacques Piccard and U.S. naval officer Don Walsh in the bathyscaphe *Trieste* in 1960, which dove to a depth of 35,798 feet (10,914 m). Humans have not visited a trench since the 1960s.

Historical observations of TEMPERATURE and SALINITY in the deep trenches have shown that temperature increases with depth in the trench by perhaps a Celsius degree or more from top to bottom, whereas salinity is almost constant. This situation results in minimum water column stability. It is the result of adiabatic (without gain or loss of heat) compression within the trench. Gravity is weaker directly over a trench, which results in lower sea surface height above. The changes in gravity over oceanic features such as trenches and ridges have made it possible to map the BATHYMETRY of the ocean floor from orbiting satellites.

Further Reading

Erickson, Jon. *Marine Geology: Exploring the New Frontiers of the Ocean.* New York: Facts On File, 2003.

trim Trim refers to the relation of the draft of a vessel at the bow and stern. To improve the trim, the vessel operator may shift passengers, equipment, and cargo or make mechanical adjustments to the power unit or trim tabs based on the roll and pitch axes of a vessel. In other words, the vessel operator improves performance by eliminating or reducing listing or heeling and raising or lowering the bow. These adjustments are especially important in head, following, or quartering seas to avoid pitch poling (surfing down a wave such that the bow dives under the preceding wave and the wave crest pitches the stern over the bow), broaching (the sudden uncontrolled motion of a vessel so that the hull is broadside to the seas), and parametric rolling motions. Parametric equations describe rolling as a combination of translational and rotational motion. The phenomena of parametrically induced roll may occur when the ship's length is comparable to the WAVELENGTH of the sea conditions and when there is a physical dependence among the ship's speed, the wavelength, and the vessel's natural rolling frequency. The amount of wetted surface area on the hull affects speed and fuel consumption. If the bow of the hull is raised, the wetted area of the hull is reduced, which improves efficiency. This reduction in friction is why planing hull boats travel faster than displacement hull boats of the same size. In hydrographic surveying, many vessels have a deep V-hull design since they need to maintain an even trim throughout a wide speed range. Another stable hull type gaining increasing popularity is the small waterplane twin-hull design. Many vessels include stabilizer fins to control the pitch and roll of the vessel and allow adjustment of the heel and trim of the ship while under way.

Marine scientists use boats that have outboard or inboard/outboard (stern drive) power units in shallow water. These vessels are typically equipped with power trims, which raise or lower the drive unit. Hydraulic trim tabs (rectangular control flaps that project parallel to the surface of the water) may also be installed in the stern (along the outside bottom of the transom). They redirect water and provide an upper force at the stern of the boat. When unable to balance heavy equipment, trim tabs can be lowered on the listing side to bring the boat level. By changing the running position of the engine drive unit, the COXSWAIN raises or lowers the bow. There may also be associated changes in steering since thrust current is being modified. The term *trimming in* lowers the bow and may improve ride quality in choppy water. In head seas, trimming in allows the hull of the boat to better absorb the impact of the waves. Neutral trimming means that the propeller shaft is parallel to the surface of the water. "Trimming out" lifts the bow and increases speed and clearance in shallow water. In following seas or when running an inlet, the trim tabs should be fully retracted. The operator must avoid letting the bow dig into the water, which may lead to capsizing.

There are many commercial schools that teach boat handling. Most academic organizations require marine scientists to complete courses with either the U.S. Power Squadron or the Coast Guard Auxiliary along with a practical check-out cruise prior to using research boats for science.

See also BOAT; HYDROGRAPHIC SURVEY.

Further Reading
Maloney, Elbert S. *Chapman Piloting and Seamanship.* New York: Hearst Books/Sterling, 2003.

National Water Safety Program. U.S. Army Corps of Engineers. Available online. URL: http://watersafety.usace. army.mil. Accessed February 3, 2007.
United States Coast Guard Auxiliary home page. Available online. URL: http://nws.cgaux.org/index.html. Accessed February 3, 2007.
United States Power Squadron. Available online. URL: http://www.usps.org/newpublic2/index.html. Accessed February 3, 2007.

tsunami   A seismic sea wave generated by sudden displacements in the seafloor resulting from landslides or volcanic activity is called a tsunami. It is a Japanese word meaning "harbor wave." These natural hazards are common along the RING OF FIRE, coasts found along the Pacific plate associated with a convergent margin where Pacific seafloor plunges under or subducts beneath continental plates. For example, undersea earthquakes at SUBDUCTION ZONES near the Japanese islands, Kamchatka peninsula, or the Aleutian islands could generate a tsunami able to strike coastal areas in the Hawaiian Islands. Once established, a tsunami is a series of long waves. Their WAVELENGTH may be on the order of hundreds of miles (km) and wave heights may be only a few inches (cm) in deep water. Since tsunami wavelengths are very long compared to the depth, they are treated as shallow-water waves with wave speeds that can be determined from the following simple equation:

$$c = (g \times h)^{\frac{1}{2}}, \text{ where}$$

c = wave speed,
g = acceleration due to gravity or 9.8 m/s^2, and
h = depth of the water.

The speed of a tsunami varies as the square root of the water depth. Thus, a tsunami generated within a TRENCH and propagating across an ABYSSAL PLAIN at an approximate depth of 3.4 miles (5.5 km) would travel at a speed of 519.6 miles per hour (232.3 m/s). As the tsunami shoals, its wave height increases and wavelength decreases. The first part of the wave reaching the BEACH is the wave trough and appears as a rapidly receding sea. The arrival of the wave crest results in a sudden increase in sea level as the run-up floods low-lying coastal areas. Run-up is the maximum vertical height that the sea surface reaches

(opposite page) **A)** A tsunami can be generated when a section of the seafloor subsides. **B)** Tsunamis propagate across the ocean as a shallow water wave of low height and very long length at speeds approaching that of a jetliner. **C)** A tsunami reaching shallow water slows and the wave height builds. **D)** At the shoreline, the wave breaks and large volumes of seawater are carried inland, sweeping everything before them.

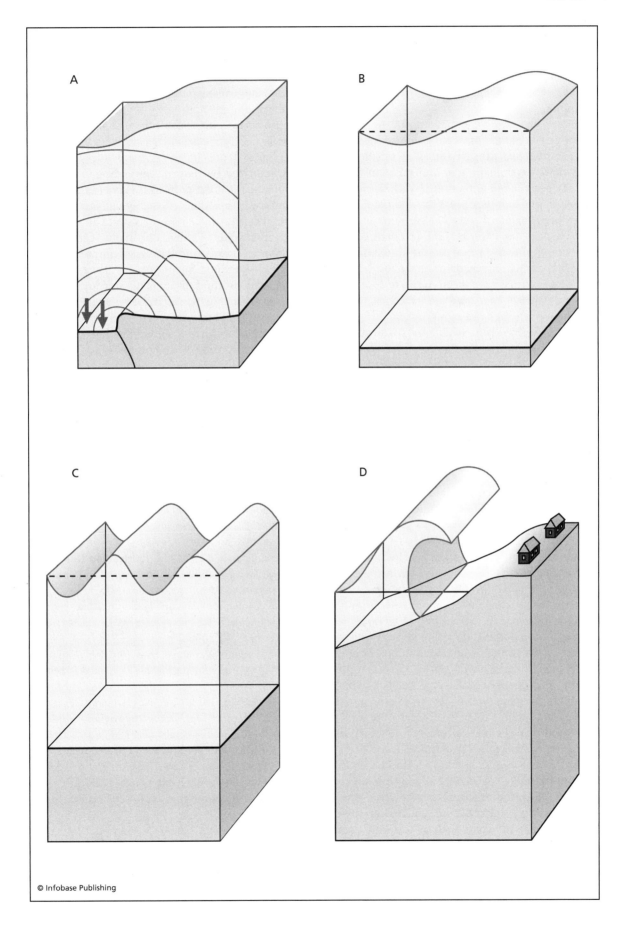

during a tsunami with respect to MEAN SEA LEVEL. The advancing wave is similar to a rapid, very strong, and exaggerated tide and is sometimes mistakenly called a tidal wave. Loss of life and damage to property results from strong currents, floating debris, and flooding.

Tsunami warning systems are used to prepare people in coastal areas of a possible tsunami. One component of a warning system is a seismograph that measures earthquake size and position on the Earth's surface. Other components include systems of anchored seafloor bottom pressure gauges and linked moored surface BUOYS. Linkages are made by sea cable or acoustic modems. Buoys outfitted with satellite communication equipment transmit real-time water level fluctuation data to tsunami warning centers. Algorithms have been written to rapidly analyze time histories of WATER LEVEL FLUCTUA-TIONS where warnings are issued when differences between observed pressure (water level) and the predicted (water level) exceed the prescribed threshold for a particular location. Marine scientists are able to utilize all available information from seismometers, pressure gauges, WAVE BUOYS, and weather stations to distinguish a tsunami from phenomena such as tides and ship wakes.

A tsunami is a very powerful destructive force to people and property living in low-lying coastal areas. On December 26, 2004, a megathrust earthquake off the west coast of northern Sumatra, Indonesia, caused a tsunami, which propagated across the INDIAN OCEAN. This earthquake epicenter was near Indonesia's Aceh province and was recorded as a magnitude 9.0 by seismologists. The magnitude helps characterize the seismic energy released by each earthquake. For this large quake, the resultant tsunami waves, which radiate outward in all directions from the underwater disturbance, were reported to be 50 feet (15 m) high across the Indonesian islands and the shores of Sri Lanka, India, Thailand and other countries such as Somalia, with more than 230,000 reportedly dead or missing. Coastal impacts and the magnitude of human suffering are still being investigated, utilizing satellite imagery. Images were used to prioritize and plan the relief effort by foreign nations that came to the rescue of tsunami victims. Before-and-after satellite images highlight a flood area that extends approximately 1.24-mile (2 km) inland from the shoreline. This Indian Ocean tsunami was particularly devastating because unsuspecting people living and working near the coast such as farmers, fishermen, and tourists were caught in the run-up of approximately six tsunami waves that flooded lowlands, washed commercial vessels hundreds of feet (up to a 100 m) inland, collapsed bungalos, knocked

down power lines, and carried cars into the ocean. Survivors were miraculously able to cling to vegetation or were able to reach high ground that afforded protection from the surging waters. Tragically, beachgoers who investigated rapidly receding waters as the tsunami trough approached shore were trapped as the powerful and turbulent crest moved across the beach.

The Atlantic basin is seismically passive and is characterized by seafloor spreading. Therefore, the only way for a tsunami to be generated would be a submarine landslide, possibly at the shelf break, that abruptly deforms and vertically displaces the overlying water. Other less likely scenarios would involve a significant meteorite impact that could force a tsunami. Such impacts have occurred as evidenced by the Barringer Crater and cosmogenic sediments such as iridium found at the Cretaceous-Tertiary (K-T) boundary, which contributed to mass extinctions of the dinosaurs. The Barringer Meteorite Crater is a mile-wide hole that is 570 feet (173.7 m) deep in the middle of the arid sandstone of the Arizona desert.

Further Reading

International Tsunami Information Center. Available online. URL: http://www.tsunamiwave.info. Accessed December 31, 2006.

National Geophysical Data Center, National Satellite and Information Service, National Oceanic and Atmospheric Administration. "Tsunami Data at NGDC." Available online. URL: http://www.ngdc.noaa.gov/seg/hazard/tsu.shtml. Accessed December 31, 2006.

Nichols, Charles R., David L. Porter, and Robert G. Williams. *Recent Advances and Issues in Oceanography.* Westport, Conn.: Greenwood, 2003.

NOAA Center for Tsunami Research home page. Available online. URL: http://nctr.pmel.noaa.gov. Accessed December 31, 2006.

Pacific Tsunami Warning Center. Available online. URL: http://www.prh.noaa.gov/ptwc. Accessed December 31, 2006.

University of Southern California Tsunami Research Center. Available online. URL: http://www.usc.edu/dept/tsunamis. Accessed December 31, 2006.

West Coast and Alaska Tsunami Warning Center, National Weather Service, National Oceanic and Atmospheric Administration. Available online. URL: http://wcatwc.arh.noaa.gov/main.htm. Accessed December 31, 2006.

tugboats (tugs)   Tugboats (or tugs) are equipped with very powerful propulsion systems for towing or pushing larger vessels. Tugboats are normally designed for inland (harbors and rivers) or open ocean missions. Inland tugs might be involved with ship

A tug pushes a barge into position for loading and unloading. *(NOAA)*

docking and moving BARGES up and down a waterway, while open ocean tugs might be involved in rescue, salvage, and pipeline construction. Tugboat crews may range in size from approximately three to 10 members consisting of deckhands, seamen, engineers, a mate (second in command), and a master (or captain). In the United States, the captain holds a Coast Guard license as Master of Towing Vessel, which is obtained after considerable experience, including earning an apprentice mate license and then work as a mate (or pilot in inland waters).

See also DOCKING PILOT; HARBOR PILOT; MARINE TRADES; WORKBOAT.

Further Reading

North River Tugboat Museum. Available online. URL: http://tugmuseum.com. Accessed August 7, 2007.

turbidity The degree of transparency based on the amount of particulate matter that is suspended in water is turbidity. Particulates are small pieces of material (such as SEDIMENTS) floating in the water that cause incident light rays to scatter. Cloudy water in an ESTUARY has high turbidity, while clear water near a CORAL REEF has low turbidity. In the field, marine scientists may use a SECCHI DISK to measure the clarity of water or a transmissometer. A black-and-white Secchi disk is lowered into the water until it cannot be seen, and then the depth of the disk is measured. The transmissometer is an instrument that may be fixed to a MOORING to assess turbidity by measuring the amount of beam attenuation between a light source and a light detector. When measuring turbidity using a nephelometer, light is passed through a sample and the amount of light deflected (usually at a 90-degree angle) is then measured. This turbidity unit is the nephelometric turbidity units (NTU). In the United States, the ENVIRONMENTAL PROTECTION AGENCY has defined the light source and formal geometry (detector angle to the incident light beam) for NTU measurements, especially to assess drinking water.

Further Reading

International Organization for Standardization. "Water quality—determination of turbidity: Geneva, Switzerland, International Organization for Standardization," IOS 7027, 1999, 10 p.

U.S. Environmental Protection Agency. "Methods for the Determination of Inorganic Substances in Environmental Samples, Method 180.1, EPA/600/R-93/100." U.S.

Environmental Protection Agency, Cincinnati, Ohio, August 1993.

U.S. Geological Survey. "National field manual for the collection of water-quality data: U.S. Geological Survey Techniques of Water-Resources Investigations," Book 9, Chaps. A1-A9. Various dates. Available online. URL: http://pubs.water.usgs.gov/twri9A. Accessed August 7, 2007.

turbidity currents   Turbidity currents are high-speed CURRENTS of SEDIMENT-laden water cascading down the CONTINENTAL SLOPE into the CONTINENTAL RISE. They typically occur in SUBMARINE CANYONS, eroding the canyon floor and walls, and transporting sediments from the continental shelf to the continental rise. The momentum of these currents is often so high that they extend far out onto the ABYSSAL PLAINS, leaving sediments in fanlike deposits.

Turbidity currents, also called submarine landslides and slumps, usually occur from catastrophic events, such as earthquakes, volcanic eruptions, or extremely intense storms. They can cause large TSUNAMIS that can severely damage neighboring coasts, such as the slump-initiated tsunami that devastated the coast of Papua New Guinea in July 1998.

Although there is no agreement about the speed attained by turbidity currents, many geologists believe it may be as great as 10 times the speed of strong ocean currents.

The sediments deposited by the turbidity currents show a gradation from heavy sediments, which are dropped first on the continental slope, to fine sediments, which can be carried far out on to the abyssal plain. These sediments are called "turbidites." Geologists use the presence of turbidites to locate ancient submarine canyons.

Turbidity currents flow down slope because the water near the bottom contains large quantities of sediment, put into suspension by a seismic or major meteorological disturbance, making it denser than the surrounding water, and so is accelerated downward by gravity. Turbidity currents might be compared to KATABATIC WINDS in the atmosphere, which are dense "waterfalls" of cold air pouring down the mountain slopes in locations such as Antarctica.

The frequency of occurrence of turbidity currents is not known, but they are believed to be rather infrequent events, maybe recurring on the order of tens of years. They are thought to have occurred more frequently during the ice ages when many of the world's continental shelves were above sea level, and the sediments would have been much closer to the edge of the slope. Hence, it would have been much easier for the sediments to flow over the edges down onto the continental rise and the abyssal plain.

turbulence   Chaotic fluctuations in a fluid are called turbulence. At the molecular level, random velocity

A turbidity current (undersea landslide) roars down the continental slope to deposit continental sediments on the continental rise and abyssal plain.

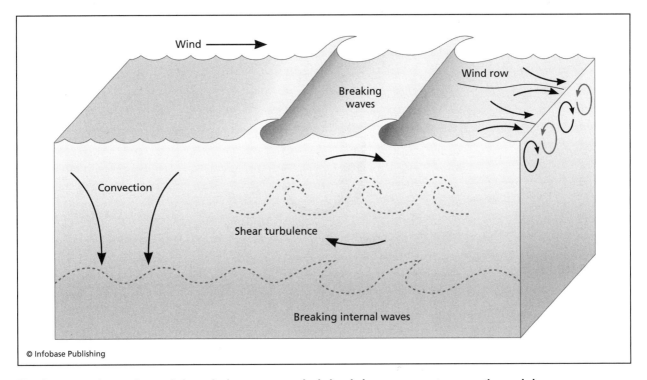

The phenomena that produce turbulence in the upper ocean include wind, waves, currents, convection, and shear.

fluctuations distort and confuse the flow. Turbulence in the atmosphere and ocean is important in mixing. Winds aloft distribute particulates such as water vapor and smoke, but irregular whirls or EDDIES of air can pose a serious effect on the motion of aircraft. In the ocean, marine scientists actively research the structure and evolution of upper ocean mixing as well as the development of eddies and vortices.

Turbulence in the upper ocean usually develops from the SHEAR that the wind stress exerts on the surface. A gentle, steady wind would produce a smooth LAMINAR FLOW, but in reality, the winds are usually gusting and produce turbulent flow. While laminar flow can be described in terms of molecular VISCOSITY coefficients, turbulent flow is usually described in terms of "eddy coefficients," which indicate the magnitude of the transfer of heat, salt, and momentum of bulk masses of water. Turbulent flow is characterized by large Reynolds stresses. Turbulence is a very important process in the cascade of energy in the ocean from the large-scale ocean cur-

rents down through waves and eddies of ever-decreasing size, to the smallest scales of millimeters where the energy is eventually dissipated as heat. Turbulence is also produced by the stress of ocean currents along the bottom and by breaking INTERNAL WAVES within the water column.

Practical studies of turbulence are a component of fluid dynamics. Numerical modelers are also using the Navier-Stokes equations to study complicated fluid flows on computers. These equations were developed by French engineer Claude Louis Marie Henri Navier (1785–1836) and Irish mathematician George Gabriel Stokes (1819–1903).

See also CHAOS; FLUID MECHANICS; NUMERICAL MODELING.

Further Reading

Center for Turbulence Research, Stanford University, Stanford, California. Available online. URL: http://www.stanford.edu/group/ctr/. Accessed August 7, 2007.

U

umbrella species Umbrella species require a large area for survival, and their health ensures the success of other flora and fauna from the same habitat. Umbrella species in the marine environment might include steelhead (*Oncorhyncus mykiss*), the sea-run form of rainbow trout and the southern sea otter (*Enhydra lutris*). Anadromous fish such as steelhead migrate from salt water to freshwater to spawn. After two or three years in the PACIFIC OCEAN, steelhead are known to enter rivers such as the Columbia River and swim approximately 400 miles (644 km) upriver to spawn. Adult steelhead gather near the mouths of rivers and streams, detect the discharge, and enter heading upstream to swim strongly against heavy CURRENTS to suitable spawning habitat. They have spectacular migrations and are sometimes seen jumping over waterfalls. The part-time river, part-time seagoing steelhead tends to inhabit cool waters in the Northern Hemisphere, and their presence indicates a healthy river or stream ECOSYSTEM. Sea otters are small marine mammals inhabiting regions near the shores of the Kurile and Aleutian Islands and along the coasts of Alaska, Washington, Oregon, and California. They are fond of kelp beds and may travel 0.6–1.2 miles per day (1–2 km) to feed. Sea otter populations may decline in nearshore waters that are polluted with contaminants such as anti-fouling paints and pesticides. Sea otters consume crustaceans such as crabs, echinoderms such as sea urchins, MOLLUSKS such as abalones, clams, and octopi, and their presence is often an indicator of a healthy kelp ecosystem. Safeguarding umbrella species also protects the needs of other species such those inhabiting steelhead spawning grounds or kelp forests. Besides protecting other species and having large ranges, umbrella species of note are easily observed, well studied, migratory, and persistent. Conservationists may pick a specific umbrella species in order to safeguard other species that depend on the same habitat.

See also ANADROMOUS; MARINE MAMMALS.

Further Reading

NatureServe Explorer home page. Available online. URL: http://www.natureserve.org/explorer. Accessed October 17, 2007.

The Otter Project. Available online. URL: http://www.otterproject.org. Accessed October 25, 2006.

undercurrent An undercurrent flows beneath another CURRENT usually at a different speed or in nearly the opposite direction. These currents are particularly important in transporting WATER MASSES and delivering nutrients, such as carbon dioxide (naturally absorbed by the ocean), nitrate (results from the recycling of organic materials), phosphate (released into the ocean through geological and biological processes) and iron (results from soil transported and deposited through currents and winds). Examples include the equatorial undercurrents, the Davidson Current, and the Gunther Current. The equatorial undercurrent is approximately 656 feet (200 m) thick and 0.25 miles (400 m) wide and travels at speeds up to 2.9 knots (1.5 m/s).

See also EQUATORIAL UNDERCURRENT; THERMOHALINE CIRCULATION.

Further Reading

The Agulhas Undercurrent Experiment, Rosenstiel School of Marine and Atmospheric Science. Available online. URL: http://www.rsmas.miami.edu/personal/lbeal/current.html. Accessed July 18, 2007.

undertow *See* RIP CURRENT.

upwelling Upwelling is the vertical movement of nutrient-rich subsurface water of lower TEMPERATURE toward the surface of the ocean. Upwelling is known to occur in the coastal ocean such as the western coastlines of North America (California) and South America (Peru), when waters are driven offshore by prevailing alongshore winds. Since the deep water is usually much colder than the surface water, the air-sea temperature gradient acts to cool the lowest layers of the atmosphere, and often the adjacent shoreline and islands. Vagn Walfrid Ekman (1874–1954) developed the theory, which described how upwelling occurs when winds blow parallel to the shore. There is a net volume transport of water 90° to the right of the wind in the Northern Hemisphere, and 90° to the left of the wind in the Southern Hemisphere. Thus, winds blowing parallel to coastlines with the water on the right result in a net transport of warmed coastal water offshore, to be replaced by colder bottom water. In the Southern Hemisphere, the directions are just reversed, with upwelling occurring when the ocean is to the left of the wind. At the equator, especially in the eastern PACIFIC OCEAN, equatorial upwelling occurs in which deep waters move in to replace surface waters that move away from the equator in response to trade winds that are blowing from the east to west. Regions where this process is occurring are known as "cold islands." The vertical velocity of the upwelling water is usually quite slow, on the order of 3.3 feet (1 m) or less per day.

Seasonal upwelling may also occur especially in response to land breezes. As the land cools and warm air over the water rises, offshore winds blow water away from the coast, which is replaced by cool deep waters. Small-scale coastal upwelling of short duration can take place off the northern portion of the Outer Banks of North Carolina, when the winds are persistently from the southwest (parallel to the shoreline, with the ocean to the right of the wind. Upwelling is an important process for ocean productivity since it brings nutrient-rich deep waters to the PHOTIC ZONE, to replace nutrients consumed by the PLANKTON and marine animals up the food chain. For this reason, considerable effort has been devoted in the oceanographic community to developing computer models of upwelling over varying time and space scales, as well as to explore the connections between upwelling and climate.

Perhaps the most notable upwelling zone is the region of the PERU CURRENT that sustains one of the most productive fisheries in the world (except during El Niño years when the prevailing winds fail and upwelling is reduced or ceases; *see* EL NIÑO), as well as the CALIFORNIA CURRENT. The BENGUELA CURRENT off the Atlantic coast of southern Africa is also highly productive.

See also EKMAN SPIRAL; EKMAN SPIRAL AND TRANSPORT; THERMOHALINE CIRCULATION.

Further Reading

Allaby, Michael. *Encyclopedia of Weather and Climate,* rev. ed. New York: Facts On File, 2007.

Johnson, John A., and Nicole Rockliff. "Shelf Break Circulation Processes." In *Baroclinic Processes on Continental Shelves,* edited by Christopher N.K. Mooers. Washington, D.C.: American Geophysical Union, 1986.

Knauss, John A. *Introduction to Physical Oceanography,* 2nd ed. Upper Saddle River, N.J.: Pearson/Prentice Hall, 1996.

National Environmental Satellite, Data, and Information Service. "Investigating the Ocean." Available online. URL: http://www.science-house.org/resources/ee.htm. Accessed August 19, 2008.

V

vector A vector is a physical quantity that has both magnitude and direction and is represented by a set of values. Scalar quantities, in contrast, have only magnitude and can be represented by a single value. Displacement, velocity, acceleration, force, and momentum are examples of vector quantities. TEMPERATURE, SALINITY, DENSITY, and PRESSURE are examples of scalar quantities. Vectors are represented graphically by arrows. The length of the arrow is proportional to the magnitude of the vector, and the direction is generally given in units of direction, such as degrees or radians, or as points of the compass. Vectors in printed material are usually indicated by boldface type, as is the convention here.

Environmental scientists depict vectors according to geographic direction, where 0° is due north and rotation is clockwise. A vector can be specified as a magnitude and direction, such as 10 knots (5.2 m/s) at 30°, or in terms of x- and y-coordinates. The system used in meteorology and oceanography is that of standard map coordinates, with north toward the top of the paper, and east to the right of the paper, thus making south down and west to the left. For example, the vector mentioned above has a north component of 5.0 knots (2.6 m/s) and an east component of 8.7 knots (4.5 m/s). The perpendicular (orthogonal) components completely specify the vector. The system in mathematics and computer science is slightly different, where 0° is due east, and rotation to 360° is counterclockwise.

Vectors can be added by putting the tail of the second vector against the head of the first vector (the beginning vector from the origin of coordinates), the tail of the third vector against the head of the second vector, the tail of the fourth vector against the head of the third vector, and so forth. The vector sum is obtained by drawing the "resultant" vector from the origin to the head of the last vector. Vector methods are extensively used in the navigation of aircraft and ships. For example, if the vectors represent displacement of a ship after successive movements, the vector from the origin (the point of departure) to the head of the final vector represents the most recent position of the ship. In meteorology and oceanography, a sequence of vectors (usually in time) is called a "progressive vector diagram," or "provec," and is often used to represent successive measurements of current velocity. For progressive vector diagrams, the x- and y-axis, which are in velocity units, are converted to distance units. Provecs are very useful in plotting the movements of drifting BUOYS.

A vector can be subtracted from another by simply reversing its direction and then adding as above. A vector can be multiplied by a scalar, which simply contracts or expands the length of the vector. A vector is represented in three-dimensional space by specifying the three angles between its direction and the x-, y-, and z-axes. It can be represented by its components (projections) along the three axes in terms of its direction cosines, such as $v \cdot \cos(\alpha)$, $v \cdot \cos(\beta)$, $v \cdot \cos(\gamma)$, where v is the magnitude of the vector (say, the velocity), and α, β, and γ are the angles of the vector with respect to the coordinate axes. The reader is referred to textbooks on physics and advanced calculus for the algebra and calculus of vector operations.

In marine science, CURRENTS are normally provided as a speed (such as knots or m/sec) and a heading given in degrees measured clockwise from true north (direction). Speed is the magnitude of the velocity vector. Another way to describe currents involves separating the flow into its horizontal components that are measured along perpendicular axes. In this case, the "u" component represents horizontal flow in the east-west direction while the "v" component represents flow in the north-south direction. So, in a CURRENT METER record the u-components are computed by multiplying speed times the sine of the direction (speed · sine (direction)) and the v components are computed by multiplying speed times the cosine of the direction (speed · cosine (direction)). Directions are in degrees from true north, where directions to the west and directions to the south are negative.

Vectors also represent the strengths of vector fields, such as electric, magnetic, and gravitational fields. They are often used to represent average ocean currents in, say, one degree squares. Before averaging, the vectors must be represented in Cartesian component form. For example, in two dimensions, the x-, and y-components are added together, and then the resultant vector is given by taking the square root of the sums of the squares of the components (the Pythagorean theorem), while the resultant direction is given by the inverse tangent of the sum of the y-components divided by the sum of the x-components.

Vectors can represent scalar fields as well. For example, if the observations of PRESSURE, plotted on a grid, such as a weather map, are isoplethed (drawing the contour lines of the isobars, at, say, every 25 millibars), the strength and direction of the PRESSURE GRADIENT at any point is determined by constructing a vector perpendicular to the ISOBARS, whose magnitude represents the pressure difference divided by the distance over which the difference occurs. If the pressure difference between two isobars is Δp over a distance Δn, then the pressure gradient is $\Delta p/\Delta n$. Large pressure gradients correspond to strong winds; small pressure gradients to light winds.

Vectors can represent rotations of rigid bodies, or of fluids, as in VORTICITY. For example, the rotation of the Earth as seen above the North Pole is represented by a vector Ω along the Earth's axis, whose length represents the rotation rate $\Omega = 7.29 \times 10^{-5}/s$ (radians per second). As a reminder, motions on a rotating Earth will appear to deflect to the right in the Northern Hemisphere (see CORIOLIS FORCE). Vorticity is represented by a vector along the axis of the circulation, whose length is proportional to the magnitude of the vorticity.

Vectors are used in wave motion to represent the direction of propagation of the wave by the wave number **k**, which is equal in magnitude to 2 pi (π) divided by the wavelength, or $k = 2\pi/\lambda$. The wave number is inversely proportional to the wavelength, so that small waves have high wave numbers, and long waves have small wave numbers.

In geology, vectors represent the direction of stresses, which may be forcing blocks of rock apart, such as in strike-slip faults. Earth scientists such as geographers, cartographers, and hydrographic surveyors use a vector graphics structure involving the storage of geographic information with x- and y-coordinates within a computer database. These systems are called geographic information systems. Vector features are represented on a digital map as points, lines, and polygons. On a nautical chart, point features might represent navigation aids such as buoys, lines might represent depth contours and shorelines, and polygons may represent marshes, rivers, and bays.

Further Reading
The Math Forum. Available online. URL: http://mathforum. org. Accessed February 12, 2007.

velocimeter A velocimeter is an instrument for the in situ measurement of the speed of sound in seawater by transmission of high-frequency acoustic pulses across the face of the instrument as it is being raised or lowered from the deck of a research vessel. An electrical cable is connected from the instrument to an electronic recording system on the deck of the supporting research vessel. Prior to the development of velocimeters in the 1960s, the speed of sound had to be computed from measurements of TEMPERATURE, SALINITY, and depth (PRESSURE).

A popular type of velocimeter uses the "sing around" principle to measure the travel time of sound between transducers that are part of a "sing around" circuit, in which an electrical impulse energizes a transducer that transmits an ultrasonic pulse across a known distance above the face of the velocimeter to a receiving transducer, which in turn converts the sonic pulse to an electrical pulse that travels the circuit and transmits another sonic pulse and so on, at a high frequency. Since the speed of sound in seawater is much slower than the speed of the electrons in the electrical circuit, the rate of transmission of pulses is determined by the speed of sound and is recorded as an electronic frequency. The output of the instrument is thus a nearly linear function giving the speed of sound as measured by the frequency of the pulses in the "sing around" circuit.

Today, sound velocimeters achieve accuracies on the order of 0.03 feet/second (0.01 m/s) at depths from the surface to the bottom of the ocean. Velocimeters can be deployed in various configurations. They can be hull-mounted on surface vessels or submarines or can be deployed on fixed platforms, BUOYS, AUTONOMOUS UNDERWATER VEHICLES (AUVs), and REMOTELY OPERATED VEHICLES (ROVs). Velocimeters are also available in expendable form, in which the sound velocity signal can be transmitted via a wire to a surface vessel or submarine during its descent. The wire breaks at the end of the cast, in a manner similar to the expendable bathythermograph (XBT). The XBTs measure temperature by means of a THERMISTOR mounted in a weighted expendable body where wire from the thermistor, which is connected to a recorder on the ship, unspools as the body descends. Air-expendable velocimeters can be deployed from aircraft or helicopters.

Applications of sound velocimeters include sonar operation, bathymetric mapping studies and surveys, underwater communication systems, geophysical prospecting, and fisheries studies of ambient noise.

See also ACOUSTIC DOPPLER CURRENT PROFILER; OCEANOGRAPHIC INSTRUMENTS; SONAR.

Further Reading

Cold Regions Research and Engineering Laboratory, U.S. Army Corps of Engineers. "Fact Sheet: A Doppler Velocimeter for Monitoring Groundwater Flow." Available online. URL: http://www.crrel.usace.army.mil/techpub/factsheets/pdfs/003.pdf. Accessed January 25, 2007.

NOAA Ocean Explorer. "Equipment Used on the Davidson Seamount Cruise." Available online. URL: http://www.oceanexplorer.noaa.gov/Explorations/06davidson/background/instruments/instrume nts.html. Accessed October 17, 2007.

Odom Hydrographic Systems, Inc. "Digibar-Pro Profiling sound Velocimeter Operation Manual." Available online. URL: http://www.odomhydrographic.com. Accessed January 23, 2007.

Venturi effect   The Venturi effect describes "choked flow" or velocity increases as a fluid is forced through a constriction. The flow must increase owing to conservation of energy as pressure changes as the flow moves through a smaller space. The Venturi effect is named after Italian physicist Giovanni Battista Venturi (1746–1822). Earth scientists observe this phenomena as increased wind speed and decreased pressure as air flows through a mountain pass or as increased current (jets) as tides propagate through a coastal inlet. In both of these examples, flows accelerate to conserve volume. Marine scientists interested in predicting wind speeds and currents must understand variables such as pass or inlet dimensions (channel lengths, cross sectional area, and depths), water level differences, and surface roughness. A Venturi tube is a practical instrument exploiting the Venturi effect. The concept involves inferring fluid velocity by measuring pressure differences at the inlet (inward pressure) and throat (constricted pressure) as fluids pass through the specially designed tube. Venturi tubes look like cylindrical pipes with a narrow constriction and are frequently used to study the flow of liquids and gases.

See also COASTAL INLET; FLUID MECHANICS; HYDRAULIC CURRENT; PRESSURE.

vertical datum   The zero surface to which elevations or heights are referred to in surveying is called a vertical datum. It is called a tidal datum when defined in terms of a certain phase of the TIDE such as mean lower low water. Relationships between these tidal datums and geodetic vertical datums such as North American Vertical Datum of 1988 (NAVD 88) are maintained by the National Geodetic Survey (NGS) of the NATIONAL OCEANIC AND ATMOSPHERIC ADMINISTRATION. The NAVD 88 is a geodetic reference for elevations, created by the NGS and published in 1991. It is based on fieldwork prior to 1929 as well as surveys as recent as 1988. Features on a map are referenced to a specific map datum. MEAN SEA LEVEL is used as a standard in determining land elevation or sea depths.

See also DATUM; HYDROGRAPHIC SURVEY; CHARTS, NAUTICAL; PERMANENT CURRENTS; WATER LEVEL FLUCTUATION.

Further Reading

American Congress on Surveying and Mapping. Available online. URL: http://www.acsm.net/error.html. Accessed December 24, 2006.

Center for Operational Oceanographic Products and Services, National Ocean Service, National Oceanic and Atmospheric Administration. "NOAA Tides and Currents." Available online. URL: http://tidesandcurrents.noaa.gov. Accessed December 24, 2006.

Continuously Operating Reference Stations (CORS), National Geodetic Survey, National Ocean Service, National Oceanic and Atmospheric Administration. Available online. URL: http://www.ngs.noaa.gov/CORS. Accessed December 24, 2006.

Federal Geographic Data Committee. Available online. URL: http://www.fgdc.gov/international. Accessed July 19, 2007.

National Geodetic Survey, National Ocean Service, National Oceanic and Atmospheric Administration. Available online. URL: http://www.ngs.noaa.gov. Accessed December 24, 2006.

NOAA, GEBCO: General Bathymetric Chart of the Oceans. Available online. URL: http://www.ngdc.noaa.gov/mgg/gebco/gebco.html. Accessed December 24, 2006.

vibrocorer A vibrocorer is an oceanographic instrument that takes geological core samples by vibrating a coring tube down into the ocean sediments or wetland soils. The coring tube consists of a core barrel encasing a PVC core liner in which the actual core sample is extruded. The core samples are typically as long as 66 feet (20 m) and on the order of four inches (10 cm) wide. Typical depths of operation range from 164 to 328 feet (50 to 100 m), but larger depths can be achieved by armoring the power cables and other equipment.

The vibrocorer is made up of a coring tube, a control box, a motor, and a vibrating head. It may include a weight stand, weights, and frame to support the instrument in such a way as to keep the coring tube vertical during sampling operations. Depending on platform or size, a vibrocorer may be powered by gasoline, hydraulic, or electrical motors. Regardless of size, a vibrating head is usually clamped to the vertical coring tube, which is often made of steel or aluminum. Gravity aided by the vibrating energy drives the core tube into SEDIMENT containing the geological history of that point. When the insertion is completed, the vibrocorer is turned off, and the tube is extracted with the aid of hoist equipment such as an A-frame on the stern of a research vessel. Core catchers (springlike steel fingers) are often used to keep sediments from falling out of the tube.

The performance of vibrocorers depends on the characteristics of the sediments. The best results are obtained in unconsolidated, water-logged, heterogeneous sediments that give way as the core tube vibrates. Vibrocorers are especially useful in such situations as operating from SUBMERSIBLES or REMOTELY OPERATED VEHICLES (ROVs) where a long distance of fall from a surface support vessel or platform to achieve penetration with a gravity corer (PISTON CORER) is not available. In wetland applications, a disadvantage to vibracoring relates to hauling the bulky equipment into a marsh or mudflat. The vibrocorer generally requires one person to operate the engine and several technicians to support the core tubes. Vibrocorers are traditionally employed in oil and gas explorations, predredging surveys, pipeline route surveys, scientific experiments, and coastal resource evaluations. Other means to collect ocean sediments are gravity, box, and piston corers. Wetland soils are often collected with can corers and GRAB SAMPLERS.

Further Reading

Massachusetts Institute of Technology. "Atlantis 1: Core Sampler." Available online. URL: http://web.mit.edu/12.000/www/m2005/al/Robotics/csampler.htm. Accessed January 23, 2007.

Rossfelder Corp. "Selecting a Vibrocorer." Available online. URL: http://www.rossfelder.com/selecting.html. Accessed January 23, 2005.

viscosity The measure of resistance to flow (internal friction) in a fluid is called viscosity. To understand viscosity, consider the speeds of small spheres passing through containers of molasses and salt water. Molasses is much more viscous than brine as evidenced by the slowly sinking sphere in the molasses container. The faster the sphere falls, the lower the viscosity. Thus, a fluid with high viscosity will strongly resist flow. In the marine realm, the viscosity of water generally increases with decreasing temperature. In the atmosphere and ocean, viscosity also becomes important near boundaries such as the seafloor and sea surface, where one sees strong velocity gradients. Velocity gradients are damped as a consequence of viscosity and will be zero at solid boundaries.

See also FLUID MECHANICS; NUMERICAL MODELING; TEMPERATURE.

vorticity Vorticity is a vector measure describing the tendency of a fluid to spin about an axis. One might put a small paddle wheel in a tub and observe the paddle wheel's motion. The rotation of the paddle wheel is a measure of vorticity. In Cartesian coordinates, the axis may be horizontal or vertical; in meteorology and oceanography, the spin about a vertical axis is of major importance. The vorticity is partly a result of the motion of fluid particles relative to adjacent particles. This motion is known as velocity shear (*see* SHEAR). The faster particles tend to curve around the slower-moving particles. In the atmosphere and the oceans, friction with obstructing topographic features such as mountains or islands and friction with the ground or the bottom results in velocity shears, which generate vorticity.

In defining vorticity, marine scientists consider many variables such as the CORIOLIS FORCE, all the directional components of the wind or current, the angular speed of the Earth, PRESSURE, volume, and latitude. Vorticity can be described as planetary, relative, potential, or absolute vorticity. The sum of planetary and relative vorticity is absolute vorticity. "Planetary vorticity" is produced by the counterclockwise rotation of the Earth as seen from above the North Pole. Its magnitude is equal to the Coriolis parameter, usually denoted by the letter f, where $f = 2\Omega\sin\varphi$, where Ω is the Earth's angular velocity, 7.29×10^{-5} /sec, and φ is the geographic latitude (*see* CORIOLIS FORCE). As seen from above, counterclockwise rotation has been labeled positive vorticity, while clockwise rotation is called negative vorticity. "Relative or local vorticity with respect to the ground is denoted by ζ. The sum of the planetary and relative vorticities, the absolute vorticity $(f + \zeta)$ must be constant from the law of conservation of angular momentum in physics. For example, if the JET STREAM (polar front) in the Northern Hemisphere is diverted northward, the planetary vorticity increases, so the relative vorticity must decrease, turning the air southward. The southward flow decreases the planetary vorticity, and thus the relative vorticity must increase, turning the flow northward again. These oscillations in the jet stream or any other air or water current are called ROSSBY WAVES and explain the long planetary waves observed in air and water flowing zonally (parallel to lines of latitude).

Relative vorticity is due to winds and currents. This effect can be thought of as a result of the horizontal shear. Even the great circulating ocean basin-wide GYRES are driven by the wind stress and, in particular, the ability of the wind to supply torque, equivalent to relative vorticity, to the ocean surface. As an example, consider the difference in velocity across the coastal ocean between western and eastern boundary currents such as the KUROSHIO and CALIFORNIA CURRENTS, respectively. The average velocity of the Kuroshio Current is 1.5 knots (77 cm/s) northward and the California Current is 0.4 knots (21 cm/s) southward.

In large-scale current models that describe such circulations, oceanographers usually employ the potential vorticity that includes the expansion or contraction of the fluid column thickness. The definition of potential vorticity is $(\zeta+f)/Z$, where Z is the thickness of the fluid layer of vorticity $(\zeta+f)$. It can be shown that in the absence of strong frictional forces the potential vorticity must be conserved, which implies that the relative vorticity must change if a water column deepens or shoals, or if the water column changes latitude. The potential vorticity is an important concept in explaining the physics of the observed flows of ocean currents that change position and orientation, and in formulating numerical circulation models.

In the atmosphere, as air converges into a low-pressure area, the horizontal area occupied by the air decreases, creating velocity shears, and so its tendency to spin increases. Diverging air, on the other hand, as from a high-pressure system, results in decreasing vorticity. The condition of surface convergence and increasing vorticity are used by weather forecasters as warning signs that the possibility for vortex generation is increasing, and a tornado watch may be issued.

Beside Rossby waves in the jet stream, manifestations of vorticity include vortices in the atmosphere such as dust devils, tornadoes and water spouts at small scales, and the polar vortex of rotating air over the Antarctic continent at large scales. In the oceans, vorticity induced by current shear is seen in meanders and EDDIES of the GULF STREAM, and in the equatorially-trapped Rossby waves that frequently perturb the great zonal equatorial currents.

See also FLUID MECHANICS; NUMERICAL MODELING.

Further Reading

Allaby, Michael. *Encyclopedia of Weather and Climate,* rev. ed. New York: Facts On File, 2007.

Colling, Angela. *Ocean Circulation,* 2nd ed. Milton Keynes, U.K.: Butterworth/Heinemann, 2001.

Stewart, R. R. *Introduction to Physical Oceanography.* Department of Oceanography, Texas A&M University, College Station, Texas. Available online. URL: http://oceanworld.tamu.edu/resources/ocng_textbook/contents.html. Accessed October 4, 2006.

W

warm-water vent A warm-water vent is a component of a hydrothermal system where water is generally free of smoke particulates and emerges from the seafloor at TEMPERATURES ranging between 50° and 77°F (10° and 25°C). The source of heat for this water is the hot magma rising in CONVECTION currents in the Earth's mantle. Warm-water vents are also known as diffuse vents. The less dense hydrothermal water escapes from cracks and fissures into surrounding bottom waters that range between 34° and 37°F (1° and 3°C). Since the warm water is exiting over a large area, the warm-water vents do not build the sulfide chimneys that usually characterize the hydrothermal vents known as black smokers. Warm-water vents were first discovered and explored by marine scientists during 1977 near the Galápagos Islands in approximately 1,367 fathoms (2,500 m) of water. This particular vent field is located where the 86° west meridian intersects with the east to west trending boundary of the Cocos and Nazca plates. The scientists used the deep submergence vessel *Alvin,* a three-person research submarine that can dive to depths of approximately three miles (4,500 m) to explore the hydrothermal vents. Scientists aboard ships use CONDUCTIVITY-TEMPERATURE-DEPTH measurements known as CTD casts to find the warm hydrothermal waters.

See also HYDROTHERMAL VENTS.

Further Reading

BIODEEP, Biotechnologies from the Deep. Available online. URL: http://www.geo.unimib.it/BioDeep/Project.html. Accessed February 1, 2007.

Neptune, University of Washington. Available online. URL: http://www.neptune.washington.edu. Accessed February 1, 2007.

University of Delaware College of Marine Studies and Sea Grant College Program. "Extreme 2000: Voyage to the Deep." Available online. URL: http://www.ocean.udel.edu/deepsea. Accessed February 1, 2007.

Vents Program, Pacific Marine Environmental Laboratory, National Oceanic and Atmospheric Administration. Available online. URL: http://www.pmel.noaa.gov/vents. Accessed February 1, 2007.

Woods Hole Oceanographic Institution. "Dive and Discover, Expeditions to the Seafloor." Available online. URL: http://www.divediscover.whoi.edu/vents/index.html. Accessed February 1, 2007.

water level fluctuations Water level fluctuations reflect the changing amount of water in reservoirs such as an ESTUARY, where changes result from recharge, discharge, basin changes, and physical forces. The total amount of water on Earth is conserved as it moves through the oceans, land, and the atmosphere in what is called the HYDROLOGICAL CYCLE. Levels within these reservoirs are important for reasons associated with water supply for drinking, agriculture, construction, and preventing hazards such as flooding.

At any one time, approximately 97.5 percent of the Earth's water is pooled in the ocean, 2.4 percent on land, and less than 1 percent in the atmosphere. Water evaporates from the oceans and other water bodies and is extracted for industrial and agricultural purposes. Meteorological forces such as precipitation and rainfall infiltration into the ground recharge water into subsurface reservoirs called aquifers and surface water bodies such as rivers and lakes. On land, groundwater moves slowly through the soil and quickly as runoff through streams and rivers.

The volumetric flow through a river is called discharge and may be estimated by taking the area of a cross section of the river and multiplying it by the average current speed. Other meteorological forces include freshets from winter melt of snow and ice and severe storms such as HURRICANES and typhoons, which cause storm surge. The Earth's vegetation also collects water through a process called transpiration, which includes the absorption of water through plant roots and evaporation through the leaves. When the normally permeable ground becomes fully saturated, the potential for flooding increases. Anthropogenic changes to water transportation include changing discharge through the construction of dams and wetland reclamation projects.

Coastal water levels may change periodically in synchronization with the tide-producing forces that are exerted by the Moon and Sun. In general, the extreme range of this horizontal movement of water from low to high tide depends on location and the phase of the Moon. Other factors such as soils fully saturated through extreme rainfall, storm surge, and extreme river discharge can combine to flood coastal or low-lying regions. In contrast, structures such as the Hoover dam can reduce river flow and affect discharges such as the Colorado River, which once flowed into the Gulf of California, Sea of Cortez, in Mexico. Groundwater is one of the major sources for water supply in many communities. Freshwater from land drainage mixes with ocean waters in estuaries; the mixed water is called brackish water.

Because of differences in topography, climate, and populations, there are local and regional variations in the amount of water that is available for human use. Groundwater collects in the aquifers over many years through infiltration and groundwater flow recharge. Some is replenished through rainwater infiltration. Sustainable use of groundwater involves the balance between withdrawal of groundwater and replenishment through recharge. Faster withdrawal rates would lead to a fall in the water table and finally depletion of groundwater. In barrier islands, a freshwater aquifer may overlie a saltwater aquifer, and the thickness of the freshwater aquifer may be modified by tides, interdunal drainage, and transpiration.

In conclusion, water level fluctuations are caused by a combination of forces. They are measured by various types of instruments such as pressure gauges and tide gauges. Navigators know they must pay close attention to water level variations in navigation channels, turning basins, and anchorages in order to avoid running aground. Tide predictions are based entirely on astronomical forces, so they do not report the total rise or fall of coastal waters. In addition to astronomical forces, water levels may rise or fall in synchronization with meteorological phenomena such as winds, precipitation, watershed discharge, severe storms, waves, and icing. Long-term sea levels may also rise or fall due to adjustments in land elevations from subsidence due to the extraction of underground fluids, expansion of the land from the last deglaciation, or the expansion of seawater due to GLOBAL WARMING.

See also BARRIER ISLANDS; OCEANOGRAPHIC INSTRUMENTS; STORM SURGE; TIDE; TIDE GAUGE; WAVE AND TIDE GAUGE.

Further Reading

Hydrologic Information Center, National Weather Service, National Oceanic and Atmospheric Administration. Available online. URL: http://www.nws.noaa.gov/oh/hic. Accessed February 2, 2007.

The National Water Level Program (NWLP) and the National Water Level Observation Network (NWLON), Center for Operational Oceanographic Products and Services (CO-OPS). Available online. URL: http://tides andcurrents.noaa.gov/nwlon.html. Accessed February 2, 2007.

The San Diego State University. *Center for Inland Waters: A New Resource for Students, Teachers, Researchers, Decision Makers, and the General Public.* Available online. URL: http://www.sci.sdsu.edu/salton/index. html. Accessed February 2, 2007.

water mass A water mass is a homogeneous body of ocean water having a common formation history as evidenced by temperature, salinity, and density signatures, and characterized by a particular set of physical and chemical parameters. Unique composition may help to identify a water mass where sentinels may be factors such as dissolved oxygen content, chemical isotopes, or nutrient load. A water mass should not be confused with the conceptual region of water from the sea surface to seafloor, which is called the water column.

Water masses are generally indicated on a temperature-salinity (T-S) diagram, on which lines of equal density appear as parameters. This method of data analysis was first developed by Norwegian oceanographer B. Helland-Hansen (1877–1957) in 1916. The T-S diagram is the most commonly used thermodynamic diagram for seawater. Near-surface values of temperature and salinity are generally not included on the T-S plot because of seasonal variability. Water masses such as ANTARCTIC BOTTOM WATER, Antarctic Intermediate Water, Black Sea water, Circumpolar Deep Water, European Mediterranean Water, Indian Central Water, North Atlantic Deep Water, and Red Sea Water are found within specific locations in the ocean owing to density. Water masses are distinct from surrounding waters and will mix with other water masses over time due to processes such as CONVECTION, DIFFUSION, advection (the physical movement of water) and TURBULENCE. Water types represent a single point on a temperature and salinity diagram, while a water mass reflects a curve along a specific region of the temperature and salinity diagram. From the curve connecting two water types, the amount of mixing between them can be deduced. Water masses are real physical entities with a measurable volume. The range of values of temperature, salinity, and density can be determined from the T-S diagram. For the most part, water masses are formed only at the sea surface, where temperature and salinity can be dramatically altered by atmospheric conditions, such as extremely cold temperatures. As seawater freezes, the salt is expelled from the ice, increasing the salinity of the surrounding water. This process is responsible for production of the North Atlantic Deep Water mass, south of Greenland and Labrador. Once the water becomes extremely dense, it sinks to the level of like density, or to the bottom of the ocean floor if denser than the surrounding water. The most dense major water mass in the world ocean is the Antarctic Bottom Water, which fills the South Atlantic and part of the North Atlantic basins, and mixes with other water masses to fill the bottom of the Pacific and Indian Ocean basins.

Because specific water masses can be identified over great regions of the globe, it follows that mixing along near-horizontal surfaces in the ocean is much greater than vertical mixing, except near the poles. In the Tropics, the highly stratified water and strong PYCNOCLINE severely restrict vertical exchange of properties. Raymond B. Montgomery (1928–88) showed in the 1950s that mixing tends to occur along surfaces of constant potential density, called isentropic (σ_θ) surfaces (approximated by σ_t surfaces; *see* DENSITY). This type of mixing requires a minimum of energy to drive it, and so is very widespread. By plotting the temperature and salinity distribution on σ_t surfaces, it is possible to infer the movement of water masses. An alternative means was developed earlier by German oceanographer Georg Wüst (1890–1977), called the "core method." In this method, the temperature and salinity data along oceanographic sections are used to plot T-S diagrams from which the extreme values of temperature or salinity can be extracted to show the dilution of tongues of, for example, high salinity water, which can then be traced back to its source such as the Mediterranean Water spilling over the sill into the Atlantic in the Strait of Gibraltar. This particular water mass flows across the entire North Atlantic basin.

The study of the transport and mixing of water masses was greatly facilitated by the hydrogen bomb tests of the 1960s, which caused a 30-fold increase in the amount of tritium that would occur naturally in the surface layers of the ocean. In the polar oceans, where strong vertical convection occurs, tritium is found in the deep layers of the ocean; in the Tropics, where strong stratification suppresses vertical convection, little or no tritium is found. By this means, and by the ever-growing archive of data obtained from CONDUCTIVITY-TEMPERATURE-DEPTH (CTD) PROBES, the water masses of the ocean are being mapped with greater and greater accuracy.

See also SALINITY; TEMPERATURE.

Further Reading

Knauss, John A. *Introduction to Physical Oceanography*, 2nd ed. Upper Saddle River, N.J.: Pearson/Prentice Hall, 1996.

Pickard, George L., and William J. Emery. *Descriptive Physical Oceanography: An Introduction*. Oxford: Pergamon Press, 1990.

waterspouts Tornadolike features above the surface of the ocean are called waterspouts. They are funnel-shaped clouds extending from the base of cumulus clouds to the water surface. In fact, if they do not reach all the way to the surface of the water, they are known as "funnel clouds." They are generally smaller than tornadoes, ranging from about 16 to 328 feet (5–100 m) across. They generally have rotating wind velocities up to about 90 miles per hour (150 km/h), although winds in especially well developed ones are believed to be as high as 180 miles per hour (300 km/h). Although they contain much less energy than tornadoes, they can damage boats. Boaters are advised to travel as fast as possible in a direction perpendicular to the path of the waterspout to avoid what could be a dangerous encounter.

A waterspout is visible because as the water vapor that evaporates from the surface rises, it is cooled adiabatically (without gain or loss of heat) below the dew point and condenses. The more unstable the lower atmosphere is, the more favorable are the conditions for a waterspout. The rotating winds circulate counterclockwise in the Northern Hemisphere and clockwise in the Southern Hemisphere. Waterspouts terminate when rain begins to fall from the associated clouds, cooling the upward-moving warm moist air that provides it with energy.

Waterspouts generally occur in the Tropics, where warm surface water favors the convective heating essential to their development. They are frequently seen in the United States around the Florida Keys, with perhaps 400 to 500 waterspouts per year.

Waterspouts are the marine equivalent of tornadoes, but they usually do not cause much damage. *(NOAA)*

They have been observed on the U.S. Atlantic coast as far north as CHESAPEAKE BAY, and occasionally on the Great Lakes.

See also CLOUDS; CONVECTION.

water type   A water type in marine science refers to a completely homogeneous water mass of a single TEMPERATURE and SALINITY. A water type would be represented on a temperature-salinity diagram as a single point. Water types are generally visualized as forming at the sea surface under conditions of uniform temperature and salinity, especially in the Arctic and Antarctic regions. In reality, water types are rarely observed because they immediately begin mixing and form water masses. Water types are a useful theoretical concept in explaining how the water masses are formed and are distributed throughout the world's oceans.

See also WATER MASS.

wave and tide gauge   A wave and tide gauge is an in situ oceanographic instrument that measures gauge pressure, which is used to record oscillations in water level in response to wave-generating forces such as winds, astronomical motions of the Sun and Moon, and earthquakes. Gauge pressure is equal to the absolute pressure minus atmospheric pressure. Atmospheric pressure is typically about 14.7 pounds per square inch (1,013.5 mb), but changes with altitude and weather events. For this reason, in most marine science applications, a series of barometric pressure readings may be subtracted from the series to determine pressure at the location of the senor. Most precision wave and tide gauges use state-of-the-art silicon semiconductor strain gauge pressure sensors that are contained within an oil-filled chamber. The gauge's sensors (pressure, THERMISTOR, conductivity) and electronics (processor, clock, batteries, memory) are then secured within a pressure-rated housing, especially for deep-sea applications, where a titanium case is used. Less expensive wave and tide gauges may incorporate a Bourdon tube to measure pressure. Named after its inventor, Eugène Bourdon (1808–84), the radius of the small c-shaped tube changes slightly in response to pressure. The greater the pressure the further the radius or tip displaces, which is used to mechanically determine the pressure. After the wave and tide gauge is calibrated and set-up parameters are loaded, the instrument may be lowered to the seafloor, fixed to a piling, or bolted to a mounting frame for deployment. Set-up parameters may account for a standard atmosphere and govern sampling rates, especially when users want to measure shorter period waves. Most wave and tide gauges measure wave periods between three seconds (0.3 Hz) to 30 seconds (0.033 Hz) with a sampling rate around 2 Hz and depending on the manner in which pressure fluctuations are averaged, output sequential readings of wave height, wave period and tides. Wave statistics are computed for the time-based pressure measurements, including means, standard deviations, moments about the mean, and the frequency spectrum, computed by means of the Fast Fourier Transform (*see* STATISTICS). The pressure records can be converted to wave heights by a mathematical transformation, but information on the higher frequency waves is lost because the water acts as a low-pass filter; the higher the frequency of the waves, the greater the attenuation with depth. To recover the low-frequency wave height information, an inverse filter transfer function is applied. TIDES, which are very long period waves are determined by continuously integrating pressure sensor output over short intervals to average out wind-driven waves. Many studies to predict tides rely on consistent water level readings every six minutes over time periods from at least 29 days to more than 19 years. Depths for bottom-mounted instruments may also be calculated by a formula such as the following:

measured pressure = pressure of water + atmospheric pressure, where

pressure of water = density of water (ρ_w) x depth, atmospheric pressure is determined from a barometer, and

ρ_w is either 62.4 pounds per cubic feet (1,000 kg/m³) for freshwater or computed from TEMPERATURE and SALINITY sensors in the ocean. Since seawater contains about 35 pounds of salt per 1,000 pounds of water, it is approximately 3.5 percent heavier than freshwater.

A wave and tide gauge could be employed to assess the set-up of water by prevailing winds. For example, if an offshore wind has been blowing for several days and the dock master on a lake is concerned about the depth near the gas pump for an incoming yacht, then he or she should check the wave and tide gauge readings. If the measured pressure is 8.6 pounds per square inch and the barometric pressure is 14.7 pounds per square inch, then the above relationship yields the following:

8.6 pounds per square inch = 62.4 pounds per cubic feet × 1 square foot/144 square inches × depth + 14.7 pounds per square inch.

By rearranging this equation

depth = 2.3 feet square inches per pound (8.6 pounds per square inch - 14.7 pounds per square inch) = -14.03 feet.

Pressure at the 14-foot (4.3 m) depth in freshwater would be less than the gauge pressure at the same depth in seawater since the density of freshwater is less than the density of seawater. The density of seawater near the surface is approximately 64.1 pounds per cubic feet (1,027 kg/m³). These types of calculations are handled automatically in a properly configured wave and tide gauge.

Measuring waves and tides is important for many navigation and construction activities. Wave and tide gauges support installed real-time reporting networks such as the U.S. Coast Guard's Vessel Traffic and Information Systems, long-term water level monitoring stations, DREDGING operations, platform leveling, and the navigation of underwater vehicles (AUTONOMOUS UNDERWATER and REMOTELY OPERATED VEHICLES) and tow bodies (trawls and side-scan SONAR). There are various types of wave and tide gauges available to marine scientists and engineers. Some provide software that simplifies the generation of products that highlight parameters such as significant wave height, wave steepness, high water, and low water.

See also OCEANOGRAPHIC INSTRUMENTS; PREDICTION; PRESSURE; WAVE PREDICTION.

wave buoy A wave buoy is a floating instrument platform used for measuring oscillations of the surface and ranging in diameter from several inches (approximately 5 cm) for air deployment to approximately 40 feet (12 m) for ship deployment. They may be moored or free floating and may be outfitted as directional wave buoys measuring wave heights, wave periods, and wave directions, or nondirectional wave buoys simply measuring wave height and wave period. Most wave buoys are anchored to the seafloor with specialized MOORING systems and measure wave heights and wave directions using accelerometers and a COMPASS. The accelerometers and compass are mounted on a gravity-stabilized platform inside a watertight sphere to measure the buoy's vertical and horizontal accelerations. Raw sensor data is processed on board the buoy to gain a vertical displacement or wave height. The data processing usually involves applying a Fast Fourier Transform (FFT) to convert the data from the temporal to the frequency domain. The function is similar to the convergent trigonometric series formulated by French mathematician Joseph Fourier (1768–1830). Wave energies with their associated frequencies allow the computation of parameters such as significant wave height and average wave period. Computed wave information is transmitted by packet radio to the shore station receiver as a radio signal. Final processing and wave product development is completed, and information is usually disseminated through a Web site. Coastal scientists and managers may use this information to monitor the type and variability of wave conditions, which helps short-term and long-term investigations of natural coastal processes, including erosion and accretion. HARBOR PILOTS, port operators, and NAVAL ARCHITECTS may use wave data to support ship maneuver, plan coastal construction, or assess ride quality of a high-speed vessel.

See also BUOYS; TIME SERIES ANALYSIS.

Further Reading

National Data Buoy Center, National Oceanic and Atmospheric Administration. Available online. URL: http://www.ndbc.noaa.gov. Accessed November 10, 2006.

Wave Buoy Survey, The Johns Hopkins University Applied Physics Laboratory. Available online. URL: http://fermi.jhuapl.edu/usmc. Accessed November 10, 2006.

wavelength   Wavelength is the distance from one wave crest to the next, or, more generally, the distance between like points on the wave profile, points having the same phase. Wavelength can also be defined as the distance for which a periodic wave repeats itself. Periodic waves obey the relationship c = L/T, where c is the phase speed of the wave, L is the wavelength, and T is the period. If L is in units of meters, T is in seconds, then c will be in meters per second. This is just the wave equivalent of the time-distance-speed relationship: speed = distance traveled/time.

For theoretical sinusoidal waves of the form $\eta = A\cos(kx - \omega t)$ in deep water, the waves obey the dispersion relationship $\omega^2 = gk$, where ω = angular frequency of the wave, that is, $2\pi f = 2\pi/T$; g = the acceleration of gravity; and k = the wave number $2\pi/L$. When solved for the wavelength, this relationship yields $L = (g/2\pi)T^2$. Since the constant $g/2\pi$ is 1.56 m/s^2 for the wavelength in meters, and $g/2\pi$ is 5.12 ft/s^2 for the wavelength in feet, there is a simple relationship connecting wavelength and wave period.

However, in the case of the real world ocean, this relationship is only a rough approximation, and works best for long, low swells from a distant storm, which approximate sine waves. Waves in rough, storm-driven seas are much more complex (see GRAVITY WAVES).

Observers on ships often estimate wavelength at sea by measuring the distance between the wave crests as the crests pass specific points on the ship's hull when it is moving in the predominant wave direction. If the ship is moving, a correction must be made for the Doppler effect. In general, wavelength must be estimated statistically by means of photography, laser profiling, or by deployment of an array of probes measuring the wave height.

It has long been known that water waves entering shallow water undergo modification so that the wavelength is shortened and the height is increased, while the period is constant. This effect is responsible for some waves such as TSUNAMIS to grow to menacing heights as they approach the shoreline.

See also WAVES.

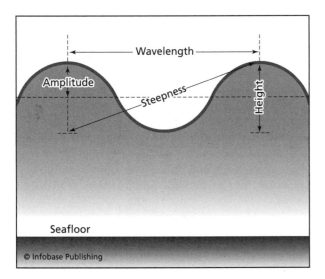

Waves are described in terms of wavelength—the distance between one crest or trough and the next (or any two points of constant phase), the amplitude—half the height, and steepness—the angle between the horizontal and a line drawn from a point level with a trough and directly beneath a crest to the top of the adjacent crest.

wave prediction   Forecasting the future condition of the ocean's irregular surface in terms of its roughness, including the calculation of wave heights, periods, and directions, is known as wave prediction. Waves are oscillations with crests and troughs, where wave height may be viewed as the distance from the trough (bottom of the wave) to the crest (top of the wave), and WAVELENGTH as the distance from one crest to the next (see WAVES). The period of the wave is the time that it takes for two successive crests to pass a fixed point. Because there is a distribution of wave heights and periods and variability in wave heights, they may be quantified in different manners. For example, wave processes may be observed by eye, measured mechanically with sensors such as accelerometers, sensed remotely by RADAR or SONAR, or modeled mathematically. The amplitudes and associated frequencies of the waves are generally used to describe their propagation on the ocean surface. Waves transform through processes such as shoaling, refraction, diffraction, reflection, and friction. Waves moving from deep water to shallow water change form as they feel the bottom and begin to transfer energy, which will be released completely as heat and sound upon breaking as SURF on the BEACH. As the wave shoals, wave speed and wavelengths decrease, while the period remains the same. The height or wave energy increases until the wave becomes unstable and breaks. As the long crested wave approaches a beach, the portion of the wave in shallow water moves slower than the deeper portion, causing the wave to bend. In addition, wave heights change along the wave crest as waves encounter obstacles such as a peninsula or BREAKWATER, causing the wave to turn into the sheltered region.

Wave models consider the physics of generating forces such as winds and possibly bathymetric effects to forecast wave properties within individual grids or transects (see NUMERICAL MODELING). An example operational model that requires only winds as input is the Wave Action Model or WAM. Another model, which considers shallow water processes in the coastal zone, is Simulating Waves Nearshore, or SWAN. The model domain is developed from a nautical chart or bathymetric data and takes the form of a detailed matrix, mesh, or grid. Wind data may be derived from weather stations or surface weather charts. Numerical model output of wave parameters is obtained for each cell within the grid by considering a combination of winds, bottom depths, and CURRENTS. In most cases, a spectrum is used to define the range of wave amplitudes or energy in terms of frequency for each cell in the grid. Some models may also look at the transfer of energy along a transect perpendicular to the shore, which is actually a depth profile. Wave forecasts are produced from spectral data that is computed from wind and possibly wave data that is incorporated into the model, usually at specific time increments. The model considers energy from wind forcing, the manner in which waves interact with one another, and dissipation. Real data from WAVE BUOYS may be assimilated into a model to make local corrections by algorithms that modify model output. For example, sea surface elevations estimated from a model could be adjusted toward the observed values of wave heights and wave periods.

In some cases, wave heights may be generalized to an empirical scale called SEA STATE, averaged into a statistical measurement called "significant wave height," or forecasted as height contours from a numerical model. Sea state refers to the height, period, and character of well-developed wind waves on a particular scale such as the World Meteorological Organization's Table 3700, where sea state is represented as Code 0 for glassy (flat), Code 1 for rippled (waves are 0 to .33 feet or 0 to 0.1 m), Code 2 for wavelets (waves are .33 to 1.64 feet or 0.1 to 0.5 m), Code 3 for slight (1.64 to 4.1 feet or 0.5 to 1.25 m), Code 4 for moderate (waves are 4.1 to 8.2 feet or 1.25 to 2.5 m), Code 5 for rough (waves are 8.2 to 13.12 feet or 2.5 to 4 m), Code 6 for very rough (waves are 13.12 to 19.69 feet or 4 to 6 m), Code 7 for high (waves are 19.69 to 29.53 feet or 6 to 9 m), Code 8 for very high (waves are 29.53 to 45.93 feet or 9 to 14 m), and Code 9 for phenomenal waves (waves are > 45.93 feet or 14m). Waves from a particular region may be described in tables highlighting the percentage of dominant wave heights recorded by specific periods such as five seconds or less, six or seven seconds, eight or nine seconds, 10 or 11 seconds to greater than 21 seconds. The period of the wave may be computed by dividing the wavelength by the wave speed. Significant wave height (H1/3) is the average of the highest one-third of the wave heights over a given time period. From a spectral computation, the zero-moment wave height

(Hmo) is equal to four times the standard deviation of the sea surface elevations, and the peak wave period (T_p) is the wave period associated with the maximum energy that is observed in the spectrum.

Wave predictions forecast the manner in which energy will be transported across the water as waves in response to astronomical, meteorological, and geological forces such as the orbits of the Earth and Moon about the Sun, winds blowing across the ocean, and tectonic activity such as earthquakes. Knowing how waves impact the seafloor and beaches is important to understanding processes such as erosion and accretion in the nearshore region. Waves also impact commercial shipping, recreational boaters, and coastal structures such as piers and waterfront property. Wave steepness, the ratio of wave height to wavelength, is the wave parameter most likely to capsize an unwary boat. Wave predictions are an essential element to any plan aimed at protecting life and property at sea and near the shore. Wave forecasts are used to route ships, and hindcasts are used to assess damages following a storm or maritime mishap (*see* HINDCASTING). Such information may be used to close beaches when conditions are unsafe for recreational activities such as swimming. Improving the measurement and forecasting of waves is a major activity by government organizations such as the NATIONAL OCEANIC AND ATMOSPHERIC ADMINISTRATION's National Data Buoy Center or the United States Army Corps of Engineers' Coastal Engineering Research Center and academic institutions such as the Delft Technical University in the Netherlands.

See also FIELD RESEARCH FACILITY; FORECASTING; MODEL; MODEL BASIN; PREDICTIONS.

Further Reading

Coastal Data Information Program. Available online. URL: http://cdip.ucsd.edu. Accessed November 8, 2006.

Computational and Information Systems Laboratory's Research Data Archive, Data for Atmospheric and Geosciences Research, the University Corporation for Atmospheric Research. Available online. URL: http://dss.ucar.edu. Accessed October 17, 2007.

Delft Technical University, Surface Waves Nearshore. Available online. URL: http://www.fluidmechanics.tudelft.nl/swan/index.htm. Accessed November 9, 2006.

National Data Buoy Center, National Oceanic and Atmospheric Administration. Available online. URL: http://www.ndbc.noaa.gov. Accessed November 9, 2006.

National Oceanographic Data Center, National Oceanic and Atmospheric Administration. Available online. URL: http://www.nodc.noaa.gov. Accessed November 9, 2006.

U.S. Army Corps of Engineers, Engineer Research and Development Center, Waterways Experiment Station,

Coastal and Hydraulics Laboratory. Available online. URL: http://chl.erdc.usace.army.mil. Accessed November 9, 2006.

World Meteorological Organization. Available online. URL: http://www.wmo.ch/. Accessed November 9, 2006.

waves   Waves are caused by perturbations on the boundary between two media, within a medium, or within a field or fields of force. Perturbations are disturbances or secondary influences that cause changes such as a puff of wind blowing over the ocean to cause capillary waves or "cat's paws." The medium such as the ocean is the material through which the physical or mechanical wave travels. Usually, waves transport energy and momentum but not matter. Energy is the capacity to do work or the product of displacement and force in the direction of the displacement. Momentum is the product of mass and velocity. Matter is anything that has mass and resists changes in rotational or translational motion (inertia).

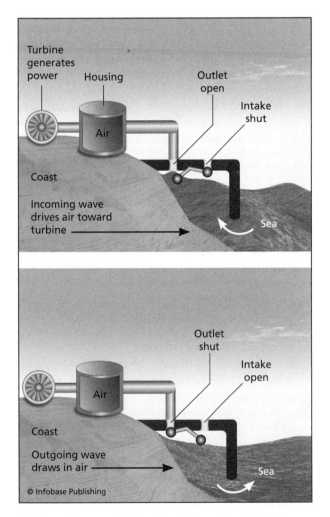

One type of proposed system (the air-trapping system) for generating electrical power from ocean wind waves

Mechanical waves, including waves on ropes, drumheads, sound waves, water waves, and SEISMIC WAVES generated by earthquakes, require a medium in which to move or propagate. Electromagnetic waves, including radio, microwaves, infrared, visible light, ultraviolet light, X-rays, and gamma rays, do not require a medium and can propagate in the near-vacuum of interstellar space. Waves exhibit both phase velocity, or the velocity of the wave profile, and group velocity, or the velocity of a wave packet or group of waves. The speed of mechanical waves depends on the nature of the medium; thus, sound waves in air travel at about 1,086 feet per second (331 m/s), sound waves in seawater travel at about 5,030 feet per second (1,533 m/s), and sound waves in an iron bar travel at about 19,522 feet per second (5,950 m/s). The speed of electromagnetic waves is constant in a vacuum, at close to 186,000 miles per second (300,000 km/s). The wave disturbance can be longitudinal, that is, in the direction of propagation. The molecules of the medium move back and forth as the disturbance passes through them, as is the case for sound waves, waves in an iron bar, and some seismic waves (P-waves). The disturbance can also be transverse, that is, perpendicular to the direction of propagation, in which the molecules of the medium move up and down, such as waves on a rope, a drumhead, and all electromagnetic waves. Water waves exhibit features of both longitudinal and transverse waves. The water particles move in circular orbits as the wave passes. In deep water, the waves are exponentially attenuated with depth such that the radius of the water particles gets smaller with depth, until the disturbance disappears. This is the reason that submarines are much less affected by storms than surface vessels. In shallow water, the circular motion of the waves becomes increasingly more elliptical as the depth increases, until it is linear, or a back-and-forth motion just above the bottom.

The phase speeds of water waves depend on their WAVELENGTH: a property known as dispersion. The long waves travel faster than the shorter waves; thus, the long SWELL waves will outrun the shorter waves generated in a storm and become harbingers of a storm's approach. Light waves are dispersive when traveling through glass; this is how a prism separates white light into its constituent colors, each of a specific wavelength. Waves can occur as an infinitely long train of waves, such as light waves from a galaxy at the far end of the universe; as a short burst of waves, such as seismic waves from an earthquake; or as a single wave, such as a soliton water wave. Waves experience a restoring force, which acts to draw the disturbance back to the undisturbed or equilibrium position. For an oscillating mass on the end of a spring, the spring tension provides the restoring force; for very small water waves, the restoring force is capillarity (*see* CAPILLARY WAVES); for ripples and larger waves, gravity provides the restoring force (*see* GRAVITY WAVES), for the very long Rossby waves, the restoring force is provided by the Earth's rotation through the CORIOLIS FORCE (*see* ROSSBY WAVES).

Waves are described in physics by their amplitude, or height above the undisturbed reference level, their period, or duration of an up-and-down or back-and-forth motion, their frequency, or the number of complete motions in a unit of time (referred to as a cycle; the frequency is the reciprocal of the period); and their wavelength, or the distance from one specific point on a wave to the equivalent point on the adjacent wave, such as the distance from one crest to the next, or from one trough to the next for water waves. The phase of a wave is its position with respect to a reference point, or with respect to neighboring waves. Waves can occur as a single disturbance at a single frequency, such as the waves produced when a tuning fork is struck, or there can be a multitude of frequencies of waves, such as those produced on the ocean surface by the wind. The energy per unit time in a wave is the *power,* and the power per unit area traversed by the wave is the intensity of the wave. The SI (International System) units for these quantities are as follows: amplitude in meters, period in seconds, frequency in cycles per second, known as Hertz (Hz), wavelength in meters, phase in units of angle, or radians, energy in joules, power in watts, and intensity in watts per square meter.

The most common mathematical model of waves used in introductory physics courses is the simple harmonic wave $y(x,t)$ expressed in SI units as:

$$y(x,t) = A \sin 2\pi (x/\lambda - t/T),$$

where $y(x,t)$ is the wave height in meters, x is distance in meters, λ is the wavelength in meters, t is the time in seconds, T is the period of the wave in seconds (the reciprocal of the frequency f), and A is the amplitude in meters. The arguments of the function $y(x,t)$ indicate that waves are functions of both space and time. Such waves are called propagating or progressive waves. The speed of the wave (c) is determined by dividing the wavelength by the time for one wavelength to pass a given point (the period); that is,

$$c = \lambda/T.$$

By the principle of superposition, simple harmonic waves can be added together to model almost any complex wave form.

The device or phenomenon producing the waves is called the source. A pebble tossed into a pond is the source of waves spreading out in a circular pattern on the surface; an explosion is the source of sound waves spreading out in air. Depending on the geometry of the source and the environment, waves can be spherical, cylindrical, or planar. Surfaces of constant phase (where the motion is everywhere the same with respect to a reference surface) are called wavefronts; VECTORS in the direction of propagation, perpendicular to the wavefronts, are called rays. An example of wavefronts would be the crests of ocean swells as they approach a BEACH. Imaginary lines in the direction of motion of the wavefronts (and hence, perpendicular to them), are the rays. When waves propagate from one medium to another of different properties, wave energy is reflected, as in light striking a slab of glass, and refracted, or bent into the glass. Thus, a pencil in a glass of water appears to be bent at the air-water interface. Spear fishermen on shore or in a boat must compensate for the effect of refraction to hit their prey. In the case of water waves, changes in the depths of the water have the same effect as changes in the refractive properties of the medium. Waves in deep water approaching a beach at an angle are refracted toward the beach, so that when they arrive at the beach, their crests are generally parallel to the shoreline. Where two wave trains combine, they can interfere with each other such that when they are in phase, their crests and troughs are aligned; their amplitudes add, and there is constructive interference. When they are out of phase, the crests of one wave occur at the same positions as the troughs of the other wave, their amplitudes subtract, and there is destructive interference. Thus, an interference pattern consists of a series of bands of rougher and smoother water. When a train of plane waves strikes a small slit of dimensions on the same order as the wavelength, circular wavelets are emitted from the side of the slit opposite the incident wave, a phenomenon called diffraction. Diffraction patterns can be seen in water waves when an incoming wave train strikes an opening in a barrier such as a seawall. Seen from above, the wave crests fan out from the opening in circular arcs. Were it not for the diffraction of sound waves, humans would not be able to hear around corners and through open doors and windows if they were not on a direct line with the sound source. Light waves, on the other hand, are not diffracted in passing through windows because the size of the window is much greater than the wavelength of the light. Light waves can be diffracted as they pass over the thin edges of razor blades. The diffraction grating, a laboratory device with thousands of narrow slits, is designed to produce diffraction of light waves.

Not all waves propagate. Waves can be excited by a person plucking a guitar string, by the wind blowing on a skyscraper, by water sloshing up and down in a bathtub, and by strong winds followed by calm over a lake or bay. Such water waves are called standing waves. Standing waves result from the interference of two progressive waves of the same amplitude traveling in opposite directions. These oscillations can be either forced, as in the case of the wind blowing over the bay; or free, when the wind stops blowing and the oscillations continue on their own. These oscillations are called seiches (*see* SEICHE). Eventually, friction will cause the wave motion to dampen out and cease until the next storm event causes a displacement of the medium.

Standing waves can be modeled mathematically as a fundamental, in which the whole object or boundary oscillates up and down in phase, and higher harmonics, in which a number of standing oscillations or loops separated by nodes are formed. The number of loops increases with the number of the harmonic. The first harmonic is the fundamental; the second harmonic is twice the frequency of the fundamental; the third harmonic is three times the frequency of the fundamental, and so forth. In real physical systems, the number of harmonics needed to approximate the oscillations mathematically is usually fairly small. This analysis, which has wide application in many fields including the study of the TIDES, is called harmonic analysis. The mathematics has its roots in Fourier series and Fourier transform analysis.

The wave spectrum, or distribution of either amplitude or energy with frequency, is perhaps the single most important analytical tool used by scientists to study waves. The most familiar spectrum is that of sunlight passing through a prism and splitting into its constituent colors, as demonstrated by Sir Isaac Newton (1642–1727) in 1666. Each color represents a different wavelength of light. From the known speed of the light in the prism, the frequencies of the constituents can also be determined. In a similar fashion, sound can be passed into an instrument called a spectrum analyzer, which displays the energy of the sounds present at each frequency. The spectrum of waves on the surface of the ocean is usually estimated from TIME SERIES of wave height at one or more locations, and presented as a plot of wave energy versus frequency. By this means, the wave characteristics can be used by mariners, marine architects, emergency responders, and many others on or near the sea. The wave spectrum is a fundamental tool in wave height forecasting. Instruments to measure water wave characteristics have undergone great improvement since scientific measurements of wave heights began in the 19th century. Today, water waves can be studied from the shore, from ships, BUOYS, platforms, aircraft, and spacecraft (*see* OCEANOGRAPHIC INSTRUMENTS).

Further Reading
Kingsland, Rosemary. *Savage Seas.* New York: TV Books, L.L.C., 1999.
Open University Course Team. *Waves, Tides, and Shallow-Water Processes,* 2nd ed. Oxford: Butterworth-Heinemann, 1999.
Serway, Raymond A., and John W. Jewett, Jr. *Physics for Scientists and Engineers,* 6th ed. Belmont, Calif.: Thomson/Brooks/Cole, 2004.

wave spectrum The wave spectrum results from transforming time domain signals of water level to the frequency domain. Wave spectra identify the distribution of wave energy with frequency or period (the period is the reciprocal of the frequency) over a specific sampling period or location. Wave spectra may be used in WAVE BUOY measurements to differentiate between wind waves and SWELL.

Wave spectra estimated from directional wave buoys or wave arrays show the distribution of wave energy with both frequency and direction.

Numerous theoretical and empirical studies were made to develop a realistic wave spectrum. The first practical wave spectrum that made a realistic forecasting system immediately possible was developed in the early 1950s by Gerhard Neumann (1911–97). Since that time, a great many spectral forms of increasing sophistication have been developed. The wave spectrum is an essential component of numerical wave prediction models.

See also SEA STATE; TIME SERIES ANALYSIS; WAVE PREDICTION.

Further Reading
Bishop, Joseph M. *Applied Oceanography.* New York: Wiley-Interscience, 1984.
Kinsman, Blair. *Wind Waves: Their Generation and Propagation on the Ocean Surface.* Englewood, N.J.: Prentice Hall, 1965.

wave staff A wave staff is a capacitance or resistance type water level sensor also called a "staff gauge" with a timing device to measure the water surface height due to waves ranging from high-frequency CAPILLARY WAVES to low-frequency TIDES. Changes in water level result in changes in the capacitance or resistance of an immersed probe that produces time-varying voltage changes in a recording circuit. These surface-piercing wave staffs are mounted vertically on piers and typically range in length from three to

30 feet (1 to 10 m). The electronic package remains dry at the top of the pier, while the staff portion is directed at a 90-degree angle into the fluid. Instrument output is a relative measure of the air-sea interface height. An array of wave staffs could be employed to capture phase and thus directional information from the various WATER LEVEL FLUCTUATIONS.

Wave staffs and poles are especially suitable in shallow water and high-traffic locations where surface buoys are not suitable. There are also field expedient methods employed to measure waves with a staff or pole. Some wave staffs have simple graduations for measuring wave heights in conjunction with a timing device for the measurement of wave periods. In the deep sea, surface waves are sometimes measured with a wave staff mounted on a spar buoy. An attached disk suspended below the depth of wave action minimizes staff oscillations on such a system. Wave heights are recorded using visual, electrical, or acoustical sensors.

Wave staffs are important instruments to measure waves, tides, and even shallow water phenomena such as RIP CURRENTS and wave runup. Investigators studying shallow-water processes may use other sensors in combination with a wave staff. For example, a pressure sensor directly beneath a wave staff would provide useful data to study the manner in which waves break. The U.S. Army Corps of Engineers uses instruments such as a Baylor Staff Gauge and pressure sensors attached to the FIELD RESEARCH FACILITY pier to provide long-term data on water level fluctuations occurring along a straight BARRIER ISLAND beach in Duck, North Carolina.

Further Reading

National Data Buoy Center: Center of Excellence in Marine Technology. National Oceanic and Atmospheric Administration. Available online. URL: http://www.ndbc.noaa.gov. Accessed October 17, 2007.

U.S. Army Corps of Engineers. "Welcome to the Field Research Facility (FRF)." Available online. URL: http://www.frf.usace.army.mil/frf.shtml. Accessed October 17, 2007.

Weddell Sea The Weddell Sea is a marginal sea of the SOUTHERN OCEAN. By definition, the Southern Ocean includes the southern portions of the ATLANTIC, INDIAN, and PACIFIC OCEANS from the 60° south latitude (parallel) to Antarctica. The Weddell Sea forms the southern most tip of the Atlantic Ocean. It was named after James Weddell (1787–1834), a British shipmaster who discovered the sea in 1823 while hunting seals. The Weddell Sea is located southeast of South America and is bounded by the Scotian Sea, the Larsen, Filchner and Ronne Ice Shelf, and the Maud Rise. Ice shelves are a thick mass of ice extending from the polar coast. ICEBERGS leaving the Weddell Sea may be transported by the ANTARCTIC CIRCUMPOLAR CURRENT or "West Wind Drift." Cracks in the ice called leads may form as a result of the rise and fall of tides. Cracks caused by tides may be miles (km) long and as wide as 1.5 feet (45 cm). Cracks and holes in the ice are important areas for animals such as snow petrels (*Pagrodama nivea*) that feed on small crustaceans and as a breathing hole for seals. Air-sea interactions in the Weddell Sea contribute to the formation of WATER MASSES such as ANTARCTIC BOTTOM WATER (AABW) and Weddell Sea Bottom Water (WSBW). Water masses are identified by their temperature and salinity ranges. AABW is considered to be the densest water in the open ocean and is associated with TEMPERATURES ranging from 32° to 30.6°F (0° to -0.8°C) and SALINITIES from 34.6 to 34.7 psu. WSBW is found beneath AABW in the seas surrounding Antarctica with temperatures ranging from 30.6° to 34.5°F (-0.8° to 1.4°C) and salinities of 34.65 psu. The Weddell Sea is known for its rich marine resources which include krill (*Euphausa superba*), penguins such as the King penguin (*Aptenodytes patagonicus*) and Emperor penguin (*Aptenodytes forsteri*), seals such as the Weddell seal (*Leptonychotes weddellii*) and Crabeater seal (*Lobodon carcinophagus*), and whales such as humpback whale (*Megaptera novaeangliae*), southern right whale (*Eubalaena australis*), and minke whale (*Balaenoptera acutorostrata*).

See also ANTARCTICA; MARGINAL SEA.

Further Reading

Cool Antarctica. Available online. URL: http://www.cool antarctica.com/Antarctica%20fact%20file/antarctica%20fact%20file%20index.htm. Accessed November 7, 2006.

wetland A land area that is inundated by surface water or groundwater and supports plants and animals especially adapted for life in saturated soil conditions is called a wetland. Wetlands often form the boundary between terrestrial and aquatic systems. The characteristics of a wetland vary in accordance to geographic location, climatic patterns, and anthropogenic factors. Coastal wetlands include swamps, marshes, and bayous, while inland wetlands include vernal pools and bogs. They are important in maintaining clean water bodies, preventing floods, controlling erosion, and are often nursery areas for many marine animals. The loss of wetlands is often caused by natural disasters such as HURRICANES and droughts or anthropogenic factors such as land reclamation, levees, and invasive species.

Wetlands are known for submergent plants such as coon tail (*Ceratophyllum demersum*), two leaf

Preserving wetlands is essential to preserving the health of coasts.   *(NOAA)*

milfoil (*Myriophyllum heterophyllum*) and wild celery (*Vallisneria americana*) that live underwater, emergents such as broadleaf arrowhead (*Sagittaria latifolia*), skunk cabbage (*Symplocarpus foetidus*), and wild rice (*Zizania aquatica*) that grow out of the water, and floating plants such as cow lily (*Nuphar lutea*), giant duckweed (*Spirodela polyrhiza*), and American lotus (*Nelumbo lutea*). These plants are also called hydrophytes and may be found in wetlands with hydric soils such as marshes or swamps and areas without hydric soils such as impoundments or excavations. Hydric soils are associated with a water table that equals 0.0 feet and little oxygen during the growing season. The water table is the top of the water surface in the saturated portions of water-bearing geological formations (aquifers).

Cypress trees (*Taxodium distichum and Taxodium ascendens*) that are found in swamps have a root system that extends above the water, which helps provide support in the soft hydric soils. These specialized roots or knees may also help supply oxygen to the roots. Wetland soils may consist of fine particles, silts, and clays. They tend to be low in permeability and have little oxygen. Hydric soils contain high concentrations of anaerobic bacteria

that break down organic material and produce methane (CH_4), which is also known as swamp gas. In some cases, wetland soils contain more mineral matter (sand, silt, clay, or loam) than organic matter, and the saturated conditions result in the reduction of iron and other elements. The resultant gray, greenish gray, or bluish gray sediments are called gleyed soils.

In tropical areas (between approximately 25° north and 25° south latitude) with low wave action such as along portions of the western and eastern coast of Africa, the eastern coast of Australia, the islands, gulfs, and bays common to Southeast Asia, and the Americas near warm ATLANTIC OCEAN, Caribbean Sea, and GULF OF MEXICO waters, wetlands take the form of mangrove forests and swamps. Mangroves impede water flow and enhance the deposition of SEDIMENT. They thrive in the intertidal zone near estuaries, along coastal regions behind CORAL REEFS, and are associated with flat gradients and muddy bottoms. The intertidal is an interface between the land and water and is defined as the region between extreme high and extreme low tides. The mangrove swamp is a coastal wetland that offers refuge and nursery grounds for many species of finfish and shellfish. Mangrove strands are also prime nesting and migratory sites for hundreds of bird species. They have been credited with helping to reduce loss of life and the destruction of property by natural hazards such as floods.

Another class of coastal wetland includes salt marshes that may be associated with tidal inlets, shifting channels, and mudflats. Salt marshes develop along the shores of estuaries where freshwater mixes with salt ocean water. They are low-energy areas that are alternatively flooded and drained through the action of tides. Plants cannot survive in high-energy environments where wave slamming would uproot vegetation. In the low marsh area between low and mean high tide, one finds rooted emergent vascular macrophytes such as cordgrass (*Spartina alterniflora*), while in the high marsh area between mean high tide and the spring tide one finds marsh-hay cordgrass (*Spartina patens*) and saltgrass (*Distichlis spicata*). Toward the edge of the marsh, black grass (*Juncus gerardii*) predominates.

Freshwater marshes occur in ponds and slow-moving streams, while riparian wetlands occur along the banks of steams, rivers, and lakes. In Louisiana and Mississippi, branches of the Mississippi River and lakes in lowland areas are called bayous. They are sluggish watercourses flowing through the Mississippi River delta or neighboring the Gulf of Mexico. These low-SALINITY waters are known for their American alligators (*Alligator Mississippiensis*), crawfish (*Orconectes blacki, Procambarus viaeviridis, Fallicambarus dissitus, Faxonella creaseri*), blue crabs (*Callinectes sapidus*), and shrimp (*Penaeus setiferus* and *Penaeus aztecus*). Bogs tend to have soils composed of peat, and mosses predominate. Vernal pools occur in small depressions underlain by dense, impenetrable claypan soils that allow water to accumulate in winter and spring. A claypan is slowly permeable and overlaid with nonclay materials. The pools may support annual plants, which flower as the collected water from precipitation begins to evaporate.

Due to their importance as the transition between land and water, wetlands in the United States have been protected since passage of the Clean Water Act during 1972. Regulatory authority for the discharge of fill or dredged material into wetlands rests with the U.S. Army Corps of Engineers. Thus, any development to change the course of a stream or to build a dam requires a permit from the Army Corp of Engineers. Organizations such as the ENVIRONMENTAL PROTECTION AGENCY are also actively involved in wetland restoration, the renewing of natural wetlands that have been lost or degraded and reclaiming their functions and values as vital ecosystems.

See also ESTUARY; MANGROVE SWAMP.

Further Reading
National Estuary Program, U.S. Environmental Protection Agency. Available online. http://www.epa.gov/owow/estuaries. Accessed November 6, 2006.
North Carolina State University. "Plant Fact Sheets." Available online. URL: http://www.ces.ncsu.edu/depts/hort/consumer/factsheets. Accessed November 3, 2006.
U.S. Environmental Protection Agency. "Wetlands." Available online. URL: http://www.epa.gov/owow/wetlands. Accessed November 3, 2006.
Wetlands Research and Technology Center, U.S. Army Engineer Research and Development Center, Engineer Research and Development Center, Waterways Experiment Station. Available online. URL: http://el.erdc.usace.army.mil/wrtc. Accessed November 10, 2006.

winch   This instrument is also known as a drum in a steel spool connected to a power source that facilitates the reeling or unreeling of rope, wire, or cables. Winch stations are fixed to the vessel's working decks. In marine science, the drum is usually loaded with steel wire or jacketed electrical cables (sea cables) that connect to OCEANOGRAPHIC INSTRUMENTS to relay data or measurements to the mother ship or a data acquisition system installed in a BUOY.

Specific DECK GEAR such as the capstan, a windlass, and many other forms of winches are essential for the safe lifting of critical loads on a vessel. A winch normally stores wire and cable, while a capstan and windless are usually used to restrain and manipulate the anchor chain on a ship and the anchor line on a boat. The term *capstan,* which dates back to the era of tall ships, is used to weigh anchors with a vertical cylinder or barrel. A windlass consists of a horizontal cylinder, which is mechanically rotated, in

Winches are the workhorses of deployments and retrievals of ocean measurement systems. *(USGS)*

order to lift and lower heavy objects such as an ANCHOR. Other forms of winches are commonly found aboard vessels to handle rigging such as heaving in the ship's MOORING lines, raising the sails, furling the jib, and trimming the sheets.

Winches in combination with one of the ship's cranes may be used to lift loads such as hauling in an instrument mooring or in combination with an A-frame on the stern to deploy a SUBMERSIBLE. The submersible is slowly lifted off the deck, and the A-frame is rotated out behind the vessel while the winch pays out cable until the submersible is safely lowered to the water. A-frame and winch or crane and winch combinations are commonly used by oceanographic research vessels to deploy instrument moorings, trawls, sediment cores, towing equipment such as side-scan SONARS and REMOTELY OPERATED VEHICLES, and to make CTD (CONDUCTIVITY-TEMPERATURE-DEPTH) casts. REMOTELY OPERATED VEHICLES are called ROVs and are actually unmanned submersibles used to collect oceanographic data. They are connected to an ROV OPERATOR aboard a research platform by a communication and power tether. The tether is usually maintained on a drum or spool. Several marine science technicians from the science crew might help steady and direct loads (oceanographic instruments fixed to a mooring cable) with several tag lines while one marine technician directs winch or crane operators that are part of the ship's crew. The winch operator carefully moves a combination of levers, pedals, and throttles in order to regulate the speed of winch drums in response to visible or audible directions or by observing dial indicators or cable marks.

Other more routine examples of winches include their use on WORKBOATS. Using an appropriately sized trailer and tow vehicle may be the most efficient means to gain access to numerous sites around an ESTUARY or along the coastal ocean with a workboat. Boats on a trailer are deployed from a boat ramp where scientists may cruise to nearby island research stations, marine protected areas, visit a larger research vessel or platform, or navigate to the inland or coastal site to make observations (collect data). Trailer winches are forward mounted on a trailer to assist in launching or retrieving a workboat from a boat trailer. The winch line that is wound around the drum is unreeled and attached to a hull-mounted steel bow eye so that the vessel can be efficiently lowered or pulled from the water in a controlled manner. Workboats are small research vessels with deck gear such as winches to fully support the execution of planned scientific investigations. A hydraulic winch or hand winch and a davit might be used aboard a workboat to pull in fixed gear such as traps or nets. A davit is a small crane

used to hoist a load. Two davits are traditionally seen on the stern of yachts to hoist a dinghy, and one davit is used on the hauling side of a lobster boat to hoist lobster pot (traps) trawls. Two or more lobster pots (usually five to 10) attached by a ground line define a lobster trawl and are hauled onboard with the aid of a hydraulic winch (pot hauler). Buoy lines at either end of the trawl connect the set of pots to a marker float at the surface. Individual pots may be attached to the ground line with a bridle and a short piece of line called a gangion.

Woods Hole Oceanographic Institution (WHOI)

Woods Hole Oceanographic Institution (WHOI) is the world's largest private, nonprofit ocean research, engineering, and educational organization. WHOI scientists roam the oceans of the world to gather data and test hypotheses and theories of the biology, chemistry, geology, physics, and climatology of the Earth. The primary mission of WHOI is "to develop and effectively communicate a fundamental understanding of the processes and characteristics governing how the oceans function and how they interact with the Earth as a whole." Woods Hole takes a multidisciplinary approach to problem solving and has recruited top scientists and engineers from all over the world to implement its staff. It has provided great public service by distributing the results of its research to the public and to policy makers, and it has fostered the applications of its technology developments.

Woods Hole was founded in 1930 as the premier East Coast oceanographic laboratory. It is located on Cape Cod at Woods Hole, Massachusetts. It's first major research vessel, a sailing ship called the R/V *Atlantis* became famous throughout the ports of the world for representing the best in ocean science. Today, Woods Hole maintains a fleet of four world-class research vessels for investigating and solving the mysteries of the seas, as well as many smaller vessels. The institution has a staff of about 135 scientists, 155 technicians, 60 marine personnel, and 130 students. It maintains close ties with the U.S. Navy, NOAA, the Massachusetts Institute of Technology, and many other academic institutions and federal agencies.

Since its inception, Woods Hole has made major contributions to marine science, engineering, and technology. Its SUBMERSIBLE, the *Alvin,* is widely known for such feats as the recovery of a lost hydrogen bomb off the coast of Spain and the discovery of WARM-WATER VENTS in the Mid-Atlantic Ridge. Research Scientist Robert D. Ballard is famous for his discovery of the wreck of the White Star liner *Titanic.* Besides basic research, WHOI maintains educational programs for undergraduate and graduate students. Undergraduate programs include Minority Fellowship and Summer Student Fellowships specially designed for college juniors or seniors who are studying in any of the fields of science or engineering such as biology, chemistry, engineering, geology, geophysics, mathematics, meteorology, physics, oceanography, and marine policy. Graduate students study at both the Massachusetts Institute of Technology (MIT) and the WHOI. In order to help recent Ph.D. recipients with an opportunity to hone their research skills, WHOI offers several temporary postdoctoral opportunities. Postdoctoral appointees may perform a significant role in a principal investigator's research and augment the role of graduate faculty in providing research instruction to graduate students.

Further Reading

Ballard, Robert D. *The Eternal Darkness: A Personal History of Deep-sea Exploration.* Princeton, N.J.: Princeton University Press, 2000.

WHOI at a glance. Available online. URL: http://whoi.edu/page.do?/pid=8117. Accessed August 7, 2007.

workboat   A workboat is a motorized vessel usually less than 65 feet (19.8 m) in length that is used for marine operations. They may transport scientists or people in the marine trades to work sites such as field stations (data retrieval sites) or platforms that are located on inland waters such as canals, rivers, and ESTUARIES and the coastal ocean. Depending on size, workboats may provide a cabin with accommodations for a science team, small crew, a galley, potable water, and a head for several days of at-sea operations. The galley is the boat's kitchen, and the head is the toilet and may include showers. Boats less than 26 feet (8 m) tend to be used for daylong operations. Workboats are specially designed to meet marine requirements ranging from collecting sea truth data for aerial surveys and the conduct of HYDROGRAPHIC SURVEYS to completing science missions requiring

This NOAA workboat can perform many tasks in surveying. *(NOAA)*

BUOY and MOORING deployments, trawling to collect fish, grab sampling to collect SEDIMENTS, tagging fish species such as sharks, collecting water samples for chemical analysis, and deploying a variety of oceanographic instruments. A hydrographic survey is conducted to collect bathymetric data to make or update a nautical chart. Commercial workboats are used to support coastal construction, dive operations, fishing, piloting, and patrolling. Unusual designs such as liftboats have hydraulically actuated legs that descend to the seafloor and lift the vessel completely out of the water to provide a stable platform for construction, drilling, or diving. Workboats normally feature ample deck space to handle equipment such as buoys and instrument moorings. They may be outfitted with deck handling gear such as cranes, WINCHES, and lifts. Onboard electronic systems usually include RADAR, DEPTH FINDER, the GLOBAL POSITIONING SYSTEM, radios, and a satellite phone. Workboats are usually classified according to their function such as crewboat, lobster boat, pilot boat, pipeline barge, shrimp boat, side trawler, tender, towboat, and TUGBOAT. Tugboats have powerful engines and are used to help move barges and ships in confined harbor areas. A vessel primarily used to push barges along a waterway such as the Mississippi River may be called a towboat or pushboat. Barges are classified by their function and capability. For example, a barge with a crane is called a crane barge, while a deck barge carries cargo and materials for coastal construction projects. Some barges are self-powered.

See also BARGE; BATHYMETRY; BOAT; CHARTS, NAUTICAL.

The U.S. Army Corps of Engineers' workboat *Gannet* can perform many tasks to maintain coastal waterways.   *(U.S. Army Corps of Engineers)*

Further Reading

American Society of Naval Engineers. Available online. URL: http://www.navalengineers.org. Accessed November 2, 2006.

xenolith Xenoliths are fragments of foreign rock that are captured within a larger mass of molten rock that cools before it reaches the surface. Molten rock that is still underground is called magma. The xenoliths must be older since these usually small pieces of rock were locked in as the magma cooled. The larger mass is an intrusive igneous rock meaning that the magma cools and crystallizes before it reaches the surface of the Earth. Examples of intrusive rocks are gabbro, diorite, and granite. Xenoliths are unrelated to the surrounding igneous material and provide marine geologists with an opportunistic sample from the deep Earth. Xenoliths may provide information on the nature of the mantle where no material has been exposed by tectonic activity. The mantle lies from 12.4 to 1,802 miles (20 to 2,900 km) beneath the crust and has two parts, a solid upper part and a liquid lower part owing to very high TEMPERATURES and PRESSURE. It deforms slowly in a plastic manner.

Geophysicists use SEISMIC WAVES to study the mantle as well as receive clues from xenoliths that have been extracted from cores such as those obtained through the OCEAN DRILLING PROGRAM. Seismic waves slow down in hot material and speed up in cold, causing refraction, or bending, of the waves, which is detected by an array of receiving seismometers. Much of what is known about the Earth's mantle and core has been learned from the study of seismic waves generated by earthquakes.

The two common types of xenoliths are peridotites and eclogites. Marine geologists believe that peridotite (an igneous rock composed chiefly of olivine [magnesium iron silicate]), is the most common rock found in the upper mantle of the Earth, and it is the primary reservoir for naturally occurring diamonds and chromium ore. The Kimberley district of South Africa is famous for its coarse-grained peridotites and diamonds. Eruptions within the mantle push peridotites into the crust as vertical columns of rock, which are also called kimberlite pipes. Diamonds are the transparent form of pure carbon, the hardest mineral, and are found in peridotes and eclogites. Eclogites are dense and coarse-grained rocks that either form as basaltic rock subducts (*see* SUBDUCTION ZONE) into TRENCHES or from magma that crystallizes and cools within regions near the upper mantle. Basalt is a hard, fine-grained, and dark volcanic rock rich in iron (Fe) and magnesium (Mg). Xenoliths range from sand-grain sizes to the size of a soccer ball.

Further Reading

Hawaiian Volcano Observatory, U.S. Geological Survey. Available online. URL: http://hvo.wr.usgs.gov/index.html. Accessed July 19, 2007.

Kusky, Timothy. *Encyclopedia of Earth Science.* New York: Facts On File, 2005.

Lapidus, Dorothy Farris, and Donald Robert Coates. *The Facts On File Dictionary of Geology and Geophysics.* New York: Facts On File, 1987.

Smith, Jacqueline. *The Facts On File Dictionary of Earth Science,* rev. ed. New York: Facts On File, 2006.

U.S. Geological Survey. Available online. URL: http://www.usgs.gov/. Accessed August 27, 2006.

Y

yaw The oscillation of a vessel about the vertical axis or deviation from a straight course is called yaw. Vessels may experience yaw in the presence of WAVES, which is controlled by moving the rudder or engines in the case of outboard motors. In aircraft, yaw occurs when the nose of the plane turns left or right to the direction of its motion and is controlled by the rudders. Other important vessel motions include pitch (rising and falling of the bow and stern) and roll (transverse oscillations). Survey vessels must measure and record vessel's motions using the ship's navigation systems. The vessel motion is then removed from the output data of OCEANOGRAPHIC INSTRUMENTS that measure parameters such as depths or CURRENTS.

See also BOAT; BUOYS.

year class strength Year class strength is a term used by many fisheries scientists to describe the number of animals in a cohort, meaning the number of animals that are of the same age in a stock. The stock is the population that is being harvested. Fisheries scientists collect physical and biological data to develop indexes that are indicative of year class strength. Physical data may be obtained by OCEANOGRAPHIC INSTRUMENTS to identify important hydrographical factors such as sea surface TEMPERATURES (°F or °C) and SALINITIES (PSU) that are preferred by certain species, particular stages of the TIDES producing water level changes that may flood or dry critical habitat, seasonal storms that trigger spawning, extensive nutrient-rich river or estuarine plumes that retain migrating species at the front, the presence of HYDRAULIC CURRENTS (knots or m/s) near a COASTAL INLET that may transport passive drifters into or out of the ocean, and estuarine circulation that transports young fish to shallow water nursery areas. Biological data includes information such as length, weight, sexual maturity, and age for individual fish sampled from a trap or net. Physical and biological data is analyzed to determine fish habits and to make important recruitment estimates. Fisheries managers evaluate recruitment by determining the numbers of new fish that will be available to the exploitable part of the population by growth from among smaller size categories.

Numerous researchers at locations such as Oregon Inlet, North Carolina, have measured hydrological variables, such as water quality, WATER LEVEL FLUCTUATIONS and transport, which influence recruitment of commercially important finfish such as striped bass (*Morone saxatilis*) and red drum (*Sciaenops ocellatus*) common to an area such as the Carolina Capes. The marine scientists evaluate how annual and seasonal flows and water levels inside and outside of the coastal inlet are related to year class strength of the captured larvae from either side of the coastal inlet. Some species of concern include demersal fish such as Atlantic croaker (*Micropogonius undulates*), spot (*Leiostomus zanthurus*), southern flounder (*Paralichthys lethostigma*), and pelagic fish such as Atlantic menhaden (*Brevoortia tyrannus*). Many of these fish are harvested for human consumption, or processed into products such as fishmeal or fish oil, and are an important food source to many predators (marine fish, birds, and mammals). Hydrometeorological data such as precipitation, winds, and river flow data may be obtained from the National Weather Service and U.S. Geological Survey, respectively. Estuarine and coastal ocean measurements are obtained from bottom-mounted ACOUSTIC DOPPLER CURRENT PROFILERS (ADCPS), instrumented moorings, electromagnetic

CURRENT METERS fixed to structures such as bridges, and water level gauges sited in the ocean and inside the ESTUARY. Fisheries biologists study samples of fish larvae caught in nets (seines, push nets, and trawls) to assess relationships between hydrological variables and year class strength. Age and growth rates for larvae, juveniles, and adults may be derived from otolith analysis. An otolith, similar to the inner ear in mammals, is a small crystal of calcium carbonate ($CaCO_3$) that is found in the head of bony fishes. Some long-term investigations may involve exposing captured fish to solutions of strontium chloride hexahydrate ($SrCl_2 \cdot 6H_2O$), which stains the otolith. Patterns on the otolith can then be studied with a scanning electron microscope equipped with a BACKSCATTER electron detector.

Many of the above species migrate from estuaries such as the Pamlico Sound onto the CONTINENTAL SHELF to spawn. Adult species are often caught in pound nets as they migrate into or out of an estuary. After the eggs have hatched in offshore waters, fish larvae are advected inshore to estuaries by wind-driven currents. The young fish may reach the coast and consolidate at an ebb current plume and then ride into their estuarine nursery when the current shifts to exaggerated flood currents. These events are in synchronization with storms, which pile water along the coast and lower water levels inside the inlet. Such nontidal flooding currents transport the fish larvae well into the estuary, helping to retain the fish larvae in the nursery.

The biological and physical data are used to assess stock status and to prepare fishery management plans. The Regional Fishery Management Councils write the plans, which are then implemented by the National Marine Fisheries Service. For some species such as southern flounder (*Paralichthys lethostigma*), declines have been observed and attributed to overfishing from COMMERCIAL FISHING, especially as bycatch, nursery habitat changes, and jetties, which disrupt the transport of larvae from offshore spawning grounds to estuarine settlement sites.

See also FISHERIES RECRUITMENT; NET.

Further Reading

Mangone, Gerard J. *Concise Marine Almanac.* New York: Van Nostrand Reinhold, 1986.

Mid-Atlantic Fisheries Management Council. Available online. URL: http://www.mafmc.org/mid-atlantic/mafmc.htm. Accessed August 27, 2006.

National Marine Fisheries Service. Available online. URL: http://www.nmfs.noaa.gov/. Accessed August 27, 2006.

Yellow Sea The Yellow Sea is a MARGINAL SEA that the Yellow River (Huang He) of China empties into (Huanghai). It is part of the PACIFIC OCEAN, located between China and the Korean Peninsula. It receives its name from the sand particles that color its water from the 1.6 billion tons of SEDIMENTS that wash into it from the Yellow River. The peak discharge is in the summer and has important effects on the Yellow Sea's SALINITY and hydrography. This sediment load is huge. It is over seven times larger than that which drains into the Gulf of Tonkin. The Yellow Sea's major inlets are Korea Bay, Bohai, and Liadong Gulf. It is also considered to be the northern part of the East China Sea. It has a maximum depth of about 492 feet (150 m), which is shallow for such a large coastal sea and covers an area of about 3.3 million square miles (1,200,000 km²).

The Yellow Sea's surficial sediments, morphology, and shallow structure of its deeper parts reflect the sedimentary input from the rivers that feed it, its post-glacial history, and the effects of local oceanographic conditions. Not only is there a seasonal variation in the sediment load, but also the flux of sediments has varied greatly since the last glacial maximum. About 10,000 years ago the Southwest Asian MONSOON system developed, which greatly increased the amount of runoff. Human impacts from deforestation and agricultural practices can be found in sediment records from about 3,000 years ago. Anthropogenic impact on these waters is presently large and continues to grow.

The Yellow Sea has marked seasonal variations and supports substantial populations of FISH and MARINE MAMMALS. It has both cold-water and warm-water species of fish. The Yellow Sea has many mammal species, including whales, finless porpoise, seals, and sea lions. The commercial fish stocks were heavily fished in the 1960s, which has a significant negative effect on the present ecosystem.

China and the two Koreas have massive populations (10 percent of the world's population) living in the Yellow Sea drainage basin. The pollution problems they share include industrial wastewater, agricultural runoff including pesticides, oil from ships and oil exploration. In addition, the Yellow Sea waters are a highway for international shipping.

The Battle of the Yellow Sea was a major naval battle in the Russo-Japanese war. In February 1904, the Japanese launched an attack on the Russian naval forces in Port Arthur. This battle resulted in the loss of a number of Russian warships that were destroyed by land bombardment. The Yellow Sea remains a very important military area separating China and the Koreas.

Further Reading

Large Marine Ecosystems of the World portal. Available online. URL: http://www.lme.noaa.gov/Portal. Accessed October 17, 2007.

Z

zebra mussel   The zebra mussel is a small fresh and brackish water invertebrate known to scientists as *Dreissena polymorpha,* having two hinged shells with a stripe pattern. These bivalves are filter feeders, similar to clams, oysters, and scallops (*see* MOLLUSKS). Their gills are used to trap food particles (PHYTOPLANKTON and small ZOOPLANKTON) as well as respiration. Zebra mussels are native to eastern Europe and western Asia and are commonly found in the Black, Caspean, and Azov Seas. They are a nonindegenious aquatic species found in all the Great Lakes and many navigable rivers in the eastern United States. They are preyed upon by birds such as the American coot (*Fulica americana*) and herring gull (*Larus argentatus*); waterfowl such as canvasback (*Aythya valisineria*), greater scaup (*Aythya marila*), goldeneye (*Bucephala clangula*), lesser scaup (*Aythya afinis*), mallards (*Anas platyrhynchos*), oldsquaw (*Clangula hyemalis*), tufted duck (*Aythya fuligula*), white-winged scoter (*Melanitta fusca*); finfish such as freshwater drum (*Aplodinotus grunniens*), river redhorse (*Moxostoma carinatum*), common carp (*Cyprinus carpio*), bluegill (*Lepomis macrochirus*), quillback (*Carpiodes cyprinus*) and flathead catfish (*Pylodictis olivaris*); and crustaceans such as the blue crab (*Callinectes sapidus*).

Despite predation, zebra mussels reproduce at rapid rates, and the larval forms, called veligers, are zooplankton for two to five weeks until they settle to hard benthic surfaces in clumps. They may grow to two inches (5 cm) and have a life span of approximately six years. Water currents, draining of outboard engine cooling water, and release of bilge water may spread larvae. The spread of this INVASIVE SPECIES to North America is thought to have occurred sometime around 1985 as a result of ballast water introductions into the lower Great Lakes from seagoing vessels. Zebra mussels were first discovered in 1988 in Lake St. Clair, Michigan, located between Lakes Huron and Erie. Other possible methods of spread include aquarium releases and accidental releases from AQUACULTURE facilities. Zebra mussels are considered a nuisance species since they clog piping systems, interfere with mechanical systems such as locks, and damage vessel hulls and engines. Locks are chambered structures on a waterway that are closed off with gates for the purpose of raising or lowering the water level within a lock chamber so that vessels can move from one elevation to another along the waterway. Natural predation has not controlled the spread of the infamous zebra mussel, which may smother native species. Strategies such as filtration, desiccation from water drawdowns, chemical applications, heat treatment, and physical removal have been used to combat the spread of this mollusk. Coastal zone managers advocate improvements in habitat conditions for natural predators so that the predator's population will increase at a gradual rate. For example, American shad (*Alosa sapidissima*), which feed on PLANKTON, are a potential predator for zebra mussel veligers. Fisheries biologists have discovered that shad do not feed during upstream spawning migration, but resume after spawning during their downstream migration back to the ocean. Locations such as the Patuxent Research Refuge in Laurel, Maryland, have installed fish ladders, a series of stepped pools that enable ANADROMOUS fish to migrate up the Patuxent River despite dams. Another important strategy to slow the spread of zebra mussels is the cleaning of boats and equipment before and after use with steam, salt water, or a chlorine solution.

Further Reading
Great Lakes Sport Fishing Council home page. Available online. URL: http://www.great-lakes.org/exotics.html. Accessed October 17, 2007.
NAS-Nonindigenous Aquatic Species, U.S. Geological Survey. Available online. URL: http://nas.er.usgs.gov. Accessed October 22, 2006.
National Invasive Species Information Center, National Agricultural Library. "Zebra Mussel Profile." Available online. URL: http://www.invasivespeciesinfo.gov/aquatics/zebramussel.shtml. Accessed October 17, 2007.

zinc (Zn) Zinc is a brittle bluish white inorganic solid with a crystalline structure. Zinc is an essential mineral for growth in all plants and animals. People obtain zinc from mixed animal and plant sources. Oysters, a filter-feeding bivalve MOLLUSK common to ESTUARIES, may contain more zinc per serving than any other food. Zinc is also a metal since it conducts heat and electricity and can be formed into sheets or foils. Zinc is used commercially as a dust, or powder, in granules, or pieces, as stringlike shavings, or in the form of shot. On the periodic table, zinc has atomic number 30, which means that there are 30 protons found in the nucleus of a zinc atom. The atom is the smallest particle retaining an element's chemical properties. Subatomic particles are electrons, protons, and neutrons. The protons and neutrons or nucleons make up the atomic nucleus, which is surrounded by an electron cloud.

Zinc has many important commercial applications from purifying water to protecting steel ships from corrosion. Granules of zinc have been used to remove chlorine, heavy metals, and bacteria from water. During Roman times, zinc was used with copper to make the alloy brass. Zinc along with other metals such as tin (Sn), copper (Cu), and silver (Ag) is a component to make solder that is free of lead (Fe). Solder is an alloy with a low melting point and is used with soldering irons to join metallic items such as copper wires. Zinc is used to coat carbon steel and cast iron to prevent rust and corrosion. This might involve submerging a steel gun barrel into a phosphoric acid (H_3PO_4) solution whose additional ingredient is zinc or manganese. Zinc is also used to coat or galvanize iron or steel to prevent corrosion by making it less reactive to air and water. Galvanizing methods, after Italian physicist Luigi Galvani (1737–98), include dipping material into a vat of molten zinc, electroplating, and spraying. Electroplating involves coating objects such as steel bolts or fasteners with zinc. The fasteners are placed into a container of a corrosive solution such as zinc ($ZnCl_2$) and ammonium chloride (NH_4Cl). The fasteners are connected to an electrical circuit, forming the cathode (negative) of the circuit, while an electrode typically of zinc forms the anode (positive). When an electrical current is passed through the circuit, metal ions in the solution take up excess electrons at the fasteners. The result is a layer of zinc on the fasteners. Zinc sprays fuse zinc to metal surfaces.

In the marine environment, it is important to understand galvanic processes. For example, metal components on a boat or MOORING that are physically or electrically connected and immersed in seawater become a battery. For example, an aluminum (Al) propeller could be attached to a stainless steel shaft. Electrical currents are induced that flow between the aluminum propeller and the stainless steel shaft. Electrons making up the current are supplied by one of the metals, which then pass metal ions to the seawater. This loss or galvanic corrosion may destroy engine components or the integrity of a mooring. Sailors and marine technicians compensate by adding a third piece of metal or sacrificial anode into the circuit to protect exposed stainless steel such as the propeller shaft or an instrument frame. Sacrificial anodes are usually zinc since they are quick to give up electrons to counteract the effects of galvanic corrosion. The bare zinc is usually mounted directly to the metal being protected.

Further Reading
The American Zinc Association. Available online. URL: http://www.zinc.org/. Accessed August 23, 2006.
The International Zinc Association. Available online. URL: http://www.iza.com/. Accessed August 23, 2006.

zooplankton Zooplankton are generally small, weakly swimming, nonphotosynthesizing organisms such as daphnia and copepods that move to and fro with tides and currents. Daphnia or water fleas are small freshwater crustaceans that eat ALGAE and are forage for finfish such as trout. Copepods are also small crustaceans and a major component of the zooplankton community. Plankton are classified as synthesizing (PHYTOPLANKTON) or nonsynthesizing (zooplankton) organisms rather than plants or animals because some phytoplankton are able to hunt for their food when photosynthetic activity is low, such as during periods of cloudy weather, when light levels are low. In common usage, however, phytoplankton usually refer to plant plankton, and zooplankton to animal plankton.

Zooplankton are found in both freshwater and saltwater environments. They are mostly microscopic heterotrophs, including bacteria, protozoans, and the larval stages of an assortment of larger animals. For example, the larval forms of many fish and shellfish are in the zooplankton community. Fish larvae may swim one or two body lengths per second, making them passive drifters when compared to much stronger bottom currents, tidal currents, and wind drift currents. Physical oceanographers focused on sampling WATER MASSES and measuring currents often work with fisheries biologists focused on collecting eggs, larvae, juveniles, and adults when trying to determine spawning and migration paths for specific finfish and shellfish.

Plankton may be categorized according to size, and many zooplankton are further classified by developmental stage. Plankton measuring less than two micrometers in size are called picoplankton, between two and 20 micrometers are called nanoplankton, between 20 and 200 micrometers are called microplankton, between 0.2 and 20 millimeters are called mesoplankton, between 20 and 200 millimeters are called macroplankton, and greater than 200 millimeters are called megaplankton. Zooplankton that metamorphose from larvae to adults such as worms, mollusks, crustaceans, corals, echinoderms, or fish, are called meroplankton. They are temporary members of the zooplankton community. Holoplankton are in the plankton community for their entire life cycle and include pteropods (snails), chaetognaths (worms), larvaceans (small transparent animals related to sea squirts), siphonophores (jellyfish and sea anemones), and copepods (small crustaceans). Some of the largest zooplankton include jellyfish, sea nettles, and comb jellies. The Portuguese man-of-war (*Physalia physalis*) consists of a surface float or bladder for propulsion and long strings of stinging tentacles and is a hazard to swimmers in many parts of the world.

Copepods are a key group in the trophic levels of the ocean, forming the base of the food chain of the carnivores. They are about 0.2 mm to 20 mm in length, and often have numerous appendages used for feeding on phytoplankton and organic particulate matter, and for propulsion. Another tiny crustacean includes approximately 86 species of euphausiids, the primary food source for many fish, MARINE BIRDS, seals, and baleen whales. They look like small shrimp, and are usually from 0.8 to two inches (2 to 5 cm) in length. In the waters of the SOUTHERN OCEAN surrounding Antarctica, they are called krill (*Euphausia superba*), and are found in great abundance, attracting large numbers of whales. During the late 19th and early 20th centuries, great numbers of whaling ships reduced these whale populations to the point where many, such as the blue whale, are endangered (*see* MARINE MAMMALS).

The decapods include crabs, lobsters, and shrimp. Crabs and lobsters are zooplankton only in their larval stages, but some species of shrimp may be classified as holoplanktonic because of their limited swimming ability. There are several thousand species of shrimp, and they are of great economic value in terms of the commercial catch and AQUACULTURE potential. The micro-crustaceans known as fairy (*Eubranchipus vernalis*), brine (*Artemia salina*), clam (*Cyzicus mexicanus*) and tadpole (*Lepidurus packardi*) shrimp are actually branchiopods. Most zooplankton whether larvae or holoplankton are filter feeders and use their appendages to strain bacteria, algae, and other fine particles from the water. Predators, which may include larger zooplankton, feed on smaller zooplankton.

See also PLANKTON.

Further Reading

COPEPOD: the Global Plankton Database. Available online. URL: http://www.st.nmfs.gov/plankton. Accessed October 17, 2007.

Davis, Richard A., Jr. *Oceanography: An Introduction to the Marine Environment*, 2nd ed. Dubuque, Iowa: Wm. C. Brown, 1991.

zooxanthellae Zooxanthellae are single-celled yellow-brown ALGAE related to dinoflagellates that live symbiotically in the tissues of animals such as reef-forming corals or cnidarians, giant clams (*Tridacna gigas*), and nudibranchs. Dinoflagellates are a type of PHYTOPLANKTON and an important part of the food chain. They use their flagella to move through the water column. The zooxanthella are found in the open ocean, and upon entering a host (coral, sponge, clam) they lose their motility by dropping their flagella. Sponges are suspension feeders that derive benefit from zooxanthellae, especially in producing their calcium carbonate ($CaCO_3$) shells. Species such as *Cliona vastifica*, a boring sponge, contain zooxanthella. These sponges bore into old shell and help fragment the shells into shell hash and calcareous sands. Hosts with zooxanthellae such as clams must expose their zooxanthellae-laden tissues toward sunlight. Giant clams when disturbed by an inquisitive SCUBA diver or hungry finfish may retract their mantle with photosynthesizing zooxanthellae and then close their shell valves. Nudibranchs are shell-less snails in the mollusk family that obtain zooxanthellae by foraging on cnidarians. Carbon (C) compounds derived from photosynthesis occurring within the animal's tissues are used by the host. Cnidarians use the carbon to make calcium carbonate ($CaCO_3$) shells that builds reefs in warm tropical waters. This symbiotic relationship helps the zooxanthellae and host recycle nutrients and survive in the nutrient-poor reef setting. The loss of the zooxanthellae due to TEMPERATURE changes or increased TURBIDITY causes white areas to appear within a reef, a phenomenon called CORAL BLEACHING.

Further Reading

The Cnidaria home page, University of California at Irvine. Available online, URL: http://www.ucihs.uci.edu/biochem/steele/default.html. Accessed August 25, 2006.

MacRae, Andrew. "Dinoflagellates." Deptartment of Geology and Geophysics, University of Calgary, Canada. Available online, URL: http://www.geo.ucalgary.ca/~macrae/palynology/dinoflagellates/dinoflagellates.html. Accessed August 25, 2006.

NOVA. Science in the News, Australian Institute of Marine Science. "Coral bleaching—Will Global Warming Kill the Reefs?" Available online. URL: http://www.science.org.au/nova/076/076key.htm. Accessed October 17, 2007.

APPENDIX I
Marine Science Further Resources

Marine science is a dynamic field where advances are being made frequently. Issues are being studied and solutions are being delivered by a range of professionals from technologists building AUTONOMOUS UNDERWATER VEHICLES in their garages to seasoned university researchers working with governance to mitigate natural disasters. In order to provide a relevant *Encyclopedia of Marine Science,* numerous print and electronic resources proved invaluable in developing content of sufficient detail to familiarize the reader on the diversity of principles, topics, and ideas that make up marine science. Several resources, such as textbooks and handbooks, were used for many of the entries; Web sites provided the most recent and specialized information. Many of the listed print and electronic resources are suitable for writing research papers or learning a particular topic. In order to deepen understandings about marine science topics provided in this encyclopedia, the reader can investigate the works suggested in "Further Reading" or use some of the basic references listed below.

PRINT RESOURCES

Abel, Daniel C., and Robert L. McConnell. *Environmental Issues In Oceanography,* 2nd ed. Upper Saddle River, N.J.: Prentice Hall, 2002.

Promotes critical thinking on topics ranging from natural resources to population dynamics. Provides a good complement to other references focused on geology, physics, chemistry, biology and ocean processes. A companion Web site may be found at URL: http://wps.prenhall.com/esm_abel_issues-ocean_2. Accessed October 17, 2007.

Bishop, Joseph M. *Applied Oceanography.* New York: Wiley-Interscience, 1984.

An introduction to the field of applied oceanography using basic techniques of physical oceanography to solve real world problems.

Bowditch, Nathaniel. *The American Practical Navigator: An Epitome of Navigation.* Bethesda, Md.: National Geospatial-Intelligence Agency, 2002.

Serves as an encyclopedia of navigation and as a valuable handbook on oceanography and meteorology, and contains useful tables and a maritime glossary.

Charton, Barbara. *A to Z of Marine Scientists.* New York: Facts On File, 2003.

Profiles approximately 140 male and female marine scientists from the beginning of recorded history to modern times.

———. *The Facts On File Dictionary Of Marine Science.* New York: Facts On File, 2008.

Defines more than 3,000 terms across topics such as biology, water chemistry, marine ecosystems, and physical features of the oceans, such as reefs, coastlines, and tides.

Colling, Angela, ed. *Ocean Circulation,* 2nd ed. Milton Keynes, U.K.: The Open University, Butterworth/Heinemann, 2001.

Addresses ocean-atmosphere interactions, global fluxes, surface circulation, and climatic systems with clearly written descriptions and numerous color charts, figures, tables, and images.

Cousteau, Jacques-Yves. *The Ocean World.* New York: Abradale Press/Harry N. Abrams, 1985.

Presents colorfully illustrated written introductions to major oceanographic topics, with many spectacular undersea color photographs.

Davis, Richard A. *Oceanography: An Introduction to the Marine Environment,* 2nd ed. Dubuque, Iowa: Wm. C. Brown, 1991.

Describes the nature of different marine environments.

Dean, Cornelia. *Against the Tide: The Battle for America's Beaches.* New York: Columbia University Press, 1999.

A discussion of engineering efforts to protect America's beaches written by the science editor of the *New York Times.*

Duxbury, Alyn C., Alison B. Duxbury, and Keith A. Sverdrup. *An Introduction to the World's Oceans.* Boston: McGraw-Hill, 2000.

Defines biological, chemical, geological, and physical properties of the world's oceans.

Emery, William J., and Richard E. Thomson. *Data Analysis Methods in Physical Oceanography.* New York: Elsevier Science, 2001.

Identifies approaches to the collection, analysis, and presentation of marine data. Discussions cross traditional descriptive statistics and time series analysis techniques and move into innovations such as wavelets.

Erickson, Jon. *Marine Geology: Exploring the New Frontiers of the Ocean.* New York: Facts On File, 2003.

Examines the origins of the Earth and interrelationship among land features, water processes, and life-forms.

Field, John G., Gotthilf Hempel, and Colin P. Summerhayes. *Oceans 2020: Science, Trends, and the Challenge of Sustainability.* Washington, D.C.: Island Press, 2002.

Discusses social issues, public policy, and ocean management, especially impacts from climate change and overfishing.

Garrison, Tom S. *Oceanography, An Invitation to Marine Science.* Belmont, Calif.: Wadsworth, 1993.

Provides a multidisciplinary overview on marine science topics.

Hsu, Shih-Ang. *Coastal Meteorology.* San Diego, Calif.: Academic Press, Inc., 1988.

Describes the coastal zone as a system involving complex interactions among the atmosphere, ocean, and land.

Knauss, John A. *Introduction to Physical Oceanography,* 2nd ed. Upper Saddle River, N.J.: Prentice Hall/Pearson, 1996.

A graduate level introduction focused on motion in the ocean.

Kunzig, Robert. *Mapping the Deep: The Extraordinary Story of Ocean Science.* New York: Norton, 2000.

A fast-moving account of ocean science and discovery for both marine specialists and general readers.

Kusky, Timothy M. *Encyclopedia of Earth Science.* New York: Facts On File, 2006.

General descriptions of more than 700 topics in hydrology, structural geology, petrology, isotope geology, geochemistry, geomorphology, and oceanography.

Mangone, Gerard J. *Concise Marine Almanac.* New York: Van Nostrand Reinhold, 1986.

A general reference discussing a variety of topics from the dimensions of oceans and marginal seas and available marine resources to the approximate numbers of fishing vessels and gear operated by selected maritime nations.

McCutcheon, Scott, and Bobbi McCutcheon. *The Facts On File Marine Science Handbook.* New York: Facts On File, 2004.

Discusses general topics from acoustics to zooplankton with more than 50 black-and-white photographs and 70 line illustrations.

Mullison, Jerry, Humfrey Melling, H. Paul Freitag, and William Johns. "Climate variability research aided by moored current profilers." In *Sea Technology* 45, no. 2: 17–27. Arlington, Va.: Compass, 2004.

Highlights the use of moored instrumentation to study ocean circulation and climatic patterns. Analysis of data gained during research cruises provides a basis for the study of large-scale wind-driven circulation, dynamics of western boundary currents, global thermohaline circulation and climate variability.

Nichols, Charles R., David L. Porter, and Robert G. Williams. *Recent Advances and Issues in Oceanography.* Westport, Conn.: Greenwood, 2003.

Summarizes several multidisciplinary and integrated oceanographic research programs. Details innovations in ocean observing and provides some biographical sketches for selected United States marine scientists.

Pickard, George L., and William J. Emery. *Descriptive Physical Oceanography: An Introduction,* 5th ed. Oxford: Pergamon Press, 1990.

Introduces the reader to physical properties of seawater, processes influencing the distribution of heat, salt and density in the ocean, and describes the ocean's water masses.

Scavia, Donald, Michael P. Sissenwine, H. Lee Dantzler, Michael J. McPhaden, and Paul F. Moersdorf. "NOAA's ocean observing programs." In *Oceanography* 16, 61–67. Rockville, Md.: The Oceanography Society, 2003.

Identifies selected ocean-observing programs and efforts to integrate those systems into an international network of ocean-observing systems.

Seibold, Eugen, and Wolfgang H. Berger. *The Sea Floor: An Introduction to Marine Geology,* 3rd ed. Berlin: Springer, 1996.

A brief survey of the most important topics in marine geology, with numerous diagrams and photographs explaining the text.

Seidov, Dan, Bernd J. Haupt, and Mark Maslin, eds. *The Oceans and Rapid Climate Change: Past, Present, and Future.* Washington, D.C.: American Geophysical Union, 2001.

Highlights the role of observations and modeling to assess the different components of climate change. Points are made through discussions of paleoclimatic and paleoceanographic reconstructions with the modeling of past and future, colder and warmer climate systems.

Serway, Raymond A., and John W. Jewett, Jr. *Physics for Scientists and Engineers,* 6th ed. Belmont, Calif.: Brooks/Cole-Thomson Learning, 2004.

Discusses classical physics with accompanying media and online resources. Provides excellent detail on Newtonian mechanics, waves, and optics, using calculus. Introduces the reader to new developments and discoveries in physics such as remote sensing.

Thurman, Harold V., and Elizabeth A. Burton. *Introductory Oceanography,* 9th ed. Upper Saddle River, N.J.: Prentice Hall, 2001.

Provides a good summary of the multidisciplinary field of oceanography with broad information on biological, chemical, geological, and physical oceanography.

U.S. Army Corps of Engineers. *Coastal Geology.* EM 1110-2-1810, U.S. Army Corps of Engineers, Vicksburg, Miss., January 31, 1995.

Provides an overview of the science of landforms, structures, rocks, and sediments within the coastal zone. Exposes the reader to data sources and study methods applicable to coastal geological field investigations.

Wright, John, Angela Colling, and Dave Park. *Waves, Tides and Shallow-Water Processes.* Prepared for the Open University Course Team, 2nd ed. Milton Keynes, U.K.: The Open University, Butterworth/Heinemann, 2002.

Up-to-date information on the waves and tides of the ocean, with numerous color illustrations and photographs.

WEB RESOURCES

Glossaries, Online Publications, and Databases

AlgaeBase, Martin Ryan Institute. Available online. URL: http://www.algaebase.org/. Accessed July 31, 2007.

Database of information on algae, especially seaweeds. Repository includes information on sea grasses even though they are flowering plants.

American Meteorological Society (AMS), Glossary of Meteorology. Available online. URL: http://amsglossary.allenpress.com/glossary. Accessed November 16, 2006.

Offers authoritative definitions of more than 12,000 important meteorology and related science terms.

Baum, Steven K. *Glossary of Oceanography and the Related Geosciences with References.* Available online. URL: http://stommel.tamu.edu/~baum/paleo/ocean/ocean.html. Accessed October 4, 2006.

Provides terms and concepts used in oceanography and related fields along with links to online sites that are directly related to many of the entries.

Blackhart, Kristan, David G. Stanton, and Allen M. Shimada. "NOAA Fisheries Glossary." NOAA Technical Memorandum NMFS-F/SPO-69, Office of Science and Technology. Available online. URL: http://www.st.nmfs.gov/st4/documents/FishGlossary.pdf. Accessed October 24, 2007.

Lists definitions for technical terms used by fisheries scientists and managers, especially in programs relevant to protected species, enforcement, and habitat conservation.

Bowditch, Nathaniel. *The American Practical Navigator.* Document NVPUB9V1, Bethesda, Md.: National Geospatial-Intelligence Agency, 2002 bicentennial ed. Available online. URL: http://164.214.12.145/pubs. Accessed January 15, 2007.

Serves as an encyclopedia of navigation and as a valuable handbook on oceanography and meteorology, and contains useful tables and a maritime glossary.

Campbell, David, and Joni Lawrence. MarineBio.org, Inc. Available online. URL: http://marinebio.org. Accessed January 7, 2007.

Serves online news, facts, and photos of marine life.

Central Intelligence Agency. *The World Factbook.* Available online. URL: https://www.cia.gov/cia/publications/factbook/index.html. Accessed October 4, 2006.

Provides national-level information including maps on countries, territories, dependencies, and natural resources.

Chapman, Rick. *The Practical Oceanographer: A Guide to Working At-Sea.* Available online. URL: http://fermi.jhuapl.edu/book/index.html. Accessed October 4, 2006.

Recommends research tasks and procedures for chief scientists, principal investigators, postdoctoral students, graduate students, and undergraduates.

Coastal and Marine Geology Program, U.S. Geological Survey National Center. Available online. URL: http://marine.usgs.gov/. Accessed July 31, 2007.

Describes the geology of coastal and marine systems to help ensure the wise use and protection of the nation's coastal and offshore resources.

Coastal Inlets Research Program, Coastal and Hydraulics Laboratory, U.S. Army Engineer Research and Development Center. Available online. URL: http://cirp.wes.army.mil/cirp/cirp.html. Accessed July 31, 2007.

Investigates federal inlets and entrance channels and responds to authorizations for channel deepening and new channels. Researchers are advancing knowledge and developing predictive technology that supports navigation channel reliability.

Col, Jeananda. "Enchanted Learning." Available online. URL: http://www.EnchantedLearning.com. Accessed October 4, 2006.

Displays simple picture dictionaries on a variety of topics including the natural sciences.

DUCKDATA, Patuxent Wildlife Research Center. Available online. URL: http://www.pwrc.usgs.gov/library/duckdata/. Accessed July 31, 2007.

Compiles scientific literature on the ecology, conservation, and management of North American waterfowl and their wetland habitats.

Froese, Rainer, and Daniel Pauly, eds. FishBase home page. Available online. URL: www.fishbase.org. Accessed December 21, 2006.

Provides general users with access to authoritative biological information on all fish.

Humboldt State University Library. "Research Tools for Searching the Oceanography Literature." Available online. URL: http://library.humboldt.edu/~rls/resocean.htm. Accessed October 4, 2006.

Offers guidance on reference materials that define terminology, provides topical overviews, lists factual information, summarizes current research, and explains standard methods and procedures.

International Association for the Physical Sciences of the Ocean (IAPSO). "The International System of Units (SI) in Oceanography: Papers in marine sciences, no. 45." Available online. URL: http://unesdoc.unesco.org/images/0006/000650/065031eb.pdf. Accessed January 15, 2007.

Identifies physical quantities, units and symbols, and SI units and basic rules for the field of physical sciences of the ocean.

LarvalBase, LarvalBase Project, Leipniz-Institut for Marine Science. Available online. URL: http://www.larvalbase.org. Accessed August 4, 2007.

Provides information on fish larvae from scientific literature, the Internet, and aquaculturists.

Levinton, Jeffrey S. "Glossary of Marine Biology." Available online. URL: http://life.bio.sunysb.edu/marinebio/glossary.html. Accessed October 24, 2007.

Displays an alphabetical listing of marine biology terms.

National Aeronautics and Space Administration. "Destination Earth." Available online. URL: http://earth.nasa.gov. Accessed January 15, 2007.

Reports on research projects that require remote sensing from and into space.

National Environmental Methods Index. Available online. URL: http://www.nemi.gov. Accessed October 4, 2006.

Documents methods and protocols for monitoring water, sediment, air, and tissues.

National Institute of Standards and Technology. "Physics Laboratory: Physical Reference Data, 1994." Available online. URL: http://physics.nist.gov/PhysRefData/contents.html.

Gives information ranging from values of basic constants to procedures to express uncertainty in measurement.

National Oceanic and Atmospheric Administration. "NOAA Celebrates 200 Years of Science, Service, and Stewardship." Available online. URL: http://celebrating200years.noaa.gov/. Accessed August 6, 2007.

Describes historical agencies that were the forerunners of NOAA offices and programs from their origins at the beginning of the 19th century to today.

National Oceanographic Data Center. *World Ocean Atlas 2001.* Available online. URL: http://www.nodc.noaa.gov/OC5/WOA01/pr_woa01.html. Accessed October 4, 2006.

Provides statistics for variables such as in situ temperature, salinity, dissolved oxygen, apparent oxygen utilization, percent oxygen saturation, dissolved inorganic nutrients (phosphate, nitrate, and silicate), chlorophyll at standard depth levels, and plankton biomass sampled from 0 to 200 meters.

———. "World Ocean Database Select and Search." Available online. URL: http://www.nodc.noaa.gov/OC5/SELECT/dbsearch/dbsearch.html. Accessed January 15, 2007.

Allows the user to search World Ocean Database 2005, a set of climatologic fields containing in situ temperature, salinity, dissolved oxygen, Apparent Oxygen Utilization (AOU), percent oxygen saturation, phosphate, silicate, and nitrate at standard depth levels for annual, seasonal, and monthly compositing periods for the World Ocean.

National Ocean Service. "Tide and Current Glossary." Available online. URL: http://tidesandcurrents.noaa.gov/publications/glossary2.pdf. Accessed October 4, 2006.

Describes important terms relevant to horizontal and vertical movements of water that are in response to astronomical forces.

National Sea Grant Library home page. Available online. URL: http://nsgd.gso.uri.edu/searchguide.html. Accessed October 24, 2007.

Archives Sea Grant–funded documents (CD-ROMs, videos, DVDs, and scientific reports) relevant to oceanography, marine education, aquaculture, fisheries, aquatic nuisance species, coastal hazards, seafood safety, limnology, coastal zone management, marine recreation, and maritime law.

National Weather Service. "National Weather Service Glossary." Available online. URL: http://www.nws.noaa.gov/glossary. Accessed October 24, 2007.

Standardizes weather-related terms that may be either heard or used by weather observers and, in particular, storm spotters.

NOAA Central Library, National Oceanic and Atmospheric Administration. *WINDandSEA: The Oceanic and Atmospheric Sciences Internet Guide.* Available online. URL: http://www.lib.noaa.gov/docs/wind/windandsea.html. Accessed January 15, 2007.

Provides more than 1,000 links to science and policy sites common to the National Oceanic and Atmospheric Administration.

Office of Naval Research. "Science and Technology Focus." Available online. URL: http://www.onr.navy.mil/focus/. Accessed October 24, 2007.

Disseminates crafted science and technology facts through links to educational resources for both teachers and students.

Oliver, James K., Marco Noordeloos, Yusri bin Yusuf, Moi Khim Tan, Nasir Bin Nayan, Calvin Foo, and Fadhilatul Shahriyah. *ReefBase: A Global Information System on Coral Reefs.* Available online. URL: http://www.reefbase.org. Accessed August 3, 2007.

Identifies the location, status, threats, monitoring, and management of coral reef resources in more than 100 countries and territories.

Physical Oceanography Program. "Ocean Motion and Surface Currents." Available online. URL: http://oceanmotion.org/. Accessed August 6, 2007.

Describes the surface circulation of Earth's ocean using satellite and numerical model output while highlighting topics relevant to navigation, weather, climate, natural hazards, and marine resources.

Pidwirny, Michael. "PhysicalGeography.Net." Available online. URL: http://www.physicalgeography.net/home.html. Accessed August 5, 2007.

Links to an online textbook of physical geography and a physical geography glossary containing approximately 1,800 definitions.

Soil Science Society of America. "Glossary of Soil Science Terms." Available online. URL: https://www.soils.org/sssagloss/. Accessed August 3, 2007.

Provides definitions, conversion factors, tables, figures, and several appendices relevant to soil science.

Stewart, Robert H. *Our Ocean Planet: Oceanography in the 21st Century.* Available online. URL: http://oceanworld.tamu.edu/resources/oceanography-book/contents2.htm. Accessed October 4, 2006.

Discusses climatic issues such as El Niño, natural hazards such as tsunamis, coastal problems such as pollution, and natural resources such as fisheries.

———. *Introduction to Physical Oceanography.* Available online. URL: http://oceanworld.tamu.edu/resources/ocng_textbook/contents.html. Accessed October 4, 2006.

A college and graduate school level text on oceanography, meteorology, and ocean engineering.

Texas A&M University. *OceanWorld.* Available online. URL: http://oceanworld.tamu.edu. Accessed January 7, 2007.

Provides rich information for students and teachers on a variety of marine science topics such as weather, icebergs, fisheries, coral reefs, waves, and currents.

Toppan, Andrew. *World Navies Today.* Bath, Maine: Toppan. Available online. URL: http://www.hazegray.org/worldnav. Accessed January 15, 2007.

Provides information on seagoing military forces found throughout the world, especially fleets of ships belonging to a particular country. Vessels are broken down in accordance to types such as surface combatants, submarines, and auxiliaries.

Tortell, Philip, and Larry Awosika. "Oceanographic Survey Techniques and Living Resources Assessment Methods, Intergovernmental Oceanographic Commission Manuals and Guides No. 32, Paris, France: United Nations Educational, Scientific and Cultural Organization, 1996." Available online. URL: http://ioc.unesco.org/iocweb/ioc pub/iocpdf/m032.pdf. Accessed October 4, 2006.

Guides marine scientists on the uses of navigational systems and measurement techniques to quantify marine features and processes.

United Nations Foundation. *United Nations Atlas of the Oceans.* Available online. URL: http://www.oceansatlas. org. Accessed January 6, 2007.

Presents images, statistics, and encyclopedic information pertaining to the sustainable development of the oceans.

United States Army Corps of Engineers. "CIRP, Coastal Inlets Research Program." Available online. URL: http:// cirp.wes.army.mil/cirp/cirp.html. Accessed December 30, 2006.

Provides information on research and development tools, which reduce costs associated with stabilizing or dredging federal inlets and channels. Products include predictive numerical models of waves, currents, sediment transport, morphology change, and water-structure interaction, a coastal inlets database, and technical publications describing both applied and basic research advances. CIRP periodically hosts technology-transfer workshops, and training with CIRP technologies.

———. "Coastal Engineering Manual: Part I, II, III, IV, V, and Appendix A, EM 1110-2-1100, CECW-EW." Available online. URL: http://www.usace.army.mil/ publications/eng-manuals/em.htm. Accessed October 4, 2006.

Explains tools and procedures to plan, design, construct, and maintain coastal projects. Includes the basic principles of coastal processes, methods for computing coastal planning and design parameters, and guidance on how to formulate and conduct studies in support of coastal flooding, shore protection, and navigation projects.

United States Naval Observatory, Astronomical Applications Department (AAD). "Phases of the Moon and Percent of the Moon Illuminated." Washington, D.C.: AAD. Available online. URL: http://aa.usno.navy.mil/ faq/docs/moon_phases.html. Accessed January 15, 2007.

Provides pictures, animations, and comments regarding primary phases of the Moon, fraction of the Moon illuminated and complete Sun and Moon data for a particular day.

U.S. Geological Survey, Coastal and Marine Geology Program. "Marine Realms Information Bank." Available online. URL: http://mrib.usgs.gov. Accessed January 15, 2007.

Classifies, integrates, and facilitates access to scientific information about the oceans and the adjacent parts of the atmosphere and solid Earth, as well as to the people, techniques, and organizations involved in marine science.

———. "Water Resources of the United States." Available online. URL: http://water.usgs.gov/. Accessed August 5, 2007.

Disseminates science-based information on water resources of the United States.

University of Rhode Island, Office of Marine Programs. *Census of Marine Life.* Available online. URL: http:// www.coml.org/coml.htm. Accessed August 5, 2007.

Reports results of multinational scientific collaborations (more than 80 nations) to catalog as many species of the world's oceans as possible, and assess their diversity, distribution, and abundance.

Water Encyclopedia. "Science and Issues." Available online. URL: http://www.waterencyclopedia.com. Accessed October 4, 2006.

Defines key concepts, provides essays, and offers links for topics ranging from acid mine drainage to women in water sciences.

Water Quality Association. "Glossary of Terms." Available online. URL: http://www.wqa.org/glossary.cfm. Accessed July 31, 2007.

Provides short definitions of words ranging from atomic absorption spectroscopy to zooplankton.

Professional Associations Relevant to Marine Science

Acoustical Society of America, Melville, NY. Available online. URL: http://asa.aip.org. Accessed July 31, 2007.

An organization of more than 7,000 men and women who work in acoustics and have interests in acoustical standards, measurement procedures, and criteria for determining the effects of noise and vibration.

American Academy of Underwater Sciences, Dauphin Island Sea Lab, Dauphin Island, AL. Available online. URL: http://www.aaus.org. Accessed August 6, 2007.

A body of marine scientists interested in developing, reviewing and revising standards for safe scientific diving certification and the safe operation of scientific diving programs.

American Fisheries Society, Bethesda, MD. Available online. URL: http://www.fisheries.org. Accessed July 31, 2007.

Fisheries scientists organized and focused on promoting the conservation, development, and wise use of the fisheries.

American Geophysical Union, Washington, DC. Available online. URL: http://www.agu.org. Accessed July 31, 2007.

A body of more than 45,000 scientists focused on the scientific study of Earth and its environment in space and to pass results to the public. Efforts to advance the various geophysical disciplines are made through scientific discussion, publication, and dissemination of information.

American Malacological Society. Available online. URL: http://www.malacological.org. Accessed August 7, 2007.

Members promote the scientific study of mollusks through the presentation and discussion of research results. The society publishes *The American Malacological Bulletin* and *AMS News.*

American Shore and Beach Preservation Association, Fort Myers, FL. Available online. URL: http://www.asbpa.org/. Accessed August 5, 2007.

A group dedicated to preserving, protecting and enhancing the beaches, shores and other coastal resources of America. This association publishes *Shore & Beach,* a quarterly, peer-reviewed journal.

American Society of Civil Engineers, ASCE World Headquarters, Reston, VA. Available online. URL: http://www.asce.org/asce.cfm. Accessed August 5, 2007.

The oldest national engineering society in the United States with more than 140,000 members worldwide. The society's publications include *Journal of Environmental Engineering, Journal of Hydrologic Engineering,* and *Journal of Water Resources Planning and Management.*

American Society of Limnology and Oceanography, Waco, TX. Available online. URL: http://aslo.org. Accessed July 31, 2007.

A society for the promotion of limnology, oceanography, and related sciences. It is best known for its publications such as *Limnology and Oceanography, Limnology and Oceanography: Methods,* and *Limnology and Oceanography Bulletin.*

Association of State Wetland Managers, Windham, ME. Available online. URL: http://www.aswm.org. Accessed July 31, 2007.

A nonprofit membership organization focused on protection and management of wetland resources, and promoting the application of sound science to wetland management efforts.

The Australian Hydrographers' Association, Cooma, New South Wales, Australia. Available online. URL: http://www.aha.net.au. Accessed July 31, 2007.

An organization promoting educational and technical standards within the field of hydrography. This group hosts the National Australian Hydrographers' Association Conference and publishes *Australasian Hydrographer,* a quarterly journal.

Australian Marine Sciences Association, Kilkivan Queensland, Australia. Available online. URL: http://www.amsa.asn.au. Accessed July 31, 2007.

A national nonprofit organization dedicated to promoting marine science and coordinating discussion and debate of marine issues in Australia.

British Marine Life Study Society, Shoreham-by-Sea, Sussex, UK. Available online. URL: www.glaucus.org.uk. Accessed July 31, 2007.

Members of this society are interested in studying and protecting marine fauna and flora that inhabit littoral and coastal regions common to the British Isles.

Canadian Meteorological and Oceanographic Society, Ottawa, Ontario, Canada. Available online. URL: http://www.cmos.ca. Accessed July 31, 2007.

The national society of individuals dedicated to advancing atmospheric and oceanic sciences and related environmental disciplines in Canada.

Chinese Society for Oceanology and Limnology, Qingdao, People's Republic of China. Available online. URL: http://csol.qdio.ac.cn/#. Accessed July 31, 2007.

The Chinese Society of Oceanology and Limnology publishes the *Chinese Journal of Oceanology and Limnology,* quarterly. The society is affiliated with the Chinese-American Oceanic and Atmospheric Association in Silver Spring, Maryland.

Cousteau Society, Hampton, VA. Available online. URL: http://www.cousteau.org. Accessed July 31, 2007.

The Cousteau Society is a membership organization focused on educating people to understand and protect the hydrosphere for the well-being of future generations. Operations involve expeditions around the globe in fresh, brackish, and saltwater environments.

The Crustacean Society, Lawrence, KS. Available online. URL: http://www.vims.edu/tcs. Accessed July 31, 2007.

The mission of the Crustacean Society is to advance the study of all aspects of the biology of the Crustacea. Publications include the *Journal of Crustacean Biology* and *The Ecdysiast,* a newsletter.

Estuarine Research Federation, Port Republic, MD. Available online. URL: http://www.erf.org. Accessed July 31, 2007.

This organization provides the framework to advance scientific understandings of estuarine and coastal ecosystems worldwide through scientific exchanges and technical publications such as *Estuaries and Coasts*. One goal involves using marine science and education to promote estuarine and coastal stewardship.

The German Society for Marine Research (Deutsche Gesellschaft für Meeresforschung), Hamburg, Germany. Available online. URL: http://www.meeresforschung.de/DGM/de/news/dgmnews_de.html. Accessed July 31, 2007.

This society facilitates the exchange of information and opinions concerning marine research topics.

Indian Underwater Robotics Society, Uttar Pradesh, India. Available online. URL: http://www.iurs.org. Accessed July 31, 2007.

A society committed to education and research and development in the field of robotics, especially autonomous underwater robotic technologies.

Institute of Marine Engineering, Science & Technology, London, UK. Available online. URL: http://www.imarest.org. Accessed July 31, 2007.

A body of more than 16,000 members worldwide focused on staying current on recent marine technology advances and issues. The society publishes trade magazines such as *Marine Engineers Review* and journals such as the *Journal of Offshore Technology* and *The Marine Scientist*.

International Federation of Hydrographic Societies, Plymouth, UK. Available online. URL: http://www.hydrographicsociety.org. Accessed July 31, 2007.

A global organization promoting the science of hydrographic surveying as well as recognizing standards of education and training for those engaged in or pursuing careers in hydrographic work.

International Society for Reef Studies, Lawrence, KS. Available online. URL: http://www.fit.edu/isrs. Accessed July 31, 2007.

The society produces and distributes scientific knowledge pertinent to coral reefs, both living and fossil. The society holds conferences and meetings and publishes the journal *Coral Reefs* and the newsletter *Reef Encounter*.

International Society of Protistologists, Tuscaloosa, AL. Available online. URL: http://www.uga.edu/protozoa. Accessed July 31, 2007.

An organization of scientists worldwide who perform research on protists, single-celled eukaryotic organisms. Society publications include *The Journal of Eukaryotic Microbiology* and *Stentor*, an electronic newsletter.

Japanese Coral Reef Society, Tokyo, Japan. Available online. URL: http://wwwsoc.nii.ac.jp/jcrs/english. Accessed July 31, 2007.

A society focused on coral reef conservation and management in Japan and publishing the journal *Galaxea* and quarterly newsletters.

The Korean Society of Oceanography, Seoul, Korea. Available online. URL: http://ksocean.or.kr. Accessed July 31, 2007.

A national society that hosts professional meetings and publishes the *Journal of the Korean Society of Oceanography*.

Marine Technology Society, Columbia, MD. Available online. URL: http://www.mtsociety.org, Accessed July 31, 2007.

An international organization of scientists, policy makers, engineers, technologists, and students endeavoring to study and explore the ocean and to create a broader understanding of the importance of marine science. The society publishes the *Marine Technology Society Journal*, a bi-monthly newsletter, and conference proceedings

National Marine Educators Association, Ocean Springs, MS. Available online. URL: http://www.marine-ed.org. Accessed July 31, 2007.

A professional organization delivering aquatic, marine, and ocean science education nationally and regionally through more than 1,100 members and 17 state chapters. Members receive *CURRENT: The Journal of Marine Education* and *NMEA News*.

New Zealand Marine Sciences Society, Christchurch, New Zealand. Available online. URL: http://nzmss.rsnz.org. Accessed July 31, 2007.

A society fostering understanding and appreciation for marine science. The society provides exchanges of information within the marine science community, encourages and assists marine science students and young scientists, and provides advice to government on marine policy issues.

Oceanographic Society of Japan, Tokyo, Japan. Available online. URL: http://wwwsoc.nii.ac.jp/kaiyo/mt3. Accessed July 31, 2007.

Membership society founded in 1941 to assure the growth of oceanography as a science. Holds workshops and seminars, publishes the scientific periodical *The Journal of Oceanography*, and presents awards for the encouragement of oceanographic research.

The Oceanography Society, Rockville, MD. Available online. URL: http://www.tos.org. Accessed July 31, 2007.

Founded in 1988 to disseminate knowledge of oceanography and its application through research and education. Publishes the journal *Oceanography*, holds meetings to exchange knowledge and promote communication among oceanographers, and gives awards in recognition of distinguished research in and contributions to oceanography.

Scottish Association for Marine Science, Dunstaffnage Marine Laboratory, Oban, Argyll, UK. Available online. URL: http://www.sams.ac.uk. Accessed July 31, 2007.

An internationally renowned marine research establishment focused on applying multidisciplinary and integrated scientists to answer questions relevant to the marine realm from Scottish coastal waters to the Arctic Ocean.

La Société Hydrotechnique de France, Paris, France. Available online. URL: http://www.shf.asso.fr/index2.htm. Accessed July 31, 2007.

Supports the progress and the development of knowledge and scientific culture in all the fields of water resources and sciences. Produces the journal *Hydroelectric Power* and sponsors a bi-annual gathering of hydraulicians, mechanics, hydrologists, geologists, and meteorologists.

Society for Indian Ocean Studies, New Delhi, India. Available online. URL: http://www.sios-india.org. Accessed July 31, 2007.

Founded in 1987 to promote the study of problems common to the Indian Ocean. Issues of importance are connected with its history, geography, living and nonliving resources, legal regime, as well as strategic, scientific, technical, social, and economic aspects.

Society for Marine Mammalogy, San Francisco, CA. Available online. URL: http://www.marinemammalogy.org. Accessed July 31, 2007.

Evaluates and promotes the educational, scientific and managerial advancement of marine mammal science. Publishes the *Journal of Marine Mammal Science* and the society's newsletter for use by its members.

Western Indian Ocean Marine Science Association, Zanzibar, Tanzania. Available online. URL: http://www. wiomsa.org. Accessed July 31, 2007.

Promotes the educational, scientific and technological development of all aspects of marine sciences throughout the Western Indian Ocean region, including research activities, publications, and training.

APPENDIX II
Selected Major Oceans, Seas, Gulfs, and Bays Worldwide

Name	Area		Average Width		Maximum Depth	
	MILES²	KM²	MILES	KM	FATHOMS	METERS
Adriatic Sea	50,599	131,050	110	175	724	1,324
Aegean Sea	83,000	214,000	186	299	164	300
Arctic Ocean	5.427 x 10⁶	14.056 x 10⁶	circumpolar		2,980	5,450
Atlantic Ocean	29.638 x 10⁶	76.762 x 10⁶	4,101	6,600	4,705	8,605
Baltic Sea	149,000	386,000	120	193	251	459
Barents Sea	542,000	1.405 x 10⁶	650	1,050	328	600
Bay of Bengal	838,614	2.172 x 10⁶	1,069	1,720	2902	5,258
Bay of Fundy	115,831	300,000	62	100	117	214
Beaufort Sea	184,000	476,000	385	620	2,560	4,682
Bering Sea	884,900	2.292 x 10⁶	1,233	1,984	2,610	4,773
Black Sea	178,000	461,000	380	610	1,208	2,210
Cardigan Bay	1,940	5,024	36	60	44	80
Caribbean Sea	1.049 x 10⁶	2.178 x 10⁶	404	650	3,798	6,946
Chaleur Bay	2,455	6,358	25	40	34	62
Chesapeake Bay	2,317	6,000	62	100	32	58
Coral Sea	581,084	1.505 x 10⁶	746	1,200	2,822	5,160
East China Sea	479,925	1.243 x 10⁶	497	800	1,486	2,718
Flores Sea	53,465	138,474	149	240	2,800	5,121
Greenland Sea	327,822	849,056	497	800	2,018	3,690
Gulf of Guinea	591,895	1.533 x 10⁶	1,181	1,900	3,479	6,363
Gulf of Mexico	595,756	1.543 x 10⁶	124	200	2,203	4,029
Indian Ocean	26.470 x 10⁶	68.556 x 10⁶	4,350	7,000	3,969	7,258
Ionian Sea	48,494	125,600	171	275	2,717	4,968
Irish Sea	40,155	104,000	150	240	96	175
Kara Sea	303,090	785,000	186	300	328	600
Labrador Sea	422,010	1.093 x 10⁶	482	775	2,297	4,200

Marmara Sea	4,300	11,137	37	60	810	1,482
Mediterranean Sea	1.145 x 10⁶	2.966 x 10⁶	311	500	2,533	4,632
Mollucca Sea	74,807	193,750	155	250	2,242	4,100
North Sea	220,001	569,800	342	550	400	731
Norwegian Sea	407,724	1.056 x 10⁶	482	775	2,171	3,970
Okhotsk Sea	589,964	1.528 x 10⁶	404	650	1,843	3,371
Pacific Ocean	60.061 x 10⁶	155.557 x 10⁶	8,202	13,200	5,966	10,911
Persian Gulf	93,051	241,000	35	56	93	170
Philippine Sea	1,767,576	4.578 x 10⁶	932	1,500	4,374	8,000
Red Sea	170,001	440,300	219	352	1,167	2,134
Ross Sea	193,978	502,400	482	775	2,078	3,800
Sargasso Sea	2,000,009	5.180 x 10⁶	1,553	2,500	2,400	4,389
Scotia Sea	1,745,954	4.522 x 10⁶	621	1,000	3,488	6,378
Sea of Japan	501,933	1.300 x 10⁶	497	800	2,168	3,965
Solomon Sea	109,112	282,600	249	400	5,468	10,000
South China Sea	895,400	2.319 x 10⁶	597	960	2,641	5,016
Southern Ocean	7.848 x 10⁶	20.327 x 10⁶	circumpolar		3,956	7,235
Sulu Sea	6,178	16,000	233	375	3,336	6,100
Tasman Sea	1,212,361	3.140 x 10⁶	1367	2,200	2,880	5,267
Timor Sea	124,146	321,536	298	480	1,969	3,600
Weddell Sea	436,450	1.130 x 10⁶	1,429	2,300	2,641	4,830
Yellow Sea	480,002	1.243 x 10⁶	398	640	50	91

NOTES:

Average widths are provided for oceans and marginal seas. The width of the mouth is provided for gulfs and bays.

Area, width, and depth measurements are approximate and useful in comparing some of the water bodies that make up the oceans, an area of approximately 525.508 x 10⁶ mi² (361.059 x 10⁶ km²).

For additional information, see entries on the specific ocean or the entry on MARGINAL SEAS. (Source: Adapted from *CIA World Factbook, Concise Marine Almanac, National Geographic Atlas of the World*, and *Introduction to Physical Oceanography*.)

APPENDIX III
Classification of Marine Organisms

By Biological Taxonomy	By Lifestyle
Monera (bacteria and blue-green algae)	**Plankton** (passively floating or drifting organisms) bacterioplankton phytoplankton zooplankton holoplankton (Spend whole lifecycle in plankton) meroplankton (live in plankton only during a few stages of lifecycle)
Protista (single-celled organisms with a nucleus) Major examples include: foraminifera dinoflagellates coccolithophores diatoms	
Fungi (decomposers found growing on sea grasses, protozoans, driftwood, corals, sea foam, and many other substrates)	**Nekton** (freely swimming organisms) **Invertebrates** crustaceans mollusks (squid) **Vertebrates** fish reptiles birds **Mammals** whales pinnipeds sea cows
Metaphytae (plants that grow attached to the seafloor) green, red, and brown seaweeds flowering, seed-bearing plants mangroves	
Metazoa (Animalia, i.e., all the multi-cellular animals) **Invertebrates** crustaceans mollusks (squid) **Vertebrates** fish (Chondrichthyes and Osteichthyes) reptiles birds mammals: whales pinnipeds sea cows	**Benthos** (organisms living attached to, on, or in the seafloor) epifauna infauna

Note: Classification systems, regardless of how complex, are only a guide, especially since there are many exceptions. Some organisms such as barnacles may have different classifications based on their life cycle. For more detailed information, see the Ocean Biogeographic Information System, or OBIS, online at URL: http://iobis.org.

OBIS was established during 1999 as part of the Census of Marine Life program.

APPENDIX IV

Periodic Table of the Elements

Numbers in parentheses are atomic mass numbers of most stable isotopes.

| | 57
La
138.9055 | 58
Ce
140.115 | 59
Pr
140.908 | 60
Nd
144.24 | 61
Pm
(145) | 62
Sm
150.36 | 63
Eu
151.966 | 64
Gd
157.25 | 65
Tb
158.9253 | 66
Dy
162.500 | 67
Ho
164.9303 | 68
Er
167.26 | 69
Tm
168.9342 | 70
Yb
173.04 |

Lanthanides / Actinides rows

© Infobase Publishing

The Chemical Elements

Element	Element	Atomic number	Element	Element	Atomic number
actinium	Ac	89	molybdenum	Mo	42
aluminum	Al	13	neodymium	Nd	60
americium	Am	95	neon	Ne	10
antimony	Sb	51	neptunium	Np	93
argon	Ar	18	nickel	Ni	28
arsenic	As	33	niobium	Nb	41
astatine	At	85	nitrogen	N	7
barium	Ba	56	nobelium	No	102
berkelium	Bk	97	osmium	Os	76
beryllium	Be	4	oxygen	O	8
bismuth	Bi	83	palladium	Pd	46
bohrium	Bh	107	phosphorus	P	15
boron	B	5	platinum	Pt	78
bromine	Br	35	plutonium	Pu	94
cadmium	Cd	48	polonium	Po	84
calcium	Ca	20	potassium	K	19
californium	Cf	98	praseodymium	Pr	59
carbon	C	6	promethium	Pm	61
cerium	Ce	58	protactinium	Pa	91
cesium	Cs	55	radium	Ra	88
chlorine	Cl	17	radon	Rn	86
chromium	Cr	24	rhenium	Re	75
cobalt	Co	27	rhodium	Rh	45
copper	Cu	29	rubidium	Rb	37
curium	Cm	96	ruthenium	Ru	44
darmstadtium	Ds	110	rutherfordium	Rf	104
dubnium	Db	105	samarium	Sm	62
dysprosium	Dy	66	scandium	Sc	21
einsteinium	Es	99	seaborgium	Sg	106
erbium	Er	68	selenium	Se	34
europium	Eu	63	silicon	Si	14
fermium	Fm	100	silver	Ag	47
fluorine	F	9	sodium	Na	11
francium	Fr	87	strontium	Sr	38
gadolinium	Gd	64	sulfur	S	16
gallium	Ga	31	tantalum	Ta	73
germanium	Ge	32	technetium	Tc	43
gold	Au	79	tellurium	Te	52
hafnium	Hf	72	terbium	Tb	65
hassium	Hs	108	thallium	Tl	81
helium	He	2	thorium	Th	90
holmium	Ho	67	thulium	Tm	69
hydrogen	H	1	tin	Sn	50
indium	In	49	titanium	Ti	22
iodine	I	53	tungsten	W	74
iridium	Ir	77	ununbium	Uub	112
iron	Fe	26	ununhexium	Uuh	116
krypton	Kr	36	ununoctium	Uuo	118
lanthanum	La	57	ununpentium	Uup	115
lawrencium	Lr	103	ununquadium	Uuq	114
lead	Pb	82	ununseptium	Uus	117
lithium	Li	3	ununtrium	Uut	113
lutetium	Lu	71	unununium	Uuu	111
magnesium	Mg	12	uranium	U	92
manganese	Mn	25	vanadium	V	23
meitnerium	Mt	109	xenon	Xe	54
mendelevium	Md	101	ytterbium	Yb	70
mercury	Hg	80	yttrium	Y	39
			zinc	Zn	30
			zirconium	Zr	40

APPENDIX V

The Geologic Timescale

Era	Period	Epoch	Age (millions of years)	First Life-forms	Geology
	Quaternary	Holocene	0.01		
Cenozoic		Pleistocene	3	Humans	Ice age
		Pliocene	11	Mastodons	Cascades
		Neogene			
		Miocene	26	Saber-toothed tigers	Alps
	Tertiary	Oligocene	37		
		Paleogene			
		Eocene	54	Whales	
		Paleocene	65	Horses, Alligators	Rockies
	Cretaceous		135		
				Birds	Sierra Nevada
Mesozoic	Jurassic		210	Mammals	Atlantic
				Dinosaurs	
	Triassic		250		
	Permian		280	Reptiles	Appalachians
	Pennsylvanian		310		Ice age
				Trees	
	Carboniferous				
Paleozoic	Mississippian		345	Amphibians	Pangaea
				Insects	
	Devonian		400	Sharks	
	Silurian		435	Land plants	Laursia
	Ordovician		500	Fish	
	Cambrian		544	Sea plants	Gondwana
				Shelled animals	
			700	Invertebrates	
Proterozoic			2500	Metazoans	
			3500	Earliest life	
Archean			4000		Oldest rocks
			4600		Meteorites

APPENDIX VI
Selected Listing of Marine Laboratories

Laboratory	Research Thrusts	Web Address	Location
Playa Unión Photobiology Station	Photobiology (the study of interactions of light with living organisms)	Available online. URL: http://www.efpu.org.ar/. Accessed on August 14, 2007.	Argentina
Australian Institute of Marine Science	Tropical ocean and marine technology	Available online. URL: http://www.aims.gov.au/. Accessed on August 14, 2007.	Australia
Bellairs Research Institute	Coral reef studies	Available online. URL: http://www.mcgill.ca/bellairs/. Accessed on August 14, 2007.	Barbados
Perry Institute for Marine Science	Conservation, ecosystems, and biodiversity	Available online. URL: http://www.perryinstitute.org/. Accessed on August 14, 2007.	Bahamas
Bermuda Institute of Ocean Sciences	Marine biology and ocean observations	Available online. URL: http://www.bbsr.edu/. Accessed on August 14, 2007.	Bermuda
Huntsman Marine Science Center	Marine resources	Available online. URL: http://www.huntsmanmarine.ca/. Accessed on August 14, 2007.	Canada
Instituto Antartico Chileno	Polar science	Available online. URL: http://www.inach.cl/InachWebNeo/index.aspx?channel=6139. Accessed on August 14, 2007.	Chile
Marine Biology Station, Puntarenas	Sustainable development	Available online. URL: http://www.una.ac.cr/biol/unaluw/ebm/. Accessed on August 14, 2007.	Costa Rica
Institute of Oceanography and Fisheries	Oceanographic research and data center	Available online. URL: http://www.izor.hr/izor.html. Accessed on August 14, 2007.	Croatia
Marine Biological Laboratory	Fish physiology and ecology	Available online. URL: http://www.mbl.ku.dk/. Accessed on August 14, 2007.	Denmark
Institute for Tropical Marine Ecology	Marine biology	Available online. URL: http://www.itme.org/. Accessed on August 14, 2007.	Dominica

Charles Darwin Research Station	Conservation	Available online. URL: http://www.darwinfoundation.org/. Accessed on August 14, 2007.	Equador
Estonian Marine Institute	Marine biology	Available online. URL: http://www.sea.ee/. Accessed on August 14, 2007.	Estonia
Archipelago Research Institute	Baltic Sea studies	Available online. URL: http://www.utu.fi/erill/saarmeri/. Accessed on August 14, 2007.	Finland
Observatoire Océanologique de Banyuls-sur-mer	Biochemical processes	Available online. URL: http://www.obs-banyuls.fr/. Accessed on August 14, 2007.	France
Alfred Wegener Institute	Polar science	Available online. URL: http://www.awi.de/de/. Accessed on August 14, 2007.	Germany
Hellenic Center for Marine Research	Marine science	Available online. URL: http://www.hcmr.gr/index.html. Accessed on August 14, 2007.	Greece
Marine Research Institute	Marine biology	Available online. URL: http://www.hafro.is/. Accessed on August 14, 2007	Iceland
Marine Institute	Marine science	Available online. URL: http://www.marine.ie/home/. Accessed on August 14, 2007.	Ireland
Israel Oceanographic and Limnological Research	Hydrogeology, biogeochemical cycles, and environmental monitoring	Available online. URL: http://www.ocean.org.il/. Accessed on August 14, 2007.	Israel
SACLANT Undersea Research Center	Underwater acoustics	Available online. URL: http://www.saclantc.nato.int/. Accessed on August 14, 2007.	Italy
Mutsu Institute for Oceanography	Paleoenvironmental reconstruction	Available online. URL: http://www.jamstec.go.jp/jamstec-e/mutu/. Accessed on August 14, 2007.	Japan
Korea Ocean Research and Development Institute	Marine technologies	Available online. URL: http://www.kordi.re.kr/eng2006/bin/main.asp. Accessed on August 14, 2007.	Korea
National Center for Marine Sciences	Living resources	Available online. URL: http://www.cnrs.edu.lb/research/marinesciences.html. Accessed on August 14, 2007.	Lebanon
Marine Environment Laboratory	Chemistry and marine pollution	Available online. URL: http://www-naweb.iaea.org/naml/. Accessed on August 14, 2007.	Monaco
Cawthron Instititute	Marine ecology	Available online. URL: http://www.cawthron.org.nz/. Accessed on August 14, 2007.	New Zealand
Norwegian Institute for Water Research	Use and protection of water bodies	Available online. URL: http://www.niva.no/. Accessed on August 14, 2007.	Norway
Marine Science Institute	Marine conservation	Available online. URL: http://www.msi.upd.edu.ph/web/. Accessed on August 14, 2007.	Philippines

(continues)

(continued)

Laboratory	Research Thrusts	Web Address	Location
P. P. Shirshov Institute of Oceanology	Oceanography	Available online. URL: http://www.sio.rssi.ru/. Accessed on August 14, 2007.	Russia
University Marine Biological Station Millport	Fishery-related biology	Available online. URL: http://www.gla.ac.uk/ Acad/Marine/. Accessed on August 14, 2007.	Scotland
JLB Smith Institute of Ichthyology	Fisheries biology	Available online. URL: http://cdserver2.ru.ac.za/ cd/011120_1/Aqua/Ichthyology/Ichthyology/JLB. htm. Accessed on August 14, 2007.	South Africa
Marine Zoology Unit	Conservation biology	Available online. URL: http://www.uv.es/cava-nilles/zoomarin/index.htm. Accessed on August 14, 2007.	Spain
Kristineberg Marine Research Station	Marine ecology	Available online. URL: http://www.kmf.gu.se/. Accessed on August 14, 2007.	Sweden
Institute of Marine Affairs	Marine science and technology	Available online. URL: http://www.ima.gov.tt/. Accessed on August 14, 2007.	Trinidad and Tobago
Marine Hydrophysical Institute	Physics and chemistry	Available online. URL: http://www.mhi.iuf.net/. Accessed on August 14, 2007.	Ukraine
Plymouth Marine Laboratory	Marine ecosystems	Available online. URL: http://www.pml.ac.uk/. Accessed on August 14, 2007.	United Kingdom
Duke University Marine Laboratory	Ocean and estuarine science	Available online. URL: http://www.nicholas. duke.edu/marinelab/. Accessed on August 14, 2007.	United States-East Coast
Mote Marine Laboratory	Ocean science and fisheries	Available online. URL: http://www.mote.org/. Accessed on August 14, 2007.	United States-Gulf Coast
Hopkins Marine Station	Marine biology	Available online. URL: http://www-marine.stan-ford.edu/. Accessed on August 14, 2007.	United States-West Coast

Note: For additional information on marine laboratories see organizations such as the National Association of Marine Laboratories (available online at URL: http://www.mbl.edu/naml/), the Mediterranean Science Commission (available online at URL: http://www.ciesm.org/), or the Intergovernmental Oceanographic Commission of the United Nations Educational, Scientific and Cultural Organization (UNESCO) (available online at URL: http://ioc.unesco.org/iocweb/index.php). All accessed August 14, 2007.

APPENDIX VII
Modern Map of the World's Oceans

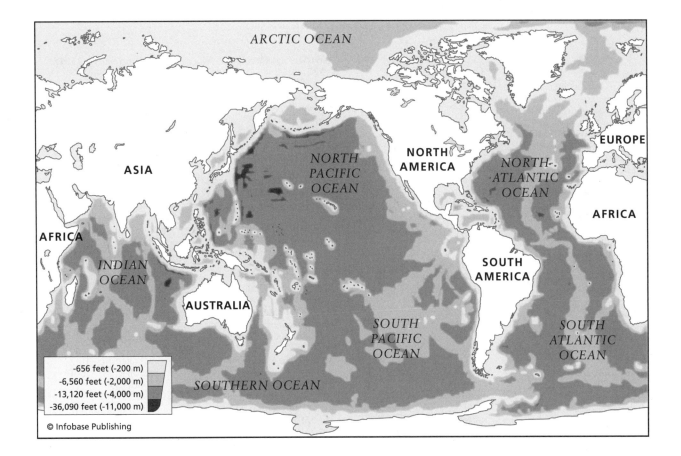

INDEX

Note: **Boldface** page numbers indicate main entries. *Italic* page numbers indicate illustrations.

thermal pollution from
308
types 169–170
wastewater 170
nuclear weapons
Marshall Islands tests 44
radioactive waste disposal
306–307
nudibranchs 345, 575
Nuestra Señora de Atocha
archaeological site 287
numerical (computer)
modeling 382–384. *See also*
Fourier analysis
and chaos theory 85
in climate forecasting
90–91
in current forecasting
361
in geophysics 396–397
of halocline 473
uses 344
in weather forecasting
178
NURP1 462
nutria 251, 298, 401
nutrient cycles 384, 384–385.
See also carbon cycle;
nitrogen cycle
nutrient regeneration
385–386
NWS. *See* National Weather
Service

O

oaks, on dunes 50
observatories. *See* Lamont-
Doherty Earth Observatory;
Longterm Ecosystem
Observatory at 15m;
long-term environmental
observatory; Martha's
Vineyard Coastal
Observatory; ocean
observatory; Sargasso
Sea Ocean/Atmosphere
Observatory
Occupational Safety and
Health Administration
(OSHA) 536, 537
ocean(s). *See also* Arctic
Ocean; Atlantic Ocean;
Indian Ocean; Pacific
Ocean; Southern Ocean
average depth 52, 388
heat budget 213–214
pH of 426
salinity 471, 471
Ocean Alliance drill platform
410
ocean basins 387, 387–390,
389
ocean conditions, forecasting
of 179, 361, 382
Ocean Drilling Program
(ODP) 345, 390–391, 391,
404, 570
Ocean Dumping Act of 1972
540
ocean engineers 391–392
ocean floor, research on
187–188

Oceanic and Atmospheric
Research (OAR)
laboratories 355, 357
oceanic crust. *See also* mid-
ocean spreading ridges;
plate tectonics; subduction
zone
age of 279, 339, 435
density of 252, 339,
344–345
depth of sediment on 436
and isostatic equilibrium
252–253
thickness of 252, 279,
334, 387, 401
Oceanic Engineering Society
322
oceanic province 150
oceanic zone 392
ocean observatory 392–394,
393
maritime commerce and
304
Ocean Observatory
Initiative 327
Oceanographer of the Navy
337
oceanographers
biological 62–63, 288,
400, 401, 440
chemical 86–87, 176,
184, 400, 401
fisheries management
201
geological 187–188, 400,
401
physical 184, 205, 395,
402, 428–430, 430
professional associations
22, 402–403
oceanographic instruments
394–400
digital, value of 400
in situ sensors 392
for life-forms study 395
for physical properties
study 397–399
for seafloor study
396–397
for seawater study
395–396
testing and maintenance
395
oceanographic libraries 361
oceanographic moorings,
anchoring of 27–28, 28
oceanography 400–402
acoustical 18–19
history 400
subjects of study 22, 400
subspecialties 400,
401–402
Oceanography Society, The
133, 402–403
Oceanography Society
Fellows Program 402
ocean sciences
disciplines included in 21
subjects of study 21
Oceans Conference, The 244
Ocean Star drill platform
410
ocean sunfish 205

ocean surface current radar.
See OSCR
Ocean.US 202, 393, 403
ocean wind waves 195
Octopoda 82. *See also*
octopus
octopus
blue-ringed 82
characteristics 81, 82
classification of 80, 81,
345
commercial importance
82
Giant Pacific 82
Octopus Hole Conservation
Area 311
ODP. *See* Ocean Drilling
Program
Odyssey (AUV) 395
Office of Naval Research
(ONR) 403, 403–404
Mobile Offshore Base
project 49
Naval Research
Laboratory 362
research 364
research funding 87, 244,
303, 327, 429–430
Office of Response and
Restoration (NOAA) 321
offshore drilling 404–408,
405. *See also* Deep-Sea
Drilling Project; oil
production platforms
in Antarctica 31, 33
in Arctic 33
contour currents and
110
diving and 144, 255
drill platform design 391
drill platform types 405,
406
eddies and 145, 152
exploration 406–407,
434–435
in Gulf of Mexico 145,
203, 275, 406
history 406
internal waves and 247
in North Sea 33
offshore region 98
Offshore Sea-Barge (SeaBee)
106
offshore structures, design
of 391
offshore zone, of barrier
island 50
Ohio class submarines 100
oil. *See also* offshore drilling;
oil production platforms
formation of 33, 408
natural releases of 406
offshore leases 152
in Persian Gulf 424
as pollutant 306
reserves, in Barents Sea
381
spill response 72–73, 179,
309, 319–322, 407
spills 165, 319, 320–321
transport of 406
Oil Pollution Act of 1990
319

oil production platforms
408–412
design and construction
408
old, disposal of 408
as research platforms 409
selected examples
410–412
types 405, 406, 408
oil shale 406
Oil Spill Liability Trust Fund
319
Okhotsk Current 122
Oleander, CMV 62
Oliver Hazard Perry class
100
olive trees 334
Olympic Coast National
Marine Sanctuary 318
OMEGA navigation system
368
OML. *See* oxygen minimum
layer
ONR. *See* Office of Naval
Research
ooze 412–413
calcareous 75, 390, 412,
486
definition of 75, 486
formation 99, 209
and petroleum formation
78
radiolarian 209
siliceous 390, 412, 486
OPEC (Organization of
the Petroleum Exporting
Countries) 406
open coast beach 413
open roadstead 441
Operation Castle 44
Operation Crossroads 44
Operation Desert Storm. *See*
Gulf War (1991)
Operation Iraqi Freedom 218
Operation UNIFIED
ASSISTANCE 337
Opler, Paul A. 158
optical beam transmittance
271
optical properties of water 34
optimal ship track planning
304, 402, 413–414, 518
Optimum Track Ship Routing
(OTSR) 12
orangemouth corvina 272
Ordovician period 12, 113
Oregon State University 258
organic compounds
definition of 427, 447
in food chain 427, 447
Organization of the Petroleum
Exporting Countries
(OPEC) 406
Ortelius, Abraham 435
orthophosphates 384
oscillators, Quality (Q) factor
450–451, 451
OSCR (ocean surface current
radar) 125, 399, 454
OSHA. *See* Occupational
Safety and Health
Administration
osprey 25, 288